THE TRANS-NEPTUNIAN SOLAR SYSTEM

THE TRANS-NEPTUNIAN SOLAR SYSTEM

Edited by

DINA PRIALNIK

Department of Geosciences, Tel Aviv University, Tel Aviv, Israel

M. ANTONIETTA BARUCCI

LESIA, Paris Observatory, PSL, CNRS, Sorbonne University, University Paris-Diderot, Paris, France

LESLIE A. YOUNG

Southwest Research Institute, Boulder, CO, United States

ELSEVIER

Elsevier
Radarweg 29, PO Box 211, 1000 AE Amsterdam, Netherlands
The Boulevard, Langford Lane, Kidlington, Oxford OX5 1GB, United Kingdom
50 Hampshire Street, 5th Floor, Cambridge, MA 02139, United States

Library of Congress Cataloging-in-Publication Data
A catalog record for this book is available from the Library of Congress

British Library Cataloguing-in-Publication Data
A catalogue record for this book is available from the British Library

ISBN: 978-0-12-816490-7

Publisher: Candice Janco
Acquisition Editor: Marisa LaFleur
Editorial Project Manager: Hilary Carr
Production Project Manager: Prem Kumar Kaliamoorthi
Cover Designer: Christian Bilbow

Typeset by SPi Global, India

Contents

Contributors

Michele T. Bannister Astrophysics Research Center, School of Mathematics and Physics, Queen's University Belfast, Belfast, United Kingdom

M. Antonietta Barucci LESIA, Paris Observatory, PSL, CNRS, Sorbonne University, University Paris-Diderot, Paris, France

James G. Bauer Department of Astronomy, University of Maryland, College Park, MD, United States

Felipe Braga-Ribas Federal University of Technology, Paraná (UTFPR), Curitiba, PR; Observatório Nacional/MCTIC, Rio de Janeiro; LIneA, Rio de Janeiro, Brazil

Adrián Brunini National Council of Scientific and Technical Research, Buenos Aires; National University of Southern Patagonia, Caleta Olivia Academic Unit, Santa Cruz, Argentina

Julio I.B. Camargo LIneA; Observatório Nacional/MCTIC, Rio de Janeiro, Brazil

Manfred Cuntz Department of Physics, University of Texas at Arlington, Arlington, TX, United States

Audrey Delsanti Aix Marseille Université, CNRS, LAM (Laboratoire d'Astrophysique de Marseille) UMR 7326, Marseille, France

Josselin Desmars Paris Observatory, PSL Research University, CNRS, Sorbonne University, Univ. Paris Diderot, Sorbonne Paris City, LESIA, Meudon, France

Rudolf Dvorak Institute of Astronomy, University of Vienna, Vienna, Austria

Heather E. Elliott Southwest Research Institute, San Antonio, TX, United States

Julio A. Fernández Department of Astronomy, Faculty of Sciences, Universidad de la República, Montevideo, Uruguay

Sonia Fornasier LESIA, Paris Observatory,PSL University, CNRS, Université de Paris, Sorbonne University, Meudon Pricipal Cedex, France

William M. Grundy Lowell Observatory, Flagstaff, AZ, United States

Aurélie Guilbert-Lepoutre LGL-TPE, UMR 5276, CNRS, Claude Bernard Lyon 1 University, ENS Lyon, Villeurbanne Cedex, France

Aurélie Guilbert-Lepoutre LGL-TPE, UMR 5276/CNRS, Université de Lyon, Université Claude Bernard Lyon 1, ENS Lyon, Villeurbanne Cedex, France

Bryan J. Holler Space Telescope Science Institute, Baltimore, MD, United States

Robert E. Johnson University of Virginia, Charlottesville, VA, United States

J.J. Kavelaars Department of Physics and Astronomy, University of Victoria; Herzberg Astronomy and Astrophysics Research Center, National Research Council of Canada, Victoria, BC, Canada

Samantha M. Lawler Herzberg Astronomy and Astrophysics Research Center, National Research Council of Canada, Victoria, BC, Canada

Rodrigo Leiva Department of Space Studies, Southwest Research Institute, Boulder, CO, United States

Emmanuel Lellouch LESIA, Paris Observatory,PSL University, CNRS, Université de Paris, Sorbonne University, Meudon Pricipal Cedex, France

Birgit Loibnegger Institute of Astronomy, University of Vienna, Vienna, Austria

Robin Métayer LGL-TPE, UMR 5276, CNRS, Claude Bernard Lyon 1 University, ENS Lyon, Villeurbanne Cedex, France

Frederic Merlin LESIA, Paris Observatory, PSL, CNRS, Sorbonne University, University Paris-Diderot, Paris, France

Alessandro Morbidelli Laboratory Lagrange, UMR7293, University of Nice Sophia-Antipolis, CNRS, Nice, France

Maryame El Moutamid Carl Sagan Institute; Center for Astrophysics and Planetary Science, Cornell University, Ithaca, NY, United States

Thomas Müller Max Planck Institute for Extraterrestrial Physics, Garching, Germany

David Nesvorný Southwest Research Institute, Boulder, CO, United States

Francis Nimmo Department of Earth and Planetary Sciences, University of California, Santa Cruz, CA, United States

Keith S. Noll Goddard Space Flight Center, Greenbelt, MD, United States

José L. Ortiz Instituto de Astrofisica de Andalucia (CSIC), Granada, Spain

Nuno Peixinho CITEUC—Center for Earth and Space Research of the University of Coimbra, Geophysical and Astronomical Observatory of the University of Coimbra, Coimbra, Portugal

Noemí Pinilla-Alonso Arecibo Observatory, Arecibo, Puerto Rico; Florida Space Institute, UCF, Orlando, FL, United States

Simon P. Porter Southwest Research Institute, Boulder, CO, United States

Dina Prialnik Department of Geosciences, Tel Aviv University, Tel Aviv, Israel

Stefan Renner Paris Observatory, CNRS UMR 8028, Lille University, Lille Observatory, IMCCE, Lille, France

Françoise Roques Paris Observatory, PSL Research University, CNRS, Sorbonne University, Univ. Paris Diderot, Sorbonne Paris City, LESIA, Meudon, France

Pablo Santos-Sanz Instituto de Astrofísica de Andalucía (CSIC), Granada, Spain

Cory Shankman Department of Physics and Astronomy, University of Victoria, Victoria, BC, Canada

Bruno Sicardy Paris Observatory, PSL Research University, CNRS, Sorbonne University, Univ. Paris Diderot, Sorbonne Paris City, LESIA, Meudon, France

Romina P. Di Sisto Facultad de Ciencias Astronómicas y Geofísicas, UNLP; Instituto de Astrofísica de La Plata, CCT—CONICET—Universidad Nacional de La Plata, La Plata, Argentina

John R. Spencer Southwest Research Institute, Boulder, CO, United States

John A. Stansberry Space Telescope Science Institute, Baltimore, MD, United States

S. Alan Stern Southwest Research Institute, Boulder, CO, United States

Stephen C. Tegler Department of Physics and Astronomy, Northern Arizona University, Flagstaff, AZ, United States

Audrey Thirouin Lowell Observatory, Flagstaff, AZ, United States

Chadwick A. Trujillo Department of Physics and Astronomy, Northern Arizona University, Flagstaff, AZ, United States

Anne Verbiscer University of Virginia, Charlottesville, VA, United States

Mark C. Wyatt Institute of Astronomy, University of Cambridge, Cambridge, United Kingdom

Leslie A. Young Southwest Research Institute, Boulder, CO, United States

Preface

The Trans-Neptunian region of the solar system first came into focus in 1930, with the discovery of Pluto, designated at the time as the ninth planet. Then, about 20 years later, Gerard Kuiper postulated the existence of a belt of small objects extending beyond the orbit of Neptune to about 50 AU. Another 30 years or so elapsed before this belt was named after him, when dynamical simulations showed that it could constitute a source of short-period comets. It is known since as the Kuiper belt, or sometimes as the Edgeworth-Kuiper belt, since a few years before Kuiper, Kenneth Edgeworth put forward the hypothesis that the solar nebula must have extended beyond the planetary orbits, a region that should be occupied by a large number of relatively small objects. The second object beyond Neptune, Pluto's moon Charon, was discovered in 1978 by James Christy. Finally, in 1992, the first Kuiper belt object was discovered by David Jewitt and Jane Luu, and designated 1992 QB_1 (now, 15760 *Albion*). Since then, the number of Kuiper belt object discoveries has grown exponentially and the acronym KBO (Kuiper belt object) has gained wide usage; there are now well over a thousand known KBOs, divided among different classes. The number of KBOs with diameters over 100 km is thought to exceed 100,000.

With the turn of the millennium came the discoveries of large KBOs: first, Varuna and Quaoar, then Sedna, Eris, Haumea, and others. Since Eris is very similar in size and larger in mass than Pluto, the definition of a planet was reconsidered by the International Astronomical Union and as a result, at the General Assembly of 2006, Pluto was reclassified by majority vote as a Kuiper belt dwarf planet. This was just after the first space mission to Pluto, *New Horizons*, was launched. Ten years later, the existence of another planet in the Trans-Neptunian region was proposed, inferred from gravitational effects on the orbits of other KBOs. The planet, dubbed Planet 9, has not yet been discovered, but in the search for it, other very distant objects have been detected. Since all these later objects have orbits way beyond the original Kuiper belt, Trans-Neptunian objects (TNOs) is a more appropriate name for them, which has gained popularity in reference to the entire population—observed or not yet discovered—beyond the orbit of Neptune. This is the story of the Trans-Neptunian solar system in a nutshell.

The extent of research devoted to TNOs since 1992 has been enormous, resulting in hundreds of papers, numerous conferences and workshops, and culminating with the *New Horizons* space mission to the Kuiper belt and beyond, which is still ongoing. The close-up images of Pluto and Charon, taken in 2015, revolutionized our views of these distant worlds. The first close-up image of a KBO beyond Pluto, 2014 MU_{69} (nicknamed *Ultima Thule*), was taken by *New Horizons* in January 2019, just as the chapters of our book were being written.

The book is largely based on the invited review talks of a conference held in the spring of 2018 in the beautiful town of Coimbra in Portugal, but the chapters extend beyond that, in an attempt to cover and summarize the current research and also to look into the future. Since the last comprehensive review of the subject, published in 2007 (*The solar system beyond Neptune*, eds.: M.A. Barucci, H. Boehnhardt, D.P. Cruikshank, and A. Morbidelli, University of Arizona Press), we have seen the flyby of a dwarf-planet system and a small contact binary, the discovery of additional Sednoids and detached objects with their orbital characteristics, the results of large observing programs, such as the *TNOs are Cool* survey with Herschel and Spitzer, the Outer Solar System Origins Survey (OSSOS) and Colors of the Outer Solar System Origins Survey (Col-OSSOS) with CFHT, and much more. All these warrant a new, fresh look at the Trans-Neptunian region.

Between a comprehensive introduction and an epilogue, the book is divided into five parts, each addressing a different subject. Part I covers the dynamics of the formation and evolution of the Trans-Neptunian populations, and also discusses the possible existence of additional distant, massive planets. Part II deals with the properties of TNOs: their surface composition, thermal emission, and in a more speculative manner, their interior structure, volatile evolution, and temporary atmospheres. Binaries and multiple systems, as well as rings, are discussed in Part III, including both physical and dynamical properties, and theories of their formation. Part IV considers relations to other populations in our solar system, and also looks at extrasolar planetary systems. Finally, the bright prospects for the future of TNO studies are addressed in Part V, including physical studies enabled by space-based observatories, the expected impact of new large surveys on the discovery rate of TNOs, and the promise of future occultations in the age of the Gaia astrometric catalog. We hope and expect that these reviews will be useful for both researchers and students in many different fields, astronomers, geologists, dynamicists, and modelers alike.

We wish to thank all the authors who have contributed to this book. All the chapters included in this volume underwent a full refereeing process. We express our thanks to the reviewers who provided valuable assistance in improving the manuscripts by insightful comments and suggestions: Susan Benecchi, Hermann Boehnhardt, Luke Dones, Julio Fernandez, Marc Fouchard, Olivier Hainaut, David Jewitt, Lynne Jones, Samantha Lawler, Emmanuel Lellouch, Javier Licandro, Brenda Matthews, Mario Melita, Cathy Olkin, Alex Parker, Kelsi Singer, and Junichi Watanabe.

Many ideas presented in this book are new and not yet fully confirmed by observation or theory. Hence, we still have a long way to go and as the study of the Trans-Neptunian solar system progresses, quoting Kenneth Edgeworth's words of seventy years ago, "Helpful ideas must be retained, faulty ideas must be eliminated and gaps in the resulting theory must be carefully filled." (Edgeworth, K.E., 1949. *The origin and evolution of the solar system.* Mon. Not. R. Astron. Soc. 109, 600–609.)

Dina Prialnik
M. Antonietta Barucci
Leslie A. Young

CHAPTER

1

Introduction: The Trans-Neptunian belt—Past, present, and future

Julio A. Fernández

Department of Astronomy, Faculty of Sciences, Universidad de la República, Montevideo, Uruguay

1.1 The solar system beyond Neptune: The search for planet X

The discovery of Neptune in 1846 stimulated the search for other more distant planets, motivated by the allegedly irregularities in the motion of Uranus that required another more distant planet besides Neptune. Percival Lowell devoted several years to search for this hypothetical planet that he called planet X until his death in 1916. The search was restarted at the Lowell Observatory in Flagstaff, Arizona, in 1929 by a young observer Clyde Tombaugh who succeeded in discovering Pluto on February 18, 1930. He continued his search for other Trans-Neptunian (TN) planets for another 13 years but none appeared. It seemed that the inventory of planets was complete, so Pluto marked the outer edge of the solar system. A detailed chronicle of Pluto's discovery can be found in Davies et al. (2008).

Right after its discovery, Pluto was hailed as the ninth planet with an uncertain mass but that, in principle, could be as large as the Earth's, depending on its albedo and bulk density. Yet, it showed significant differences with the other planets that moved on regular, near-circular, and near-coplanar orbits. Pluto's orbit was indeed so eccentric that it crossed that of Neptune, and its inclination was of about 17 degrees, far higher than those of the other planets. With such an orbit the puzzling question was why Pluto had not been ejected by Neptune's perturbations throughout the solar system lifetime. Lyttleton (1936) suggested that Pluto was a former Neptune's satellite that moved with Tritton in direct regular orbits around the planet. Tidal friction brought their orbits close together allowing close encounters between these satellites. One of such encounters led to the ejection of Pluto and the reversal of Tritton's orbit. The ejected Pluto remained in an orbit around the Sun crossing that of Neptune.

The Trans-Neptunian Solar System.
https://doi.org/10.1016/B978-0-12-816490-7.00001-1

Where Lyttleton's hypothesis proved to be true, it would have exposed a curious situation: the changing role of Pluto during its lifetime, being at the beginning a "satellite" to become later a "planet."

The puzzle about the stability of Pluto's orbit was finally solved by Cohen and Hubbard (1965), who carried out numerical computations for 120,000 year showing that the orbit of Pluto librates relative to that of Neptune around the 2:3 commensurability, in such a way that Pluto can never approach Neptune to less than 18 AU.

Nicholson and Mayall (1931) estimated Pluto's mass at about two-thirds that of the Earth from the presumed perturbations that this planet caused on Neptune and Uranus. Later Kuiper (1950) measured Pluto's diameter with the 200-in. Palomar telescope, finding an apparent diameter of $0''.20 \pm 0''.01$ or 0.46 of the Earth's diameter which would imply a rather low visual albedo of 0.17. Yet Cruikshank et al. (1976) found evidence for methane frost on Pluto's surface suggesting a much higher albedo than previously thought, probably above 0.4. Therefore, Pluto should be much smaller, perhaps even smaller than the Moon. In 1977, a large moon was discovered around Pluto by Christy and Harrington (1978) that was later named Charon. The motion of the Pluto-Charon system around their center of mass allowed to estimate its mass. The result of the combined mass was about 0.0017 $M_{\oplus}$ (Earth's mass), thus confirming that Pluto was really very small, much smaller than the Moon. New astrometric measurements of Pluto, Charon, and the four other satellites discovered later (Styx, Nix, Kerberos, and Hydra) allowed to refine Pluto's mass to 0.0022 $M_{\oplus}$ (Buie et al., 2006; Stern et al., 2018).

1.2 Early cosmogonic ideas

The discovery of Pluto prompted Leonard (1930) to speculate about the existence of a population of "ultra-Neptunian" and "ultra-Plutonic" planets on the basis that the Sun's gravitational field extends far beyond Pluto's orbit. But the first thorough cosmogonic model was developed by Edgeworth (1938, 1943, 1949) who assumed that the early Sun was surrounded by a vast disk of meteorites that extended beyond the orbits of the planets, retaking the idea of a Laplacian disk where planets formed. While in the inner regions of the disk densities were high enough to allow the formation of big planets by gravitational collapse, in the outer edge of the disk densities were too low for this to happen. He went on to argue that the particulate fluid became turbulent favoring the agglomeration of dust particles within eddies into comet-size bodies, described as 10-km size heap of gravel loosely held together by its own gravity. The total number of comets would have been of about 200–2000 million, amounting to a total mass of a few tenths $M_{\oplus}$ scattered over a region between $\sim 10^{10}$ km and $3-4 \times 10^{10}$ km ($\sim$70 to $\sim$200–300 AU). Edgeworth conjectured that this vast reservoir was the source of the observed comets. It is interesting to mention that Edgeworth's ideas were recently taken by Johansen et al. (2007) who invoked turbulence to explain two opposite effects. On one side, the inhibition of particle sedimentation in the midplane of the protoplanetary disk prevented densities from growing to the point of triggering gravitational collapse into kilometer-size planetesimals. On the other side, turbulence can create transient high-pressure regions that can grow by an order of magnitude by a streaming instability driven by the relative flow of gas and solids. These dense regions can collapse into planetesimals from comet to dwarf-planet size bodies.

In an independent work, Kuiper (1951) also discussed a model of solar system formation in which he envisaged that beyond Neptune densities were too low for the solid material to collect into a single massive body, leaving instead a swarm of planetesimals spread in a ring between ~30 and 50 AU. Some years later, and seemingly independent from the previous authors, Cameron (1962) argued that in a massive solar nebula a large mass of nebular gas was shed beyond Neptune, so a large amount of solid material must remain there as comets. Inspired by the work of Cameron (1962), Whipple (1964) developed further the idea of a stable belt of comets beyond the orbit of Neptune. In order to account for observed disturbances in the motion of Neptune's latitude, he estimated that the belt should have a mass of 10–20 $M_\oplus$ if it were located at 40–50 AU from the Sun. Later Standish (1993) showed that such disturbances in Neptune's motion were spurious so Whipple's result was invalidated. Whipple and colleagues tried another approach by using 1P/Halley as a natural probe since its aphelion lies right in the putative belt. No perturbations were detected on 1P/Halley's motion which allowed the authors to set an upper limit of 0.5 $M_\oplus$ if the belt was located near the invariable plane at 40 AU, or 1.3 $M_\oplus$ at 50 AU (Hamid et al., 1968). In this early stage, Whipple was probably the one who did the most to disseminate the idea of a TN belt in a quite explicit way as it is shown by his sketch of the putative belt (Fig. 1.1). It is interesting to mention that Whipple did not believe that this belt could contribute appreciably to the observable comets, rather he thought on a belt dynamically frozen in the outer edge of the planetary system beyond the reach of planetary perturbations. In the same sense, Delsemme (1973) argued that because belt comets had circular orbits, they could not be the source of the visible comets which must come from the Oort cloud.

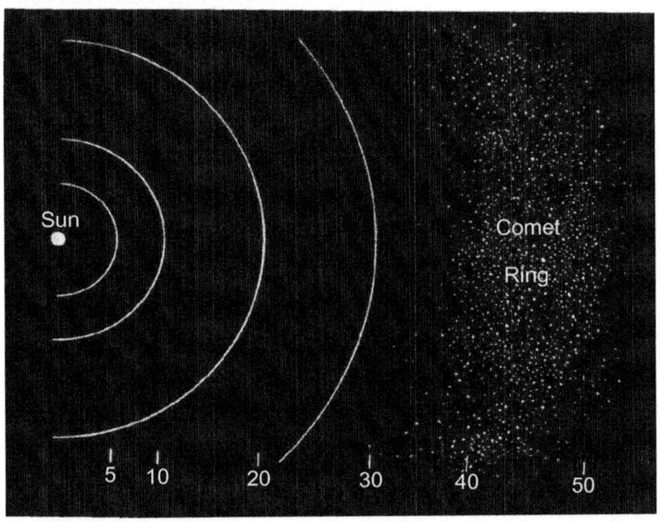

FIG. 1.1 Whipple's concept of a Trans-Neptunian ring or belt beyond the reach of planetary perturbations (Whipple, 1972).

1.3 The Jupiter family comet connection

Until the 1970s, the discussion about the existence of a TN belt was entirely theoretical since there was no observational evidence supporting it. For instance, the concept of the Oort cometary cloud was developed to explain the anomalous excess of near-parabolic orbits (semimajor axes $a > 10^4$ AU) among the observed long-period comets (LPCs). From this observation, Oort (1950) inferred that a vast spherically distributed comet reservoir should exist at several 10^4 AU subject to stellar perturbations. But there seemed to be no observed comet population in the Earth's vicinity coming from a flat source beyond Neptune. If this were the case, how could we learn about a putative belt safe from planetary or stellar perturbations, and thus dynamically frozen?

At the end of the 1970s, I became interested in the problem of the evolution of comets and their transfer from long-period orbits to short-period orbits. The standard theory at that time was that short-period comets (SPCs) were the end product of the dynamical evolution of Oort cloud comets captured by Jupiter (the classical Laplace's hypothesis), with the difference that the single-encounter capture was substituted by the multiple-step capture by Jupiter and the other Jovian planets (Kazimirchak-Polonskaya, 1967; Everhart, 1972). I found rather puzzling to have an evolutionary path connecting both populations since LPCs have a more or less isotropic distribution of orbital inclinations whereas SPCs show a flat distribution strongly concentrated to the ecliptic plane. How was it possible to pass from an isotropic distribution to a flat distribution of inclinations? Everhart (1972) tried to solve the problem by assuming that SPCs were the end product of the evolution of Oort cloud comets through multiple encounters with Jupiter. By setting an upper limit of 2000 revolutions to the evolution, he found that captures to short-period orbits occurred for comets within the region of perihelion distances $4 < q < 6$ AU and inclinations $i < 9$ degrees. The cut at 2000 revolutions seemed to be tailored to allow only low-i Oort cloud comets to reach short-period orbits. But what would happen if we extended the evolution to a greater number of revolutions? In that case, many high-inclination and retrograde comets would leak into short-period orbits in conflict with the observations. One could argue that the limit of 2000 revolutions had a physical explanation in terms of the finite physical lifetimes, but this argument had no quantitative basis.

Vaghi (1973) pointed out to another inconsistency in Everhart's hypothesis: the Tisserand parameter with respect to Jupiter, T_J, which is a quasiinvariant of the comet's motion, must fulfill the condition $T_J \leq 2\sqrt{2} \simeq 2.82$ for comets coming from a parabolic flux, while about 40% of SPCs have $T_J > 2\sqrt{2}$, so they could not have come, in principle, from original parabolic orbits. Vaghi's objection is illustrated in Fig. 1.2, where we plot the Tisserand parameters of different comet populations: SPCs, LPCs, and intermediate-period or Halley-type comets (HTCs), versus their orbital energies (represented by the reciprocal of the original semimajor axis, $1/a$, as a proxy). Note that the figure plots JFCs with orbital periods $P < 20$ years, while SPCs in Everhart's nomenclature were those with $P < 13$ years (reciprocal semimajor axes $\gtrsim 0.18\ \mathrm{AU}^{-1}$). This is a minor difference that does not change the arguments given here. We can see that SPCs (open circles at the upper right corners of the panels) are clearly detached from the remaining populations represented by filled circles. Another problem came from the capture efficiency from Everhart's scheme, Joss (1973) found that the rate of captures into short-period orbits would be 40,000 times too low to keep the current observed population of SPCs in steady state. To overcome this shortcoming, Delsemme (1973) argued about the existence of a population of 30,000–100,000 intermediate-period, low-inclination comets with perihelia come between $4 < q < 6$ AU as the immediate source of SPCs. Yet Delsemme did not explain how this population was formed and maintained. The ultimate question was: what was the origin of the flat, intermediate-period comet source?

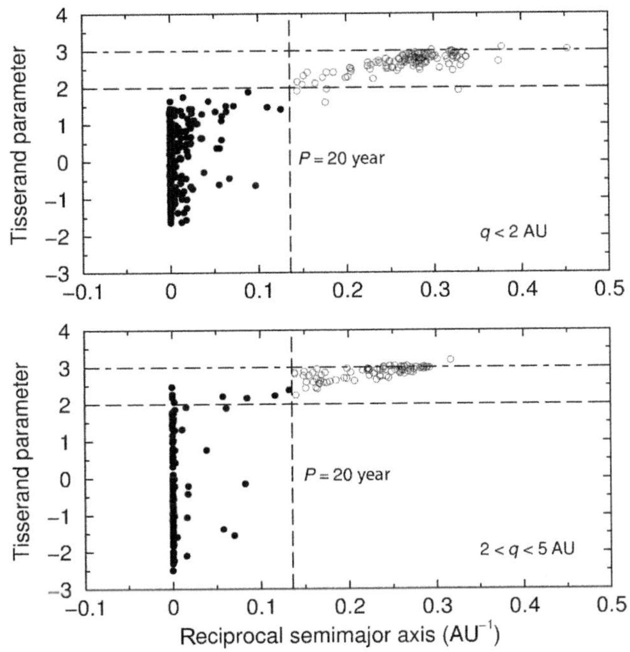

FIG. 1.2 Tisserand parameter T_J versus reciprocal semimajor axis for comets with $q < 2$ AU (*upper panel*) and $2 < q < 5$ AU (*bottom panel*). Jupiter family comets with periods $P < 20$ years (*open circles*) have in general $T_J > 2$, and a significant fraction $T_J > 2\sqrt{2}$, while intermediate-period comets and long-period comets (*filled circles*) have in most cases $T_J < 2$ (Fernández, 2005).

We rechecked the efficiency of the transfer process from near-parabolic to short-period orbits by considering Everhart's capture window and also the contribution from captures by the other Jovian planets (Fernández, 1980). In agreement with Joss's conclusion that the capture efficiency would be extremely low, we found that for every captured SPC, 300 near-parabolic comets would be ejected per year, which would give $\sim 1.3 \times 10^{12}$ throughout the solar system lifetime at a steady loss rate. This is an enormous wastage of matter, one order of magnitude greater than the Oort cloud population estimated by Oort. By applying Okham's razor principle of choosing a theory that requires the smaller number of hypothesis, we thought: "Why not to assume that SPCs come from a flat source?" After reading the works by Kuiper and Whipple, we came to the conclusion that the TN belt should be a more suitable source of SPCs. This idea was developed in a paper (Fernández, 1980) where from a numerical model we showed that weak perturbations by Pluto-size or Moon-size bodies could drive some TN objects (TNOs) near Neptune's orbit. Thus, contrary to what Whipple (1972) and Delsemme (1973) thought of a dynamically frozen TN belt, we found that a steady leakage of bodies to Neptune's gravitational influence zone could take place. Once under the control of Neptune, the bodies would be either ejected or handed down to the next planet inside (Uranus) and so on until ending up under the gravitational control of Jupiter. The reasoning was that if for each one of the Jovian planets half of the bodies were handed down to the next planet inside (or to the inner planetary region in the case of Jupiter) and the other half were ejected (see, e.g., Everhart, 1977), the efficiency of the transfer process from the TN belt to short-period orbits would be: $(1/2)^4 = 0.0625$. This back-of-the-envelope estimate turned out to be too low, later numerical simulations by Duncan et al. (1988) rose the transfer efficiency to 0.17, and the Duncan et al.'s (1995) estimate rose again the ratio to 0.34. The transfer efficiency turns out to be two orders of magnitude greater than that from an isotropic flux of near-parabolic comets.

1.4 The naming controversy

The TN belt did not have a proper name until Duncan et al. (1988) coined the term "Kuiper belt" that was rapidly adopted by the scientific community at large. The name seems to be arbitrary and unfair with other researchers. Regrettably, Edgeworth's pioneer work on a TN belt of comets was overlooked until Bailey and Stagg (1990) mentioned Edgeworth's (1949) paper. McFarland (1996) advocated strongly in favor of recognizing Edgeworth's important contribution to the subject, which led Davis and Farinella (1997) to coin the name "Edgeworth-Kuiper" belt. Even though for a time both names coexisted in the literature, the latter has faded with time, perhaps because it is just simpler to use a single name. But the story is more complex since there were other important players. As we saw before, Whipple—inspired by Cameron's work—was possibly the one who best tried to convey the idea of a TN belt in a more explicit way and to constrain its mass (at least the article by Whipple (1972) was the first to call the writer's attention about this subject).

But, after all, why does the belt have to have a proper name? For instance, does the asteroid belt have a proper name? No, it does not. It could have been named, for example, the Piazzi belt (after Ceres's discoverer) or Olbers belt (after the first one that proposed a theory about the origin of asteroids) but, with good criterion, it was left unnamed thus preventing potential controversies. In summary, if we choose a certain name, surely we will be unfair with others, so why not to just simply adopt the neutral term "Trans-Neptunian belt" (TN) and Trans-Neptunian object (TNO)? These will be the terms that we will be using in the rest of the chapter.

1.5 The discovery

In the late 1980s, the situation was ripe to search for TNOs inspired in the previous theoretical and computational studies. Kowal (1989) searched 6400 square degrees photographically to approximately $m_V = 20$ which led to the discovery of the Centaur 2060 Chiron in 1977 but no TNOs. Other deep searches were carried out by Luu and Jewitt (1988) and Levison and Duncan (1990), again with negative results. It was a matter of perseverance, access to the medium- and large-size telescopes, and better CCD cameras (that at the beginning of the 1990s were replacing the old photographic plates) to detect these elusive bodies. Observing with the 2.2 m-telescope of Mauna Kea, Hawaii, Jewitt and Luu (1993) finally succeeded in finding a couple of TNOs: 15,670 Albion and a few months later 1993 FW. Both objects were found to have low-eccentricity orbits and semimajor axes of $a = 43.82$ AU and $a = 44.07$ AU, right at the distances where the TN belt was expected to be.

Once an observing campaign was successful, it was just a matter of time to find new members. This was what actually happened, so the population of discovered TNOs has been steadily growing until reaching a number of 2183 at the end of 2017 (see Fig. 1.3). The discovery rate of TNOs has been particularly high during the period 1999–2005 thanks to the *Deep Ecliptic Survey* a program carried out from Kitt Peak National Observatory and Cerro Tololo Inter-American Observatory (Elliot et al., 2005) that yielded more than 300 TNOs and Centaurs with well-determined orbits, and again during 2013–17 thanks to the Outer Solar System Origins Survey (OSSOS) program that used the Canada-France-Hawaii Telescope, surveying 155.3 square degrees of sky near the invariable plane to a limiting (red) magnitude $m(R) = 25.2$ providing more than 800 discoveries (e.g., Bannister et al., 2018).

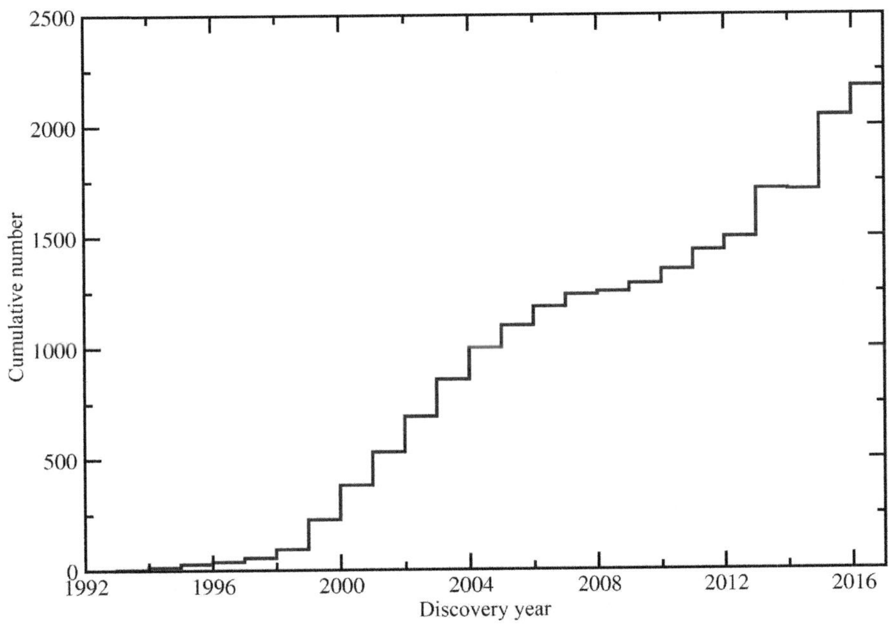

FIG. 1.3 The discovery rate of TNOs with perihelion distances $q > 30$ AU during 1992–2017. *Data taken from JPL Solar System Dynamics.*

1.6 Dynamical structure and transfer mechanisms

The increasing computer power prompted several authors to carry out massive numerical integrations to build dynamical maps of stable and unstable regions in the TN region (Torbett, 1989; Torbett and Smoluchowski, 1990; Levison, 1991; Levison and Duncan, 1993; Holman and Wisdom, 1993). These early results showed that such a belt had a complex structure, shaped by numerous secular and mean-motion resonances (MMRs) that generated chaos in the motion of

TNOs inducing high eccentricities. As a result, some TNOs may become Neptune-crossers or approachers with their consequent removal from the belt, causing a strong dynamical erosion for semimajor axes $a \gtrsim 42$ AU (Fig. 1.4). Early orbit computations of the first discovered TNOs showed that about 40% moved in Pluto-like orbits, that is, in the 2:3 MMR with Neptune. They have been called *Plutinos* owing to their smaller sizes as compared to Pluto's. Most Plutinos have eccentricities and inclinations higher than the nonresonant TNOs in such a way that their perihelia approach or fall inside Neptune's orbit. However, as it happens with Pluto, the resonance protects them from having close encounters with Neptune. Duncan et al. (1995) carried out massive numerical computations of fictitious bodies for the solar system age showing that for $e \gtrsim 0.1$ and $a \gtrsim 42$ AU, the only stable orbits were those in the 2:3, 3:4, 4:5, and 5:6 MMRs. No stable orbits were found for $a \gtrsim 34$ AU because the overlapping of MMRs causes large-scale chaos, and Neptune's proximity increases the instability of bodies located there (Morbidelli et al., 1995). The slow diffusion of TNOs located at the borders of chaotic zones causes a slow leakage of TNOs to Neptune-crossing orbits which feeds a transient population of bodies in the outer planetary region that were called *Centaurs* (Stern and Campins, 1996). A fraction of them will end up in the inner planetary region, the majority as JFCs and a small fraction as HTCs (Fernández et al., 2016).

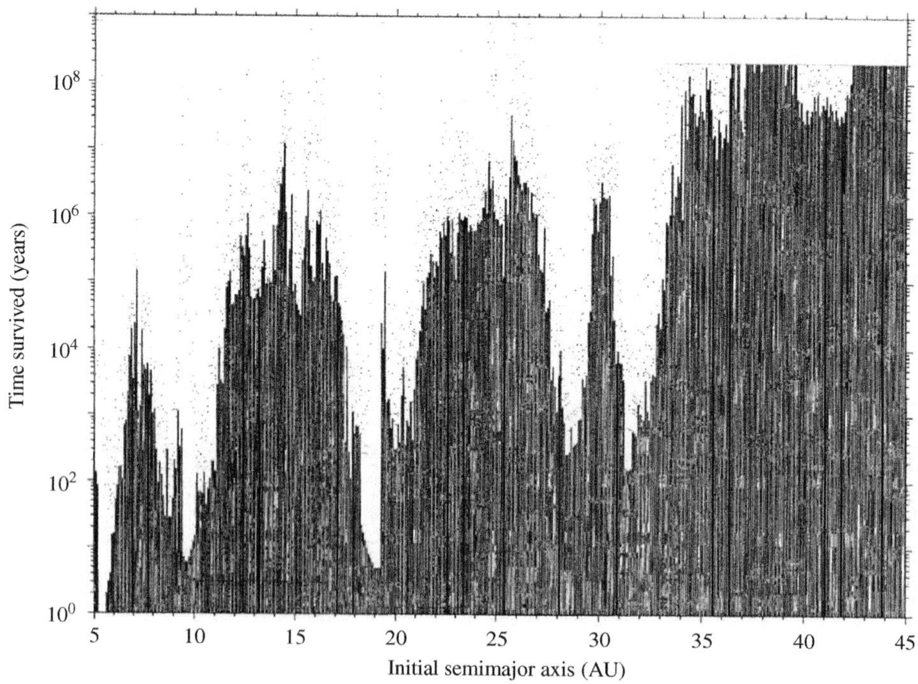

FIG. 1.4 Dynamical erosion in the planetary zone and the TN belt depicted through the survival times of test particles with semimajor axes covering the range $5 < a < 45$ AU. It is seen that for $a > 43$ AU the particles survive for the whole integration period of 200 Myr (Holman and Wisdom, 1993).

Jewitt et al. (1998) characterized three different dynamical classes: (1) Classical TNOs, which occupy low-eccentricity ($e < 0.25$) orbits with semimajor axes $41 \gtrsim a \gtrsim 46$ AU, they are estimated to constitute ~70% of the observed population. They are subdivided according to their inclinations into the "cold" ($i < 5$ degrees) and "hot" ($i > 5$ degrees) populations. We will see below that both populations show distinct features. (2) Resonant TNOs, which occupy MMRs with Neptune, in particular the 2:3 ($a \approx 39.4$ AU) and comprise ~20% of the known objects. (3) Scattered disk objects (SDOs), which possess the most extreme orbits, with perihelion distances $30 \leq q \leq 40$ AU, median semimajor axis $a \sim 90$ AU and eccentricities $e \approx 0.6$, and comprise about 10% of the known TNOs. Yet, there is a strong bias against the detection of SDOs since they are on average more distant and therefore fainter. The actual population of SDOs should therefore be substantially larger than the observed one, probably of similar size or slightly larger than the classical (cold + hot) population (Trujillo et al., 2001; Adams et al., 2014). Because SDOs approach Neptune near perihelion, they are subject to weak perturbations by this planet that excite their eccentricities to higher and higher values. It is then likely that an important fraction of them will end up in the Oort cloud (Fernández et al., 2004).

Besides the dynamical classes described earlier, we have also found *detached* SDOs with $q >$ 40 AU which are too far away for planetary perturbations to have excited their eccentricities by the classical diffusion mechanism. Yet the Kozai mechanism (coupling between e and i when ω librates around a certain value) can force high q up to values around 70 AU (Gomes et al., 2005). Fig. 1.5 shows the distribution of all the discovered TNOs in the parametric

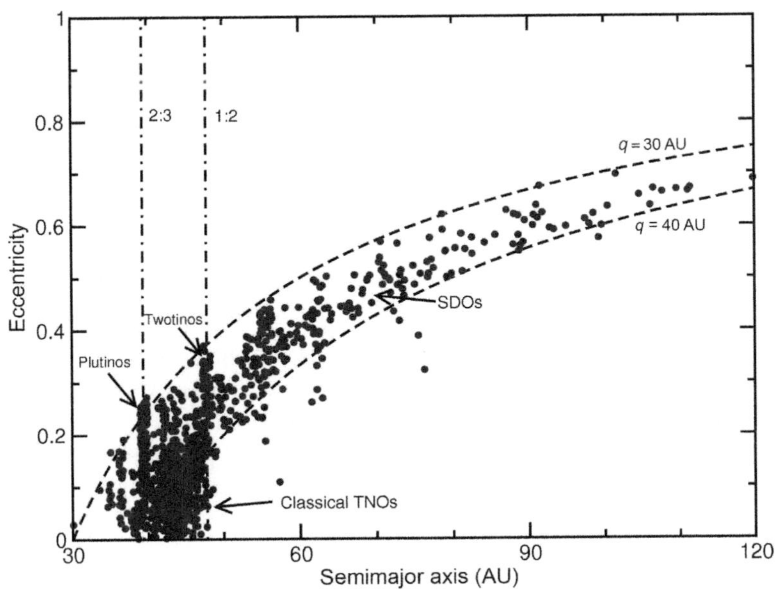

FIG. 1.5 Distribution of the discovered TNOs with $q > 30$ AU in the parametric plane eccentricity versus semimajor axis. We can see several groupings: classical TNOs (cold and hot), TNOs in the 2:3 and 1:2 MMRs (Plutinos and Twotinos), and SDOs. We can also see some detached TNOs. There are some distant TNOs with $a > 120$ AU that are not plotted in the figure. *Data taken from JPL Solar System Dynamics.*

plane (a, e). Besides the TNOs heavily concentrated in the 2:3 MMR ($a = 39.4$ AU), there is a quite notorious concentration around the 1:2 MMR at $a = 47.7$ AU. The OSSOS survey has enriched the observed sample of TNOs, in particular highlighting some other clusterings of TNOs around other MMRs, such as 3:4 ($a = 36.4$ AU), 4:7 ($a = 43.7$ AU), and 2:5 ($a = 55.4$ AU) (Bannister et al., 2018).

1.7 Size distribution and massive TNOs

Early results showed that the cumulative luminosity function (CLF) could fit to a linear relation

$$\log \Sigma(m_R) = C + \alpha m_R \tag{1.1}$$

where $\Sigma(m_R)$ is the density of TNOs per square degree near the ecliptic brighter than the apparent (red) magnitude m_R, and C and α are constants. Trujillo et al. (2001) derived for the CLF a slope: $\alpha = 0.63 \pm 0.06$.

By making appropriate corrections for heliocentric and geocentric distances, we can convert the CLF into a cumulative size distribution (CSD) which will follow a power law of the kind

$$N_R(> R) \propto R^{-s} \tag{1.2}$$

where $N_R(> R)$ is the number of objects with radii greater than R. Bearing in mind that α is related to the exponent s through $s = 5\alpha$ we obtain $s = 3.15 \pm 0.3$ if we take the value of α derived by Trujillo et al. The conversion from luminosity to radius implies to adopt an average albedo for the TNOs. Typical geometric albedos of all TNOs and Centaurs are about 0.07–0.08 (Stansberry et al., 2008), though there is a large dispersion among individual objects and albedos seem to be correlated with color. Lacerda et al. (2014) distinguished the two groups: one with dark neutral surfaces and albedos ~0.05, and other with bright red surfaces and higher albedos ~0.15.

More recently it was found that the CLF is better represented by a broken power law for both the hot population (hot classicals + SDOs) and cold classicals (Fuentes et al., 2010; Adams et al., 2014; Fraser et al., 2014). Yet the slopes and the break magnitude, H_b, differ for hot and cold TNOs. Fraser et al. (2014) obtained bright-end slopes: $\alpha_{1c} = 1.5^{+0.4}_{-0.2}$ and a break magnitude $H_{bc} = 6.9^{+0.1}_{-0.2}$ (radius $R_{bc} \simeq 105$ km) for the cold population, and $\alpha_{1h} = 0.87^{+0.07}_{-0.2}$ and $H_{bh} = 7.7^{+1.0}_{-0.5}$ (radius $R_{bh} \simeq 72$ km) for the hot population. The radii R_{bc} and R_{bh} are derived by adopting an average albedo $p = 0.07$. As regards the faint-end slopes both populations show a similar value $\alpha_2 \sim 0.2$. The masses of the hot and cold populations are estimated to be ~0.01 and 3×10^{-4} $M_\oplus$, respectively. How much mass is enclosed in the scattered disk and the detached scattered disk is still very uncertain given the high degree of incompleteness of the observed sample and the difficulty to estimate biases.

The size distribution we observe today must arise from the combination of two processes: (1) The primordial distribution of planetesimals in the protoplanetary disk and their growth to massive bodies via runaway and oligarchic accretion, and (2) their later collisional evolution. From theoretical models Schlichting et al. (2013) found that collisional evolution has shaped the size distribution for radii $R \gtrsim 30$ km while for greater sizes it was keptprimordial. It is then possible to associate the bimodal CLF found before with two different evolutionary stages: one primordial and one collisionally evolved.

Most of the largest TNOs seem to have been detected by now, some of them rivaling in size with Pluto, as are the cases of Haumea with an estimated diameter $D \simeq 1600$ km, Makemake with $D \simeq 1430$ km, and Eris with a size very similar to Pluto's and slightly more massive. All the largest TNOs belong to the hot population (either the hot classical or the SD). Schwamb et al. (2014) estimated that the inventory of large classical TNOs brighter than apparent (red) magnitude $m(R) \sim 19.5$ is nearly complete. This roughly corresponds to an object of $D \sim 750$ km of albedo $\sim$0.1 at $r = 42$ AU. This of course does not rule out the possibility that massive TNOs of Pluto size or larger are lurking at distances greater than 100 AU.

At the other end of the size distribution, we have small comet-size TNOs beyond our current capabilities of detection. The most meaningful constraint on the small TNOs population (diameters $D \simeq 1-10$ km) comes from the Taiwanese-American Occultation Survey (TAOS) based on serendipitous occultations of stars by small TNOs. From the lack of observed occultations, the TAOS team (Zhang et al., 2013) found an upper limit of $s = 2.34-2.82$ for the exponent of the CSD, which is consistent with a decrease in the slope at the faint end of the size distribution found by Fraser et al. (2014). The 1–10 km size range is relevant since it roughly corresponds to the size range of the observed JFCs which show a power-law CSD with an estimated $s \simeq 1.9-2.7$ (Lamy et al., 2004; Tancredi et al., 2006; Fernández et al., 2013). It is also possible that bodies coming from the TN belt will experience sublimation, outbursts and splittings, during their journey to the inner planetary region, leaving two or more daughter comets that will change the CSD of the incoming small TNOs. Therefore, the CSD of JFCs does not have to match that observed in small TNOs. One of the striking features of the population of near-Earth JFCs is the scarcity of comets at the faintest end (roughly in the size range $\sim$0.1–0.5 km) (Fernández and Morbidelli, 2006), Is such a scarcity an original property of the population in the TN belt? or Is it due to their depletion as they approach to the Sun? This is no doubt an interesting topic whose solution will depend on expanding capabilities to investigate the subkilometer size population at larger heliocentric distances.

1.8 Is Pluto a planet? Discussion of its status and redefinition of planet

In ancient times, the concept of planet was very clear: It applied to those celestial bodies moving among the stars. The word itself comes from the Greek and means "wandering star." In the Copernican heliocentric universe, planets were those bodies moving around the Sun known at that time, namely Mercury, Venus, Earth, Mars, Jupiter, and Saturn. The discovery of Uranus in 1781 and Neptune in 1846 did not need any reformulation of the classical definition of planet. The search for a planet between the orbits of Mars and Jupiter, as predicted by the Titius-Bode law, led to the discovery of Ceres in 1801. It was immediately hailed as the eighth planet, but its tiny size and the quick discovery in a row of three other bodies in the same interplanetary region, Pallas, Juno, and Vesta between 1801 and 1807, led astronomers to conclude that rather than a single planet, the region between Mars and Jupiter was populated by a swarm of small bodies that were called "asteroids" alluding to their star-like appearance as seen through the telescope.

The fact that Pluto, discovered in 1930, stood unchallenged as the sole known object of the TN region for more than six decades (with the satellite Charon since 1977) helped to cement the idea that it was the "ninth planet," despite successive estimates of its size showing that it was extremely small, even smaller than the Moon. One of the unexpected consequences of the discovery of a TN population was to challenge the status of Pluto as a planet. The first discovered TNOs were too small in comparison to Pluto to promote any revision, but the situation changed when massive TNOs were discovered, and the crisis finally erupted with the discovery of Eris rivaling in size with Pluto itself. So, what to do? Should Eris, Makemake, Haumea, and others be added to the list of planets? or Should Pluto and the other Pluto-sized TNOs be classified as a new class of objects, following the historical path of Ceres?

To be useful, any classification in science must be supported by some theoretical background. In the case of the definition of planet, once that the old descriptions as a wandering star or a body that circles the Sun proved to be insufficient, a new theoretical framework was required. From a cosmogonic perspective, the concept of planet should be associated to the end products of the accretion process in the protoplanetary disk. The process is not fully efficient: A fraction of the mass in a certain accretion zone will remain unaccreted and be either ejected or remain in dynamically stable niches as comets, asteroids, TNOs, or trojans. In the sketch of Fig. 1.6, we depict in a simple way the passage from an initial population of kilometer-size planetesimals to a planet through intermediate stages. Planetesimals accrete until the most massive members reach the stage of runaway accretion in which their gravitational

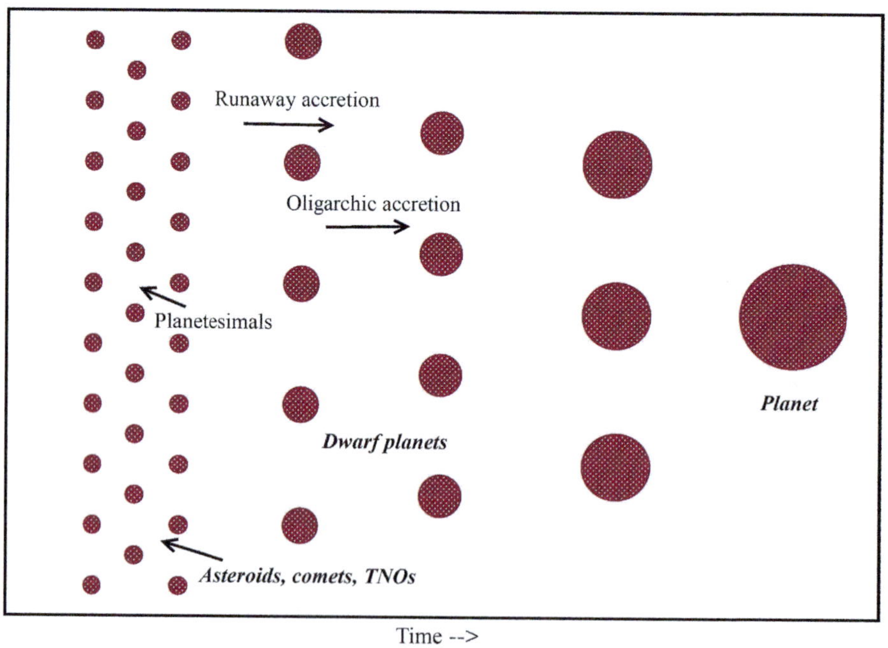

FIG. 1.6 Sketch showing the different stages of accretion in the solar nebula: From kilometer-size planetesimal to a planet as the final product. The process left a residual population that survived until now with the names indicated at the bottom of the figure.

fields become strong enough to significantly increase their collision cross-sections, so their masses start to detach from those of the rest of the planetesimals. When runaway bodies become sufficiently massive, it is the gravitational scattering of background planetesimals which dominates the random velocity evolution of the population in the accretion zone rather than the interactions among planetesimals. The accretion cross-section of massive runaway bodies will decrease with an increase of the random velocity, and then the accretion rate starts to slow down. The more massive the runaway bodies are, the greater the random velocities, and thus the smaller the cross-section, so the massive runaway bodies will end up with similar masses orderly distributed in the disk. The subsequent accretion mode receives the name of *oligarchic accretion* (Kokubo and Ida, 1998; Thommes et al., 2004). The oligarchs dominate the field of relative velocities. There is accretion via collision and also dispersal of the residual matter, leaving as a final product a set of planets in regular orbits, perhaps with the ejection of one or more massive protoplanets.

The crisis triggered by the discovery of Eris and other large TNOs generated an interesting debate that reached its climax during the IAU General Assembly in Prague in 2006. From a cosmogonical point of view, it made sense to reserve the category of *planet* to the final product of accretion in a given accretion zone, leaving the products of the initial and intermediate stages of the accretion process (planetesimals, runaway bodies, oligarchs) for other categories of objects. Contrary to what cosmogonical reasons would have advised, an ad hoc committee appointed by the IAU tried to push for a lax definition that kept Pluto in the planet category. In essence such definition said that a planet is a body in hydrostatic equilibrium circling the Sun, and that it is not a satellite. Talking in cosmogonic terms, it was like putting in the same bag planets, oligarchs, and even runaway bodies. Facing this proposal, a group of astronomers attending the Prague General Assembly presented an alternative definition that took into account cosmogonic principles as described earlier. In the alternative definition, a new cosmogonic-dynamic requirement was added to qualify as a planet: The body must have been able to clean its influence zone of planetesimals and competing embryo planets during its accretion. While the lax definition would open the door to a myriad of planet candidates in the solar system that fulfill the criterion of hydrostatic equilibrium (e.g., Tancredi and Favre, 2008), the more restricted definition would constrain the number of planets to eight, from Mercury to Neptune, corresponding to the end products of the accretion process in their respective influence zones.

At the closing assembly both definitions were subject to vote, gathering the second one a clear majority, thus leaving Pluto and the other planet candidates of the first option in an unspecified category. They were certainly larger than most comets, asteroids, and TNOs. Finally, bodies fulfilling the first criterion, but not the second one were categorized as "dwarf planets." They were the oligarchs and even runaway bodies that remained unaccreted in dynamically stable niches. In retrospect, it is amazing the hot debate generated by the issue of planet definition that spilled over the academy reaching the public at large, few times seen in the history of astronomy or of science in general, and in which the issue was finally settled through a show of hands more typical of a political debate than of a scientific one. This is particularly astonishing for laymen that consider science as objective, cool, describing the "reality," and far away from human passions and interests. For the general public, the episode has come to be known as the "demotion" of Pluto. Certainly, Pluto was the "victim" of the astronomers' success at discovering TNOs.

1.9 TNOs today: Current picture and new challenges

Our current view depicts TNOs as icy bodies characteristic of the outer solar system, a mixture of water ice, other volatiles, and minerals. Water ice has been detected on the surface of some TNOs in the near-infrared region (e.g., Brown et al., 1999). Other more volatile substances: CH_4, N_2, CO, and NH_3 have been detected in the Pluto-Charon system (Stern et al., 2018). In massive TNOs in eccentric orbits, CH_4 and N_2 may circulate between the atmosphere and the surface on which they are deposited as a thin frost layer giving a high albedo. The largest TNOs (diameters $\gtrsim$ 1200 km) show indeed clear signatures of volatile retention and they seem to be coated by a frost layer of methane, molecular nitrogen and in some cases carbon monoxide (Schaller and Brown, 2007). Their albedos are quite high going from about 9% for Quaoar to about 86% for Eris. Another process that might contribute to raise the albedo is impact gardening and cryovolcanism that leave exposed fresh unirradiated ice (Jewitt and Luu, 2004).

The icy surfaces of TNOs are subject to chemical alteration by cosmic-ray bombardment that can dissociate water molecules leading to a hydrogen depletion. Cosmic rays can also trigger the formation of "ultrared" material, constituted by carbon-rich, complex organic molecules like tholins that are synthetic macromolecular compounds produced by the irradiation of gaseous or solid mixtures of hydrocarbons and water. The ultrared material can be removed by collisions that leave exposed fresh ice, thus leading to changes in the surface color (Luu and Jewitt, 1996). It is then to be expected that TNOs will show different colors and this is what is actually observed. What is interesting and rather surprising is that colors are related to the dynamical class: cold classical TNOs (inclinations $i < 5$ degrees) show red surface colors, while hot classical TNOs ($i > 5$ degrees), high-i Plutinos and SDOs show a wide range of colors going from red to gray. Tegler et al. (2008) argue that objects formed at $\lesssim$40 AU are usually gray, while objects formed at greater distances tend to be red. They explain the color difference in terms of the sublimation of CH_4 or its release from clathrates closer to the Sun, leaving a colorless water ice crust and depriving the surface of a substance like the CH_4 able to form red organic compounds.

The bulk density of TNOs is a key parameter to understand their chemical composition and physical structure. A pure-ice body will have a density of 1 g cm^{-3} (or slightly higher due to the gravitational compression of the material). An ice/rock mixture would give densities 2–2.5 g cm^{-3}, as are the cases of Pluto and Eris. Bodies growing to sizes of several hundred kilometers will produce at an early stage enough internal heat from colliding planetesimals and radioactivity to melt water ice. This solid-liquid phase transition will reduce porosity to negligible values, thus increasing the bulk density of the body to values of 2–2.5 g cm^{-3} typical of porosity-free, rock/ice mixtures (Brown, 2013). In order to determine bulk densities, it is necessary to measure both size and mass. Dynamical masses can be determined in binary pairs, while sizes can be estimated from occultations of stars or from observations in the IR from space-based observatories like *Spitzer* and *Herschel*.

As regards some open problems of current active research we can mention:

1.9.1 In situ formation versus implantation

One key question to learn about formation, evolution, and transport of matter in the early solar system is to estimate what are the fractions of matter in the TN belt that are primordial and implanted from inner regions. Current theories of planet formation suggest that the Jovian planets underwent migration during their formation. In particular, Uranus and Neptune would have experienced a significant outward migration (Fernández and Ip, 1984; Tsiganis et al., 2005). In the course of their migration, these outer planets would have also pushed outward via resonant coupling swarms of planetesimals that ended up in the TN region, most of them in MMRs with Neptune (Malhotra, 1995). This push-out mechanism of planetesimals would have originated a high-inclination TNO population with moderate eccentricities, the source of the current hot population (Gomes, 2003). A slow migration of Neptune from ~26 to 27 AU to its current location would have produced a hot population with the observed inclination distribution (Nesvorný, 2015). From numerical simulations Nesvorný (2015) derived a capture efficiency in the classical disk of $\sim 2-4 \times 10^{-4}$ for each initial particle, which would require a massive transplanetary disk of planetesimals at <30 AU of about 20 $M_\oplus$.

The scattering of planetesimals by Neptune would have destroyed wide binary TNOs, thus explaining their scarcity in the hot population (Parker and Kavelaars, 2010). In a more extreme view, Levison and Morbidelli (2003) and Levison et al. (2008) argued that the cold classicals were also implanted from a region <30 AU. However, the existence of a large fraction of wide binary pairs in the cold disk argues against their implantation from an inner region which would have implied close interactions with Neptune. The cold classical belt could be the remnant of the primordial population that survived the high eccentric phase of Neptune's migration before settling in its current low-eccentricity orbit (Batygin et al., 2011). All the implantation models require a massive disk outside the early Neptune of ~20 $M_\oplus$ that abruptly ends at ~30 AU, condition that it is required in order to stop the outward migration of Neptune beyond its current distance. The question is: What caused the disk to end so abruptly? Could it have been possible that the mass in the cold disk was much larger at the beginning and suffered collisional attrition that left only a small fraction of its primordial content? Stern and Campins (1996) found that the primordial disk must have been two orders of magnitude more massive than the current one in order to form 100-km size bodies at distances ~45 AU. The question is how was it possible to lose most of the mass. Morbidelli and Valsecchi (1997) conjectured that Neptune scattered Earth-sized bodies that passed through the TN belt exciting higher eccentricities and inclinations of TNOs, thus favoring mutual collisions that ground down most of them to dust and meteoroids.

1.9.2 The outer edge of the belt

Is there a sharp cutoff in the number density of classical TNOs at $r = 48$ AU? The fact that the edge is very close to the 1:2 MMR with Neptune ($a = 47.7$ AU) might suggest that it is related to the push-out of bodies drove by Neptune's migration (Levison and Morbidelli, 2003; Levison et al., 2008). But this mechanism will not work if cold classical objects formed in situ. The problem will vanish if there is not such a sharp cutoff at about 48 AU. In this regard, the OSSOS, a wide-field imaging program carried out between 2013 and 2017, provided accurate data of more than 800 TNO discoveries showing that the cold disk should extend at least 1 AU beyond the 1:2 MMR boundary (Bannister et al., 2016, 2018), thus putting into question the reality of such a sharp cutoff. If this were the case, could it be possible to have an extended cold disk well beyond 48 AU populated by small TNOs ($\gtrsim$100 km)? The rationale for setting an upper size limit to distant TNOs is that the low densities of the protoplanetary disk at distances

>50 AU would have slowed down or impeded altogether the growth of planetesimals to larger sizes.

1.9.3 The detached population

We can divide this population into two groups: (1) That formed by planetesimals scattered by Neptune from <30 to >40 AU where they were trapped in high-amplitude MMRs, as discussed by, for example, Nesvorný (2015), then the Kozai mechanism and other secular resonances worked for decreasing their eccentricities, that is, raising their q's above 40 AU. This mechanism could have been efficient to raise perihelia up to ~70 AU (Gomes et al., 2005), and able to create a numerous population in MMRs of type 1: k with $q \simeq 50-70$ AU and $a \simeq 100-300$ AU (Saillenfest et al., 2017). (2) Those bodies with $q \gtrsim 70$ AU, like Sedna, for which the previous mechanism did not work. The source was the same: These bodies were scattered by the migrating Neptune from <30 AU but with the difference that they were not trapped in MMRs, so they could continue evolving with ever larger semimajor axes until they fell under the gravitational influence of Galactic tides and passing stars that raised their perihelia (Fernández et al., 2004). If the planetesimal disk encountered by the migrating Neptune had a mass of 20 $M_{\oplus}$, as estimated by Nesvorný (2015), we can conjecture that a significant mass worth of the order of one $M_{\oplus}$ could have been trapped in the detached disk/inner Oort cloud, including planet-size bodies. In third regard, the discovery of the detached TNO 2013 SY_{99} with a perihelion distance $q = 50.0$ AU and a semimajor axis $a \simeq 733$ AU led Bannister et al. (2017) to suggest the existence of a numerous detached population whose perihelia have been lifted to $a \sim 40-50$ AU and their semimajor axes to 1000–2000 AU by the interplay between the weak Galactic tides and planetary perturbations.

1.9.4 Planet nine

Is there any planet-size body lurking in the region between the TN belt and the Oort cloud? Such body would have escaped detection because it is too far and then too faint. Furthermore, it would be located in a region dynamically safe from planetary perturbations and external perturbers (passing stars and Galactic tides), so it could have been there since the early solar system when the transport of matter was substantial. The possible existence of planet-size bodies at distances ~10^3 AU is supported by sound theoretical arguments. For instance, the current architecture of the solar system can be better reproduced if a fifth Jovian planet is included in the models, which was ejected or placed in the Oort cloud (Nesvorný, 2011). The planet-sized companion could have been snatched from other stellar system or as a free-floating body while the Sun was in its birth star cluster. Furthermore, if, as discussed before, Neptune scattered ~20 $M_{\oplus}$ worth of material, one or more of the scattered bodies could have been of the size of the Earth or Mars. They could have been trapped in the inner Oort cloud by perturbations from cluster stars that lifted their perihelia beyond the planetary region.

Trujillo and Sheppard (2014) found a pronounced clustering of the argument of perihelion ω of distant TNOs ($q > 30$ AU and $a > 150$ AU) around $\omega \approx 0$ degrees. They attributed this clustering to the presence of a massive perturber at about 250 AU that force librations of ω around zero via the Kozai mechanism. In the same direction, Batygin and Brown (2016) argue that a highly eccentric ($e \sim 0.6$), massive planet ($\gtrsim 10\,M_\oplus$) with semimajor axis $a \sim 700$ AU can drive such apsidal alignment of interior TNOs and also cluster the longitude of the ascending node Ω, and thus the longitude of perihelion $\varpi = \omega + \Omega$. Yet from numerical integrations of all the observed distant TNOs, the Jovian planets, and a hypothetical Planet 9, Shankman et al. (2017a) failed to reproduce the simultaneous clustering of ω, Ω, and ϖ, and they also found that the distant TNOs should spend over half of the time in high inclination or retrograde orbits in conflict with the observed ones that show low-i orbits. Furthermore, from the OSSOS survey Shankman et al. (2017b) discovered eight new distant TNOs for which the ω clustering is severely weakened or absent, calling into question the previous claim by Trujillo and Sheppard (2014) of the ω clustering and hence of the necessity of Planet Nine. The problem continues to be open for discussion.

1.9.5 Binary TNOs

A significant fraction of TNOs are found in binary or multiple systems. What is very interesting is that among the 100-km size TNOs the fraction of binary pairs in the cold population is about 30% (with separations >0.06 arcsec and magnitude difference <2), while they are nearly absent in the hot population (Noll et al., 2008), suggesting a different origin for both populations. As regards the size ratio between primary and secondary, Brunini and Zanardi (2016) argue that binaries with diameter ratio $D_{SEC}/D_{PRIM} \gtrsim 0.3$ are not observed, though it may be due to selection effects rather than to an intrinsic property. They found that the ratio of initial wide binaries would have decreased by ~30% due to a combination of Kozai cycles, tidal friction, and collisions with TNOs. Therefore, it is possible that most TNOs formed as binaries. The discovery of a blue-colored population of binary pairs in the cold classical disk, led Fraser et al. (2017) to conclude that they were implanted from an interior region at ~38 AU and that in order to survive migration, the initial rate of binary pairs should have been nearly 100%.

From accretion theory, the assemblage of a swarm of planetesimals may lead to the formation of pairs rather than a single object due to the excess of angular momentum residing in the accumulated mass (Tancredi and Fernández, 1991; Nesvorný et al., 2010). Yet, a common origin cannot explain the different percentages of rock by mass between Pluto and Charon, and probably the other small satellites (McKinnon et al., 2017; Stern et al., 2018). Such scenario in which the primary is accompanied by a cohort of icier bodies is more compatible with a mega-collision between two partially differentiated precursors with a mass ratio ~0.1 (Canup, 2011). Mega-collisions in a more crowded primordial environment can thus account for most of the binary or multiple systems we observe today.

1.9.6 Bulk densities and rock/ice mass ratios

There is a correlation between bulk density and size suggesting that different porosities, rather than different rock/ice ratios are responsible for the variations in the bulk density. Thus small TNOs (less than a few hundred kilometers in diameter) are found to have densities of about $0.6\,\mathrm{g\,cm^{-3}}$, thus implying porosities of 50% or larger. The material gets more compacted as the size and self-gravity increase, so bodies with diameters ~1000 km or larger will have bulk densities of 2–2.5 g cm^{-3}, as are the cases of Quaoar, Haumea, Pluto, and Eris. The transition from a porous to a porosity-free rock/ice material is found to occur at a diameter ~700 km, where a bulk density of 0.8 g cm^{-3} is found for the binary TNO 2002 UX_{25} (Brown, 2013), while for the binary Varda-Ilmarë, Grundy et al. (2015) found a bulk density of 1.24 g cm^{-3} for the system and a diameter of ~720 km for the largest member Varda. Brown (2013) argues that asteroids can support high porosities (of about 50%) up to diameters ~350 km. He also states that rocky material is able to support porosities under compression better than icy material. Therefore, if we translate these results to the rocky/icy TNOs we should expect that the transition from porous to porosity-free bodies should occur at ~350 km at most. Why then do we observe TNOs with twice the size (~10 times more massive) with such low bulk densities, suggesting high porosities? One possibility is that some TNOs are icier than others, so we need much less porosity to have low bulk densities. Undoubtedly, we need to determine many more bulk densities, mainly in the range of 300–800 km to learn more about rock/ice compositions and porosities.

1.9.7 Cryovolcanism and interior oceans

One of the most exciting discoveries of the last few decades was the existence of glaciers and cryovolcanism able to erode and smooth cratered surfaces, as the examples of Europa, Enceladus, and now Pluto show (McKinnon et al., 2017). More interesting, these activity strongly suggests the existence of interior oceans where the energy to keep water in liquid state can come from tides or the radioactive decay of radioisotopes contained in the rocks. Given its importance for astrobiology, the question of the existence of interior oceans in Pluto and other large TNOs is very relevant and target of future explorations.

1.10 Concluding remarks

We have reviewed the history of the TN belt since the discovery of Pluto until present and how it shaped our current ideas of the architecture of the solar system. In less than a century, we have greatly expanded the frontiers of the known solar system. The *New Horizons* mission reached the Pluto-Charon system in July 2015 allowing for the first time to learn about the geophysics, tectonics, cryovolcanism, and interior structure of a TNO (for a summary of *New Horizons* results see, e.g., McKinnon et al., 2016, 2017; Stern et al., 2018). *New Horizons* is heading to TNO 2014 MU_{69} for a historical flyby on January 1, 2019. As our body of knowledge grew, new questions emerged on key aspects to understand the accretion processes in the outer solar system and the transport of matter outward drove by the migration of Uranus and Neptune. The existence of the TN belt, mainly of its hot population, may be the definitive proof that Neptune sustained a substantial migration (of at least several AU) during its final accretion stage. The discovery of massive TNOs obliged us to discuss the definition of planet, and what should be the requirements to reach such a status. Will we find TNOs much larger than Pluto? Earth's size or even larger? This is a very exciting question awaiting response. We are still at the beginning of the exploration of the TN region so we look forward to new and important discoveries that will very likely revolutionize our current view of how the solar system formed and evolved.

Acknowledgments

The authors gratefully acknowledge financial support from Project CSIC Grupo I+D 831725 Planetary Sciences.

References

Adams, E.R., Gulbis, A.A.S., Elliot, J.L., Benecchi, S.D., Buie, M.W., Trilling, D.E., Wasserman, L.H., 2014. De-biased populations of Kuiper belt objects from the deep ecliptic survey. Astron. J. 148, 55.

Bailey, M.E., Stagg, C.R., 1990. The origin of short-period comets. Icarus 86, 2–8.

Bannister, M.T., Kavelaars, J.J., Petit, J.M., et al., 2016. The Outer Solar System Origins Survey. I. Design and first-quarter discoveries. Astron. J. 152, 70. https://doi.org/10.3847/0004-6256/152/3.70.

Bannister, M.T., Shankman, C., Volk, K., et al., 2017. OSSOS. V. Diffusion in the orbit of a high-perihelion distant solar system object. Astron. J. 153, 262. https://doi.org/10.3847/1538-3881/aa6db5.

Bannister, M.T., Gladman, B.J., Kavelaars, J.J., et al., 2018. OSSOS. VII. 800+ Trans-Neptunian objects—the complete data release. Astrophys. J. Suppl. Ser. 236, 18. https://doi.org/10.3847/1538-4365/aab77a.

Batygin, K., Brown, M.E., 2016. Evidence for a distant giant planet in the solar system. Astron. J. 151, 22.

Batygin, K., Brown, M.E., Fraser, W.C., 2011. Retention of a primordial cold classical Kuiper belt in an instability-driven model of solar system formation. Astrophys. J. 738, 13.

Brown, M.E., 2013. The density of mid-sized Kuiper belt object 2002 UX25 and the formation of the dwarf planets. Astrophys. J. Lett. 778, L34.

Brown, R.H., Cruikshank, D.P., Pendleton, Y., 1999. Water ice on Kuiper belt object 1996 TO66. Astrophys. J. 519, L101–L104.

Brunini, A., Zanardi, M., 2016. Dynamical and collisional evolution of Kuiper belt binaries. Mon. Not. R. Astron. Soc. 455, 4487–4497.

Buie, M.W., Grundy, W.M., Young, E.F., Young, L.A., 2006. Orbit and photometry of Pluto's satellites. Astron. J. 132, 190–192.

Cameron, A.G.W., 1962. The formation of the Sun and planets. Icarus 1, 13–69.

Canup, R.M., 2011. On a giant impact origin of Charon, Nyx, and Hydra. Astron. J. 141, 35.

Christy, J.W., Harrington, R.S., 1978. The satellite of Pluto. Astron. J. 83, 1005–1008.

Cohen, C.J., Hubbard, E.C., 1965. Libration of the close approaches of Pluto to Neptune. Astron. J. 70, 10–13.

Cruikshank, D.P., Pilcher, C.B., Morrison, D., 1976. Pluto—evidence for methane frost. Science 194, 835–837.

Davies, J.K., McFarland, J., Bailey, M.E., Marsden, B.G., Ip, W.H., 2008. The early development of ideas concerning the Trans-Neptunian region. In: Boehnhardt, H., Cruikshank, D.P., Morbidelli, A. (Eds.), The Solar System Beyond Neptune. University of Arizona Press, Tucson, AZ, pp. 11–23.

Davis, D.R., Farinella, P., 1997. Collisional evolution of Edgeworth-Kuiper belt objects. Icarus 125, 50–60.

Delsemme, A.H., 1973. Origin of the short-period comets. Astron. Astrophys. 29, 377–381.

Duncan, M., Quinn, T., Tremaine, S., 1988. The origin of short-period comets. Astrophys. J. 328, L69–L73.

Duncan, M.J., Levison, H.F., Budd, S.M., 1995. The dynamical structure of the Kuiper belt. Astron. J. 110, 3073–3081.

Edgeworth, K.E., 1938. The Evolution of the Solar System. Trustees of the National Library of Ireland. Unpublished manuscript Ms. No. 16869/47 & /48, 226 pp.

Edgeworth, K.E., 1943. The evolution of our planetary system. J. Br. Astron. Soc. 53, 181–188.

Edgeworth, K.E., 1949. The origin and evolution of the solar system. Mon. Not. R. Astron. Soc. 109, 600–609.

Elliot, J.L., Kern, S.D., Clancy, K.B., et al., 2005. The deep ecliptic survey: a search for Kuiper belt objects and Centaurs. II. Dynamical classification, the Kuiper belt plane, and the Oore population. Astron. J. 129, 1117–1162.

Everhart, E., 1972. The origin of short-period comets. Astrophys. Lett. 10, 131–135.

Everhart, E., 1977. The evolution of comet orbits as perturbed by Uranus and Neptune. In: Delsemme, A.H. (Ed.), Comets, Asteroids, Meteorites: Interrelations, Evolution and Origins. University of Toledo, Toledo, OH, pp. 99–104.

Fernández, J.A., 1980. On the existence of a comet belt beyond Neptune. Mon. Not. R. Astron. Soc. 192, 481–491.

Fernández, J.A., 2005. Comets. Nature, Dynamics, Origin, and Their Cosmogonical Relevance. Springer, Dordrecht.

Fernández, J.A., Ip, W.H., 1984. Some dynamical aspects of the accretion of Uranus and Neptune: the exchange of orbital angular momentum with planetesimals. Icarus 58, 109–120.

Fernández, J.A., Morbidelli, A., 2006. The population of faint Jupiter family comets near the Earth. Icarus 185, 211–222.

Fernández, J.A., Gallardo, T., Brunini, A., 2004. The scattered disk population as a source of Oort cloud comets: evaluation of its current and past role in populating the Oort cloud. Icarus 172, 372–381.
Fernández, Y.R., Kelley, M.S., Lamy, P.L., et al., 2013. Thermal properties, sizes, and size distribution of Jupiter-family cometary nuclei. Icarus 226, 1138–1170.
Fernández, J.A., Gallardo, T., Young, J.D., 2016. The end states of long-period comets and the origin of Halley-type comets. Mon. Not. R. Astron. Soc. 461, 3075–3088.
Fraser, W.C., Brown, M.E., Morbidelli, A., Parker, A., Batygin, K., 2014. The absolute magnitude distribution of Kuiper belt objects. Astrophys. J. 782, 100.
Fraser, W.C., Bannister, M.T., Pike, R.E., et al., 2017. All planetesimals born near the Kuiper belt formed as binaries. Nat. Astron. 1, 88.
Fuentes, C.I., Holman, M.J., Trilling, D.E., Protopapas, P., 2010. Trans-Neptunian objects with Hubble Space Telescope ACS/WFC. Astrophys. J. 722, 1290–1302.
Gomes, R.S., 2003. The origin of the Kuiper belt high inclination population. Icarus 161, 404–418.
Gomes, R.S., Gallardo, T., Fernández, J.A., Brunini, A., 2005. On the origin of the high-perihelion scattered disk: the role of the Kozai mechanism and mean motion resonances. Celest. Mech. Dyn. Astron. 91, 109–129.
Grundy, W.M., Porter, S.B., Benecchi, S.D., Roe, H.G., Noll, K.S., Trujillo, C.A., Thirouin, A., Stansberry, J.A., Barker, E., Levison, H.F., 2015. The mutual orbit, mass, and density of the large Trans-Neptunian binary system Varda and Ilmarë. Icarus 257, 130–138.
Hamid, S.E., Marsden, B.G., Whipple, F.L., 1968. Influence of a comet belt beyond Neptune on the motions of periodic comets. Astron. J. 73, 727–729.
Holman, M.J., Wisdom, J., 1993. Dynamical stability in the outer Solar System and the delivery of short period comets. Astron. J. 105, 1987–1999.
Jewitt, D.C., Luu, J.X., 1993. Discovery of the candidate Kuiper belt object 1992 QB_1. Nature 362, 730–732.
Jewitt, D.C., Luu, J.X., 2004. Crystalline water ice on the Kuiper belt object (50000) Quaoar. Nature 432, 731–733.
Jewitt, D., Luu, J.X., Trujillo, C., 1998. Large Kuiper belt objects: the Mauna Kea 8K CCD survey. Astron. J. 115, 2125–2135.
Johansen, A., Oishi, J.S., MacLow, M.M., Klahr, H., Henning, T., Youdin, A., 2007. Rapid planetesimal formation in turbulent circum stellar discs. Nature 448, 1022–1025.
Joss, P.C., 1973. On the origin of short-period comets. Astron. Astrophys. 25, 271–273.
Kazimirchak-Polonskaya, E.I., 1967. Evolution of short-period comet orbits from 1660 to 2060, and the role of the outer planets. Sov. Astron. AJ 11, 349–365.
Kokubo, E., Ida, S., 1998. Oligarchic growth of protoplanets. Icarus 131, 171–178.
Kowal, C., 1989. A solar system survey. Icarus 77, 118–123.
Kuiper, G.B., 1950. The diameter of Pluto. Publ. Astron. Soc. Pacif. 62, 133–137.
Kuiper, G.B., 1951. On the origin of the solar system. In: Hynek, J.A. (Ed.), Astrophysics. McGraw-Hill, New York, NY, pp. 357–424.
Lacerda, P., Fornasier, S., Lellouch, E., et al., 2014. The albedo-color diversity of Trans-Neptunian objects. Astrophys. J. Lett. 793, L2.
Lamy, P.L., Toth, I., Fernández, Y.R., Weaver, H.A., 2004. The sizes, shapes, albedos, and colors of cometary nuclei. In: Festou, M.C., Keller, H.U., Weaver, H.A. (Eds.), Comets II. University of Arizona Press, Tucson, AZ, pp. 223–264.
Leonard, F.C., 1930. The new planet Pluto. Astron. Soc. Pacif. Leaflet 30, 121–124.
Levison, H.F., 1991. The long-term dynamical behavior of small bodies in the Kuiper belt. Astron. J. 102, 787–794.
Levison, H.F., Duncan, M.J., 1990. A search for proto-comets in the outer regions of the solar system. Astron. J. 100, 1669–1675.
Levison, H.F., Duncan, M.J., 1993. The gravitational sculpting of the Kuiper belt. Astrophys. J. 406, L35–L38.
Levison, H.F., Morbidelli, A., 2003. The formation of the Kuiper belt by the outward transport of bodies during Neptune's migration. Nature 426, 419–421.
Levison, H.F., Morbidelli, A., VanLaerhoven, C., Gomes, R., Tsiganis, K., 2008. Origin of the structure of the Kuiper belt during a dynamical instability in the orbits of Uranus and Neptune. Icarus 196, 258–273.
Luu, J.X., Jewitt, D.C., 1988. A two-part search for slow-moving objects. Astron. J. 95, 1256–1262.
Luu, J.X., Jewitt, D.C., 1996. Color diversity among the Centaurs and Kuiper belt objects. Astron. J. 112, 2310–2318.
Lyttleton, R.A., 1936. On the possible results of an encounter of Pluto with the Neptunian system. Mon. Not. R. Astron. Soc. 97, 108–115.

Malhotra, R., 1995. The origin of Pluto's orbit: implications for the solar system beyond Neptune. Astron. J. 110, 420–429.
McFarland, J., 1996. Kenneth Essex Edgeworth—Victorian polymath and founder of the Kuiper belt? Vistas Astron. 40, 343–354.
McKinnon, W.B., Moore, J.M., Spencer, J.R., et al., 2016. The Pluto-Charon system revealed: geophysics, activity, and origins. In: 47th LPSC, LPI Contribution No. 1903, p. 1995.
McKinnon, W.B., Stern, S.A., Weaver, H.A., et al., 2017. Origin of the Pluto-Charon system: constraints from the new horizons flyby. Icarus 287, 2–11.
Morbidelli, A., Valsecchi, G.B., 1997. Neptune scattered planetesimals could have sculpted the primordial Edgeworth-Kuiper belt. Icarus 128, 464–468.
Morbidelli, A., Thomas, F., Moons, M., 1995. The resonant structure of the Kuiper belt and the dynamics of the first five Trans-Neptunian objects. Icarus 118, 322–340.
Nesvorný, D., 2011. Young solar system's fifth giant planet? Astrophys. J. Lett. 742, L22.
Nesvorný, D., 2015. Evidence for slow migration of Neptune from the inclination distribution of Kuiper belt objects. Astron. J. 150, 73.
Nesvorný, D., Youdin, A.N., Richardson, D.C., 2010. Formation of Kuiper belt binaries by gravitational collapse. Astron. J. 140, 785–793.
Nicholson, S.B., Mayall, N.U., 1931. Positions, orbit and mass of Pluto. Astrophys. J. 73, 1–12.
Noll, K.S., Grundy, W.M., Stephens, D.C., Levison, H.F., Kern, S.D., 2008. Evidence for two populations of classical Trans-Neptunian objects: the strong inclination dependence of classical binaries. Icarus 194, 758–768.
Oort, J.H., 1950. The structure of the cloud of comets surrounding the solar system and a hypothesis concerning its origin. Bull. Astron. Inst. Neth. 11, 91–110.
Parker, A.H., Kavelaars, J.J., 2010. Destruction of binary minor planets during Neptune scattering. Astrophys. J. Lett. 722, L204–L208.
Saillenfest, M., Fouchard, M., Tommei, G., Valsecchi, G.B., 2017. Study and application of the resonant secular dynamics beyond Neptune. Celest. Mech. Dyn. Astron. 127, 477–504.
Schaller, E.L., Brown, M.E., 2007. Volatile loss and retention on Kuiper belt objects. Astrophys. J. 659, L61–L64.
Schlichting, H.E., Fuentes, C.I., Trilling, D.E., 2013. Initial planetesimal sizes and the size distribution of small Kuiper belt objects. Astron. J. 146, 36.
Schwamb, M.E., Brown, M.E., Fraser, W.C., 2014. The small numbers of large Kuiper belt objects. Astron. J. 147, 2.
Shankman, C., Kavelaars, J.J., Lawler, S.M., Gladman, B.J., Bannister, M.T., 2017a. Consequences of a distant massive planet on the large semimajor axis Trans-Neptunian objects. Astron. J. 153, 63.
Shankman, C., Kavelaars, J.J., Bannister, M.T., et al., 2017b. OSSOS VI. Striking biases in the detection of large semimajor axis Trans-Neptunian objects. Astron. J 154, 50. https://doi.org/10.3847/1538-3881/aa7aed.
Standish Jr., E.M., 1993. Planet X—no dynamical evidence in the optical observations. Astron. J. 105, 2000–2006.
Stansberry, J., Grundy, W., Brown, M., Cruikshank, D., Spencer, J., Trilling, D., Margot, J.L., 2008. Physical properties of Kuiper belt and centaur objects: constraints from the Spitzer space telescope. In: Boehnhardt, H., Cruikshank, D.P., Morbidelli, A. (Eds.), The Solar System Beyond Neptune. University of Arizona Press, Tucson, AZ, pp. 161–179.
Stern, S.A., Campins, H., 1996. Chiron and the Centaurs: escapees from the Kuiper belt. Nature 382, 507–510.
Stern, S.A., Grundy, W.M., McKinnon, Wm.B., Weaver, H.A., Young, L.A., 2018. The Pluto system after New Horizons. Ann. Rev. Astron. Astrophys. 56, 357–392.
Tancredi, G., Favre, S., 2008. Which are the dwarfs in the solar system? Icarus 195, 851–862.
Tancredi, G., Fernández, J.A., 1991. The angular momentum of the Pluto-Charon system: considerations about its origin. Icarus 93, 298–315.
Tancredi, G., Fernández, J.A., Rickman, H., Licandro, J., 2006. Nuclear magnitudes and the size distribution of Jupiter family comets. Icarus 182, 527–549.
Tegler, S.C., Bauer, J.M., Romanishin, W., Peixinho, N., 2008. Colors of Centaurs. In: Boehnhardt, H., Cruikshank, D.P., Morbidelli, A. (Eds.), The Solar System Beyond Neptune. University of Arizona Press, Tucson, AZ, pp. 105–114.
Thommes, E.W., Duncan, M.J., Levison, H.F., 2004. Oligarchic growth of giant planets. Icarus 161, 431–455.
Torbett, M.V., 1989. Chaotic motion in a comet disk beyond Neptune: the delivery of short-period comets. Astron. J. 98, 1477–1481.
Torbett, M.V., Smoluchowski, R., 1990. Chaotic motion in a primordial comet disk beyond Neptune and comet influx to the solar system. Nature 345, 49–50.

Trujillo, C.A., Sheppard, S.S., 2014. A Sedna-like body with a perihelion of 80 astronomical units. Nature 507, 471–474. https://doi.org/10.1038/nature13156.
Trujillo, C.A., Jewitt, D.C., Luu, J.X., 2001. Properties of the Trans-Neptunian belt: statistics from the Canada-France-Hawaii telescope survey. Astron. J. 122, 457–473.
Tsiganis, K., Gomes, R., Morbidelli, A., Levison, H.F., 2005. Origin of the orbital architecture of the giant planets of the solar system. Nature 435, 459–461.
Vaghi, S., 1973. The origin of Jupiter's family of comets. Astron. Astrophys. 24, 107–110.
Whipple, F.L., 1964. Evidence for a comet belt beyond Neptune. Proc. Natl. Acad. Sci. USA 51, 711–718.
Whipple, F.L., 1972. The origin of comets. In: Chebotarev, G.A., Kazimirchak-Polonskaya, E.I., Marsden, B.G. (Eds.), The Motion, Evolution of Orbits, and Origin of Comets. IAU Symp. 45. Reidel, Dordrecht, pp. 401–408.
Zhang, Z.W., Lehner, M.J., Wang, J.H., et al., 2013. The TAOS project: results from seven years of survey data. Astron. J. 146, 14.

Further reading

Stern, S.A., 1996. On the collisional environment accretion time scales, and architecture of the massive, primordial Kuiper belt. Astron. J. 112, 1203–1211.

PART I

Dynamics and evolution

CHAPTER

2

Kuiper belt: Formation and evolution

Alessandro Morbidelli[a], *David Nesvorný*[b]

[a]Laboratory Lagrange, UMR7293, University of Nice Sophia-Antipolis, CNRS, Nice, France
[b]Southwest Research Institute, Boulder, CO, United States

2.1 Introduction

The Kuiper belt provides a very rich set of clues to unveil how the solar system formed and evolved. The very existence of objects so far from the Sun is significant. The outer edge of the classical Kuiper belt, together with the extremely low mass of the cold population, suggests that the Kuiper belt represents a relic of planetesimal formation at the maximal distance where such formation process worked. The size distribution of the Kuiper belt objects (KBOs), their colors, density, and binarity provides important constraints on this process. Section 2.2 will detail on this.

The complex orbital structure of the Kuiper belt, with at least five distinct components partially overlapping (cold classicals, hot classicals, resonant objects, scattered disk, and fossilized scattered disk—see Gladman et al., 2008 and Chapter 1 for formal definitions) shows that a great deal of dynamical sculpting occurred that profoundly transformed the pristine disk of planetesimals. Reconstructing the origin of the Kuiper belt orbital structure provides key information on the complex dynamical evolution of the giant planet system, as described in Section 2.3. Understanding this evolution also provides predictions on the existence of other KBOs outside of the Kuiper belt, such as the Oort cloud objects, Trojans and irregular satellites of the giant planets, and even some asteroids, as reviewed in Section 2.4.

Finally, the orbital evolution is coupled with the collisional evolution. The existence of delicate features, like wide binaries, provides constraints on how intense the collisional evolution was. This, in turn, has implication for the size below which KBOs are more likely to be collisional fragments rather than primordial planetesimals. This will be discussed in Section 2.5

Since the time of the first KBO book in 2008, our knowledge of the Kuiper belt and our understanding of its formation and evolution have significantly evolved. The five dynamical classes had already been identified back in 2008, but their quantitative characterization has been improved since, allowing for more constraining tests of the models of the belt's primordial sculpting (Petit et al., 2011; Lawler et al., 2018).

The Trans-Neptunian Solar System.
https://doi.org/10.1016/B978-0-12-816490-7.00002-3

The new models of planetesimal formation based on the streaming instability (Youdin and Goodman, 2005; Johansen et al., 2007) had just started to emerge but their relevance for the Kuiper belt had not yet been understood. It was nevertheless noted that planetesimal formation by gravitational instability proposed by Goldreich and Ward (1973), dismissed in Weidenschilling (1995) and resurrected by Youdin and Shu (2002) with the concept of a critical solids/gas density ratio could change completely the picture set by the collisional coagulation scenario. Now a lot more is known on these instability processes and comparisons with the Kuiper belt properties can be done, as discussed in Section 2.2.

The size distribution of the Kuiper belt was poorly known below $D \sim 100$ km; the existence of a knee was still debated (Petit et al., 2008). Thus, it was difficult to assess whether the KBO population shows evidence for a collisional equilibrium, like the asteroid belt (Kenyon et al., 2008). Today deep surveys (Fraser et al., 2014) and crater counts on Pluto and Charon provide information about the size distribution of KBOs down to about a couple of kilometers in diameter (Robbins et al., 2017), so the possibility of a collisional equilibrium can be tested (Section 2.5).

The notion that a phase of orbital instability of the giant planets (a.k.a. the Nice model—Tsiganis et al., 2005; Gomes et al., 2005a; Morbidelli et al., 2007) should have occurred and played a major role in sculpting the Kuiper belt was already well established (Morbidelli et al., 2008), but the reproduction of the main structures of the Kuiper belt was still sketchy (Levison et al., 2008). Moreover, the possible evolutions of the giant planets during the instability were much less constrained than they are today (Nesvorný and Morbidelli, 2012).

With the model of a giant planet instability, it was already understood that some KBOs should survive today very far from their parent reservoir, as Trojans of Jupiter (Morbidelli et al., 2005) or irregular satellites of the giant planets (Nesvorný et al., 2007). These capture models are now much more precisely tested. Moreover, it has become clear that the Oort cloud and primitive D- or P-type asteroids are also related to the KBO population (Section 2.4).

This chapter will provide a review of our current understanding of the origin and evolution of the Kuiper belt, focusing on the advances achieved in the last decade.

2.2 Accretion of KBOs

In 2008, it was generally believed that KBOs (and large asteroids as well) formed by progressive collisional coagulation of primordial subkilometer planetesimals. The origin of these small planetesimals was unexplained. It is in fact known that dust coagulation is effective in building particles up to a few centimeters in size, but this process is then blocked by a number of effective barriers (the so-called bouncing, fragmentation, and drift barriers; see Weidenschilling, 1977; Brauer et al., 2008; Güttler et al., 2009; Birnstiel et al., 2016). The review by Kenyon et al. (2008) extensively described the statistical coagulation-fragmentation codes employed to simulate the growth of KBOs from subkilometer planetesimals to their final size distribution. Relatively little work has been done on collisional coagulation since (e.g., Schlichting and Sari, 2011; Kenyon and Bromley, 2012; Schlichting et al., 2013; Shannon et al., 2016).

Although the results somewhat differ from one coagulation model to the other, depending, for instance, on the assumptions on the strength of planetesimals and on the evolution of Neptune, the general trends are well summarized in Fig. 2.1, taken from Kenyon and Bromley (2004a). In essence, a handful of Pluto-size objects are formed, from which follows a steep single-slope size distribution down to small sizes unobserved with current telescopes. The currently observed KBO size distribution is different and sketched by the red lines in the figure. It is not a single power law: It has a "knee" at about $R_{\text{knee}} \sim 50$ km where it turns from shallow (for $R < R_{\text{knee}}$) to steep (for $R > R_{\text{knee}}$) and an "ankle" at $R_{\text{ankle}} \sim 300$–$400$ km where it turns to being very shallow (for $R > R_{\text{ankle}}$). Obviously, this last feature is only visible in the hot Kuiper belt/scattered disk, because the cold belt does not contain objects with $R > R_{\text{ankle}}$. The model by Schlichting et al. (2013) works better than the one presented in the figure in that it features a turnover in the size distribution around R_{knee}. To do so, however, it predicts huge waves in the size distribution at around ~10 km in size, which are ruled out by the crater size distribution on Pluto and Charon, as it will be shown later (see Section 2.5, which will also discuss the size distribution at kilometer scale).

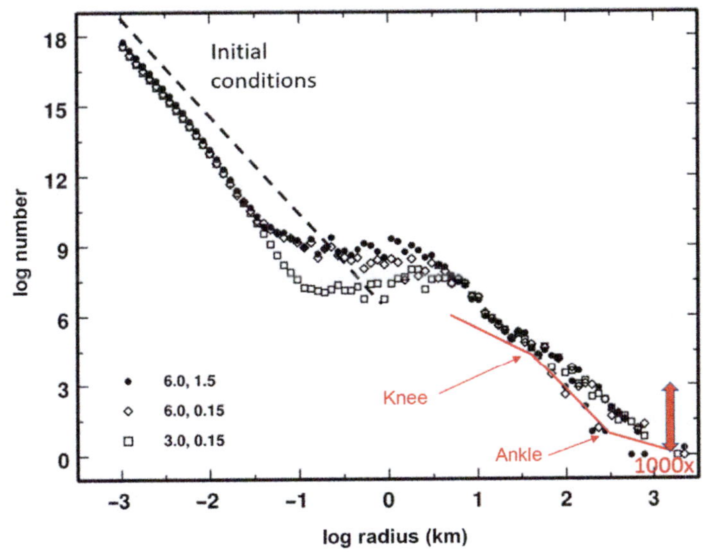

FIG. 2.1 The initial (*dashed line*) and final (*open and filled symbols*, from three simulations assuming different strength parameters) size distributions in the Kuiper belt in the collisional coagulation model of Kenyon and Bromley (2004a)—taken from their Fig. 9. The *red lines* sketch the currently observed size distribution. The *vertical arrow* indicates that the primordial distribution needs to be scaled up from the current one by a factor ~1000. Coagulation models, therefore, fail to produce the original size distribution of the Trans-Neptunian disk.

Nevertheless, the main issue with coagulation models is not related to the observed broken power law. By reproducing roughly the current number of objects, these models fail to reproduce the *original* number of objects in the Trans-Neptunian disk. As it will be discussed in Section 2.3, there is ample evidence that the original disk population was 100–1000 times more massive than the current scattered (Duncan and Levison, 1997; Brasser and Morbidelli, 2013; Nesvorný et al., 2017) and hot populations (Nesvorný, 2015b), respectively, and that the number of Pluto-size objects was 1000–4000 (Stern, 1991; Nesvorný and Vokrouhlický, 2016). In a nutshell, all collisional coagulation models fail to produce enough big objects by orders of magnitude.

Given the difficulty of collisional coagulation models in producing a large number of big objects, as well as the problem of the origin of the initial subkilometer planetesimals assumed in these models that we mentioned at the beginning of this section, a large body of work has been developed in the last 13 years on the possibility to produce large bodies directly from dust aggregates (dubbed "pebbles") via the so-called streaming instability.

Although originally discovered as a linear instability (Youdin and Goodman, 2005; Jacquet et al., 2011), this instability in the nonlinear regime generates even more powerful effects, which can be qualitatively explained as follows. The key factors are the speed difference between gas (in a slightly sub-Keplerian rotation) and solid particles and the back reaction of solids onto the gas. Thus, the differential speed causes gas drag onto the particles and the friction exerted from the particles back onto the gas accelerates the gas and diminishes its difference from the Keplerian speed. Consequently, if there is a small overdensity of particles, the local gas is in a less sub-Keplerian rotation than elsewhere; this, in turn, reduces the local headwind on the particles, which therefore drift more slowly toward the star. As a result, an isolated particle located farther away in the disc, feeling a stronger headwind and drifting faster toward the star, eventually joins this overdensity region. This enhances the local density of particles and reduces further its radial drift. It is easy to see that this process drives a positive feedback, that is, instability, where the local density of particles increases exponentially with time.

The particle clumps generated by the streaming instability, if dense enough, can become self-gravitating and contract to form planetesimals. Numerical simulations of the streaming instability process (Johansen et al., 2015; Simon et al., 2016) show that planetesimals of a variety of sizes can be produced, but those that carry most of the final total mass are those of ~100 km in size. Thus, these models suggest that planetesimals form (at least preferentially) big, in stark contrast with the collisional coagulation model in which planetesimals would grow progressively from subkilometer objects by pair-wise collisions. The fact that the size of 100 km is prominent (i.e., the "knee") in the observed size-frequency distributions of both asteroids and KBOs is a first important support for the streaming instability model.

An attempt to reproduce the size distribution of the Kuiper belt starting from the size distribution produced by the streaming instability model has been done in Lambrechts and Morbidelli (2016) and is illustrated in Fig. 2.2. The planetesimals produced in the streaming instability (marked "initial distribution" in the figure) have initially a phase of collisional coagulation, which brings the largest objects to acquire a size of 300–400 km in radius. These objects then start to efficiently accrete individual pebbles, here assumed to be 0.4 mm in size, as the latter migrate radially in the disk. The acceleration in growth due to this process of *pebble accretion* (Lambrechts and Johansen, 2012, 2014) produces the "ankle" and the terminal, shallow part of the observed size distribution. The total number of objects, about 1000 times larger than at the present time for any size, is also successfully reproduced. Thus, for the first time we have a satisfactory model for the primordial size distribution in the Trans-Neptunian disk, although it should be noticed that the values of the model parameters (pebble size and flux, disk turbulence and aspect ratio, and mass of the initial planetesimals produced by the streaming instability) have been chosen to achieve a best fit.

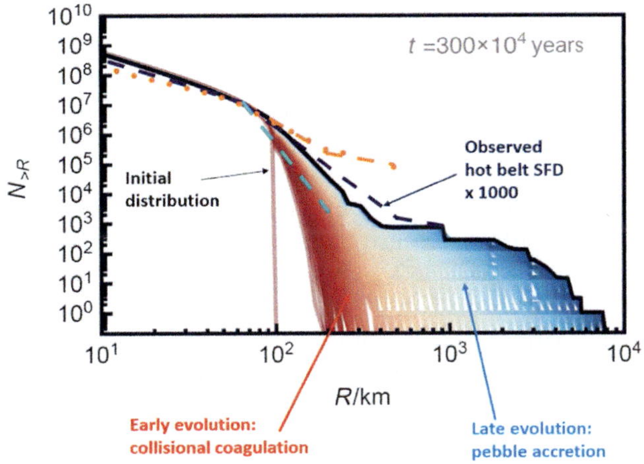

FIG. 2.2 The evolution of the size distribution in the Trans-Neptunian disk, centered at 25 AU, over 3 Myr, starting from the initial size distribution predicted in the streaming instability model of Johansen et al. (2015) and progressing via mutual collisions and pebble accretion. Here the initial mass of planetesimals was 25 Earth masses. Pebbles had a size of 0.4 mm. The disk's aspect ratio, that is the thickness of the disk of gas relative to the stellar distance, was 5% and its turbulence was described by the α-prescription of Shakura and Sunyaev (1973) with $\alpha = 10^{-5}$. *From Lambrechts, M., Morbidelli, A., 2016. Reconstructing the size distribution of the small body population in the Solar System. AAS/Division for Planetary Sciences Meeting Abstracts No. 48, ID.105.08.*

A strong support for the streaming instability model comes from Kuiper belt binaries. A large fraction of the cold population KBOs is binary. These binaries are typically made of objects of similar size and identical colors (see Noll et al., 2008 and Chapter 9 for reviews). Accounting for the fact that wide binaries can be dissociated by collisions (Petit and Mousis, 2004; Nesvorný et al., 2011), it is possible that nearly all initial cold population KBOs were binary (Fraser et al., 2017). The discovery that Ultima Thule—a member of the cold population that looked like a single object from Hubble Space Telescope (HST) observations (Benecchi et al., 2018)—is in reality a contact binary (Stern et al., 2019), strengthens this conclusions, suggesting that planetesimals are born twins, although with a range of possible separations. The fact that the fraction of binary KBOs is much lower in the hot population does not contradict this interpretation: The deficiency of hot KBO binaries can be understood because these bodies should have undergone close encounters with Neptune in the past, dissociating most of the binaries (Parker and Kavelaars, 2010, see Section 2.3.2.1). The same is true for Centaurs, which are objects currently undergoing planetary encounters: Only tight binaries can survive (e.g., the Typhon/Echidna binary; Noll et al., 2006). Comet-size objects, instead, even if they had formed as binaries, would have been dissociated or collapsed as bilobed objects by a combination of collisions and planetary encounters (Nesvorný et al., 2018b, see also Section 2.5). In addition, the two components of a binary comet may suffer differential torques produced by their possible unequal activity or be affected by the drag exerted by the gas in the coma formed during the activity phase. These effects may dissociate the binary or make the two components collapse into a contact binary object (Belton et al., 2018). In summary, only the cold population can be diagnostic of an initial high binary fraction of planetesimals.

Nesvorný et al. (2010) showed that the formation of a binary is the natural outcome of the gravitational collapse of the clump of pebbles formed in the streaming instability if the angular momentum of the clump is large. Their simulations show that the typical semimajor axes, eccentricities, and size ratios of the observed binaries are well reproduced by the model. The identity in colors between the two components (Benecchi et al., 2009) is a natural consequence of the fact that both components are made of the same material. This is a big strength of the model because such color identity cannot be explained in any capture or collisional scenario, given the observed intrinsic difference in colors between any random pair of KBOs (even within the cold population).

Additional evidence for the formation of equal-size KBO binaries by streaming instability is provided by the spatial orientation of binary orbits. Grundy et al. (2019) have recently tabulated KBO binaries with known orbits. For about 20 of these binaries in the cold classical Kuiper belt, the ambiguity between the true orbit and its mirror through the sky plane has been broken. They show a broad distribution of binary inclinations with $\simeq$80% of prograde orbits ($i_b < 90$ degrees) and $\simeq$20% of retrograde orbits ($i_b > 90$ degrees). To explain these observations, Nesvorný et al. (2019) analyzed new high-resolution simulations and determined the angular momentum vector of the gravitationally bound clumps produced by the streaming instability. Because the orientation of the angular momentum vector is approximately conserved during collapse (Nesvorný et al., 2010), the distribution obtained from these simulations can be compared with known binary inclinations. The comparison shows that the model and observed distributions are indistinguishable from each other (Nesvorný et al., 2019). This clinches an argument in favor of the planetesimal formation by the streaming instability and binary formation by gravitational collapse. For comparison, Goldreich et al. (2002) proposed that binaries formed by capture during the coagulation growth of KBOs. Their L^2s mechanism predicts retrograde orbits with $i_b \sim 180$ degrees and their L^3 mechanism predicts a 3:2 preference for retrograde orbits (Schlichting and Sari, 2008). These capture models can, therefore, be ruled out.

Davidsson et al. (2016) developed a hybrid model of KBO formation. In their model, large KBOs are formed by the streaming instability as described earlier. Small, comet-size objects, however, are formed after gas removal, by collisional coagulation of the remaining, not-yet accreted, pebbles. The motivation for invoking this bimodal accretion process is that the chemical analysis of comet 67P/CG by the mission ESA/Rosetta revealed that the internal temperature of the comet should never have exceeded 70 K. This suggests that comets formed late, otherwise they would have been heated significantly by radioactive decay of short-lived nuclei, like ^{26}Al (Prialnik et al., 2009). If comets had formed after gas removal, that is, at least several millions of years after time 0, this would not have been the case. The problem is that, after gas removal, the velocity dispersion of pebbles, stirred by the presence of large KBOs, grows relatively fast in time. In Davidsson et al. model, the average random velocity in the primordial disk (from 15 to 30 AU) is $\sim$40 m s^{-1} during the first 25 Myr. Pebbles are assumed to coagulate during mutual collisions, but at these speeds, this is far from obvious. Moreover, the model produces "only" 4×10^{10} comets with a diameter $D > 2$ km, which is somewhat anemic compared to expectations. In fact, dynamical models need about 2×10^{11} objects with $D > 2$ km originally in the Trans-Neptunian disk to obtain a scattered disk and an Oort cloud sufficiently populated to explain the currently observed fluxes of short-period and long-period comets (Nesvorný et al., 2017; Brasser and Morbidelli, 2013; Volk and Malhotra, 2008; Duncan and Levison, 1997).

An implicit assumption in Davidsson et al. is that the streaming instability would form objects early (so vulnerable to radioactive decay). Hence, the requirement to form comets at a late time prompts to search for an alternative model of comet formation. But this assumption may not be correct. From the observational point of view, there are indications that also large bodies did not heat up substantially. Comet Hale-Bopp, one of the largest comets ever observed, with a diameter perhaps of 70 km, is also one of the richest in super-volatiles such as CO (Biver et al., 1996). Had this comet suffered substantial heating, these super-volatiles would have been rapidly lost. Moreover, many KBOs with $D < 600$ km have bulk densities smaller than $1\,g\,cm^{-3}$ (Brown, 2013), suggesting an undifferentiated structure which, again, points to a low amount of internal heating, hence late formation. If big KBOs formed by the streaming instability, this suggests that the instability occurred relatively late in the lifetime of the disk, not early.

Indeed, from the modeling side, it has been shown that triggering the streaming instability requires a ratio between the surface densities of solids and gas which is 2–5 times the solar ratio (Johansen et al., 2009; Yang et al., 2017), the exact value depending on parameters, such as the particle size. This enhanced ratio could be achieved locally by piling up radially drifting particles in some specific ring of the disk (Drążkowska et al., 2016; Ida and Guillot, 2016; Schoonenberg and Ormel, 2017; Drążkowska and Alibert, 2017). Or it could be achieved globally in the disk at a much later time when a substantial fraction of the gas was removed by photoevaporation (Carrera et al., 2017). This predicts a dichotomy of formation ages of planetesimals. In the inner solar system, it is confirmed by the measurement of formation ages of meteorites, which show a clear dichotomy between iron meteorites, daughter products of planetesimals accreted in the first $\sim 10^5$ years (Kleine et al., 2005) and chondrites, which formed 2–3 Myr later (Villeneuve et al., 2009). In the outer solar system, there is evidence that planetesimals formed very early in Jupiter's region, given that this planet reached ~ 20 Earth masses within 1 Myr (Kruijer et al., 2014), which argues for the pileup of solids at/near the snowline (Ida and Guillot, 2016; Schoonenberg and Ormel, 2017; Drążkowska and Alibert, 2017) while, as said earlier, KBOs formed late. If the streaming instability in the Trans-Neptunian disk occurred late, then comets may have formed by the streaming instability as well, either directly (Blum et al., 2017) or indirectly, namely as collisional fragments of larger objects formed by the streaming instability (see Section 2.5). In either case, we think that there is no need of invoking a specific mechanism for the formation of comets as in Davidsson et al. (2016).

We cannot fail noticing, however, that there is a contradiction between the idea of a late formation of KBOs triggered by gas removal and the results of Fig. 2.2. If it takes really 3 Myr of collisional evolution and pebble accretion to evolve the original size distribution to the one currently observed for large objects, as the figure suggests, it is unlikely that the streaming instability that formed the first KBOs was triggered by photoevaporation of the gas. In fact, pebble accretion requires the presence of gas, so the gas had to remain for a long time after the formation of the first KBOs. This important issue deserves further and more detailed investigation. Perhaps the streaming instability directly produced bodies as large as 300–400 km in radius, so that a long phase of collisional evolution was not needed before that pebble accretion could operate.

An additional indication of how the streaming instability might have operated comes from the structure of the Kuiper belt. As it will be discussed in Section 2.3, the cold component of the Kuiper belt formed locally and never had a large mass (probably never exceeding a tenth of an

Earth mass). That population ended within 47 AU with no planetesimal disk beyond this limit. Instead, inside of 30–35 AU, the disk of planetesimals carried a considerable mass, perhaps 20–30 Earth masses. This is required to form a populated enough hot Kuiper belt and scattered disk, but also to drive the giant planets to their current orbits at the end of their dynamical instability. All this suggests that the streaming instability never occurred beyond ~45 AU, probably because the threshold solid/gas density could never be achieved; between ~30 and 45 AU the streaming instability could occur, but only sporadically, forming a population of KBOs (the cold population) with a small total mass; within ~30 AU the streaming instability was efficient, converting pebbles into tens of Earth masses of planetesimals; and near the snowline the streaming instability was not just efficient, but could operate early on, despite the large gas density, forming the precursors of the giant planet cores. This view can be qualitatively understood considering that pebbles drift radially due to gas drag (the outer disk is depleted in solid material and the inner disk in enriched, with a pileup formed near the snowline; Ida and Guillot, 2016; Schoonenberg and Ormel, 2017; Drążkowska and Alibert, 2017) and that the streaming instability is triggered when the solid/gas ratio exceeds a threshold value.

2.3 Dynamical sculpting of the Kuiper belt

It was already clear in 2008 that the structure of the Kuiper belt is intimately related to the dynamical instability that the giant planets should have experienced after the removal of gas from the protoplanetary disk (Morbidelli et al., 2008). But, as we commented in Section 2.1, at the time the planets' instability had not yet been fully characterized and, consequently, the reproduction of the Kuiper belt structure was, at best, qualitative (Levison et al., 2008). Thus, below we start reviewing the major advances achieved since 2008 in our understanding of the planets' evolution and of their instability, then we focus on how the main structural features of the Kuiper belt can be understood quantitatively in that framework.

2.3.1 Giant planet instability

The idea of a giant planet instability has its roots in the work by Thommes et al. (2002) and was then formalized in the so-called Nice model (Tsiganis et al., 2005; Gomes et al., 2005a). While Thommes et al. aimed simply at demonstrating that Uranus and Neptune could have formed closer to the Sun, the motivation of the Nice model was more profound and still valid today. The planets should have formed on coplanar and circular orbits. Their interaction with the gas and with the planetesimals should have contributed in damping their eccentricities and inclinations if any. So, why are the planets on moderately eccentric and inclined orbits today? Only two mechanisms can excite the eccentricity and the inclination of the planets: resonance crossing (which, typically, affects only the eccentricities) and mutual close encounters. Hence the need for a phase of instability.

In the original version of the Nice model published in 2005, the initial orbits of the planets were chosen ad hoc, conveniently close to the 1:2 mean-motion resonance between Jupiter and Saturn to trigger an instability during their planetesimal-driven migration

(Tsiganis et al., 2005). But contemporary studies on planet migration in gas-dominated protoplanetary disks soon showed that these initial conditions were not realistic. Instead, because Jupiter should have migrated toward the Sun more slowly than Saturn and these two planets together should have migrated more slowly than Uranus and Neptune, the giant planets should have been captured in mutual mean-motion resonances during the gas-disk phase and be in a multiresonant chain when the gas was removed (Morbidelli et al., 2007). Thus, the original Nice model was abandoned in favor of a new model with initial planetary orbits taken from the output of hydrodynamical simulations of planet migration (Morbidelli et al., 2007; Levison et al., 2011). Nevertheless, the original multiresonant configuration is not univocally determined (e.g., it depends on the density of the disk of gas and its aspect ratio).

A detailed exploration of the outcome of the giant planet instability depending on the initial configuration was done in Nesvorný and Morbidelli (2012) (NM12 hereafter).

NM12 defined four criteria to measure the overall success of their simulations. First of all, the final planetary system must have four giant planets (criterion A) with orbits that resemble the present ones (criterion B), that is, with final semimajor axes within 20% to their present values, and the final eccentricities and inclinations not larger than 0.11 and 2 degrees, respectively. NM12 also required that the proper eccentricity of Jupiter e_{55}, which is the most difficult to excite due to the planet's largest mass, is at least half of its current value, that is, is larger than 0.022 (criterion C). Moreover, the period ratio between Saturn and Jupiter was required to evolve from <2.1 to >2.3 in $\ll 1$ Myr (criterion D), because previous work had shown that this is necessary to preserve the inner asteroid belt (Morbidelli et al., 2010) and the terrestrial planet system (Brasser et al., 2009) if it already existed at the time of the giant planet instability. NM12 relaxed the condition that only the four known giant planets existed at the end of the gas-disk phase. They also tested configurations with five and six planets, with the sum of the masses of the rogue planets comparable to that of Neptune. We will refer below to Uranus, Neptune, and the rogue planet(s) generically as "ice giants." All planets were initially in multiresonant configurations for the reasons explained earlier, but different period ratios were tested.

Fig. 2.3 shows an example of a successful simulation starting from a five-planet configuration that satisfied all four criteria. The instability happened in this case about 6 Myr after the start of the simulation. Before the instability, the three ice giants slowly migrated by scattering planetesimals. The instability was triggered when the inner ice giant crossed an orbital resonance with Saturn and its eccentricity was pumped up. Following that the ice giant had encounters with all other planets and was ejected from the solar system by Jupiter. Jupiter was scattered inward and Saturn outward during the encounters, with period ratio P_6/P_5 moving from ~1.7 to ~2.4 in less than 10^5 years (right panel of Fig. 2.3). The orbits of Uranus and Neptune became excited as well, with Neptune reaching an eccentricity of 0.15 just after the instability. The orbital eccentricities were subsequently damped by interactions with the planetesimal disk (dynamical friction). Uranus and Neptune, propelled by the planetesimal-driven migration, reached their current orbits some 100 Myr after the instability. The final eccentricities of Jupiter and Saturn were 0.031 (with a proper value $e_{55} = 0.030$) and 0.058, respectively. For comparison, the mean eccentricities of the real planets today are 0.046 and 0.054.

NM12 found (see also Nesvorný, 2011; Batygin et al., 2012) that the dynamical evolution is typically violent and leads to the ejection of at least one ice giant from the solar system. Planet ejection could be avoided if the mass of the planetesimal disk exceeds 50 Earth masses, but a massive disk would lead to excessive dynamical damping and would drive the planets too

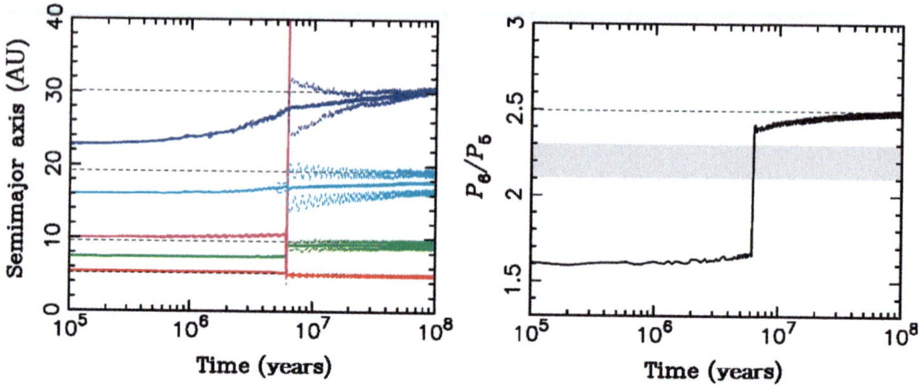

FIG. 2.3 Orbital histories of the giant planets from NM12. Five planets were started in the (3:2, 4:3, 2:1, 3:2) resonant chain, and $M_{\text{disk}} = 20$ Earth masses. *Left panel*: The semimajor axes (*solid lines*) and perihelion and aphelion distances (*dashed lines*) of each planet's orbit. The *horizontal dashed lines* show the semimajor axes of planets in the present solar system. The final orbits obtained in the model, including e_{55}, are a good match to those in the present solar system. *Right panel*: The period ratio P_6/P_5. The *dashed line* shows $P_6/P_5 = 2.49$ corresponding to the present orbits of Jupiter and Saturn. The *shaded area* approximately denotes the zone where secular resonances with the terrestrial planets and asteroids occur. These resonances are not activated, because the period ratio "jumps" over the *shaded area* as Jupiter and Saturn scatter off of the ejected ice giant.

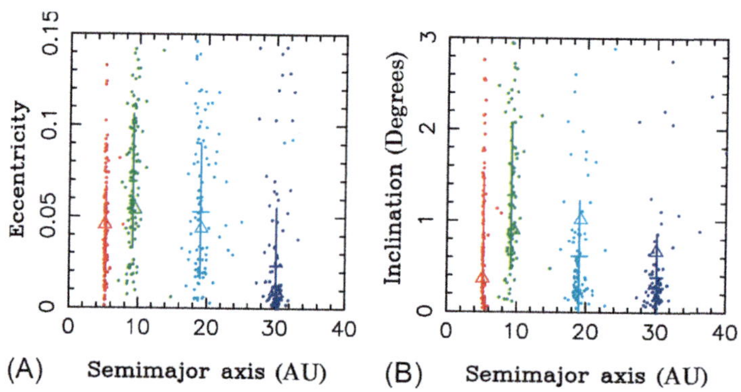

FIG. 2.4 Final planetary orbits obtained in 500 simulations with five outer planets started in the (3:2, 3:2, 2:1, 3:2) resonant chain and $M_{\text{disk}} = 20\ M_\oplus$ (panel A: eccentricity vs. semimajor axis; panel B: inclination vs. semimajor axis). The mean orbital elements were obtained by averaging the osculating orbital elements over the last 10 Myr of the simulation. Only the systems ending with four outer planets are plotted here (*dots*). The *bars* show the mean and standard deviation of the model distribution of orbital elements. The mean orbits of real planets are shown by *triangles*. Colors *red*, *green*, *turquoise*, and *blue* correspond to Jupiter, Saturn, Uranus, and Neptune, respectively.

far from each other. Thus, the dynamical simulations starting with a resonant system of four giant planets have a very low success rate. In fact, NM12 did not find any case that would satisfy all four criteria in nearly 3000 simulations of the four planet case. Consequently, either the solar system followed an unusual evolution path (<1/3000 probability to satisfy criteria A–D), or some constraints are misunderstood, or there were originally more than four planets in the outer solar system.

Better results were obtained in NM12 when the solar system was assumed to have five giant planets initially and one ice giant was ejected into interstellar space by Jupiter (Fig. 2.3). The best results were obtained when the ejected planet was placed into the external 3:2 or 4:3 resonance with Saturn and the mass of the planetesimal disk was 15–20 Earth masses. The range of possible outcomes is rather broad in this case (Fig. 2.4), indicating that the present solar system is neither a typical nor expected result for a given initial state, and occurs, in best cases, with a ~5% probability (as defined by the NM12 success criteria). Notice that if it is assumed that each of the four NM12 criteria is satisfied in 50% of cases, and the success statistics are uncorrelated, the expected overall success rate is $0.5^4 = 0.063$, that is, 6.3%.

2.3.2 Reproducing the main structures of the Kuiper belt

Because in all versions of the Nice model, the massive portion of the planetesimal disk had to have an edge around 30 AU, Levison et al. (2008) assumed that no planetesimals existed originally beyond this limit. The current Kuiper belt was empty. They showed that during the dynamical instability, a fraction of the disk's planetesimal could be implanted into the Kuiper belt, and they proposed to identify the hot population with the bodies implanted from the inner part of the disk, more violently scattered by the giant planets, and the cold population with the bodies implanted from the outer part, more gently scattered. However, the predicted distribution of orbital inclinations of the hot population was found to be narrower than the one inferred from observations while the orbital excitation of the cold population was too high. The distinction between hot and cold populations was not sharp enough also in terms of initial location in the disk, and therefore it could not explain the observed differences in spectral/color properties. Moreover, Parker and Kavelaars (2010) demonstrated that most KBO binaries would not have survived the encounters with Neptune required to transport the cold population to its current location. Finally, the resonant populations were overpopulated in Levison et al. (2008). Clearly, the Levison et al.'s paper was a pioneer in attempting a global reconstruction of the Kuiper belt in the framework of the giant planet instability, but it needed to be improved. Armed with a better understanding of the evolutions of the giant planets that are compatible with the planets' current orbits (see Section 2.3.1), the origin of the complex structure of the Kuiper belt has been revised completely, as we review later for each of its five components.

2.3.2.1 *The hot population*

The successful simulations of the Nice model like that presented in Fig. 2.3 show that Neptune can move away from the other planets by planetesimal-driven migration by several AUs before the instability happens. Nesvorný (2015b) realized that this preinstability migration phase, which did not exist in the model considered by Levison et al. (2008), could be key for the appropriate excitation of the disk prior to its implantation in the hot Kuiper belt. Thus, in a series of simulations, he started Neptune at different positions between 20 and 30 AU and migrated it into the disk on an e-folding timescale τ between 1 and 100 Myr, to test the dependence of the results on the migration range and timescale. He found that reproducing the inclination distribution of the hot Kuiper belt requires that Neptune's migration was slow (with an e-folding timescale $\tau \geq 10$ Myr) and long range (initial $a_{\mathrm{Neptune}} < 25$ AU). The models

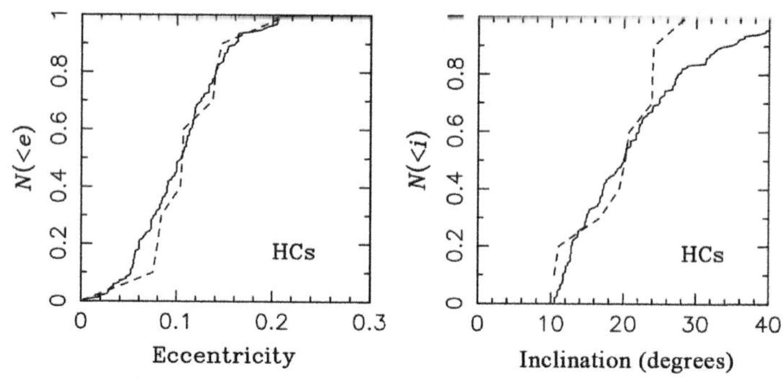

FIG. 2.5 The cumulative distribution of eccentricities (*left*) and inclinations (*right*) for the hot Kuiper belt. The *dashed lines* show the actual CFEPS detections. The *solid lines* show the distributions of model bodies, detected by the CFEPS simulator. The comparison is done only for $i > 10$ degrees to avoid any potential contamination of the detection statistics from the cold population. *From Nesvorný, D., 2015. Evidence for slow migration of Neptune from the inclination distribution of Kuiper belt objects. Astron. J. 150, 73.*

with $\tau < 10$ Myr do not satisfy the inclination constraint, because there is not enough time for dynamical processes to raise inclinations.

Fig. 2.5 shows the eccentricity and inclination distribution of the population implanted at $i > 10$ degrees in a simulation with Neptune starting at 24 AU and migrating with $\tau = 30$ Myr. For a quantitative comparison with the observations, it is important to have an accurate estimate of the observational biases. For this reason, only the objects detected by the CFEPS survey has been considered because of the availability of an accurate survey simulator (Petit et al., 2011). By simulating the detection of synthetic objects generated according to the model, the observational bias is applied to the model distribution. The resulting distribution can then be directly compared with the observed one. The agreement between model and observations is remarkable, as shown in Fig. 2.5. Unfortunately, it is not possible to extend the comparison to the objects detected by other surveys because of the lack of quantitative knowledge of the corresponding biases.

Only a small fraction ($\sim 10^{-3}$) of the disk planetesimals became implanted into the Kuiper belt in this kind of simulations. The implantation process is basically the one already described by Gomes (2003, 2011) and Gomes et al. (2005b). Namely, the objects are first scattered by Neptune, which pushes them outward in semimajor axis and increases their inclination. Then, they are captured in some mean-motion resonance. The complex secular dynamics in the resonance decrease the objects' eccentricity and raise their perihelion distance, so that Neptune is no longer able to have a close encounter with them. The continued migration of Neptune (and hence of its resonances) can finally drop these objects off resonance so that they remain permanently trapped in the Kuiper belt region, preserving the inclination they previously acquired.

The slow migration of Neptune required to reproduce the hot population represents an important clue about the original mass of the outer disk. For example, in the NM12 planetary migration/instability model where the Trans-Neptunian disk extends from ~23 to 30 AU, $\tau \geq 10$ Myr implies $M_{disk} \sim 15$–20 Earth masses.

2.3.2.2 *The cold population*

Parker and Kavelaars (2010) showed that the transport of the members of the cold Kuiper belt from smaller heliocentric distances via close encounters with Neptune, as proposed by Levison et al. (2008), would have dissociated most of the KBO binaries. A resonant transport during Neptune's migration, not involving close encounters with the planet had also been proposed (Levison and Morbidelli, 2003). But the point remains that the sharp differences in color (Trujillo and Brown, 2002) and size distributions (Fraser et al., 2010, 2014) between cold and hot KBOs suggest a different origin for the two populations. If the cold population had been transported from within 30 AU as advocated in Levison et al. (2008), given that this portion of the disk is also the source of the hot population (see Section 2.3.2.1), these differences would be difficult to understand. Thus, over the years, the idea that the cold population formed in situ became more and more popular in the Kuiper belt science community.

Even if the cold belt formed in situ, the preservation of small eccentricities and inclinations is far from trivial, if Neptune experienced a phase of large eccentricity during the giant planet instability (Batygin et al., 2011; Ribeiro de Sousa et al., 2018). But Neptune's eccentricity and inclination are never large in the NM12 models (i.e., $e_{\mathrm{Neptune}} < 0.15$ and $i_{\mathrm{Neptune}} < 2$ degrees) so that there is no excessive orbital excitation in the >40 AU region, where the cold population is expected to have formed.

On the other hand, simulations that do not overexcite the cold population also lead to a limited dynamical depletion, of only a factor of ~2 (Nesvorný, 2015a). This implies that the total mass of sizeable objects in the 42–47 AU region, currently of 3×10^{-4} Earth masses (Fraser et al., 2014), was always small. It has been proposed that the cold population originally may have been significantly more massive and have lost most of its mass by collisional grinding (Pan and Sari, 2005), but the presence of loosely bound binaries places a strong constraint on how much mass can be removed by collisions (Petit and Mousis, 2004; Nesvorný et al., 2011). As we said in Section 2.2, the existence of a population of large bodies with a total small mass may suggest that the streaming instability in this frontier region of the solar system could occur only sporadically.

A particularly puzzling feature of the cold population is the so-called kernel, a concentration of orbits with $a = 44$ AU, $e \sim 0.05$, and $i < 5$ degrees (Petit et al., 2011). This feature can either be interpreted as a sharp edge beyond which the number density of the cold population drops (i.e., the original outer edge of the cold population) or as a genuine concentration of bodies. If it is the latter, the kernel can be explained if Neptune's migration was interrupted by a discontinuous change of Neptune's semimajor axis when Neptune reached ~27.7 AU (Petit et al., 2011; Nesvorný, 2015a), presumably due to the giant planet instability (see Fig. 2.3). Before the discontinuity happened, planetesimals located at ~40 AU were swept into the Neptune's 1:2 resonance and were carried with the migrating resonance outward (Levison and Morbidelli, 2003). The 1:2 resonance was at ~44 AU when Neptune reached ~27.7 AU. If Neptune's semimajor axis changed by a fraction of an AU at this point, perhaps because it had a close encounter with another planet, the 1:2 population would have been released at ~44 AU and would remain there to this day. The orbital distribution of bodies produced in this model provides a good match to the orbital properties of the kernel (Nesvorný, 2015a).

Fig. 2.6 shows the resulting cold Kuiper belt in one of the simulations of Nesvorný (2015a), featuring a Neptune's jump at 27.8 AU. The comparison with the observations of the CFEPS

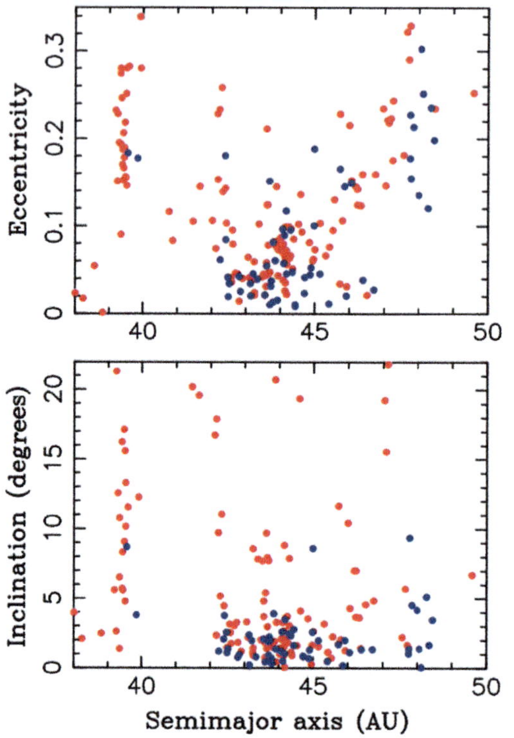

FIG. 2.6 The distribution of eccentricities (*top*) and inclinations (*bottom*) versus semimajor axis for the KBOs detected by CFEPS (*red*) and those obtained in the model and detected by the CFEPS simulator. The apparent mismatch in inclination is due to the fact that here only the local population is modeled, so there is no hot population generated in the simulations. *From Nesvorný, D., 2015. Jumping Neptune can explain the Kuiper belt Kernel. Astron. J. 150, 68.*

survey is done properly, by passing the model population through the CFEPS simulator. There are observed objects at inclinations larger than 5 degrees, that are absent in the model. This is an additional indication that the hot Kuiper belt does not have a local origin, but rather comes from below 30 AU, as discussed in Section 2.3.2.1. Taking this consideration into account and restricting the comparison to $i < 5$ degrees, the match between the model and the observations is excellent.

2.3.2.3 *The resonant populations*

The existence of resonant populations particularly the group of Plutinos in the 2:3 resonance with Neptune triggered the studies of the planetesimal-driven migration of this planet (Malhotra, 1993, 1995). These studies have been important from a perspective of history of science, because they have been the first to break the paradigm of in situ formation of the planets (see also Fernandez and Ip, 1984).

These works proposed that the resonant objects had been captured from a dynamically cold planetesimal disk swept by the resonances during Neptune's migration. But it was soon realized that the results of the model were at odds with observations. The resonant populations have an inclination distribution similar to that of the hot population, while they were expected to share one of the cold population (Hahn and Malhotra, 2005). Their colors (Sheppard, 2012) and binary fraction (Noll et al., 2008) are also closer to the hot population properties.

First Gomes (2003), then Levison et al. (2008), showed that objects can also be captured into resonance from the population of planetesimals scattered by Neptune, and hence explained why resonant objects are more closely related to the hot population and the scattered disk than the cold population. Nesvorný (2015b), while reproducing the inclination distribution of the hot population, also reproduced very accurately the inclination distribution of the Plutinos. These simulations showed that not only the 2:3 resonance is filled with objects, but also other resonances, such as the 1:2, 2:5, 1:3, etc.

Models with smooth migration of Neptune invariably predict excessively large resonant populations (e.g., Hahn and Malhotra, 2005; Nesvorný, 2015b), while observations show that the nonresonant orbits are in fact the most common (e.g., the classical belt population is ~2–4 times larger than Plutinos in the 2:3 resonance; Gladman et al., 2012). This problem can be resolved if Neptune's migration was grainy, as expected if Neptune scattered massive planetesimals. The grainy migration acts to destabilize resonant bodies with large libration amplitudes, a fraction of which ends up on stable nonresonant orbits. Thus, the nonresonant/resonant ratio obtained with the grainy migration is up to ~10 times higher for the range of parameters investigated in Nesvorný and Vokrouhlický (2016), than in a model with smooth migration. The best fit to observations was obtained when it was assumed that the outer planetesimal disk below 30 AU contained 1000–4000 Pluto-mass objects. The combined mass of Pluto-class objects in the original disk was thus ~2–8 Earth masses, which represents 10%–50% of the estimated planetesimal disk mass.

2.3.2.4 The scattered disk

The scattered disk is the Kuiper belt structure that was understood first (Duncan and Levison, 1997). Hence, models of the scattered disk origin have not significantly evolved over the last decade. The scattered disk is also the least sensitive structure to the actual details of the past planets' dynamics. In fact, the scattered disk is what remains today of the population of objects scattered by Neptune since the beginning of the solar system and which have not found a stable parking orbit. Duncan and Levison (1997) showed that this surviving population accounts for approximately 1% of the original population scattered by the planet. This fraction may appear surprisingly high (e.g., it is 10 times larger than the fraction captured on stable orbits in the hot population; see Section 2.3.2.1). This is due to the fact that many of the scattered disk objects are temporarily trapped in mean-motion resonances with Neptune and therefore can live on nonencountering orbits for long time, before going back to scattered dynamics.

More modern simulations, conducted in the framework of the Nice model (Brasser and Morbidelli, 2013; Nesvorný et al., 2017), confirmed that the current scattered disk comprises 0.5%–1% of the original planetesimals in the Trans-Neptunian disk.

Therefore, this population is quite massive, compared to the other components of the Kuiper belt, as confirmed by observations (Trujillo et al., 2000). Possibly, only the fossilized scattered disk contains more objects. But, because the scattered disk is intrinsically unstable and its population keeps decaying today, it dominates the flux of objects toward the giant planet region (Centaurs) and the inner solar system (Jupiter-family comets [JFCs]). Therefore, it can be considered as the reservoir of these populations.

Several works have reproduced the orbital distribution of JFCs from the flux of objects from the scattered disk, provided that an appropriate physical lifetime is assumed for the comets (Duncan and Levison, 1997; Volk and Malhotra, 2008; Brasser and Morbidelli, 2013; Nesvorný et al., 2017). These works concluded that the scattered disk should contain $\sim 2 \times 10^9$ objects with $D > 2$–3 km to explain the JFC population currently observed. Given the surviving fraction in the scattered disk, this implies that the original Trans-Neptunian disk contained about 2×10^{11} of these objects (Nesvorný et al., 2017).

2.3.2.5 *The fossilized scattered disk*

The orbital structure of the Kuiper belt shows a population of objects that follows approximately the (a, e) orbital distribution of the scattered disk, but have a larger perihelion distances, so that they are out of reach from Neptune's scattering action. Their orbital distribution suggests that these bodies have been transported to a large semimajor axis by encounters with Neptune. This means that they have been part of the scattered disk in the past but they are not part of it today. Hence the name fossilized scattered disk. This name, however, is not unique. Gladman et al. (2008) used "Detached Population" instead.

The observations suggest the existence of two subcomponents of the fossilized scattered disk. One is present at all semimajor axes. It is basically the extension of the hot population to $a > 50$ AU. The perihelion distances are typically within $\sim$45 AU. Its origin is most likely related to (i) capture in mean-motion resonance of scattered disk objects, (ii) increase in perihelion distance due to the secular dynamics in the resonance, and (iii) drop-off resonance while Neptune (and its resonances) was still migrating (Gomes et al., 2005b; Gomes, 2011). We have already described this process for the origin of the hot population. In fact, in this scenario, there is basically no difference between the hot population and this component of the fossilized scattered disk, apart from the arbitrarily chosen semimajor axis divide at 50 AU. Lawler et al. (2019) concluded that, to have enough objects dropping off resonance, Neptune's migration should have been grainy and slow, in agreement to what was already concluded for the hot and resonant populations (see Sections 2.3.2.1 and 2.3.2.3).

There seems to be a gap in perihelion distance between the two subgroups of the fossilized scattered disk (Trujillo and Sheppard, 2014). In fact, the other subpopulation of the fossilized scattered disk has perihelion distances $q > 60$ AU. Three objects are currently known in this group: Sedna, 2012VP$_{113}$ and 2015TG$_{387}$. They are sometimes called "Sednoids." Given that these bodies are very big and can only be observed during a tiny arc of their orbital motion, they may represent the visible tip of the iceberg. In other words, the Sednoids may be the most massive component of the Trans-Neptunian population.

The fact that no bodies with perihelion distances comparable to those of the Sednoids have been observed on orbits with semimajor axis smaller than 200 AU, despite the shorter orbital periods would make their discovery more probable, suggests that the population with very large q exists only on wide orbits. This, in turn, suggests a specific trapping mechanism, operating only at large distances.

This consideration immediately calls to mind the effect of passing stars and, more generally, the external potential due to the Galactic environment of the Sun. In fact, the current Galactic environment can raise the perihelion distance of objects, detaching them from the giant planets, if their orbits have $a \gtrsim 10{,}000$ AU (Dones et al., 2004). This is related to the formation of the Oort cloud, as we will see in Section 2.4.3. However, if the Sun formed in a stellar cluster, as most stars do, the external perturbations would have been stronger in the early times and they could have raised the perihelion distance of bodies already at a few hundred AUs in the semimajor axis. Thus, since the discovery of Sedna, it has been proposed that Sednoids are objects ejected from the giant planet region when the Sun was still in its natal cluster (Morbidelli and Levison, 2004; Kenyon and Bromley, 2004b; Brasser et al., 2006, 2008, 2012). In alternative, it has been conjectured that these objects have been captured from the disk surrounding another star which had a close encounter with the Sun (Morbidelli and Levison, 2004; Kenyon and Bromley, 2004b; Jílková et al., 2015). More recently, the planet IX hypothesis (Trujillo and Sheppard, 2014; Batygin and Brown, 2016) has opened the possibility that Sednoids have been trapped onto their current orbits by the perturbations exerted by this planet. However, the origin of the planet in first place would still be due to the same mechanisms: either ejection from the giant planet region and perturbations from the cluster or capture from another star.

We end stressing that, if the formation of the Sednoids is due to the effects of a stellar cluster, the origin of these objects has to predate the giant planet instability. In fact, it is not possible to form the Sednoids and the Oort cloud at the same time. In a cluster such as the one required to capture the Sednoids, the Oort cloud would be unbound (Brasser et al., 2006, 2008). Because the Oort cloud exists, and the giant planet instability is the last event that scattered planetesimals around, it is necessary that the Sun was already in a Galactic environment similar to the current one when the giant planets became unstable. Thus, presumably the ejection of the Sednoids occurred as the giant planets were still forming. Because gas was also present at that time and gas drag prevents the ejection of comet-size bodies, this leads to the prediction that the Sednoid population is made only of big objects (Brasser et al., 2007).

2.4 Relationships with other populations of small bodies

The dispersal of the original Trans-Neptunian planetesimal disk during the giant planet instability allowed some objects to be captured in regions of the solar system quite far from the Kuiper belt. These objects, strictly parented to those of the hot Kuiper belt, resonant populations, and scattered disk comprise today the Trojan populations of Jupiter and Neptune, the irregular satellites populations of all four giant planets and the Oort cloud, but can also be found in the main asteroid belt. We focus on each of these subpopulations as follows.

2.4.1 Trojan populations

Jupiter and Neptune have Trojan bodies, that is, small bodies leading or trailing the planet's position on its orbit. Saturn and Uranus do not have any Trojans but this is not surprising because their coorbital regions are unstable on long timescales (Nesvorný and Dones, 2002). For a long time, only the Trojans of Jupiter have been known; those of Neptune have started to be discovered only recently (Sheppard and Trujillo, 2010). It was supposed that Trojans are a local population of planetesimals that got captured in coorbital motion with the planet when the latter grew in mass (see Marzari et al., 2002). But this scenario cannot explain why the observed orbital inclinations of Jupiter's Trojans can reach very large values (now also true for Neptune's Trojans).

Morbidelli et al. (2005) have been the first to show that the Trojans of Jupiter could have been captured during the instability of the giant planets. Their model explained the large orbital inclination of the Trojans but was developed in the framework of the original version of the Nice model, then abandoned in 2007. As it was relying on a specific feature of the first Nice model, namely Saturn's crossing of the 1:2 resonance with Jupiter, it could not be applied to the new version of the model (Morbidelli et al., 2007; Nesvorný and Morbidelli, 2012) and it could also not be applied to explain the origin of Neptune's Trojans.

The capture of Trojans of both giant planets has been explained in the framework on the new Nice model by Nesvorný et al. (2013) and Gomes and Nesvorný (2016), respectively. In essence, in both cases, when the considered planet had its last significant jump in semimajor axis (due to a close encounter with another planet) the planetesimals that happened to have semimajor axes similar to the postencounter planet's semimajor axis got captured in the new coorbital zone.

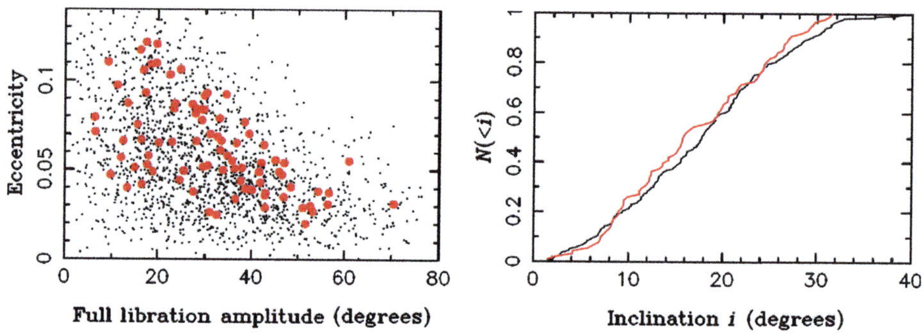

FIG. 2.7 *Left*: The distribution of eccentricity versus libration amplitude of observed Trojans (*black dots*) and planetesimals captured in the simulation (*red*). *Right*: Comparison between the observed (*black*) and simulated (*red*) cumulative inclination distributions. *From Nesvorný, D., Vokrouhlický, D., Morbidelli, A., 2013. Capture of Trojans by jumping Jupiter. Astrophys. J. 768, 45.*

In principle, the planet's orbital jump can produce Trojans from any source reservoir that populated the planet's vicinity at the time of the jump. But, at the time of the giant planet instability, the overwhelmingly dominant source was without doubts the Trans-Neptunian disk, undergoing dynamical dispersal. Using this source, the model reproduces in a quantitative manner the observed distribution of Jupiter's Trojans in terms of libration amplitude, eccentricity and inclination (Fig. 2.7). The model can also potentially explain the observed asymmetry in the number of objects larger than a given size between the leading and trailing Jupiter's Trojan populations. In fact, the capture process is not symmetric: One of the two populations can be partially depleted, depending on the specific trajectory of the rogue planet during the close encounter with Jupiter that led to Trojans' capture. Simulations show asymmetries up to a factor of 50%, sometimes in favor of one population, sometimes the other. The observed asymmetry is of ~30% in favor of the leading population.

Because the dispersed Trans-Neptunian disk was also the source of the hot population (see Section 2.3.2.1), this model leads to the strong prediction that Trojans and the hot population should have indistinguishable size and color distributions (Morbidelli et al., 2009).

For the size distribution, this prediction was confirmed by the dedicated study of Fraser et al. (2014) for Jupiter's Trojans. The observed match can be hardly considered a coincidence, because the slope of the common distribution is very different from those of other populations, such as the asteroids or the cold Kuiper belt.

For colors, however, it is observed that Jupiter's Trojans cover only part of the color distribution of the hot population. The reddest colors of the hot KBOs (those with $B-R \gtrsim 1.5$) are missing among the Jovian Trojans. Wong and Brown (2016) explained this difference by assuming that the Trojans were resurfaced (e.g., by sublimation of near-surface volatiles) as they moved from the Trans-Neptunian disk (where surface temperatures are ~50 K) to Jupiter's orbit (~125 K). But a recent work (Jewitt, 2018) shows that also Neptune's Trojans are missing the reddest objects (a first red object has been found by (Lin et al., 2019)). If this is confirmed when more Neptune's Trojans are discovered and characterized, it will invalidate the Wong and Brown's explanation because the temperature of Neptune's Trojans is about the same as that of the KBOs.

In this case, the explanation has to be dynamical. While the hot Kuiper belt population samples the entire disk dispersed by Neptune during its radial migration, the Trojans of Jupiter and Saturn just sample the dispersed population at the time of the last orbital jump suffered by these planets. If the original Trans-Neptunian disk was characterized by a radial color gradient, with the grayer colors in the inner part and the redder ones in the outer part (Hainaut et al., 2012; Wong and Brown, 2016), the absence of very red colors among the Trojans implies that the last orbital jumps of the planets occurred before that the very red component of the Trans-Neptunian disk had started to be dispersed. Numerical quantitative models are needed to test whether this is compatible with a jump of Neptune's orbit when the planet was at ~28 AU, as discussed in Section 2.3.2.2. One has to check whether the portion of the Trans-Neptunian disk dispersed *after* this event could be a sufficient source of the red component of the hot population and whether these objects would have the correct inclination dispersion in the end. If the result were negative, the implication would be that Neptune was significantly closer to the Sun than 28 AU when its last orbital jump happened. In this case, the kernel of the cold Kuiper belt could not be related to Neptune's jump and most likely reveals the original outer edge of the cold population.

2.4.2 Irregular satellites

The irregular satellites of the giant planets are bodies that orbit the planet with inclined and eccentric trajectories, in sharp contrast with the coplanar and quasicircular orbits of the regular satellite systems.

The irregular satellite populations of the four giant planets are fairly similar to each other if the orbital semimajor axes are rescaled relative to the Hill radius of the planet (Jewitt and Sheppard, 2005). This led these authors to reject capture scenarios based on planet's growth or gas drag in the primordial planetary envelopes because in this case the vastly different accretional histories of gas giants (Jupiter and Saturn) and ice giants (Uranus and Neptune) would imply very different irregular satellite populations. Instead, Jewitt and Sheppard favored unspecified mechanisms of conservative dynamics because the capability of a planet to deflect an object scales with its Hill radius.

This surprising intuition for a team of observers was confirmed by Nesvorný et al. (2007), again in the framework of the Nice model. They showed that irregular satellites could be captured from the background planetesimal population generated by the dispersal of the Trans-Neptunian disk during planetary encounters. More specifically, capture happened when the trajectory of a background planetesimal approached a pair of planets in a mutual close encounter. In this three-body interaction, the planetesimal may end up on a bound orbit around one of the planets, where it remains permanently trapped once the planets move away from each other.

Modeling this mechanism in detail, Nesvorný et al. (2014) found that the orbital distribution of bodies captured during planetary encounters provides a good match to the observed distribution of the irregular satellites around all four giant planets.

Given the capture probabilities of irregular satellites in the numerical simulations and the expected size distribution in the original Trans-Neptunian disk (deduced from the size distribution of the hot Kuiper belt and of the Trojans, multiplied by the inverse of their capture probabilities), the model predicts within a factor of 2 the size of the largest irregular satellite around each giant planet (Nesvorný et al., 2007). The overall size distribution of irregular satellites, however, is much shallower than that of the hot Kuiper belt population. This discrepancy was explained by Bottke et al. (2010) by showing that the irregular satellite populations are strongly collisionally evolved. Due to the extremely high relative velocities between prograde and retrograde satellites, collisions lead to the super-catastrophic disruption of the targets. The fragments are therefore very small so that the collisional evolution leads to an equilibrium distribution with a slope of ~ -2 only for $D \lesssim 1$ km. Above this threshold, satellites are only destroyed, never regenerated, so that the resulting size distribution is extremely shallow (exponent of the cumulative size distribution larger than -1), as observed.

Because the captures of irregular satellites and of Trojans are coeval because they both occurring during the planetary close encounters, this model predicts that the color distributions of these two populations of bodies should be indistinguishable. This prediction has been validated by Graykowski and Jewitt (2018), who compared the irregular satellite colors with the Jovian Trojan color distribution, finding no evidence for a significant difference.

2.4.3 The Oort cloud

The main mechanism of formation of the Oort cloud was understood long ago (Shoemaker and Wolfe, 1984; Duncan et al., 1987) and was reviewed in a very detailed manner in Dones et al. (2004). In essence, planetesimals scattered by the giant planets onto orbits with semimajor axis $\gtrsim$10,000 AU can have their perihelion distance increased enough to avoid a new close encounter with the planetary system during the subsequent revolutions. The perihelion-increase process is due to the tidal force exerted by the distribution of mass in the galaxy. Passages of nearby stars randomize the orbital distribution of the objects, giving to the Oort cloud its characteristic spherical structure. From time to time, the action of the Galactic tide and of passing stars can decrease the perihelion distance of an object to <5 AU in less than an orbital period: In this case, a new long-period comet is born (Oort, 1950).

Brasser et al. (2007) pointed out that gas drag prevents comet-size objects to be ejected onto orbits with semimajor axes typical of the Oort cloud. This is the case even if the comets are scattered off the disk's plane (where most of the gas is supposed to be) and if the disk itself is fairly small (with an outer edge of 50–100 AU). The implication of this result is very strong. Unlike what was previously supposed (Duncan et al., 1987; Dones et al., 2004) comets could not be sent to the Oort cloud during the formation phase of the giant planets, because the latter necessarily happened when there was still a significant amount of gas in the disk. Therefore, comets must have been sent to the Oort cloud only after gas removal. But, after gas removal, planetesimals were mostly beyond the giant planets and therefore they could only be scattered during the giant planet instability and the migration of Neptune, that is, during the same phase that led to the formation of the hot Kuiper belt and of the scattered disk and to the capture of Trojans and irregular satellites.

However, it is far from obvious that the dispersal of the Trans-Neptunian disk during giant planet instability and migration could produce at the same time a scattered disk and an Oort cloud that are sufficiently populated to sustain the observed fluxes of Jupiter-family and long-period comets. This has eventually been verified with numerical simulations in Brasser and Morbidelli (2013) and Nesvorný et al. (2017). However, this result can be achieved only assuming that Oort cloud comets are intrinsically brighter (i.e., more active per unit surface) than JFCs, as independently advocated by Sosa and Fernández (2011) by looking at the strength of nongravitational accelerations. Using the Sosa and Fernández (2011) conversion from comet brightness to nuclear size, the population estimates for the scattered disk and the Oort cloud are well reproduced in the Nice-model simulations, suggesting that both structures formed in the same event from the same parent population.

This conclusion leads to a strong prediction: Although comets can be quite diverse from each other (because they sample a disk originally extended from ~15 to ~30 AU), statistically the populations of JFCs (coming from the scattered disk) and long-period comets (coming from the Oort cloud) should be the same, given their common ultimate parent reservoir (the Trans-Neptunian disk). No systematic differences indeed appear in the $^{13}C/^{12}C$ and $^{15}N/^{14}N$ ratios between these two classes of comets (Jehin et al., 2008).

However, for quite some time this prediction seemed to be at odds with the observed systematic difference in D/H ratio between long-period comets and JFCs. But the ESA/Rosetta mission found that the JFC 67P-CG has a D/H ratio larger than any long-period comet measured to date (Altwegg et al., 2015), suggesting that the previously claimed systematic difference was just apparent and due to small number statistics. Another apparent systematic difference between the two populations of comets is that carbon-depleted comets are only observed among JFCs (A'Hearn et al., 1995). This could be due to the facts that the observed carbon-bearing species are very volatile and that JFCs have suffered many more revolutions through the inner solar system than long-period comets, so it is not necessarily diagnostic of an original compositional difference between the two types of comets.

2.4.4 Primitive asteroids

Levison et al. (2009) pointed out that during the giant planet instability a fraction of the planetesimals coming from the Trans-Neptunian disk should have been implanted in the main asteroid belt and in the 3:2 resonance with Jupiter (Hilda population). By analogy with the spectral properties of Jupiter's Trojans, classified as P- and D-type, Levison et al. proposed that all P- and D-type asteroids in the main belt and Hilda populations are dormant comets parented with the hot Kuiper belt.

The implantation process has been revisited by Vokrouhlický et al. (2016) in the framework of the most successful realizations of the Nice model of Nesvorný and Morbidelli (2012). They found a mean probability of $\sim 5 \times 10^{-6}$ for each Trans-Neptunian disk body to be implanted into the asteroid belt. From the expected size distribution of the original Trans-Neptunian disk, this capture probability leads to a number of large P-/D-type bodies in the belt ($D > 150$ km) consistent with observations. However, it predicts a significant excess over the estimated population of smaller P-/D-types because the size distribution of the Trans-Neptunian disk was steeper than that of the main asteroid belt. This problem can be solved invoking the massive collisional destruction of small P-/D-type bodies, which is possible if they are more fragile than indigenous asteroids (Levison et al., 2009).

Unlike Levison et al. (2009) and Vokrouhlický et al. (2016) observed implantation of Trans-Neptunian objects also in the inner asteroid belt ($a < 2.5$ AU), where indeed some D-type asteroids are observed (DeMeo and Carry, 2014).

2.5 Collisional evolution

Given the dynamical history described in Section 2.3, it is clear that the KBOs suffered two stages of collisional evolution. The first stage was in the Trans-Neptunian disk, very massive but with moderate dynamical excitation, and the second phase was during the dispersal of the disk and the implantation of some of its members in the various structures of the Kuiper belt. Given that the scattered disk is still decaying in population, there is no sharp transition between this second phase and today's environment, so both have to be treated together.

Let us start our discussion from this second stage, because it is unavoidable that it happened and the associated dynamical evolution is now well characterized (see Section 2.3). The first stage is instead more uncertain, because the duration of the massive disk is poorly known.

The collisional evolution during the dispersal of the Trans-Neptunian disk till today has been modeled in Jutzi et al. (2017), from the dynamical simulations of Brasser and Morbidelli (2013). They showed that the effects of collisions are fairly moderate. The bodies with a catastrophic disruption probability of ~ 1 are those with $D \sim 4$ km. This means that bodies larger than 10 km (namely all visible KBOs) have been relatively unaffected. It is unlikely that the size distribution of the Kuiper belt populations has significantly evolved above this size limit. For comet-size bodies (few kilometers), though, the conclusion is the opposite. Most of the original comet-size planetesimals (if they existed—see Section 2.2 on the possible preferential formation of "big" objects in the streaming instability process) should have been destroyed. The survivors should have suffered multiple reshaping collisions (Jutzi et al., 2017).

This suggests that the comets that we see today, or at least their morphological structures (e.g., the bilobed shapes), are not primordial. Bilobed comets should be either the product of subcatastrophic collisions (Jutzi and Benz, 2017) or fragments from catastrophic collisions (Schwartz et al., 2018).

The collisional evolution of KBOs during the massive disk phase was addressed in Morbidelli and Rickman (2015). They found, not surprisingly, that the collisional activity is very intense in this phase, given the large amount of bodies in the disk (about 1000 times the current population in the hot Kuiper belt; see Section 2.3.2.1). The orbital eccentricities and inclinations were smaller than those characterizing the current Kuiper belt, but there was nevertheless a significant excitation due to the distant perturbations from Neptune and internal stirring from the Pluto-mass objects (whose number is estimated to have been ~1000; see Section 2.3.2.3). Given that the orbital periods were shorter than those of the current Kuiper belt, Morbidelli and Rickman estimated collisional velocities in the range 0.25–1 $\mathrm{km\,s^{-1}}$ (for comparison, the current collisional velocity for bodies with $a \sim 42$ AU, $e \sim 0.1$, and $i = 10$ degrees is $\sim 1\,\mathrm{km\,s^{-1}}$). Thus, collisions were clearly erosive and disruptive if the projectiles were large enough.

The key question is then how long the disk remained in this state before getting dispersed by the giant planet instability and the migration of Neptune. When the original Nice model was proposed, it was suggested that the giant planet instability occurred ~4 Gyr ago (namely ~550 Myr after gas removal from the disk), so to explain the origin of the so-called late heavy bombardment (LHB) of the Moon (Gomes et al., 2005a; Bottke et al., 2012). In fact, at the time the most common interpretation of the LHB was that it corresponded to a sudden increase in the flux of impactors of the Earth-Moon system (Tera et al., 1974; Ryder, 2002). Thus, there was the need for a dynamical mechanism destabilizing reservoirs of small bodies after hundreds of Myr of relative stability: A late giant planet instability seemed perfect for that. But the evidence for an impact spike got weaker with the assessment of older impact events (Norman and Nemchin, 2014) and a reconsideration of the effects of impact age resetting (Boehnke and Harrison, 2016). Finally, it was shown that the LHB can be explained as the tail of a bombardment ongoing since the time of terrestrial planet formation and declining over time if one relaxes the constraint on the total amount of mass accreted by the Moon that had been (possibly incorrectly) deduced from the concentration of highly siderophile elements in the lunar mantle (Morbidelli et al., 2018). Thus there is no need that the Trans-Neptunian disk remained massive for 550 Myr.

Nesvorný et al. (2018a) provided an upper limit of ~100 Myr for the dispersal of the Trans-Neptunian disk. Their considerations are based on the existence of a relatively wide binary object (650 km separation) among the largest Jupiter's Trojans: Patroclus-Menoetius. They first verified that this binary would have had a probability of ~70% to survive through the typical close encounters with the giant planets that transported objects from the Trans-Neptunian disk to the Jupiter's coorbital region. This supports the idea that this binary, like all other Trojans, was captured from the Trans-Neptunian disk. Then, they measured the rate at which similar binaries would have been dissociated by collisions in such disk. They concluded that the probability that Patroclus-Menoetius survived in the disk drops below 10% if the disk lasted more than 100 Myr and below 7×10^{-5} if the disk lasted more than 400 Myr. Thus, it is very unlikely that the giant planet instability and the dispersal of the Trans-Neptunian disk occurred at the time of the LHB.

Clement et al. (2018) suggested that the giant planet instability occurred very early, within 10 Myr from gas photoevaporation. In this case, most of the solid mass would have been removed from Mars' feeding zone while the planet was growing, thus explaining why this planet remained small.

At the opposite end of the spectrum, Marty et al. (2017) argued for a significantly later instability, possibly around 100 Myr. By analyzing the xenon isotope composition of comet 67P-CG, they found that the original xenon in the Earth's atmosphere (called U-Xe), for long time of unknown origin, is just the mixture of cometary Xe with meteoritic Xe. There is, however, no cometary Xe in the Earth's mantle. Only the meteoritic Xe can be found. This suggests that the cometary bombardment, associated with the giant planet instability and dispersal of the Trans-Neptunian disk occurred toward the end of the Earth's formation, possibly even after the Moon-forming event and the crystallization of the terrestrial magma ocean. As the Moon-forming event is dated at ~40–100 Myr, this result argues for a comparable lifetime of the Trans-Neptunian disk.

Thus, it is fair to say that the lifetime of the Trans-Neptunian disk is unknown and therefore it is not possible to assess a priori how much collisional evolution was suffered by the KBO population prior to its implantation in the Kuiper belt.

However, thanks to the New Horizons mission (Stern et al., 2015), we may possibly turn the argument around. The mission allowed counting craters down to a few kilometers in size on Pluto and Charon (Robbins et al., 2017; Singer et al., 2019). The observed crater size-frequency distribution can be turned into a projectile (i.e., KBO) size-frequency distribution using appropriate scaling laws (roughly relating the crater size to 10 times the projectile size). This deduced size-frequency distribution for small KBOs can then be smoothly branched with the size distribution of large KBOs determined by telescopic surveys from Earth. This is done in Fig. 2.8. The figure compares the resulting size distribution of KBOs with the size distribution of main belt asteroids, which is well known down to subkilometer sizes. Both are very similar for $D \gtrsim 2$ km.

Below $D \sim 2$ km there are discordant results. Singer et al. (2019) claim that there is a paucity of craters on Pluto and Charon below ~13 km in diameter, which would imply a sharp turnover to a very shallow size distribution of KBOs smaller than 1–2 km, as illustrated in Fig. 2.8. Robbins et al. (2017), however, did not see this turnover, at least on the surface units imaged at the highest resolution (see Fig. 11 of Robbins et al. and compare it with Fig. 2B of Singer et al.). The reason for this discrepancy within the New Horizons team is not clear to us. The issue of the size distribution at kilometer scale becomes even more complicated with the result by Arimatsu et al. (2019) reporting the serendipitous detection by two telescopes of a stellar occultation by a Trans-Neptunian object likely of 2.6 km in diameter at 30–50 AU. The statistical analysis of this detection implies a population of $D > 2$ km bodies (red dot in Fig. 2.8) at least an order of magnitude larger than that suggested by Pluto-Charon's crater record, but consistent with the population estimate deduced by another claimed stellar occultation detection by Schlichting et al. (2012) (red losange in Fig. 2.8).

The issue is important and is related to both formation (see Section 2.2) and collisional evolution models. If there is really a deficit of subkilometer objects as claimed by Singer et al., this means that very few small objects formed originally (consistent with streaming instability models) and that the size distribution of KBOs is not at collisional equilibrium (otherwise the small objects would have been produced in the collisional cascade, even if originally

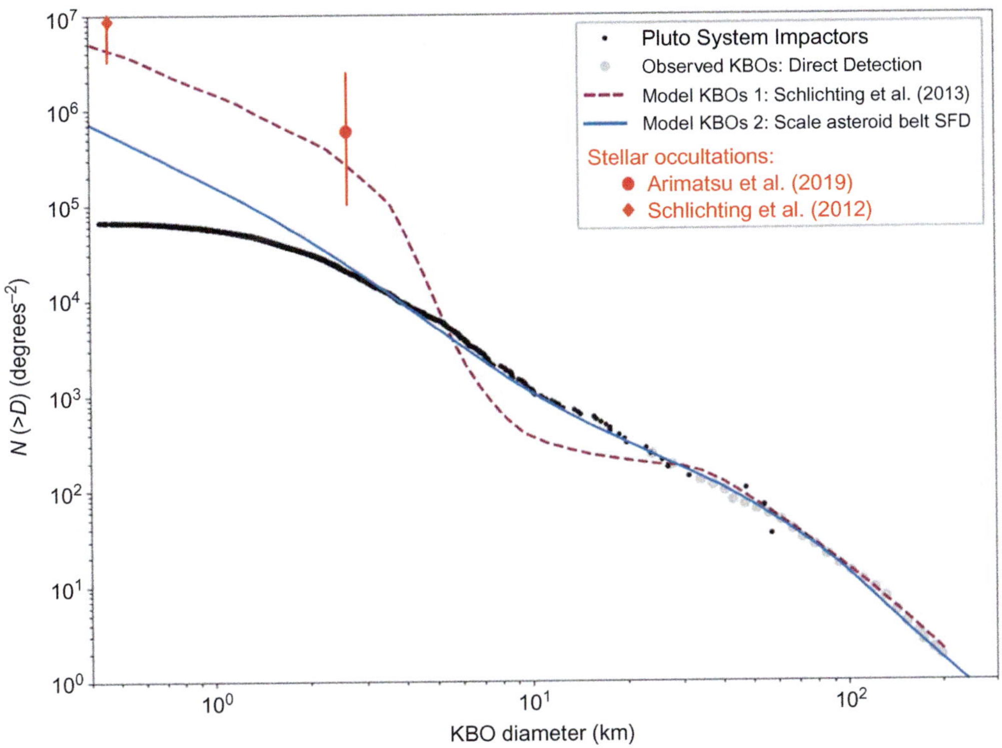

FIG. 2.8 The *black dots* show the size distribution of small KBOs determined by crater counting on Pluto and Charon. The *gray dots* show the size distribution of large KBOs determined by ground-based surveys. Both distributions have been translated into a density of bodies per sky unit area near the ecliptic. For comparison, the *blue curve* shows the size distribution of main belt asteroids, vertically rescaled. The *dashed magenta curve* shows the size distribution predicted in the coagulation model of Schlichting et al. (2013) (see Section 2.2). The *red dots* and *vertical error bars* show the sky densities of bodies larger than 500 m and 2.6 km in diameter as inferred by serendipitous stellar occultations (Schlichting et al., 2012; Arimatsu et al., 2019). *Adapted from Singer, K.N., et al., 2019. Impact craters on Pluto and Charon indicate a deficit of small Kuiper belt objects. Science 363, 955. Courtesy of A. Parker.*

absent). If instead, the kilometer-sized bodies overnumber the power-law extrapolation from 30 to 100 km, as claimed by Schlichting et al. and Arimatsu et al., this would suggest that planetesimals formed preferentially at these sizes as postulated in the collisional coagulation models. However, we have already commented in Section 2.2 that collisional coagulation models fail to reproduce the large-size end of the primordial Trans-Neptunian disk.

Our tentative interpretation of these discordant observations is that the Kuiper belt populations have a normal, power-law size distribution similar to that of the asteroid belt. The crater record on Pluto and Charon is mostly dominated by bodies in the hot and cold components of the Kuiper belt (Greenstreet et al., 2015), while the occultation technique is mostly sensitive to scattered disk objects. There may be an order of magnitude ratio between the populations in the scattered disk and in the hot/cold Kuiper belt, thus explaining the apparent discordance in the detected population densities. Notice that 90% of the scattered disk population at any given time is within 100 AU (Duncan and Levison, 1997), so this is consistent with a claimed detection distance possibly of 50 AU (Arimatsu et al., 2019). Of course, this tentative interpretation needs to be confirmed through quantitative tests using an occultation survey simulator. At the time of writing, these tests have not yet been done.

Because of all these uncertainties, let us base our consideration on the size distribution of objects between 10 and 100 km, that is now quite safely assessed thanks to the New Horizons observations and telescopic surveys (see Fig. 2.8). In this range, the size distribution of the main asteroid belt for $D \lesssim 100$ km is known to be the result of collisional equilibrium (Bottke et al., 2005). Thus, the similarity with the KBO size distribution suggests that the KBO population is also at collisional equilibrium below this size limit. Nesvorný and Vokrouhlický (2019) adopted an initial size distribution from the streaming instability simulations (Simon et al., 2016), and modeled the collisional evolution of the massive planetesimal disk at 20–30 AU (Fig. 2.9A). Their target was to obtain the size distribution of dynamically hot KBOs and Jupiter Trojans (see Section 2.4.1; Fraser et al., 2014). They found that it is indeed possible that the initially rounded size distribution from the streaming instability collisionally evolved to match the distribution of hot KBOs and Jupiter's Trojans. There is a trade-off between the timescale of this evolution and the assumed strength of KBOs, with stronger (weaker) disruption laws requiring longer (shorter) timescale. For instance, the result shown in Fig. 2.9A has been obtained assuming a specific energy of disruption for KBOs equal to that computed by Benz and Asphaug (1999) for strong ice, divided by three. If the specific energy for disruption is reduced

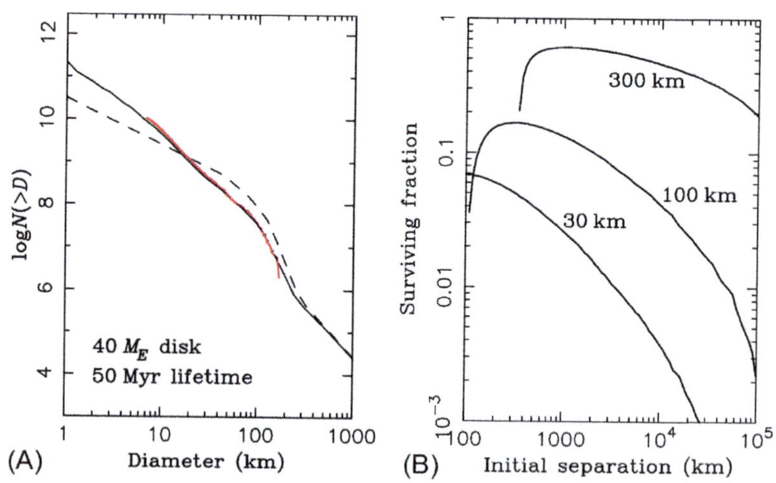

FIG. 2.9 (A) Collisional evolution of the outer planetesimal disk at 20–30 AU. The *dashed line* shows the initial size distribution that was adopted from streaming instability simulations (Simon et al., 2016). It corresponds to the initial disk mass $M_{\text{disk}} = 40$ Earth masses. The *solid line* shows the size distribution after $t_{\text{disk}} = 50$ Myr of collisional grinding when $M_{\text{disk}} = 20$ Earth masses. The *red line* is the size distribution of known Jupiter Trojans (incomplete for $D < 10$ km) scaled up by their implantation efficiency (Nesvorný et al., 2013). (B) Collisional survival of equal-size binaries in the disk. The *lines* show the results for $t_{\text{disk}} = 50$ Myr and $R_1 + R_2 = 30, 100$, and 300 km, where R_1 and R_2 are the effective radii of two binary components. *From Nesvorný, D., Vokrouhlický, D., 2019. Binary survival in the outer solar system. Icarus 331, 49–61.*

by a factor of 10, the time required to acquire the observed size distribution decreases to 10 Myr. Thus, the disk lifetime thus cannot be inferred precisely from these simulations alone.

Nesvorný and Vokrouhlický (2019) and also evaluated the effect of collisions on equal-size binaries and found that the survival depends on the binary size and separation (Fig. 2.9B). For size/separation of equal-size binaries found in the cold Kuiper belt, the survival probability in the original, massive portion of the Trans-Neptunian disk is 1%–10%. Remembering that this part of the disk is the progenitor of the hot population (Section 2.3.2.1), this explains why the equal-size binary fraction among hot KBOs is relatively small. Moreover, the existence of a large fraction of binaries in the cold population reinforces the idea that the disk beyond 40 AU was never massive (see Sections 2.2 and 2.3.2.2).

This result suggests that (i) the Trans-Neptunian disk lasted ~50 Myr, which is consistent with the upper bound of ~100 Myr of Nesvorný et al. (2018a) and the indication of a post-Earth-formation cometary bombardment of Marty et al. (2017) and (ii) that the KBO population below 100 km in diameter has been strongly affected by collisions. In particular, the vast majority of cometary-size objects in the scenario would be fragments of larger objects, in a collisional cascade scenario.

This result raises the question of whether catastrophic fragmentation could be consistent with the high porosity and low-temperature chemistry observed for 67P-CG and other comets. Schwartz et al. (2018) showed that only a tiny fraction of the comet's material would have been heated by more than a few degrees by the catastrophic collision that generated it. A larger fraction of the parent body might have been substantially heated, but the heated material typically is not incorporated in macroscopic fragments. An analogous result was found for the porosity. It should be noted, however, that the collisions in Schwartz et al. (2018) are just above the catastrophic limit and the parent bodies considered are quite small: about 7 km in diameter. Catastrophic collisions of parent bodies of tens of kilometers in size have not yet been simulated, so we do not know whether the same results concerning heating and compaction apply. Simulations of this kind can provide constraints on how big the progenitors of the collisional cascade leading to comet-sized objects could be.

2.6 Conclusions

This review chapter has been divided into three parts. The initial part concerns the formation of KBOs; the central part concerns the origin of the different subcomponents of the Kuiper belt, as well as the relationships with other population of objects, now distant from the Kuiper belt (e.g., Trojans, irregular satellites). The final part discusses the collisional evolution of the KBOs coupled to their dynamical evolution.

The central part is the most consolidated one. Unlike the situation in 2008, now all the components of the Kuiper belt and the related distant populations are reproduced on quantitative grounds using a single model. In this model, there are five giant planets on initially resonant and compact orbits, surrounded by a massive disk of planetesimals extended up to ~30 AU, with a low-mass extension to ~45 AU. Neptune migrates through the disk, dispersing its inner part and eventually triggering instability in the planets' dynamics. At the instability one planet is ejected, all planets have encounters with at least another planet, which leads to orbital jumps. The final interaction of the planets with the dispersed disk stabilizes the planetary system and brings the planets to their current orbits.

The power of this model has been to reproduce the observations in great details. Observers often ask for theoretical predictions, in additions to explanations of what is already known. This model implies several predictions, some of which already confirmed. It predicted that the size distributions of Jupiter's Trojans and of the hot Kuiper belt should be the same (Morbidelli et al., 2009). This has been confirmed in Fraser et al. (2014). From the expected no systematic differences between Jupiter-family and Oort cloud comets, it predicted the existence of JFCs with large D/H ratio, which has been confirmed by the ESA/Rosetta mission (Altwegg et al., 2015). It predicts that, with the exception of the cold population and the Sednoids, all other subpopulations of the Kuiper belt, the Trojans, the irregular satellites, and the Oort cloud are parented; this can be checked by looking at spectral/color distributions, once irradiation and alteration effects are properly understood (prediction already validated for the Jovian Trojans—irregular satellites connection by Graykowski and Jewitt, 2018). It predicts a deficit of small bodies in the Sednoid population compared to all other Trans-Neptunian populations; this prediction can be tested with serendipitous stellar occultation experiments. Finally, it predicts that future surveys, such as the Large Synoptic Survey Telescope (LSST) will not find anything really new in the Kuiper belt structure but will just confirm and consolidate what is already known and explained. Only beyond 50 AU, where our current knowledge of the real population is sketchy, the LSST will unveil possible new features, but the model makes clear predictions of what will be found there (Nesvorný et al., 2016).

Given the current understanding of the orbital structure of the Kuiper belt, other expected contributions of LSST will be:

- Improving the characterization of the "kernel" of the cold Kuiper belt. This will help understanding if this structure is related to the sudden jump of Neptune's orbit, implying that the giant planet instability happened once Neptune had already reached ~28 AU (see Section 2.3.2.2).
- Discovering/characterizing more Neptune's Trojans. The color distribution of Trojans, compared to the color distribution of the hot Kuiper belt can be diagnostic of the original color distribution in the Trans-Neptunian disk and of the location of Neptune at the time of its orbital jump (see Section 2.4.1).
- Understanding whether there is really a distant giant planet in the solar system. In addition to potentially discovering the planet itself, the LSST will find a number of new fossilized scattered disk objects. With enough statistics, it will become clear if there is an anomaly in orbital orientations that requires the existence of a distant planet to be explained.
- Constraining the original stellar cluster in which the Sun formed. By finding more Sednoids and characterizing their distribution in perihelion distance versus semimajor axis, it will become possible to constrain quite precisely the density of the natal stellar cluster of the solar system and even its lifetime.

The origin of KBOs is not yet as clear as the orbital sculpting of the Kuiper belt. The realization that the original planetesimal population in the Trans-Neptunian disk was about 1000 times more populated than the current Kuiper belt defies the predictions of collisional coagulation models. The streaming instability model, possibly combined with subsequent pebble accretion, seems to be more successful. The contrast between the massive disk within 30 AU and a low-mass disk in the 40–45 AU region also seems to suggest that large planetesimals formed in an instability process, which could act sporadically or frequently depending on local conditions. Nevertheless, the streaming instability model has to be improved on quantitative grounds to allow a pertinent comparison with the observations. Similarly, the fact that most objects of the cold population are binaries points to the gravitational collapse of rotating clouds of pebbles, as expected in the streaming instability model. The statistics of prograde/retrograde fractions of the equal-size binaries also support planetesimal formation by streaming instability. The streaming instability model predicts that tight, equal-size binaries (similar to Patroclus-Menoetius), not easily dissociated during planetary encounters, should be found in the hot population when observations reach the needed resolution. Verifying this prediction would also serve as a further confirmation that Jupiter Trojans have been captured from the original Trans-Neptunian disk.

The subsequent collisional evolution of KBOs depends crucially on the duration of the Trans-Neptunian disk before its dynamical dispersal by the giant planet instability. This time could range from $\sim$0 to 100 Myr, with profound implications on how collisionally evolved the KBO population is. It will take time to clarify this issue. On one hand, insight on the timing of the giant planet instability can be obtained by constraining the timing of the cometary bombardment of the Earth, given that some distinct cometary isotope signatures are now known. The paper by Marty et al. (2017) clearly establishes that a cometary bombardment existed on Earth, but the interpretation that the absence of cometary xenon in the Earth interior implies a post-Earth-formation bombardment has to be substantiated with geochemical models. On the other hand, achieving a better understanding of the size distribution produced in the streaming instability and comparing it with the observed size distribution can lead to an assessment of the importance of the collisional evolution. The images of Ultima Thule taken during the flyby the New Horizons mission should eventually extend the size range on which the KBO size distribution is confidently determined from crater counting, thus solving the current controversy between Robbins et al. (2017), Singer et al. (2019), and Arimatsu et al. (2019). This would help constraining the collisional evolution of the Kuiper belt population. Unfortunately, at the time of writing of this review, the required high-resolution images are not yet available.

In summary, although the Kuiper belt is no longer *Terra Incognita*, there is still room for new exploration and discoveries. Like the orbital structure of the Kuiper belt was determinant to constrain quite precisely the past orbital dynamics of the planets, its size distribution will provide crucial information to theorists to constrain formation models and to understand how collisional evolved the KBO population is.

References

A'Hearn, M.F., Millis, R.C., Schleicher, D.O., Osip, D.J., Birch, P.V., 1995. The ensemble properties of comets: results from narrowband photometry of 85 comets, 1976–1992. Icarus 118, 223–270.

Altwegg, K., et al., 2015. 67P/Churyumov-Gerasimenko, a Jupiter family comet with a high D/H ratio. Science 347, 1261952.

Arimatsu, K., Tsumura, K., Usui, F., Shinnaka, Y., Ichikawa, K., Ootsubo, T., Kotani, T., Wada, T., Nagase, K., Watanabe, J., 2019. A kilometre-sized Kuiper belt object discovered by stellar occultation using amateur telescopes. Nat. Astron. 3, 301–306.

Batygin, K., Brown, M.E., 2016. Evidence for a distant giant planet in the solar system. Astron. J. 151, 22.
Batygin, K., Brown, M.E., Fraser, W.C., 2011. Retention of a primordial cold classical Kuiper belt in an instability-driven model of solar system formation. Astrophys. J. 738, 13.
Batygin, K., Brown, M.E., Betts, H., 2012. Instability-driven dynamical evolution model of a primordially five-planet outer solar system. Astrophys. J. 744, L3.
Belton, M.J.S., Zou, X.-D., Li, J.-Y., Asphaug, E., 2018. On the origin of internal layers in comet nuclei. Icarus 314, 364–375.
Benecchi, S.D., Noll, K.S., Grundy, W.M., Buie, M.W., Stephens, D.C., Levison, H.F., 2009. The correlated colors of transneptunian binaries. Icarus 200, 292–303.
Benecchi, S.D., Porter, S., Buie, M., Zangari, A., Verbiscer, A., Noll, K., Stern, S.A., Spencer, J., Parker, A., 2018. The HST lightcurve of (486958) 2014 MU_{69}. arXiv e-prints arXiv:1812.04758.
Benz, W., Asphaug, E., 1999. Catastrophic disruptions revisited. Icarus 142, 5–20.
Birnstiel, T., Fang, M., Johansen, A., 2016. Dust evolution and the formation of planetesimals. Space Sci. Rev. 205, 41–75.
Biver, N., Rauer, H., Despois, D., Moreno, R., Paubert, G., Bockelée-Morvan, D., Colom, P., Crovisier, J., Gérard, E., Jorda, L., 1996. Substantial outgassing of CO from comet Hale-Bopp at large heliocentric distance. Nature 380, 137–139.
Blum, J., et al., 2017. Evidence for the formation of comet 67P/Churyumov-Gerasimenko through gravitational collapse of a bound clump of pebbles. Mon. Not. R. Astron. Soc. 469, S755–S773.
Boehnke, P., Harrison, T.M., 2016. Illusory late heavy bombardment. Proc. Natl. Acad. Sci. USA 113, 10802–10806.
Bottke, W.F., Durda, D.D., Nesvorný, D., Jedicke, R., Morbidelli, A., Vokrouhlický, D., Levison, H., 2005. The fossilized size distribution of the main asteroid belt. Icarus 175, 111–140.
Bottke, W.F., Nesvorný, D., Vokrouhlický, D., Morbidelli, A., 2010. The irregular satellites: the most collisionally evolved populations in the solar system. Astron. J. 139, 994–1014.
Bottke, W.F., Vokrouhlický, D., Minton, D., Nesvorný, D., Morbidelli, A., Brasser, R., Simonson, B., Levison, H.F., 2012. An Archaean heavy bombardment from a destabilized extension of the asteroid belt. Nature 485, 78–81.
Brasser, R., Morbidelli, A., 2013. Oort cloud and scattered disc formation during a late dynamical instability in the Solar System. Icarus 225, 40–49.
Brasser, R., Duncan, M.J., Levison, H.F., 2006. Embedded star clusters and the formation of the Oort Cloud. Icarus 184, 59–82.
Brasser, R., Duncan, M.J., Levison, H.F., 2007. Embedded star clusters and the formation of the Oort cloud. II. The effect of the primordial solar nebula. Icarus 191, 413–433.
Brasser, R., Duncan, M.J., Levison, H.F., 2008. Embedded star clusters and the formation of the Oort cloud. III. Evolution of the inner cloud during the Galactic phase. Icarus 196, 274–284.
Brasser, R., Morbidelli, A., Gomes, R., Tsiganis, K., Levison, H.F., 2009. Constructing the secular architecture of the solar system II: the terrestrial planets. Astron. Astrophys. 507, 1053–1065.
Brasser, R., Duncan, M.J., Levison, H.F., Schwamb, M.E., Brown, M.E., 2012. Reassessing the formation of the inner Oort cloud in an embedded star cluster. Icarus 217, 1–19.
Brauer, F., Dullemond, C.P., Henning, T., 2008. Coagulation, fragmentation and radial motion of solid particles in protoplanetary disks. Astron. Astrophys. 480, 859–877.
Brown, M.E., 2013. The density of mid-sized Kuiper belt object 2002 UX25 and the formation of the dwarf planets. Astrophys. J. 778, L34.
Carrera, D., Gorti, U., Johansen, A., Davies, M.B., 2017. Planetesimal formation by the streaming instability in a photoevaporating disk. Astrophys. J. 839, 16.
Clement, M.S., Kaib, N.A., Raymond, S.N., Walsh, K.J., 2018. Mars' growth stunted by an early giant planet instability. Icarus 311, 340–356.
Davidsson, B.J.R., et al., 2016. The primordial nucleus of comet 67P/Churyumov-Gerasimenko. Astron. Astrophys. 592, A63.
DeMeo, F.E., Carry, B., 2014. Solar system evolution from compositional mapping of the asteroid belt. Nature 505, 629–634.
Dones, L., Weissman, P.R., Levison, H.F., Duncan, M.J., 2004. Oort cloud formation and dynamics. In: Festou, M.C., Keller, H.U., Weaver, H.A. (Eds.), Comets II. University of Arizona Press, Tucson, pp. 153–174.
Drążkowska, J., Alibert, Y., 2017. Planetesimal formation starts at the snow line. Astron. Astrophys. 608, A92.

Drążkowska, J., Alibert, Y., Moore, B., 2016. Close-in planetesimal formation by pile-up of drifting pebbles. Astron. Astrophys. 594, A105.
Duncan, M.J., Levison, H.F., 1997. A scattered comet disk and the origin of Jupiter family comets. Science 276, 1670–1672.
Duncan, M., Quinn, T., Tremaine, S., 1987. The formation and extent of the solar system comet cloud. Astron. J. 94, 1330–1338.
Fernandez, J.A., Ip, W.-H., 1984. Some dynamical aspects of the accretion of Uranus and Neptune—the exchange of orbital angular momentum with planetesimals. Icarus 58, 109–120.
Fraser, W.C., Brown, M.E., Schwamb, M.E., 2010. The luminosity function of the hot and cold Kuiper belt populations. Icarus 210, 944–955.
Fraser, W.C., Brown, M.E., Morbidelli, A., Parker, A., Batygin, K., 2014. The absolute magnitude distribution of Kuiper belt objects. Astrophys. J. 782, 100.
Fraser, W.C., et al., 2017. All planetesimals born near the Kuiper belt formed as binaries. Nat. Astron. 1, 0088.
Gladman, B., Marsden, B.G., Vanlaerhoven, C., 2008. Nomenclature in the outer solar system. In: Barucci, M.A., Boehnhardt, H., Cruikshank, D.P., Morbidelli, A. (Eds.), Solar System Beyond Neptune. University of Arizona Press, Tucson, 592 pp.
Gladman, B., et al., 2012. The resonant Trans-Neptunian populations. Astron. J. 144, 23.
Goldreich, P., Ward, W.R., 1973. The formation of planetesimals. Astrophys. J. 183, 1051–1062.
Goldreich, P., Lithwick, Y., Sari, R., 2002. Formation of Kuiper-belt binaries by dynamical friction and three-body encounters. Nature 420, 643–646.
Gomes, R.S., 2003. The origin of the Kuiper belt high-inclination population. Icarus 161, 404–418.
Gomes, R.S., 2011. The origin of TNO 2004 XR_{190} as a primordial scattered object. Icarus 215, 661–668.
Gomes, R., Nesvorný, D., 2016. Neptune trojan formation during planetary instability and migration. Astron. Astrophys. 592, A146.
Gomes, R., Levison, H.F., Tsiganis, K., Morbidelli, A., 2005a. Origin of the cataclysmic late heavy bombardment period of the terrestrial planets. Nature 435, 466–469.
Gomes, R.S., Gallardo, T., Fernández, J.A., Brunini, A., 2005b. On the origin of the high-perihelion scattered disk: the role of the Kozai mechanism and mean motion resonances. Celest. Mech. Dyn. Astron. 91, 109–129.
Graykowski, A., Jewitt, D., 2018. Colors and shapes of the irregular planetary satellites. Astron. J. 155, 184.
Greenstreet, S., Gladman, B., McKinnon, W.B., 2015. Impact and cratering rates onto Pluto. Icarus 258, 267–288.
Grundy, W.M., et al., 2019. Mutual orbit orientations of Transneptunian binaries. Icarus (in press).
Güttler, C., Blum, J., Zsom, A., Ormel, C.W., Dullemond, C.P., 2009. The first phase of protoplanetary dust growth: the bouncing barrier. Geochim. Cosmochim. Acta Suppl. 73, A482.
Hahn, J.M., Malhotra, R., 2005. Neptune's migration into a stirred-up Kuiper belt: a detailed comparison of simulations to observations. Astron. J. 130, 2392–2414.
Hainaut, O.R., Boehnhardt, H., Protopapa, S., 2012. Colours of minor bodies in the outer solar system. II. A statistical analysis revisited. Astron. Astrophys. 546, A115.
Ida, S., Guillot, T., 2016. Formation of dust-rich planetesimals from sublimated pebbles inside of the snow line. Astron. Astrophys. 596, L3.
Jacquet, E., Balbus, S., Latter, H., 2011. On linear dust-gas streaming instabilities in protoplanetary discs. Mon. Not. R. Astron. Soc. 415, 3591–3598.
Jehin, E., Manfroid, J., Hutsemékers, D., Cochran, A., Zucconi, J.-M., Schulz, R., Arpigny, C., 2008. Carbon and nitrogen isotopic ratios in comets. In: Asteroids, Comets, Meteors 2008, vol. 1405, p. 8339.
Jewitt, D., 2018. The Trojan color conundrum. Astron. J. 155, 56.
Jewitt, D., Sheppard, S., 2005. Irregular satellites in the context of planet formation. Space Sci. Rev. 116, 441–455.
Jílková, L., Zwart, S.P., Pijloo, T., Hammer, M., 2015. How Sedna and family were captured in a close encounter with a solar sibling. Mon. Not. R. Astron. Soc. 453, 3157–3162.
Johansen, A., Oishi, J.S., Low, M.-M.M., Klahr, H., Henning, T., Youdin, A., 2007. Rapid planetesimal formation in turbulent circum stellar disks. Nature 448, 1022–1025.
Johansen, A., Youdin, A., Low, M.-M.M., 2009. Particle clumping and planetesimal formation depend strongly on metallicity. Astrophys. J. 704, L75–L79.
Johansen, A., Low, M.-M.M., Lacerda, P., Bizzarro, M., 2015. Growth of asteroids, planetary embryos, and Kuiper belt objects by chondrule accretion. Sci. Adv. 1, 1500109.

Jutzi, M., Benz, W., 2017. Formation of bi-lobed shapes by sub-catastrophic collisions. A late origin of comet 67P's structure. Astron. Astrophys. 597, A62.
Jutzi, M., Benz, W., Toliou, A., Morbidelli, A., Brasser, R., 2017. How primordial is the structure of comet 67P? Combined collisional and dynamical models suggest a late formation. Astron. Astrophys. 597, A61.
Kenyon, S.J., Bromley, B.C., 2004a. The size distribution of Kuiper belt objects. Astron. J. 128, 1916–1926.
Kenyon, S.J., Bromley, B.C., 2004b. Stellar encounters as the origin of distant Solar System objects in highly eccentric orbits. Nature 432, 598–602.
Kenyon, S.J., Bromley, B.C., 2012. Coagulation calculations of Icy planet formation at 15–150 AU: a correlation between the maximum radius and the slope of the size distribution for Trans-Neptunian objects. Astron. J. 143, 63.
Kenyon, S.J., Bromley, B.C., O'Brien, D.P., Davis, D.R., 2008. Formation and collisional evolution of Kuiper belt objects. In: The Solar System Beyond Neptune. University of Arizona Press, Tucson, AZ, pp. 293–313.
Kleine, T., Mezger, K., Palme, H., Scherer, E., Münker, C., 2005. Early core formation in asteroids and late accretion of chondrite parent bodies: evidence from ^{182}Hf-^{182}W in CAIs, metal-rich chondrites, and iron meteorites. Geochim. Cosmochim. Acta 69, 5805–5818.
Kruijer, T.S., Touboul, M., Fischer-Gödde, M., Bermingham, K.R., Walker, R.J., Kleine, T., 2014. Protracted core formation and rapid accretion of protoplanets. Science 344, 1150–1154.
Lambrechts, M., Johansen, A., 2012. Rapid growth of gas-giant cores by pebble accretion. Astron. Astrophys. 544, A32.
Lambrechts, M., Johansen, A., 2014. Forming the cores of giant planets from the radial pebble flux in protoplanetary discs. Astron. Astrophys. 572, A107.
Lambrechts, M., Morbidelli, A., 2016. Reconstructing the size distribution of the small body population in the Solar System. AAS/Division for Planetary Sciences Meeting Abstracts No. 48, ID.105.08.
Lawler, S.M., Kavelaars, J.J., Alexandersen, M., Bannister, M.T., Gladman, B., Petit, J.-M., Shankman, C., 2018. OSSOS: X. How to use a survey simulator: statistical testing of dynamical models against the real Kuiper belt. Front. Astron. Space Sci. 5, 14.
Lawler, S.M., et al., 2019. OSSOS: XIII. Fossilized resonant dropouts tentatively confirm Neptune's migration was grainy and slow. Astron. J. 157, 253.
Levison, H.F., Morbidelli, A., 2003. The formation of the Kuiper belt by the outward transport of bodies during Neptune's migration. Nature 426, 419–421.
Levison, H.F., Morbidelli, A., Van Laerhoven, C., Gomes, R., Tsiganis, K., 2008. Origin of the structure of the Kuiper belt during a dynamical instability in the orbits of Uranus and Neptune. Icarus 196, 258–273.
Levison, H.F., Bottke, W.F., Gounelle, M., Morbidelli, A., Nesvorný, D., Tsiganis, K., 2009. Contamination of the asteroid belt by primordial Trans-Neptunian objects. Nature 460, 364–366.
Levison, H.F., Morbidelli, A., Tsiganis, K., Nesvorný, D., Gomes, R., 2011. Late orbital instabilities in the outer planets induced by interaction with a self-gravitating planetesimal disk. Astron. J. 142, 152.
Lin, H.W., et al., 2019. Evidence for color dichotomy in the primordial Neptunian Trojan population. Icarus 321, 426.
Malhotra, R., 1993. The origin of Pluto's peculiar orbit. Nature 365, 819–821.
Malhotra, R., 1995. The origin of Pluto's orbit: implications for the solar system beyond Neptune. Astron. J. 110, 420.
Marty, B., et al., 2017. Xenon isotopes in 67P/Churyumov-Gerasimenko show that comets contributed to Earth's atmosphere. Science 356, 1069–1072.
Marzari, F., Scholl, H., Murray, C., Lagerkvist, C., 2002. Origin and evolution of Trojan asteroids. In: Bottke, Jr., W.F., Cellino, A., Paolicchi, P., Binzel, R.P. (Eds.), Asteroids III. University of Arizona Press, Tucson, pp. 725–738.
Morbidelli, A., Levison, H.F., 2004. Scenarios for the origin of the orbits of the Trans-Neptunian objects 2000 CR_{105} and 2003 VB_{12} (Sedna). Astron. J. 128, 2564–2576.
Morbidelli, A., Rickman, H., 2015. Comets as collisional fragments of a primordial planetesimal disk. Astron. Astrophys. 583, A43.
Morbidelli, A., Levison, H.F., Tsiganis, K., Gomes, R., 2005. Chaotic capture of Jupiter's Trojan asteroids in the early solar system. Nature 435, 462–465.
Morbidelli, A., Tsiganis, K., Crida, A., Levison, H.F., Gomes, R., 2007. Dynamics of the giant planets of the solar system in the gaseous protoplanetary disk and their relationship to the current orbital architecture. Astron. J. 134, 1790–1798.
Morbidelli, A., Levison, H.F., Gomes, R., 2008. The dynamical structure of the Kuiper belt and its primordial origin. In: The Solar System Beyond Neptune. University of Arizona Press, Tucson, AZ, pp. 275–292.

Morbidelli, A., Levison, H.F., Bottke, W.F., Dones, L., Nesvorný, D., 2009. Considerations on the magnitude distributions of the Kuiper belt and of the Jupiter Trojans. Icarus 202, 310–315.
Morbidelli, A., Brasser, R., Gomes, R., Levison, H.F., Tsiganis, K., 2010. Evidence from the asteroid belt for a violent past evolution of Jupiter's orbit. Astron. J. 140, 1391–1401.
Morbidelli, A., Nesvorný, D., Laurenz, V., Marchi, S., Rubie, D.C., Elkins-Tanton, L., Wieczorek, M., Jacobson, S., 2018. The timeline of the lunar bombardment: revisited. Icarus 305, 262–276.
Nesvorný, D., 2011. Young solar system's fifth Giant planet? Astrophys. J. 742, L22.
Nesvorný, D., 2015a. Jumping Neptune can explain the Kuiper belt Kernel. Astron. J. 150, 68.
Nesvorný, D., 2015b. Evidence for slow migration of Neptune from the inclination distribution of Kuiper belt objects. Astron. J. 150, 73.
Nesvorný, D., Dones, L., 2002. How long-lived are the hypothetical Trojan populations of Saturn, Uranus, and Neptune? Icarus 160, 271–288.
Nesvorný, D., Morbidelli, A., 2012. Statistical study of the early solar system's instability with four, five, and six giant planets. Astron. J. 144, 117.
Nesvorný, D., Vokrouhlický, D., 2016. Neptune's orbital migration was Grainy, not smooth. Astrophys. J. 825, 94.
Nesvorný, D., Vokrouhlický, D., 2019. Binary survival in the outer solar system. Icarus 331, 49–61.
Nesvorný, D., Vokrouhlický, D., Morbidelli, A., 2007. Capture of Irregular Satellites during planetary encounters. Astron. J. 133, 1962–1976.
Nesvorný, D., Youdin, A.N., Richardson, D.C., 2010. Formation of Kuiper belt binaries by gravitational collapse. Astron. J. 140, 785–793.
Nesvorný, D., Vokrouhlický, D., Bottke, W.F., Noll, K., Levison, H.F., 2011. Observed binary fraction sets limits on the extent of collisional grinding in the Kuiper belt. Astron. J. 141, 159.
Nesvorný, D., Vokrouhlický, D., Morbidelli, A., 2013. Capture of Trojans by jumping Jupiter. Astrophys. J. 768, 45.
Nesvorný, D., Vokrouhlický, D., Deienno, R., 2014. Capture of irregular satellites at Jupiter. Astrophys. J. 784, 22.
Nesvorný, D., Vokrouhlický, D., Roig, F., 2016. The orbital distribution of Trans-Neptunian objects beyond 50 AU. Astrophys. J. 827, L35.
Nesvorný, D., Vokrouhlický, D., Dones, L., Levison, H.F., Kaib, N., Morbidelli, A., 2017. Origin and evolution of short-period comets. Astrophys. J. 845, 27.
Nesvorný, D., Vokrouhlický, D., Bottke, W.F., Levison, H.F., 2018a. Evidence for very early migration of the solar system planets from the Patroclus-Menoetius binary Jupiter Trojan. Nat. Astron. 2, 878–882.
Nesvorný, D., Parker, J., Vokrouhlický, D., 2018b. Bi-lobed shape of comet 67P from a collapsed binary. Astron. J. 155, 246.
Nesvorný, D., Li, R., Youdin, N.Y., Simon, J.B., Grundy, W.M., 2019. Trans-Neptunian binaries provide evidence for planetesimal formation by the streaming instability. Nat. Astron. 3, 808.
Noll, K.S., Levison, H.F., Grundy, W.M., Stephens, D.C., 2006. Discovery of a binary Centaur. Icarus 184, 611–618.
Noll, K.S., Grundy, W.M., Chiang, E.I., Margot, J.-L., Kern, S.D., 2008. Binaries in the Kuiper belt. In: The Solar System Beyond Neptune. University of Arizona Press, Tucson, AZ, pp. 345–363.
Norman, M.D., Nemchin, A.A., 2014. A 4.2 billion year old impact basin on the Moon: U-Pb dating of zirconolite and apatite in lunar melt rock 67955. Earth Planet. Sci. Lett. 388, 387–398.
Oort, J.H., 1950. The structure of the cloud of comets surrounding the Solar System and a hypothesis concerning its origin. Bull. Astron. Inst. 11, 91–110.
Pan, M., Sari, R., 2005. Shaping the Kuiper belt size distribution by shattering large but strengthless bodies. Icarus 173, 342–348.
Parker, A.H., Kavelaars, J.J., 2010. Destruction of binary minor planets during Neptune scattering. Astrophys. J. 722, L204–L208.
Petit, J.-M., Mousis, O., 2004. KBO binaries: how numerous were they? Icarus 168, 409–419.
Petit, J.-M., Kavelaars, J.J., Gladman, B., Loredo, T., 2008. Size distribution of multikilometer transneptunian objects. In: The Solar System Beyond Neptune. University of Arizona Press, Tucson, AZ, pp. 71–87.
Petit, J.-M., et al., 2011. The Canada-France Ecliptic plane survey-full data release: the orbital structure of the Kuiper belt. Astron. J. 142, 131.
Prialnik, D., Sarid, G., Rosenberg, E.D., Merk, R., 2009. Thermal and chemical evolution of comet nuclei and Kuiper belt objects. In: Origin and Early Evolution of Comet Nuclei 147.
Ribeiro de Sousa, R., Gomes, R., Morbidelli, A., Vieira Neto, E., 2018. Dynamical effects on the classical Kuiper Belt during the excited-Neptune model. arXiv e-prints arXiv:1808.02146.

Robbins, S.J., et al., 2017. Craters of the Pluto-Charon system. Icarus 287, 187–206.
Ryder, G., 2002. Mass flux in the ancient Earth-Moon system and benign implications for the origin of life on Earth. J. Geophys. Res. (Planets) 107, 5022-1.
Schlichting, H.E., Sari, R., 2008. The ratio of retrograde to prograde orbits: a test for Kuiper belt binary formation theories. Astrophys. J. 686, 741–747.
Schlichting, H.E., Sari, R., 2011. Runaway growth during planet formation: explaining the size distribution of large Kuiper belt objects. Astrophys. J. 728, 68.
Schlichting, H.E., Ofek, E.O., Sari, R., Nelan, E., Gal-Yam, A., Wenz, M., Muirhead, P., Javanfar, N., Livio, M., 2012. Measuring the Abundance of sub-kilometer-sized Kuiper belt objects using stellar occultations. Astrophys. J. 761, 150.
Schlichting, H.E., Fuentes, C.I., Trilling, D.E., 2013. Initial planetesimal sizes and the size distribution of small Kuiper belt objects. Astron. J. 146, 36.
Schoonenberg, D., Ormel, C.W., 2017. Planetesimal formation near the snowline: in or out? Astron. Astrophys. 602, A21.
Schwartz, S.R., Michel, P., Jutzi, M., Marchi, S., Zhang, Y., Richardson, D.C., 2018. Catastrophic disruptions as the origin of bilobate comets. Nat. Astron. 2, 379–382.
Shakura, N.I., Sunyaev, R.A., 1973. Black holes in binary systems. Observational appearance. Astron. Astrophys. 24, 337–355.
Shannon, A., Wu, Y., Lithwick, Y., 2016. Forming the cold classical Kuiper belt in a light disk. Astrophys. J. 818, 175.
Sheppard, S.S., 2012. The color differences of Kuiper belt objects in resonance with Neptune. Astron. J. 144, 169.
Sheppard, S.S., Trujillo, C.A., 2010. Detection of a trailing (L5) Neptune Trojan. Science 329, 1304.
Shoemaker, E.M., Wolfe, R.F., 1984. Evolution of the Uranus-Neptune planetesimal swarm. Lunar Planet. Sci. Conf. 15, 780–781.
Simon, J.B., Armitage, P.J., Li, R., Youdin, A.N., 2016. The mass and size distribution of planetesimals formed by the streaming instability. I. The role of self-gravity. Astrophys. J. 822, 55.
Singer, K.N., et al., 2019. Impact craters on Pluto and Charon indicate a deficit of small Kuiper belt objects. Science 363, 955.
Sosa, A., Fernández, J.A., 2011. Masses of long-period comets derived from non-gravitational effects—analysis of the computed results and the consistency and reliability of the non-gravitational parameters. Mon. Not. R. Astron. Soc. 416, 767–782.
Stern, S.A., 1991. On the number of planets in the outer solar system—evidence of a substantial population of 1000-km bodies. Icarus 90, 271–281.
Stern, S.A., et al., 2015. The Pluto system: initial results from its exploration by New Horizons. Science 350, 292.
Stern, S.A., et al., 2019. Initial results from the new Horizons exploration of 2014 MU_{69}, a small Kuiper belt object. Science 364 (6441), Article ID: aaw9771.
Tera, F., Papanastassiou, D.A., Wasserburg, G.J., 1974. Isotopic evidence for a terminal lunar cataclysm. Earth Planet. Sci. Lett. 22, 1–21.
Thommes, E.W., Duncan, M.J., Levison, H.F., 2002. The formation of Uranus and Neptune among Jupiter and Saturn. Astron. J. 123, 2862–2883.
Trujillo, C.A., Brown, M.E., 2002. A correlation between inclination and color in the classical Kuiper belt. Astrophys. J. 566, L125–L128.
Trujillo, C.A., Sheppard, S.S., 2014. A Sedna-like body with a perihelion of 80 astronomical units. Nature 507, 471–474.
Trujillo, C.A., Jewitt, D.C., Luu, J.X., 2000. Population of the scattered Kuiper belt. Astrophys. J. 529, L103–L106.
Tsiganis, K., Gomes, R., Morbidelli, A., Levison, H.F., 2005. Origin of the orbital architecture of the giant planets of the solar system. Nature 435, 459–461.
Villeneuve, J., Chaussidon, M., Libourel, G., 2009. Homogeneous distribution of ^{26}Al in the solar system from the Mg isotopic composition of Chondrules. Science 325, 985.
Vokrouhlický, D., Bottke, W.F., Nesvorný, D., 2016. Capture of Trans-Neptunian planetesimals in the main asteroid belt. Astron. J. 152, 39.
Volk, K., Malhotra, R., 2008. The scattered disk as the source of the Jupiter family comets. Astrophys. J. 687, 714–725.
Weidenschilling, S.J., 1977. Aerodynamics of solid bodies in the solar nebula. Mon. Not. R. Astron. Soc. 180, 57–70.
Weidenschilling, S.J., 1995. Can gravitational instability form planetesimals? Icarus 116, 433–435.

Wong, I., Brown, M.E., 2016. A hypothesis for the color bimodality of Jupiter trojans. Astron. J. 152, 90.
Yang, C.-C., Johansen, A., Carrera, D., 2017. Concentrating small particles in protoplanetary disks through the streaming instability. Astron. Astrophys. 606, A80.
Youdin, A.N., Goodman, J., 2005. Streaming instabilities in protoplanetary disks. Astrophys. J. 620, 459–469.
Youdin, A.N., Shu, F.H., 2002. Planetesimal formation by gravitational instability. Astrophys. J. 580, 494–505.

CHAPTER

3

Perspectives on the distribution of orbits of distant Trans-Neptunian objects

J.J. Kavelaars[a,b], *Samantha M. Lawler*[b], *Michele T. Bannister*[c], *Cory Shankman*[a]

[a]Department of Physics and Astronomy, University of Victoria, Victoria, BC, Canada [b]Herzberg Astronomy and Astrophysics Research Center, National Research Council of Canada, Victoria, BC, Canada [c]Astrophysics Research Centre, School of Mathematics and Physics, Queen's University Belfast, Belfast, United Kingdom

3.1 Biases in the detection of distant solar system objects

The Kuiper belt is over 4.5-billion-km distant from the Earth-bound observer, with the most distant Trans-Neptunian objects (TNOs) known being three times further away still. The challenge of detecting objects at these great distances should not be underestimated. The Sun's light reflected on solar system bodies at a distance r is dimmed by the factor r^{-4}, greatly exaggerating our sensitivity to nearby objects in comparison to more distant objects. The volume of the Solar neighborhood that a survey is sensitive to, its *detection volume*, is, at minimum, limited in radial extent. The strength of the r^{-4} observational bias is frequently underappreciated when attempting to interpret the distributions of objects detected by a particular survey.

TNOs on orbits with moderate-to-large eccentricities also present a distorted view of the population. Eccentric TNOs occupy a range of Solar distances during their orbits, resulting in a time-variable r^{-4} flux bias. A TNO may only spend a small fraction of its orbital period within the detection volume of a particular survey. The larger the semimajor axis, the larger the eccentricity needed to bring the object within the detection volume, and the smaller the fraction of the orbit for which that object remains in the detection volume.

The Trans-Neptunian Solar System.
https://doi.org/10.1016/B978-0-12-816490-7.00003-5

Limited telescopic resources add another layer of complexity to the problem of quantifying the biases inherent in the detected sample of TNOs. One can only detect objects in the part of the sky where one looks. This observer direction bias imposes a relation between the nodal angle (Ω), argument of pericenter (ω), mean anomaly ($\mathcal{M}$), and inclination (i) of the orbits that can be detected. This coupling of multiple angles can be difficult to conceptualize. For illustration, consider a discovery survey whose fields all straddle the ecliptic plane. In those surveys, orbits that are inclined to the ecliptic plane with an inclination that exceeds the latitude sensitivity of the survey will only be detectable when looking toward their nodes. Combine the forced detection at the node with the preferential discovery of objects near their pericenter q, caused by the r^{-4} flux bias discussed earlier, and such a survey will find that the detected sample of TNOs on inclined orbits all have arguments of pericenter ω near 0 and 180 degrees. This effect is well known, and is described here to remind the reader of the basic processes at work. When attempting to maximize the science return of scarce observing time, observing fields in specific areas of sky inherently induces biases in the detected sample, and some of these biases may be difficult to recognize.

In Fig. 3.1, we present the orbital distribution of a subset of the TNOs reported to the Minor Planet Center (MPC) as of October 15, 2018: All TNOs with $a > 150$ AU and $q > 30$ AU. The vertical line at 1000 AU *roughly* indicates the more distant phase space where the effects of Galactic tides and stellar passages become important (e.g., Kaib and Quinn, 2009). The horizontal dotted curve indicates the zone below which outward diffusion (chaotic scattering) from the Kuiper belt is significant. Between that and the dot-dashed curve indicates the zone where inward diffusion from the inner Oort cloud occurs (see Section 3.3 and Bannister et al. 2017 for details). Also shown in this figure is the background color coding estimating the size of the intrinsic population that would be needed to detect a single object on a given (a, q) orbit, in a survey that also detected one object with $a \sim 150$ AU and $q \sim 30$ AU (assuming a size-frequency distribution of TNOs $\Sigma N = 10^{0.5(H-H_o)}$). From this figure, we can see that the present low detection rate in the $q > 60$ AU, $a > 1000$ AU orbits is only a weak constraint on the size of that population as our ability to see into this zone is quite limited. In order to detect one TNO in this (a, q) range, a survey that detected one TNO at low-a and low-q would require a couple hundred times as many TNOs with obits in the large (a, q) zone. We would need hundreds of TNO detections in the low-(a, q) zone just to rule out a uniform distribution in this phase space. These numbers are in basic agreement with more careful computation provided elsewhere (e.g., Sheppard et al., 2019) and are given here to guide the reader's understanding of the influence of orbit and flux bias in the detected sample. An important consideration is that to use the biases in Fig. 3.1 to aid in understanding the structure of the Trans-Neptunian region, we need to know the full range of orbits, including those with $a < 200$ AU and $q < 40$ AU, that were detected in a given survey. We can then use the relative sensitivity to scale between the regions.

Given the sample of TNOs that are publicly known, one is tempted to pursue mechanisms to debias the observed sample. Two classes of approaches are common. In the first, the characteristics of the survey itself are used to determine the efficiency of detections of various orbits, this approach has been employed in numerous project (e.g., Trujillo et al., 2000; Elliot et al., 2005; Petit et al., 2011; Adams et al., 2014; Bannister et al., 2018). A weakness of this approach is that each project has taken somewhat different approaches to documenting their characterization and this makes combining data sets, to enhance statistical power, difficult.

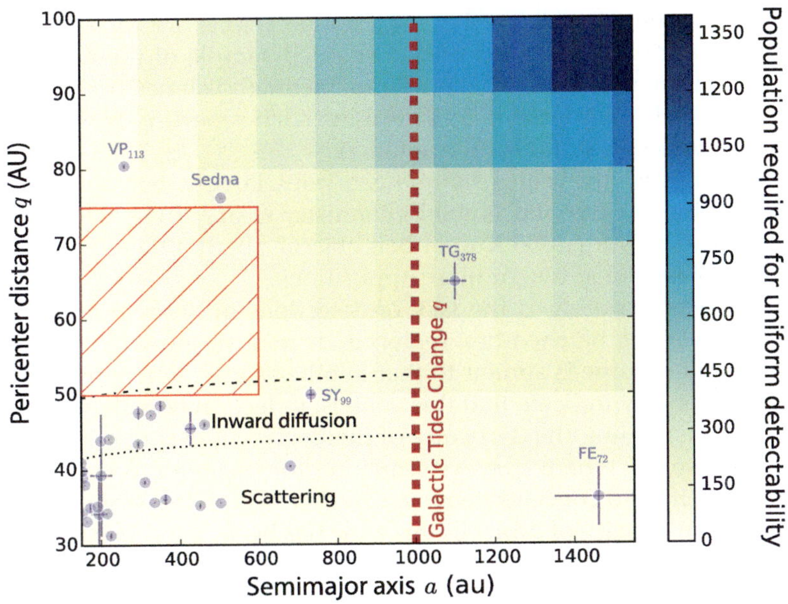

FIG. 3.1 Orbital distribution of known TNOs with observed arcs longer than 10 months in the MPC Database as of October 2018, on orbits with $a > 150$ AU and $q > 30$ AU in a–q (*blue dots*); a few extremes are noted by name, uncertainty in orbital parameters extracted from orbit (Bernstein and Khushalani, 2000). The *background grid of colored boxes* indicates the population of TNOs needed in each (a, q) bin for detection of that orbit to have similar probability if there is a single object in the $a = 150$ and $q = 30$ bin; see text for details. Horizontal curves approximate the boundary below which inward diffusion from $a > 1000$ AU is significant (*dot-dashed line*; Bannister et al., 2017), and the boundary below which outward diffusion (chaotic scattering) in a is significant (*dotted line*). The *thick dashed vertical line* roughly indicates where Galactic tides become a notable long-term influence on the pericenters of orbits. The *red hatched box* indicates a region of (a, q) orbits that currently has no TNO detections, although detectability is actually easier here than for the two known highest-q TNOs, Sedna, and VP_{113}.

Recently, Brown (2017), and continuing in Brown and Batygin (2019), have implemented procedures that attempt to "self-characterize" the detection biases present in the MPC reported sample of KBOs. This approach has the advantage that it combines together a larger data set but has some disadvantages also.

Using the public catalog to attempt to debias the detected sample makes at least two implicit assumptions that are not true: That the sky coverage of a survey is well represented by the detections in that survey and that all objects detected by a survey are report.

Using the public catalog provides no capacity to know where surveys did not detect objects. As an example, consider the Canada France Ecliptic Plane Survey High Latitude Component (Petit et al., 2017). CFEPS-HiLat imaged 700 square degrees of sky searching for TNOs on high-inclination orbits but only reported the detection of 24 objects. This survey provides constraints on the number of objects that can be on highly inclined orbits, a constraint that is not visible if one uses the known objects as probes of the locations of sky that have been surveyed.

An additional weakness is that the use of the detected sample to determine the characterization assumes that there is no reporting bias (i.e., all objects detected are tracked and reported), but this is known not to be the case. Some projects are forced, by the nature of resource restrictions, to only report and track a selected sample of their detections (such as only reporting the detection of objects beyond some defined distance from the observer; e.g., Sheppard et al., 2019), it is impossible to determine such reporting bias from the detected sample and can lead to significant misinterpretation.

The desire for a statistically useful sample can now, largely, be achieved by using the sample provided by the "OSSOS Ensemble" (Bannister et al., 2018), which includes orbits and detection circumstances for 1086 Kuiper belt objects, about 40% of the currently known population of KBOs observed at two or more oppositions.

One point that is worth noting is the lack of detections in the region shown by the red hatched box in Fig. 3.1. The hatched-box region is devoid of known TNOs—however, our sensitivity to orbits in this zone is similar to that in other zones, where in contrast a number of detections exist. This may indicate that this zone is, indeed, relatively underpopulated, an important point in constraining the dynamics of this region of the solar system. Or this may be a region of phase space, where surveys searching for distant TNOs have simply culled their detections in an effort to concentrate resources on tracking the large pericenters, distant members. The apparent paucity of orbits in this zone has been noted previously (e.g., Trujillo and Sheppard, 2014; Bannister et al., 2018), and is discussed further in Section 3.2.

3.2 Potential mechanisms forming the orbits of high-pericenter TNOs

In the recent literature, there has been much discussion of the handful of known large-a, high-q so-called "extreme TNOs," which are defined with a variety of orbital selection criteria,[1] sometimes as having orbits with $a > 250$ AU (alternatively sometimes $a > 150$ AU) and perihelia detached from active interaction with Neptune of $q \gtrsim 37$ AU (e.g., Kiss et al., 2013; Sheppard et al., 2016; de la Fuente Marcos and de la Fuente Marcos, 2016; Shankman et al., 2017a; Bannister et al., 2017; Becker et al., 2018). This definition is vague, reflecting the current lack of understanding of the population, and may not necessarily be dynamically distinguishing. Batygin et al. (2019) provide a lengthy review of the "Planet 9" hypothesis and provide some dynamical considerations on where the boundary should be drawn when considering dynamics that could be induced by a large external planet. They draw the boundaries at $q > 30$ AU and $a > 250$ AU. As of October 2018, 17 TNOs with multiopposition orbits reported in the MPC database have $150 < a < 1000$ AU with $q > 37$ AU, mostly with pericenters in the range $37 < q < 50$ AU. Two of these TNOs have much larger pericenter distances, with $q > 75$ AU: Sedna ($q = 76.19 \pm 0.03$ AU, $a = 507 \pm 10$ AU; Brown et al., 2004) and 2012 VP_{113} ($q = 80.3^{+1.2}_{-1.6}$ AU, $a = 266^{+26}_{-17}$ AU; Trujillo and Sheppard, 2014).[2]

[1] We note that definitions that make use of orbital elements for such orbits must use the barycentric not the heliocentric orbit, due to the long-term effect of Jupiter on these distant orbits. However, heliocentric orbital elements are what is provided by the most frequently used databases of orbits, the MPC Database and JPL Horizons. Uncertainties in the orbit fit to the measured TNO astrometry should also be considered.

[2] Barycentric orbits after Bannister et al. (2017).

It has been long recognized that the orbital distribution of TNOs can be used to understand the past dynamical history of the solar system, particularly the outer giant planets (e.g., Malhotra, 1993; Levison et al., 2008). For a recent review of this topic, see Nesvorný (2018). The basic origin scenario is that TNOs formed in a dynamically cold disk of planetesimals, which was largely disrupted when Neptune migrated, placing TNOs into classes based on orbits that display unique dynamical behaviors (see Gladman et al., 2008). High-q TNOs with $a < 1000$ AU are difficult to explain in this framework: They never approach Neptune closely enough to receive the strong dynamical kicks needed to change their orbits, their large eccentricities preclude in situ formation, and their orbits are not large enough to be affected significantly by Galactic tides.

In a sign of a vigorously active area of theoretical investigation, the high-pericenter TNOs have recently spawned a flurry of studies to explain these dynamically interesting orbits. While attention has been focused on the hypothesis that an undiscovered distant giant planet can be used to explain properties of large-a, high-q orbits (see Batygin et al., 2019, for a recent review). While the Planet 9 hypothesis provides a compelling explanation for various orbital characteristics, a planet presently orbiting in the distant solar system is by no means the only theory being advanced. We highlight several proposed classes of theories that can raise TNO orbital pericenters.

Simulations that include perturbations by passing stars at various times in solar system history produce high-pericenter TNOs on large-a orbits. Stellar perturbations have been considered to raise perihelia of native TNOs, while the Sun is still in the denser stellar environment of its birth cluster with lower relative velocities (Fernández and Brunini, 2000; Kenyon and Bromley, 2004; Morbidelli and Levison, 2004; Brasser et al., 2006, 2012; Kaib and Quinn, 2008; Brasser and Schwamb, 2015; Pfalzner et al., 2018), by field stars during the Sun's postcluster orbits of the Galaxy, and possible radial migration (Kaib et al., 2011). Certain geometries of stellar fly-by may permit capture of TNOs from passing star systems (Kenyon and Bromley, 2004; Morbidelli and Levison, 2004; Jílková et al., 2015; see also Levison et al., 2010 for Oort comet capture). All these simulations have many degrees of freedom related to the stellar mass, and the distance and geometry of a flyby, which require an abundance of known high-q TNOs to well confine their possible parameter space.

An intriguing new theoretical mechanism is being explored by simulations that take into account the self-gravitational influence of great densities of small TNOs. Madigan and McCourt (2016) have shown an "inclination instability" within a massive planetesimal disk can produce high-q orbits. A disk that begins as axisymmetric eccentric orbits like that of the scattering disk will increase in inclination while lowering its orbital eccentricities, forming an asymmetric cone (Madigan et al., 2018). The inclination instability effect requires substantial mass in planetesimals in the distant solar system for it to initiate; Madigan et al. (2018) infer about half an Earth mass at hundreds of AU. Sefilian and Touma (2019) have further explored this mechanism and find that a massive disk would induce clustering in orbital angles and argue that massive disk may be more likely than expected as such disks are common around other stars. Batygin et al. (2019), however, expect that such a massive disk unlikely to have remained in place for the age of solar system, making this explanation of clustering unlikely to be correct. Bounds on the size distribution of distant TNOs and thus the mass at large a remain limited, but will constrain this theory tightly in the future.

Two mechanisms can produce many high-q TNOs solely from the known planets of the solar system. First, chaotic diffusion in semimajor axis caused by weak gravitational kicks from Neptune can cause minor planets to migrate from the inner Oort cloud to large-a, high-q TNO orbits (Duncan et al., 1987; Kaib et al., 2009; Bannister et al., 2017); we discuss this further in Section 3.3. Second, high-q orbits may be produced during Neptune's migration. TNOs that are captured into Neptune's mean-motion resonances (MMRs) experience Kozai oscillations inside the MMR. As Neptune migrates outward, they may drop out of the resonance at high-q, where the resonance is narrower. The TNO is then "fossilized" on a dynamically detached, long-term stable orbit (Gomes, 2003). Several nonresonant, stable TNOs have been discovered on high-q orbits near strong resonances, lending observational support to this theory (Pike et al., 2015; Lawler et al., 2018b). Including dwarf planets in migration simulations (cf. "grainy" migration; Nesvorný and Vokrouhlický 2016), as required from the size distribution of the initial disk, is a recent refinement that is still being explored, but appears important. Grainy simulations show that the mode and timescale of Neptune's migration affects the distribution of these high-q resonant dropouts (Nesvorný et al., 2016; Kaib and Sheppard, 2016). While simulations of scattering TNO capture into MMRs show that this can be effective for raising pericenters as high as $q \gtrsim 70$ AU, this can only happen for scattering TNOs that already have large inclinations prior to resonant capture (Gallardo et al., 2012), and while grainy migration can modify the inclination distribution (Muñoz-Gutiérrez et al., 2018), it may not be sufficient and so this may not be the emplacement mechanism for Sedna and VP_{113} at $i < 25$ degrees, despite their locations near low-order, distant resonances.

A separate class of hypotheses invokes planetary-mass bodies to raise pericenters. The possible presence of an undiscovered massive distant planet has been discussed extensively in the literature from the early days of Kuiper belt discoveries to the present (Gladman et al., 2002; Brown et al., 2004; Gladman, 2005; Lykawka and Mukai, 2008; Soares and Gomes, 2013; Trujillo and Sheppard, 2014). Many recent simulations have shown that a distant massive planet would be quite effective at raising the pericenters of large-a TNOs (Batygin and Brown, 2016; Shankman et al., 2017b; Lawler et al., 2017; Li et al., 2018, see Fig. 3.2). However, other aspects of the observed Kuiper belt are solidly inconsistent with this particular planetary scenario (Lawler et al., 2017; Shankman et al., 2017a,b; Nesvorný et al., 2017). Simulations that include one or more "rogue planets," with masses similar to Mars or Earth that are ejected after orbiting in the Kuiper belt region for a few hundred million years, are also very successful at lifting pericenters for large-a TNOs (Gladman and Chan, 2006; Silsbee and Tremaine, 2018).

Fig. 3.2 shows the TNO orbits resulting from a rogue planet simulation where a two Earth mass planet started at $a = 35$ AU and $q = 30$ AU, and was ejected by interactions with the giant planets after 200 Myr (from Gladman and Chan, 2006). The TNO test particles are then integrated for 1 Gyr in the presence of the four giant planets, resulting in the distribution shown (the clustering results from cloning of particles, prior to the 1 Gyr simulation). Also shown for comparison in Fig. 3.2 is a TNO emplacement simulation from Lawler et al. (2017) that includes an eccentric distant planet. For comparison, the same real TNOs are plotted as in Fig. 3.1. We note that this is provided as an indicative rather than a fully quantitative comparison, as survey characterizations do not exist for all of the discovery surveys of this ensemble of known TNOs, and thus cannot be applied to appropriately bias the simulation outcomes. While both the rogue planet model and distant giant planet model manage to produce TNOs at very high-pericenter distances, even higher q than 2012 VP_{113} and Sedna,

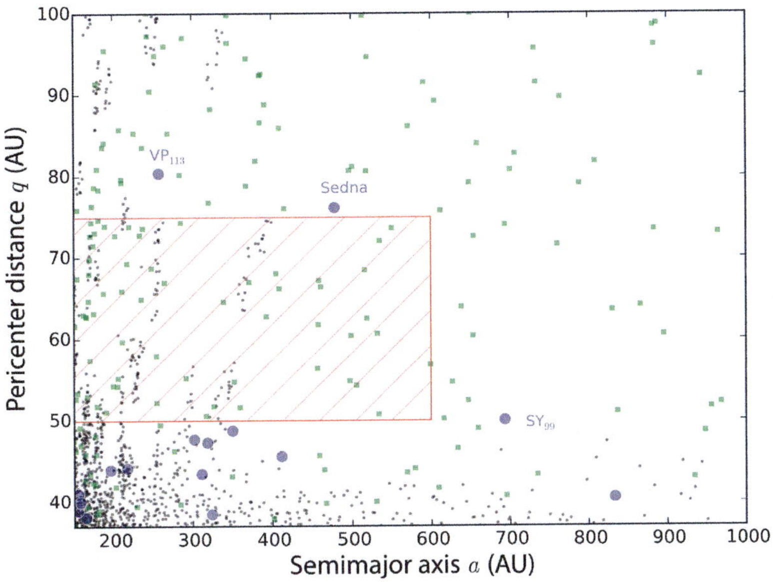

FIG. 3.2 Orbital distribution of simulated TNOs resulting from the rogue planet simulation in Gladman and Chan (2006), cloned and integrated for an additional Gyr (*small black points*), and simulated TNOs resulting from an emplacement simulation of Lawler et al. (2017) including an eccentric distant giant planet (*green squares*). As in Fig. 3.1, *large blue dots* show real TNOs and the *red hatched box* indicates the region of (a, q) orbits yet without real TNO detections. Both the rogue planet model and the additional planet model are able to produce TNOs with very high q values, even higher than 2012 VP_{113} and Sedna, and both models also produce TNOs inside the hatched box. If either of these models represents reality, Fig. 3.1 shows that it is very unlikely that there would be zero detections of TNOs inside the box, due to the greater detectability of TNOs in that box as compared with 2012 VP_{113} and Sedna.

both models also produce many TNOs inside the red hatched box, which contains zero real TNO detections to date. The TNO distribution produced by either of these models predicts that there should be more easily detectable TNOs inside the red box, and it is unlikely that the highest-q TNOs would be detected without any TNOs detected in the 50 AU $< q <$ 70 AU range. With either of these models, 2012 VP_{113} and Sedna remain hard-to-explain outliers.

All of these hypotheses for lifting pericenters of distant TNOs have associated simulations modeling a small-body population, produced with many degrees of freedom. The critical test for each hypothesis is how well it reproduces the observed TNO population. Several of these models are currently providing population outcomes at the level of detail necessary for testing against the observed TNO population (cf. Fig. 3.2); others are still maturing toward that critical point.

3.3 Diffusion and motion of large semimajor axes orbits

The nuanced effects of gravitational perturbation from the planets extend over remarkably wide spatial scales and timescales for large semimajor axes orbits, in ways not seen in the inner solar system. As initially suggested by Duncan et al. (1987), each distant encounter of Neptune by a TNO on a near-parabolic $a \gtrsim 100$ AU orbit with a perihelion exterior to Neptune will produce an energy change in the TNO's orbit, even for high-q orbits. The effect of the energy change at each perihelion passage is a change in the size of the orbit's semimajor axis, while the orbit's perihelion stays constant. The weak kicks by Neptune at the TNO's perihelion change its orbital a on a timescale $\propto a^{-1/2}$. As the semimajor axis changes can be modeled as a random walk, with the orbit either becoming larger or decreasing in size with each passage, this change in orbital dimensions for large-a detached TNOs is an example of dynamical diffusion. It is an effect that occurs purely under the gravitational influence of the known planets.

Bannister et al. (2017) showed that diffusion is a substantive effect over Gyr for the large-a detached (high-q) TNO orbits. This investigation was prompted by the discovery of 2013 SY_{99} in the course of the OSSOS survey, on an orbit with $q = 50.0$ AU, $a = 733 \pm 42$ AU (noted in Fig. 3.1). Perihelia passages for orbits as large and distant as SY_{99}'s are only every 20 kyr, thus the energy walk is slow and requires the passage of Gyr to show changes in orbital semimajor axis. The semimajor axis of SY_{99} can change by a factor of two over the age of the solar system, due to the semimajor axis diffusion of 100 AU or more that it experiences on Gyr timescales—despite being fully 20 AU separate at perihelia passages with Neptune's orbit. There were hints of the presence of diffusion in the earlier studies of large TNO orbits: Gladman et al. (2002) found diffusive chaos when examining the orbit of 2001 CR_{105} (the first known member of the extreme orbit group), Sheppard and Trujillo (2016) noted semimajor axis mobility in the orbit of their discovery 2013 FT_{28} ($q = 43.47 \pm 0.08$ AU, $a = 295 \pm 7$ AU), while Gallardo et al. (2012) and Brasser and Schwamb (2015) saw diffusion in their modeling of subsamples of extreme TNO orbital phase space. Integrations of the then-known $45 < q < 50$ AU TNOs, with $180 < a < 300$ AU, in the presence of the giant planets showed that they exhibit diffusive semimajor axis behavior (Bannister et al., 2017). Like Sedna and 2012 VP_{113}, these orbits are within the placid $a \lesssim 1000$ AU region where they are isolated from the Gyr-timescale influence of perturbations by the Galactic tide (e.g., Brasser and Schwamb, 2015).

The evaluation of the largest minor planet orbits cannot take place in isolation. Orbits at several thousand AU start to experience the effects of the Galactic tide, in the inner fringe of the Oort cloud (Dones et al., 2004). The population density of the $a \sim 2000$ AU inner Oort cloud region is presently largely mysterious. Long-period comets are sourced from several tens of thousands of AU, where the influence of the Galactic tide is dominant (e.g., Dones et al., 2004; Vokrouhlický et al., 2019). The apocenters of scattering disk member orbits can extend into the inner Oort region, such as that of 2014 FE_{72} ($q = 36.3 \pm 0.1$ AU, $a = 1505 \pm 540$ AU[3]; noted in Fig. 3.1; Sheppard and Trujillo 2016). While other scattering members have more significant evolution of their orbits, like that of 2006 SQ_{372} ($q = 24.2$ AU, $a = 796$ AU; Kaib et al. 2009), while still spending some fraction of their orbit in the inner Oort region.

Such large-a scattering orbits as FE_{72} provide a conceptual link between the scattering disk and the inner Oort cloud, both past and present. The emplacement of the scattering disk and its subsequent decay under encounters with Neptune require many millions of minor planets to have been placed on exceptionally large-a orbits (Gladman, 2005; Levison et al., 2006).

[3] Heliocentric JPL Horizons elements from a 1511 day arc, computed June 11, 2018.

The combination of the existence of diffusion in so many of the large-a, high-q TNOs, and the way in which the scattering disk overlaps with the inner Oort cloud led Bannister et al. (2017) to the proposal of a mechanism for populating this region, which follows entirely from known physics and the existence of the known planets:

> An object scatters outward in the initial emplacement of the scattering disk, pushing the orbital semimajor axis into the inner fringe of the Oort cloud. At a semimajor axis of a thousand or more AU, Galactic tides couple and torque out the orbit's perihelion. Once an object is orbiting with $q = 50$ AU and $a \sim 1000-2000$ AU, it diffuses to a lower-a orbit via planetary energy kicks. A reservoir population of objects must then exist that cycles under diffusion with $q = 40-50$ AU and $a \sim 1000-2500$ AU.

The scattering-to-diffusion scenario made a prediction for future large-a discoveries:

> Our scenario for forming 2013 SY99's orbit does show that for an inner Oort cloud object with q lifted to $\gtrsim$55 AU, diffusion will be too weak to retract the semimajor axis. Thus, future discoveries with $q \sim 60$ AU should have $a \gtrsim 1000$ AU.

The next year, Sheppard et al. (2019) reported the discovery of the first TNO with perihelion intermediate between SY_{99} and the two very high-q TNOs (Sedna and VP_{113}): 2015 TG_{387} has $q = 65 \pm 1$ AU. This TNO has $a = 1190 \pm 70$ AU (Sheppard et al., 2019), in line with the scattering-to-diffusion scenario. Under the scenario outlined earlier, its orbit is fossilized. Sheppard et al. (2019) show that in the configuration of the known solar system, TG_{387}'s orbit is presently stable, with Galactic tides cycling q on very long (Gyr) timescales. Potentially, orbits like TG_{387}'s could return to the more actively diffusing part of the proposed cycling population. If stellar fly-bys are also modeled, TG_{387}'s orbit can have its perihelion driven to lower values of $q \sim 50-55$ AU by the combination of tidal and stellar perturbations. In this case, TG_{387}'s orbit becomes more actively altering, diffusing in a on order of a hundred AU or more (Sheppard et al., 2019).

The scattering-to-diffusion scenario has an inherent limit to the most distant perihelion orbit it can explain: The kicks from Neptune that permit diffusion to a lower semimajor axis eventually become too weak. Thus, diffusion does not explain the $q \sim 80$ AU orbits of VP_{113} and Sedna. However, it provides an interesting possibility that explains well the orbits of the remainder of the currently known large-a high-q TNOs in the solar system as we know it.

3.4 Dynamical effects expected to be imprinted on the distant Kuiper belt by the presence of an additional massive planet

In this section, we discuss the results of published n-body simulations that take into account the strong pericenter-raising effects of an additional distant planet (e.g., Lawler et al., 2017). The presence or the absence of a massive, distant planet results in very different orbital distributions for large-a TNOs. It was originally proposed by Trujillo and Sheppard (2014) that an apparent clustering in ω for the six known high-q TNOs at the time could be explained by an undiscovered planet in the distant solar system. This theory was expanded on by Batygin and Brown (2016), who proposed that certain orbits for this distant planet will cause large-a, high-q TNOs to have their orbits physically aligned, so the longitude of the ascending node ω, the argument of pericenter Ω, and the longitude of pericenter ϖ (where $\varpi = \Omega + \omega$) will remain confined for all time. As more high-q TNOs have been discovered, the statistical strength of clustering in all of these orbital angles has grown weaker; through modifications to which orbital a/q cuts are applied, some continue to argue (e.g., Brown and Batygin, 2019) that a clustering signal remains in one or more orbital angles. Other simulations have highlighted dynamical effects that a massive distant planet would have on this detached TNO population that were not highlighted in the initial published theories (e.g., Shankman et al., 2017b).

Lawler et al. (2017) used n-body simulations to create a Kuiper belt analog in the presence of a distant massive planet and the four known giant planets, focusing on realistically creating the scattering TNOs using the method of Kaib et al. (2011), including Galactic tides and stellar flybys. These simulations also demonstrated that a distant massive planet will take an initially dynamically cold distribution of TNOs and raise pericenters and inclinations on Gyr timescales while creating the scattering disk. The resulting TNO distributions from these five-planet emplacement simulations were then compared with a control simulation that included just the known planets (Kaib et al., 2011), which have been shown to reproduce the orbital properties of the scattering TNOs at all a (Shankman et al., 2013; Lawler et al., 2018c). The five-planet simulations easily produce a large population of high-q TNOs, but simultaneously produce a wide distribution of inclinations, including a large fraction of retrograde scattering and detached TNOs. Although substantive in size, Lawler et al. (2017) conclude that such orbits would not be strongly detectable in current surveys.

Shankman et al. (2017b) showed that the same inclination-raising mechanism will cause all (then) known high-q TNOs to flip to retrograde inclinations on Gyr timescales, thus there should be a nearly equal number of retrograde as prograde high-q TNOs. They also showed that with such a broad inclination distribution, the detection of just one of these objects, Sedna, requires a massive number of TNOs on similar a and q orbits spread over a range of inclination, implying a total mass of order tens of Earth masses on such obits. Batygin et al. (2019) also discuss this effect and find it provides a reasonable explanation for known highly inclined TNOs but do not provide detectable population estimates as these would be highly dependent on particulars of the model and not well constrained by current observations. Li et al. (2018) showed inclination flipping will continue to occur for a moderate-eccentricity ($e \gtrsim 0.4$) and near-coplanar distant planet, in a similar mechanism to the near-coplanar flip induced in a hierarchical three-body system (Li et al., 2014). Dynamical simulations of the newly discovered high-q TNO described in Sheppard et al. (2019), 2015 TG_{387}, agree with the existence of inclination flipping; a large fraction of clones of 2015 TG_{387} in simulations that include a distant giant planet flip to retrograde orbits on Gyr timescales.

While observational constraints on a large retrograde population are currently weak due to their large predicted distances (Lawler et al., 2017, 2018a), these simulations imply that if there is a giant distant planet, the inclination distribution of high-q, large-a TNOs should be nearly isotropic (though Li et al. 2018 find some substructure will occur for $a \lesssim 300$ AU). As yet, there remains little evidence of such a dynamically hot inclination distribution. The highest-q known TNOs both have $i < 25$ degrees. Promisingly, the highest-i TNO yet known, 2015 BP_{519} has $i = 54$ degrees; however, it is on an orbit actively interacting with Neptune ($q = 35.25 \pm 0.08$ AU, $a = 449.0 \pm 0.5$ AU; Becker et al. 2018). In additional, the masses required for detection of even one high-q isotropic TNO are worryingly high (Shankman et al., 2017b).

An additional giant planet is one possible way to explain the orbits of high-q TNOs, but some of the other effects it would have on the orbits of TNOs do not appear to agree with observations. The science driver behind this latest cycle of additional giant planet simulations was initially proposed was to explain the apparent simultaneous clustering of the three orbital angles (Ω, ω, and ϖ) of these high-q TNOs, and here we must discuss the complicated and unintuitive biases that are introduced by surveys of this observationally challenging TNO population.

3.5 Detectability of orbital effects

All observational surveys contain biases. By understanding and carefully keeping track of as many biases as possible, one can understand which types of detections (in this case, which types of orbits) were most unlikely in a survey, and thus which classes of objects represent larger populations than a survey's raw number of detections naively suggest. Accounting for the fraction of time that a given TNO is visible on its orbit and the survey's sky coverage are the biggest effects, and attempts have been made to quantify and account for biases in several TNO surveys at this level (e.g., Schwamb et al., 2010; Adams et al., 2014).

The OSSOS Ensemble of surveys (Petit et al., 2011, 2017; Alexandersen et al., 2016; Bannister et al., 2018) was specifically designed with bias characterization as a top priority, resulting in Survey Simulator software (Petit et al., 2018) that allows TNO orbital distribution models to be forward biased by all the characteristics of the survey, including sky pointing for each survey block, magnitude limits, detection efficiencies, and chip gaps. The OSSOS Ensemble of surveys also took great pains to track every single TNO that was detected, using careful orbital measurements over 5 months in each discovery year and recovery over >3 oppositions (Bannister et al., 2018), so there is no bias in orbit type, unlike other surveys which preferentially do not track low-q TNOs, or have a high rate of lost TNOs. Because of this, TNOs that were detected as part of the OSSOS Ensemble can be analyzed statistically, and a degree of de-biasing can be achieved for each subpopulation, measuring orbital properties and size distributions (Lawler et al., 2018a). The information needed to perform analysis using the OSSOS Ensemble is freely available (see Bannister et al., 2018) and given that this sample represents about 40% of the currently known TNOs with reliable orbits, the reader is encouraged to consider this particular sample in examinations of the TNO orbital structure.

3.5.1 Biases in the angle of pericenter detection in the large-*q* large-*a* TNO sample

Much has been made of the alignment of the pericenter angles of large-a TNOs. When all of the high-q TNOs detected in the OSSOS Ensemble are analyzed separately from all other known high-q TNOs, the distribution of orbital angles ω, Ω, and ϖ are consistent with a uniform distribution (Shankman et al., 2017a; Bannister et al., 2018). Brown and Batygin (2019) examine the OSSOS Ensemble along, using the characterization information provided with the sample, and also find that, while the sample is not inconsistent with alignment, the sample does not require to be drawn from an aligned distribution. To restate this a different way, when a uniform distribution of high-q, large-a orbits is forward biased by the OSSOS survey pointings and detection efficiencies, it produces sets of high-q, large-a simulated detections that are statistically indistinguishable from the real survey detections. The high-q, large-a TNOs detected by the OSSOS ensemble of surveys show no evidence for orbital clustering in any of the three orbital angles (Ω, ω, and ϖ).

Many of the high-q, large-a TNOs discovered to date are from surveys that have not yet reported their pointing history or tracking fraction, so one cannot statistically test the populations in the same way as the OSSOS detections, but we can make some assumptions about telescope pointing in order to test the biases that are likely present in some of these surveys. In Fig. 3.3, we present the current sample of such orbits (as of October 1, 2018) to allow some examination of that sample. The figure presents the sample of six high-q TNO orbits that created the original speculation (red points), the next six TNOs detected (gray points), and the most recently discovered TNOs in blue. The feature that originally drew the attention of Trujillo and Sheppard (2014) was the detection of objects with ω near 0 degrees and a complete lack of detections with ω near 180 degrees. Flux bias in the detected sample causes most detections to be of TNOs near the pericenters of their orbits (i.e., with mean anomaly $\mathcal{M}$ near 0 degrees = 360 degrees). This is coupled with the habit of conducting TNO searches in fields that predominantly straddle the sky location of the ecliptic plane, which forces most discovered TNO orbits to have ω near 0 degrees and 180 degrees (as described in Section 3.1). Although the preference for angles near $\omega = 0$ degrees is clearly present in the early sample, there were no detections found near $\omega = 180$ degrees, which is a puzzling feature of the sample.

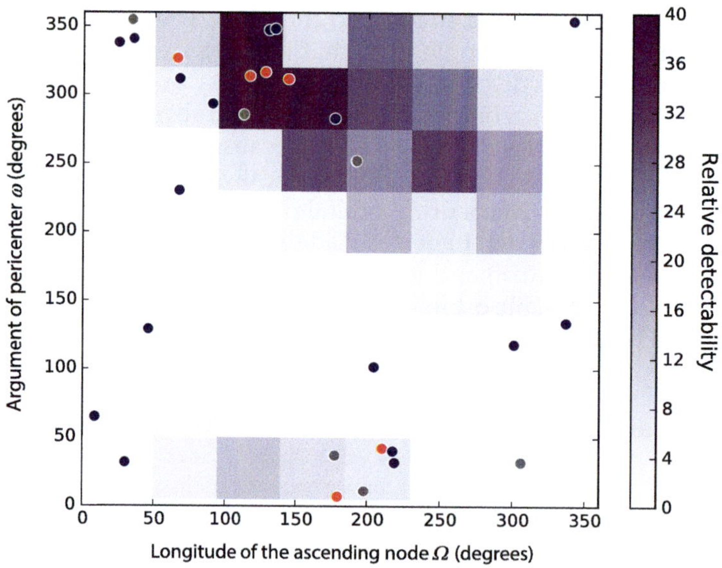

FIG. 3.3 Dots show distribution of known TNOs (observed arcs longer than 10 months) with $a > 150$ AU and $q > 30$ AU in (Ω, ω). *Red dots* indicate the first six such known TNOs, *gray dots* the next six, and the *blue dots* the most recently discovered TNOs. The lack of objects with values of ω near 180 degrees is not an easy bias to disentangle. The *grid of colored boxes* indicates the number of objects a mock survey would have detected on a grid of Ω, ω values, sampled from the measured a, q, and i elements of known TNOs, simulating a southern hemisphere survey. The most detectable area of (Ω, ω)-space (*darkest purple squares*) occurs where most of the first known high-q TNOs are.

In Fig. 3.3, we also give, on a grid of (ω, Ω) values, the relative number of detections one might expect at the given Ω and ω values when drawing from a sample that is uniformly distributed but with a, q, and i sampled from the known TNOs. For simplicity, we assume a flux-limited survey focusing on fields south of the ecliptic and observing in September, October, November, February, and March (when the best weather conditions occur in the mountains of Chile). From the grid of numbers, we can see that there are parts of the (Ω, ω) space where these orbits are much more strongly detectable than others.

This hypothetical survey is examined as a thought experiment to alert the reader to the complexity of the bias interactions. The (Ω, ω) alignment first reported is now largely washed out by the increased sample size (see blue points in Fig. 3.3), but there continues to be a paucity of detections near $\omega = 180$ degrees. Without detailed knowledge of the pointing history and careful measuring of a survey's detection and tracking efficiency through the various seasons of observations, interpretation of Fig. 3.3 is problematic at best. Regardless of the distribution's physical reality, there are as yet no described dynamical processes that keep ω values away from 180 degrees, and accepting that the ω distribution is most likely due to observational biases is the only supportable explanation.

Subsequent to the claim of an alignment of ω values, possible alignment in the longitude of pericenter ($\varpi = \Omega + \omega$) has become a popular point of discourse, the appeal being that one can conceive of physical processes that might align the values of ϖ (e.g., Batygin and Brown, 2016), making this a plausibly physical structure. However, one must consider that the observationally biased alignment that exists within the raw detected distribution of ω values propagates forward into a clustering ϖ, as the values of Ω are not uncorrelated, and ϖ is defined as the sum of the two angles Ω and ω (see Fig. 3.3). Thus, although there are good proposed physical mechanisms to cause a clustering or alignment of ϖ, the clustering of the observed values of ϖ is contaminated by the same observational biases discussed in the previous paragraph.

3.6 Summary and conclusions

There are several dynamical effects under active theoretical development to explain the observed high-q TNOs. The newest announced high-q TNO, 2015 TG_{387}, perfectly falls into the (a, q) range predicted to be affected by chaotic diffusion as described in Bannister et al. (2017). Among the less-explored dynamical mechanisms, rogue planets appear to create distributions of high-q TNOs that match observations reasonably well (Lawler et al., 2018a; Silsbee and Tremaine, 2018). Precursor simulations like that of Gladman and Chan (2006) should be revisited in light of the new high-q TNO discoveries to date. These types of simulations produce Sedna-like TNOs without a substantial retrograde TNO population at large a. Raising perihelia to the values of the highest-q TNOs, Sedna and VP_{113}, remains a challenge to several of the other proposed mechanisms, though stellar fly-bys remain a promising route. Rogue planet models, however, fail to provide a hole in the pericenter distribution as appears to be present in the detected sample. Indeed, the authors are not aware of any models that reproduce this feature.

While a distant massive planet is effective at raising pericenters, it also substantially raises inclinations, and current surveys have not yet reported abundant high-i TNOs. The authors of this chapter have already reported some of the problematic orbital evolution effects that an additional massive planet in the outer solar system would create. In those works, we found that the alignment of orbits caused by a massive external planet are not particularly strong (Shankman et al., 2017a) and the signature of such an alignment would be difficult to detect in the current sample of known TNOs (Lawler et al., 2017). Thus, our expectation is that at present there is *not* strong evidence of a massive external perturber.

The lack of TNO detections inside the red box in Figs. 3.1 and 3.2, however, provides an intriguing possibility. We may be able to exclude the existence of such a planet with present published TNO data sets. There are no TNOs reported with pericenters between 50 and 75 AU and semimajor axis interior to 1000 AU. Indeed, other authors have already remarked on the absence of such orbits (e.g., Trujillo and Sheppard, 2014; Bannister et al., 2017). Recall that in Fig. 3.1 the grid of colored boxes provides some measure of the inverse probability of detection of particular orbits, given a survey. A survey that might have detected a TNO at $a \sim 500$ AU and $q \sim 75$ AU is actually more likely to have detected objects with similar a but smaller values of q. The same is true of the other $q > 75$ AU detections: The lower-q but similar a detections are always more likely. Thus, the lack of detections in the $50 < q < 75$ AU range may be indicating that there is really an absence of TNOs on orbits in this range. This strongly contradicts models of orbital evolution that include an additional planet, as the gravitational action of such an object would cause TNOs to be distributed across a range of q values at any given moment (Fig. 3.2; Shankman et al., 2017b; Lawler et al., 2017). Thus, if the lack of objects in the 50 AU $< q <$ 70 AU range is real, the hypothesized external planet can be excluded.[4] In most physical situations, multiple effects play at any given point in time. Perhaps, we should be cautious of requiring reduction to a single mechanism to produce all the complexity of the distant TNO populations across the solar system's history.

Acknowledgments

The authors thank Brett Gladman (UBC) for useful discussions during the preparation of this manuscript.

References

Adams, E.R., Gulbis, A.A.S., Elliot, J.L., Benecchi, S.D., Buie, M.W., Trilling, D.E., Wasserman, L.H., 2014. De-biased populations of Kuiper belt objects from the deep ecliptic survey. Astron. J. 148, 55. https://doi.org/10.1088/0004-6256/148/3/55.

Alexandersen, M., Gladman, B., Kavelaars, J.J., Petit, J.-M., Gwyn, S.D.J., Shankman, C.J., Pike, R.E., 2016. A carefully characterized and tracked Trans-Neptunian survey: the size distribution of the Plutinos and the number of Neptunian Trojans. Astron. J. 152, 111. https://doi.org/10.3847/0004-6256/152/5/111.

Bannister, M.T., Shankman, C., Volk, K., Chen, Y.-T., Kaib, N., Gladman, B.J., Jakubik, M., Kavelaars, J.J., Fraser, W.C., Schwamb, M.E., Petit, J.-M., Wang, S.-Y., Gwyn, S.D.J., Alexandersen, M., Pike, R.E., 2017. OSSOS. V. Diffusion in the orbit of a high-perihelion distant solar system object. Astron. J. 153, 262. https://doi.org/10.3847/1538-3881/aa6db5.

Bannister, M.T., Gladman, B.J., Kavelaars, J.J., Petit, J.-M., Volk, K., Chen, Y.-T., Alexandersen, M., Gwyn, S.D.J., Schwamb, M.E., Ashton, E., Benecchi, S.D., Cabral, N., Dawson, R.I., Delsanti, A., Fraser, W.C., Granvik, M., Greenstreet, S., Guilbert-Lepoutre, A., Ip, W.-H., Jakubik, M., Jones, R.L., Kaib, N.A., Lacerda, P., Van Laerhoven,

[4] There may be some very specialized orbital configurations of a distant planet that preserve the emptiness of this q zone. As of this writing, none have been proposed.

C., Lawler, S., Lehner, M.J., Lin, H.W., Lykawka, P.S., Marsset, M., Murray-Clay, R., Pike, R.E., Rousselot, P., Shankman, C., Thirouin, A., Vernazza, P., Wang, S.-Y., 2018. OSSOS. VII. 800+ Trans-Neptunian objects—the complete data release. Astrophys. J. Suppl. Ser. 236, 18. https://doi.org/10.3847/1538-4365/aab77a.

Batygin, K., Brown, M.E., 2016. Evidence for a distant giant planet in the solar system. Astron. J. 151, 22. https://doi.org/10.3847/0004-6256/151/2/22.

Batygin, K., Adams, F.C., Brown, M.E., Becker, J.C., 2019. The planet nine hypothesis. Phys. Rep. 805, 1–53. https://doi.org/10.1016/j.physrep.2019.01.009.

Becker, J.C., Khain, T., Hamilton, S.J., Adams, F.C., Gerdes, D.W., Zullo, L., Franson, K., Millholland, S., Bernstein, G.M., Sako, M., Bernardinelli, P., Napier, K., Markwardt, L., Lin, H.W., Wester, W., Abdalla, F.B., Allam, S., Annis, J., Avila, S., Bertin, E., Brooks, D., Carnero Rosell, A., Carrasco Kind, M., Carretero, J., Cunha, C.E., D'Andrea, C.B., da Costa, L.N., Davis, C., De Vicente, J., Diehl, H.T., Doel, P., Eifler, T.F., Flaugher, B., Fosalba, P., Frieman, J., García-Bellido, J., Gaztanaga, E., Gruen, D., Gruendl, R.A., Gschwend, J., Gutierrez, G., Hartley, W.G., Hollowood, D.L., Honscheid, K., James, D.J., Kuehn, K., Kuropatkin, N., Maia, M.A.G., March, M., Marshall, J.L., Menanteau, F., Miquel, R., Ogando, R.L.C., Plazas, A.A., Sanchez, E., Scarpine, V., Schindler, R., Sevilla-Noarbe, I., Smith, M., Smith, R.C., Soares-Santos, M., Sobreira, F., Suchyta, E., Swanson, M.E.C., Walker, A.R., DES Collaboration, 2018. Discovery and dynamical analysis of an extreme Trans-Neptunian object with a high orbital inclination. Astron. J. 156, 81. https://doi.org/10.3847/1538-3881/aad042.

Bernstein, G., Khushalani, B., 2000. Orbit fitting and uncertainties for Kuiper belt objects. Astron. J. 120, 3323–3332. https://doi.org/10.1086/316868.

Brasser, R., Schwamb, M.E., 2015. Re-assessing the formation of the inner Oort cloud in an embedded star cluster-II. Probing the inner edge. Mon. Not. R. Astron. Soc. 446, 3788–3796. https://doi.org/10.1093/mnras/stu2374.

Brasser, R., Duncan, M.J., Levison, H.F., 2006. Embedded star clusters and the formation of the Oort cloud. Icarus 184 (1), 59–82.

Brasser, R., Schwamb, M.E., Lykawka, P.S., Gomes, R.S., 2012. An Oort cloud origin for the high-inclination, high-perihelion Centaurs. Mon. Not. R. Astron. Soc. 420, 3396–3402. https://doi.org/10.1111/j.1365-2966.2011.20264.x.

Brown, M.E., 2017. Observational bias and the clustering of distant eccentric Kuiper belt objects. Astron. J. 154, 65. https://doi.org/10.3847/1538-3881/aa79f4.

Brown, M.E., Batygin, K., 2019. Orbital clustering in the distant solar system. Astron. J. 157, 62. https://doi.org/10.3847/1538-3881/aaf051.

Brown, M.E., Trujillo, C., Rabinowitz, D., 2004. Discovery of a candidate inner Oort cloud planetoid. Astrophys. J. 617, 645–649. https://doi.org/10.1086/422095.

de la Fuente Marcos, C., de la Fuente Marcos, R., 2016. Finding planet nine: apsidal anti-alignment Monte Carlo results. Mon. Not. R. Astron. Soc. 462, 1972–1977. https://doi.org/10.1093/mnras/stw1778.

Dones, L., Weissman, P.R., Levison, H.F., Duncan, M.J., 2004. Oort cloud formation and dynamics. In: Festou, M.C., Keller, H.U., Weaver, H.A. (Eds.), Comets II. University of Arizona Press, San Francisco, pp. 153–174.

Duncan, M., Quinn, T., Tremaine, S., 1987. The formation and extent of the solar system comet cloud. Astron. J. 94, 1330–1338.

Elliot, J.L., Kern, S.D., Clancy, K.B., Gulbis, A.A.S., Millis, R.L., Buie, M.W., Wasserman, L.H., Chiang, E.I., Jordan, A.B., Trilling, D.E., Meech, K.J., 2005. The deep ecliptic survey: a search for Kuiper belt objects and Centaurs. II. Dynamical classification, the Kuiper belt plane, and the core population. Astron. J. 129, 1117–1162. https://doi.org/10.1086/427395.

Fernández, J.A., Brunini, A., 2000. The buildup of a tightly bound comet cloud around an early Sun immersed in a dense Galactic environment: numerical experiments. Icarus 145, 580–590. https://doi.org/10.1006/icar.2000.6348.

Gallardo, T., Hugo, G., Pais, P., 2012. Survey of Kozai dynamics beyond Neptune. Icarus 220, 392–403. https://doi.org/10.1016/j.icarus.2012.05.025.

Gladman, B., 2005. The Kuiper belt and the Solar system's comet disk. Science 307, 71–75. https://doi.org/10.1126/science.1100553.

Gladman, B., Chan, C., 2006. Production of the extended scattered disk by rogue planets. Astrophys. J. 643, L135–L138. https://doi.org/10.1086/505214.

Gladman, B., Holman, M., Grav, T., Kavelaars, J., Nicholson, P., Aksnes, K., Petit, J.-M., 2002. Evidence for an extended scattered disk. Icarus 157, 269–279. https://doi.org/10.1006/icar.2002.6860.

Gladman, B., Marsden, B.G., Vanlaerhoven, C., 2008. Nomenclature in the outer solar system. In: The Solar System Beyond Neptune. University of Arizona Press, Tucson, AZ, pp. 43–57.

Gomes, R.S., 2003. The origin of the Kuiper belt high-inclination population. Icarus 161, 404–418. https://doi.org/10.1016/S0019-1035(02)00056-8.

Jílková, L., Portegies Zwart, S., Pijloo, T., Hammer, M., 2015. How Sedna and family were captured in a close encounter with a solar sibling. Mon. Not. R. Astron. Soc. 453, 3157–3162. https://doi.org/10.1093/mnras/stv1803.

Kaib, N.A., Quinn, T., 2008. The formation of the Oort cloud in open cluster environments. Icarus 197, 221–238. https://doi.org/10.1016/j.icarus.2008.03.020.

Kaib, N.A., Quinn, T., 2009. Reassessing the source of long-period comets. Science 325, 1234. https://doi.org/10.1126/science.1172676.

Kaib, N.A., Sheppard, S.S., 2016. Tracking Neptune's migration history through high-perihelion resonant Trans-Neptunian objects. Astron. J. 152, 133. https://doi.org/10.3847/0004-6256/152/5/133.

Kaib, N.A., Becker, A.C., Jones, R.L., Puckett, A.W., Bizyaev, D., Dilday, B., Frieman, J.A., Oravetz, D.J., Pan, K., Quinn, T., Schneider, D.P., Watters, S., 2009. 2006 SQ 372: a likely long-period comet from the inner Oort cloud. Astrophys. J. 695 (1), 268–275.

Kaib, N.A., Roškar, R., Quinn, T., 2011. Sedna and the Oort Cloud around a migrating Sun. Icarus 215, 491–507. https://doi.org/10.1016/j.icarus.2011.07.037.

Kenyon, S.J., Bromley, B.C., 2004. Stellar encounters as the origin of distant solar system objects in highly eccentric orbits. Nature 432, 598–602. https://doi.org/10.1038/nature03136.

Kiss, C., Szabó, G., Horner, J., Conn, B.C., Müller, T.G., Vilenius, E., Sárneczky, K., Kiss, L.L., Bannister, M., Bayliss, D., Pál, A., Góbi, S., Verebélyi, E., Lellouch, E., Santos-Sanz, P., Ortiz, J.L., Duffard, R., Morales, N., 2013. A portrait of the extreme solar system object 2012 DR_{30}. Astron. Astrophys. 555, A3. https://doi.org/10.1051/0004-6361/201321147.

Lawler, S.M., Shankman, C., Kaib, N., Bannister, M.T., Gladman, B., Kavelaars, J.J., 2017. Observational signatures of a massive distant planet on the scattering disk. Astron. J. 153, 33. https://doi.org/10.3847/1538-3881/153/1/33.

Lawler, S.M., Kavelaars, J.J., Alexandersen, M., Bannister, M.T., Gladman, B., Petit, J.-M., Shankman, C., 2018a. OSSOS: X. How to use a survey simulator: statistical testing of dynamical models against the real Kuiper belt. Front. Astron. Space Sci. 5, 14. https://doi.org/10.3389/fspas.2018.00014.

Lawler, S.M., Pike, R.E., Kaib, N., Alexandersen, M., Bannister, M.T., Chen, Y.-T., Gladman, B., Gwyn, S., Kavelaars, J.J., Petit, J.-M., Volk, K., 2018b. OSSOS: XIII. Fossilized resonant dropouts imply Neptune's migration was grainy and slow. Astron. J. 157, 235.

Lawler, S.M., Shankman, C., Kavelaars, J.J., Alexandersen, M., Bannister, M.T., Chen, Y.-T., Gladman, B., Fraser, W.C., Gwyn, S., Kaib, N., Petit, J.-M., Volk, K., 2018c. OSSOS. VIII. The transition between two size distribution slopes in the scattering disk. Astron. J. 155, 197. https://doi.org/10.3847/1538-3881/aab8ff.

Levison, H.F., Duncan, M.J., Dones, L., Gladman, B.J., 2006. The scattered disk as a source of Halley-type comets. Icarus 184, 619–633. https://doi.org/10.1016/j.icarus.2006.05.008.

Levison, H.F., Morbidelli, A., Van Laerhoven, C., Gomes, R., Tsiganis, K., 2008. Origin of the structure of the Kuiper belt during a dynamical instability in the orbits of Uranus and Neptune. Icarus 196, 258–273. https://doi.org/10.1016/j.icarus.2007.11.035.

Levison, H.F., Duncan, M.J., Brasser, R., Kaufmann, D.E., 2010. Capture of the Sun's Oort cloud from stars in its birth cluster. Science 329, 187–190. https://doi.org/10.1126/science.1187535.

Li, G., Naoz, S., Kocsis, B., Loeb, A., 2014. Eccentricity growth and orbit flip in near-coplanar hierarchical three-body systems. Astrophys. J. 785, 116. https://doi.org/10.1088/0004-637X/785/2/116.

Li, G., Hadden, S., Payne, M., Holman, M.J., 2018. The secular dynamics of TNOs and planet nine interactions. Astron. J. 156, 263. https://doi.org/10.3847/1538-3881/aae83b.

Lykawka, P.S., Mukai, T., 2008. An outer planet beyond Pluto and the origin of the Trans-Neptunian belt architecture. Astron. J. 135, 1161–1200. https://doi.org/10.1088/0004-6256/135/4/1161.

Madigan, A.-M., McCourt, M., 2016. A new inclination instability reshapes Keplerian discs into cones: application to the outer solar system. Mon. Not. R. Astron. Soc. 457, L89–L93. https://doi.org/10.1093/mnrasl/slv203.

Madigan, A.-M., Zderic, A., McCourt, M., Fleisig, J., 2018. On the dynamics of the inclination instability. Astron. J. 156, 141. https://doi.org/10.3847/1538-3881/aad95c.

Malhotra, R., 1993. The origin of Pluto's peculiar orbit. Nature 365, 819–821. https://doi.org/10.1038/365819a0.

Morbidelli, A., Levison, H.F., 2004. Scenarios for the origin of the orbits of the Trans-Neptunian objects 2000 CR_{105} and 2003 VB_{12} (Sedna). Astron. J. 128, 2564–2576. https://doi.org/10.1086/424617.

Muñoz-Gutiérrez, M.A., Peimbert, A., Pichardo, B., 2018. The contribution of Dwarf planets to the origin of low-inclination comets by the replenishment of mean motion resonances in debris disks. Astron. J. 156, 108. https://doi.org/10.3847/1538-3881/aad4f8.

Nesvorný, D., 2018. Dynamical evolution of the early solar system. Ann. Rev. Astron. Astrophys. 56, 137–174. https://doi.org/10.1146/annurev-astro-081817-052028.

Nesvorný, D., Vokrouhlický, D., 2016. Neptune's orbital migration was Grainy, not smooth. Astrophys. J. 825, 94. https://doi.org/10.3847/0004-637X/825/2/94.

Nesvorný, D., Vokrouhlický, D., Roig, F., 2016. The orbital distribution of Trans-Neptunian objects beyond 50 AU. Astrophys. J. Lett. 827, L35. https://doi.org/10.3847/2041-8205/827/2/L35.

Nesvorný, D., Vokrouhlický, D., Dones, L., Levison, H.F., Kaib, N., Morbidelli, A., 2017. Origin and evolution of short-period comets. Astrophys. J. 845, 27. https://doi.org/10.3847/1538-4357/aa7cf6.

Petit, J.-M., Kavelaars, J.J., Gladman, B.J., Jones, R.L., Parker, J.W., Van Laerhoven, C., Nicholson, P., Mars, G., Rousselot, P., Mousis, O., Marsden, B., Bieryla, A., Taylor, M., Ashby, M.L.N., Benavidez, P., Campo Bagatin, A., Bernabeu, G., 2011. The Canada-France ecliptic plane survey—full data release: the orbital structure of the Kuiper belt. Astron. J. 142, 131. https://doi.org/10.1088/0004-6256/142/4/131.

Petit, J.-M., Kavelaars, J.J., Gladman, B.J., Jones, R.L., Parker, J.W., Bieryla, A., Van Laerhoven, C., Pike, R.E., Nicholson, P., Ashby, M.L.N., Lawler, S.M., 2017. The Canada-France Ecliptic Plane Survey (CFEPS)—high-latitude component. Astron. J. 153, 236. https://doi.org/10.3847/1538-3881/aa6aa5.

Petit, J.-M., Kavelaars, J.J., Gladman, B., Alexandersen, M., 2018. OSS: OSSOS Survey Simulator. ASCL at Michigan Technological University, Houghton, MI.

Pfalzner, S., Bhandare, A., Vincke, K., Lacerda, P., 2018. Outer solar system possibly shaped by a stellar fly-by. Astrophys. J. 863, 45. https://doi.org/10.3847/1538-4357/aad23c.

Pike, R.E., Kavelaars, J.J., Petit, J.M., Gladman, B.J., Alexandersen, M., Volk, K., Shankman, C.J., 2015. The 5:1 Neptune resonance as probed by CFEPS: dynamics and population. Astron. J. 149, 202. https://doi.org/10.1088/0004-6256/149/6/202.

Schwamb, M.E., Brown, M.E., Rabinowitz, D.L., Ragozzine, D., 2010. Properties of the distant Kuiper belt: results from the Palomar Distant Solar System Survey. Astrophys. J. 720, 1691–1707. https://doi.org/10.1088/0004-637X/720/2/1691.

Sefilian, A.A., Touma, J.R., 2019. Shepherding in a self-gravitating disk of Trans-Neptunian objects. Astron. J. 157, 59. https://doi.org/10.3847/1538-3881/aaf0fc.

Shankman, C., Gladman, B.J., Kaib, N., Kavelaars, J.J., Petit, J.M., 2013. A possible divot in the size distribution of the Kuiper belt's scattering objects. Astrophys. J. Lett. 764, L2. https://doi.org/10.1088/2041-8205/764/1/L2.

Shankman, C., Kavelaars, J.J., Bannister, M.T., Gladman, B.J., Lawler, S.M., Chen, Y.-T., Jakubik, M., Kaib, N., Alexandersen, M., Gwyn, S.D.J., Petit, J.-M., Volk, K., 2017a. OSSOS. VI. Striking biases in the detection of large semimajor axis Trans-Neptunian objects. Astron. J. 154, 50. https://doi.org/10.3847/1538-3881/aa7aed.

Shankman, C., Kavelaars, J.J., Lawler, S.M., Gladman, B.J., Bannister, M.T., 2017b. Consequences of a distant massive planet on the large semimajor axis Trans-Neptunian objects. Astron. J. 153, 63. https://doi.org/10.3847/1538-3881/153/2/63.

Sheppard, S.S., Trujillo, C., 2016. New extreme Trans-Neptunian objects: toward a super-Earth in the outer solar system. Astron. J. 152, 221. https://doi.org/10.3847/1538-3881/152/6/221.

Sheppard, S.S., Trujillo, C., Tholen, D.J., 2016. Beyond the Kuiper belt edge: new high perihelion Trans-Neptunian objects with moderate semimajor axes and eccentricities. Astrophys. J. Lett. 825, L13. https://doi.org/10.3847/2041-8205/825/1/L13.

Sheppard, S.S., Trujillo, C.A., Tholen, D.J., Kaib, N., 2019. A new high perihelion Trans-Plutonian inner Oort cloud object: 2015 TG_{387}. Astron. J. 157, 139. https://doi.org/10.3847/1538-3881/ab0895.

Silsbee, K., Tremaine, S., 2018. Producing distant planets by mutual scattering of planetary embryos. Astron. J. 155, 75. https://doi.org/10.3847/1538-3881/aaa19b.

Soares, J.S., Gomes, R.S., 2013. Comparison of forming mechanisms for Sedna-type objects through an observational simulator. Astron. Astrophys. 553, A110. https://doi.org/10.1051/0004-6361/201219840.

Trujillo, C.A., Sheppard, S.S., 2014. A Sedna-like body with a perihelion of 80 astronomical units. Nature 507, 471–474. https://doi.org/10.1038/nature13156.

Trujillo, C.A., Jewitt, D.C., Luu, J.X., 2000. Population of the Scattered Kuiper belt. Astrophys. J. Lett. 529, L103–L106. https://doi.org/10.1086/312467.

Vokrouhlický, D., Nesvorný, D., Dones, L., 2019. Origin and evolution of long-period comets. Astron. J. 157, 181. https://doi.org/10.3847/1538-3881/ab13aa.

CHAPTER

4

Observational constraints on an undiscovered giant planet in our solar system

Chadwick A. Trujillo

Department of Physics and Astronomy, Northern Arizona University, Flagstaff, AZ, United States

4.1 Introduction

The idea of an undiscovered giant planet present in our solar system was rekindled several years ago by Trujillo and Sheppard (2014), who used it to explain the fact that the most distant known Trans-Neptunian objects (TNOs) in the solar system had similar arguments of perihelion. These arguments of perihelion would randomize on timescales of a few hundred million years due to the action of the giant planets without the presence of some external perturber, namely an undetected giant planet of some several Earth masses. This spawned a number of dynamical investigations assessing the validity of this possibility (de la Fuente Marcos and de la Fuente Marcos, 2014; Iorio, 2014; de la Fuente Marcos et al., 2015; Gomes et al., 2015) as well as formation scenarios that could create such an object (Bromley and Kenyon, 2014; Kenyon and Bromley, 2015). Most notably Batygin and Brown (2016a) performed extensive dynamical simulations coupled with analytic theory to suggest that an undiscovered giant planet would create not just a similarity in the argument of perihelion, but a constraint in physical space of the most distant objects. In addition, Batygin and Brown (2016b) found that such a planet could also create highly inclined TNOs whose origins have been in doubt for some time (Gladman et al., 2009). Batygin and Brown (2016a,b) combined with an assessment

The Trans-Neptunian Solar System.
https://doi.org/10.1016/B978-0-12-816490-7.00004-7

of the observational constraints on the discovery of such a planet (Brown and Batygin, 2016) renewed and expanded interest in the distant giant planet hypothesis. Trujillo and Sheppard (2014) found that not only were the arguments of perihelion clustered, but also that the discovered objects were not symmetric in the sky in terms of ecliptic longitude. They suggested that further discoveries (or null detections outside of the favored autumn region of the sky) would be needed to confirm this sky location clustering.

Concurrent with and subsequent to Batygin and Brown (2016a), several works attempted to place further observational constraints on the location of the planet by wide-field direct observation using ground-based telescopes (Sheppard and Trujillo, 2016b; Sheppard et al., 2016) and space-based telescopes (Kuchner et al., 2017; Meisner et al., 2017, 2018). Dynamical studies attempted to constrain the planet's orbit by its inferred action on the Cassini spacecraft (Fienga et al., 2016; Holman and Payne, 2016b) as well as the planet's effects on comets (Medvedev et al., 2017; Guliyev and Guliyev, 2017; Nesvorný et al., 2017). Many works examined the effect that a distant giant planet would have on the outermost TNOs. Studies were done on the secular and resonant dynamics of the distant TNOs (Beust, 2016; Batygin and Morbidelli, 2017; Saillenfest et al., 2017; Hadden et al., 2018; Bailey et al., 2018; Cáceres and Gomes, 2018). An astrometric study of the planet's effects on Pluto and the distant TNOs was also conducted (Holman and Payne, 2016a). The possibility of the distant TNOs being in mean-motion resonance with the planet was examined (Malhotra et al., 2016; de la Fuente Marcos and de la Fuente Marcos, 2016a). A few studies attempted to make an observable prediction of the location of the planet (de la Fuente Marcos and de la Fuente Marcos, 2016b,c; Millholland and Laughlin, 2017; Trujillo, 2020).

The effects of the planet on the most distant known TNOs have also been considered (de la Fuente Marcos et al., 2016; Shankman et al., 2017b; Becker et al., 2017) as well as its effects on the orbit of the Earth (Zeebe, 2017) and the other planets (Iorio, 2017), specifically high inclination objects (Chen et al., 2016; Becker et al., 2018) and a high aphelion object (Bannister et al., 2017b). In addition, the planet's effects on the protoplanetary disk were examined (Lawler et al., 2017) as well as constraints that the planet's orbit could place on the early solar system (Khain et al., 2018a).

Formation scenarios have been considered to explain how such a planet could form in our solar system (Kenyon and Bromley, 2016; Bromley and Kenyon, 2016; Wyatt et al., 2017; Silsbee and Tremaine, 2018; Batygin et al., 2019) and possibly be ejected (Li and Adams, 2016; Parker et al., 2017) or originate as a captured exoplanet (Mustill et al., 2016). Some works also considered the effects of stellar encounters on the planet (Martínez-Barbosa et al., 2017; Pfalzner et al., 2018). In addition, the physical characteristics of the planet have been explored by studying models of the known giant planets with differing masses and solar distances (Ginzburg et al., 2016; Fortney et al., 2016; Linder and Mordasini, 2016; Levi et al., 2017). The presence of a distant giant planet has also been suggested as the reason for the obliquity of the Sun (Bailey et al., 2016; Lai, 2016; Gomes et al., 2017).

Despite the large number of works discussing the hypothesized giant planet detectability, properties, gravitational effects, and formation mechanisms, questions remain about the veracity of its existence. In this work, we discuss the observational evidence for a distant undiscovered giant planet, discuss the biases that may affect those observations, and examine some of the observational constraints that can be placed on the hypothesized planet's location.

4.2 Observational evidence for the planet

Here, we review the observational evidence for the presence of an undiscovered giant planet in our solar system. Most TNO orbits can be fully explained by the action of Neptune and the known giant planets. In addition, Galactic tides become important for the most distant objects as they start to approach the Oort cloud, objects with semimajor axes $a \gtrsim 1000$–2000 AU (Dones et al., 2004; Gladman et al., 2008; Kaib et al., 2011; Bannister et al., 2017b; Sheppard et al., 2019). The most distant objects in the solar system, here referred as the extreme TNOs (ETNOs) and the inner Oort cloud objects (IOCs) (see specific orbital definition in Section 4.2.1), appear to currently have an orbital alignment that cannot be explained nor maintained by the known mass in the solar system. Thus, there must be some other activity or process in the outer solar system to explain the aligned orbits. Throughout this work, we use the following nomenclature to specify the Keplerian orbit of an object: semimajor axis a; eccentricity e; inclination to the ecliptic i; argument of perihelion ω, which is the angle from the ecliptic plane to the perihelion point; and longitude of ascending node Ω, which is the angle from the ecliptic/celestial plane crossing to the point in the orbit where an object first moves northward of the ecliptic. There is also the derived angle called longitude of perihelion $\bar{\omega} = \omega + \Omega$. Despite its name, $\bar{\omega}$ is not truly representative of the actual ecliptic longitude of perihelion since ω and Ω are angles in different planes if $i \neq 0$ (Brown and Batygin, 2016). The position of the object at a given epoch is described by the true anomaly f, which is the angle from perihelion to the object. Historically this is sometimes instead described by the mean anomaly M, which is the fraction of the orbital period that has elapsed since perihelion, expressed as an angle.

4.2.1 Argument of perihelion

The first suggestion of evidence for a distant giant planet was made by Trujillo and Sheppard (2014) who observed that the argument of perihelion of the most distant known TNOs (those with semimajor axes $a > 150$ AU) appeared to cluster near an argument of perihelion $\omega \sim 0$ degrees for objects with perihelion greater than $q > 30$ AU (i.e., exterior to Neptune's orbit) and were not related to observational bias. These specific limits of ($q > 30$ AU and $a > 150$ AU) were chosen somewhat arbitrarily based on the appearance of the trend at the time and were not based on dynamical analysis. One might expect that a perihelion significantly larger than $q > 30$ AU or semimajor axis significantly greater than $a > 150$ AU might be needed to minimize the influence of Neptune, with true limits only ascertained based on the exact orbit and mass of the hypothesized planet. This was clear even before an undiscovered giant planet was posited as Gomes et al. (2008) suggest that a combination of mean-motion resonances and Kozai resonances can create objects with perihelia as high as 60 AU given the action of Neptune and the Sun alone. In terms of simulations specific to the giant planet hypothesis, Batygin and Brown (2016a) suggest that larger values are needed for a and specific objects might need to be excluded based on the many Neptune mean-motion resonances that extend outward from the region. They specifically consider all objects with semimajor axis $a > 250$ AU and $q > 30$ AU as tracers of the planet trend and 150 AU $< a < 250$ AU on a case-by-case basis depending on dynamical stability simulations of Neptune's gravitational effects. The choice of $q > 30$ AU is probably also not ideal since

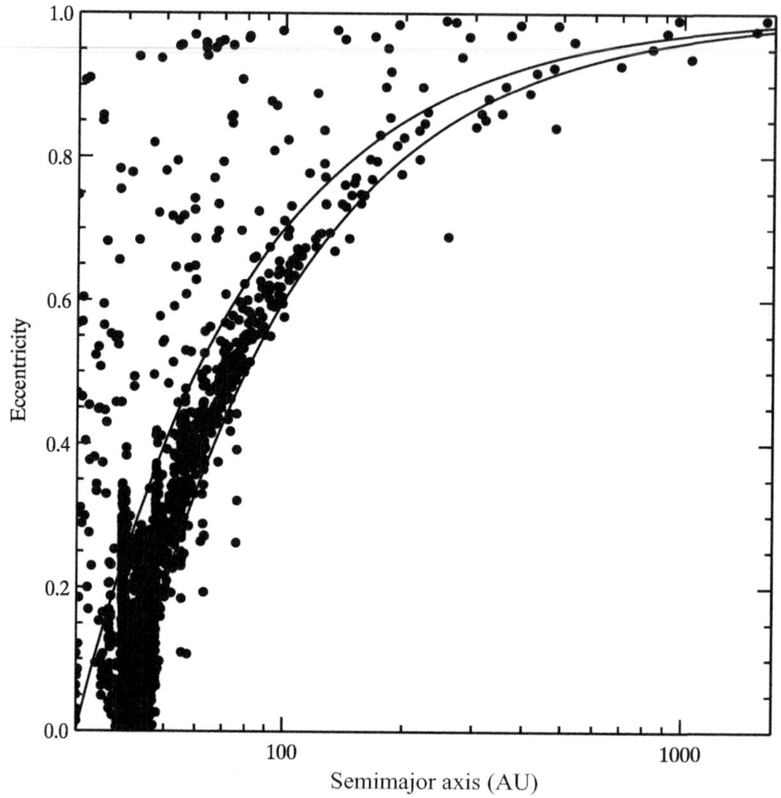

FIG. 4.1 Eccentricity versus semimajor axis for all multiopposition minor planets with semimajor axes greater than 30 AU. Two lines bound perihelion of 30 and 40 AU. Orbits within these perihelion bounds and with semimajor axes greater than about 50 AU can be created by gravitational interactions with Neptune. The objects with perihelion beyond 40 AU appear to be more affected by the hypothesized giant planet than Neptune.

there are numerous objects that are scattered by Neptune out to $q = 40$ AU, as depicted in Fig. 4.1. It is clear that there is a large population of objects scattered to large semimajor axes for $q < 40$ AU. Thus, it seems reasonable that 40 AU might be a valid lower limit for the perihelion of the objects most influenced by the hypothesized planet.

Nomenclature for these types of objects has not been established and it might not be reasonable to establish nomenclature without a more thorough understanding of whether the hypothesized planet exists and the strength of the Galactic tide. Using the terminology of Gladman et al. (2008), the objects most relevant to the hypothesized giant planet would be a subset of the "Detached TNOs" ($e > 0.24$ and not affected by Neptune) and IOCs ($a > 2000$ AU). Such a definition does not specifically split the detached TNOs into different groups around 150 AU $< a <$ 250 AU, where the apparent transition regime appears to be for objects that are affected more by the hypothesized planet than Neptune. For the IOCs, currently, there are no known objects with $a > 2000$ AU, but the concept behind this definition is that objects with very distant orbits begin to be affected by the Galactic tide. Since the strength of the Galactic tide is not well known, this suggests that the true transition to inner Oort cloud is also difficult to define. However, Sheppard et al. (2019) show that 2015 TG_{387} is likely to be marginally affected by the Galactic tide, so it may truly be the first IOC. We also note that upon its discovery, Sedna was suggested to be an IOC as well due to its unusually high perihelion (Brown et al., 2004).

Due to these complexities and the anticipation that these definitions will change as understanding increases, for the purposes of this work, we simply refer to the objects more affected by the hypothesized giant planet than Neptune as ETNOs ($a > 150$ AU and $q > 40$ AU) and suggest that the three objects with the highest perihelion be called IOCs ($a > 150$ AU and $q > 65$ AU, Sedna, 2012 VP_{113} and 2015 TG_{387}). These are all listed in Table 4.1. A thorough review of the dynamical significance of the IOCs, with their unusually high perihelia, is beyond the scope of this work as there are many works discussing their formation. However, it is worth noting that all formation scenarios for the IOCs require significant mass well beyond Neptune in the nascent solar system to raise the perihelia of these objects beyond ~40 AU. This is covered in more detail in Chapter 2. The object 2014 FE_{72}, which has an extremely large semimajor axis $a \approx 2100$ AU, might be considered a candidate IOC object. However, due to its low perihelion $q \approx 36.3$ AU, it is affected by both Neptune and the Galactic tide and Sheppard et al. (2019) find that its orbit is likely unstable and it will be ejected from the solar system within a few hundred mega years.

Several very distant ETNOs and IOCs have been found since the argument of perihelion trend was identified several years ago. We note that the argument of perihelion trend recognized several years ago continues today as depicted in Fig. 4.2. The overall trend has broadened with the addition of 2015 TG_{387}; however, the 11 objects with $q > 40$ AU and $a > 250$ AU all have arguments of perihelion in the range -98 degrees $< \omega < 118$ degrees. In the absence of observational bias, this is a decidedly nonrandom distribution. The probability of finding 11 objects drawn from a uniform distribution with -98 degrees $< \omega < 118$ degrees is $\left(\frac{216\text{ degrees}}{360\text{ degrees}}\right)^{11} = 0.0036$, which is a 2.9σ effect assuming Gaussian statistics. For the 15 $a > 150$ AU objects, all fall within -129 degrees $< \omega < 118$ degrees, the probability of which is $\left(\frac{247\text{ degrees}}{360\text{ degrees}}\right)^{15} = 0.0035$, again 2.9σ assuming Gaussian statistics. This is a rather simplistic method for calculating statistical significance which relies on specific choices of limits for ω and ignores observational bias. A more thorough treatment would perhaps employ a Monte-Carlo approach, such as that done for Sheppard et al. (2019) for assessing $\bar{\omega}$ statistical significance or an end-to-end estimate of all known biases such as that used by Shankman et al. (2017a), Khain et al. (2018b), and Sheppard et al. (2019). We will discuss observational bias in more detail in Section 4.3. However, here we note that one might expect a distribution favoring $\omega = 0$ degrees when observing near the ecliptic since by definition $\omega = 0$ degrees objects come to perihelion in the ecliptic and eccentric objects are most likely to be found near perihelion. If this were the case, one would expect to find objects in equal numbers with $\omega = 180$ degrees since those objects also come to perihelion in the ecliptic, but pass through perihelion in the opposite manner to the $\omega = 0$ degrees objects (i.e., moving from $+z$ to $-z$ rather than $-z$ to $+z$ in heliocentric Cartesian coordinates). Both the $\omega = 0$ degrees and the 180 degrees objects are prograde, so it is very difficult to envision an observational bias that would affect the $\omega = 180$ degrees objects but not the $\omega = 0$ degrees objects; such a bias would be needed to

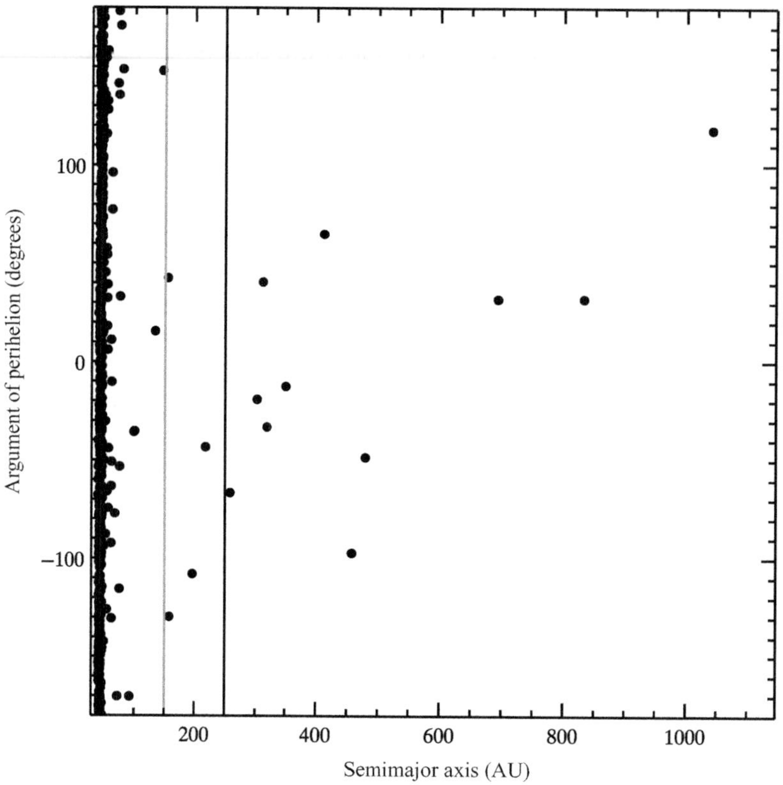

FIG. 4.2 Argument of perihelion versus semimajor axis for all multiopposition minor planets with perihelion greater than 40 AU. Two semimajor axis lines are indicated at 150 AU (*gray*) and 250 AU (*black*). Objects with semimajor axes less than the lines appear to have a random distribution of argument of perihelion. Objects beyond these lines appear to be clustered around $\omega \sim 0$ degrees, while no objects are near $\omega \sim 180$ degrees.

create the observed distribution from an underlying uniform one. Shankman et al. (2017a) show that their survey does not have such a bias and suggest the same should be true for other ecliptic surveys. They point out that off-ecliptic surveys can be biased in ω, but Sheppard and Trujillo (2016b) analyze this bias and show it is not significant in their survey which covers a very large sky area in both hemispheres. Sheppard and Trujillo (2016b) do suggest that biases may exist for $\bar{\omega}$, but this is shown to be minimal with the additional sky area surveyed in Sheppard et al. (2019). Thus, even without a detailed knowledge of observational biases, the observed trend seems compellingly interesting. In Section 4.2.2, we discuss the physical alignment of the ETNO/IOC orbits with respect to the hypothesized giant planet.

4.2.2 Orbit alignment

The choice of argument of perihelion to parameterize the giant planet's action is somewhat arbitrary since there are few physical mechanisms that would affect argument of perihelion but not other orbit aspects. In fact, as detailed by Batygin and Brown (2016a), the true orbital alignment is in physical space, relating to the plane of the objects as well as their perihelion location with respect to the hypothesized giant planet. Thus, a measure of the true orbital alignment of ETNOs/IOCs must include a mixture of inclination and longitude of ascending node, which together define the orbit plane, as well as the argument of perihelion which is needed to determine the perihelion location. Eccentricity and semimajor axis of course also play a role in determining the mean-motion resonant strength of an object with the planet.

It is difficult to visualize this orbital alignment as it is most evident in three-dimensional (3D) space. To illustrate this, we have created a 3D geometrically accurate model. We have made this by creating a custom JavaScript shape generator for the publicly available Autodesk Tinkercad software which is commonly used to define volumes for 3D printing.[1] A plan view of the orbits of the most distant ETNOs/IOCs is depicted in Fig. 4.3. From the figure, and an examination of the model in three dimensions, it becomes apparent that the true orbital alignment is a combination of orbit plane and longitude of perihelion.

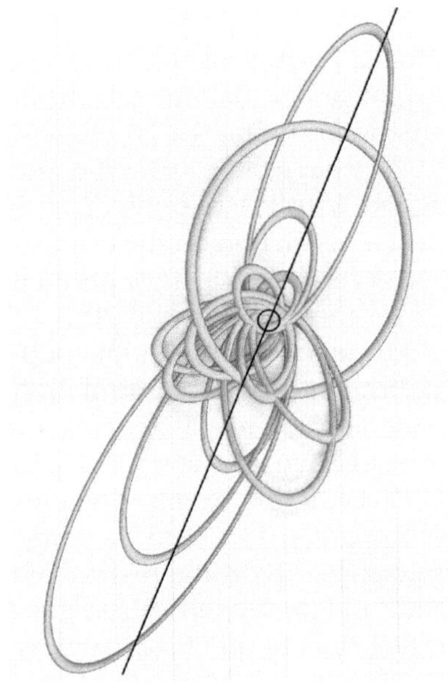

FIG. 4.3 Plan view of the most distant ETNO/IOC orbits with exaggerated orbit thickness. These are the same objects as depicted in Fig. 4.2 ($a > 150$ AU and $q > 40$ AU) with the addition of Neptune (orbit marked by *black circle*) and a hypothesized giant planet (the least eccentric object in the figure). The faint (*bold*) grid lines are about 15 AU (150 AU) apart and demarcate a translucent ecliptic plane. Right ascension of 0 hours is to the right. Orbit arcs which are below the ecliptic plane have grid lines visible above them. It is apparent that most objects have aphelion within about 45 degrees of a common straight line (in *black*) passing through the Sun.

[1] A 3D printable model is available at: https://www.tinkercad.com/things/geKbAcTJrF2.

Here, we introduce an alignment metric m to quantify the alignment relationship, which is similar to that used by Batygin and Brown (2016a) and produces similar results. We apply our metric to $q > 40$ AU objects and to a variety of semimajor axes. This metric can be applied to any subset of objects to assess whether they are aligned in a common way. It is based on two unit vectors computed from the population of n objects: (1) the mean orbit plane normal unit vector $\bar{\mathbf{p}}$ and (2) the mean longitude of perihelion unit vector $\bar{\bar{\omega}}$. The ecliptic latitude β and longitude λ of the orbit plane normal in spherical coordinates are first determined from inclination i and longitude of ascending node Ω via $\beta = 90$ degrees $- i$ and $\lambda = \Omega - 90$ degrees. These values are then converted to rectangular coordinates to determine the mean orbit plane normal unit vector and subsequent dot products. Computing the population metric m requires computing the average deviation of the population. To do so we use the following formula involving the kth object's orbit normal unit vector $\mathbf{p}_k$ and longitude of perihelion $\bar{\omega}_k$ unit vector via the following equation involving vector dot products:

$$m = \frac{1}{n}\sum_{k} |\mathbf{p_k} \cdot \bar{\mathbf{p}}| \times \sum_{k} |\bar{\omega}_\mathbf{k} \cdot \bar{\bar{\omega}}| \tag{4.1}$$

The metric has a maximum value of 1 when all objects in a population are perfectly aligned or antialigned with the mean vector and a theoretical minimum of 0 which would never be reached by any physical population. The reason to include both aligned and antialigned objects is because both types of objects have been observed and are predicted by theory. Most of the known ETNOs are antialigned with the hypothesized planet in terms of longitude of perihelion; in other words, they reach perihelion on the opposite side of the solar system from the hypothesized planet. However, several objects have perihelion similar to the hypothesized planet. This was first observed with the discovery of 2013 FT_{28} and Sheppard and Trujillo (2016a,b) suggested that this could be an additional stable population of objects.

Rather than considering the metric for a perfect population, we consider a perhaps more realistic example population where all objects are 20 degrees from the mean normal unit vector and mean $\bar{\omega}$; such a population would have $m = \cos^2(20 \text{ degrees}) = 0.88$. Below, we find that the actual distribution of objects exceeds this example value for large a. We stress that the above metric is purely geometric in nature. It would be better to assess alignment based on a dynamical understanding of the long-term orbital history of the objects taking into account resonance behavior of both Neptune and the hypothesized giant planet. Such a method would require dynamically integrating the orbits of all objects and assessing their membership in a uniform way, and to implement correctly, would require knowledge of (or at a minimum an assumption of) the hypothesized planet's orbit and mass.

We compute the alignment metric m for all $q > 40$ AU objects as a function of minimum semimajor axis a_{min} in Fig. 4.4. It shows a low value for small a_{min}, due to the random nature of the orbits of typical TNOs near Neptune. But for most $a_{\text{min}} \gtrsim 150$ AU, the value is large. To assess the statistical significance of the alignment metric m we used a Monte-Carlo method to create a random sample of mock ETNO/IOC orbits by drawing independent ω, Ω, and i randomly from true orbits in the semimajor axis interval 50 AU $< a < a_{\text{min}}$. In case the ETNOs and IOCs are not aligned, the m statistics drawn from the random sample should match those with $a_{\text{min}} < a$, a similar method to Batygin and Brown (2016a). This procedure includes only the scattered TNOs in the random sample and excludes the classical and most of the

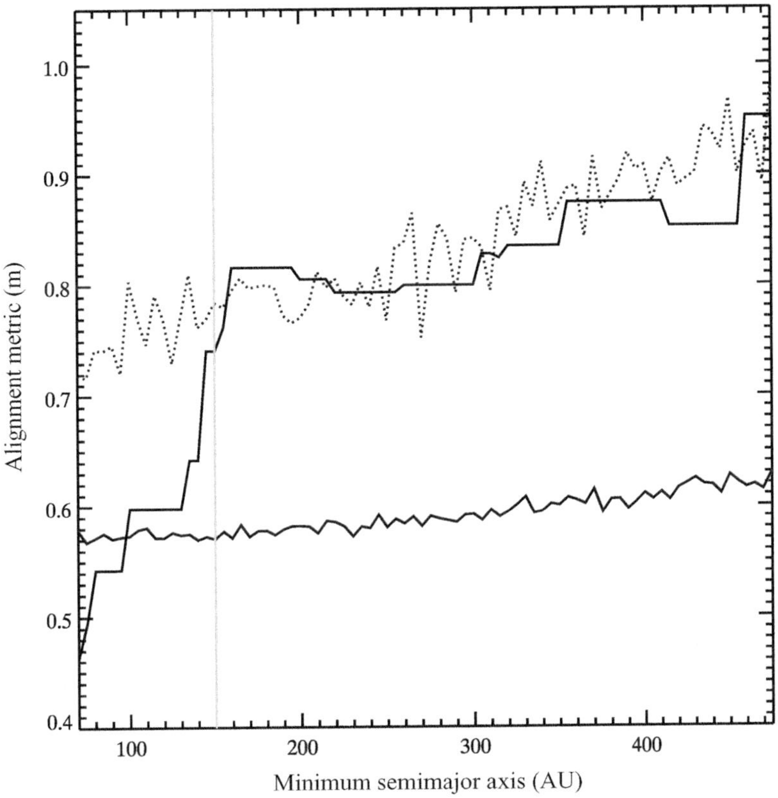

FIG. 4.4 Our alignment metric m as a function of minimum semimajor axis (*black*). The *blue line* is the mean alignment metric for a random distribution of Ω and ω with i drawn from the 50 AU $< a < a_{min}$ AU bodies (i.e., those not exhibiting alignment) as described in the text. The 3σ upper limit to m for the random bodies is shown as a *dotted line*. This illustrates that the most distant ETNOs/IOCs are aligned; a trend that is near a statistically significant level for most a_{min} greater than about 150 AU (delineated with a *vertical gray line*).

resonant TNOs as their i distribution is dissimilar to the ETNOs/IOCs. From these hundreds of m statistics computed from the Monte-Carlo random orbital distribution, we compute the 3σ maximum for m (the 99.73 percentile highest m value), which is plotted as a dotted blue line. This dotted line represents an upper limit to the possible m that could be created by chance. Both the m for the real objects and the random sample slowly increase at large a_{min} due to the reduction in sample size (i.e., it is easier to produce an aligned random sample with small numbers of objects). The m for the real objects falls around this line for $a_{min} \gtrsim 150$ AU, which suggests that the trend is borderline statistically significant and not particularly sensitive to the choice of minimum semimajor axis. This $\sim 3\sigma$ statistical significance is similar but somewhat lower than that found by Batygin and Brown (2016a, 3–8σ) and de la Fuente Marcos and de la Fuente Marcos (2017, 3–7σ), but our methodologies and samples differ somewhat. For reference, the orbital elements of the currently known $q > 40$ AU and $a > 150$ AU ETNOs/IOCs are shown in Table 4.1. The previous discussion is only valid in the case that observational bias is not a major issue for most of the ETNOs/IOCs. We consider the effects of observational bias in Section 4.3.

4.3 Survey methodology and observational bias

We have previously discussed the observation that the orbits of the most distant ETNOs/IOCs exhibit a statistically significant alignment. However, most of these orbits are very eccentric and for the most part the objects are only bright enough to be discovered near perihelion. Thus, the possibility must be considered that there could be an observational bias in the longitude of perihelion discovery statistics which could translate to an apparent alignment where none exists. We consider these both at the survey level and at the population level below.

4.3.1 Small sky area surveys

The survey methodology chosen greatly affects the discovery statistics of the ETNOs/IOCs discovered by that survey. Surveys which search a small sky area ($\lesssim$200 square degrees), patches of sky at specific ecliptic longitudes (such as opposition at only one time of the year) or only observe near the ecliptic can lead to biases which favor objects like the observed distribution of ETNOs and IOCs. One of the most studied recent examples of this effect is the Outer Solar System Origins Survey (OSSOS). The OSSOS used the Canada France Hawaii Telescope MegaPrime camera (Boulade et al., 2003) and pipeline (Magnier and Cuillandre, 2004) to search eight patches of sky each ~21 square degrees in extent over several years to discover and characterize the orbits of about 800 outer solar system objects (Bannister et al., 2014, 2018). The OSSOS examined several subpopulations of TNOs in detail (Shankman et al., 2016; Bannister et al., 2016, 2017b; Lawler et al., 2018b; Volk et al., 2016, 2018) and included color information with a follow-up survey (Pike et al., 2017; Bannister et al., 2017a, Col-OSSOS).

Most relevant to the observed ETNO/IOC orbital alignment is the very detailed study of observational biases (Shankman et al., 2017a) which were characterized with the OSSOS observational simulator (Lawler et al., 2018a). For narrow-field surveys, such detail in survey simulations allows biases to be extremely well-understood. Since the OSSOS was able to characterize an unprecedented number of TNOs, over 830, and track them to produce reliable orbits, this leads to excellent constraints on the underlying population. The OSSOS found objects that are generally consistent with the aforementioned clustering in ω for the ETNOs/IOCs. However, after a careful examination of the observational biases involved, Shankman et al. (2017a) demonstrate there is no evidence for intrinsic clustering in the ω, Ω, or $\bar{\omega}$ distributions in the OSSOS and they cannot reject the null hypothesis that the ETNO/IOC orbits are drawn from a uniform sample.

When the biases for the OSSOS are considered, Shankman et al. (2017a) find that their observed objects are consistent with a uniform one which differs from other works, such as Sheppard and Trujillo (2016b) and Sheppard et al. (2019) who report a trend as described later. The solution to this apparent conflict is most likely to find more ETNOs and IOCs and characterize the survey biases carefully to determine if the underlying distribution of objects departs from the uniform (see Section 4.3.4). Despite the biases, we reiterate that no object with a semimajor axis $a > 150$ AU and $q > 40$ degrees has been found with an argument of perihelion within $\omega = 180 \pm 40$ degrees while about a dozen have been found with an argument of perihelion $\omega = 0 \pm 120$ degrees. There is no known bias that would account for

TABLE 4.1 The currently known multiopposition minor planets and dwarf planets with perihelion greater than 40 AU and semimajor axis greater than 150 AU are listed with designation (Desig.) and provisional designation (Prov. desig.).

Desig.	Prov. desig.	q (AU)	a (AU)	e	i (degrees)	ω (degrees)	Ω (degrees)	ϖ (degrees)	M (degrees)	Discoverers
(541132)	2015 TG_{387}	65.018	1042	0.938	11.7	118.0	300.8	58.8	359.4	Tholen, Sheppard, and Trujillo
	2015 RX_{245}	45.597	412	0.889	12.1	65.1	8.6	73.7	358.0	OSSOS
	2015 KH_{163}	40.013	158	0.747	27.1	230.4	67.6	298.0	353.5	OSSOS
	2015 KG_{163}	40.498	833	0.951	14.0	32.2	219.1	251.3	360.0	OSSOS
	2014 SR_{349}	47.691	302	0.842	18.0	340.9	34.8	15.7	357.7	Sheppard and Trujillo
	2013 SY_{99}	50.019	694	0.928	4.2	32.1	29.5	61.6	359.3	OSSOS
	2013 RA_{109}	45.986	458	0.899	12.4	262.8	104.7	7.5	0.4	Dark Energy Survey
	2013 FT_{28}	43.454	312	0.861	17.3	40.5	217.8	258.3	357.3	Sheppard
	2012 VP_{113}	80.389	258	0.689	24.1	293.5	90.7	24.2	3.6	Sheppard and Trujillo
	2010 GB_{174}	48.746	351	0.861	21.6	347.4	130.8	118.2	3.7	NGVS
(505478)	2013 UT_{15}	43.968	197	0.776	10.7	252.1	192.0	84.1	353.8	OSSOS
(496315)	2013 GP_{136}	41.031	155	0.735	33.5	42.5	210.7	253.2	356.7	OSSOS
(474640)	2004 VN_{112}	47.296	319	0.852	25.6	326.8	66.0	32.8	0.6	ESSENCE
(148209)	2000 CR_{105}	44.175	218	0.797	22.8	316.7	128.3	85.0	6.1	Deep Ecliptic Survey
(90377) Sedna	2003 VB_{12}	76.164	479	0.841	11.9	311.5	144.3	95.8	358.0	Caltech Wide Area Sky Survey

Notes: Orbital elements are epoch April 27, 2019 and are defined by perihelion q, semimajor axis a, eccentricity e, inclination i, argument of perihelion ω, longitude of ascending node Ω, longitude of perihelion ϖ, and mean anomaly M. All numeric data are from the Minor Planet Center. Discoverers are listed by surname or survey: Outer Solar System Origins Survey (OSSOS) (Bannister et al., 2014), Dark Energy Survey (DES) (Dark Energy Survey Collaboration et al., 2016), Next Generation Virgo Cluster Survey (NGVS) (Chen et al., 2013), Equation of State: SupErNovae trace Cosmic Expansion (ESSENCE) (Miknaitis et al., 2007; Becker et al., 2008), the Deep Ecliptic Survey (Elliot et al., 2005), and the Caltech Wide Area Sky Survey (Trujillo and Brown, 2003).

favoring $\omega \sim 0$ degrees over $\omega \sim 180$ degrees as the only difference in these two types of orbits is that they cross the ecliptic plane moving in opposite vertical directions (i.e., up vs. down) with respect to the ecliptic. For example, the OSSOS is similarly sensitive to both (Shankman et al., 2017a, Fig. 4.3) while Sheppard and Trujillo (2016b, Fig. 11) are actually somewhat more sensitive to the unobserved $\omega \sim 180$ degrees population.

4.3.2 Large sky area surveys

The best way to remove biases from the ETNO/IOC population is to observe at all times of the year and in a relatively uniform ecliptic longitude and to observe away from the ecliptic. The two surveys which have done this are the DES (Dark Energy Survey Collaboration et al., 2016) and the long-term observational work of Trujillo and Sheppard (2014), Sheppard and Trujillo (2016b), and Sheppard et al. (2016, 2019). The DES has serendipitously found detached TNOs as well as other interesting outer solar system objects (Gerdes et al., 2016, 2017; Becker et al., 2018). This survey covers about 2500 square degrees north of declination −40 degrees and predominantly south of the celestial equator. Since the original DES survey was designed to avoid the ecliptic, most of the objects found in the DES are high inclination and have different observational biases from an ecliptic survey. There are biases, however, since the DES observes primarily in the fall sky (Hamilton and Gerdes, 2017). In the literature, most of the DES works focus on specific objects of dynamical interest, such as objects with similar orbits (Khain et al., 2018b). Rather than try to aggregate the findings of the DES works, which could be the subject of an independent work and would require comprehensive analysis such as that discussed in Hamilton et al. (2018), we focus here primarily on the recent work of Sheppard et al. (2019) which builds upon the statistics from Trujillo and Sheppard (2014) and Sheppard and Trujillo (2016b).

Whereas Sheppard and Trujillo (2016b) surveyed over 1000 square degrees to $r \sim 24$, additional observations have allowed Sheppard et al. (2019) to incorporate data from over 2100 square degrees surveyed to $24 < r < 25.5$ using both the Dark Energy Camera (DECam) (Flaugher et al., 2015) and Subaru Hyper Suprime-Cam (HSC) (Miyazaki et al., 2012). The survey methodology employed by Sheppard et al. (2019) was to survey off the ecliptic since nearly all known ETNOs/IOCs have an inclination in the range 10 degrees $\lesssim i \lesssim$ 35 degrees and across many ecliptic longitudes to reduce bias. They are sensitive to objects out to 1000–2000 AU but limit orbital refinement to objects with current heliocentric distance beyond 50 AU. This minimum heliocentric distance limit is a practical limit because orbital tracking of all discovered objects would require a factor ~10–50 more telescope time since typically no two discovered objects are in the same telescopic field of view and most detected objects are nearby TNOs, not ETNOs/IOCs. This does introduce bias, but it is characterized in their survey simulations, and as illustrated in Table 4.1, half of the ETNOs/IOCs they have found have perihelia less than 50 AU. Sheppard et al. (2019) find that their survey is relatively unbiased in longitude of perihelion. The results of their observational bias simulation are shown in Fig. 4.5 which demonstrates that the biases in their survey are smaller than many other surveys. Specifically, Fig. 4.5 shows the expected distribution of longitude of perihelion, $\bar{\omega}$, for ETNOs/IOCs in the hypothetical case that the underlying distribution is uniform. Although the bias does vary by about a factor of ~2 peak to peak, this is considerably betterthan narrower surveys such as the OSSOS which vary by a factor of ~10 peak to peak (Shankman et al., 2017a, Fig. 4.3).

Since only a few of the known ETNOs/IOCs were found in the survey (Sheppard et al., 2019), the bias corrected significance of the orbital alignment using their survey alone is in the range of 2.2σ to 2.5σ, depending on the exact choice of a and q cutoffs used. So more ETNOs/IOCs are still needed in their sample to bring the statistics to the level of 3σ. They note that if the most conservative definition of ETNOs/IOCs is used with $q > 45$ AU and $a > 250$ AU, which selects eight objects from all surveys, ignoring biases, the trend is significant at the $\sim 4\sigma$ level.

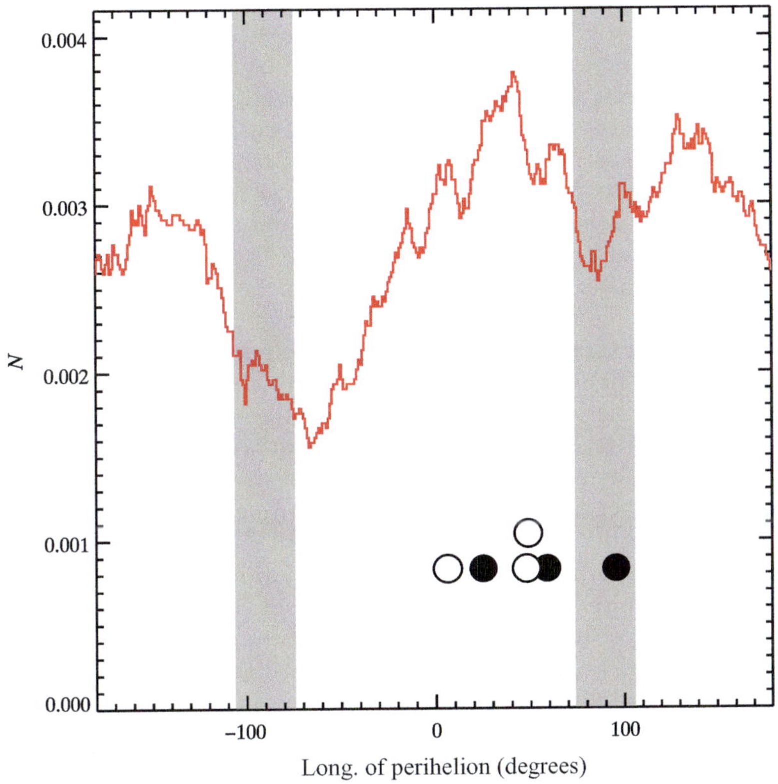

FIG. 4.5 The longitude of perihelion bias from Sheppard et al. (2019) assuming their survey parameters and an ETNO/IOC orbital distribution distributed uniformly around the sky. N represents a normalized histogram of the number of detected objects at a given longitude of perihelion $\bar{\omega}$ assuming a uniform angular distribution of ETNOs/IOCs. The ecliptic longitude of the Galactic plane (±15 degrees) is shown in *gray*. The bias due to the northern Galactic plane (the fall sky) is not especially strong. The most biased areas and the least biased areas differ by about a factor 2 suggesting that objects can be found with any longitude of perihelion. The longitude of perihelion of the objects with the most distant perihelia (the IOCs) and thus those most likely to exhibit bias, Sedna, 2012 VP_{113} and 2015 TG_{387} are shown by *solid circles* and their discovery locations are shown by *open circles*. The mean number of expected objects with 0 degrees $< \bar{\omega} <$ 100 degrees, those suggested by Batygin and Brown (2016a) as the aligned objects, is about 10% greater than those outside this range, yet none of the 15 known ETNO/IOC objects from any survey have longitude of perihelion 120 degrees $< \bar{\omega} <$ 250 degrees.

4.3.3 Statistical significance of observed orbital alignment

The best assessment of observational biases would encompass all surveys. Brown (2017) specifically looked at this using a simulation method to quantify observational biases of objects based purely on discovery data publicly available from the Minor Planet Center. The method relies on the assumption that if an object was discovered at a particular sky location and magnitude, similar objects should have been possible to discover that are brighter than the discovered object and at a similar location. There are some inherent limitations to this technique in that it makes assumptions about other survey characteristics such as the minimum and maximum heliocentric distance range, which vary from survey to survey. However, this method is powerful as it allows the observational biases of all surveys to be assessed without detailed knowledge of the survey parameters such as sensitivity and sky area covered. This is possible since the survey parameters such as sensitivity and sky area are in effect encoded into the survey discovery statistics themselves. The analysis found that while a few IOC/ETNO objects are discovered with longitude of perihelion near where they would be expected due to bias, this was confined to a few objects. Brown (2017) find that the probability of the longitude of perihelion clustering could be derived from a uniform distribution to be 1.2%, which assuming Gaussian statistics is the equivalent of 2.5σ, which is similar to the statistical significance found by Sheppard et al. (2019). Brown (2017) further suggests that because an orbital pole clustering is also observed, in the case they are independent, the probability of these two trends being observed from a uniform sample is 0.025% which is the equivalent of 3.7σ in the Gaussian case. The independence of orbit pole, which relies on Ω and i, and longitude of perihelion, which relies on Ω and ω has not been well-demonstrated. Although there have been shown to be strong biases in these parameters, especially in small-area surveys (Shankman et al., 2017a), their interdependence has not been well-studied. The calculation of longitude of perihelion relies on the sum of Ω and ω and the orbit pole is related to Ω and i. One might expect these values to not be completely independent of one another especially if there are biases in Ω. However, considering a distribution of objects with a single Ω value, any longitude of perihelion and orbit pole vector can be chosen since there are still two degrees of freedom left in spherical coordinates, i and ω. Thus, it seems possible that the orbit pole and longitude of perihelion are independent of one another although this may be less true in very biased samples. Brown and Batygin (2019) recently quantified the observational bias in the longitude of perihelion and orbital pole position and found it to be significant at the 99.8% confidence level.

4.3.4 The need for additional distant TNOs

Regardless of the significance of the orbital alignment of the ETNOs/IOCs, the numbers of objects are small (of order ~10) and the significance reported in the literature are in the 2σ to 4σ range. It would be highly beneficial for additional ETNOs and IOCs to be discovered. Ten additional ETNOs/IOCs would boost nearly all the statistical significance reported above the 3σ level, especially if the survey discovering the objects was relatively unbiased. It is especially important to survey areas outside of the known clustering region, which puts objects near perihelion in the fall sky. Such a survey could take several forms which would be beneficial assuming observational bias was well-characterized: a narrow-field survey similar to the OSSOS (Bannister et al., 2018) could work if it were sensitive to objects outside the clustering region, a wide-field survey such as the work of Sheppard et al. (2019) which does target nonclustered objects, and another might be the Large Synoptic Survey Telescope (LSST).

We expect that other small-area surveys will be undertaken in coming years, some similar to analysis of existing data such as Khain et al. (2018b) and likely other surveys targeting the ecliptic similar to the OSSOS (Bannister et al., 2018). Sheppard et al. (2019) have found four of the ETNOs/IOCs listed in Table 4.1 in about 2000 square degrees for a rate of 1 per 500 square degrees in a relatively uniform survey. Their survey coverage has increased significantly with the recent inclusion of HSC in their survey area to about 500 square degrees per year (comparing the 1000 square degrees surveyed between Sheppard and Trujillo (2016b) and Sheppard et al. (2019)). Thus, a similar survey could find ~10 more ETNOs/IOCs with an additional 5000 square degrees which might take 5 years to complete. Due to the ETNO/IOC orbital alignment, discovery in the autumn sky is favored, so if resources were only deployed then, the discovery rate could be approximately doubled to 1/250 square degrees. This would of course lead to additional observational bias which might not be a prudent survey design. Regardless, it seems likely that such a survey could be completed with ~10 additional ETNOs/IOCs discovery before LSST's first year of survey observations are completed in 2023.

Since the LSST should cover essentially the entire sky to magnitude ~24, it has the possibility of finding every ETNO/IOC that is brighter than this magnitude. The ETNOs/IOCs have a thick observed inclination distribution 10 degrees $\lesssim i \lesssim$ 30 degrees, so they could inhabit around 20000 square degrees of sky in the uniformly distributed case. However, due to orbital alignment, the effective sky area available for ETNO/IOC discovery should be reduced by a factor ~2 down to 10000 square degrees. Since the ETNOs/IOCs are favored in the autumn sky when the nighttime ecliptic is in the northern hemisphere, only about 5000 square degrees of prime ETNO/IOC survey area might be available to LSST. LSST's discovery rate could be as high as 1/200 square degrees in this area of sky. Thus, one might expect the LSST to detect about 25 ETNOs/IOCs some of which might have already been known. Despite the fact that many might be detections of previously known objects, the well-characterized biases of such a uniform survey as the LSST would add a significant amount of leverage to understanding the underlying ETNO/IOC populations.

4.4 Action of the giant planet

Until now, we have only considered that there is an orbital alignment for the distant ETNOs/IOCs, but have not explored what can cause or sustain such an alignment. If this orbital alignment was created by some infrequent or one time stochastic process early in the solar system, the action of the currently known giant planets should precess the orbits of the known ETNOs/IOCs to randomize them on a timescale of ~1 billion years. The precession rate for a body can be found using Lagrange's Planetary Equations (e.g., Murray and Dermott, 1998, and references therein). It is a simple matter to simulate this for any solar system body and even distant bodies such as the IOCs will precess at different rates depending on their orbits (Trujillo and Sheppard, 2014).

Placing a large mass in the outer solar system can maintain orbital alignment and constrain the orbital parameters of distant ETNOs/IOCs, as was demonstrated by dynamical integration by Trujillo and Sheppard (2014) and other works. However, it was not until recently that a semianalytical theory for the constraining action was developed (Batygin and Morbidelli, 2017, and references therein). It should be noted that the confinement of orbits is not strong and that objects will drift in and out of confinement with time. This, combined with the fact that specific planet orbit choices have a large effect on the stability of the ETNOs/IOCs has led to some question about the strength of the hypothesized giant planet action (Shankman et al., 2017b; Lawler et al., 2017). In fact Trujillo (2020) found that of the ~2500 different hypothesized planet orbits studied, only a small fraction (~1%) of simulations produced the most favorable results for the majority of ETNOs/IOCs. Thus, orbital confinement of the current observed population appears to be very sensitive to the specific planet parameters chosen. We demonstrate that the orbital evolution of the longitude of perihelion of the known ETNOs/IOCS can be confined by a hypothesized giant planet in Fig. 4.6. This figure illustrates how the longitude of perihelion of the currently known ETNOs/IOCs has evolved historically over time based on the four giant planets and a hypothesized undiscovered giant planet. It shows that although not all ETNOs/IOCs are confined for the entire simulation, many are at any given time and at no time are more than ~2 objects out of confinement.

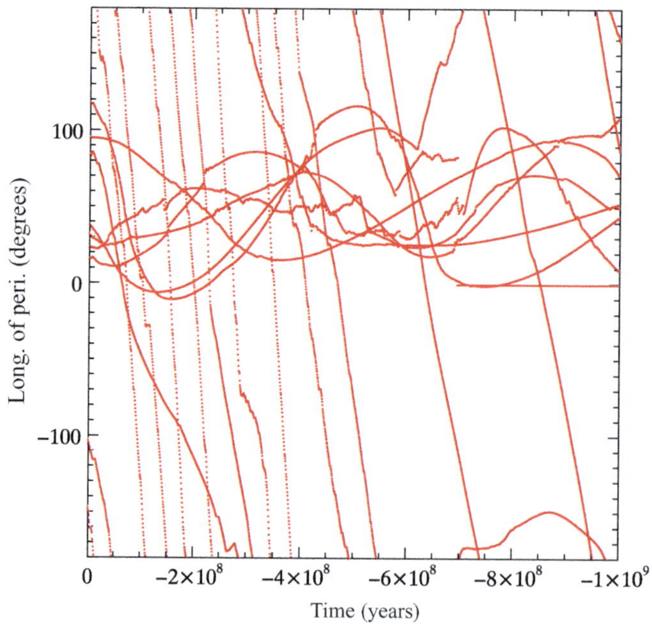

FIG. 4.6 The orbital evolution of the longitude of perihelion of the ETNOs/IOCs due to the presence of an undiscovered giant planet. The dynamical simulation pictured was one of the best simulations performed by Trujillo (2020) using only the $q > 40$ AU and $a > 150$ AU objects known at the time, a total of nine objects. Although all ETNOs/IOCs are not fully constrained by the planet, most are for most of the simulation and only a single object is ejected from the system. Without the hypothesized planet, the orbital evolution of the angular parameters would cycle uniformly through 360 degrees on differing timescales due to the gravitational effects of the four known giant planets and would quickly randomize.

4.4.1 Alternative explanations

It seems very likely that more explanations will be found for the ETNO/IOC orbital alignment including alternatives to the hypothesized giant planet. Historically, this has been true. For instance, both Urbain Le Verrier and John Couch Adams have been credited for predicting the existence of Neptune in 1846 prior to its discovery by analyzing orbital anomalies of Uranus in one of the greatest mathematical achievements in modern observational astronomy. However in 1859, to explain Mercury's perihelion precession, Le Verrier hypothesized a planet interior to Mercury's orbit, an effect that is now attributed to general relativity. To date there have been few dynamical studies of ETNO/IOC alignment processes besides that of a hypothesized planet. One possibility is the effect of a self-gravitating disk, which has incongruously been shown both to produce effects similar to a planet (Madigan and McCourt, 2016; Madigan et al., 2018; Sefilian and Touma, 2019) and to not significantly alter the final planetary orbits after the epoch of planet formation (Fan and Batygin, 2017). It seems likely that this should and will be studied in more detail in the future as should other alternative explanations.

4.5 Finding the planet

Under the assumption at the hypothesized giant planet exists, then it would be most useful to know if there are dynamical constraints that could yield clues to its orbit. Certainly there are broad constraints due to the apsidal and mean-motion resonances with the ETNOs/IOCs (Batygin and Brown, 2016a; Batygin et al., 2019) and the highly inclined TNOs (Batygin and Brown, 2016b). Brown and Batygin (2016) combined these dynamical constraints with known observational surveys that would be likely to find the planet, such as Panoramic Survey Telescope and Rapid Response System (Pan-STARRS) (Chambers et al., 2016) which might detect the planet if it were near perihelion. With these constraints, Brown and Batygin (2016) were able to narrow down the prime search region to about 30 × 70 degrees, or about 2100 square degrees near right ascension $\alpha = 5$ hours and declination $\delta = 0$ degrees. This is quite a large location given that the two best survey instruments for discovery are likely to be the DECam and HSC, due to their combination of lightgrasp and depth. DECam and HSC have field of view of about 3 square degrees and about 1.75 square degrees, respectively. Assuming 10-hour nights, 5-minute exposures, and a minimum of two images required to see the hypothesized planet's $\sim 2''$ day^{-1} apparent motion, it would require about 14 nights to survey this region assuming perfect conditions and no observing overheads. Given the realities of observing, it is likely that a full month of telescope time would be needed to cover the region in a dedicated survey, which is a very large allocation. This also assumes that the other prior surveys were indeed sensitive to detecting the hypothesized planet. Historically, there have been many cases where newly discovered objects were previously imaged by observers but not recognized until later, so the assumption that the planet could be serendipitously found in a survey not tailored to very distant objects may be risky.

4.5.1 Mean-motion resonances and the planet location

One of the main difficulties of locating the planet using to dynamical clues is that apsidal resonances like those demonstrated by Batygin and Brown (2016a) do not confine the true anomaly of the object and thus one cannot determine where in its orbit the hypothesized planet might be located. This is not true of mean-motion resonances, which do constrain the location of the object. Pluto, for instance, comes to aphelion when near Neptune's ecliptic longitude due to its presence in Neptune's 3:2 mean-motion resonance (Cohen and Hubbard, 1964; Malhotra, 1996). Thus, if the ETNOs/IOCs inhabit mean-motion resonances with the hypothesized planet, they could constrain the hypothesized planet's location. Malhotra et al. (2016) pointed out that some of the known ETNOs/IOCs could be in mean-motion resonance based on their distribution of semimajor axes. Interestingly, such a mean-motion resonance effect would be a correlation completely independent of any orbital alignment correlation relying on orbit plane or longitude of perihelion. Millholland and Laughlin (2017) built on the idea that the ETNOs/IOCs may be in mean-motion resonance with the hypothesized planet by identifying planet semimajor axes that would maximize the number of the ETNOs/IOCs in mean-motion resonance and conducted dynamical simulations to constrain the orbit. They compute the most favorable planet orbits based on the clustering in argument of perihelion and the presence of ETNOs/IOCs in antialignment with the planet. They find a wide sky area favorable for the hypothesized planet, about 100 $\times$ 45 degrees or 4500 square degrees. A similar technique was used by Trujillo (2020), although using a different metric based on number of objects retained throughout the simulation and orbit pole location. Trujillo (2020) also used a more constrained set of orbital parameters as a basis for the investigation. While the simulated planets spanned a sky area of $\sim$10,000 square degrees, the most favorable hypothesized planet locations were in a $\sim$100 square degree patch of sky near 3 hours $< \alpha <$ 6 hours and $\delta = 0.56\alpha - 36$ degrees$\pm$5 degrees, as shown in Fig. 4.7. Such a small area of sky can be surveyed in several clear nights with either DECam or HSC. Both Trujillo (2020) and Brown and Batygin (2016) suggest that the planet should be between magnitude 22.5 and 25–26 for most cases, so this is within the range of brightness available to DECam and HSC. Interestingly, simulations that Trujillo (2020) performed to estimate orbital parameters of a hypothesized planet stable with the known ETNOs/IOCs but without knowledge of 2015 TG_{387} also showed 2015 TG_{387} to be stable upon its discovery (Sheppard et al., 2019).

4.5.2 Limitations of dynamics-based observational constraints

The main limitation of the mean-motion resonance technique is that there is no guarantee that the currently known ETNOs/IOCs are indeed in mean-motion resonance with the hypothesized planet. Indeed, Bailey et al. (2018) specifically explore this possibility by simulating the action of a hypothesized planet on a suite of test particles. They find that high-order resonances are numerous and overlapping in the region inhabited by the currently known ETNOs and IOCs. Thus, the test particles do not inhabit a single low-order resonance as is the case for Neptune and the Resonant TNOs, but move between low-order and high-order resonances throughout their simulations. Thus, the utility of observational constraints based on mean-motion resonances may be in question. Thus, if apsidal resonances dominate, the planet orbit may be constrained but its location in that orbit may not, and the difficulty of finding the planet could increase by a factor of $\sim$50.

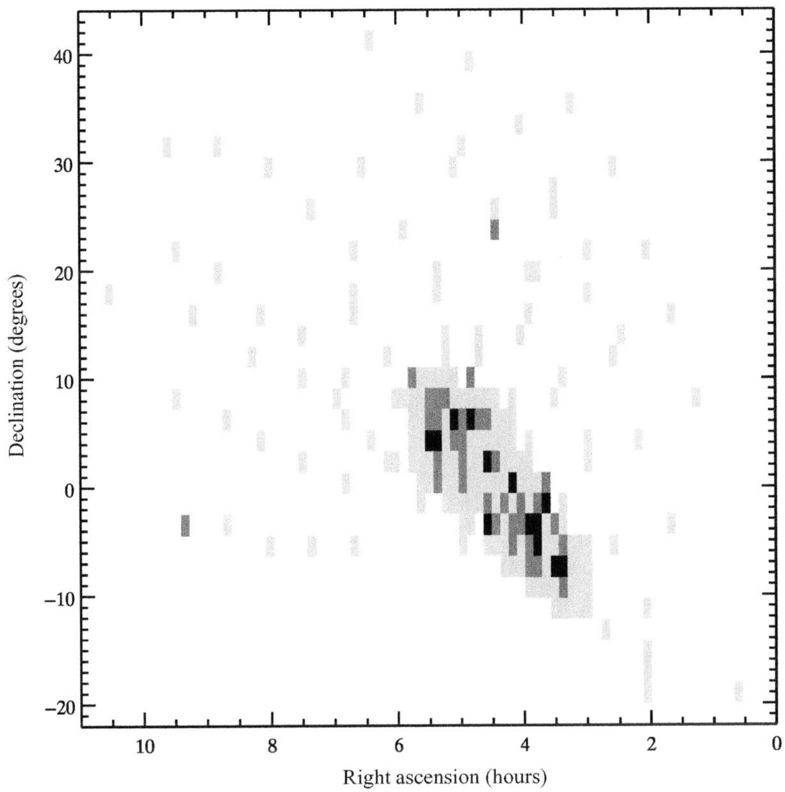

FIG. 4.7 The current sky locations of the hypothesized giant planets simulated by Trujillo (2020) are shown. Sky locations where unsuccessful hypothesized planets were tested are shown as *light gray rectangles*. A high fraction of successful hypothesized planets (*dark gray rectangles*) and a high fraction of very successful planets (*black rectangles*) are also shown. The most favorable region for the hypothesized planet is in a ~100 square degree patch near right ascension 4.5 hours and declination 0 degrees. For reference, the range of current epoch orbital parameters providing a good fit are $a = 721.933$ AU, $e = 0.55 \pm 0.02$, $i = 28.75 \pm 1$ degrees, $\omega = 125–145$ degrees, $\bar{\omega} = 235–250$ degrees, $M = 130–215$ degrees with a mass of 10 ± 1 Earth masses.

Another limitation to dynamical constraints of the hypothesized planet's location may be that the ETNOs/IOCs are not dynamically stable. Most dynamical investigations using the known ETNOs/IOCs either explicitly or implicitly assume that they are stable for the age of the solar system. However, the ETNOs/IOCs that we see today may be transitory in nature, similar to the Comets or the Centaurs, which have lifetimes much shorter than the solar system but are drawn from a distant and more stable population (Fernandez, 1980; Tiscareno and Malhotra, 2003). Since the current populations of ETNOs/IOCs are thought to be similar to the TNO population in the stable case (Sheppard et al., 2019), this suggests that the transitory population could be an order of magnitude larger or more. If this were the case, then even stability arguments may be of questionable utility and it is possible that a very wide range of planet orbits could be plausible. Hope is not lost, however, because even very transitory populations such as the Jupiter Family Comets do provide information about the perturber (Nesvorný et al., 2017).

4.6 Summary

We have reviewed the observational evidence for a hypothesized giant planet in the distant reaches of our solar system and have discussed possible bias effects that could generate the observed orbital alignment of the ETNOs/IOCs. Several of the 14 $q > 40$ AU and $a > 150$ AU ETNOs/IOCs that exhibit the most obvious orbital alignment have been found in surveys that are known to have strong biases. Thus, the true statistical significance of the orbital alignment remains difficult to compute, but in the literature ranges from 2σ to 4σ range (i.e., 95%–99.99% confidence). In the most conservative case, it seems that the ETNO/IOC alignment deserves more study. Excepting direct detection of the hypothesized planet itself, the best way to improve the significance of the orbital alignment to certainty is through the discovery of more distant ETNOs/IOCs. With typical telescope time allocations, statistical significance of the orbital alignment could be raised to the unequivocal level in the next few years prior to the LSST first light. Additional ETNOs/IOCs would also provide stronger dynamical constraints on the hypothesized planet orbit. LSST will certainly provide a large amount of information about the true distribution of the ETNOs/IOCs and constraints on the hypothesized planet especially since its observational biases should be exceptionally well-characterized.

Specific predictions for the location of the hypothesized planet in the sky also range from ~100 to ~1000 square degrees depending on the method used. These methods may have pitfalls that are not fully understood, so again, the best hope for direct detection of the planet is likely to be the discovery of additional ETNOs/IOCs so that specific planet hypotheses can be created, tested, and confirmed or ruled out. There remains a wide gap between the most studied explanation for the observed orbital distribution—that a giant planet exists undiscovered in our solar system—and alternative explanations which to date have only been studied in a few works. Overall, much remains to be discovered and investigated related to the minor, dwarf, and giant planets inhabiting the most distant regions of our solar system, the region beyond 50 AU.

Acknowledgments

The author thanks Scott Sheppard for helpful suggestions for inclusion in this work. The author thanks an anonymous referee for providing a very detailed report which significantly improved this work. This work was funded in part by NASA Grants NNX17AK35G, 80NSSC18K1006, and the State of Arizona Technology and Research Initiative Program.

References

Bailey, E., Batygin, K., Brown, M.E., 2016. Solar obliquity induced by planet nine. Astron. J. 152, 126. https://doi.org/10.3847/0004-6256/152/5/126.

Bailey, E., Brown, M.E., Batygin, K., 2018. Feasibility of a resonance-based planet nine search. Astron. J. 156, 74. https://doi.org/10.3847/1538-3881/aaccf4.

Bannister, M., Kavelaars, J., Gladman, B.J., Petit, J., Gwyn, S., Chen, Y., Alexandersen, M., Outer Solar System Origins Survey Collaboration, 2014. First discoveries from OSSOS—the outer solar system origins survey. In: Muinonen, K., Penttilä, A., Granvik, M., Virkki, A., Fedorets, G., Wilkman, O., Kohout, T. (Eds.), Asteroids, Comets, Meteors—Book of Abstracts. University of Helsinki, Helsinki, Finland. ISBN 978-952-10-8962-6. Published at http://www.helsinki.fi/acm2014/pdf-material/ACM2014.pdf.

Bannister, M.T., Alexandersen, M., Benecchi, S.D., Chen, Y.-T., Delsanti, A., Fraser, W.C., Gladman, B.J., Granvik, M., Grundy, W.M., Guilbert-Lepoutre, A., Gwyn, S.D.J., Ip, W.-H., Jakubik, M., Jones, R.L., Kaib, N., Kavelaars, J.J., Lacerda, P., Lawler, S., Lehner, M.J., Lin, H.W., Lykawka, P.S., Marsset, M., Murray-Clay, R., Noll, K.S., Parker, A., Petit, J.-M., Pike, R.E., Rousselot, P., Schwamb, M.E., Shankman, C., Veres, P., Vernazza, P., Volk, K., Wang, S.-Y., Weryk, R., 2016. OSSOS. IV. Discovery of a dwarf planet candidate in the 9:2 resonance with Neptune. Astron. J. 152, 212. https://doi.org/10.3847/0004-6256/152/6/212.

Bannister, M.T., Schwamb, M.E., Fraser, W.C., Marsset, M., Fitzsimmons, A., Benecchi, S.D., Lacerda, P., Pike, R.E., Kavelaars, J.J., Smith, A.B., Stewart, S.O., Wang, S.-Y., Lehner, M.J., 2017a. Col-OSSOS: colors of the interstellar planetesimal 1I/'Oumuamua. Astrophys. J. Lett. 851, L38. https://doi.org/10.3847/2041-8213/aaa07c.

Bannister, M.T., Shankman, C., Volk, K., Chen, Y.-T., Kaib, N., Gladman, B.J., Jakubik, M., Kavelaars, J.J., Fraser, W.C., Schwamb, M.E., Petit, J.-M., Wang, S.-Y., Gwyn, S.D.J., Alexandersen, M., Pike, R.E., 2017b. OSSOS. V. Diffusion in the orbit of a high-perihelion distant solar system object. Astron. J. 153, 262. https://doi.org/10.3847/1538-3881/aa6db5.

Bannister, M.T., Gladman, B.J., Kavelaars, J.J., Petit, J.-M., Volk, K., Chen, Y.-T., Alexandersen, M., Gwyn, S.D.J., Schwamb, M.E., Ashton, E., Benecchi, S.D., Cabral, N., Dawson, R.I., Delsanti, A., Fraser, W.C., Granvik, M., Greenstreet, S., Guilbert-Lepoutre, A., Ip, W.-H., Jakubik, M., Jones, R.L., Kaib, N.A., Lacerda, P., Van Laerhoven, C., Lawler, S., Lehner, M.J., Lin, H.W., Lykawka, P.S., Marsset, M., Murray-Clay, R., Pike, R.E., Rousselot, P., Shankman, C., Thirouin, A., Vernazza, P., Wang, S.-Y., 2018. OSSOS. VII. 800+ Trans-Neptunian objects—the complete data release. Astrophys. J. Suppl. 236, 18. https://doi.org/10.3847/1538-4365/aab77a.

Batygin, K., Brown, M.E., 2016a. Evidence for a distant giant planet in the solar system. Astron. J. 151, 22. https://doi.org/10.3847/0004-6256/151/2/22.

Batygin, K., Brown, M.E., 2016b. Generation of highly inclined Trans-Neptunian objects by planet nine. Astrophys. J. Lett. 833, L3. https://doi.org/10.3847/2041-8205/833/1/L3.

Batygin, K., Morbidelli, A., 2017. Dynamical evolution induced by planet nine. Astron. J. 154, 229. https://doi.org/10.3847/1538-3881/aa937c.

Batygin, K., Adams, F.C., Brown, M.E., Becker, J.C., 2019. The planet nine hypothesis. Phys. Rep. 805, 1. https://doi.org/10.1016/j.physrep.2019.01.009.

Becker, A.C., Arraki, K., Kaib, N.A., Wood-Vasey, W.M., Aguilera, C., Blackman, J.W., Blondin, S., Challis, P., Clocchiatti, A., Covarrubias, R., Damke, G., Davis, T.M., Filippenko, A.V., Foley, R.J., Garg, A., Garnavich, P.M., Hicken, M., Jha, S., Kirshner, R.P., Krisciunas, K., Leibundgut, B., Li, W., Matheson, T., Miceli, A., Miknaitis, G., Narayan, G., Pignata, G., Prieto, J.L., Rest, A., Riess, A.G., Salvo, M.E., Schmidt, B.P., Smith, R.C., Sollerman, J., Spyromilio, J., Stubbs, C.W., Suntzeff, N.B., Tonry, J.L., Zenteno, A., 2008. Exploring the outer solar system with the ESSENCE supernova survey. Astrophys. J. Lett. 682, L53–L56.

Becker, J.C., Adams, F.C., Khain, T., Hamilton, S.J., Gerdes, D., 2017. Evaluating the dynamical stability of outer solar system objects in the presence of planet nine. Astron. J. 154, 61. https://doi.org/10.3847/1538-3881/aa7aa2.

Becker, J.C., Khain, T., Hamilton, S.J., Adams, F.C., Gerdes, D.W., Zullo, L., Franson, K., Millholland, S., Bernstein, G.M., Sako, M., Bernardinelli, P., Napier, K., Markwardt, L., Lin, H.W., Wester, W., Abdalla, F.B., Allam, S., Annis, J., Avila, S., Bertin, E., Brooks, D., Carnero Rosell, A., Carrasco Kind, M., Carretero, J., Cunha, C.E., D'Andrea, C.B., da Costa, L.N., Davis, C., De Vicente, J., Diehl, H.T., Doel, P., Eifler, T.F., Flaugher, B., Fosalba, P., Frieman, J., García-Bellido, J., Gaztanaga, E., Gruen, D., Gruendl, R.A., Gschwend, J., Gutierrez, G., Hartley, W.G., Hollowood, D.L., Honscheid, K., James, D.J., Kuehn, K., Kuropatkin, N., Maia, M.A.G., March, M., Marshall, J.L., Menanteau, F., Miquel, R., Ogando, R.L.C., Plazas, A.A., Sanchez, E., Scarpine, V., Schindler, R., Sevilla-Noarbe, I., Smith, M., Smith, R.C., Soares-Santos, M., Sobreira, F., Suchyta, E., Swanson, M.E.C., Walker, A.R., DES Collaboration, 2018. Discovery and dynamical analysis of an extreme Trans-Neptunian object with a high orbital inclination. Astron. J. 156, 81. https://doi.org/10.3847/1538-3881/aad042.

Beust, H., 2016. Orbital clustering of distant Kuiper belt objects by hypothetical planet 9. Secular or resonant? Astron. Astrophys. 590, L2. https://doi.org/10.1051/0004-6361/201628638.

Boulade, O., Charlot, X., Abbon, P., Aune, S., Borgeaud, P., Carton, P.-H., Carty, M., Da Costa, J., Deschamps, H., Desforge, D., Eppellé, D., Gallais, P., Gosset, L., Granelli, R., Gros, M., de Kat, J., Loiseau, D., Ritou, J.L., Roussé, J.Y., Starzynski, P., Vignal, N., Vigroux, L.G., 2003. MegaCam: the new Canada-France-Hawaii Telescope wide-field imaging camera. In: Iye, M., Moorwood, A.F.M. (Eds.), Instrument Design and Performance for Optical/Infrared Ground-Based Telescopes, Proc. SPIE, vol. 4841, pp. 72–81.

Bromley, B.C., Kenyon, S.J., 2014. The fate of scattered planets. Astrophys. J. 796, 141. https://doi.org/10.1088/0004-637X/796/2/141.

Bromley, B.C., Kenyon, S.J., 2016. Making planet nine: a scattered giant in the outer solar system. Astrophys. J. 826, 64. https://doi.org/10.3847/0004-637X/826/1/64.

Brown, M.E., 2017. Observational bias and the clustering of distant eccentric Kuiper belt objects. Astron. J. 154, 65. https://doi.org/10.3847/1538-3881/aa79f4.

Brown, M.E., Batygin, K., 2016. Observational constraints on the orbit and location of planet nine in the outer solar system. Astrophys. J. Lett. 824, L23. https://doi.org/10.3847/2041-8205/824/2/L23.

Brown, M.E., Batygin, K., 2019. Orbital clustering in the distant solar system. Astron. J. 157, 62. https://doi.org/10.3847/1538-3881/aaf051.

Brown, M.E., Trujillo, C., Rabinowitz, D., 2004. Discovery of a candidate inner Oort cloud planetoid. Astrophys. J. 617, 645–649. https://doi.org/10.1086/422095.

Cáceres, J., Gomes, R., 2018. The influence of planet nine on the orbits of distant TNOs: the case for a low-perihelion planet. Astron. J. 156, 157. https://doi.org/10.3847/1538-3881/aad77a.

Chambers, K.C., Magnier, E.A., Metcalfe, N., Flewelling, H.A., Huber, M.E., Waters, C.Z., Denneau, L., Draper, P.W., Farrow, D., Finkbeiner, D.P., Holmberg, C., Koppenhoefer, J., Price, P.A., Saglia, R.P., Schlafly, E.F., Smartt, S.J., Sweeney, W., Wainscoat, R.J., Burgett, W.S., Grav, T., Heasley, J.N., Hodapp, K.W., Jedicke, R., Kaiser, N., Kudritzki, R.-P., Luppino, G.A., Lupton, R.H., Monet, D.G., Morgan, J.S., Onaka, P.M., Stubbs, C.W., Tonry, J.L., Banados, E., Bell, E.F., Bender, R., Bernard, E.J., Botticella, M.T., Casertano, S., Chastel, S., Chen, W.-P., Chen, X., Cole, S., Deacon, N., Frenk, C., Fitzsimmons, A., Gezari, S., Goessl, C., Goggia, T., Goldman, B., Grebel, E.K., Hambly, N.C., Hasinger, G., Heavens, A.F., Heckman, T.M., Henderson, R., Henning, T., Holman, M., Hopp, U., Ip, W.-H., Isani, S., Keyes, C.D., Koekemoer, A., Kotak, R., Long, K.S., Lucey, J.R., Liu, M., Martin, N.F., McLean, B., Morganson, E., Murphy, D.N.A., Nieto-Santisteban, M.A., Norberg, P., Peacock, J.A., Pier, E.A., Postman, M., Primak, N., Rae, C., Rest, A., Riess, A., Riffeser, A., Rix, H.W., Roser, S., Schilbach, E., Schultz, A.S.B., Scolnic, D., Szalay, A., Seitz, S., Shiao, B., Small, E., Smith, K.W., Soderblom, D., Taylor, A.N., Thakar, A.R., Thiel, J., Thilker, D., Urata, Y., Valenti, J., Walter, F., Watters, S.P., Werner, S., White, R., Wood-Vasey, W.M., Wyse, R., 2016. The Pan-STARRS1 surveys. ArXiv e-prints. arXiv:1612.05560.

Chen, Y.-T., Kavelaars, J.J., Gwyn, S., Ferrarese, L., Côté, P., Jordán, A., Suc, V., Cuillandre, J.-C., Ip, W.-H., 2013. Discovery of a new member of the inner Oort cloud from the Next Generation Virgo Cluster Survey. Astrophys. J. Lett. 775, L8.

Chen, Y.-T., Lin, H.W., Holman, M.J., Payne, M.J., Fraser, W.C., Lacerda, P., Ip, W.-H., Chen, W.-P., Kudritzki, R.-P., Jedicke, R., Wainscoat, R.J., Tonry, J.L., Magnier, E.A., Waters, C., Kaiser, N., Wang, S.-Y., Lehner, M., 2016. Discovery of a new retrograde Trans-Neptunian object: hint of a common orbital plane for low semimajor axis, high-inclination TNOs and Centaurs. Astrophys. J. Lett. 827, L24. https://doi.org/10.3847/2041-8205/827/2/L24.

Cohen, C.J., Hubbard, E.C., 1964. Libration of Pluto-Neptune. Science 145, 1302–1303. https://doi.org/10.1126/science.145.3638.1302.

Dark Energy Survey Collaboration, Abbott, T., Abdalla, F.B., Aleksić, J., Allam, S., Amara, A., Bacon, D., Balbinot, E., Banerji, M., Bechtol, K., Benoit-Lévy, A., Bernstein, G.M., Bertin, E., Blazek, J., Bonnett, C., Bridle, S., Brooks, D., Brunner, R.J., Buckley-Geer, E., Burke, D.L., Caminha, G.B., Capozzi, D., Carlsen, J., Carnero-Rosell, A., Carollo, M., Carrasco-Kind, M., Carretero, J., Castander, F.J., Clerkin, L., Collett, T., Conselice, C., Crocce, M., Cunha, C.E., D'Andrea, C.B., da Costa, L.N., Davis, T.M., Desai, S., Diehl, H.T., Dietrich, J.P., Dodelson, S., Doel, P., Drlica-Wagner, A., Estrada, J., Etherington, J., Evrard, A.E., Fabbri, J., Finley, D.A., Flaugher, B., Foley, R.J., Fosalba, P., Frieman, J., García-Bellido, J., Gaztanaga, E., Gerdes, D.W., Giannantonio, T., Goldstein, D.A., Gruen, D., Gruendl, R.A., Guarnieri, P., Gutierrez, G., Hartley, W., Honscheid, K., Jain, B., James, D.J., Jeltema, T., Jouvel, S., Kessler, R., King, A., Kirk, D., Kron, R., Kuehn, K., Kuropatkin, N., Lahav, O., Li, T.S., Lima, M., Lin, H., Maia, M.A.G., Makler, M., Manera, M., Maraston, C., Marshall, J.L., Martini, P., McMahon, R.G., Melchior, P., Merson, A., Miller, C.J., Miquel, R., Mohr, J.J., Morice-Atkinson, X., Naidoo, K., Neilsen, E., Nichol, R.C., Nord, B., Ogando, R., Ostrovski, F., Palmese, A., Papadopoulos, A., Peiris, H.V., Peoples, J., Percival, W.J., Plazas, A.A., Reed, S.L., Refregier, A., Romer, A.K., Roodman, A., Ross, A., Rozo, E., Rykoff, E.S., Sadeh, I., Sako, M., Sánchez, C., Sanchez, E., Santiago, B., Scarpine, V., Schubnell, M., Sevilla-Noarbe, I., Sheldon, E., Smith, M., Smith, R.C., Soares-Santos, M., Sobreira, F., Soumagnac, M., Suchyta, E., Sullivan, M., Swanson, M., Tarle, G., Thaler, J., Thomas, D., Thomas, R.C., Tucker, D., Vieira, J.D., Vikram, V., Walker, A.R., Wechsler, R.H., Weller, J., Wester, W., Whiteway, L., Wilcox, H., Yanny, B., Zhang, Y., Zuntz, J., 2016. The Dark Energy Survey: more than dark energy—an overview. Mon. Not. R. Astron. Soc. 460, 1270–1299. https://doi.org/10.1093/mnras/stw641.

de la Fuente Marcos, C., de la Fuente Marcos, R., 2014. Extreme Trans-Neptunian objects and the Kozai mechanism: signalling the presence of Trans-Plutonian planets. Mon. Not. R. Astron. Soc. 443, L59–L63. https://doi.org/10.1093/mnrasl/slu084.

de la Fuente Marcos, C., de la Fuente Marcos, R., 2016a. Commensurabilities between ETNOs: a Monte Carlo survey. Mon. Not. R. Astron. Soc. 460, L64–L68. https://doi.org/10.1093/mnrasl/slw077.

de la Fuente Marcos, C., de la Fuente Marcos, R., 2016b. Finding planet nine: a Monte Carlo approach. Mon. Not. R. Astron. Soc. 459, L66–L70. https://doi.org/10.1093/mnrasl/slw049.

de la Fuente Marcos, C., de la Fuente Marcos, R., 2016c. Finding planet nine: apsidal anti-alignment Monte Carlo results. Mon. Not. R. Astron. Soc. 462, 1972–1977. https://doi.org/10.1093/mnras/stw1778.

de la Fuente Marcos, C., de la Fuente Marcos, R., 2017. Evidence for a possible bimodal distribution of the nodal distances of the extreme Trans-Neptunian objects: avoiding a Trans-Plutonian planet or just plain bias? Mon. Not. R. Astron. Soc. 471, L61–L65. https://doi.org/10.1093/mnrasl/slx106.

de la Fuente Marcos, C., de la Fuente Marcos, R., Aarseth, S.J., 2015. Flipping minor bodies: what comet 96P/Machholz 1 can tell us about the orbital evolution of extreme Trans-Neptunian objects and the production of near-Earth objects on retrograde orbits. Mon. Not. R. Astron. Soc. 446, 1867–1873. https://doi.org/10.1093/mnras/stu2230.

de la Fuente Marcos, C., de la Fuente Marcos, R., Aarseth, S.J., 2016. Dynamical impact of the planet nine scenario: N-body experiments. Mon. Not. R. Astron. Soc. 460, L123–L127. https://doi.org/10.1093/mnrasl/slw078.

Dones, L., Weissman, P.R., Levison, H.F., Duncan, M.J., 2004. Oort cloud formation and dynamics. In: Festou, M.C., Keller, H.U., Weaver, H.A. (Eds.), Comets II. University of Arizona Press, Tucson, AZ, pp. 153–174.

Elliot, J.L., Kern, S.D., Clancy, K.B., Gulbis, A.A.S., Millis, R.L., Buie, M.W., Wasserman, L.H., Chiang, E.I., Jordan, A.B., Trilling, D.E., Meech, K.J., 2005. The deep ecliptic survey: a search for Kuiper belt objects and Centaurs. II. Dynamical classification, the Kuiper belt plane, and the core population. Astron. J. 129, 1117–1162. https://doi.org/10.1086/427395.

Fan, S., Batygin, K., 2017. Simulations of the solar system's early dynamical evolution with a self-gravitating planetesimal disk. Astrophys. J. Lett. 851, L37. https://doi.org/10.3847/2041-8213/aa9f0b.

Fernandez, J.A., 1980. On the existence of a comet belt beyond Neptune. Mon. Not. R. Astron. Soc. 192, 481–491. https://doi.org/10.1093/mnras/192.3.481.

Fienga, A., Laskar, J., Manche, H., Gastineau, M., 2016. Constraints on the location of a possible 9th planet derived from the Cassini data. Astron. Astrophys. 587, L8. https://doi.org/10.1051/0004-6361/201628227.

Flaugher, B., Diehl, H.T., Honscheid, K., Abbott, T.M.C., Alvarez, O., Angstadt, R., Annis, J.T., Antonik, M., Ballester, O., Beaufore, L., Bernstein, G.M., Bernstein, R.A., Bigelow, B., Bonati, M., Boprie, D., Brooks, D., Buckley-Geer, E.J., Campa, J., Cardiel-Sas, L., Castander, F.J., Castilla, J., Cease, H., Cela-Ruiz, J.M., Chappa, S., Chi, E., Cooper, C., da Costa, L.N., Dede, E., Derylo, G., DePoy, D.L., de Vicente, J., Doel, P., Drlica-Wagner, A., Eiting, J., Elliott, A.E., Emes, J., Estrada, J., Fausti Neto, A., Finley, D.A., Flores, R., Frieman, J., Gerdes, D., Gladders, M.D., Gregory, B., Gutierrez, G.R., Hao, J., Holland, S.E., Holm, S., Huffman, D., Jackson, C., James, D.J., Jonas, M., Karcher, A., Karliner, I., Kent, S., Kessler, R., Kozlovsky, M., Kron, R.G., Kubik, D., Kuehn, K., Kuhlmann, S., Kuk, K., Lahav, O., Lathrop, A., Lee, J., Levi, M.E., Lewis, P., Li, T.S., Mandrichenko, I., Marshall, J.L., Martinez, G., Merritt, K.W., Miquel, R., Muñoz, F., Neilsen, E.H., Nichol, R.C., Nord, B., Ogando, R., Olsen, J., Palaio, N., Patton, K., Peoples, J., Plazas, A.A., Rauch, J., Reil, K., Rheault, J.-P., Roe, N.A., Rogers, H., Roodman, A., Sanchez, E., Scarpine, V., Schindler, R.H., Schmidt, R., Schmitt, R., Schubnell, M., Schultz, K., Schurter, P., Scott, L., Serrano, S., Shaw, T.M., Smith, R.C., Soares-Santos, M., Stefanik, A., Stuermer, W., Suchyta, E., Sypniewski, A., Tarle, G., Thaler, J., Tighe, R., Tran, C., Tucker, D., Walker, A.R., Wang, G., Watson, M., Weaverdyck, C., Wester, W., Woods, R., Yanny, B., DES Collaboration, 2015. The dark energy camera. Astron. J. 150, 150. https://doi.org/10.1088/0004-6256/150/5/150.

Fortney, J.J., Marley, M.S., Laughlin, G., Nettelmann, N., Morley, C.V., Lupu, R.E., Visscher, C., Jeremic, P., Khadder, W.G., Hargrave, M., 2016. The hunt for planet nine: atmosphere, spectra, evolution, and detectability. Astrophys. J. Lett. 824, L25. https://doi.org/10.3847/2041-8205/824/2/L25.

Gerdes, D.W., Jennings, R.J., Bernstein, G.M., Sako, M., Adams, F., Goldstein, D., Kessler, R., Hamilton, S., Abbott, T., Abdalla, F.B., Allam, S., Benoit-Lévy, A., Bertin, E., Brooks, D., Buckley-Geer, E., Burke, D.L., Capozzi, D., Carnero Rosell, A., Carrasco Kind, M., Carretero, J., Cunha, C.E., D'Andrea, C.B., da Costa, L.N., DePoy, D.L., Desai, S., Dietrich, J.P., Doel, P., Eifler, T.F., Fausti Neto, A., Flaugher, B., Frieman, J., Gaztanaga, E., Gruen, D., Gruendl, R.A., Gutierrez, G., Honscheid, K., James, D.J., Kuehn, K., Kuropatkin, N., Lahav, O., Li, T.S., Maia, M.A.G., March, M., Martini, P., Miller, C.J., Miquel, R., Nichol, R.C., Nord, B., Ogando, R., Plazas, A.A., Romer, A.K., Roodman, A., Sanchez, E., Santiago, B., Schubnell, M., Sevilla-Noarbe, I., Smith, R.C., Soares-Santos, M., Sobreira,

F., Suchyta, E., Swanson, M.E.C., Tarlé, G., Thaler, J., Walker, A.R., Wester, W., Zhang, Y., DES Collaboration, 2016. Observation of two new L4 Neptune Trojans in the dark energy survey supernova fields. Astron. J. 151, 39. https://doi.org/10.3847/0004-6256/151/2/39.

Gerdes, D.W., Sako, M., Hamilton, S., Zhang, K., Khain, T., Becker, J.C., Annis, J., Wester, W., Bernstein, G.M., Scheibner, C., Zullo, L., Adams, F., Bergin, E., Walker, A.R., Mueller, J.H., Abbott, T.M.C., Abdalla, F.B., Allam, S., Bechtol, K., Benoit-Lévy, A., Bertin, E., Brooks, D., Burke, D.L., Carnero Rosell, A., Carrasco Kind, M., Carretero, J., Cunha, C.E., da Costa, L.N., Desai, S., Diehl, H.T., Eifler, T.F., Flaugher, B., Frieman, J., García-Bellido, J., Gaztanaga, E., Goldstein, D.A., Gruen, D., Gschwend, J., Gutierrez, G., Honscheid, K., James, D.J., Kent, S., Krause, E., Kuehn, K., Kuropatkin, N., Lahav, O., Li, T.S., Maia, M.A.G., March, M., Marshall, J.L., Martini, P., Menanteau, F., Miquel, R., Nichol, R.C., Plazas, A.A., Romer, A.K., Roodman, A., Sanchez, E., Sevilla-Noarbe, I., Smith, M., Smith, R.C., Soares-Santos, M., Sobreira, F., Suchyta, E., Swanson, M.E.C., Tarle, G., Tucker, D.L., Zhang, Y., DES Collaboration, 2017. Discovery and physical characterization of a large scattered disk object at 92 AU. Astrophys. J. Lett. 839, L15. https://doi.org/10.3847/2041-8213/aa64d8.

Ginzburg, S., Sari, R., Loeb, A., 2016. Blackbody radiation from isolated Neptunes. Astrophys. J. Lett. 822, L11. https://doi.org/10.3847/2041-8205/822/1/L11.

Gladman, B.J., Marsden, B.G., Vanlaerhoven, C., 2008. Nomenclature in the outer solar system. In: Barucci, M.A., Boehnhardt, H., Cruikshank, D.P., Morbidelli, A., Dotson, R. (Eds.), The Solar System Beyond Neptune. University of Arizona Press, Tucson, AZ, pp. 43–57.

Gladman, B.J., Kavelaars, J., Petit, J.-M., Ashby, M.L.N., Parker, J., Coffey, J., Jones, R.L., Rousselot, P., Mousis, O., 2009. Discovery of the first retrograde Trans-Neptunian object. Astrophys. J. Lett. 697, L91–L94. https://doi.org/10.1088/0004-637X/697/2/L91.

Gomes, R.S., Fernández, J.A., Gallardo, T., Brunini, A., 2008. The scattered disk: origins, dynamics, and end states. In: Barucci, M.A., Boehnhardt, H., Cruikshank, D.P., Morbidelli, A., Dotson, R. (Eds.), The Solar System Beyond Neptune. University of Arizona Press, Tucson, AZ, pp. 259–273.

Gomes, R.S., Soares, J.S., Brasser, R., 2015. The observation of large semi-major axis Centaurs: testing for the signature of a planetary-mass solar companion. Icarus 258, 37–49. https://doi.org/10.1016/j.icarus.2015.06.020.

Gomes, R., Deienno, R., Morbidelli, A., 2017. The inclination of the planetary system relative o the solar equator may be explained by the presence of planet 9. Astron. J. 153, 27. https://doi.org/10.3847/1538-3881/153/1/27.

Guliyev, R., Guliyev, A., 2017. On the relationship between long-period comets and large Trans-Neptunian planetary bodies. In: Brescia, M., Djorgovski, S.G., Feigelson, E.D., Longo, G., Cavuoti, S. (Eds.), Astroinformatics, IAU Symposium, vol. 325, pp. 263–265.

Hadden, S., Li, G., Payne, M.J., Holman, M.J., 2018. Chaotic dynamics of Trans-Neptunian objects perturbed by planet nine. Astron. J. 155, 249. https://doi.org/10.3847/1538-3881/aab88c.

Hamilton, S., Gerdes, D.W., 2017. Detection bias for Trans-Neptunian objects on highly elliptical orbits with the dark energy survey. In: AAS/Division for Planetary Sciences Meeting Abstracts No. 49.

Hamilton, S., Napier, K., Gerdes, D., Lin, H.-W., 2018. A combined study of extreme Trans-Neptunian objects from three surveys and implications for planet nine. In: AAS/Division for Planetary Sciences Meeting Abstracts No. 50.

Holman, M.J., Payne, M.J., 2016a. Observational constraints on planet nine: astrometry of Pluto and other Trans-Neptunian objects. Astron. J. 152, 80. https://doi.org/10.3847/0004-6256/152/4/80.

Holman, M.J., Payne, M.J., 2016b. Observational constraints on planet nine: Cassini range observations. Astron. J. 152, 94. https://doi.org/10.3847/0004-6256/152/4/94.

Iorio, L., 2014. Planet X revamped after the discovery of the Sedna-like object 2012 VP_{113}? Mon. Not. R. Astron. Soc. 444, L78–L79. https://doi.org/10.1093/mnrasl/slu116.

Iorio, L., 2017. Preliminary constraints on the location of the recently hypothesized new planet of the Solar System from planetary orbital dynamics. Astrophys. Space Sci. 362, 11. https://doi.org/10.1007/s10509-016-2993-8.

Kaib, N.A., Roškar, R., Quinn, T., 2011. Sedna and the Oort cloud around a migrating Sun. Icarus 215, 491–507. https://doi.org/10.1016/j.icarus.2011.07.037.

Kenyon, S.J., Bromley, B.C., 2015. Formation of super-Earth mass planets at 125-250 AU from a solar-type star. Astrophys. J. 806, 42. https://doi.org/10.1088/0004-637X/806/1/42.

Kenyon, S.J., Bromley, B.C., 2016. Making planet nine: pebble accretion at 250–750 AU in a gravitationally unstable ring. Astrophys. J. 825, 33. https://doi.org/10.3847/0004-637X/825/1/33.

Khain, T., Batygin, K., Brown, M.E., 2018a. The generation of the distant Kuiper belt by planet nine from an initially broad perihelion distribution. Astron. J. 155, 250. https://doi.org/10.3847/1538-3881/aac212.

Khain, T., Becker, J.C., Adams, F.C., Gerdes, D.W., Hamilton, S.J., Franson, K., Zullo, L., Sako, M., Napier, K., Lin, H.W., Markwardt, L., Bernardinelli, P., Abbott, T.M.C., Abdalla, F.B., Annis, J., Avila, S., Bertin, E., Brooks, D., Carnero Rosell, A., Carrasco Kind, M., Carretero, J., Cunha, C.E., da Costa, L.N., Davis, C., De Vicente, J., Desai, S., Diehl, H.T., Doel, P., Eifler, T.F., Flaugher, B., Frieman, J., García-Bellido, J., Gruen, D., Gruendl, R.A., Gutierrez, G., Hartley, W.G., Hollowood, D.L., Honscheid, K., James, D.J., Krause, E., Kuehn, K., Kuropatkin, N., Lahav, O., Maia, M.A.G., Menanteau, F., Miquel, R., Nord, B., Ogando, R.L.C., Plazas, A.A., Romer, A.K., Sanchez, E., Scarpine, V., Schindler, R., Schubnell, M., Sevilla-Noarbe, I., Smith, M., Soares-Santos, M., Sobreira, F., Suchyta, E., Swanson, M.E.C., Tarle, G., Vikram, V., Walker, A.R., Wester, W., Zhang, Y., DES Collaboration, 2018b. Dynamical analysis of three distant Trans-Neptunian objects with similar orbits. Astron. J. 156, 273. https://doi.org/10.3847/1538-3881/aaeb2a.

Kuchner, M.J., Faherty, J.K., Schneider, A.C., Meisner, A.M., Filippazzo, J.C., Gagné, J., Trouille, L., Silverberg, S.M., Castro, R., Fletcher, B., Mokaev, K., Stajic, T., 2017. The first brown Dwarf discovered by the backyard worlds: planet 9 citizen science project. Astrophys. J. Lett. 841, L19. https://doi.org/10.3847/2041-8213/aa7200.

Lai, D., 2016. Solar obliquity induced by planet nine: simple calculation. Astron. J. 152, 215. https://doi.org/10.3847/0004-6256/152/6/215.

Lawler, S.M., Shankman, C., Kaib, N., Bannister, M.T., Gladman, B.J., Kavelaars, J.J., 2017. Observational signatures of a massive distant planet on the scattering disk. Astron. J. 153, 33. https://doi.org/10.3847/1538-3881/153/1/33.

Lawler, S.M., Kavelaars, J.J., Alexandersen, M., Bannister, M.T., Gladman, B.J., Petit, J.-M., Shankman, C., 2018a. OSSOS: X. How to use a survey simulator: statistical testing of dynamical models against the real Kuiper belt. Front. Astron. Space Sci. 5, 14. https://doi.org/10.3389/fspas.2018.00014.

Lawler, S.M., Shankman, C., Kavelaars, J.J., Alexandersen, M., Bannister, M.T., Chen, Y.-T., Gladman, B.J., Fraser, W.C., Gwyn, S., Kaib, N., Petit, J.-M., Volk, K., 2018b. OSSOS. VIII. The transition between two size distribution slopes in the scattering disk. Astron. J. 155, 197. https://doi.org/10.3847/1538-3881/aab8ff.

Levi, A., Kenyon, S.J., Podolak, M., Prialnik, D., 2017. H-atmospheres of icy super-Earths formed in situ in the outer solar system: an application to a possible planet nine. Astrophys. J. 839, 111. https://doi.org/10.3847/1538-4357/aa6ba6.

Li, G., Adams, F.C., 2016. Interaction cross sections and survival rates for proposed solar system member planet nine. Astrophys. J. Lett. 823, L3. https://doi.org/10.3847/2041-8205/823/1/L3.

Linder, E.F., Mordasini, C., 2016. Evolution and magnitudes of candidate planet nine. Astron. Astrophys. 589, A134. https://doi.org/10.1051/0004-6361/201628350.

Madigan, A.-M., McCourt, M., 2016. A new inclination instability reshapes Keplerian discs into cones: application to the outer Solar System. Mon. Not. R. Astron. Soc. 457, L89–L93. https://doi.org/10.1093/mnrasl/slv203.

Madigan, A.-M., Zderic, A., McCourt, M., Fleisig, J., 2018. On the dynamics of the inclination instability. Astron. J. 156, 141. https://doi.org/10.3847/1538-3881/aad95c.

Magnier, E.A., Cuillandre, J.-C., 2004. The elixir system: data characterization and calibration at the Canada-France-Hawaii telescope. Publ. ASP 116, 449–464. https://doi.org/10.1086/420756.

Malhotra, R., 1996. The phase space structure near Neptune resonances in the Kuiper belt. Astron. J. 111, 504. https://doi.org/10.1086/117802.

Malhotra, R., Volk, K., Wang, X., 2016. Corralling a distant planet with extreme resonant Kuiper belt objects. Astrophys. J. Lett. 824, L22. https://doi.org/10.3847/2041-8205/824/2/L22.

Martínez-Barbosa, C.A., Jílková, L., Portegies Zwart, S., Brown, A.G.A., 2017. The rate of stellar encounters along a migrating orbit of the Sun. Mon. Not. R. Astron. Soc. 464, 2290–2300. https://doi.org/10.1093/mnras/stw2507.

Medvedev, Y.D., Vavilov, D.E., Bondarenko, Y.S., Bulekbaev, D.A., Kunturova, N.B., 2017. Improvement of the position of planet X based on the motion of nearly parabolic comets. Astron. Lett. 43, 120–125. https://doi.org/10.1134/S1063773717020037.

Meisner, A.M., Bromley, B.C., Nugent, P.E., Schlegel, D.J., Kenyon, S.J., Schlafly, E.F., Dawson, K.S., 2017. Searching for planet nine with coadded WISE and NEOWISE-reactivation images. Astron. J. 153, 65. https://doi.org/10.3847/1538-3881/153/2/65.

Meisner, A.M., Bromley, B.C., Kenyon, S.J., Anderson, T.E., 2018. A 3π search for planet nine at 3.4 μm with WISE and NEOWISE. Astron. J. 155, 166. https://doi.org/10.3847/1538-3881/aaae70.

Miknaitis, G., Pignata, G., Rest, A., Wood-Vasey, W.M., Blondin, S., Challis, P., Smith, R.C., Stubbs, C.W., Suntzeff, N.B., Foley, R.J., Matheson, T., Tonry, J.L., Aguilera, C., Blackman, J.W., Becker, A.C., Clocchiatti, A., Covarrubias, R., Davis, T.M., Filippenko, A.V., Garg, A., Garnavich, P.M., Hicken, M., Jha, S., Krisciunas, K., Kirshner, R.P.,

Leibundgut, B., Li, W., Miceli, A., Narayan, G., Prieto, J.L., Riess, A.G., Salvo, M.E., Schmidt, B.P., Sollerman, J., Spyromilio, J., Zenteno, A., 2007. The ESSENCE supernova survey: survey optimization, observations, and supernova photometry. Astrophys. J. 666, 674–693. https://doi.org/10.1086/519986.
Millholland, S., Laughlin, G., 2017. Constraints on planet nine's orbit and sky position within a framework of mean-motion resonances. Astron. J. 153, 91. https://doi.org/10.3847/1538-3881/153/3/91.
Miyazaki, S., Komiyama, Y., Nakaya, H., Kamata, Y., Doi, Y., Hamana, T., Karoji, H., Furusawa, H., Kawanomoto, S., Morokuma, T., Ishizuka, Y., Nariai, K., Tanaka, Y., Uraguchi, F., Utsumi, Y., Obuchi, Y., Okura, Y., Oguri, M., Takata, T., Tomono, D., Kurakami, T., Namikawa, K., Usuda, T., Yamanoi, H., Terai, T., Uekiyo, H., Yamada, Y., Koike, M., Aihara, H., Fujimori, Y., Mineo, S., Miyatake, H., Yasuda, N., Nishizawa, J., Saito, T., Tanaka, M., Uchida, T., Katayama, N., Wang, S.-Y., Chen, H.-Y., Lupton, R., Loomis, C., Bickerton, S., Price, P., Gunn, J., Suzuki, H., Miyazaki, Y., Muramatsu, M., Yamamoto, K., Endo, M., Ezaki, Y., Itoh, N., Miwa, Y., Yokota, H., Matsuda, T., Ebinuma, R., Takeshi, K., 2012. Hyper Suprime-Cam. In: Ground-Based and Airborne Instrumentation for Astronomy IV, Proc. SPIE, vol. 8446, p. 84460Z.
Murray, C.D., Dermott, S.F., 1998. Solar System Dynamics. Cambridge University Press, Cambridge, UK.
Mustill, A.J., Raymond, S.N., Davies, M.B., 2016. Is there an exoplanet in the Solar System? Mon. Not. R. Astron. Soc. 460, L109–L113. https://doi.org/10.1093/mnrasl/slw075.
Nesvorný, D., Vokrouhlický, D., Dones, L., Levison, H.F., Kaib, N., Morbidelli, A., 2017. Origin and evolution of short-period comets. Astrophys. J. 845, 27. https://doi.org/10.3847/1538-4357/aa7cf6.
Parker, R.J., Lichtenberg, T., Quanz, S.P., 2017. Was planet 9 captured in the Sun's natal star-forming region? Mon. Not. R. Astron. Soc. 472, L75–L79. https://doi.org/10.1093/mnrasl/slx141.
Pfalzner, S., Bhandare, A., Vincke, K., Lacerda, P., 2018. Outer solar system possibly shaped by a stellar fly-by. Astrophys. J. 863, 45. https://doi.org/10.3847/1538-4357/aad23c.
Pike, R.E., Fraser, W.C., Schwamb, M.E., Kavelaars, J.J., Marsset, M., Bannister, M.T., Lehner, M.J., Wang, S.-Y., Alexandersen, M., Chen, Y.-T., Gladman, B.J., Gwyn, S., Petit, J.-M., Volk, K., 2017. Col-OSSOS: z-band photometry reveals three distinct TNO surface types. Astron. J. 154, 101. https://doi.org/10.3847/1538-3881/aa83b1.
Saillenfest, M., Fouchard, M., Tommei, G., Valsecchi, G.B., 2017. Study and application of the resonant secular dynamics beyond Neptune. Celest. Mech. Dyn. Astron. 127, 477–504. https://doi.org/10.1007/s10569-016-9735-7.
Sefilian, A.A., Touma, J.R., 2019. Shepherding in a self-gravitating disk of Trans-Neptunian objects. Astron. J. 157, 59. https://doi.org/10.3847/1538-3881/aaf0fc.
Shankman, C., Kavelaars, J., Gladman, B.J., Alexandersen, M., Kaib, N., Petit, J.-M., Bannister, M.T., Chen, Y.-T., Gwyn, S., Jakubik, M., Volk, K., 2016. OSSOS. II. A sharp transition in the absolute magnitude distribution of the Kuiper belt's scattering population. Astron. J. 151, 31. https://doi.org/10.3847/0004-6256/151/2/31.
Shankman, C., Kavelaars, J.J., Bannister, M.T., Gladman, B.J., Lawler, S.M., Chen, Y.-T., Jakubik, M., Kaib, N., Alexandersen, M., Gwyn, S.D.J., Petit, J.-M., Volk, K., 2017a. OSSOS. VI. Striking biases in the detection of large semimajor axis Trans-Neptunian objects. Astron. J. 154, 50. https://doi.org/10.3847/1538-3881/aa7aed.
Shankman, C., Kavelaars, J.J., Lawler, S.M., Gladman, B.J., Bannister, M.T., 2017b. Consequences of a distant massive planet on the large semimajor axis Trans-Neptunian objects. Astron. J. 153, 63. https://doi.org/10.3847/1538-3881/153/2/63.
Sheppard, S.S., Trujillo, C., 2016a. 2013 FT28. Minor Planet Electron. Circ 2016-Q39.
Sheppard, S.S., Trujillo, C., 2016b. New extreme Trans-Neptunian objects: toward a super-Earth in the outer solar system. Astron. J. 152, 221. https://doi.org/10.3847/1538-3881/152/6/221.
Sheppard, S.S., Trujillo, C., Tholen, D.J., 2016. Beyond the Kuiper belt edge: new high perihelion Trans-Neptunian objects with moderate semimajor axes and eccentricities. Astrophys. J. Lett. 825, L13. https://doi.org/10.3847/2041-8205/825/1/L13.
Sheppard, S.S., Trujillo, C.A., Tholen, D.J., Kaib, N., 2019. A new high perihelion Trans-Plutonian inner Oort cloud object: 2015 TG_{387}. Astron. J. 157, 139. https://doi.org/10.3847/1538-3881/ab0895.
Silsbee, K., Tremaine, S., 2018. Producing distant planets by mutual scattering of planetary embryos. Astron. J. 155, 75. https://doi.org/10.3847/1538-3881/aaa19b.
Tiscareno, M.S., Malhotra, R., 2003. The dynamics of known Centaurs. Astron. J. 126, 3122–3131. https://doi.org/10.1086/379554.
Trujillo, C.A., 2020. Dynamical constraints on the sky location of an undiscovered giant planet in the solar system. Astrophys. J. (submitted).

Trujillo, C.A., Brown, M.E., 2003. The Caltech wide area sky survey. Earth Moon Planets 92, 99–112. https://doi.org/10.1023/B:MOON.0000031929.19729.a1.

Trujillo, C.A., Sheppard, S.S., 2014. A Sedna-like body with a perihelion of 80 astronomical units. Nature 507, 471–474. https://doi.org/10.1038/nature13156.

Volk, K., Murray-Clay, R., Gladman, B.J., Lawler, S., Bannister, M.T., Kavelaars, J.J., Petit, J.-M., Gwyn, S., Alexandersen, M., Chen, Y.-T., Lykawka, P.S., Ip, W., Lin, H.W., 2016. OSSOS III—resonant Trans-Neptunian populations: constraints from the first quarter of the outer solar system origins survey. Astron. J. 152, 23. https://doi.org/10.3847/0004-6256/152/1/23.

Volk, K., Murray-Clay, R.A., Gladman, B.J., Lawler, S.M., Yu, T.Y.M., Alexandersen, M., Bannister, M.T., Chen, Y.-T., Dawson, R.I., Greenstreet, S., Gwyn, S.D.J., Kavelaars, J.J., Lin, H.W., Lykawka, P.S., Petit, J.-M., 2018. OSSOS. IX. Two objects in Neptune's 9:1 resonance—implications for resonance sticking in the scattering population. Astron. J. 155, 260. https://doi.org/10.3847/1538-3881/aac268.

Wyatt, M.C., Bonsor, A., Jackson, A.P., Marino, S., Shannon, A., 2017. How to design a planetary system for different scattering outcomes: giant impact sweet spot, maximizing exocomets, scattered discs. Mon. Not. R. Astron. Soc. 464, 3385–3407. https://doi.org/10.1093/mnras/stw2633.

Zeebe, R.E., 2017. Numerical solutions for the orbital motion of the solar system over the past 100 Myr: limits and new results. Astron. J. 154, 193. https://doi.org/10.3847/1538-3881/aa8cce.

PART II

Properties and structure

CHAPTER

5

Surface composition of Trans-Neptunian objects

M. Antonietta Barucci, Frederic Merlin

LESIA, Paris Observatory, PSL, CNRS, Sorbonne University, University Paris-Diderot, Paris, France

5.1 Introduction

Knowledge of the properties of the Trans-Neptunian population is essential in understanding the processes involved in the early solar nebula as well as in other planetary disks. After 25 years from the first discovery (Pluto excluded) of a Trans-Neptunian object (TNO) and after the NASA/New Horizons mission that visited with a great success the Pluto Charon system, we start to have a better knowledge of dynamical and physical properties of this population.

These objects remain very faint and the majority of them have today a large photometric data available and moderate-quality spectroscopy even if observed by large ground-based telescopes (8–10 m size) and few of them by space telescopes as Hubble and Spitzer. Recently, a large survey has been obtained by the mid-far infrared space telescope Herschel (Chapter 7) determining albedo properties for more than 130 objects which allowed a better interpretation of the surface composition.

The composition of these objects remains one of the most important information to define the physical and chemical conditions in the early solar system. The material of these objects did not experienced high temperature and pressure processes. The interpretation of the composition is strongly connected to the competition between composition, space weathering effects, and rejuvenating processes. Modeling the obtained spectra, their surface composition is interpreted with the presence of different ices (H_2O, CH_4, CH_3OH, C_2H_6, CO, NH_3, etc.) and possibly contribution of silicates and refractory carbonaceous compounds are to interpret the spectral slope and their albedo. Their surfaces have been affected by energetic phenomena which implied modification on molecular composition of their surfaces. Several

The Trans-Neptunian Solar System.
https://doi.org/10.1016/B978-0-12-816490-7.00005-9

specific laboratory experiments have been carried out on different compounds, irradiated by energetic particles to simulate the space environment (e.g., Hudson et al., 2008). The cosmic-ray bombardment and solar wind alter the surface causing irreversible physical and chemical changes on the surface of these ice bodies.

Most of them remain too small and too far to be observed even with the largest telescopes. Short period comets, coming from the Trans-Neptunian population can allow a better understanding the composition of objects coming from external solar system. Rosetta mission gave an extraordinary amount of information on the comet 67P/Churyumov-Gerasimenko (Barucci and Fulchignoni, 2017) and this help in constraining the properties of the accretion region of all these objects.

5.2 Techniques

The TNOs are the most challenging objects to observe in the solar system and it remains still difficult to study them by spectroscopy. Photometry is, from the beginning of the discovery of these objects, the technique available to observe a large number of them that allowed to have a first indication on the surface properties and capable to give a global view.

5.2.1 Spectrophotometry, taxonomy

As soon as TNOs were discovered, many surveys started to investigate the photometric properties with different filters to study the colors and spectral gradient of TNOs and Centaurs and have a first information on the surface properties. In fact, broadband photometry represents the simplest observing technique to study these faint objects and color data provide information of the global surface reflectance of the objects. The objective was to study them by statistical purposes and to compare the color properties with the different dynamical groups to investigate how TNOs were formed and how their surfaces have evolved.

Color properties and color-color diagrams were analyzed by many authors starting by Tegler and Romanishin (1998), who analyzing the BVR colors of 13 objects found two groups. With the increasing of data, clustering was less obvious, nevertheless clustering was found when considered different dynamical populations. Many others analyzed a large number of objects (order of hundreds) showing differences in colors and correlations with diameters, and orbital parameters, trying to interpret the entire population and a complete review is presented by Doressoundiram et al. (2008). The main results are the strong color diversity and the fact that classical low inclination is redder than the other populations. The found cluster of objects similar in color (red) and dynamical properties, the so-called cold classical objects (classical population at small inclinations beyond 40 AU), coincides with the dynamically cold class defined by Gomez (2003), which are associated with the more pristine objects. This color-dynamical correlation presents the intriguing idea that the surfaces of TNOs contain information on more than composition, but as well hold the key to understanding the dynamical processes that dispersed the protoplanetesimal disk and populating the Kuiper belt region.

A more recent and complete work using albedo data obtained by Herschel Space observatory and the available colors on different dynamical populations (detailed discussion is presented by Chapter 7) was done by Lacerda et al. (2014). They conclude that their finding on the color-albedo separation could be an evidence for a compositional discontinuity in the young solar system. Based on the plot "visible spectral slope—albedo" for 109 objects, Lacerda et al. (2014) distinguished two main surface types of TNOs, composed of dark neutral and bright red objects. The objects located in dynamical stable orbits within the classical dynamical region and beyond follow to the bright red group, believed to be formed in the outer solar system.

The interpretation of different colors remains unclear, even if it is associated with different compositions and/or different surface evolutions (Table 5.1). In fact, spectrophotometry can provide only limited constraints on the surface composition as photometry can be influenced

TABLE 5.1 Names of the observed TNOs and Centaurs divided by taxonomical classes.

BB group: name	Perihelion distance (q)	First ice detection
2060 Chiron	8.5	H_2O (Luu et al., 2000)
15874 1996 TL_{66}	35	H_2O (Barucci et al., 2011)
19308 1996 TO_{66}	38.4	H_2O (Brown et al., 1999)
24835 1995 SM_{65}	37.5	H_2O (Barkume et al., 2008)
55636 2002 TX_{300}	38	H_2O (Barkume et al., 2008)
90482 Orcus	30.3	H_2O–NH_3 (Barucci et al., 2008)
120178 2003 OP_{32}	38.8	H_2O (Barkume et al., 2008)
134340 Pluto	29.7	H_2O–N_2–CH_4–CO (Owen et al., 1993)
136108 Haumea	34.5	H_2O (Barkume et al., 2008)
136199 Eris	38.5	CH_4 (Brown et al., 2005)
145451 2005 RM_{43}	35.1	H_2O (Barucci et al., 2011)
145453 2005 RR_{43}	37.2	H_2O (Barucci et al., 2011)
208996 2003 AZ_{84}	32.3	H_2O (Merlin et al., 2010b)
2003 UZ_{413}	30.6	H_2O (Barucci et al., 2011)
BR group: name	q	**Ice detection**
8405 Asbolus	6.8	– (Romon-Martin et al., 2002)
10199 Chariklo	13	H_2O (Guilbert et al., 2009)
29981 1999 TD_{10}	12.3	H_2O (Barkume et al., 2008)
32532 Thereus	8.5	H_2O (Guilbert et al., 2009)
42355 Typhon	17.5	H_2O (Guilbert et al., 2009)
47932 2000 GN_{171}	28.3	– (Guilbert et al., 2009)
52872 Okyrhoe	5.8	– (Dotto et al., 2003)

Continued

TABLE 5.1 Names of the observed TNOs and Centaurs divided by taxonomical classes—cont'd

IR group: name	*q*	Ice detection
54598 Bienor	13.2	H_2O (Guilbert et al., 2009)
63252 2001 BL_{41}	6.9	– (Doressoundiram et al., 2003)
73480 2002 PN_{34}	13.4	H_2O (DeMeo et al., 2010b)
90568 2004 GV_9	38.8	– (Guilbert et al., 2009)
95626 2002 GZ_{32}	18	H_2O (Barucci et al., 2011)
120061 2003 CO_1	10.9	– (Barucci et al., 2011)
136472 Makemake	37.9	$CH_4 + C_2H_6 + C_2H_4$ + other (Licandro et al., 2006; Brown et al., 2007a, 2015)
2007 VH_{305}	8.2	H_2O (Barucci et al., 2011)
19521 Chaos	40.7	– (Barkume et al., 2008)
20000 Varuna	40.5	H_2O (Licandro et al., 2001)
26375 1999 DE_9	32.2	H_2O (Jewitt and Luu, 2001)
28978 Ixion	29.9	H_2O (Guilbert et al., 2009)
33340 1998 VG_{44}	29.4	– (Barkume et al., 2008)
38628 Huya	28.5	H_2O (Alvarez-Candal et al., 2007)
55565 2002 AW_{197}	41	– (Guilbert et al., 2009)
119951 2002 KX_{14}	37	– (Guilbert et al., 2009)
120132 2003 FY_{128}	37	– (Guilbert et al., 2009)
144897 2004 UX_{10}	37.4	H_2O (Barucci et al., 2011)
174567 2003 MW_{12}	39.3	– (Barucci et al., 2011)
2003 QW_{90}	40.7	H_2O (Guilbert et al., 2009)
RR group: name	*q*	**Ice detection**
5145 Pholus	8.7	H_2O, CH_3OH + other (Cruikshank et al., 1998; Barucci et al., 2011; Brown, 2000)
15789 1993 SC	32	Hydrocarbons (Brown et al., 1997; Jewitt and Luu, 2001)
15875 1996 TP_{66}	26.3	– (Barkume et al., 2008)
31824 Elatus	7.2	H_2O + other (Bauer et al., 2002; Dalle Ore et al., 2015)
42301 2001 UR_{163}	37	– (Barkume et al., 2008)
44594 1999 OX_3	17.6	– (Barucci et al., 2011)
47171 1999 TC_{36}	30.6	H_2O + other (Dotto et al., 2003; Merlin et al., 2005; Barkume et al., 2008)
50000 Quaoar	41.8	H_2O, CH_4, C_2H_6, $NH_3 \cdot H_2O$ + other (Guilbert et al., 2009; Jewitt and Luu, 2004; Barucci et al., 2015)

TABLE 5.1 Names of the observed TNOs and Centaurs divided by taxonomical classes—cont'd

RR group: name	q	Ice detection
55576 Amycus	15.2	H_2O + other (Barucci et al., 2011; Dalle Ore et al., 2015)
55637 2002 UX_{25}	36.8	– (Barucci et al., 2011)
55638 2002 VE_{95}	28	H_2O, CH_3OH + other (Barkume et al., 2008; Barucci et al., 2012; Dalle Ore et al., 2015)
60558 Echeclus	5.8	– (Guilbert et al., 2009)
66652 1999 RZ_{253}	40	CH_4? (Barkume et al., 2008; Dalle Ore et al., 2015)
79360 1997 CS_{29}	43.4	– (Grundy et al., 2005)
83982 Crantor	14	H_2O (Guilbert et al., 2009)
90377 Sedna	90	H_2O, CH_4, C_2H_6, N_2+ other (Trujillo et al., 2005; Barucci et al., 2010; Dalle Ore et al., 2015)
145452 2005 RN_{43}	40.6	– (Guilbert et al., 2009)
2002 KY_{14}	8.6	H_2O (Barucci et al., 2011)
2007 UM_{126}	10.1	– (Barucci et al., 2011)
2008 FC_{76}	10.2	– (Barucci et al., 2011)

Notes: Perihelion distance and detected ices are reported with references of the first detection.

by scattering effects, but colors are very useful to classify objects and can be used for statistical purposes. Barucci et al. (2001, 2005) were the first that analyzed the TNO population as was done on asteroids to define the first tentative taxonomy. Using multivariate statistical analysis of 51 objects with BVRIJ colors, they found four classes of objects defined as RR group (including the reddest objects), BB group (containing objects with neutral colors), and two intermediary groups RI and BR (containing objects with intermediate behaviors). With the increase of data Fulchignoni et al. (2008) analyzed 67 objects with four color indices (B-V, V-R, V-I, and V-J) as in the previous works by Barucci et al. and 55 objects designed by six color indices (B-V, V-R, V-I, V-H, and V-K) with two different multivariate statistical methods confirming the four defined groups which indicate differences in the surface nature of these objects. They extended the taxonomy to a total of 135 objects even when only three colors (B-V, V-R, and V-I) were available. They studied the distribution with the dynamical classes and orbital elements. The RR objects have low inclinations while Centaurs show a clear bimodality (BR or RR taxonomy). With the increase of colors in the visible and near infrared, Belskaya et al. (2015) reanalyzed again the available sample updating the classification to 258 TNOs and Centaurs and obtained the four classes of objects confirming the previous classification into the BB, BR, IR, and RR taxonomic groups. The increasing accuracy of color measurements allowed to better separate the classes. They also used the albedo obtained by Herschel Space Observatory, nevertheless it does not seem to have an impact on the classification except for the separation of a subgroup of the brightest bodies inside the BB group. Using the albedo as an additional independent parameter, Belskaya et al. (2015) obtained the same classes of objects. The only difference is a separation of subgroup of brightest bodies inside the BB group which is associated with the largest TNOs Pluto, Eris, Makemake, Haumea, and the members of the Haumea family. Color observations continue on all discovered objects. A new survey named COL-OSSOS, covering 4-year program to have large number of colors, was started. At the end, a total of 400 hours observations on Gemini-North telescope will be performed to observe optical and NIR colors of all targets in the outer solar system brighter than magnitude 23.5 in V to obtain homogeneous and consistent data.

Merlin et al. (2017) analyzing more than 40 spectra of TNOs and Centaurs, generated general mean typical spectra for each of the four defined classes to allow each new observation, even if not available in the previous used colors/filters, to be easily classify in the different groups. They confirmed the increase content of H_2O ice from IR and BR (at similar level) to RR and BB. All the spectra of the objects following the BB group present some ice content in their surface. They also showed that the RR and IR mean spectra exhibit an absorption feature at wavelength above about 2.3 μm, that could be an indicator of hydrocarbons and/or methanol present at the surface of these ultrared objects, as presented by Dalle Ore et al. (2015).

5.2.2 Spectroscopy

The investigation of the composition of any object from remote observations is possible only by spectroscopy technique. From ground-based observations, the available wavelength is generally from 0.4 to 2.4 μm, which allows the characterization of the major mineral phases and ices present in the TNOs (Table 5.1). The interpretation is connected with the presence of diagnostic features on the spectra. These spectra probe only the upper surface layers, no deeper than few centimeters.

The visible spectra obtained are generally featureless but confirm the spectral gradient found with the colors going from neutral (flat spectrum) to very red (reflectance increasing with the wavelength) making some of these objects among the reddest objects of the solar system. The reddening in the spectral slope could be associated with the presence of particular chemical compounds (Section 5.4) or be associated with the presence of organic or other irradiated compounds (Section 5.5). Among possible absorption features present in the visible range, we could find faint bands associated with the presence of aqueous altered mineral such as phyllosilicates which have been detected on several objects, as well as the presence of CH_4. The difficulty to confirm the shallow features is often connected with the quality (the noise) of the spectra.

The near-infrared spectra covering 1–2.4 μm range represent the most diagnostic region to detect the presence of ices and volatiles. Indeed, most of the absorption features due to combination of harmonic modes (e.g., stretching, bending modes) are located in this wavelength range. Pluto and Triton, a possible captured TNO, were the first objects observed, due to their brightness, showing clear absorption features in this wavelength range owing to CH_4, N_2, and CO (Brown et al., 1995; Cruikshank et al., 1993, 1997). The first spectra of new discovered TNOs were dramatically noisy, but as soon as large objects have been discovered, the quality increased in particular when the largest telescopes were used like Keck, Gemini, Subaru, and VLT-ESO. At today more than 60 objects have been observed in the near infrared, but the quality of the data remains poor as the visibility of these distant objects remains at the limit of available ground instrumentations. More than half of the observed objects show signatures of ices on their surface. The relative abundance of volatile content is varying on surface of these objects and their presence has been easily explained by a model of volatile loss and retention on the surface of these objects in function of their size and temperature/location (Schaller and Brown, 2007). Amorphous carbon, olivines, pyroxenes, tholins, and other types of organics have been used to fit the spectral behavior of the spectra, but without clear features, the real presence of these materials is difficult to be confirmed.

The mid-infrared is well suited to investigate the surface composition of these objects. The ISO satellite demonstrated the utility on brighter asteroids, while Spitzer with the increased sensibility, open the door to new exploration of the more distant objects. The spectral region (2.5–40 μm) may be used to study the fundamental vibrations and associated rotational-vibrational structure. H_2O and CH_3OH ices also show features in this wavelength range. Some constraints on the grain size can also be deduced. Unfortunately, due to the faintness of these objects, only few Centaurs were observed showing emissivity peaks at around 10 and 20 μm to detect the presence of small grain silicates. For about 30 objects the broad band fluxes at 3.55 and 4.5 μm have been obtained and for few of them also at 5.8 μm (Emery et al., 2006). The measured reflectance used as a proxy to extend the near-infrared spectra obtained by ground (Emery et al., 2007; Dalle Ore et al., 2015; Barucci et al., 2015) represented an important constraint in discriminating the presence of the different ices. New more precise detection of albedo (see Chapter 7) allowed more precise spectral modeling.

5.3 Surface modeling

Beyond the detection of spectral features, mainly associated with the presence of absorption bands of ices or minerals, spectral modeling could be intended to provide deeper insights on the physical and chemical properties of the surface of these atmosphereless bodies.

5.3.1 Scattering theories and requirements

From the different analytical approaches, the model developed by Hapke (1981, 1993) has been the widest used to constrain the physical and chemical properties of icy surfaces. The main approach relies on a component to describe the albedo or the reflectance singly scattered light by an average regolith grain that could be pure or composed by a mixture of different chemical materials. It is computed as the ratio of the average scattering coefficient of the medium to the average extinction coefficient of the medium depending on the external and internal surface-scattering coefficients and the absorption coefficient, which are related to the microscopic quantities (the optical constants) and the size of the particle. The other components, also taken into account, are the one to describe how photons are multiply scattered among and from an aggregate volume of average grains (analytical simplification of Chandrasekhar's 1960, or more rigorous treatment for strongly anisotropic grains by Hapke, 2002), the one to describe the effect of macroscopic surface texture and porosity (see Hapke, 1984, 2008) and finally the one to describe the effect of the phase function (initially approximated by a simple approximation of a first-order Legendre polynomial and then improved using expression provided by the Henyey-Greenstein function; Henyey and Greenstein, 1941). The most recent and promising attempt to validate these models were then done by Shepard and Helfenstein (2007) and Helfenstein and Shepard (2011) from laboratory measurements. While these tests suggest that refinements are real for low-to-medium albedo surface, the correction has still little effect on high-albedo surfaces.

Several requirements are needed to better investigate the surface composition of planetary objects from the use of a radiative transfer model. The first crucial requirement is the measurement of the albedo at a given wavelength, fundamental to constrain the concentration of ices, carbons, silicates, or organics, which can have widely differing reflectivities. The second parameter is the presence of clear absorption features. They are not only mandatory to clearly identify a chemical compound on the surface, but they also allow convergence of the model into a given pure or diluted state and could possibly be used for ice thermometry (see, e.g., Merlin et al., 2018 for the case of Triton). Such particular investigation is usually restricted to the case of the brightest objects, covered by huge amounts of ice. Despite the common absence of deep and wide absorption features, the visible range is also useful since effects of mixtures and alteration, for instance, due to space weathering (see Section 5.5), are mainly detectable in this wavelength range and give strong constraints on the properties of possible silicates and hydrocarbons.

5.3.2 Limits of the models

Despite several refinements to overcome deficiencies and improve its physical realism, this kind of model has yet to be fully validated for retrieving accurate chemical and physical properties of planetary surfaces. Optical constants are the basis of the modeling, which could generate synthetic spectra considering several kind of mixtures (intimate, geographical or intramolecular). Each optical constant is derived from laboratory measurements for a given molecule at a given state (pure, diluted, isolated), temperature, and phase (amorphous and crystalline). Extended investigations considering the effects of all these parameters are rare and focused on a restricted set of ices such as water ice (Grundy and Schmitt, 1998), and methane diluted or not in nitrogen (Quirico and Schmitt, 1997a; Grundy et al., 2002). Optical constants retrieved from laboratory also depend on the properties of the samples, such as the dispersive degree of the media that could differ to the reality. In these conditions, it is difficult to completely simulate the local conditions of the analyzed surfaces and the retrieved physical properties are still only indicative. In addition to that, models are usually applied in case of unresolved objects considering simple surface (composed with a restricted number of spatial units), which is not likely considering resolved images of the entire Pluto system as seen by the New Horizons mission (e.g., Weaver et al., 2016).

5.4 Surface composition

Absorption features have been reported from a data sample composed of merely 50 objects, allowing detection of ices and traces of aqueous alteration. At the contrary, several TNOs exhibit almost featureless spectra, and attempt to derive surface composition is difficult.

5.4.1 Ice detections

Observation of the surface of Centaurs and TNOs mainly led to the detection of several ices. Water ice is the most detected ice (Barucci et al., 2008) possibly in its amorphous or crystalline form (see Grundy and Schmitt, 1998). Methane ice has been detected on the biggest TNOs so far. Ammonia ice has been detected on the surface of Charon, for instance (see Dumas et al., 2001), and we suspect that ammonia is probably in its hydrated state rather than in its pure state on the surface of this body (see Merlin et al., 2010a). Nitrogen has been detected on the surface and atmosphere of Triton and Pluto (see, e.g., DeMeo et al., 2010a; Quirico and Schmitt, 1997a). Heavier hydrocarbons (Merlin et al., 2010a,b) and carbon dioxide (e.g., Jewitt and Luu, 2004; Barucci et al., 2015; Merlin et al., 2018) have been reported for the objects 50,000 Quaoar, Pluto, or Triton. Carbon monoxide has been detected for both Pluto and Triton (Quirico and Schmitt, 1997b). Alcohol compounds, such as methanol, have been detected on the surface of a few objects, even if its presence needs confirmation (see, e.g., Cruikshank et al., 1997; Merlin et al., 2012).

5.4.2 Aqueous alteration

A few TNOs spectra appear to show spectral features related to the presence of aqueous altered materials on their surfaces (De Bergh et al., 2004; Alvarez-Cantal et al., 2008; Fornasier et al., 2009). Its detection might indicate the presence of water in the original bodies, that were enough heated during their early evolutionary phases to develop the alteration at their surface. Hydrous altered materials seem to be present in comets and hydrous silicates have been detected in interplanetary dust particles and micrometeorites. As suggested by Jarvis and Vilas (2000), if such altered material is confirmed on TNOs, a mechanism to produce aqueous alteration at large heliocentric distances must be defined. How aqueous alteration could have occurred at such low temperatures far from the Sun is not well understood. Large TNOs may have been subjected to significant radiogenic heating (Luu and Jewitt, 2002), but it remains to be seen if this would have been sufficient to enable aqueous alteration of the anhydrous material. Another way to heat a TNO is through impacts, but the heat and the duration of the heating episode have to be sufficient to enable aqueous alteration (e.g., Kerridge and Bunch, 1979). It cannot be excluded that hydrated minerals could have been formed directly in the early solar nebula.

5.4.3 Ambiguous cases

Many spectra appear featureless when observed at low signal-to-noise level, suggesting, at least, that the surface of these objects is mainly depleted in (fresh) icy mantle. Even in this case, spectral models could help to possibly point out on the presence of hydrocarbons-like material for red and moderately bright targets, such as tholins or kerogene, displaying a continuous red spectral slope from the visible to the beginning of the near infrared. For gray and dark objects, with spectra comparable to those of the dead comets and the Jovian Trojans (see Dotto et al., 2011), interpretation of their surface content is more difficult. Ambiguous cases also appear when multiobservations lead to clear different results indicating possible heterogeneity (see, e.g., Barucci et al., 2006). This shows the necessity to investigate the surface of these objects at different rotational phases and at different times, as seen in the case of the Centaur Chariklo, surrounded by a ring observed with different tilt angles during the last 15 years (Braga-Ribas et al., 2014).

5.4.4 Physical constrains on the retrieved chemical compounds

Spectral models applied on spectra obtained at relative high spectral resolution, ranging from 1000 to 10,000, could enable the determination of the ice temperature (e.g., as shown for H_2O by Grundy and Schmitt, 1998, for CH_4 by Grundy et al., 2002, or for N_2 by Quirico and Schmitt, 1997a). The study of the peak location and shape of the CH_4 absorption bands have been used to state on its diluted form in nitrogen in different cases (see, e.g., Licandro et al., 2006 for Eris, Tegler et al., 2007 and Lorenzi et al., 2015 for Makemake). Laboratory experiments (Quirico and Schmitt, 1997a) confirmed that the spectral band shifted blueward indicates that methane is in solid solution with nitrogen. Such investigation could also give strong constraints on the mixing state or stratification level of the ices on the subsurface (see, e.g., Merlin et al., 2009 for the object Eris or Merlin et al., 2018 for the object Triton).

5.5 Space weathering

The color diversity of TNOs and Centaurs populations was initially explained as the resulting competition between the color change produced by the cosmic-ray bombardment or/and UV-photolysis on fresh icy surfaces, and the resurfacing mechanisms, such as microscopic to kilometric scale impacts or cometary like activity (Luu and Jewitt, 1996). Other hypotheses to explain the color diversity are also proposed, such as the role of the SO_2 snow line (Wong and Brown, 2017), suggesting that the search of a unique and complete scenario at work for all objects is probably unrealistic.

5.5.1 Irradiation of the surface

Using magnetohydrodynamic (MHD) model, Cooper et al. (2003) showed that the dose and properties of ions and particles are related to the heliocentric distance, so the surface of the objects from different populations should evolve differently since ions and particles at different energies do not probe the same layers of a surface and do not react similarly with the exposed chemical content. The solar protons affect only the upper layers, modifying very quickly the color of the material in the visible range, while the Galactic cosmic rays penetrate more deeply and affect a thick layer of many meters. Moore et al. (1983) or more recently Brunetto et al. (2006) demonstrated from laboratory experiments that icy objects in the outer solar system may have grown an irradiation red mantle, produced by cosmic ion irradiation of simple hydrocarbons and/or alcohols. In particular, they have reproduced two different processes in their experiments: starting from pure C-rich frozen species, they induced strong reddening of the V-NIR spectra, connected to the formation of a refractory residue; conversely, starting from a complex hydrocarbon structure such as asphaltite or kerite, they induced spectral flattening, connected to a modification in the structure of the sample. Amorphization of the crystalline water ice is observed from ion irradiation at low temperature (Mastrapa and Brown, 2006).

Such experiments allow us to constrain the implications of the irradiation in the evolution of the surface of the atmosphereless members of the solar system and give some hints on the formation, stability, and detectability of different chemical species.

5.5.2 Resurfacing processes

The competition between cosmic ray reddening and collisions produces a continuous modification of the instantaneous color of the surface. As previously stated, the solar protons affect only the upper layers modifying very fast the color of the material. The Galactic cosmic rays penetrate more deeply and affect a thick layer of many meters, making necessary more energetic collisions to give fresh ices to the surface. As a consequence of the previous point, two different collisional regimes are expected from the collisional resurfacing model and Gil-Hutton (2002) considers that a thick mantle could survive in the largest TNOs because the resurfacing time scale in the depth regime for these objects is longer than the time needed to form the mantle, and that it is possible that TNOs in different regions of the belt are affected by projectile populations truncated at a different radius. At lower scale, submicron iron has been found in the Saturn system by the Cassini Cosmic Dust Analyzer. The origin of the iron may be from meteoritic dust falling into the Saturn system, and may indicate such physical process operates in the Saturn system as it does in other parts of the solar system. In this particular case, we see implication of submicron meteorite on the water ice surface of Iapetus, with different levels of reddening and darkening (Clark et al., 2012). The last resurfacing process is the activation of cometary like activity, reported for several Centaurs which could potentially lead to color variation, blanketing the subsurface with fresher excavated material. Jewitt (2009) finds that the "active Centaurs" in the studied sample have perihelia smaller than the inactive Centaurs (median 5.9 AU vs. 8.7 AU), and smaller than the median perihelion distance computed for all known Centaurs (12.4 AU). This suggests that their activity is thermally driven, but most of the objects are too cold for activity at the observed levels to originate via the sublimation of crystalline water ice. At the opposite, solid carbon monoxide and carbon dioxide are so volatile that they should drive activity in Centaurs at much larger distances than observed. Moreover, Melita and Licandro (2012) found that Centaurs following the Gray group color had experienced cometary activity, released volatiles in the past, and consequently their surfaces are covered by a cometary-color dust mantle.

5.6 Discussion and conclusion

The composition of TNOs is different and as shown in Fig. 5.1, some objects show spectra with clear signatures, others are featureless and exhibit different spectral slopes.

The largest TNOs as Eris, Pluto, Makemake, Sedna, and Quaoar, for which three of them (Eris, Pluto, and Makemake) follow in the official category of dwarf planets as defined by International Astronomical Union (IAU), show a clear detection of CH_4. The abundance is not the same as well the spectral slope.

Makemake, as brightest object, has been most easily studied. Brown et al. (2007a) suggested a surface with large grains of methane, clear presence of ethane (product of irradiated methane),

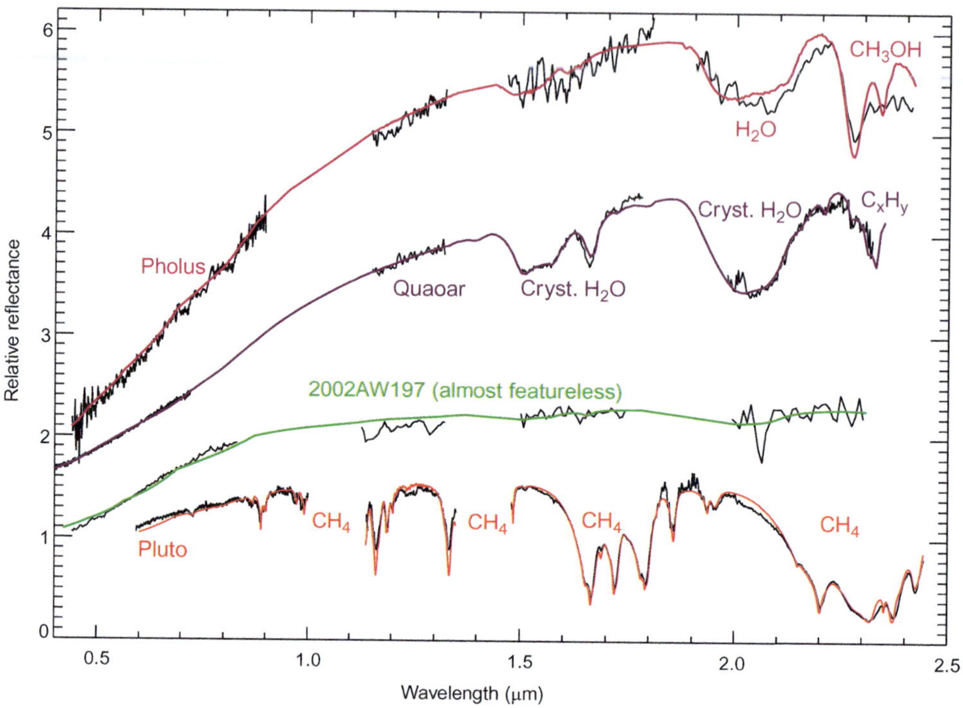

FIG. 5.1 Spectra of several objects associated with their best fit models. The spectra of Pholus and Quaoar both exhibit strong absorption bands of water ice and additional absorption features near 2.3 μm, attributable to methanol or complex hydrocarbons (*CxHy*), respectively. The spectrum of Pluto is well reproduced by a model including large amount of methane, in pure and diluted state in nitrogen, while the fit of the almost featureless spectrum of 2002 AW197 lets us assume a few amount a water ice only.

and tholin-like material with no evidence of N_2. Brown et al. (2015), with new high S/N spectra, presented detection of solid ethylene and acetylene, which are clear evidence of irradiation processes of methane. Following the model of Schaller and Brown (2007) large objects can retain volatiles as they are massive enough to prevent escape or cold enough to prevent vapor escape of frosts.

Eris and Pluto are the biggest objects in the known population. The near-infrared spectra of Eris are very similar to that of Pluto even if the weak band of N_2 (at 2.15 μm) has not been identified on Eris. Merlin et al. (2009) confirmed the possible presence of N_2 as a consequence of the band shift of CH_4 suggesting a stratification of the diluted methane ice and pure ice, as well the presence of small and large particles of methane ice. The visible properties of Eris also differ from those of Pluto which is much red with much more variation with rotational period than Eris. The high albedo and the lack of rotational variations on Eris allow us to suggest a surface dominate by seasonal cycling atmosphere.

Sedna is one of the reddest object among the more distant object accessible to investigation and well observed. The observed spectra show clear heterogeneity. Spectral modeling has been performed using the various sets of data in the visible and near infrared, including Spitzer observations at photometry at 3.6 and 4.5 μm to better constrain the composition (Barucci et al., 2010). The spectra can be modeled with organic materials (triton and titan tholin), serpentine, and H_2O ice in significant amounts, with CH_4, N_2, and C_2H_6 in varying trace amounts (Barucci et al., 2010).

Quaoar could be associated with the list of dwarf planets. Also in this case recent new observations (Barucci et al., 2015) made by the most sophisticated instruments and analyzed including the Spitzer photometry at longer wavelength allowed to confirm the presence of H_2O, CH_4, and C_2H_6, as previously reported (Dalle Ore et al., 2009), along with indication of the possible presence of $NH_3 \cdot H_2O$. This detection supports the hypothesis of possible presence of cryovolcanism, even if heterogeneity has not been detected on the 40% of the observed surface.

The fourth included in the dwarf planets by IAU, *Haumea*, is one of the most fascinating objects among TNOs with satellites and rings (see Chapter 11). Although its large dimension would allow us to predict following the model by Schaller and Brown (2007) to retain volatiles, its surface is dominated of nearly pure water ice as well its moons and the dynamical family objects. All these are possible product of a giant impact onto proto-Haumea. The pure water ice surface of Haumea is due to the fact that they exposed layers of the interior pure ice as well the satellites and family members are fragments of the interior (Brown et al., 2007b). The fact that they are still uncontaminated by more dark dust, impactors or irradiated hydrocarbons is still a surprise. All members contain crystalline water (showing the band at 1.65 μm) and also this is surprising as irradiation studies suggest that crystalline water become amorphous.

Abundance of water ice has been studied in details by Barucci et al. (2011) and Brown et al. (2012) showing that all the dynamical class objects contains similar amount of water, while smaller objects (except Haumea family members) contain less water ice. Ammonia seems present in some intermediate size objects like Orcus or Charon. Many small class objects also show water on their surface as for example the TNO 2002 VE95 (Barucci et al., 2006) which shows spectra with H_2O on their surface. Smaller Centaurs also present similar spectral trend compared to those of TNOs, as in the case of Pholus which spectrum is very red, similar to that of Sedna (see Fig. 5.1) with clear methanol presence. Many red objects exist in the TNOs and Centaurs population. Dalle Ore et al. (2015) analyzed the composition of the ultrared objects concluding they contains methanol/hydrocarbon ices and they may be formed in the outer part of the solar system. Small TNOs are difficult to be observed by spectroscopy and their properties are not yet well known.

In Fig. 5.2 have been reported all the observed objects in function of their heliocentric distance (except Sedna) and their albedo and diameter. The different taxonomic classes have also been reported to better analyze the trend in function of the different properties. The presence of ice seems equally distributed at all sizes and at all heliocentric distances. Ice has been detected so far on the surface of all objects with diameter larger than about 700 km.

The BB taxonomic class is present essential on TNOs and only in one Centaur and all show presence of ices, while the RR class is distributed equally on objects at all heliocentric distances. Concerning the intermediate classes, it seems to have more BR class objects in the Centaurs, while more IR in the TNOs.

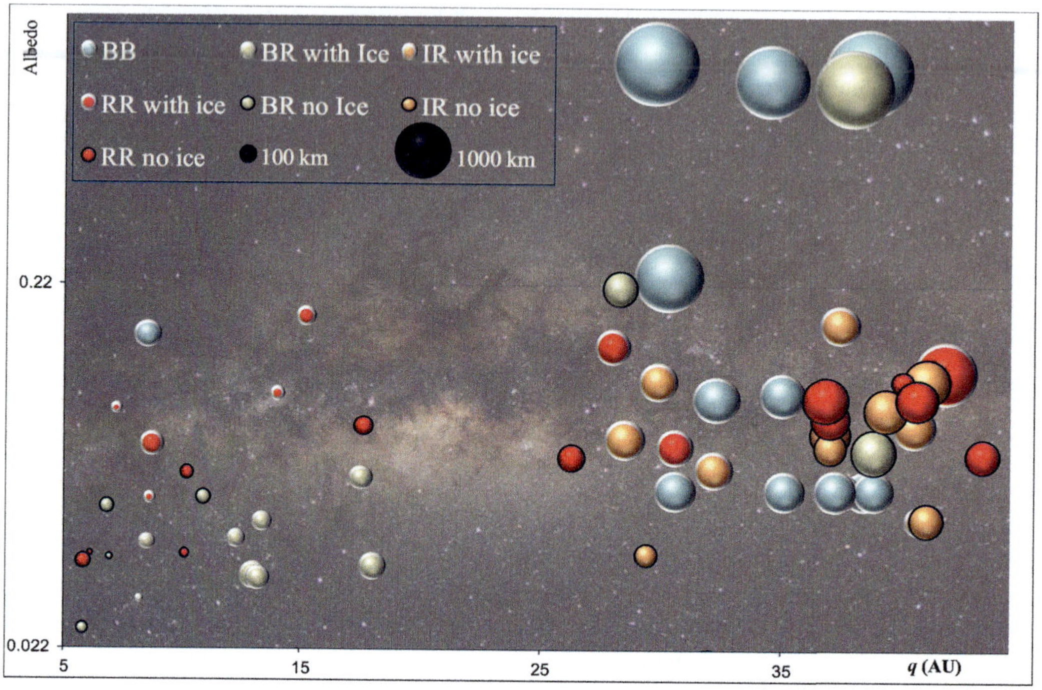

FIG. 5.2 Representation of icy and no icy surface content bodies depending on their taxonomical group, their size, and dispatched as a function of their perihelion distance and estimated albedo in V band.

The color properties of TNOs (except for Centaurs and classical) do not show any connection with their color-dynamical properties, implying that local conditions do not have primary influence in their colors, while the identical colors for binary objects (see Chapter 9) argue more on the fact that they could be primordial. To better understand if TNO surface can be modified when they enter in the Centaurs location needs more deep analyses on both populations.

Spectroscopy and photometry can give an idea on the surface composition, but to understand the real composition, we need to investigate on the bulk composition. The density is an important parameter to estimate the ratio ice/rock, which is available approximately only for few objects. The Trans-Neptunian population was expected to be relatively homogeneous, growing in the same location of the solar system, but we discovered a huge diversity in composition and internal properties (see Chapter 8).

Silicates seem also present in the surface of these objects, but more deep observations are needed in particular on far infrared. At the present, we are at the limit of the visibility from the ground. We will improve our knowledge with the new telescope generations of 30–40 m. The JWST, supposed to launch in 2021, will allow for sure to make a big step in the improvement of knowledge of these objects as well as the observations of one or two TNOs by NASA/New Horizons mission. The analysis of the data collected by the NASA fly-by on January 1, 2019 of Ultima Thule (2014 MU_{69}) with about 30 km in diameter and reddish surface, will improve our knowledge on the small size TNOs.

References

Alvarez-Candal, A., Barucci, M.A., Merlin, F., et al., 2007. A search for rotational variations on Trans-Neptunian objects. Astron. Astrophys. 475, 369.
Alvarez-Cantal, A., Fornasier, S., Barucci, M.A., et al., 2008. Visible spectroscopy of the new ESO large program on transneptunian objects and Centaurs. Astron. Astrophys. 487, 741.
Barkume, K.M., Brown, M.E., Schaller, E.L., 2008. Near-infrared spectra of Centaurs and Kuiper belt objects. Astron. J. 135, 55.
Barucci, M.A., Fulchignoni, M., 2017. Major achievements of the Rosetta mission in connection with the origin of the solar system. Astron. Astrophys. Rev. 25, 3.
Barucci, M.A., Fulchignoni, M., Birlan, M., et al., 2001. Analysis of Trans-Neptunian and centaur colours: continuous trend or grouping? Astron. Astrophys. 371, 1150.
Barucci, M.A., Belskaya, I., Fulchignoni, M., Birlan, M., 2005. Taxonomy of Centaurs and transneptunians objects. Astron. J. 130, 1291.
Barucci, M.A., Merlin, F., Dotto, E., et al., 2006. TNO surface ices. Observations of the TNO 55638 (2002 VE_{95}). Astron. Astrophys. 455, 725.
Barucci, M.A., Brown, M.E., Emery, J.P., Merlin, F., 2008. Composition and surface properties of transneptunian objects and Centaurs. In: The Solar System Beyond Neptune. University of Arizona Press, Tucson, AZ, pp. 143–160.
Barucci, M.A., Dalle Ore, C.M., Alvarez-Cantal, A., et al., 2010. Sedna: investigation of surface compositional variation. Astron. J. 140, 2095.
Barucci, M.A., Alvarez-Candal, A., Merlin, F., et al., 2011. New insights on ices in centaur and transneptunian populations. Icarus 214, 297.
Barucci, M.A., Merlin, F., Perna, D., et al., 2012. The extra red Plutino (55638) 2002 VE95. Astron. Astrophys. 539, A152.
Barucci, M.A., Dalle Ore, C.M., Perna, D., et al., 2015. (50000) Quaoar: surface composition variability. Astron. Astrophys. 584, A107.
Bauer, J.M., Meech, K.J., Fernández, Y.R., et al., 2002. Observations of the centaur 1999 UG_5: evidence of a unique outer solar system surface. Publ. Astron. Soc. Pac. 114, 1309.
Belskaya, I., Barucci, M.A., Fulchignoni, M., et al., 2015. Updated taxonomy of transneptunian objects and Centaurs: influence of Albedo. Icarus 250, 482.
Braga-Ribas, F., Sicardy, B., Ortiz, J.L., et al., 2014. A ring system detected around the centaur (10199) Chariklo. Nature 508, 72.
Brown, M.E., 2000. Near infrared spectroscopy of Centaurs and irregular satellites. Astron. J. 119, 977.
Brown, R.H., Cruikshank, D.P., Veverka, J., et al., 1995. Surface composition and photometric properties of Triton. In: Cruikshank, D.P. (Ed.), Neptune and Triton. University of Arizona Press, Tucson, AZ, p. 1249.
Brown, R.H., Cruikshank, D.P., Pendleton, Y.J., Veeder, G.J., 1997. Surface composition of Kuiper belt object 1993 SC. Science 276, 937.
Brown, R.H., Cruikshank, D.P., Pendleton, Y., et al., 1999. Water ice on Kuiper belt object 1996 TO_{66}. Astron. J. 519, L101.
Brown, M.E., Bouchez, A.H., Rabinowitz, D., et al., 2005. Keck observatory laser guide star adaptive optics discovery and characterization of a satellite to the large Kuiper belt object 2003 EL_{61}. Astrophys. J. 635, L97.
Brown, M.E., Barkume, K.M., Blake, G.A., et al., 2007a. Methane and ethane on the bright Kuiper belt object 2005 FY9. Astron. J. 133, 284.
Brown, M.E., Barkume, K.M., Ragozzine, D., et al., 2007b. A collisional family of icy objects in the Kuiper belt. Nature 446, 294.
Brown, M.E., Schaller, E.L., Fraser, W.C., 2012. Water ice in the Kuiper belt. Astron. J. 143, 146.
Brown, M.E., Schaller, E.L., Blake, G.A., 2015. Irradiation products on dwarf planet Makemake. Astron. J. 149, 105.
Brunetto, R., Baruci, M.A., Dotto, E., Strazzulla, G., 2006. Ion irradiation of frozen methanol, methane, and benzene, linking to the colors of Centaurs and Trans-Neptunian objects. Astrophys. J. 644, 646.
Chandrasekhar, S., 1960. Radiative Transfer. Dover, New York, NY.
Clark, R.N., Cruikshank, D.P., Jaumann, R., et al., 2012. The surface composition of Iapetus: mapping results from Cassini VIMS. Icarus 218, 831.
Cooper, J.F., Christian, E.R., Richardson, J.D., Wang, C., 2003. Proton irradiation of Centaurs, Kuiper belt, and Oort cloud objects at plasma to cosmic ray energy. Earth Moon Planet. 92, 261.
Cruikshank, D.P., Roush, T.L., Owen, T.C., et al., 1993. Ices on the surface of triton. Science 261, 742.

Cruikshank, D.P., Rouch, T.L., Bartholomew, M.J., et al., 1997. The surfaces of Pluto and Charon. In: Stern, S.A., Tholen, D.J. (Eds.), Pluto and Charon. University of Arizona Press, Tucson, AZ, p. 221.

Cruikshank, D.P., Roush, T.L., Bartholomew, M.J., et al., 1998. The composition of centaur 5145 Pholus. Icarus 135, 389.

Dalle Ore, C.M., Barucci, M.A., Emery, J.P., et al., 2009. Composition of kbp (50000) Quaoar. Astron. Astrophys. 501, 349.

Dalle Ore, C.M., Barucci, M.A., Emery, J.P., et al., 2015. The composition of "ultra-red" TNOs and Centaurs. Icarus 252, 311.

De Bergh, C., Boehnhardt, H., Barucci, M.A., et al., 2004. Aqueous altered silicates in the surface of two Plutinos? Astron. Astrophys. 416, 791.

DeMeo, F.E., Dumas, C., de Bergh, C., et al., 2010a. A search for ethane on Pluto and Triton. Icarus 208, 412.

DeMeo, F.E., Barucci, M.A., Merlin, F., et al., 2010b. A spectroscopic analysis of Jupiter-coupled object (52872) Okyrhoe, and TNOs (90482) Orcus and (73480) 2002 PN_{34}. Astron. Astrophys. 521, 35.

Doressoundiram, A., Tozzi, G.P., Barucci, M.A., et al., 2003. ESO large programme on Trans-Neptunian objects and Centaurs: spectroscopic investigation of centaur 2001 BL_{41} and TNOs (26181) 1996 GQ_{21} and (26375) 1999 DE_9. Astron. J. 125, 2721.

Doressoundiram, A., Boehnhardt, H., Tegler, S.C., Trujillo, C., 2008. Color properties and trends of the transneptuian objects. In: Barucci, M.A., Boehnhardt, H., Cruikshank, D.P., Morbidelli, A. (Eds.), The Solar System Beyond Neptune. University of Arizona Press, Tucson, AZ, pp. 91–104.

Dotto, E., Barucci, M.A., Leyrat, C., et al., 2003. Unveiling the nature of 10199 Chariklo: near-infrared observations and modeling. Icarus 162, 408.

Dotto, E., Emery, J.P., Barucci, M.A., et al., 2011. De Troianis: The Trojans in the Planetary System. University of Arizona Press, Tucson, AZ, pp. 383–396.

Dumas, C., Terrile, R.J., Brown, R.H., et al., 2001. Hubble space telescope NICMOS spectroscopy of Charon's leading and trailing hemispheres. Astron. J. 121, 1163.

Emery, J.P., Dalle Ore, C., Cruikshank, D.P., et al., 2006. Reflectances of Kuiper Belt Objects at Lambda > 2.5 Microns. American Geophysical Union, Abstract id: P13C-0183.

Emery, J.P., Dalle Ore, C., Cruikshank, D.P., et al., 2007. Ices on (90377) Sedna: confirmation and compositional constraints. Astron. Astrophys. 466, 395.

Fornasier, S., Barucci, M.A., de Bergh, C., et al., 2009. Visible spectroscopy of the new ESO large programme on Trans-Neptunian objects and Centaurs: final results. Astron. Astrophys. 508, 457.

Fulchignoni, M., Belskaya, I., Barucci, M.A., et al., 2008. Transneptunian objects taxonomy. In: Barucci, M.A., Boehnhardt, H., Cruikshank, D.P., Morbidelli, A. (Eds.), The Solar System Beyond Neptune. University of Arizona Press, Tucson, AZ, pp. 181–192.

Gil-Hutton, R., 2002. Color diversity among Kuiper belt objects: the collisional resurfacing model revisited. Planet. Space Sci. 50, 57.

Gomez, R.S., 2003. The origin of the Kuiper belt high-inclination population. Icarus 161, 404.

Grundy, W.M., Schmitt, B., 1998. The temperature-dependent near-infrared absorption spectrum of hexagonal H_2O ice. J. Geophys. Res. 103, 25809.

Grundy, W.M., Schmitt, B., Quirico, E., 2002. The temperature-dependent spectrum of methane ice I between 0.7 and 5 μm and opportunities for near-infrared remote thermometry. Icarus 155, 486.

Grundy, W.M., Buie, M.W., Spencer, J.R., 2005. Near-infrared spectrum of low-inclination classical Kuiper belt object (79360) 1997 CS_{29}. Astron. J. 130, 1299.

Guilbert, A., Alvarez-Candal, A., Merlin, F., et al., 2009. ESO-large program on TNOs: near-infrared spectroscopy with Sinfoni. Icarus 201, 272.

Hapke, B., 1981. Bidirectional reflectance spectroscopy. 1. Theory. J. Geophys. Res. 86, 565.

Hapke, B., 1984. Bidirectional reflectance spectroscopy. III—correction for macroscopic roughness. Icarus 59, 41.

Hapke, B., 1993. Theory of Reflectance and Emittance Spectroscopy. Cambridge University Press, Cambridge.

Hapke, B., 2002. Bidirectional reflectance spectroscopy. 5. The coherent Bacjscatter opposition effect and anisotropic scattering. Icarus 157, 523.

Hapke, B., 2008. Bidirectional reflectance spectroscopy. 6. Effects of porosity. Icarus 195, 918.

Helfenstein, P., Shepard, M.K., 2011. Testing the Hapke photometric model: improved inversion and the porosity correction. Icarus 215, 83.

Henyey, C., Greenstein, J., 1941. Diffuse radiation in the galaxy. Astron. J. 93, 70.

Hudson, R.L., Palumbo, M.E., Strazulla, G., et al., 2008. Laboratory studies of the chemistry of transneptunian object surface materials. In: Barucci, M.A., Boehnhardt, H., Cruikshank, D.P., Morbidelli, A. (Eds.), The Solar System Beyond Neptune. University of Arizona Press, Tucson, AZ, pp. 507–524.

Jarvis, K.S., Vilas, F., 2000. Searching for clues to the composition of Trans-Neptunian objects. BAAS 32, 1641.

Jewitt, D.C., 2009. The active Centaurs. Astron. J. 137, 4296.

Jewitt, D.C., Luu, J.X., 2001. Colors and spectra of Kuiper belt objects. Astron. J. 122, 2099.

Jewitt, D.C., Luu, J.X., 2004. Crystalline water ice on the Kuiper belt object (50000) Quaoar. Nature 432, 731.

Kerridge, J.F., Bunch, T.E., 1979. Aqueous activity on asteroids—evidence from carbonaceous meteorites. In: Gehrels, T. (ed.), pp. 745–764.

Lacerda, P., Fornasier, S., Lellouch, E., et al., 2014. The Albedo-color diversity of transneptunian objects. Astrophys. J. Lett. 793, L2.

Licandro, J., Oliva, E., Di Martino, M., 2001. NICS-TNG infrared spectroscopy of Trans-Neptunian objects 2000 EB173 and 2000 WR106. Astron. Astrophys. 373, L29.

Licandro, J., Grundy, W.M., Pinilla-Alonso, N., et al., 2006. Visible spectroscopy of 2003 UB313: evidence for N_2 ice on the surface of the largest TNO? Astron. Astrophys. 445, L35.

Lorenzi, V., Pinilla-Alonso, N., Licandro, J., 2015. Rotationally resolved spectroscopy of dwarf planet (136472) Makemake. Astron. Astrophys. 577, A86.

Luu, J.X., Jewitt, D.C., 1996. Color diversity among the Centaurs and Kuiper belt objects. Astron. J. 112, 2310.

Luu, J.X., Jewitt, D.C., 2002. Kuiper belt objects: Relics from the accretion disk of the Sun. A. R. Astron. Astrophys. 40, 63.

Luu, J.X., Jewitt, D.C., Trujillo, C., 2000. Water ice in 2060 Chiron and its implications for Centaurs and Kuiper belt objects. Astrophys. J. 531, L151.

Mastrapa, R.M.E., Brown, R.H., 2006. Ion irradiation of crystalline H_2O-ice: effect on the 1.65 μm band. Icarus 183, 207.

Melita, M.D., Licandro, J., 2012. Links between the dynamical evolution and the surface color of the Centaurs. Astron. Astrophys. 539, A144.

Merlin, F., Barucci, M.A., Dotto, E., et al., 2005. Search for surface variations on TNO 47171 and Centaur 32532. Astron. Astrophys. 444, 977.

Merlin, F., Alvarez-Candal, A., Delsanti, A., et al., 2009. Stratification of methane ice on Eris' surface. Astron. J. 137, 315.

Merlin, F., Barucci, M.A., de Bergh, C., et al., 2010a. Chemical and physical properties of the variegated Pluto and Charon surfaces. Icarus 210, 930.

Merlin, F., Barucci, M.A., de Bergh, C., et al., 2010b. Surface composition and physical properties of several Trans-Neptunian objects from the Hapke scattering theory and Shkuratov model. Icarus 208, 945.

Merlin, F., Quirico, E., Barucci, M.A., de Bergh, C., 2012. Methanol ice on the surface of the Centaur Pholus and TNOs 2002 VE95 and 2004 TY364. Astron. Astrophys. 544, A20.

Merlin, F., Hromakina, T., Perna, D., et al., 2017. Taxonomy of Trans-Neptunian objects and Centaurs as seen from spectroscopy. Astron. Astrophys. 604, A86.

Merlin, F., Lellouch, E., Quirico, E., Schmitt, B., 2018. Triton's surface ices: distribution, temperature and mixing state from VLT/SINFONI observations. Icarus 314, 274.

Moore, M.H., Donn, B., Khanna, R., A'Hearn, M.F., 1983. Studies of proton-irradiated cometary-type ice mixtures. Icarus 54, 388.

Owen, T.C., Roush, T.L., Cruikshank, D.P., et al., 1993. Surface ices and the atmospheric composition of Pluto. Science 261, 745.

Quirico, E., Schmitt, B., 1997a. Near-infrared spectroscopy of simple hydrocarbons and carbon oxides diluted in solid N_2 and as pure ices: implications for Triton and Pluto. Icarus 127, 354.

Quirico, E., Schmitt, B., 1997b. A spectroscopic study of co diluted in N_2 ice: applications for Triton and Pluto. Icarus 128, 181.

Romon-Martin, J., Barucci, M.A., de Bergh, C., et al., 2002. Observations of centaur 8405 Asbolus: searching for water ice. Icarus 160, 59.

Schaller, E.L., Brown, M.E., 2007. Volatile loss and retention on Kuiper belt objects. Astrophys. J. 659, 61.

Shepard, M.K., Helfenstein, P., 2007. A test of the Hapke photometric model. J. Geophys. Res. 112, E03001.

Tegler, S.C., Romanishin, W., 1998. Two distinct populations of Kuiper-belt objects. Nature 392, 49.
Tegler, S.C., Grundy, W.M., Romanishin, W., et al., 2007. Optical spectroscopy of the large Kuiper belt objects 136472 (2005 FY9) and 136108 (2003 EL61). Astron. J. 133, 526.
Trujillo, C.A., Brown, M.E., Rabinowitz, D.L., Geballe, T.R., 2005. Near-infrared surface properties of the two intrinsically brightest minor planets : (90377) Sedna and (90482) Orcus. Astrophys. J. 627, 1057.
Weaver, H.A., Buie, M.W., Buratti, B.J., et al., 2016. The small satellites of Pluto as observed by New Horizons. Science 351, 0030.
Wong, I., Brown, M.E., 2017. The bimodal color distribution of small Kuiper belt objects. Astron. J. 153, 145.

CHAPTER

6

Volatile evolution and atmospheres of Trans-Neptunian objects

Leslie A. Young[a], *Felipe Braga-Ribas*[b], *Robert E. Johnson*[c]

[a]Southwest Research Institute, Boulder, CO, United States [b]Federal University of Technology, Paraná, Curitiba (UTFPR), PR, Brazil [c]Federal University of Technology, Paraná (UTFPR), Curitiba, PR, Brazil

6.1 Introduction

Several bodies in the Trans-Neptunian region have volatiles on their surfaces that have significant vapor pressures at the temperatures of the outer solar system: CH_4, N_2, and CO (with CO only detected on Pluto and Triton). When present, volatiles may raise significant atmospheres around these bodies at some times during the orbit of these Trans-Neptunian objects (TNOs). These would be vapor-pressure supported atmospheres, where the main atmospheric species exists also as a surface ice, and the surface pressure is a very sensitive function of that ice's temperature. Mars, Pluto, Triton, and, to some extent, Io are examples of vapor-pressure supported atmospheres. If, as expected, Eris, Makemake, and other TNOs are occasionally in this class at some time in their orbit, then vapor-pressure supported atmospheres would be more numerous than terrestrial atmospheres (Venus, Earth, and Titan), or gas giants. Rather than an oddity, vapor-pressure supported atmospheres may be the most common style of atmosphere in our solar system. These atmospheres are characterized by large seasonal pressure variations, and global transport of volatiles across the surface.

Vapor-pressure supported atmospheres may have been important in the evolution of the outer solar system. All the TNOs should have formed with some measure of these species, but not all TNOs have volatiles detected on their surface. The explanation is tied to the gaseous phase through atmospheric escape.

The atmospheres of Pluto and Triton have been extensively studied, by spacecraft, ground-based occultations and spectroscopy, and modeling. Great effort has been made to search for atmospheres around other TNOs. To date, only upper limits have been placed.

The Trans-Neptunian Solar System.
https://doi.org/10.1016/B978-0-12-816490-7.00006-0

6.2 Spectral evidence of N_2, CO, and CH_4 on the surfaces of TNOs

Methane ice has been reported or inferred on the surfaces Triton, Pluto, Eris, Makemake, Quaoar, Varuna, Sedna, and 2007 OR_{10}. CO and N_2 ices have been directly detected on Pluto and Triton. N_2 ice has been inferred from its effect on the CH_4-dominated spectra of Eris and Makemake, and possibly Quaoar and Sedna. We detail the evidence for the presence of such ices below and summarize it in Table 6.1. See de Bergh et al. (2013), Brown (2008, 2012), and Grasset et al. (2017) for earlier reviews, and Chapter 5 for more general discussion of the composition of TNO surfaces. Despite attempts to characterize surface compositions with photometric systems (Trujillo et al., 2011; Dalle Ore et al., 2015), spectral resolution of at least 500 seems to be needed for definitive detections of volatile ices, and we focus on spectral data in this chapter.

Triton: We consider Triton in the discussion of volatiles on TNOs, even though it is a moon of Neptune, because it is thought to be a captured TNO (McKinnon and Leith, 1995) and, more frivolously, its orbit clearly crosses that of Neptune. Its surface and atmospheric properties

TABLE 6.1 Volatile ices reported or suggested on TNOs.

Body	CH_4 ice	N_2 ice	CO ice
Triton	Multiple CH_4 bands directly detected	N_2 band directly detected. N_2-rich ice inferred from CH_4 band shifts	Multiple CO bands directly detected
(134340) Pluto	Multiple CH_4 bands directly detected	N_2 band directly detected. N_2-rich ice inferred from CH_4 band shifts	Multiple CO bands directly detected
(136199) Eris	Multiple CH_4 bands directly detected	N_2-rich ice inferred from CH_4 band shifts	No spectral evidence
(136472) Makemake	Multiple CH_4 bands directly detected	N_2-rich ice inferred from CH_4 band shifts	No spectral evidence
(50000) Quaoar	Multiple CH_4 bands possibly directly detected	N_2 band possibly directly detected. N_2-rich ice inferred from CH_4 band shifts	No spectral evidence
(225088) 2007 OR_{10}	CH_4 inferred by analogy with Quaoar	No spectral evidence	No spectral evidence
(90377) Sedna	Weak evidence longward of 2.2 μm; more work is needed	Weak evidence reported	No spectral evidence
(20000) Varuna	Weak evidence at 2.3 μm; more work is needed	No spectral evidence	No spectral evidence
(90482) Orcus	Possible absorption at 2.2 μm	No spectral evidence	No spectral evidence

are usefully compared to those of Pluto and other large TNOs. The volatiles N_2 and CH_4 were measured in Triton's atmosphere by Voyager during its 1989 flyby with its ultraviolet (UV) spectrometer (Broadfoot et al., 1989). Three relatively volatile ices, N_2, CH_4, and CO, and two nonvolatile ices, CO_2 and H_2O, were detected on Triton's surface from ground-based near-IR spectroscopy (Cruikshank et al., 1993, 2000). The spectral profile of the 2.15-μm N_2 absorption band is temperature dependent, and transitions from very broad for the higher-temperature hexagonal β-N_2 phase above 35.61 K to very narrow for the colder cubic α-N_2 phase. Noting this, Quirico et al. (1999) showed that the N_2 ice was in the warmer β phase on Triton. In addition, the spectrum of CH_4 ice diluted in N_2 is shifted toward shorter wavelengths with respect to pure CH_4 ice (Quirico and Schmitt, 1997; Protopapa et al., 2015). Therefore, it was concluded that solid CH_4 on Triton exists predominantly diluted in N_2 (Quirico et al., 1999), with subsequent observations suggesting the presence of some pure CH_4 ice (Merlin et al., 2018). These ices are not uniformly distributed across Triton's surface, as seen from the variation of its spectrum with subobserver longitude as the body rotates (Grundy and Young, 2004; Grundy et al., 2010; Holler et al., 2016). These studies suggest that the nonvolatiles H_2O and CO_2 may dominate the terrains nearest Triton's current summer pole, that the more volatile species, N_2 and CO, are colocated, and that the CH_4 distribution and mixing state may vary with depth and longitude. Furthermore, there is evidence for isolated areas of CH_4 diluted in N_2 that is too fine grained to allow direct detection of the inherently weak 2.15-μm N_2 band, where the presence of the N_2 is revealed only by its effect of shifting the CH_4 spectrum. Further clues to the distribution of Triton's ices comes from a long-time base of observations and Triton's changing aspect. Infrared spectra from 1986 to 1992—approaching Triton's summer solstice in 2000—show a dramatic decrease in Triton's 2.2–2.4 μm CH_4 band (Brown et al., 1995), while post-solstice spectra 2002–14 show a mild increase in CH_4 band depth (Holler et al., 2016), suggesting a combination of changing aspect and perhaps active volatile transport. Voyager flew no infrared spectrometer, and so it is likely that these speculations will remain unconfirmed until a mission to Neptune/Triton can fly an infrared spectrometer.

Pluto: Pluto's surface has been studied with disk-integrated ground-based spectra in the visible, near-infrared, and mid-infrared (see the review by Cruikshank et al., 2015) and at geologically relevant spatial scales by the LEISA-infrared spectrometer and MVIC color imager on NASA's New Horizons spacecraft (see Chapter 12). The N_2 ice on Pluto is concentrated in a large, deep, N_2-filled basin called Sputnik Planitia, near the equator and opposite Charon, in which some CO and CH_4 are diluted in solid solution with large grained or annealed β-N_2. N_2-rich ice of similar composition is also seen at mid-latitudes, 35–55°N (Protopapa et al., 2017; Schmitt et al., 2017). It is now understood that Pluto has more CH_4 than Triton overall. Areas of CH_4-rich ice was first inferred from ground-based spectra, and were mapped with the LEISA spectrometer to be predominately (i) at a northern cap, north of ~55°N, (ii) in a band from 20° to 35°N, and (iii) on the area named Tartarus Dorsae on the eastern terminator limb of the encounter hemisphere near 220–250°E, 10°S to 30°N.

Eris: The visible and near-infrared spectrum of Eris clearly shows strong CH_4 absorption features (Brown et al., 2005; see review by Brown, 2008 and analysis by Tegler et al., 2012). Absorption at 1.684 μm indicates pure CH_4 or CH_4-rich ice (Dumas et al., 2007), and the spectral shifts are much smaller than for either Triton or Pluto spectra. The depth, shape, and the precise wavelengths of the CH_4 features on Eris can be modeled to derive grain size, dilution state, and stratification of CH_4 ice, even without the direct detection of the N_2 feature (e.g., Merlin et al., 2009). The issue of stratification—that is, whether the CH_4 concentration state is constant or varying with depth—is particularly interesting because it relates to the evolution of Eris's ices over its 561-year orbit, which in turn is determined by the order in which different species freeze out onto the surface each orbit after perihelion. Different studies have disagreed on the stratification, variously concluding that the dilute CH_4 lies above pure CH_4 (Licandro et al., 2006a), below pure CH_4 (Abernathy et al., 2009), sandwiched between two layers of pure CH_4 (Merlin et al., 2009), or that the stoichiometry of CH_4:N_2 is constant with depth (Tegler et al., 2012). Clearly, better data and continued modeling is needed on this question. Eris's surface temperature is almost certainly below the α-β phase transition for N_2 (Sicardy et al., 2011), given its bright geometric albedo $p_V = 0.96$. Since the α-N_2 absorption feature is weaker and much narrower than that of β-N_2, it has eluded direct detection on Eris.

Makemake: Very strong CH_4 absorption is seen in the visible and infrared spectrum of Makemake with spectral shifts to shorter wavelengths, but with smaller shifts than are seen on Eris, Pluto, or Triton (Licandro et al., 2006b; Brown et al., 2007; Tegler et al., 2007, 2008; Lorenzi et al., 2015; Perna et al., 2017). This suggests that although some CH_4 is in solution with N_2, N_2 itself is not as prevalent on Makemake as on the previous three bodies. At visible wavelengths, Tegler et al. (2007) detected four absorption features between 0.54 and 0.62 μm that they attributed to CH_4, but ice-phase laboratory spectra of these features are lacking. The nonvolatile irradiation products of CH_4 are also seen on Makemake (Brown et al., 2015).

Quaoar: The spectrum of Quaoar is dominated by H_2O ice, analogous to H_2O-rich Charon. Jewitt and Luu (2004) interpreted absorption near 1.65 μm as crystalline water ice and absorption near 2.2 μm as ammonia hydrate, and reported no detected CH_4 absorption. Improved SNR at 2.2–2.4 μm strengthened the case for CH_4 in the spectrum (Schaller and Brown, 2007b), which, if real, was nevertheless subtle when superimposed on the significant H_2O absorption. Guilbert et al. (2009) claim a marginal detection ($2 \pm 2\%$ deep) of the 1.724 μm feature due to CH_4 ice, which they say adds weight to the attribution of CH_4 ice for the 2.2 μm feature. Dalle Ore et al. (2009) model the near-IR spectrum plus photometry at 3.6 and 4.5 μm using CH_4 to explain both the 2.2 μm absorption and the mid-IR photometry, with no need for ammonia hydrate, and propose that N_2 ice in the β-phase may cover 20% of the surface, as seen by relatively subtle effects near 2.15 μm and by photometry at 3.6 and 4.5 μm. Barucci et al. (2015) investigate the 1.67-μm CH_4 band, and N_2 implied by the shifts in the 1.67-μm CH_4 band, but measuring this shift is complicated by the broad 1.65-μm crystalline H_2O band. In short, on Quaoar strong H_2O bands complicate the definitive detection and analysis of CH_4, and the derived presence of N_2 from CH_4 shifts.

2007 OR_{10}: The large TNO 2007 OR_{10} (Pál et al., 2016) shows an extremely red visible spectrum, and its near-IR spectrum shows water-ice absorption (Brown et al., 2011). 2007 OR_{10} shares both traits with Quaoar. This similarity with Quaoar is indirect evidence that CH_4 may also be present on 2007 OR_{10} (Brown et al., 2011), if one accepts both the equivalence of surface type and the evidence of CH_4 on Quaoar. Both Quaoar and 2007 OR_{10} are near the CH_4 retention line of Schaller and Brown (2007a) (see Section 6.4).

Sedna: Distant Sedna, currently at ~85 AU, has a spectrum that is difficult to interpret. Barucci et al. (2005) reported hints of a Triton-like spectrum with β-N_2 (an unexpected N_2

phase at Sedna's distance) and CH_4, from spectra with a spectral resolution of only 100 in the near-IR, while Trujillo et al. (2005) reported a featureless spectrum, from spectra at a binned resolution of 215 near 2.15 μm. Near-infrared spectra were analyzed in combination with Spitzer photometry at 3.6 and 4.5 μm (Barucci et al., 2010; Emery et al., 2007). No diagnostic CH_4 absorption bands were evident, but a decrease in albedo longward of 2.2 μm was interpreted as due to CH_4 and C_2H_6 (ethane), or possibly serpentine, and the addition of N_2 was also consistent with the Spitzer photometry (Barucci et al., 2010). The surface appears heterogeneous (Barucci et al., 2010).

Varuna: The spectrum of Varuna is in general well explained by various nonvolatiles, such as water ice, olivine, pyroxene, tholin, and amorphous carbon as a darkening agent. The strong CH_4 absorption bands at 1.67, 1.72, 1.8, or 2.21 μm are not detected in the current best spectra (Lorenzi et al., 2014), limiting the amount of CH_4 ice to less than 10% of the surface. Lorenzi et al. (2014) discuss how the addition of a small amount methane ice slightly improves the match to the data, but as their data have a spectral resolution of only 50, further work is critically needed to confirm or constrain CH_4 on Varuna.

Orcus: The water-dominated near-infrared spectrum of Orcus (de Bergh et al., 2005; Fornasier et al., 2004; Trujillo et al., 2005) has an absorption feature near 2.2 μm, compatible with absorption by CH_4, NH_3, or $NH_4{}^+$ (Barucci et al., 2008; Delsanti et al., 2010; Carry et al., 2011; DeMeo et al., 2010). Because of the lack of other CH_4 absorptions, specifically those near 1.67 and 1.72 μm, it is likely that CH_4 is not the main cause of the 2.2-μm absorption feature.

6.3 Volatile-supported atmospheres

For a TNO with little or no atmosphere, the equilibrium surface temperature, T_0, will depend on its rotation rate and thermal inertia (i.e., fast vs. slow rotator), latitude (λ), subsolar latitude (λ_{sol}), and time of day (e.g., hour angle of the Sun, h). The simplest approach is to assume thermal emission balances absorbed insolation, and that the atmosphere is nearly transparent, in which case the equilibrium surface temperature is

$$\epsilon \sigma T_0^4 = \overline{\mu(\lambda, \lambda_{sol}, h)} \left(\frac{S_{1AU}(1 - A)}{\eta R^2} \right) \tag{6.1}$$

where ϵ is the emissivity, σ is the Stefan-Boltzmann constant, S_{1AU} is the normal solar insolation at 1 AU (1367 W m^{-2}), A is the bolometric Bond albedo, and R is the heliocentric distance in AU. $\overline{\mu(\lambda, \lambda_{sol}, h)}$ is the average cosine of the incidence angle for the problem at hand: For the subsolar temperature on a slow rotator, T_{ss}, $\overline{\mu} = 1$; for the equatorial latitude for fast rotator, $\overline{\mu} = 1/\pi$; for a single characteristic equilibrium temperature balancing the global average of the insolation, T_{eq}, $\overline{\mu} = 1/4$. η is the so-called "beaming factor" (e.g., Spencer, 1990; Harris, 1998; Müller et al., 2010; Lellouch et al., 2013), which can raise the temperature for a rough surface (or lower it for a body with high thermal inertia in some models of thermal emission). Many of the smaller TNOs have $A \approx 5\%$, while the largest TNOs have $A \approx 60\%$

–80% (Lellouch et al., 2013). Thus, the temperatures relevant for TNOs range from $T_{ss} \approx 70\,K$ for $\epsilon = 1$, $\eta = 1$, $A = 5\%$ at $R = 30\,AU$ down to $T_{eq} \approx 20\,K$ for $\epsilon = 1$, $\eta = 1$, $A = 80\%$ at 90 AU. Detached objects at R much greater 90 AU will have even colder temperatures. Eq. (6.1) ignores latent heat of sublimation and internal heat flux, and the conduction of heat into or from the subsurface is empirically included in the beaming factor.

The surface pressure of an atmosphere over a pure ice in thermodynamic equilibrium (the sublimation pressure, $p_s(T_0)$) is a function only of the ice temperature (Fray and Schmitt, 2009). Of the three volatiles seen in the outer solar system, N_2, CO, and CH_4, N_2 is by far the most volatile (Fig. 6.1). TNOs with volatiles show a mix of ices: CH_4 or CO diluted in N_2-rich ice; N_2 or CO diluted in CH_4-rich ice; or pure CH_4 ices. The mixtures present complicated surface-ice interaction, including lag deposits and the influence of CH_4-rich warm patches (see review by Trafton et al., 1998). Trafton (2015) and Tan and Kargel (2018) have more recent work on the mixtures, including the exciting conclusion that the CH_4 and N_2 partial pressures above a mixture of CH_4 saturated in N_2-rich ice plus N_2 saturated in CH_4 ice is independent of the bulk N_2:CH_4 ice ratio, but these have not been tested under laboratory conditions. Moreover, as seen in the compilation by Fray and Schmitt (2009), laboratory measurements for the pure ices only exist for temperatures above 54.78 K for CO and above 48.15 K for CH_4; and the only laboratory data below 35.4 K for N_2 were published in 1960 (Borovik et al., 1960).

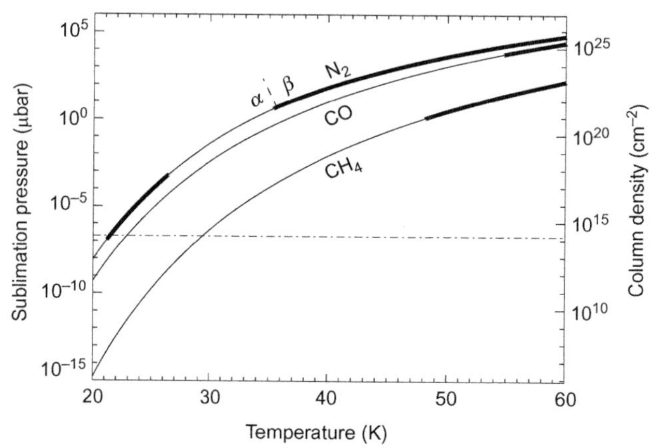

FIG. 6.1 Sublimation pressure (*left axis*) for N_2, CO, and CH_4 (Fray and Schmitt, 2009; FS09). The column density (*right axis*) is calculated for a body with bulk density ρ of $1.9\,g\,cm^{-3}$ and a surface radius r of 500 km for N_2 or CO ($\mu = 28$) For a given pressure, the column density scales as $1/(\mu\,\rho\,r_0)$. Thus, the plotted column density (*right axis*) also applies for $\rho = 1.9\,g\,cm^{-3}$ and a surface radius r of 875 km for CH_4 ($\mu = 16$). *Thick lines* show the ranges of temperatures at or below 60 K from laboratory measurements included in FS09: For N_2, 21.20–26.40 K (Borovik et al., 1960), 35.40–59.17 K (Frels et al., 1974), and 54.78–61.70 K (Giauque and Clayton, 1933); for CO, 54.78–68.07 K (Shinoda, 1969); and for CH_4, 48.15–77.65 K (Tickner and Lossing, 1951) and 53.15–90.66 K (Armstrong et al., 1955). The α-β phase transition for N_2 is indicated (35.61 K). The *dot-dashed line* indicates the rough transition between ballistic and collisional atmospheres, for $\rho = 1.9\,g\,cm^{-3}$ and $r_0 = 500\,km$.

The surface pressure is the weight of the column of gas, and so is closely related to the column density at the surface, N_0

$$p_s(T_0) \approx mg_0N_0 \approx mg_0H_0n_0 \tag{6.2}$$

where m is the mass of molecule, and g_0 is surface gravity, given by $g_0 = GM/r_0^2$, where G is the gravitational constant, $M = \rho(4/3)\pi r_0^3$ is the TNO mass, ρ is the TNO bulk density, and r_0 is the surface radius. $H = kT_0/mg_0$ is the pressure scale height at the surface, k is Boltzmann's constant, and n_0 is the local number density at the surface. The relation is only approximate for small bodies, because gravity decreases with altitude.

For small column densities, the atmosphere can be described as a surface-bounded exobase, since if $N_0\,\sigma_{eff} \ll 1$, where σ_{eff} is the effective collision cross-section, then an escaping molecule is not likely to suffer a collision on its way out. This condition is equivalent to a large Knudsen number, Kn, the ratio of the mean free path between collisions, $l_{mfp} = 1/(n_0\,\sigma_{eff})$, to a characteristic length scale. The scale in question depends on the problem at hand; taking the scale height for the length scale (Zhu et al., 2014), $Kn = 1$ is one classic definition of an exobase. At the surface:

$$Kn_0 = \frac{l_{mfp}}{H_0} = \frac{1}{N_0\sigma_{eff}} \tag{6.3}$$

σ_{eff} is $\sqrt{2}$ times larger than the collisional cross-section, σ_{coll} (Johnson et al., 2015). Since $\sigma_{coll} \approx 0.46 \times 10^{-14}$ for CH_4 (Atkins and de Paula, 2009) or $\sigma_{coll} \approx 0.43 \times 10^{-14}$ for N_2 (Kaye and Laby, 1973), with some dependence on temperature, atmospheres with N_0 greater than $\sim 1.6 \times 10^{14}$ molecule cm^{-2} can be considered collisional. This is achieved for N_2 or CH_4 at extremely small surface pressures (Fig. 6.1), for example, $\sim 3 \times 10^{-8}$ μbar for CH_4 on a $\rho = 1.6\,\mathrm{g\,cm^{-3}}$, 100-km radius body, or $\sim 8 \times 10^{-7}$ μbar for N_2 on a $\rho = 2.5\,\mathrm{g\,cm^{-3}}$, 1000-km radius body. Extrapolating to these small pressures from the pressures measured in the lab is questionable, but application of the Fray and Schmitt (2005) compilations predicts that the transition to continuum atmospheres happens near temperatures of ~29 K for CH_4, and ~21 K for N_2, depending weakly on the TNO radius and density. For larger column densities, the atmosphere becomes opaque to UV radiation. This transition occurs for Ly-α at $N_0 \sim 10^{18}$ molecule cm^{-2} ($T_0 \sim 26$ K) for N_2 with 3% CH_4 gaseous molar mixing ratio, or $N_0 \sim 3 \times 10^{16}$ molecule cm^{-2} ($T_0 \sim 32$ K) for pure CH_4 (Johnson et al., 2015).

The Jeans parameter at the surface, or ratio of potential energy to thermal energy, is a measure of how tightly bound the atmosphere is, and is given by

$$\lambda = \frac{gm}{rkT} = \frac{U}{kT} = \frac{r}{H} = \frac{GMm}{r^2} \tag{6.4}$$

The Jeans parameter at the surface, λ_0, varies from ~0.1 (unbound) for CH_4 gas on 100-km radius objects at 70 K to ~100 (bound) for N_2 gas on 1000-km radius objects at 20 K. Although the composition surely varies from object to object, assuming each TNO has N_2 on its surface it can be seen from Fig. 6.2 that due to the large range of temperatures, the TNO atmospheres can have a large range of λ_0 and N_0, with implications for atmospheric escape, seasonal variation, and detectability.

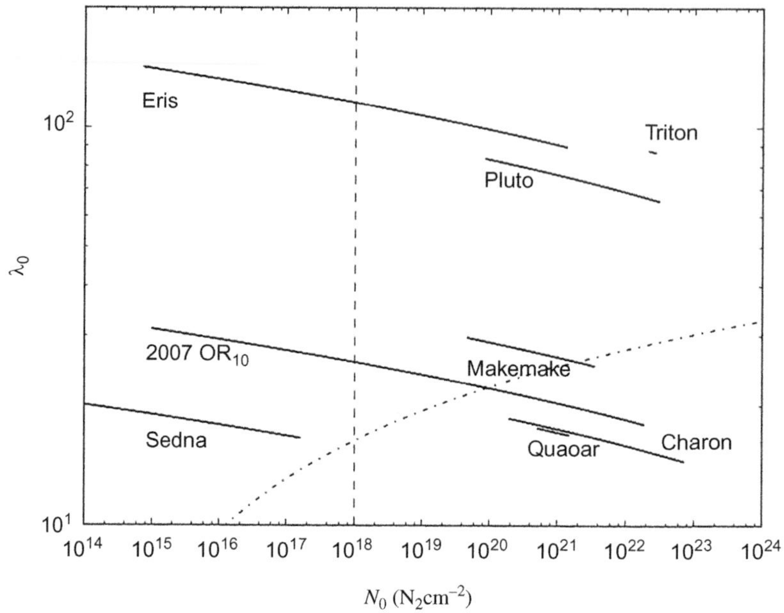

FIG. 6.2 Range of surface Jeans parameter, λ_0, versus total column density, N_0, over the orbits of a number of TNOs assuming an N_2 atmosphere (Johnson et al., 2015) using $\varepsilon = 0.8$ and $A = 0.67$, appropriate for an early TNO with a bright surface. For Makemake, 2007 OR_{10}, and Sedna we used $\rho = 1.8\,\mathrm{g\,cm^{-3}}$. Sedna's values are shown only near perihelion as its aphelion ($N_0 \ll 10^{14}\,\mathrm{cm^{-2}}$) is off-scale. *Dashed line*: $N_0 = 10^{18}\,N_2\,\mathrm{cm^{-2}}$; to the right the CH_4 and N_2 components are sufficient so that escape is mainly driven by solar heating of the atmosphere. *Dotted-dashed line*: Surface-heating-induced escape rate is smaller than Jeans formula below this line (warmer surface or lower surface gravity) and greater above this line. *From Johnson, R.E., Oza, A., Young, L.A., Volkov, A.N., Schmidt, C., 2015. Volatile loss and classification of Kuiper belt objects. Astrophys. J. 809, 43. Reproduced by permission of the AAS.*

6.4 Expected volatile retention

In general, only the largest TNOs have had volatiles detected or suspected on their surfaces (Table 6.1). This is not merely an observational effect (i.e., because higher-quality spectra are more easily obtained on larger TNOs), but is linked to the escape of their initial inventory of volatiles (Schaller and Brown, 2007a; Levi and Podolak, 2009; Johnson et al., 2015). Volatile escape can be driven by heating of the surface by solar visible radiation and by absorption of solar radiation in the atmosphere. The relative importance of these processes depends on how tightly bound the atmosphere is, parameterized by the surface Jeans parameter λ_0, and its UV opacity, parameterized by its surface column density, N_0. Since the work of Schaller and Brown (2007a) and Levi and Podolak (2009), there has been progress in estimating the escape rate in the transition to the fluid regime for transparent atmospheres (Volkov et al., 2011a,b), and the role of atmospheric heating on escape (Johnson et al., 2013a,b, 2015), as well as new observations relating to initial volatile inventories (Glein and Waite, 2018) and the complexity of escape at Pluto (Gladstone and Young, 2019).

In order to provide a link between the presence of volatiles, the bulk properties, and the orbits of TNOs, Schaller and Brown (2007a) considered sublimation-induced escape directly from the TNO surface, and updated this work in Brown et al. (2011). Starting with the equilibrium surface temperature (e.g., Eq. 6.1 with $\overline{\mu} = 1/4$ and $\eta = 1$), they used the Jeans expression for escape from an exobase, evaluated at the conditions of the surface (subscripted here as SJ for surface Jeans):

$$\Phi_{SJ} = 4\pi r_0^2 n_0 (\bar{v}/4)(1 + \lambda_0)\exp(-\lambda_0) \tag{6.5}$$

where Φ is the total escape rate of volatiles in molecule s^{-1}, and $\bar{v} = \sqrt{8kT/\pi m}$ is the mean molecular speed. Using these expression for T_0 and Φ, they divided the TNOs into those that would likely keep their volatiles and those that likely lost their initial volatile inventory over the age of the solar system (Fig. 6.3).

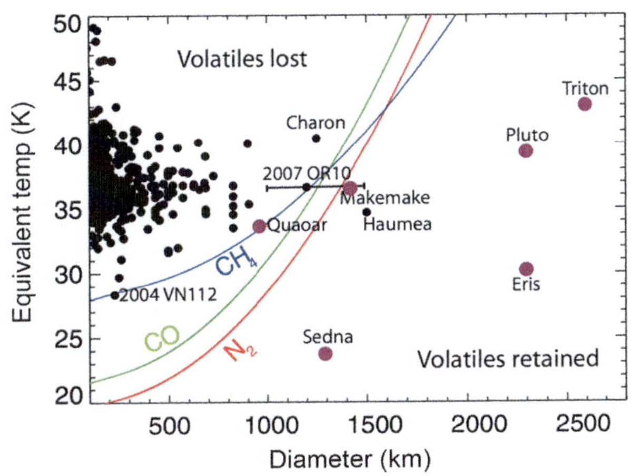

FIG. 6.3 Plot of volatile retention and loss in the Kuiper belt. Using the surface-Jeans formulation for atmospheric escape, objects to the left of the CH_4, CO, and N_2 lines lose surface volatiles over the age of the solar system. *From Brown, M.E., Burgasser, A.J., Fraser, W.C., 2011. The surface composition of large Kuiper belt object 2007 OR$_{10}$. Astrophys. J. 738, L26, updated from Schaller, E.L., Brown, M.E., 2007. Volatile loss and retention on Kuiper belt objects. Astrophys. J. 659, L61–L64. Reproduced by permission of the AAS.*

The surface-Jeans estimate of the escape rate is roughly accurate if the atmosphere is noncollisional, $N_0 \lesssim 10^{14}$ molecule cm^{-2} (Fig. 6.4), which holds for very distant TNOs. However, Eq. (6.5) is problematic, even for atmospheres that are transparent to solar heating. Levi and Podolak (2009) subsequently used a hydrodynamics model that transitioned to Jeans escape. Volkov et al. (2011a,b) used a molecular kinetic model, the direct simulation Monte Carlo (DSMC) method (Bird, 1994), to calculate the surface-heated escape rate, Φ_S, from a single-component atmosphere for a range of surface values of T_0 and N_0, and then expanded this range to very thick atmospheres by coupling iteratively to a fluid model in Johnson et al. (2015), thereby covering the full range of escape due to surface heating. Note that Volkov et al. (2011a,b) use the *radial* Knudsen number, scaling the mean free path by the surface radius, which is appropriate for small bodies with extended atmospheres (i.e., small λ_0). We denote that here as Kn_0^r (r for radial) and relate it to Eq. (6.3) through $Kn_0^r = l_{mfp}/r_0 = Kn_0/\lambda_0$. Johnson et al. (2015) fit an analytic expression to the numerical results of Volkov et al. (2011a,b) to find a correction to the surface-Jeans flux that depends on the surface values of the Knudsen number and Jeans parameter. We use Kn_0^r here to allow direct comparison with Volkov et al. (2011a,b) and Johnson et al. (2015).

$$\Phi_S = \Phi_{SJ}/[(Kn_0^r)^{0.09} + \lambda_0^{2.55} \exp(-\lambda_0)/(70Kn_0^r)] \tag{6.6}$$

For cold atmospheres and large bodies (high gravity), the escape rate driven only by the surface temperature is throttled for a bound atmosphere by the exp($-\lambda_0$) term in Eq. (6.5), (Fig. 6.4). This can be overcome by direct absorption of solar radiation in the atmosphere. Assuming ~2%–3% CH_4 in the more volatile N_2 background gas, Johnson et al. (2015) calculated that the column of gas that is sufficient for the UV and EUV to be primarily absorbed in the atmosphere is $N_0 > N_C$, where $N_C \approx 10^{18}$ molecule cm^{-2} is the minimum column for UV absorption. Prior to the New Horizons encounter, models of Pluto's atmospheric loss coupled a fluid simulation to a DSMC molecular kinetic simulation (e.g., Erwin et al., 2013; Tucker et al., 2012) to describe the UV/EUV absorption versus depth as well as the escape from the exobase region. In their Pluto modeling, the simulated loss rate was matched reasonably well by the simple so-called energy-limited escape model, in which the gravitational energy lost by the escaping molecules is set equal to the heating produced by the absorbed UV radiation. The limits to the applicability of that model were explored in Johnson et al. (2013a,b) and subsequently applied to other TNOs (Johnson et al., 2015). Ly-α dominates the absorbed UV flux if CH_4 is optically thick. Since scattered interplanetary Ly-α and stellar flux contribute to the UV flux in the Trans-Neptunian region, the energy-limited flux falls off more slowly than R^2. Eq. (6.7) gives the expression adopted by Johnson et al. (2015) for the escape due to UV/EUV heating in the upper atmosphere, Φ_U, scaled to the detailed rate calculated for Pluto, Φ_P

$$\Phi_U = \Phi_P(\rho_P/\rho)[(30/R)^2 + 0.09] \tag{6.7}$$

where ρ_P and ρ are the density of Pluto and the TNO, R is the heliocentric distance in AU, and Φ_P is the escape rate at Pluto due to the direct solar UV flux. The second term in the brackets accounts for the so-called interplanetary UV flux (Gladstone, 1998) ignored in those papers but which becomes important at very large R. Johnson et al. (2015), based on Erwin et al. (2013) and Tucker et al. (2012), adopted $\rho_P = 2.05\,\mathrm{g\,cm^{-3}}$, and $\Phi_P = 120\,\mathrm{kg\,s^{-1}}/m_{N_2} = 2.6 \times 10^{27}\,\mathrm{N_2\,s^{-1}}$. Johnson et al. (2015) combined surface and upper atmospheric heating by restricting Φ_U to $N_0 > N_C$ and choosing the larger of Φ_S and Φ_U (Fig. 6.4). Fig. 6.4 only gives a rough guide for ranges in which global models of the two escape processes dominate. Johnson et al. (2015) integrated the combined atmospheric loss rates over the lifetime of individual TNOs (Table 2 in Johnson et al., 2015, which did not include Orcus and Varuna). Besides Charon, which has likely lost its initial volatiles, they concluded that for the objects studied only Makemake, Quaoar, and 2007 OR_{10} likely lost a large fraction of their volatiles, primarily due to short wavelength absorption in their upper atmospheres, while Pluto, Triton, and Sedna had retained most of theirs. While similar conclusions were reached by Schaller and Brown (2007a), the fraction of volatiles retained by Pluto, Triton, and Sedna differed.

The picture created by these simulations has been altered by two recent spacecraft measurements. One is our more recent understanding of Pluto's atmosphere based on the New Horizons observations (Gladstone and Young, 2019). Prior to the flyby, the expected escape rate was ~[0.4–4] × $10^{27}\,\mathrm{N_2\,s^{-1}}$ (Zhu et al., 2014), consistent with energy-limited escape. The escape rate that is derived from the observed density and temperature profile is much

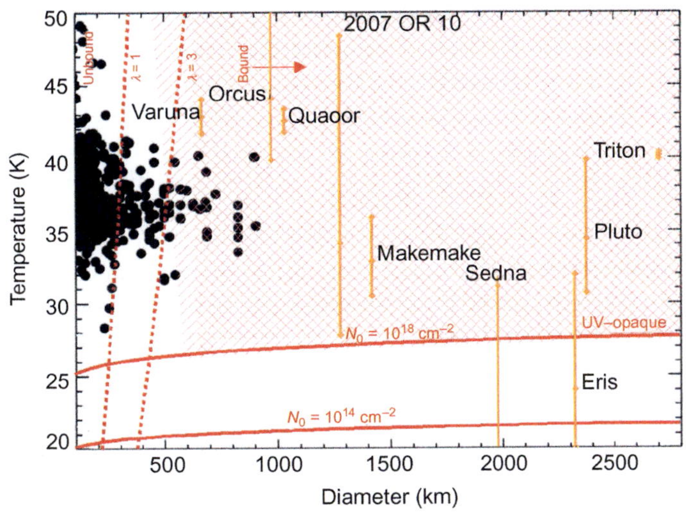

FIG. 6.4 Diagram of escape regimes. *Yellow*: The nine bodies in Table 6.1 have been plotted: *Solid vertical lines* show the variation in the average equilibrium temperatures between aphelion and perihelion (Eq. 6.1), using diameters and albedos from the *tnosarecool* database (i.e., updated from Brown et al., 2011 and Johnson et al., 2015). *Black*: Diameters and temperatures for other bodies are taken from the retention plot of Brown et al. (2011), as plotted in Fig. 6.3. *Solid red*: Lines of constant column density at the surface for two critical values: for $N_0 = 10^{14}\ cm^{-3}$ defining the classical surface-bounded exobase, and $N_0 = 10^{18}\ cm^{-3}$ for an N_2 atmosphere with 3% CH_4 to be opaque to Ly-α. *Dashed red*: Lines of constant Jeans parameter for two critical values dividing the unbound and bound atmosphere (Volkov et al., 2011b). *Hashed red*: Atmospheric heating dominates the loss rate in the shaded region for (adapted from the shaded regions in Fig. 3 of Johnson et al., 2015) $A = 0.1$ (*single hash*) and $A = 0.67$ (*cross-hatched*). Outside the hashed area, the surface-heated escape rate for warm, small bodies is much smaller than the surface-Jeans estimate, but the surface-heated escape rate for cold, large bodies is slightly larger than the surface-Jeans estimate (cf. *dot-dashed line* in Fig. 6.1).

lower: (3–7) $\times\ 10^{22}\ N_2\ s^{-1}$ and (4–8) $\times\ 10^{25}\ CH_4\ s^{-1}$ (Young et al., 2018). Applying these new observations has been confounded by the fact that the principal cooling agent in Pluto's upper atmosphere, which also compressed the extent of Pluto's atmosphere, is still uncertain (Gladstone and Young, 2019). Pluto's current escape rate must be better understood before volatile loss can be calculated for Pluto at other epochs, or for other bodies. The second is new constraints on the initial inventory of volatiles. While Schaller and Brown (2007a) and Johnson et al. (2015) used an N_2 to H_2O mass ratio of 2%, Glein and Waite (2018) use the mixing ratio of N_2 measured in the coma of 67P (Rubin et al., 2015) to derive an N_2 to H_2O mass ratio of only [0.7–6] $\times\ 10^{-4}$.

6.5 Variation of atmospheres over an orbit

Because the sublimation pressures depend exponentially on the temperatures of the volatile ices, the gases surrounding volatile-bearing TNOs vary with heliocentric distance and subsolar latitude, and possibly time of day and latitude. This was initially modeled for Triton and

Pluto (see reviews by Spencer et al., 1997; Yelle et al., 1995). Since those reviews, trends of increasing atmospheric pressure for both Triton and Pluto were observed using the technique of stellar occultation, with an increase by factors of 2 and 3, respectively (Elliot et al., 1998, 2000a,b, 2003a; Olkin et al., 1997, 2015; Meza et al., 2019; see Section 6.6). The new time base of atmospheric observations and the New Horizons flyby of Pluto inspired new models of seasonal variation (e.g., Young, 2012, 2013, 2017; Hansen et al., 2015; Olkin et al., 2015), including general circulation models (e.g., Forget et al., 2017) and evolution of atmospheres on the timescale of millions of years (e.g., Bertrand and Forget, 2016; Bertrand et al., 2018).

When N_2 was discovered on the surface of Eris, authors speculated that volatiles on TNOs, especially N_2, could raise temporary atmospheres near perihelion (e.g., Dumas et al., 2007). This was generalized in Stern and Trafton (2008), and applied numerically to the known or suspected volatile-bearing TNOs by Young and McKinnon (2013). When thinking about atmospheres on TNOs, it is useful to distinguish three types: global, collisional, and ballistic. For global sublimation-supported atmospheres, such as Mars or current-day Pluto and Triton, volatiles sublime from areas of higher insolation, and recondense on areas of lower insolation, transporting latent heat as well as mass (Trafton, 1984; Ingersoll, 1990; also see reviews by Spencer et al., 1997; Yelle et al., 1995; Stern and Trafton, 2008). As long as the volatiles can be effectively transported, the surface pressures and the volatile ice temperatures will be nearly constant across the surface. Sublimation winds transport mass from latitudes of high insolation to low insolation. Trafton (1984) showed that pressures stay within 10% across the surface if the sublimation winds (v) are less than 7.2% of the sound speed (v_s). The sublimation wind speeds can be found by conservation of mass; the mass per time crossing a given latitude equals the integral of the net deposition from that latitude to the pole. The wind speeds depend on the subsolar latitude (Trafton, 1984), if we consider diurnally averaged insolation; higher wind speeds are needed to transport volatiles pole to pole (high subsolar latitudes) than equator to pole (low subsolar latitudes). For an "ice ball" uniformly covered in volatiles, the maximum sublimation wind speed, v, can be expressed as

$$vmN_0 = \xi Sr/L \tag{6.8}$$

where $S = S_{1AU}(1 - A)/R^2 = 4\varepsilon\sigma T_{avg}^4$ is the absorbed normal insolation, and L is the latent heat of sublimation, in energy per mass. ξ in Eq. (6.8) is a numerical factor accounting for the subsolar latitude, λ_{Sun}. We calculated ξ numerically, following the prescription of Young (1993). From these calculations, ξ is well approximated by a cubic expression

$$\xi(\lambda_{Sun}) \approx 0.044 + 0.148(\lambda_{Sun}/90°) + 0.4012(\lambda_{Sun}/90°)^2 - 0.296(\lambda_{Sun}/90°)^3 \tag{6.9}$$

For a 400–1400-km radius body uniformly covered with CH_4 ice to have a global atmosphere, the pressure needs to be greater than ~17–295 nbar for polar illumination (3.1×10^{19} to 1.6×10^{20} cm^{-2}, 41.0–45.6 K; Fig. 6.5), or 1.9–33 nbar for equatorial illumination (3.6×10^{18} to 1.8×10^{19} cm^{-2}, 38.1–42.0 K). For N_2, the pressures are similar, so the temperatures are lower: 14–244 nbar for polar (1.5×10^{19} to 7.4×10^{19} cm^{-2}, 29.1–31.9 K) or 2–28 nbar for equatorial (1.8×10^{18} to 68.5×10^{18} cm^{-2}, 27.2–29.7 K). N_0 increases slightly faster than r because both S and N_0 increase with temperature; p_0 increases even faster, slightly faster than r^2, because of its dependence on g_0 (Eq. 6.2).

The temperatures in Fig. 6.5 are highly simplified. Seasonal thermal inertia can be important, even at the long timescales in the outer solar system. More significantly, bodies are unlikely to be uniformly covered in volatiles. For example, much of the N_2 on Pluto is located in the basin known as Sputnik Planitia (Moore et al., 2016), and Triton's N_2 may be perennially confined to the southern hemisphere (Moore and Spencer, 1990).

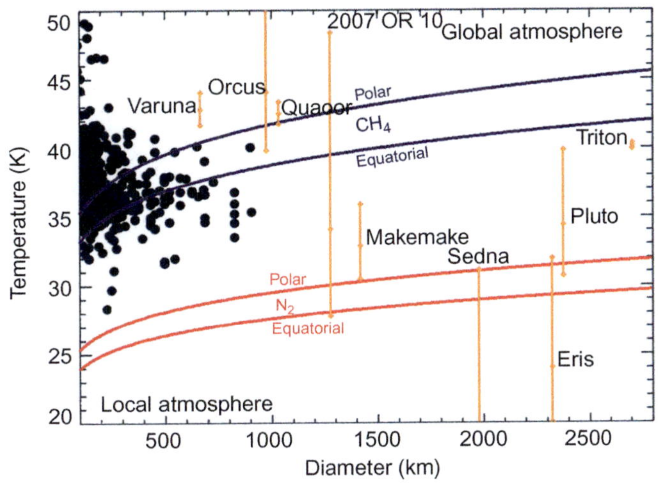

FIG. 6.5 In a global atmosphere, pressures are high enough so that the sublimation winds can keep surface pressures and volatile ice temperatures nearly uniform over the globe. Minimum temperatures for a global atmosphere are plotted, assuming a surface uniformly covered with CH_4 ice (*blue*) or N_2 ice (*red*), for two extreme subsolar latitudes. Several bodies would have global atmospheres at some portion of their orbit if covered with N_2 ice (as reported on Triton, Pluto, Eris, Makemake, and possibly Quaoar). A much smaller number of bodies would have global atmospheres at some portion of their orbit if covered with only CH_4 ice (as reported possibly on Varuna, Orcus, Quaoar, Sedna, and 2007 OR_{10}). Thermal inertia and nonuniform volatile ice coverage can change the range of temperatures actually achieved. Temperatures and diameters are as in Fig. 6.4.

Nonglobal atmospheres will vary with location and time of day, but may still be collisional, if the column density is greater than $\sim 10^{14}$ cm^{-2} for either N_2 or CH_4. Io is a classic example of a local atmosphere that is collisional around the subsolar point, and demonstrates some of the processes that are active in even these thin atmospheres. Atmospheric chemistry can occur even in these local, tenuous atmospheres (Wong and Smyth, 2000). Supersonic winds certainly flow and transport volatiles, even if they are not effective at equalizing pressures and temperatures (e.g., Walker et al., 2012). Recently, Hofgartner et al. (2018) used the Ingersoll et al. (1985) meteorological model developed for Io study the transport of N_2 on Eris at aphelion, when it is a local, collisional atmosphere, and found significant transport of N_2 ice. Even for more tenuous "surface-bounded exospheres," the loss of volatiles can modify landforms (see review by Mangold, 2011). For example, sublimation erosion may lead to the narrow divides between craters on Hyperion (Howard et al., 2012) or redeposition on the crater rims on Callisto, where the convex summits see less of the warm surface than do concave crater interiors, and are therefore local cold traps (Howard and Moore, 2008).

6.6 Detections of or limits on atmospheres by stellar occultation

The technique of stellar occultation is one of the most powerful ways to search for atmospheres around these bodies. While infrared absorption or radio emission have detected low column densities of CO and CH_4 on Triton (Lellouch et al., 2010) or CO and HCN on Pluto (Lellouch et al., 2017), these are bright targets. Stellar occultations depend on the brightness of the occulted star, and also study the target's size, shape, and the presence or nature of rings or jets (see Chapter 19; Elliot and Olkin, 1996; Elliot and Kern, 2003; Santos-Sanz et al., 2016). In an occultation, the starlight is refracted through a bending angle that increases in magnitude roughly in proportion to the line-of-sight column density (N_{los}). This leads to differential refraction, or a divergence of the refracted rays (cf. Elliot and Olkin, 1996), which dims the occulted starlight according to the scale height H and the column density. Some insight can be gained (Fig. 6.6) from the approximate, analytic expression for the relative stellar flux (e.g., Elliot and Young, 1992)

$$\varphi \approx \left|1 - \theta \tfrac{\Delta}{H}\right|^{-1} \left|1 + \theta \tfrac{\Delta}{r}\right|^{-1} e^{-\sigma_{ext} N_{los}} \tag{6.10}$$

where θ is the bending angle, given by $\theta \approx -\nu_{STP} N_{los}/n_{STP} H$, ν_{STP} is the refractivity at standard temperature and pressure (2.9×10^{-4} and 4.4×10^{-4} for N_2 and CH_4 at 0.7 μm), n_{STP} is Loschmidt's constant ($2.6868 \times 10^{19}\,cm^{-3}$), and Δ is the target-observer distance. As before, r is the radius (distance from target center) and H is the atmospheric scale height. N_{los} is the line-of-sight column density (molecule per area), which is larger than N, the vertical

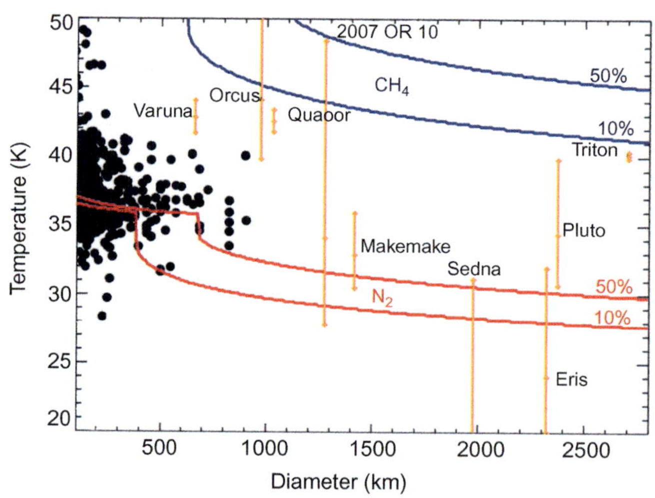

FIG. 6.6 Detectability of TNO atmospheres by ground-based stellar occultation, under the assumption that the atmosphere is isothermal, and both atmosphere and surface are at the plotted temperature. The *blue* (CH_4) and *red* (N_2) detectability limits are for 50% or 10% drop in stellar flux for a surface-grazing ray. Near 35 K for N_2 and 50 K for CH_4, the surface-grazing ray causes a central flash (Eq. 6.10, second term), which leads to a discontinuity in the derived limits; the plotted limits are somewhat conservative near this discontinuity Temperatures and diameters are as in Fig. 6.4.

column density, by the unitless factor $\sqrt{2\pi\lambda}$ (e.g., 8–25 for $\lambda \approx$ 10–100). σ_{ext} is the extinction cross-section (area per molecule). The first term represents the decrease due to the divergence (defocusing) of rays perpendicular to the limb, and halves the starlight when $\theta\Delta \approx -H$. The second term represents the increase due to refocusing parallel to the limb, leading to a "central flash" near the center of the shadow. The final term represents extinction, which becomes important when $N_{los} \approx 1/\sigma_{ext}$ (e.g., $N_{los} \approx 10^{26}$ molecule cm^{-2} for Rayleigh scattering at visible wavelengths, or $N_{los} \approx 10^{17}$ molecule cm^{-2} for typical cross-sections in the extreme UV). Refraction is typically dominant over extinction for Earth-based stellar occultations of TNOs, and extinction dominates for spacecraft UV occultations.

Light curves from model atmospheres can be calculated analytically for some idealized atmospheres (e.g., isothermal or $T \propto r^{\beta}$, $\beta \ll 1$; Elliot and Young, 1992). However, the presence of CH_4 or its by-products (including photochemical haze) can heat up the atmosphere by 10 s of K (Yelle and Lunine, 1989; Zhang et al., 2017). For more complex atmospheres, synthetic light curves are calculated under the assumption of geometric optics/ray tracing (Sicardy et al., 1999 and references therein) or wave optics/Fresnel diffraction (French and Gierasch, 1976). Standard model fitting can then be used to extract the geometric edge and the refractivity of the atmosphere at the surface. For high-quality data (as has been obtained for Pluto and Triton), light curves can be inverted to extract temperature, pressure, and number-density profiles (e.g., Elliot et al., 2003b). Derived surface pressures (or upper limits) from occultation light curves can be compared to the sublimation pressures for the atmospheric molecules under consideration, as expected from its surface equilibrium temperature (Fray and Schmitt, 2009).

Synthetic light curves versus shadow radius are plotted in Fig. 6.7 for an example TNO with an N_2 or CH_4dominated atmosphere (see caption for model details). Very thin atmospheres (10–100 s of nanobar, or ~10^{19}–10^{20} molecule cm^{-2}) cause only a small drop of flux very close to the surface, requiring very high signal-to-noise ratio for detection. Denser atmospheres (a few μbar, or ~10^{20} molecule cm^{-2}) cause a gradual drop in the star flux at a significant distance from the object's surface, so they will be easier to detect. For denser atmospheres, the ray that grazes the TNO surface is refracted inward significantly toward smaller shadow radii (bent by 94–112 km for 1 μbar CH_4 or N_2 atmospheres). For the 10 μbar curve in Fig. 6.7, the surface-grazing ray is refracted past the shadow center. The bottom flux never reaches zero and much of the light curve is the sum of the near and far limb contributions. In the example plotted, the limb-grazing ray is bent by 900–1180 km for a CH_4 or N_2 atmospheres, leading to the discontinuities seen at ±400 or ±680 km for the CH_4- or the N_2-dominated atmospheres. When the star crosses the center of the object as seen from Earth, it causes a prominent central flash for the 10 μbar atmosphere, as the atmosphere acts as a lens, converging the starlight from all parts of the atmosphere to the observer.

Stellar occultations by Pluto, Charon, and Triton have been observed since the 1980s.

Pluto: A single-chord occultation from 1985 was observed under extremely difficult circumstances (Brosch, 1995). The definitive discovery of Pluto's atmosphere was from the 1988 stellar occultation (Hubbard et al., 1988; Millis et al., 1993). Observations in 2002 showed a doubling of the pressure since 1988 (Sicardy et al., 2003; Elliot et al., 2003a). Many high-quality observations since have revealed the continued changes in Pluto's atmosphere, its thermal structure, and waves (Young et al., 2008; Toigo et al., 2010; Sicardy et al., 2016; Dias-Oliveira et al., 2015; Pasachoff et al., 2017; Meza et al., 2019). The New Horizons radio and solar occultations

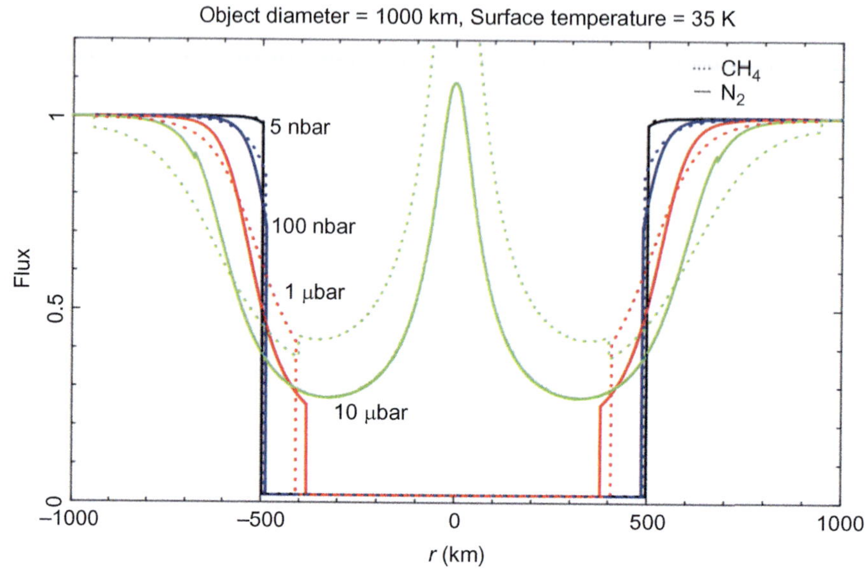

FIG. 6.7 Synthetic light curve models of normalized stellar flux versus shadow radius for an object with $r_0 = 500$ km, density 1.9 g cm^{-3}, $\Delta = 40$ AU, with a surface temperature of $T_0 = 35$ K and a 5 K km^{-1} gradient to a 100 K upper atmosphere. *Continuous lines* are for an N_2-dominated atmosphere with surface pressures and column densities of $p_0 = 5$ nbar, $N_0 = 4.0 \times 10^{18}$ molecule cm^{-2} (*black*); 100 nbar, 8.1×10^{19} molecule cm^{-2} (*blue*); 1 μbar, 8.1×10^{20} molecule cm^{-2} (*red*), and 10 μbar, 8.1×10^{21} molecule cm^{-2} (*green*). *Dashed lines* are for a CH_4-dominated atmosphere with surface pressures and column densities of 0.005 μbar, 7.1×10^{18} molecule cm^{-2} (*black*); 0.1 μbar, 1.4×10^{20} molecule cm^{-2} (*blue*); 1 μbar, 1.4×10^{21} molecule cm^{-2} (*red*); and 10 μbar, 1.4×10^{22} molecule cm^{-2} (*green*).

revealed N_2 in vapor-pressure equilibrium with the N_2-rich ice, an overabundance of gaseous CH_4 possibly explained by CH_4-rich patches, hazes, and photochemical products, a cold upper atmosphere, and an atmospheric escape rate that was dominated by CH_4 and was much smaller than expected (see Chapter 12; Gladstone and Young, 2019).

Triton: Triton, like Pluto, was visited by a spacecraft, and has had several high-quality occultations, notably the November 4, 1997 Triton occultation observed from HST (Elliot et al., 1998). Occultations in the 1990s show a doubling of surface pressure on decadal timescales compared to the 1989 flyby by Voyager 2 (Elliot et al., 1998, 2000a,b; Young et al., 2002). An occultation from 2017 suggests that the increase has peaked (Oliveira et al., 2018; Person et al., 2018).

Charon: Prior to the New Horizons flyby, limits on Charon's atmosphere were set to be <50 nbar (1σ) from stellar occultation (Gulbis et al., 2006), and the New Horizons UV occultation set limits of <1.4 pbar (Stern et al., 2017).

Since 2009, stellar occultations have been successfully used to study other TNOs, and the search for an atmospheric signature was carried for a few of them. Most of the largest bodies have already observed during a stellar occultation event and no atmosphere around a TNO has been found so far (other than Pluto and ex-TNO Triton). Upper limits on the presence of putative atmospheres were obtained and they are summarized in Table 6.2.

TABLE 6.2 Results on atmospheres from stellar occultations.

Name	Equivalent radius (km)	Surface pressure	References
Triton	1350	3–11 μbar	Elliot et al. (2000b)
Pluto	1188	4–12.7 μbar	Sicardy et al. (2016)
Eris	1163	<1 nbar[a]	Sicardy et al. (2011)
Haumea	816	<10 nbar[a]	Ortiz et al. (2017)
Makemake	715	<12 nbar[a]	Ortiz et al. (2012)
Charon	606	<1.4 nbar[a]	Stern et al. (2017)
Quaoar	555	<21 nbar[a]	Braga-Ribas et al. (2013)
Sedna	445	Inconclusive	Braga-Ribas et al. (2014b)

[a] *$1-\sigma$ upper limits.*

Eris: A stellar occultation of a $V = 17.1$ star by Eris was observed on November 6, 2010 (Sicardy et al., 2011). Eris, the second-biggest TNO after Pluto, was at 95.7 AU and had a surface temperature estimated at 30 K. With methane and nitrogen detected on its surface, it was a good candidate to have retained volatiles. The occultation allowed a constraint on the presence of an isothermal N_2 atmosphere to an upper limit of 1 nbar at the surface, with similar limits for CH_4 and argon. The occultation revealed that Eris is one of the brightest objects of the solar system, with a geometric albedo of $p_V = 0.96$ $(+0.09/-0.04)$, which may be caused by a collapsed atmosphere, as discussed earlier. As the subsolar temperature of Eris can reach 35 K, the authors mentioned that a local subsolar atmosphere of N_2 may exist, but it would freeze to undetectable values at the limb. It is suggested that, due to the eccentric orbit of Eris, when it approaches its perihelion at 37.8 AU, it may develop a global atmosphere of 2 μbar.

Haumea: An observation of a stellar occultation by Haumea on February 21, 2017 revealed that it also possesses a dense ring of 70 km in width (Ortiz et al., 2017), making it the second small solar system object, after (10199) Chariklo (Braga-Ribas et al., 2014a), known to have a ring. Upper limits (1σ) were obtained for the atmosphere of $p_0 < 10$ nbar (CH_4) or $p_0 < 3$ nbar (N_2). This is consistent with the lack of spectral evidence of CH_4 or N_2 on Haumea's surface. Haumea's ring, two moons, and the Haumea dynamical family suggest that volatiles were lost via a disruptive collision (e.g., Schaller and Brown, 2007b).

Makemake: A Makemake occultation of a $V = 18.5$ star on April 23, 2011 was detected from seven telescopes in Chile (Ortiz et al., 2012). The preferred solution for the projected shape is an ellipsoid with axes of 1430 ± 9 and 1502 ± 45 km. Constraining the shape to a circular cross-section gives a radius of 1430 ± 9 km, and is also consistent with the observations. The preferred (elliptical) solution gives a geometric albedo at V of $p_V = 0.77 \pm 0.03$. This is a bright albedo, and is consistent with surface freshening by a periodic atmosphere. No global atmosphere was detected, with limits on a global atmosphere of <12 nbar (1σ). This is somewhat surprising, as Makemake was a candidate for a global atmosphere, especially given the report of N_2 ice from CH_4 band shifts. The conclusion is that Makemake has lost almost all of its N_2 gas, or that the atmosphere near the limb has frozen to undetectable values (e.g., for a pole-on orientation or a slow rotation/low thermal inertia). Ortiz et al. (2012) also modeled the effect of a local atmosphere on the occultation light curves, motivated by some points close to centrality with an elevated flux a few σ above the noise, and placed loose limits on a local atmosphere of $p_0 < 30\,\mu$bar.

Quaoar: A large TNO ($r_0 = 550$ km) with CH_4 reported on its surface, Quaoar was a candidate to possess a thin atmosphere. It was observed crossing in front of a star on May 4, 2011, from which upper limits on the presence of an atmosphere were derived (Braga-Ribas et al., 2013). Being at a distance of 42.4 AU, with a geometric albedo of 0.109, a surface equilibrium temperature of 40 K is expected. Considering a CH_4 dominant atmosphere, reaching 102 K above 10 km from the surface, an upper limit of $p_0 < 21$ nbar (1σ) was obtained. As N_2 would have a vapor pressure of 66–176 μbar at 40–42 K, the data rule out this scenario.

Smaller TNOS: Additional TNOs have been observed during stellar occultations, and no clear signals of any atmosphere were detected. For these events, there are no published upper limits, because the decrease in stellar flux was abrupt (rather than dimming by atmospheric refraction), and the data did not have high enough quality (cadence or signal-to-noise ratio) for any constraints on a thin atmosphere. The larger bodies (equivalent radii $r_0 > 200$ km) so probed include Varuna ($r_0 = 568$ km; Sicardy et al., 2010), 2003 VS_2 ($r_0 = 274$ km; Benedetti-Rossi et al., 2019), 2002 TC_{302} ($r_0 = 499$ km; Santos-Sanz, 2017), 2003 AZ_{84} ($r_0 = 382$ km; Dias-Oliveira et al., 2017), and 2007 UK_{126} ($r_0 = 319$ km; Benedetti-Rossi et al., 2016).

6.7 Future research

The earlier discussion suggests some critical directions for future research.

Surface compositions:

- The spectra of Pluto and Triton change on the timescale of decades, and are related to both changing viewing geometry and possibly real changes in the volatile distribution or physical state of volatile ices (e.g., grain size). The changing compositions of Triton (visual magnitude $V = 13.4$) and Pluto ($V = 14.2$) can and should continue to be tracked with moderate ground-based telescopes, to decouple these effects.
- For other large TNOs, new understanding of the surface stratification of CH_4 and N_2 on Eris ($V = 18.7$) needs higher spectral resolution and SNR than previous spectra, as does the confirmation of or constraints on CH_4 on Quaoar ($V = 18.7$), 2007 OR_{10} ($V = 21.3$), Sedna ($V = 20.8$), Varuna ($V = 20.1$), and Orcus ($V = 19.2$). Some progress is likely to come from the new generation of infrared detectors (e.g., NIRES on Keck, or NIHTS on DCT). JWST's spectral sensitivity from 1 to 5 μm will be very powerful for measuring the volatiles and other species on the surfaces of these and other TNOs (Parker et al., 2016; see Chapters 5, 12, 18, and 19). Both the Thirty Meter Telescope and the European Extremely Large Telescope have planned near-infrared integral field spectrometers as first light instruments, and their various white papers emphasize how they will revolutionize TNO spectroscopy.

Laboratory work:

- The vapor pressures of CO and CH_4 should be measured at TNO temperatures of 20–40 K, and the vapor pressure of N_2 should be remeasured in a modern lab. Also, recent work on the vapor pressures of mixed volatile ices (Trafton, 2015; Tan and Kargel, 2018) should be tested under laboratory conditions.

- Measurements of the spectrum of CH_4 ice in the visible, both pure and in solution with N_2, are needed for quantitative analysis of Makemake's CH_4 features 0.54–0.62 μm.

Atmospheric retention and escape:

- The most critical work is a better model of why the escape rate at Pluto in 2015 ($\sim 5 \times 10^{22}$ N_2 s^{-1} and $\sim 6 \times 10^{25}$ CH_4 s^{-1}) is so much lower than the energy-limited rate expected preencounter ($\sim 2 \times 10^{27}$ N_2 s^{-1}), and how that model can be applied to Pluto at other seasons, and to other TNOs.
- Current DSMC work on volatile retention only models a single species, but the observed mixed surfaces with both N_2 and CH_4 suggest new DSMC work with multiple species. Similarly, local atmospheres (e.g., dome-like atmospheres over the subsolar point) can be more completely modeled.

Seasonal transport:

- Models such as Hofgartner et al. (2018) can be used to investigate the transition between global and collisional atmospheres.

Atmospheric searches:

- Higher SNR occultations with higher cadence can help the sensitivity, and can help break ambiguity in the cases of suspected local atmospheres. Accurate star positions are now given by the Gaia catalog, but ephemeris errors are nearly always much greater than the object's apparent angular size. Astrometric positions are needed for the targeted TNOs. One source of positions can be the big and deep surveys like the Large Synoptic Sky Survey that is expected to begin observations in 2022, and will help to update TNOs ephemerides and so improve stellar occultation predictions (Camargo et al., 2018).

Acknowledgments

Leslie Young was supported in part by the NASA OPR Grant NNX14AO45G "Connecting Present and Past KBOs." Felipe Braga Ribas acknowledges CNPq Grant 309578/2017-5 and the funding from the European Research Council under the European Communities H2020 (2014–20/ERC Grant Agreement No. 669416 "LUCKY STAR"). Apurva Oza was invaluable in the calculations behind the work in Fig. 6.4. Thanks to Bryan Holler for discussions on Section 6.2, and Larry Trafton for discussion relating to the requirements for a global atmosphere. Mike Brown graciously allowed the reproduction of his figure from Brown et al. (2011), which provided some level of coherence and clarity to our subsequent figures. This chapter was greatly improved by a thorough review by Emmanuel Lellouch.

References

Abernathy, M.R., Tegler, S.C., Grundy, W.M., Licandro, J., Romanishin, W., Cornelison, D., et al., 2009. Digging into the surface of the icy dwarf planet Eris. Icarus 199, 520–525.

Armstrong, G.T., Brickwedde, F.G., Scott, R.B., 1955. Vapor pressures of methane. J. Res. Natl. Bur. Stand. 55 (1), 39–52.

Atkins, P., de Paula, J., 2009. Elements of Physical Chemistry, fifth ed. Oxford University Press, Oxford.

Barucci, M.A., Cruikshank, D.P., Dotto, E., Merlin, F., Poulet, F., Dalle Ore, C.M., et al., 2005. Is Sedna another Triton? Astron. Astrophys. 439, L1–L4.

Barucci, M.A., Merlin, F., Guilbert, A., de Bergh, C., Alvarez-Candal, A., Hainaut, O., et al., 2008. Surface composition and temperature of the TNO Orcus. Astron. Astrophys. 479, L13–L16.

Barucci, M.A., Dalle Ore, C.M., Alvarez-Candal, A., de Bergh, C., Merlin, F., Dumas, C., et al., 2010. (90377) Sedna: investigation of surface compositional variation. Astron. J. 140, 2095–2100.

Barucci, M.A., Dalle Ore, C.M., Perna, D., Cruikshank, D.P., Doressoundiram, A., Alvarez-Candal, A., et al., 2015. (50000) Quaoar: surface composition variability. Astron. Astrophys. 584, A107.

Benedetti-Rossi, G., Sicardy, B., Buie, M.W., Ortiz, J.L., Vieira-Martins, R., Keller, J.M., et al., 2016. Results from the 2014 November 15th multi-chord stellar occultation by the TNO (229762) 2007 UK_{126}. Astron. J. 152, 156.

Benedetti-Rossi, G., Santos-Sanz, P., Ortiz, J.L., Assafin, M., Sicardy, B., Morales, N., et al., 2019. The Trans-Neptunian object (84922) 2003 VS2 through stellar occultations. https://ui.adsabs.harvard.edu/abs/2019arXiv190806645B/abstract.

Bertrand, T., Forget, F., 2016. Observed glacier and volatile distribution on Pluto from atmosphere-topography processes. Nature 540, 86–89.

Bertrand, T., Forget, F., Umurhan, O.M., Grundy, W.M., Schmitt, B., Protopapa, S., et al., 2018. The nitrogen cycles on Pluto over seasonal and astronomical timescales. Icarus 309, 277–296.

Bird, G.A., 1994. Molecular Gas Dynamics and the Direct Simulation of Gas Flows. Clarendon, Oxford.

Borovik, E.S., Grishin, S.F., Grishina, E.Y., 1960. The vapor pressure of nitrogen and hydrogen at low pressures. Sov. Phys. Tech. Phys. (Engl. Transl.) 5 (5), 506–511.

Braga-Ribas, F., Sicardy, B., Ortiz, J.L., Lellouch, E., Tancredi, G., Lecacheux, J., 2013. The size, shape, albedo, density, and atmospheric limit of transneptunian object (50000) Quaoar from multi-chord stellar occultations. Astrophys. J. 773, 26.

Braga-Ribas, F., Sicardy, B., Ortiz, J.L., Snodgrass, C., Roques, F., Vieira-Martins, R., 2014a. A ring system detected around the Centaur (10199) Chariklo. Nature 508, 72–75.

Braga-Ribas, F., Vieira-Martins, R., Assafin, M., Camargo, J.I.B., Sicardy, B., Ortiz, J.L., 2014b. Stellar occultations by transneptunian and centaurs objects: results from more than 10 observed events. Revista Mexicana Astron. Astrofisica Conf. Ser. 44, 3.

Broadfoot, A.L., Atreya, S.K., Bertaux, J.L., Blamont, J.E., Dessler, A.J., Donahue, T.M., et al., 1989. Ultraviolet spectrometer observations of Neptune and Triton. Science 246, 1459–1466.

Brosch, N., 1995. The 1985 stellar occultation by Pluto. Mon. Not. R. Astron. Soc. 276, 571–578.

Brown, M.E., 2008. The largest Kuiper belt objects. In: The Solar System Beyond Neptune. University of Arizona Press, Tucson, AZ, pp. 335–344.

Brown, M.E., 2012. The compositions of Kuiper belt objects. Annu. Rev. Earth Planet. Sci. 40, 467–494.

Brown, R.H., Cruikshank, D.P., Veverka, J., Helfenstein, P., Eluszkiewicz, J., 1995. Surface composition and photometric properties of Triton. In: Cruikshank, D.P. (Ed.), Neptune and Triton. University of Arizona Press, Tucson, AZ, pp. 991–1030.

Brown, M.E., Trujillo, C.A., Rabinowitz, D.L., 2005. Discovery of a planetary-sized object in the scattered Kuiper belt. Astrophys. J. 635, L97–L100.

Brown, M.E., Barkume, K.M., Blake, G.A., Schaller, E.L., Rabinowitz, D.L., Roe, H.G., et al., 2007. Methane and ethane on the bright Kuiper belt object 2005 FY_9. Astron. J. 133, 284–289.

Brown, M.E., Burgasser, A.J., Fraser, W.C., 2011. The surface composition of large Kuiper belt object 2007 OR_{10}. Astrophys. J. 738, L26.

Brown, M.E., Schaller, E.L., Blake, G.A., 2015. Irradiation products on dwarf planet Makemake. Astron. J. 149, 105.

Camargo, J.I.B., Desmars, J., Braga-Ribas, F., Vieira-Martins, R., Assafin, M., Sicardy, B., et al., 2018. The future of stellar occultations by distant solar system bodies: perspectives from the Gaia astrometry and the deep sky surveys. Planet. Space Sci. 154, 59–62.

Carry, B., Hestroffer, D., DeMeo, F.E., Thirouin, A., Berthier, J., Lacerda, P., et al., 2011. Integral-field spectroscopy of (90482) Orcus-Vanth. Astron. Astrophys. 534, A115.

Cruikshank, D.P., Roush, T.L., Owen, T.C., Geballe, T.R., de Bergh, C., Schmitt, B., et al., 1993. Ices on the surface of Triton. Science 261, 742–745.

Cruikshank, D.P., Schmitt, B., Roush, T.L., Owen, T.C., Quirico, E., Geballe, T.R., et al., 2000. Water ice on Triton. Icarus 147, 309–316.

Cruikshank, D.P., Grundy, W.M., DeMeo, F.E., Buie, M.W., Binzel, R.P., Jennings, D.E., et al., 2015. The surface compositions of Pluto and Charon. Icarus 246, 82–92.

Dalle Ore, C.M., Barucci, M.A., Emery, J.P., Cruikshank, D.P., Dalle Ore, L.V., Merlin, F., et al., 2009. Composition of KBO (50000) Quaoar. Astron. Astrophys. 501, 349–357.

Dalle Ore, C.M., Barucci, M.A., Emery, J.P., Cruikshank, D.P., de Bergh, C., Roush, T.L., et al., 2015. The composition of "ultra-red" TNOs and centaurs. Icarus 252, 311–326.

de Bergh, C., Delsanti, A., Tozzi, G.P., Dotto, E., Doressoundiram, A., Barucci, M.A., 2005. The surface of the transneptunian object 90482 Orcus. Astron. Astrophys. 437, 1115–1120.

de Bergh, C., Schaller, E.L., Brown, M.E., Brunetto, R., Cruikshank, D.P., Schmitt, B., 2013. The ices on transneptunian objects and centaurs. Astrophys. Space Sci. Library 356, 107.

Delsanti, A., Merlin, F., Guilbert-Lepoutre, A., Bauer, J., Yang, B., Meech, K.J., 2010. Methane, ammonia, and their irradiation products at the surface of an intermediate-size KBO? A portrait of Plutino (90482) Orcus. Astron. Astrophys. 520, A40.

DeMeo, F.E., Barucci, M.A., Merlin, F., Guilbert-Lepoutre, A., Alvarez-Candal, A., Delsanti, A., et al., 2010. A spectroscopic analysis of Jupiter-coupled object (52872) Okyrhoe, and TNOs (90482) Orcus and (73480) 2002 PN_{34}. Astron. Astrophys. 521, A35.

Dias-Oliveira, A., Sicardy, B., Lellouch, E., Vieira-Martins, R., Assafin, M., Camargo, J.I.B., et al., 2015. Pluto's atmosphere from stellar occultations in 2012 and 2013. Astrophys. J. 811, 53.

Dias-Oliveira, A., Sicardy, B., Ortiz, J.L., Braga-Ribas, F., Leiva, R., Vieira-Martins, R., et al., 2017. Study of the Plutino object (208996) 2003 AZ_{84} from stellar occultations: size, shape, and topographic features. Astron. J. 154, 22.

Dumas, C., Merlin, F., Barucci, M.A., de Bergh, C., Hainault, O., Guilbert, A., Vernazza, P., Doressoundiram, A., 2007. Surface composition of the largest dwarf planet 136199 Eris (2003 UB_{313}). Astron. Astrophys. 471, 331–334.

Elliot, J.L., Kern, S.D., 2003. Pluto's atmosphere and a targeted-occultation search for other bound KBO atmospheres. Earth Moon Planets 92, 375–393.

Elliot, J.L., Olkin, C.B., 1996. Probing planetary atmospheres with stellar occultations. Annu. Rev. Earth Planet. Sci. 24, 89–124.

Elliot, J.L., Young, L.A., 1992. Analysis of stellar occultation data for planetary atmospheres. I—model fitting, with application to Pluto. Astron. J. 103, 991–1015.

Elliot, J.L., Hammel, H.B., Wasserman, L.H., Franz, O.G., McDonald, S.W., Person, M.J., et al., 1998. Global warming on Triton. Nature 393, 765–767.

Elliot, J.L., et al., 2000a. The prediction and observation of the 1997 July 18 stellar occultation by Triton: more evidence for distortion and increasing pressure in Triton's atmosphere. Icarus 148, 347–369.

Elliot, J.L., Strobel, D.F., Zhu, X., Stansberry, J.A., Wasserman, L.H., Franz, O.G., 2000b. Note: The thermal structure of Triton's middle atmosphere. Icarus 143, 425–428.

Elliot, J.L., Ates, A., Babcock, B.A., Bosh, A.S., Buie, M.W., Clancy, K.B., et al., 2003a. The recent expansion of Pluto's atmosphere. Nature 424, 165–168.

Elliot, J.L., Person, M.J., Qu, S., 2003b. Analysis of stellar occultation data. II. Inversion, with application to Pluto and Triton. Astron. J. 126, 1041–1079.

Emery, J.P., Dalle Ore, C.M., Cruikshank, D.P., Fernández, Y.R., Trilling, D.E., Stansberry, J.A., 2007. Ices on (90377) Sedna: confirmation and compositional constraints. Astron. Astrophys. 466, 395–398.

Erwin, J., Tucker, O.J., Johnson, R.E., 2013. Hybrid fluid/kinetic modeling of Pluto's escaping atmosphere. Icarus 226, 375–384.

Forget, F., Bertrand, T., Vangvichith, M., Leconte, J., Millour, E., Lellouch, E., 2017. A post-New Horizons global climate model of Pluto including the N_2, CH_4 and CO cycles. Icarus 287, 54–71.

Fornasier, S., Dotto, E., Barucci, M.A., Barbieri, C., 2004. Water ice on the surface of the large TNO 2004 DW. Astron. Astrophys. 422, L43–L46.

Fray, N., Schmitt, B., 2009. Sublimation of ices of astrophysical interest: a bibliographic review. Planet. Space Sci. 57, 2053–2080.

Frels, W., Smith, D.R., Ashworth, T., 1974. Vapour pressure of nitrogen below the triple point. Cryogenics 14, 3–7.

French, R.G., Gierasch, P.J., 1976. Diffraction calculation of occultation light curves in the presence of an isothermal atmosphere. Astron. J. 81, 445–451.

Giauque, W.F., Clayton, J.O., 1933. The heat capacity and entropy of nitrogen. Heat of vaporization. Vapor pressures of solid and liquid. The reaction $1/2\ N_2 + 1/2\ O_2 = NO$ from spectroscopic data. J. Am. Chem. Soc. 55 (12), 4875–4889.

Gladstone, G.R., 1998. Lyman-α photolysis in the Primitive Solar Nebula. Final Report for NASA Grant NAG5-4290.

Gladstone, G.R., Young, L.A., 2019. New Horizons observations of the atmosphere of Pluto. Annu. Rev. Earth Planet. Sci. 47, 119–140.
Glein, C.R., Waite, J.H., 2018. Primordial N_2 provides a cosmochemical explanation for the existence of Sputnik Planitia, Pluto. Icarus 313, 79–92.
Grasset, O., Castillo-Rogez, J., Guillot, T., Fletcher, L.N., Tosi, F., 2017. Water and volatiles in the outer solar system. Space Sci. Rev. 212, 835–875.
Grundy, W.M., Young, L.A., 2004. Near-infrared spectral monitoring of Triton with IRTF/SpeX I: establishing a baseline for rotational variability. Icarus 172, 455–465.
Grundy, W.M., Young, L.A., Stansberry, J.A., Buie, M.W., Olkin, C.B., Young, E.F., 2010. Near-infrared spectral monitoring of Triton with IRTF/SpeX II: spatial distribution and evolution of ices. Icarus 205, 594–604.
Guilbert, A., Alvarez-Candal, A., Merlin, F., Barucci, M.A., Dumas, C., de Bergh, C., et al., 2009. ESO-large program on TNOs: near-infrared spectroscopy with SINFONI. Icarus 201, 272–283.
Gulbis, A.A.S., Elliot, J.L., Person, M.J., Adams, E.R., Babcock, B.A., Emilio, M., et al., 2006. Charon's radius and atmospheric constraints from observations of a stellar occultation. Nature 439, 48–51.
Hansen, C.J., Paige, D.A., Young, L.A., 2015. Pluto's climate modeled with new observational constraints. Icarus 246, 183–191.
Harris, A.W., 1998. A thermal model for near-Earth asteroids. Icarus 131, 291–301.
Hofgartner, J.D., Buratti, B.J., Hayne, P.O., Young, L.A., 2018. Ongoing resurfacing of KBO Eris by volatile transport in local, collisional, sublimation atmosphere regime. Icarus. https://doi.org/10.1016/j.icarus.2018.10.028.
Holler, B.J., Young, L.A., Grundy, W.M., Olkin, C.B., 2016. On the surface composition of Triton's southern latitudes. Icarus 267, 255–266.
Howard, A.D., Moore, J.M., 2008. Sublimation-driven erosion on Callisto: a landform simulation model test. Geophys. Res. Lett. 35, L03203.
Howard, A.D., Moore, J.M., Schenk, P.M., White, O.L., Spencer, J., 2012. Sublimation-driven erosion on Hyperion: topographic analysis and landform simulation model tests. Icarus 220, 268–276.
Hubbard, W.B., Hunten, D.M., Dieters, S.W., Hill, K.M., Watson, R.D., 1988. Occultation evidence for an atmosphere on Pluto. Nature 336, 452–454.
Ingersoll, A.P., 1990. Dynamics of Triton's atmosphere. Nature 344, 315–317.
Ingersoll, A.P., Summers, M.E., Schlipf, S.G., 1985. Supersonic meteorology of Io—sublimation-driven flow of SO_2. Icarus 64, 375–390.
Jewitt, D.C., Luu, J., 2004. Crystalline water ice on the Kuiper belt object (50000) Quaoar. Nature 432, 731–733.
Johnson, R.E., Volkov, A.N., Erwin, J.T., 2013a. Erratum: "Molecular-kinetic simulations of escape from the ex-planet and exoplanets: criterion for transonic flow". Astrophys. J. 779, L30.
Johnson, R.E., Volkov, A.N., Erwin, J.T., 2013b. Molecular-kinetic simulations of escape from the ex-planet and exoplanets: criterion for transonic flow. Astrophys J. Lett. 768, L4.
Johnson, R.E., Oza, A., Young, L.A., Volkov, A.N., Schmidt, C., 2015. Volatile loss and classification of Kuiper belt objects. Astrophys. J. 809, 43.
Kaye, G.W.C., Laby, T.H., 1973. Tables of Physical and Chemical Constants, fourteenth ed. Longman Scientific & Technical, London.
Lellouch, E., de Bergh, C., Sicardy, B., Ferron, S., Käufl, H.U., 2010. Detection of CO in Triton's atmosphere and the nature of surface-atmosphere interactions. Astron. Astrophys. 512, L8.
Lellouch, E., Santos-Sanz, P., Lacerda, P., Mommert, M., Duffard, R., Ortiz, J.L., et al., 2013. "TNOs are Cool": a survey of the Trans-Neptunian region. IX. Thermal properties of Kuiper belt objects and centaurs from combined Herschel and Spitzer observations. Astron. Astrophys. 557, A60.
Lellouch, E., Gurwell, M., Butler, B., Fouchet, T., Lavvas, P., Strobel, D.F., et al., 2017. Detection of CO and HCN in Pluto's atmosphere with ALMA. Icarus 286, 289–307.
Levi, A., Podolak, M., 2009. Cornoa-like atmospheric escape from KBOs. Icarus 202, 681–693.
Licandro, J., Grundy, W.M., Pinilla-Alonso, N., Leisy, P., 2006a. Visible spectroscopy of 2003 UB_{313}: evidence for N_2 ice on the surface of the largest TNO? Astron. Astrophys. 458, L5–L8.
Licandro, J., Pinilla-Alonso, N., Pedani, M., Oliva, E., Tozzi, G.P., Grundy, W.M., 2006b. The methane ice rich surface of large TNO 2005 FY_9: a Pluto-twin in the Trans-Neptunian belt? Astron. Astrophys. 445, L35–L38.
Lorenzi, V., Pinilla-Alonso, N., Licandro, J., Dalle Ore, C.M., Emery, J.P., 2014. Rotationally resolved spectroscopy of (20000) Varuna in the near-infrared. Astron. Astrophys. 562, A85.

Lorenzi, V., Pinilla-Alonso, N., Licandro, J., 2015. Rotationally resolved spectroscopy of dwarf planet (136472) Makemake. Astron. Astrophys. 577, A86.

Mangold, N., 2011. Ice sublimation as a geomorphic process: a planetary perspective. Geomorphology 126, 1–17.

McKinnon, W.B., Leith, A.C., 1995. Gas drag and the orbital evolution of a captured Triton. Icarus 118, 392–413.

Merlin, F., Alvarez-Candal, A., Delsanti, A., Fornasier, S., Barucci, M.A., DeMeo, F.E., et al., 2009. Stratification of methane ice on Eris' surface. Astron. J. 137, 315–328.

Merlin, F., Lellouch, E., Quirico, E., Schmitt, B., 2018. Triton's surface ices: distribution, temperature and mixing state from VLT/SINFONI observations. Icarus 314, 274–293.

Meza, E., Sicardy, B., Assafin, M., Ortiz, J.L., Bertrand, T., Lellouch, E., et al., 2019. Lower atmosphere and pressure evolution on Pluto from ground-based stellar occultations, 1988–2016. Astron. Astrophys. 625, A42.

Millis, R.L., et al., 1993. Pluto's radius and atmosphere—results from the entire 9 June 1988 occultation data set. Icarus 105, 282.

Moore, J.M., Spencer, J.R., 1990. Koyaanismuuyaw: the hypothesis of a perennially dichotomous Triton. Geophys. Res. Lett. 17, 1757–1760.

Moore, J.M., McKinnon, W.B., Spencer, J.R., Howard, A.D., Schenk, P.M., Beyer, R.A., et al., 2016. The geology of Pluto and Charon through the eyes of New Horizons. Science 351, 1284–1293.

Müller, M., Marchis, F., Emery, J.P., Harris, A.W., Mottola, S., Hestroffer, D., et al., 2010. Eclipsing binary Trojan asteroid Patroclus: thermal inertia from Spitzer observations. Icarus 205, 505–515.

Oliveira, J.M., Sicardy, B., Meza, E., Desmars, J., Lecacheux, J., Assafin, M., et al., 2018. Stellar occultation of Triton on October 5th, 2017. In: European Planetary Science Congress EPSC2018-172.

Olkin, C.B., Elliot, J.L., Hammel, H.B., Cooray, A.R., McDonald, S.W., Foust, J.A., et al., 1997. The thermal structure of Triton's atmosphere: results from the 1993 and 1995 occultations. Icarus 129, 178–201.

Olkin, C.B., Young, L.A., Borncamp, D., Pickles, A., Sicardy, B., Assafin, M., et al., 2015. Evidence that Pluto's atmosphere does not collapse from occultations including the 2013 May 04 event. Icarus 246, 220–225.

Ortiz, J.L., Sicardy, B., Braga-Ribas, F., Alvarez-Candal, A., Lellouch, E., Duffard, R., et al., 2012. Albedo and atmospheric constraints of dwarf planet Makemake from a stellar occultation. Nature 491, 566–569.

Ortiz, J.L., Santos-Sanz, P., Sicardy, B., Benedetti-Rossi, G., Bérard, D., Morales, N., et al., 2017. The size, shape, density and ring of the dwarf planet Haumea from a stellar occultation. Nature 550, 219–223.

Pál, A., Kiss, C., Müller, T.G., Molnár, L., Szabó, R., Szabó, G.M., et al., 2016. Large size and slow rotation of the Trans-Neptunian object (225088) 2007 OR_{10} discovered from Herschel and K2 observations. Astron. J. 151, 117.

Parker, A., Pinilla-Alonso, N., Santos-Sanz, P., Stansberry, J., Alvarez-Candal, A., Bannister, M., et al., 2016. Physical characterization of TNOs with the James Webb Space Telescope. Publ. Astron. Soc. Pac. 128, 018010.

Pasachoff, J.M., Babcock, B.A., Durst, R.F., Seeger, C.H., Levine, S.E., Bosh, A.S., et al., 2017. Pluto occultation on 2015 June 29 UTC with central flash and atmospheric spikes just before the New Horizons flyby. Icarus 296, 305–314.

Perna, D., Hromakina, T., Merlin, F., Ieva, S., Fornasier, S., Belskaya, I., et al., 2017. The very homogeneous surface of the dwarf planet Makemake. Mon. Not. R. Astron. Soc. 466, 3594–3599.

Person, M.J., Schindler, K., Bosh, A.S., Wolf, J., Levine, S.E., Zuluaga, C.A., et al., 2018. Airborne and ground observations of the stellar occultation by Triton on 5 October 2017. In: AAS/Division for Planetary Sciences Meeting Abstracts No. 416, 10.

Protopapa, S., Grundy, W.M., Tegler, S.C., Bergonio, J.M., 2015. Absorption coefficients of the methane-nitrogen binary ice system: implications for Pluto. Icarus 253, 179–188.

Protopapa, S., Grundy, W.M., Reuter, D.C., Hamilton, D.P., Dalle Ore, C.M., Cook, J.C., et al., 2017. Pluto's global surface composition through pixel-by-pixel Hapke modeling of New Horizons Ralph/LEISA data. Icarus 287, 218–228.

Quirico, E., Schmitt, B., 1997. Near-infrared spectroscopy of simple hydrocarbons and carbon oxides diluted in solid N_2 and as pure ices: implications for Triton and Pluto. Icarus 127, 354–378.

Quirico, E., Douté, S., Schmitt, B., de Bergh, C., Cruikshank, D.P., Owen, T.C., et al., 1999. Composition, physical state, and distribution of ices at the surface of Triton. Icarus 139, 159–178.

Rubin, M., Altwegg, K., Balsiger, H., Bar-Nun, A., Berthelier, J.-J., Bieler, A., et al., 2015. Molecular nitrogen in comet 67P/Churyumov-Gerasimenko indicates a low formation temperature. Science 348, 232–235.

Santos-Sanz, P., 2017. Physical characterization of Kuiper belt objects from stellar occultations and thermal measurements. In: European Planetary Science Congress 11, EPSC2017-97.

Santos-Sanz, P., French, R.G., Pinilla-Alonso, N., Stansberry, J., Lin, Z.-Y., Zhang, Z.-W., et al., 2016. James Webb Space Telescope observations of stellar occultations by solar system bodies and rings. Publ. Astron. Soc. Pac. 128, 018011.

Schaller, E.L., Brown, M.E., 2007a. Volatile loss and retention on Kuiper belt objects. Astrophys. J. 659, L61–L64.
Schaller, E.L., Brown, M.E., 2007b. Detection of methane on Kuiper belt object (50000) Quaoar. Astrophys. J. 670, L49–L51.
Schmitt, Philippe, S., Grundy, W.M., Reuter, D.C., Côte, R., Quirico, E., et al., 2017. Physical state and distribution of materials at the surface of Pluto from New Horizons LEISA imaging spectrometer. Icarus 287, 229–260.
Shinoda, T., 1969. Vapor pressure of carbon monoxide in condensed phases. Bull. Chem. Soc. Jpn 42, 1815–1820.
Sicardy, B., Ferri, F., Roques, F., Lecacheux, J., Pau, S., Brosch, N., et al., 1999. The structure of Titan's stratosphere from the 28 SGR occultation. Icarus 142, 357–390.
Sicardy, B., Widemann, T., Lellouch, E., Veillet, C., Cuillandre, J.-C., Colas, F., et al., 2003. Large changes in Pluto's atmosphere as revealed by recent stellar occultations. Nature 424, 168–170.
Sicardy, B., Colas, F., Maquet, L., Vachier, F., Doressoundiram, A., Roques, F., et al., 2010. The 2010, February 19 stellar occultation by Varuna. Bull. Am. Astron. Soc. 42, 23.11.
Sicardy, B., Ortiz, J.L., Assafin, M., Jehin, E., Maury, A., Lellouch, E., et al., 2011. A Pluto-like radius and a high albedo for the dwarf planet Eris from an occultation. Nature 478, 493–496.
Sicardy, B., Talbot, J., Meza, E., Camargo, J.I.B., Desmars, J., Gault, D., et al., 2016. Pluto's atmosphere from the 2015 June 29 ground-based stellar occultation at the time of the new Horizons Flyby. Astrophys. J. 819, L38.
Spencer, J.R., 1990. A rough-surface thermophysical model for airless planets. Icarus 83, 27–38.
Spencer, J.R., Stansberry, J.A., Trafton, L.M., Young, E.F., Binzel, R.P., Croft, S.K., 1997. Volatile transport, seasonal cycles, and atmospheric dynamics on Pluto. In: In: Pluto and Charon. University of Arizona Press, Tucson, AZ, pp. 435–473.
Stern, S.A., Trafton, L.M., 2008. On the atmospheres of objects in the Kuiper belt. In: The Solar System Beyond Neptune. University of Arizona Press, Tucson, AZ, pp. 365–380.
Stern, S.A., Kammer, J.A., Gladstone, G.R., Steffl, A.J., Cheng, A.F., Young, L.A., et al., 2017. New Horizons constraints on Charon's present day atmosphere. Icarus 287, 124–130.
Tan, S.P., Kargel, J.S., 2018. Solid-phase equilibria on Pluto's surface. Mon. Not. R. Astron. Soc. 474, 4254–4263.
Tegler, S.C., Grundy, W.M., Romanishin, W., Consolmagno, G.J., Mogren, K., Vilas, F., 2007. Optical spectroscopy of the large Kuiper belt objects 136472 (2005 FY9) and 136108 (2003 EL61). Astron. J. 133, 526–530.
Tegler, S.C., Grundy, W.M., Vilas, F., Romanishin, W., Cornelison, D.M., Consolmagno, G.J., 2008. Evidence of N_2-ice on the surface of the icy dwarf planet 136472 (2005 FY9). Icarus 195, 844–850.
Tegler, S.C., Grundy, W.M., Olkin, C.B., Young, L.A., Romanishin, W., Cornelison, D.M., et al., 2012. Ice mineralogy across and into the surfaces of Pluto, Triton, and Eris. Astrophys. J. 751, 76.
Tickner, A.W., Lossing, F.P., 1951. The measurement of low vapor pressures by means of a mass spectrometer. J. Phys. Chem. 55 (5), 733–740.
Toigo, A.D., Gierasch, P.J., Sicardy, B., Lellouch, E., 2010. Thermal tides on Pluto. Icarus 208, 402–411.
Trafton, L.M., 1984. Large seasonal variations in Triton's atmosphere. Icarus 58, 312–324.
Trafton, L.M., 2015. On the state of methane and nitrogen ice on Pluto and Triton: implications of the binary phase diagram. Icarus 246, 197–205.
Trafton, L.M., Matson, D.L., Stansberry, J.A., 1998. Surface/atmosphere interactions and volatile transport (Triton, Pluto and Io). In: Solar System Ices. Kluwer Academic Publishers, Dordrecht. p. 773.
Trujillo, C.A., Brown, M.E., Rabinowitz, D.L., Geballe, T.R., 2005. Near-infrared surface properties of the two intrinsically brightest minor planets: (90377) Sedna and (90482) Orcus. Astrophys. J. 627, 1057–1065.
Trujillo, C.A., Sheppard, S.S., Schaller, E.L., 2011. A photometric system for detection of water and methane ices on Kuiper belt objects. Astrophys. J. 730, 105.
Tucker, O.J., Erwin, J.T., Deighan, J.I., Volkov, A.N., Johnson, R.E., 2012. Thermally driven escape from Pluto's atmosphere: a combined fluid/kinetic model. Icarus 217, 408–415.
Volkov, A.N., Johnson, R.E., Tucker, O.J., Erwin, J.T., 2011a. Thermally driven atmospheric escape: transition from hydrodynamic to jeans escape. Astrophys. J. 729, L24.
Volkov, A.N., Tucker, O.J., Erwin, J.T., Johnson, R.E., 2011b. Kinetic simulations of thermal escape from a single component atmosphere. Phys. Fluids 23, 066601-16.
Walker, A.C., Moore, C.H., Goldstein, D.B., Varghese, P.L., Trafton, L.M., 2012. A parametric study of Io's thermophysical surface properties and subsequent numerical atmospheric simulations based on the best fit parameters. Icarus 220, 225–253.
Wong, M.C., Smyth, W.H., 2000. Model calculations for Io's atmosphere at eastern and western elongations. Icarus 146, 60–74.

Yelle, R.V., Lunine, J.I., 1989. Evidence for a molecule heavier than methane in the atmosphere of Pluto. Nature 339, 288–290.

Yelle, R.V., Lunine, J.I., Pollack, J.B., Brown, R.H., 1995. Lower atmospheric structure and surface-atmosphere interactions on Triton. In: Neptune and Triton. University of Arizona Press, Tucson, AZ, pp. 1031–1105.

Young, E.F., 1993. An Albedo Map and Frost Model of Pluto (Ph.D. thesis). Massachusetts Institute of Technology.

Young, L.A., 2012. Volatile transport on inhomogeneous surfaces: I—analytic expressions, with application to Pluto's day. Icarus 221, 80–88.

Young, L.A., 2013. Pluto's seasons: new predictions for New Horizons. Astrophys. J. 766, L22.

Young, L.A., 2017. Volatile transport on inhomogeneous surfaces: II. Numerical calculations (VT3D). Icarus 284, 443–476.

Young, L., McKinnon, W.B.. 2013. Atmospheres on volatile-bearing Kuiper belt objects. In: AAS/Division for Planetary Sciences Meeting Abstracts No. 45, ID507.02.

Young, E.F., French, R.G., Young, L.A., Ruhland, C.R., Buie, M.W., Olkin, C.B., et al., 2008. Vertical structure in Pluto's atmosphere from the 2006 June 12 stellar occultation. Astron. J. 136, 1757–1769.

Young, L.A., Stark, G., Vervack, R.J., 2002. Triton's atmosphere in 1989: new lab data, new profiles. In AAS/Division for Planetary Sciences Meeting Abstracts No. 34, ID21.04.

Young, L.A., Kammer, J.A., Steffl, A.J., Gladstone, G.R., Summers, M.E., Strobel, D.F., et al., 2018. Structure and composition of Pluto's atmosphere from the New Horizons solar ultraviolet occultation. Icarus 300, 174–199.

Zhang, X., Strobel, D.F., Imanaka, H., 2017. Haze heats Pluto's atmosphere yet explains its cold temperature. Nature 551, 352–355.

Zhu, X., Strobel, D.F., Erwin, J.T., 2014. The density and thermal structure of Pluto's atmosphere and associated escape processes and rates. Icarus 228, 301–314.

CHAPTER

7

Trans-Neptunian objects and Centaurs at thermal wavelengths

Thomas Müller[a], *Emmanuel Lellouch*[b], *Sonia Fornasier*[b]

[a]Max Planck Institute for Extraterrestrial Physics, Garching, Germany [b]LESIA, Paris Observatory, PSL University, CNRS, Université de Paris, Sorbonne University, Meudon Pricipal Cedex, France

7.1 Introduction

Thermal infrared (IR) observations are crucial for the physical and thermal characterization of distant objects which are typically too small to be resolved even by the largest ground or space telescopes. The thermal emission measurements allow one to determine accurate radiometric size and albedo information, but they also put constraints on the object's thermal properties and spin-axis orientations. Albedos derived from thermal data are important to interpret broadband colors and NIR spectral data. The submillimeter/millimeter (submm/mm)-range emission originates from a few millimeters below the surface and can be used to determine the object's long-wavelength emissivity. A recent new approach is to combine multidata information (occultations, thermal, high-resolution imaging, etc.) to perform more sophisticated physical and thermal characterization of Trans-Neptunian objects (TNOs) and Centaurs, and to constrain properties which are otherwise not accessible, for example, the object's spin-axis orientation. At the same time, the multidisciplinary approach allows one to improve and extend model techniques, for example, the verification of radiometric sizes via occultation results. A comparison between radiometric and occultation size values is given in Chapter 19.

The Trans-Neptunian Solar System.
https://doi.org/10.1016/B978-0-12-816490-7.00007-2

However, the thermal emission from TNOs and Centaurs[1] is difficult to detect, mainly due to their large distance from the Sun (and the observer) and their low surface temperatures.[2]

We give an overview of the existing thermal emission measurements for TNOs and Centaurs (Section 7.2), the modeling concepts (Section 7.3), and we present the derived radiometric properties (Sections 7.4 and 7.5). The chapter is concluded by a short outlook (Section 7.6).

7.2 Thermal data for TNOs and Centaurs

The thermal IR covers approximately the wavelength range from about 5 to 300 μm, often subdivided into the mid-IR (MIR) range up to about 30 μm (ground-based N-/Q-band, SOFIA,[3] WISE/NEOWISE,[4] AKARI,[5] Spitzer[6]-IRS/IRAC) and the far-IR (FIR) at wavelengths up to several hundred micron (Herschel[7]-PACS, Spitzer-MIPS), followed by the submm range below 1 mm wavelength (e.g., Herschel-SPIRE; CSO[8]; ALMA[9] bands 7–10), and the mm range (e.g., IRAM,[10] ALMA bands 1–6, NOEMA[11]).

The first thermal measurements for a Centaur go back to ground-based IR and IRAS measurements of Chiron (Lebofsky et al., 1984; Sykes and Walker, 1991; Campins et al., 1994), followed by millimeter observations (Jewitt and Luu, 1992; Altenhoff and Stumpff, 1995). Pholus (Howell et al., 1992; Davies et al., 1993) and Chariklo (Jewitt and Kalas, 1998) were next, with solid detections mainly in Q-band at ~20 μm. A first overview of the physical characteristics of TNOs and Centaurs (Davies, 2000) listed sizes for eight Centaurs (three based on thermal measurements, other sizes are estimated assuming an albedo of 4%) and no TNO sizes (Pluto was still considered as planet at that time).

The Pluto-Charon system had the first thermal detections by IRAS in 1983 (Aumann and Walker, 1987; Sykes et al., 1987; Tedesco et al., 1987) and a decade later at submm/mm (Stern et al., 1993; Jewitt, 1994). The characterization of the system's IR and millimeter variability and the corresponding modeling followed: Sykes (1999) using reanalyzed IRAS data, Lellouch

[1] We use the term "Trans-Neptunian object (TNO)" or "(Edgeworth-) Kuiper belt object (KBO)" to describe objects with orbits in the range 30–50 AU, including also the scattered-disk objects (SDO) with high eccentricities and inclinations. The Centaurs are closer and partly active objects with perihelia between Jupiter and Neptune.

[2] Blackbody radiation from TNOs and Centaurs peaks in the FIR between 50 and 100 μm, a range which is only accessible from outside the Earth's atmosphere.

[3] See https://www.sofia.usra.edu/.

[4] See https://neowise.ipac.caltech.edu/.

[5] See https://www.ir.isas.jaxa.jp/AKARI/.

[6] See http://www.spitzer.caltech.edu/.

[7] See http://sci.esa.int/herschel/.

[8] See http://www.cso.caltech.edu/.

[9] See http://www.almaobservatory.org/.

[10] See http://www.iram-institute.org/.

[11] See http://iram-institute.org/EN/noema-project.php.

et al. (2000a) interpreting dedicated ISO-ISOPHOT measurements, and Lellouch et al. (2000b) by searching for variability in Pluto's millimeter-wave emission. ISO also contributed with low-SNR FIR detections of two TNOs: 15789 (1993 SC) and 15874 (1996 TL66) (Thomas et al., 2000). In the years after, Varuna was measured in the submm (Jewitt et al., 2001) and mm (Lellouch et al., 2002), the Centaurs Asbolus and Chiron (Fernández et al., 2002) were detected at MIR, Chiron and Chariklo also at FIR/submm/mm (Altenhoff et al., 2001; Groussin et al., 2004), the TNOs 55565 (2002 AW197) (Margot et al., 2002), 47171 Lempo (1999 TC36) (Altenhoff et al., 2004), and Eris (Bertoldi et al., 2006) at radio wavelengths. However, many attempts to measure the TNO thermal emission produced only upper limits (e.g., Altenhoff et al., 2004; Brown et al., 2004; Ortiz et al., 2004). This changed dramatically with the availability of the Spitzer Space Telescope: Stansberry et al. (2004) reported Spitzer-MIPS FIR detections of 14 KBOs and 8 Centaurs, Grundy et al. (2005) covered 20 TNOs, followed by several other Spitzer-related TNO or Centaur projects (Cruikshank et al., 2005, 2006; Stansberry et al., 2006; Grundy et al., 2007, 2008). Stansberry et al. (2008) summarized the Spitzer-provided constraints on the physical properties of 47 KBOs and Centaurs. In the years after, the Spitzer-based studies focused either on individual TNOs or smaller samples (e.g., Brucker et al., 2009; Stansberry et al., 2012). Four Centaurs and seven TNOs were also detected with AKARI (T. Sekiguchi, private communication); however, no fluxes or radiometric results were published so far.

The next big step in thermal observations of Centaurs and TNOs came with Herschel's large Open Time Key Project on "TNOs are Cool: A survey of the Trans-Neptunian region with Herschel" (Müller et al., 2009) which produced more than 20 publications. The Herschel data (partly also Spitzer data) allowed to interpret the thermal emission of almost 170 TNOs and Centaurs (Müller et al., 2010, 2018; Lellouch et al., 2010, 2013, 2016; Lim et al., 2010; Barucci et al., 2012; Mommert et al., 2012; Pál et al., 2012, 2015, 2016; Santos-Sanz et al., 2012, 2017; Vilenius et al., 2012, 2014, 2018; Fornasier et al., 2013; Kiss et al., 2013, 2016, 2019; Duffard et al., 2014; Lacerda et al., 2014; Marton et al., 2015; Kovalenko et al., 2017).

In parallel, the WISE project performed an all-sky survey at MIR wavelengths and detected the brightest Centaurs (Bauer et al., 2013).

Since Spitzer and Herschel, the situation has become much more difficult and only a dozen TNOs have been observed in the submm/mm range, mainly by ALMA (Moullet et al., 2011; Gerdes et al., 2017; Brown and Butler, 2017, 2018; Lellouch et al., 2017).

7.3 Radiometric techniques

7.3.1 Models to interpret thermal measurements

Radiometry is a powerful technique to derive size, albedo, and thermal properties from thermal IR measurements. The technique consists in the exploitation and interpretation of thermal data (see Section 7.2) in combination with data in the visible. The visible data are mainly obtained from the ground and include the object's absolute magnitude and estimates for the phase integral q. The most accurate radiometric properties are determined when objects are observed close to their thermal emission peaks, preferentially shortwards and longwards of the peak. For Centaurs and TNOs, these peaks are located in the FIR regime between ≈50 and 100 μm (accessible from space only).

There are different thermal models to obtain size and albedo of atmosphereless small bodies: Simple models like the standard thermal model (STM), the fast rotating thermal model, or the near-Earth asteroid thermal model (NEATM) or more complex thermophysical models (TPMs) are used (see Delbo' et al., 2015 for a detailed discussion). For Centaurs and TNOs, mostly the NEATM (Harris, 1998) and the TPM (Lagerros, 1996, 1997, 1998; Müller and Lagerros, 1998, 2002) are applied. In cases where the object is lacking shape or spin properties, the NEATM is often used. The beaming factor η is then either fitted to multiband thermal measurements or a default value is applied. The STM/NEATM concepts can also handle elongated shapes (Brown, 1985; Lellouch et al., 2017; Vilenius et al., 2018), but have severe limitations when thermal measurements cover different aspect angles.[12]

In cases where the rotation period is known, the TPM concepts have the advantage that in addition to the standard size-albedo solution, also the spin-axis and thermophysical parameters can be constrained (Fornasier et al., 2013; Kiss et al., 2019; Müller et al., 2018). Pál et al. (2012) applied STM and TPM techniques to the Herschel measurements of (90377) Sedna and 2010 EK139. The derived size-albedo values agree within ~10%. Pál et al. (2012) also show that NEATM (fitted beaming factor) and TPM (assuming equator-on viewing geometry) approaches lead to the same size-albedo conclusions for (225088) 2007 OR_{10}. However, the discovery of a satellite (Kiss et al., 2017) suggests a nearly pole-on viewing geometry for 2007 OR_{10}. The recalculated TPM size (Kiss et al., 2019) is 18% smaller than the NEATM best-η solution, but agrees within 5% when assuming a default $\eta = 1.2$ in the radiometric calculations, indicating that NEATM solutions require a proper error handling for the fits of the beaming factors.

The TPM concept has advantages when objects have known spin or shape properties, or in cases where multiepoch thermal data for different observing geometries (like for Centaurs observed over a wide range of heliocentric distances or aspect angles) are available (e.g., Kiss et al., 2013). Recently, stellar occultations by Centaurs and TNOs revealed ring systems, elongated shapes, constraints on possible atmospheres, and very accurate size information (see Chapter 19). Here, the TPM applications allow one to determine the object's thermal properties and testing of the spin-axis orientation (Kiss et al., 2013; Müller et al., 2018). For multiwavelength thermal data which include also submm/mm observations, a wavelength-dependent emissivity within the TPM code is necessary to explain the observed fluxes (Fornasier et al., 2013; Lellouch et al., 2017).

The accuracy of the radiometrically derived size and albedo values depends very much on the model input values: A good-quality H-magnitude is needed for deriving the object's albedo. A typical 0.1 mag error for H translates into a ~10% error in geometric albedo. The thermal flux error is crucial for the final size accuracy, but here the beaming parameter (or thermal inertia and surface roughness in case of the TPM) and the object's emissivity are also relevant. In general, the final quality of the radiometric size benefits from high-quality multiepoch and multiwavelengths observations. It is also important to mention that radiometric solutions for high-albedo dwarf planets like Pluto, Eris, and Makemake suffer from unknowns for the object's surface scattering properties, expressed in the phase integral

[12] The aspect angle is defined such that it equals 0 degrees when we observe the North pole and it equals 180 degrees when we observe the South pole.

(Stansberry et al., 2008). Using the formula $q = 0.336p_V + 0.479$, with q being the phase integral and p_V the object's geometric V-band albedo, by Brucker et al. (2009) solves only part of the problem since individual objects can deviate significantly from that relation (Verbiscer et al., 2018). To account for a wide range of scattering properties for icy surfaces, Müller et al. (2018) used a phase integral of $q = 0.65 \pm 0.20$ for Haumea which has a geometric albedo of $p_V = 0.51$ (Ortiz et al., 2017). For higher-albedo objects, the uncertainty in q might be even larger, for lower-albedo objects the phase integral is less relevant and a default value of 0.39 (Bowell et al., 1989) can be used.

7.3.2 Satellite thermal emission

Almost all large TNOs have satellites (Parker et al., 2016b; Kiss et al., 2017) and these satellites can contribute to the observed thermal emission measurements. At MIR and FIR, it was so far not possible to resolve the satellite and the main-body emission with the current technical limitations. Only in the submm/mm range, it is possible to detect large satellites around dwarf planets with the high spatial resolution of the ALMA array. Brown and Butler (2018) examined the spatially resolved Orcus-Vanth and Eris-Dysnomia systems and found low albedos similar to other TNOs of similar size and different from the main body's properties. It is therefore reasonable to estimate the thermal emission of satellites by using adequate beaming and albedo properties from published statistical analysis of TNO samples. However, the thermal emission estimate for satellites remains uncertain. This has to be considered in the radiometric analysis in cases where primaries and secondaries cannot be seen directly (e.g., for 2007 OR_{10}: see discussion in Kiss et al., 2019; for Haumea: see discussion in Müller et al., 2018).

7.3.3 Ring thermal emission

Rings have been detected around Chariklo (Braga-Ribas et al., 2014) and Haumea (Ortiz et al., 2017) and are suspected on other bodies (e.g., Ortiz et al., 2015). Lellouch et al. (2017) estimated the contribution of rings to thermal emission, using a simplified version of a model for Saturn's rings. In this model, the only source of energy for ring particles is absorbed solar radiation, but mutual shadowing—as seen both from the Sun and the observer—and optical depth effects are taken into consideration. The model further assumes that ring particles have a bolometric and spectral emissivity of unity. By analogy with Saturn's rings, the latter assumption is likely valid up to $\sim$200 μm, but the spectral emissivity could decrease at longer wavelengths if ring particles are made of water ice. Model free parameters are the ring radius, width, opacity, and Bond albedo (related to the I/F reflectivity). The relative importance of ring contribution to solid body emission is of course strongly geometry dependent, but was found to have a minor effect, for example, affecting the equivalent diameter determination by at most 5% for Chariklo. In that sense, for ellipsoid bodies, including/omitting rings are much less important than properly accounting for the varying pole orientation and hence projected surface of the object.

7.4 Albedos, sizes, and densities

Sizes, albedos, and in many cases beaming factors, derived from Spitzer, WISE, and Herschel thermal data, are available for 178 TNOs and Centaurs and listed in Table 7.1. Trans-Neptunian bodies show a huge diversity both in albedo and size (Fig. 7.1, top). Measured diameters range from a few tens of kilometers to ~2000 km but the distribution of sizes is

TABLE 7.1 Absolute magnitude used in the modeling and derived radiometric diameters, geometric albedos, and beaming factor for TNOs and Centaurs.

Object	Dyn. class	H_v	Diam. (km)	p_v	η	Refs.
2000 GM_{137}	Centaur	14.36 ± 0.38	$8.6^{+1.5}_{-1.5}$	$0.043^{+0.026}_{-0.016}$	[a]$1.20^{+0.35}_{-0.35}$	a
2004 QQ_{26}	Centaur	9.53 ± 0.36	$79.0^{+19.0}_{-19.0}$	$0.044^{+0.039}_{-0.014}$	[a]$1.20^{+0.35}_{-0.35}$	a
2013 AZ_{60}	Centaur/SDO	10.45 ± 0.10	$62.3^{+5.3}_{-5.3}$	$0.029^{+0.006}_{-0.006}$	$1.70^{+0.90}_{-0.90}$	b
(2060) Chiron	Centaur	5.92 ± 0.20	$215.6^{+9.9}_{-9.9}$	$0.167^{+0.037}_{-0.030}$	$0.95^{+0.09}_{-0.10}$	c
(5145) Pholus	Centaur	7.68 ± 0.28	$99.0^{+15.0}_{-14.0}$	$0.155^{+0.076}_{-0.049}$	$0.77^{+0.16}_{-0.16}$	a
(7066) Nessus	Centaur	9.51 ± 0.22	$57.0^{+17.0}_{-14.0}$	$0.086^{+0.075}_{-0.034}$	[a]$1.20^{+0.35}_{-0.35}$	a
(8405) Asbolus	Centaur	9.13 ± 0.25	$85.0^{+8.0}_{-9.0}$	$0.056^{+0.019}_{-0.015}$	$0.97^{+0.14}_{-0.18}$	a
(10199) Chariklo	Centaur	7.40 ± 0.25	$238.0^{+10.0}_{-10.0}$	$0.042^{+0.005}_{-0.005}$	$1.12^{+0.14}_{-0.14}$	c, d
(10370) Hylonome	Centaur	9.51 ± 0.08	$74.0^{+16.0}_{-16.0}$	$0.051^{+0.030}_{-0.017}$	$1.29^{+0.31}_{-0.31}$	a
(31824) Elatus	Centaur	10.40 ± 0.09	$49.8^{+10.4}_{-9.8}$	$0.049^{+0.028}_{-0.016}$	[a]$1.20^{+0.35}_{-0.35}$	a
(32532) Thereus	Centaur	9.40 ± 0.16	$62.0^{+3.0}_{-3.0}$	$0.083^{+0.016}_{-0.013}$	$0.87^{+0.08}_{-0.08}$	a
(52872) Okyrhoe	Centaur	11.07 ± 0.10	$35.0^{+3.0}_{-3.0}$	$0.056^{+0.012}_{-0.010}$	$0.71^{+0.12}_{-0.13}$	a
(52975) Cyllarus	Centaur	9.02 ± 0.15	$56.0^{+21.0}_{-18.0}$	$0.139^{+0.157}_{-0.064}$	[a]$1.20^{+0.35}_{-0.35}$	a
(54598) Bienor	Centaur	7.57 ± 0.34	$198.0^{+6.0}_{-7.0}$	$0.043^{+0.016}_{-0.012}$	$1.58^{+0.07}_{-0.07}$	a
(55576) Amycus	Centaur	8.27 ± 0.11	$104.0^{+8.0}_{-8.0}$	$0.083^{+0.016}_{-0.015}$	$1.00^{+0.12}_{-0.13}$	a
(60558) Echeclus	Centaur	9.78 ± 0.14	$64.6^{+1.6}_{-1.6}$	$0.052^{+0.007}_{-0.007}$	$0.86^{+0.04}_{-0.04}$	a
(63252) 2001 BL_{41}	Centaur	11.34 ± 0.21	$34.6^{+6.6}_{-6.1}$	$0.043^{+0.028}_{-0.014}$	[a]$1.20^{+0.35}_{-0.35}$	a
(83982) Crantor	Centaur	9.03 ± 0.16	$59.0^{+11.0}_{-12.0}$	$0.121^{+0.064}_{-0.038}$	[a]$1.20^{+0.35}_{-0.35}$	a
(95626) 2002 GZ_{32}	Centaur	7.37 ± 0.10	$237.0^{+8.0}_{-8.0}$	$0.037^{+0.004}_{-0.004}$	$0.97^{+0.05}_{-0.07}$	a
(119315) 2001 SQ_{73}	Centaur	9.15 ± 0.11	$90.0^{+23.0}_{-20.0}$	$0.048^{+0.030}_{-0.018}$	[a]$1.20^{+0.35}_{-0.35}$	a
(119976) 2002 VR_{130}	Centaur	11.26 ± 0.39	$24.4^{+5.4}_{-4.6}$	$0.093^{+0.066}_{-0.036}$	[a]$1.20^{+0.35}_{-0.35}$	a
(120061) 2003 CO_1	Centaur	9.07 ± 0.05	$94.0^{+5.0}_{-5.0}$	$0.049^{+0.005}_{-0.006}$	$1.23^{+0.12}_{-0.11}$	a
(136204) 2003 WL_7	Centaur	8.75 ± 0.16	$105.0^{+6.0}_{-7.0}$	$0.053^{+0.010}_{-0.010}$	$1.02^{+0.07}_{-0.05}$	a
(145486) 2005 UJ_{438}	Centaur	11.14 ± 0.32	$16.0^{+1.0}_{-2.0}$	$0.256^{+0.097}_{-0.076}$	$0.34^{+0.09}_{-0.08}$	a
(248835) 2006 SX_{368}	Centaur	9.45 ± 0.11	$76.0^{+2.0}_{-2.0}$	$0.052^{+0.007}_{-0.006}$	$0.87^{+0.04}_{-0.06}$	a

TABLE 7.1 Absolute magnitude used in the modeling and derived radiometric diameters, geometric albedos, and beaming factor for TNOs and Centaurs—cont'd

Object	Dyn. class	H_v	Diam. (km)	p_v	η	Refs.
(250112) 2002 KY_{14}	Centaur	10.37 ± 0.07	$47.0^{+3.0}_{-4.0}$	$0.057^{+0.011}_{-0.007}$	[a]$1.20^{+0.35}_{-0.35}$	a
(281371) 2008 FC_{76}	Centaur	9.44 ± 0.10	$68.0^{+6.0}_{-7.0}$	$0.067^{+0.017}_{-0.011}$	[a]$1.20^{+0.35}_{-0.35}$	a
(447178) 2005 RO_{43}	Centaur	7.34 ± 0.51	$194.0^{+10.0}_{-10.0}$	$0.056^{+0.036}_{-0.021}$	$1.12^{+0.05}_{-0.08}$	a
2000 CN_{105}	Cold classical	5.20 ± 0.30	$247.0^{+63.0}_{-40.0}$	$0.151^{+0.070}_{-0.059}$	[a]$1.20^{+0.35}_{-0.35}$	e
2001 QS_{322}	Cold classical	6.91 ± 0.68	$186.0^{+99.0}_{-24.0}$	$0.095^{+0.531}_{-0.060}$	[a]$1.20^{+0.35}_{-0.35}$	f
2001 RZ_{143}	Cold classical	6.69 ± 0.13	$140.0^{+39.0}_{-33.0}$	$0.191^{+0.066}_{-0.045}$	$0.75^{+0.23}_{-0.19}$	g
2001 XR_{254}	Cold classical	6.05 ± 0.15	$221.0^{+41.0}_{-71.0}$	$0.136^{+0.168}_{-0.044}$	[a]$1.20^{+0.35}_{-0.35}$	f
2003 QR_{91}[b]	Cold classical	6.55 ± 0.56	$280.0^{+27.0}_{-30.0}$	$0.054^{+0.035}_{-0.028}$	$1.20^{+0.10}_{-0.12}$	f
2003 WU_{188}	Cold classical	5.96 ± 0.64	<220	>0.15	$1.20^{+0.35}_{-0.35}$	f
(66652) Borasisi[b]	Cold classical	6.121 ± 0.07	$163.0^{+32.0}_{-66.0}$	$0.236^{+0.438}_{-0.077}$	$0.77^{+0.19}_{-0.47}$	f
(79360) Sila-Nunam	Cold classical	5.56 ± 0.04	$343.0^{+42.0}_{-42.0}$	$0.090^{+0.027}_{-0.017}$	$1.36^{+0.21}_{-0.19}$	g
(88611) Teharonhiawako	Cold classical	6.00 ± 0.13	$220.0^{+41.0}_{-44.0}$	$0.145^{+0.086}_{-0.045}$	$1.08^{+0.30}_{-0.28}$	f
(119951) 2002 KX_{14}	Cold classical	4.86 ± 0.10	$455.0^{+27.0}_{-27.0}$	$0.097^{+0.014}_{-0.013}$	$1.79^{+0.16}_{-0.15}$	g
(120181) 2003 UR_{292}	Cold classical	7.40 ± 0.40	$136.0^{+16.0}_{-26.0}$	$0.105^{+0.081}_{-0.033}$	[a]$1.20^{+0.35}_{-0.35}$	f
(135182) 2001 QT_{322}	Cold classical	7.29 ± 0.67	$159.0^{+30.0}_{-47.0}$	$0.085^{+0.424}_{-0.052}$	[a]$1.20^{+0.35}_{-0.35}$	f
(275809) 2001 QY_{297}	Cold classical	5.86 ± 0.31	$229.0^{+22.0}_{-108.0}$	$0.152^{+0.439}_{-0.035}$	$1.52^{+0.22}_{-0.92}$	f
(385266) 2001 QB_{298}	Cold classical	6.10 ± 0.30	$196.0^{+71.0}_{-53.0}$	$0.167^{+0.162}_{-0.082}$	[a]$1.20^{+0.35}_{-0.35}$	i
(385437) 2003 GH_{55}	Cold classical	6.43 ± 0.12	$178.0^{+21.0}_{-56.0}$	$0.150^{+0.182}_{-0.031}$	[a]$1.20^{+0.35}_{-0.35}$	f
(469438) 2002 GV_{31}	Cold classical	6.10 ± 0.60	<180.0	>0.019	[a]$1.20^{+0.35}_{-0.35}$	f
(469514) 2003 QA_{91}[b]	Cold classical	5.76 ± 0.63	$260.0^{+30.0}_{-36.0}$	$0.130^{+0.119}_{-0.075}$	$0.83^{+0.10}_{-0.15}$	f
(469705) 2005 EF_{298}	Cold classical	6.40 ± 0.50	$174.0^{+27.0}_{-32.0}$	$0.160^{+0.130}_{-0.070}$	[a]$1.20^{+0.35}_{-0.35}$	g
(508869) 2002 VT_{130}	Cold classical	5.80 ± 0.30	$324.0^{+57.0}_{-68.0}$	$0.097^{+0.098}_{-0.049}$	[a]$1.20^{+0.35}_{-0.35}$	i
1996 TS_{66}	Hot classical	6.50 ± 0.05	$159.0^{+44.0}_{-46.0}$	$0.179^{+0.173}_{-0.070}$	$0.75^{+0.21}_{-0.27}$	f
2001 KA_{77}	Hot classical	5.64 ± 0.12	$310.0^{+170.0}_{-60.0}$	$0.099^{+0.052}_{-0.056}$	$2.52^{+0.18}_{-0.83}$	g
2001 QC_{298}	Hot classical	6.26 ± 0.32	$303.0^{+27.0}_{-30.0}$	$0.061^{+0.027}_{-0.017}$	$0.99^{+0.08}_{-0.10}$	f
2001 QD_{298}	Hot classical	6.71 ± 0.17	$233.0^{+27.0}_{-63.0}$	$0.067^{+0.062}_{-0.014}$	[a]$1.26^{+0.35}_{-0.35}$	g
2002 GH_{32}	Hot classical	6.58 ± 0.28	<180	>0.13	[a]$1.26^{+0.35}_{-0.35}$	f
(19308) 1996 TO_{66}	Hot classical	4.81 ± 0.14	<330.0	>0.200	[a]$1.74^{+0.17}_{-0.17}$	j
(19521) Chaos	Hot classical	5.00 ± 0.06	$600.0^{+140.0}_{-130.0}$	$0.050^{+0.030}_{-0.016}$	$2.20^{+1.20}_{-1.10}$	g
(20000) Varuna	Hot classical	3.76 ± 0.035	$668.0^{+154.0}_{-86.0}$	$0.127^{+0.040}_{-0.042}$	$2.18^{+1.04}_{-0.49}$	h
(24835) 1995 SM_{55}	Hot classical	4.49 ± 0.035	<280.0	>0.360	[a]$1.74^{+0.17}_{-0.17}$	j
(35671) 1998 SN_{165}	Hot classical	5.707 ± 0.085	$393.0^{+39.0}_{-38.0}$	$0.060^{+0.019}_{-0.013}$	[a]$1.23^{+0.35}_{-0.35}$	g

Continued

TABLE 7.1 Absolute magnitude used in the modeling and derived radiometric diameters, geometric albedos, and beaming factor for TNOs and Centaurs—cont'd

Object	Dyn. class	H_V	Diam. (km)	p_V	η	Refs.
(50000) Quaoar[b]	Hot classical	2.73 ± 0.06	$1073.6^{+37.9}_{-37.9}$	$0.127^{+0.010}_{-0.009}$	$1.73^{+0.08}_{-0.08}$	c
(55565) 2002 AW_{197}	Hot classical	3.57 ± 0.05	$768.0^{+39.0}_{-38.0}$	$0.112^{+0.012}_{-0.011}$	$1.29^{+0.13}_{-0.10}$	f
(55636) 2002 TX_{300}	Hot classical	3.37 ± 0.05	$323.0^{+95.0}_{-37.0}$	$0.760^{+0.180}_{-0.450}$	$1.80^{+0.50}_{-0.90}$	j
(55637) 2002 UX_{25}[b]	Hot classical	3.87 ± 0.02	$697.2^{+23.0}_{-24.5}$	$0.107^{+0.008}_{-0.008}$	$1.07^{+0.05}_{-0.05}$	h
(78799) 2002 XW_{93}	Hot classical	5.40 ± 0.70	$565.0^{+71.0}_{-73.0}$	$0.038^{+0.043}_{-0.025}$	$0.79^{+0.27}_{-0.24}$	g
(86177) 1999 RY_{215}	Hot classical	7.235 ± 0.093	$263.0^{+29.0}_{-37.0}$	$0.039^{+0.012}_{-0.007}$	[a]$1.20^{+0.35}_{-0.35}$	g
(90568) 2004 GV_9	Hot classical	4.23 ± 0.10	$680.0^{+34.0}_{-34.0}$	$0.077^{+0.008}_{-0.008}$	$1.93^{+0.09}_{-0.07}$	g
(120178) 2003 OP_{32}	Hot classical	4.10 ± 0.07	$274.0^{+47.0}_{-25.0}$	$0.540^{+0.110}_{-0.150}$	[a]$1.74^{+0.17}_{-0.17}$	j
(120347) Salacia[b]	Hot classical	4.25 ± 0.05	$901.0^{+45.0}_{-45.0}$	$0.044^{+0.004}_{-0.004}$	$1.16^{+0.03}_{-0.03}$	l
(136108) Haumea[b]	Hot classical	0.43 ± 0.11	$1239.5^{+68.7}_{-57.8}$	$0.804^{+0.062}_{-0.095}$	$0.95^{+0.33}_{-0.26}$	c
(136472) Makemake	Hot classical	0.14 ± 0.05	$1430.0^{+9.0}_{-9.0}$	$0.770^{+0.020}_{-0.020}$	$2.29^{+0.46}_{-0.40}$	h
(138537) 2000 OK_{67}	Hot classical	6.47 ± 0.13	$164.0^{+33.0}_{-45.0}$	$0.169^{+0.159}_{-0.052}$	[a]$1.20^{+0.35}_{-0.35}$	f
(145452) 2005 RN_{43}	Hot classical	3.89 ± 0.05	$679.0^{+55.0}_{-73.0}$	$0.107^{+0.029}_{-0.018}$	[a]$1.20^{+0.35}_{-0.35}$	g
(145453) 2005 RR_{43}	Hot classical	4.13 ± 0.08	$300.0^{+43.0}_{-34.0}$	$0.440^{+0.120}_{-0.100}$	[a]$1.74^{+0.17}_{-0.17}$	j
(148780) Altjira[b]	Hot classical	6.44 ± 0.14	$331.0^{+51.0}_{-187.0}$	$0.043^{+0.183}_{-0.009}$	$1.62^{+0.24}_{-0.83}$	f
(174567) Varda[b]	Hot classical	3.61 ± 0.05	$792.0^{+91.0}_{-84.0}$	$0.102^{+0.024}_{-0.020}$	$0.84^{+0.28}_{-0.22}$	f
(182934) 2002 GJ_{32}	Hot classical	6.16 ± 0.13	$416.0^{+81.0}_{-78.0}$	$0.035^{+0.019}_{-0.011}$	$2.05^{+0.38}_{-0.36}$	f
(202421) 2005 UQ_{513}	Hot classical	3.87 ± 0.14	$498.0^{+63.0}_{-75.0}$	$0.202^{+0.084}_{-0.049}$	[a]$1.27^{+0.35}_{-0.35}$	g
(230965) 2004 XA_{192}	Hot classical	4.42 ± 0.63	$339.0^{+120.0}_{-95.0}$	$0.260^{+0.340}_{-0.150}$	$0.62^{+0.79}_{-0.49}$	f
(307251) 2002 KW_{14}	Hot classical	5.88 ± 0.11	$161.0^{+35.0}_{-40.0}$	$0.310^{+0.281}_{-0.094}$	[a]$1.20^{+0.35}_{-0.35}$	f
(307261) 2002 MS_4	Hot classical	4.00 ± 0.60	$934.0^{+47.0}_{-47.0}$	$0.051^{+0.036}_{-0.022}$	$1.06^{+0.06}_{-0.06}$	g
(307616) 2003 QW_{90}	Hot classical	5.00 ± 0.30	$401.0^{+63.0}_{-48.0}$	$0.084^{+0.026}_{-0.022}$	[a]1.11	e
(416400) 2003 UZ_{117}	Hot classical	5.23 ± 0.15	$222.0^{+57.0}_{-42.0}$	$0.290^{+0.160}_{-0.110}$	[a]$1.74^{+0.17}_{-0.17}$	j
(444030) 2004 NT_{33}	Hot classical	4.74 ± 0.11	$423.0^{+87.0}_{-80.0}$	$0.125^{+0.069}_{-0.039}$	$0.69^{+0.460}_{-0.32}$	f
(469306) 1999 CD_{158}	Hot classical	5.35 ± 0.67	<310.0	>0.130	[a]$1.20^{+0.35}_{-0.35}$	j
(469615) 2004 PT_{107}	Hot classical	6.33 ± 0.11	$400.0^{+45.0}_{-51.0}$	$0.032^{+0.011}_{-0.007}$	[a]$1.53^{+0.35}_{-0.35}$	f
2001 KD_{77}	Plutino	6.42 ± 0.08	$232.3^{+40.5}_{-39.4}$	$0.089^{+0.044}_{-0.027}$	[a]$1.20^{+0.35}_{-0.35}$	k
2002 XV_{93}	Plutino	5.42 ± 0.46	$549.2^{+21.7}_{-23.0}$	$0.040^{+0.020}_{-0.015}$	$1.24^{+0.06}_{-0.06}$	k
2003 UT_{292}	Plutino	6.85 ± 0.68	$185.6^{+17.9}_{-18.0}$	$0.067^{+0.068}_{-0.034}$	[a]$1.20^{+0.35}_{-0.35}$	k
(15820) 1994 TB	Plutino	7.934 ± 0.354	$85.0^{+36.0}_{-28.0}$	$0.172^{+0.258}_{-0.097}$	$1.260^{+0.97}_{-0.65}$	h
(15875) 1996 TP_{66}	Plutino	7.51 ± 0.09	$154.0^{+28.8}_{-33.7}$	$0.074^{+0.063}_{-0.031}$	[a]$1.20^{+0.35}_{-0.35}$	k
(28978) Ixion	Plutino	3.83 ± 0.04	$617.0^{+19.0}_{-20.0}$	$0.141^{+0.011}_{-0.011}$	$0.91^{+0.04}_{-0.06}$	h

TABLE 7.1 Absolute magnitude used in the modeling and derived radiometric diameters, geometric albedos, and beaming factor for TNOs and Centaurs—cont'd

Object	Dyn. class	H_v	Diam. (km)	p_v	η	Refs.
(33340) 1998 VG$_{44}$	Plutino	6.67 ± 0.04	$248.0^{+43.0}_{-41.0}$	$0.063^{+0.026}_{-0.017}$	$1.55^{+0.58}_{-0.38}$	h
(38628) Huya[b]	Plutino	5.04 ± 0.03	$458.0^{+9.2}_{-9.2}$	$0.083^{+0.004}_{-0.004}$	$0.93^{+0.02}_{-0.02}$	c
(47171) 1999 TC$_{36}$[b]	Plutino	5.41 ± 0.10	$393.1^{+25.2}_{-26.8}$	$0.079^{+0.013}_{-0.011}$	$1.10^{+0.07}_{-0.08}$	k
(47932) 2000 GN$_{171}$	Plutino	6.45 ± 0.34	$147.1^{+20.7}_{-17.8}$	$0.215^{+0.093}_{-0.070}$	$1.11^{+0.24}_{-0.21}$	k
(55638) 2002 VE$_{95}$	Plutino	5.70 ± 0.06	$249.8^{+13.5}_{-13.1}$	$0.149^{+0.019}_{-0.016}$	$1.40^{+0.12}_{-0.11}$	k
(84719) 2002 VR$_{128}$	Plutino	5.58 ± 0.37	$448.5^{+42.1}_{-43.2}$	$0.052^{+0.027}_{-0.018}$	[a]$1.20^{+0.35}_{-0.35}$	k
(84922) 2003 VS$_{2}$	Plutino	4.11 ± 0.38	$523.0^{+35.1}_{-34.4}$	$0.147^{+0.063}_{-0.043}$	$1.57^{+0.30}_{-0.23}$	k
(90482) Orcus[b]	Plutino	2.31 ± 0.03	$958.4^{+22.9}_{-22.9}$	$0.231^{+0.018}_{-0.011}$	$0.97^{+0.05}_{-0.02}$	c
(120216) 2004 EW$_{95}$	Plutino	6.69 ± 0.35	$291.1^{+20.3}_{-25.9}$	$0.044^{+0.021}_{-0.015}$	[a]$1.20^{+0.35}_{-0.35}$	k
(120348) 2004 TY$_{364}$	Plutino	4.52 ± 0.07	$512.0^{+37.0}_{-40.0}$	$0.107^{+0.020}_{-0.015}$	$1.55^{+0.15}_{-0.10}$	h
(133067) 2003 FB$_{128}$	Plutino	6.80 ± 0.30	$186.0^{+27.0}_{-29.0}$	$0.074^{+0.035}_{-0.021}$	[a]$1.20^{+0.35}_{-0.35}$	e
(144897) 2004 UX$_{10}$	Plutino	4.75 ± 0.16	$398.1^{+32.6}_{-39.3}$	$0.141^{+0.044}_{-0.031}$	[a]$1.20^{+0.35}_{-0.35}$	k
(175113) 2004 PF$_{115}$	Plutino	4.54 ± 0.25	$468.2^{+38.6}_{-49.1}$	$0.123^{+0.043}_{-0.033}$	[a]$1.20^{+0.35}_{-0.35}$	k
(208996) 2003 AZ$_{84}$[b]	Plutino	3.74 ± 0.08	$727.0^{+61.9}_{-66.5}$	$0.107^{+0.023}_{-0.016}$	$1.05^{+0.19}_{-0.15}$	k
(307463) 2002 VU$_{130}$	Plutino	5.47 ± 0.83	$252.9^{+33.6}_{-31.3}$	$0.179^{+0.202}_{-0.103}$	[a]$1.20^{+0.35}_{-0.35}$	k
(450265) 2003 WU$_{172}$	Plutino	6.30 ± 0.30	$312.0^{+0.0}_{-0.0}$	$0.039^{+0.000}_{-0.000}$	[a]$2.60^{+0.35}_{-0.35}$	e
(455502) 2003 UZ$_{413}$	Plutino	5.70 ± 0.30	$670.0^{+84.0}_{-82.0}$	$0.070^{+0.022}_{-0.015}$	[a]$1.20^{+0.35}_{-0.35}$	e
(469372) 2001 QF$_{298}$	Plutino	5.43 ± 0.07	$408.2^{+40.2}_{-44.9}$	$0.071^{+0.020}_{-0.014}$	[a]$1.20^{+0.35}_{-0.35}$	k
(469987) 2006 HJ$_{123}$	Plutino	5.32 ± 0.66	$216.4^{+29.7}_{-34.2}$	$0.281^{+0.259}_{-0.152}$	[a]$1.20^{+0.35}_{-0.35}$	k
(126154) 2001 YH$_{140}$	Med res	5.80 ± 0.20	$349.0^{+81.0}_{-81.0}$	$0.080^{+0.050}_{-0.050}$	[a]$1.20^{+0.35}_{-0.35}$	m
2002 GP$_{32}$	Out res	6.90 ± 0.30	$201.0^{+25.0}_{-29.0}$	$0.091^{+0.061}_{-0.024}$	[a]$1.20^{+0.35}_{-0.35}$	e
(26308) 1998 SM$_{165}$	Out res	6.02 ± 0.08	$291.0^{+22.0}_{-26.0}$	$0.083^{+0.018}_{-0.013}$	$1.55^{+0.11}_{-0.10}$	h
(26375) 1999 DE$_{9}$	Out res	5.16 ± 0.05	$311.0^{+29.0}_{-32.0}$	$0.163^{+0.041}_{-0.026}$	$0.71^{+0.10}_{-0.11}$	h
(42301) 2001 UR$_{163}$	Out res	4.10 ± 0.30	$352.0^{+85.0}_{-53.0}$	$0.209^{+0.082}_{-0.074}$	[a]$1.20^{+0.35}_{-0.35}$	e
(82075) 2000 YW$_{134}$	Out res	4.88 ± 0.05	<500	>0.08	[a]$1.20^{+0.35}_{-0.35}$	m
(84522) 2002 TC$_{302}$	Out res	4.17 ± 0.10	$584.1^{+105.6}_{-88.0}$	$0.115^{+0.047}_{-0.033}$	$1.09^{+0.37}_{-0.25}$	c
(119979) 2002 WC$_{19}$[b]	Out res	4.88 ± 0.07	$348.0^{+45.0}_{-45.0}$	$0.167^{+0.052}_{-0.037}$	$1.12^{+0.16}_{-0.17}$	h
(143707) 2003 UY$_{117}$	Out res	5.70 ± 0.30	$247.0^{+30.0}_{-29.0}$	$0.126^{+0.039}_{-0.028}$	[a]$1.20^{+0.35}_{-0.35}$	e
(225088) 2007 OR$_{10}$[b]	Out res	2.34 ± 0.05	$1252.0^{+43.0}_{-42.0}$	$0.13^{+0.01}_{-0.01}$	$1.20^{+0.35}_{-0.35}$	l
(308379) 2005 RS$_{43}$	Out res	5.00 ± 0.30	$271.0^{+45.0}_{-40.0}$	$0.193^{+0.071}_{-0.053}$	[a]$1.20^{+0.35}_{-0.35}$	e
(469505) 2003 FE$_{128}$	Out res	6.30 ± 0.30[c]	$157.0^{+60.0}_{-7.0}$	[c]$0.167^{+0.085}_{-0.072}$	[a]$1.20^{+0.35}_{-0.35}$	f
(471143) 2010 EK$_{139}$	Out res	3.80 ± 0.10	$433.0^{+63.0}_{-64.0}$	$0.297^{+0.113}_{-0.078}$	$0.60^{+0.33}_{-0.25}$	h

Continued

TABLE 7.1 Absolute magnitude used in the modeling and derived radiometric diameters, geometric albedos, and beaming factor for TNOs and Centaurs—cont'd

Object	Dyn. class	H_v	Diam. (km)	p_v	η	Refs.
2012 DR_{30}	SDO	7.04 ± 0.35	$188.0^{+9.4}_{-9.4}$	$0.076^{+0.031}_{-0.025}$	$0.81^{+0.074}_{-0.062}$	n
(15874) 1996 TL_{66}	SDO	5.39 ± 0.12	$339.0^{+20.0}_{-20.0}$	$0.110^{+0.021}_{-0.015}$	$1.15^{+0.08}_{-0.05}$	o
(26181) 1996 GQ_{21}	SDO	5.20 ± 0.30	$349.0^{+43.0}_{-49.0}$	$0.127^{+0.043}_{-0.026}$	[a]$1.20^{+0.35}_{-0.35}$	i
(29981) 1999 TD_{10}	SDO	8.93 ± 0.30	$103.7^{+13.6}_{-13.5}$	$0.044^{+0.014}_{-0.010}$	$1.64^{+0.32}_{-0.31}$	p
(42355) Typhon[b]	SDO	7.72 ± 0.04	$185.0^{+7.0}_{-7.0}$	$0.044^{+0.003}_{-0.003}$	$1.48^{+0.07}_{-0.07}$	o
(44594) 1999 OX_3	SDO	7.72 ± 0.09	$135.0^{+13.0}_{-12.0}$	$0.081^{+0.018}_{-0.015}$	$1.04^{+0.27}_{-0.22}$	h
(48639) 1995 TL_8[b]	SDO	5.29 ± 0.06	$244.0^{+82.0}_{-63.0}$	$0.231^{+0.189}_{-0.102}$	$1.38^{+0.80}_{-0.49}$	h
(65489) Ceto[b]	SDO	6.54 ± 0.06	$281.0^{+11.0}_{-11.0}$	$0.056^{+0.006}_{-0.006}$	$1.04^{+0.05}_{-0.05}$	o
(73480) 2002 PN_{34}	SDO	8.66 ± 0.03	$112.0^{+7.0}_{-7.0}$	$0.049^{+0.006}_{-0.006}$	$1.02^{+0.07}_{-0.09}$	o
(82158) 2001 FP_{185}	SDO	6.39 ± 0.07	$332.0^{+31.0}_{-24.0}$	$0.046^{+0.007}_{-0.007}$	$1.23^{+0.24}_{-0.19}$	o
(127546) 2002 XU_{93}	SDO	8.11 ± 0.10	$164.0^{+9.0}_{-9.0}$	$0.038^{+0.004}_{-0.004}$	$1.12^{+0.05}_{-0.08}$	o
(309239) 2007 RW_{10}	SDO	6.39 ± 0.61[c]	$247.0^{+30.0}_{-30.0}$	[c]$0.083^{+0.068}_{-0.039}$	[a]$1.20^{+0.35}_{-0.35}$	o
(40314) 1999 KR_{16}	Detached	6.24 ± 0.15	$232.0^{+34.0}_{-36.0}$	$0.105^{+0.049}_{-0.027}$	[a]$1.20^{+0.35}_{-0.35}$	j
(90377) Sedna	Detached	1.83 ± 0.05	$906.0^{+314.0}_{-258.0}$	$0.410^{+0.390}_{-0.190}$	$0.72^{+0.78}_{-0.54}$	h
(120132) 2003 FY_{128}	Detached	5.09 ± 0.09	$460.0^{+21.0}_{-21.0}$	$0.079^{+0.010}_{-0.010}$	$1.07^{+0.08}_{-0.08}$	o
(136199) Eris[b]	Detached	−1.12 ± 0.03	$2326.0^{+12.0}_{-12.0}$	$0.960^{+0.040}_{-0.040}$	$0.87^{+0.26}_{-0.41}$	h
(145480) 2005 TB_{190}	Detached	4.40 ± 0.11	$464.0^{+62.0}_{-62.0}$	$0.148^{+0.051}_{-0.036}$	[a]$1.20^{+0.35}_{-0.35}$	o
(229762) 2007 UK_{126}	Detached	3.69 ± 0.10	$599.0^{+77.0}_{-77.0}$	$0.167^{+0.058}_{-0.038}$	[a]$1.20^{+0.35}_{-0.35}$	o
(303775) 2005 QU_{182}	Detached	3.80 ± 0.32	$416.0^{+73.0}_{-73.0}$	$0.328^{+0.160}_{-0.109}$	[a]$1.20^{+0.35}_{-0.35}$	o
(470316) 2007 OC_{10}	Detached	5.43 ± 0.10	$309.0^{+37.0}_{-37.0}$	$0.127^{+0.040}_{-0.028}$	[a]$1.20^{+0.35}_{-0.35}$	o
(330759) 2008 SO_{218}	Centaur	12.8 ± 0.3	11.8 ± 0.4	0.097 ± 0.017	0.823 ± 0.046	q
2008 JS_{14}	Centaur	13.2 ± 0.3	14.5 ± 1.8	0.044 ± 0.019	1.046 ± 0.186	q
2010 CR_{140}	Centaur	15.5 ± 0.3	7.5 ± 1.4	0.020 ± 0.01	1.111 ± 0.283	q
2010 HU_{20}	Centaur	13.0 ± 0.3	10.513 ± 1.1	0.101 ± 0.024	0.976 ± 0.162	q
2010 LG_{61}	Centaur	18.5 ± 0.3	0.89 ± 0.19	0.089 ± 0.056	1.00 ± 0.400	q
2010 OR_1	Centaur	16.2 ± 0.3	3.25 ± 0.64	0.055 ± 0.013	0.831 ± 0.146	q
2010 OM_{101}	Centaur	17. ± 0.3	3.12 ± 0.17	0.029 ± 0.005	1.054 ± 0.105	q
2010 PO_{58}	Centaur	14.5 ± 0.3	8.88 ± 0.63	0.035 ± 0.007	0.915 ± 0.093	q
167P/CINEOS	Centaur	9.7 ± 0.3	66.17 ± 22.9	0.053 ± 0.019	0.8 ± 0.360	q
29P/Schwassm.-W. 1	Centaur	9.0 ± 0.3	46.0 ± 13.0	0.033 ± 0.015	0.64 ± 0.29	q
(148975) 2001 XA_{255}	Centaur	11.2 ± 0.3	37.7 ± 10.5	0.041 ± 0.014	0.703 ± 0.186	q
(309139) 2006 XQ_{51}	Centaur	9.8 ± 0.3	39.1 ± 15.7	0.139 ± 0.058	0.8 ± 0.456	q

TABLE 7.1 Absolute magnitude used in the modeling and derived radiometric diameters, geometric albedos, and beaming factor for TNOs and Centaurs—cont'd

Object	Dyn. class	H_v	Diam. (km)	p_v	η	Refs.
(310071) 2010 KR_{59}	Centaur	7.7 ± 0.3	110.06 ± 30.82	0.121 ± 0.037	0.8 ± 0.324	q
(309737) 2008 SJ_{236}	Centaur	12.2 ± 0.3	17.7 ± 1.5	0.074 ± 0.021	0.800 ± 0.110	q
(328884) 2010 LJ_{109}	Centaur	10.1 ± 0.3	44.2 ± 3.8	0.083 ± 0.021	0.748 ± 0.103	q
(332685) 2009 HH_{36}	Centaur	10.6 ± 0.3	33.0 ± 2.8	0.078 ± 0.018	0.739 ± 0.095	q
(342842) 2008 YB_3	Centaur	9.5 ± 0.3	67.1 ± 1.0	0.062 ± 0.012	0.839 ± 0.012	q
(346889) 2009 QV_{38}	Centaur	11.8 ± 0.3	23.2 ± 9.5	0.062 ± 0.049	0.8 ± 0.389	q
2007 VH_{305}	Centaur	11.6 ± 0.3	23.8 ± 8.0	0.070 ± 0.036	0.8 ± 0.384	q
2008 HY_{21}	Centaur	12.1 ± 0.3	24.0 ± 1.5	0.044 ± 0.010	1.22 ± 0.094	q
2010 BL_4	Centaur	11.9 ± 0.3	15.7 ± 3.2	0.114 ± 0.052	0.8 ± 0.333	q
2010 ES_{65}	Centaur	11.8 ± 0.3	26.9 ± 7.9	0.049 ± 0.024	0.8 ± 0.28	q
2010 FH_{92}	Centaur	11.7 ± 0.3	28.0 ± 0.6	0.047 ± 0.007	0.730 ± 0.023	q
2010 RM_{64}	Centaur	11.0 ± 0.3	21.0 ± 2.0	0.159 ± 0.048	0.85 ± 0.144	q
2010 TH	Centaur	9.2 ± 0.3	69.9 ± 24.2	0.078 ± 0.033	0.8 ± 0.363	q
2011 MM_4	Centaur	9.3 ± 0.3	63.7 ± 6.2	0.083 ± 0.024	0.841 ± 0.119	q
2005 VJ_{119}	SDO	10.6 ± 0.3	28.5 ± 6.9	0.126 ± 0.060	0.8 ± 0.30	q
2010 BK_{118}	SDO	10.2 ± 0.3	46.4 ± 1.8	0.068 ± 0.013	0.821 ± 0.043	q
2010 GW_{64}	SDO	14.9 ± 0.3	6.42 ± 0.38	0.047 ± 0.012	0.795 ± 0.075	q
2010 GW_{147}	SDO	13.2 ± 0.3	15.9 ± 0.7	0.037 ± 0.006	0.869 ± 0.056	q
2010 JH_{124}	SDO	14.6 ± 0.3	7.04 ± 0.74	0.052 ± 0.024	0.959 ± 0.164	q
C/2010 KW_7 (WISE)	SDO	15.5 ± 0.3	4.87 ± 0.22	0.047 ± 0.011	0.75 ± 0.06	q
2010 WG_9	SDO	8.1 ± 0.3	112.7 ± 61.9	0.074 ± 0.080	0.8 ± 0.423	q
(336756) 2010 NV_1	SDO	10.5 ± 0.3	44.2 ± 8.0	0.057 ± 0.030	0.661 ± 0.168	q
C/2011 KP_{36} (Spacewatch)	Centaur	9.4 ± 0.3	55.1 ± 19.4	0.101 ± 0.062	0.8 ± 0.4	q
Pluto	Plutino	−0.7	2376.6 ± 3.2	0.52 ± 0.03 (0.08–1.0)		r, s, t
Charon	Plutino	1.0	1212.0 ± 1.0	0.41 ± 0.02 (0.11–0.73)		t, u

Notes: The individual fluxes and the criteria for the dynamical classification for the objects are available in the TNOs are Cool public database: http://public-tnosarecool.lesia.obspm.fr/Published-observations.html. "Med/Out res" refers to resonants beside Plutinos, not including Haumea and family members. Geometric albedo and Absolute magnitude are given in the V filter, except for the two objects 2003 FE_{128} and 2007 RW_{10}.
[a] *Fixed η.*
[b] *Binary systems.*
[c] *Absolute magnitude and albedo are in the R band.*
References: (a) Duffard et al. (2014); (b) Pál et al. (2015); (c) Fornasier et al. (2013); (d) Fornasier et al. (2014); (e) Lacerda et al. (2014); (f) Vilenius et al. (2014); (g) Vilenius et al. (2012); (h) Lellouch et al. (2013); (i) Mommert (2013) PhD thesis; (j) Vilenius et al. (2018); (k) Mommert et al. (2012); (l) Kiss et al. (2019); (m) Müller et al. (2010); (n) Kiss et al. (2013); (o) Santos-Sanz et al. (2012); (p) Stansberry et al. (2008); (q) Bauer et al. (2013); (r) Nimmo et al. (2017); (s) Buratti et al. (2015); (t) Buratti et al. (2017); (u) Stern et al. (2018); in cases where the uncertainty for H-mag is unknown, we assumed 0.3 mag.

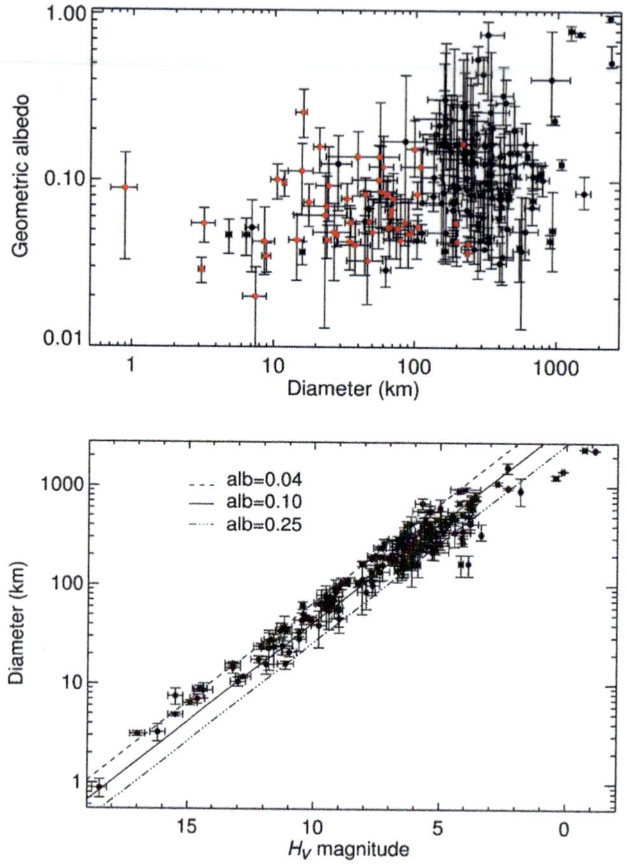

FIG. 7.1 *Top*: Geometric albedo versus diameter for 170 TNOs (in *black*) and Centaurs (in *red*) from Herschel, Spitzer, and WISE data (see Table 7.1, but excluding objects with upper/lower limit estimates). The selection bias due to distance is clearly visible as small sizes are measured only for closer bodies, that is, Centaurs. *Bottom*: Diameter versus H_v magnitude for 170 TNOs and Centaurs. Absolute magnitude with an albedo of 10% is a good proxy for size for the majority of TNOs.

strongly affected by discovery and selection biases in the Spitzer/Herschel/WISE sample. Eleven objects have diameters larger than 900 km: Charon, Orcus, Quaoar, Salacia, 2002 MS_4, 2007 OR_{10}, Sedna and the four so-called dwarf planets Pluto, Eris, Makemake, and Haumea. Geometric albedos p_V vary over a factor of ~25–30 from $p_V = 3\%–4\%$ for the darkest objects to $p_V \sim 50\%–90\%$ for the brightest (see Table 7.1). The latter category includes (1) volatile-rich bodies (Eris, Makemake, Pluto), where the high albedo likely results from seasonally variable resurfacing processes (deposition of fresh ice); (2) Haumea and some of its collisional family members, covered by pure H_2O ice likely excavated from Haumea's mantle. Excluding these bodies, the vast majority of TNOs have albedos ranging from 4% to 25%, with a median (mean and standard deviation) of 8% (10±6%), see Table 7.2. As routinely assumed, H_v magnitude is a good proxy for size (Fig. 7.1, bottom): Using a 10% albedo ensures an error on the diameter

TABLE 7.2 Average albedo values for different dynamic groups.

Dynamic group	n	Median	Mean	σ	$\sigma/\sqrt{n}$	min/max	Comment
Cold classical	17	0.136	0.132	0.045	0.0109	0.054 0.236	All
Hot classical	32	0.107	0.195	0.225	0.0397	0.032 0.804	All
	26	0.084	0.102	0.072	0.0141	0.032 0.310	Without Makemake, Haumea, and family
Plutino	25	0.089	0.114	0.064	0.0128	0.039 0.281	Without Pluto/Charon
Outer resonant	11	0.163	0.148	0.071	0.0215	0.049 0.297	All[a]
Detached	8	0.167	0.291	0.294	0.1040	0.079 0.960	All
	7	0.148	0.195	0.125	0.0471	0.079 0.410	Without Eris
SDO	20	0.057	0.075	0.046	0.0103	0.037 0.231	All
Haumea family	9		median $0.48^{+0.28}_{-0.18}$				Vilenius et al. (2018)
Centaur	55	0.056	0.074	0.043	0.0058	0.020 0.256	All
All TNOs and Centaurs	170	0.083	0.126	0.138	0.0106	0.020 0.960	All
	160	0.083	0.099	0.062	0.0049	0.020 0.328	Without P/C, E, M, S, H, Hfam[b]

Notes: The number of objects in a given sample (n), the median albedo value, the mean and standard deviation (σ), the standard error of the mean ($\sigma/\sqrt{n}$), the minimum and maximum values (min/max) in the sample, and comments on the sample: "All" refers to all entries in Table 7.1 with diameter and albedo solutions, excluding the objects with only upper/lower limit solutions. In cases of dwarf planets within a given dynamic group, we list the values with and without these special objects.

[a] *2001 YH140 is in a 3:5 resonance with Neptune and not considered here.*

[b] *Pluto/Charon, Eris, Makemake, Sedna, Haumea, and Haumea family members.*

lower than a factor of 1.6. Fig. 7.1 (top) further shows that except for the very bright objects, there is no obvious correlation between albedo and diameter, the only trend being (except for one outlier, 2005 UJ_{438}, whose radiometric solution may be flawed by coma activity) the lack of objects with albedos >0.2 at diameters <100 km. In Table 7.2, we present the median, mean, standard deviation σ, the error of the mean ($\sigma/\sqrt{n}$), and the minimum and maximum albedo values for each dynamic group. In the calculations, we excluded the objects which have only upper/lower-limit radiometric solutions. For the hot classicals and the Detached categories, we present these values with and without the dwarf planets and the Haumea family. We also looked at the weighted mean albedos which are typically below the median values. But the albedo errors are closely connected to the H-magnitude errors which are coming from different sources, including rough estimates or affected by unknown opposition effects and different handling of light curve amplitudes. It is therefore problematic to work with error-weighted albedos.

Twenty-six binary systems with known system masses have equivalent diameter values, yielding density estimates. Density values are reported in Sicardy et al. (2011), Mommert et al. (2012), Ortiz et al. (2012, 2017), Santos-Sanz et al. (2012), Vilenius et al. (2012, 2014), Brown (2013), Fornasier et al. (2013), Lellouch et al. (2013), Brown and Butler (2017), Dias-Oliveira et al. (2017), Kovalenko et al. (2017), Leiva et al. (2017), Kiss et al. (2019), and Stern et al. (2018). The derived bulk densities span a wide range, from below that of water ice to that of nearly pure rock (Fig. 7.2). TNOs smaller than 400 km have density lower than 1 g cm^{-3}, as already noticed (for 1999 TC_{36}; Stansberry et al., 2006), and this implies both a small rock-to-ice ratio and a high porosity. Possible outliers are the cold classical Borasisi and the 2:1 resonant body 2002 WC_{19}; those may have higher densities but uncertainties are huge. For comparison, precise measurement of the 67P/Churyumov-Gerasimenko nucleus by the Rosetta mission yields a density of 537.8±0.7 kg m^{-3}, and a large porosity (70%–80%, Sierks et al., 2015; Pätzold et al., 2016; Preusker et al., 2017). Vilenius et al. (2014) also found that the sizes of the binaries components for objects smaller than 400 km are not significantly different from each other.

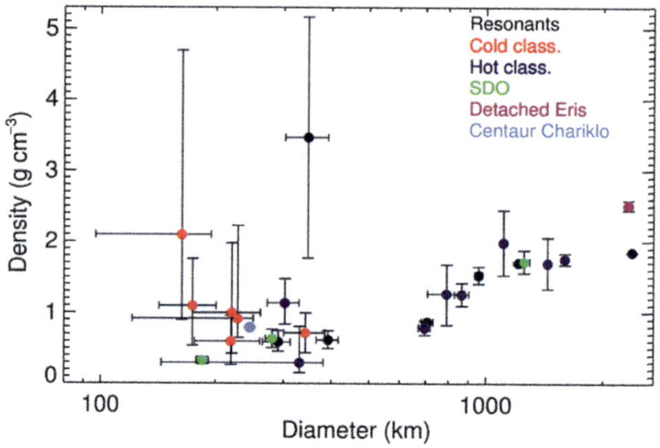

FIG. 7.2 Bulk density versus the effective diameter for the observed TNOs. *Red symbols*: Cold classicals; *blue symbols*: Hot classicals; *black symbols*: resonants; *green symbols*: SDO; *magenta symbol*: detached. Density values are from the references given in the text, as well as a description of the outliers. We also add Pluto and Charon from New Horizons mission results.

The general increase of density with object size (Fig. 7.2) may be partly explained by gravitational self-compaction, reducing the macroporosity. However, this process alone seems unlikely to explain the factor-of-five variation of density from small to large bodies, so that large KBOs may truly be rock-richer than small ones. Thus accreting large rock-rich KBOs from smaller, rock-poor bodies is not a viable scenario (Brown, 2013). Dwarf planets may have larger rock-to-ice ratios possibly due to different formation location/times compared with smaller ones. Barr and Schwamb (2016) note that among dwarf planets, those with large moons tend to have smaller densities than those with small moons.[13] They propose that low-velocity collisions between undifferentiated primordial dwarf planets make large planet/moon pairs, in which both bodies retain their original compositions, while higher velocity collisions between differentiated dwarf planets could yield rock-enriched primaries with small ice-rich satellites.

The median albedos increase from ~5% to 6% for Centaurs and SDO to ~8%–10% for hot classicals and Plutinos, to ~14% for cold classicals, and to ≳15% for detached objects (see

[13] However, Fig. 2 in Barr and Schwamb (2016) suggests that the correlation between density and satellite mass ratio is not strong.

Table 7.2). Within the classicals, the cold and hot populations have a clearly distinct albedo cumulative probability distributions. Given the clustering of objects in two groups in the color-albedo diagram (Lacerda et al., 2014), and the hypothesis that colors reflect the formation distance (Brown et al., 2011), this indicates that cold and hot classicals likely formed in different regions of the solar system. Particularly remarkable is the similarity of the albedo distribution of Centaurs and SDO (median $p_V = 5\%–6\%$; Lacerda et al., 2014; Table 7.2). It is consistent with an origin of Centaurs in the scattered disk (Volk and Malhotra, 2008), and may suggest that these transitioning objects do not change surface properties when entering the Giant Planet region, although Jewitt (2009) and Melita and Licandro (2012) have advocated that the color bimodality of Centaurs is caused by differential thermal processing.

7.4.1 Classical population

Classicals—as well as the other dynamical classes—are defined according to the Gladman et al. (2008) dynamical classification. Vilenius et al. (2012, 2014) analyzed 44 classical TNOs, whose sizes range from ~130 to 930 km, from the Herschel and Spitzer thermal data. Classicals are distinguished between the cold and hot populations according to inclination with a limit at 5 degrees.

Vilenius and coworkers found that the cold and hot classical populations are distinct in terms of size and albedo. First, they have different averaged albedos, as previously noticed (Grundy et al., 2005; Brucker et al., 2009): The cold classicals have a median albedo of 14% (mean: 13±5% on a sample of 17 objects), higher than that of the hot population with 8% (mean 10 ± 7% from a sample of 26 objects), excluding the Haumea family members. Second, they have different (debiased) size distributions: cold classicals are smaller than 400 km, while the hot ones have a wider size distribution, and diameters up to ~900 km (excluding Makemake). The cumulative size distribution is also clearly different, having a much steeper slope ($q = 5.1 \pm 1.1$) for cold than for hot ($q = 2.3 \pm 0.1$) classicals, evaluated for $160 < D < 280$ km and $100 < D < 500$ km, respectively. Similarly, Fraser et al. (2014) found that dynamically quiescent ("cold") KBOs have a steeper bright-end slope than excited ("hot") KBOs, although the distinction vanishes in the faint end of the distribution.

The fact that the cold and hot populations are distinct in terms of spectral slope (Tegler et al., 2000; Doressoundiram et al., 2008), number of binary systems (Noll et al., 2008; Chapter 9), size and albedo (Vilenius et al., 2014), strengths the hypothesis that these two subpopulations had different dynamical histories and therefore formed in different regions of the planetary disk. These results are consistent with the Nice model (Batygin et al., 2011) indicating that hot classicals formed closer to the Sun compared to their present location and migrated during the early solar system evolution, while the cold population formed in situ.

7.4.2 Resonant

Thirty-seven resonant TNOs were investigated in the thermal wavelengths. Most of them (25) are Plutinos populating the 3:2 resonance with Neptune (Table 7.1). The Plutinos sizes range from about 85 to 730 km for the largest body investigated (2003 AZ_{84}), except the dwarf planet Pluto. Their albedos range from 4% to 28%, as reported in Mommert et al. (2012)

and Lellouch et al. (2013), with a median (mean) value $p_V = 9\%(11 \pm 6\%)$ comparable to that of Centaurs, Jupiter family comets, and other TNOs, excepting the detached and cold classicals. Mommert et al. (2012) found that the Plutino size distribution is reproduced using a cumulative power law with $q = 2$ at sizes ranging from 120 to 400 km and $q = 3$ at larger sizes. They did not find any correlation between the different physical parameters of the Plutinos, except a clear anticorrelation between eccentricity and diameter, likely caused by a discovery bias. Among the 18 Plutinos investigated in their study, 6 have evidence of water ice and show an higher albedo (>11%) than the average Plutino value.

The investigated outer resonants are brighter ($p_V = 16\%$ median and $15 \pm 7\%$ mean values) and bigger than Plutinos, very likely because of the discovery and selection biases. Their size ranges from ~150 to 1200 km for the largest body investigated, 2007 OR_{10} (Pál et al., 2016; Kiss et al., 2019), located in the 3:10 mean motion resonance with Neptune.

7.4.3 Detached/SDO population

The thermal properties were determined for 20 SDOs and 8 detached Trans-Neptunians (Table 7.1). Santos-Sanz et al. (2012) analyzed 15 SDO/detached object. They report that SDO are smaller than 400 km (excluding 2007 OR_{10}, which is also classified as outer resonant), while detached objects are larger than 250 km, the larger size of the measured detached bodies being again a discovery/selection bias. Excluding the dwarf planet Eris, the two populations have different median (mean) albedos: 6% (8±5%) for the SDOs, and 15% (20±13%) for the detached bodies. Santos-Sanz et al. (2012) and Stansberry et al. (2008) proposed that larger objects have higher albedo because they may more easily retain volatiles. A possible positive correlation between albedo, size, and perihelion distance was interpreted as a consequence of increased volatile sublimation and/or space weathering effects at low heliocentric distances, both of which leading to surface darkening.

7.4.4 Centaurs

Size, albedo, and thermal properties were derived for 55 Centaurs combining the Spitzer, Herschel (and in some cases WISE) observations (Bauer et al., 2013; Duffard et al., 2014; Tegler et al., 2016; Romanishin and Tegler, 2018). Centaurs have a median (mean) albedo of 6% (7 ± 4%). Most of the Centaurs (92%) are smaller than ~120 km in diameter, and they are thought to be fragments from collisions of larger parent bodies (Pan and Sari, 2005). Regarding sizes, Duffard et al. (2014) claim a lack of objects with sizes between 120 and 190 km, while the largest two, 2002 GZ_{32} and Chariklo, reach $D \sim 240$ km. They further report a size dichotomy between red Centaurs, which are all small (<120 km) and gray ones, which can reach ~240 km.

7.4.5 Haumea family

Haumea is in a weak 7:12 resonance with Neptune (Ragozzine and Brown, 2007) and it is hypothesized to be the parent body of a TNO collisional family (Brown et al., 2007). Vilenius et al. (2018) found that the thermally detected Haumea family members have high albedos in the range ~0.3–0.8 (median $p_V = 0.48^{+0.28}_{-0.18}$), indicative for the presence of water ice. Deep water ice signatures are actually detected spectroscopically on Haumea, one of its moons (Hi'iaka) and several of the family members (see Brown et al., 2012). Also the cumulative size distribution for sizes in the range 175–300 km is steeper ($q = 3.2^{+0.7}_{-0.4}$) than for dynamical interlopers[14] with $D < 500$ km.

7.4.6 Colors/albedo correlations

A comprehensive study of the color-albedo distribution for 109 TNOs and Centaurs with available visible spectral slopes or colors and albedo values (see Table 7.1) was presented by Lacerda et al. (2014). They found that excluding the dwarf planet and the Haumea family members, the Trans-Neptunian population is globally split in two clusters: the dark neutral and the bright red objects. The dark neutral group includes spectrally gray objects, having an average spectral slope $S \sim 10\%/100$ nm and values $<18\%/100$ nm, and low albedo values, clustered at $p_V \sim 5\%$. The red and bright objects have $18 < S < 58\%/100$ nm, and albedos $>6\%$, with a higher median albedo value (around 15%). The dwarf planets and Haumea family members represent a third group of objects, having high albedo and small spectral slope values resulting from the presence of surface volatiles or exposed water ice.

The detailed analysis of spectral slope versus albedo for the different dynamical populations shows that Centaurs, Plutinos, hot classicals, and SDOs have objects in both groups, while cold classicals, detached, and outer resonants have only bright red TNOs (Lacerda et al., 2014). These results agree with conclusions from previous studies (that lacked the albedo information), finding that dynamically excited TNO populations are composed of two main types of surfaces (Fraser and Brown, 2012; Bauer et al., 2013). This agrees with the scenario (Malhotra, 1995; Gladman et al., 2002; Batygin et al., 2011) that cold classicals, detached, and outer resonants formed in distant regions from the Sun and that their bright red surfaces may be related to heliocentric-distance-dependent fractionation of surface volatiles with different sublimation temperatures (Brown et al., 2011; Fraser and Brown, 2012; Wong and Brown, 2016), with methanol and/or hydrogen sulfide playing a key role.[15] Conversely, hot classicals, Plutinos, Centaurs, and SDOs formed closer to the Sun and in a wider range of heliocentric distances, from 20 AU to about 48 AU (Petit et al., 2011), resulting in a wider range of physical properties, both in term of surface colors and albedo.

The color-albedo splitting in two groups is particularly prominent for Centaurs (Bauer et al., 2013; Duffard et al., 2014; Tegler et al., 2016; Romanishin and Tegler, 2018) with median albedos of ~5% for the dark-neutral group and ~8.4% for the bright red, with the red group also having lower mean inclinations. An early explanation of the Centaur color bimodality in terms of collisional resurfacing has been invalidated by Thébault and Doressoudiram (2003). Thermal processing has also been invoked (Melita and Licandro, 2012) whereby red objects tend to spend shorter amount of times at small heliocentric distances; but in this case, Pholus would

14 Interlopers (as defined by Ragozzine and Brown, 2007) belong to the same dynamical cluster as family members but lack the H_2O spectral features.

15 We note that CH_3OH has been detected on a few objects, but H_2S has not so far (see Chapter 5).

be a striking counter example. Romanishin and Tegler (2018) find that gray Centaurs have albedos not significantly different from those of the Trojans, consistent with a common origin.

7.5 Thermal and emissivity properties

7.5.1 Thermal inertia

In addition to size, shape, and albedo, physical properties accessible to thermal measurements include thermal inertia, surface roughness, and surface emissivity. Thermal inertia Γ is related to thermal conductivity κ through a relation that also involves the material density ρ and heat capacity C ($\Gamma = \sqrt{\kappa \rho C}$). The latter two parameters are usually unknown but as pointed out by Delbo' et al. (2015), their plausible range of variation is much smaller than that for κ—which can vary by order of magnitudes between fine grain regolith and compact rock/ices. This justifies that Γ, the most directly accessible variable in a TPM, be used as a proxy for κ, itself an indicator on the processes contributing to thermal conduction (intragrain, intergrain, or radiation-assisted heat transfer, e.g., Gundlach and Blum, 2013; Ferrari and Lucas, 2016).

7.5.1.1 *Ensemble properties*

The unambiguous determination of an object's thermal inertia requires good-quality spin and shape information (e.g., Hanuš et al., 2018). Rotation periods have been measured for ~150 objects, but less than 10% of them have shapes and pole direction available. Therefore, the simple NEATM model still remains the default approach, when dealing with "large" samples. Although the typical NEATM accuracy is 15% on diameter (Wolters and Green, 2009), NEATM-based biases for asteroids occur mostly at large-phase angle, a problem that is not relevant for TNOs/Centaurs. The NEATM-derived diameters usually agree with equivalent diameters from stellar occultation values to within much better accuracy (see Chapter 19). The beaming factor is then a proxy for the combined effect of thermal inertias, spin state, and surface roughness. In rare cases where the thermal inertia is estimated, it characterizes the response of the object to the diurnal cycle and pertains to a layer comparable to the diurnal skin depth.

Lellouch et al. (2013) compiled (or rederived) 85 beaming factors (η) values for TNOs/Centaurs to assess ensemble thermal inertia properties for the population. They found that: (i) Beaming factors range from values <1 to ~2.5 (close to the maximum value expected in the limit of high thermal inertia for an equator-on object); (ii) η values >2 are lacking at small (<30 AU) heliocentric distances; (iii) beaming factors lower than 1 occur frequently (~40% of the objects), implying that surface roughness effects are important. Based on a statistical description of the bodies' spin properties and surface roughness, they determined that all these trends could be explained by a mean thermal inertia of $\Gamma = (2.5 \pm 0.5)\,\mathrm{J\,m^{-2}\,s^{-0.5}\,K^{-1}}$ (SI), with evidence for a decrease of Γ with increasing heliocentric distance r_h, corresponding to a power law $\Gamma \sim r_h^{-(1.0-1.7)}$. Another finding was that high-albedo objects have preferentially lower thermal inertias.

Such thermal inertias are considerably smaller than for other solar system icy bodies (e.g., Γ = 50–70 for Jupiter's satellites, 5–20 SI for Saturn's (Howett et al., 2010 and references therein), Γ = 10–25 for Pluto/Charon). They are also two to three orders of magnitude smaller than values for compact ices, typically 2000–2600 SI (resp. 700–160 SI) for crystalline (resp. amorphous) ice at 150–30 K. Surface porosity and granularity strongly reduce the thermal conductivity, and are likely to be the cause of the measured low thermal inertias. Regarding the temperature dependence of the thermal inertia, Lellouch et al. (2013) noted that if thermal conduction is dominantly radiation-assisted (i.e., $\kappa \propto T^3$), Γ is expected to be nearly proportional to T^2, that is, r_h^{-1}. Based on a model of grain contact and radiative conduction, Ferrari and Lucas (2016) could quantitatively explain the low Γ values and their heliocentric distance if amorphous ice is present at centimeter depths below a thin layer of crystalline ice. The smaller value of Γ for "the mean TNO" compared, for example, to Pluto/Charon does not necessarily imply different regolith properties. Rather it may be related to different diurnal skin depths, which for equal density/thermal properties, is typically 4–5 times shallower on a 8-hour rotating TNO than on the 6.39-day period Pluto/Charon. This interpretation implies that the thermal inertia increases with depth, which is confirmed by recent estimates of the seasonal thermal inertia (Bertrand and Forget, 2016).

7.5.1.2 Pluto/Charon and other prominent TNOs

Thermal inertias of Pluto (Γ = 16–26 SI) and Charon (Γ = 9–14 SI) have been estimated based on unresolved thermal light curves of the system, observed by ISO, Spitzer, and Herschel (Lellouch et al., 2000a, 2011, 2016), covering altogether 20–500 μm. The pre-New Horizons Pluto model was described with three surface units (N_2 ice, CH_4 ice, and H_2O-tholin) and the determined thermal inertia globally applies to the non-N_2 ice (i.e., the nonisothermal) regions. With a lower thermal inertia (and a lower Bond albedo, e.g., Charon's dayside surface temperature (estimated to 55.4 ± 2.6 K by Lellouch et al., 2011) must be warmer than Pluto's, a result consistent with the spatially-separated measurements of the cm/mm/submm brightness temperature ($T_B = 31-33$ K for Pluto vs. 42–45 K for Charon; Butler et al., 2015, 2019), although these values are not direct surface temperatures, being depressed by long-wavelength spectral emissivity effects. The seemingly higher thermal inertia for Pluto versus Charon may result from a significant contribution of atmospheric conduction within a porous upper surface (Spencer et al., 1992; Lellouch et al., 2000b).

At Pluto, the seasonal skin depth is 120 times larger than the diurnal skin depth for a given set of thermal properties. Determining the seasonal thermal inertia requires measurements over seasonal timescales, or at least a temperature value in the polar night (Leyrat et al., 2016). But an object's subsurface seasonal inertia also controls volatile cycles and the distribution of ices at a point in time (e.g., Young, 2013). Bertrand and Forget (2016) found that Pluto's pressure cycle, as well as the accumulation of N_2 and CO ices in the low-latitude Sputnik Planitia basin, can be explained by a seasonal thermal inertia $\Gamma_{seasonal}$ = (500–1500) SI, with 800 SI as the preferred value, corresponding to an annual skin depth of ~40 m.

Although Charon lacks an atmosphere, a somewhat similar reasoning may be used to constrain its seasonal thermal inertia. Grundy et al. (2016) interpret the red coloration of Charon's poles as due to winter cold-trapping of methane gas escaping from Pluto's atmosphere and captured by Charon, followed by photolytic processing of the material into more complex and less volatile molecules. For methane to be cold-trapped, Charon's winter poles temperatures must fall below an estimated 25 K. This condition is met for a seasonal thermal inertia $\Gamma_{seasonal} = 2.5\text{–}40$ SI (i.e., comparable to the diurnal value), but conversely it implies some upper limit to $\Gamma_{seasonal}$, and it is likely that a value as high as the Pluto one would violate the constraint. These results (i) confirm the vertical variation of Pluto's thermal inertia with depth, (ii) suggest that Charon's seasonal thermal inertia is smaller than Pluto's, as is the case for the diurnal Γ.

For more details on the Pluto-Charon system and the interpretation of New Horizons measurements see also Chapters 12 and 13.

Dwarf planet (136138) Haumea is the only other TNO with a definitely detected thermal light curve (Lellouch et al., 2010; Lockwood et al., 2014; Santos-Sanz et al., 2017; Müller et al., 2018). Shape information is also available from high-quality light curve observations (e.g., Lacerda et al., 2008), and most recently from stellar occultation, which also provides the pole orientation from the detection of a ring system (Ortiz et al., 2017). Adopting a shape model from Lacerda et al. (2008), Santos-Sanz et al. (2017) inferred an extremely small thermal inertia ($\Gamma < 0.5$ SI), essentially constrained by the lack of temporal shift between the optical and thermal light curves, and modeling the light curve amplitude then required a phase integral $q > 0.73$. After the occultation results, Müller et al. (2018) revisited the problem by (i) rereducing the Herschel data; (ii) using the occultation-light curve derived 3D size-spin-shape solution; and (iii) estimating and correcting for the contribution of satellites and rings to the thermal flux. They reported a thermal inertia in the range of 2–15 SI, with $\Gamma = 5$ SI as the most likely value. The higher value compared to the previous estimate results from the combination of (i) a strongly subdued light curve amplitude in the latest version of data reduction; (ii) the larger effective diameter and lower albedo yielded by the occultation results; and (iii) the neglect of satellite/rings fluxes in the former study. In spite of the large difference, the new value for Γ is in light with the gross picture of low thermal inertia for the TNO population. Further reanalysis for Haumea might still be warranted in the light of the newly available value for q (0.45 from Verbiscer et al., 2018).

Thermal inertia results for other individual TNOs, derived from TPMs exploring parameter space (diameter, thermal inertia, roughness, and spin state), are scarce. They are summarized in Table 7.3.

7.5.2 Emissivity

A factor impacting surface/subsurface temperatures calculated either by NEATM or TPM is the so-called bolometric emissivity (ϵ_b). ϵ_b is defined as the Planck-function weighted average of the spectral emissivity $\epsilon(\lambda)$, where the latter is related to the spectral directional hemispheric reflectivity.[16] In addition, for a given surface/subsurface temperature profile, the radiation locally emitted at some wavelength depends on $\epsilon(\lambda)$. Many studies on asteroids have shown evidence for subdued fluxes at long wavelengths (submm/mm) compared to model expectations. This behavior is usually termed "emissivity effect," but quantitative estimates

[16] In asteroid science, ϵ_b is routinely taken as 0.9 on the basis that it is a typical value for silicates, given their MIR albedo. As extensively discussed by Myhrvold (2018), this assumption and other approximations of the NEATM are questionable, but we do not dwell into this here.

TABLE 7.3 Thermal inertia values for individual TNOs.

Object	Thermal inertia (SI)
Chiron	3–10[a], 0–10[b]
Chariko	3–30[a], 1.5–10[b]
Quaoar	2–10[a]
Orcus	0.4–2[a]
Bienor	6–20[b]
2007 UK_{126}	0.7–10[c]
2013 AZ_{60}	$\gtrsim$50[d]
2007 OR_{10}	1–10[e]
Haumea	0–0.5[f], 2–15[g]
Pluto	16–26[h]
Charon	9–14[h]
Galilean satellites	50–70[i]
Saturnian satellites	5–20[i]

[a] *Fornasier et al. (2013).*
[b] *Lellouch et al. (2017).*
[c] *Schindler et al. (2017).*
[d] *Pál et al. (2015).*
[e] *Kiss et al. (2019).*
[f] *Santos-Sanz et al. (2017).*
[g] *Müller et al. (2018).*
[h] *Lellouch et al. (2011, 2016).*
[i] *Howett et al. (2010).*

of this "emissivity" depend on how much physics is put in the reference model. For example, the increasing transparency of ices toward long wavelengths (see e.g., Mishima et al., 1983 for H_2O ice) implies that radiation progressively originates from the subsurface. As dayside temperatures decrease with depth, this could lead one to misinterpret an apparent decline of the T_B with λ in terms of a lower "physical" emissivity. The mixing of horizontally variable surface temperatures has the same effect of decreasing T_B with λ (until the Rayleigh-Jeans limit is hit). Nonetheless, spectrally declining emissivities may also have physical causes. These include (i) reflection of upward-propagating thermal radiation at the surface (often described by Fresnel reflection for a dielectric surface); (ii) particle scattering, which produces an emissivity minimum for particle sizes a comparable to $\lambda/4\pi$; (iii) volume scattering, a process in which subsurface inhomogeneities or voids on scales comparable or larger than the wavelength in weakly absorbing medium cause multiple internal reflections; and (iv) thermal inertia increase with depth over the first approximate centimeters into the surface, as observed for the Moon regolith (Keihm, 1984), Mercury (Mitchell and de Pater, 1994), or asteroid 21 Lutetia (Gulkis et al., 2012).

These effects are still poorly characterized for TNOs. For Pluto/Charon, system-averaged T_B decline from ~53 K at 20 μm to ~35 K at 500 μm (Lellouch et al., 2016) and level out at this value across the mm/cm range (Butler et al., 2015, 2019, submitted to AJ). Lellouch et al. (2016) showed that 35 K is lower than any expected temperature for the dayside surface/subsurface of Pluto and Charon, and inferred a "true" spectral emissivity decreasing steadily from 1 at 20–25 μm to ~0.7 at 500 μm. This kind of behavior is usually not observed in asteroids (when proper allowance is made for subsurface sounding, see Keihm et al., 2013), but is found in several outer solar system icy surfaces (e.g., Muhleman et al., 1991; Ostro et al., 2006; Janssen et al., 2009; Le Gall et al., 2014; Ries et al., 2015) and in various kinds of ice and snow on Earth (Hewison and English, 1999). Lellouch et al. (2016) concluded that the combination of a high dielectric constant ($\epsilon_r = 3-5$) and a considerable surface material transparency (typical penetration depth ~1 cm at 500 μm) was responsible for the effect.

Similar studies, albeit less detailed, have been performed on a dozen other TNO or Centaurs. This includes (i) Herschel/SPIRE 250–500 μm measurements of nine objects strongly detected with Herschel/PACS (Fornasier et al., 2013); (ii) ALMA observations of four objects at 0.87 and 1.30 mm (Brown and Butler, 2017) and of six other at 1.30 mm only (Lellouch et al., 2017), complemented by ancient and less accurate but useful data from IRAM and JCMT. The SPIRE measurements generally did not detect the objects at 500 μm (and sometimes not at 350 or even 250 μm), implying an emissivity decline at wavelengths >200 μm, but the derived emissivity curves are very uncertain. The much more accurate ALMA data show long-wavelength emissivity effects for all objects except Makemake, and lead to a mean *relative* radio emissivity ϵ_r (i.e., rationed to the bolometric emissivity) of $\epsilon_r = 0.70 \pm 0.13$. Lellouch et al. (2017) demonstrated the importance of including proper shape models and accounting for variable spin orientation to infer these emissivities. No correlation was found between the radio emissivity and other (semi)physical parameters such as diameter, color, composition, beaming factor, albedo, subsolar temperature, although a possible trend of increasing emissivity with grain size was suggested. A surprising result from Brown and Butler (2017) is sometimes marked (but not in consistent directions) spectral variation of emissivity for some objects over 0.87–1.3 mm. This is unexpected because in the framework of scattering effects, the emissivity minimum at $\lambda \sim 4\pi a$ is expected to be broad due to particle size distribution.

7.6 Outlook

The Spitzer, WISE, and Herschel missions provided a wealth of thermal data which was the key for the determination of thermophysical (sample-)properties of Centaurs and TNOs. But with the end of the Herschel mission in 2013 it is not possible right now to detect these objects close to their thermal emission peak in the FIR regime. Individual objects can in principle be measured at MIR (e.g., with SOFIA, possibly also from ground), but only the brightest ones are detectable and with very little scientific gain over the existing observations from Spitzer and Herschel. In the submm/mm wavelength range, the situation is more comfortable (e.g., with ALMA, IRAM), but measurements of a significant sample remain very time consuming. In addition, the unknowns about the objects' emissivity make the interpretation more challenging. The next big step in thermal emission observations of TNOs is expected to come with JWST.[17] Norwood et al. (2016) stated that MIRI[18] will be capable to characterize the thermal emission of TNOs in the 18–25 μm filters. Parker et al. (2016a) expect that the MIRI measurements will resolve ambiguities in the TPM studies and constrain thermal inertia and roughness of TNO surfaces. JWST will also see the short-wavelengths thermal emission excessively related to hotter terrains (low-albedo regions) or originating from dark satellites. In combination with existing Spitzer and Herschel measurements, the JWST thermal emission measurements will lead to a significant refinement of the radiometric size-albedo solutions and possibly provide evidence for seasonal volatile transport (e.g., on Sedna). A major contribution to the thermal studies of Centaurs and TNOs is expected to come with the Space Infrared Telescope for Cosmology and Astrophysics (SPICA). It will cover the FIR regime, but with much better sensitivity compared to Spitzer or Herschel. However, the launch of this JAXA/ESA-proposed mission would not be before 2032.

Acknowledgments

The research leading to these results has received funding from the European Union's Horizon 2020 Research and Innovation Program, under Grant Agreement No. 687378.

References

Altenhoff, W.J., Stumpff, P., 1995. Size estimate of "asteroid" 2060 Chiron from 250 GHz measurements. Astron. Astrophys. 293, L41–L42.

Altenhoff, W.J., Menten, K.M., Bertoldi, F., 2001. Size determination of the Centaur Chariklo from millimeter-wavelength bolometer observations. Astron. Astrophys. 366, L9–L12.

Altenhoff, W.J., Bertoldi, F., Menten, K.M., 2004. Size estimates of some optically bright KBOs. Astron. Astrophys. 415, 771–775.

Aumann, H.H., Walker, R.G., 1987. IRAS observations of the Pluto-Charon system. Astron. J. 94, 1088–1091.

Barr, A.C., Schwamb, M.E., 2016. Interpreting the densities of the Kuiper belt's dwarf planets. MNRAS 460, 1542–1548.

Barucci, M.A., Merlin, F., Perna, D., Alvarez-Candal, A., Müller, T., Mommert, M., et al., 2012. The extra red Plutino (55638) 2002 VE95. Astron. Astrophys. 539, A152.

Batygin, K., Brown, M.E., Fraser, W.C., 2011. Retention of a primordial cold classical Kuiper belt in an instability-driven model of solar system formation. Astrophys. J. 738, 13.

Bauer, J.M., Grav, T., Blauvelt, E., Mainzer, A.K., Masiero, J.R., Stevenson, R., et al., 2013. Centaurs and scattered disk objects in the thermal infrared: analysis of WISE/NEOWISE observations. Astrophys. J. 773, 22.

Bertoldi, F., Altenhoff, W., Weiss, A., Menten, K.M., Thum, C., 2006. The Trans-Neptunian object UB313 is larger than Pluto. Nature 439, 563–564.

Bertrand, T., Forget, F., 2016. Observed glacier and volatile distribution on Pluto from atmosphere-topography processes. Nature 540, 86–89.

Bowell, E., Hapke, B., Domingue, D., Lumme, K., Peltoniemi, J., Harris, A.W., 1989. Application of photometric models to asteroids. In: Asteroids II; Proceedings of the Conference, Tucson, AZ, March 8–11, 1988. University of Arizona Press, pp. 524–556.

Braga-Ribas, F., Sicardy, B., Ortiz, J.L., Snodgrass, C., Roques, F., Vieira-Martins, R., et al., 2014. A ring system detected around the Centaur (10199) Chariklo. Nature 508, 72–75.

[17] See James Webb Space Telescope: https://jwst.stsci.edu.

[18] See MIR Instrument: https://jwst.stsci.edu/instrumentation/miri

Brown, R.H., 1985. Ellipsoidal geometry in asteroid thermal models: the standard radiometric model. Icarus 64, 53–63.

Brown, M.E., 2013. The density of mid-sized Kuiper belt object 2002 UX25 and the formation of the dwarf planets. Astrophys. J. Lett. 778, L34.

Brown, M.E., Butler, B.J., 2017. The density of mid-sized Kuiper belt objects from ALMA thermal observations. Astron. J. 154, 19.

Brown, M.E., Butler, B.J., 2018. Medium-sized satellites of large Kuiper belt objects. Astron. J. 156 (4), 164.

Brown, M.E., Trujillo, C., Rabinowitz, D., 2004. Discovery of a candidate inner Oort cloud planetoid. Astrophys. J. 617, 645–649.

Brown, M.E., Barkume, K.M., Ragozzine, D., Schaller, E.L., 2007. A collisional family of icy objects in the Kuiper belt. Nature 446, 294–296.

Brown, M.E., Schaller, E.L., Fraser, W.C., 2011. A hypothesis for the color diversity of the Kuiper belt. Astrophys. J. 739, L60.

Brown, M.E., Schaller, E.L., Fraser, W.C., 2012. Water ice in the Kuiper belt. Astron. J. 143, 146.

Brucker, M.J., Grundy, W.M., Stansberry, J.A., Spencer, J.R., Sheppard, S.S., Chiang, E.I., Buie, M.W., 2009. High albedos of low inclination classical Kuiper belt objects. Icarus 201, 284–294.

Buratti, B.J., Hicks, M.D., Dalba, P.A., Chu, D., O'Neill, A., Hillier, J.K., et al., 2015. Photometry of Pluto 2008–2014: evidence of ongoing seasonal volatile transport and activity. Astrophys. J. Lett. 804, L6.

Buratti, B.J., Hofgartner, J.D., Hicks, M.D., Weaver, H.A., Stern, S.A., Momary, T., et al., 2017. Global albedos of Pluto and Charon from LORRI New Horizons observations. Icarus 287, 207–217.

Butler, B.J., Gurwell, M., Lellouch, E., Moullet, A., Moreno, R., Bockelee-Morvan, D., et al., 2015. Long wavelength observations of thermal emission from Pluto and Charon with ALMA. In: DPS Meeting No. 47, Id.210.04.

Butler, B.J., Gurwell, M.A., Lellouch, E., Fouchet, T., Lis, D., Moreno, R., et al., 2019. Emission from Pluto and Charon at long wavelengths: observations using ALMA, SMA, and VLA. Astron. J. (submitted).

Campins, H., Telesco, C.M., Osip, D.J., Rieke, G.H., Rieke, M.J., Schulz, B., 1994. The color temperature of (2060) Chiron: a warm and small nucleus. Astron. J. 108, 2318–2322.

Cruikshank, D.P., Stansberry, J.A., Emery, J.P., Fernández, Y.R., Werner, M.W., Trilling, D.E., et al., 2005. The high--Albedo Kuiper belt object (55565) 2002 AW197. Astrophys. J. 624, L53–L56.

Cruikshank, D.P., Stansberry, J.A., Emery, J.P., van Cleve, J., Fernández, Y.R., Werner, M.W., et al., 2006. Solar System observations with Spitzer Space Telescope. In: Armus, L., Reach, W.T. (Eds.), The Spitzer Space Telescope: New Views of the Cosmos, Proceedings ASP Conference Series 357, November 9–12, 2004 in Pasadena, California, USA. Astronomical Society of the Pacific, San Francisco, CA, p. 23.

Davies, J.R., 2000. Physical characteristics of Trans-Neptunian objects and centaurs. In: Fitzsimmons, A., Jewitt, D., West, R.M. (Eds.), Minor Bodies in the Outer Solar System: Proceedings of the ESO Workshop Held at Garching, Germany, November 2–5, 1998. Springer-Verlag, p. 9.

Davies, J.R., Spencer, J., Sykes, M., Tholen, D., Green, S., 1993. (5145) Pholus. IAU Circ., No. 5698, #2 (1993).

Delbo', M., Mueller, M., Emery, J.P., Rozitis, B., Capria, M.T., 2015. Asteroid thermophysical modeling. In: Michel, P., Francesca, E.D., Bottke, W.F. (Eds.), Asteroids IV. University of Arizona Press, Tucson, AZ, pp. 107–128.

Dias-Oliveira, A., Sicardy, B., Ortiz, J.L., Braga-Ribas, F., Leiva, R., Vieira-Martins, R., et al., 2017. Study of the Plutino object (208996) 2003 AZ_{84} from Stellar Occultations: size, shape, and topographic features. Astron. J. 154, 22.

Doressoundiram, A., Boehnhardt, H., Tegler, S.C., Trujillo, C., 2008. Color properties and trends of the transneptunian objects. In: Barucci, M.A., Boehnhardt, H., Cruikshank, D., Morbidelli, A. (Eds.), The Solar System Beyond Neptune. University of Arizona Press, Tucson, AZ, pp. 91–104.

Duffard, R., Pinilla-Alonso, N., Santos-Sanz, P., Vilenius, E., Ortiz, J.L., Müller, T., et al., 2014. "TNOs are Cool": a survey of the transneptunian region: a Herschel-PACS view of 16 Centaurs. Astron. Astrophys. 564, A92.

Fernández, Y.R., Jewitt, D.C., Sheppard, S.S., 2002. Thermal properties of Centaurs Asbolus and Chiron. Astron. J. 123, 1050–1055.

Ferrari, C., Lucas, A., 2016. Low thermal inertias of icy planetary surfaces. Evidence for amorphous ice? Astron. Astrophys. 588, 14.

Fornasier, S., Lellouch, E., Müller, T., Santos-Sanz, P., Panuzzo, P., Kiss, C., et al., 2013. TNOs are Cool program: combined observations of 9 Centaurs and TNOs with PACS and SPIRE instruments onboard the Herschel Space Telescope. Astron. Astrophys. 555, A15.

Fornasier, S., Lazzaro, D., Alvarez-Candal, A., Snodgrass, C., Tozzi, G.P., Carvano, J.M., et al., 2014. The Centaur 10199 Chariklo: investigation into rotational period, absolute magnitude, and cometary activity. Astron. Astrophys. 568, L11.

Fraser, W.C., Brown, M.E., 2012. The Hubble Wide Field Camera 3 Test of surfaces in the outer solar system: the compositional classes of the Kuiper belt. Astron. J. 749, 33.

Fraser, W.C., Brown, M.E., Morbidelli, A., Parker, A., Batygin, K., 2014. The absolute magnitude distribution of Kuiper belt objects. Astrophys. J. 782, 100.

Gerdes, D.W., Sako, M., Hamilton, S., Zhang, K., Khain, T., Becker, J.C., et al., 2017. Discovery and physical characterization of a large scattered disk object at 92 AU. Astrophys. J. Lett. 839, L15.

Gladman, B., Holman, M., Grav, T., Kavelaars, J., Nicholson, P., Aksnes, K., et al., 2002. Evidence for an extended scattered disk. Icarus 157, 269–279.

Gladman, B., Marsden, B.G., VanLaerhoven, C., 2008. Nomenclature in the outer Solar System. In: Barucci, M.A., Boehnhardt, H., Cruikshank, D., Morbidelli, A. (Eds.), The Solar System Beyond Neptune. University of Arizona Press, Tucson, AZ, pp. 43–57.

Groussin, O., Lamy, P., Jorda, L., 2004. Properties of the nuclei of Centaurs Chiron and Chariklo. Astron. Astrophys. 413, 1163–1175.

Grundy, W.M., Noll, K.S., Stephens, D.C., 2005. Diverse albedos of small Trans-Neptunian objects. Icarus 176, 184–191.

Grundy, W.M., Stansberry, J.A., Noll, K.S., Stephens, D.C., Trilling, D.E., Kern, S.D., et al., 2007. The orbit, mass, size, albedo, and density of (65489) Ceto/Phorcys: a tidally-evolved binary Centaur. Icarus 191, 286–297.

Grundy, W.M., Noll, K.S., Virtanen, J., Muinonen, K., Kern, S.D., Stephens, D.C., et al., 2008. (42355) Typhon Echidna: scheduling observations for binary orbit determination. Icarus 197, 260–268.

Grundy, W.M., Cruikshank, D.P., Gladstone, G.R., Howett, C.J.A., Lauer, T.R., Spencer, J.R., et al., 2016. The formation of Charon's red poles from seasonally cold-trapped volatiles. Nature 539, 65–68.

Gulkis, S., Keihm, S., Kamp, L., Lee, S., Hartogh, P., Crovisier, J., et al., 2012. Continuum and spectroscopic observations of asteroid (21) Lutetia at millimeter and submillimeter wavelengths with the MIRO instrument on the Rosetta spacecraft. Planet. Space Sci. 66, 31–42.

Gundlach, B., Blum, J., 2013. A new method to determine the grain size of planetary regolith. Icarus 223, 479–492.

Hanuš, J., Delbo', M., Ďurech, J., Alí-Lagoa, V., 2018. Thermophysical modeling of main-belt asteroids from wise thermal data. Icarus 309, 297–337.

Harris, A.W., 1998. A thermal model for near-Earth asteroids. Icarus 131, 291–301.

Hewison, T.J., English, S.J., 1999. Airborne retrievals of snow and ice surface emissivity at millimeter wavelengths. IEEE Trans. Geosci. Remote Sens. 37, 1871–1879.

Howell, E., Marcialis, R., Cutri, R., Nolan, M., Lebofsky, L., Sykes, M., 1992. 1992 AD, IAU Circ., No. 5449, #2.

Howett, C.J.A., Spencer, J.R., Pearl, J., Segura, M., 2010. Thermal inertia and bolometric Bond albedo values for Mimas, Enceladus, Tethys, Dione, Rhea and Iapetus as derived from Cassini/CIRS measurements. Icarus 206, 573–593.

Janssen, M.A., Lorenz, R.D., West, R., Paganelli, F., Lopes, R.M., Kirk, R.L., et al., 2009. Titan's surface at 2.2-cm wavelength imaged by the Cassini RADAR radiometer: calibration and first results. Icarus 200, 222–239.

Jewitt, D., 1994. Heat from Pluto. Astron. J. 107, 372–378.

Jewitt, D., 2009. The active Centaurs. Astron. J. 137, 4296–4312.

Jewitt, D., Kalas, P., 1998. Thermal observations of Centaur 1997 CU26. Astrophys. J. 499, L103–L106.

Jewitt, D., Luu, J., 1992. Submillimeter continuum observations of 2060 Chiron. Astron. J. 104, 398–404.

Jewitt, D., Aussel, H., Evans, A., 2001. The size and albedo of the Kuiper-belt object (20000) Varuna. Nature 411, 446–447.

Keihm, S.J., 1984. Interpretation of the lunar microwave brightness temperature spectrum—feasibility of orbital heat flow mapping. Icarus 60, 568–589.

Keihm, S., Kamp, L., Gulkis, S., Hofstadter, M., Lee, S., Janssen, M., Choukroun, M., 2013. Reconciling main belt asteroid spectral flux density measurements with a self-consistent thermophysical model. Icarus 226, 1086–1102.

Kiss, C., Szabó, G., Horner, J., Conn, B.C., Müller, T.G., Vilenius, E., Sarneczky, K., et al., 2013. A portrait of the extreme solar system object 2012 DR30. Astron. Astrophys. 555, A3.

Kiss, C., Pál, A., Farkas-Takács, A.I., Szabó, G.M., Szabó, R., Kiss, L.L., Molnár, L., et al., 2016. Nereid from space: rotation, size and shape analysis from K2, Herschel and Spitzer observations. MNRAS 457, 2908–2917.

Kiss, C., Marton, G., Farkas-Takács, A., Stansberry, J., Müller, T., Vinkó, J., et al., 2017. Discovery of a satellite of the large Trans-Neptunian object (225088) 2007 OR_{10}. Astrophys. J. Lett. 838, L1.

Kiss, C., Marton, G., Parker, A.H., Grundy, W., Farkas-Takács, A., Stansberry, J., et al., 2019. The mass and density of the dwarf planet (225088) 2007 OR 10. arXiv e-prints, arXiv:1903.05439.

Kovalenko, I.D., Doressoundiram, A., Lellouch, E., Vilenius, E., Müller, T., Stansberry, J., 2017. "TNOs are Cool": a survey of the Trans-Neptunian region. XIII. Statistical analysis of multiple Trans-Neptunian objects observed with Herschel space observatory. Astron. Astrophys. 608, A19.

Lacerda, P., Jewitt, D., Peixinho, N., 2008. High-precision photometry of extreme KBO 2003 EL61. Astron. J. 135, 1749–1756.

Lacerda, P., Fornasier, S., Lellouch, E., Kiss, C., Vilenius, E., Santos-Sanz, P., et al., 2014. TNOs are Cool: a survey of the transneptunian region. XII. The albedo-color diversity of transneptunian objects. Astrophys. J. Lett. 793, L2.

Lagerros, J.S.V., 1996. Thermal physics of asteroids. I. Effects of shape, heat conduction and beaming. Astron. Astrophys. 310, 1011–1020.

Lagerros, J.S.V., 1997. Thermal physics of asteroids III. Irregular shapes and albedo variegations. Astron. Astrophys. 325, 1226–1236.

Lagerros, J.S.V., 1998. Thermal physics of asteroids IV. Thermal infrared beaming. Astron. Astrophys. 332, 1123–1132.

Le Gall, A., Leyrat, C., Janssen, M.A., Keihm, S., Wye, L.C., West, R., et al., 2014. Iapetus' near surface thermal emission modeled and constrained using Cassini RADAR Radiometer microwave observations. Icarus 241, 221–238.

Lebofsky, L.A., Tholen, D.J., Rieke, G.H., Lebofsky, M.J., 1984. 2060 Chiron: visual and thermal infrared observations. Icarus 60, 532–537.

Leiva, R., Sicardy, B., Camargo, J.I.B., Ortiz, J.L., Desmars, J., Bérard, D., et al., 2017. Size and shape of Chariklo from multi-epoch Stellar occultations. Astron. J. 154, 159.

Lellouch, E., Laureijs, R., Schmitt, B., Quirico, E., de Bergh, C., Crovisier, J., Coustenis, A., 2000a. Pluto's non-isothermal surface. Icarus 147, 220–250.

Lellouch, E., Paubert, G., Moreno, R., Schmitt, B., 2000b. Note: Search for variations in Pluto's millimeter-wave emission. Icarus 147, 580–584.

Lellouch, E., Moreno, R., Ortiz, J.L., Paubert, G., Doressoundiram, A., Peixinho, N., 2002. Coordinated thermal and optical observations of Trans-Neptunian object (20000) Varuna from Sierra Nevada. Astron. Astrophys. 391, 1133–1139.

Lellouch, E., Kiss, C., Santos-Sanz, P., Müller, T.G., Fornasier, S., Groussin, O., et al., 2010. TNOs are Cool: a survey of the Trans-Neptunian region. II. The thermal lightcurve of (136108) Haumea. Astron. Astrophys. 518, L147.

Lellouch, E., Stansberry, J., Emery, E., Grund, W., Cruikshank, D.P., 2011. Thermal properties of Pluto's and Charon's surfaces from Spitzer observations. Icarus 214, 701–716.

Lellouch, E., Santos-Sanz, P., Lacerda, P., Mommert, M., Duffard, R., Ortiz, J.L., et al., 2013. "TNOs are Cool": a survey of the Trans-Neptunian region. IX. Thermal properties of Kuiper belt objects and centaurs from combined Herschel and Spitzer observations. Astron. Astrophys. 557, A60.

Lellouch, E., Santos-Sanz, P., Fornasier, S., Lim, T., Stansberry, J., Vilenius, E., et al., 2016. The long-wavelength thermal emission of the Pluto-Charon system from Herschel observations. Evidence for emissivity effect. Astron. Astrophys. 588, 15.

Lellouch, E., Moreno, R., Müller, T., Fornasier, S., Santos-Sanz, P., Moullet, A., et al., 2017. The thermal emission of centaurs and Trans-Neptunian objects at millimeter wavelengths from ALMA observations. Astron. Astrophys. 608, 21.

Leyrat, C., Lorenz, R.D., Le Gall, A., 2016. Probing Pluto's underworld: ice temperatures from microwave radiometry decoupled from surface conditions. Icarus 268, 50–55.

Lim, L.T., Stansberry, J., Müller, T.G., Mueller, M., Lellouch, E., Kiss, C., et al., 2010. "TNOs are Cool": a survey of the Trans-Neptunian region. III. Thermophysical properties of 90482 Orcus and 136472 Makemake. Astron. Astrophys. 518, L148.

Lockwood, A.C., Brown, M.E., Stansberry, J., 2014. The size and shape of the oblong dwarf planet Haumea. Earth Moon Planet. 111, 127–137.

Malhotra, R., 1995. The origin of Pluto's orbit: implications for the Solar System Beyond Neptune. Astron. J. 110, 420–429.

Margot, J.L., Trujillo, C., Brown, M.E., Bertoldi, F., 2002. The size and albedo of KBO 2002 AW197. In: DPS Meeting No. 34, Id.17.03; BAAS 34, p. 871.

Marton, G., Kiss, C., Balog, Z., Lellouch, E., Verebélyi, E., Klaas, U., 2015. Search for signatures of dust in the Pluto-Charon system using Herschel/PACS observations. Astron. Astrophys. 579, L9.

Melita, M.D., Licandro, J., 2012. Links between the dynamical evolution and the surface color of the Centaurs. Astron. Astrophys. 539, A144.

Mishima, O., Klug, D.D., Whalley, E., 1983. The far-infrared spectrum of ice Ih in the range 8–25 cm^{-1}. Sound waves and difference bands, with application to Saturn's rings. J. Chem. Phys. 78, 6399–6404.

Mitchell, D.L., de Pater, I., 1994. Microwave imaging of Mercury's thermal emission at wavelengths from 0.3 to 20.5 cm. Icarus 110, 2–32.

Mommert, M., 2013. Remnant Planetesimals and Their Collisional Fragments: Physical Characterization From Thermal Infrared Observations (Ph.D. thesis). Freie Universität Berlin.

Mommert, M., Harris, A.W., Kiss, C., Pál, A., Santos-Sanz, P., Stansberry, J., et al., 2012. "TNOs are Cool": a survey of the Trans-Neptunian region. V. Physical characterization of 18 Plutinos using Herschel-PACS observations. Astron. Astrophys. 541, 17.

Moullet, A., Lellouch, E., Moreno, R., Gurwell, M., 2011. Physical studies of Centaurs and Trans-Neptunian objects with the Atacama Large Millimeter Array. Icarus 213, 382–392.

Muhleman, D.O., Berge, G.L., 1991. Observations of Mars, Uranus, Neptune, Io, Europa, Ganymede, and Callisto at a wavelength of 2.66 MM. Icarus 92, 263–272.

Müller, T.G., Lagerros, J.S.V., 1998. Asteroids as far-infrared photometric standards for ISOPHOT. Astron. Astrophys. 338, 340–352.

Müller, T.G., Lagerros, J.S.V., 2002. Asteroids as calibration standards in the thermal infrared for space observatories. Astron. Astrophys. 381, 324–339.

Müller, T.G., Lellouch, E., Böhnhardt, H., Stansberry, J., Barucci, A., Crovisier, J., et al., 2009. TNOs are Cool: a survey of the transneptunian region. Earth Moon Planet. 105, 209–219.

Müller, T.G., Lellouch, E., Stansberry, J., Kiss, C., Santos-Sanz, P., Vilenius, E., et al., 2010. "TNOs are Cool": a survey of the Trans-Neptunian region. I. Results from the Herschel science demonstration phase (SDP). Astron. Astrophys. 518, L146.

Müller, T., Kiss, C., Alí-Lagoa, V., Ortiz, J.L., Lellouch, E., Santos-Sanz, P., et al., 2018. Haumea's thermal emission revisited in the light of the occultation results. arXiv e-prints, arXiv:1811.09476.

Myhrvold, N., 2018. Asteroid thermal modeling in the presence of reflected sunlight. Icarus 303, 91–113.

Nimmo, F., Umurhan, O., Lisse, C.M., Bierson, C.J., Lauer, T.R., Buie, M.W., et al., 2017. Mean radius and shape of Pluto and Charon from New Horizons images. Icarus 287, 12–29.

Noll, K.S., Grundy, W.M., Stephens, D.C., Levison, H.F., Kern, S.D., 2008. Evidence for two populations of classical transneptunian objects: the strong inclination dependence of classical binaries. Icarus 194, 758–768.

Norwood, J., Hammel, H., Milam, S., Stansberry, J., Lunine, J., Chanover, N., et al., 2016. Solar System observations with the James Webb Space Telescope. Publ. Astron. Soc. Pac. 128 (960), 025004.

Ortiz, J.L., Sota, A., Moreno, R., Lellouch, E., Biver, N., Doressoundiram, A., et al., 2004. A study of Trans-Neptunian object 55636 (2002 TX_{300}). Astron. Astrophys. 420, 383–388.

Ortiz, J., Sicardy, B., Braga-Ribas, F., Alvarez-Candal, A., Lellouch, E., Duffard, R., et al., 2012. Albedo and atmospheric constraints of dwarf planet Makemake from a stellar occultation. Nature 491, 566–569.

Ortiz, J.L., Duffard, R., Pinilla-Alonso, N., Alvarez-Candal, A., Santos-Sanz, P., Morales, N., et al., 2015. Possible ring material around centaur (2060) Chiron. Astron. Astrophys. 576, A18.

Ortiz, J.L., Santos-Sanz, P., Sicardy, B., Benedetti-Rossi, G., Bérard, D., Morales, N., et al., 2017. The size, shape, density and ring of the dwarf planet Haumea from a stellar occultation. Nature 550, 219–223.

Ostro, S.J., West, R.D., Janssen, M.A., Lorenz, R.D., Zebker, H.A., Black, G.J., et al., 2006. Cassini RADAR observations of Enceladus, Tethys, Dione, Rhea, Iapetus, Hyperion, and Phoebe. Icarus 183, 479–490.

Pál, A., Kiss, C., Müller, T.G., Santos-Sanz, P., Vilenius, E., Szalai, N., et al., 2012. "TNOs are Cool": a survey of the Trans-Neptunian region. VII. Size and surface characteristics of (90377) Sedna and 2010 EK139. Astron. Astrophys. 541, L6.

Pál, A., Kiss, C., Horner, J., Szakats, R., Vilenius, E., Müller, T., et al., 2015. Physical properties of the extreme Centaur and super-comet candidate 2013 AZ60. Astron. Astrophys. 583, A93.

Pál, A., Kiss, C., Müller, T., Molnar, L., Szabo, R., Szabo, G.M., et al., 2016. Large size and slow rotation of the Trans-Neptunian object (225088) 2007 OR_{10}. Discovered from Herschel and K2 observations. Astron. J. 151, 117.

Pan, M., Sari, R., 2005. Shaping the Kuiper belt size distribution by shattering large but strengthless bodies. Icarus 173, 342–348.

Parker, A., Pinilla-Alonso, N., Santos-Sanz, P., Stansberry, J., Alvarez-Candal, A., Bannister, M., et al., 2016a. Physical characterization of TNOs with the James Webb Space Telescope. Publ. Astron. Soc. Pac. 128 (959), 018010.

Parker, A.H., Buie, M.W., Grundy, W.M., Noll, K.S., 2016b. Discovery of a Makemakean moon. Astrophys. J. Lett. 825, L9.

Pätzold, M., Andert, T., Hahn, M., Asmar, S.W., Barriot, J.P., Bird, M.K., 2016. A homogeneous nucleus for comet 67P/Churyumov-Gerasimenko from its gravity field. Nature 530, 63–65.
Petit, J.M., Kavelaars, J.J., Gladman, B.J., Jones, R.L., Parker, J.W., Van Laerhoven, C., et al., 2011. The Canada-France ecliptic plane survey full data release: the orbital structure of the Kuiper belt. Astron. J. 142, 131.
Preusker, F., Scholten, F., Matz, K.D., Roatsch, T., Hviid, S.F., Mottola, S., et al., 2017. The global meter-level shape model of comet 67P/Churyumov-Gerasimenko. Astron. Astrophys. 607, L1.
Ragozzine, D., Brown, M.E., 2007. Candidate members and age estimate of the family of Kuiper belt object 2003 EL61. Astron. J. 134, 2160–2167.
Ries, P.A., Janssen, M., 2015. A large-scale anomaly in Enceladus' microwave emission. Icarus 257, 88–102.
Romanishin, W., Tegler, S.C., 2018. Albedos of Centaurs, Jovian Trojans, and Hildas. Astron. J. 156, 19.
Santos-Sanz, P., Lellouch, E., Fornasier, S., Kiss, C., Pal, A., Müller, T.G., 2012. TNOs are Cool: a survey of the Trans-Neptunian region. IV. Size/albedo characterization of 15 scattered disk and detached objects observed with Herschel space observatory-PACS. Astron. Astrophys. 541, A92.
Santos-Sanz, P., Lellouch, E., Groussin, O., Lacerda, P., Müller, T.G., Ortiz, J.L., et al., 2017. "TNOs are Cool": a survey of the Trans-Neptunian region. XII. Thermal light curves of Haumea, VS2 and 2003 AZ_{84} with Herschel/PACS, Astron. Astrophys. 604, A95.
Schindler, K., Wolf, J., Bardecker, J., Olsen, A., Müller, T., Kiss, C., et al., 2017. Results from a triple chord stellar occultation and far-infrared photometry of the Trans-Neptunian object (229762) 2007 UK126. Astron. Astrophys. 600, A12.
Sicardy, B., Ortiz, J.L., Assafin, M., Jehin, E., Maury, A., Lellouch, E., et al., 2011. A Pluto-like radius and a high albedo for the dwarf planet Eris from an occultation. Nature 478, 493–496.
Sierks, H., Barbieri, C., Lamy, P.L., Rodrigo, R., Koschny, D., Rickman, H., et al., 2015. On the nucleus structure and activity of comet 67P/Churyumov-Gerasimenko. Science 347, aaa1044.
Spencer, J.R., Moore, J.M., 1992. The influence of thermal inertia on temperatures and frost stability on Triton. Icarus 99, 261–272.
Stansberry, J.A., Cruikshank, D.P., Grundy, W.M., Emery, J.P., Osip, D.J., Fernandez, Y.R., et al., 2004. Far-IR photometry of centaurs and Kuiper belt objects with Spitzer. In: DPS Meeting No. 36, Id.43.01; BAAS, 36, p. 1173.
Stansberry, J.A., Grundy, W.M., Margot, J.L., Cruikshank, D.P., Emery, J.P., Rieke, G.H., et al., 2006. The albedo, size, and density of binary Kuiper belt object (47171) 1999 TC36. Astrophys. J. 643, 556–566.
Stansberry, J., Grundy, W., Brown, M., Cruikshank, D., Spencer, J., Trilling, D., Margot, J.-L., 2008. Physical properties of Kuiper belt and Centaur objects: constraints form the Spitzer Space Telescope. In: Barucci, M.A., Boehnhardt, H., Cruikshank, D., Morbidelli, A. (Eds.), The Solar System Beyond Neptune. University of Arizona Press, Tucson, AZ, pp. 161–179.
Stansberry, J.A., Grundy, W.M., Mueller, M., Benecchi, S.D., Rieke, G.H., Noll, K.S., et al., 2012. Physical properties of Trans-Neptunian binaries (120347) Salacia-Actaea and (42355) Typhon-Echidna. Icarus 219, 676–688.
Stern, S.A., Weintraub, D.A., Festou, M.C., 1993. Evidence for a low surface temperature on Pluto from millimeter-wave thermal emission measurements. Science 261, 1713–1716.
Stern, S.A., Grundy, W.M., McKinnon, W.B., Weaver, H.A., Young, L.A., 2018. The Pluto system after new horizons. Annu. Rev. Astron. Astrophys. 56, 357–392.
Sykes, M.V., 1999. IRAS survey-mode observations of Pluto-Charon. Icarus 142, 155–159.
Sykes, M.V., Walker, R.G., 1991. Constraints on the diameter and albedo of 2060 Chiron. Science 251, 777–780.
Sykes, M.V., Cutri, R.M., Lebofsky, L.A., Binzel, R., 1987. IRAS serendipitous survey observations of Pluto and Charon. Science 237, 1336–1340.
Tedesco, E.F., Veeder, G.J., Dunbar, R.S., Lebofsky, L.A., 1987. IRAS constraints on the sizes of Pluto and Charon. Nature 327, 127–129.
Tegler, S.C., Romanishin, W., 2000. Extremely red Kuiper-belt objects in near-circular orbits beyond 40 AU. Nature 407, 979–981.
Tegler, S.C., Romanishin, W., Consolmagno, G., 2016. Two color populations of Kuiper belt and Centaur objects and the smaller orbital inclinations of red centaur objects. Astron. J. 152, 210.
Thébault, P., Doressoudiram, A., 2003. Colors and collision rates within the Kuiper belt: problems with the collisional resurfacing scenario. Icarus 162, 27–37.
Thomas, N., Eggers, S., Ip, W.H., Lichtenberg, G., Fitzsimmons, A., Jorda, L., et al., 2000. Observations of the Trans-Neptunian objects 1993 SC and 1996 TL66 with the infrared space observatory. Astrophys. J. 534, 446–455.

Verbiscer, A., Porter, S.B., Benecchi, S., Kavelaars, J.J., Weaver, H.A., Spencer, J.R., et al., 2018. Photometric properties of distant KBOs observed by New Horizons LORRI at moderate and high phase angles. In: The Transneptunian Solar System, March 26–29, 2018, Coimbra, Portugal.

Vilenius, E., Kiss, C., Mommert, M., Müller, T., Santos-Sanz, P., Pál, A., et al., 2012. "TNOs are Cool": a survey of the Trans-Neptunian region. VI. Herschel/PACS observations and thermal modelling of 19 classical Kuiper belt objects. Astron. Astrophys. 541, A94.

Vilenius, E., Kiss, C., Müller, T., Mommert, M., Santos-Sanz, P., Pál, A., et al., 2014. "TNOs are Cool": a survey of the Trans-Neptunian region. X. Analysis of classical Kuiper belt objects from Herschel and Spitzer observations. Astron. Astrophys. 564, A35.

Vilenius, E., Stansberry, J., Müller, T., Mueller, M., Kiss, C., Santos-Sanz, P., et al., 2018. "TNOs are Cool": a survey of the Trans-Neptunian region. XIV. Size/albedo characterization of the Haumea family observed with Herschel and Spitzer. Astron. Astrophys. 618, A136.

Volk, K., Malhotra, R., 2008. The scattered disk as the source of the Jupiter family comets. Astrophys. J. 687, 714–725.

Wolters, S.D., Green, S.F., 2009. Investigation of systematic bias in radiometric diameter determination of near-Earth asteroids: the night emission simulated thermal model (NESTM). MNRAS 400, 204–218.

Wong, I., Brown, M.E., 2016. A hypothesis for the color bimodality of Jupiter Trojans. Astron. J. 152, 90.

Young, L.A., 2013. Pluto's seasons: new predictions for New Horizons. Astrophys. J. Lett. 766, L22.

CHAPTER

8

Internal structure and cryovolcanism on Trans-Neptunian objects

Aurélie Guilbert-Lepoutre[a], *Dina Prialnik*[b], *Robin Métayer*[a]

[a]LGL-TPE, UMR 5276, CNRS, Claude Bernard Lyon 1 University, ENS Lyon, Villeurbanne Cedex, France [b]Department of Geosciences, Tel Aviv University, Tel Aviv, Israel

8.1 Introduction

The Trans-Neptunian population includes objects with sizes ranging from a few kilometers to a few 10^3 km (Petit et al., 2008), but most of our knowledge relates to large-size (~2000 km) and mid-size (500–1000 km) objects, which are more easily observable. The bulk of the population, that is, the smaller objects, are thought to have undergone less processing than their larger counterparts, and any processing would be preferentially at their surface. In contrast, large- and mid-size Trans-Neptunian objects (TNOs), like planets, could have been strongly altered by geological activity that modified both their internal structure and their surface.

8.1.1 Cryovolcanism

Similar to volcanism on the Earth, which plays an important role in reshaping the surface of our planet, cryovolcanism may have acted in shaping the largest TNOs as we observe them today. Cryovolcanism is a process by which material from the internal or subsurface layers of an icy object reaches the surface as liquids or vapors, where they would otherwise be frozen (Geissler, 2015). Active cryovolcanism is known to occur on moons of the giant planets, and past activity has been recently evidenced in the Pluto-Charon system by the NASA/New Horizons probe (Stern et al., 2015). Because the liquid material is emplaced from below the surface, cryovolcanism is generally associated with the occurrence of melts inside an object. Hence, evidence for cryovolcanism can be regarded as evidence for conditions favorable to the physical and chemical differentiation of icy bodies.

The Trans-Neptunian Solar System.
https://doi.org/10.1016/B978-0-12-816490-7.00008-4

8.1.2 Observational motivation

With time, several observational features have been proposed as suggestive of past activity, or at least of past surface renewal.

The first was the detection of crystalline water ice at the surface of Charon and Quaoar (Brown and Calvin, 2000; Buie and Grundy, 2000; Dumas et al., 2001; Jewitt and Luu, 2004), later observed at the surface of many other TNOs (Barkume et al., 2008; Guilbert et al., 2009; Brown, 2012). The reason for the presence of large quantities of crystalline water ice at the surface of TNOs remains to be fully understood. Whether water ice is formed amorphous or crystalline, once crystalline water ice is present at the surface of an atmosphere-less object, it is subject to irradiation, which is effective in converting it to amorphous water ice in several million years (Cooper et al., 2003; Mastrapa and Brown, 2006; Cook et al., 2007). Therefore, there are at least two reasons why crystalline water ice may be observed at the surface of TNOs today. First, it may require some replenishment process, including cryovolcanism, bringing fresh material to the surface (e.g., Cook et al., 2007). Second, as indicated by laboratory experiments, an equilibrium between the amorphous and crystalline phases may be reached at temperatures higher than 30 K (Leto and Baratta, 2003; Mastrapa and Brown, 2006; Zheng et al., 2009). Holler et al. (2017a) showed that at a surface temperature typical of TNOs, this equilibrium could be reached in less than 1 Gyr, and the amount of crystalline water ice would increase with temperature.

The second observational evidence for internal processing was the detection of ammonia and ammonia hydrates, first at the surface of Charon (Brown and Calvin, 2000; Buie and Grundy, 2000; Dumas et al., 2001; Cook et al., 2007; Merlin et al., 2010a; Cruikshank et al., 2015; DeMeo et al., 2015; Holler et al., 2017a), then Orcus (Barucci et al., 2008; Delsanti et al., 2010; DeMeo et al., 2010; Carry et al., 2011). Similar to crystalline water ice, ammonia and its hydrates should be destroyed at the surface in about 20 Myr in the radiation environment of TNOs (Cooper et al., 2003; Cook et al., 2007). Contrarily to water ice, for which conversion between amorphous to crystalline phases is possible, ammonia is not undergoing any phase transition under irradiation, but is rather destroyed by the process (Moore et al., 2007). Therefore, in order to be present at the surface, ammonia needs to be replenished from a subsurface reservoir. For Charon, no mechanism other than the recently localized emplacement at the surface by cryovolcanism has been able to explain the observations (Cook et al., 2007; McKinnon et al., 2008; Desch et al., 2009) at the time these were done. However, we now know that though cryovolcanism did shape parts of the surface of Charon, it must have occurred very early in the history of the satellite, as the surface is estimated to be as old as 4 Gyr (Singer et al., 2019). Alternate explanations, which do not require the presence of liquids inside TNOs, have been proposed, such as the diffusion of ammonia through a layer of water ice, resulting in the hydration of ammonia (Cruikshank et al., 2005; Holler et al., 2017a).

In this chapter, we review the current knowledge on the thermal processing of TNOs, in particular, the possible occurrence of liquid water and its consequences on the internal structure and on the rejuvenation of the surface (Section 8.2). We then review the most recent modeling results obtained for Charon, and show how the evidence for its past geological activity obtained by the NASA/New Horizons probe can be used to study TNOs in general (Section 8.3). Finally, we review the observed physical properties that may be used to constrain the actual processing of TNOs and propose ways to improve our understanding of their differentiation state, their internal structure, and the production of liquid water in the outer solar system (Section 8.4). We conclude (Section 8.5) by indicating future lines of research.

8.2 Evolution of large- and mid-size TNOs

Early spectral observations mentioned earlier have rapidly prompted studies of the thermal evolution of TNOs. If the presence of crystalline water ice or ammonia hydrates does not require past cryovolcanism, it nonetheless implies some sort of replenishment from subsurface material. This, in turn, implies the presence of liquid water and differentiation. Ammonia in particular can play a significant role in the differentiation of icy objects since it depresses the melting point of the ice mixture and lowers the thermal conductivity and the viscosity of melts. Eventually, it can allow for a liquid phase to be preserved in TNOs above a rocky core and below a thermally insulating crust of unprocessed material (Desch et al., 2009). The early studies have been reviewed by Coradini et al. (2008), McKinnon et al. (2008), and Prialnik et al. (2008). We thus refer to these reviews for details, and summarize the findings later. Given the uncertainties in TNOs' thermophysical properties, and the unknowns regarding their formation mechanism, time or location, the goal of these works has been to identify key characteristics of the thermal evolution of TNOs, and to propose statistically significant outcomes for the overall population.

8.2.1 Early processing: Accretion and short-lived radiogenic heating

The effect of early radiogenic heating remains largely unknown, since it depends not only on the formation epoch but also on the formation timescale of TNOs, hence on the formation mechanism, which may not be unique to all TNOs, given the range of heliocentric distances involved for forming the entire TNO population. Radiogenic heating during accretion was investigated by Merk and Prialnik (2003, 2006). Their model considered a simple composition of mixed amorphous water ice and dust and allowed for crystallization and melting of ice, though no gas or liquid flow was taken into account. Depending on free initial parameters such as initial composition, size, or heliocentric distance, pristine compositions could be largely retained. However, the authors showed that some TNOs might have undergone significant processing, including crystallization of amorphous water ice, and production of liquid water. Interestingly, the potential outcomes included objects which could hold liquid layers in their innermost parts while retaining a pristine uppermost layer of amorphous water ice.

8.2.2 Chemical differentiation: Sublimation and crystallization

Regardless of the effects of early heating, long-term radiogenic heating should be inevitable. In general, radiogenic heating results in a temperature gradient from the warm interior to the cold surface, in contrast to solar heating, which results in a warm surface relative to the interior.

De Sanctis et al. (2001), Choi et al. (2002), or McKinnon (2002), for example, studied the effect of radiogenic heating, in particular with respect to the survival of volatile species such as CO or CO_2. They showed that TNOs' surface layers remain cold enough to retain volatiles in icy form, although CO—and by analogy, all volatiles sublimating below ~50 K initially present as pure ices—should be lost at the surface regardless of radiogenic heating (De Sanctis et al., 2001; Choi et al., 2002), due to insolation alone.

Radiogenic heating may also trigger crystallization at various depths. In general, large-scale crystallization is expected to occur, especially for the larger objects. Crystallization is accompanied by the migration of released volatiles and their possible recondensation in appropriately cold regions of the nucleus. The outcome depends on many free parameters that are extremely temperature dependent (Prialnik, 2000). Furthermore, these processes have similar characteristic timescales and hence they are expected to create a complex pattern of chemical differentiation, with highly volatile species closer to the surface, and less volatile ones layered toward the interior in order of volatility.

8.2.3 Physical differentiation: Internal ocean formation

Large objects are expected to have undergone more internal processing than their smaller counterparts, simply because they may attain higher internal temperatures due to radiogenic heating. Indeed, central temperatures high enough to reach the water-ammonia, water-methanol, or water-methanol-ammonia eutectics can be obtained in objects larger than 500 km (McKinnon et al., 2008, and references therein). The larger the object, the larger the internal volume reaching the temperature thresholds required for producing a liquid phase. Such a liquid phase should be buoyant with respect to the surrounding solid material made of water

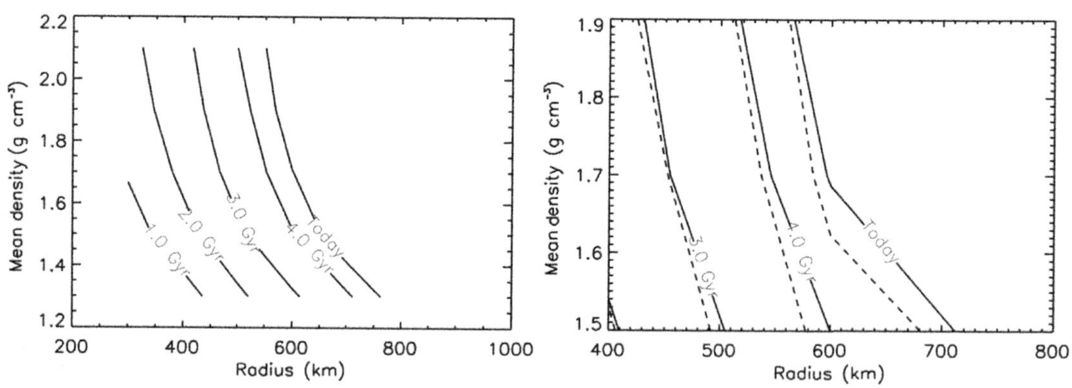

FIG. 8.1 Times after formation at which all liquid freezes within a TNO with initial ammonia content $X = 1\%$ (*left*) and 5% (*right*). TNOs with mean densities and radii that place them to the right of the rightmost curve potentially can retain subsurface liquid to the present day. The *dashed curves* on the right represent models where convection has been shut off. *From Desch, S.J., Cook, J.C., Doggett, T.C., Porter, S.B., 2009. Thermal evolution of Kuiper belt objects, with implications for cryovolcanism. Icarus 202, 694–714. https://doi.org/10.1016/j.icarus.2009.03.009.*

ice and rock, and is expected to percolate upward, possibly making its way up to the surface and erupt (Stevenson, 1982).

Hussmann et al. (2006) studied the formation of internal oceans within TNOs. In general, they found that in order to sustain an internal ocean, the composition should include some sort of antifreeze, like ammonia, methanol, or salts, so as to overcome solid-state convection. The key to the production of internal oceans, according to their modeling results, was the amount of antifreeze. The effect of ammonia on the internal structure of medium-size TNOs (Desch et al., 2009) is shown in Fig. 8.1. For objects larger than ~900–1000 km, it is entirely possible that the water ice melting point is reached even when no ammonia, methanol or other antifreeze is involved, and even without the contribution of early radiogenic heating.

More recently, Malamud and Prialnik (2015) studied the effect of serpentinization on the thermal evolution of TNOs. They applied their model to Charon, Orcus, and Salacia, and showed that the final internal structure should be differentiated, with an inner rocky core, and an outer ice-rich mantle, the degree of differentiation and exact final structure depending on the mass of the object (see Fig. 8.2).

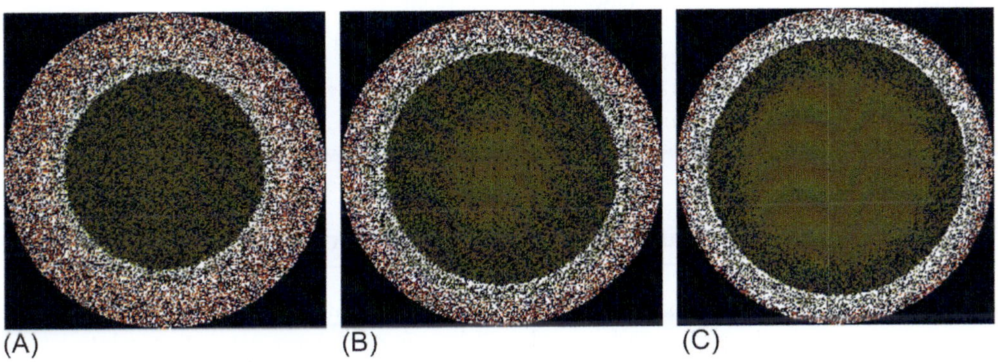

FIG. 8.2 Compositional cross-section of the final differentiated state of three TNO models after 4.6 Gyr of evolution. The color coding is: *black* (pores); *white* (crystalline ice); *pink* (amorphous ice); *brown* (unprocessed rock); and *olive* (processed rock). The initial models differ only in mass, while all other parameters physical properties and composition are identical. The different densities of these objects, as derived from observations, may be uniquely determined by their different masses. (A) Salacia, (B) Orcus, (C) Charon. *From Malamud, U., Prialnik, D., 2015. Modeling Kuiper belt objects Charon, Orcus and Salacia by means of a new equation of state for porous icy bodies. Icarus 246, 21–36. https://doi.org/10.1016/j.icarus.2014.02.027.*

In summary, large- and mid-size TNOs are expected to have a complex internal structure, resulting from chemical stratification and physical differentiation, either or both. Finally, it is entirely possible that the larger objects may have experienced cryovolcanic events, which should be reflected in their surface properties (such as composition and albedo). Table 8.1 summarizes the processes and factors that are bound to affect the internal structure and evolution of TNOs. All of them have been studied by different groups, but no model to date includes all of these factors.

TABLE 8.1 Processes and factors affecting internal and surface structure.

Process/factor	Effect
Crystallization of amorphous ice	Heat and volatile release; thermal conductivity change
Serpentinization	Heat release; density and thermal conductivity change
Dehydration	Heat absorption; density and thermal conductivity change
Sublimation/condensation	Heat absorption/release; porosity change; composition change
Self-gravity	Compaction; settling; differentiation; energetic changes
Accretion	Size change; heat release
Water flow	Differentiation; ocean formation
Volatile flow	Composition alteration; formation of atmosphere
Cryovolcanism	Resurfacing; formation of geological structures
Irradiation by energetic protons	Surface composition changes (amorphization)
Ammonia	Lowering freezing point
Porosity	Lowering material strength and thermal conductivity
Short-lived radionuclides	Intense heat release during early evolution
Long-lived radionuclides	Slow heating during long-term evolution

8.3 Evolution of the Pluto-Charon system

The aforementioned studies have highlighted the fact that any evolutionary track—from entirely pristine to heavily processed—is plausible, given the uncertainties related to TNO formation, and the lack of constraints on many of their thermophysical properties. For the large- and mid-size TNOs, it is thus relevant to consider each object individually, with its own thermal and collisional history. In this regard, the Pluto-Charon system, recently studied in great detail by the NASA/New Horizons probe (see Chapters 13 and 14), is most informative on the processes responsible for shaping TNOs as we observe them today. Pluto, being one of the largest TNOs, was expected to be at least partially differentiated, and some studies even predicted its complete differentiation, accompanied by the presence of liquid today (McKinnon et al., 2008; Robuchon and Nimmo, 2011; Barr and Collins, 2015). Images provided by the NASA/New Horizons probe (Stern et al., 2015) have brought the evidence for such past geological activity. A variety of geological features can be observed, including impact craters, tectonics, signs of cryovolcanism, or surface-atmosphere interactions (Stern et al., 2015; McKinnon et al., 2016; Trowbridge et al., 2016; Trilling, 2016; Buhler and Ingersoll, 2018).

8.3.1 Pre-New Horizons: Thermal history of Charon after formation

Ever since its discovery, Charon has been considered as too small to sustain any cryovolcanic activity, or even to differentiate (Stern, 1989; Durand-Manterola, 2003; Hussmann et al.,

2006; Schubert et al., 2010). Therefore, there was no reason to suspect that any geological activity would be seen on Charon. However, as stated in Section 8.1, near-infrared spectra of Charon's observed separately from Pluto showed the presence of crystalline water and ammonia or ammonia hydrates are clearly identified (Brown and Calvin, 2000; Buie and Grundy, 2000; Dumas et al., 2001; Cook et al., 2007; Merlin et al., 2010a; Cruikshank et al., 2015; DeMeo et al., 2015; Holler et al., 2017a).

A series of dedicated models, with an increasing degree of complexity, has been developed before New Horizons to study Charon's thermal history in link with possible cryovolcanism (Desch et al., 2009; Neveu et al., 2015; Desch, 2015; Malamud and Prialnik, 2015). We note again that alternate explanations (which do not require the presence of liquids inside TNOs) have been proposed to explain the presence of ammonia at its surface, such as the diffusion of ammonia through a layer of water ice, resulting in the hydration of ammonia (Cruikshank et al., 2005; Holler et al., 2017a). New Horizons taught us that though cryovolcanism did shape parts of the surface of Charon, it must have occurred very early in the history of the satellite, as the surface is estimated to be as old as 4 Gyr (Singer et al., 2019). Also, the link between ammonia and cryovolcanic features is not so clear, and could require impacts excavating subsurface material (Grundy et al., 2016; Dalle Ore et al., 2018). Today, Charon does not display any sign of active cryovolcanism (Moore et al., 2016), but its surface clearly shows evidence for past processing (Moore et al., 2016; Beyer et al., 2017). Therefore, now that New Horizons has provided the evidence for surface renewal processes such as tectonics and cryovolcanism, these models are of particular significance for studying other TNOs with intermediate sizes similar to Charon's.

8.3.2 Post-New Horizons: Forming the Pluto-Charon system

The new constraints brought by New Horizons allow for even more detailed models of the internal structure and evolution of Charon to be developed. For example, McKinnon et al. (2017) argued that in order to achieve the density contrast observed between Pluto and Charon without any initial compositional difference, the entire icy shell of Pluto needed to melt. Bierson et al. (2018, see Fig. 8.3) tried to explain it by constraining the evolution of the internal porosity. They showed that porosity variations alone cannot explain the density contrast, and that Pluto needed to have accreted more silicates than Charon. This appears to be consistent with a formation scenario in which Pluto formed quickly enough to lose a significant fraction of its original water ice content. The matter of how the system formed is still not clear though. Stern et al. (2015) suggested that the impactors at the origin of the Pluto-Charon system should have been moderately differentiated or not differentiated at all, a scenario in which Charon tends to form more or less intact from its precursor, rather than from the disk of debris resulting from the collision (Canup, 2005, 2011). However, Desch (2015) and Desch and Neveu (2017) argue that Charon should instead have formed from the collisional debris disk, and that it accreted not only material from the precursor icy mantle but also from the undifferentiated inner parts of the progenitors. Such progenitors are in this framework formed within 5 Myr after CAI formation in order to sustain early radiogenic heating discussed in Section 8.2.

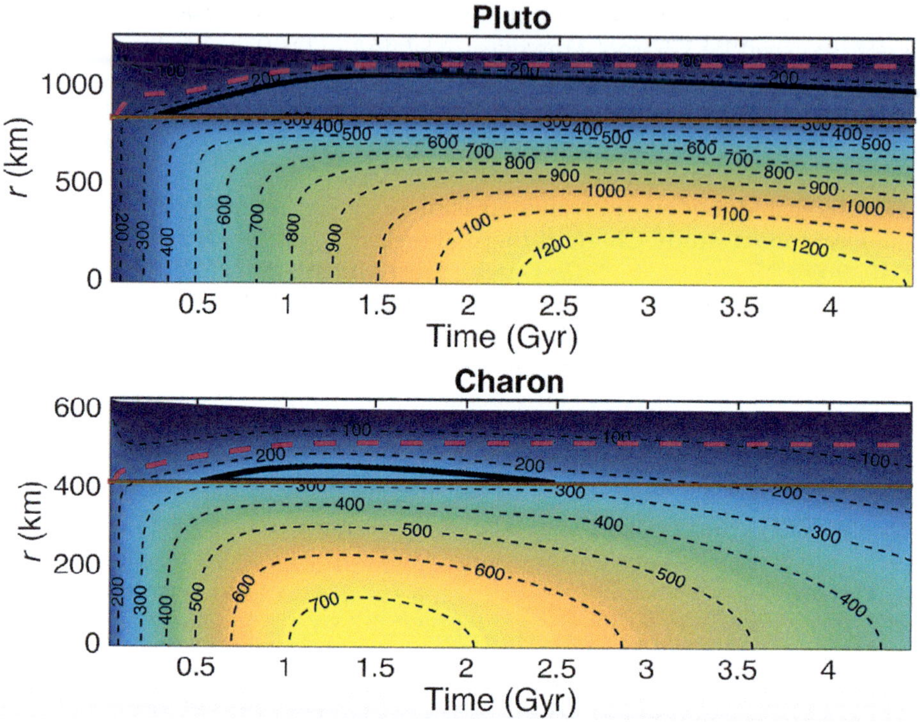

FIG. 8.3 Thermal histories of Pluto and Charon, assuming an initial porosity of 0.3, and a silicate mass fraction of 0.69 and density of 3500 kg m^{-3}. Temperature is contoured at a 100 K interval. The *horizontal brown line* is the top of the silicate core, the *dashed magenta line* contours the bottom of the porous layer, and the *thick black line* is the top of the ocean layer. In this model, Pluto forms a substantial ocean that persists to the present day, while Charon forms an ocean 40 km thick that then refreezes. Charon maintains a larger porous layer than Pluto due to lower temperatures and pressure in the ice mantle. At the end of evolution, Pluto and Charon have radii of 1191 and 605.2 km, respectively. *From Bierson, C.J., Nimmo, F., McKinnon, W.B., 2018. Implications of the observed Pluto-Charon density contrast. Icarus 309, 207–219. https://doi.org/10.1016/j.icarus.2018.03.007.*

8.3.3 Evolution, geological activity, and internal structure of Charon

Desch and Neveu (2017) proposed an even more detailed model in which Pluto and Charon formed from the collision of two partially differentiated precursors. Charon and the other small satellites would form from the collisional debris disk. Then, as described in Section 8.2, heating due to the decay of long-lived radionuclides, assisted by the antifreeze effect of ammonia, allowed the material to reach the temperature threshold for melting the ice, thus creating a subsurface ocean that eventually refroze 1.7–2.5 Gyr ago. The freezing of that ocean should have caused extensional stresses at the origin of Serenity Chasma, and should have led to widespread resurfacing (perhaps explaining the younger age of Vulcan Planum). Subsequently, radiogenic heat would have been able to build up to the point of creating a second ocean which refroze 0.5–1.7 Gyr ago. The freezing enabled cryovolcanism, resulting in the creation of Kubric Mons (Desch and Neveu, 2017).

In addition, Malamud et al. (2017) highlighted the role that serpentinization may play in causing the extensional tectonics detected on Charon, observed by Beyer et al. (2017). In this model, the tectonic belt observed at the surface of Charon would be due to contraction and expansion episodes which are related to ice compaction, differentiation, and chemical reactions in the interior of Charon. The authors propose a complex history, resulting in a stratified internal structure, which also includes the formation and freezing of a subsurface ocean. The strength of their previous model (Malamud and Prialnik, 2015), also applied to Orcus and Salacia, was to explain the density contrasts between those three objects with a similar set of initial properties. While the scenarios described by Desch and Neveu (2017) and Malamud et al. (2017) need to be confirmed and may not be unique in explaining the features observed at the surface of Charon, they deserve attention since the same processes may take place on other TNOs with properties similar to Charon's.

8.4 Constraining cryovolcanism and the internal structure of TNOs

From the breadth of thermophysical properties involved in the numerical calculations of thermal evolution, as described in Sections 8.2 and 8.3, key characteristics can be identified, which should help constrain the actual evolutionary track of each TNO.

8.4.1 Density

Density, more than the mass or the size of an object, is crucial for understanding the formation and evolution of large- and mid-size TNOs (Brown and Butler, 2017). It is one of the most important measurements for determining the first-order structure and composition of any object. Many of the large objects do have satellites (Table 8.2; Brown et al., 2006a), hence the mass of the system can be determined to assess the density of the primary object. Though no correlation could initially be found between the density and the size of a TNO (Brown, 2008), we now know that a trend exists, where larger objects tend to be denser than the smaller ones (McKinnon et al., 2017). This trend may be understood by considering the combined thermal and collisional history of these systems. Indeed, large-scale collisions could have played a role in changing the density of large objects, especially if those were differentiated beforehand (Brown et al., 2007; Brown, 2008, 2013b). Barr and Schwamb (2016) argued that TNOs may fall in two categories: systems with large satellites (like Pluto-Charon and Orcus-Vanth) could be formed from the low-speed collision between undifferentiated (or moderately differentiated) objects, while systems with small satellites (like Eris-Dysnomia) could be formed from the high-speed collision between differentiated objects. Further constraints on the density of TNOs and their satellites are obviously desirable for understanding the formation and early evolution of these worlds.

8.4.2 Shape

The stratified internal structure of TNOs might be reflected in their general shape (due to their rotation), and may thus be constrained without any dedicated space mission or in situ

TABLE 8.2 General characteristics of large- and mid-size TNOs.

Number	Name	Satellite	ϱ (g cm^{-3})	D (km)	Surface
134340	Pluto[a]	Styx Nix Kerberos Hydra	1.854 ± 0.006	2374 ± 8	H_2O, N_2, CH_4, CO, C_2H_6, evidence for complex geology: craters, tectonics, mountains, etc.
		Charon	1.702 ± 0.021	1212 ± 6	H_2O, NH_3, $NH_4{}^+$, CH_4, NH_3:H_2O, evidence for tectonics and surface renewal
136199	Eris	Dysnomia	2.52 ± 0.05	**2326 ± 12**	CH_4, CO, N_2
136472	Makemake	Yes[b]	–	**1430 ± 9** **1434 ± 14**	CH_4 partially diluted in N_2, C_2H_6, C_2H_2; C_2H_4, C_3H_8
136108	Haumea	Hi'iaka Namaka	$2.551^{+0.115}_{-0.010}$	$1922 \times 1536 \times 998$ **$2322 \times 1704 \times 1026$**	H_2O, almost pure
225088	2007 OR_{10}	Yes	–	1142^{+647}_{-467} 1535^{+75}_{-225}	H_2O, CH_4?, CH_3OH?
50000	Quaoar	Weywot	2.13 ± 0.29 **1.99 ± 0.46** **$<1.82 \pm 0.28$**	1070 ± 38 **1110 ± 5** **$>1138 \pm 25$**	H_2O, CH_4, C_2H_6, NH_3:H_2O?, N_2?, CO?
90377	Sedna	?	–	906^{+314}_{-258} 1420^{+110}_{-180}	H_2O, CH_4, C_2H_6?
90482	Orcus	Vanth	2.47 ± 0.89 $1.65^{+0.35}_{-0.24}$ $1.53^{+0.15}_{-0.13}$	917 ± 25 906 ± 72 910^{+50}_{-40}	H_2O, NH_3?, CH_4, NH_3:H_2O?, $NH_4{}^+$?, C_2H_6?
134340	Pluto[a]	Styx Nix Kerberos Hydra	1.854 ± 0.006	2374 ± 8	H_2O, N_2, CH_4, CO, C_2H_6, evidence for complex geology: craters, tectonics, mountains, etc.
120347	Salacia	Actaea	$1.29^{+0.29}_{-0.23}$ 1.26 ± 0.16	829 ± 30 866 ± 37	Inconclusive

TABLE 8.2 General characteristics of large- and mid-size TNOs—cont'd

Number	Name	Satellite	ϱ (g cm^{-3})	*D* (km)	Surface
174567	Varda	Ilmarë	$1.27^{+0.41}_{-0.44}$	705^{+81}_{-75}	Inconclusive
			$1.24^{+0.50}_{-0.35}$	722^{+82}_{-76}	
	2013 FY_{27}	Yes	–	740^{+85}_{-90}	–
				724 ± 64	
208996	2003 AZ_{84}[c]	Yes	*0.87 ± 0.01*	**764 ± 6**	H_2O, CH_3OH?
				692 ± 12	
				670 ± 34	
55637	2002 UX_{25}	Yes	*0.80 ± 0.13*	659 ± 38	Inconclusive
				695^{+30}_{-29}	

[a] *Values for the Pluto/Charon system are from the NASA/New Horizons mission (Stern et al., 2015).*
[b] *Parker et al. (2016) noted that Makemake has a satellite but the tracking of the orbit was too sparse to provide a reliable estimate of the system's mass (currently estimated at* 3.15×10^{21} *kg, leading to a density of 2.1).*
[c] *Plutino 2003 AZ_{84} may have a nonspherical shape as revealed by stellar occultations, which provide the a size of (470 ± 20) × (383 ± 10) × (245 ± 8) km (Dias-Oliveira et al., 2017), equivalent to a spherical radius of 383 ± 6 km, comparable to the thermal measurements (364 ± 33 km; Mommert et al., 2012). The two solutions given in this table are related to the presence of a surface feature, which can be interpreted as a canyon or a depression.*
Notes: Diameters given with a gray background are from the "TNOs are Cool" key program on ESA/Herschel for the primary object (available at http://public-tnosarecool.lesia.obspm.fr/Published-results.html). Measurements obtained with different techniques are given for comparison, in particular stellar occultation results are given in bold. References are given in Table 8.3 to keep the table readable.
Mid-size object 2015 RR_{245} has not been included in this table due to few constraints existing on its physical parameters. From an absolute magnitude of 3.6 ± 0.1 and an assumed albedo of 12%, its diameter is estimated at ~670 km (Bannister et al., 2016).

measurements. The hydrostatic equilibrium shape may indicate whether an object is uniform or stratified. Rambaux et al. (2017) studied the shape of TNOs Makemake, Quaoar, Orcus, Salacia, and Sedna. They showed that differences between the equatorial and polar axes for a homogeneous and a heterogeneous internal structure could be significant enough to be measured by ground-based techniques like stellar occultations. For instance, they predict that for their best case examples, Makemake and Salacia, variations would be of the order of 40 km between a uniform internal structure and a structure with two layers. This difference would further increase if an additional layer is considered (Rambaux personal communication). For the other TNOs studied, the differences are of the order of the size of topographic features possibly present at their surface.

8.4.3 Geomorphological features and surface composition

Cryovolcanic activity may be identified from specific geomorphological features, thermal anomalies, or spectroscopic features related to the composition of cryolava. Since it is unlikely that thermal anomalies can be identified in a near future, given the spatial resolution required, we must turn to surface composition and geomorphological features which can be seen from ground- or space-based telescopes, in order to constrain past activity. Interestingly, stellar

TABLE 8.3 References corresponding to Table 8.1.

Number	Name	References
134340	Pluto/Charon	Values from the NASA/New Horizons mission (Stern et al., 2015)
136199	Eris	Sicardy et al. (2011), Brown et al. (2005, 2006b), Dumas et al. (2007), Merlin et al. (2009), Santos-Sanz et al. (2012), Tegler et al. (2010, 2012), and Alvarez-Candal et al. (2011)
136472	Makemake	Ortiz et al. (2012), Brown (2013a), Licandro et al. (2006), Brown et al. (2007, 2015), Parker et al. (2016), Lim et al. (2010), Lellouch et al. (2017), and Lorenzi et al. (2015)
136108	Haumea	Lacerda and Jewitt (2007), Barkume et al. (2006), Trujillo et al. (2007), Merlin et al. (2007), Dumas et al. (2011), Pinilla-Alonso et al. (2009), Lellouch et al. (2010), Lockwood et al. (2014), Ortiz et al. (2017), and Rabinowitz et al. (2006)
225088	2007 OR_{10}	Kiss et al. (2017), Santos-Sanz et al. (2012), Pál et al. (2016), Brown et al. (2011, 2012), and Holler et al. (2017b)
50000	Quaoar	Brown and Butler (2017), Braga-Ribas et al. (2013), Davis et al. (2014), Fornasier et al. (2013), Jewitt and Luu (2004), Schaller and Brown (2007), Guilbert et al. (2009), Dalle Ore et al. (2009), and Barucci et al. (2015)
90377	Sedna	Lellouch et al. (2013), Lykawka and Mukai (2005), Trujillo et al. (2005), Emery et al. (2007), and Barucci et al. (2010)
90482	Orcus	Carry et al. (2011), Brown and Butler (2017, 2018), Fornasier et al. (2004, 2013), Thirouin et al. (2014), de Bergh et al. (2005), Trujillo et al. (2005), Barucci et al. (2008), Delsanti et al. (2010), and DeMeo et al. (2010)
120347	Salacia	Fornasier et al. (2013) and Brown and Butler (2017)
	2013 FY_{27}	Sheppard et al. (2018)
174567	Varda	Vilenius et al. (2014) and Grundy et al. (2015)
208996	2003 AZ_{84}	Brown and Suer (2007), Dias-Oliveira et al. (2017), Thirouin et al. (2014), Guilbert et al. (2009), Merlin et al. (2010b), and Trujillo et al. (2011)
55637	2002 UX_{25}	Brown and Suer (2007), Brown and Butler (2017), Vilenius et al. (2014), and Lellouch et al. (2017)

occultations of Plutino 2003 AZ_{84} observed by Dias-Oliveira et al. (2017) reveal a topographic feature along the object's limb that can be interpreted as a canyon-like feature, perhaps similar to that observed at close range on the surface of Charon.

Surface composition alone may not be sufficient to offer a diagnostic for past activity, but the presence of volatiles easily destroyed under irradiation may help identify objects for which surface renewal is likely. Although the presence of methane may effectively hide it, ammonia (or hydrates) has been detected at the surface of Charon, Orcus (see Table 8.2), and possibly Quaoar. In addition, methanol (another antifreeze) has been observed at the surface of some smaller objects, like Pholus or 2002 VE_{95} (Cruikshank et al., 1998; Barucci et al., 2006). Merlin et al. (2012) considered the formation and destruction processes of methanol, and suggested that at least part of the surface of these objects is younger than the age of the solar system. McKinnon et al. (2008) discussed the fact that although it would be tempting to attribute the presence of methanol on the surface to cryovolcanism, these bodies are significantly smaller than any prediction discussed in Section 8.2. Dedicated studies would thus be required to address this specific issue.

8.5 Where should we go from here?

The key question we face when coming to investigate the internal structure of TNOs is whether surface and bulk properties, which may be obtained from observations, are sufficient for inferring internal properties and more ambitiously, the evolutionary course of a TNO from formation to present day. If a robust and comprehensive evolutionary model could be constructed and if the initial parameters were known, the answer to this question would be in the affirmative. The problem is that the physical characteristics of these objects are poorly known and theories about formation scenarios vary. Therefore, observations performed to date can only weakly constrain interior models. In the near future, promising observations will be performed to advance this field. Stellar occultations in the ESA/Gaia era can be expected to allow precise constraints on the shapes of TNOs, and reveal possible topographic features at their surface. JWST/NIRSpec observations will access the 3–5 μm wavelength range, where the fundamental absorption features of ices can be observed, thus allowing for the composition of possible cryolava to be identified (Metayer et al., 2019). In addition, extreme adaptive optics systems, such as the ESO/VLT SPHERE instrument (Beuzit et al., 2008), are now able to image binaries and multiple systems, providing us with new reliable constraints on TNOs' densities.

Modeling the evolution of mid-size TNOs is strongly dependent on the details of the model itself, such as the many free parameters, or the thermal, physical, and chemical processes accounted for. Studies summarized in this chapter have taught us that cryovolcanism is enabled by radiogenic heating, aided by the presence of ammonia, methanol, and other antifreeze compounds like salts. In this regard, advance can be achieved by further constraining the effect of CH_3OH, for example, which has been poorly studied so far, or the role of geochemical reactions, like serpentinization. The complexity of the problem stems from the fact that each object may need to be treated as a unique individual world. Eventually, however, consistent initial parameters will be required for the overall population. We may not reach as detailed empirical constraints for each large- and mid-size TNO as for Pluto or Charon, but we should be able to model a large enough number of well-observed objects, using similar processes and initial properties. Thus, thermal evolution modeling remains a powerful tool for studying TNOs in general and cryovolcanism in particular, which should lead to significant progress in our understanding of how the entire TNO population formed and evolved.

References

Alvarez-Candal, A., Pinilla-Alonso, N., Licandro, J., Cook, J., Mason, E., Roush, T., Cruikshank, D.P., Gourgeot, F., et al., 2011. The spectrum of (136199) Eris between 350 and 2350 nm: results with X-Shooter. Astron. Astrophys. 532, A130. https://doi.org/10.1051/0004-6361/201117069.

Bannister, M.T., Alexandersen, M., Benecchi, S.D., Chen, Y.-T., Delsanti, A., Fraser, W.C., Gladman, B.J., Granvik, M., et al., 2016. OSSOS. IV. Discovery of a dwarf planet candidate in the 9:2 resonance with Neptune. Astron. J. 152, 212. https://doi.org/10.3847/0004-6256/152/6/212.

Barkume, K.M., Brown, M.E., Schaller, E.L., 2006. Water ice on the satellite of Kuiper belt object 2003 EL_{61}. Astrophys. J. Lett. 640, L87–L89. https://doi.org/10.1086/503159.

Barkume, K.M., Brown, M.E., Schaller, E.L., 2008. Near-infrared spectra of centaurs and Kuiper belt objects. Astron. J. 135, 55–67. https://doi.org/10.1088/0004-6256/135/1/55.

Barr, A.C., Collins, G.C., 2015. Tectonic activity on Pluto after the Charon-forming impact. Icarus 246, 146–155. https://doi.org/10.1016/j.icarus.2014.03.042.

Barr, A.C., Schwamb, M.E., 2016. Interpreting the densities of the Kuiper belt's dwarf planets. Mon. Not. R. Astron. Soc. 460, 1542–1548. https://doi.org/10.1093/mnras/stw1052.

Barucci, M.A., Merlin, F., Dotto, E., Doressoundiram, A., de Bergh, C., 2006. TNO surface ices. Observations of the TNO 55638 (2002 VE_{95}). Astron. Astrophys. 455, 725–730. https://doi.org/10.1051/0004-6361:20064951.

Barucci, M.A., Merlin, F., Guilbert, A., de Bergh, C., Alvarez-Candal, A., Hainaut, O., Doressoundiram, A., Dumas, C., et al., 2008. Surface composition and temperature of the TNO Orcus. Astron. Astrophys. 479, L13–L16. https://doi.org/10.1051/0004-6361:20079079.

Barucci, M.A., Dalle Ore, C.M., Alvarez-Candal, A., de Bergh, C., Merlin, F., Dumas, C., Cruikshank, D.P., 2010. (90377) Sedna: investigation of surface compositional variation. Astron. J. 140, 2095–2100. https://doi.org/10.1088/0004-6256/140/6/2095.

Barucci, M.A., Dalle Ore, C.M., Perna, D., Cruikshank, D.P., Doressoundiram, A., Alvarez-Candal, A., Dotto, E., Nitschelm, C., 2015. (50000) Quaoar: surface composition variability. Astron. Astrophys. 584, A107. https://doi.org/10.1051/0004-6361/201526119.

Beuzit, J.-L., Feldt, M., Dohlen, K., Mouillet, D., Puget, P., Wildi, F., Abe, L., Antichi, J., et al., 2008. SPHERE: a "Planet Finder" instrument for the VLT. In: Ground-Based and Airborne Instrumentation for Astronomy II, Proc. SPIE, vol. 7014, p. 701418.

Beyer, R.A., Nimmo, F., McKinnon, W.B., Moore, J.M., Binzel, R.P., Conrad, J.W., Cheng, A., Ennico, K., et al., 2017. Charon tectonics. Icarus 287, 161–174. https://doi.org/10.1016/j.icarus.2016.12.018.

Bierson, C.J., Nimmo, F., McKinnon, W.B., 2018. Implications of the observed Pluto-Charon density contrast. Icarus 309, 207–219. https://doi.org/10.1016/j.icarus.2018.03.007.

Braga-Ribas, F., Sicardy, B., Ortiz, J.L., Lellouch, E., Tancredi, G., Lecacheux, J., Vieira-Martins, R., Camargo, J.I.B., et al., 2013. The size, shape, albedo, density, and atmospheric limit of transneptunian object (50000) Quaoar from multi-chord stellar occultations. Astrophys. J. 773, 26. https://doi.org/10.1088/0004-637X/773/1/26.

Brown, M.E., 2008. The largest Kuiper belt objects. In: Barucci, M.A., Boehnhardt, H., Cruikshank, D.P., Morbidelli, A. (Eds.), The Solar System Beyond Neptune. University of Arizona Press, Tucson, pp. 335–344.

Brown, M.E., 2012. The compositions of Kuiper belt objects. Annu. Rev. Earth Planet. Sci. 40, 467–494. https://doi.org/10.1146/annurev-earth-042711-105352.

Brown, M.E., 2013a. On the size, shape, and density of dwarf planet Makemake. Astrophys. J. Lett. 767 (1), Article ID: L7.

Brown, M.E., 2013b. The density of mid-sized Kuiper belt object 2002 UX25 and the formation of the dwarf planets. Astrophys. J. Lett. 778 (2), Article ID: L34.

Brown, M.E., Butler, B.J., 2017. The density of mid-sized Kuiper belt objects from ALMA thermal observations. Astron. J. 154, 19. https://doi.org/10.3847/1538-3881/aa6346.

Brown, M.E., Butler, B.J., 2018. Medium-sized satellites of large Kuiper belt objects. Astron. J. 156, 164. https://doi.org/10.3847/1538-3881/aad9f2.

Brown, M.E., Calvin, W.M., 2000. Evidence for crystalline water and ammonia ices on Pluto's satellite Charon. Science 287, 107–109. https://doi.org/10.1126/science.287.5450.107.

Brown, M.E., Suer, T.-A., 2007. Satellites of 2003 AZ_84, (50000), (55637), and (90482). Int. Astron. Union Circ. 8812, 1.

Brown, M.E., Trujillo, C.A., Rabinowitz, D.L., 2005. Discovery of a planetary-sized object in the scattered Kuiper belt. Astrophys. J. Lett. 635, L97–L100. https://doi.org/10.1086/499336.

Brown, M.E., van Dam, M.A., Bouchez, A.H., Le Mignant, D., Campbell, R.D., Chin, J.C.Y., Conrad, A., Hartman, S.K., et al., 2006a. Satellites of the largest Kuiper belt objects. Astrophys. J. Lett. 639, L43–L46. https://doi.org/10.1086/501524.

Brown, M.E., Schaller, E.L., Roe, H.G., Rabinowitz, D.L., Trujillo, C.A., 2006b. Direct measurement of the size of 2003 UB313 from the Hubble Space Telescope. Astrophys. J. Lett. 643, L61–L63. https://doi.org/10.1086/504843.

Brown, M.E., Barkume, K.M., Blake, G.A., Schaller, E.L., Rabinowitz, D.L., Roe, H.G., Trujillo, C.A., 2007. Methane and ethane on the bright Kuiper belt object 2005 FY9. Astron. J. 133, 284.

Brown, M.E., Burgasser, A.J., Fraser, W.C., 2011. The surface composition of large Kuiper belt object 2007 OR_{10}. Astrophys. J. Lett. 738, L26. https://doi.org/10.1088/2041-8205/738/2/L26.

Brown, M.E., Schaller, E.L., Fraser, W.C., 2012. Water ice in the Kuiper belt. Astron. J. 143, 146. https://doi.org/10.1088/0004-6256/143/6/146.

Brown, M.E., Schaller, E.L., Blake, G.A., 2015. Irradiation products on dwarf planet Makemake. Astron. J. 149, 105. https://doi.org/10.1088/0004-6256/149/3/105.

Buhler, P.B., Ingersoll, A.P., 2018. Sublimation pit distribution indicates convection cell surface velocities of ~10 cm per year in Sputnik Planitia, Pluto. Icarus 300, 327–340. https://doi.org/10.1016/j.icarus.2017.09.018.

Buie, M.W., Grundy, W.M., 2000. The distribution and physical state of H_2O on Charon. Icarus 148, 324–339. https://doi.org/10.1006/icar.2000.6509.

Canup, R.M., 2005. A giant impact origin of Pluto-Charon. Science 307, 546–550. https://doi.org/10.1126/science.1106818.

Canup, R.M., 2011. On a giant impact origin of Charon, Nix, and Hydra. Astron. J. 141, 35. https://doi.org/10.1088/0004-6256/141/2/35.

Carry, B., Hestroffer, D., DeMeo, F.E., Thirouin, A., Berthier, J., Lacerda, P., Sicardy, B., Doressoundiram, A., et al., 2011. Integral-field spectroscopy of (90482) Orcus-Vanth. Astron. Astrophys. 534, A115. https://doi.org/10.1051/0004-6361/201117486.

Choi, Y.-J., Cohen, M., Merk, R., Prialnik, D., 2002. Long-term evolution of objects in the Kuiper belt zone—effects of insolation and radiogenic heating. Icarus 160, 300–312. https://doi.org/10.1006/icar.2002.6976.

Cook, J.C., Desch, S.J., Roush, T.L., Trujillo, C.A., Geballe, T.R., 2007. Near-infrared spectroscopy of Charon: possible evidence for cryovolcanism on Kuiper belt objects. Astrophys. J. 663, 1406–1419. https://doi.org/10.1086/518222.

Cooper, J.F., Christian, E.R., Richardson, J.D., Wang, C., 2003. Proton irradiation of centaur, Kuiper belt, and Oort cloud objects at plasma to cosmic ray energy. Earth Moon Planets 92, 261–277. https://doi.org/10.1023/B:MOON.0000031944.41883.80.

Coradini, A., Capria, M.T., de Sanctis, M.C., McKinnon, W.B., 2008. The structure of Kuiper belt bodies: link with comets. In: Barucci, M.A., Boehnhardt, H., Cruikshank, D.P., Morbidelli, A. (Eds.), The Solar System Beyond Neptune. University of Arizona Press, Tucson, pp. 243–256.

Cruikshank, D.P., Roush, T.L., Bartholomew, M.J., Geballe, T.R., Pendleton, Y.J., White, S.M., Bell, J.F., Davies, J.K., et al., 1998. The composition of centaur 5145 Pholus. Icarus 135, 389–407. https://doi.org/10.1006/icar.1998.5997.

Cruikshank, D.P., Owen, T.C., Dalle Ore, C.M., Geballe, T.R., Roush, T.L., de Bergh, C., Sandford, S.A., Poulet, F., et al., 2005. A spectroscopic study of the surfaces of Saturn's large satellites: H_2O ice, tholins, and minor constituents. Icarus 175, 268–283. https://doi.org/10.1016/j.icarus.2004.09.003.

Cruikshank, D.P., Grundy, W.M., DeMeo, F.E., Buie, M.W., Binzel, R.P., Jennings, D.E., Olkin, C.B., Parker, J.W., et al., 2015. The surface compositions of Pluto and Charon. Icarus 246, 82–92. https://doi.org/10.1016/j.icarus.2014.05.023.

Dalle Ore, C.M., Barucci, M.A., Emery, J.P., Cruikshank, D.P., Dalle Ore, L.V., Merlin, F., Alvarez-Candal, A., de Bergh, C., et al., 2009. Composition of KBO (50000) Quaoar. Astron. Astrophys. 501, 349–357. https://doi.org/10.1051/0004-6361/200911752.

Dalle Ore, C.M., Protopapa, S., Cook, J.C., Grundy, W.M., Cruikshank, D.P., Verbiscer, A.J., et al., 2018. Ices on Charon: distribution of H_2O and NH_3 from new horizons LEISA observations. Icarus 300, 21–32. https://doi.org/10.1016/j.icarus.2017.08.026.

Davis, A.B., Pasachoff, J.M., Babcock, B.A., Person, M.J., Zuluaga, C.A., Bosh, A.S., Levine, S., Naranjo, O.A., et al., 2014. Observation and analysis of a single-chord stellar occultation by Kuiper belt object (50000) Quaoar. In: American Astronomical Society Meeting Abstracts No. 223.

de Bergh, C., Delsanti, A., Tozzi, G.P., Dotto, E., Doressoundiram, A., Barucci, M.A., 2005. The surface of the transneptunian object 90482 Orcus. Astron. Astrophys. 437, 1115–1120. https://doi.org/10.1051/0004-6361:20042533.

De Sanctis, M.C., Capria, M.T., Coradini, A., 2001. Thermal evolution and differentiation of Edgeworth-Kuiper belt objects. Astron. J. 121, 2792–2799. https://doi.org/10.1086/320385.

Delsanti, A., Merlin, F., Guilbert-Lepoutre, A., Bauer, J., Yang, B., Meech, K.J., 2010. Methane, ammonia, and their irradiation products at the surface of an intermediate-size KBO? A portrait of Plutino (90482) Orcus. Astron. Astrophys. 520, A40. https://doi.org/10.1051/0004-6361/201014296.

DeMeo, F.E., Barucci, M.A., Merlin, F., Guilbert-Lepoutre, A., Alvarez-Candal, A., Delsanti, A., Fornasier, S., de Bergh, C., 2010. A spectroscopic analysis of Jupiter-coupled object (52872) Okyrhoe, and TNOs (90482) Orcus and (73480) 2002 PN_{34}. Astron. Astrophys. 521, A35. https://doi.org/10.1051/0004-6361/201014042.

DeMeo, F.E., Dumas, C., Cook, J.C., Carry, B., Merlin, F., Verbiscer, A.J., Binzel, R.P., 2015. Spectral variability of Charon's 2.21-μm feature. Icarus 246, 213–219. https://doi.org/10.1016/j.icarus.2014.04.010.

Desch, S.J., 2015. Density of Charon formed from a disk generated by the impact of partially differentiated bodies. Icarus 246, 37–47. https://doi.org/10.1016/j.icarus.2014.07.034.

Desch, S.J., Neveu, M., 2017. Differentiation and cryovolcanism on Charon: a view before and after New Horizons. Icarus 287, 175–186. https://doi.org/10.1016/j.icarus.2016.11.037.

Desch, S.J., Cook, J.C., Doggett, T.C., Porter, S.B., 2009. Thermal evolution of Kuiper belt objects, with implications for cryovolcanism. Icarus 202, 694–714. https://doi.org/10.1016/j.icarus.2009.03.009.

Dias-Oliveira, A., Sicardy, B., Ortiz, J.L., Braga-Ribas, F., Leiva, R., Vieira-Martins, R., Benedetti-Rossi, G., Camargo, J.I.B., et al., 2017. Study of the Plutino object (208996) 2003 AZ_{84} from stellar occultations: size, shape, and topographic features. Astron. J. 154, 22. https://doi.org/10.3847/1538-3881/aa74e9.

Dumas, C., Terrile, R.J., Brown, R.H., Schneider, G., Smith, B.A., 2001. Hubble space telescope NICMOS spectroscopy of Charon's leading and trailing hemispheres. Astron. J. 121, 1163–1170. https://doi.org/10.1086/318747.

Dumas, C., Merlin, F., Barucci, M.A., de Bergh, C., Hainault, O., Guilbert, A., Vernazza, P., Doressoundiram, A., 2007. Surface composition of the largest dwarf planet 136199 Eris (2003 UB_{313}). Astron. Astrophys. 471, 331–334. https://doi.org/10.1051/0004-6361:20066665.

Dumas, C., Carry, B., Hestroffer, D., Merlin, F., 2011. High-contrast observations of (136108) Haumea. A crystalline water-ice multiple system. Astron. Astrophys. 528, A105. https://doi.org/10.1051/0004-6361/201015011.

Durand-Manterola, H.J., 2003. Internal structure of planetary icy bodies. In: EGS-AGU-EUG Joint Assembly, p. 7389.

Emery, J.P., Dalle Ore, C.M., Cruikshank, D.P., Fernández, Y.R., Trilling, D.E., Stansberry, J.A., 2007. Ices on (90377) Sedna: confirmation and compositional constraints. Astron. Astrophys. 466, 395–398. https://doi.org/10.1051/0004-6361:20067021.

Fornasier, S., Dotto, E., Barucci, M.A., Barbieri, C., 2004. Water ice on the surface of the large TNO 2004 DW. Astron. Astrophys. 422, L43–L46. https://doi.org/10.1051/0004-6361:20048004.

Fornasier, S., Lellouch, E., Müller, T., Santos-Sanz, P., Panuzzo, P., Kiss, C., Lim, T., Mommert, M., et al., 2013. TNOs are Cool: a survey of the Trans-Neptunian region. VIII. Combined Herschel PACS and SPIRE observations of nine bright targets at 70–500 μm. Astron. Astrophys. 555, A15. https://doi.org/10.1051/0004-6361/201321329.

Geissler, P.E., 2015. Cryovolcanism in the outer solar system. In: Sigurdsson, H. (Ed.), The Encyclopedia of Volcanoes, pp. 763–776.

Grundy, W.M., Porter, S.B., Benecchi, S.D., Roe, H.G., Noll, K.S., Trujillo, C.A., Thirouin, A., Stansberry, J.A., et al., 2015. The mutual orbit, mass, and density of the large transneptunian binary system Varda and Ilmarë. Icarus 257, 130–138. https://doi.org/10.1016/j.icarus.2015.04.036.

Grundy, W.M., Binzel, R.P., Buratti, B.J., Cook, J.C., Cruikshank, D.P., Dalle Ore, C.M., et al., 2016. Surface compositions across Pluto and Charon. Science 351, aad9189. https://doi.org/10.1126/science.aad9189.

Guilbert, A., Alvarez-Candal, A., Merlin, F., Barucci, M.A., Dumas, C., de Bergh, C., Delsanti, A., 2009. ESO-large program on TNOs: near-infrared spectroscopy with SINFONI. Icarus 201, 272–283. https://doi.org/10.1016/j.icarus.2008.12.023.

Holler, B.J., Young, L.A., Buie, M.W., Grundy, W.M., Lyke, J.E., Young, E.F., Roe, H.G., 2017a. Measuring temperature and ammonia hydrate ice on Charon in 2015 from Keck/OSIRIS spectra. Icarus 284, 394–406. https://doi.org/10.1016/j.icarus.2016.12.003.

Holler, B.J., Young, L.A., Bus, S.J., Protopapa, S., 2017b. Methanol ice on Kuiper belt objects 2007 OR_{10} and Salacia: implications for formation and dynamical evolution. Eur. Planet. Sci. Congr. 11, EPSC2017-330.

Hussmann, H., Sohl, F., Spohn, T., 2006. Subsurface oceans and deep interiors of medium-sized outer planet satellites and large Trans-Neptunian objects. Icarus 185, 258–273. https://doi.org/10.1016/j.icarus.2006.06.005.

Jewitt, D.C., Luu, J., 2004. Crystalline water ice on the Kuiper belt object (50000) Quaoar. Nature 432, 731–733. https://doi.org/10.1038/nature03111.

Kiss, C., Marton, G., Farkas-Takács, A., Stansberry, J., Müller, T., Vinkó, J., Balog, Z., Ortiz, J.-L., et al., 2017. Discovery of a satellite of the large Trans-Neptunian object (225088) 2007 OR_{10}. Astrophys. J. Lett. 838, L1. https://doi.org/10.3847/2041-8213/aa6484.

Lacerda, P., Jewitt, D.C., 2007. Densities of solar system objects from their rotational light curves. Astron. J. 133, 1393. https://doi.org/10.1086/511772.

Lellouch, E., Kiss, C., Santos-Sanz, P., Müller, T.G., Fornasier, S., Groussin, O., Lacerda, P., Ortiz, J.L., et al., 2010. "TNOs are Cool": a survey of the Trans-Neptunian region. II. The thermal lightcurve of (136108) Haumea. Astron. Astrophys. 518, L147. https://doi.org/10.1051/0004-6361/201014648.

Lellouch, E., Santos-Sanz, P., Lacerda, P., Mommert, M., Duffard, R., Ortiz, J.L., Müller, T.G., Fornasier, S., et al., 2013. "TNOs are Cool": a survey of the Trans-Neptunian region. IX. Thermal properties of Kuiper belt objects and Centaurs from combined Herschel and Spitzer observations. Astron. Astrophys. 557, A60. https://doi.org/10.1051/0004-6361/201322047.

Lellouch, E., Moreno, R., Müller, T., Fornasier, S., Santos-Sanz, P., Moullet, A., Gurwell, M., Stansberry, J., et al., 2017. The thermal emission of Centaurs and Trans-Neptunian objects at millimeter wavelengths from ALMA observations. Astron. Astrophys. 608, A45. https://doi.org/10.1051/0004-6361/201731676.

Leto, G., Baratta, G.A., 2003. Ly-alpha photon induced amorphization of IC water ice at 16 Kelvin. Effects and quantitative comparison with ion irradiation. Astron. Astrophys. 397, 7–13. https://doi.org/10.1051/0004-6361:20021473.

Licandro, J., Pinilla-Alonso, N., Pedani, M., Oliva, E., Tozzi, G.P., Grundy, W.M., 2006. The methane ice rich surface of large TNO 2005 FY_9: a Pluto-twin in the Trans-Neptunian belt? Astron. Astrophys. 445, L35–L38. https://doi.org/10.1051/0004-6361:200500219.

Lim, T.L., Stansberry, J., Müller, T.G., Mueller, M., Lellouch, E., Kiss, C., Santos-Sanz, P., Vilenius, E., et al., 2010. "TNOs are Cool": a survey of the Trans-Neptunian region. III. Thermophysical properties of 90482 Orcus and 136472 Makemake. Astron. Astrophys. 518, L148. https://doi.org/10.1051/0004-6361/201014701.

Lockwood, A.C., Brown, M.E., Stansberry, J., 2014. The size and shape of the oblong dwarf planet Haumea. Earth Moon Planets 111, 127–137. https://doi.org/10.1007/s11038-014-9430-1.

Lorenzi, V., Pinilla-Alonso, N., Licandro, J., 2015. Rotationally resolved spectroscopy of dwarf planet (136472) Makemake. Astron. Astrophys. 577, A86. https://doi.org/10.1051/0004-6361/201425575.

Lykawka, P.S., Mukai, T., 2005. Higher albedos and size distribution of large transneptunian objects. Planet. Space Sci. 53, 1319–1330. https://doi.org/10.1016/j.pss.2005.06.004.

Malamud, U., Prialnik, D., 2015. Modeling Kuiper belt objects Charon, Orcus and Salacia by means of a new equation of state for porous icy bodies. Icarus 246, 21–36. https://doi.org/10.1016/j.icarus.2014.02.027.

Malamud, U., Perets, H.B., Schubert, G., 2017. The contraction/expansion history of Charon with implications for its planetary-scale tectonic belt. Mon. Not. R. Astron. Soc. 468, 1056–1069. https://doi.org/10.1093/mnras/stx546.

Mastrapa, R.M.E., Brown, R.H., 2006. Ion irradiation of crystalline H_2O-ice: effect on the 1.65-μm band. Icarus 183, 207–214. https://doi.org/10.1016/j.icarus.2006.02.006.

McKinnon, W.B., 2002. On the initial thermal evolution of Kuiper belt objects. In: Warmbein, B. (Ed.), Asteroids, Comets, and Meteors: ACM 2002, vol. 500. ESA Special Publication, pp. 29–38.

McKinnon, W.B., Prialnik, D., Stern, S.A., Coradini, A., 2008. Structure and evolution of Kuiper belt objects and dwarf planets. In: Barucci, M.A., Boehnhardt, H., Cruikshank, D.P., Morbidelli, A. (Eds.), The Solar System Beyond Neptune. University of Arizona Press, Tucson, pp. 213–241.

McKinnon, W.B., Nimmo, F., Wong, T., Schenk, P.M., White, O.L., Roberts, J.H., Moore, J.M., Spencer, J.R., et al., 2016. Convection in a volatile nitrogen-ice-rich layer drives Pluto's geological vigour. Nature 534, 82–85. https://doi.org/10.1038/nature18289.

McKinnon, W.B., Stern, S.A., Weaver, H.A., Nimmo, F., Bierson, C.J., Grundy, W.M., Cook, J.C., Cruikshank, D.P., et al., 2017. Origin of the Pluto-Charon system: constraints from the new horizons flyby. Icarus 287, 2–11. https://doi.org/10.1016/j.icarus.2016.11.019.

Merk, R., Prialnik, D., 2003. Early thermal and structural evolution of small bodies in the Trans-Neptunian zone. Earth Moon Planets 92, 359–374. https://doi.org/10.1023/B:MOON.0000031952.89891.a4.

Merk, R., Prialnik, D., 2006. Combined modeling of thermal evolution and accretion of Trans-Neptunian objects—occurrence of high temperatures and liquid water. Icarus 183, 283–295. https://doi.org/10.1016/j.icarus.2006.02.011.

Merlin, F., Guilbert, A., Dumas, C., Barucci, M.A., de Bergh, C., Vernazza, P., 2007. Properties of the icy surface of the TNO 136108 (2003 EL{61}). Astron. Astrophys. 466, 1185–1188. https://doi.org/10.1051/0004-6361:20066866.

Merlin, F., Alvarez-Candal, A., Delsanti, A., Fornasier, S., Barucci, M.A., DeMeo, F.E., de Bergh, C., Doressoundiram, A., et al., 2009. Stratification of methane ice on Eris' surface. Astron. J. 137, 315–328. https://doi.org/10.1088/0004-6256/137/1/315.

Merlin, F., Barucci, M.A., de Bergh, C., DeMeo, F.E., Alvarez-Candal, A., Dumas, C., Cruikshank, D.P., 2010a. Chemical and physical properties of the variegated Pluto and Charon surfaces. Icarus 210, 930–943. https://doi.org/10.1016/j.icarus.2010.07.028.

Merlin, F., Barucci, M.A., de Bergh, C., Fornasier, S., Doressoundiram, A., Perna, D., Protopapa, S., 2010b. Surface composition and physical properties of several Trans-Neptunian objects from the Hapke scattering theory and Shkuratov model. Icarus 208, 945–954. https://doi.org/10.1016/j.icarus.2010.03.014.

Merlin, F., Quirico, E., Barucci, M.A., de Bergh, C., 2012. Methanol ice on the surface of minor bodies in the solar system. Astron. Astrophys. 544, A20. https://doi.org/10.1051/0004-6361/201219181.

Metayer, R., Guilbert-Lepoutre, A., Ferruit, P., Merlin, F., Holler, B.J., Cabral, N., et al., 2019. JWST/NIRSpec prospects on transneptunian objects. Front. Astron. Space Sci. 6, A8. https://doi.org/10.3389/fspas.2019.00008.

Mommert, M., Harris, A.W., Kiss, C., Pál, A., Santos-Sanz, P., Stansberry, J., Delsanti, A., Vilenius, E., et al., 2012. TNOs are Cool: a survey of the Trans-Neptunian region. V. Physical characterization of 18 Plutinos using Herschel-PACS observations. Astron. Astrophys. 541, A93. https://doi.org/10.1051/0004-6361/201118562.

Moore, M.H., Ferrante, R.F., Hudson, R.L., Stone, J.N., 2007. Ammonia water ice laboratory studies relevant to outer Solar System surfaces. Icarus 190, 260–273. https://doi.org/10.1016/j.icarus.2007.02.020.

Moore, J.M., McKinnon, W.B., Spencer, J.R., Howard, A.D., Schenk, P.M., Beyer, R.A., Nimmo, F., Singer, K.N., et al., 2016. The geology of Pluto and Charon through the eyes of new horizons. Science 351, 1284–1293. https://doi.org/10.1126/science.aad7055.

Neveu, M., Desch, S.J., Shock, E.L., Glein, C.R., 2015. Prerequisites for explosive cryovolcanism on dwarf planet-class Kuiper belt objects. Icarus 246, 48–64. https://doi.org/10.1016/j.icarus.2014.03.043.

Ortiz, J.L., Sicardy, B., Braga-Ribas, F., Alvarez- Candal, A., Lellouch, E., Duffard, R., Pinilla-Alonso, N., Ivanov, V.D., et al., 2012. Albedo and atmospheric constraints of dwarf planet Makemake from a stellar occultation. Nature 491, 566–569. https://doi.org/10.1038/nature11597.

Ortiz, J.L., Santos-Sanz, P., Sicardy, B., Benedetti-Rossi, G., Bérard, D., Morales, N., Duffard, R., Braga-Ribas, F., et al., 2017. The size, shape, density and ring of the dwarf planet Haumea from a stellar occultation. Nature 550, 219–223. https://doi.org/10.1038/nature24051.

Pál, A., Kiss, C., Müller, T.G., Molnár, L., Szabó, R., Szabó, G.M., Sárneczky, K., Kiss, L.L., 2016. Large size and slow rotation of the Trans-Neptunian object (225088) 2007 OR_{10} discovered from Herschel and K2 observations. Astron. J. 151, 117. https://doi.org/10.3847/0004-6256/151/5/117.

Parker, A.H., Buie, M.W., Grundy, W.M., Noll, K.S., 2016. Discovery of a Makemakean Moon. Astrophys. J. Lett. 825, L9. https://doi.org/10.3847/2041-8205/825/1/L9.

Petit, J.-M., Kavelaars, J.J., Gladman, B., Loredo, T., 2008. Size distribution of multikilometer transneptunian objects. In: Barucci, M.A., Boehnhardt, H., Cruikshank, D.P., Morbidelli, A. (Eds.), The Solar System Beyond Neptune. University of Arizona Press, Tucson, pp. 71–87.

Pinilla-Alonso, N., Brunetto, R., Licandro, J., Gil-Hutton, R., Roush, T.L., Strazzulla, G., 2009. The surface of (136108) Haumea (2003 EL{61}), the largest carbon-depleted object in the Trans-Neptunian belt. Astron. Astrophys. 496, 547–556. https://doi.org/10.1051/0004-6361/200809733.

Prialnik, D., 2000. Physical characteristics of distant comets. In: Fitzsimmons, A., Jewitt, D., West, R.M. (Eds.), Minor Bodies in the Outer Solar System ESO Astrophysics Symposia (European Southern Observatory). Springer, Berlin, Heidelberg, p. 33.

Prialnik, D., Sarid, G., Rosenberg, E.D., Merk, R., 2008. Thermal and chemical evolution of comet nuclei and Kuiper belt objects. Space Sci. Rev. 138, 147–164. https://doi.org/10.1007/s11214-007-9301-4.

Rabinowitz, D.L., Barkume, K., Brown, M.E., Roe, H., Schwartz, M., Tourtellotte, S., Trujillo, C., 2006. Photometric observations constraining the size, shape, and albedo of 2003 EL61, a rapidly rotating, Pluto-sized object in the Kuiper belt. Astrophys. J. 639, 1238–1251. https://doi.org/10.1086/499575.

Rambaux, N., Baguet, D., Chambat, F., Castillo-Rogez, J.C., 2017. Equilibrium shapes of large Trans-Neptunian objects. Astrophys. J. Lett. 850, L9. https://doi.org/10.3847/2041-8213/aa95bd.

Robuchon, G., Nimmo, F., 2011. Thermal evolution of Pluto and implications for surface tectonics and a subsurface ocean. Icarus 216, 426–439. https://doi.org/10.1016/j.icarus.2011.08.015.

Santos-Sanz, P., Lellouch, E., Fornasier, S., Kiss, C., Pál, A., Müller, T.G., Vilenius, E., Stansberry, J., et al., 2012. "TNOs are Cool": a survey of the Trans-Neptunian region. IV. Size/albedo characterization of 15 scattered disk and detached objects observed with Herschel-PACS. Astron. Astrophys. 541, A92. https://doi.org/10.1051/0004-6361/201118541.

Schaller, E.L., Brown, M.E., 2007. Detection of methane on Kuiper belt object (50000) Quaoar. Astrophys. J. Lett. 670, L49–L51. https://doi.org/10.1086/524140.

Schubert, G., Hussmann, H., Lainey, V., Matson, D.L., McKinnon, W.B., Sohl, F., Sotin, C., Tobie, G., et al., 2010. Evolution of icy satellites. Space Sci. Rev. 153, 447–484. https://doi.org/10.1007/s11214-010-9635-1.

Sheppard, S.S., Fernandez, Y.R., Moullet, A., 2018. The albedos, sizes, colors, and satellites of dwarf planets compared with newly measured dwarf planet 2013 FY27. Astron. J. 156, 270. https://doi.org/10.3847/1538-3881/aae92a.

Sicardy, B., Ortiz, J.L., Assafin, M., Jehin, E., Maury, A., Lellouch, E., Hutton, R.G., Braga-Ribas, F., et al., 2011. A Pluto-like radius and a high albedo for the dwarf planet Eris from an occultation. Nature 478, 493–496. https://doi.org/10.1038/nature10550.

Singer, K.N., McKinnon, W.B., Gladman, B., Greenstreet, S., Bierhaus, E.B., Stern, S.A., et al., 2019. Impact craters on Pluto and Charon indicate a deficit of small Kuiper belt objects. Science 363, 955–959. https://doi.org/10.1126/science.aap8628.

Stern, S.A., 1989. Pluto—comments on crustal composition, evidence for global differentiation. Icarus 81, 14–23. https://doi.org/10.1016/0019-1035(89)90121-8.

Stern, S.A., Bagenal, F., Ennico, K., Gladstone, G.R., Grundy, W.M., McKinnon, W.B., Moore, J.M., Olkin, C.B., et al., 2015. The Pluto system: initial results from its exploration by New Horizons. Science 350, aad1815. https://doi.org/10.1126/science.aad1815.

Stevenson, D.J., 1982. Volcanism and igneous processes in small icy satellites. Nature 298, 142–144. https://doi.org/10.1038/298142a0.

Tegler, S.C., Cornelison, D.M., Grundy, W.M., Romanishin, W., Abernathy, M.R., Bovyn, M.J., Burt, J.A., Evans, D.E., et al., 2010. Methane and nitrogen abundances on Pluto and Eris. Astrophys. J. 725, 1296–1305. https://doi.org/10.1088/0004-637X/725/1/1296.

Tegler, S.C., Grundy, W.M., Olkin, C.B., Young, L.A., Romanishin, W., Cornelison, D.M., Khodadadkouchaki, R., 2012. Ice mineralogy across and into the surfaces of Pluto, Triton, and Eris. Astrophys. J. 751, 76. https://doi.org/10.1088/0004-637X/751/1/76.

Thirouin, A., Noll, K.S., Ortiz, J.L., Morales, N., 2014. Rotational properties of the binary and non-binary populations in the Trans-Neptunian belt. Astron. Astrophys. 569, A3. https://doi.org/10.1051/0004-6361/201423567.

Trilling, D.E., 2016. The surface age of Sputnik Planum, Pluto, must be less than 10 million years. PLoS ONE 11, e0147386. https://doi.org/10.1371/journal.pone.0147386.

Trowbridge, A.J., Melosh, H.J., Steckloff, J.K., Freed, A.M., 2016. Vigorous convection as the explanation for Pluto's polygonal terrain. Nature 534, 79–81. https://doi.org/10.1038/nature18016.

Trujillo, C.A., Brown, M.E., Rabinowitz, D.L., Geballe, T.R., 2005. Near-infrared surface properties of the two intrinsically brightest minor Planets: (90377) Sedna and (90482) Orcus. Astrophys. J. 627, 1057–1065. https://doi.org/10.1086/430337.

Trujillo, C.A., Brown, M.E., Barkume, K.M., Schaller, E.L., Rabinowitz, D.L., 2007. The surface of 2003 EL_{61} in the near-infrared. Astrophys. J. 655, 1172–1178. https://doi.org/10.1086/509861.

Trujillo, C.A., Sheppard, S.S., Schaller, E.L., 2011. A photometric system for detection of water and methane ices on Kuiper belt objects. Astrophys. J. 730, 105. https://doi.org/10.1088/0004-637X/730/2/105.

Vilenius, E., Kiss, C., Müller, T., Mommert, M., Santos-Sanz, P., Pál, A., Stansberry, J., Mueller, M., et al., 2014. "TNOs are Cool": a survey of the Trans-Neptunian region. X. Analysis of classical Kuiper belt objects from Herschel and Spitzer observations. Astron. Astrophys. 564, A35. https://doi.org/10.1051/0004-6361/201322416.

Zheng, W., Jewitt, D., Kaiser, R.I., 2009. On the state of water ice on Saturn's Moon Titan and implications to icy bodies in the outer solar system. J. Phys. Chem. A 113, 11174–11181. https://doi.org/10.1021/jp903817y.

PART III

Multiple systems

CHAPTER

9

Trans-Neptunian binaries (2018)

Keith S. Noll[a], William M. Grundy[b], David Nesvorný[c], Audrey Thirouin[b]

[a]Goddard Space Flight Center, Greenbelt, MD, United States [b]Lowell Observatory, Flagstaff, AZ, United States [c]Southwest Research Institute, Boulder, CO, United States

9.1 Overview

At the time of the Coimbra conference in early 2018 four decades had elapsed since the discovery of the first Trans-Neptunian binary (TNB) companion—Charon, almost two decades since the first discoveries of additional TNBs (e.g., Veillet et al., 2002), and 10 years since the last major review of this topic (Noll et al., 2008a). In this chapter, we start from the foundation of this earlier work and review new developments in the field that have occurred in the last decade.

In 2008, many of the themes relating to TNBs had already been established and the basic outlines of our current understanding of the Trans-Neptunian solar system and the role of binaries within that population had begun to take shape. Since that time there has been continued progress in the identification of binaries, determination of mutual orbits, and understanding of system mass, density, rotational state, component colors, and mutual events. Formation models have shifted away from capture scenarios in favor of models that invoke direct gravitational collapse—binaries offer tests of such models while also constraining subsequent dynamical evolution. In 2018, the study of TNBs remains one of the most active areas of progress in understanding the solar system beyond Neptune.

The Trans-Neptunian Solar System.
https://doi.org/10.1016/B978-0-12-816490-7.00009-6

9.2 Inventory

9.2.1 Direct imaging

The discovery of new binaries has continued over the last decade, with the majority of new systems being discovered by the Hubble Space Telescope (HST) as was also true in the previous decade. At the time of the last review, 43 binary systems were known (Noll et al., 2008a); that number has now doubled to 86 Trans-Neptunian objects (TNOs) with one or more companions (see compilation[1] maintained by Johnston, 2018).

TNBs have been found in every dynamical population of the Kuiper belt as shown in Fig. 9.1. As discussed in more detail here, this includes a large number of binaries in the cold classical population (nonresonant, not scattering, and low inclination) and significant numbers also in the dynamically hot populations—that is, objects in mean-motion resonance with Neptune, objects that are scattered by Neptune, detached objects with large perihelia, and high inclination (hot) classicals (Elliot et al., 2005; Gladman et al., 2008). Large snapshot surveys with HST have historically been the most productive source of binary discovery accounting for approximately 80% of discoveries. Snapshot observations have been an efficient way to observe a significant number of solar system objects with HST; however, the reduced exposure times dictated by this observing strategy favored brighter objects and observations in a single broad filter. In the last decade, searches with HST have been focused more on follow-up of distinct subpopulations underrepresented in earlier surveys. Overall, HST has observed more than 500 TNOs, a significant fraction of the total Trans-Neptunian population with well-known orbits.

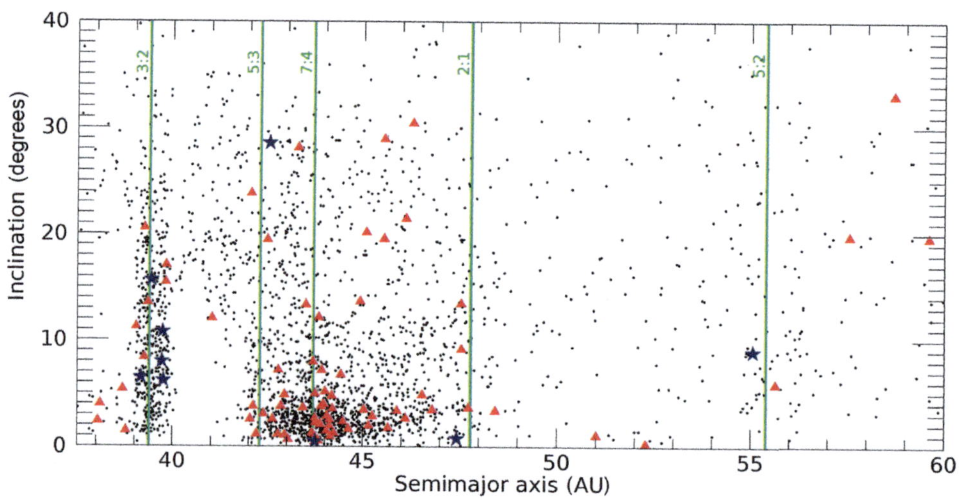

FIG. 9.1 TNOs with semimajor axes between 38 and 60 AU (*black points*) are shown in this plot of semimajor axis versus inclination. Resolved TNBs are shown as *red triangles* while possible contact binaries identified from light curve analysis are shown as *blue stars*.

[1] See http://www.johnstonsarchive.net/astro/asteroidmoons.html.

9.2.2 Light curves

Light curve evidence for elongated bodies, close binaries, and/or bilobed objects (contact binaries) is found in most small body populations including near-Earth objects (Benner et al., 2015), main belt asteroids (Agarwal et al., 2017), trojans (Ryan et al., 2017), comets (Harmon et al., 2010; Nesvorný et al., 2018), and TNOs (Sheppard and Jewitt, 2004; Lacerda et al., 2014; Thirouin and Sheppard, 2017, 2018; Thirouin et al., 2014, 2017). For a subset of objects that are large and/or close enough to the Earth or a spacecraft direct imaging or radar data is available to confirm their bilobed/contact-binary shape. However, for the majority of objects with similar light curve signatures, the detailed shape is indeterminate. It is nevertheless clear that such objects are quite common across the solar system. The Trans-Neptunian population is no exception with early estimates of the fraction of such objects at about 30% (Sheppard and Jewitt, 2004; Lacerda, 2011).

Potential close/contact binaries in the Kuiper belt are not directly resolvable due to the small angular separation of the system's components but candidates can be identified by rotational light curves (Thirouin et al., 2017) or by occultations. In the case of the Plutino 2001 QG_{298}, Sheppard and Jewitt (2004) argued that the large 1.14 mag amplitude light curve coupled with the 13.77-hour (double-peaked) rotation period was incompatible with an equilibrium fluid body with the required axial ratio. They proposed instead that 2001 QG_{298} is a likely contact binary. The decreased amplitude of 0.7 mag observed in 2010 was compatible with the change in viewing geometry of a contact-binary-like system (Lacerda, 2011).

A second example of a TNO with a large light curve amplitude (0.85 mag) is 2003 SQ_{317} (Lacerda et al., 2014). This object is of special interest because it is a possible member of the Haumea collisional family. The authors noted the possibility of distinguishing between different hydrostatic equilibrium shape models by monitoring the light curve amplitude over multiple years, a technique that can separate elongated ellipsoids and contact-binary shapes given the right alignment and sufficient time baseline

Six additional contact-binary candidates have recently been identified from a survey using *Lowell's Discovery Channel Telescope* and the *Magellan Telescope* (Thirouin et al., 2017; Thirouin and Sheppard, 2017, 2018; Fig. 9.2). Preliminary estimates from this survey suggest that as many as ~40%–50% of Plutinos but only ~10%–25% of cold classicals are candidate contact binaries (Thirouin and Sheppard, 2019). This difference, if confirmed with larger numbers of objects, may be an important clue to the dynamical evolution of these populations.

Finally, although it is routine in the literature to refer to the objects we have described as candidate contact binaries, it is worth pointing out that light curves are insensitive to concavities and that even rubble piles have sufficient strength that they cannot be treated as hydrostatic fluid bodies (Harris and Warner, 2018). On the other hand, the contact-binary shape of the cold classical (486958) 2014 MU_{69}, first revealed by stellar occultation (Buie et al., 2019) and dramatically confirmed by the New Horizons fly-by (Porter et al., 2019) stands as an existence proof and strong validation of both indirect shape determination methods and the models that predict a plentitude of such binaries. The lack of a detectable light curve for this object (Benecchi et al., 2019) is a reminder that the light curve technique is sensitive to the pole orientation and that many other similar hidden binaries may exist. Continued investigation of the Trans-Neptunian contact-binary population promises to be a productive observational and theoretical enterprise.

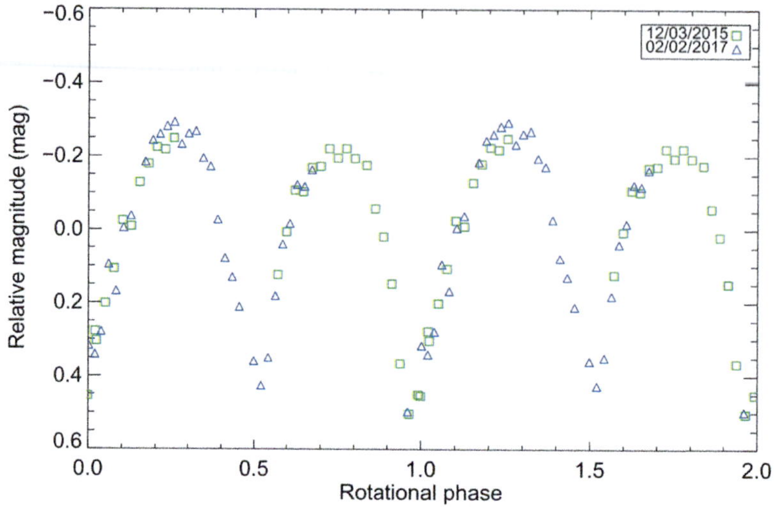

FIG. 9.2 The light curve of 2004 TT_{357} shows a large amplitude and a narrow V-shaped minimum that is characteristic of highly elongated bodies including bilobed configurations. The rotational period of 7.79 hours rules out a hydrostatic fluid shape as a possible source of this elongation, leaving a "contact" binary as a possibility (Thirouin et al., 2017).

9.3 Binary frequency

One of the most frequently asked and seemingly simple questions about TNBs is what fraction of the population they represent. However, any such question must be qualified by knowledge of observational limitations and biases, possible differences between populations, possible size dependencies, knowledge of the separation distribution and regions of orbital stability as well other confounding factors. Any attempt to simply divide the number of detected companions by the number of objects observed will yield a result that is nearly devoid of significance. Following we explore some of the different aspects of determining binary frequency.

While the first TNBs were found from ground-based telescopic observations, the vast majority of known binaries have been discovered by HST (Noll et al., 2008a). Only a small fraction of known binaries are ever widely enough separated that they can be observed from ground-based telescopes without adaptive optics (AO), even in the best of seeing. And AO systems must use a laser guide star with stellar appulse because the targets are too faint for wavefront correction. HST detections have heterogeneous instrumental resolution and observational S/N that must be considered when attempting to estimate the frequency of binaries.

In addition to instrumental considerations, the heterogeneity of the sample itself must be considered. The angular size of the Hill sphere is independent of distance for objects with equal mass. But the wide range in albedo (Stansberry et al., 2008) and differing heliocentric eccentricities result in Hill radii that vary depending on an object's density, albedo, and orbit. The largest TNOs may retain volatile ices on their surfaces that can cause unusually high albedos. Among smaller TNOs, higher albedos are found among the cold classicals than among other dynamical classes (Brucker et al., 2009; Vilenius et al., 2012, 2014). Thus, magnitude-limited surveys will, in general, have a biased sensitivity to secondaries when comparing different classes of objects. We consider two different examples that illustrate issues in appropriately determining binary frequency.

9.3.1 Binary fraction in the cold classicals

One of the earliest published results on the binary fraction in the Kuiper belt populations was the observation that the fraction of binaries in the cold classical population appeared to be greater than that of the dynamically excited populations and, more specifically, the so-called hot classicals (Stephens and Noll, 2006; Noll et al., 2008b) which overlap the cold classicals in semimajor axis space, but have a higher inclination (Levison and Stern, 2001; Brown, 2001). This observation is one of a group of apparently unique features of the cold classicals that along with their color (Tegler and Romanishin, 2000) and the presence of wide binaries (Parker and Kavelaars, 2010) has been used to argue that the cold classicals are a relatively undisturbed remnant of the primordial planetesimal disk.

But the high binary fraction in the cold classicals is subject to a surprising complication that must be considered carefully. Fig. 9.3 plots 110 cold classicals ($i \leq 5.5$ degrees) observed with HST in two different ways. When binned by the absolute magnitude of the unresolved system (i.e., the magnitude reported by ground-based discovery observations), there is a large excess of cold classical binaries for $6.15 > H > 4.95$ mag (Noll et al., 2014). The hot classicals do not show a similar behavior. This effect can be understood as being due to the extreme steepness of the magnitude-frequency distribution (MFD) of the cold classicals (Fraser et al., 2014) resulting in the brightest objects being dominated by binaries. Because the observed targets are heavily biased to the brightest available members of the population, the apparent binary fraction can be magnified. However, even at fainter system magnitudes, $H > 6.15$ mag, the binary frequency among cold classicals is ~20% and exceeds that found in the hot classicals in the same magnitude range.

The higher albedo for cold classicals means that the mass of a cold classical is likely to be lower than a hot classical of the same absolute magnitude. Because the Hill radius scales with the system mass, the likelihood of finding a secondary at the same separation in terms of Hill radius is, therefore, lower for a cold classical than for a hot classical of the same absolute magnitude. Thus, the high fraction of cold classical binaries appears to be a robust feature of the population.

When the same data are plotted as a function of the absolute magnitude of the primary, there is no excess among the largest objects and the frequency is roughly constant down to $H \sim 8$ mag. Crudely debiasing by scaling all binaries to the faintest absolute magnitude bin and removing undetectable secondaries does not significantly alter this result. Although the dip from $H \sim 6.5$–7.0 appears significant, Nesvorný et al. (2011) used the lack of a sustained drop off at even fainter magnitudes to constrain collisional grinding among cold classicals. Further observations of cold classicals, especially fainter ones, are needed to better understand this structure.

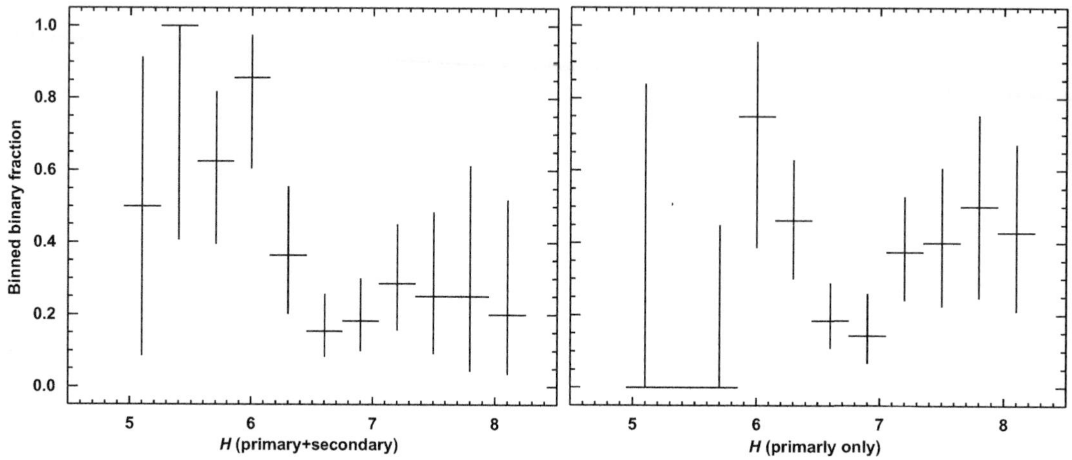

FIG. 9.3 The binned binary fraction of cold classical KBOs observed with HST is plotted as a function of the overall system magnitude (*left*) and by the brightness of the largest component (*right*). The large excess of binaries among the brightest systems (*left*) is the product of an extremely steep MFD that causes the brightest cold classicals to be dominated by near-equal mass binaries (Noll et al., 2014). When plotted by the magnitude of the largest component (*right*), the lack of a drop-off at faint magnitudes is evidence that the observed slope change in the MFD is not due to collisional grinding (Nesvorný et al., 2011).

9.3.2 Resonant binaries

Binaries found in dynamically excited populations, including hot classicals, scattered disk, and Resonant populations, are potentially diagnostic indicators of the dynamical perturbations experienced by these objects over their histories. Noll et al. (2012) examined binaries in populations in mean-motion resonance with Neptune where a total of 76 Plutinos and 96 objects in other resonances had been observed with HST. As shown in Fig. 9.4, the lack of binary Plutinos at low inclination is particularly striking, with a binary fraction of only $5^{+6}_{-2}\%$ for $i < 12$ degrees and $5 \leq H \leq 8$ mag. In this same data set, there are no binary Plutinos with $i < 5.5$ degrees. This can be compared to $27^{+16}_{-9}\%$ found for the 2:1 resonance with $i < 12$ degrees in the same absolute magnitude range, a value similar to that found for the cold classicals ($29^{+7}_{-6}\%$). This result appears to be consistent with the predicted behavior of low-*e* migration of Neptune and coherent capture into resonances (Murray-Clay and Schlichting, 2011) if the 2:1 resonance swept over the cold classicals, but the 3:2 resonance did not.

9.4 Mutual orbits

Orbits of TNBs have mostly been obtained from multiepoch imaging with sufficiently high spatial resolution to separate the two objects. The widest binaries are separated by a few arcsec, and can be resolved even in natural seeing, but most are separated by much less, requiring AO or the HST.

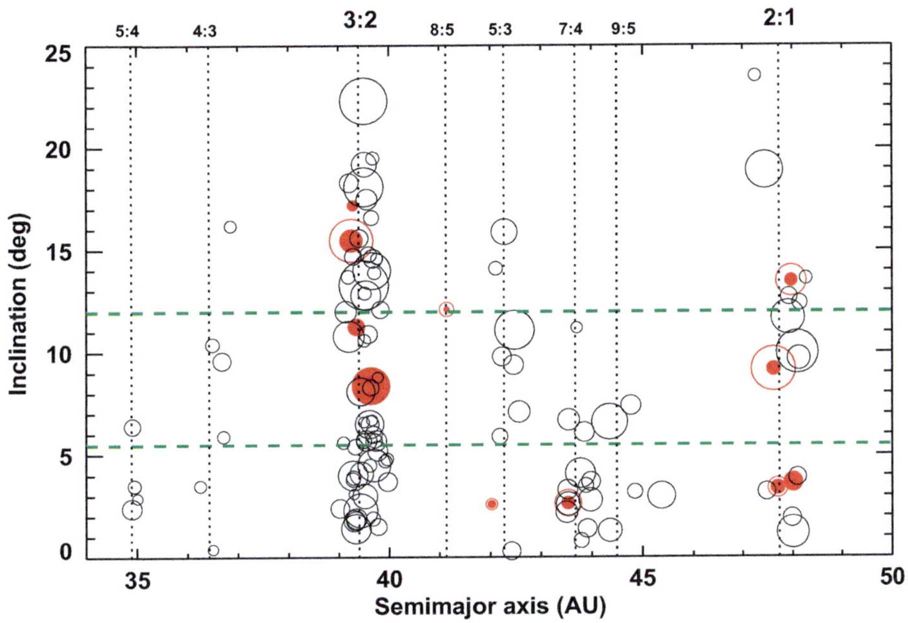

FIG. 9.4 Resonant Trans-Neptunians with absolute magnitudes $5 \leq H \leq 8$ mag observed by HST as of 2012 are shown by *circular symbols*. The size of the symbol scales inversely with absolute magnitude. Objects identified as binaries are shown in *red* with the primary represented by an *open circle* and the secondary shown as a *filled circle*. Resonances with identified objects are shown along the top of the figure. The *green dashed horizontal lines* at $i = 5.5$ degrees and $i = 12$ degrees mark two proposed "crossover" boundaries of the cold and hot classical inclination distributions. Objects captured by low-e resonance sweeping should have inclinations mostly below these bounds, as described in the text.

Resolved observations at four epochs can, in principle, determine the seven parameters required to fully describe a Keplerian binary orbit (total mass and six orbital elements), but in practice this is a function of how well the observations sample the orbit. The sampling can be improved by using the accumulating collection of data to optimize the timing of each subsequent observation (e.g., Grundy et al., 2008). The mirror ambiguity of an orbit projected onto the sky-plane can eventually be resolved through changing observation geometry, but it takes multiple years for sufficient parallax to accumulate. In some cases, the similarity in brightness between the two components makes it impossible to tell which of them is which from one epoch to the next—more observations are then required to yield a unique solution. Despite these complications, the last decade has seen significant progress. Of the 86 TNBs/multiples known, the orbit is fully known for 37. For 11 more, the period, semimajor axis, and eccentricity are known, but the sky-plane mirror ambiguity remains to be broken. These orbital parameters are collected in Table 9.1

TABLE 9.1 Mutual orbital parameters of TNBs.

Object	*P* (days)	*a* (km)	*e*	Incl.[a] (degrees)	M_{sys} (10^{18} kg)	a/r_H[b]	Refs.
26308 1998 SM_{165}	130.154	11,374	0.4732	75.46	6.88	0.02425	Margot et al. (2004) and Grundy et al. (2011)
42355 Typhon—Echidna	18.9815	$15\underline{8}$	0.507	54	0.87	0.0115	Grundy et al. (2008) and Stansberry et al. (2012)
50000 Quaoar—Weywot	12.431	$133\underline{3}$	0.021	–	$12\underline{1}$	0.00364	Fraser et al. (2013) and Brown and Butler (2017)
55637 2002 UX_{25}	8.3095	$47\underline{5}$	0.18	65.0	123	0.00314	Brown and Butler (2017) and Brown (2013)
58534 Logos—Zoe	309.9	$82\underline{2}$	0.546	74.2	0.458	0.0326	Margot et al. (2004), Grundy et al. (2011), and Noll et al. (2004b)
60458 2000 CM_{114}	24.825	2500	0.03	56	2.00	0.00667	Grundy et al. (2019a)
65489 Ceto—Phorcys	9.560	$18\underline{5}$	0.008	–	5.5	0.0071	Grundy et al. (2007)
66652 Borasisi—Pabu	46.289	$45\underline{3}$	0.470	49.4	3.44	0.00916	Noll et al. (2004a)
79360 Sila—Nunam	12.51006	$27\underline{7}$	0.026	123.3	10.8	0.00351	Grundy et al. (2012)
88611 Teharonhiawako—Sawiskera	828.8	$27,\underline{6}00$	0.249	127.6	2.44	0.0581	Grundy et al. (2011) and Osip et al. (2003)
90482 Orcus—Vanth	9.53915	9000	0.0009	106.7	635	0.004311	Brown and Butler (2017), Grundy et al. (2019a), and Brown et al. (2010)
119979 2002 WC_{19}	8.403	$40\underline{9}$	0.20	–	77	0.00329	Grundy et al. (2019a)
120347 Salacia—Actaea	5.49388	$57\underline{2}$	0.010	41.4	492	0.00233	Stansberry et al. (2012) and Grundy et al. (2019a)
123509 2000 WK_{183}	30.913	$23\underline{7}$	0.014	–	1.10	0.00655	Grundy et al. (2011)
134860 2000 OJ_{67}	22.0585	$22\underline{7}$	0.012	–	1.90	0.00523	Grundy et al. (2009)

136199 Eris—Dysnomia	15.78590	3727	0.0062	78.5	16,470	0.004724	Grundy et al. (2011) and Brown and Schaller (2007)
148780 Altjira	139.56	9900	0.344	25.4	3.95	0.0182	Grundy et al. (2011)
160091 2000 OL_{67}	347.1	7800	0.24	–	0.32	0.035	Grundy et al. (2019a)
160256 2002 PD_{149}	1675	26,800	0.588	21.9	0.54	0.099	Grundy et al. (2019a)
174567 Varda	5.7506	481	0.022	–	266	0.00232	Grundy et al. (2015)
229762 G!kún||'hòmdímà	11.3147	604	0.024	32.3	136	0.00383	Grundy et al. (2019b)
275809 2001 QY_{297}	138.118	996	0.418	161.0	4.10	0.01855	Grundy et al. (2011, 2019a)
341520 Mors—Somnus	972.2	2099	0.1494	24.3	0.776	0.0955	Sheppard et al. (2012)
364171 2996 JZ_{81}	1500	3300	0.85	11	1.3	0.090	Parker et al. (2011)
385446 Manwë	110.18	667	0.563	49.1	1.94	0.0167	Grundy et al. (2014)
469514 2003 QA_{91}	10.1089	159	0.02	–	3.1	0.00318	Grundy et al. (2019a)
469705 Ķágára	128.11	7700	0.69	11.2	2.2	0.0178	Grundy et al. (2019a)
508788 2000 CQ_{114}	220.48	694	0.095	44.4	0.545	0.0253	Grundy et al. (2019a)
508869 2002 VT_{130}	30.761	303	0.019	–	2.3	0.0067	Grundy et al. (2019a)
1998 WW_{31}	587.3	2262	0.819	51.7	2.66	0.04839	Grundy et al. (2019a) and Veillet et al. (2002)
1999 OJ_4	84.115	331	0.368	57.4	0.405	0.01448	Grundy et al. (2009, 2019a)
1999 RT_{214}	126.50	3400	0.30	23	0.19	0.0179	Grundy et al. (2019a)
2000 CF_{105}	3900	34,300	0.33	–	0.22	0.166	Parker et al. (2011)
2000 QL_{251}	56.449	499	0.489	134.1	3.09	0.01097	Grundy et al. (2009, 2019a)

Continued

TABLE 9.1 Mutual orbital parameters of TNBs—cont'd

Object	*P* (days)	*a* (km)	*e*	Incl.[a] (degrees)	M_{sys} (10^{18} kg)	a/r_H[b]	Refs.
2001 QC_{298}	19.2287	$38\underline{1}$	0.334	73.7	11.9	0.00502	Margot et al. (2004) and Grundy et al. (2011)
2001 QW_{322}	$62\underline{8}$	$102{,}\underline{1}00$	0.464	152.8	2.14	0.223	Parker et al. (2011)
2001 XR_{254}	125.58	$93\underline{1}$	0.556	21.1	4.06	0.01687	Grundy et al. (2009, 2019a)
2002 XH_{91}	371.1	$22{,}\underline{4}00$	0.71	39	6.5	0.0359	Grundy et al. (2019a)
2003 QY_{90}	309.6	$85\underline{5}$	0.663	51.4	0.52	0.0320	Grundy et al. (2011)
2003 TJ_{58}	137.68	$38\underline{3}$	0.516	63	0.236	0.0187	Grundy et al. (2011, 2019a)
2003 UN_{284}	$31\underline{8}$	$5\underline{4}00$	0.38	23.0	1.27	0.145	Parker et al. (2011)
2004 PB_{108}	97.020	10,400	0.438	83.2	9.5	0.0148	Grundy et al. (2009, 2019a)
2005 EO_{304}	$3\underline{6}00$	$6\underline{9}00$	0.21	–	2.0	0.155	Parker et al. (2011)
2006 BR_{284}	1500	$25{,}\underline{4}00$	0.275	54.1	0.576	0.0884	Parker et al. (2011)
2006 CH_{69}	$14\underline{2}$	$2\underline{7}00$	0.896	133	0.8	0.081	Parker et al. (2011)

[a] *Inclination between the mutual orbit and the heliocentric orbit, where known.*
[b] *Semimajor axis in units of Hill radii, a measure of how tightly bound the orbit is.*
[c] *The first uncertain decimal in each figure is either underlined or the last decimal shown.*

9.5 Properties derived from orbits

9.5.1 Mass and density

The total mass of a binary system can be calculated once the period and semimajor axis of the binary orbit are known. Mass is a fundamental property, providing valuable constraints on dynamics, as well as composition and internal structure, via the bulk density computed by dividing the mass by the estimated total volume of the bodies in the system. Object sizes, necessary to derive density, have been challenging to obtain. Observations at thermal infrared wavelengths can constrain radii to about ±10% precision, translating to a ±30% uncertainty in volume and density. As shown in Fig. 9.5, TNBs with system masses, M_{sys}, of $\sim 10^{19}$ kg and less, corresponding roughly to component diameters of $d < 300$ km, have very low densities, mostly less than 100 kg m^{-3}. More massive objects, $M_{sys} > 10^{21}$ kg, $d > 900$ km have higher densities, generally greater than 1500 kg m^{-3}. In between, diverse densities are seen among the mid-sized bodies. Brown (2013) argued that the objects in the 300–900-km-diameter-size range are too large to have retained much internal porosity, and suggested that their lower densities result from predominately ice-rich compositions, with more rock-rich compositions in the larger bodies accounting for their higher densities. However, a plot of density versus mass (Fig. 9.5) shows a consistent trend suggesting that a gradual reduction in pore space with increasing mass can explain the variation in density without resorting to ad hoc assumptions about composition. Looking ahead, stellar occultations enabled by the GAIA astrometric catalog have the potential to provide much more accurate sizes, and consequently better densities that will allow more detailed study of the densities of TNBs.

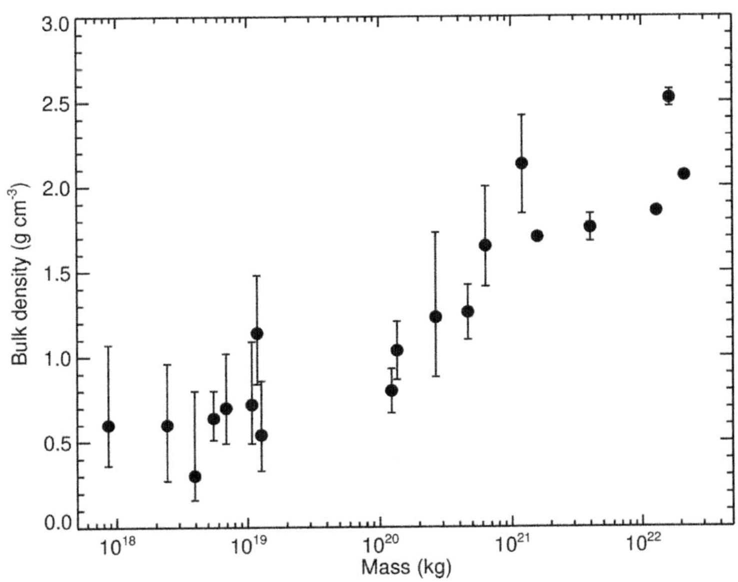

FIG. 9.5 Densities of TNOs versus mass show a smooth increasing trend over the 10^{19}–10^{22} kg mass range. (Triton and Charon are included because they fall in this mass range and have well-determined densities.)

9.5.2 Orbital eccentricity, inclination, and separation

The orbital properties of binaries can provide information about their formation circumstances as well as their subsequent dynamical evolution. For instance, circular orbits occur mostly among tightly bound binaries (Fig. 9.6), which is not a surprise, since these are the systems most prone to tidal evolution. Wide binaries with appreciable inclinations between their mutual orbits and heliocentric orbits are subject to the Kozai-Lidov effect (Kozai, 1962; Lidov, 1962) where the gravity of the distant Sun perturbs the mutual orbit, causing an oscillation between e and i. Grundy et al. (2011) argued that the broad peak of eccentricities around $e = 0.5$ was consistent with a population undergoing these oscillations. Because wide binaries are easier to discover and characterize than tight ones, it is likely that the tight circular orbits are underrepresented among the known orbits relative to the wider more eccentric ones.

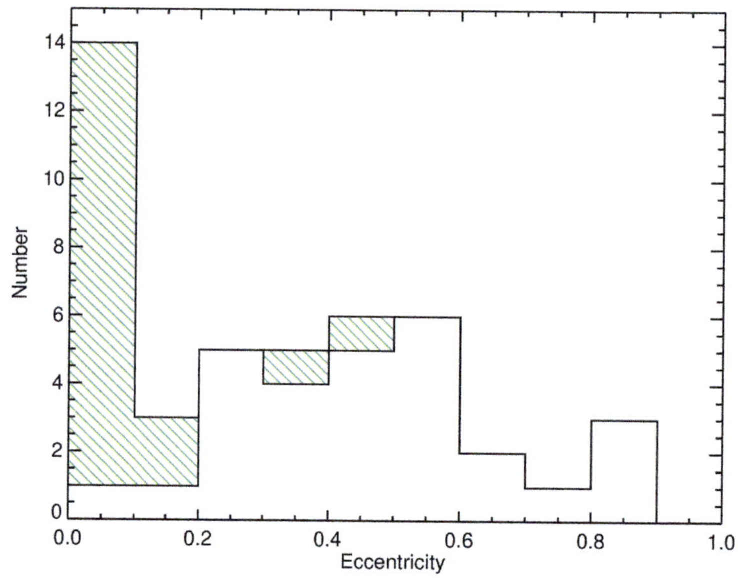

FIG. 9.6 Distribution of eccentricities of known binary orbits. Close binaries, that is, those with $a/r_H < 0.1$, are indicated by *green* hatching. Only one of the nearly circular orbits ($e < 0.1$) is not a close binary. At higher eccentricities most binaries are wider, although four close binaries also have eccentricities higher than 0.1.

The observation that there are considerably more prograde orbits than retrograde ones (Fig. 9.7; Grundy et al., 2009, 2011) is contrary to expectation from various capture scenarios (e.g., Goldreich et al., 2002; Schlichting and Sari, 2008; Kominami et al., 2011) that favor retrograde orbits, or, at most equal proportions of the two. Collisions should also not produce an excess of prograde systems Gravitational collapse from locally concentrated regions of the nebula, however, might preferentially produce prograde binaries (Nesvorný et al., 2010; Grundy et al., 2019a).

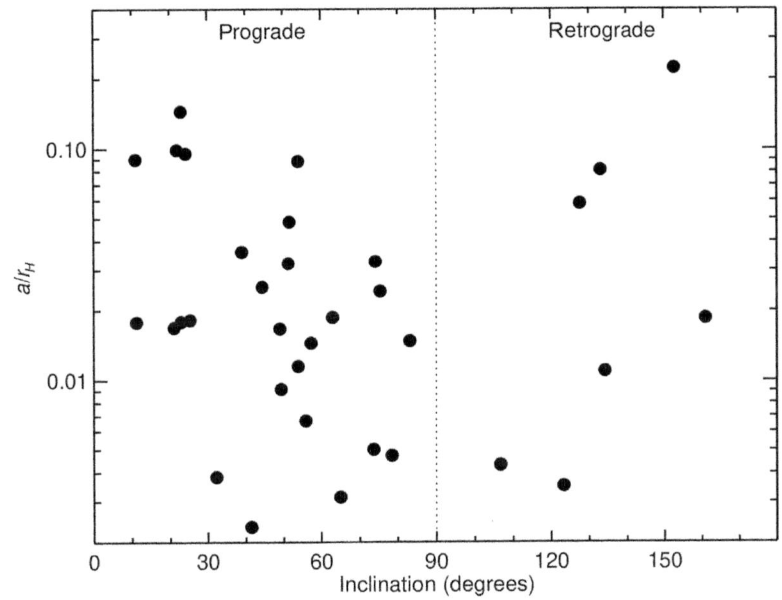

FIG. 9.7 Prograde orbits vastly outnumber retrograde orbits, except among the widest binaries. There is also a shortage of inclinations near 90 degrees among the wider binaries.

The widest binaries, those with a/r_H greater than about 0.05, do not show a strong prograde-retrograde asymmetry but do lack orbits with inclinations near 90 degrees. These highly inclined orbits are the ones that undergo the largest amplitude Kozai oscillations, leading to times of especially high eccentricity that could accelerate tidal evolution (e.g., Perets and Naoz, 2009; Naoz et al., 2010; Porter and Grundy, 2012), converting them into much tighter binaries, or perhaps leading to their becoming unbound.

A plot of a/r_H versus the excitation of the heliocentric orbit shows that most of the TNBs on more perturbed orbits are tightly bound (Fig. 9.8). This may be because the process of perturbing them into their present-day heliocentric orbits also dissociated most of the less tightly bound binaries. An exception is Mors-Somnus, a 3:2 resonant object (Sheppard et al., 2012) that is also a wide binary. Its existence suggests that nondestructive paths to excited heliocentric orbits, such as resonance sweeping, may have affected TNO populations.

9.5.3 Mutual events and occultations

Two additional channels for obtaining physical information that have begun to be exploited are mutual events and occultations. Mutual events occur twice in each TNB orbit when the binary orbit plane and the line of sight to Earth are nearly aligned. The duration of the individual mutual events and mutual event season depend on the particulars of the binary orbit, but are on the order of hours for individual events and approximate years for the season. With >40 TNB orbits known, it is reasonable to expect that one or two TNBs will have observable mutual events per decade.

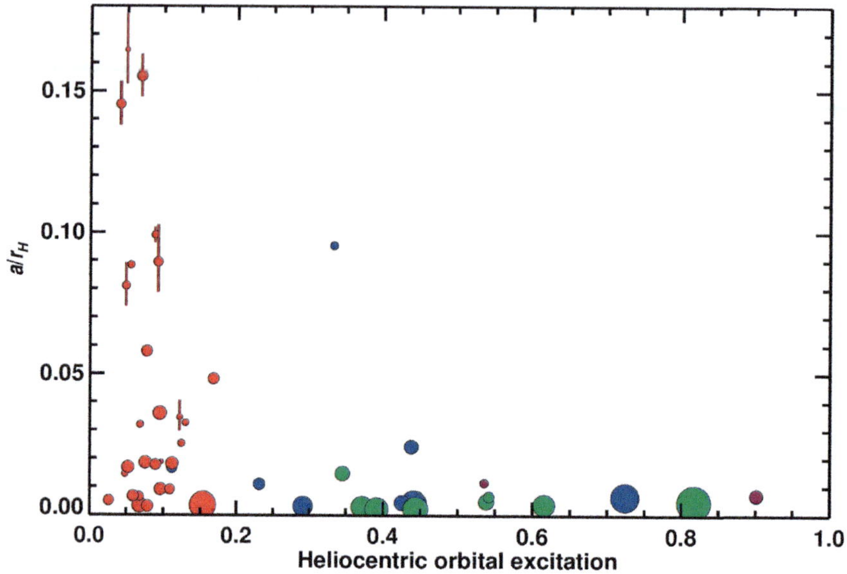

FIG. 9.8 Binary orbital looseness, as measured by a/r_H, versus the excitation of the heliocentric obit, as measured by $\sqrt{\sin(i_{\odot})^2 + e_{\odot}^2}$, where $i_{\odot}$ and $e_{\odot}$ are the inclination and eccentricity of the heliocentric orbit. Colors indicate dynamical class: *red* for classical TNOs in low eccentricity, low inclination orbits, *blue* for objects in mean-motion resonance with Neptune, and *green* for scattered disk objects. The two *purple* objects are Centaurs, effectively scattered disk objects that have been perturbed into short-lived orbits crossing those of one or more of the giant planets.

Mutual events of the binary (79360) Sila-Nunam were predicted by Grundy et al. (2012). Benecchi et al. (2014) successfully observed one such event in February 2013 from multiple telescope facilities. Grundy et al. (2014) also predicted mutual events for (385446) Manwë-Thorondor in 2015–17. Attempts to observe Manwë-Thorondor with the SOAR telescope in August 2016 were reported by Rabinowitz et al. (2016), but final analysis remains pending. Fabrycky et al. (2008) and Ragozzine and Brown (2009) described a series of mutual events involving Haumea and one of its satellites, Namaka and attempted to observe one event using HST. Results of this observation were inconclusive. Fully realizing the potential of TNB mutual events will require more significant investments of large telescope observing time than has been available up to now.

Occultations by TNBs are another avenue of study that has only begun to be exploited. Sickafoose et al. (2019) report a ground-based occultation of Vanth, the satellite of Orcus, leading to diameter determination of 443±10 km assuming a spheroidal body. The occultation-derived contact-binary shape of (486958) 2014 MU_{69} (Buie et al., 2019) is another poignant example of the potential of occultation observations. The availability of precise stellar astrometric data from Gaia (e.g., Brown et al., 2018) promises to greatly expand the opportunities for similar observations in the future.

9.6 Colors

One of the most significant results to emerge from the study of TNBs has been the near identity of component colors (Noll et al., 2008a; Benecchi et al., 2009). This identity coupled with the wide range of colors of TNBs puts strong constraints on both formation mechanisms and subsequent color evolution. Capture models are limited to homogeneous regions and/or times within the larger potential color range seen in the current Kuiper belt. Subsequent physical evolution is also limited; in particular, nondisruptive collisions do not appear to result in color changes which would be seen as rotational color variability and/or variations in component colors.

Exceptions to binary component color identity among the largest TNOs and their satellites are attributable to highly volatile ices that are retained on the larger body, but are not stable on smaller components (e.g., Parker et al., 2016). More generally, it is possible that the satellites of the largest TNOs may have originated from impacts (e.g., Leinhardt et al., 2010), rather than by coaccretion or similar coformation mechanisms (see Section 9.6) and thus could have more global compositional differences.

Sheppard et al. (2012) noted that the wide Plutino binary, Mors-Somnus, has a very red color and argued for capture from the cold classical population based on the combination of this color and the near-equal component sizes. Fraser et al. (2017) used a similar color-based argument to leverage the neutral colors of several wide cold classical binaries into a more global conclusion on the compositional segregation of the protoplanetary disk and the ubiquity of binaries. However, the wide range of colors found in all Trans-Neptunian populations (e.g., Peixinho et al., 2015) greatly limits the use of color as an identifier capable of linking an individual object to a parent population. A much larger data set will be required in order to test these and similar ideas.

9.7 Formation scenarios

TNBs are an important tracer of planetesimal formation in the outer solar system and understanding how binaries formed is a critical test of planetesimal formation models. For example, if formed by capture (e.g., Goldreich et al., 2002), they constrain the dynamical conditions in the early protoplanetary disk (Schlichting and Sari, 2008). However, as noted earlier, the matching colors of binary components and prograde orbital orientations argue against this formation model.

An appealing alternative is a model in which planetesimals formed by the streaming instability (SI; Youdin and Goodman, 2005). The results of detailed hydrodynamical simulations (Johansen et al., 2009; Simon et al., 2017; Li et al., 2018) show that the SI clumps become gravitationally bound and result in a characteristic planetesimal size of $\sim$100 km (depending on the details of disk parameters; Johansen et al., 2009).

The existing SI simulations also indicate that the collapsing clouds have vigorous rotation and their initial angular momentum typically exceeds, by a large margin, that of a critically rotating Jacobian ellipsoid with density $\rho = 1000\,\mathrm{kg\,m^{-3}}$. This large excess of angular momentum that must be lost can produce a binary (Nesvorný et al., 2010, hereafter NYR10) with most of the angular momentum being deposited into the binary orbit.

NYR10 studied the collapse stage and Fig. 9.8 summarizes some results of the gravitational collapse. Equal-size KBO binaries ($R_2/R_1 > 0.5$, where R_1 and R_2 are the radii of primary and secondary components) form in the simulations if the initial clumps contract below the Hill radius ($R \leq 0.6\, R_H$) and/or have fast rotation ($\omega \geq 0.5\, \omega_c$). This is roughly consistent with the results of the SI simulations where the clumps are small $R \sim 0.1 R_H$ and rotate as fast as they could. The gravitational collapse simulations also predict large separations between binary components ($\sim$1000 to $\sim 10^5$ km) and a broad distribution of binary eccentricities. Depending on the local disk conditions, the gravitational collapse may be capable of producing up to 100% binary fraction (for $\omega \geq 0.5\omega_c$). Because the binary components form from the local composition mix, the components should have identical compositions and colors, consistent with observations (see Section 9.5).

Another argument can be based on the inclination distribution of binary orbits (see Section 9.4). The SI simulations can be used to extract the orientation of the angular momentum vectors of individual clumps. NYR10 showed that the initial angular momentum vectors are a good proxy for the final orientation of binary orbits, allowing the SI results to be directly compared with observed binary orbits (Table 9.1); the results show a strikingly good match (Fig. 9.9).

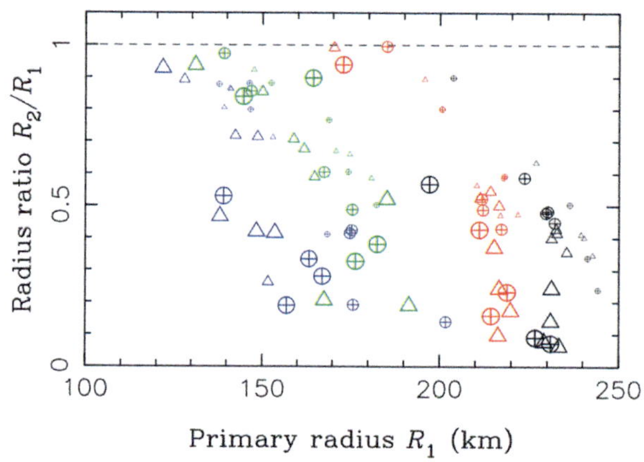

FIG. 9.9 The plot shows the sizes of binary components obtained in 96 simulations of gravitational collapse. Different initial conditions were used in each of these simulations. The initial rotation was assumed to be prograde (*triangles*) or retrograde (*crossed circles*) with respect to the orbital motion. The initial size of clumps, R, was set to be a fraction of the Hill radius at 30 AU, R_H (0.4, 0.6, and 0.8 R_H as indicated by the symbol size). The initial rotation was assumed to be a fraction of the critical frequency Ω_c (see NYR10 for definition): 0.1 (*black*), 0.25 (*red*), 0.5 (*green*), and 0.75 Ω_c (*blue*). For each setup, we performed four different simulations with slightly altered distributions of $N = 10^5$ superparticles.

9.8 Future observations and summary

The entire field of study of Trans-Neptunian space stands on the cusp of a major revolution in data availability to be brought about by new surveys that will greatly expand the number of known TNOs. Because these surveys will be carried out by ground-based telescopes with limited angular resolution, a proportionally large jump in the number of known binaries will come about as well, but more slowly as the result of painstaking follow ups with advanced AO or space-based telescopes. More binary orbits will also be determined, both from already-known binaries and yet-to-be discovered systems. Taken in whole on the timescale of the next decade, these developments promise improved statistics that will bring some of the conclusions reached so far into sharper focus.

Perhaps the greatest potential for the study of TNBs lies in the further observations of TNO light curves and of stellar occultations. Both of these techniques make it possible to explore close-in binaries—a population that cannot be reached by direct imaging. These efforts should be bolstered by the incontrovertible evidence delivered by the New Horizons mission that a heretofore underappreciated population of primordial binary objects awaits our study.

Finally, the expansion of our understanding of TNBs opens the door to comparisons with other related population including the Jupiter and Neptune Trojans, Centaurs, and comets. At least one of the targets of the Lucy mission, the Patroclus-Menoetius binary, has properties that look very much like those of many of the known TNBs (e.g., Buie et al., 2015). Answering whether this signals a genetic link or is simply a matter of chance will rely both on spacecraft data and a deep understanding of the binary populations of the outer solar system.

Acknowledgments

William M. Grundy and Keith S. Noll gratefully acknowledge support from the National Aeronautics and Space Administration (NASA)/ESA Hubble Space Telescope programs 13404, 13668, 13692, and 15233. Support for these programs was provided by the NASA through grants from the Space Telescope Science Institute, operated by the Association of Universities for Research in Astronomy, Inc. under NASA contract NAS 5-26555. Additional support was provided to William M. Grundy through NASA Keck PI Data Awards, administered by the NASA Exoplanet Science Institute. David Nesvorný's work is supported by NASA's Emerging Worlds. Audrey Thirouin is partly supported by the National Science Foundation, grant number AST-1734484. Audrey Thirouin also acknowledges Scott Sheppard for his contribution to the contact-binary search and characterization.

References

Agarwal, J., Jewitt, D., Mutchler, M., Weaver, H., Larson, S., 2017. A binary main-belt comet. Nature 549, 357.

Benecchi, S.D., Noll, K.S., Grundy, W.M., Buie, M.W., Stephens, D.C., Levison, H.F., 2009. The correlated colors of Trans-Neptunian binaries. Icarus 200, 292–303.

Benecchi, S.D., Noll, K.S., Thirouin, A., Ryan, E., Grundy, W.M., Verbiscer, A., et al., 2014. The UT 7/8 February 2013 Sila-Nunam mutual event Spiandspi future predictions. Icarus 229, 423–427.

Benecchi, S.D., Porter, S., Buie, M.W., Zangari, A.M., Verbiscer, A.J., Noll, K.S., et al., 2019. The HST lightcurve of (486958) 2014 MU_{69}. Icarus (in press). https://doi.org/10.1016/j.icarus.2019.01.023.

Benner, L.A.M., Busch, M.W., Giorgini, J.D., Taylor, P.A., Margot, J.L., 2015. Radar observations of near-Earth and main-belt asteroids. In: Michel, P., DeMeo, F.E., Bottke, W.F. (Eds.), Asteroids IV. University of Arizona Press, Tucson, AZ, pp. 165–182.

Brown, M.E., 2001. The inclination distribution of the Kuiper belt. Astron. J. 121, 2804–2814.

Brown, M.E., 2013. The density of mid-sized Kuiper belt object 2002 UX_{25} and the formation of the dwarf planets. Astrophys. J. Lett. 778, L34.1-5.

Brown, M.E., Butler, B.J., 2017. The density of mid-sized Kuiper belt objects from alma thermal observations. Astron. J. 154, 19.1-7.

Brown, M.E., Schaller, E.L., 2007. The mass of dwarf planet Eris. Science 316, 1585.

Brown, M.E., Ragozzine, D., Stansberry, J., Fraser, W.C., 2010. The size, density, and formation of the Orcus-Vanth system in the Kuiper belt. Astron. J. 139, 2700–2705.

Brown, A.G.A., Vallenari, A., Prusti, T., de Bruijne, H.J., Babusiaux, C., Bailer-Jones, C.A.L., et al., 2018. Gaia data release 2: summary of the contents and survey properties. Astron. Astrophys. 616, A1.

Brucker, M.J., Grundy, W.M., Stansberry, J.A., Spencer, J.R., Sheppard, S.S., Chiang, E.I., Buie, M.W., 2009. High albedos of low inclination classical Kuiper belt objects. Icarus 201, 284–294.

Buie, M.W., Olkin, C.B., Merline, W.J., Walsh, K.J., Levison, H.F., Timerson, B., et al., 2015. Size and shape from stellar occultation observations of the double Jupiter trojan Patroclus and Menoetius. Astron. J. 149, 113.

Buie, M.W., Porter, S.B., Tamblyn, P., Terrell, D., Verbiscer, A.J., Keeney, B., et al., 2019. Stellar occultation results for (486958) 2014 MU_{69}: a pathfinding effort for the new Horizons flyby. In: 50th Lunar and Planetary Science Conference, The Woodlands, Texas, ID 3120.

Elliot, J., Kern, S.D., Clancy, K.B., Gulbis, A.A.S., Millis, R.L., Buie, M.W., et al., 2005. The deep ecliptic survey: a search for Kuiper belt objects and centaurs. II. Dynamical classification, the Kuiper belt plane, and the core population. Astron. J. 129, 1117–1162.

Fabrycky, D.C., Ragozzine, D., Brown, M.E., Holman, M.J., 2008. Mutual events of (136108) 2003 EL_{61} and S/2005 (136108) 2. Int. Astron. Union Circ. 8949, 1.

Fraser, W.C., Batygin, K., Brown, M.E., Bouchez, A., 2013. The mass, orbit, and tidal evolution of the Quaoar-Weywot system. Icarus 222, 357–363.

Fraser, W.C., Brown, M.E., Morbidelli, A., Parker, A., Batygin, K., 2014. The absolute magnitude distribution of Kuiper belt objects. Astrophys. J. 742, 100F.

Fraser, W.C., Bannister, M.T., Pike, R.E., Marsset, M., Schwamb, M.E., Kavelaars, J.J., et al., 2017. All planetesimals born near the Kuiper belt formed as binaries. Nat. Astron. 1, 0088.

Gladman, B., Marsden, B.G., VanLaerhoven, C., 2008. Nomenclature in the outer solar system. In: Barucci, M.A., Boehnhardt, H., Cruikshank, D.P., Morbidelli, A. (Eds.), The Solar System Beyond Neptune. University of Arizona Press, Tucson, AZ, pp. 43–57.

Goldreich, P., Lithwick, Y., Sari, R., 2002. Formation of Kuiper-belt binaries by dynamical friction and three-body encounters. Nature 420, 643–646.

Grundy, W.M., Stansberry, J.A., Noll, K.S., Stephens, D.C., Trilling, D.E., Kern, S.D., Spencer, J.R., Cruikshank, D.P., Levison, H.F., 2007. The orbit, mass, size, Albedo, and density of (65489) Ceto/Phorcys: a tidally-evolved binary centaur. Icarus 191, 286–297.

Grundy, W.M., Noll, K.S., Virtanen, J., Muinonen, K., Kern, S.D., Stephens, D.C., et al., 2008. (42355) Typhon Echidna: scheduling observations for binary orbit determination. Icarus 197, 260–268.

Grundy, W.M., Noll, K.S., Buie, M.W., Benecchi, S.D., Stephens, D.C., Levison, H.F., 2009. Mutual orbits and masses of six transneptunian binaries. Icarus 200, 627–635.

Grundy, W.M., Noll, K.S., Nimmo, F., Roe, H.G., Buie, M.W., Porter, S.B., et al., 2011. Five new and three improved mutual orbits of transneptunian binaries. Icarus 213, 678–692.

Grundy, W.M., Benecchi, S.D., Rabinowitz, D.L., Porter, S.B., Wasserman, L.H., Skiff, B.A., et al., 2012. Mutual events in the cold classical transneptunian binary system Sila and Nunam. Icarus 220, 74–83.

Grundy, W.M., Benecchi, S.D., Porter, S.B., Noll, K.S., 2014. The orbit of transneptunian binary Manwë and Thorondor and their upcoming mutual events. Icarus 237, 1–8.

Grundy, W.M., Porter, S.B., Benecchi, S.D., Roe, H.G., Noll, K.S., Trujillo, C.A., Thirouin, A., Stansberry, J.A., Barker, E., Levison, H.F., 2015. The mutual orbit, mass, and density of the large transneptunian binary system Varda and Ilmarë. Icarus 257, 130–138.

Grundy, W.M., Noll, K.S., Roe, H.G., Buie, M.W., Porter, S.B., et al., 2019a. Mutual orbit orientations of transneptunian binaries. Icarus (in press). https://doi.org/0.1016/j.icarus.2019.03.035.

Grundy, W.M., Noll, K.S., Buie, M.W., Benecchi, S.D., Ragozzine, D., Roe, H.G., 2019b. The mutual orbit, mass, and density of transneptunian binary G!kún||'hòmdímà (229762 2007 UK_{126}). Icarus (in press). https://doi.org/10.1016/j.icarus.2018.12.037.

Harmon, J.K., Nolan, M.C., Giorgini, J.D., Howell, E.S., 2010. Radar observations of 8P/Tuttle: a contact-binary comet. Icarus 207, 499–502.

Harris, A.W., Warner, B., 2018. Asteroid lightcurves: can't tell a contact binary from a brick. In: AAS/Division for Planetary Sciences Meeting No. 50, ID.414.03.

Johansen, A., Youdin, A., Low, M.M., 2009. Particle clumping and planetesimal formation depend strongly on metallicity. Astrophys. J. 704, L75–L79.

Johnston, W.R., 2018. Binary minor planets compilation C2.0 (PDS4). In: NASA Planetary Data System.
Kominami, J.D., Makino, J., Daisaka, H., 2011. Binary formation in planetesimal disks I. Equal mass planetesimals. Publ. Astron. Soc. Jpn 63, 1331–1344.
Kozai, Y., 1962. Secular perturbations of asteroids with high inclination and eccentricity. Astron. J. 67, 591–598.
Lacerda, P., 2011. A change in the light curve of Kuiper belt contact binary (139775) 2001 QG_{298}. Astron. J. 142, 90.
Lacerda, P., McNeill, A., Peixinho, N., 2014. The unusual Kuiper belt object 2003 SQ_{317}. Mon. Not. R. Astron. Soc. 437, 3824–3831.
Leinhardt, Z.M., Marcus, R.A., Stewart, S.T., 2010. The formation of a collisional family around the dwarf planet Haumea. Astrophys. J. 714, 1789–1799.
Levison, H.F., Stern, A.S., 2001. On the size dependence of the inclination distribution of the main Kuiper belt. Astron. J. 121, 1730–1735.
Li, R., Youdin, A.N., Simon, J.B., 2018. On the numerical robustness of the streaming instability: particle concentration and gas dynamics in protoplanetary disks. Astrophys. J. 862, 14.
Lidov, M.L., 1962. The evolution of orbits of artificial satellites of planets under the action of gravitational perturbations of external bodies. Planet. Space Sci. 9, 719–759.
Margot, J.L., Brown, M.E., Trujillo, C.A., Sari, R., 2004. Hst observations of Kuiper belt binaries. Bull. Am. Astron. Soc. 36, 1081.
Murray-Clay, R.A., Schlichting, H.E., 2011. Using Kuiper belt binaries to constrain Neptune's migration history. Astrophys. J. 730, 132.
Naoz, S., Perets, H.B., Ragozzine, D., 2010. The observed orbital properties of binary minor planets. Astrophys. J. 719, 1775–1783.
Nesvorný, D., Youdin, A.N., Richardson, D.C., 2010. Formation of Kuiper belt binaries by gravitational collapse. Astron. J. 140, 785–793.
Nesvorný, D., Vokrouhlickyì, D., Bottke, W.F., Noll, K., Levison, H.F., 2011. Observed binary fraction sets limits on the extent of collisional grinding in the Kuiper belt. Astron. J. 141, 159.
Nesvorný, D., Parker, J., Vokrouhlickyì, D., 2018. Bi-lobed shape of comet 67P from a collapsed binary. Astron. J. 155, 246.
Noll, K.S., Stephens, D.C., Grundy, W.M., Griffin, I., 2004a. The orbit, mass, and Albedo of (66652) 1999 RZ_{253}. Icarus 172, 402–407.
Noll, K.S., Stephens, D.C., Grundy, W.M., Osip, D., Griffin, I., 2004b. The orbit and Albedo of Trans-Neptunian binary (58534) 1997 CQ_{29}. Astron. J. 128, 2547–2552.
Noll, K.S., Grundy, W.M., Chiang, E.I., Margot, J.L., Kern, S.D., 2008a. Binaries in the Kuiper belt. In: Barucci, M.A., Boehnhardt, H., Cruikshank, D.P., Morbidelli, A. (Eds.), The Solar System Beyond Neptune. University of Arizona Press, Tucson, AZ, pp. 345–363.
Noll, K.S., Grundy, W.M., Stephens, D.C., Levison, H.F., Kern, S.D., 2008b. Evidence for two populations of classical transneptunian objects: the strong inclination dependence of classical binaries. Icarus 194, 758–768.
Noll, K.S., Grundy, W.M., Schlichting, H.E., Murray-Clay, R.A., Benecchi, S.D., 2012. Resonant transneptunian binaries: evidence for slow migration of Neptune. In: AAS/Division for Planetary Sciences Meeting No. 44, ID.405.07.
Noll, K.S., Parker, A.W., Grundy, W.M., 2014. All bright Cold Classicals are binary. In: AAS/Division for Planetary Sciences Meeting No. 46, ID.507.05.
Osip, D.J., Kern, S.D., Elliot, J.L., 2003. Physical characterization of the binary Edgeworth-Kuiper belt object 2001 QT297. Earth Moon Planets 92, 409–421.
Parker, A.H., Kavelaars, J.J., 2010. Destruction of binary minor planets during Neptune scattering. Astrophys. J. 722, L204–L208.
Parker, A.H., Kavelaars, J.J., Petit, J.M., Jones, L., Gladman, B., Parker, J., 2011. Characterization of seven ultra-wide transneptunian binaries. Astrophys. J. 743, 1.
Parker, A.H., Buie, M.W., Grundy, W.M., Noll, K.S., 2016. Discovery of a Makemakean moon. Astrophys. J. 825, L9.
Peixinho, N., Delsanti, A., Doressoundiram, A., 2015. Reanalyzing the visible colors of Centaurs and KBOs: what is there and what we might be missing. Astron. Astrophys. 577, A35.
Perets, H.B., Naoz, S., 2009. Kozai cycles, tidal friction, and the dynamical evolution of binary minor planets. Astrophys. J. 699, L17–L21.
Porter, S.B., Grundy, W.M., 2012. KCTF evolution of Trans-Neptunian binaries: connecting formation to observation. Icarus 220, 947–957.

Porter, S.B., Bierson, C.J., Umurhan, O., Beyer, R.A., Lauer, T.R., Buie, M.W., et al., 2019. A contact binary in the Kuiper belt: the shape and pole of (486958) 2014 MU_{69}. In: 50th Lunar and Planetary Science Conference, The Woodlands, Texas, ID 1611.

Rabinowitz, D.L., Benecchi, S.D., Grundy, W.M., Thirouin, A., Verbiscer, A.J., 2016. Observations of mutual eclipses by the binary Kuiper belt object Manwë-Thorondor. In: AAS/Division for Planetary Sciences Meeting No. 48, ID.120.10.

Ragozzine, D., Brown, M.E., 2009. Orbits and masses of the satellites of the dwarf planet Haumea (2003 EL_{61}). Astron. J. 137, 4766–4776.

Ryan, E.L., Sharkey, B.N.L., Woodward, C.E., 2017. Trojan asteroids in the Kepler campaign 6 field. Astrophys. J. 153, 116.

Schlichting, H.E., Sari, R., 2008. The ratio of retrograde to prograde orbits: a test for Kuiper belt binary formation theories. Astrophys. J. 686, 741–747.

Sheppard, S.S., Jewitt, D., 2004. Extreme Kuiper belt object 2001 QG_{298} and the fraction of contact binaries. Astrophys. J. 127, 3023–3033.

Sheppard, S.S., Ragozzine, D., Trujillo, C., 2012. 2007 TY_{430}: a cold classical Kuiper belt type binary in the Plutino population. Astron. J. 143, 58.1-13.

Sickafoose, A.A., Bosh, A.S., Levine, S.E., Zuluaga, C.A., Genade, A., Schindler, K., et al., 2019. A stellar occultation by Vanth, a satellite of (90482) Orcus. Icarus 319, 657–668.

Simon, J.B., Armitage, P.J., Youdin, A.N., Li, R., 2017. Evidence for Universality in the initial planetesimal mass function. Astrophys. J. 847, L12.

Stansberry, J.A., Grundy, W.M., Brown, M.E., Cruikshank, D.P., Spencer, J.R., Trilling, D.E., et al., 2008. Physical properties of Kuiper belt and centaur objects: constraints from the Spitzer Space Telescope. In: Barucci, M.A., Boehnhardt, H., Cruikshank, D.P., Morbidelli, A. (Eds.), The Solar System Beyond Neptune. University of Arizona Press, Tucson, AZ, pp. 161–179.

Stansberry, J.A., Grundy, W.M., Mueller, M., Benecchi, S.D., Rieke, G.H., Noll, K.S., Buie, M.W., Levison, H.F., Porter, S.B., Roe, H.G., 2012. Physical properties of transneptunian binaries (120347) Salacia-Actaea and (42355) Typhon-Echidna. Icarus 219, 676–688.

Stephens, D.C., Noll, K.S., 2006. Detection of six trans-Neptunian binaries with NICMOS: a high fraction of binaries in the cold classical disk. Astron. J. 131 (2), 1142–1148.

Tegler, S., Romanishin, W., 2000. Extremely red Kuiper-belt objects in near-circular orbits beyond 40 AU. Nature 407, 979–981.

Thirouin, A., Sheppard, S.S., 2017. A possible dynamically cold classical contact binary: (126719) 2002 CC_{249}. Astron. J. 154, 241.

Thirouin, A., Sheppard, S.S., 2018. The Plutino population: an abundance of contact binaries. Astron. J. 155, 248.

Thirouin, A., Sheppard, S.S., 2019. Lightcurves and rotational properties of the pristine cold classical Kuiper belt objects. Astron. J. (submitted).

Thirouin, A., Sheppard, S.S., Noll, K.S., 2017. 2004 TT_{357}: a potential contact binary in the Trans-Neptunian belt. Astrophys. J. 844, 135.

Thirouin, A., Noll, K.S., Ortiz, J.L., Morales, N., 2014. Rotational properties of the binary and non-binary populations in the Trans-Neptunian belt. Astron. Astrophys. 569, A3.

Veillet, C., Parker, J.W., Griffin, I., Marsden, B., Doressoundiram, A., Buie, M., Tholen, D.J., Connelley, M., Holman, M.J., 2002. The binary Kuiper-belt object 1998 WW_{31}. Nature 416, 711–713.

Vilenius, E., Kiss, C., Mommert, M., Müller, T., Santos-Sanz, P., Pal, A., et al., 2012. "TNOs are Cool": a survey of the Trans-Neptunian region. VI. Herschel/PACS observations and thermal modeling of 19 classical Kuiper belt objects. Astron. Astrophys. 541, A94.1-17.

Vilenius, E., Kiss, C., Müller, T., Mommert, M., Santos-Sanz, P., Pal, A., et al., 2014. "TNOs are Cool": a survey of the Trans-Neptunian region X. Analysis of classical Kuiper belt objects from Herschel and Spitzer observations. Astron. Astrophys. 564, A35.1-18.

Youdin, A.N., Goodman, J., 2005. Streaming instabilities in protoplanetary disks. Astrophys. J. 620, 459–469.

Further reading

Parker, A.H., Kavelaars, J.J., 2012. Collisional evolution of ultra-wide transneptunian binaries. Astrophys. J. 744, 139.1-14.

CHAPTER

10

Trans-Neptunian binary formation and evolution

Adrián Brunini[a,b]

[a]National University of Southern Patagonia, Caleta Olivia Academic Unit, Santa Cruz, Argentina [b]National Council of Scientific and Technical Research, Buenos Aires, Argentina

10.1 Introduction

The fraction of binary objects in the Trans-Neptunian (TN) region is large, ranging from $29^{+7}_{-6}\%$ in the cold classical population to $5.5^{+4}_{-2}\%$ for all other classes combined (Noll et al., 2002; Kern and Elliot, 2006; Lin et al., 2010). The distribution of their orbits is an important tracer of the conditions prevailing during the formation of the outer solar system and its subsequent evolution (Goldreich et al., 2002; Weidenschilling, 2002; Funato et al., 2004; Petit and Mousis, 2004; Astakhov et al., 2005; Nazzario et al., 2007; Parker and Kavelaars, 2010; Nesvorný et al., 2010, 2011; Kominami et al., 2011; Fraser et al., 2017).

Much theoretical work was done and a considerable amount of observational information collected since the publication of the last review on Trans-Neptunian binaries (TNBs) (Noll et al., 2008a). The known population of TNB increased by almost a 100%, and the Trans-Neptunian objects (TNOs) with fully or partially known orbits passed from 13 to ~40 systems (Grundy et al., 2019).

Space-based observations can discover closely spaced binaries with separations $\gtrsim$0.013 (Trujillo and Brown, 2002; Brown and Trujillo, 2002; Noll et al., 2002; Kern and Elliot, 2006; Thirouin et al., 2017), while ground-based observations are sensitive only to wider separations of a fraction of the seeing disk, depending on the instrumental pixel scale and atmospheric effects (Parker, 2012). Up to date, most binaries were discovered with the Hubble Space Telescope (Grundy et al., 2008, 2009, 2011, 2012, 2014, 2015; Noll et al., 2002; Fuentes et al., 2010), thus our current capacity to resolve binaries in the TN region range from minimum separations of ~300 km at 30 AU (the Neptune Trojan region) to ~500 km at 48 AU (the 2:1

The Trans-Neptunian Solar System.
https://doi.org/10.1016/B978-0-12-816490-7.00010-2

resonance). Therefore, there is a bias in the observed sample that should be taken into account when analyzing the orbital distribution of TNB.

The distribution of the projected separation in the sky at discovery carries information on the distribution of true orbital separation, and thus of the semimajor axes of TNB (Grundy et al., 2008). To illustrate this, in Fig. 10.1 we show the correlation between separation and orbital semimajor axis a for TNB.

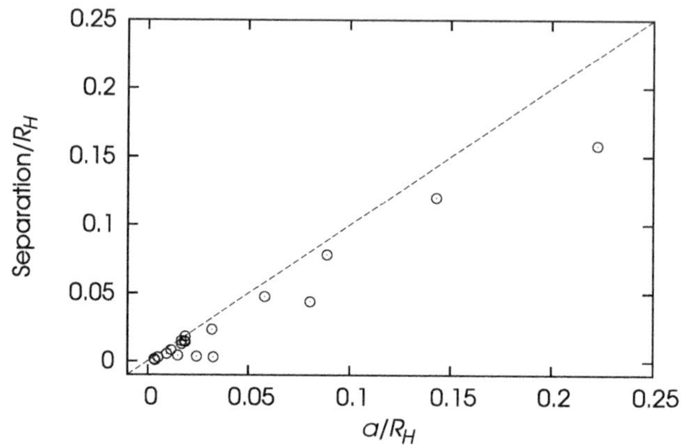

FIG. 10.1 Separation of TNB at discovery versus semimajor axis.

This was computed as follows: For each one of the TNB with fully known orbit (Grundy et al., 2019), we obtained 1000 simulated projected separations on the sky plane, considering that at the time of discovery the secondary was placed at random on its orbit (mean anomaly at random in the interval $[0-2\pi]$). We see that the semimajor axes are slightly underestimated by the observed separations, but the discrepancy is only noticeable for the widest orbits. Therefore, the distribution of separations at discovery is representative of the distribution of semimajor axes (Grundy et al., 2008), which we show in Fig. 10.2.

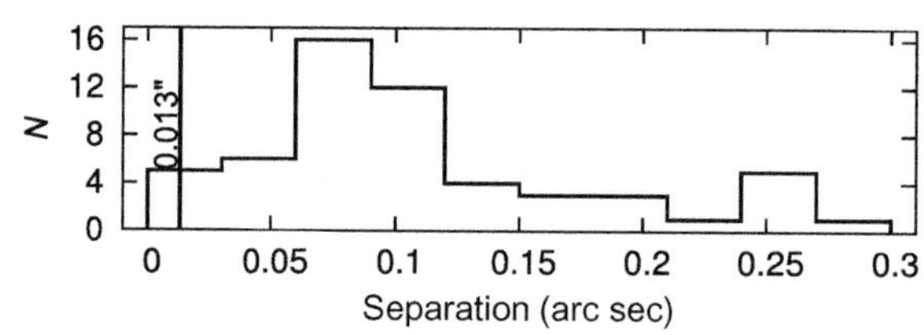

FIG. 10.2 Distribution of separations at discovery.

The drop in the number of very tight binaries is attributed to limitations in separating something so close in the 100-km-sized object range (Thirouin et al., 2017) which is the size of most known TNB. There are no known systems with $D_{SEC}/D_{PRIM} \leq 0.3$ (except dwarf planets with satellites). Such systems, with so small a size ratio, would have a flux difference between the components of ~3 magnitudes. A typical survey of the TN region reaches a magnitude $r \approx 24.5$ (Alexandersen et al., 2016; Elliot et al., 2005; Petit et al., 2008). A primary for such a small secondary component would have $r \approx 21.5$. There are not many known Kuiper belt objects (KBOs) with this magnitude, and therefore the lack of binaries with small secondary to primary size ratio in the observed sample might be interpreted as an observational bias rather than a primordial characteristic of the population.

From the point of view of orbital stability of a binary against external perturbations, the semimajor axes are usually scaled with the binary Hill radius (the distance to the second Lagrange point in the restricted three-body problem), defined as

$$R_H = a_\odot(1 - e_\odot)\left(\frac{M_{bin}}{3M_\odot}\right)^{1/3}, \tag{10.1}$$

where $a_\odot$ and $e_\odot$ are the semimajor axis and eccentricity of the binary heliocentric orbit, respectively, $M_{bin} = M_{prim} + M_{sec}$ is the combined mass of the primary and secondary components and $M_\odot$ is the mass of the Sun.

As shown in Fig. 10.3, the sample of binaries with well-determined orbits (Grundy et al., 2019) exhibits a tendency for the wider orbits to have higher eccentricities.

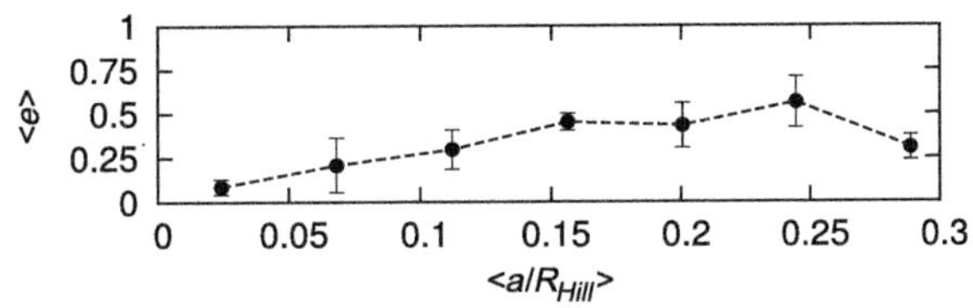

FIG. 10.3 Mean eccentricity versus mean semimajor axis for TNB with well-known orbits.

The orbital inclinations of TNB are difficult to obtain unambiguously. When the geometry between the observer and the binary object is essentially unchanged over the course of the observations, the actual orbit is astrometrically indistinguishable from its reflection on the sky plane (Descamps, 2005). As the heliocentric motion of a TNB is very slow, this mirror ambiguity can only be broken by the parallactic angle arising from the relative motion of the Earth and the binary. Therefore, a long-term astrometric campaign is needed in order to determine unambiguously the orbital inclinations.

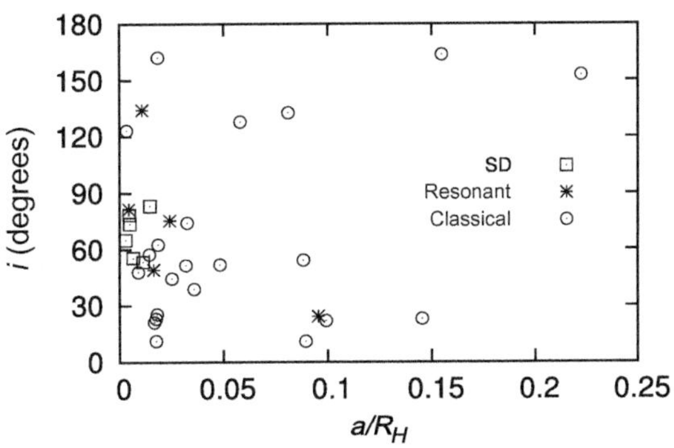

FIG. 10.4 Orbital inclination versus semimajor axis of TNB in the different populations.

Fig. 10.4 shows that tight binaries have a clear preference for direct orbits. Although the number of ultrawide binaries is still small to draw a definitive conclusion, this tendency is attenuated when we go to large separations where, in addition, only orbits with low inclinations are observed (Grundy et al., 2019).

Two important issues in relation to the orbital distribution of TNB can be summarized as follows:

- How the different processes involved in the dynamical evolution of TNB sculpted their current orbital distribution?
- Is it possible to unveil from what we see today some clues about the formation mechanism of TNB?

Throughout this review, we will try to advance on these two points, keeping in mind that the orbital information at hand to constrain the origin and the dynamical evolution of TNB, although increased during the last years, comes still from a relatively small sample. Consequently, we warn the reader that everything we may say from now on could change as more observational data will be collected, and more accurate orbits computed.

10.2 Dynamical mechanisms driving the orbital evolution of Trans-Neptunian binaries

10.2.1 Kozai-Lydov oscillations

The torque exerted by the Sun when the planes of the binary mutual orbit and that of the heliocentric orbit of the binary center of mass are not aligned, causes a periodic oscillation of the orbital elements of the binary (Kozai, 1962; Lidov, 1962), known as Kozai-Lydov cycles (KLC). KLC have been suggested as a mechanism affecting the mutual orbits of TNB (Perets and Naoz, 2009) and further applied to model their evolution by Porter and Grundy (2012) and Brunini (2014).

If $a_{\odot} \gg a_{bin}$, where a_{bin} is the binary semimajor axis, and if in addition, the orientation of the heliocentric orbit can be considered constant during the cycle, the binary orbital eccentricity and inclination oscillate in such a way that the z-component of the orbital angular momentum

$$L_z = \sqrt{a_{bin}(1 - e_{bin}^2)} \cos i \qquad (10.2)$$

is conserved (Kiseleva and Eggleton, 1999; Fabrycky and Tremaine, 2007). In the KLC secular theory, there are also two other conserved quantities: The binary semimajor axis a_{bin} (a natural consequence of the time averaging procedure) and

$$H = -2 - 3e_{bin}^2 + (3 + 12e_{bin}^2 - 15e_{bin}^2 \cos^2 \omega_{bin}) \sin^2 i, \qquad (10.3)$$

where ω_{bin} is the argument of the pericenter of the binary orbit and i_{bin} is the mutual orbital inclination of the binary with respect to the fixed plane of the heliocentric orbit.

There is a critical inclination i_c = 39 degrees · 23 that cannot be crossed. If the initial inclination is smaller than the i_C, the variation of the orbital elements is rather small. But if the initial orbital inclination is $i_0 > i_c$, KLC induce large oscillations in the eccentricity of the binary. Kozai (1962) proved that if $i_0 > i_c$, a binary on an initially circular orbit can reach a maximum eccentricity given by

$$e_{bin}(\max) = \sqrt{1 - \frac{5}{3}\cos^2(i_0)}, \tag{10.4}$$

which could be very large.

In addition, the amplitude of the oscillation does not depend on the distance to the Sun, but on the characteristics of the heliocentric orbit. The oscillation timescale T_K is given by (Innanen et al., 1997; Fabrycky and Tremaine, 2007)

$$\frac{T_K}{P_{bin}} \simeq \frac{2}{3\pi}\left(\frac{P_\odot}{P_{bin}}\right)^2 (1 - e_\odot^2)^{3/2}, \tag{10.5}$$

where $P_\odot$ is the period of the binary heliocentric orbit, and P_{bin} is the period of the mutual orbit. For the population of known TNB, T_K ranges from 10^3 to 10^6 years.

10.2.2 Tidal interaction

The dissipative force due to tides raised on binary components is significant only at very short binary separations (Dziembowski, 1967; Kiseleva and Eggleton, 1997; Fabrycky and Tremaine, 2007; Perets and Naoz, 2009; Porter and Grundy, 2012). Tidal evolution depends on the internal structure and composition of the objects. In tidal theories, these characteristics are represented by two quantities: One quantity is the second tidal love number, which is a dimensionless parameter measuring the rigidity of the body and the susceptibility to change its shape in response to the tidal potential (Jeffreys, 1925). For monolithic solid bodies, the uncertainty is small due to our knowledge of the orbits of the outer solar system satellites, as long as we accept that their structure and composition are similar to that of TNB (Gladman et al., 1996). In this case, the second tidal love number is given by

$$K_L = \frac{3}{2}\left(1 + \frac{19\mu_r R}{2GM_{bin}\rho}\right), \tag{10.6}$$

with the rigidity $\mu_r = 4 \times 10^9\,\mathrm{N\,m^{-2}}$. Here R is the radius of the object. For rubble piles, theoretical considerations and laboratory experiments set values for the tidal second love number (Goldreich and Sari, 2009)

$$K_L = R/10^5\,\mathrm{km}. \tag{10.7}$$

The other quantity is the fraction of tidal energy lost as heat per binary orbit. The inverse of this quantity, per radian, is known as Q, the *quality factor*. Typical values of the quality factor are (Goldreich and Soter, 1966) $Q = 100$ for solid bodies, and $Q = 10$ for rubble piles. Although they are related, no simple relation exists between the quality factor and the time lag between the maximum of the tidal potential and the tidal bulge in each body, except for a few particular cases (Leconte et al., 2010).

For a binary with both components of equal size, the timescale of tidal evolution of the mutual orbit T_F is approximately given by (Porter and Grundy, 2012)

$$\frac{T_F}{P_{bin}} \simeq \frac{1}{3}\frac{Q}{K_L}\frac{a_{bin}}{R}. \tag{10.8}$$

The tidal friction timescale depends strongly on the separation of the binary. For a binary component with $Q = 10$ and $R = 50$ km,

$$\frac{T_F}{P_{bin}} \simeq 10^3 \left(\frac{a_{bin}}{R}\right)^5. \tag{10.9}$$

In the case of very eccentric orbits, the tidal interaction acts as kicks at pericenter (Hut, 1981). Therefore, the important parameter to measure the relevance of tidal forces is the pericentric distance rather than the semimajor axis.

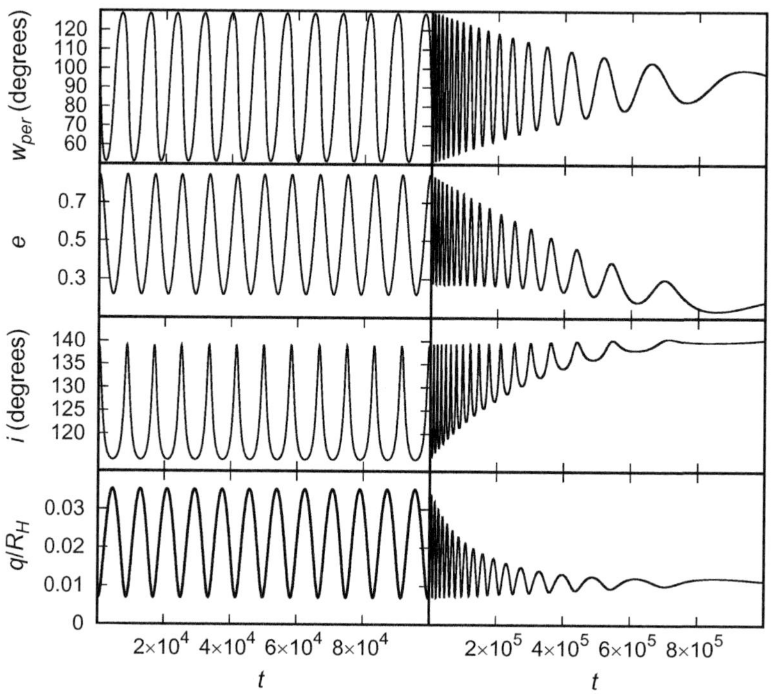

FIG. 10.5 Dynamical evolution a typical TNB subject to Kozai cycles and with and without tidal friction operating.

Depicted in Fig. 10.5 is an example of the evolution of a TNB subject to KLC with and without tidal friction. In this example, the binary is composed of two rubble piles of 100 km. Every time the binary is at pericenter, tidal interaction tends to circularize the orbit. The system ends up on a frozen, much less elliptical orbit, on a timescale of $\sim 10^6$ years. As the eccentricity gets lower, the pericenter goes beyond the zone where tidal friction operates. Tides do not always operate in this way. Tidal friction evolution also depends on the relation between the diurnal rotation and the orbital motion (i.e., if the tidal bulge precedes or supersedes the position in the orbit of the companion).

The spin period and obliquity of objects are also affected by tides. The timescale of evolution of the rotation rate T_{spin}, for the case of a binary with two spherical equal components on circular orbit, is approximately given by (Hut, 1981)

$$\frac{T_{spin}}{T_F} \simeq \frac{1}{25}\frac{P_{spin}}{P_{orb}}\left(\frac{a_{bin}}{R}\right)^2. \tag{10.10}$$

At separations larger than 5R, the evolution of the rotation rate is slower than the orbital evolution. For $a_{bin} \geq 16R$, the rotation rate evolution is at least one order of magnitude slower than the orbital evolution. Tidal synchronization is only achieved after a long period of evolution, and for very tight systems. Therefore, contact binaries are expected to be in synchronous rotation state (Thirouin et al., 2017). For two equal size spherical objects with density $\rho = 1\,\mathrm{g\,cm^{-3}}$, the spin period (the orbital period if they were in contact) would be of ~6.6 hours.

For highly eccentric orbits, the rotation rate evolves to match the orbital angular velocity at pericenter (the so-called "quasisynchronization state"; Hut, 1981). The orbital behavior in this case is very complex. Certain combinations of spin period, orbital eccentricity and obliquity, can produce decoupling of TNB, forming pairs of single objects with slow rotation rate (Brunini and Zanardi, 2016).

10.2.3 Nonsphericity of the bodies

TNOs of 100 km size are probably not large enough to have reached hydrostatic equilibrium (Stevenson, 2009). Nonsphericity of one or both components of TNB leads to a noncentral force field that causes an apsidal precession of the mutual orbit (Greenberg, 1981). If this precession is too fast, libration of the argument of the pericenter forced by the Sun is no longer possible, inhibiting KLC (Porter and Grundy, 2012).

The quadrupole potential is characterized by J_2, the quadrupolar moment that, for an axisymmetric body, is defined as (Hubbard, 1984)

$$J_2 = \frac{(C-A)}{MR_{eq}^2},$$

where C and A are the moments of inertia about the polar and equatorial radii R_{eq}, respectively. We do not have much information about the internal structure of TNB or about their shape. However, we can make some speculations from the amplitude of the observed light curves of TNOs, which may be related to the shape of an axisymmetric spheroid. The mean value of the sample with observed light curve is about $R_{pol}/R_{eq} = 0.77$ (Benecchi and Sheppard, 2013; Barucci et al., 2004), and is similar to the mean shape found for the irregular satellites of the giant planets (Graykowski and Jewitt, 2018). J_2 is related to the shape of an homogeneous axisymmetric nonrotating spheroid as $J_2 = (2/3)(1 - R_{pol}/R_{eq})$ (Danby, 1962), which would have a mean value of ~0.15.

The timescale of orbital evolution due to the quadrupolar potential is approximately given by (Campanella et al., 2013)

$$T_{J_2} \approx \frac{2}{3}\left(\frac{a}{R}\right)^2\frac{1}{J_2}. \tag{10.11}$$

Even for the large value of $J_2 = 0.15$, the effect of the quadrupolar component is only appreciable at a small mutual distance. Assuming the mutual orbit and the heliocentric orbit to be both circular, and taking into account Eqs. (10.5), (10.11), the precession rate supersedes the timescale of KLC if (Nicholson et al., 2008)

$$\frac{a_{bin}}{R} \lesssim 145 \left\{ \frac{\rho}{1\,\mathrm{g\,cm^{-3}}} \left(\frac{a_\odot}{45\,\mathrm{AU}} \right)^3 \frac{J_2}{0.1} (1 + R_{sec}/R_{prim}) \right\}^{1/5}. \tag{10.12}$$

Inside this distance, KLC are no longer driving the precession of the orbit, and eccentricity oscillations stop. Even so, tidal friction continues operating.

10.2.4 Collisional evolution

TNBs are immersed in a sea of small objects, and therefore subject to collisional evolution (Petit and Mousis, 2004; Nesvorný et al., 2011; Parker, 2012; Brunini and Zanardi, 2016). At each collision, the primary or the secondary suffer several changes: If the collision is catastrophic, the target (one of the components of the binary) is disrupted, and probably the binary ceases to exist as such. If not, the target is eroded, its size, mass, and moment of inertia change, but most importantly, the collision imparts a certain impulse, which changes the mutual orbit and the spin vector. Following Petit and Mousis (2004), an impact of a projectile of radius R_i at a relative velocity v_{rel} can unbind the binary if

$$R_i = R \left(\frac{0.62}{v_{rel}} \right)^{1/3} \left(\frac{GM_{bin}}{a_{bin}} \right)^{1/6}, \tag{10.13}$$

where R is the radius of the target component of the TNB. It is clear that the separation of the binary is an important factor, with the ultrawide TNB being the most affected. But the size of the target is important as well. Given that the typical impact velocity for TN populations is of the order of $1\,\mathrm{km\,s^{-1}}$ (Dell'Oro et al., 2013) it is found that impacts of small objects of size $D_i \geq 0.1 D_t$ could disrupt a binary (Nesvorný et al., 2011). Ultrawide TNB are easily unbound by collisions with objects in the 1–5-km radius range, and the existence of these systems constrains the number of small impactors that exist in the present TN space (Nesvorný et al., 2011).

Not always can an impact unbind the orbit, but important orbital changes are possible. The velocity vector v of the binary orbit will change by

$$\delta v = f \frac{m_i}{m_t} v_i, \tag{10.14}$$

where m_i is the mass of the impactor, m_t is the target mass, and v_i is the relative velocity of impact, while f is a factor ($f < 1$) taking into account the linear momentum carried away by the escaping ejecta (Benz and Asphaug, 1999; Dell'Oro and Cellino, 2007). If the impacts are randomly oriented, on average, the change in orbital elements of the binary, after many small impacts, is given by (Nesvorný et al., 2011)

$$\frac{\langle (\delta a_{bin})^2 \rangle}{a_{bin}^2} = f^2 \frac{4}{3} \left(\frac{\delta v}{v} \right)^2,$$

$$\langle (\delta e_{bin})^2 \rangle = f^2 \frac{5}{6} (1 - e_{bin}^2) \left(\frac{\delta v}{v} \right)^2,$$

$$\langle(\delta \cos i_{bin})^2\rangle = f^2 \frac{2 + 3e_{bin}^2}{12(1 - e_{bin}^2)} \left(\frac{\delta v}{v}\right)^2,$$

where v is the orbital velocity of the binary.

Besides the spin, the obliquity of the target also changes, affecting the tidal evolution. Modeling this change is a very difficult task. Yanagisawa and Hasegawa (2000) conducted high-velocity impact experiments to study the total angular impulse that a target acquires by a collision. They have found that it is possible to write

$$\delta L = \psi L_i \cos \theta, \tag{10.15}$$

where δL is the angular momentum acquired by a spherical target during the collision, L_i denotes the preimpact momentum of the projectile, θ is the incident angle with respect to the normal surface of the target, and ψ is a factor measuring the efficiency of the impulse transfer. They also found that ψ depends on the target material, but is rather independent of the impact relative velocity. For basaltic targets, $\psi = 0.4$. For sand targets, Gault and Schultz (1986) found $\psi \sim 0.7$. It is expected that ψ should be even larger for icy objects, because the efficiency of angular momentum transfer increases with the projectile penetration (Yanagisawa and Itoi, 1994).

10.2.5 Close encounters with the giant planets

During early stages of the outer solar system formation, some populations of TNB were subject to the effects of close encounters with the giant planets (Tsiganis et al., 2005). The same situation is applicable to binary Centaurs during their journey through the giant planets region.

The typical change of the orbital elements of a binary, after a close encounter with a planet of mass M_p, averaging over all encounter geometries, is approximately given by (Fang and Margot, 2012)

$$\frac{\Delta a_{bin}}{a_{bin}} \simeq 1.48 \sqrt{\frac{G}{M_{bin}} \frac{M_p a_{bin}^{3/2}}{v_{rel} b^2}},$$

$$\Delta e_{bin} \simeq 1.28 \frac{\Delta a_{bin}}{a_{bin}},$$

$$\Delta i_{bin} \simeq 0.51 \frac{\Delta a_{bin}}{a_{bin}},$$

where v_{rel} is encounter velocity and b is the encounter closest approach distance. These expressions are valid under certain conditions, and should be used with caution, but nevertheless, they show that the outcome of a close encounter is rather sensitive to several parameters: the binary separation, the relative velocity, and the closest approach distance to the planet.

To illustrate the dynamical effect, we carried out a series of close encounter experiments between synthetic binaries and the giant planets. The initial binary orbits were generated in a similar way as in Parker and Kavelaars (2010) in order to cover the range of known classical Kuiper belt binaries (Naoz et al., 2010; Grundy et al., 2009, 2019; Petit et al., 2008; Veillet et al.,

2002). The procedure is as follows: a/R_H is generated from a uniform distribution between 0.01 and 0.25. Then a is generated at random with uniform distribution between 2×10^3 and 1.2×10^5 km. Given a and a/R_H, M_{bin} is determined. If M_{bin} lies outside the range 10^{16}–10^{19} kg, a is recomputed. The density of the binary was set $\rho = 1\,\mathrm{g\,cm^{-3}}$. Both components were assumed of equal mass. The orbital eccentricity was generated with a uniform distribution between 0 and 0.90, and all initial angular parameters (including inclination) were uniformly distributed between 0 and 180 degrees. The relative velocity of the encounters was sampled from the distributions found in the close encounters of the Centaur simulation of Di Sisto and Brunini (2007). At the beginning, the center of mass of the binary was placed at random on the Hill sphere of the planet, assumed with its present orbit. The encounter was integrated with a high-precision numerical Bullirsch-Stöer routine (Press et al., 2002) until the binary left the Hill sphere or the system was considered destroyed. This happens if the binary becomes unbound, the mutual semimajor axis is larger than R_H, or the binary components collide.

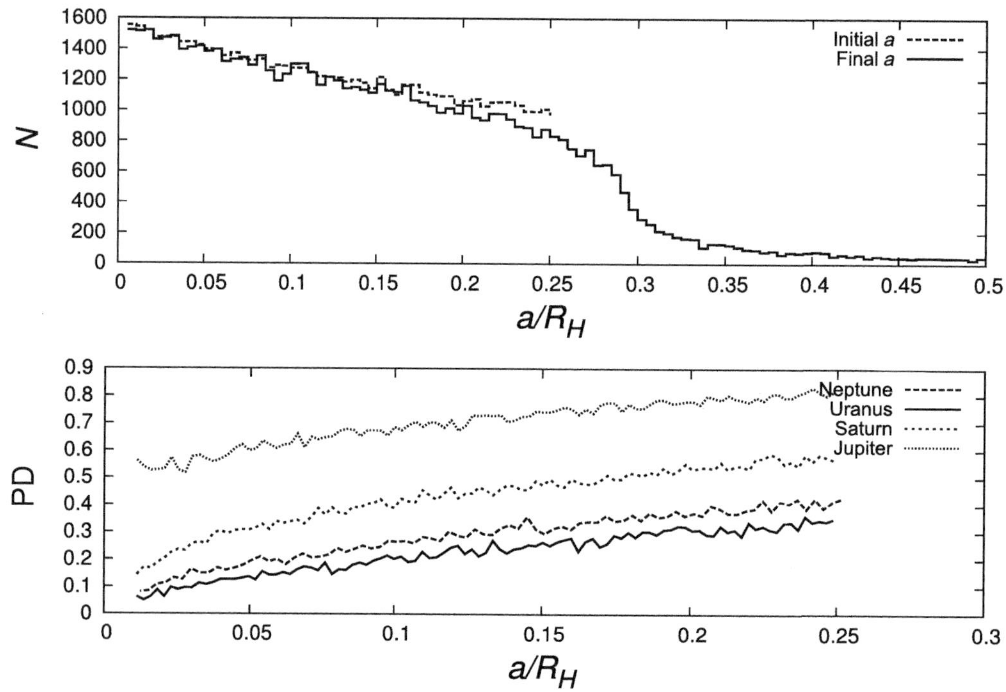

FIG. 10.6 *Upper panel*: Mean effect of one single close encounter with Neptune (see main text for explanation). *Lower panel*: Fraction of binary destruction (PD) against initial separation during a single close encounter with the giant planets.

In the upper panel of Fig. 10.6, we show the mean effect of one single close encounter with Neptune. After one single close encounter, the number of objects within a given bin of initial a drops with the separation. We can also observe that the distribution of a the binaries have after the encounters shows the classical tail at large values (Everhart, 1969; Weidenschilling,

1975). Each binary whose orbit moves to the tail of large a during the first close encounter is susceptible to be much more easily disrupted during the next one. Therefore, the curve of the fraction of binary survival against separation becomes more and more steep as successive close encounters take place. In the lower panel, we show the result of similar numerical experiments for the case of the four giant planets. As expected, close encounters with Saturn and Jupiter are much more critical for the survival of the binaries, whereas Uranus and Neptune only affect substantially ultrawide binaries. This fact was already shown by Parker and Kavelaars (2010) for the particular case of scattering histories that result in KBOs being implanted in the cold classical Kuiper belt (CCKB), showing that ultrawide binaries in the cold Kuiper belt probably formed in situ.

Close encounters also change the heliocentric orbit of the binary center of mass. Therefore, they can have different consequences for binary survival: They could change the binary orbit, disrupt the binary as such, eject the binary from the solar system, or place it into a different minor body reservoir, but still conserving its binary nature. We wish to note that possibly many binary objects are at present residing in the Oort cloud, but this possibility has not yet been investigated.

10.3 All together now

10.3.1 Kozai-cycle tidal friction

The first extensive exploration of the action of Kozai-cycle tidal friction (KCTF) and oblateness perturbation on TNB was done by Porter and Grundy (2012). They found that these effects can significantly transform the orbits of TNB. Most of their synthetic TNB (~90%) survive for the age of the solar system as such. A considerable fraction of the surviving population decays to very tight circular orbits, some of them probably ending up as a contact binaries. For these very tight systems, there is a clear tendency to pole alignment and tidal synchronization. As expected from the secular theory, all resulting systems preserve their initial prograde/retrograde orbit.

Collisional evolution of TNB was investigated by Nesvorný et al. (2011). They found that the observed break in the slope of the size distribution function of the present KBOs at $R_0 \approx$ 25–50 km (Fraser et al., 2014) could be explained by an intense primordial collisional grinding. However, such a strong collisional grinding required to produce the break would imply the existence of a strong gradient of the binary fraction with size and separation. As the present observational data do not show such a gradient, they conclude that the break in the slope at R_0 is not produced by a collisional process, and it should be a fossil remnant of the KBO formation process.

Parker (2012) also performed simulations of the collisional evolution, but applied to ultrawide TNB. The collisional lifetimes of these populations constrain the size distribution function of KBO at small radii ($R < 1$ km), suggesting that the slope in the size distribution function at small size cannot be too steep, probably with a slope of $q < 3.5$, in accordance to the slope of $q \sim 2$ found by Fraser et al. (2014). They also conclude that the current observed population of ultrawide binaries does not come from a primordial population of tight binaries widened by collisional processes. In their simulation model, ultrawide binaries generated in this way would have high orbital eccentricities ($e \gtrsim 0.8$).

The first attempt to include KCTF and collisional evolution together was done by Brunini and Zanardi (2016) (BZ16). They showed that both effects are strongly linked. As shown in Fig. 10.7, a collisional process cannot be considered as a pure random walk diffusion of the orbit. In this particular case, a collision excites the eccentricity, reducing the pericentric distance within the tidal friction operative limit. One (or a few) more collisions would probably unbind the binary if tidal friction were not operating, but on the contrary, the binary ends up with a very tight separation, slowly circularizing its orbit. Further collisions cannot change this trend when the orbit becomes very tight.

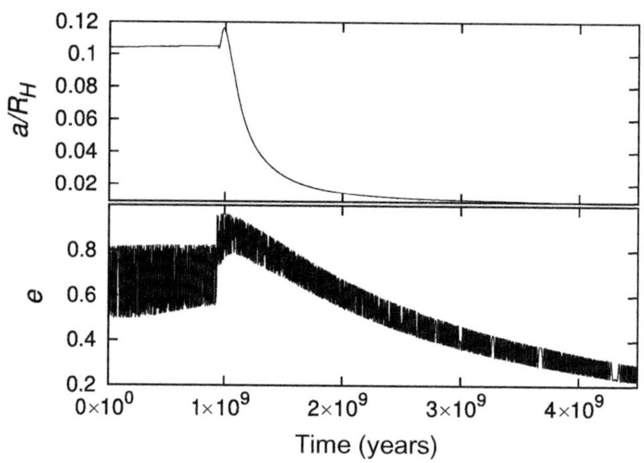

FIG. 10.7 Typical behavior of a TNB subject to KCTF + collisional evolution.

Many of the synthetic binaries (more than 50%) do not survive the entire simulation. Some are decoupled mainly by the action of repeated collisions that increase the orbital eccentricity up to ejection, or by a mutual collision of both components. Nevertheless, the fraction of unbound objects is much smaller than that reported by Nesvorný et al. (2011) or Parker (2012). This is because, as already mentioned, the collisional process cannot be treated as a pure random walk process. From Eq. (10.9), it becomes evident that if a wide system has $a_{bin} = 15{,}000$ km, an eccentricity of $e \approx 0.8$ is enough for tidal friction to operate. Consequently, any conclusion drawn from simulations including collisions, planetary encounters, or any other kind of physical process affecting the binary eccentricity and not including tidal friction, should be taken with extreme caution. When tidal friction is included in the model, the TNB can experience several kinds of behavior: in some cases, the binary starts its life as a wide system, with moderate eccentricity. The pericentric distance never reaches sufficiently small values for tidal friction to operate and the collisional evolution is well represented by a random walk of the orbital elements. Other systems evolve like the one shown in Fig. 10.8. In some other cases, the binary undergoes successive collisions and it is gradually transformed from a wide system into an ultrawide system. The survivors have a distribution of eccentricities and inclinations resembling the observed ones.

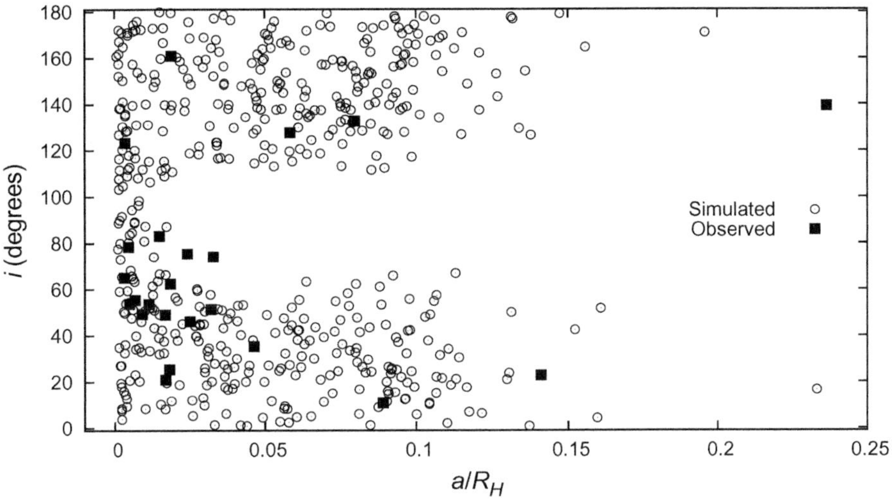

FIG. 10.8 Comparison between the observed inclinations and the ones obtained by Brunini and Zanardi (2016).

The different physical effects are quite evident in Fig. 10.8: Tidal friction creates many circular and tight orbits (perhaps contact binaries). KLC depletes orbits with large inclinations, but at short range, the flattening factor counteracts the Kozai oscillation. The eccentricity trend with increasing separations is consistent with the known binary systems, but BZ16 found an over production of wide and ultrawide systems.

On the hand, as a natural by-product of the collisional decoupling of binaries, there is some tendency for an increasing binary fraction with the relative size of the components. Small secondaries are hard to detect for this class of objects, but the lack of detection should be due in part to collisional evolution, because, for a given primary size, a binary with a small secondary is less gravitationally bound, and easier to separate.

Ultrawide binaries are the most useful for imposing constraints on primordial characteristics of the TN region (Parker, 2012), because they are more sensitive to the size distribution of the impactors and the original mass in the region. BZ16 observed that in their initial conditions there were already wide binaries, but most of them were unbound by successive collisions, as demonstrated in the previous papers (Parker, 2012). The remaining fraction of initially primordial wide + ultrawide binaries is low. But at the same time, collisions generated new wide and ultrawide binaries from the tight population. As a conclusion, only half of the present population of wide and ultrawide binaries would be primordial. The other half is generated from an original tight population. Their orbital characteristics are quite similar to the observed ones, especially the distribution of inclinations. Starting from a uniform distribution of i, a nonuniform final distribution is obtained, where low inclinations prevail, explaining one of the main characteristics of the ultrawide population.

About 50% of the binaries are decoupled during evolution. Each decoupled binary forms two new single objects. Considering at the beginning a 100% binary fraction, BZ16 end up with two-thirds of the population as single objects and one-third as binaries. This binary fraction is similar to the observed one, suggesting that the planetesimals in the TN region could have been born as binaries. If this is the case, any binary forming mechanism should be very efficient.

10.3.2 Facing planetary encounters

For most of the populations of TNB not originated in situ, we have to consider planetary encounters and resonance sweeping.

10.3.2.1 Centaurs

Most binary Centaurs end up being disrupted before reaching the orbit of Jupiter. The most common mechanism of disruption is close encounters with a giant planet.

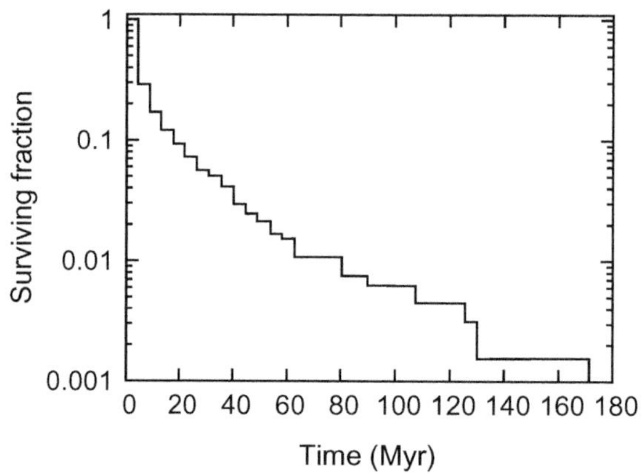

FIG. 10.9 Lifetime of binary Centaurs.

Fig. 10.9 shows the lifetime distribution of binaries as Centaurs for one of the simulations by Brunini (2014). The lifetime as binary Centaurs is very short: After 10^6 years, the population drops to 50%. Only a very tight binary survives its entire life, around 72 Myr, as a Centaur (Di Sisto and Brunini, 2007). Taking into account these surviving rates, and considering a 10% binary fraction in the source (the scattered disk), a small fraction of no more than 2% of binaries is expected to be in the Centaur population. In addition, only contact binaries could survive encounters with Saturn and Jupiter. Therefore, we may speculate that only tight binaries should be found in the population of Centaurs and only very tight or contact binaries, in the region between Saturn and Jupiter. Similarly, only very tight or contact binaries should be found among Jupiter family comets. In some cases, the contact binary end state is achieved when a close encounter with a giant planet affects the pericentric distance and enables the operation of tidal friction, which is the mechanism leading to the contact binary end state. We note that the distribution shown in Fig. 10.9 underestimates the binary lifetime, because the simulation stops for those achieving the contact binary end state.

The effect of successive encounters, like as in the case of collisions, cannot be treated as a random walk process.

10.3.2.2 *Neptune Trojans and resonant populations*

The dynamical evolution of TNB in the different resonances of the TN space has not been studied, with the exception of the Neptune Trojan region, where in fact, binary objects were not found yet. Brunini (2017) investigated the evolution of a fictitious population of binaries in this region. The collisional evolution was not included, because it is not so intense today as to deserve being considered in the dynamical model. The behavior of binaries located at a resonance is affected by possible similarities between the KLC period and the resonance libration period, as is the case for binary Neptune Trojans (Brunini, 2017). It is important to include the perturbing planet in the model, although the perturbation is much smaller than the one of the Sun.

For binary Neptune Trojans, Brunini (2017) found that more than 50% of the original binaries survive for 4.5×10^9 years. The orbital distribution of the resulting population, not being excited by mutual collisions, resembles that found for the CCKB region in the simulations by Porter and Grundy (2012). One interesting result is that nearly 50% of the fictitious Neptune Trojan binaries survive, and that ~80% of the survivors end up as very tight circular binaries. Therefore, contact binaries should be abundant in the Neptune Trojan population (if binaries indeed exist in this population!). In fact, an overabundance of potential contact binaries has been reported in the Plutino population (Thirouin and Sheppard, 2018). Interestingly, this is in agreement with what has been found in the Trojan population. Nevertheless, a complete dynamical exploration of binary Plutinos has not been done yet, nor of other resonant populations.

10.4 Formation mechanisms

In this section, we will review briefly the different formation mechanisms of TNB proposed so far. If we accept that the primordial binary fraction should be large, near 100%, as proposed by Fraser et al. (2017) and as suggested by the results of BZ16, the formation mechanism of TNB should be very efficient. Several formation routes have been proposed, but only two fulfill this condition: capture mechanisms and gravitational collapse. They are conceptually very different: Capture mechanisms assume interaction of already formed planetesimals, whereas gravitational collapse assumes that planetesimals form directly as binaries (or multiple) objects. In what follow, we will describe each one of them. An origin based on a collision between two large objects is not addressed here, as it is applicable only to satellites of large TNOs, which are not included in this chapter.

10.4.1 Capture mechanisms

One way to form binary pairs is by means of a gravitational capture of two single objects (Weidenschilling, 2002; Goldreich et al., 2002; Funato et al., 2004; Astakhov et al., 2005; Lee et al., 2007). Once planetesimals have formed through the classical process of pebble accretion (Safronov, 1972), close encounters of couples of them should be very common. If during close encounters some dynamical mechanism causes enough kinetic energy loss, the encountering pair could be transformed into a bound binary system.

There are three ways to lose energy during a fly-by:

(i) A third interacting object ejected during the close encounter carrying away the excess energy (the L3 mechanism), as proposed by Goldreich et al. (2002).
(ii) Dynamical friction with the sea of small objects, removing kinetic energy, known as the L2s mechanism (Goldreich et al., 2002).
(iii) A collision of one of the objects with a third small object (Weidenschilling, 2002).

Weidenschilling (2002) proposed a formation scenario involving a collision between two bodies inside the Hill sphere of a third. However, the Hill sphere radius given by Eq. (10.1), applied to two equal sized TNB on circular heliocentric orbit can be expressed as

$$\frac{R_H}{R} = \frac{a_\odot}{R_\odot}\left(\frac{2\rho}{3\rho_\odot}\right)^{1/3}, \tag{10.16}$$

where $R_\odot$ and $\rho_\odot$ are the radius and mean density of the Sun, respectively. As the term in bracket in Eq. (10.16) is of order unity, at 45 AU the Hill sphere radius is $\sim 10^4$ times the radii of the binary components (Goldreich et al., 2002). Therefore, encounters are many orders of magnitude more frequent than collisions. Nevertheless, option (iii) could be an effective mechanism for creating the small moons of large KBOs. Astakhov et al. (2005) and Lee et al. (2007) suggested that the most promising candidates to form a binary by the L2s and L3 mechanisms is by means of objects experiencing a long-lasting capture as transient satellites.

Schlichting and Sari (2008b) have shown that the relevance of transitory binary capture is not significant for binary formation in the TN region. They estimate that the frequency of binary formation by both mechanisms is

$$\frac{F_{L3}}{F_{L2s}} \sim 0.005, \tag{10.17}$$

as long as the relative velocity of objects during the formation of the Kuiper belt is in the shear dominated regime ($v_{rel} \ll \Omega R_H$, where Ω is the orbital frequency of the heliocentric orbit). This is supported by a series of arguments, including the existence the Haumea collisional family (Brown and Trujillo, 2002) and the ubiquity of small satellites around large KBOs (Green, 2006; Parker et al., 2016; Showalter et al., 2011, 2012; Weaver et al., 2006). Therefore, option (ii) seems to be the most efficient, and perhaps the only one providing the possibility of a large binary fraction (Schlichting and Sari, 2008a,b) within the frame of the capture hypothesis.

But the L2s mechanism presents some problems: The sea of small objects should be massive enough as to enable dynamical friction during the short encounter times, and the relative velocity of the big objects (the components of the binary that is forming) with respect to the sea of small objects should be very small (0.2 m s^{-1} or less). The primordial size distribution should be rather bimodal. It remains as a matter of investigation if this is possible.

An additional problem with capture mechanisms is that they favor retrograde orbits, because retrograde encounters are much more long lasting than direct ones (Brunini, 1996). As the loss of energy must occur during the encounter, there is more chance to produce a retrograde binary than a direct one. As in the observed sample there is not such an asymmetry in orbit orientation, the problem is then finding a way to eliminate the excess of retrograde binaries. This is still an open question.

10.4.2 Gravitational instability of clumps: The streaming instability

The other proposed mechanism that we will discuss here resembles the old idea of the disk instability hypothesis (Safronov, 1972; Goldreich and Ward, 1973), where the self-gravitation of local dense clumps of dust grains starts a gravitational collapse, forming planetesimals very fast. The gravitational instability can be prevented by a small amount of turbulence in the gas disk (Weidenschilling, 1980; Cuzzi et al., 1993). However, particles can also clump in a turbulent environment (Johansen et al., 2006). Theoretically, there are several hydrodynamic processes able to produce the self-gravitating clumps (Johansen et al., 2014). They include long-lived gaseous vortices (Barge and Sommeria, 1995; Meheut et al., 2012), pressure shocks (Johansen et al., 2009, 2011; Simon et al., 2012), or accumulation of dust particles by a runaway radial drift, which is known as the *streaming instability* (Youdin and Goodman, 2005; Johansen et al., 2007; Bai and Stone, 2010; Simon et al., 2016, 2017; Carrera et al., 2017). This mechanism has proved to be effective at producing dense clumps of particles and then forming planetesimals on very short timescales. Nesvorný et al. (2010) explored the possibility of forming binary objects by this mechanism, by performing a series of numerical simulations. They show that the streaming instability may be very efficient in forming binaries with components of similar size, with size ratios ranging from 0.3 to 1, and with most systems having $R_{sec}/R_{prim} > 0.7$. Correlated colors of both binary components (Benecchi et al., 2009) are a natural consequence of this mechanism, because both objects form with material of the same local clump. The orbital parameters of the formed binaries present a wide range of orbital parameters, with semimajor axes between 10^3 to several 10^5 km and most eccentricity values below 0.6, which are in agreement with the range of orbital parameters of the known population. The prograde-to-retrograde ratio of binaries formed by this mechanism is correlated with the orientation of the angular momentum vector of the collapsing clump. Therefore, it is not clear whether this mechanism could form retrograde systems (Schlichting and Sari, 2008b). This is because the angular momentum of the clumps should be essentially oriented in the same way as that of the protoplanetary disk. The computed orbits of the know population of tight TNB show a clear tendency for direct orbits. Nevertheless, there are ~10% of retrograde systems (Grundy et al., 2019). If we accept that TNB form on direct orbits, we would need to invoke an additional mechanism to flip the orbital inclinations after formation.

Nesvorný et al. (2010) also found that the accretion timescale depends on the initial density of solids within the collapsing cloud. Therefore, at different heliocentric distances, binary formation by this mechanism should proceed at different rates.

The most recent hydrodynamic simulation of planetesimal formation by the streaming instability (Simon et al., 2016, 2017) has a resolution fine enough as to produce hundreds of gravitationally bound clumps whose masses are in the range of 30–300-km-sized planetesimals with a rotation where the centrifugal force is of the order of the cloud's self-gravity (Grundy et al., 2019). This is the necessary condition for producing binaries or multiple systems. Moreover, the clumps have direct rotation with a considerable spread of obliquity, just as the TNB data show (Grundy et al., 2019). This last point opens a new question which is addressed in the following section.

10.4.3 Orbital flip after formation

If we accept that there was a unique formation mechanism giving rise to a population of TNB all orbiting in a given sense (direct or retrograde), the most probable situation—as most TNB are at present orbiting in the direct orbits—is a flip from direct to retrograde sense. Otherwise, most TNB should have suffered an orbital flip—an unlikely situation. A few possible flipping mechanisms will be summarized in what follows.

One possible way to flip from direct to retrograde orbit (or vice versa) is by angular momentum transfer during collisional evolution.

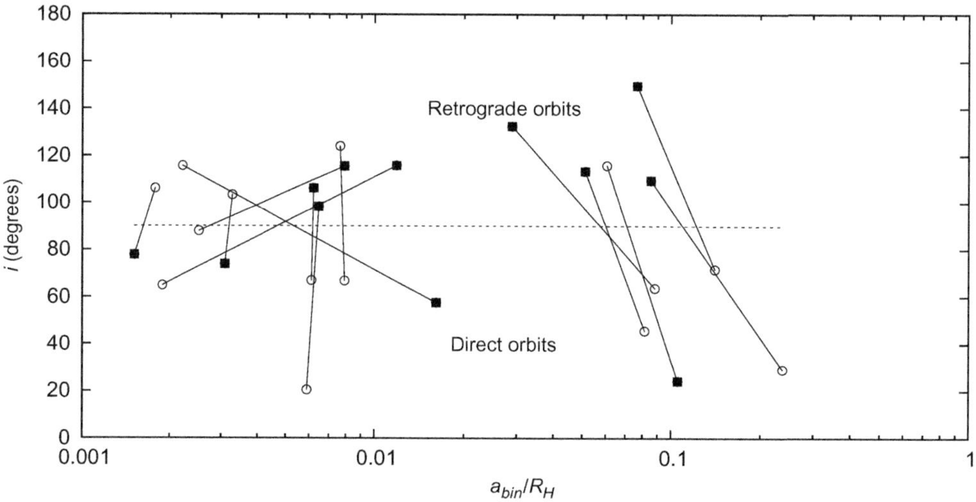

FIG. 10.10 Inclination flips by collisional evolution.

For example, the simulations of BZ16 started with 800 binaries with uniform distribution of inclinations between 0 and 180 degrees. On average, half the systems were originally in direct orbits. They found 12 cases of orbital swap from retrograde to prograde state or vice versa. The initial and final orbital states after the inclination swap are shown in Fig. 10.10. Therefore, we can speculate that from an initially direct population of 400 binaries (half the whole initial population in BZ16), out of the surviving population of 160 binaries (a fraction of ~40%), only eight systems swap from direct to retrograde orbit. Indeed, 8 of 160 represents a small fraction of retrograde end states originating from a direct population. Nevertheless, these simulations did not include the intense primordial collisional period that could change this number.

A sequence of small impulses cannot produce the desired effect because it would result in a smooth random walk evolution of the inclination. This would force the orbital inclination to pass through the Kozai-Lydov forbidden region, where the eccentricity increases up to the unbinding of the system.

An alternative, not explored yet, is the interaction between a mean motion resonance of the heliocentric orbit with the orbit of Neptune, and Kozai-Lidov oscillation. For the case of Neptune Trojans, there is a clear overlap of the libration period in the 1:1 resonance and the

KLC period that modifies the dynamical evolution of the binaries (Brunini, 2017). Other mean motion resonances could have similar effects. Pluto, for example, has a period of libration close to 20,000 years, which is within the range of the KLC periods for TNB. If a binary is at resonance at some moment of its life (i.e., during the migration period of Neptune), the question is whether the overlap of the periods of libration could cause an inclination flip.

As to collisions, one possible vehicle of change is a close encounters with a giant planet. Close encounters change both the heliocentric orbit and the binary orbit. Some particular encounter geometries do not change the orbital eccentricity, but change the inclination (Collins and Sari, 2008). Brunini (2019) has recently shown that this is a certain possibility, but it only applies to TNB populations that suffered close encounters with Neptune.

10.4.4 The wide population of blue binaries

Fraser et al. (2017) reported resolved photometry of both the primary and the secondary components of 23 TN binaries obtained with the Hubble Space Telescope. They found that the color of both components is almost identical, and within the range of those reported for apparently single TNOs belonging to the same population. These observations reinforce the conclusion that the colors of TNOs are primordial, being a signature of the place where they have formed. They also reported numerical simulations of dynamical implantation of binaries in the CKB, showing that objects originally placed at 35–40 AU are temporarily swept into the 2:1 mean motion resonance with Neptune, and transported outward during Neptune's migration. A small fraction was deposited into the CCKB region, if Neptune experienced a jump in semimajor axis at the end of the migration period (Nesvorný and Vokrouhlický, 2016). A very small fraction of 0.5% of wide binaries survived this journey. This is an interesting result, because it gives a natural explanation for the wide blue binaries found in that region (Fraser et al., 2017). All the known blue binaries are wide systems. There is no known tight blue binary in the CCKB. It is difficult to understand though, why only wide binaries formed at 35–40 AU, since tight binaries should have survived the sweeping mechanisms much more easily than wide binaries. Some explanation were proposed, based on the fact that, if binaries formed through the streaming instability mechanism, the separation of the formed binaries would depend on the mass of the aggregates, which in turns depends on the density profile of the primordial nebula and hence on the distance to the Sun. But for the moment, much more work should be done in order to clarify this question. If the sweeping mechanism is correct in explaining the presence of the blue wide binaries in the CCKB, almost all the planetesimals formed in the region just outside Neptune's primordial orbit should have been born as binary or multiple systems—a result consistent with that found in the simulations of BZ16 for the TNOs formed in situ.

10.5 Conclusions and perspectives

In the last decade, there was a huge advance in modeling the dynamical evolution of TNB. Contrasting these models with observations is crucial, thus, collecting more information is mandatory. Surveys designed to search for binaries in the resonant populations are required, especially Trojans of Neptune and 2:1 resonance, where only tight binaries are at present known. It is also necessary to increase the number of binaries with known orbits. The distribution of inclinations of TNB is crucial in order to determine which formation mechanism is viable and whether postformation mechanisms able to flip the orbital planes are efficient. The original distribution of orbital elements is difficult to constrain, because several of the mechanisms involved in the dynamical evolution of TNB, such as collisions, tides, and close encounters with the planets, are essentially of irreversible nature. Therefore, linking the current orbital elements of TNB to the orbits they had upon formation is a difficult task.

References

Alexandersen, M., Gladman, B., Kavelaars, J.J., Petit, J.-M., Gwyn, S.D.J., Shankman, C.J., Pike, R.E., 2016. A carefully characterized and tracked Trans-Neptunian survey: the size distribution of the Plutinos and the number of Neptunian Trojans. Astron. J. 152, 111.

Astakhov, S.A., Lee, E.A., Farrelly, D., 2005. Formation of Kuiper-belt binaries through multiple chaotic scattering encounters with low-mass intruders. Mon. Not. R. Astron. Soc. 360, 401–415.

Bai, X.-N., Stone, J.M., 2010. Dynamics of solids in the midplane of protoplanetary disks: implications for planetesimal formation. Astrophys. J. 722, 1437–1459.

Barge, P., Sommeria, J., 1995. Did planet formation begin inside persistent gaseous vortices? Astron. Astrophys. 295, L1–L4.

Barucci, M.A., Doressoundiram, A., Cruikshank, D.P., 2004. Surface characteristics of transneptunian objects and centaurs from photometry and spectroscopy. In: Comets II, vol. 647.

Benecchi, S.D., Sheppard, S.S., 2013. Light curves of 32 large transneptunian objects. Astron. J. 145, 124.

Benecchi, S.D., Noll, K.S., Grundy, W.M., Buie, M.W., Stephens, D.C., Levison, H.F., 2009. The correlated colors of transneptunian binaries. Icarus 200, 292–303.

Benz, W., Asphaug, E., 1999. Catastrophic disruptions revisited. Icarus 142, 5–20.

Brown, M.E., Trujillo, C.A., 2002. (26308) SM_165. Int. Astron. Union Circ. 7807, 1.

Brunini, A., 1996. On the satellite capture problem capture and stability regions for planetary satellites. Celest. Mech. Dyn. Astron. 64, 79–92.

Brunini, A., 2014. On the dynamical evolution and end states of binary centaurs. Mon. Not. R. Astron. Soc. 437, 2297–2302.

Brunini, A., 2017. Origin of slow-rotating minor bodies by tidal decoupling of binary pairs. Mon. Not. R. Astron. Soc. 465, 3949–3955.

Brunini, A., 2019. Flipping the orbital planes of primordial Trans-Neptunian binaries by close encounters with Neptune during the planetary instability migration phase of the outer solar system. Mon. Not. R. Astron. Soc. 483, 5042–5044.

Brunini, A., Zanardi, M., 2016. Dynamical and collisional evolution of Kuiper belt binaries. Mon. Not. R. Astron. Soc. 455, 4487–4497.

Campanella, G., Nelson, R.P., Agnor, C.B., 2013. Possible scenarios for eccentricity evolution in the extrasolar planetary system HD 181433. Mon. Not. R. Astron. Soc. 433, 3190–3207.

Carrera, D., Gorti, U., Johansen, A., Davies, M.B., 2017. Planetesimal formation by the streaming instability in a photoevaporating disk. Astrophys. J. 839, 16.

Collins, B.F., Sari, R., 2008. Lévy flights of binary orbits due to impulsive encounters. Astron. J. 136, 2552–2562.

Cuzzi, J.N., Dobrovolskis, A.R., Champney, J.M., 1993. Particle-gas dynamics in the midplane of a protoplanetary Nebula. Icarus 106, 102.

Danby, J., 1962. Fundamentals of Celestial Mechanics. MacMillan, New York, NY.

Dell'Oro, A., Cellino, A., 2007. The random walk of main belt asteroids: orbital mobility by non-destructive collisions. Mon. Not. R. Astron. Soc. 380, 399–416.

Dell'Oro, A., Campo Bagatin, A., Benavidez, P.G., Alemañ, R.A., 2013. Statistics of encounters in the Trans-Neptunian region. Astron. Astrophys. 558, A95.

Descamps, P., 2005. Orbit of an astrometric binary system. Celest. Mech. Dyn. Astron. 92, 381–402.

Di Sisto, R.P., Brunini, A., 2007. The origin and distribution of the centaur population. Icarus 190, 224.
Dziembowski, W., 1967. Internal structure and evolutionary changes in binaries: tidal friction in close binary systems. In: Dommanget, J. (Ed.), On the Evolution of Double Stars 105. Computes Rondus. Communications Serie B, No. 17.
Elliot, J.L., et al., 2005. The deep ecliptic survey: a search for Kuiper belt objects and centaurs. II. Dynamical classification, the Kuiper belt plane, and the core population. Astron. J. 129, 1117–1162.
Everhart, E., 1969. Close encounters of comets and planets. Astron. J. 74, 735.
Fabrycky, D., Tremaine, S., 2007. Shrinking binary and planetary orbits by Kozai cycles with tidal friction. Astrophys. J. 669, 1298–1315.
Fang, J., Margot, J.-L., 2012. Binary asteroid encounters with terrestrial planets: timescales and effects. Astron. J. 143, 25.
Fraser, W.C., et al., 2017. All planetesimals born near the Kuiper belt formed as binaries. Nat. Astron. 1, 0088.
Fraser, W.C., Brown, M.E., Morbidelli, A., Parker, A., Batygin, K., 2014. The absolute magnitude distribution of Kuiper belt objects. Astrophys. J. 782, 100.
Fuentes, C.I., Holman, M.J., Trilling, D.E., Protopapas, P., 2010. Trans-Neptunian Objects with Hubble Space Telescope. Astrophys. J. 722, 1290.
Funato, Y., Makino, J., Hut, P., Kokubo, E., Kinoshita, D., 2004. The formation of Kuiper-belt binaries through exchange reactions. Nature 427, 518–520.
Gault, D.E., Schultz, P.H., 1986. Oblique impact: projectile ricochet, concomitant ejecta and momentum transfer. Meteoritics 21, 368.
Gladman, B.J., Burns, J.A., Duncan, M., Lee, P., Levison, H.F., 1996. The exchange of impact ejecta between terrestrial planets. Science 271, 1387–1392.
Goldreich, P., Sari, R., 2009. Tidal evolution of rubble piles. Astrophys. J. 691, 54–60.
Goldreich, P., Soter, S., 1966. Q in the solar system. Icarus 5, 375–389.
Goldreich, P., Ward, W.R., 1973. The formation of planetesimals. Astrophys. J. 183, 1051–1062.
Goldreich, P., Lithwick, Y., Sari, R., 2002. Formation of Kuiper-belt binaries by dynamical friction and three-body encounters. Nature 420, 643–646.
Graykowski, A., Jewitt, D., 2018. Colors and shapes of the irregular planetary satellites. Astron. J. 155, 184.
Green, D.W.E., 2006. (134340) Pluto, (136199) Eris, and (136199) Eris I (Dysnomia). Int. Astron. Union Circ. 8747, 1.
Greenberg, R., 1981. Apsidal precession of orbits about an oblate planet. Astron. J. 86, 912–914.
Grundy, W.M., Noll, K.S., Virtanen, J., Muinonen, K., Kern, S.D., Stephens, D.C., Stansberry, J.A., Levison, H.F., Spencer, J.R., 2008. (42355) Typhon echidna: scheduling observations for binary orbit determination. Icarus 197, 260–268.
Grundy, W.M., Noll, K.S., Buie, M.W., Benecchi, S.D., Stephens, D.C., Levison, H.F., 2009. Mutual orbits and masses of six transneptunian binaries. Icarus 200, 627–635.
Grundy, W.M., Noll, K.S., Nimmo, F., Roe, H.G., Buie, M.W., Porter, S.B., Benecchi, S.D., Stephens, D.C., Levison, H.F., Stansberry, J.A., 2011. Five new and three improved mutual orbits of transneptunian binaries. Icarus 213, 678–692.
Grundy, W.M., et al., 2012. Mutual events in the Cold Classical transneptunian binary system Sila and Nunam. Icarus 220, 74–83.
Grundy, W.M., Benecchi, S.D., Porter, S.B., Noll, K.S., 2014. The orbit of transneptunian binary Manwë and Thorondor and their upcoming mutual events. Icarus 237, 1–8.
Grundy, W.M., Porter, S.B., Benecchi, S.D., Roe, H.G., Noll, K.S., Trujillo, C.A., Thirouin, A., Stansberry, J.A., Barker, E., Levison, H.F., 2015. The mutual orbit, mass, and density of the large transneptunian binary system Varda and Ilmarë. Icarus 257, 130–138.
Grundy, W.M., Noll, K.S., Roe, H.G., Buie, M.W., Porter, S.B., Parker, A.H., Nesvorný, D., Levison, H.F., Benecchi, S.D., Stephens, D.C., Trujillo, C.A., 2019. Mutual orbit orientations of transneptunian binaries. Icarus (in press).
Hubbard, W.B., 1984. Planetary Interiors, vol. 1984. Van Nostrand Reinhold Co., New York, NY, p. 343.
Hut, P., 1981. Tidal evolution in close binary systems. Astron. Astrophys. 99, 126–140.
Innanen, K.A., Zheng, J.Q., Mikkola, S., Valtonen, M.J., 1997. The Kozai mechanism and the stability of planetary orbits in binary star systems. Astron. J. 113, 1915.
Jeffreys, H., 1925. On the dynamical theory of the tides. Geophys. J. 1, 244–246.
Johansen, A., Henning, T., Klahr, H., 2006. Dust sedimentation and self-sustained Kelvin-Helmholtz turbulence in protoplanetary disk midplanes. Astrophys. J. 643, 1219–1232.
Johansen, A., Oishi, J.S., MacLow, M.-M., Klahr, H., Henning, T., Youdin, A., 2007. Rapid planetesimal formation in turbulent circumstellar disks. Nature 448, 1022–1025.

Johansen, A., Youdin, A., MacLow, M.-M., 2009. Particle clumping and planetesimal formation depend strongly on metallicity. Astrophys. J. 704, L75–L79.

Johansen, A., Klahr, H., Henning, T., 2011. High-resolution simulations of planetesimal formation in turbulent protoplanetary discs. Astron. Astrophys. 529, A62.

Johansen, A., Blum, J., Tanaka, H., Ormel, C., Bizzarro, M., Rickman, H., 2014. The multifaceted planetesimal formation process. In: Protostars and Planets VI, pp. 547–570.

Kern, S.D., Elliot, J.L., 2006. The frequency of binary Kuiper belt objects. Astrophys. J. Lett. 643, L57–L60.

Kiseleva, L.G., Eggleton, P.P., 1997. Effects of Tidal Friction in non-isolated close binary stars. In: The Combination of Theory, Observations, and Simulation for the Dynamics of Stars and Star Clusters in the Galaxy" 23rd Meeting of the IAU, Joint Discussion 15, Kyoto, Japan.

Kiseleva, L.G., Eggleton, P.P., 1999. Can combination of "Kozai effect" and tidal friction produce close stellar and planetary orbits? In: Henrard, J., Ferraz-Mello, S. (Eds.), Impact of Modern Dynamics in Astronomy. Springer, Dordrecht, pp. 409–410.

Kominami, J.D., Makino, J., Daisaka, H., 2011. Binary formation in planetesimal disks. I. Equal mass planetesimals. Publ. Astron. Soc. Jpn 63, 1331–1344.

Kozai, Y., 1962. Secular perturbations of asteroids with high inclination and eccentricity. Astron. J. 67, 579.

Leconte, J., Chabrier, G., Baraffe, I., Levrard, B., 2010. Is tidal heating sufficient to explain bloated exoplanets? Consistent calculations accounting for finite initial eccentricity. Astron. Astrophys. 516, A64.

Lee, E.A., Astakhov, S.A., Farrelly, D., 2007. Production of Trans-Neptunian binaries through chaos-assisted capture. Mon. Not. R. Astron. Soc. 379, 229–246.

Lidov, M.L., 1962. The evolution of orbits of artificial satellites of planets under the action of gravitational perturbations of external bodies. Planet. Space Sci. 9, 719–759.

Lin, H.-W., Kavelaars, J.J., Ip, W.-H., Gladman, B.J., Petit, J.M., Jones, R.L., Parker, J.W., 2010. On the detection of two new Trans-Neptunian binaries from the CFEPS Kuiper belt survey. Publ. Astron. Soc. Pac. 122, 1030.

Meheut, H., Keppens, R., Casse, F., Benz, W., 2012. Formation and long-term evolution of 3D vortices in protoplanetary discs. Astron. Astrophys. 542, A9.

Naoz, S., Perets, H.B., Ragozzine, D., 2010. The observed orbital properties of binary minor planets. Astrophys. J. 719, 1775–1783.

Nazzario, R.C., Orr, K., Covington, C., Kagan, D., Hyde, T.W., 2007. Kuiper binary formation. Adv. Space Res. 40, 280–283.

Nesvorný, D., Vokrouhlický, D., 2016. Neptune's orbital migration was grainy, not smooth. Astrophys. J. 825, 94.

Nesvorný, D., Youdin, A.N., Richardson, D.C., 2010. Formation of Kuiper belt binaries by gravitational collapse. Astron. J. 140, 785–793.

Nesvorný, D., Vokrouhlický, D., Bottke, W.F., Noll, K., Levison, H.F., 2011. Observed binary fraction sets limits on the extent of collisional grinding in the Kuiper belt. Astron. J. 141, 159.

Nicholson, P.D., Cuk, M., Sheppard, S.S., Nesvorný, D., Johnson, T.V., 2008. Irregular satellites of the giant planets. In: Barucci, A., Boehnhardt, H., Cruikshank, D., Morbidelli, A. (Eds.) The Solar System Beyond Neptune. University of Arizona Press, Tucson, AZ, pp. 411–424.

Noll, K.S., Stephens, D.C., Grundy, W.M., Millis, R.L., Spencer, J., Buie, M.W., Tegler, S.C., Romanishin, W., Cruikshank, D.P., 2002. Detection of two binary Trans-Neptunian objects, 1997 CQ_{29} and 2000 CF_{105}, with the Hubble Space Telescope. Astron. J. 124, 3424–3429.

Noll, K.S., Grundy, W.M., Chiang, E.I., Margot, J.-L., Kern, S.D., 2008a. Binaries in the Kuiper belt. In: The Solar System Beyond Neptune. University of Arizona Press, Tucson, AZ, pp. 345–363.

Parker, A.H., 2012. Discovery and characterization of Trans-Neptunian binaries in large-scale surveys. In: Arenou, F., Hestroffer, D. (Eds.) Proceedings of the workshop "Orbital Couples: Pas de Deux in the Solar System and the Milky Way". Held at the Observatoire de Paris, 10–12 October 2011, pp. 19–23.

Parker, A.H., Kavelaars, J.J., 2010. Destruction of binary minor planets during Neptune scattering. Astrophys. J. 722, L204–L208.

Parker, A.H., Buie, M.W., Grundy, W.M., Noll, K.S., 2016. Discovery of a Makemakean moon. Astrophys. J. 825, L9.

Perets, H.B., Naoz, S., 2009. Kozai cycles, tidal friction, and the dynamical evolution of binary minor planets. Astrophys. J. 699, L17–L21.

Petit, J.-M., Mousis, O., 2004. KBO binaries: how numerous were they? Icarus 168, 409–419.

Petit, J.M., Kavelaars, J.J., Gladman, B., Loredo, T., 2008. Size distribution of multikilometer transneptunian objects. In: The Solar System Beyond Neptune. University of Arizona Press, Tucson, AZ, pp. 71–87.

Porter, S.B., Grundy, W.M., 2012. KCTF evolution of Trans-Neptunian binaries: connecting formation to observation. Icarus 220, 947–957.

Press, W.H., Teukolsky, S.A., Vetterling, W.T., Flannery, B.P., 2002. Numerical Recipes in C++: The Art of Scientific Computing. ISBN: 0521750334.

Safronov, V.S., 1972. Evolution of the Protoplanetary Cloud and Formation of the Earth and Planets (Trans.) Israel Program for Scientific Translations. Keter Publishing House, Jerusalem (Israel) (in Russian).

Schlichting, H.E., Sari, R., 2008a. The ratio of retrograde to prograde orbits: a test for Kuiper belt binary formation theories. Astrophys. J. 686, 741–747.

Schlichting, H.E., Sari, R., 2008b. Formation of Kuiper belt binaries. Astrophys. J. 673, 1218–1224.

Showalter, M.R., Hamilton, D.P., Stern, S.A., Weaver, H.A., Steffl, A.J., Young, L.A., 2011. New satellite of (134340) Pluto: S/2011 (134340) 1. Int. Astron. Union Circ. 9221, 1.

Showalter, M.R., Weaver, H.A., Stern, S.A., Steffl, A.J., Buie, M.W., Merline, W.J., Mutchler, M.J., Soummer, R., Throop, H.B., 2012. New satellite of (134340) Pluto: S/2012 (134340) 1. Int. Astron. Union Circ. 9253, 1.

Simon, J.B., Armitage, P.J., Beckwith, K., 2012. Simulations of protoplanetary disk turbulence: connecting theory and observations. American Astronomical Society Meeting Abstracts 219, Article ID: 337.13.

Simon, J.B., Armitage, P.J., Li, R., Youdin, A.N., 2016. The mass and size distribution of planetesimals formed by the streaming instability. I. The role of self-gravity. Astrophys. J. 822 (55), 1–18.

Simon, J.B., Armitage, P.J., Youdin, A.N., Li, R., 2017. Evidence for universality in the initial planetesimal mass function. Astrophys. J. Lett. 847, L12.1-6.

Stevenson, D.J., 2009. Planetary Structure and Evolution. Available from: www.gps.caltech.edu/classes/ge131.

Thirouin, A., Sheppard, S.S., 2018. The Plutino population: an abundance of contact binaries. Astron. J. 155, 248.

Thirouin, A., Sheppard, S.S., Noll, K.S., 2017. 2004 Tt_{357}: a potential contact binary in the Trans-Neptunian belt. Astrophys. J. 844, 135.

Trujillo, C.A., Brown, M.E., 2002. A correlation between inclination and color in the classical Kuiper belt. Astrophys. J. 566, L125–L128.

Tsiganis, K., Gomes, R., Morbidelli, A., Levison, H.F., 2005. Origin of the orbital architecture of the giant planets of the solar system. Nature 435, 459–461.

Veillet, C., Parker, J.W., Griffin, I., Marsden, B., Doressoundiram, A., Buie, M., Tholen, D.J., Connelley, M., Holman, M.J., 2002. The binary Kuiper-belt object 1998 WW31. Nature 416, 711–713.

Weaver, H.A., Stern, S.A., Mutchler, M.J., Steffl, A.J., Buie, M.W., Merline, W.J., Spencer, J.R., Young, E.F., Young, L.A., 2006. Discovery of two new satellites of Pluto. Nature 439, 943–945.

Weidenschilling, S.J., 1975. Close encounters of small bodies and planets. Astron. J. 80, 145.

Weidenschilling, S.J., 1980. Dust to planetesimals—settling and coagulation in the solar Nebula. Icarus 44, 172–189.

Weidenschilling, S.J., 2002. On the origin of binary transneptunian objects. Icarus 160, 212–215.

Yanagisawa, M., Hasegawa, S., 2000. Momentum transfer in oblique impacts: implications for asteroid rotations. Icarus 146, 270–288.

Yanagisawa, M., Itoi, T., 1994. Impact fragmentation experiments of porous and weak targets. In: 75 Years of Hirayama Asteroid Families: The Role of Collisions in the Solar System History, vol. 63, p. 243.

Youdin, A.N., Goodman, J., 2005. Streaming instabilities in protoplanetary disks. Astrophys. J. 620, 459–469.

Further reading

Brunini, A., 2018. Dynamical evolution of a fictitious population of binary Neptune Trojans. Mon. Not. R. Astron. Soc. 474, 3912–3920.

Fedkin, A.V., Grossman, L., Ciesla, F.J., Simon, S.B., 2012. Mineralogical and isotopic constraints on chondrule formation from shock wave thermal histories. Geochim. Cosmochim. Acta 87, 81–116.

Noll, K.S., Grundy, W.M., Stephens, D.C., Levison, H.F., Kern, S.D., 2008b. Evidence for two populations of classical transneptunian objects: the strong inclination dependence of classical binaries. Icarus 194, 758–768.

Thirouin, A., Noll, K.S., Ortiz, J.L., Morales, N., 2014. Rotational properties of the binary and non-binary populations in the Trans-Neptunian belt. Astron. Astrophys. 569, A3.

CHAPTER

11

The dynamics of rings around Centaurs and Trans-Neptunian objects

Bruno Sicardy[a], *Stefan Renner*[b], *Rodrigo Leiva*[c], *Françoise Roques*[a], *Maryame El Moutamid*[d,e], *Pablo Santos-Sanz*[f], *Josselin Desmars*[a]

[a]Paris Observatory, PSL Research University, CNRS, Sorbonne University, Univ. Paris Diderot, Sorbonne Paris City, LESIA, Meudon, France [b]Paris Observatory, CNRS UMR 8028, Lille University, Lille Observatory, IMCCE, Lille, France [c]Department of Space Studies, Southwest Research Institute, Boulder, CO, United States [d]Center for Astrophysics and Planetary Science, Cornell University, Ithaca, NY, United States [e]Carl Sagan Institute, Cornell University, Ithaca, NY, United States [f]Instituto de Astrofísica de Andalucía (CSIC), Granada, Spain

11.1 Introduction

Trans-Neptunian objects (TNOs) are very remote and faint objects. They orbit at some 30–48 AU from the Sun and have magnitudes typically fainter than 18, thus requiring large telescopes. They subtend small angular diameters, for example, at most 100 milli-arcsec (mas) or so for the largest one, Pluto, and less than 35 mas for other large ones like Eris and Makemake. Centaur objects that are thought to be dynamically related to TNOs, orbit closer to us between Jupiter and Saturn, but they are small (300 km at most) and thus faint too and no more than 25 mas in angular size as seen from Earth. Classical imaging, either using adaptive optics or space telescopes, cannot provide any accurate details about their sizes, shapes, and surroundings.

In that context, stellar occultations are a precious tool, in particular to detect rings around small bodies, see also Chapter 19. Out of about 25 TNOs or Centaurs probed up to now using occultations, two proved to be surrounded by rings, providing a rather high frequency of about 8% for the presence of rings around those remote bodies. First in 2013, when two narrow and

The Trans-Neptunian Solar System.
https://doi.org/10.1016/B978-0-12-816490-7.00011-4

dense rings were discovered around Chariklo (Braga-Ribas et al., 2014; Sicardy et al., 2018), and then in 2017, when a ring was spotted around the dwarf planet Haumea (Ortiz et al., 2017). Physical parameters of those rings are listed in Table 11.1. Note that dense and narrow material has also been detected around the second largest known Centaur, Chiron. It could be a ring system alike that of Chariklo (Ortiz et al., 2015) (in which case the frequency of rings would be about 12%), or a bound cometary dust shell (Ruprecht et al., 2015).

This detection rate is statistically high, an augury for more ring detections around other bodies. Those new rings were the first to be detected elsewhere than around the four giant planets, and they may outnumber them soon. Moreover, the occultation technique now greatly benefits from the Gaia catalogs (Gaia Collaboration et al., 2016, 2018). Combining astrometric measurements of the occulting bodies, past stellar occultations and the Gaia catalogs, accuracies at the level of 10 mas (corresponding to some 200 km at the object) can now be reached for predicting these events, and even better for some bodies like Chariklo or Pluto, for which several occultations have been detected (Desmars et al., 2017, 2019).

Those recent (and possibly future) detections launched a new wave of interests on ring studies. First, because they can tell us something about the history of the object they surround and second, because they reside in a dynamical environment that is specific to small bodies, thus revealing some hitherto unstudied mechanisms.

11.2 Rings around irregular bodies

Currently, little is known about the nature, origin, and evolution of rings around small bodies like Chariklo or Haumea (Sicardy et al., 2018). Note that Haumea's ring composition is still unknown (Ortiz et al., 2017), while silicates, tholins, and water ice may be present in Chariklo's rings (Duffard et al., 2014; Table 11.1). However, this latter point is debated because the suspected water ice feature in Chariklo's ring spectrum presented in Duffard et al. (2014) is close to a gap in the spectral coverage of the instrument. Also, it could be that the water ice feature is caused by Chariklo itself, and that the rings are dark (like those of Uranus; Leiva et al., 2017) and spectroscopically featureless. The importance of water ice for forming rings has been suggested, predicting ring existence in the heliocentric range 8–20 AU (Hedman, 2015). This is at odd with the discovery of Haumea's ring, currently at 50 AU from the Sun, and with perihelion distance larger than 35 AU. In any case, this opens the interesting question of whether rings should be composed of water ice (and why), a possibility that would favor ring presences in the TNO realm rather than in the asteroid Main Belt.

Those objects are angularly so small that detection of putative shepherd satellites near the rings is extremely challenging (Sicardy et al., 2015). The chaotic nature of Chariklo's orbit suggests a scenario where that body has close encounter(s) with a giant planet, thus triggering ring formation through tidal disruption (Hyodo et al., 2016). However, the probability of such encounters being small (Araujo et al., 2016; Wood et al., 2017), this route to ring formation seems unlikely. Satellite orbital evolution followed by disruption has also been invoked to form Chariklo's rings, as well as a possible impact onto the body by an external body (Melita et al., 2017). But again such events are rare, and the satellite tidal disruption scenario should

TABLE 11.1 Chariklo and Haumea's rings: Physical parameters.

Parameters	Chariklo	Haumea
Main body		
Rotation period, T_{rot} (hours) (Fornasier et al., 2014; Lellouch et al., 2010)	7.004	3.915341
Mass M (kg) (Leiva et al., 2017; Ragozzine and Brown, 2009)	6.3×10^{18}	4.006×10^{21}
Rotational parameter q[a]	0.226	0.268
Semiaxes $A \times B \times C$ (km) (Leiva et al., 2017; Ortiz et al., 2017)	$157 \times 139 \times 86$[b]	$1161 \times 852 \times 513$
Reference radius R[c] (km)	115	712
Elongation parameter[d] ϵ	0.20	0.61
Oblateness parameter[d] f	0.55	0.76
Rings		
Pole position (degrees) α_p	151.25 ± 0.50	285.1 ± 0.5[e]
(equatorial J2000) δ_p	41.48 ± 0.22	-10.6 ± 1.2
Width (km) (Bérard et al., 2017; Ortiz et al., 2017)	C1R: ~5–7	~70
	C2R: ~1–3	
Apparent transmission[f] (Bérard et al., 2017; Sicardy et al., 2018; Ortiz et al., 2017)	C1R: ~0.5	~0.5
	C2R: ~0.9	
Visible reflectivity $(I/F)_V$ (Duffard et al., 2014)	0.07	Unknown
Composition[g] (Duffard et al., 2014; Ortiz et al., 2017)	20% Water ice	Unknown
	40%–70% Silicates	
	10%–30% Tholins	
	Some amorphous carbon	
Orbital radii (km) (Braga-Ribas et al., 2014; Ortiz et al., 2017)	390 (CR1)	2287
	405 (C2R)	
Corotation radius a_{cor} (km)	189	1104
Outer 2/4 resonance radius[h] (km)	300	1752
Outer 2/6 resonance radius[h] (km)	392	2285
Classical Roche limit a_{Roche}[i] (km)	~280	~2420

[a] *Eq. (11.5).*
[b] *Assuming a Jacobi equilibrium shape. Other solutions are possible (Leiva et al., 2017).*
[c] *Eq. (A.2).*
[d] *Eq. (A.4).*
[e] *The ring is coplanar with the orbit Haumea's main satellite Hi'iaka (Ortiz et al., 2017).*
[f] *Fraction of light transmitted by the rings, as seen in the plane of the sky at present epoch.*
[g] *Still debated, see text.*
[h] *Using Eqs. (11.10), (A.6).*
[i] *Using Eq. (11.19) and assuming fluid particles ($\gamma = 0.85$) with icy composition ($\rho' = 1000\,kg\,m^{-3}$). This is probably unrealistic, see text.*

be supported by more observational evidences, like the presence of a retinue of small satellites around Chariklo. As mentioned earlier, this is very challenging. Concerning Haumea, a violent collisional history (Brown et al., 2007) plus the coplanarity of the ring orbit with that of its main satellite Hi'iaka orbit suggest cogenetic origins.

More dynamical works sought to constrain the mass of Chariklo's main ring C1R, using the fact that it may be eccentric, and should be maintained as such through self-gravity, like those of Uranus (Pan and Wu, 2016). However, the eccentricity of CR1 is not well constrained (Bérard et al., 2017), especially because the ring precession rate is not known (Sicardy et al., 2018). Thus, the similarity between Chariklo and Uranus' rings orbital structure has yet to be proved, and inferences on the ring mass should be taken with caution.

Numerical studies using collisional codes have proved useful to describe the local behavior of Chariklo's rings. But one of these codes does not include the nonspherical shape of Chariklo (Michikoshi and Kokubo, 2017), while the other one does consider it, but without rotation (Gupta et al., 2018), an important ingredient of ring dynamics, see following.

Considering the small amount of information currently available on rings around small objects, we focus here on some more basic and generic processes. Let us first recall that a collisional disk around a central body dissipates energy, while conserving its total angular momentum $\mathbf{H}$. Assuming that the body rotates around its principal axis of largest inertia (i.e., that it does not wobble), and after averaging its potential $U(\mathbf{r})$ over the azimuth, one obtains that the average angular momentum $\langle \mathbf{H} \rangle_z$ of a particle projected on the rotation z-axis of the body is conserved. As a consequence, the system tends to flatten in the plane perpendicular to $\langle \mathbf{H} \rangle_z$ (called the equatorial plane herein), minimizing its energy while conserving $\langle \mathbf{H} \rangle_z$. The competing effects of pressure (stemming from the particle velocity dispersion c), rotation and self-gravity lead to an equilibrium where the local Toomre's parameter Q is a few times unity:

$$Q = \frac{c\kappa}{\pi G \Sigma_0} >\sim 1, \tag{11.1}$$

where κ is the particle radial epicyclic frequency, G is the gravitational constant, and Σ_0 is the disk surface density. Let us consider a broad disk of radius r, vertical thickness h, and total mass M_D surrounding the body. Noting that $Q \sim 1$, $GM \sim r^3\kappa^2$, $h \sim c/\kappa$, and $M_\mathrm{D} \sim \pi r^2 \Sigma_0$, we obtain

$$\frac{h}{r} \sim \frac{M_\mathrm{D}}{M}. \tag{11.2}$$

Only crude mass estimations for the current masses M_r of Chariklo and Haumea's rings are possible. Since the orbital periods of ring particles (some 10–20 hours) are of same order as for Saturn's ring A, the frequencies κ are similar. Consequently, the local kinematic conditions in all those rings are expected to be comparable (Sicardy et al., 2018). This suggests that Chariklo's ring C1R, Haumea's ring, and Saturn's ring A share the same surface density, some $\Sigma_0 \sim 500-1000\,\mathrm{kg\,m^{-2}}$ (Colwell et al., 2009). Table 11.1 then yields a ring mass estimate $M_\mathrm{r} \sim 10^{13}\,\mathrm{kg}$ for C1R, equivalent to an icy body of radius $\sim$1 km, corresponding to very small fractions of Chariklo's mass, $M_\mathrm{r}/M_\mathrm{C} \sim 10^{-6}$ and angular momentum, $H_\mathrm{r}/H_\mathrm{C} \sim 10^{-5}$. The same exercise for Haumea provides $M_\mathrm{r}/M_\mathrm{H} \sim 10^{-7}$ (equivalent to an icy body of radius $\sim$5 km) and $H_\mathrm{r}/H_\mathrm{H} \sim 10^{-6}$.

Noting that broad disks around Chariklo and Haumea would have masses about 100 times that of the current rings, we have from Eq. (11.2) $h/r \sim M_D/M \sim 10^{-5}$–$10^{-4}$, that is, h of some 10 m, corresponding to a very flat disk. To keep the problem at its simplest, we will assume the body is not wobbling as mentioned earlier that any departure of the body from axisymmetry is purely sectoral, that is, depends only on longitude, not latitude.

In that context, small objects like Chariklo or Haumea appear quite different from the ring-bearing giant planets, size put apart. In fact, they may support much larger nonaxisymmetric terms in their gravitational potentials. For instance, a topographic feature of height or depth z represents a mass anomaly $\mu \sim (z/2R)^3$ relative to the body. Considering $R = 130$ km, typical of Chariklo, and $z \sim 10$ km (not excluded for such a body; Leiva et al., 2017), we obtain $\mu \sim 10^{-4}$. This is comparable to Titan's mass relative to Saturn. Such a body placed on Saturn's equator would have devastating effects on the rings. In fact, the Saturnian satellite Janus is able to sharply truncate the outer edge of Saturn's ring A with a mere $\mu \sim 3 \times 10^{-9}$, while mass anomalies inside the planet have at most $\mu \sim 10^{-12}$ (Hedman and Nicholson, 2014). The differences between small bodies and giant planets are exacerbated when their shapes are considered. For instance, Haumea is extremely elongated (elongation $\epsilon \sim 0.6$, see Table 11.1 and Eq. A.1), as is possibly Chariklo too ($\epsilon \sim 0.2$). The mass anomalies stored in the bulges of those bodies are of order ϵ, which still increases the difference between the essentially axisymmetric giant planets and small bodies.

We focus here on the coupling between the central body and a surrounding collisional disk. We will not discuss the very origin of this disk, as discussed earlier. We feel that a good understanding of the ring dynamics around a nonspherical body is a necessary preliminary step before investigating further the origin of those rings.

11.3 Potential of a nonaxisymmetric body

As mentioned earlier, the body is assumed to rotate without wobbling. Consequently, each of its volume elements executes a circular motion at constant angular speed, noted Ω herein ($T_{\rm rot} = 2\pi/\Omega$ being the rotation period). The problem is simplified further by assuming that departure from sphericity is described either by a mass anomaly (under the form of a topographic feature) lying on the equator of a homogeneous sphere of mass M and radius R, or by a homogeneous ellipsoid of principal semiaxes $A > B > C$ rotating around C. In that context, a collisional disk, once settled in the equatorial plane of the body, does not feel any perpendicular force that would excite off-plane motions.

The vector $\mathbf{r}$ denotes the position vector of the particle (counted from the center of the body), with $r = \|\mathbf{r}\|$. The angle L is the true longitude of the particle (counted from an arbitrary origin) and $L' = \Omega t$ measures the orientation of the body through its mass anomaly or its longest axis A, t being the time. By symmetry, the expansion of the potential then contains only cosine terms, with the generic form

$$U(\mathbf{r}) = \sum_{m=-\infty}^{+\infty} U_m(r) \cos(m\theta), \tag{11.3}$$

where $\theta = L - L'$. The integer m is called the azimuthal number.

In the mass anomaly case, $U(\mathbf{r})$ can be split into a Keplerian term, $-GM/R$, and terms depending on μ (see Sicardy et al., 2019 for details), so that

$$U(\mathbf{r}) = -\frac{GM}{r} - \frac{GM}{2R}\mu\left\{\left[\sum_{m=-\infty}^{+\infty} b_{1/2}^{(m)}(r/R)\cos(m\theta)\right] - 2q\left(\frac{r}{R}\right)\cos(\theta)\right\}, \tag{11.4}$$

where G is the constant of gravitation and $b_{1/2}^{(m)}(r/R)$ are the classical Laplace coefficients. The dimensionless coefficient

$$q = \frac{R^3\Omega^2}{GM} \tag{11.5}$$

measures the rotational stability of the body, so that $q < 1$ for bodies of interest.

In the case of a homogeneous ellipsoid, *only even values of m are allowed*, ensuring the invariance of the potential under a rotation of π radians. As seen later, this will affect the orders of the resonances. Posing $m = 2p$, we have

$$U(\mathbf{r}) = \sum_{p=-\infty}^{+\infty} U_{2p}(r)\cos(2p\theta). \tag{11.6}$$

The calculation of $U_{2p}(r)$ is detailed in Appendix A.1. Defining the sequence $S_{|p|}$ recursively as

$$S_{|p|+1} = 2\frac{(|p|+1/4)(|p|+3/4)}{(|p|+1)(|p|+5/2)} \times S_{|p|} \text{ with } S_0 = 1, \tag{11.7}$$

and turning back to $m = 2p$, we obtain at lowest order in the elongation ϵ

$$U(\mathbf{r}) = -\frac{GM}{r}\sum_{m=-\infty}^{+\infty}\left(\frac{R}{r}\right)^{|m|} S_{|m/2|}\epsilon^{|m/2|}\cos(m\theta) \quad (m \text{ even}). \tag{11.8}$$

Eqs. (11.4), (11.8) define the potentials caused by a mass anomaly and by a homogeneous ellipsoid, respectively. Note that $U_m(r)$ has an exponential behavior with m, that is, $|U_m(r)| \propto K^{|m|}$, where $K < 1$ is a constant depending on the problem under study (Sicardy et al., 2019). Thus, the strengths of the resonances rapidly decrease as $|m|$ increases.

11.4 Resonances around nonaxisymmetric bodies

A particle moving in the equatorial plane of a body has two degrees of freedom. They are associated with two fundamental frequencies: The mean motion n and the radial epicyclic frequency κ, whose calculations are outlined in Appendix A.2. However, the nonaxisymmetric terms $\cos(m\theta)$ ($m \neq 1$) in Eq. (11.3) are time dependent through $L' = \Omega t$, introducing another degree of freedom. For a rigid body rotating at constant angular speed Ω along a fixed axis, this dependence can be eliminated by using the variable $\theta = L - L'$ instead of L. This is equivalent to writing the equations of motion in a frame corotating with the body. The two fundamental frequencies of the autonomous system so obtained are now $n - \Omega$ and κ. Two kinds of resonances happen, one being the corotation resonance

$$n = \Omega, \tag{11.9}$$

also referred to as the synchronous orbit.

The other kind of resonances (called sectoral, see later) occurs for

$$j\kappa = m(n - \Omega), \tag{11.10}$$

where j and m are integers. Those two kinds of resonances are examined in turn.

11.4.1 Corotation resonance

The corotation region, where $n \sim \Omega$, can in principle host ring-arc material. The particle motion is then best studied by reexpressing the potential $U(\mathbf{r})$ in a frame corotating with the body, that is,

$$V(\mathbf{r}) = U(\mathbf{r}) - \frac{\Omega^2 r^2}{2}. \tag{11.11}$$

Level curves of $V(\mathbf{r})$ are shown in Fig. 11.1. Note the presence of two stable points C_2 and C_4, reminiscent of the Lagrange points L_4 and L_5, while C_1 and C_3 are reminiscent of the unstable Lagrange point L_3.

The points C_2 and C_4 being maxima of potential, they are unstable against dissipative collisions, even if dynamically stable. With the expected values of μ (some 10^{-5} for Chariklo), the librating (or "tadpole") orbits around C_2 and C_4 are so narrow (a few kilometers) that spreading time scales for removing the particles from that region is very short, a few years (Sicardy et al., 2019). An elongated body with $\epsilon = 0.2$ creates much wider corotation zones, and thus much longer spreading time scales (some 10^4 years). However, another problem appears, as the points C_2 and C_4 are linearly unstable for

$$\left(4\Omega^2 + V_{xx} + V_{yy}\right)^2 \leq V_{xx}V_{yy} - V_{xy}^2, \tag{11.12}$$

where the indices x and y stand for partial derivatives (Murray and Dermott, 2000).

Using Eq. (11.4) and Chariklo's parameters, this equation provides a critical value of about 0.04 for μ, above which C_2 and C_4 are linearly unstable. This is unrealistically large, as this it requires a topographic feature of several tens of kilometers. Turning to the case of an elongated body, and limiting the expansion in Eq. (11.8) to the term $m = 2$, Eq. (11.12) implies that C_2 and C_4 are unstable for $\epsilon > \epsilon_{\text{crit}} \sim 0.06/q^{2/3}$, that is, $\epsilon_{\text{crit}} \sim 0.16$ in the case of Chariklo ($q = 0.226$). Using the nominal value of Chariklo's elongation ($\epsilon = 0.20$), this shows that C_2 and C_4 are unstable for Chariklo, as illustrated in Fig. 11.1.

In summary, the corotation region does not appear as a viable environment to host ring-arc material.

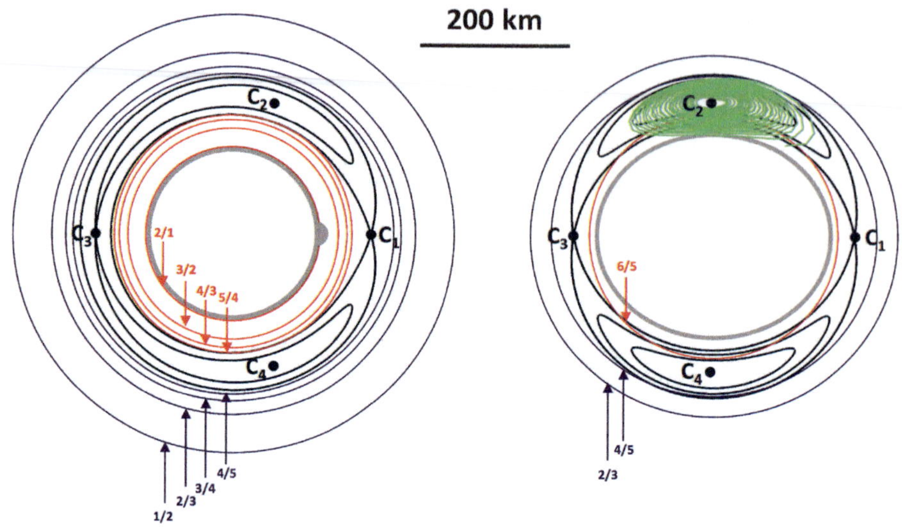

FIG. 11.1 Structure of corotation potential. *Black lines*: Levels curves of the potential $V(\mathbf{r})$ in the corotating frame (Eq. 11.11). The parameters are those of Chariklo (Table 11.1), the bar at the top providing the distance scale. *Left*: Case of a mass anomaly on a spherical body ($R \sim 130$ km) and $\mu = 2.5 \times 10^{-2}$. This is much larger than Chariklo's topographic features (height $z \sim 5$ km, $\mu \sim 10^{-5}$; Leiva et al., 2017) in order to better show the level curves around the stable points C_2 and C_4. Note the presence of two unstable points C_1 and C_3. A few inner $m/(m-1)$ Lindblad resonances radii (LRs, see Section 11.5.1) are plotted in *red* ($m = 2, \ldots, 5$), while outer LRs are plotted in *blue* ($m = -1, \ldots, -4$). *Right*: The same with an homogeneous ellipsoid with semiaxes $A \times B \times C = 157 \times 139 \times 86$ km and elongation $\epsilon = 0.2$. Only LRs with m even are now present, with an example for $m = 6$ (*red*) and $m = -2, -4$ (*blue*). The *green line* shows the motion of a particle starting at C_2. The trajectory is linearly unstable because ϵ is larger than the critical value $\epsilon_{\text{crit}} = 0.16$, see text. *Adapted from Sicardy, B., et al., 2019. Ring dynamics around non-axisymmetric bodies. Nat. Astron. 3, 147.*

11.4.2 Sectoral resonances

The potential $U(\mathbf{r})$ in Eq. (11.3) can be expanded in a Fourier series containing harmonics $j\kappa - m(n - \Omega)$ of the two frequencies κ and $n - \Omega$. Without loss of generality, we can take $j \geq 0$, while m is either positive or negative. Other conventions are found in the literature, for example, with m positive and j positive or negative. Our choice is motivated by the fact that it avoids the tedious use of the symbol $\pm$ in the equations.

The resonances studied here will be qualified as sectoral because they stem only from azimuth-dependent parts of the potential, with no dependence on latitude. In other words, we do not consider here tesseral harmonics of the potential, classically studied in the dynamics of artificial satellites of Earth, or associated with oscillation modes of giant planets (Marley and Porco, 1993; Marley, 2014). We avoid the terminology "spin-orbit resonances," that usually describes the response of a rotating secondary to the torque exerted by the primary, thus introducing confusions. In fact, sectoral resonances are akin to mean motion resonances between a ring particle and an equatorial satellite, the role of the satellite being played by the nonaxisymmetric terms of the potential.

The position vector of the particle, $\mathbf{r}$, can be expressed in terms of the orbital elements a, e, λ, and ϖ, that is, the semimajor axis, orbital eccentricity, mean longitude, and pericentric longitude of the particle, respectively. The relevant resonance angle associated with the resonance defined in Eq. (11.10) is

$$\phi_{m/(m-j)} = [mL' - (m-j)\lambda - j\varpi]/j, \tag{11.13}$$

recalling that $L' = \Omega t$. Note the presence of the dividing factor j in the expression earlier. This is needed to obtain the proper choice of canonical variables in the Hamiltonian describing the resonance (Peale, 1986).

The expression of $\phi_{m/(m-j)}$ satisfies the d'Alembert's relation, that is, it is invariant under a change of origin, since $m-(m-j)-j=0$. Moreover, the d'Alembert's characteristics state that the amplitude of the term containing $\phi_{m/(m-j)}$ is of lowest order $e^{|j|}$ ($= e^j$ since $j \geq 0$ by convention). For that reason, the integer j is called herein the order of the resonance. In summary

$$U(\mathbf{r}) = U(a, e, \lambda, \varpi) = \sum_{m=-\infty}^{+\infty} \sum_{j=0}^{+\infty} A_{m,j}(a)e^j \cos\left(\phi_{m/(m-j)}\right), \tag{11.14}$$

where $A_{m,j}(a)$ depends on the problem under study. The case $j=1$ (LRs) will be detailed in Section 11.5.1.

Condition (11.10) means that after j synodic periods $2\pi/(n-\Omega)$, the particle completes exactly m radial oscillations, which excites the particle orbital eccentricity (El Moutamid et al., 2014). Noting that the apsidal precession rate is $\dot{\varpi} = n - \kappa$, Eq. (11.10) reads

$$\frac{n-\dot{\varpi}}{\Omega-\dot{\varpi}} = \frac{m}{m-j}, \tag{11.15}$$

meaning that in a frame rotating at rate $\dot{\varpi}$, the particle completes m revolutions, while the body completes $m-j$ rotations. Since usually $\dot{\varpi} \ll n, \Omega$, we have

$$\frac{n}{\Omega} \sim \frac{m}{m-j}, \tag{11.16}$$

loosely referred to as a "$m/(m-j)$" resonance.

Without loss of generality, we may assume that $\Omega > 0$. If $n > 0$ (and then also $\kappa > 0$), the particle has a prograde motion around the body. Then, $m > 0$ implies that the resonance occurs inside the corotation radius, and will be qualified as an inner resonance. Conversely, $m < 0$ will correspond to outer resonances. Note that we may have $n < 0$ (with $\kappa < 0$) that describes a "retrograde resonance," in which the particle and the body move in the opposite directions. From $j > 0$, retrograde resonances always require $m > 0$.

11.4.3 Resonance order

We define the order of a resonance as the lowest power in eccentricity present in the corresponding resonant term, for example, the integer j in Eq. (11.14). For a given n/Ω, the order depends on the symmetry of the potential.

For instance, due to its invariance under a π-rotation, an ellipsoid creates only terms with *even* azimuthal numbers $m = 2p$ (Eq. 11.6), so that $n/\Omega = 2p/(2p-j)$. Thus, to get $n/\Omega = 1/2$, one must take $m = 2p = -2$ and $j = 2$, corresponding to a second-order resonance. In the case of a mass anomaly, this ratio is obtained with $m = -1$ and $j = 1$, corresponding to a

TABLE 11.2 Resonance orders around an elongated body.

Azimuthal number	$m = +2$	$m = +4$	$m = +6$	$m = +8$
$j = 1$ (LRs)	$e\epsilon$ [2/1]	$e\epsilon^2$ [4/3]	$e\epsilon^3$ [6/5]	$e\epsilon^4$ [8/7]
$j = 2$	Apsidal	* * **	$e^2\epsilon^3$ [6/4]	* * **
$j = 3$	$e^3\epsilon$ [−2/1]	$e^3\epsilon^2$ [4/1]	* * **	$e^3\epsilon^4$ [8/5]
$j = 4$ ...	$e^4\epsilon$ [−1]	Apsidal	$e^4\epsilon^3$ [6/2]	* * **
Azimuthal number	$m = -2$	$m = -4$	$m = -6$	$m = -8$
$j = 1$ (LRs)	$e\epsilon$ [2/3]	$e\epsilon^2$ [4/5]	$e\epsilon^3$ [6/7]	$e\epsilon^4$ [8/9]
$j = 2$	$e^2\epsilon$ [2/4]	* * **	$e^2\epsilon^3$ [6/8]	* * **
$j = 3$	$e^3\epsilon$ [2/5]	$e^3\epsilon^2$ [4/7]	* * **	$e^3\epsilon^4$ [8/11]
$j = 4$ ...	$e^4\epsilon$ [2/6]	* * **	$e^4\epsilon^3$ [6/10]	* * **

*Notes: Each entry shows the leading terms $e^j\epsilon^{|m/2|}$ of a few $n/\Omega \sim m/(m-j)$ sectoral resonances around an elongated body. The fractions in brackets are the corresponding ratios n/Ω. The ratios that have absolute values larger (resp. smaller) than one correspond to inner (resp. outer) resonances. The ratios that have positive (resp. negative) values correspond to prograde (resp. retrograde) resonances. Because only the leading terms are considered here, some resonances are not replicated (symbols * * **). For instance, the $m = -4, j = 2$ resonance (corresponding to the second-order ($e^2\epsilon^2$) resonance $n/\Omega = 4/6$) is already considered in the $m = -2, j = 1$ case, corresponding to the 2/3 first-order ($e\epsilon$) LR.*

first-order resonance. More physically, one can see an ellipsoid as a body with two opposite mass anomalies. The resonant term $\propto e\cos(2\lambda - L' - \varpi)$ created by one bulge is canceled out by the term $\propto e\cos[2\lambda - (L' + \pi) - \varpi]$ created by the other bulge. This leaves the place for the second-order term $\propto e^2\cos(4\lambda - 2L' - 2\varpi)$, which is now invariant under $L' \to L' + \pi$.

For this reason, we use the notation 1/2 in the case of a mass anomaly, and 2/4 in the case of an elongated body, to enhance their respective second and fourth orders. In the same vein, a mass anomaly creates a second-order resonance $n/\Omega = 1/3$ with $m = -1, j = 2$, while an elongated body creates a fourth-order resonance $n/\Omega = 2/6$ ($m = -2, j = 4$) that have respective potentials $\propto e^2$ and $\propto e^4$. The resonance orders, both in orbital eccentricity and in elongation parameter, are listed in Table 11.2.

11.5 Lindblad resonances

11.5.1 Definition

The strongest resonances in Eq. (11.10) correspond to the lowest order, $j = 1$. They are called Lindblad eccentric resonances, or simply LRs.[1] Thus, an LR corresponds to, recalling that m can be positive or negative:

$$\kappa = m(n - \Omega). \tag{11.17}$$

LRs have been extensively studied in the frame of Galactic and ring dynamics. One reason is that they are the strongest ones for a given m. Another reason is that the resonant streamlines do not cross themselves, contrarily to the cases $j > 1$. This avoids singularities in the

[1] Lindblad inclined resonances are not considered here, as no off-plane forces are present.

hydrodynamical equations, and allows an analytical treatment of the disk response to the resonant forcing.

A more general way to define an LR is in fact $\kappa = m(n - \Omega_p)$, where Ω_p is the pattern speed of the forcing potential. In our case (a nonwobbling body), $\Omega_p = \Omega$ because each volume element of the body moves at constant speed along a circular path. This is not true for a perturbation caused by a satellite with orbital eccentricity e_s. In that case, besides the fundamental pattern speed n_s (the satellite mean motion), there is an infinity of other pattern speeds $n_s + \kappa_s/(m_s - 1)$, where κ_s is the satellite epicyclic frequency and m_s is an integer. The associated perturbing potential is then of order $ee_s^{m_s-1}$, and the order of the LR is said to be m_s, *although it remains of first order in e*. For instance, the second-order 5:3 Mimas LR in Saturn's A ring is in fact a 4:3 first-order LR with one of the harmonics of Mimas' potential.

11.5.2 Torques

Each $m/(m-1)$ LR exerts a torque Γ_m onto the disk (Sicardy et al., 2019),

$$\Gamma_m = \text{sign}(\Omega - n) \left(\frac{4\pi^2 \Sigma_0}{3n} \right) \frac{(GM)^2}{\Omega R^2} \mathcal{A}_m^2, \tag{11.18}$$

where $\mathcal{A}_m$ is a dimensionless coefficient that measures the strength of the resonance, see Appendix A.3. This coefficient depends on the coefficient $U_m(r)$ that appears in Eqs. (11.4), (11.8) and its derivative, see Eqs. (A.7), (A.8).

The sign$(\Omega - n)$ factor shows that the material inside the corotation radius receives a negative torque and thus is pushed onto the body, while the material outside of the corotation is pushed outwards, and more precisely, outside the outermost 1/2 LR. The timescales for these migrations are short, a few years only for bodies with the typical elongations ϵ of Chariklo and Haumea (Table 11.1). Even a spherical body with a unique topographic feature of height 5 km sitting on its equator can clear the entire zone between its surface and the outer 1/2 LR in less than 1 Myr (Sicardy et al., 2019).

This model shows that rings can exist around a nonaxisymmetric body only if the latter rotates fast enough. This is because the 1/2 LR at radius $a_{1/2}$ must lie inside the Roche limit a_{Roche} of the body, to prevent the ring from accreting into satellites. One can estimate

$$a_{\text{Roche}} \sim \left(\frac{3}{\gamma} \right)^{1/3} \left(\frac{M}{\rho'} \right)^{1/3}, \tag{11.19}$$

where ρ' is the density of the ring particles and γ is a factor describing the particle shape (Tiscareno et al., 2013). From Kepler's third law, we have $a_{1/2} = 2^{2/3} a_{\text{cor}}$ (where a_{cor} is the corotation radius) so that thus the condition $a_{1/2} < a_{\text{Roche}}$ reads

$$\gamma \rho' < \frac{3}{4} \frac{\Omega^2}{G}. \tag{11.20}$$

Although γ is poorly known, a nominal value of $\gamma \sim 1.6$ can be used,[2] representing particles filling their lemon-shaped Roche lobes (Tiscareno et al., 2013). A value $\rho' \sim 450\,\text{kg}\,\text{m}^{-3}$ stems from the typical density of the small Saturnian satellites (Thomas, 2010), thus a good proxy of

[2] The classical Roche limit is obtained with $\gamma = 0.85$ and particles with same density as the central body.

ring particles. Eq. (11.20) finally yields $T_{\rm rot} <\sim 7$ hours, a condition met by both Chariklo and Haumea (Table 11.1).

11.6 Beyond the first order

11.6.1 Streamline self-crossings

A lowest order in eccentricity, a particle moves at distance $\rho = a[1 - e\cos(M)]$ from the body, where $M = \lambda - \varpi$ is the mean anomaly. Eq. (11.13) shows that $M = (m/j)\theta + \phi_{m/(m-j)}$, where $\theta = L - L' \sim \lambda - L'$ is the longitude of the particle as observed in a frame corotating with the body. All the particles in a given resonant streamline share the same resonant angle $\phi_{m/(m-j)}$, so that its polar equation is

$$\rho(\theta) = a\left[1 - e\cos\left(\frac{m}{j}\theta + \phi_{m/(m-j)}\right)\right], \tag{11.21}$$

showing that $\phi_{m/(m-j)}$ defines the orientation of the streamline relative to the body. To simplify the expressions used hereafter, we take without loss of generality $\phi_{m/(m-j)} = 0$, so that $\rho(\theta) = a[1 - e\cos(m\theta/j)]$, noting that e depends on a.

For a LR ($j = 1$), the polar equation $\rho(\theta) = a[1 - e\cos(m\theta)]$ shows that the resonance forces a m-lobbed pattern on the streamline that propagates as a m-armed spiral wave in the disk. In this case, the function $\rho(\theta)$ is single-valued, and a given streamline does not have self-crossing points.[3] For $j > 1$, self-crossing usually occurs because the same value of θ (mod 2π) may correspond to different values of $\rho(\theta)$, providing a multivalued function. To see that, let us first divide m and j by their greatest common divider, *so that m and j are now relatively prime*. Consider the sequence of j angles $\theta_0 = \theta, \theta_1 = \theta + 2\pi, \ldots \theta_k = \theta + 2k\pi, \ldots (0 \le k < j)$. They correspond to the same angular position of a particle in the rotating frame. Since m and j are relatively prime, $k(m/j)$ cannot be an integer for $0 < k < j$. Thus, all the angles $(m/j)\theta_k$, taken two by two, are different (mod 2π), and so are in general the corresponding values of the radius $\rho(\theta_k)$, except at self-crossing points. Consequently, the streamline exhibits j distinct braids, each defined by a given $k = 0, \ldots, j-1$. Note that the streamline eventually closes onto itself for $k = j$, since $(m/j)\theta_j = (m/j)\theta_0 + 2m\pi$.

The existence of several braids means that the streamline must cross itself for $j > 1$, see, for example, Fig. 11.2. To count the number of self-crossing points, let us first show that the streamline is invariant by a rotation of $2\pi/m$ in the rotating frame. Then, it is enough to count the crossing points for $\theta \in [0, 2\pi/|m|[$ and multiply the result by $|m|$.

The equation of the streamline, once rotated by $2\pi/m$, is $\rho'(\theta) = \rho(\theta - 2\pi/m) = a\{1 - e\cos[(m\theta - 2\pi)/j]\}$. Thus, $\rho'(\theta_k) = a\{1 - e\cos[(m\theta + (km-1)2\pi)/j]\}$. From Bachet-Bézout theorem, and because m and j are relatively prime, there exists two integers x and y such that $mx + jy = 1$. Thus, $\rho'(\theta_k) = a\{1 - e\cos[m(\theta + 2(k-x)\pi)/j]\} = a\{1 - e\cos[m(\theta + 2k'\pi)/j]\}$ after posing $k' = k - x$. Thus, $\rho'(\theta_k)$ takes the same values as $\rho(\theta_k)$, except for a circular permutation on the indices $k = 0, \ldots, j-1$, showing the invariance of the streamline under a rotation of $2\pi/m$. In other

[3] However, *adjacent* streamlines may cross if the local gradient of eccentricity de/da is large enough.

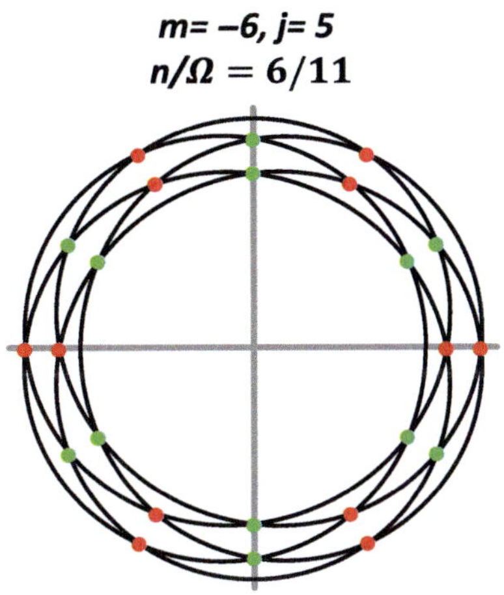

FIG. 11.2 Example of self-crossing resonant streamlines. *Left*: A periodic streamline with $m = -6, j = 5$, corresponding to $n/\Omega = m/(m-j) = 6/11$, observed in a corotating frame. Note the presence of $j = 5$ braids and $|m|(j-1) = 24$ self-crossing points. The difference between the *red* and *green points* is discussed in Fig. 11.3. *Right*: The case $m = -5, j = 6$ ($n/\Omega = 5/11$). The streamline has now $j = 6$ braids and $|m|(j-1) = 25$ self-crossing points.

words, the streamline can be divided in identical sectors, each of angular aperture $2\pi/|m|$ (e.g., $360/6 = 60$ degrees in Fig. 11.2, left panel, and $360/5 = 72$ degrees in the right panel).

Let us now consider θ varying in $[0, 2\pi/m[$, corresponding to the first sector of the streamline. Let us pose $M_k = m(\theta + 2k\pi)/j$, with $k = 0, \ldots, j-1$. As θ goes from 0 to $2\pi/m$, M_k goes from $2\pi(mk)/j$ to $2\pi(mk)/j + 2\pi/j$ and $\rho(M_k) = a[1 - e\cos(M_k)]$ then describes one of the j braids of the streamline. The extremities of each angular interval $[2\pi(mk)/j, 2\pi(mk)/j + 2\pi/j[$ can be distributed on the unit circle (Fig. 11.3). Because m and j are relatively prime, this defines exactly j points on the circle for $0 \leq k < j$. The self-crossing points then correspond to those angles that have the same $\cos(M_k)$, that is, to the points that have the same projection onto the x-axis (Fig. 11.3). This occurs at $2(j-1)$ angular positions along the circle, corresponding to $j-1$ self-crossing points as θ varies in $[0, 2\pi/m[$, and thus a total of $|m|(j-1)$ self-crossing points in the entire streamline, see examples in Fig. 11.2.

Note that this result is valid only if m and j are relatively prime. However, as discussed in the previous section, it is desirable to enhance the order of the resonance by using nonrelatively prime integers. For instance, the notation 2/6 resonance ($m = -2, j = 4$) enhances its fourth-order nature in the case of an elongated body, but its irreductible version 1/3 ($m = -1, j = 2$) must be used to calculate the number of self-crossing points of the corresponding streamline, $|-1| \times (2-1) = 1$.

In summary, a $m/(m-j)$ resonance is of order j, as long as m and j have not yet been reduced to their relatively prime version. Once m and j are reduced to their relatively prime version, the resonant streamline has

- $|m|$ identical sectors, each of extension $(2\pi/|m|)$;
- j braids; and
- $|m|(j-1)$ self-crossing points.

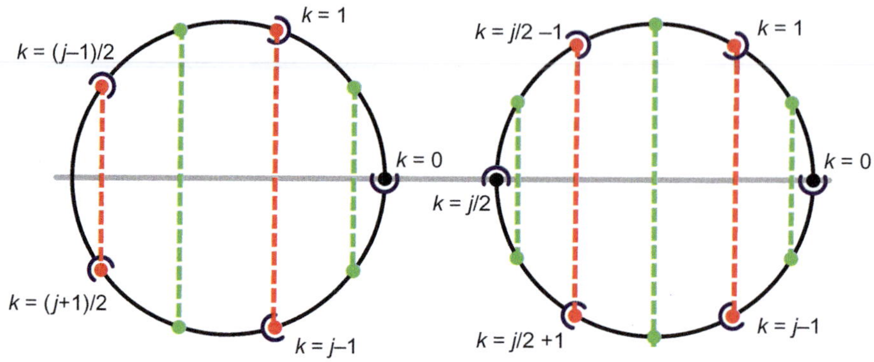

FIG. 11.3 Counting the self-crossing points of a resonant streamline. Consider a $m/(m-j)$ resonance of order j (with m and j relatively prime) and divide the unit circle in j identical arcs with opening angles $2\pi/j$. The points with angular positions $M_k = m(\theta + 2k\pi/|m|)/j$ $(0 \leq k < j-1)$ then move simultaneously along each arc, from its closed extremity (*dots*) to its open extremity (*blue semicircles*), as θ varies in the interval $[0, 2\pi/|m|[$. Those points have a common projection onto the x-axis (*gray line*) for a given θ at the *red* or *green points*. *Left panel*: An example with $j=5$ odd. There are $j-1=4$ *red points* and $j-1=4$ *green points*, that is, a total of $2j-2=8$ points that provides $j-1=4$ self-crossing points in the streamline, defined by the four *vertical dashed lines*. *Right panel*: The same, but with $j=6$ even. There are now $j-2=4$ *red points* and $j=6$ *green points*, that is, again a total of $2j-2=10$ points defining $j-1=5$ self-crossing points. It can be shown that this result is general, that is, that the number of self-crossing points is always $j-1$ as θ moves in the sector $0 \leq \theta < 2\pi/|m|$.

Resonances are often noted $n/\Omega = p/q$ in the literature, instead of $m/(m-j)$. Then, it is of order $|p-q|$ before p and q have been reduced. After reduction to their relatively prime version, the results previously show that the resonant streamline has $|p|(|p-q|-1)$ self-crossing points.

11.6.2 Phase portraits of 1/2 and 1/3 resonances

We present here examples of phase portraits corresponding to outer sectoral resonances. We focus on the 1/2 and 1/3 resonances caused by a mass anomaly, and on the 2/4 and 2/6 sectoral resonances caused by an elongated body. This stems from the fact that the 1/2 is the outermost LR where the classical torque formula (11.5.2) applies, while the 2/4 resonance, although not an LR, has still a significant effect on the ring (Sicardy et al., 2019). Moreover, the 1/3 and 2/6 resonances are close to the locations of both Chariklo and Haumea's rings (Table 11.1), and thus deserve special attention.

We do not care at this stage about the particular numerical values of μ or ϵ, but rather about the topologies of the phase portraits and their consequences on a ring. Phase portraits are displayed in Fig. 11.4, using the formalism of Murray and Dermott (2000). Each panel shows the level curves of the resonant Hamiltonian, for a given Jacobi constant. Note that while the left panel can only be reproduced in the case of a mass anomaly (1/2 resonance), the central panel can be obtained either with the 1/3 resonance with a mass anomaly, or with the 2/4 resonance with an elongated body. Similarly, the rightmost portrait can be obtained either with a 2/6 resonance with an elongated body, or with a 1/5 resonance with a mass anomaly.

The topology of each portrait provides interesting hints about the ring response. Collisions actually tend to damp the forced eccentricities, that is, to attract the eccentricity vector (X, Y)

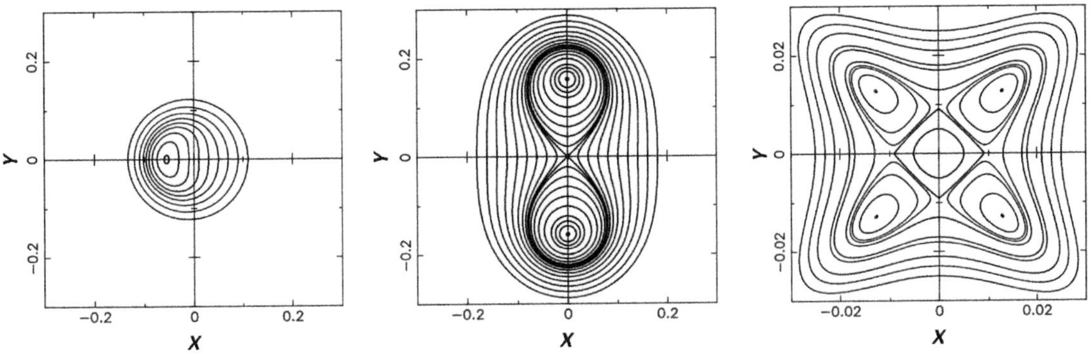

FIG. 11.4 Phase portraits of first-, second-, and fourth-order resonances. The trajectories of the eccentricity vector $(X, Y) = (e\cos\phi_{m/(m-j)}, e\sin\phi_{m/(m-j)})$ are shown for various resonance orders. *Left*: Representative phase portrait of the first-order resonance $n/\Omega = 1/2$, $\phi_{1/2} = 2\lambda - L' - \varpi$, caused by a mass anomaly. *Center*: Phase portrait close to a second-order resonance, either caused by a mass anomaly with $n/\Omega = 1/3$, $\phi_{1/3} = (3\lambda - L' - 2\varpi)/2$, or by an elongated body with $n/\Omega = 2/4$, $\phi_{2/4} = (4\lambda - 2L' - 2\varpi)/2$. *Right*: Phase portrait close to the fourth-order resonance caused by, for instance, an elongated body with $n/\Omega = 2/6$, $\phi_{2/6} = (6\lambda - 2L' - 4\varpi)/4$.

in Fig. 11.4 toward the origin $X = Y = 0$. Conversely, the resonance tends to force that vector to move around a stable fixed point of the phase portrait.

Those antagonist effects may result in an equilibrium. In the case of a first-order resonance (Fig. 11.4, left panel), (X, Y) stabilizes somewhere between the origin and the fixed point, but not along the OX-axis. This creates the torques that are discussed in Section 11.5.2. The case of a second-order resonance (middle panel) is different because the origin is an unstable hyperbolic point, at least near the resonance. As particles are driven toward the origin, the resonance forces it to move around one of the two stable points, that is, to acquire large orbital eccentricities. It is again expected that (X, Y) stabilizes somewhere between the origin and one of the stable points, but not along the OY-axis. To our knowledge, no work has been done yet to describe the torque that appears in this case. In the case of a fourth-order resonance (right panel), it can be shown that the origin is always a stable equilibrium point. Thus, it is anticipated that the excited streamlines near such resonances damp down to circular orbits, so that these resonances have eventually little effects on a collisional disk. However, this still has to be tested, for example, using collisional codes, and remembering that more subtle effects that are not considered here may show up, like viscous effects between neighbor streamlines.

11.7 Rings and satellite formation

Rings around small bodies are an interesting route toward satellite formation. The resonant mechanisms described here show that nonaxisymmetric bodies tend to push the initial disk material surrounding the body to more external regions. In fact, binarity or multiplicity is common among TNOs (Noll et al., 2008; Fraser et al., 2017). Although not so common (Richardson and Walsh, 2006), binary asteroids tend to be better known due to their proximity.

For TNOs, dynamical models are usually invoked to explain the formation of binaries with similar masses and wide separation. Gravitational interactions between two TNOs can lead to binary formation provided that energy is dissipated to bind the pair (Nesvorný et al., 2010).

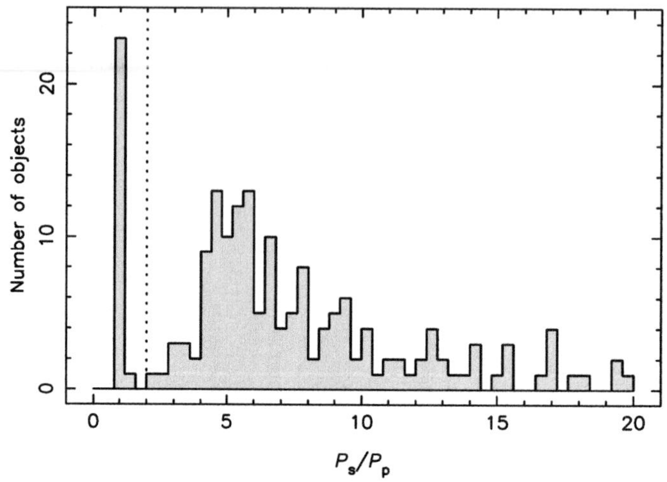

FIG. 11.5 Orbital periods of satellites compared to spin of primaries. The orbital periods P_S of 179 satellites known around multiple asteroids and TNOs have been normalized to the rotation period P_P of their respective primaries. The resulting histogram shows a peak near unity, corresponding to tidally evolved synchronous systems. The 1/2 outer resonance is shown as a *dotted line*. The increase of satellite number beyond that resonance may be a result of the repulsive torques at outer LRs discussed in this chapter. *From Sicardy, B., et al., 2019. Ring dynamics around non-axisymmetric bodies. Nat. Astron. 3, 146–153, Supplement Figure 1.*

This dissipation could involve dynamical friction with small bodies (Goldreich et al., 2002), or the intervention of a third body (Schlichting and Sari, 2008). For small asteroids (less than 10 km in size) close to the Sun, YORP fission may be a route to satellite formation. For other asteroids, collisional origin may be a good explanation (Walsh and Jacobson, 2015).

Fig. 11.5 shows the distribution of satellite orbital periods normalized to the primary rotation period,[4] P_S/P_P (i.e., the inverse of the ratio n/Ω considered earlier). Apart from the peak at $P_S/P_P = 1$, due to tidal locking into synchronous orbit, an interesting feature of the histogram is the absence of satellites with P_S/P_P between 1 and 2. This might be the footprint of an ancient collisional disks (whatever their origin) that suffered a strong coupling with the central body though sectoral resonances, and thus moved outside the outermost 1/2 resonance to finally accrete into a satellite. Note finally that in light of our results, the formation of satellites outside the 1/2 resonance is favored for slow rotators ($P_P > 7$ hours), for which the Roche limit will be inside the aforementioned resonance.

11.8 Conclusions

Studies of rings around small bodies are just at their beginning. Future stellar occultations can pin down further the physical parameters of the already known systems (width, optical depth, and orbital parameters) and possibly reveal new rings around other Centaurs and TNOs.

[4] See http://www.johnstonsarchive.net/astro/astmoontable.html, as of April 2018.

However, direct imaging remain difficult, owing to the very small angular spans of those rings, less than 100 mas for the material detected so far around Chariklo, Haumea, and Chiron. Meanwhile, spectroscopy is challenging too, due to the proximity of the central body that prevents a clear separation of the ring contribution from the total flux, although some long-term observations may disentangle to two (Duffard et al., 2014).

Consequently, basic questions such as the existence of small moons (shepherding or not the rings), prograde or retrograde nature of the ring particle motion, or extended dust sheets around the main rings remain unanswered. Future large telescopes, such as the European Extremely Large Telescope or the Thirty Meter Telescope may have sufficient angular resolution to resolve these features.

At the other end of methodologies, theoretical approaches will help understand better the rich dynamical environments of nonaxisymmetric bodies. In contrast to what happens around giant planets, the strong coupling between the spin of the body and the particle mean motions is expected to create previously unknown processes in the rings. In that context, collisional codes should be most useful to test some of the mechanisms that are expected to shape a collisional disk around a small body.

Appendix

A.1 Potential of a homogeneous ellipsoid

We consider the expansion of the potential created by a homogeneous ellipsoid of semiaxes $A > B > C$, see details in Sicardy et al. (2019), Balmino (1994), and Boyce (1997). The coefficients $U_{2p}(r)$ in Eq. (11.6) can be expanded in power of R/r,

$$U_{2p}(r) = -\frac{GM}{r} \sum_{l=|p|}^{+\infty} \left(\frac{R}{r}\right)^{2l} Q_{2l,2|p|}, \tag{A.1}$$

where the reference radius R is given by

$$\frac{3}{R^2} = \frac{1}{A^2} + \frac{1}{B^2} + \frac{1}{C^2}, \tag{A.2}$$

and

$$Q_{2l,2|p|} = \frac{3}{2^{l+2|p|}(2l+3)} \frac{(2l+2|p|)!(2l-2|p|)!l!}{(l+|p|)!(l-|p|)!(2l+1)!} \times \sum_{i=0}^{\mathrm{int}\left(\frac{l-|p|}{2}\right)} \frac{1}{16^i} \frac{\epsilon^{|p|+2i}}{(|p|+i)!i!} \frac{f^{l-|p|-2i}}{(l-|p|-2i)!}. \tag{A.3}$$

The dimensionless parameters ϵ and f measure the elongation and oblateness of the body, respectively:

$$\epsilon = \frac{A^2 - B^2}{2R^2} \quad \text{and} \quad f = \frac{A^2 + B^2 - 2C^2}{4R^2}. \tag{A.4}$$

For bodies close to spherical, $(A-B)/R \ll 1$, we retrieve the classical definition of oblateness, $f \sim (A-C)/A$ and elongation, $\epsilon \sim (A-B)/A$.

The coefficient $Q_{2l,2|p|}$ is of order $(\epsilon f)^l$. For order of magnitude considerations, it is enough to consider only the term of lowest order in Eq. (A.1), that is, $l = |p|$. Defining the sequence $S_{|p|} = Q_{2|p|,2|p|}/\epsilon^{|p|}$ and turning back to $m = 2p$, we obtain the expressions in Eqs. (11.7), (11.8).

A.2 Mean motion and epicyclic frequency

The mean motion and epicyclic frequency of a ring particle can be obtained using the classical formulae

$$n^2(r) = \frac{1}{r}\frac{dU_0(r)}{dr} \quad \text{and} \quad \kappa^2(r) = \frac{1}{r^3}\frac{d(r^4 n^2)}{dr}, \tag{A.5}$$

where $U_0(r)$ is the azimuthally averaged potential. Using the previous results for a homogeneous ellipsoid, and keeping the lowest-order term in R/r, we have $U_0(r) \sim -(GM/r)\left[1 + (f/5)(R/r)^2\right]$, so that

$$n^2(r) \sim \frac{GM}{r^3}\left[1 + \frac{3f}{5}\left(\frac{R}{r}\right)^2\right] \quad \text{and} \quad \kappa^2(r) \sim \frac{GM}{r^3}\left[1 - \frac{3f}{5}\left(\frac{R}{r}\right)^2\right]. \tag{A.6}$$

A.3 Lindblad resonance strengths

The coefficient $\mathcal{A}_m$ defining the strength of the $m/(m-1)$ LRs are given in Sicardy et al. (2019). In the case of the mass anomaly of relative mass μ sitting at the equator of a sphere of radius R, it reads (where a is the semimajor axis at the resonance)

$$\mathcal{A}_m = \left\{\left[m + \frac{a}{2}\frac{d}{da}\right] b_{1/2}^{(m)}\left(\frac{a}{R}\right) + q\left(\frac{a}{2R}\right)\delta_{(m,-1)}\right\}\mu. \tag{A.7}$$

In the case of an elongated body, it is

$$\mathcal{A}_m = \left[2m - (|m| + 1)\right] S_{|m/2|}\left(\frac{R}{a}\right)^{|m|+1} \epsilon^{|m/2|}, \tag{A.8}$$

where the reference radius is given by Eq. (A.2).

Acknowledgments

The work leading to this results has received funding from the European Research Council under the European Community's H2020 2014-20 ERC Grant Agreement No. 669416 "Lucky Star." Pablo Santos-Sanz acknowledges financial support by the European Union's Horizon 2020 Research and Innovation Program, under Grant Agreement No. 687378 (SBNAF).

References

Araujo, R.A.N., Sfair, R., Winter, O.C., 2016. The rings of Chariklo under close encounters with the giant planets. Astrophys. J. 824, 80. https://doi.org/10.3847/0004-637X/824/2/80.

Balmino, G., 1994. Gravitational potential harmonics from the shape of an homogeneous body. Celest. Mech. Dyn. Astron. 60, 331–364. https://doi.org/10.1007/BF00691901.

Bérard, D., Sicardy, B., Camargo, J.I.B., Desmars, J., Braga-Ribas, F., Ortiz, J.-L., Duffard, R., Morales, N., Meza, E., Leiva, R., Benedetti-Rossi, G., Vieira-Martins, R., Gomes Júnior, A.-R., Assafin, M., Colas, F., Dauvergne, J.-L., Kervella, P., Lecacheux, J., Maquet, L., Vachier, F., Renner, S., Monard, B., Sickafoose, A.A., Breytenbach, H., Genade, A., Beisker, W., Bath, K.-L., Bode, H.-J., Backes, M., Ivanov, V.D., Jehin, E., Gillon, M., Manfroid, J., Pollock, J., Tancredi, G., Roland, S., Salvo, R., Vanzi, L., Herald, D., Gault, D., Kerr, S., Pavlov, H., Hill, K.M., Bradshaw, J., Barry, M.A., Cool, A., Lade, B., Cole, A., Broughton, J., Newman, J., Horvat, R., Maybour, D., Giles, D., Davis, L., Paton, R.A., Loader, B., Pennell, A., Jaquiery, P.-D., Brillant, S., Selman, F., Dumas, C., Herrera, C., Carraro, G., Monaco, L., Maury, A., Peyrot, A., Teng-Chuen-Yu, J.-P., Richichi, A., Irawati, P., De Witt, C., Schoenau, P., Prager, R., Colazo, C., Melia, R., Spagnotto, J., Blain, A., Alonso, S., Román, A., Santos-Sanz, P., Rizos, J.-L., Maestre, J.-L., Dunham, D., 2017. The structure of Chariklo's rings from stellar occultations. Astron. J. 154, 144. https://doi.org/10.3847/1538-3881/aa830d.

Boyce, W., 1997. Comment on a formula for the gravitational harmonic coefficients of a triaxial ellipsoid. Celest. Mech. Dyn. Astron. 67, 107–110. https://doi.org/10.1023/A:1008224825003.

Braga-Ribas, F., Sicardy, B., Ortiz, J.L., Snodgrass, C., Roques, F., Vieira-Martins, R., Camargo, J.I.B., Assafin, M., Duffard, R., Jehin, E., Pollock, J., Leiva, R., Emilio, M., Machado, D.I., Colazo, C., Lellouch, E., Skottfelt, J., Gillon, M., Ligier, N., Maquet, L., Benedetti-Rossi, G., Gomes, A.R., Kervella, P., Monteiro, H., Sfair, R., El Moutamid, M., Tancredi, G., Spagnotto, J., Maury, A., Morales, N., Gil-Hutton, R., Roland, S., Ceretta, A., Gu, S.-H., Wang, X.-B., Harpsøe, K., Rabus, M., Manfroid, J., Opitom, C., Vanzi, L., Mehret, L., Lorenzini, L., Schneiter, E.M., Melia, R., Lecacheux, J., Colas, F., Vachier, F., Widemann, T., Almenares, L., Sandness, R.G., Char, F., Perez, V., Lemos, P., Martinez, N., Jørgensen, U.G., Dominik, M., Roig, F., Reichart, D.E., Lacluyze, A.P., Haislip, J.B., Ivarsen, K.M., Moore, J.P., Frank, N.R., Lambas, D.G., 2014. A ring system detected around the Centaur (10199) Chariklo. Nature 508, 72–75. https://doi.org/10.1038/nature13155.

Brown, M.E., Barkume, K.M., Ragozzine, D., Schaller, E.L., 2007. A collisional family of icy objects in the Kuiper belt. Nature 446, 294–296. https://doi.org/10.1038/nature05619.

Colwell, J.E., Nicholson, P.D., Tiscareno, M.S., Murray, C.D., French, R.G., Marouf, E.A., 2009. The structure of Saturn's rings. In: Dougherty, M.K., Esposito, L.W., Krimigis, S.M. (Eds.), Saturn From Cassini-Huygens, Springer, Dordrecht, pp. 375–412.

Desmars, J., Camargo, J., Berard, D., Sicardy, B., Leiva, R., Vieira-Martins, R., Braga-Ribas, F., Assafin, M., Rossi, G., Chariklo Occultations Team, Rio Group, Lucky Star Occultation Team, Granada Occultation Team, 2017. Predictions of stellar occultations by TNOs/Centaurs using Gaia. In: AAS/Division for Planetary Sciences Meeting Abstracts No. 49, ID.216.03.

Desmars, J., Meza, E., Sicardy, B., 2019. Pluto's ephemeris from ground-based stellar occultations. Astron. Astrophys. 625, A43. https://doi.org/10.1051/0004-6361/201834958.

Duffard, R., Pinilla-Alonso, N., Ortiz, J.L., Alvarez-Candal, A., Sicardy, B., Santos-Sanz, P., Morales, N., Colazo, C., Fernández-Valenzuela, E., Braga-Ribas, F., 2014. Photometric and spectroscopic evidence for a dense ring system around Centaur Chariklo. Astron. Astrophys. 568, A79. https://doi.org/10.1051/0004-6361/201424208.

El Moutamid, M., Sicardy, B., Renner, S., 2014. Coupling between corotation and Lindblad resonances in the presence of secular precession rates. Celest. Mech. Dyn. Astron. 118, 235–252. https://doi.org/10.1007/s10569-014-9533-z.

Fornasier, S., Lazzaro, D., Alvarez-Candal, A., Snodgrass, C., Tozzi, G.P., Carvano, J.M., Jiménez-Teja, Y., Silva, J.S., Bramich, D.M., 2014. The Centaur 10199 Chariklo: investigation into rotational period, absolute magnitude, and cometary activity. Astron. Astrophys. 568, L11. https://doi.org/10.1051/0004-6361/201424439.

Fraser, W.C., Bannister, M.T., Pike, R.E., Marsset, M., Schwamb, M.E., Kavelaars, J.J., Lacerda, P., Nesvorný, D., Volk, K., Delsanti, A., Benecchi, S., Lehner, M.J., Noll, K., Gladman, B., Petit, J.-M., Gwyn, S., Chen, Y.-T., Wang, S.-Y., Alexandersen, M., Burdullis, T., Sheppard, S., Trujillo, C., 2017. All planetesimals born near the Kuiper belt formed as binaries. Nat. Astron. 1, 0088. https://doi.org/10.1038/s41550-017-0088.

Gaia Collaboration, Brown, A.G.A., Vallenari, A., Prusti, T., de Bruijne, J.H.J., Mignard, F., Drimmel, R., Babusiaux, C., Bailer-Jones, C.A.L., Bastian, U., et al., 2016. Gaia Data Release 1. Summary of the astrometric, photometric, and survey properties. Astron. Astrophys. 595, A2. https://doi.org/10.1051/0004-6361/201629512.

Gaia Collaboration, Brown, A.G.A., Vallenari, A., Prusti, T., de Bruijne, J.H.J., Babusiaux, C., Bailer-Jones, C.A.L., Biermann, M., Evans, D.W., Eyer, L., et al., 2018. Gaia Data Release 2. Summary of the contents and survey properties. Astron. Astrophys. 616, A1. https://doi.org/10.1051/0004-6361/201833051.

Goldreich, P., Lithwick, Y., Sari, R., 2002. Formation of Kuiper-belt binaries by dynamical friction and three-body encounters. Nature 420, 643–646. https://doi.org/10.1038/nature01227.

Gupta, A., Nadkarni-Ghosh, S., Sharma, I., 2018. Rings of non-spherical, axisymmetric bodies. Icarus 299, 97–116. https://doi.org/10.1016/j.icarus.2017.07.012.

Hedman, M.M., 2015. Why are dense planetary rings only found between 8 AU and 20 AU? Astrophys. J. Lett. 801, L33. https://doi.org/10.1088/2041-8205/801/2/L33.

Hedman, M.M., Nicholson, P.D., 2014. More Kronoseismology with Saturn's rings. Mon. Not. R. Astron. Soc. 444, 1369–1388. https://doi.org/10.1093/mnras/stu1503.

Hyodo, R., Charnoz, S., Genda, H., Ohtsuki, K., 2016. Formation of Centaurs' rings through their partial tidal disruption during planetary encounters. Astrophys. J. Lett. 828, L8. https://doi.org/10.3847/2041-8205/828/1/L8.

Leiva, R., Sicardy, B., Camargo, J.I.B., Ortiz, J.-L., Desmars, J., Bérard, D., Lellouch, E., Meza, E., Kervella, P., Snodgrass, C., Duffard, R., Morales, N., Gomes-Júnior, A.R., Benedetti-Rossi, G., Vieira-Martins, R., Braga-Ribas, F., Assafin, M., Morgado, B.E., Colas, F., De Witt, C., Sickafoose, A.A., Breytenbach, H., Dauvergne, J.-L., Schoenau, P., Maquet, L., Bath, K.-L., Bode, H.-J., Cool, A., Lade, B., Kerr, S., Herald, D., 2017. Size and shape of Chariklo from multi-epoch stellar occultations. Astron. J. 154, 159. https://doi.org/10.3847/1538-3881/aa8956.

Lellouch, E., Kiss, C., Santos-Sanz, P., Müller, T.G., Fornasier, S., Groussin, O., Lacerda, P., Ortiz, J.L., Thirouin, A., Delsanti, A., Duffard, R., Harris, A.W., Henry, F., Lim, T., Moreno, R., Mommert, M., Mueller, M., Protopapa, S., Stansberry, J., Trilling, D., Vilenius, E., Barucci, A., Crovisier, J., Doressoundiram, A., Dotto, E., Gutiérrez, P.J., Hainaut, O., Hartogh, P., Hestroffer, D., Horner, J., Jorda, L., Kidger, M., Lara, L., Rengel, M., Swinyard, B., Thomas, N., 2010. "TNOs are Cool": a survey of the Trans-Neptunian region. II. The thermal lightcurve of (136108) Haumea. Astron. Astrophys. 518, L147. https://doi.org/10.1051/0004-6361/201014648.

Marley, M.S., 2014. Saturn ring seismology: looking beyond first order resonances. Icarus 234, 194–199. https://doi.org/10.1016/j.icarus.2014.02.002.

Marley, M.S., Porco, C.C., 1993. Planetary acoustic mode seismology—Saturn's rings. Icarus 106, 508. https://doi.org/10.1006/icar.1993.1189.

Melita, M.D., Duffard, R., Ortiz, J.L., Campo-Bagatin, A., 2017. Assessment of different formation scenarios for the ring system of (10199) Chariklo. Astron. Astrophys. 602, A27. https://doi.org/10.1051/0004-6361/201629858.

Michikoshi, S., Kokubo, E., 2017. Simulating the smallest ring world of Chariklo. Astrophys. J. Lett. 837, L13. https://doi.org/10.3847/2041-8213/aa6256.

Murray, C.D., Dermott, S.F., 2000. Solar System Dynamics. Cambridge University Press, Cambridge.

Nesvorný, D., Youdin, A.N., Richardson, D.C., 2010. Formation of Kuiper belt binaries by gravitational collapse. Astron. J. 140, 785–793. https://doi.org/10.1088/0004-6256/140/3/785.

Noll, K.S., Grundy, W.M., Chiang, E.I., Margot, J.-L., Kern, S.D., 2008. Binaries in the Kuiper belt. In: Barucci, M.A., Boehnhardt, H., Cruikshank, D.P., Morbidelli, A., Dotson, R. (Eds.), The Solar System Beyond Neptune. University of Arizona Press, Tucson, AZ, pp. 345–363.

Ortiz, J.L., Duffard, R., Pinilla-Alonso, N., Alvarez-Candal, A., Santos-Sanz, P., Morales, N., Fernández-Valenzuela, E., Licandro, J., Campo Bagatin, A., Thirouin, A., 2015. Possible ring material around centaur (2060) Chiron. Astron. Astrophys. 576, A18. https://doi.org/10.1051/0004-6361/201424461.

Ortiz, J.L., Santos-Sanz, P., Sicardy, B., Benedetti-Rossi, G., Bérard, D., Morales, N., Duffard, R., Braga-Ribas, F., Hopp, U., Ries, C., Nascimbeni, V., Marzari, F., Granata, V., Pál, A., Kiss, C., Pribulla, T., Komžík, R., Hornoch, K., Pravec, P., Bacci, P., Maestripieri, M., Nerli, L., Mazzei, L., Bachini, M., Martinelli, F., Succi, G., Ciabattari, F., Mikuz, H., Carbognani, A., Gaehrken, B., Mottola, S., Hellmich, S., Rommel, F.L., Fernández-Valenzuela, E., Campo Bagatin, A., Cikota, S., Cikota, A., Lecacheux, J., Vieira-Martins, R., Camargo, J.I.B., Assafin, M., Colas, F., Behrend, R., Desmars, J., Meza, E., Alvarez-Candal, A., Beisker, W., Gomes-Junior, A.R., Morgado, B.E., Roques, F., Vachier, F., Berthier, J., Mueller, T.G., Madiedo, J.M., Unsalan, O., Sonbas, E., Karaman, N., Erece, O., Koseoglu, D.T., Ozisik, T., Kalkan, S., Guney, Y., Niaei, M.S., Satir, O., Yesilyaprak, C., Puskullu, C., Kabas, A., Demircan, O., Alikakos, J., Charmandaris, V., Leto, G., Ohlert, J., Christille, J.M., Szakáts, R., Takácsné Farkas, A., Varga-Verebélyi, E., Marton, G., Marciniak, A., Bartczak, P., Santana-Ros, T., Butkiewicz-Bąk, M., Dudziński, G., Alí-Lagoa, V., Gazeas, K., Tzouganatos, L., Paschalis, N., Tsamis, V., Sánchez-Lavega, A., Pérez-Hoyos, S., Hueso, R., Guirado, J.C., Peris, V., Iglesias-Marzoa, R., 2017. The size, shape, density and ring of the dwarf planet Haumea from a stellar occultation. Nature 550, 219–223. https://doi.org/10.1038/nature24051.

Pan, M., Wu, Y., 2016. On the mass and origin of Chariklo's rings. Astrophys. J. 821, 18. https://doi.org/10.3847/0004-637X/821/1/18.

Peale, S.J., 1986. Orbital resonances, unusual configurations and exotic rotation states among planetary satellites. In: Burns, J.A., Matthews, M.S. (Eds.), IAU Colloq. 77: Some Background About Satellites, pp. 159–223.

Ragozzine, D., Brown, M.E., 2009. Orbits and masses of the satellites of the dwarf planet Haumea (2003 EL61). Astron. J. 137, 4766–4776. https://doi.org/10.1088/0004-6256/137/6/4766.

Richardson, D.C., Walsh, K.J., 2006. Binary minor planets. Annu. Rev. Earth Planet. Sci. 34, 47–81. https://doi.org/10.1146/annurev.earth.32.101802.120208.

Ruprecht, J.D., Bosh, A.S., Person, M.J., Bianco, F.B., Fulton, B.J., Gulbis, A.A.S., Bus, S.J., Zangari, A.M., 2015. 29 November 2011 stellar occultation by 2060 Chiron: symmetric jet-like features. Icarus 252, 271–276. https://doi.org/10.1016/j.icarus.2015.01.015.

Schlichting, H.E., Sari, R., 2008. Formation of Kuiper belt binaries. Astrophys. J. 673, 1218–1224. https://doi.org/10.1086/524930.

Sicardy, B., Buie, M.W., Benedetti-Rossi, G., Braga-Ribas, F., Bueno de Camargo, J.I., Duffard, R., Ortiz, J.L., Gratadour, D., Dumas, C., 2015. Constraints on Chariklo's rings from HST and VLT observations. In: AAS/Division for Planetary Sciences Meeting Abstracts No. 47, ID.104.01.

Sicardy, B., El Moutamid, M., Quillen, A.C., Schenk, P.M., Showalter, M.R., Walsh, K., 2018. Rings beyond the giant planets. In: Tiscareno, M.S., Murray, C.D. (Eds.), Planetary Ring Systems. Properties, Structure, and Evolution, Cambridge University Press, Cambridge, pp. 135–154.

Sicardy, B., et al., 2019. Ring dynamics around non-axisymmetric bodies. Nat. Astron. 3, 146–153.

Thomas, P.C., 2010. Sizes, shapes, and derived properties of the Saturnian satellites after the Cassini nominal mission. Icarus 208, 395–401. https://doi.org/10.1016/j.icarus.2010.01.025.

Tiscareno, M.S., Hedman, M.M., Burns, J.A., Castillo-Rogez, J., 2013. Compositions and origins of outer planet systems: insights from the Roche critical density. Astrophys. J. Lett. 765, L28. https://doi.org/10.1088/2041-8205/765/2/L28.

Walsh, K.J., Jacobson, S.A., 2015. Formation and evolution of binary asteroids. In: Patrick, M., Francesca, E.D., William, F.B. (Eds.), Asteroids IV. University of Arizona Press, Tucson, AZ, pp. 375–393.

Wood, J., Horner, J., Hinse, T.C., Marsden, S.C., 2017. The dynamical history of Chariklo and its rings. Astron. J. 153, 245. https://doi.org/10.3847/1538-3881/aa6981.

CHAPTER

12

The Pluto system after New Horizons

John R. Spencer[a], *William M. Grundy*[b], *Francis Nimmo*[c], *Leslie A. Young*[a]

[a]Southwest Research Institute, Boulder, CO, United States [b]Lowell Observatory, Flagstaff, AZ, United States [c]Department of Earth and Planetary Sciences, University of California, Santa Cruz, CA, United States

12.1 Knowledge of Pluto before New Horizons

Physical studies of Pluto began with determination of its light curve, establishing its rotation period and the presence of albedo features on its surface (Walker and Hardie, 1955). Spectroscopic studies found CH_4 on its surface (Cruikshank et al., 1976), and the discovery of Charon (Christy and Harrington, 1978) allowed determination of the system mass. In the 1980s, mutual eclipses and occultations between Pluto and Charon allowed determination of the diameters of both bodies, initial mapping of their albedo features (Buie et al., 1992), and established that Charon's surface was dominated by water ice (Marcialis et al., 1987). Pluto's atmosphere was discovered by stellar occultation (Elliot et al., 1989). The atmospheric scale height (Yelle and Lunine, 1989), and detection of nitrogen, in addition to CH_4 and CO on the surface (Owen et al., 1993), established that the atmosphere was dominantly N_2. The high amplitude of the light curve, and direct imaging of the surface with the Hubble Space Telescope (HST), revealed very high contrast albedo features on Pluto's surface (Buie et al., 2010), and rotationally resolved spectroscopy revealed large longitudinal variations in surface composition (Grundy et al., 2013). Thermal observations also revealed large surface temperature contrasts (Lellouch et al., 2000). Additional stellar occultations between 2002 and 2015 revealed, surprisingly, that Pluto's atmosphere was increasing in density as Pluto receded from the Sun (Elliot et al., 2007; Dias-Oliveira et al., 2015). Two additional small moons, Nix and Hydra, were discovered in 2005 (Weaver et al., 2006), followed by the smaller moons Styx and Kerberos in 2011 and 2012 (Showalter and Hamilton, 2015) (Table 12.1).

Meanwhile, starting in 1992, the discovery of additional objects beyond Pluto, including many that shared Pluto's 3:2 mean-motion orbital resonance with Neptune, and some that approached Pluto itself in size, finally revealed Pluto's true place in the solar system.

The Trans-Neptunian Solar System.
https://doi.org/10.1016/B978-0-12-816490-7.00012-6

TABLE 12.1 Bodies in the Pluto system.

Body	Highest-resolution NH imaging ($km\,pix^{-1}$)	Radius (km)	Density ($kg\,m^{-3}$)	Geometric albedo	Rotation period (days)	Spin pole obliquity (degrees)	Orbit relative to barycenter: Semimajor axis	Eccentricity	Inclination (degrees)	Period (days)	Period relative to Charon
Pluto	0.07	1188[a]	1854[a]	0.57[c]	6.387[d]	0	2125[d,e]	0.0000[d]	0.00[d]	6.3872[d]	1.00
Charon	0.15	606[a]	1701[a]	0.36[c]	6.387[d]	0	17,448[d,e]	0.0000[d]	0.00[d]	6.3872[d]	1.00
Styx	3.13	16 × 9 × 8[b]	–	0.65[b]	3.24[b]	82[b]	42,656[f]	0.0058[f]	0.81[f]	20.1616[f]	3.16
Nix	0.30	50 × 35 × 33[b]	–	0.56[b]	1.83[b]	131[b]	48,694[f]	0.0020[f]	0.13[f]	24.8546[f]	3.89
Kerberos	1.97	19 × 10 × 9[b]	–	0.56[b]	5.31[b]	96[b]	57,783[f]	0.0033[f]	0.39[f]	32.1676[f]	5.04
Hydra	1.14	65 × 45 × 25[b]	–	0.83[b]	0.43[b]	117[b]	64,738[f]	0.0059[f]	0.24[f]	38.2018[f]	5.98

References: [a]*Nimmo et al. (2017);* [b]*Weaver et al. (2016);* [c]*Buie et al. (1997), adjusted to true radii;* [d]*Buie et al. (2012);* [e]*Brozovic et al. (2015); and* [f]*Showalter and Hamilton (2015).*

12.2 The New Horizons encounter with Pluto

The New Horizons mission was launched on January 19, 2006, and encountered the Pluto system on July 14, 2015. Its payload (Table 12.2; Weaver et al., 2008) provided a scientifically rich portrait of the Pluto system.

TABLE 12.2 New Horizons payload.

Instrument	Angular resolution	Field of view	Wavelength/ energy range	Best resolution at Pluto (km)	Notes
Ralph/MVIC	19.8 μrad pix^{-1}	99.0 mrad	400–550 nm; 540–700 nm; 860–910 nm; 860–910 nm; 400–975 nm	0.33	Color + Pan pushbroom camera, Pan framing camera
Ralph/LEISA	60.8 μrad pix^{-1}	15.6 × 15.6 mrad	1.25–2.5 μm	2.83	Near-IR imaging spectrometer: $\lambda/\Delta\lambda = 240$
LORRI	4.96 μrad pix^{-1}	5.08 × 5.08 mrad	350–850 nm	0.07	Framing camera
Alice	1700 μrad pix^{-1}	35 × 35 mrad; 1.7 × 70 mrad	47–188 nm	23	UV imaging spectrometer: spectral resolution 0.18 nm
REX	20 mrad	20 mrad	4.2 cm	240	Doppler tracking of uplink DSN signals; Passive radiometry
SWAP	200 × 10 degrees	200 × 10 degrees	0.25–7.5 keV	N/A	Low-energy plasma spectrometer
PEPSSI	25 × 12 degrees	160 × 12 degrees	1–1000 keV	N/A	High-energy plasma spectrometer
SDC	N/A	180 × 180 degrees	Particle mass $> 10^{-12}$ g	N/A	Dust detector

The encounter was designed to satisfy multiple constraints (Young et al., 2008). Flying through Pluto and Charon's shadows enabled Earth and solar occultation studies of Pluto's atmosphere at radio and UV wavelengths, and placed upper limits on Charon's atmosphere. This geometry also enabled observations of Pluto's highest-contrast hemisphere near closest approach. Closest approach distance to Pluto, 12,500 km, was a compromise between maximizing spatial resolution and probing the atmosphere and plasma interaction, and providing observations at a wide range of viewing geometries.

Approach observations of Pluto began in January 2015, with periodic LORRI images for light curve studies and optical navigation. By May 2015, the resolution of the LORRI images exceeded that of the best HST images. From 64 to 13 days before closest approach, a series of deep image mosaics were obtained to search for potentially hazardous rings, or new moons capable of generating hazardous debris, though no new moons or other hazards were found.

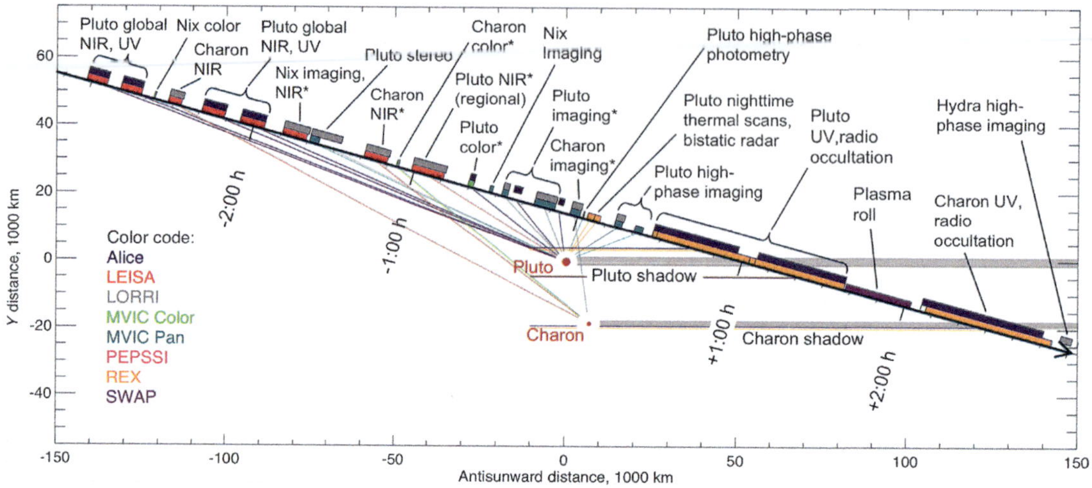

FIG. 12.1 The Pluto encounter sequence, with instruments identified for each observation: Frequently more than one instrument was used at a time. Highest-resolution observations are indicated with an *asterisk*. Lines extending from observation midpoints along the trajectory indicate instrument pointing (except for Nix and Hydra observations, which pointed out of the plane of this graphic). *NIR*, near infrared, *UV*, ultraviolet. The spacecraft traveled from upper left to lower right. In situ observations with PEPSSI, SWAP, and SDC were made throughout the flyby.

The close approach sequence is shown in Fig. 12.1, and the best resolution obtained on Pluto by the different instruments is shown in Table 12.2. Fig. 12.2 shows the best global color images of Pluto and Charon. Departure observations focused first on Alice and REX occultations by Pluto and Charon, and then high phase remote sensing of Pluto, Charon, and the small satellites, continuing for 2 weeks after the flyby corresponding to roughly two rotation periods of Pluto and Charon. Data downlink was completed in Fall 2016. Note that many of the feature names referred to in this chapter are informal.

FIG. 12.2 The best color images of Pluto (*left*) and Charon (*right*). Colors are enhanced relative to those visible to the human eye. Pluto north is up. Light-colored Sputnik Planitia occupies most of the center of the Pluto image.

12.3 Pluto

12.3.1 Interior

The interior of Pluto remains poorly understood. The New Horizons spacecraft provided an accurate radius measurement (1188 km; Nimmo et al., 2017), allowing determination of Pluto's density (1854 kg m^{-3}). This value suggests that Pluto is roughly two-thirds rock and one-third ices (McKinnon et al., 2017). Although there are no measurements of Pluto's moment of inertia, it is generally assumed that Pluto is differentiated. This is partly because ices dominate the surface spectra (Grundy et al., 2016a) but also because an undifferentiated Pluto would have experienced high-pressure ice phase transformations, leading to surface compression—which is not observed in its surface geological record (McKinnon et al., 2017). Pluto's high rock fraction and relatively large size mean that radioactive heat production would be sufficient to melt a conductive ice shell and maintain a global, subsurface ocean to the present day (Hammond et al., 2016; Bierson et al., 2018). However, if the ice shell were convective, heat would be removed rapidly enough that an ocean would never develop (Robuchon and Nimmo, 2011). Therefore, if there is a present-day ocean, its presence would suggest a cold and rigid ice shell, with little communication between the ocean and the surface.

A freezing ocean would generate extensional surface stresses, which are a likely explanation for the large and apparently young normal faults observed on Pluto's surface (Moore et al., 2016). Freezing would also pressurize the ocean, perhaps helping produce the putative "cryovolcanic" features observed (Moore et al., 2016). The location of the nitrogen-filled Sputnik Planitia basin near the anti-Charon point suggests that it caused true polar wander, and thus represents a mass excess (Keane et al., 2016; Nimmo et al., 2016). One plausible way of generating such a mass excess is if the ice shell beneath the basin were thinned and replaced with denser water (Nimmo et al., 2016). The ice shell must be cold to avoid rapid relaxation of the basin; this might suggest a cold, ammonia-rich ocean beneath. Fig. 12.3 shows a plausible interior structure.

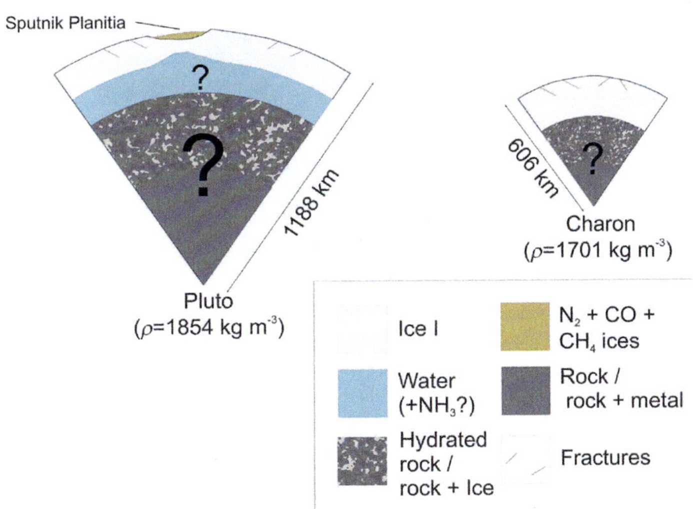

FIG. 12.3 Cartoons of plausible interior structure for Pluto and Charon. Cross-sections are to scale, except that the depth of Sputnik Planitia has been exaggerated for clarity.

12.3.2 Surface geology

Pluto's geology is remarkably varied (Figs. 12.2 and 12.4; Moore et al., 2016), and shows us that the expected few $mW\,m^{-2}$ of radiogenic heat is capable of powering ongoing varied geological activity on volatile-rich ice worlds. Impact crater densities vary from near saturation to negligible (Singer et al., 2019), implying surface ages from ~4 Ga to <10 Ma (Moore et al., 2016). The oldest preserved feature on Pluto's surface is the basin containing Sputnik Planitia (Fig. 12.5), probably an ancient impact basin (Moore et al., 2016). Much of the encounter hemisphere west of Sputnik is covered by moderately cratered smooth planes of unknown origin, cut by graben suggesting global expansion (Fig. 12.4A), and by scarps that may result from sublimation erosion (Moore et al., 2017). North and east of Sputnik the terrain is much more complex, with what appear to be several generations of mantling material, several kilometers thick with steep-sided pits and marginal scarps (Fig. 12.4B; Howard et al., 2017a; Umurhan et al., 2017). Extensive valley erosion is seen in many places (Fig. 12.4B and F). A distinctive high-standing methane-rich unit is extensively eroded, probably by sublimation, into penitente-like "bladed terrain" (Moore et al., 2018).

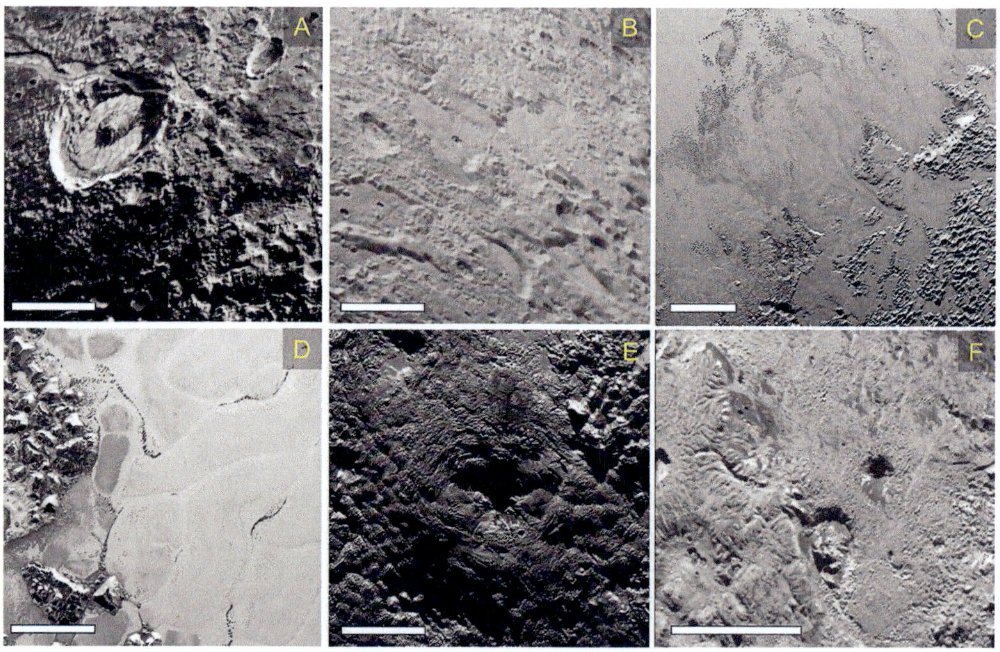

FIG. 12.4 Illustration of the diversity of terrain types on Pluto: Coordinates are those of the image centers. (A) Elliot, a relatively pristine impact crater with a central peak, an apparent nitrogen ice deposit occupying its floor, and a graben cutting part of its rim (141 E, 12 N). (B) Eroded mantles northeast of Sputnik Planitia (195 E, 59 N); (C) glacial flow into Sputnik Planitia from the highlands to the east (193 E, 2 N); (D) convection cells, and neighboring chaotic mountain blocks, near the west margin of Sputnik Planitia (174 E, 6 N); (E) Wright Mons, a structure of unknown origin, possibly a cryovolcanic construct (172 E, 20 S); and (F) dendritic channels and other erosional features southwest of Sputnik Planitia (158 E, 20 S). Portions of the MVIC image taken near close approach, with a resolution of 330 m $pixel^{-1}$. Scale bar is 50 km long in all panels.

The most remarkable known terrains on Pluto occur in and around Sputnik Planitia. The Planitia itself appears to be the surface of a nitrogen-rich deposit, several kilometers thick, partially filling the original impact basin. All of the Planitia except the southernmost part features polygonal cells strongly suggesting active convection (Fig. 12.4D): Convection of solid N_2 is plausible for expected radiogenic heat fluxes (McKinnon et al., 2016). Clusters of hills at the margins of the convection cells appear to be buoyant material rafted to those locations by the convection (White et al., 2017). The northern margin is depressed (Fig. 12.5) and shows evidence of flow toward the margin, perhaps driven by sublimation of the northern margin (Bertrand et al., 2018a; Protopapa et al., 2017). The southern portion is marked by complex patterns of pits, probably sublimation generated. The west margin of Sputnik features rugged mountain blocks up to 6.5 km high with a chaotic jumbled appearance (Fig. 12.4D; Schenk et al., 2018a). It is likely that these are fragments of the water ice bedrock, perhaps undermined and transported by denser, more fluid, nitrogen ice. On the east side of Sputnik, active N_2 glaciers (Fig. 12.4C) flow down from pitted highlands into the Planitia, perhaps resupplied by nitrogen condensed out of the atmosphere (Howard et al., 2017b). To the southwest of Sputnik, a pair of large mountains with central depressions, Wright Mons (Fig. 12.4E) and Piccard Mons (175 E, 35 S), may be cryovolcanic constructs, unlike any seen elsewhere in the solar system.

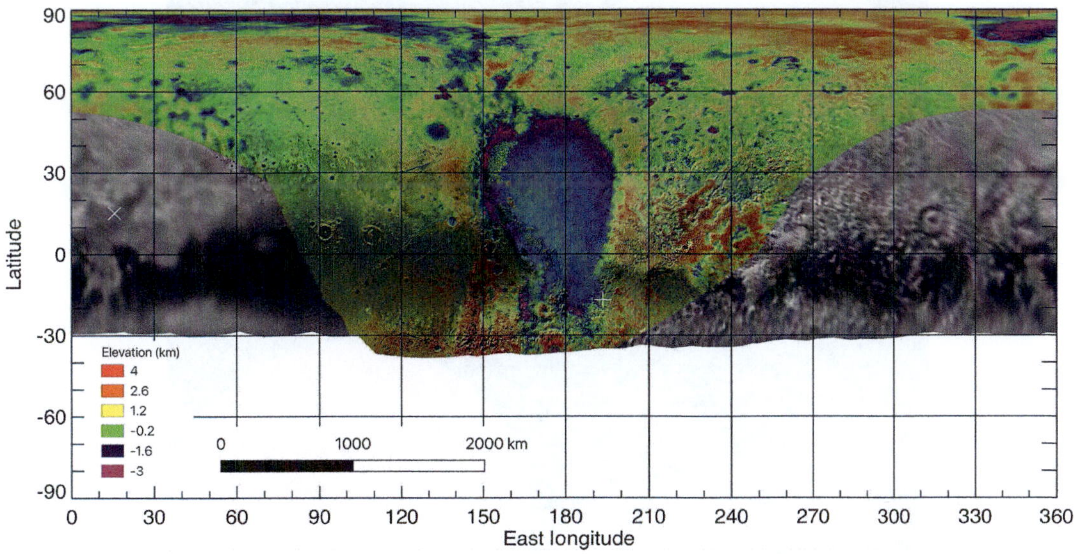

FIG. 12.5 Global map of Pluto, colored by elevation using stereo images (Schenk et al., 2018a). Sputnik Planitia is the depression near the center of the image. "+" and "×" mark the locations of the radio occultation ingress and egress, respectively; see Fig. 12.8.

12.3.3 Surface chemistry

Pluto's surface materials can be divided into two categories: Volatile ices that are mobilized through sublimation and condensation and nonvolatile materials that are inert at Pluto surface temperatures (30–60 K). The volatile ices are N_2, CO, and CH_4 (Fig. 12.6). N_2 is the most volatile and dominates Pluto's lower atmosphere and the seasonal migration of ice, driven by changing patterns of sunlight. New Horizons detected N_2 at mid-northern latitudes, where it preferentially occupies low-lying areas like the floors of craters and valleys where atmospheric pressure is highest (Fig. 12.6C; Schmitt et al., 2017). N_2 ice is mostly absent from the north pole, currently experiencing continuous summer sunlight. It partly fills the Sputnik Planitia basin, along with CO and CH_4, both of which are soluble in N_2 ice (Protopapa et al., 2017). The distribution of CH_4 ice (Fig. 12.6D) contrasts with that of N_2, due to varying relative abundances of the CH_4- and N_2-rich phases of the N_2/CH_4 solid solution. CH4 is abundant at high northern latitudes, and also at high elevations at lower latitudes, such as the rims of craters and crests of mountains, and in the bladed terrain (Earle et al., 2018; Moore et al., 2018). Being much less volatile, CH_4 migrates more slowly, on the million-year timescales of the pole precessional cycle (e.g., Bertrand et al., 2018b).

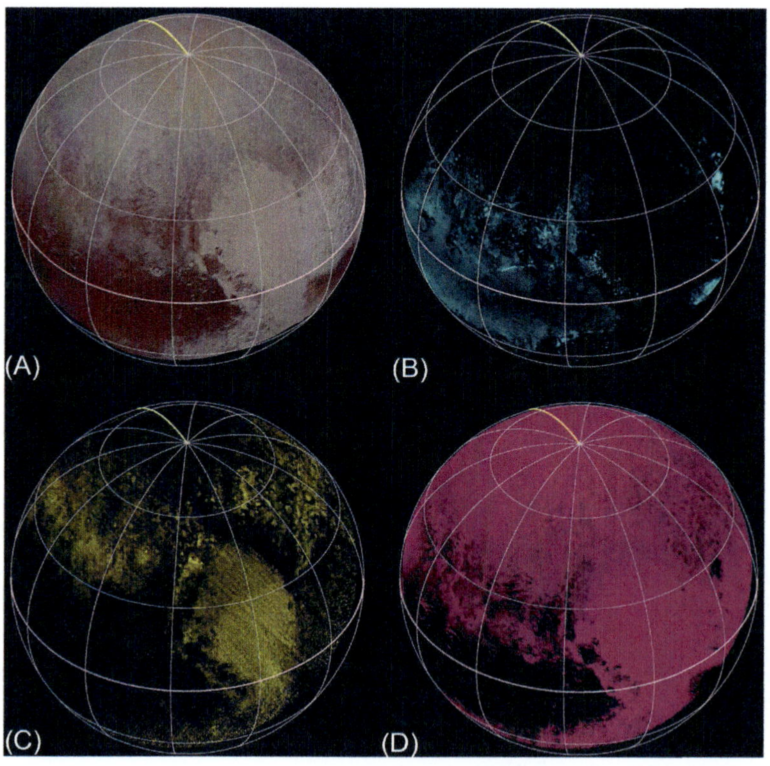

FIG. 12.6 Distribution of color (approximately natural) (A), and absorption band strengths for H_2O (B), N_2 (C), and CH_4 (D) ices on the encounter hemisphere of Pluto. The *yellow line* marks the Charon-facing meridian.

The inert bedrock underlying Pluto's surface is H_2O ice. It is exposed at the surface in several places, mostly at mid-to-low latitudes (Fig. 12.6B). It appears in the floors of numerous craters, in rugged mountains within and adjacent to Sputnik Planitia, and in the broken belt of dark red maculae typified by Cthulhu (20 E–150 E, 20 S–10 N) and Krun (190 E–220 E, 25 S–5 S) that encircles Pluto along its equator (e.g., Grundy et al., 2016a; Cook et al., 2019). It is mostly associated with dark, reddish regions, but there are a few regions showing comparatively clean, bright, H_2O ice, most notably in Supay Facula (215 E, 25 N, Fig. 12.7). The red coloration (Fig. 12.6A) mostly originates in complex organic chemistry in Pluto's upper atmosphere, where solar ultraviolet radiation drives processing of CH_4 and N_2 into complex reddish macromolecules known as tholins. These form aerosol particles that eventually settle to the surface (e.g., Grundy et al., 2018). Additional radiolytic and photolytic chemistry could occur directly in Pluto's surface ices (e.g., Cruikshank et al., 2015). Smaller, but still inert molecules such as C_2H_6 and CH_3OH have been tentatively detected in eastern Cthulhu (Cook et al., 2019). They could be produced from CH_4 plus other Pluto materials through energetic radiation.

12.3.4 Surface/atmosphere interactions

Like Mars and Triton, Pluto has an atmosphere whose density is controlled by vapor-pressure equilibrium with surface ices, and which is global in extent, because subsonic winds are able to transport gas from regions of net sublimation to regions of net condensation (Trafton and Stern, 1983). The transport of latent heat by wind means that the N_2 ice surface is virtually isothermal regardless of local insolation. The ice temperature, and thus (by vapor-pressure equilibrium) the atmospheric pressure, is controlled by the global balance between absorbed insolation, thermal radiation of the nitrogen ice, and subsurface heat flow, and thus by the spatial distribution, albedo, thermal inertia, and emissivity of the ice, as well as the distribution of incident sunlight. Because about a meter of ice can be transported between hemispheres over a seasonal cycle, complex feedback is possible between ice distribution and atmospheric pressure over a Pluto year and longer (precessional) timescales (Hansen and Paige, 1996; Young, 2013; Bertrand et al., 2018a,b). Stellar occultations show a tripling in atmospheric pressure between 1988 and 2013, consistent with many model predictions (Olkin et al., 2015). These models also predict that pressure will soon begin to fall, though the degree of atmospheric collapse at aphelion is still unclear. Because condensation is enhanced at low elevation due to higher atmospheric pressure, there is a tendency for nitrogen to migrate to depressions, which likely explains the presence and probable long-term stability of kilometers of nitrogen in the basin of Sputnik Planitia (Fig. 12.7; Bertrand et al., 2018a). The permanent reservoir of N_2 in Sputnik Planitia has a major influence on the seasonal and longer-term distribution of volatiles on Pluto, and the variations in atmospheric pressure (Bertrand and Forget, 2016; Bertrand et al., 2018a). For simple static volatile models, perihelion surface pressure could exceed Martian values ($\sim$6 mbar) at times in the precessional cycle when solstice coincides with perihelion (Stern et al., 2017). Fig. 12.7 shows how the compositional provinces seen in Fig. 12.6 may be understood in terms of Pluto's volatile cycles.

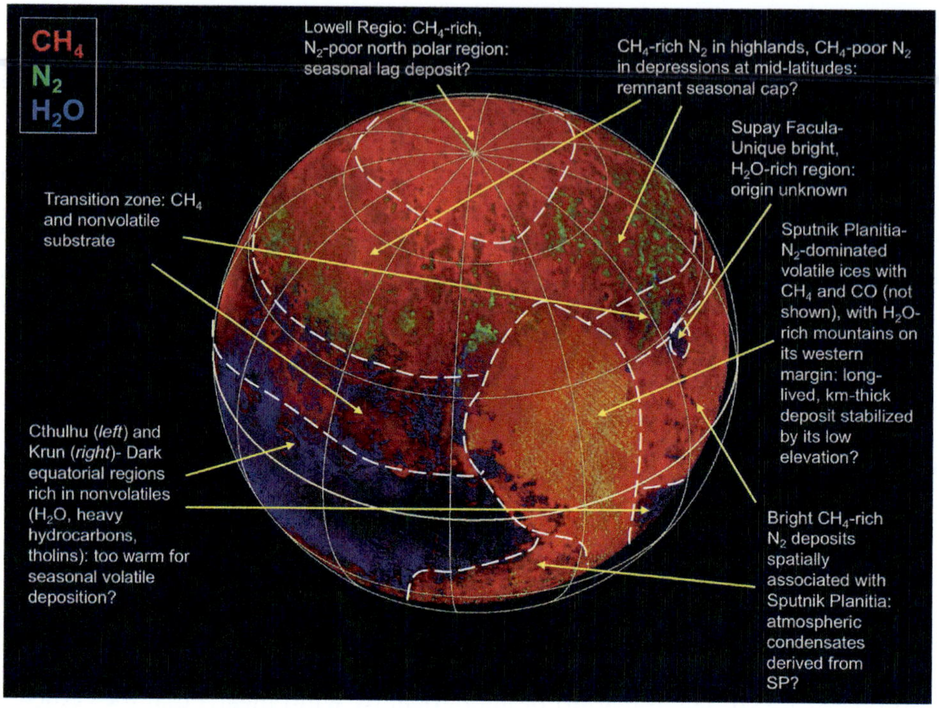

FIG. 12.7 Composite image of the distribution of CH_4, N_2, and H_2O on Pluto (from Fig. 12.6), annotated to show the major compositional provinces and tentative interpretations.

12.3.5 Atmospheric structure and composition

The pressure and temperature of Pluto's atmosphere (Fig. 12.8) was probed by New Horizons at altitudes below 100 km by the REX radio occultation (Hinson et al., 2017) and at altitudes near 1000 km by the Alice ultraviolet solar occultation (Young et al., 2018). It is constrained at intermediate altitudes by ground-based stellar occultations (Sicardy et al., 2016), and hydrostatic equilibrium (see Young et al., 2018). The temperature is surprisingly cold at the highest altitudes measured by New Horizons; CH_4, CO, HCN, C_2H_6, atmospheric hazes, and H_2O may all be important in the energy budget and temperature of Pluto's atmosphere (Strobel and Zhu, 2017; Yelle et al., 2017; Zhang et al., 2017; Krasnopolsky, 2018). Thermal conduction via a strong thermal gradient connects the warm middle atmosphere to the colder surface (Hinson et al., 2017). The surface pressure at 1189.9 km radius was $11.5 \pm 0.7\ \mu$bar in 2015. Over the N_2-filled Sputnik Planitia basin, at lower elevation (radius 1187.4 km), the surface pressure is $12.8 \pm 0.7\ \mu$bar, in vapor-pressure equilibrium with N_2 ice at 37.1 K (Fray and Schmitt, 2009).

From the Alice ultraviolet solar occultation (Young et al., 2018), we find: (1) the atmosphere is primarily N_2 below 1200 km altitude above the surface, (2) the CH_4 mixing ratio increases from 0.3% at 80 km to 25% at 1200 km altitude, and should dominate over N_2 at the exobase, at

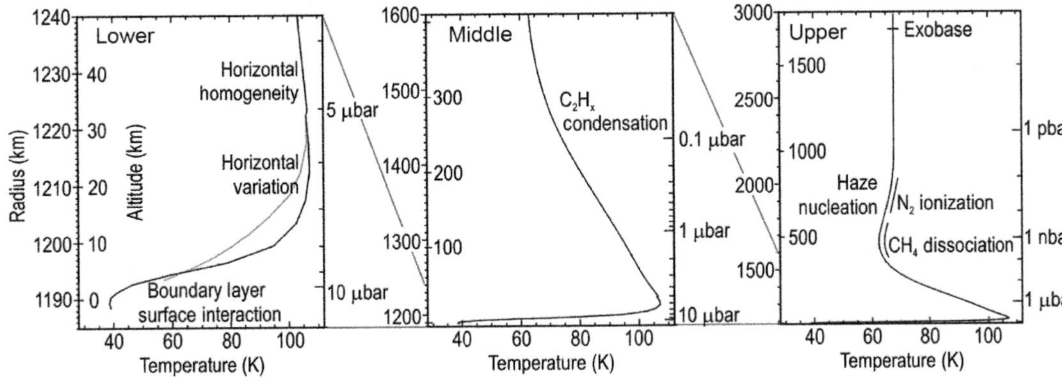

FIG. 12.8 Overview of Pluto's vertical atmospheric structure at the time of the New Horizons flyby, at three different magnifications. Below 30 km, *black* indicates the dawn atmosphere over the southern tip of Sputnik Planitia (radio ingress, "+" symbol in Fig. 12.5), while *gray* indicates the dusk atmosphere over an uncertain terrain type near the center of the Charon-facing hemisphere (radio egress, "×" symbol in Fig. 12.5). Above 30 km, ingress and egress are similar in both the radio and UV occultations. The surface at radio ingress was at planetary radius = 1187.4 ± 3.6 km, while at egress was at radius = 1192.4 ± 3.6 km. The altitude scale is relative to a reference radius of = 1190 km. *Modified from Young, L.A., et al., 2018. Structure and composition of Pluto's atmosphere from the New Horizons solar ultraviolet occultation. Icarus 300, 174–199.*

~1700 km altitude, and (3) the simple hydrocarbons C_2H_2, C_2H_4, and C_2H_6 have mixing ratios near 0.1% at ~550 km altitude, decreasing to $\sim 1\times10^{-6}$, 8×10^{-6}, and 4×10^{-6} at 200 km altitude for C_2H_4, C_2H_2, and C_2H_6, respectively (Young et al., 2018), likely due to condensation on haze (Cheng et al., 2017). The Atacama Large Millimeter Array measured CO with a mixing ratio of 5×10^{-4}, and HCN supersaturated in Pluto's atmosphere with a mixing ratio of $\sim 1.5 \times 10^{-5}$ to 4×10^{-5} that varies with altitude (Lellouch et al., 2017).

12.3.6 Solar wind interaction

Due to the unexpectedly cold upper atmosphere, the escape rate derived from analysis of the Alice data is 100 times smaller than preencounter estimates (6×10^{25} CH_4 s^{-1} (Young et al., 2018) vs. the expected 3.5×10^{27} N_2 s^{-1} (Zhu et al., 2014)), which leads to lower mass loading and a smaller obstacle to the solar wind. Moreover, the solar proton flux at the time of the flyby was unusually strong, about four times larger than the predicted mean (Bagenal et al., 2016), further shrinking the interaction region. Analysis of data from the SWAP instrument (McComas et al., 2016), supported by the PEPSSI instrument (Bagenal et al., 2016), paints a picture of a small bow shock at the time of the flyby, only 4.5 Pluto radii (R_P) from Pluto center, and a wake or tail that extends at least 100 R_P. New Horizons entered the bow shock region in its flanks, and detected a 20% slowing of the solar wind. Plutogenic species fill the wake of the interaction region: PEPSSI detected suprathermal particles (Bagenal et al., 2016), and SWAP detected heavy ions, likely CH_4^+ (Zirnstein et al., 2016).

12.4 Charon

12.4.1 Interior

Charon is about half the size of Pluto (Nimmo et al., 2017) and about 10% less dense (1701 kg m^{-3}), suggesting it is about three-fifths rock (Fig. 12.3). Like Pluto, the observed surface ices and lack of compression suggest a differentiated interior. Modeling shows that although Charon could have developed a subsurface ocean via radioactive heating midway through its history, such an ocean would not survive to the present day (Desch and Neveu, 2017; Bierson et al., 2018). The imaged hemisphere of Charon shows a wide range of extensional tectonic features (Figs. 12.2 and 12.9), implying a total strain of order 1% (Beyer et al., 2017). This strain is what one would expect from refreezing of an ocean ~40 km deep, consistent with thermal evolution models. Charon also has extensive smooth plains that apparently postdate extensional scarp formation, perhaps indicative of an episode of cryovolcanism and/or crustal overturn (Beyer et al., 2019).

The lower density of Charon compared to Pluto is unlikely to be due to differences in porosity alone (Bierson et al., 2018), and indicates a slightly lower rock mass fraction. As with Pluto, Charon's rock core most likely underwent significant hydrothermal alteration (Desch and Neveu, 2017; Malamud et al., 2017). Charon also differs from Pluto in that it may have experienced an early episode of tidal heating and stress generation if its orbit was initially noncircular (Rhoden et al., 2015).

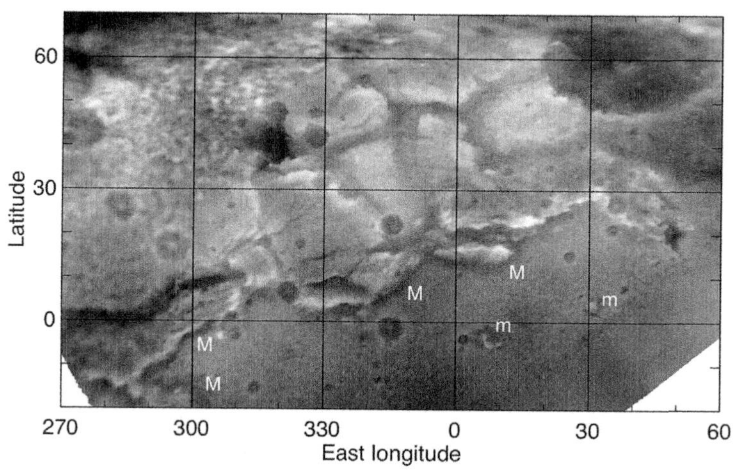

FIG. 12.9 Topography of the best-imaged part of Charon, showing the contrast between the smooth plains of Vulcan Planitia (*lower right*) and the polygonally fractured rugged terrain (*upper left*). Total displayed elevation range is −6 km (*black*) to +6 km (*white*). "M" indicates examples of the moat at the north and west margin of Vulcan Planitia, and "m" indicates the moated mountains within Vulcan. *Modified from Schenk, P., et al., 2018. Breaking up is hard to do: global cartography and topography of Pluto's mid-sized icy moon Charon from New Horizons. Icarus 315, 124–145.*

12.4.2 Surface geology

Charon's surface is on average older than that of Pluto, as expected for a smaller, less volatile-rich world. Crater counts suggest that most of the surface has an age of ~4 Ga (Moore et al., 2016). Two dominant terrain types are visible in the New Horizons images (Figs. 12.2 and 12.9). The oldest, occupying more than half of the encounter hemisphere, is a heavily cratered rugged terrain broken into polygonal blocks up to several 100 km in size by canyons up to at least 12 km deep (Beyer et al., 2017; Schenk et al., 2018b), providing evidence for global expansion likely due to freezing of a subsurface ocean. The other terrain type, which has similar crater density but is younger based on superposition relationships, is a smooth plains unit embaying the rugged terrain (Beyer et al., 2019). The margins of the plains units abutting the rugged terrain form a "moat" that is depressed by 1–3 km ("M," Fig. 12.9), and several mountains partially embayed by the smooth plains show similar moats ("m," Fig. 12.9, also visible near the terminator in Fig. 12.2). These margins suggest that the flows had high viscosity, emplaced either in the solid state or as a liquid/solid mixture (Beyer et al., 2019).

12.4.3 Surface chemistry

New Horizons spectral observations show ubiquitous H_2O ice across Charon (Grundy et al., 2016a). A band at 1.65 μm indicates the ice is predominantly crystalline rather than amorphous. Exposure to space radiation amorphizes cold H_2O ice over time. If warmed, radiation-damaged ice can recrystallize, but Charon's surface temperature was thought to be too cold for that, triggering speculation about recent cryovolcanic activity (Cook et al., 2007). The New Horizons images show no evidence of recent cryovolcanism, casting doubt on the extrapolation from laboratory to geological timescales. A band at 2.21 μm is seen across Charon (Dalle Ore et al., 2018). It had been attributed to ammonia or ammonia hydrate, but these materials are likewise thought to be readily destroyed by space radiation, so some other, more stable ammoniated compound is suspected (e.g., Cook et al., 2018). Much higher concentrations in the ejecta of a few craters suggest a localized subsurface reservoir. Charon's dark polar caps are among its most visually striking features. They are hypothesized to slowly form through cold-trapping of CH_4 escaping from Pluto's atmosphere onto Charon's frigid winter pole, where it is radiolytically processed into heavier and less volatile hydrocarbons and ultimately dark, reddish, tholin-like material (Grundy et al., 2016b).

12.5 Small satellites

12.5.1 Dynamics

The orbits of Pluto's four small satellites (Table 12.1; Fig. 12.10) are almost coplanar with Charon's orbit, and are close to, but not exactly in, the 1:3, 1:4, 1:5, and 1:6 mean-motion resonances with the orbit of Charon. Their rotation is nonsynchronous with their orbits, and is much faster than that of Pluto or Charon. The fastest period, that of Hydra (10.3 hours), is typical for small bodies in heliocentric orbit, suggesting that it simply is too far from Pluto for tides to have slowed its rotation—the fact that the closer satellites spin more slowly may indicate that they have been subject to some tidal despinning. The obliquities of the spin axes are extreme, perhaps having been tilted by resonances between spin precession and mean motion (Quillen et al., 2017).

FIG. 12.10 Montage of New Horizons images of Pluto's small satellites, to scale, compared to the limb of Charon.

12.5.2 Physical properties

The masses of the small moons have not been directly determined, though there is hope that mutual perturbations can eventually provide useful information. Albedos are higher than that of Charon, which is surprising if, as is likely, they formed from the same circum-Pluto debris disk following the Charon-forming impact. Their shapes are irregular, and both Kerberos and Hydra have a bi-lobed shape suggesting formation from the coalescence of two smaller bodies. Craters are visible on Hydra and Nix, with a density slightly higher than that of Charon (Weaver et al., 2016). The largest crater imaged on Nix is associated with both bright spots and dark reddish regions, due either to excavation of contrasting subsurface material or contamination by the impactor.

12.5.3 Surface chemistry

New Horizons near-infrared spectroscopy of Nix, Hydra, and Kerberos revealed deep H_2O ice absorption bands, consistent with clean coarse-grained ice (Cook et al., 2018). Styx is likely similar, as it shares the others' high albedos. As on Charon, the 1.65 μm band indicates the ice is predominantly crystalline. These small bodies are clearly not cryovolcanically active, so some other mechanism must keep their ice crystalline. Nix and Hydra have prominent 2.21 μm absorptions tentatively attributed to the same ammoniated species as on Charon (Cook et al., 2018).

12.6 System origin and evolution

The high Charon/Pluto mass ratio is reminiscent of the Earth-Moon system. As with the latter system, the leading explanation for Pluto-Charon is that Charon was formed during a collision between Pluto and another KBO (Canup, 2011). Although Charon would have initially orbited closer to Pluto, tidal dissipation within the two bodies drove them apart until they reached their current, doubly synchronous state (Dobrovolskis et al., 1997). Assuming that the small satellites also formed during the giant impact, the outward evolution of Charon also presumably pushed these bodies out into their present near-resonant configuration (Showalter and Hamilton, 2015), although how this happened in detail is not currently understood (e.g., Walsh and Levison, 2015). The survival of the small satellites also places constraints on how the Pluto-Charon system as a whole migrated outward from the Sun in response to Neptune's outward motion (Pires et al., 2015).

If proto-Pluto had been fully differentiated, then Charon would consist almost entirely of ice, which is not the case (Canup, 2011). On the other hand, a completely undifferentiated Pluto is hard to reconcile with the apparently icy nature of the four small moons (at least at their surfaces—density constraints are currently lacking). So the most likely proto-Pluto structure is a partially differentiated body, with an icy shell but a rock-ice mixture in the interior (McKinnon et al., 2017).

An only partially differentiated proto-Pluto suggests that it cannot have formed while ^{26}Al was still live, that is, within the first 3 Myr of solar system history (McKinnon et al., 2017). Such a timescale makes it difficult to have formed the Pluto-Charon system via pebble accretion (Johansen et al., 2015), which requires nebular gas to still be present. Pebble accretion also has difficulty in explaining the (apparently) different bulk compositions of the small satellites compared to Pluto and Charon. Conversely, forming Pluto and Charon via more conventional coagulation (Kenyon and Bromley, 2012) does not suffer from these problems and currently seems the more likely explanation.

12.7 Conclusions

The New Horizons data have revealed the Pluto system to be even more diverse and spectacular than anticipated. We already have at least plausible explanations for many aspects of the system, including the high Charon/Pluto mass ratio; the origin and nature of Sputnik Planitia; the distribution of many of Pluto's other compositional provinces; the composition and structure of Pluto's lower atmosphere; the extensional features and smooth plains of Charon; and Charon's dark poles. However, many mysteries remain, such as the origin of many of Pluto's peculiar endogenic and erosional features; the low temperature of Pluto's upper atmosphere; and the origin and orbits of its small satellites. There is much still to learn from continued analysis of existing data, from future telescopic observations and, eventually, from future missions to the Pluto system.

References

Bagenal, F., et al., 2016. Pluto' interaction with its space environment: solar wind, energetic particles, and dust. Science 351, aad9045.

Bertrand, T., Forget, F., 2016. Observed glacier and volatile distribution on Pluto from atmosphere-topography processes. Nature 540, 86.

Bertrand, T., et al., 2018a. The nitrogen cycles on Pluto over seasonal and astronomical timescales. Icarus 309, 277.

Bertrand, T., Forget, F., Umurhan, O.M., Moore, J.M., Young, L.A., et al., 2018b. The methane cycles on Pluto over seasonal and astronomical timescales. Icarus 329, 148–165.

Beyer, R.A., Nimmo, F., McKinnon, W.B., Moore, J.M., et al., 2017. Charon tectonics. Icarus 287, 161–174.
Beyer, R.A., et al., 2019. The nature and origin of Charon's smooth plains. Icarus 323, 16–32.
Bierson, C.J., Nimmo, F., McKinnon, W.B., 2018. Implications of the observed Pluto-Charon density contrast. Icarus 309, 207–219.
Brozovic, M., Showalter, M.R., Jacobson, R.A., Buie, M.W., 2015. The orbits and masses of satellites of Pluto. Icarus 246, 317.
Buie, M.W., Tholen, D.J., Horne, K., 1992. Albedo maps of Pluto and Charon-initial mutual event results. Icarus 97, 211.
Buie, M.W., Tholen, D.J., Wasserman, L.H., 1997. Separate lightcurves of Pluto and Charon. Icarus 125, 233.
Buie, M.W., Grundy, W.M., Young, E.F., Young, L.A., Stern, S.A., 2010. Pluto and Charon with the Hubble Space Telescope. II. Resolving changes on Pluto's surface and a map for Charon. Astron. J. 139, 1128–1143.
Buie, M.W., Tholen, D.J., Grundy, W.M., 2012. The orbit of Charon is circular. Astron. J. 144, 15.
Canup, R.M., 2011. On a giant impact origin of Charon, Nix, and Hydra. Astron. J. 141, 9.
Cheng, A.F., et al., 2017. Haze in Pluto's atmosphere. Icarus 290, 112–133.
Christy, J.W., Harrington, R.S., 1978. The satellite of Pluto. Astron. J. 83, 1005–1008.
Cook, J.C., Desch, S.J., Roush, T.L., Trujillo, C.A., Geballe, T.R., 2007. Near-infrared spectroscopy of Charon: possible evidence for cryovolcanism on Kuiper belt objects. Astrophys. J. 663, 1406–1419.
Cook, J.C., Dalle Ore, C.M., Protopapa, S., Binzel, R.P., Cartwright, R., et al., 2018. Composition of Pluto's small satellites: analysis of New Horizons spectral images. Icarus 315, 30–45.
Cook, J.C., Dalle Ore, C.M., Protopapa, S., Binzel, R.P., Cruikshank, D.P., et al., 2019. The distribution of H_2O, CH_3OH, and hydrocarbon-ices on Pluto: analysis of New Horizons spectral images. Icarus 331, 148–169.
Cruikshank, D.P., Pilcher, C.B., Morrison, D., 1976. Pluto: evidence for methane frost. Science 194, 835–837.
Cruikshank, D.P., Grundy, W.M., DeMeo, F.E., Buie, M.W., Binzel, R.P., et al., 2015. The surface compositions of Pluto and Charon. Icarus 246, 82–92.
Dalle Ore, C.M., Protopapa, S., Cook, J.C., Grundy, W.M., Cruikshank, D.P., et al., 2018. Ices on Charon: distribution of H_2O and NH_3 from New Horizons LEISA observations. Icarus 300, 21–32.
Desch, S.J., Neveu, M., 2017. Differentiation and cryovolcanism on Charon: a view before and after New Horizons. Icarus 287, 175–186.
Dias-Oliveira, A., et al., 2015. Pluto's atmosphere from stellar occultations in 2012 and 2013. Astrophys. J. 811, 53.
Dobrovolskis, A.R., Peale, S.J., Harris, A.W., 1997. Dynamics of the Pluto-Charon binary. In: Stern, S.A., Tholen, D.J. (Eds.), Pluto and Charon. University of Arizona Press, Tucson, AZ, p. 159.
Earle, A.M., Grundy, W., Howett, C.J.A., Olkin, C.B., Parker, A.H., et al., 2018. Methane distribution on Pluto as mapped by the New Horizons Ralph/MVIC instrument. Icarus 314, 195–209.
Elliot, J.L., Dunham, E.W., Bosh, A.S., et al., 1989. Pluto's atmosphere. Icarus 77, 148.
Elliot, J.L., et al., 2007. Changes in Pluto's atmosphere: 1988–2006. Astron. J. 134, 1.
Fray, N., Schmitt, B., 2009. Sublimation of ices of astrophysical interest: a bibliographic review. Planet. Space Sci. 57, 2053–2080.
Grundy, W.M., Olkin, C.B., Young, L.A., Buie, M.W., Young, E.F., 2013. Near-infrared spectral monitoring of Pluto's ices: spatial distribution and secular evolution. Icarus 223, 710.
Grundy, W.M., Binzel, R.P., Buratti, B.J., Cook, J.C., Cruikshank, D.P., et al., 2016a. Surface compositions across Pluto and Charon. Science 351, 1283.
Grundy, W.M., Cruikshank, D.P., Gladstone, G.R., Howett, C.J.A., Lauer, T.R., et al., 2016b. Formation of Charon's red poles from seasonally cold-trapped volatiles. Nature 539, 65–68.
Grundy, W.M., Bertrand, T., Binzel, R.P., Buie, M.W., Buratti, B.J., et al., 2018. Pluto's haze as a surface material. Icarus 314, 232–245.
Hammond, N.P., Barr, A.C., Parmentier, E.M., 2016. Recent tectonic activity on Pluto driven by phase changes in the ice shell. Geophys. Res. Lett. 43, 6775–6782.
Hansen, C.J., Paige, D.A., 1996. Seasonal nitrogen cycles on Pluto. Icarus 120, 247.
Hinson, D.P., et al., 2017. Radio occultation measurements of Pluto's neutral atmosphere with New Horizons. Icarus 290, 96–111.
Howard, A.D., et al., 2017a. Pluto: pits and mantles on uplands north and east of Sputnik Planitia. Icarus 293, 218.
Howard, A.D., et al., 2017b. Present and past glaciation on Pluto. Icarus 287, 287.
Johansen, A., MacLow, M., Lacerda, P., Bizzarro, M., 2015. Growth of asteroids, planetary embryos, and Kuiper belt objects by chondrule accretion. Sci. Adv. 1, Article ID: 1500109.

Keane, J.T., Matsuyama, I., Kamata, S., Steckloff, J.K., 2016. Reorientation and faulting of Pluto due to volatile loading within Sputnik Planitia. Nature 540, 90–93.

Kenyon, S.J., Bromley, B.C., 2012. Coagulation calculations of icy planet formation at 15–150 AU: a correlation between the maximum radius and the slope of the size distribution for Trans-Neptunian objects. Astron. J. 143, 63.

Krasnopolsky, V.A., 2018. Some problems in interpretation of the New Horizons observations of Pluto's atmosphere. Icarus 301, 152–154.

Lellouch, E., Laureijs, R., Schmitt, B., Quirico, E., de Bergh, C., Crovisier, J., Coustenis, A., 2000. Pluto's non-isothermal surface. Icarus 147, 220–250.

Lellouch, E., et al., 2017. Detection of CO and HCN in Pluto's atmosphere with ALMA. Icarus 286, 289–307.

Malamud, U., Perets, H.B., Schubert, G., 2017. The contraction/expansion history of Charon with implications for its planetary-scale tectonic belt. Mon. Not. R. Astron. Soc. 468, 1056–1069.

Marcialis, R.L., Rieke, G.H., Lebofsky, L.A., 1987. The surface composition of Charon: tentative identification of water ice. Science 237, 1349.

McComas, D.J., et al., 2016. Pluto's interaction with the solar wind. J. Geophys. Res. (Space Phys.) 121, 4232.

McKinnon, W.B., et al., 2016. Convection in a volatile nitrogen-ice-rich layer drives Pluto's geological vigour. Nature 534, 82.

McKinnon, W.B., Stern, S.A., Weaver, H.A., Nimmo, F., et al., 2017. Origin of the Pluto-Charon system: constraints from the New Horizons flyby. Icarus 287, 2–11.

Moore, J.M., et al., 2016. The geology of Pluto and Charon through the eyes of New Horizons. Science 351, 1284.

Moore, J.M., et al., 2017. Sublimation as a landform-shaping process on Pluto. Icarus 287, 320.

Moore, J.M., Howard, A.D., Umurhan, O.M., White, O.L., Schenk, P.M., et al., 2018. Bladed terrain on Pluto: possible origins and evolution. Icarus 300, 129–144.

Nimmo, F., et al., 2016. Reorientation of Sputnik Planitia implies a subsurface ocean on Pluto. Nature 540, 94–96.

Nimmo, F., et al., 2017. Mean radius and shape of Pluto and Charon from New Horizons images. Icarus 287, 12.

Olkin, C.B., et al., 2015. Evidence that Pluto's atmosphere does not collapse from occultations including the 2013 May 04 event. Icarus 246, 220.

Owen, T.C., Roush, T.L., Cruikshank, D.P., et al., 1993. Surface ices and the atmospheric composition of Pluto. Science 261, 745.

Pires, P., Winter, S.M.G., Gomes, R.S., 2015. The evolution of a Pluto-like system during the migration of the ice giants. Icarus 246, 330–338.

Protopapa, S., Grundy, W.M., Reuter, D.C., Hamilton, D.C., Dalle Ore, C.M., et al., 2017. Pluto's global surface composition through pixel-by-pixel Hapke modeling of New Horizons Ralph/LEISA data. Icarus 287, 218–228.

Quillen, A.C., Nichols-Fleming, F., Chen, Y.-Y., Noyelles, B., 2017. Obliquity evolution of the minor satellites of Pluto and Charon. Icarus 293, 94.

Rhoden, A.R., Henning, W., Hurford, T.A., Hamilton, D.P., 2015. The interior and orbital evolution of Charon as preserved in its geologic record. Icarus 246, 11–20.

Robuchon, G., Nimmo, F., 2011. Thermal evolution of Pluto and implications for surface tectonics and a subsurface ocean. Icarus 216, 426–439.

Schenk, P., et al., 2018a. Basins, fractures and volcanoes: global cartography and topography of Pluto from New Horizons. Icarus 314, 400–433.

Schenk, P., et al., 2018b. Breaking up is hard to do: global cartography and topography of Pluto's mid-sized icy moon Charon from New Horizons. Icarus 315, 124–145.

Schmitt, B., Philippe, S., Grundy, W.M., Reuter, D.C., Côte, R., et al., 2017. Physical state and distribution of materials at the surface of Pluto from New Horizons LEISA imaging spectrometer. Icarus 287, 229–260.

Showalter, M.R., Hamilton, D.P., 2015. Resonant interactions and chaotic rotation of Pluto's small moons. Nature 522, 45.

Sicardy, B., et al., 2016. Pluto's atmosphere from the 2015 June 29 ground-based stellar occultation at the time of the New Horizons flyby. Astrophys. J. 819, L38.

Singer, K., et al., 2019. Impact craters on Pluto and Charon reveal a deficit of small Kuiper belt objects. Science 363, 955–959.

Stern, S.A., et al., 2017. Past epochs of significantly higher pressure atmospheres on Pluto. Icarus 287, 47.

Strobel, D.F., Zhu, X., 2017. Comparative planetary nitrogen atmospheres: density and thermal structures of Pluto and Triton. Icarus 291, 55–64.

Trafton, L., Stern, S.A., 1983. On the global distribution of Pluto's atmosphere. Astrophys. J. 267, 872.
Umurhan, O.M., et al., 2017. Modeling glacial flow on and onto Pluto's Sputnik Planitia. Icarus 287, 301.
Walker, M.F., Hardie, R.H., 1955. A photometric determination of the rotational period of Pluto. Pub. Astron. Soc. Pac. 67, 224–231.
Walsh, K.J., Levison, H.F., 2015. Formation and evolution of Pluto's small satellites. Astron. J. 150, 11.
Weaver, H.A., et al., 2006. Discovery of two new satellites of Pluto. Nature 439, 943–945.
Weaver, H.A., Gibson, W.C., Tapley, M.B., Young, L.A., Stern, S.A., 2008. Overview of the New Horizons science payload. Space Sci. Rev. 140, 75.
Weaver, H.A., et al., 2016. The small satellites of Pluto as observed by New Horizons. Science 351, Aae0030.
White, O.L., et al., 2017. Geological mapping of Sputnik Planitia on Pluto. Icarus 287, 261.
Yelle, R.V., Lunine, J.I., 1989. Evidence for a molecule heavier than methane in the atmosphere of Pluto. Nature 339, 288.
Yelle, R., Koskinen, T., Lavvas, P., 2017. Thermal structure of Pluto's lower atmosphere. In: European Planetary Science Congress, vol. 11, p. 923.
Young, L.A., 2013. Pluto's seasons: new predictions for New Horizons. Astrophys. J. 766, L22.
Young, L.A., et al., 2008. New Horizons: anticipated scientific investigations at the Pluto system. Space Sci. Rev. 140, 93.
Young, L.A., et al., 2018. Structure and composition of Pluto's atmosphere from the New Horizons solar ultraviolet occultation. Icarus 300, 174–199.
Zhang, X., Strobel, D.F., Imanaka, H., 2017. Haze heats Pluto's atmosphere yet explains its cold temperature. Nature 551, 352–355.
Zhu, X., Strobel, D.F., Erwin, J.T., 2014. The density and thermal structure of Pluto's atmosphere and associated escape processes and rates. Icarus 228, 301–314.
Zirnstein, E.J., et al., 2016. Interplanetary magnetic field sector from solar wind around Pluto (SWAP) measurements of heavy ion pickup near Pluto. Astrophys. J. 823, L30.

PART IV

Relations with other populations

CHAPTER

13

Pluto and Charon as templates for other large Trans-Neptunian objects

William M. Grundy
Lowell Observatory, Flagstaff, AZ, United States

13.1 Introduction

Following its discovery in 1930, Pluto was seen by planetary scientists as a lone oddball, distinct from both the rocky planets and from the giant planets. Imagination of what its geology might be like was initially limited by lack of experience with spacecraft exploration of comparable bodies. The 1989 Voyager II flyby of the Neptune system changed that providing a key point of reference. Voyager's instruments revealed Neptune's largest satellite Triton to be a spectacularly complex world with a wealth of geological features, some of them bearing little resemblance to landforms seen elsewhere in the solar system and some pointing to recent or even ongoing geologic activity (e.g., Smith et al., 1989; Croft et al., 1995; Schenk and Zahnle, 2007). For a generation of planetary scientists, Triton served as a template for the surprises Pluto might hold.

Triton and Pluto are similar in many ways. They are comparable in size, with diameters of 2707 and 2377 km for Triton and Pluto, respectively. Their bulk densities are similar too, being intermediate between the bulk densities of water ice and of rock, more rock rich than the mid-sized icy satellites of Saturn and Uranus. Like Pluto, and unlike other large icy satellites of the giant planets, Triton is thought to have originated in the outer protoplanetary nebula prior to its capture into orbit around Neptune (e.g., McKinnon, 1984; Lunine, 1993; Agnor and Hamilton, 2006; Nogueira et al., 2011). Triton and Pluto both have thin, nitrogen-dominated atmospheres (e.g., Hubbard et al., 1988; Tyler et al., 1989), along with the same volatile ices on their surfaces: Nitrogen (N_2), methane (CH_4), and carbon monoxide (CO) (e.g., Cruikshank

The Trans-Neptunian Solar System.
https://doi.org/10.1016/B978-0-12-816490-7.00013-8

et al., 1993; Owen et al., 1993; Cruikshank, 2005), mobile on seasonal timescales (e.g., Trafton et al., 1998). Of course, there were differences, too, making Triton an imperfect template. For instance, Triton's orbit around Neptune offers a source of internal heating via tidal dissipation, unavailable to Pluto. And Triton's location within Neptune's magnetosphere puts it in a different charged particle environment from Pluto's setting in the solar wind. Nonetheless, Triton's exotic and enigmatic geology pointed to the diversity of features that might be expected at Pluto, and helped spur interest in sending a spacecraft to explore it (Neufeld, 2014).

NASA's New Horizons spacecraft flew through the Pluto system in July 2015, sending back a wealth of data from its seven scientific instruments (Stern et al., 2015). This chapter will show data from two of the instruments. The LOng-Range Reconnaissance Imager (LORRI) is a panchromatic 1k × 1k charge coupled device (CCD) detector mounted on an 8″ telescope (Cheng et al., 2008). The Ralph instrument consists of two components, the Multispectral Visible Imaging Camera (MVIC) and the Linear Etalon Imaging Spectral Array (LEISA; Reuter et al., 2008). MVIC provides wide field of view CCD imaging in four-color filter channels plus panchromatic channels. LEISA performs spectral imaging at wavelengths from 1.25 to 2.5 μm at a spectral resolution ($\lambda/\Delta\lambda$) of about 240.

This chapter will examine the Pluto system's distinctive suite of features revealed by New Horizons and consider how its distinct types of activity might provide templates for revelations from future exploration of other large trans-neptunian bodies.

13.2 Powering planetary activity

Planetary activity requires the presence of materials that can be mobilized along with the energy to mobilize them. The primary source of energy in the Pluto system is sunlight, despite Pluto's orbiting at a mean distance 39 times further from the Sun than Earth does. Pluto's version of the solar "constant" ranges from 1.52 W m^{-2} at perihelion to 0.57 W m^{-2} at aphelion. Additional energy is available from the decay of radioactive isotopes in Pluto's interior, chiefly ^{232}Th and ^{238}U, with smaller contributions from ^{235}U and ^{40}K which would have been more important early in Pluto's history. If these radionuclides are present in chondritic abundances in rock that accounts for 65.5% of Pluto's total mass (and 59% of Charon's), their decay would contribute about 0.003 W m^{-2} at Pluto's surface and about half that at Charon's surface (McKinnon et al., 2017). These energy sources are minuscule compared to their counterparts on Earth (1361 W m^{-2} for the terrestrial solar constant and about 0.09 W m^{-2} geothermal heat from a combination of radioisotope decay and residual heat from Earth's formation and differentiation; Davies and Davies, 2010). Yet they are sufficient to make Pluto geologically active thanks to the presence on Pluto of abundant volatile materials N_2, CH_4, and CO. These form weakly bonded solid ices that are readily mobilized with modest energy inputs. The following sections review some of Pluto's more striking geological features enabled by the mobilization of these volatile materials, and then consider a few of Charon's features.

13.3 Pluto

The diversity of processes shaping Pluto's surface is hinted at by the topography and color views in Fig. 13.1. Many distinct landforms are produced.

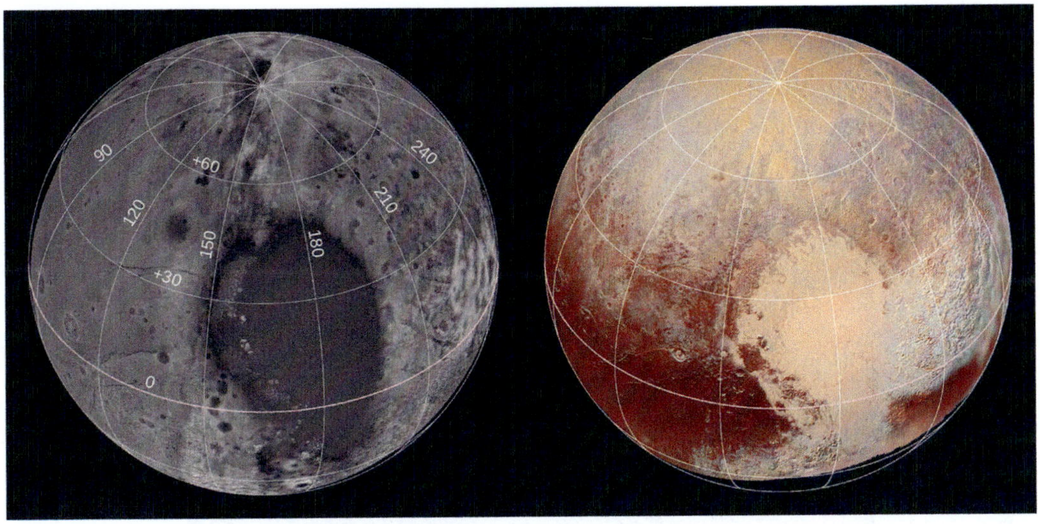

FIG. 13.1 Topography (*left*) and color mosaic (*right*) of Pluto in orthographic projection centered on the encounter hemisphere. The topographic data (Schenk et al., 2018a) are scaled from 4 km below Pluto's 1188.3 km mean radius (*black*) to 10 km above the mean radius (*white*). The color image is made from three of the four MVIC filters: 400–550 nm displayed in *blue*, 540–700 nm displayed in *green*, and 780–975 nm displayed in *red*.

The bladed terrain of Tartarus Dorsa[1] (Fig. 13.2) stands out as a prominent high-altitude region in the topographic map near the eastern edge of the encounter hemisphere. Broad swells are separated by deep troughs. Perched atop the swells are narrower blades, oriented roughly north-south and spaced from 3 to 7 km apart. The blades are up to 1 km in height, with the largest of them occurring near the crests of the broader swells (Moore et al., 2016, 2018). Few if any craters cut into the bladed terrain, suggesting a recent or ongoing formation process (Moore et al., 2018). The composition of the blades appears to be dominated by CH_4 ice, the least volatile of Pluto's three volatile ices (Grundy et al., 2016a; Schmitt et al., 2017; Earle et al., 2018). Bladed terrain shows a distinctive coloration in MVIC images, and that color extends much further to the east along the equator in lower-resolution approach images (Olkin et al., 2017), suggesting that bladed terrain is much more widespread on the nonencounter hemisphere of Pluto. The blades are hypothesized to be analogous to terrestrial penitentes (Moores et al., 2017; Moore et al., 2018). Penitentes are jagged residual features formed as a thick deposit of snow or ice sublimates away, with sunlight controlling where sublimation occurs, possibly in concert with wind. How the thick layer of CH_4 ice from which the blades were carved came to be deposited in the first place remains unclear, but could be related to the several million year "mega-season" cycles of Pluto's obliquity and longitude of perihelion (e.g., Dobrovolskis and Harris, 1983; Earle and Binzel, 2015; Bertrand et al., 2018).

[1] Pluto system feature names used here include a mix of informal and officially approved names.

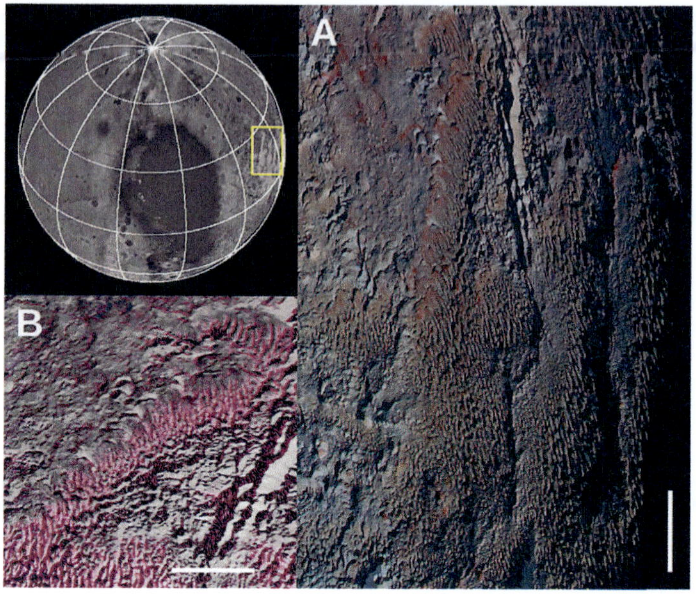

FIG. 13.2 Bladed terrain location indicated by *yellow box* on the elevation map from Fig. 14.1. (A) Oblique MVIC color image and (B) CH_4 ice absorption map (Earle et al., 2018) highlighted in *magenta* on the panchromatic base map (Schenk et al., 2018a). Scale bars are 50 km.

Elsewhere at low latitudes, CH_4 ice can be seen occupying high-altitude sites such as the crests of mountains and the rims of craters (Fig. 13.3). It is unclear if these outcrops are the last remnants of a once-widespread deposit, or the result of preferential deposition of CH_4 frost on high and thus cold regions (e.g., Grundy et al., 2016a; Moore et al., 2017). If the latter, that could suggest an alternate formation scenario for the bladed terrain, that the blades are built up through deposition of CH_4 at high altitudes rather than eroded through sunlight-driven sublimation. Or perhaps both processes work in tandem or alternating with Pluto's climate cycles.

At Pluto surface temperatures N_2 ice is much less rigid than CH_4 ice. Under the influence of Pluto's weak gravity (6.3% of Earth's) it flows downhill like water ice does in valley glaciers on Earth (Umurhan et al., 2017). Several currently active valley glaciers can be seen in Tombaugh Regio, the large, bright feature on Pluto's encounter hemisphere (Fig. 13.4). These flow westward, transporting N_2 ice from higher elevations in eastern Tombaugh Regio down into the large basin of Sputnik Planitia, carving valleys. The source of the N_2 ice feeding these glaciers appear to be seasonal cycles of sublimation, atmospheric transport, and condensation (e.g., Bertrand et al., 2018). N_2 glaciers on Pluto have several key differences from H_2O glaciers on Earth. First, if N_2 were to melt within the glacier, the liquid is less dense than N_2 ice, so it will percolate upward toward the cold surface of the glacier, where it would refreeze or perhaps erupt as jets or geysers. H_2O liquid is more dense than ice, so any liquid that forms in a terrestrial glacier drains downward toward the base of the glacier, where it can help lubricate the glacier's progress, and also help erode the bedrock through freeze-thaw cycles.

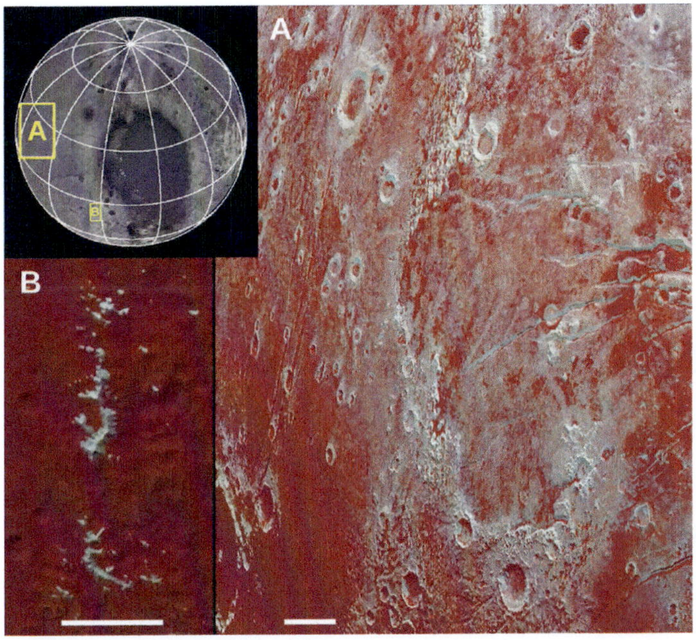

FIG. 13.3 MVIC color images of bright CH_4 ice on the uplands of Piri Rupes and numerous crater rims in Vega Terra (A) and on peaks of Enrique Montes (B). Scale bars are about 50 km.

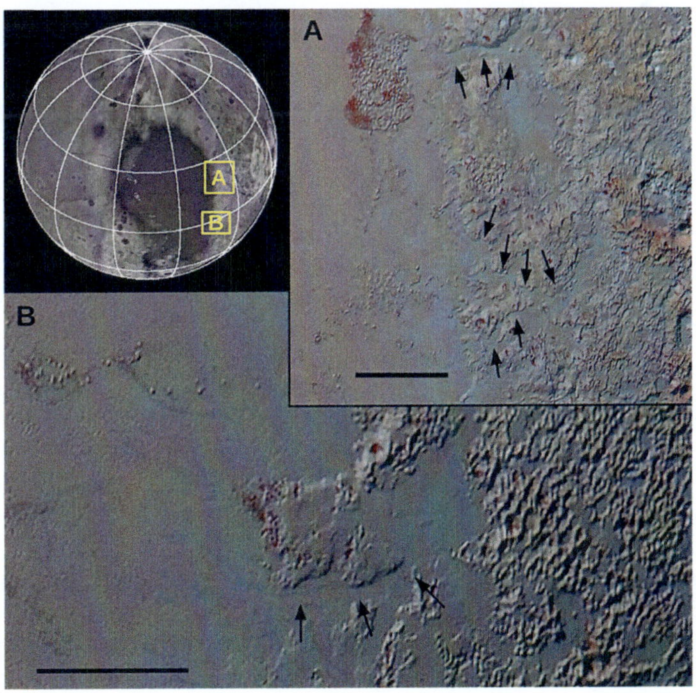

FIG. 13.4 MVIC color images of valley glaciers flowing into Sputnik Planitia from the east. Scale bars are 50 km.

Terrestrial rocks are much denser than H_2O ice, so glaciers on Earth generally flow over the rocky substrate. But on Pluto, the bedrock is mostly H_2O ice, which is slightly less dense than N_2 ice. This density inversion suggests that boulders of H_2O ice bedrock could rise up through the glacier, and that glaciers might exploit fractures to flow down into the ground.

Additional channels can be seen in various locations across Pluto (Fig. 13.5), exhibiting a range of morphologies (Moore et al., 2016; Howard et al., 2017a). It is unclear whether these result from past epochs of glacial erosion, or from subglacial or subsurface liquid flow, or even from surface liquid erosion during climatic extremes of temperature and pressure (e.g., Stern et al., 2017).

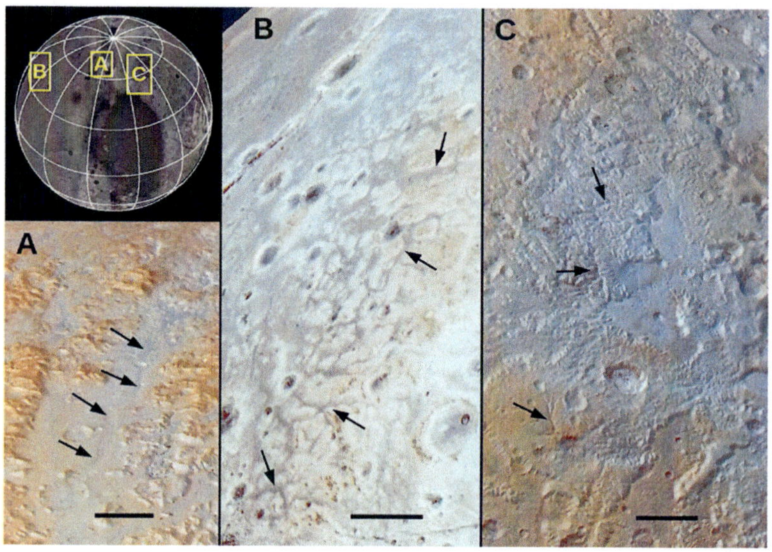

FIG. 13.5 MVIC color images of channels with a variety of morphologies. Scale bars are 50 km.

Pluto's largest glacier by far occupies Sputnik Planitia, the ~1000 km wide ancient impact basin that forms the western half of Tombaugh Regio. Sputnik is partly filled with N_2 ice, along with small amounts of CO and CH_4 (Stern et al., 2015; Moore et al., 2016; Protopapa et al., 2017). This deposit is thick, probably more than a kilometer deep, and N_2 ice is a good thermal insulator, so the meager heat from Pluto's interior is sufficient to drive solid-state convective overturn (McKinnon et al., 2016; Trowbridge et al., 2016). This churning continually refreshes the surface, keeping it free of impact craters. The nearly level surface is marked by a distinctive pattern of polygonal convection cells some 10 s of kilometers across (White et al., 2017). Warm ice rises in the centers of the cells and cold ice sinks along the margins (Fig. 13.6). This type of solid-state convective overturn is not seen in terrestrial glaciers, despite the much greater energy available from geothermal heat, and is unlike anything seen elsewhere in the solar system so far.

Rugged mountains jut up from the western side of Sputnik Planitia (Fig. 13.7), including Al-Idrisi, Zheng He, Baré, Hillary, and Tenzing Montes. These are thought to be blocks of

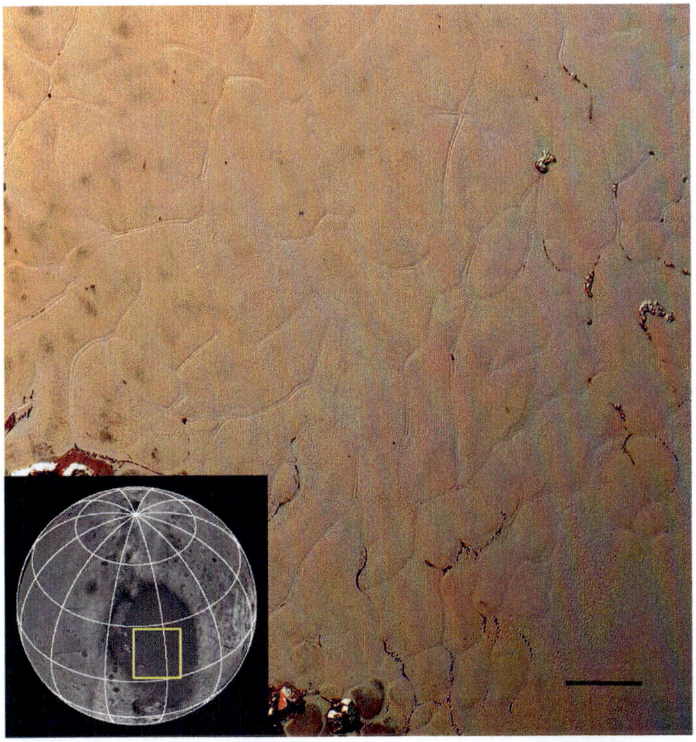

FIG. 13.6 Contrast enhanced MVIC color image of convection cells in Sputnik Planitia. Scale bar is 50 km.

Pluto's crust that were buoyantly supported by the N_2 ice glacier, enabling them to shift and rotate (Stern et al., 2015; Moore et al., 2016). The density contrast between H_2O and N_2 ices is small, so if these mountains are made of water ice, only the tips of the bergs should protrude above the N_2 surface. That seems inconsistent with their close proximity to one another. Possible explanations include the level of the N_2 ice having receded, exposing more of the roots of the mountains, or their being partly composed of much lower density CH_4 ice that would float higher in the N_2 glacier.

Wright Mons and Piccard Mons are a pair of large, circular structures towering over the landscape in the south of the encounter hemisphere (Fig. 13.8). With their deep central pits and peculiarly hummocky texture, these features have been identified as cryovolcanoes (Moore et al., 2016). The paucity of impact craters marring their flanks suggests they were active relatively recently in geological time. The details of how the volcanoes are powered and what materials erupt from them remain unknown. Since liquid H_2O is more dense than H_2O ice, something else would have to be influencing it to make it erupt at the surface. One possibility is dissolved gases that come out of solution as the liquid rises, creating expanding bubbles (e.g., Neveu et al., 2015). Another is compression of a residual liquid pocket as an interior H_2O ocean expands on freezing. Other more exotic scenarios have been discussed, including subsurface circulation of more volatile fluids involving N_2, CO, or CH_4 that could perhaps blast local H_2O

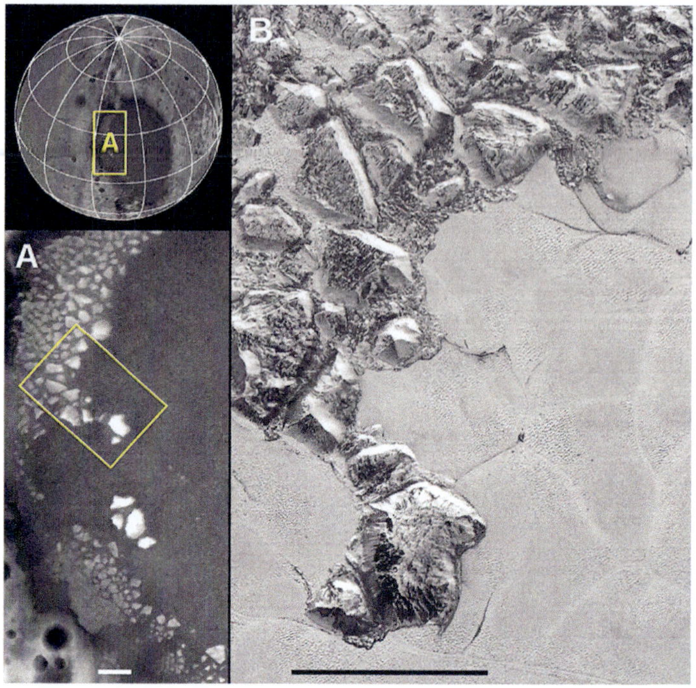

FIG. 13.7 Mountain blocks rising out of Sputnik Planitia. (A) Topographic map scaled from −2.5 km (*black*) to +2.5 km (*white*) relative to Pluto's mean radius of 1188.3 km (Nimmo et al., 2017; Schenk et al., 2018a). (B) High-resolution LORRI panchromatic image showing details of the region in the *yellow box* from (A). Scale bars are 50 km.

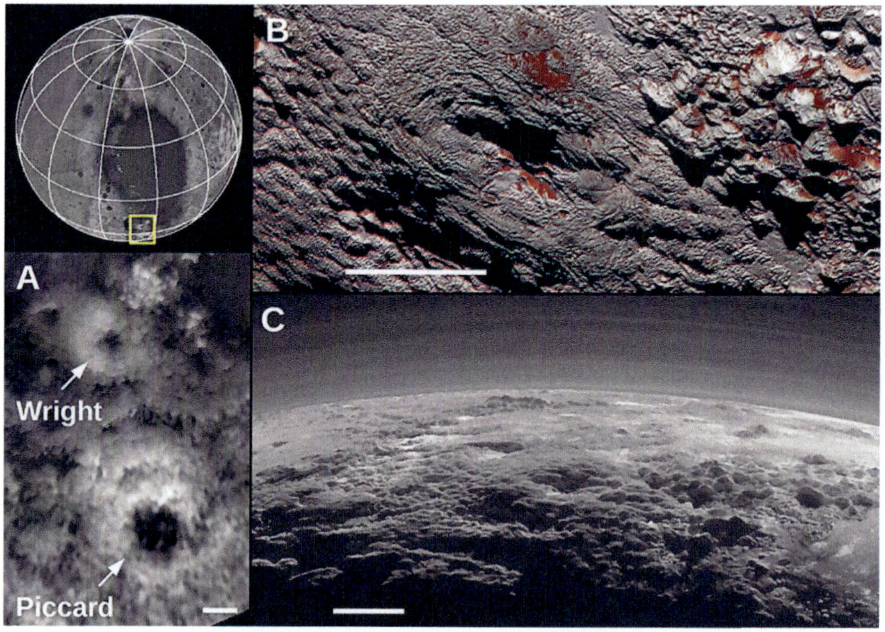

FIG. 13.8 Cryovolcanoes Wright and Piccard Mons. (A) Topographic map scaled from −3 km (*black*) to +5 km (*white*) relative to Pluto's mean radius of 1188.3 km (Nimmo et al., 2017; Schenk et al., 2018a). (B) MVIC color view of Wright Mons. (C) High-phase oblique MVIC panchromatic view of Wright Mons and nearby mountains, showing atmospheric haze layers above the limb. Scale bars are approximately 50 km.

bedrock out as they explosively escape from below the surface. Other small pits, and clusters of pits on Pluto might be the result of eruptive processes, too (Howard et al., 2017b; see Fig. 13.9). Some of these are surrounded by relatively clean H_2O ice (Schmitt et al., 2017; Cook et al., 2019). Since H_2O ice is inert in Pluto's surface environment, exposed H_2O ice should soon become coated by the volatile ices, along with the dark organic haze that is continually settling out of Pluto's atmosphere (Cheng et al., 2017; Grundy et al., 2018), implying relatively recent or even ongoing activity.

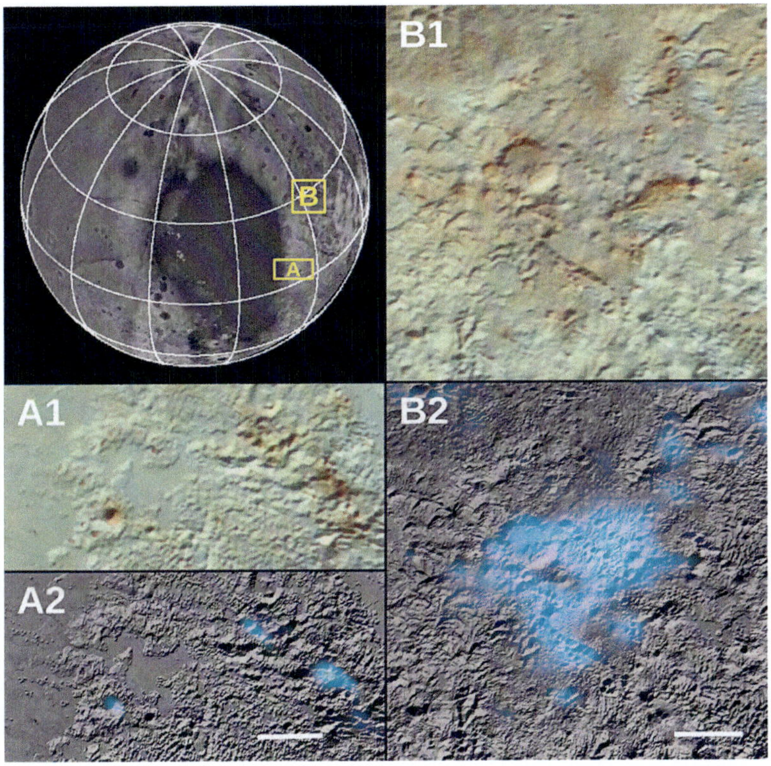

FIG. 13.9 Pits, sometimes occurring in clusters or chains in eastern Tombaugh Regio. A1 and B1 panels show MVIC color. Some pits are associated with H_2O ice deposits (Schmitt et al., 2017; Cook et al., 2019), as indicated by the *blue overlay* on the base map in A2 and B2. The H_2O ice is often, but not always associated with reddish coloration in the MVIC color images. The large crater in (B) is Pulfrich, and the associated H_2O-rich region is Supay Facula. Scale bars are 50 km.

13.4 Charon

Charon's surface lacks the abundant volatile ices that enable much of Pluto's geological activity. Its surface gravity, just 2.9% of Earth's, is insufficient to prevent molecules like N_2 and CH_4 from escaping to space (e.g., Schaller and Brown, 2007a). Abundant impact craters attest to an ancient age for Charon's surface. Yet there is also clear evidence for past activity that could provide useful hints to the sorts of geology we might find on comparably sized trans-neptunian objects. Charon's activity involved H_2O, a material that is far less volatile than those enabling ongoing activity on Pluto, but is thought to have benefited from the presence of various antifreezes such as ammonia, alcohol, and salts, as well as higher abundances of radioactive elements leading to greater internal heat in the past (e.g., Desch and Neveu, 2017).

A striking feature of Charon's encounter hemisphere (Fig. 13.10) is the dichotomy between the battered-looking northern hemisphere (Oz Terra), and the smoother southern hemisphere (Vulcan Planitia), separated by a belt of rugged cliffs and canyons of tectonic origin (e.g., Moore et al., 2016; Beyer et al., 2017; Schenk et al., 2018b). Vulcan Planitia's surface appears to have been flooded by a vast cryolava eruption. Several features suggest that it was an extremely viscous flow, including mountains surrounded by moats, as well as a moat along the northern perimeter of the planitia, suggesting a thick slurry or even glacier-like solid-state flow, rather than a liquid. One scenario involves breakup and foundering of crustal blocks into a warm, softer mantle which oozed up around the blocks, submerging some of them completely, but leaving others protruding through in places.

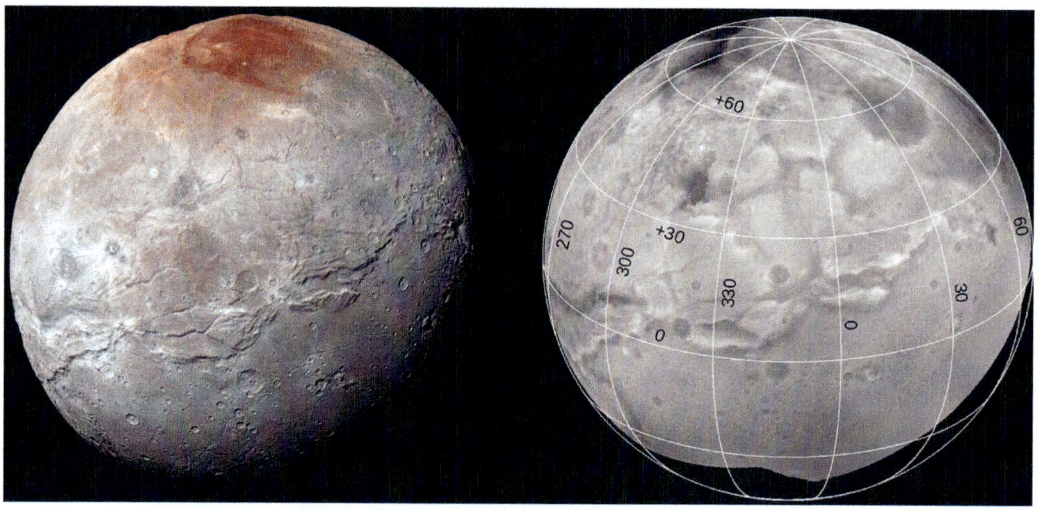

FIG. 13.10 Charon MVIC color (*left*) and topographic map (*right*) of the encounter hemisphere, scaled from −10 km (*black*) to +5 km (*white*) relative to Charon's mean radius of 606.0 km (Nimmo et al., 2017; Schenk et al., 2018b).

Oz Terra has dramatic tectonic features, too. They are hard to discern in the images, but stand out in the topographic map (Fig. 13.10) as a network of deep chasms separating high-standing blocks. This region of crust evidently did not completely founder as in Vulcan Planitia. Its more degraded appearance suggests the blocks in Oz Terra formed during an earlier episode. Freezing of an internal ocean is a probable explanation, since as the water freezes, it expands. The resulting extensional tectonics created horsts separated by grabens, similar to those seen on the Uranian Satellite Ariel (e.g., Beyer et al., 2017; Schenk et al., 2018b). Models show that after Charon's internal ocean freezes, radiogenic heat from the core is less able to escape to the surface, leading to rising internal temperatures, and under some circumstances, a second melting of the lower mantle, followed by a second freezing (e.g.,Desch and Neveu, 2017). A second thaw and freeze could explain the much less eroded appearance of Vulcan Planitia and the tectonic belt separating it from Oz Terra. Hints of even more recent activity on Charon can be seen near the terminator as fields of small pits and clusters of small hills that appear to postdate the resurfacing of Vulcan Planitia (Fig. 13.11). These could be related to escape of volatiles, perhaps from the submerged crustal blocks. Volatiles escaping from Charon's interior could perhaps have become cold trapped in the near surface environment before being mobilized again later when blocks of crust were overrun by warmer mantle material.

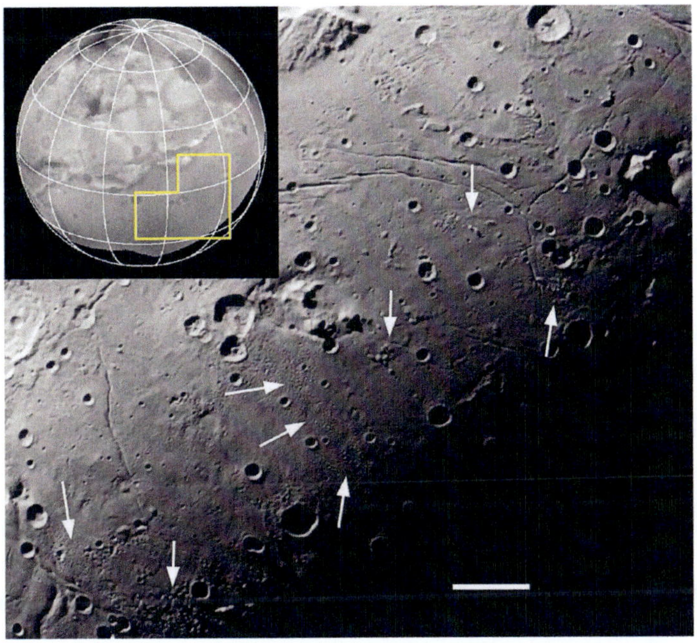

FIG. 13.11 Clusters of small pits and hills near Charon's terminator in an MVIC panchromatic image. Kubrick Mons, one of the moated mountains, is visible at *upper right*. Scale bar is 50 km.

Another striking feature is Charon's dark, reddish pole, Mordor Macula (Fig. 13.10). This feature (along with its counterpart on the opposite pole that is barely discernible in Plutoshine images) is hypothesized to form through a process that is still operating today, with no obvious analog elsewhere in the solar system (Grundy et al., 2016b). Gas, chiefly CH_4, escapes from Pluto and streams radially outward. Some of it passes close enough to Charon to form an extremely thin, transiently trapped atmosphere around Charon. Charon's winter pole goes without sunlight for more than a century, becoming so cold that CH_4 gas freezes out there as ice. Although the winter pole is oriented away from the Sun, it is still exposed to enough ultraviolet radiation (e.g., Gladstone et al., 2018) to convert some of the CH_4 ice into heavier hydrocarbons that remain behind after the Sun rises again, and these hydrocarbons are eventually further processed into dark, reddish, tholin-like material. Over the age of the solar system, estimates suggest some 10 s of centimeters of this material could be produced, more than enough to paint Charon's poles, even as it is mixed down into the H_2O ice substrate via impact gardening.

13.5 Expectations for other large TNOs

Other large TNOs have abundant volatile ices. Eris and Makemake exhibit deeper CH_4 ice absorption bands than Pluto does (e.g., Licandro et al., 2006; Dumas et al., 2007; Brown et al., 2007). Subtle shifts in the wavelengths of their CH_4 bands suggest the presence of abundant N_2 too (Tegler et al., 2008, 2010). Their high albedos and abundant volatiles led Grundy and Umurhan (2017) to pose the question of whether Eris and Makemake might be "Sputnik planets" covered with thick volatile ice deposits, convectively overturning to keep their surfaces refreshed. Eris has a higher mass, density, and thus rock content than Pluto, resulting in greater internal energy from decay of long-lived radionuclides. Makemake probably has a smaller internal energy budget. Or could they be "bladed planets" with globally distributed bladed terrain? Pluto's bladed terrain has deep CH_4 ice absorptions (e.g., Earle et al., 2018) reminiscent of the spectra of Eris and Makemake.

Volatile ices have been more tentatively reported on other large TNOs, such as Sedna and Quaoar (Schaller and Brown, 2007b; Barucci et al., 2010, 2015). Their much more subtle absorption features suggest far less volatile material in their surface environments, and presumably correspondingly less geological activity. Considering the gradual escape of volatiles to space (e.g., Schaller and Brown, 2007a), could what is currently present be the last remnants of once abundant supplies? That idea raises questions about what sorts of features might be left behind after volatiles had played a key role in creating landforms before being lost to space. Could Triton's cantaloupe terrain be the remnant of a once vast glacier of convecting volatile ice? Triton's cantaloupe cells have similar spatial scales to the convective cells in Sputnik Planitia. What might a planetary surface look like after all the volatile ice in bladed terrain has gone away? Perhaps wind-blown debris that had accumulated between the blades would remain behind, leaving a more subdued pattern of parallel ridges. Valleys carved by volatile ice glaciers could certainly be expected to scar the landscape long after the glaciers had sublimated away and H_2O ice edifices constructed by cryovolcanic eruptions could be expected to persist, too.

Numerous even smaller dwarf planets exist in the Kuiper belt, having long since lost whatever primordial endowments of volatile ices they accreted with. But Charon teaches us that even these smaller bodies could have experienced activity in the past that may have dramatically reshaped their geology. Haumea and Orcus both show strong H_2O ice spectral features like Charon (Fornasier et al., 2004; Trujillo et al., 2007), suggestive of extensive melting and resurfacing. Smaller, darker dwarf planets might never have been resurfaced by molten H_2O, maintaining cold, primordial surfaces. But if their deep interiors ever became warm enough to collapse their initially high primordial porosities, or even melt and partially differentiate, we can expect them to have shrunk, leaving a telltale signature of compressional tectonics in their surface geology.

Acknowledgments

This work was supported by NASA's New Horizons project.

References

Agnor, C.B., Hamilton, D.P., 2006. Neptune's capture of its moon Triton in a binary-planet gravitational encounter. Nature 441, 192–194.

Barucci, M.A., Dalle Ore, C.M., Alvarez-Candal, A., de Bergh, C., Merlin, F., et al., 2010. (90377) Sedna: investigation of surface compositional variation. Astron. J. 140, 2095–2100.

Barucci, M.A., Dalle Ore, C.M., Perna, D., Cruikshank, D.P., Doressoundiram, A., Alvarez-Candal, A., Dotto, E., Nitschelm, C., 2015. (50000) Quaoar: surface compositional variability. Astron. Astrophys. 584, A107.1-7.

Bertrand, T., Forget, F., Umurhan, O.M., Grundy, W.M., Schmitt, B., et al., 2018. The nitrogen cycles on Pluto over seasonal and astronomical timescales. Icarus 309, 277–296.

Beyer, R.A., Nimmo, F., McKinnon, W.B., Moore, J.M., Binzel, R.P., et al., 2017. Charon tectonics. Icarus 287, 161–174.

Brown, M.E., Barkume, K.M., Blake, G.A., Schaller, E.L., Rabinowitz, D.L., et al., 2007. Methane and ethane on the bright Kuiper belt object 2005 FY_9. Astron. J. 133, 284–289.

Cheng, A.F., Weaver, H.A., Conard, S.J., Morgan, M.F., Barnouin-Jha, O., et al., 2008. Long-range reconnaissance imager on New Horizons. Space Sci. Rev. 140, 189–215.

Cheng, A.F., Summers, M.E., Gladstone, G.R., Strobel, D.F., Young, L.A., et al., 2017. Haze in Pluto's atmosphere. Icarus 290, 112–133.

Cook, J.C., Dalle Ore, C.M., Protopapa, S., Binzel, R.P., Cruikshank, D.P., et al., 2019. The distribution of H_2O, CH_3OH, and hydrocarbon ices on Pluto: analysis of New Horizons spectral images. Icarus 331, 148–169.

Croft, S.K., Kargel, J.S., Kirk, R.L., Moore, J.M., Schenk, P.M., Strom, R.G., 1995. The geology of Triton. In: Cruikshank D.P. (Ed.), Neptune and Triton. University of Arizona Press, Tucson, AZ, pp. 879–947.

Cruikshank, D.P., 2005. Triton, Pluto, centaurs, and transneptunian bodies. Space Sci. Rev. 116, 421–439.

Cruikshank, D.P., Roush, T.L., Owen, T.C., Geballe, T.R., de Bergh, C., et al., 1993. Ices on the surface of Triton. Science 261, 742–745.

Davies, J.H., Davies, D.R., 2010. Earth's surface heat flux. Solid Earth 1, 5–24.

Desch, S.J., Neveu, M., 2017. Differentiation and cryovolcanism on Charon: a view before and after New Horizons. Icarus 287, 175–186.

Dobrovolskis, A.R., Harris, A.W., 1983. The obliquity of Pluto. Icarus 55, 231–235.

Dumas, C., Merlin, F., Barucci, M.A., de Bergh, C., Hainault, O., et al., 2007. Surface composition of the largest dwarf planet 136199 Eris (2003 UB_{313}). Astron. Astrophys. 471, 331–334.

Earle, A.M., Binzel, R.P., 2015. Pluto's insolation history: latitudinal variations and effects on atmospheric pressure. Icarus 250, 405–412.

Earle, A.M., Grundy, W., Howett, C.J.A., Olkin, C.B., Parker, A.H., et al., 2018. Methane distribution on Pluto as mapped by the New Horizons Ralph/MVIC instrument. Icarus 314, 195–209.

Fornasier, S., Dotto, E., Barucci, M.A., Barbieri, C., 2004. Water ice on the surface of the large TNO 2004 DW. Astron. Astrophys. 422, L43–L46.

Gladstone, G.R., Pryor, W.R., Stern, S.A., Ennico, K., Olkin, C.B., et al., 2018. The Lyman-α sky background as observed by New Horizons. Geophys. Res. Lett. 45, 8022–8028.

Grundy, W.M., Umurhan, O.M., 2017. Are Makemake and Eris Sputnik planets? In: American Astronomical Society, DPS Meeting No. 49.

Grundy, W.M., Binzel, R.P., Buratti, B.J., Cook, J.C., Cruikshank, D.P., et al., 2016a. Surface compositions across Pluto and Charon. Science 351, 1283.

Grundy, W.M., Cruikshank, D.P., Gladstone, G.R., Howett, C.J.A., Lauer, T.R., et al., 2016b. Formation of Charon's red poles from seasonally cold-trapped volatiles. Nature 539, 65–68.

Grundy, W.M., Bertrand, T., Binzel, R.P., Buie, M.W., Buratti, B.J., et al., 2018. Pluto's haze as a surface material. Icarus 314, 232–245.

Howard, A.D., Moore, J.M., Umurhan, O.M., White, O.L., Anderson, R.S., et al., 2017a. Present and past glaciation on Pluto. Icarus 287, 287–300.

Howard, A.D., Moore, J.M., White, O.L., Umurhan, O.M., Schenk, P.M., et al., 2017b. Pluto: pits and mantles on uplands north and east of Sputnik Planitia. Icarus 293, 218–230.

Hubbard, W.B., Hunten, D.M., Dieters, S.W., Hill, K.M., Watson, R.D., 1988. Occultation evidence for an atmosphere on Pluto. Nature 336, 452–454.

Licandro, J., Grundy, W.M., Pinilla-Alonso, N., Leisy, P., 2006. Visible spectroscopy of 2003 UB_{313}: evidence for N_2 ice on the surface of the largest TNO? Astron. Astrophys. 458, L5–L8.

Lunine, J.I., 1993. Triton, Pluto, and the origin of the solar system. Science 261, 697–698.
McKinnon, W.B., 1984. On the origin of Triton and Pluto. Nature 311, 355–358.
McKinnon, W.B., Nimmo, F., Wong, T., Schenk, P.M., White, O.L., et al., 2016. Convection in a volatile nitrogen-ice-rich layer drives Pluto's geological vigour. Nature 534, 82–85.
McKinnon, W.B., Stern, S.A., Weaver, H.A., Nimmo, F., Bierson, C.J., et al., 2017. Origin of the Pluto-Charon system: constraints from the New Horizons flyby. Icarus 287, 2–11.
Moore, J.M., McKinnon, W.B., Spencer, J.R., Howard, A.D., Schenk, P.M., et al., 2016. The geology of Pluto and Charon through the eyes of New Horizons. Science 351, 1284–1293.
Moore, J.M., Howard, A.D., Umurhan, O.M., White, O.L., Schenk, P.M., et al., 2017. Sublimation as a landform-shaping process on Pluto. Icarus 287, 320–333.
Moore, J.M., Howard, A.D., Umurhan, O.M., White, O.L., Schenk, P.M., et al., 2018. Bladed terrain on Pluto: possible origins and evolution. Icarus 300, 129–144.
Moores, J.E., Smith, C.L., Toigo, A.D., Guzewich, S.D., 2017. Penitentes as the origin of the bladed terrain of Tartarus Dorsa on Pluto. Nature 541, 188–190.
Neufeld, M.J., 2014. First mission to Pluto: policy, politics, science, and technology in the origins of New Horizons, 1989–2003. Hist. Stud. Nat. Sci. 44, 234–276.
Neveu, M., Desch, S.J., Glein, C.R., 2015. Prerequisites for explosive cryovolcanism on dwarf planet-class Kuiper belt objects. Icarus 246, 48–64.
Nimmo, F., Umurhan, O., Lisse, C.M., Bierson, C.J., Lauer, T.R., et al., 2017. Mean radius and shape of Pluto and Charon from New Horizons images. Icarus 287, 12–29.
Nogueira, E., Brasser, R., Gomes, R., 2011. Reassessing the origin of Triton. Icarus 214, 113–130.
Olkin, C.B., Spencer, J.R., Grundy, W.M., Parker, A.H., Beyer, R.A., et al., 2017. The global color of Pluto from New Horizons. Astron. J. 154, 258.1-13.
Owen, T.C., Roush, T.L., Cruikshank, D.P., Elliot, J.L., Young, L.A., et al., 1993. Surface ices and atmospheric composition of Pluto. Science 261, 745–748.
Protopapa, S., Grundy, W.M., Reuter, D.C., Hamilton, D.C., Dalle Ore, C.M., et al., 2017. Pluto's global surface composition through pixel-by-pixel Hapke modeling of New Horizons Ralph/LEISA data. Icarus 287, 218–228.
Reuter, D.C., Stern, S.A., Scherrer, J., Jennings, D.E., Baer, J.W., et al., 2008. Ralph: a visible/infrared imager for the New Horizons Pluto/Kuiper belt mission. Space Sci. Rev. 140, 129–154.
Schaller, E.L., Brown, M.E., 2007a. Volatile loss and retention on Kuiper belt objects. Astrophys. J. 659, L61–L64.
Schaller, E.L., Brown, M.E., 2007b. Detection of methane on Kuiper belt object (50000) Quaoar. Astrophys. J. 670, L49–L51.
Schenk, P.M., Zahnle, K., 2007. On the negligible surface age of Triton. Icarus 192, 135–149.
Schenk, P.M., Beyer, R.A., McKinnon, W.B., Moore, J.M., Spencer, J.R., et al., 2018a. Basins, fractures, and volcanoes: global cartography and topography of Pluto from New Horizons. Icarus 314, 400–433.
Schenk, P.M., Beyer, R.A., McKinnon, W.B., Moore, J.M., Spencer, J.R., et al., 2018b. Breaking up is hard to do: global cartography and topography of Pluto's mid-sized icy moon Charon from New Horizons. Icarus 315, 124–145.
Schmitt, B., Philippe, S., Grundy, W.M., Reuter, D.C., Côte, R., et al., 2017. Physical state and distribution of materials at the surface of Pluto from New Horizons LEISA imaging spectrometer. Icarus 287, 229–260.
Smith, B.A., Soderblom, L.A., Banfield, D., Barnet, C., Basilevksy, A.T., et al., 1989. Voyager 2 at Neptune: imaging science results. Science 246, 1422–1449.
Stern, S.A., Bagenal, F., Ennico, K., Gladstone, G.R., Grundy, W.M., et al., 2015. The Pluto system: initial results from its exploration by New Horizons. Science 350, 292.
Stern, S.A., Binzel, R.P., Earle, A.M., Singer, K.N., Young, L.A., et al., 2017. Past epochs of significantly higher pressure atmospheres on Pluto. Icarus 287, 47–53.
Tegler, S.C., Grundy, W.M., Vilas, F., Romanishin, W., Cornelison, D.M., et al., 2008. Evidence of N_2-ice on the surface of the icy dwarf planet 136472 (2005 FY_9). Icarus 195, 844–850.
Tegler, S.C., Cornelison, D.M., Grundy, W.M., Romanishin, W., Abernathy, M.R., et al., 2010. Methane and nitrogen abundances on Pluto and Eris. Astrophys. J. 725, 1296–1305.
Trafton, L.M., Matson, D.L., Stansberry, J.A., 1998. Surface/atmosphere interactions and volatile transport (Triton, Pluto, and IO). In: Schmitt, B., de Bergh, C., Festou, M. (Eds.), Solar System Ices. Kluwer Academic, Dordrecht, pp. 773–812.
Trowbridge, A.J., Melosh, H.J., Steckloff, J.K., Freed, A.M., 2016. Vigorous convection as the explanation for Pluto's polygonal terrain. Nature 534, 79–81.

Trujillo, C.A., Brown, M.E., Barkume, K.M., Schaller, E.L., Rabinowitz, D.L., 2007. The surface of 2003 EL_{61} in the near-infrared. Astrophys. J. 655, 1172–1178.

Tyler, G.L., Sweetnam, D.N., Anderson, J.D., Borutzki, S.D., Campbell, J.K., et al., 1989. Voyager radio science observations of Neptune and Triton. Science 246, 1466–1473.

Umurhan, O.M., Howard, A.D., Moore, J.M., Earle, A.M., White, O.L., et al., 2017. Modeling glacial flow on and onto Pluto's Sputnik Planitia. Icarus 287, 301–319.

White, O.L., Moore, J.M., McKinnon, W.B., Spencer, J.R., Howard, A.D., et al., 2017. Geological mapping of Sputnik Planitia on Pluto. Icarus 287, 261–286.

CHAPTER

14

From Centaurs to comets: 40 Years

Nuno Peixinho[a], *Audrey Thirouin*[b], *Stephen C. Tegler*[c], *Romina P. Di Sisto*[d,e], *Audrey Delsanti*[f], *Aurélie Guilbert-Lepoutre*[g], *James G. Bauer*[h]

[a]CITEUC—Center for Earth and Space Research of the University of Coimbra, Geophysical and Astronomical Observatory of the University of Coimbra, Coimbra, Portugal [b]Lowell Observatory, Flagstaff, AZ, United States [c]Department of Physics and Astronomy, Northern Arizona University, Flagstaff, AZ, United States [d]Instituto de Astrofísica de La Plata, CCT—CONICET—Universidad Nacional de La Plata, La Plata, Argentina [e]Facultad de Ciencias Astronómicas y Geofísicas, UNLP, La Plata, Argentina [f]Aix Marseille Université, CNRS, LAM (Laboratoire d'Astrophysique de Marseille) UMR 7326, Marseille, France [g]LGL-TPE, UMR 5276, CNRS, Claude Bernard Lyon 1 University, ENS Lyon, Villeurbanne Cedex, France [h]Department of Astronomy, University of Maryland, College Park, MD, United States

14.1 Introduction

Centaurs are a group of solar system minor bodies loosely classified as those orbiting around the Sun between the orbits of Jupiter and Neptune that are not trapped in any mean-motion resonance with a planet. They have been discovered 15 years before the Trans-Neptunian objects (TNOs), or Kuiper belt objects, except for Pluto/Charon. First announced as "slow-moving object Kowal," on November 4, 1977, and then with the provisional designation 1977 UB, on November 14, when it was shown that a very high eccentricity orbit was not viable for it, (2060) Chiron was the first Centaur to be discovered (Kowal and Gehrels, 1977). Chiron was also precovered in photographic plates from 1941 and even from 1895 at the Boyden Observatory, South Africa (Davies, 2001).

Fernández (1980) suggested that Chiron might be a comet diffusing inwards, through planetary perturbations—from a cometary belt beyond Neptune—the by-then undetected Trans-Neptunian belt. But the orbit of Chiron was not its only unusual characteristic. Tholen et al. (1988) found that its brightness increased more than expected and, later on,

The Trans-Neptunian Solar System.
https://doi.org/10.1016/B978-0-12-816490-7.00014-X

Meech and Belton (1990) directly detected a low-surface-brightness coma for Chiron. Then, it was hypothesized that Chiron was potentially an ice-rich object, a comet nucleus 180 km across, darkened by C-class carbonaceous soil that has been scattered inward from the Oort cloud (Hartmann et al., 1990). A second idea was that Chiron is most likely a bright member of what is known today as the Trans-Neptunian belt (Duncan et al., 1988). Finally, Hahn and Bailey (1990) concluded that Chiron-type orbits are short-lived and chaotic, with an ejection half-life of $t_{1/2} \sim 1$ Myr, therefore Chiron's current orbit should be a transition between a TNO and a short-period comet.

In 1992, 7 months before the discovery of the first TNO, a second Centaur was announced: 1992 AD, now known as (5145) Pholus (Scotti et al., 1992). No sign of coma was identified, but Pholus became the reddest known object in the solar system (Hainaut et al., 1992).

Given their close link, Centaurs have been mostly studied together with TNOs. Previous to this book, the most extensive review of this field was the book *The Solar System Beyond Neptune* (Barucci et al., 2008), preceded by works as the *Trans-Neptunian Objects and Comets* (Jewitt et al., 2008) and the pioneering *The first decadal review of the Edgeworth-Kuiper belt* (Davies and Barrera, 2004). Other reviews, such as Nesvorný (2018), Ferrari (2018), Luu and Jewitt (2002), and Jewitt (1999), are also of highly recommended reading. In this chapter, we will propose a brief overview of the current state of the art regarding composition, rotational and physical properties, and activity regarding the Centaur population as well as their links to the Trans-Neptunian and comet populations.

14.2 Centaurs as progeny of TNOs

Levison and Duncan (1997) and Duncan and Levison (1997) found that a belt of TNOs on rather circular and low inclined orbits—presently known as: classical TNOs—would create a dynamically excited population through scattering interactions with Neptune, hence called the scattered disk objects (SDOs), and that population should be the main source of Centaurs. Integrating different models for that scattered population for 4.5 Gyr provides injection rates of $\sim(1-5) \times 10^{-10}$ SDOs/year and an estimated number of Centaurs larger than ~1 km of the order of 10^6–10^8 (Di Sisto and Brunini, 2007; Volk and Malhotra, 2008). Plutinos, the TNOs in the 3:2 mean-motion resonance with Neptune, should also be a source of Centaurs (Morbidelli, 1997; Yu and Tremaine, 1999), contributing ~3 times less than SDOs (Di Sisto and Brunini, 2007), and the 2:1 resonant TNOs might also contribute (Tiscareno and Malhotra, 2009; de Elía et al., 2008). Although very secondary as source, every ~60–200 years a kilometer-sized Neptune Trojan enters the Centaurs region (Horner and Lykawka, 2010). Jupiter Trojans were also found to be a minor source of Centaurs (Di Sisto et al., 2019).

The mean lifetime of Centaurs is strongly dependent on their definition, because lifetime correlates with perihelion distance and semimajor axis and with the initial inclination of the progenitors, and can be as short as few Myr up to tens of Myr (Tiscareno and Malhotra, 2003; Di Sisto and Brunini, 2007; Volk and Malhotra, 2008; Bailey and Malhotra, 2009).

Interesting is the very low probability for a high inclination Centaur, or even a retrograde one, originating from the Trans-Neptunian region (Volk and Malhotra, 2013) and the suggestion that a heterogeneous group of up to 200 Centaurs with diameters $D \gtrsim 150$ km, inclinations

$i > 70$ degrees, and perihelia $q > 15$ AU should exist, originating mainly in the Oort cloud (Brasser et al., 2012; de la Fuente Marcos and de la Fuente Marcos, 2014).

Another very interesting aspect of Centaurs is the existence of binaries. From the 81 known Trans-Neptunian binaries, (42355) Typhon/Echidna and (65489) Ceto/Phorcys cross the orbit of Neptune. Brunini (2014) concludes that a small population of contact binary Centaurs may exist and Araujo et al. (2018) find that Typhon/Echidna could survive close encounters with the planets and even reach the terrestrial planets region as a binary.

14.3 Centaurs as progenitors of Jupiter-family comets

On early days, the expression short-period comets ($P < 200$ years) was frequently used and the Trans-Neptunian belt was thought to be their main source. Centaurs were merely a transient phase. The usage of short-period comets was replaced by the more concrete Jupiter-family comets (JFCs) and Halley-type comets (HTCs)—Chiron and Encke type are also used—using not the orbital period as a criterion but the Tisserand parameter relative to Jupiter, T_J (Tisserand, 1896; Kresák, 1972; Levison, 1996). JFCs are those with $2 \lesssim T_J \lesssim 3$.

It is interesting to notice that the studies regarding the source of JFCs have been made, mostly, regarding the direct link between them and TNOs. Centaurs are taken as a short-lived transient phase not decoupled from this JFC-TNO link. Right after the discovery of Pluto, Leonard (1930) speculated about the existence of Trans-Neptunian and Trans-Plutonian objects but without linking them explicitly to comets. The existence of a Trans-Neptunian region as a source of JFCs was first suggested by Edgeworth (1943, 1949) and Kuiper (1951), being properly addressed much later by Fernández (1980) and Duncan et al. (1988).

Relatively little attention has been given to the dynamics of the Centaur population itself. Tiscareno and Malhotra (2003) find that one-third of the Centaurs will actually be injected into the JFC population, while two-thirds will be ejected from the solar system or will enter the Oort cloud. About 20% of them have lifetimes as Centaurs shorter than 1 Myr, while another 20% have lifetimes exceeding 100 Myr. Horner et al. (2004b) found even that about one-fifth of the Centaurs that became JFCs will become Earth-crossing objects. Some Centaurs may even become temporarily captured as Trojans of the giant planets for up to 100 kyr (Horner and Evans, 2006). Most interesting is the recent finding that active Centaurs tend to evolve to JFCs, while the inactive ones tend to evolve to HTCs (Fernández et al., 2018).

14.4 Centaurs by themselves

14.4.1 Classifying Centaurs

There is no strict definition of Centaurs, which poses some evident problems when comparing different works. The most frequent classification forces them to have both perihelion and semimajor axis between the orbits of Jupiter and Neptune ($q > 5.2$ AU, $a < 30.1$ AU) (e.g., Gladman et al., 2008). With the previous definition, the most strict, as of February 1,

2019[1] there are 292 Centaurs, of which 26 are also classified as comets. Also widely used is the Deep Ecliptic Survey (DES) definition, in which the perihelion needs to be between the orbits of Jupiter and Neptune but without any constraints regarding the semimajor axis (Elliot et al., 2005). With this definition, there are 360 Centaurs listed[2] (see Fig. 14.1).

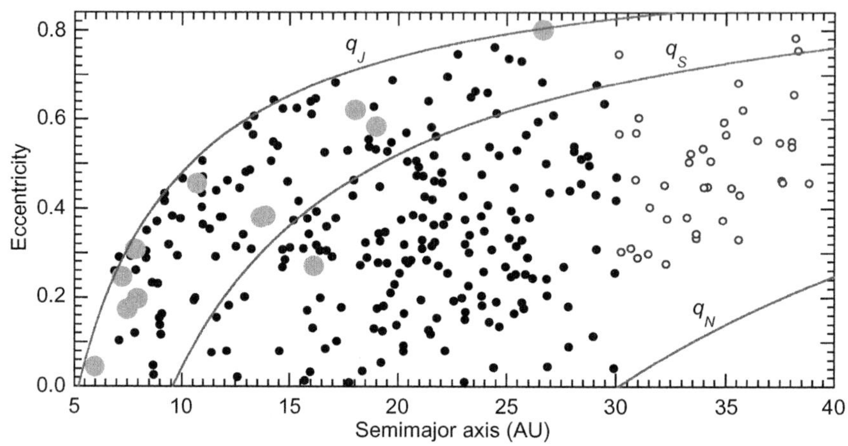

FIG. 14.1 Semimajor axis and eccentricity plot of known Centaurs. *Large gray dots* indicate known active Centaurs. *Black dots* indicate Centaurs with both perihelion and semimajor axis between those of Jupiter and Neptune, that is, the more strict definition. *White dots* indicate objects classified as Centaurs by DES alone. The *curves* where objects have perihelion equal to the semimajor axis of Jupiter, q_J, of Saturn, q_S, and of Neptune, q_N, are drawn. Several objects are out of scale.

14.4.2 Surface properties

14.4.2.1 Colors

When photometry is acquired almost simultaneously in various bands, or the eventual rotational brightness variations are negligible or averaged down, the magnitude difference between those bands defines a color. Centaurs' colors are usually measured using the Johnson UBVRJHK photometric system—from ≈3600 Å to ≈2.2 μm—or the SDSS *ugriz* system (e.g., Schwamb et al., 2018). In the most used color, Centaurs range from a neutral/gray solar-like B−R = 1.0 mag to an extraordinarily red B−R = 2.0 mag. Since complex organic molecules are known to efficiently absorb shorter wavelengths of optical light, the redder the object the richer the object might be in complex organic molecules, although other compounds are capable of doing so. Surface colors can also be used to obtain the low-resolution reflectance spectrum, at a relatively low observation time cost and constraint, at least, the spectral continuum of the studied object (see Delsanti et al., 2004; Dalle Ore et al., 2015). Two large surveys post their

[1] See https://ssd.jpl.nasa.gov/sbdb_query.cgi.

[2] See http://www.boulder.swri.edu/~buie/kbo/desclass.html.

measurements of colors for these objects online the Tegler, Romanishin, and Consolmagno's survey (TRC),[3] and the Minor Bodies of the Outer Solar System Survey (MBOSS).[4]

Early on, in the observations of Centaurs and TNOs, a controversy arose as to whether these objects exhibited a unimodal color distribution (Luu and Jewitt, 1996; Jewitt and Luu, 2001; Hainaut and Delsanti, 2002) or a bimodal color distribution (Tegler and Romanishin, 1998, 2003; Tegler et al., 2003). Models attempting to explain the two possible behaviors were very different. Luu and Jewitt (1996) put forth a model of steady radiation reddening and stochastic impacts to explain the unimodal colors of Centaurs and TNOs, but it has been shown that such simple model could not explain the color diversity as it was observed (Jewitt and Luu, 2001; Thébault and Doressoundiram, 2003; Thébault, 2003). Posterior modeling and laboratory works argue that it is, in fact, possible to reproduce the whole range of colors observed with an appropriate combination of initial albedo, meteoritic bombardment, and space weathering (Kaňuchová et al., 2012). On the other hand, Tegler et al. (2003) argue in favor of a thermally driven composition gradient in the primordial Trans-Neptunian belt that would result in bimodal colors. Brown et al. (2011) and Wong and Brown (2016) argue in favor of H_2S, rather than simple organic molecules, as the main reddening agent, where objects located beyond the H_2S sublimation line, hence not depleted of it, would suffer radiation reddening contrarily to the depleted ones that would result in gray surfaces.

We now see that Centaurs and different dynamical classes of TNOs, as well as different sizes, exhibit different colors (e.g., Tegler et al., 2008, 2016; Peixinho et al., 2012; Fraser and Brown, 2012; Wong and Brown, 2017). Consensus is growing toward a scenario where migration of outer planets disrupted the primordial disk of icy planetesimals, objects were scattered onto dynamically hot orbits and then onto Centaur orbits to present us with the bimodal color we see today. In contrast, dynamically cold TNOs were far enough away from the migrating outer planets that they were not scattered onto dynamically hot TNO orbits and so represent a third compositional class. Nonetheless, recent observations call the bimodality of Centaurs and, therefore, models to explain the bimodality into question. The 99.5% confidence level for the bimodality found by Peixinho et al. (2003) and Tegler et al. (2008) drops to 81% with today's larger sample (Tegler et al., 2016, using TRC survey data). Even reanalyzing TRC data for the 50 objects that follow Gladman et al.'s more strict definition of a Centaur, we get a confidence level of 91.6%, that is, no different conclusion. Looking at Fig. 14.2 we can see how the red population has a larger spreading today. It is interesting that Brown et al. (2011) predicted that the red population should have a broader color distribution than the gray one, given the several reddening agents with sublimation lines in the 20–35 AU region of the primordial Trans-Neptunian belt—although in their most recent paper, they claim only one species, H_2S, is serving as the reddening agent. It is, evidently, premature to rule out irradiation of simple organic molecules into complex organic molecules in favor of irradiation H_2S without further evidence, and without studying what would happen when ices are not in their pure form but in a certainly more realistic mixture with refractory material and how would that affect the sublimation lines.

[3] See http://www.physics.nau.edu/~tegler/research/survey.htm.

[4] See http://www.eso.org/~ohainaut/MBOSS.

Further B–R observations of Centaurs are essential. Confirmation of bimodal Centaur colors will bolster models like that of Wong and Brown (2016). On the other hand, refuting bimodal colors for Centaurs will require a rethinking of the important processes acting on Centaurs. The fact that red Centaurs have a smaller orbital inclination angle distribution than the neutral ones (Tegler et al., 2016), suggesting a distinct origin, is also puzzling when dynamical models say that Centaurs do not preserve their inclinations (Volk and Malhotra, 2013). Moreover, the recent discovery that Jupiter Trojans and Neptune Trojans possess similar colors, when they were presumably populated by very distinct regions of the primordial planetesimal disk, evidences that we are still missing something crucial (Jewitt, 2018).

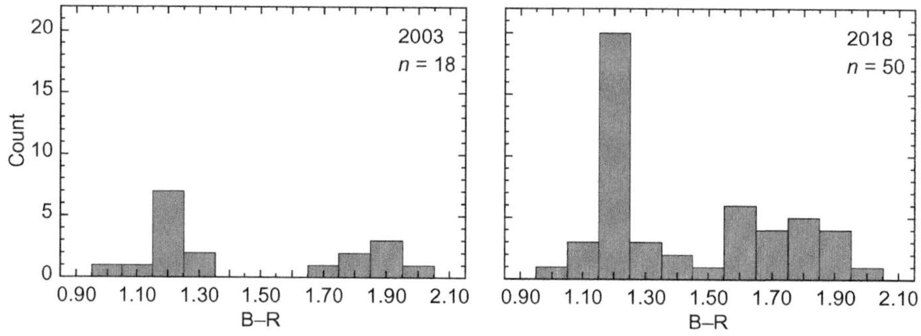

FIG. 14.2 Centaur B–R histograms as they were in 2003 and as they are in 2018 from TRC data. In the larger 2018 sample, the red population became broader and the gap between the gray and red populations became a dip. Note that we are using the more strict definition of Centaurs.

14.4.2.2 Spectra

Although surface colors provide a good proxy to the general surface properties of our objects, the dedicated investigation tool for a comprehensive surface composition study is, of course, reflectance spectroscopy. The technical limit to this method is set by the collecting power of the current large telescopes and the sensitivity of the instruments. With an 8–10 m class telescope, visible spectroscopy with a signal-to-noise ratio (SNR) above ~20 is possible only for objects with $V < 20–22$ mag. Near-infrared (NIR) spectroscopy can be performed at SNR $\gtrsim 10$ for objects with $J < 18–20$ mag. Therefore, only a subset of the currently known Centaurs can be explored with this technique.

Visible reflectance spectra ($\lambda \sim 400–700$ nm) generally show a linear increasing continuum. Most of the ices expected to be present in these regions of the solar system have absorption bands in the NIR. Signatures of water ice, H_2O, are present at 1.5, 1.65, and 2.0 μm; of methane ice, CH_4, at 1.7 and 2.3 μm; of methanol ice, CH_3OH, at 2.27 μm; and of ammonia ice, NH_3 at 2.0 and 2.25 μm. Unfortunately, observations in the region $\lambda > 3$ μm, where ices of interest have very deep signatures (including also CO and CO_2) are not feasible. However, some predictions on the surface composition have been tempted by Dalle Ore et al. (2015).

(5145) Pholus masterfully opened the era of the spectroscopic exploration of the outer solar system belt and what we know today about other objects is just a briefer story. Since Pholus, about two dozens of Centaurs were studied with spectrographs from the largest facilities available from the ground. Its first spectrum showed to be linear between 400 nm and 1 μm, being the steepest red slope detected so far on a solar system body (Fink et al., 1992; Binzel, 1992; Hoffmann et al., 1993). Such slope was incompatible with the conventional silicate or meteoritic analogs that prevailed in asteroids studies at that time, but rather compatible with a mixture of tholins (Khare et al., 1989), suggesting the presence of organic material processed by irradiation. Binzel (1992) suggested that this body has a primitive composition and is at either the beginning or the end state of its thermal evolution. Although on an unstable orbit, Pholus has been undisturbed for at least 10^5 years, that is, long enough to hold, possibly, a thin surface crust of organics due to space weathering. If Pholus just entered the Centaur region, it should start to experience heating and surface processing. In alternative, it might have experienced past cometary activity, maybe on a different orbit in that region, and rather be at its thermal end state with an insulating mantle of organic tholins. Using Hapke (1993) theory, laboratory data, and new observations, Cruikshank et al. (1998) conclude that the red color of Pholus is caused by the irradiation processing of simple molecules (CH_4, H_2O, CO, NH_3, N_2, and CH_3OH) and describe the full spectrum of Pholus with the presence of silicate olivine, carbon, a complex refractory solid (organic molecules), and a mixture of H_2O and CH_3OH (or an equivalent molecule), highlighting that this inventory includes all the basic components of comets. A nice display of Pholus NIR spectrum can be found in Barucci et al. (2011).

Spectroscopically, Centaurs seem to exhibit the same variety of features as their TNO progenitors, except for methane, seen only among the large TNOs. Water ice has been detected on (2060) Chiron, (8405) Asbolus, (10199) Chariklo, (31824) Elatus, (32532) Thereus, (52872) Okyrhoe, (55576) Amycus, and 2007 UM_{126}, and methanol has been detected on (5145) Pholus, (54598) Bienor, (83982) Crantor, and 2008 FC_{76} (see Table 14.1).

14.4.3 Rotations, shapes, and sizes

14.4.3.1 Light curves

The rotational brightness variation, commonly named *light curve*, is the periodic variation of an object brightness as a function of time due to its rotation. The light curve can be produced by various mechanisms: (i) albedo variations on the body surface, (ii) elongated shape, and/or (iii) contact/close binary (e.g., Sheppard and Jewitt, 2002; Lacerda, 2005). Therefore, light curves carry a lot of information about the rotational and physical properties of small bodies.

The first two basic parameters derived from a light curve are the rotational period of the object (P), and the peak-to-peak light curve amplitude (Δm). Assuming a triaxial object in hydrostatic equilibrium, one can estimate the lower limit to its density using the figures of equilibrium from Chandrasekhar (1987). Assuming the light curve is dominated by the shape of the object, one can estimate its elongation assuming a certain viewing angle. A light curve can also be caused by a spherical object with albedo variation(s) on its surface and so one can extract constraints. Finally, a light curve with a U- and V-shape at the maximum and minimum of brightness and $\Delta m > 0.9$ mag is likely due to a contact/close binary, and thus we can infer the binarity of the object and derive the physical characteristics of both system's components (Leone et al., 1984; Lacerda et al., 2014b; Thirouin and Sheppard, 2018).

Only ~5% of the known Centaurs have been observed for complete/partial light curves (Table 14.1) showing, nonetheless, a large variety of amplitudes and rotational periods. From Maxwellian fits the mean rotation period of Centaurs is 8.1 hours, being slower than the mean period of TNOs, that is, 8.45 hours (Thirouin et al., 2016; Thirouin and Sheppard, 2019). The mean amplitude is ~0.15 mag. Because of Centaurs relatively short orbital periods, compared to the TNOs, the viewing angle between the rotation axis and the line of sight will change significantly over the course of a couple of years, and thus the light curve amplitude will vary, as seen for Bienor and Pholus (Tegler et al., 2005a; Fernández-Valenzuela et al., 2017). In such cases, it is possible to derive the object's pole orientation. So far, no strong correlations between P and Δm and orbital parameters have been detected among Centaurs (Duffard et al., 2009). But the correlation between light curve amplitude and absolute magnitude (smaller objects are more elongated than the larger ones) noticed among TNOs is not present in the Centaur population which infer for a different evolution/collisional history (Duffard et al., 2009; Thirouin, 2013; Benecchi and Sheppard, 2013).

14.4.3.2 Size distribution

The radius of an object (in kilometer) is given by: $p_R \times R^2 = 2.24 \times 10^{16} \times 10^{0.4\times(R_{\odot}-H_R)}$, where $R_{\odot}$ is the R-filter magnitude of the Sun and p_R and H_R are, respectively, the geometric albedo and the absolute magnitude of the object in the R-filter (Russell, 1916).

Several techniques can be used to derive or constrain the albedo and absolute magnitude and by extension the size of an object. Stellar occultations can be used to derive shape, size, albedo, detection of rings and satellites, and constraints about atmosphere, such as for Chiron and Chariklo (Braga-Ribas et al., 2014; Ruprecht et al., 2015; Ortiz et al., 2015). With data from the Spitzer Space Telescope, the Herschel Space Observatory, the Wide-field Infrared Survey Explorer mission, and the Atacama Large Millimeter Array, it has been possible to perform thermal modeling in order to estimate the radiometric size and albedo (Stansberry et al., 2008; Lellouch et al., 2013, 2017; Bauer et al., 2013; Lacerda et al., 2014a; Duffard et al., 2014). Generally, Centaurs have a low albedo, of a few percent, and only a handful have an albedo up to ~20% (Duffard et al., 2014; Lacerda et al., 2014a; Lellouch et al., 2017). Also, it was noticed that the mean albedo of the red Centaurs is ~8%−12%, whereas for the gray ones is lower ~5%−6% (Bauer et al., 2013; Duffard et al., 2014).

Several surveys have been dedicated to the search of small bodies (e.g., Petit et al., 2008; Trujillo, 2008; Adams et al., 2014; Weryk et al., 2019; Sheppard et al., 2016; Bannister et al., 2018). However, these surveys are mostly focused on the search of TNOs or Near-Earth Asteroids (NEOs) and the discovery of Centaurs is mainly serendipitous. Size distributions have been estimated, but they are not debiased (Bauer et al., 2013), or used either the maximum likelihood technique (Jedicke and Herron, 1997) or the Monte Carlo simulations (Sheppard et al., 2000)—these two, however, find a similar distribution with $\alpha \sim 0.6$. Adams et al. (2014) proposed a debiased size distribution, using DES data, and inferred that the best fit for Centaurs with absolute magnitudes $7.5 < H < 11$ mag. is a power law with $\alpha = 0.42 \pm 0.02$ suggesting a knee in the size distribution around $H = 7.2$ mag. Similarly, using the OSSOS simulator, Lawler et al. (2018) confirmed that a single slope for the size distribution is not able to match the observations, and thus a break is needed at smaller sizes. Such a break is also required to explain the size distribution of the TNOs (Bernstein et al., 2004; Shankman et al., 2013).

TABLE 14.1 B–R colors of "strict" Centaurs from TRC survey and their albedos, light curves studies, and detected ices from other works.

Object	B–R	Albedo	Single peak *P* (hours)	Double peak *P* (hours)	Δm (mag)	Ices?	References
29P Schwassmann-Wachmann 1	–	$4.61^{+5.22}_{-1.90}$	–	–	–	–	S08
(2060) 1977 UB Chiron	–	$7.57^{+1.03}_{-0.87}$	–	5.9180 ± 0.0001	0.088 ± 0.003	Water	B89, O15, G16, SS08
(5145) 1992 AD Pholus	1.97 ± 0.11	$15.5^{+7.6}_{-4.9}$	–	9.98	0.15/0.60	Methanol	B92, H92, F01, T05, D14
	–	–	–	–	0.15	–	RT99
(7066) 1993 HA_2 Nessus	–	$8.6^{+7.5}_{-3.4}$	–	–	0.5	–	RT99
	–	–	–	–	<0.2	–	D98, D14
	–	–	5.68 ± 0.19	–	0.76 ± 0.08	–	K06
1994 TA	1.92 ± 0.10	–	–	–	–	–	TR00
(10370) 1995 DW_2 Hylonome	1.16 ± 0.09	$5.1^{+3.0}_{-1.7}$	–	–	–	–	TR98, RT99, D14
(8405) 1995 GO Asbolus	1.22 ± 0.05	$5.6^{+1.9}_{-1.5}$	–	8.9351 ± 0.003	0.55	Water	D98, K00, D14
	–	–	–	–	0.34	–	RT99
	–	–	–	–	0.14 ± 0.10	–	R07
(10199) 1997 CU_{26} Chariklo	1.25 ± 0.05	$5.63^{+0.76}_{-0.65}$	–	–	<0.1	Water	D98, TR98, RT99, S08
	–	$5.59^{+1.69}_{-1.17}$	–	7.004 ± 0.036	0.11	–	F14, G16
(49036) 1998 QM_{107} Pelion	1.25 ± 0.04	–	–	–	–	–	TR00
(52872) 1998 SG_{35} Okyrhoe	1.21 ± 0.02	$5.6^{+1.2}_{-1.0}$	–	16.6	0.2	Water	B03, D14, TR03
	–	–	4.68 or 6.08	–	0.07 ± 0.01	–	T10
(52975) 1998 TF_{35} Cyllarus	1.72 ± 0.12	$13.9^{+15.7}_{-6.4}$	–	–	–	–	D14, TRC03
(29981) 1999 TD_{10}	–	$4.40^{+1.42}_{-0.96}$	–	–	–	–	S08
(31824) 1999 UG_5 Elatus	1.75 ± 0.05	$4.9^{+2.8}_{-1.6}$	13.25	–	0.24	Water	D14, TR03, G01
		–	13.41 ± 0.04	–	0.102 ± 0.005	–	B02
(60558) 2000 EC_{98} Echeclus	1.39 ± 0.04	$5.2^{+0.70}_{-0.71}$	13.401	26.802	0.24 ± 0.06	–	R05, D14

Continued

TABLE 14.1 B–R colors of "strict" Centaurs from TRC survey and their albedos, light curves studies, and detected ices from other works—cont'd

Object	B–R	Albedo	Single peak P (hours)	Double peak P (hours)	Δm (mag)	Ices?	References
2000 FZ_{53}	1.17 ± 0.05	–	–	–	–	–	TRC03
2000 GM_{137}	–	$4.3^{+2.6}_{-1.6}$	–	–	–	–	D14
(54598) 2000 QC_{243} Bienor	1.15 ± 0.08	$4.3^{+1.6}_{-1.2}$	4.57 ± 0.02	9.14 ± 0.04	0.08 to 0.75	Methanol	O03, F17, D14
		–	–	9.17	0.34 ± 0.08	–	R07
		–	(4.723 or 4.594) ± 0.001	–	0.38 ± 0.02	–	K06
(63252) 2001 BL_{41}	1.21 ± 0.03	$4.3^{+2.8}_{-1.4}$	–	–	–	–	D09, TRC03
(88269) 2001 KF_{77}	1.81 ± 0.04	–	–	–	–	–	TRC03
(32532) 2001 PT_{13} Thereus	1.18 ± 0.01	$8.3^{+1.6}_{-1.3}$	4.1546 ± 0.0005	8.3091 ± 0.0001	0.16 ± 0.02	Water	O03, D14
	–	–	–	8.34	0.16	–	FD03
	–	–	–	8.34	0.34 ± 0.08	–	R07
		–	–	8.4	0.15	–	F01
(119315) 2001 SQ_{73}	1.13 ± 0.02	$4.8^{+3.0}_{-1.8}$	–	–	–	–	D14, TRC03
(148975) 2001 XA_{255}	–	–	–	–	~0.13	–	T13
	–	–	>7	>14	~0.2	–	H18
2001 XZ_{255}	1.91 ± 0.07	–	–	–	–	–	TRC03
(42355) 2002 CR_{46} Typhon	–	$5.09^{+1.24}_{-0.80}$	9.67	–	0.07 ± 0.01	–	T10
(55567) 2002 GB_{10} Amycus	1.82 ± 0.03	$8.3^{+1.6}_{-1.5}$	9.76	–	0.16 ± 0.01	Water	TRC03, T10
(83982) 2002 GO_9 Crantor	1.86 ± 0.02	$11.8^{+7.09}_{-?}$	(6.97 or 9.67) ± 0.03	–	0.14 ± 0.04	Methanol	O03, TR03, S08
	–	–	–	–	0.34	–	RT99
(95626) 2002 GZ_{32}	1.03 ± 0.04	–	–	5.80 ± 0.03	0.15 ± 0.03	–	D08
(250112) 2002 KY_{14}	1.75 ± 0.02	–	–	–	–	–	TRC16
(73480) 2002 PN_{34}	–	$4.25^{+0.83}_{-0.65}$	–	–	–	–	S08
2002 PQ_{152}	1.85 ± 0.05	–	–	–	–	–	TRC16

(523597) 2002 QX_{47}	1.08 ± 0.04	–	–	–	–	–	TRC16
(119976) 2002 VR_{130}	–	$9.3^{+6.6}_{-3.6}$	–	–	–	–	D14
(120061) 2003 CO_1	–	$4.9^{+0.5}_{-0.6}$	3.53/4.13/4.99 ± 0.01/6.30	–	0.10 ± 0.05	–	O06, D14
	–	–	4.51	–	0.06 ± 0.01	–	T10
(65489) 2003 FX_{128} Ceto	–	$7.67^{+1.38}_{-1.10}$	–	4.43 ± 0.03	0.13 ± 0.02	–	D08, S08
(136204) 2003 WL_7	1.23 ± 0.04	5.3 ± 1.0	8.24	–	0.04 ± 0.01	–	T10, TRC16, D14
2004 QQ_{26}	–	$4.4^{+3.9}_{-1.4}$	–	–	–	–	D14
(447178) 2005 RO_{43}	1.24 ± 0.03	$5.6^{+3.6}_{-2.1}$	–	–	–	–	D14, TRC16
(145486) 2005 UJ_{438}	1.64 ± 0.04	$25.6^{+9.7}_{-7.6}$	8.32	–	0.11 ± 0.01	–	T10, TRC16, D14
(248835) 2006 SX_{368}	1.22 ± 0.02	$5.2^{+0.7}_{-0.6}$	–	–	–	–	TRC16, D14
(309139) 2006 XQ_{51}	1.15 ± 0.03	–	–	–	–	–	TRC16
(341275) 2007 RG_{283}	1.26 ± 0.03	–	–	–	–	–	TRC16
2007 RH_{283}	1.15 ± 0.03	–	–	–	–	–	TRC16
2007 TK_{422}	1.22 ± 0.04	–	–	–	–	–	TRC16
(25012) 2007 UL_{126}	1.75 ± 0.02	$5.7^{+1.1}_{-0.7}$	3.56 or 4.2	7.12 or 8.4	0.11 ± 0.01	–	T10
2007 UM_{126}	1.13 ± 0.03	–	–	–	–	Water	TRC16
2007 VH_{305}	1.18 ± 0.02	–	–	–	–	–	TRC16
(281371) 2008 FC_{76}	1.60 ± 0.03	$6.7^{+1.7}_{-1.1}$	–	–	~0.1	Methanol	T13, TRC16, D14
	–	–	5.93 ± 0.05	11.86 ± 0.05	0.04 ± 0.01	–	H18
(315898) 2008 QD_4	1.20 ± 0.02	–	–	–	~0.09	–	T13, TRC16
	–	–	>7	>14	~0.15	–	H18
(309737) 2008 SJ_{238}	1.60 ± 0.02	–	–	–	–	–	TRC16

Continued

TABLE 14.1 B–R colors of "strict" Centaurs from TRC survey and their albedos, light curves studies, and detected ices from other works—cont'd

Object	B–R	Albedo	Single peak *P* (hours)	Double peak *P* (hours)	Δm (mag)	Ices?	References
(309741) 2008 UZ_6	1.52 ± 0.04	–	–	–	–	–	TRC16
(342842) 2008 YB_3	1.23 ± 0.02	–	–	–	~0.18	–	T13, TRC16
(346889) 2009 QV_{38} Rhiphonos	1.37 ± 0.02	–	–	–	–	–	TRC16
(349933) 2009 YF_7	1.18 ± 0.03	–	–	–	–	–	TRC16
2010 BK_{118}	–	–	–	–	~0.15	–	T13
(382004) 2010 RM_{64}	1.56 ± 0.02	–	–	–	–	–	TRC16
2010 TH	1.18 ± 0.03	–	–	–	–	–	TRC16
2010 TY_{53}	–	–	–	–	<0.14	–	BS13
(471339) 2011 ON_{45}	1.81 ± 0.04	–	–	–	–	–	TRC16
(449097) 2012 UT_{68}	1.68 ± 0.03	–	–	–	–	–	TRC16
(463368) 2012 VU_{85}	1.70 ± 0.07	–	–	–	–	–	TRC16
(523676) 2013 UL_{10}	1.17 ± 0.03	–	–	–	–	–	TRC16
(459865) 2013 XZ_8	1.17 ± 0.03	–	–	–	–	–	TRC16
(459971) 2014 ON_6	1.55 ± 0.03	–	–	–	–	–	TRC16

Notes: Tegler, Romanishin, and Consolmagno's (TRC) survey: http://www.physics.nau.edu/~tegler/research/survey.htm.
B89: Bus et al. (1989); B92: Buie and Bus (1992); B02: Bauer et al. (2002); B03: Bauer et al. (2003); BS13: Benecchi and Sheppard (2013); D98: Davies et al. (1998); D08: Dotto et al. (2008); D09: Duffard et al. (2009); D14: Duffard et al. (2014); F01: Farnham (2001); F14: Fornasier et al. (2014); F17: Fernández et al. (2017); FD03: Farnham and Davies (2003); G01: Gutiérrez et al. (2001); G16: Galiazzo et al. (2016); H92: Hoffmann et al. (1992); H18: Hromakina et al. (2018); K00: Kern et al. (2000); K06: Kern (2006); O03: Ortiz et al. (2003); O06: Ortiz et al. (2006); O15: Ortiz et al. (2015); R05: Rousselot et al. (2005); R07: Rabinowitz et al. (2007); RT99: Romanishin and Tegler (1999); S08: Stansberry et al. (2008); SJ02: Sheppard and Jewitt (2002); T05: Tegler et al. (2005b); T10: Thirouin et al. (2010); T13: Thirouin (2013); TR00: Tegler and Romanishin (2000); TR03: Tegler and Romanishin (2003); TRC03: Tegler et al. (2003); TRC16: Tegler et al. (2016).

14.4.4 Moons and rings

Binary and multiple systems have been found across many minor body populations. The Centaurs are no exception with two known binary systems: Ceto/Phorcys—formerly, (65489) 2003 FX_{128}—and Typhon/Echidna—formerly, (42355) 2002 CR_{46}. In the case of Ceto/Phorcys, the components have a radii of 87 and 66 km and are separated by about 1800 km and the system has a density of $\sim$1.4 g cm^{-3} (Grundy et al., 2007). Typhon/Echidna has a similar semimajor axis of $\sim$1600 km and the radii of the components are 76 and 42 km, with a density of only 0.44 g cm^{-3} (Grundy et al., 2008). Despite finding many binary systems among TNOs, only two were found among Centaurs and there is no evidence of contact binary Centaurs so far. But, it is important to keep in mind that the Centaurs have not been the topic of any large-scale survey for light curve studies.

Recently, to everyone's surprise, thanks to stellar occultations, a ring system was discovered around the largest Centaur: Chariklo (Braga-Ribas et al., 2014). Then, Ortiz et al. (2015), Ruprecht et al. (2015), and Fernández-Valenzuela et al. (2017) evaluated the possibility of ring systems around Chiron and Bienor. The formation and stability of ring systems around objects in the size scale of $\sim$100 km is still a mystery. Several hypotheses have been considered to explain the formation of these rings: debris disks from a rotational fission, debris disks due to the disruption of a satellite because of impact, and/or tidal forces among other options (Hyodo et al., 2016; Pan and Wu, 2016). Finally, their stability is also an open question (Sicardy et al., 2018).

14.4.5 Cometary activity

As already mentioned, Chiron, the first object identified as a Centaur, was found to exhibit magnitude variations on short timescales. That the first Centaur discovered was an active cometary body was no coincidence. As we saw in Section 14.3, the dynamical evidence for a link between Centaurs and JFCs is compelling.

From the $\sim$290–360 Centaurs known, depending on the classification, 29 have shown evidence of activity, that is, $\sim$8%–9%. In addition to the dynamical evidence, the Centaur population has the largest fraction of objects with confirmed activity relative to other populations, besides those of actual comet populations, evidently. Neither main-belt comets—a.k.a. active asteroids—nor NEOs approach the fraction of active bodies embedded within their dynamically defined populations that the Centaurs do. Many of the unique attributes of the Centaur population may be linked with their cometary activity and behavior, and may suggest that activity among the Centaurs is more common than the fraction observed at a given epoch. Several attempts have been made to link the bimodal color distribution with current and past activity of Centaurs as we saw in Section 14.4.2.1. A significant color-based correlation with Centaurs' albedos has been established (Stansberry et al., 2008; Tegler et al., 2008; Bauer et al., 2013; Lacerda et al., 2014a). The darker population matching the neutrally colored populations marks a convergence similar to comet surfaces. In fact, all small body populations nearer the Sun are composed entirely of dark and neutrally colored objects. A plausible explanation for the lack of red and bright surfaces among Centaurs and JFCs is the destruction of these very same red surfaces when entering the inner solar system over a lifetime comparable to that of Centaurs (Jewitt, 2002), which may even take less than 10 years when Centaurs get $q < \sim 10$ au

(Jewitt, 2015), but TNOs show that something more complex is going on. As if doubts existed still, the match between Centaur and JFC size distribution slopes (Fernández et al., 2013; Bauer et al., 2013) also evidences a probably similar origin or evolution.

In fact, technically, Chiron was not the first Centaur discovered, rather, that distinction falls to a more active body, 29P/Schwassmann-Wachmann 1, discovered in 1927. The pattern of behavior of active Centaurs varies with each Centaur. While Chiron's activity varies over large timescales (Duffard et al., 2002, and references therein) Echeclus' outbursts are shorter lived, but more periodic (Choi et al., 2006; Jaeger et al., 2011; Seccull et al., 2019), while 29P's activity is constant, though punctuated with outbursts. Jewitt (2009, 2015) identified that active Centaurs tend to have lower perihelion (see Fig. 14.1), though notable exceptions, like Chiron, exist. In fact, Fernández et al. (2018) found that active and inactive Centaurs have experienced different dynamical evolutions. They suggest that inactive Centaurs deactivated because of a larger permanence far from the Sun, while a recent drastic drop in perihelion distance could be responsible for renewing the activity in the active ones. The brightness variations of the other 26 are lower but morphological features are still identifiable (Jewitt, 2009). Brightness variations and extended morphology are not the sole indicators of possible activity. Larger bodies may exhibit temporary ring structures and dust grain atmospheres as we have seen over Chariklo and Chiron. No outburst has yet been detected on Chariklo but, as mentioned in Section 14.4.4, such kind of events may both be the origin and eventual demise of the rings. The structure surrounding Chiron could also be a manifestation of, or evolve from, a ballistic dust atmosphere (Meech and Belton, 1990).

The activity in Centaurs tends to be recurrent but, unlike standard cometary behavior, Echeclus, 29P, and Chiron all exhibit outbursts and coma decoupled from the Centaur's perihelion. The mechanism that triggers the sublimation is not clear. It may be that localized regions of surface ice are directly exposed through a combination of seasonal and diurnal variations on such bodies, rather than gases being released from the subsurface, or that volatiles may be driven from water ice via the amorphous-crystalline transition as the thermal wave reaches localized subsurface water ice reservoirs (see Prialnik, 1992; Jewitt, 2009, and Section 14.4.6). Dynamical models show that excursions of Centaurs over the inner solar system and subsequent returns to Centaur orbits are common (e.g., Horner et al., 2004a). Deriving any notion of how volatiles, or hypervolatiles, are distributed on Centaurs must accommodate the perseverance of a fraction of the ices throughout such terrestrial-planet-region excursions, possibly in subsurface reservoirs that may later be exposed.

14.4.6 Interiors

The internal structure of Centaurs is, in theory, the result of processes that occurred at three different stages of their history (see Prialnik et al., 2008; Coradini et al., 2008; Sarid and Prialnik, 2009; Guilbert-Lepoutre et al., 2015, for reviews). First, the internal structure may reflect the formation of Centaurs themselves, a process which is not fully constrained yet. Some Centaurs could be primordial aggregates of planetesimals relatively unaltered by collisions (Belton, 2014), while some others could be the outcomes of disruptive collision of larger bodies. Indeed, Stern (1996) and Davis and Farinella (1997) suggested that JFCs—therefore, Centaurs too—could be fragments from parent bodies with initial sizes up to ~100 km. The "outbursty" activity of Echeclus could be interpreted as being due to internal heterogeneities, possibly reflecting its rubble-pile nature (either primordial or collisional, see Rousselot et al., 2016). In fact, Rosenberg and Prialnik (2010) showed that internal heterogeneities could potentially be detected through the activity pattern of a comet, where erratic behaviors would be induced by blocks of varying thermophysical properties randomly distributed inside the nucleus.

The second stage of processing is related to radiogenic heating, especially their early heating. The effectiveness of ^{26}Al in heating the interior of an icy object strongly depends on the time delay between its formation and the formation of calcium-aluminum-rich inclusions (CAIs) (see Prialnik and Podolak, 1995; Prialnik et al., 2008), which remains unknown. Prialnik et al. (1987) found that early radiogenic heating by ^{26}Al could lead to the complete crystallization of amorphous ice inside comets and Centaurs, and even melting of large nuclei. Therefore, the early thermal evolution of Centaurs could have led to phase transitions and chemical differentiation occurring mostly in the core, while an outer layer might remain pristine. We should stress that the internal structure inherited from the first two stages remain speculative: the effects of accretion could be inferred from size, shape, or within the activity pattern, but most observations cannot provide us with this information. Early heating should have mostly affected the core of icy objects, which we cannot access from remote observations of their surface. In addition, the third stage, dominated by cometary activity, could erase many of the effects of the two previous stages.

Whatever stratigraphy is produced after these early stages of formation and evolution, the internal structure of Centaurs should be further modified by cometary activity. Water ice is thermodynamically stable in the giant planet region, that is, it does not sublimate, hence it cannot drive the activity of Centaurs. Jewitt (2009) suggested that crystalline-amorphous water ice transition, releasing trapped volatiles, could be the source of Centaurs activity. Guilbert-Lepoutre (2012) studied this scenario concluding that: (i) crystallization at the surface of Centaurs can be triggered at heliocentric distances as large as 16 AU; (ii) cometary activity due to crystallization could be triggered up to ~10–12 AU, which is broadly consistent with the observations of active Centaurs; and (iii) as the crystallization front propagates from the surface toward the interior of Centaurs, cometary activity should stop in a relatively short period of time—outgassing is typically sustained for hundreds to thousands of years. The simple fact that active Centaurs do exist shows that a chemical differentiation, due to thermophysical processing, is happening in the giant planet region.

14.5 Discussion

If Centaurs were discovered later than TNOs, they probably would not have been named as a class and would be just seen as low perihelion SDOs. Yet, they do not cease to surprise and puzzle us revealing themselves to be as rich as TNOs alone.

Identifying the important large-scale processes in the outer solar system is of the utmost importance to understand how the entire solar system has formed and how it has evolved. Centaurs are an essential class of minor bodies to help us do so. No longer simply because they are closer, and therefore brighter, than TNOs, allowing for more detailed observational studies, but also because they exhibit peculiar and distinct behaviors that seem to be at the core of the deep physical and chemical changes that affect icy bodies when entering the inner solar system.

The condition of transition phase between TNOs and JFCs attributed to Centaurs seems unquestionable. However, if dynamical models predict that orbital inclinations are not preserved among Centaurs, how come red Centaurs have a smaller orbital inclination angle distribution than the neutral ones? Are dynamical models that reliable? If Centaurs are scattered inwards from the Trans-Neptunian region through interaction with Neptune, how can they possibly have a different collisional history than TNOs? Indeed puzzling, but only with more data we can make it indisputable. There are no red surfaces among JFCs, therefore, red Centaurs must change their colors when becoming JFCs. Is it not a remarkable coincidence that neutral Centaurs, that should have formed beyond Jupiter in the primordial disk, do not change their colors when becoming JFCs? Too remarkable to leave us comfortable. Models for scenarios of distinct compositions over the primordial planetesimal disk of icy bodies seem increasingly more sound and refined, and yet the evidence for the two distinct color groups of Centaurs has faded. Is it not a paradox? Bearing in mind the principle of explosion, that is, from a false premise anything follows, models, however, elegant they may be, need to be evidence based. It is through observational data they can be demonstrated or refuted. Paradoxically, about 20 years ago we had a seemingly elegant model capable of explaining a continuous color distribution and colors started to appear noncontinuous, whereas today colors might be continuous and we have an apparently elegant model capable of explaining their discontinuity. Confirming or refuting is imperative. Consensus is not a demonstration.

Acknowledgments

CITEUC is funded by National Funds through FCT—Foundation for Science and Technology (project: UID/Multi/00611/2019) and FEDER—European Regional Development Fund through COMPETE 2020—Operational Program Competitiveness and Internationalization (project: POCI-01-0145-FEDER-006922). Audrey Thirouin is partly funded by Lowell Observatory funds and by the National Science Foundation, grant number AST-1734484.

References

Adams, E.R., Gulbis, A.A.S., Elliot, J.L., Benecchi, S.D., Buie, M.W., Trilling, D.E., Wasserman, L.H., 2014. De-biased populations of Kuiper belt objects from the Deep Ecliptic Survey. Astron. J. 148, 55. https://doi.org/10.1088/0004-6256/148/3/55.

Araujo, R.A.N., Galiazzo, M.A., Winter, O.C., Sfair, R., 2018. The journey of Typhon-Echidna as a binary system through the planetary region. Mon. Not. R. Astron. Soc. 476, 5323–5331. https://doi.org/10.1093/mnras/sty583.

Bailey, B.L., Malhotra, R., 2009. Two dynamical classes of Centaurs. Icarus 203, 155–163. https://doi.org/10.1016/j.icarus.2009.03.044.

Bannister, M.T., Gladman, B.J., Kavelaars, J.J., Petit, J.-M., Volk, K., Chen, Y.-T., Alexandersen, M., Gwyn, S.D.J., Schwamb, M.E., Ashton, E., Benecchi, S.D., Cabral, N., Dawson, R.I., Delsanti, A., Fraser, W.C., Granvik, M., Greenstreet, S., Guilbert-Lepoutre, A., Ip, W.-H., Jakubik, M., Jones, R.L., Kaib, N.A., Lacerda, P., Van Laerhoven, C., Lawler, S., Lehner, M.J., Lin, H.W., Lykawka, P.S., Marsset, M., Murray-Clay, R., Pike, R.E., Rousselot, P., Shankman, C., Thirouin, A., Vernazza, P., Wang, S.-Y., 2018. OSSOS. VII. 800+ Trans-Neptunian objects—the complete data release. Astrophys. J. Suppl. Ser. 236, 18. https://doi.org/10.3847/1538-4365/aab77a.

Barucci, M.A., Boehnhardt, H., Cruikshank, D.P., Morbidelli, A., Dotson, R., 2008. The Solar System Beyond Neptune. University of Arizona Press, Tucson, AZ.

Barucci, M.A., Alvarez-Candal, A., Merlin, F., Belskaya, I.N., de Bergh, C., Perna, D., DeMeo, F., Fornasier, S., 2011. New insights on ices in Centaur and Transneptunian populations. Icarus 214, 297–307. https://doi.org/10.1016/j.icarus.2011.04.019.

Bauer, J.M., Meech, K.J., Fernández, Y.R., Farnham, T.L., Roush, T.L., 2002. Observations of the Centaur 1999 UG_5: evidence of a unique outer solar system surface. Publ. Astron. Soc. Pac. 114, 1309–1321. https://doi.org/10.1086/344586.

Bauer, J.M., Meech, K.J., Fernández, Y.R., Pittichova, J., Hainaut, O.R., Boehnhardt, H., Delsanti, A.C., 2003. Physical survey of 24 Centaurs with visible photometry. Icarus 166, 195–211. https://doi.org/10.1016/j.icarus.2003.07.004.

Bauer, J.M., Grav, T., Blauvelt, E., Mainzer, A.K., Masiero, J.R., Stevenson, R., Kramer, E., Fernández, Y.R., Lisse, C.M., Cutri, R.M., Weissman, P.R., Dailey, J.W., Masci, F.J., Walker, R., Waszczak, A., Nugent, C.R., Meech, K.J., Lucas, A., Pearman, G., Wilkins, A., Watkins, J., Kulkarni, S., Wright, E.L., WISE Team, PTF Team, 2013. Centaurs and scattered disk objects in the thermal infrared: analysis of WISE/NEOWISE observations. Astrophys. J. 773, 22. https://doi.org/10.1088/0004-637X/773/1/22.

Belton, M.J.S., 2014. The size-distribution of scattered disk TNOs from that of JFCs between 0.2 and 15 km effective radius. Icarus 231, 168–182. https://doi.org/10.1016/j.icarus.2013.12.001.

Benecchi, S.D., Sheppard, S.S., 2013. Light curves of 32 large transneptunian objects. Astron. J. 145, 124. https://doi.org/10.1088/0004-6256/145/5/124.

Bernstein, G.M., Trilling, D.E., Allen, R.L., Brown, M.E., Holman, M., Malhotra, R., 2004. The size distribution of Trans-Neptunian bodies. Astron. J. 128, 1364–1390. https://doi.org/10.1086/422919.

Binzel, R.P., 1992. The optical spectrum of 5145 Pholus. Icarus 99, 238–240. https://doi.org/10.1016/0019-1035(92)90185-A.

Braga-Ribas, F., Sicardy, B., Ortiz, J.L., Snodgrass, C., Roques, F., Vieira-Martins, R., Camargo, J.I.B., Assafin, M., Duffard, R., Jehin, E., Pollock, J., Leiva, R., Emilio, M., Machado, D.I., Colazo, C., Lellouch, E., Skottfelt, J., Gillon, M., Ligier, N., Maquet, L., Benedetti-Rossi, G., Gomes, A.R., Kervella, P., Monteiro, H., Sfair, R., El Moutamid, M., Tancredi, G., Spagnotto, J., Maury, A., Morales, N., Gil-Hutton, R., Roland, S., Ceretta, A., Gu, S.-H., Wang, X.-B., Harpsøe, K., Rabus, M., Manfroid, J., Opitom, C., Vanzi, L., Mehret, L., Lorenzini, L., Schneiter, E.M., Melia, R., Lecacheux, J., Colas, F., Vachier, F., Widemann, T., Almenares, L., Sandness, R.G., Char, F., Perez, V., Lemos, P., Martinez, N., Jørgensen, U.G., Dominik, M., Roig, F., Reichart, D.E., Lacluyze, A.P., Haislip, J.B., Ivarsen, K.M., Moore, J.P., Frank, N.R., Lambas, D.G., 2014. A ring system detected around the Centaur (10199) Chariklo. Nature 508, 72–75. https://doi.org/10.1038/nature13155.

Brasser, R., Schwamb, M.E., Lykawka, P.S., Gomes, R.S., 2012. An Oort cloud origin for the high-inclination, high-perihelion Centaurs. Mon. Not. R. Astron. Soc. 420, 3396–3402. https://doi.org/10.1111/j.1365-2966.2011.20264.x.

Brown, M.E., Schaller, E.L., Fraser, W.C., 2011. A hypothesis for the color diversity of the Kuiper belt. Astrophys. J. 739, L60. https://doi.org/10.1088/2041-8205/739/2/L60.

Brunini, A., 2014. On the dynamical evolution and end states of binary Centaurs. Mon. Not. R. Astron. Soc. 437, 2297–2302. https://doi.org/10.1093/mnras/stt2045.

Buie, M.W., Bus, S.J., 1992. Physical observations of (5145) Pholus. Icarus 100, 288–294. https://doi.org/10.1016/0019-1035(92)90101-C.

Bus, S.J., Bowell, E., Harris, A.W., Hewitt, A.V., 1989. 2060 Chiron: CCD and electronographic photometry. Icarus 77, 223–238. https://doi.org/10.1016/0019-1035(89)90087-0.

Chandrasekhar, S., 1987. Ellipsoidal Figures of Equilibrium. Dover, New York, NY.

Choi, Y.-J., Weissman, P., Chesley, S., Bauer, J., Stansberry, J., Tegler, S., Romanishin, W., Consolmagno, G., 2006. Comet 174P/Echeclus. Cent. Bureau Electron. Telegr. 563, 1.

Coradini, A., Capria, M.T., de Sanctis, M.C., McKinnon, W.B., 2008. The structure of Kuiper belt bodies: link with comets. In: Barucci, M.A., Boehnhardt, H., Cruikshank, D.P., Morbidelli, A., Dotson, R. (Eds.), The Solar System Beyond Neptune. University of Arizona Press, Tucson, AZ, pp. 243–256.

Cruikshank, D.P., Roush, T.L., Bartholomew, M.J., Geballe, T.R., Pendleton, Y.J., White, S.M., Bell, J.F., Davies, J.K., Owen, T.C., de Bergh, C., Tholen, D.J., Bernstein, M.P., Brown, R.H., Tryka, K.A., Dalle Ore, C.M., 1998. The composition of centaur 5145 Pholus. Icarus 135, 389–407. https://doi.org/10.1006/icar.1998.5997.

Dalle Ore, C.M., Barucci, M.A., Emery, J.P., Cruikshank, D.P., de Bergh, C., Roush, T.L., Perna, D., Merlin, F., Dalle Ore, L.V., 2015. The composition of "ultra-red" TNOs and centaurs. Icarus 252, 311–326. https://doi.org/10.1016/j.icarus.2015.01.014.

Davies, J. (Ed.), 2001. Beyond Pluto. Cambridge University Press, Cambridge, p. 246.

Davies, J.K., Barrera, L.H. (Eds.), 2004. The First Decadal Review of the Edgeworth-Kuiper Belt. Kluwer Academic Publishers, Dordrecht (Reprinted from: Earth Moon Planets 92 (1–4), 2003).

Davies, J.K., McBride, N., Ellison, S.L., Green, S.F., Ballantyne, D.R., 1998. Visible and infrared photometry of six centaurs. Icarus 134, 213–227. https://doi.org/10.1006/icar.1998.5931.

Davis, D.R., Farinella, P., 1997. Collisional evolution of Edgeworth-Kuiper belt objects. Icarus 125, 50–60. https://doi.org/10.1006/icar.1996.5595.

de Elía, G.C., Brunini, A., di Sisto, R.P., 2008. Collisional and dynamical evolution of Plutinos. Astron. Astrophys. 490, 835–842. https://doi.org/10.1051/0004-6361:200809865.

de la Fuente Marcos, C., de la Fuente Marcos, R., 2014. Large retrograde centaurs: visitors from the Oort cloud? Astrophys. Space Sci. 352, 409–419. https://doi.org/10.1007/s10509-014-1993-9.

Delsanti, A., Hainaut, O., Jourdeuil, E., Meech, K.J., Boehnhardt, H., Barrera, L., 2004. Simultaneous visible-near IR photometric study of Kuiper belt object surfaces with the ESO/Very Large Telescopes. Astron. Astrophys. 417, 1145–1158. https://doi.org/10.1051/0004-6361:20034182.

Di Sisto, R.P., Brunini, A., 2007. The origin and distribution of the Centaur population. Icarus 190, 224–235. https://doi.org/10.1016/j.icarus.2007.02.012.

Di Sisto, R.P., Ramos, X.S., Gallardo, T., 2019. The dynamical evolution of escaped Jupiter Trojan asteroids, link to other minor body populations. Icarus 319, 828–839. https://doi.org/10.1016/j.icarus.2018.10.029.

Dotto, E., Perna, D., Barucci, M.A., Rossi, A., de Bergh, C., Doressoundiram, A., Fornasier, S., 2008. Rotational properties of Centaurs and Trans-Neptunian objects, lightcurves and densities. Astron. Astrophys. 490, 829–833. https://doi.org/10.1051/0004-6361:200809615.

Duffard, R., Lazzaro, D., Pinto, S., Carvano, J., Angeli, C., Alvarez-Candal, A., Fernández, S., 2002. New activity of Chiron: results from 5 years of photometric monitoring. Icarus 160, 44–51. https://doi.org/10.1006/icar.2002.6938.

Duffard, R., Ortiz, J.L., Thirouin, A., Santos-Sanz, P., Morales, N., 2009. Transneptunian objects and Centaurs from light curves. Astron. Astrophys. 505, 1283–1295. https://doi.org/10.1051/0004-6361/200912601.

Duffard, R., Pinilla-Alonso, N., Santos-Sanz, P., Vilenius, E., Ortiz, J.L., Mueller, T., Fornasier, S., Lellouch, E., Mommert, M., Pal, A., Kiss, C., Mueller, M., Stansberry, J., Delsanti, A., Peixinho, N., Trilling, D., 2014. "TNOs are Cool": a survey of the Trans-Neptunian region. XI. A Herschel-PACS view of 16 Centaurs. Astron. Astrophys. 564, A92. https://doi.org/10.1051/0004-6361/201322377.

Duncan, M.J., Levison, H.F., 1997. A scattered comet disk and the origin of Jupiter Family Comets. Science 276, 1670–1672. https://doi.org/10.1126/science.276.5319.1670.

Duncan, M., Quinn, T., Tremaine, S., 1988. The origin of short-period comets. Astrophys. J. 328, L69. https://doi.org/10.1086/185162.

Edgeworth, K.E., 1943. The evolution of our planetary system. J. Br. Astron. Assoc. 53, 181–188.

Edgeworth, K.E., 1949. The origin and evolution of the solar system. Mon. Not. R. Astron. Soc. 109, 600–609. https://doi.org/10.1093/mnras/109.5.600.

Elliot, J.L., Kern, S.D., Clancy, K.B., Gulbis, A.A.S., Millis, R.L., Buie, M.W., Wasserman, L.H., Chiang, E.I., Jordan, A.B., Trilling, D.E., Meech, K.J., 2005. The deep ecliptic survey: a search for Kuiper belt objects and Centaurs. II. Dynamical classification, the Kuiper belt plane, and the core population. Astron. J. 129, 1117–1162. https://doi.org/10.1086/427395.

Farnham, T.L., 2001. The rotation axis of the Centaur 5145 Pholus. Icarus 152, 238–245. https://doi.org/10.1006/icar.2001.6656.

Farnham, T.L., Davies, J.K., 2003. The rotational and physical properties of the Centaur (32532) 2001 PT_{13}. Icarus 164, 418–427. https://doi.org/10.1016/S0019-1035(03)00141-6.

Fernández, J.A., 1980. On the existence of a comet belt beyond Neptune. Mon. Not. R. Astron. Soc. 192, 481–491. https://doi.org/10.1093/mnras/192.3.481.

Fernández, Y.R., Kelley, M.S., Lamy, P.L., Toth, I., Groussin, O., Lisse, C.M., A'Hearn, M.F., Bauer, J.M., Campins, H., Fitzsimmons, A., Licandro, J., Lowry, S.C., Meech, K.J., Pittichová, J., Reach, W.T., Snodgrass, C., Weaver, H.A., 2013. Thermal properties, sizes, and size distribution of Jupiter-family cometary nuclei. Icarus 226, 1138–1170. https://doi.org/10.1016/j.icarus.2013.07.021.

Fernández, J.A., Licandro, J., Moreno, F., Sosa, A., Cabrera-Lavers, A., de León, J., Birtwhistle, P., 2017. Physical and dynamical properties of the anomalous comet 249P/LINEAR. Icarus 295, 34–45. https://doi.org/10.1016/j.icarus.2017.04.004.

Fernández, J.A., Helal, M., Gallardo, T., 2018. Dynamical evolution and end states of active and inactive Centaurs. Planet. Space Sci. 158, 6–15. https://doi.org/10.1016/j.pss.2018.05.013.

Fernández-Valenzuela, E., Ortiz, J.L., Duffard, R., Morales, N., Santos-Sanz, P., 2017. Physical properties of centaur (54598) Bienor from photometry. Mon. Not. R. Astron. Soc. 466, 4147–4158. https://doi.org/10.1093/mnras/stw3264.

Ferrari, C., 2018. Thermal properties of icy surfaces in the outer solar system. Space Sci. Rev. 214, 111. https://doi.org/10.1007/s11214-018-0546-x.

Fink, U., Hoffmann, M., Grundy, W., Hicks, M., Sears, W., 1992. The steep red spectrum of 1992 AD: an asteroid covered with organic material? Icarus 97, 145–149. https://doi.org/10.1016/0019-1035(92)90064-E.

Fornasier, S., Lazzaro, D., Alvarez-Candal, A., Snodgrass, C., Tozzi, G.P., Carvano, J.M., Jimenez-Teja, Y., Silva, J.S., Bramich, D.M., 2014. The Centaur 10199 Chariklo: investigation into rotational period, absolute magnitude, and cometary activity. Astron. Astrophys. 568, L11. https://doi.org/10.1051/0004-6361/201424439.

Fraser, W.C., Brown, M.E., 2012. The Hubble wide field camera 3 test of surfaces in the outer solar system: the compositional classes of the Kuiper belt. Astrophys. J. 749, 33. https://doi.org/10.1088/0004-637X/749/1/33.

Galiazzo, M., de la Fuente Marcos, C., de la Fuente Marcos, R., Carraro, G., Maris, M., Montalto, M., 2016. Photometry of Centaurs and Trans-Neptunian objects: 2060 Chiron (1977 UB), 10199 Chariklo (1997 CU_{26}), 38628 Huya (2000 EB_{173}), 28978 Ixion (2001 KX_{76}), and 90482 Orcus (2004 DW). Astrophys. Space Sci. 361, 212. https://doi.org/10.1007/s10509-016-2801-5.

Gladman, B., Marsden, B.G., Vanlaerhoven, C., 2008. Nomenclature in the outer solar system. In: Barucci, M.A., Boehnhardt, H., Cruikshank, D.P., Morbidelli, A., Dotson, R. (Eds.), The Solar System Beyond Neptune. University of Arizona Press, Tucson, AZ, pp. 43–57.

Grundy, W.M., Stansberry, J.A., Noll, K.S., Stephens, D.C., Trilling, D.E., Kern, S.D., Spencer, J.R., Cruikshank, D.P., Levison, H.F., 2007. The orbit, mass, size, albedo, and density of (65489) Ceto/Phorcys: a tidally-evolved binary Centaur. Icarus 191, 286–297. https://doi.org/10.1016/j.icarus.2007.04.004.

Grundy, W.M., Noll, K.S., Virtanen, J., Muinonen, K., Kern, S.D., Stephens, D.C., Stansberry, J.A., Levison, H.F., Spencer, J.R., 2008. (42355) Typhon Echidna: scheduling observations for binary orbit determination. Icarus 197, 260–268. https://doi.org/10.1016/j.icarus.2008.04.004.

Guilbert-Lepoutre, A., 2012. Survival of amorphous water ice on centaurs. Astron. J. 144, 97. https://doi.org/10.1088/0004-6256/144/4/97.

Guilbert-Lepoutre, A., Besse, S., Mousis, O., Ali-Dib, M., Höfner, S., Koschny, D., Hager, P., 2015. On the evolution of comets. Space Sci. Rev. 197, 271–296. https://doi.org/10.1007/s11214-015-0148-9.

Gutiérrez, P.J., Ortiz, J.L., Alexandrino, E., Roos-Serote, M., Doressoundiram, A., 2001. Short term variability of Centaur 1999 UG5. Astron. Astrophys. 371, L1–L4. https://doi.org/10.1051/0004-6361:20010418.

Hahn, G., Bailey, M.E., 1990. Rapid dynamical evolution of giant Comet Chiron. Nature 348, 132–136. https://doi.org/10.1038/348132a0.

Hainaut, O.R., Delsanti, A.C., 2002. Colors of minor bodies in the outer solar system. A statistical analysis. Astron. Astrophys. 389, 641–664. https://doi.org/10.1051/0004-6361:20020431.

Hainaut, O., Smette, A., Zeilinger, W., 1992. 1992 AD. Int. Astron. Union Circ. 5450.

Hapke, B., 1993. Theory of reflectance and emittance spectroscopy. In: Topics in Remote Sensing. Cambridge University Press, Cambridge.

Hartmann, W.K., Tholen, D.J., Meech, K.J., Cruikshank, D.P., 1990. 2060 Chiron: colorimetry and cometary behavior. Icarus 83, 1–15. https://doi.org/10.1016/0019-1035(90)90002-Q.

Hoffmann, M., Fink, U., Grundy, W.M., Hicks, M., 1992. Photometric and spectroscopic observations of 5145 1992 AD, In: Brahic, A., Gerard, J.-C., Surdej, J. (Eds.), Liege International Astrophysical Colloquia, vol. 30, p. 203.

Hoffmann, M., Fink, U., Grundy, W.M., Hicks, M., 1993. Photometric and spectroscopic observations of 5145 Pholus. J. Geophys. Res. 98, 7403–7407. https://doi.org/10.1029/93JE00120.

Horner, J., Evans, N.W., 2006. The capture of Centaurs as Trojans. Mon. Not. R. Astron. Soc. 367, L20–L23. https://doi.org/10.1111/j.1745-3933.2006.00131.x.

Horner, J., Lykawka, P.S., 2010. The Neptune Trojans—a new source for the Centaurs? Mon. Not. R. Astron. Soc. 402, 13–20. https://doi.org/10.1111/j.1365-2966.2009.15702.x.

Horner, J., Evans, N.W., Bailey, M.E., 2004a. Simulations of the population of Centaurs—I. The bulk statistics. Mon. Not. R. Astron. Soc. 354, 798–810. https://doi.org/10.1111/j.1365-2966.2004.08240.x.

Horner, J., Evans, N.W., Bailey, M.E., 2004b. Simulations of the population of Centaurs—II. Individual objects. Mon. Not. R. Astron. Soc. 355, 321–329. https://doi.org/10.1111/j.1365-2966.2004.08342.x.

Hromakina, T., Perna, D., Belskaya, I., Dotto, E., Rossi, A., Bisi, F., 2018. Photometric observations of nine Transneptunian objects and Centaurs. Mon. Not. R. Astron. Soc. 474, 2536–2543. https://doi.org/10.1093/mnras/stx2904.
Hyodo, R., Charnoz, S., Genda, H., Ohtsuki, K., 2016. Formation of Centaurs' rings through their partial tidal disruption during planetary encounters. Astrophys. J. 828, L8. https://doi.org/10.3847/2041-8205/828/1/L8.
Jaeger, M., Prosperi, E., Vollmann, W., Sato, H., Sostero, G., Guido, E., 2011. Comet 174P/Echeclus = (60558) Echeclus. Int. Astron. Union Circ. 9213, 2.
Jedicke, R., Herron, J.D., 1997. Observational constraints on the Centaur population. Icarus 127, 494–507. https://doi.org/10.1006/icar.1997.5696.
Jewitt, D., 1999. Kuiper belt objects. Annu. Rev. Earth Planet. Sci. 27, 287–312. https://doi.org/10.1146/annurev.earth.27.1.287.
Jewitt, D.C., 2002. From Kuiper belt object to cometary nucleus: the missing ultrared matter. Astron. J. 123, 1039–1049. https://doi.org/10.1086/338692.
Jewitt, D., 2009. The active Centaurs. Astron. J. 137, 4296–4312. https://doi.org/10.1088/0004-6256/137/5/4296.
Jewitt, D., 2015. Color systematics of Comets and related bodies. Astron. J. 150, Article ID: 201, 18 pp. https://doi.org/10.1088/0004-6256/150/6/201.
Jewitt, D., 2018. The Trojan color conundrum. Astron. J. 155, 56. https://doi.org/10.3847/1538-3881/aaa1a4.
Jewitt, D.C., Luu, J.X., 2001. Colors and spectra of Kuiper belt objects. Astron. J. 122, 2099–2114. https://doi.org/10.1086/323304.
Jewitt, D., Morbidelli, A., Rauer, H., 2008. Trans-Neptunian objects and comets. In: Saas-Fee Advanced Course 35. Swiss Society for Astrophysics and Astronomy Series. ISBN: 978-3-540-71957-1.
Kaňuchová, Z., Brunetto, R., Melita, M., Strazzulla, G., 2012. Space weathering and the color indexes of minor bodies in the outer solar system. Icarus 221, 12–19. https://doi.org/10.1016/j.icarus.2012.06.043.
Kern, S.D., 2006. A Study of Binary Kuiper Belt Objects (Ph.D. thesis). Massachusetts Institute of Technology.
Kern, S.D., McCarthy, D.W., Buie, M.W., Brown, R.H., Campins, H., Rieke, M., 2000. Compositional variation on the surface of Centaur 8405 Asbolus. Astrophys. J. 542, L155–L159. https://doi.org/10.1086/312932.
Khare, B.N., Thompson, W.R., Murray, B.G.J.P.T., Chyba, C.F., Sagan, C., Arakawa, E.T., 1989. Solid organic residues produced by irradiation of hydrocarbon-containing H_2O and H_2O/NH_3 ices: infrared spectroscopy and astronomical implications. Icarus 79, 350–361. https://doi.org/10.1016/0019-1035(89)90082-1.
Kowal, C.T., Gehrels, T., 1977. Slow-moving object Kowal. Int. Astron. Union Circ. 3129, 1.
Kresák, L., 1972. On the dividing line between cometary and asteroidal orbits, In: Chebotarev, G.A., Kazimirchak-Polonskaia, E.I., Marsden, B.G. (Eds.), The Motion, Evolution of Orbits, and Origin of Comets, IAU Symposium, vol. 45, p. 503.
Kuiper, G.P., 1951. On the origin of the solar system. Proc. Natl. Acad. Sci. USA 37, 1–14. https://doi.org/10.1073/pnas.37.1.1.
Lacerda, P., 2005. The Shapes and Spins of Kuiper Belt Objects (Ph.D. thesis). Leiden Observatory, Leiden University, Leiden, The Netherlands.
Lacerda, P., Fornasier, S., Lellouch, E., Kiss, C., Vilenius, E., Santos-Sanz, P., Rengel, M., Müller, T., Stansberry, J., Duffard, R., Delsanti, A., Guilbert-Lepoutre, A., 2014a. The Albedo-color diversity of transneptunian objects. Astrophys. J. Lett. 793, L2. https://doi.org/10.1088/2041-8205/793/1/L2.
Lacerda, P., McNeill, A., Peixinho, N., 2014b. The unusual Kuiper belt object 2003 SQ_{317}. Mon. Not. R. Astron. Soc. 437, 3824–3831. https://doi.org/10.1093/mnras/stt2180.
Lawler, S.M., Shankman, C., Kavelaars, J.J., Alexandersen, M., Bannister, M.T., Chen, Y.-T., Gladman, B., Fraser, W.C., Gwyn, S., Kaib, N., Petit, J.-M., Volk, K., 2018. OSSOS. VIII. The transition between two size distribution slopes in the scattering disk. Astron. J. 155, 197. https://doi.org/10.3847/1538-3881/aab8ff.
Lellouch, E., Santos-Sanz, P., Lacerda, P., Mommert, M., Duffard, R., Ortiz, J.L., Müller, T.G., Fornasier, S., Stansberry, J., Kiss, C., Vilenius, E., Mueller, M., Peixinho, N., Moreno, R., Groussin, O., Delsanti, A., Harris, A.W., 2013. "TNOs are Cool": a survey of the Trans-Neptunian region. IX. Thermal properties of Kuiper belt objects and Centaurs from combined Herschel and Spitzer observations. Astron. Astrophys. 557, A60. https://doi.org/10.1051/0004-6361/201322047.
Lellouch, E., Moreno, R., Müller, T., Fornasier, S., Santos-Sanz, P., Moullet, A., Gurwell, M., Stansberry, J., Leiva, R., Sicardy, B., Butler, B., Boissier, J., 2017. The thermal emission of Centaurs and Trans-Neptunian objects at millimeter wavelengths from ALMA observations. Astron. Astrophys. 608, A45. https://doi.org/10.1051/0004-6361/201731676.
Leonard, F.C., 1930. The new planet Pluto. Leaflet Astron. Soc. Pac. 1, 121.

Leone, G., Paolicchi, P., Farinella, P., Zappala, V., 1984. Equilibrium models of binary asteroids. Astron. Astrophys. 140, 265–272.

Levison, H.F., 1996. Comet taxonomy. In: Rettig, T., Hahn, J.M. (Eds.), Completing the Inventory of the Solar System, ASP Conf. Ser., vol. 107. Astronomical Society of the Pacific, pp. 173–191.

Levison, H.F., Duncan, M.J., 1997. From the Kuiper belt to Jupiter-family comets: the spatial distribution of ecliptic comets. Icarus 127, 13–32. https://doi.org/10.1006/icar.1996.5637.

Luu, J., Jewitt, D., 1996. Color diversity among the Centaurs and Kuiper belt objects. Astron. J. 112, 2310. https://doi.org/10.1086/118184.

Luu, J.X., Jewitt, D.C., 2002. Kuiper belt objects: relics from the accretion disk of the Sun. Ann. Rev. Astron. Astrophys. 40, 63–101. https://doi.org/10.1146/annurev.astro.40.060401.093818.

Meech, K.J., Belton, M.J.S., 1990. The atmosphere of 2060 Chiron. Astron. J. 100, 1323. https://doi.org/10.1086/115600.

Morbidelli, A., 1997. Chaotic diffusion and the origin of comets from the 2/3 resonance in the Kuiper belt. Icarus 127, 1–12. https://doi.org/10.1006/icar.1997.5681.

Nesvorný, D., 2018. Dynamical evolution of the early solar system. Ann. Rev. Astron. Astrophys. 56, 137–174. https://doi.org/10.1146/annurev-astro-081817-052028.

Ortiz, J.L., Gutiérrez, P.J., Casanova, V., Sota, A., 2003. A study of short term rotational variability in TNOs and Centaurs from Sierra Nevada observatory. Astron. Astrophys. 407, 1149–1155. https://doi.org/10.1051/0004-6361:20030972.

Ortiz, J.L., Gutiérrez, P.J., Santos-Sanz, P., Casanova, V., Sota, A., 2006. Short-term rotational variability of eight KBOs from Sierra Nevada Observatory. Astron. Astrophys. 447, 1131–1144. https://doi.org/10.1051/0004-6361:20053572.

Ortiz, J.L., Duffard, R., Pinilla-Alonso, N., Alvarez- Candal, A., Santos-Sanz, P., Morales, N., Fernández-Valenzuela, E., Licandro, J., Campo Bagatin, A., Thirouin, A., 2015. Possible ring material around centaur (2060) Chiron. Astron. Astrophys. 576, A18. https://doi.org/10.1051/0004-6361/201424461.

Pan, M., Wu, Y., 2016. On the mass and origin of Chariklo's rings. Astrophys. J. 821, 18. https://doi.org/10.3847/0004-637X/821/1/18.

Peixinho, N., Doressoundiram, A., Delsanti, A., Boehnhardt, H., Barucci, M.A., Belskaya, I., 2003. Reopening the TNOs color controversy: centaurs bimodality and TNOs unimodality. Astron. Astrophys. 410, L29–L32. https://doi.org/10.1051/0004-6361:20031420.

Peixinho, N., Delsanti, A., Guilbert-Lepoutre, A., Gafeira, R., Lacerda, P., 2012. The bimodal colors of Centaurs and small Kuiper belt objects. Astron. Astrophys. 546, A86. https://doi.org/10.1051/0004-6361/201219057.

Petit, J.-M., Kavelaars, J.J., Gladman, B., Loredo, T., 2008. Size distribution of multikilometer transneptunian objects. In: Barucci, M.A., Boehnhardt, H., Cruikshank, D.P., Morbidelli, A., Dotson, R. (Eds.), The Solar System Beyond Neptune. University of Arizona Press, Tucson, AZ, pp. 71–87.

Prialnik, D., 1992. Crystallization, sublimation, and gas release in the interior of a porous comet nucleus. Astrophys. J. 388, 196. https://doi.org/10.1086/171143.

Prialnik, D., Podolak, M., 1995. Radioactive heating of porous comet nuclei. Icarus 117, 420–430. https://doi.org/10.1006/icar.1995.1166.

Prialnik, D., Bar-Nun, A., Podolak, M., 1987. Radiogenic heating of comets by ^{26}Al and implications for their time of formation. Astrophys. J. 319, 993. https://doi.org/10.1086/165516.

Prialnik, D., Sarid, G., Rosenberg, E.D., Merk, R., 2008. Thermal and chemical evolution of comet nuclei and Kuiper belt objects. Space Sci. Rev. 138, 147–164. https://doi.org/10.1007/s11214-007-9301-4.

Rabinowitz, D.L., Schaefer, B.E., Tourtellotte, S.W., 2007. The diverse solar phase curves of distant icy bodies. I. Photometric observations of 18 Trans-Neptunian objects, 7 Centaurs, and Nereid. Astron. J. 133, 26–43. https://doi.org/10.1086/508931.

Romanishin, W., Tegler, S.C., 1999. Rotation rates of Kuiper-belt objects from their light curves. Nature 398, 129–132. https://doi.org/10.1038/18168.

Rosenberg, E.D., Prialnik, D., 2010. The effect of internal inhomogeneity on the activity of comet nuclei—application to Comet 67P/Churyumov-Gerasimenko. Icarus 209, 753–765. https://doi.org/10.1016/j.icarus.2010.03.037.

Rousselot, P., Petit, J.-M., Poulet, F., Sergeev, A., 2005. Photometric study of Centaur (60558) 2000 EC 98 and Trans-Neptunian object (55637) 2002 UX 25 at different phase angles. Icarus 176, 478–491. https://doi.org/10.1016/j.icarus.2005.03.001.

Rousselot, P., Korsun, P.P., Kulyk, I., Guilbert-Lepoutre, A., Petit, J.-M., 2016. A long-term follow up of 174P/Echeclus. Mon. Not. R. Astron. Soc. 462, S432–S442. https://doi.org/10.1093/mnras/stw3054.

Ruprecht, J.D., Bosh, A.S., Person, M.J., Bianco, F.B., Fulton, B.J., Gulbis, A.A.S., Bus, S.J., Zangari, A.M., 2015. 29 November 2011 stellar occultation by 2060 Chiron: symmetric jet-like features. Icarus 252, 271–276. https://doi.org/10.1016/j.icarus.2015.01.015.

Russell, H.N., 1916. On the Albedo of the planets and their satellites. Astrophys. J. 43, 173–196. https://doi.org/10.1086/142244.

Sarid, G., Prialnik, D., 2009. From KBOs to Centaurs: the thermal connection. Meteorit. Planet. Sci. 44, 1905–1916. https://doi.org/10.1111/j.1945-5100.2009.tb02000.x.

Schwamb, M.E., Fraser, W.C., Bannister, M.T., Marsset, M., Pike, R.E., Kavelaars, J.J., Benecchi, S.D., Lehner, M.J., Wang, S.-Y., Thirouin, A., Delsanti, A., Peixinho, N., Volk, K., Alexandersen, M., Chen, Y.-T., Gladman, B., Gwyn, S.D.J., Petit, J.-M., 2018. Col-OSSOS: the colours of the outer solar system origins survey. Astrophys. J. Suppl. Ser. 243, Article ID: 12, 24 pp. https://doi.org/10.3847/1538-4365/ab2194.

Scotti, J.V., Rabinowitz, D.L., Shoemaker, C.S., Shoemaker, E.M., Levy, D.H., King, T.M., Helin, E.F., Alu, J., Lawrence, K., McNaught, R.H., Frederick, L., Tholen, D., Mueller, B.E.A., 1992. 1992 AD. Int. Astron. Union Circ. 5434.

Seccull, T., Fraser, W.C., Puzia, T.H., Fitzsimmons, A., Cupani, G., 2019. 174P/Echeclus and its blue coma observed post-outburst. Astron. J. 157, 88. https://doi.org/10.3847/1538-3881/aafbe4.

Shankman, C., Gladman, B.J., Kaib, N., Kavelaars, J.J., Petit, J.M., 2013. A possible divot in the size distribution of the Kuiper belt's scattering objects. Astrophys. J. 764, L2. https://doi.org/10.1088/2041-8205/764/1/L2.

Sheppard, S.S., Jewitt, D.C., 2002. Time-resolved photometry of Kuiper belt objects: rotations, shapes, and phase functions. Astron. J. 124, 1757–1775. https://doi.org/10.1086/341954.

Sheppard, S.S., Jewitt, D.C., Trujillo, C.A., Brown, M.J.I., Ashley, M.C.B., 2000. A wide-field CCD survey for Centaurs and Kuiper belt objects. Astron. J. 120, 2687–2694. https://doi.org/10.1086/316805.

Sheppard, S.S., Trujillo, C., Tholen, D.J., 2016. Beyond the Kuiper belt edge: new high Perihelion Trans-Neptunian objects with moderate semimajor axes and eccentricities. Astrophys. J. 825, L13. https://doi.org/10.3847/2041-8205/825/1/L13.

Sicardy, B., Leiva, R., Renner, S., Roques, F., El Moutamid, M., Santos-Sanz, P., Desmars, J., 2018. Ring dynamics around non-axisymmetric bodies with application to Chariklo and Haumea. Nat. Astron. https://doi.org/10.1038/s41550-018-0616-8.

Stansberry, J., Grundy, W., Brown, M., Cruikshank, D., Spencer, J., Trilling, D., Margot, J.-L., 2008. Physical properties of Kuiper belt and centaur objects: constraints from the Spitzer Space Telescope. In: Barucci, M.A., Boehnhardt, H., Cruikshank, D.P., Morbidelli, A., Dotson, R. (Eds.), The Solar System Beyond Neptune. University of Arizona Press, Tucson, AZ, pp. 161–179.

Stern, S.A., 1996. Signatures of collisions in the Kuiper disk. Astron. Astrophys. 310, 999–1010.

Tegler, S.C., Romanishin, W., 1998. Two distinct populations of Kuiper-belt objects. Nature 392, 49–51. https://doi.org/10.1038/32108.

Tegler, S.C., Romanishin, W., 2000. Extremely red Kuiper-belt objects in near-circular orbits beyond 40 AU. Nature 407, 979–981. https://doi.org/10.1038/35039572.

Tegler, S.C., Romanishin, W., 2003. Resolution of the Kuiper belt object color controversy: two distinct color populations. Icarus 161, 181–191. https://doi.org/10.1016/S0019-1035(02)00021-0.

Tegler, S.C., Romanishin, W., Consolmagno, G.J., 2003. Color patterns in the Kuiper belt: a possible primordial origin. Astrophys. J. 599, L49–L52. https://doi.org/10.1086/381076.

Tegler, S.C., Romanishin, W., Consolmagno, G.J., Rall, J., Worhatch, R., Nelson, M., Weidenschilling, S., 2005a. The period of rotation, shape, density, and homogeneous surface color of the Centaur 5145 Pholus. Icarus 175, 390–396. https://doi.org/10.1016/j.icarus.2004.12.011.

Tegler, S.C., Romanishin, W., Consolmagno, G.J., Rall, J., Worhatch, R., Nelson, M., Weidenschilling, S., 2005b. The period of rotation, shape, density, and homogeneous surface color of the Centaur 5145 Pholus. Icarus 175, 390–396. https://doi.org/10.1016/j.icarus.2004.12.011.

Tegler, S.C., Bauer, J.M., Romanishin, W., Peixinho, N., 2008. Colors of Centaurs. In: Barucci, M.A., Boehnhardt, H., Cruikshank, D.P., Morbidelli, A., Dotson, R. (Eds.), The Solar System Beyond Neptune. University of Arizona Press, Tucson, AZ, pp. 105–114.

Tegler, S.C., Romanishin, W., Consolmagno, G.J., J., S., 2016. Two color populations of Kuiper belt and Centaur objects and the smaller orbital inclinations of red Centaur objects. Astron. J. 152, 210. https://doi.org/10.3847/0004-6256/152/6/210.

Thébault, P., 2003. A numerical check of the collisional resurfacing scenario. Earth Moon Planets 92, 233–241. https://doi.org/10.1023/B:MOON.0000031941.77871.c5.

Thébault, P., Doressoundiram, A., 2003. Colors and collision rates within the Kuiper belt: problems with the collisional resurfacing scenario. Icarus 162, 27–37. https://doi.org/10.1016/S0019-1035(03)00004-6.

Thirouin, A., 2013. Study of trans-neptunian objects using photometric techniques and numerical simulations (Ph.D. thesis). University of Granada.

Thirouin, A., Sheppard, S.S., 2018. The Plutino population: an abundance of contact binaries. Astron. J. 155, 248. https://doi.org/10.3847/1538-3881/aac0ff.

Thirouin, A., Sheppard, S.S., 2019. Lightcurves and rotational properties of the pristine cold classical Kuiper belt objects. Astron. J. 157, Article ID: 228, 12 pp. https://doi.org/10.3847/1538-3881/ab18a9.

Thirouin, A., Ortiz, J.L., Duffard, R., Santos-Sanz, P., Aceituno, F.J., Morales, N., 2010. Short-term variability of a sample of 29 Trans-Neptunian objects and Centaurs. Astron. Astrophys. 522, A93. https://doi.org/10.1051/0004-6361/200912340.

Thirouin, A., Sheppard, S.S., Noll, K.S., Moskovitz, N.A., Ortiz, J.L., Doressoundiram, A., 2016. Rotational properties of the Haumea family members and candidates: short-term variability. Astron. J. 151, 148. https://doi.org/10.3847/0004-6256/151/6/148.

Tholen, D.J., Hartmann, W.K., Cruikshank, D.P., Lilly, S., Bowell, E., Hewitt, A., 1988. (2060) Chiron. Int. Astron. Union Circ. 4554, 2.

Tiscareno, M.S., Malhotra, R., 2003. The dynamics of known Centaurs. Astron. J. 126, 3122–3131. https://doi.org/10.1086/379554.

Tiscareno, M.S., Malhotra, R., 2009. Chaotic diffusion of resonant Kuiper belt objects. Astron. J. 138, 827–837. https://doi.org/10.1088/0004-6256/138/3/827.

Tisserand, F., 1896. Traité de mécanique céleste, t. IV. In-4. Gauthier-Villars et fils.

Trujillo, C.A., 2008. Future surveys of the Kuiper belt. In: Barucci, M.A., Boehnhardt, H., Cruikshank, D.P., Morbidelli, A., Dotson, R. (Eds.), The Solar System Beyond Neptune. University of Arizona Press, Tucson, AZ, pp. 573–585.

Volk, K., Malhotra, R., 2008. The scattered disk as the source of the Jupiter family comets. Astrophys. J. 687, 714–725. https://doi.org/10.1086/591839.

Volk, K., Malhotra, R., 2013. Do Centaurs preserve their source inclinations? Icarus 224, 66–73. https://doi.org/10.1016/j.icarus.2013.02.016.

Weryk, R.J., Schunova-Lilly, E., Jedicke, R., Fraser, W.C., Volk, K., Wainscoat, R.J., 2019. The Population of Trans-Neptunian Objects Detected by Pan-STARRS1 (in preparation).

Wong, I., Brown, M.E., 2016. A hypothesis for the color bimodality of Jupiter Trojans. Astron. J. 152, 90. https://doi.org/10.3847/0004-6256/152/4/90.

Wong, I., Brown, M.E., 2017. The bimodal color distribution of small Kuiper belt objects. Astron. J. 153, 145. https://doi.org/10.3847/1538-3881/aa60c3.

Yu, Q., Tremaine, S., 1999. The dynamics of Plutinos. Astron. J. 118, 1873–1881. https://doi.org/10.1086/301045.

CHAPTER

15

On the dynamics of comets in extrasolar planetary systems

Rudolf Dvorak[a], *Birgit Loibnegger*[a], *Manfred Cuntz*[b]

[a]Institute of Astronomy, University of Vienna, Vienna, Austria [b]Department of Physics, University of Texas at Arlington, Arlington, TX, United States

15.1 Introduction

For thousands of years the unpredictability of the appearance of comets contrary to the motion of the planets, the Sun, and the Moon led astronomers to the assumption that they originate within the Earth's atmosphere. Thanks to Tycho Brahe, who measured their parallax at the end of the 16th century, it could be established that these objects originate as far away as at least four times the distance to our Moon. Using Newton's laws, Sir Edmond Halley was even able to determine the orbit of a comet and predicted its reappearance after a period of 76 years. Before that comets were thought to be heavenly omens for disaster and catastrophes on Earth. Only recently, based on spectroscopic measurements, it was found that their role in the development of life on Earth is essential: Although the Earth may have been formed with some water in the crust and the mantle already, a considerable part of it (as well as molecules of biochemical significance) has been transported onto the surface of the Earth afterward through asteroids and mainly comets. Comparing the estimated water content of comets within the Oort cloud[1] the ratio between the water in this cloud compared to the Earth's water in the oceans is estimated to be on the order of 10^4! Still, there is an ongoing discussion of whether the water in the oceans and the one observed in comets is different concerning

[1] A reservoir of billions of comets in the Kuiper belt and in the Oort cloud thousands of AU away from the Sun.

The Trans-Neptunian Solar System.
https://doi.org/10.1016/B978-0-12-816490-7.00015-1

the ratio of hydrogen to deuterium (e.g., Raymond and Izidoro, 2017).[2] Considering the large number of comets which entered the inner solar system and collided with the Earth during the last billions of years, it is worth to pursue statistical analyses for these events in extrasolar planetary systems as well.

Our chapter is structured as follows: In the next section, we explain how comets could have formed around other stars. Thereafter, we discuss the observational evidence for comets in extrasolar systems together with a list of exoplanetary systems with gas giants akin to Jupiter and Saturn. In the main section, we show the results of simulations for the three systems HD 10180, 47 UMa, and HD 141399. Finally, we discuss what kind of consequence our results would have on possible water transport to possible terrestrial planets in stellar habitable zones (HZs).

15.2 Extrasolar Oort clouds

Stern (1987) explored how many comets from a so-called extrasolar Oort cloud may have hit the Earth as well as other terrestrial planets or Jupiter. According to his results, Earth may have suffered from a cometary impact originating from an Oort cloud around a star other than the Sun every 10^7 years.[3] In comparison, in this article the impact rate for solar indigenous comets was found to be three orders of magnitude higher, which was estimated according to previously computed probability rates (Whipple, 1950).

We mention this because in this article the existence of extrasolar Oort clouds is assumed, a concept previously considered as well by other authors (e.g., Kaib and Quinn, 2008; Kaib et al., 2011; Brasser et al., 2006, 2007). In fact, the formation of the Sun was not an isolated event occurring in an interstellar cloud but took place in an extended cloud where many other stars were born as well. This process was accompanied by the formation of objects such as comets containing a large array of molecules, including many molecules of astrobiological relevance. Looking at the ingredients of known comets, as, for example, 67P/Churyumov-Gerasimenko (visited by the *Rosetta* mission), the dust consists of almost 50% organic molecules and a notable part of water.[4] Thus, huge amounts of small icy bodies were created in the early disk around a star. Due to perturbations from nearby stars and other interstellar processes, many (or most) outer objects of this extrasolar Oort cloud were ejected and lost in the interstellar space as free interstellar comets (Stern, 1987). Alternatively, they were captured by the Sun or other stars, or were able to travel to the inner part of the solar system as previously discussed.

Thus, comets should be considered a by-product of planet formation. This leads to the assumption that in most, if not all planetary system there should be a cometary activity, despite the fact that the term "exocomets" has been coined only very recently.

[2] So far, the deuterium/hydrogen (D/H) ratio has been measured in 11 different comets, covering both types, but only one shows similarity with Earth: the Jupiter-family Comet 103P/Hartley 2. However, as reported in *Science*, *Rosetta* found that the D/H ratio is more than three times higher than the values found in Earth's oceans and on Hartley 2. Hence, this question is far from being settled!

[3] In this article, the author distinguished between comet nuclei and coma.

[4] The expression of "dirty snowball" was introduced by Whipple (1950).

The dynamics of scattering of planetesimals by a planet has been investigated, for example, by Wyatt et al. (2017), who explored the scattering outcome depending on the mass and the semimajor axis of the planet. Object trajectories resulted in accretion, ejection, remaining, escape, or placement in an Oort cloud or a depleted Oort cloud analog.

Here we will present the dynamical studies pursued by our group through investigating the outcome of planet-planetesimal scattering in single star systems. Therefore, the interaction between test particles distributed in a disk about the host star and (at least) one massive planet will be explored. Statistics for sets of models will be shown. This will include studies of outwardly scattered objects from solar system analogs to the Kuiper belt or the Oort cloud. Semimajor axis, eccentricity, Perihel, Aphel, inclination, and orbital period after the integration of each test particle are determined and statistical analyses are performed.

Rappaport et al. (2018) presented observational evidence for exocomet transits in the data of the *Kepler* star KIC 3542116. The signals show a distinct asymmetric shape and duration of 1 day or shorter, and have been fitted with a simple dust-tail model. The model yields that these comets have approximately the mass of Halley's comet. The inferred speeds during the transits are in the range of 35–50 $\mathrm{km\,s^{-1}}$ for the deeper transits, and 75–90 $\mathrm{km\,s^{-1}}$ for the more shallow and narrow transits. Similar observations of short-time events in high-resolution spectra of the Ca II K line of ϕ Leo have been presented by Eiroa et al. (2016) who compared those results to the ones found for β Pic (Beust et al., 1990; Kiefer et al., 2014). Based on observational data from the Atacama Large Millimeter/submillimeter Array (ALMA), Matrà et al. (2017) presented evidence for exocometary gas released within a Kuiper belt analog at $\sim$136 AU around the 440-Myr-old Fomalhaut system. The amount of CO observed implies that it originates from exocometary ices within the belt and is consistent with observations of solar system comets.

15.3 Evidence of the existence of extrasolar comets

Welsh and Montgomery (2015) presented observations of absorption line profiles for Ca II K (3933 Å) for 15 A-type and 2 B-type stars with known debris disks. They found short-term absorption variation in the line profiles in one late B-type and four A-type stars. The observed spectra contain a rotationally broadened absorption line at 3933 Å, which is typical for circumstellar gaseous disks corotating near to the stellar radial velocity of their host star. In addition, they identified transient weak absorption lines occasionally appearing to be red-shifted or blue-shifted by tens of kilometer per second. These features are believed to be caused by falling evaporating bodies (FEBs)—analogs to comets in the solar system—which evaporate the material when approaching their host stars. As the circumstellar K-line may often vary in absorption strength itself, they applied the method of comparing the night-to-night equivalent width (EW) of each observed star in order to evaluate if the observed change is accompanied by a significant change in the main circumstellar line. They concluded that the observational results are consistent with the presence of FEBs.

Moreover, Welsh and Montgomery (2015) also discussed the systems of HD 64145, HD 56537, HD 58647 (late B-type), HD 108767, HD 9672, HD 80007, HD 109573, and β Pictoris in detail. The key findings of this chapter read as follows.

The stars hosting FEBs are on average 70 Myr younger than the stars hosting no FEBs—although one has to consider that obtaining accurate ages for A-stars is still problematic. They also could not find a statistically significant difference between stars based on their metallicity or chemical peculiarity. In the observed sample, the FEB hosting stars are of an earlier type which, as argued by the authors, could be due to stellar activity. The detection of an FEB event itself strongly depends on the viewing the angle of the disk (which must be close to edge on) and the observational time frame (i.e., lower levels of activity need longer timespans of observation). They conclude that the detection of FEB activity can also be seen as evidence for the presence of associated exoplanets. FEB absorption in the Fe I line is only detected if the similar event is observed with an EW value of >50 mÅ in the Ca II K-line profile. This relationship between the Fe I and Ca II circumstellar absorption in young disk systems is still not fully understood as discussed in more detail by Welsh and Montgomery (2016).

Rotational velocity, mid-IR excess, or chemical peculiarity does not differentiate between stars with or without FEB activity. Stellar age does not seem to be a significant factor either. While the earlier described paper investigated a sample of stars in order to arrive at conclusions about parameter revealing hints on the presence of exocomets, Marino et al. (2017) focused on ALMA observations of a single-star system, that is, η Corvi. It is 1.5-Gyr-old system with a two-component debris disk.

The underlying problem was finding a theoretical explanation for the hot dust disk, which cannot be explained by a collisional cascade in situ. However, they were able to present the analysis for the first ALMA band observations (7 runs at 340 GHz); those corresponded to observations of the continuum dust emission of η Corvi's outer belt at a wavelength of 0.88 mm. At this wavelength, the continuum is dominated by ~0.1–10.0 mm sized dust grains for which radiation forces are negligible. The outer disk is detected with a peak radius of 150 AU and a radial width of over 70 AU. In order to receive estimates for the different disk parameters, they modeled the observed values to four different disk models. The first model consisted of a simple belt with radial and vertical Gaussian mass distribution peaked at 150 AU with a full-width half-maximum (FWHM) of 44 AU fit best. The comet-like composition and short life time of the observed belt led to the conclusion that it is fed from the outer belt via scattering by a chain of planets.

They propose the following scenario as being responsible for the observations.

Volatile-rich solid material from the outer belt is scattered inwardly via a chain of planets. This icy material starts to sublimate and loses part of its volatiles at specific ice lines, thus producing the CO feature observed at ~20 AU. The authors deduced a probable mass distribution for the chain of planets responsible for the scattering, which should be close to flat between 3 and 30 Earth masses. The inwardly scattered material could also explain an in situ collisional cascade or a collision with a planet of 4–10 Earth masses located at ~3 AU (sweet spot of the system) releasing large amounts of debris and causing the asymmetric structure revealed in the observations.

In our investigation, we concentrate on the orbital dynamics of exocomets as a theoretical study. We show results of computer simulations for three systems, which are HD 10180, 47 UMa system, and HD 141399. These systems are known to host one or more Jupiter-mass planets, which are expected to change the orbits of the incoming comets in characteristic ways.

15.4 Three examples of the dynamics of extrasolar comets

15.4.1 Numerical setup

Guided by the structure of the solar system and our knowledge about planet formation, we assume that most, if not all stellar systems harboring planets might also harbor an Oort-type cloud of comets away far from the star, resulting from the scattering processes occurring during the process of planet formation. These comets are expected to be gravitationally disturbed by, for example, a passing star or by Galactic tidal forces leading to highly eccentric cometary trajectories and allowing the small objects to enter the planetary region close to the star. The underlying injection mechanisms have been studied, for example, by Fouchard et al. (2007, 2017, 2018) and Rickman et al. (2008).

We have not implemented the forces due to Galactic tides and/or passing stars in our models as the initial semimajor axes of the comets are rather small and the eccentricities are rather large. Thus, the planets in the system will play the main role in the evolution of the cometary trajectories. Furthermore, the timespan of our integrations is only 1 Myr, which is too short for the Galactic tides to significantly affect the dynamics.

We consider tens of thousands of fictitious comets assumed as massless, which we distributed evenly in a sphere with initial conditions as follows:

- 80 AU $< a <$ 200 AU with $\delta a = 10$ AU
- $0.915 < e < 0.99$ with $\delta e = 0.005$
- $0 < i < 180$ with $\delta i = 10$

As the comets have been assumed to originate from an Oort-cloud analog, they have been placed in highly eccentric orbits. The total integration time was set to 1 Myr. Whenever a comet was ejected from the system or underwent a collision, another comet was inserted with the same initial conditions as the previous comet. Since at that time of insertion the configuration of the planets is different from the previous setting, the newly inserted comet will undergo different dynamics within the system.

We track the comet's osculating elements (e.g., semimajor axes and eccentricities) and determine if it is captured into a long-period or short-periodic orbit. This happens mostly for comets entering the system in the same plane as of the planets (see Fig. 15.1). For some systems (as done for 47 UMa), we monitor the collisions in order to estimate the possible amount of water transported to an Earth-mass planet in the system's habitable zone. The ejections and close encounters of the comets with the planets in the systems are examined as well.

The integration of the equations of motion was done with the Lie-integration method with adaptive step-size control (Hanslmeier and Dvorak, 1984; Eggl and Dvorak, 2010).

The initial conditions for the comets are the same for all three planetary systems considered. As the comets with smallest semimajor axes are started at 80 AU, they are far enough outside the orbit of the outermost planet of each system. Nevertheless, the noticeably different orbits of the planets in the systems in combination with the initial conditions for the comets are expected to impact the results. A relatively large semimajor axis of the outermost planet is expected to lead to a higher possibility of interaction with the comets and may thus affect the results in a way that more collisions and/or captures might occur. As mentioned in Section 15.2, one can assume that the cometary reservoirs have been formed inside a gas cloud with no preferences of inclination relative to the planetary orbits of the systems; consequently, the initial conditions were chosen as described.

Due to the lack of knowledge on the inclination of the planets in our study, all inclinations were set to randomly small values (<1 degrees).

15.4.2 HD 10180

15.4.2.1 *The system*

HD 10180 is a Sun-like star. Its parameters are $T_{eff} = 5910\,K$, $M = 1.06\,M_{\odot}$, combined with an age of about 4.3 Gyr (Lovis et al., 2011). The system consists of at least five Neptune-mass planets and one inner Earth-mass planet (HD 10180 b). Tuomi (2012) found two statistically significant signals corresponding to two additional super-Earth-mass planets located close-in. Thus, HD 10180 is a tightly packed system with six, or possibly nine planets. The most massive planet, HD 10180 h, is located beyond the other planets at a distance of approximately 3.4 AU. For our numerical investigation, we only took the four outer planets into account. The properties of the planets are shown in Table 15.1.

TABLE 15.1 Properties of the planets in the system of HD 10180.

Name	a (AU)	e	m ($m_{Jupiter}$)
HD 10180 e	0.270	0.0260	0.0789
HD 10180 f	0.49220	0.1350	0.0752
HD 10180 g	1.4220	0.00010	0.0673
HD 10180 h	3.40	0.080	0.2026

Note: For details and updates see http://exoplanet.eu/.

15.4.2.2 *Results*

Similar to the role of Jupiter in the solar system, which is mostly responsible for the scattering of small objects (especially comets) originating from the outer parts of the system, in HD 10180 the outermost and most massive planet HD 10180 h is able to change the orbits of incoming comets most profoundly. A possible outcome of gravitational interaction of a comet with the planet is a capture of that small body in an orbit with a relatively low semimajor axis and small eccentricity. Such comets may stay in the system for longer time spans akin to the Jupiter-family or Halley-type comets in the solar system. Fig. 15.1 conveys the probability for an incoming comet to be captured in such an orbit depending on the initial conditions. The three peaks visible at about 30, 90, and 150 degrees (green lines in Fig. 15.1) deserve more detailed investigations; they are most probably due to the Kozai mechanism.

Nevertheless, the most probable resulting trajectory after a close encounter is one leading to the escape of the comet from the system. Fig. 15.2 shows the number of new comets injected into the system; this number agrees with the number of comets ejected from the system. It is found that the number of ejected comets increases for comets entering the system with relatively large initial eccentricities and relatively small initial inclinations. These initial

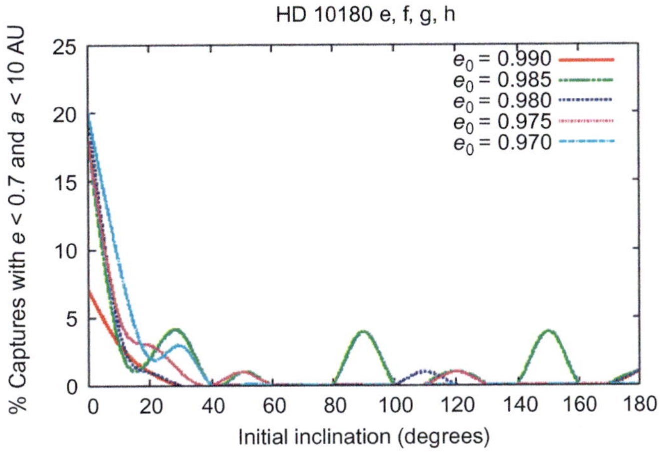

FIG. 15.1 HD 10180. We show the probability (in %) for a comet to be captured in an orbit with moderate values for the semimajor axis and eccentricity. The results depend on the comet's initial inclination and eccentricity. We find that it is more probable for comets entering the system with low initial inclination to be captured in a moderate orbit.

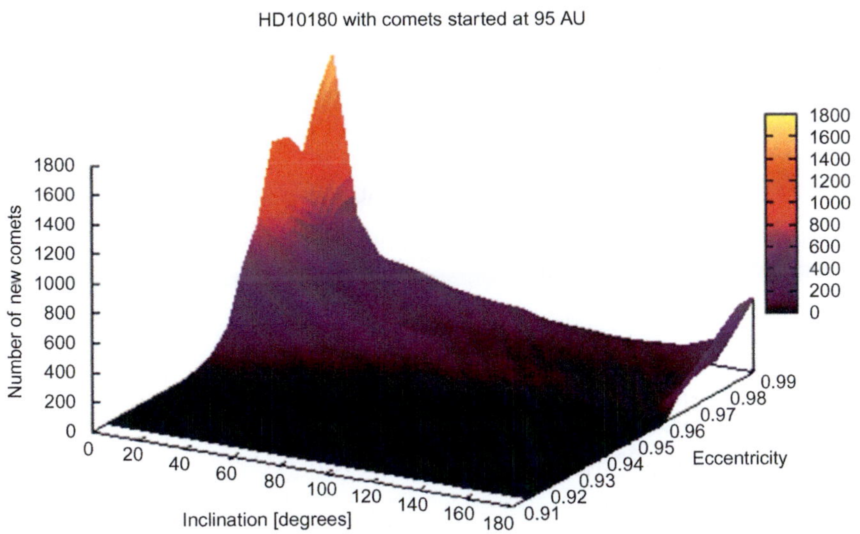

FIG. 15.2 HD 10180. The color coding shows the number of new comets inserted into the system. This number also corresponds to the number of comets ejected from the system during the duration of the integration. The higher the initial eccentricity and the lower the initial inclination, the more likely it is for a comet to get ejected.

parameters ensure that the comet is able to enter the system's inner region and experiences relatively early close encounters resulting in its escape.

The interaction of a comet with the most massive planet HD 10180 h, leading to either a stable close-in orbit, a quasistable orbit, or a cometary escape, is revealed in an example given in Fig. 3 of Loibnegger et al. (2017). Here over a period of approximately 1.5 Myr, a comet experiences a stable period. The orbit is captured at a perihelion distance of about 5 AU before it undergoes a close encounter with the most massive planet leading to pronounced changes in its orbit.

Figs. 15.3 and 15.4 show the outcome of the scattering process when leading to cometary captures. Undergoing close encounters with a planet can lead to an exchange of sufficient amounts of angular momentum that puts the comet in an orbit with a reduced semimajor axis

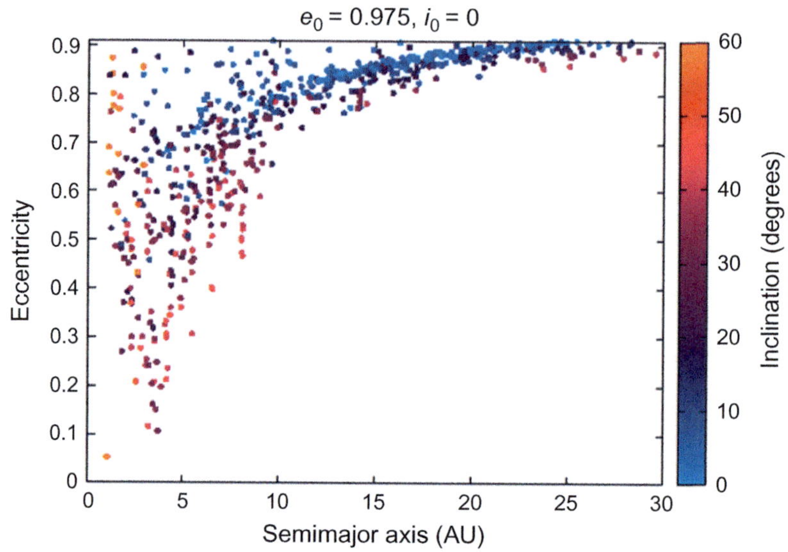

FIG. 15.3 HD 10180. Orbital properties of comets captured in the system after 1 Myr of integration. The color coding indicates the inclination at the time of the capture.

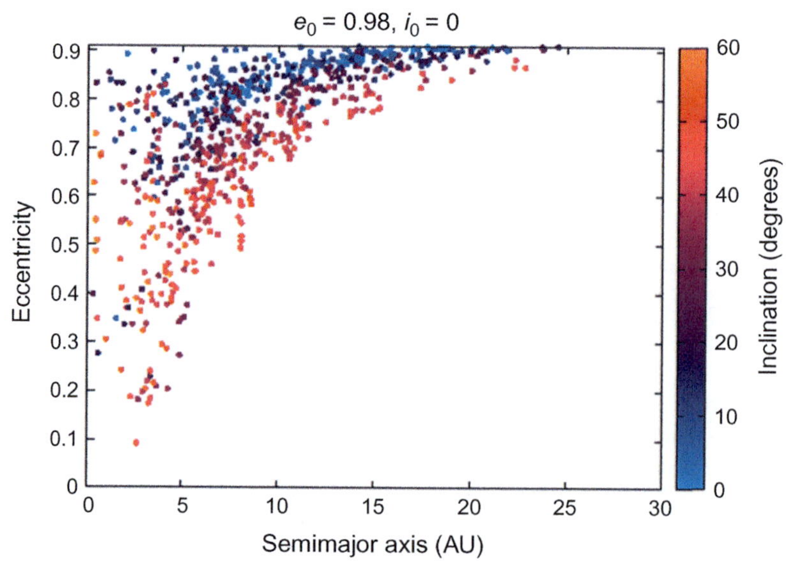

FIG. 15.4 HD 10180. Same as Fig. 15.3, but for a slightly different initial eccentricity for the comets.

and eccentricity. The figures show the cometary orbital elements after an integration time of 1 Myr. We could not find a clear distinction between long-period and short-periodic comets, as given in our solar system. However, it is evident that there are some comets captured in orbits of relatively low eccentricities ($e < 0.3$) and semimajor axes ($a < 10$ AU). The color coding shows the inclination of the orbit of the captured comets. Nevertheless, one can observe that most of the comets are captured in highly eccentric orbits, which will most likely lead to their ejection from the system after some time. Interestingly, there is a reasonable number of comets being scattered into orbits with inclinations of up to 60 degrees (value measured at the end of the integration) even for comets entering the system with an initial inclination of zero.

15.4.3 47 UMa

15.4.3.1 The system

The system of 47 UMa shows significant similarities to the solar system, although the host star is more evolved along the main sequence. The stellar parameters are: $T_{eff} = 5890$ K, $M = 1.03\ M_{\odot}$, and $R = 1.17\ R_{\odot}$, which have been determined by Henry et al. (1997) and Gonzalez (1998) with updates by Kovtyukh et al. (2003) and van Belle and von Braun (2009). Observations revealed three Jupiter-type planets with properties shown in Table 15.2. According to Kopparapu et al. (2013), the limits of the system's general HZ are given as 1.19 and 2.04 AU. The outer parts of the radiative HZ is unavailable for hosting planets due to the influence of 47 UMa b, which is responsible for a truncation of the HZ at approximately 1.6 AU.

TABLE 15.2 Properties of the planets in the system of 47 UMa.

Name	***a* (AU)**	***e***	**m ($m_{Jupiter}$)**
47 UMa b	2.1	0.032	2.53
47 UMa c	3.6	0.098	0.54
47 UMa d	11.6	0.16	1.64

Note: For details and updates see http://exoplanet.eu/.

For our simulations, besides the planets known in the 47 UMa system, we assumed additional Earth-mass planets orbiting in the HZ in order to investigate the transport of water by comets to potentially habitable terrestrial planets. These fictitious planets have been placed at 1.0, 1.25, and 1.584 AU, respectively. The latter planet is assumed to be in 3:2 resonance with 47 UMa b.

15.4.3.2 Results

With respect to 47 UMa, we focused on the possibility of water transport by comets to the system's HZ. Thus, we investigated the collisions of comets with fictitious Earth-mass planets with the known radius of the Earth. Figs. 15.5 and 15.6 show collisions of comets with those bodies. It is evident that the number of comets in each case decreases with lower distance to 47 UMa b. The Hilda-type planet (originally placed at 1.584 AU) did not undergo any collisions

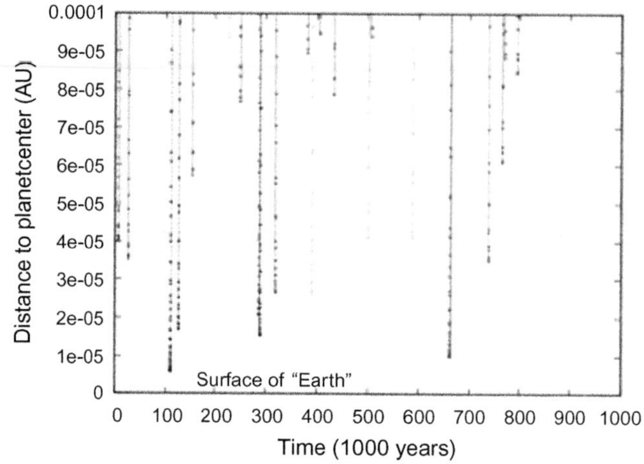

FIG. 15.5 47 UMa. The figures show cometary collisions with an Earth-mass planet placed at 1 AU. The *green line* marks the surface of the planet, assuming an Earth-like composition.

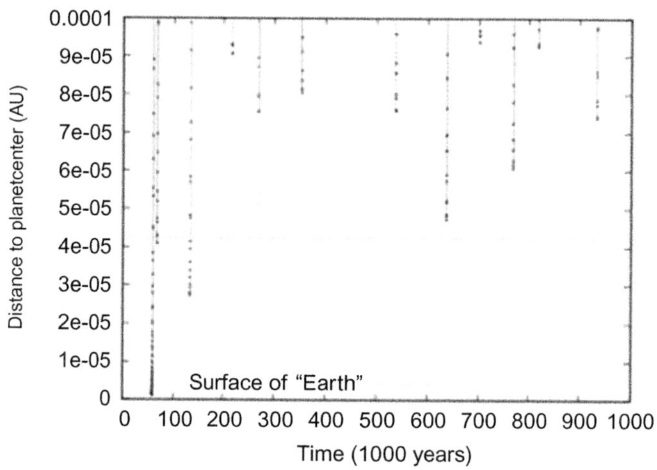

FIG. 15.6 47 UMa. Same as Fig. 15.5, but for an Earth-mass planet placed at 1.25 AU.

at all, which leads to the conclusion that an object in resonance with 47 UMa b is shielded from collisions by the massive planets.

In Fig. 15.7, we show the number of collisions of comets with all planets in the 47 UMa system. One can observe that the innermost planet 47 UMa b receives the most collisions with comets with initially high eccentricities. Only those comets are able to come as close to the star to cross 47 UMa b's orbit. The fictitious Earth-mass planet positioned further-in undergoes only very few collisions. Thus, the possibility of water transport by comets to terrestrial planets in 47 UMa's HZ is uncertain.

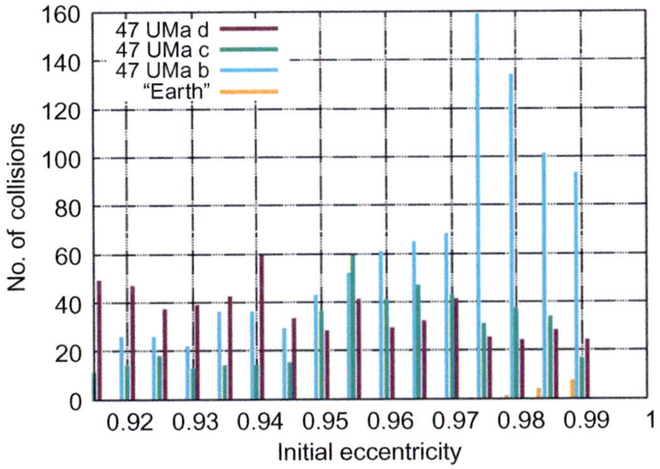

FIG. 15.7 47 UMa. The number of collisions depending on the initial eccentricity of the comets. Only comets with initially high eccentricities are able to penetrate the inner planetary system. Thus, 47 UMa b and the Earth-mass planet (originally placed at 1 AU) experience more (or only) collisions with comets of initially higher eccentricity.

Figs. 15.8 and 15.9 depict the number of comets captured in orbits with moderate values for the semimajor axis and eccentricity as a function of their initial eccentricity and inclination. A repeating pattern emerges once again: Comets with relatively high initial eccentricities are able to get closer to the star and the inner planetary region. This increases the likelihood of close encounters with one of the system planets, allowing the comets to attain orbits with relatively small semimajor axes and eccentricities. Comets which plunge into the system in the plane of

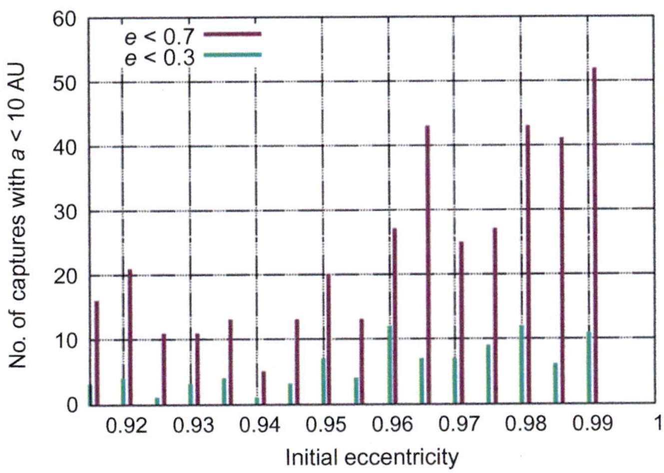

FIG. 15.8 47 UMa. The number of comets captured in orbits with moderate values for the semimajor axis and eccentricity. The higher the initial eccentricity of the comets, the more probable it is for them to interact with the planets and thus the more probable it is to be captured.

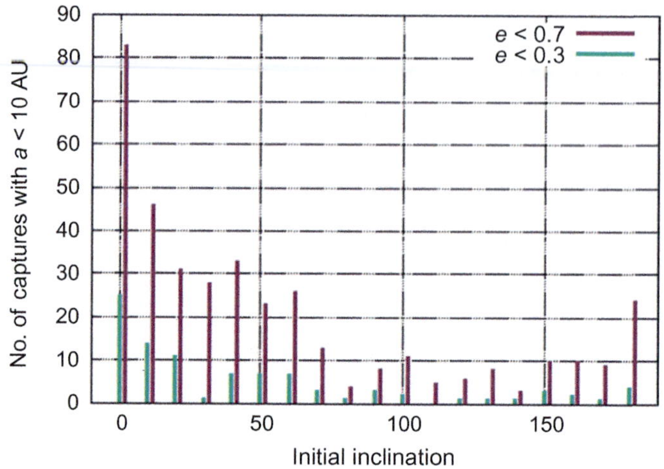

FIG. 15.9 47 UMa. Same as Fig. 15.8, but for the initial inclination of the comets. Comets entering the system in the plane have more time to interact with the planets; thus, it is more probable for them to be captured in orbits with low values for the semimajor axis and eccentricity.

the planetary orbits are also more likely to experience orbit-changing encounters caused by one of the planets. Further results have been given by Cuntz et al. (2018).

15.4.4 HD 141399

15.4.4.1 The system

HD 141399 harbors four planets. Its parameters are: $T_{\mathrm{eff}} = 5600\,\mathrm{K}$ and $M = 1.07\,M_{\odot}$; thus, this star is somewhat cooler than the Sun but slightly more massive and has a metallicity of [Fe/H] = 0.35. The properties of the known planets are given in Table 15.3. The planets of the system have masses between 0.5 and 1.4 Jupiter masses. The system's HZ extends from 1.04 and 2.34 AU (Kopparapu et al., 2013); this range corresponds to a domain between the planets HD 141399 c and HD 141399 d.

TABLE 15.3 Properties of the planets in the system of HD 141399.

Name	a (AU)	e	m (m_{Jupiter})
HD 141399 b	0.415	0.04	0.451
HD 141399 c	0.689	0.048	1.33
HD 141399 d	2.09	0.074	1.18
HD 141399 e	5.0	0.26	0.66

Note: For details and updates see http://exoplanet.eu/.

15.4.4.2 *Results*

We investigated the interaction of comets assumed to originate from an Oort cloud analog with the four system planets of HD 141399. The cometary cloud was set up as described in the previous sections. We monitored the paths of the comets on their way through the system and pursued a statistical analysis of the captures, collisions, and escapes of the comets after their interaction with the planets.

Fig. 15.10 conveys the number of collisions with each planet. One can observe that the outermost planet HD 141399 e experiences most collisions with the cometary bodies. This is accounted for with the initial eccentricity of the comets entering the system. Trajectories of comets with initial eccentricities between 0.92 and 0.99 are included in our statistics. Comets with relatively low eccentricities are unable to reach the inner planetary system without undergoing close encounters and angular momentum exchange. However, they are able to reach the orbit of HD 131499 d. Thus, the average number of comets crossing the orbit of the outermost planet is greater than the number of comets reaching the orbit of one of the planets orbiting close to the star. Interestingly, the innermost planet HD 141399 b only encounters collisions with comets in prograde orbits rather than comets with low inclinations ($i < 60$ degrees). All in all, it is more likely for comets entering the system in the ecliptic plane to collide with the planets than for comets entering the system with relatively high inclinations.

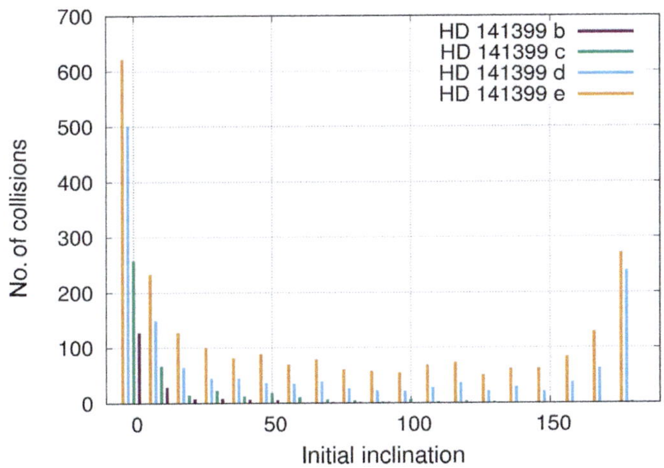

FIG. 15.10 HD 141399. We show the number of collisions with the different planets of the HD 141399 system dependent on the initial inclination of the comets. We depict the trajectories of all comets with initial eccentricity between 0.92 and 0.99 and an initial semimajor axis of 100 AU.

Fig. 15.11 depicts the number of cometary collisions depending on the initial eccentricity for comets with initial semimajor axes of 100 AU. Each bar includes the comets with all initial inclinations. One finds essentially the same behavior as given in Fig. 15.10: The outermost planet HD 141399 e encounters most collisions with the comets entering the system. Comets with initial eccentricities lower than 0.945 do not reach the orbit of even this planet; therefore, no collisions occur for those comets. The higher the eccentricity, the more collisions with the planets occur. Interestingly, for comets with initial eccentricities of $e > 0.98$, the number of collisions with the two outer planets drops while for the two inner planets that value stays

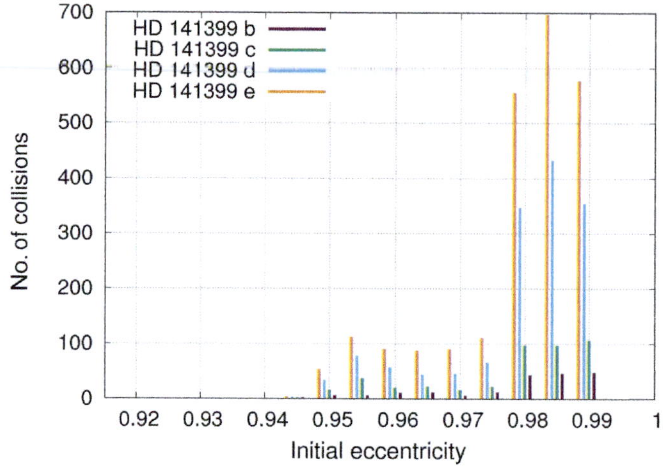

FIG. 15.11 HD 141399. Same as in Fig. 15.10, but for the initial eccentricity of the comets.

about the same. This may be due to the trajectories of the comets. Comets with high values for the eccentricity pass by the outer planets rather quickly as their perihelion is farther in. Thus, the time of interaction with the outer planets is relatively short, thus reducing the probability of a collision with one of the planets.

Another outcome based on the exchange of angular momentum between the planets and comets entering the system of HD 141399 is the capture of a comet in a moderate orbit. We define moderate orbits as orbits with $a < 10$ AU and eccentricities lower than 0.7 (violet line in Fig. 15.12). In special cases, the eccentricities are dropped to 0.3 (green lines). These orbits represent short-periodic comets. In this figure, the number of comets fulfilling this criterion is given as a function of the initial inclination of the comets. Each bar includes comets with all

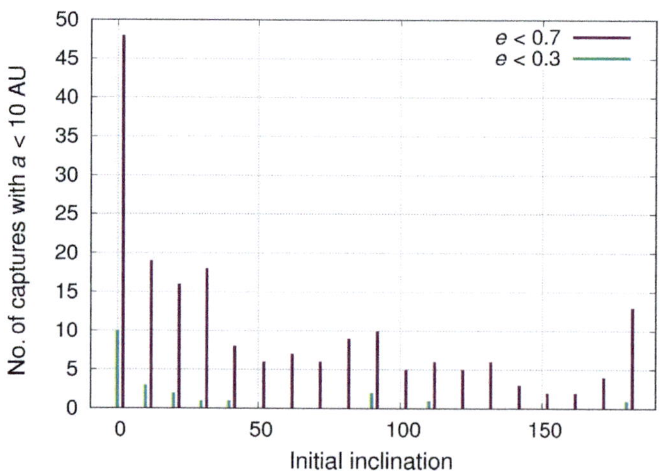

FIG. 15.12 HD 141399. Depiction of the number of captures of comets in orbits with low values for the semimajor axis and eccentricity.

initial eccentricities. It is evident that the capture of a comet in such orbits is most likely for comets entering the system in the ecliptic plane. Interestingly, it is found that even comets with the initially high inclination can be captured in such orbits. The likely underlying explanation considers the orbits of the two inner giant planets. The short-period orbits of HD 141399 b and HD 141399 c allow the planets to effectively interact with the comets entering the system from above, which makes it more likely for them to be scattered into orbits of small semimajor axes and low eccentricities.

Fig. 15.13 shows the same results but for the initial eccentricity of the comets. As expected, the number of comets captured into orbits with $a < 10$ AU and low eccentricities is higher for comets with initially high eccentricity. The same reasons apply to the probability of collisions with the planets apply here. Comets with initially high eccentricities are able to reach the planetary orbits and thus experience changes in their angular momentum resulting in capture. The same trend as given in Fig. 15.11 is here visible as well: The probability for comets with initial eccentricity $e > 0.98$ drops.

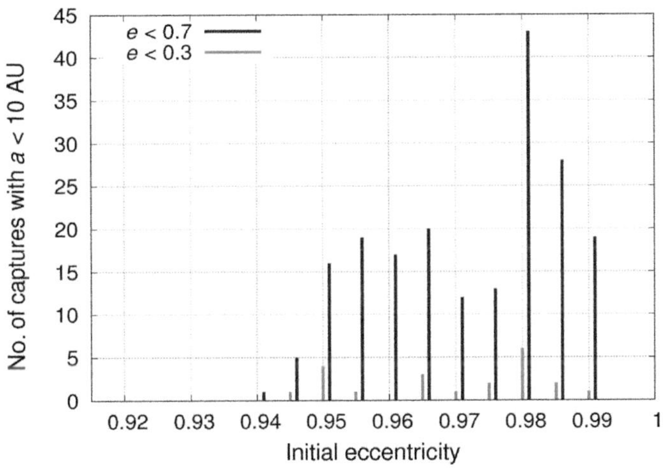

FIG. 15.13 HD 141399. Same as in Fig. 15.12, but for the initial eccentricity of the comets.

In Fig. 15.14, we depict the Delaunay elements L and G of one comet experiencing a type of stable phase before undergoing a close planetary encounter and being scattered to a chaotic orbit; this behavior occurs at about 730 kyr after the integration has been started. L represents the evolution of the semimajor axis. One can clearly identify the stable phases also visible in Fig. 15.15. Delaunay element G (i.e., the connection between a and e) confirms that behavior.

The Delaunay elements are calculated as follows:

$$L = \kappa \cdot \sqrt{a}$$
$$G = L \cdot \sqrt{1 - e^2}$$
$$H = G \cdot \cos(i)$$

Fig. 15.15 shows the evolution of the semimajor axis of the same comet as depicted in Fig. 15.14. The figure shows stable phases of the comet, which is captured in a more or less

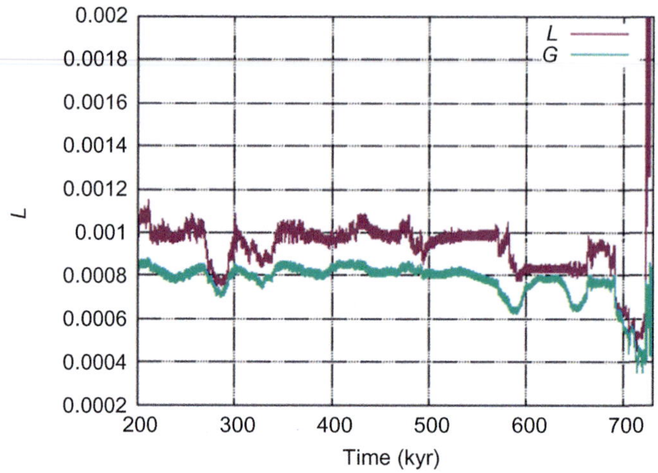

FIG. 15.14 HD 141399. Depiction of the Delaunay elements L and G of one comet experiencing a rather stable phase in a captured orbit prior to experiencing a close encounter.

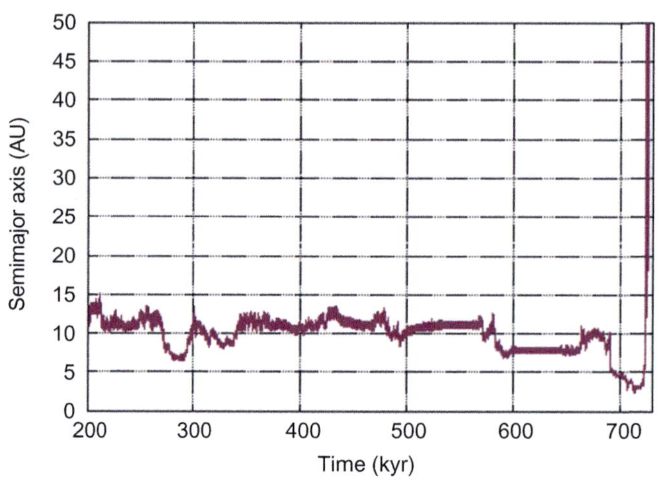

FIG. 15.15 HD 141399. Semimajor axis evolution of the comet depicted in Fig. 15.14. Phases of stability are clearly visible.

stable orbit about the host star. With κ connected to the gravitational constant and the masses involved.

15.5 Conclusion

Since very recently, we have evidence about comets in extrasolar planetary systems obtained through spectroscopic observations. The discovery of traces of objects, presumably created through stellar interaction, was first reported some 20 years ago; however, theoretical work predicting the existence of comets outside our solar system has been presented even much earlier. We scratched on the history of the discovery of exocomets since the end of the last century and discussed the role of Oort clouds. They are expected to exist around virtually every star due to their formation history in large molecular clouds. We then explored the possible dynamics of comets penetrating the inner parts of extrasolar systems. Thus, we discussed three specific examples of extrasolar systems with gas giants in order to compare the outcomes with the cometary structure in our solar system.

Our goal was twofold. First, we studied how the dynamical elements of comets arriving from large distances are altered, perhaps resulting in the formation of families of comets akin the Jupiter- and the Halley-cometary families in our solar system. Second, we planned to determine the percentages of comets colliding with the planets or being ejected from the system because of close encounters with one of the large planets. However, in the three systems, which are HD 10180, 47 UMa, and HD 141399, we could not identify the formation of such families. But we found evidence for long-term captures of comets into orbits where they spent up to millions of years in essentially stable orbits quite close to their host star.

For HD 10180, a tightly packed system with six or even nine planets (including at least five Neptune-like giants), we were able to show how the exocomets may have been captured or ejected from the system depending on their initial eccentricities and inclinations.

For 47 UMa, a system containing three gas giants, we did the same type of investigation but in addition we checked how a possible Earth-mass planet could get its water inventory from comets. Three different cases were pursued: An Earth-mass planet placed at 1 AU, an Earth-mass planet placed at 1.5 AU, and an Earth-mass planet in the 3:2 MMR with the innermost system planet (47 UMa b); the habitable zone in 47 UMa is confined approximately between ~0.9 and ~2.1 AU for this late-stage main-sequence star. The results show that there is little opportunity for the Earth-mass planets to collect water due to cometary collisions. If water is present on any of those planets (if existing), an alternate mechanism of transport is required; see, for example, Raymond and Izidoro (2017) and references therein for proposed scenarios.

HD 141399, a star hosting four massive gas giants, was the final exoplanetary system of our investigations aimed at studying cometary dynamics. In all of our numerical experiments, the outermost planet, although the least massive, suffered from many collisions, whereas the innermost giant planet had only very few ones.

As a supplementary aspect of our study, we also checked the Tisserand parameter **T** for the systems. However, no suitable information could be obtained from these values depicted in Fig. 15.16. It is well-known that in the solar system Jupiter's role is dominant (Carusi and Valsecchi, 1992; Carusi et al., 1995). However, for HD 10108, the gas giant at $a = 3.4$ AU (planet h) cannot play a dominant role due to the presence of another gas giant at $a = 1.4$ AU (planet g). Hence, no cometary family akin to the Jupiter family in the solar system can therefore occur. Furthermore, in the other systems investigated here the Tisserand parameter is not giving important information on the cometary orbits either. Although the values interior to the location of planet h in the upper panel of Fig. 15.16 show a similar signal like in the respective plot for Jupiter family comets (see Fig. 5 in Rickman, 2010), the dispersion is too large to draw meaningful conclusions. In the lower panel of Fig. 15.16, the Tisserand parameter (computed for planet g) is plotted versus the semimajor axis, but again no tendency can be seen due to the very large dispersion.

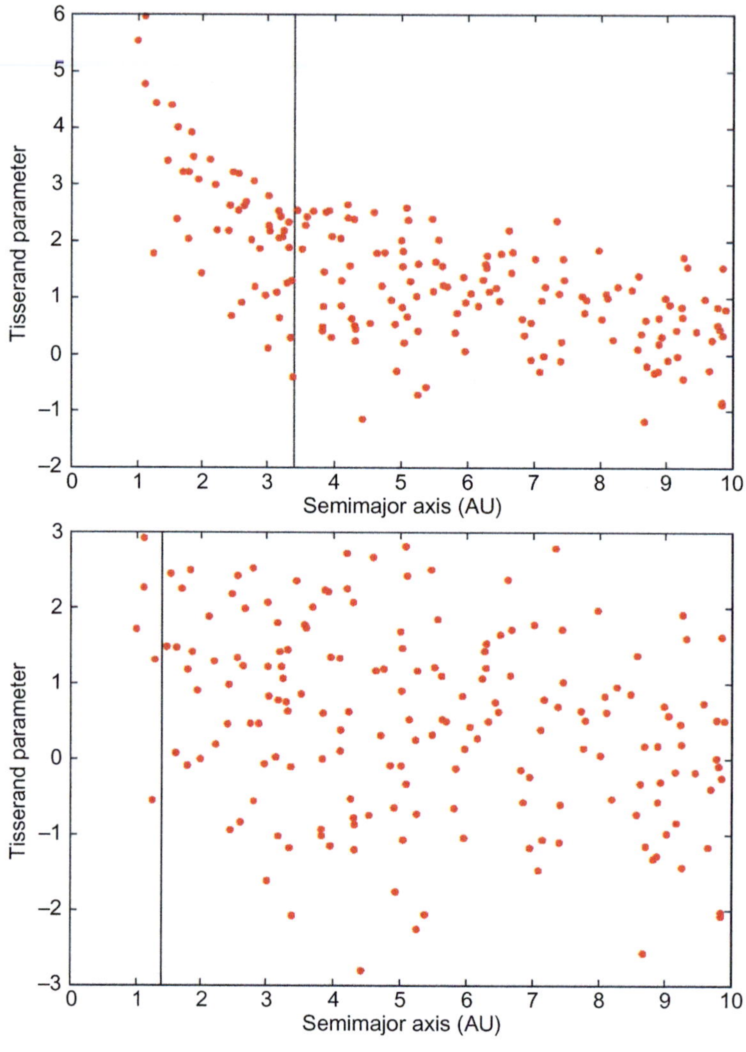

FIG. 15.16 *Upper panel*: Tisserand parameter versus semimajor axes for a gas giant at $a = 3.4$ AU pertaining to HD 10108 h. *Lower panel*: Same but for the planet HD 10108 g. The *vertical lines* in both panels depict the location of the respective planets.

In the near future, many more new observations are expected. Some of them may be indicators of comets very close to the host star. Thus, there will be a serious need for extending our numerical experiments to all systems with more than one giant planet to acquire a more complete picture on the role of exocomets in proliferating water to terrestrial planets. For that, we will carefully investigate possible terrestrial planets in the stellar HZs. Our focus will be to determine the probability of collisions between comets, expected to originate from cometary clouds at the outskirts of those systems, and terrestrial system planets, as indicated by theory or established by observations.

Acknowledgments

This research is supported by the Austrian Science Fund (FWF) through grant S11603-N16 (R. Dvorak and B. Loibnegger). Moreover, M. Cuntz acknowledges support by the University of Texas at Arlington. The computational results presented have been achieved in part using the Vienna Scientific Cluster.

References

Beust, H., Vidal-Madjar, A., Ferlet, R., Lagrange-Henri, A.M., 1990. The beta Pictoris circumstellar disk. X—numerical simulations of infalling evaporating bodies. Astron. Astrophys. 236, 202–216.

Brasser, R., Duncan, M.J., Levison, H.F., 2006. Embedded star clusters and the formation of the Oort cloud. Icarus 184, 59–82. https://doi.org/10.1016/j.icarus.2006.04.010.

Brasser, R., Duncan, M.J., Levison, H.F., 2007. Embedded star clusters and the formation of the Oort cloud. II. The effect of the primordial solar nebula. Icarus 191, 413–433. https://doi.org/10.1016/j.icarus.2007.05.003.

Carusi, A., Valsecchi, G.B., 1992. Dynamics of comets (lecture). In: Ferraz-Mello, S. (Ed.), Chaos, Resonance, and Collective Dynamical Phenomena in the Solar System, IAU Symposium, vol. 152, p. 255.

Carusi, A., Kresák, Ľ., Valsecchi, G.B., 1995. Conservation of the Tisserand parameter at close encounters of interplanetary objects with Jupiter. Earth Moon Planets 68, 71–94. https://doi.org/10.1007/BF00671499.

Cuntz, M., Loibnegger, B., Dvorak, R., 2018. Exocomets in the 47 UMa system: theoretical simulations including water transport. Astron. J. 156, 290. https://doi.org/10.3847/1538-3881/aaeac7.

Eggl, S., Dvorak, R., 2010. An introduction to common numerical integration codes used in dynamical astronomy. In: Souchay, J., Dvorak, R. (Eds.), Lecture Notes in Physics, vol. 790, Springer Verlag, Berlin, pp. 431–480.

Eiroa, C., Rebollido, I., Montesinos, B., Villaver, E., Absil, O., Henning, T., Bayo, A., Canovas, H., Carmona, A., Chen, C., Ertel, S., Iglesias, D.P., Launhardt, R., Maldonado, J., Meeus, G., Moór, A., Mora, A., Mustill, A.J., Olofsson, J., Riviere-Marichalar, P., Roberge, A., 2016. Exocomet signatures around the A-shell star phi Leonis? Astron. Astrophys. 594, L1. https://doi.org/10.1051/0004-6361/201629514.

Fouchard, M., Froeschlé, C., Rickman, H., Valsecchi, G.B., 2007. Methods for the study of the dynamics of the Oort cloud comets I: modelling the stellar perturbations. In: Benest, D., Froeschle, C., Lega, E. (Eds.), Lecture Notes in Physics, vol. 729, Springer Verlag, Berlin, p. 257.

Fouchard, M., Rickman, H., Froeschlé, C., Valsecchi, G.B., 2017. On the present shape of the Oort cloud and the flux of new comets. Icarus 292, 218–233. https://doi.org/10.1016/j.icarus.2017.01.013.

Fouchard, M., Higuchi, A., Ito, T., Maquet, L., 2018. The "memory" of the Oort cloud. Astron. Astrophys. 620, A45. https://doi.org/10.1051/0004-6361/201833435.

Gonzalez, G., 1998. Spectroscopic analyses of the parent stars of extrasolar planetary system candidates. Astron. Astrophys. 334, 221–238.

Hanslmeier, A., Dvorak, R., 1984. Numerical integration with lie series. Astron. Astrophys. 132, 203.

Henry, G.W., Baliunas, S.L., Donahue, R.A., Soon, W.H., Saar, S.H., 1997. Properties of Sun-like Stars with Planets: 51 Pegasi, 47 Ursae Majoris, 70 Virginis, and HD 114762. Astrophys. J. 474, 503–510. https://doi.org/10.1086/303451.

Kaib, N.A., Quinn, T., 2008. Assessing the threat of Oort cloud comet showers. In: AAS/Division of Dynamical Astronomy Meeting No. 39.

Kaib, N.A., Roškar, R., Quinn, T., 2011. Sedna and the Oort cloud around a migrating Sun. Icarus 215, 491–507. https://doi.org/10.1016/j.icarus.2011.07.037.

Kiefer, F., Lecavelier des Etangs, A., Boissier, J., Vidal-Madjar, A., Beust, H., Lagrange, A.-M., Hébrard, G., Ferlet, R., 2014. Two families of exocomets in the β Pictoris system. Nature 514, 462–464. https://doi.org/10.1038/nature13849.

Kopparapu, R.K., Ramirez, R., Kasting, J.F., Eymet, V., Robinson, T.D., Mahadevan, S., Terrien, R.C., Domagal-Goldman, S., Meadows, V., Deshpande, R., 2013. Habitable zones around main-sequence stars: new estimates. Astrophys. J. 765, 131. https://doi.org/10.1088/0004-637X/765/2/131.

Kovtyukh, V.V., Soubiran, C., Belik, S.I., Gorlova, N.I., 2003. High precision effective temperatures for 181 F-K dwarfs from line-depth ratios. Astron. Astrophys. 411, 559–564. https://doi.org/10.1051/0004-6361:20031378.

Loibnegger, B., Dvorak, R., Cuntz, M., 2017. Case studies of exocomets in the system of HD 10180. Astron. J. 153, 203. https://doi.org/10.3847/1538-3881/aa67ef.

Lovis, C., Ségransan, D., Mayor, M., Udry, S., Benz, W., Bertaux, J.-L., Bouchy, F., Correia, A.C.M., Laskar, J., Lo Curto, G., Mordasini, C., Pepe, F., Queloz, D., Santos, N.C., 2011. The HARPS search for southern extra-solar planets. XXVIII. Up to seven planets orbiting HD 10180: probing the architecture of low-mass planetary systems. Astron. Astrophys. 528, A112. https://doi.org/10.1051/0004-6361/201015577.

Marino, S., Wyatt, M.C., Panić, O., Matrà, L., Kennedy, G.M., Bonsor, A., Kral, Q., Dent, W.R.F., Duchene, G., Wilner, D., Lisse, C.M., Lestrade, J.-F., Matthews, B., 2017. ALMA observations of the η Corvi debris disc: inward scattering of CO-rich exocomets by a chain of 3–30 M planets? Mon. Not. R. Astron. Soc. 465, 2595–2615.

Matrà, L., MacGregor, M.A., Kalas, P., Wyatt, M.C., Kennedy, G.M., Wilner, D.J., Duchene, G., Hughes, A.M., Pan, M., Shannon, A., Clampin, M., Fitzgerald, M.P., Graham, J.R., Holland, W.S., Panić, O., Su, K.Y.L., 2017. Detection of exocometary CO within the 440 Myr old Fomalhaut belt: a similar CO + CO_2 ice abundance in exocomets and solar system comets. Astrophys. J. 842, 9. https://doi.org/10.3847/1538-4357/aa71b4.

Rappaport, S., Vanderburg, A., Jacobs, T., LaCourse, D., Jenkins, J., Kraus, A., Rizzuto, A., Latham, D.W., Bieryla, A., Lazarevic, M., Schmitt, A., 2018. Likely transiting exocomets detected by Kepler. Mon. Not. R. Astron. Soc. 474, 1453–1468. https://doi.org/10.1093/mnras/stx2735.

Raymond, S.N., Izidoro, A., 2017. Origin of water in the inner Solar System: planetesimals scattered inward during Jupiter and Saturn's rapid gas accretion. Icarus 297, 134–148. https://doi.org/10.1016/j.icarus.2017.06.030.

Rickman, H., 2010. Cometary dynamics. In: Souchay, J., Dvorak, R. (Eds.), Lecture Notes in Physics, vol. 790, Springer Verlag, Berlin, pp. 341–399.

Rickman, H., Fouchard, M., Froeschlé, C., Valsecchi, G.B., 2008. Injection of Oort cloud comets: the fundamental role of stellar perturbations. Celest. Mech. Dyn. Astron. 102, 111–132. https://doi.org/10.1007/s10569-008-9140-y.

Stern, A., 1987. Extra-solar Oort cloud encounters and planetary impact rates. Icarus 69, 185–188. https://doi.org/10.1016/0019-1035(87)90013-3.

Tuomi, M., 2012. Evidence for nine planets in the HD 10180 system. Astron. Astrophys. 543, A52. https://doi.org/10.1051/0004-6361/201118518.

van Belle, G.T., von Braun, K., 2009. Directly determined linear radii and effective temperatures of exoplanet host stars. Astrophys. J. 694, 1085–1098. https://doi.org/10.1088/0004-637X/694/2/1085.

Welsh, B.Y., Montgomery, S.L., 2015. The appearance and disappearance of exocomet gas absorption. Adv. Astron. 2015, 980323. https://doi.org/10.1155/2015/980323.

Welsh, B.Y., Montgomery, S., 2016. Exocomet circumstellar Fe I absorption in the beta Pictoris gas disk. Publ. ASP 128 (6), 064201. https://doi.org/10.1088/1538-3873/128/964/064201.

Whipple, F.L., 1950. A comet model. I. The acceleration of comet Encke. Astrophys. J. 111, 375–394. https://doi.org/10.1086/145272.

Wyatt, M.C., Bonsor, A., Jackson, A.P., Marino, S., Shannon, A., 2017. How to design a planetary system for different scattering outcomes: giant impact sweet spot, maximizing exocomets, scattered discs. Mon. Not. R. Astron. Soc. 464, 3385–3407. https://doi.org/10.1093/mnras/stw2633.

CHAPTER

16

Extrasolar Kuiper belts

Mark C. Wyatt

Institute of Astronomy, University of Cambridge, Cambridge, United Kingdom

16.1 Introduction

As the other chapters in this book testify, studies of the Kuiper belt continue to play a key role in shaping our understanding of the structure and history of the solar system. This chapter aims to provide some context to this relatively detailed information on our own planetary system, by considering how this compares to the planetary systems of other stars, in particular with regard to the component of their debris disks which might be considered analogous to the Kuiper belt. Such comparisons help to ascertain whether our system is typical or atypical, either in terms of its current architecture or in terms of its past history. They also allow a deeper understanding of processes inferred to have occurred in our own history, since those processes may be ongoing for some stars, furthermore providing evidence for how these play out in different environments.

By now it is clear that our solar system is just one of the many planetary systems, with observations of nearby stars showing that approximately half of the stars have a planetary system in which at least one of its planets may be detected in current surveys (Winn and Fabrycky, 2015). However, this seemingly abundant information is restricted to the planets that reside close to their stars (within a few AU), and information about the outer regions of planetary systems remains scarce. Indeed just a few percent of stars have planets $\gg$5 AU detectable by direct imaging (which means at least a few times more massive than Jupiter), although microlensing surveys hint that Neptune-analogues may be more common.

In contrast, our understanding of the dust content of outer planetary systems is relatively well advanced (e.g., Wyatt, 2008; Hughes et al., 2018). It is over 30 years since the far-IR satellite *IRAS* first discovered circumstellar dust orbiting nearby main sequence stars (Aumann et al., 1984). Because this dust is short-lived it was recognized that it must be continually replenished from larger planetesimals situated in a source region that was inferred

The Trans-Neptunian Solar System.
https://doi.org/10.1016/B978-0-12-816490-7.00016-3

from the dust temperature to be at 10s of AU (Backman and Paresce, 1993), a conclusion that was reinforced by imaging of the dust structures (Smith and Terrile, 1984). Stars with such dust have been called Vega-type (after the first discovery), and the circumstellar dust is referred to as a debris disk,[1] and often interpreted to arise from an extrasolar analogue to the solar system's Kuiper belt (e.g., Wyatt et al., 2003; Moro-Martín et al., 2008).

The extent to which that analogy is appropriate remains open for interpretation, but what is clear is that ~20% of stars have dust, and so presumably also planetesimals, at somewhere around 20–150 AU from their stars (Wyatt, 2008; Hughes et al., 2018). Given the aforementioned difficulty of detecting planets in these outer regions, it is usually uncertain whether this dust lies at the outer edge of a planetary system or whether there are other similarities with our own Kuiper belt, but there are clues in the dust structures. These often reinforce the solar system analogy, with the caveat that there is some anthropocentric bias in this interpretation.

This chapter starts in Section 16.2 by summarizing the various observational methods used to obtain information about these putative extrasolar Kuiper belts, then in Section 16.3 considers how our solar system fits within this context. More detail is given in Section 16.4 about specific aspects of the physical and dynamical properties of the extrasolar systems and their comparison with the solar system, before concluding in Section 16.5.

16.2 Extrasolar Kuiper belt observations

As noted in Section 16.1, most of the information about extrasolar Kuiper belts comes from observations of dust created in the destruction of their planetesimals. These observations can be roughly split into those used to discover the presence of dust using photometry (discussed in Section 16.2.1) and those that provide more detailed characterization through high-resolution imaging (discussed in Section 16.2.2). There is also an emerging area of observations of gas in the systems (see Section 16.2.3).

16.2.1 Discovery: Photometry and SED fitting

Debris disk dust is most readily detected from its thermal emission which manifests itself as the star appearing brighter than expected from purely photospheric emission at wavelengths that depend on the dust temperature, although care must be taken to ensure that potential extragalactic contamination has been removed from the photometry (e.g., Kennedy and Wyatt, 2012; Gáspár and Rieke, 2014). The emission from dust created at 10s of AU peaks at far-IR wavelengths and that is where the majority of debris disks have been discovered. The most recent surveys were those undertaken by *Herschel* (Eiroa et al., 2013; Matthews et al., 2014; Hughes et al., 2018), including an unbiased survey of the nearest several hundred stars in which dust emission was detected around 20% of stars (Thureau et al., 2014; Sibthorpe et al., 2018).

[1] Here *debris* refers to the debris left over after planet formation, which applies to any circumstellar material that is not a planet. This could include a dust or gas component left over from the protoplanetary disk, but also planetesimals and the dust and gas resulting from their destruction.

These photometric surveys are supplemented by photometry at mid-IR and sub-mm wavelengths to build up the spectral energy distribution (SED) of the dust emission, which generally shows that the emission is dominated by a single temperature and so dust at a single distance (see e.g., Fig. 16.1). Given the luminosity of the star this can be translated into a black body distance (i.e., the distance at which the dust would be if it behaved like a black body), although it is recognized that the small dust that dominates is an inefficient emitter and so this is likely an underestimate of the true location of the dust by a factor that can be several depending on the star (Booth et al., 2013; Pawellek et al., 2014). In addition, to its black body radius r_{bb}, the disk's SED is also characterized by its fractional luminosity f (i.e., the luminosity of the dust divided by that of the star), with values of 10^{-6} to 10^{-3} typical for known disks.

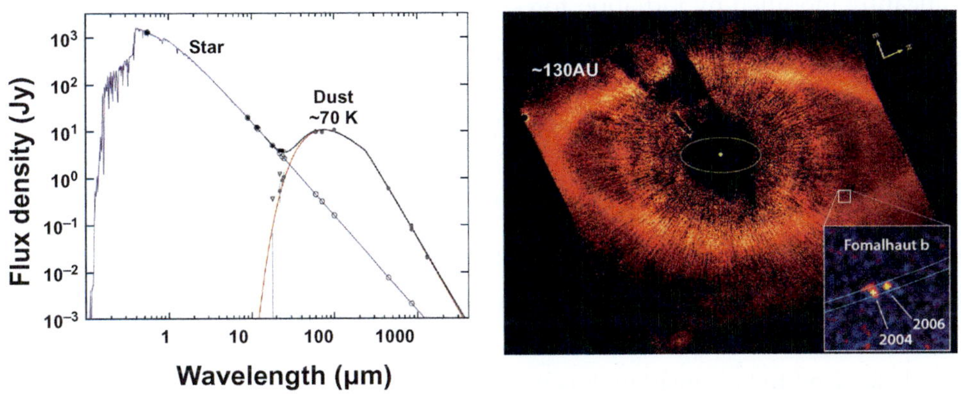

FIG. 16.1 Observations of the nearby (7.8 pc) 400 Myr A-type main sequence star Fomalhaut. *(Left)* Photometric observations show that the spectral energy distribution is comprised of two temperature components, one from the star and another from cold dust at ~70 K. *(Right)* Optical images in which the stellar emission has been subtracted using coronagraphic techniques to reveal scattered light from circumstellar dust that is distributed in a narrow ring at ~130 AU from the star (Kalas et al., 2008). The center of this ring is offset from the star implying an eccentricity of ~0.11, and a planet-like object (Fomalhaut b) is seen close to the inner edge of the ring.

For some systems, the SED has been characterized in greater detail in which case more realistic modeling is performed to constrain the dust size distribution and composition (e.g., Olofsson et al., 2012; Lebreton et al., 2012). However, the simple shape of the spectrum means that there is still value in considering the disks in terms of their two observable parameters (r_{bb} and f), particularly because there are a number of observational biases that can be readily understood within this context (Wyatt, 2008). Nevertheless, some systems have an SED that is best characterized as having two temperatures (Chen et al., 2006; Kennedy and Wyatt, 2014). In the context of an extrasolar Kuiper belt interpretation, this means that there is an additional warm component of emission that might be analogous to the solar system's zodiacal cloud. However, there remains debate about whether this component originates in an extrasolar asteroid belt analogue (Su et al., 2013), exocomets scattered in from the extrasolar Kuiper belt (Wyatt et al., 2017), or is a transient dust component perhaps caused by a recent collision (Jackson and Wyatt, 2012).

16.2.2 Characterization: Imaging

Debris disks around the nearest stars have a spatial extent of >1 arcsec which means that these can be readily resolved. The low resolution of far-IR instrumentation precludes this except in a few cases (e.g., Acke et al., 2012), but this technique has been successful at optical and near-IR wavelengths with *HST* (see Fig. 16.1), and more recently SPHERE on VLT and GPI on Gemini, where starlight scattered by the smallest sub-μm dust can be imaged (e.g., Schneider et al., 2014). In the sub-mm with ALMA (Atacama Large Millimeter/submillimeter Array), thermal emission from mm to cm-sized grains can be imaged as well (see Fig. 16.5). Given the size of dust dominating the emission at the different wavelengths, it is generally assumed that sub-mm observations trace the distribution of the parent planetesimals, while the picture is more complicated at optical wavelengths because the orbits of small dust grains are significantly modified by radiation forces. Indeed, different radial and even nonaxisymmetric structures are seen in multiwavelength imaging of the same disk providing a valuable diagnostic tool for the underlying dynamics.

Such imaging provided the first evidence that the dust is configured in a disk (rather than a spherical distribution, Smith and Terrile, 1984), and has allowed direct measurement of the radial location of the dust, its radial and vertical extents, and the presence of gaps (see Section 16.4.4), as well as providing evidence for asymmetries in the form of clumps, eccentricities, and warps (see Section 16.4.5). All of these give vital clues to the underlying dynamics and the existence of planets.

16.2.3 Gas

Debris disks used to be considered as gas-free, in contrast to the protoplanetary disks are found around young stars (<10 Myr). Recent discoveries with ALMA show that this can no longer be considered the case since there are now of order 10 main sequence stars known to exhibit CO gas emission. Many of the first gas detections are around the more massive A-type stars (Kóspál et al., 2013; Greaves et al., 2016; Moór et al., 2017), though there are a number of observational biases that facilitate the detection of gas around early-type stars, and gas is now seen around later spectral types as well (Marino et al., 2016; Matrà et al., 2019a). Most of these observations can be explained within the context of a model in which the gas in debris disks, like the dust, is the product of the destruction of planetesimals that are then required to be volatile-rich (Kral et al., 2017). An open question is whether the youngest (<50 Myr) stars retain gas from the protoplanetary disk phase (Kóspál et al., 2013; Kral et al., 2019). Observations of debris disk gas not only allow the mass and composition of gas to be measured (see Section 16.4.3) but also spatially resolved with additional kinematic information, resulting in a potentially powerful probe of the disk structure and dynamics (see Section 16.4.5, Dent et al., 2014).

16.3 An extrasolar perspective of the Kuiper belt

It is important to note that the solar system's debris disk is fainter than any known debris disk around another star. If one of the stars in the debris disk surveys hosted an exact replica of the solar system, it would only have been possible to detect the photospheric emission from the star and not that of circumstellar dust (see Figs. 16.2 and 16.3). The challenge with detecting a true Kuiper belt analogue lies in the fact that its thermal emission is outshined by more than an order of magnitude at all wavelengths by the Sun.

In fact, the level of thermal emission from dust in the Kuiper belt is not well known. Its emission has not been detected since it is masked by that of the zodiacal cloud which is much closer to the Earth, although the far-IR all-sky surveys like *IRAS*, *COBE*, and *Planck* do provide upper limits on the level of emission that can be present (e.g., Backman et al., 1995). Direct detection of dust grains in the outer solar system gives an indication of the number of dust grains present (Piquette et al., 2019), but this must be combined with dynamical models for dust production and evolution to understand the whole dust structure (Vitense et al., 2014). In doing so, the best estimate is that Kuiper belt emission peaks at ~70 μm at a flux 1% that of the Sun (Vitense et al., 2012). Even for bright stars their far-IR fluxes cannot be predicted with such accuracy, which combined with calibration uncertainties in photometric measurements, means that ≫10% excesses are required for a confident detection.

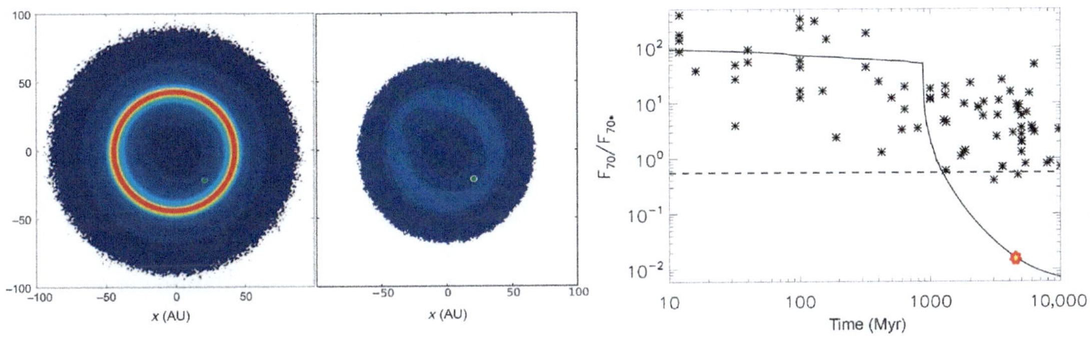

FIG. 16.2 The solar system's Kuiper belt. *(Left)* Face-on images of the number density distribution of Kuiper belt objects that have been debiased from the observed distributions. To the left is shown the full Kuiper belt which appears as a narrow ring dominated by the classical Kuiper belt objects (Matthews and Kavelaars, 2016), and to the right the contribution of resonant Kuiper belt objects that shows two clumps at ±90 degrees from Neptune (shown with a *circle*) due to the Plutino population (Lawler, 2014). *(Right)* The evolution of the 70 μm emission from dust in the Kuiper belt (shown as its level compared with that from the Sun) expected in the model of Gomes et al. (2005), and compared with the dust emission from nearby Sun-like stars (Booth et al., 2009). In this model the massive primordial Kuiper belt is comparable in brightness to those observed around nearby stars (shown with *asterisks*), but is depleted to a nondetectable level at ~800 Myr by the Late Heavy Bombardment.

If it were possible to resolve the emission from Kuiper belt dust, the structures that would be observed at long wavelengths would resemble the structures known to be present in the distribution of Kuiper belt objects (e.g., Lawler, 2014); that is, there would be a prominent ring at ~40 AU from the classical Kuiper belt (see Fig. 16.2 left), with some nonaxisymmetric structure caused by the resonant Kuiper belt objects (see Fig. 16.2 middle), and with emission extending out to larger radii from the scattered disk (which is present but hardly noticeable in Fig. 16.2 left). Given the low density of the dust distribution, the smallest dust (as traced by the shorter far-IR wavelengths) would migrate in by Poynting-Robertson drag and so fill in the 40 AU hole. However, very little dust would make it past Jupiter or even Saturn, meaning that this tenuous dust distribution would have drops in density associated with these planets (Liou and Zook, 1999; Moro-Martín and Malhotra, 2002).

There are, however, indications that the Kuiper belt was more massive in the past by several orders of magnitude. At such an epoch there would have been correspondingly more dust, resulting in detectable levels of emission (see Fig. 16.2 right, Booth et al., 2009). Thus the known debris disks could be analogues to the primordial Kuiper belt, and thus representative of systems that either never will, or have yet to undergo, large-scale instabilities. Note though that the exact timing of the depletion in the solar system remains uncertain (e.g., Gomes et al., 2005; Morbidelli et al., 2018).

16.4 Extrasolar Kuiper belt properties (and comparison with solar system)

16.4.1 Mass and radius distribution

Far-IR surveys from the past few decades have provided large quantities of information on the incidence of debris disks, as measured by the excess emission above photospheric levels at different wavelengths (typically at 24, 70, 100, and 160 μm). Surveys have been carried out for nearby stars with a variety of spectral types, distances, and ages. While main sequence star ages are usually quite inaccurate, there is sufficient information, particularly when comparing associations of young stars for which ages can be better determined, to show that disks tend to get fainter with age (Rieke et al., 2005; Su et al., 2006). As such the surveys are usually interpreted within the context of population models that take collisional erosion into account (see Section 16.4.2).

Fig. 16.3 summarizes the state of our knowledge of the debris disk population seen around nearby Sun-like (i.e., F, G, or K-type) stars. For each of the ~300 stars in the sample, the various observations (be they detections or nondetections of dust emission) have been combined to determine the best fit fractional luminosity and black body radius to their disk (see Fig. 16.1), or the upper limit on the dust luminosity for a given temperature to allow a nondetection. Particularly where disks are only detected at a single wavelength, the fitted parameters can have significant uncertainty, and so the disks are placed on this plot accordingly,[2] and the color scale shows the fraction of stars in the sample predicted to have a disk in different regions of parameter space (for more details on the method, see Sibthorpe et al., 2018). Since observational biases mean that disks of a given level could only have been detected around a fraction of the stars in the sample as shown in the contours (Wyatt, 2008), this is taken into account to get an estimate of the underlying distribution.

As noted in the figure, the distribution of known disks is bounded on the bottom edge by an observational limitation, while the top edges are set by physical considerations. For disks close to the star (<40 AU in black body radius) the maximum fractional luminosity is set by collisional erosion since most nearby stars have the same roughly Gyr age and so their disks would be expected to have undergone similar levels of collisional processing and would have

[2] More specifically, uncertainties in the photometric data mean that analysis of the SED can only determine a probability that the disk is present in a given part of fractional luminosity vs black body radius parameter space.

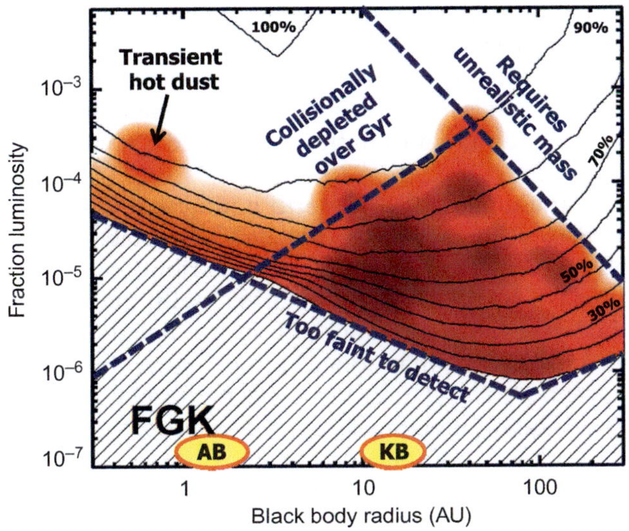

FIG. 16.3 Distribution of debris disk luminosities and radii. The color scale shows the fraction of a sample of ~300 Sun-like (FGK) stars from Phillips et al. (2010) that have dust emission as a function of their measured fractional luminosities and black body radii. The approximate locations of the asteroid belt and Kuiper belt dust are shown in *yellow*, noting that the black body radius is a factor of a few smaller than the true radius because dust emits hotter than the black body. The contours show the fraction of stars in the sample for which dust emission could have been detected, which is used to correct the observed incidence of debris to get the fraction plotted in the color scale. The *blue dashed lines* and *annotation* explain why the known disks (which are present around ~20% of stars) lie mostly in a well-defined region of parameter space. *Adapted from Sibthorpe, B., Kennedy, G.M., Wyatt, M.C., Lestrade, J.-F., Greaves, J.S., Matthews, B.C., Duchêne, G., 2018. Analysis of the Herschel DEBRIS Sun-like star sample. Mon. Not. R. Astron. Soc. 475, 3046–3064. https://doi.org/10.1093/mnras/stx3188.*

undergone more depletion the closer they are to the star (Wyatt et al., 2007b). Some disks may exhibit higher fractional luminosities if there happens to be a young star in the sample, or if there is a transient dust component (Wyatt et al., 2007a). For disks further from the star, the upper edge is set by a constraint on the mass required to produce the given fractional luminosity.

While Fig. 16.3 gives a quick visual impression of the distribution of debris disk properties, population models that also address the age dependence of the phenomenon can be used to get a more accurate parameterization for the distributions of initial masses and radii of the disks. Such models show that it is possible to fit the observations by assuming that all stars started out with a planetesimal belt, with a radius and initial mass chosen from given distributions, that mass then evolving with time due to collisional erosion (Wyatt et al., 2007b; Löhne et al., 2008; Gáspár et al., 2013; Sibthorpe et al., 2018), although it is possible that a few of the brightest disks may require unrealistic initial masses (~$1000M_\oplus$, Krivov et al., 2018). There is evidence that the disks of A-type stars tend to be brighter than their Sun-like star counterparts (Greaves and Wyatt, 2003), even accounting for the various biases (such as the on average younger ages of the more massive stars), which may be because of a difference in their planetesimal belt properties (e.g., in having more massive disks, or one made of smaller planetesimals, or lower levels of stirring). Given that many of the well known bright debris disks discussed in this review are found around A stars (like β Pic), it may be worth bearing this caveat in mind when making comparisons with the solar system. There are few disks detected around the less massive M stars, but this is likely an observational bias and there is no evidence as yet to say that their disks are any different from those of Sun-like stars (e.g., Morey and Lestrade, 2014).

Despite the success of these models, it is worth noting that they cannot make any strong conclusions about the 80% of stars that do not have detectable disks, since these would be based on assumptions about the shape of the distributions. For example, a common assumption is that the mass distribution is log-normal, with the nondetections arising from close-in disks that undergo rapid depletion and so are not detectable in the nearby star population, but it could be that the mass distribution is bimodal, with 80% of stars having no disk. Also, more recently a correlation has been found between planetesimal belt radii determined from high-resolution imaging of the sub-mm dust emission and the luminosity of the star (Matrà et al., 2018a). If this holds up as more disks are resolved, this would suggest a preferential location for the formation of planetesimal belts that could be linked to ice lines in the protoplanetary disk for example, which would provide an additional constraint on the population models (as well as significant insight into the origin of the belts).

Fig. 16.3 also includes the location of the solar system's Kuiper belt and asteroid belt (see Section 16.3). This shows that roughly 20% of stars have disks that are comparable in radius to the Kuiper belt, in having dust at temperatures that give a black body radius of 10–100 AU, but that have at least an order of magnitude more dust at fractional luminosities of 10^{-6}–10^{-3}. For the reasons noted above it is not possible to quantify where our own Kuiper belt fits into the distribution.

16.4.2 Evolution: Collisional vs dynamical erosion

As noted in Section 16.4.1, the evidence that there is any evolution in debris disk properties comes not from seeing those changes in any one system, but rather from seeing the change in properties for samples of stars with different ages. Inevitably then there remains uncertainty in how individual disks evolve. However, since it is necessary for collisions to occur to replenish the dust that is observed and that dust is lost from the system relatively quickly (compared with the age of the star), some level of collisional erosion is unavoidable. That evolution has been studied extensively both analytically and numerically, in ways that have also been applied to small body populations in the solar system. In the solar system evidence for past collisional evolution can be found in the detailed shape of the asteroid and Kuiper belt size distributions (e.g., Bottke et al., 2005; Fraser, 2009; Singer et al., 2019), whereas in extrasolar systems the evolution of the size distribution in debris disks is manifested in the evolution of disk brightness which must be inferred from the population studies.

There are several unknowns in the problem, such as the initial size distribution of the planetesimals, their level of stirring (i.e., their collision velocity), and their size dependent dispersal threshold. With these factors known, the collisional evolution would be well understood (e.g., Wyatt et al., 2011), and when coupled with models for dust optical properties that determine both its emission properties and loss due to radiation forces, this can be readily translated into an evolution in the disk's observables. In principle, this means that these fundamental physical parameters can be extracted from the observations, although in practice there are degeneracies between the parameters which make it hard to extract unambiguous information, other than that the vast majority of observations are consistent with purely steady state collisional erosion.

Indeed it is easier to identify the disk properties or processes that are incompatible with the observations, at least on a population level. For example, a slow decline in disk incidence with age that is consistent with collisional erosion means that large-scale dynamical depletions at 100s of Myr must be relatively rare (Booth et al., 2009), otherwise there would be fewer bright disks at several Gyr ages. Thus either the Kuiper belt is anomalous in this respect (i.e., if its depletion is responsible for the Late Heavy Bombardment, Gomes et al., 2005), or the depletion actually occurred much earlier on (Morbidelli et al., 2018). It is also possible to conclude that the majority of debris disks are not broad planetesimal disks (even if each is only bright at one radius at any given time, for example, Kenyon and Bromley, 2010), since broad disks are hot early on and cold later on in a manner that is incompatible with the observed 24 and 70 μm brightness evolution (Kennedy and Wyatt, 2010, see also Section 16.4.4).

However, for comparison with the Kuiper belt, it is interesting to consider what has been concluded about the largest planetesimals in debris disks and the origin and level of stirring within them. While observations are only sensitive to dust up to ~cm-size, the short collisional lifetime of such dust means that it must be continually replenished. Extrapolating the size distribution and equating the resulting collisional lifetime with age of the star shows that debris disks are replenished by planetesimals at least several km in size (Wyatt and Dent, 2002), with population models pointing to sizes up to 100 km (Gáspár et al., 2013). It has been suggested that some disks may be comprised of much smaller planetesimals, say 10–100 m, that are colliding at very low velocity so that the collisions are barely disruptive (Heng and Tremaine, 2010; Krivov et al., 2013), explaining the lack of small grains inferred to some systems (Eiroa et al., 2011), although the chance alignment of background galaxies could be confusing the interpretation of these stars (Gáspár and Rieke, 2014). Detailed observations of the halo of small grains seen to extend beyond a debris disk due to radiation pressure has the potential to constrain the level of stirring, with eccentricities inferred to be <1% in one system (Thébault and Wu, 2008). For some disks, the vertical structure has been measured with inclinations inferred to be of order a few percent (Matrà et al., 2019b; Daley et al., 2019).

The origin of the stirring is not well constrained. It could arise from large bodies embedded in the disk that excites relative velocities to around their escape velocity (Kenyon and Bromley, 2010). However, the growth of Pluto-sized objects at 100s of AU in the disk takes too long to explain large disks seen around Gyr-old stars. One possibility is that these form directly from the protoplanetary disk (rather than from the collisional growth of smaller planetesimals), or it could be that a massive disk of ~200 km planetesimals can stir itself without the need for larger bodies (Krivov and Booth, 2018). If the belts do mark the outer edge of the planetary system then those planets could stir the disk through their gravitational perturbations. This could be through overlapping resonances from a planet at the inner edge (Quillen and Faber, 2006), but could also arise from the secular perturbations of a more distant eccentric or inclined planet (Mustill and Wyatt, 2009). It is also possible that the planetesimals were born on stirred orbits, for example, if these formed in the early phases of the protoplanetary disk (Booth and Clarke, 2016).

To illustrate the range of interpretations for debris disks that do not encompass a traditional Kuiper belt-like analogy, note that in one model the planetesimals form during protoplanetary disk dispersal from the dust that is swept up by a photoevaporating gas disk (Carrera et al., 2017). While such a model does not recover the observed distribution of debris disk radii, it is worth bearing in mind that our understanding of protoplanetary disk dispersal processes is incomplete, in particular regarding the fate of the $>1M_{\oplus}$ of mm-sized dust that resides in the outer regions of the system (Wyatt et al., 2015). As such it is not inconceivable that our understanding of the origin and evolution of debris disks could need revision.

16.4.3 Planetesimal composition

A standard way of determining the composition of circumstellar material is to look for solid-state features in the emission spectrum, such as the silicate features near 10 μm. This has been successful for the few systems for which warm dust is present (e.g., Lisse et al., 2009), however, for the majority that is too cold to have significant mid-IR emission, this leaves far-IR features of which only a few were measured with *Herschel*. Nevertheless, when applied to the disk of β Pic, the 69 μm feature was used to show that 3.6% of the dust mass is in crystalline olivine grains, moreover constraining them to be relatively Mg-rich (de Vries et al., 2012).

As noted in Section 16.2.3, a new technique has been developed to probe the volatile content of the planetesimals by studying gas in the systems. Molecular gas has a short and well-defined lifetime in optically thin environments since it is destroyed by interstellar radiation, for example, CO has a lifetime of just ~120 years. Thus if a disk is observed to have a certain mass of CO, the rate at which CO is being replenished can be readily inferred. Since observations of the dust disk can also be used to determine the rate at which the refractory component is being replenished, these two rates can be used to find the volatile fraction of the planetesimals feeding the cascade. While there are some assumptions in that calculation, such as that both gas and dust are being replenished from steady state grinding, and the uncertainties in the CO mass measurement can be large if excitation conditions are not known (e.g., Matrà et al., 2015), this at least provides an order of magnitude estimate.

For now, this has been achieved for a handful of disks with conclusions being that the fraction of the planetesimals that is made up of CO and CO_2 is similar to that of solar system comets (see Fig. 16.4, Marino et al., 2016; Matrà et al., 2017b). Moreover, the same model also explains those systems for which CO is not detected (Kral et al., 2017). Thus it is plausible that all debris disks are made of volatile-rich planetesimals, and so that all have associated secondary gas disks, but that only a fraction of these has CO at detectable levels. While no other gas molecules have been detected in debris disks, this is to be expected given the lower abundance relative to CO and moreover, the even shorter lifetime of most molecules (Matrà et al., 2018b).

Atomic gas in the form of CI, CII, and OI has also been detected (e.g., Riviere-Marichalar et al., 2014; Higuchi et al., 2017), where it is inferred that this gas is the photodissociation product of CO and CO_2. These are much longer lived and survive until the atomic gas disk has undergone viscous diffusion to be accreted onto the star (Kral et al., 2016). The observed ratio of carbon to CO thus constrains the viscous diffusion timescale, with implications for the underlying physics of that accretion (Kral and Latter, 2016). For β Pic, this resulted in a

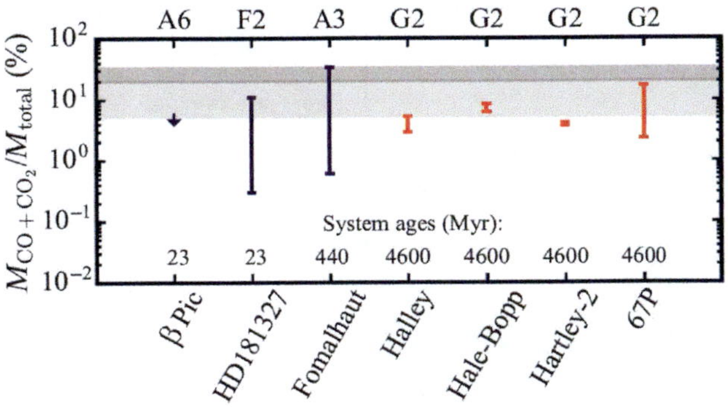

FIG. 16.4 Planetesimal volatile fraction inferred from observations of gas in debris disks compared with the same fractions for solar system comets (Matrà et al., 2017b). Volatile fraction here is defined as the fraction of the total mass that is made up of CO and CO_2.

prediction for the amount of OI that was too low compared with that observed, which was used to infer that the planetesimals were also water-rich, similar to solar system comets, since that would explain the detection (Kral et al., 2016).

While there is an emerging paradigm that has been outlined above, it is worth noting that there remain uncertainties. For example, the model does not explain the distribution of CI observed in β Pic (Cataldi et al., 2018). It was also suggested that the <50 Myr systems with protoplanetary disk levels of CO have a component of primordial gas that has yet to disperse (Kóspál et al., 2013; Moór et al., 2017). However, it was recently shown that such high levels of CO could be explained in a secondary production scenario, since if the gas production rate is particularly high, or the viscous diffusion timescale is too long, then sufficient carbon can accumulate to shield the CO from photodissociating, which then can also start to self-shield (Kral et al., 2019).

In consideration of the solar system, the points to take away are that the volatile-rich composition of our Kuiper belt may be similar for other stars and also that there may have been a tenuous atomic gas disk present early on.

16.4.4 Radial and vertical structure

As Fig. 16.1 shows, debris disks can be radially very narrow, with a fractional radial width (i.e., $\Delta r/r$) of ~10% in the example shown of the Fomalhaut disk. That width can be measured for disks that have been imaged, but more specifically requires that disks be resolved at mm wavelengths to determine the width of the planetesimal belt since the smaller dust probed at shorter wavelengths will be spread out from the parent belt by radiation forces. There is then some bias in the sense that a broad disk will have lower surface brightness than a narrow one and so will be harder to detect in high resolution imaging. Nevertheless, Fig. 16.5 shows that it is generally found that the disks are perhaps better described as rings or belts, with the fractional radial width ranging from 10% to 100%.

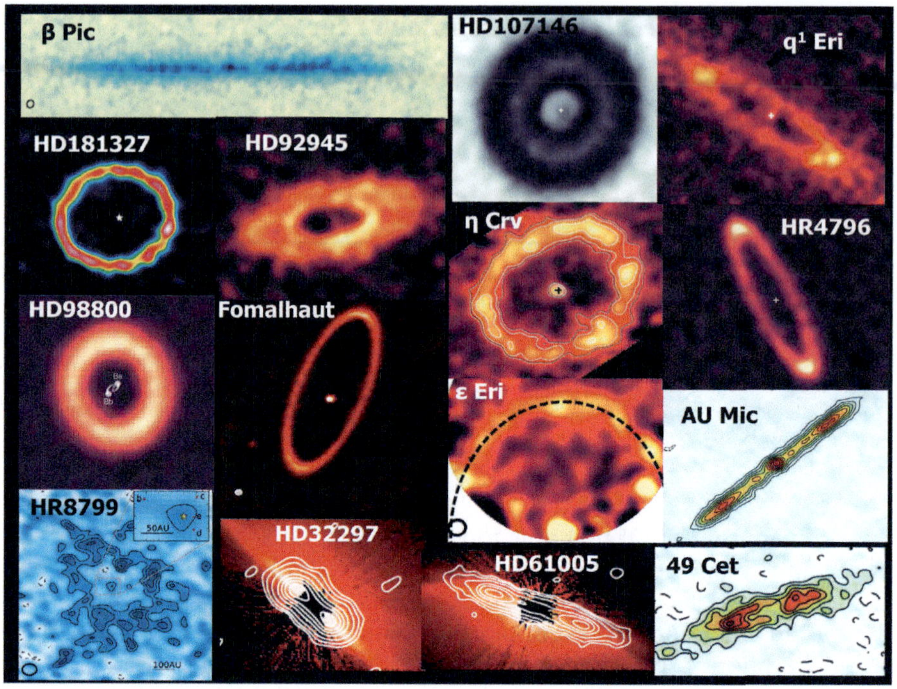

FIG. 16.5 Compilation of ALMA images of the distribution of mm-cm-sized dust in the debris disks of 15 nearby stars. Individual images are from: β Pic (Matrà et al., 2019b), HD107146 (Marino et al., 2018b), q^1 Eri (J. Lovell, private communication), HD181327 (Marino et al., 2016), HD92945 (Marino et al., 2019), η Crv (Marino et al., 2017b), HR4796 (Kennedy et al., 2018), HD98800 (Kennedy et al., 2019), Fomalhaut (MacGregor et al., 2017), ε Eri (Booth et al., 2017), AU Mic (Daley et al., 2019), HR8799 (Booth et al., 2016), HD32297 and HD61005 (MacGregor et al., 2018; Schneider et al., 2014), 49 Cet (Hughes et al., 2017).

There are some notable exceptions to this description in that some disks are broad (e.g., MacGregor et al., 2016). For example, the 61 Vir disk extends from 30 AU to beyond 150 AU (Wyatt et al., 2012; Marino et al., 2017a), a property which means that its surface brightness is so low that it is barely noticeable in ALMA imaging (but detected in the visibility data), with a radial profile that may be explained by a broad disk of ~10 km-sized planetesimals undergoing collisional erosion (Marino et al., 2017a). Other examples of broad disks include HD107146 and HD92945, which also show an annular gap in the radial profile (see Fig. 16.5, Marino et al., 2018b, 2019) that may be related to planets either embedded in the disk or nearby (Yelverton and Kennedy, 2018).

The inner edges of the debris belts have received some attention since it is noted that if the belt is truncated by dynamical interactions with a planet located at the inner edge, then the sharpness of the belt edge is indicative of the mass of the planet (e.g., Chiang et al., 2009). A more massive planet stirs up the disk to a greater extent resulting in a shallower slope. Analogy, with the solar system shows that the sharpness and location of the disk's outer edge can also be an important indicator of the system's past history (Gladman et al., 2001).

The outer edge of a debris disk is sometimes quoted as the location at which the disk becomes too faint to detect. However, it is expected that for planetesimals confined to a belt, there would be a halo of small μm-sized dust that extends beyond the belt due to radiation pressure (Strubbe and Chiang, 2006). In this case, the outer edge of the planetesimal belt would be evident as the location where there is a drop in the disk's surface brightness. However, it is now becoming clear that the extended dust components in some systems that were thought to be a halo of small dust also contain mm-sized dust (MacGregor et al., 2018). Since this dust is unlikely to be affected by radiation pressure, this calls into question whether the planetesimal disk is, in fact, broad with simply a step change in density at what had been assumed to be the outer edge of a radially confined planetesimal belt. The presence of a faint outer belt of debris in one system could support this view (Marino et al., 2016). Alternatively, the extended halo of mm-sized grains could represent an analogue of the scattered disk, that is, made of planetesimals (and the dust derived from their destruction) that are on eccentric orbits undergoing scattering with a planet located near the inner edge. Indeed the favored interpretation of the radial structure of the HR8799 disk is that it has a low eccentricity disk (i.e., a classical Kuiper belt analogue) as well as a high eccentricity population of comets in a scattered disk analogue (see Fig. 16.6, Geiler et al., 2019), an interpretation strengthened by the fact that multiple planets interior to the belt are known in this system (Marois et al., 2010).

It was already noted in Section 16.4.2 that the vertical structure of some disks has now been resolved. Furthermore, for β Pic it was found that the vertical structure was not well-described by a single Gaussian, rather it is fit well as the sum of two Gaussians that correspond to components with mean inclinations of 1 and 9 degrees (Matrà et al., 2019b). While this is a few times flatter than the solar system, there is a clear analogy with the cold and hot components of the classical Kuiper belt (see Fig. 16.6).

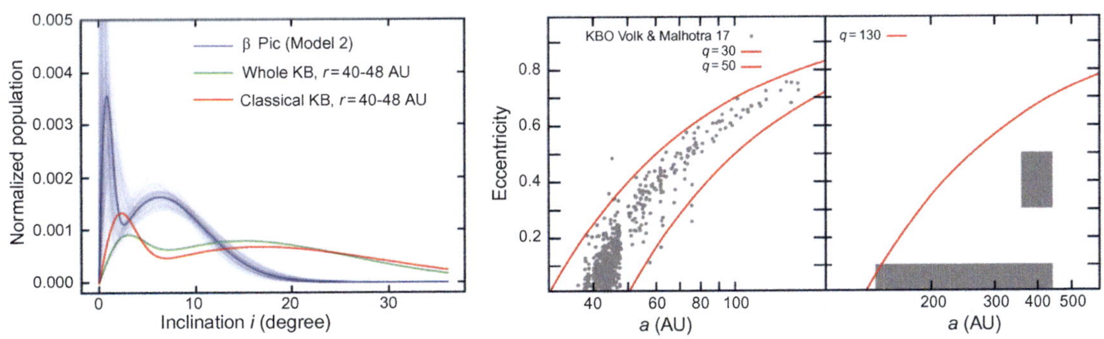

FIG. 16.6 Distributions of orbital elements inferred from ALMA images of debris disks that are analogous to dynamical populations in the Kuiper belt. *(Left)* Vertical structure in the β Pic disk (see Fig. 16.5) is best modeled as the sum of two Gaussians that imply a distribution of inclinations reminiscent of the hot and cold classical Kuiper belt (Matrà et al., 2019b). *(Right)* Radial structure in the HR8799 disk (see Fig. 16.5) includes a halo of mm-sized grains that is best modeled with a high eccentricity population potentially analogous to the scattered disk (Geiler et al., 2019).

16.4.5 Dynamical structures

Some of the radial structures discussed in Section 16.4.4 can be considered dynamical structures, in that the shape of inner and outer disk edges, and gaps within the disks, can all be related to the presence of planets. However, such an interpretation is potentially ambiguous, because this may alternatively be indicative of the initial distribution of planetesimals, in which radial structure may have been imprinted by processes ongoing during the protoplanetary disk phase (e.g., the formation of planetesimals at ice lines); that is, no planets may be required. Nonaxisymmetric structure is a clearer indicator that additional perturbations are ongoing since these would be needed to maintain such structures over long timescales.

Several of the disks that have been imaged are seen to be eccentric. This was manifested originally by a brightness asymmetry in mid-IR images of the HR4796 disk (Telesco et al., 2000), which was interpreted as arising from the pericenter side of the disk being closer to the star and so hotter and brighter (Wyatt et al., 1999). That interpretation is supported by ALMA imaging which showed that the opposite side of the disk is brighter (Kennedy et al., 2018) as expected due to the slower orbital velocity at apocenter (Pan et al., 2016). Subsequently scattered light imaging showed that the star is not exactly at the center of the disk (Milli et al., 2017), although the direction of the offset is not exactly that expected suggesting that there maybe some additional density variations around the ring (see also Löhne et al., 2017). Fomalhaut's disk (see Fig. 16.1) is perhaps a better example, being brighter and larger on the sky so that the offset and asymmetries are more clearly seen (Kalas et al., 2005; MacGregor et al., 2017), with additional azimuthal structure also evident (Kalas et al., 2013).

The origin of these eccentricities, which are of order 0.05–0.15, has not been identified, but would arise naturally from the secular perturbations of a planet on an eccentric orbit. While a planet candidate was detected orbiting near the inner edge of the Fomalhaut disk (see Fig. 16.1, Kalas et al., 2008) its eccentricity was subsequently found to put it on a disk-crossing orbit (Kalas et al., 2013) that would significantly disrupt the disk suggesting that the planet is less massive than originally thought and/or has only been put on this orbit recently (Beust et al., 2014; Tamayo, 2014; Lawler et al., 2015). The presence of an additional planet in the system that is responsible for the eccentricity is thus inferred.

More secure identification of a disk asymmetry that is caused by planetary perturbations is the warp in the β Pic disk (Mouillet et al., 1997). The change in the plane of symmetry seen in scattered light images at 80 AU in this edge-on disk (see Apai et al., 2015) was inferred to arise from the secular perturbations of a planet on an inclined orbit at $\sim$10 AU. Since that planet has subsequently been found with direct imaging (Lagrange et al., 2010), this supports the idea that disk structures can be used to identify unseen planets. That said, there remains some uncertainty in the interpretation of the vertical structure of the β Pic disk, not least because the additional clumpy structure mentioned below is situated at the same location as the warp.

Both the Vega and ϵ Eri debris disks were inferred to be clumpy based on sub-mm imaging (Holland et al., 1998; Greaves et al., 1998). Subsequent imaging gave conflicting views on the presence of the clumps (Greaves et al., 2005; Marsh et al., 2006; Hughes et al., 2012; Holland et al., 2017), but it seems likely that at least some of the clumpiness was attributable to background galaxies. The most secure evidence for clumpiness in a disk comes from β Pic for which mid-IR images show a prominent clump at $\sim$50 AU in projected separation from the star (Telesco et al., 2005). The same clump was later imaged in CO (see Fig. 16.7, Dent et al., 2014), for which the additional kinematic information showed that the clump is both radially and azimuthally broad and situated at ~85 AU distance from the star (Matrà et al., 2017a). Two models were proposed to explain the clump, either a recent collision between two Mars-sized protoplanets (Jackson et al., 2014), or the trapping of planetesimals into the 2:1 and 3:2 resonances of a $\sim 30 M_{\oplus}$ planet that migrated out from 31 to 57 AU over 23 Myr (Wyatt, 2003; Matrà et al., 2019b). The radial breadth of the clump disfavors a collisional origin leaving the favored interpretation as a structure analogous to the resonant Kuiper belt objects (the Plutinos and Twotinos). Comparison of Figs. 16.2 and 16.7 shows there are some notable differences, however, such as that the tight clump observed in the β Pic disk would require the planetesimals to have a smaller libration width than resonant objects in the Kuiper belt. This could be explained within the context of this model by the stochasticity of the migration and/or the initial orbital eccentricities (e.g., Nesvorný and Vokrouhlický, 2016; Reche et al., 2008).

16.4.6 Connection from outer to inner system

The fraction of stars with excess 10 μm emission indicative of warm dust in the region where there maybe habitable planets was long recognized to be low (Aumann and Probst, 1991) and was quantified more recently from the WISE photometric survey to find that excesses above 10% of the stellar emission only occur to $1{:}10^4$ nearby stars (Kennedy and Wyatt, 2013), although this rate is higher if considering a sample of young <100 Myr stars. This result is explained by collisional evolution (see also Fig. 16.3 and Section 16.4.1, Wyatt et al., 2007a), which suggests that the rare bright excesses are transient populations of warm dust, perhaps originating in recent collisions. However, ongoing interferometric surveys have shown that excesses at 0.1%–1% levels are more common, and present to ~20% of stars (Ertel et al., 2018), and moreover, are more prevalent (though not exclusively) around stars known to host cold outer disks (Mennesson et al., 2014).

This does not prove that the warm dust must originate somehow in the outer belt, since it could be that the conditions of a protoplanetary disk that leads to the formation of an outer cold belt also favor the formation of a massive asteroid belt (e.g., Geiler and Krivov, 2017). However, there are two mechanisms by which outer belts can replenish a warm dust population. The first is by Poynting-Robertson drag (Burns et al., 1979). While the dust population dragged in from the known outer belts is expected to be significantly depleted by mutual collisions (which is not the case in the more tenuous belt in the solar system, Wyatt, 2005), the levels of dust expected to make it in to ~1 AU are comparable to those inferred from observations in some cases (Mennesson et al., 2014; Kennedy and Piette, 2015), although the dust population may be further depleted if the dust has to cross the path of any planets (Bonsor et al., 2018). The second is by comet scattering. That is, an intervening planetary system could scatter planetesimals into the inner regions where they could sublimate or disintegrate to replenish the warm dust, much as comets in the solar system replenish the zodiacal cloud (Nesvorný et al., 2010).

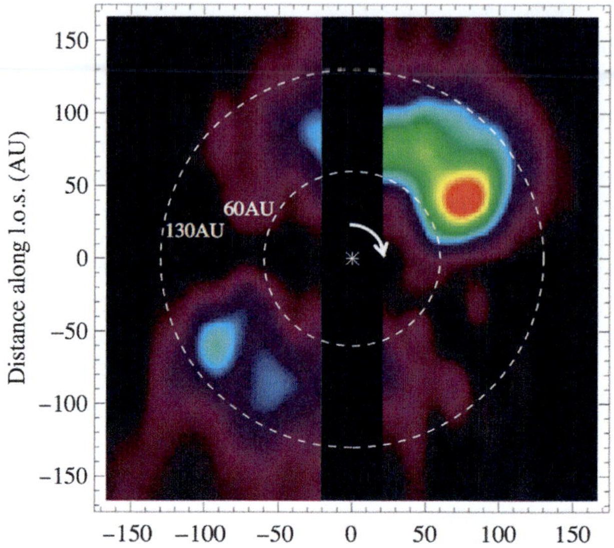

FIG. 16.7 Dynamical structure in the debris disk of β Pic that is analogous to that of resonant Kuiper belt objects. While the disk is edge-on, velocity information from CO observations allows its face-on distribution to be inferred showing that this includes two clumps of unequal brightness on opposite sides of the star (Dent et al., 2014). The best explanation is that outward migration of a planet close to the inner edge of the disk trapped volatile-rich planetesimals into its 2:1 and 3:2 resonances causing over-densities in these regions where collisions are most frequent (Wyatt, 2003). The CO lifetime is short compared with the orbital timescale meaning that it does not travel far from where it is created before being photodissociated.

For η Crv shown in Fig. 16.8, its ~10% excess from dust at ~1 AU (Smith et al., 2009; Defrère et al., 2015) cannot be explained by steady state collisions in an asteroid belt given its ~1 Gyr age (Wyatt et al., 2007a). In this case, an origin in the outer belt seen at 150 AU (Wyatt et al., 2005) is most likely, and is supported by the composition of the warm dust inferred from its infrared spectrum (Lisse et al., 2012). The warm dust is too bright to be explained by P-R drag, but inward scattering of comets is the favored scenario, particularly following the tentative detection of CO gas at ~20 AU from the star (Marino et al., 2017b). The detection of gas at a location where there is no dust is surprising, but is readily explained if this is the location where comets start sublimating on their way in before depositing their dust at ~1 AU.

The requirement to sustain a high rate of comet scattering sets constraints on the intervening planetary system, for example, since the comets should not be ejected before reaching the inner regions (Wyatt et al., 2017), but it is found that a plausible planetary system architecture is one with a chain of 3–30$M_\oplus$ planets (Marino et al., 2017b). More generally, some level of comet scattering is expected in all systems with an outer belt, with the rate determined by the processes that perturb planetesimals out of their orbits in the belt and the architecture of their planetary system (Marino et al., 2018a). Since these planetesimals may impact any planets and so deliver water or strip their atmospheres, this has significant interest for those seeking to understand the prevalence of conditions that favor the development of life.

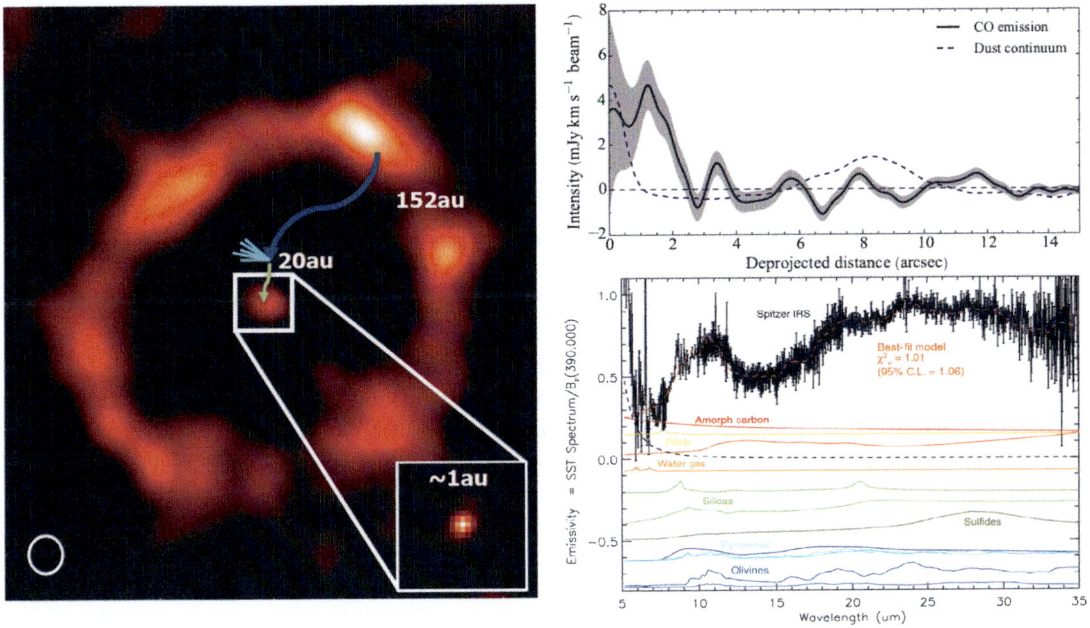

FIG. 16.8 Evidence for inward scattering of exocomets in the debris disk of the ~1 Gyr F star η Crv. *(Left)* ALMA image of its outer debris belt (and the stellar photospheric emission at the center), with inset (bottom right) of the mid-IR emission showing warm dust close to the star, and annotation to illustrate the origin of that warm dust in exocomet scattering (Marino et al., 2017b). *(Top right)* The profile of CO emission showing this peaks at ~20 AU close to where the exocomets may be expected to cross the H_2O or CO_2 ice lines (Marino et al., 2017b). *(Bottom right)* The spectrum of mid-IR emission can be fitted by a model for the dust composition that suggests it is water- and carbon-rich (Lisse et al., 2012).

16.5 Conclusions

The Sun is not unique in having either a planetary system or in having belts of planetesimals. Other stars also have what can be considered as extrasolar analogues to the Kuiper belt. However, the extent of that analogy depends on how strict the similarities must be. Certainly, other stars have belts of planetesimals at distances from their stars similar to our own Kuiper belt, but these belts contain orders of magnitude more mass, perhaps comparable to that in our primordial Kuiper belt. Since we cannot yet detect a true Kuiper belt analogue around another star we cannot yet tell how ours compares within the population, except to say that it is not in the top 20%.

There are, however, many other similarities to the Kuiper belt in the debris disks observed around other stars. For example, several have volatile-rich compositions inferred to be similar to solar system comets (from gas observations). In terms of structures, there is evidence for analogues to the hot and cold classical belt (from vertically resolved edge-on disks), to the resonant Kuiper belt object population (from clumpy disks), and to the scattered disk (from haloes of mm-sized grains). Many disks are seen to be radially narrow with sharp inner and outer edges (from resolved imaging). It is also inferred that in some systems exocomets are scattered into the inner regions (from observations of close-in dust and gas).

It is nevertheless worth bearing in mind that these similarities are based in some cases on a relatively small number of what must inevitably be considered the extremes of the debris disk population (in that they are bright enough to have been studied in sufficient detail to identify any similarity). Also, the similarities that are highlighted in the literature (and in this review) reflect to some extent our inherently anthropocentric view (e.g., some of the aforementioned features may have other interpretations). Moreover, there are also examples of disks that do not fit into a Kuiper belt-like mold (e.g., the broad disks or the eccentric rings).

These differences in the extrasolar debris disk population are to be expected, because just as we have learned by comparing the solar system to the exoplanet population, planet formation has a diverse range of outcomes of which the solar system is just one example. By studying and comparing our Kuiper belt to extrasolar debris disks, we have a different perspective on those outcomes, one which is biased toward processes that occur in the outer regions of the systems that are hard to access through exoplanet observations. What the similarities do show is that, perhaps, we share some dynamical processes in common with other stars, and by considering the solar system as one outcome we may achieve a deeper understanding of the planet formation process and the uniqueness of our own situation.

References

Acke, B., Min, M., Dominik, C., Vandenbussche, B., Sibthorpe, B., Waelkens, C., Olofsson, G., Degroote, P., Smolders, K., Pantin, E., Barlow, M.J., Blommaert, J.A.D.L., Brandeker, A., De Meester, W., Dent, W.R.F., Exter, K., Di Francesco, J., Fridlund, M., Gear, W.K., Glauser, A.M., Greaves, J.S., Harvey, P.M., Henning, T., Hogerheijde, M.R., Holland, W.S., Huygen, R., Ivison, R.J., Jean, C., Liseau, R., Naylor, D.A., Pilbratt, G.L., Polehampton, E.T., Regibo, S., Royer, P., Sicilia-Aguilar, A., Swinyard, E.T., 2012. Herschel images of Fomalhaut. An extrasolar Kuiper belt at the height of its dynamical activity. Astron. Astrophys. 540, A125. https://doi.org/10.1051/0004-6361/201118581.

Apai, D., Schneider, G., Grady, C.A., Wyatt, M.C., Lagrange, A.-M., Kuchner, M.J., Stark, C.J., Lubow, S.H., 2015. The inner disk structure, disk-planet interactions, and temporal evolution in the β Pictoris system: a two-epoch HST/STIS coronagraphic study. Astrophys. J. 800, 136. https://doi.org/10.1088/0004-637X/800/2/136.

Aumann, H.H., Probst, R.G., 1991. Search for Vega-like nearby stars with 12 micron excess. Astrophys. J. 368, 264–271. https://doi.org/10.1086/169690.

Aumann, H.H., Gillett, F.C., Beichman, C.A., de Jong, T., Houck, J.R., Low, F.J., Neugebauer, G., Walker, R.G., Wesselius, P.R., 1984. Discovery of a shell around alpha lyrae. Astrophys. J. 278, L23–L27. https://doi.org/10.1086/184214.

Backman, D.E., Paresce, F., 1993. Main-sequence stars with circumstellar solid material—the VEGA phenomenon. In: Levy, E.H., Lunine, J.I. (Eds.) Protostars and Planets III, Univ. Arizona Press, Tucson, pp. 1253–1304.

Backman, D.E., Dasgupta, A., Stencel, R.E., 1995. Model of a Kuiper belt small grain population and resulting far-infrared emission. Astrophys. J. 450, L35. https://doi.org/10.1086/309660.

Beust, H., Augereau, J.-C., Bonsor, A., Graham, J.R., Kalas, P., Lebreton, J., Lagrange, A.-M., Ertel, S., Faramaz, V., Thébault, P., 2014. An independent determination of Fomalhaut b's orbit and the dynamical effects on the outer dust belt. Astron. Astrophys. 561, A43. https://doi.org/10.1051/0004-6361/201322229.

Bonsor, A., Wyatt, M.C., Kral, Q., Kennedy, G., Shannon, A., Ertel, S., 2018. Using warm dust to constrain unseen planets. Mon. Not. R. Astron. Soc. 480, 5560–5579. https://doi.org/10.1093/mnras/sty2200.

Booth, R.A., Clarke, C.J., 2016. Collision velocity of dust grains in self-gravitating protoplanetary discs. Mon. Not. R. Astron. Soc. 458, 2676–2693. https://doi.org/10.1093/mnras/stw488.

Booth, M., Wyatt, M.C., Morbidelli, A., Moro-Martín, A., Levison, H.F., 2009. The history of the solar system's debris disc: observable properties of the Kuiper belt. Mon. Not. R. Astron. Soc. 399, 385–398. https://doi.org/10.1111/j.1365-2966.2009.15286.x.

Booth, M., Kennedy, G., Sibthorpe, B., Matthews, B.C., Wyatt, M.C., Duchêne, G., Kavelaars, J.J., Rodriguez, D., Greaves, J.S., Koning, A., Vican, L., Rieke, G.H., Su, K.Y.L., Moro-Martín, A., Kalas, P., 2013. Resolved debris discs around A stars in the Herschel DEBRIS survey. Mon. Not. R. Astron. Soc. 428, 1263–1280. https://doi.org/10.1093/mnras/sts117.

Booth, M., Jordán, A., Casassus, S., Hales, A.S., Dent, W.R.F., Faramaz, V., Matrà, L., Barkats, D., Brahm, R., Cuadra, J., 2016. Resolving the planetesimal belt of HR 8799 with ALMA. Mon. Not. R. Astron. Soc. 460, L10–L14. https://doi.org/10.1093/mnrasl/slw040.

Booth, M., Dent, W.R.F., Jordán, A., Lestrade, J.-F., Hales, A.S., Wyatt, M.C., Casassus, S., Ertel, S., Greaves, J.S., Kennedy, G.M., Matrà, L., Augereau, J.-C., Villard, E., 2017. The Northern arc of ϵ Eridani's Debris Ring as seen by ALMA. Mon. Not. R. Astron. Soc. 469, 3200–3212. https://doi.org/10.1093/mnras/stx1072.

Bottke, W.F., Durda, D.D., Nesvorný, D., Jedicke, R., Morbidelli, A., Vokrouhlický, D., Levison, H., 2005. The fossilized size distribution of the main asteroid belt. Icarus 175, 111–140. https://doi.org/10.1016/j.icarus.2004.10.026.

Burns, J.A., Lamy, P.L., Soter, S., 1979. Radiation forces on small particles in the solar system. Icarus 40, 1–48. https://doi.org/10.1016/0019-1035(79)90050-2.

Carrera, D., Gorti, U., Johansen, A., Davies, M.B., 2017. Planetesimal formation by the streaming instability in a photoevaporating disk. Astrophys. J. 839, 16. https://doi.org/10.3847/1538-4357/aa6932.

Cataldi, G., Brandeker, A., Wu, Y., Chen, C., Dent, W., de Vries, B.L., Kamp, I., Liseau, R., Olofsson, G., Pantin, E., Roberge, A., 2018. ALMA resolves C I emission from the β Pictoris debris disk. Astrophys. J. 861, 72. https://doi.org/10.3847/1538-4357/aac5f3.

Chen, C.H., Sargent, B.A., Bohac, C., Kim, K.H., Leibensperger, E., Jura, M., Najita, J., Forrest, W.J., Watson, D.M., Sloan, G.C., Keller, L.D., 2006. Spitzer IRS spectroscopy of IRAS-discovered debris disks. Astrophys. J. Suppl. 166, 351–377. https://doi.org/10.1086/505751.

Chiang, E., Kite, E., Kalas, P., Graham, J.R., Clampin, M., 2009. Fomalhaut's debris disk and planet: constraining the mass of Fomalhaut b from disk morphology. Astrophys. J. 693, 734–749. https://doi.org/10.1088/0004-637X/693/1/734.

Daley, C., Hughes, A.M., Carter, E.S., Flaherty, K., Lambros, Z., Pan, M., Schlichting, H., Chiang, E., Wyatt, M., Wilner, D., Andrews, S., Carpenter, J., 2019. The mass of stirring bodies in the AU mic debris disk inferred from resolved vertical structure. Astrophys. J. 875, 87. https://doi.org/10.3847/1538-4357/ab1074.

de Vries, B.L., Acke, B., Blommaert, J.A.D.L., Waelkens, C., Waters, L.B.F.M., Vandenbussche, B., Min, M., Olofsson, G., Dominik, C., Decin, L., Barlow, M.J., Brandeker, A., di Francesco, J., Glauser, A.M., Greaves, J., Harvey, P.M., Holland, W.S., Ivison, R.J., Liseau, R., Pantin, E.E., Pilbratt, G.L., Royer, P., Sibthorpe, B., 2012. Comet-like mineralogy of olivine crystals in an extrasolar proto-Kuiper belt. Nature 490, 74–76. https://doi.org/10.1038/nature11469.

Defrère, D., Hinz, P.M., Skemer, A.J., Kennedy, G.M., Bailey, V.P., Hoffmann, W.F., Mennesson, B., Millan-Gabet, R., Danchi, W.C., Absil, O., Arbo, P., Beichman, C., Brusa, G., Bryden, G., Downey, E.C., Durney, O., Esposito, S., Gaspar, A., Grenz, P., Haniff, C., Hill, J.M., Lebreton, J., Leisenring, J.M., Males, J.R., Marion, L., McMahon, T.J., Montoya, M., Morzinski, K.M., Pinna, E., Puglisi, A., Rieke, G., Roberge, A., Serabyn, E., Sosa, R., Stapeldfeldt, K., Su, K., Vaitheeswaran, V., Vaz, A., Weinberger, A.J., Wyatt, M.C., 2015. First-light LBT nulling interferometric observations: warm exozodiacal dust resolved within a few AU of η Crv. Astrophys. J. 799, 42. https://doi.org/10.1088/0004-637X/799/1/42.

Dent, W.R.F., Wyatt, M.C., Roberge, A., Augereau, J.-C., Casassus, S., Corder, S., Greaves, J.S., de Gregorio-Monsalvo, I., Hales, A., Jackson, A.P., Hughes, A.M., Lagrange, A.-M., Matthews, B., Wilner, D., 2014. Molecular gas clumps from the destruction of icy bodies in the β Pictoris debris disk. Science 343, 1490–1492. https://doi.org/10.1126/science.1248726.

Eiroa, C., Marshall, J.P., Mora, A., Krivov, A.V., Montesinos, B., Absil, O., Ardila, D., Arévalo, M., Augereau, J.-C., Bayo, A., Danchi, W., Del Burgo, C., Ertel, S., Fridlund, M., González-García, B.M., Heras, A.M., Lebreton, J., Liseau, R., Maldonado, J., Meeus, G., Montes, D., Pilbratt, G.L., Roberge, A., Sanz-Forcada, J., Stapelfeldt, K., Thébault, P., White, G.J., Wolf, S., 2011. Herschel discovery of a new class of cold, faint debris discs. Astron. Astrophys. 536, L4. https://doi.org/10.1051/0004-6361/201117797.

Eiroa, C., Marshall, J.P., Mora, A., Montesinos, B., Absil, O., Augereau, J.C., Bayo, A., Bryden, G., Danchi, W., del Burgo, C., Ertel, S., Fridlund, M., Heras, A.M., Krivov, A.V., Launhardt, R., Liseau, R., Löhne, T., Maldonado, J., Pilbratt, G.L., Roberge, A., Rodmann, J., Sanz-Forcada, J., Solano, E., Stapelfeldt, K., Thébault, P., Wolf, S., Ardila, D., Arévalo, M., Beichmann, C., Faramaz, V., González-García, B.M., Gutiérrez, R., Lebreton, J., Martínez-Arnáiz, R., Meeus, G., Montes, D., Olofsson, G., Su, K.Y.L., White, G.J., Barrado, D., Fukagawa, M., Grün, E., Kamp, I.,

Lorente, R., Morbidelli, A., Müller, S., Mutschke, H., Nakagawa, T., Ribas, I., Walker, H., 2013. DUst around NEarby Stars. The survey observational results. Astron. Astrophys. 555, A11. https://doi.org/10.1051/0004-6361/201321050.

Ertel, S., Defrère, D., Hinz, P., Mennesson, B., Kennedy, G.M., Danchi, W.C., Gelino, C., Hill, J.M., Hoffmann, W.F., Rieke, G., Shannon, A., Spalding, E., Stone, J.M., Vaz, A., Weinberger, A.J., Willems, P., Absil, O., Arbo, P., Bailey, V.P., Beichman, C., Bryden, G., Downey, E.C., Durney, O., Esposito, S., Gaspar, A., Grenz, P., Haniff, C.A., Leisenring, J.M., Marion, L., McMahon, T.J., Millan-Gabet, R., Montoya, M., Morzinski, K.M., Pinna, E., Power, J., Puglisi, A., Roberge, A., Serabyn, E., Skemer, A.J., Stapelfeldt, K., Su, K.Y.L., Vaitheeswaran, V., Wyatt, M.C., 2018. The HOSTS survey—exozodiacal dust measurements for 30 stars. Astron. J. 155, 194. https://doi.org/10.3847/1538-3881/aab717.

Fraser, W.C., 2009. The collisional divot in the Kuiper belt size distribution. Astrophys. J. 706, 119–129. https://doi.org/10.1088/0004-637X/706/1/119.

Gáspár, A., Rieke, G.H., 2014. The Herschel cold debris disks: confusion with the extragalactic background at 160 μm. Astrophys. J. 784, 33. https://doi.org/10.1088/0004-637X/784/1/33.

Gáspár, A., Rieke, G.H., Balog, Z., 2013. The collisional evolution of debris disks. Astrophys. J. 768, 25. https://doi.org/10.1088/0004-637X/768/1/25.

Geiler, F., Krivov, A.V., 2017. Does warm debris dust stem from asteroid belts? Mon. Not. R. Astron. Soc. 468, 959–970. https://doi.org/10.1093/mnras/stx462.

Geiler, F., Krivov, A.V., Booth, M., Löhne, T., 2019. The scattered disc of HR 8799. Mon. Not. R. Astron. Soc. 483, 332–341. https://doi.org/10.1093/mnras/sty3160.

Gladman, B., Kavelaars, J.J., Petit, J.-M., Morbidelli, A., Holman, M.J., Loredo, T., 2001. The structure of the Kuiper belt: size distribution and radial extent. Astron. J. 122, 1051–1066. https://doi.org/10.1086/322080.

Gomes, R., Levison, H.F., Tsiganis, K., Morbidelli, A., 2005. Origin of the cataclysmic late heavy bombardment period of the terrestrial planets. Nature 435, 466–469. https://doi.org/10.1038/nature03676.

Greaves, J.S., Wyatt, M.C., 2003. Some anomalies in the occurrence of debris discs around main-sequence A and G stars. Mon. Not. R. Astron. Soc. 345, 1212–1222. https://doi.org/10.1046/j.1365-2966.2003.07037.x.

Greaves, J.S., Holland, W.S., Moriarty-Schieven, G., Jenness, T., Dent, W.R.F., Zuckerman, B., McCarthy, C., Webb, R.A., Butner, H.M., Gear, W.K., Walker, H.J., 1998. A dust ring around ϵ Eridani: analog to the young solar system. Astrophys. J. 506, L133–L137. https://doi.org/10.1086/311652.

Greaves, J.S., Holland, W.S., Wyatt, M.C., Dent, W.R.F., Robson, E.I., Coulson, I.M., Jenness, T., Moriarty-Schieven, G.H., Davis, G.R., Butner, H.M., Gear, W.K., Dominik, C., Walker, H.J., 2005. Structure in the ϵ Eridani debris disk. Astrophys. J. 619, L187–L190. https://doi.org/10.1086/428348.

Greaves, J.S., Holland, W.S., Matthews, B.C., Marshall, J.P., Dent, W.R.F., Woitke, P., Wyatt, M.C., Matrà, L., Jackson, A., 2016. Gas and dust around A-type stars at tens of Myr: signatures of cometary breakup. Mon. Not. R. Astron. Soc. 461, 3910–3917. https://doi.org/10.1093/mnras/stw1569.

Heng, K., Tremaine, S., 2010. Long-lived planetesimal discs. Mon. Not. R. Astron. Soc. 401, 867–889. https://doi.org/10.1111/j.1365-2966.2009.15739.x.

Higuchi, A.E., Sato, A., Tsukagoshi, T., Sakai, N., Iwasaki, K., Momose, M., Kobayashi, H., Ishihara, D., Watanabe, S., Kaneda, H., Yamamoto, S., 2017. Detection of submillimeter-wave [C I] emission in gaseous debris disks of 49 Ceti and β Pictoris. Astrophys. J. 839, L14. https://doi.org/10.3847/2041-8213/aa67f4.

Holland, W.S., Greaves, J.S., Zuckerman, B., Webb, R.A., McCarthy, C., Coulson, I.M., Walther, D.M., Dent, W.R.F., Gear, W.K., Robson, I., 1998. Submillimetre images of dusty debris around nearby stars. Nature 392, 788–791. https://doi.org/10.1038/33874.

Holland, W.S., Matthews, B.C., Kennedy, G.M., Greaves, J.S., Wyatt, M.C., Booth, M., Bastien, P., Bryden, G., Butner, H., Chen, C.H., Chrysostomou, A., Davies, C.L., Dent, W.R.F., Di Francesco, J., Duchêne, G., Gibb, A.G., Friberg, P., Ivison, R.J., Jenness, T., Kavelaars, J., Lawler, S., Lestrade, J.-F., Marshall, J.P., Moro-Martin, A., Panić, O., Phillips, N., Serjeant, S., Schieven, G.H., Sibthorpe, B., Vican, L., Ward-Thompson, D., van der Werf, P., White, G.J., Wilner, D., Zuckerman, B., 2017. SONS: the JCMT legacy survey of debris discs in the submillimetre. Mon. Not. R. Astron. Soc. 470, 3606–3663. https://doi.org/10.1093/mnras/stx1378.

Hughes, A.M., Wilner, D.J., Mason, B., Carpenter, J.M., Plambeck, R., Chiang, H.-F., Andrews, S.M., Williams, J.P., Hales, A., Su, K., Chiang, E., Dicker, S., Korngut, P., Devlin, M., 2012. Confirming the primarily smooth structure of the Vega debris disk at millimeter wavelengths. Astrophys. J. 750, 82. https://doi.org/10.1088/0004-637X/750/1/82.

Hughes, A.M., Lieman-Sifry, J., Flaherty, K.M., Daley, C.M., Roberge, A., Kóspál, Á., Moór, A., Kamp, I., Wilner, D.J., Andrews, S.M., Kastner, J.H., Ábrahám, P., 2017. Radial surface density profiles of gas and dust in the debris disk around 49 Ceti. Astrophys. J. 839, 86. https://doi.org/10.3847/1538-4357/aa6b04.

Hughes, A.M., Duchêne, G., Matthews, B.C., 2018. Debris disks: structure, composition, and variability. Ann. Rev. Astron. Astrophys. 56, 541–591. https://doi.org/10.1146/annurev-astro-081817-052035.

Jackson, A.P., Wyatt, M.C., 2012. Debris from terrestrial planet formation: the Moon-forming collision. Mon. Not. R. Astron. Soc. 425, 657–679. https://doi.org/10.1111/j.1365-2966.2012.21546.x.

Jackson, A.P., Wyatt, M.C., Bonsor, A., Veras, D., 2014. Debris froms giant impacts between planetary embryos at large orbital radii. Mon. Not. R. Astron. Soc. 440, 3757–3777. https://doi.org/10.1093/mnras/stu476.

Kalas, P., Graham, J.R., Clampin, M., 2005. A planetary system as the origin of structure in Fomalhaut's dust belt. Nature 435, 1067–1070. https://doi.org/10.1038/nature03601.

Kalas, P., Graham, J.R., Chiang, E., Fitzgerald, M.P., Clampin, M., Kite, E.S., Stapelfeldt, K., Marois, C., Krist, J., 2008. Optical images of an exosolar planet 25 light-years from Earth. Science 322, 1345. https://doi.org/10.1126/science.1166609.

Kalas, P., Graham, J.R., Fitzgerald, M.P., Clampin, M., 2013. STIS coronagraphic imaging of Fomalhaut: main belt structure and the orbit of Fomalhaut b. Astrophys. J. 775, 56. https://doi.org/10.1088/0004-637X/775/1/56.

Kennedy, G.M., Piette, A., 2015. Warm exo-Zodi from cool exo-Kuiper belts: the significance of P-R drag and the inference of intervening planets. Mon. Not. R. Astron. Soc. 449, 2304–2311. https://doi.org/10.1093/mnras/stv453.

Kennedy, G.M., Wyatt, M.C., 2010. Are debris discs self-stirred? Mon. Not. R. Astron. Soc. 405, 1253–1270. https://doi.org/10.1111/j.1365-2966.2010.16528.x.

Kennedy, G.M., Wyatt, M.C., 2012. Confusion limited surveys: using WISE to quantify the rarity of warm dust around Kepler stars. Mon. Not. R. Astron. Soc. 426, 91–107. https://doi.org/10.1111/j.1365-2966.2012.21621.x.

Kennedy, G.M., Wyatt, M.C., 2013. The bright end of the exo-Zodi luminosity function: disc evolution and implications for exo-Earth detectability. Mon. Not. R. Astron. Soc. 433, 2334–2356. https://doi.org/10.1093/mnras/stt900.

Kennedy, G.M., Wyatt, M.C., 2014. Do two-temperature debris discs have multiple belts? Mon. Not. R. Astron. Soc. 444, 3164–3182. https://doi.org/10.1093/mnras/stu1665.

Kennedy, G.M., Marino, S., Matrà, L., Panić, O., Wilner, D., Wyatt, M.C., Yelverton, B., 2018. ALMA observations of the narrow HR 4796A debris ring. Mon. Not. R. Astron. Soc. 475, 4924–4938. https://doi.org/10.1093/mnras/sty135.

Kennedy, G.M., Matrà, L., Facchini, S., Milli, J., Panić, O., Price, D., Wilner, D.J., Wyatt, M.C., Yelverton, B.M., 2019. A circumbinary protoplanetary disk in a polar configuration. Nat. Astron. 3, 230–235. https://doi.org/10.1038/s41550-018-0667-x.

Kenyon, S.J., Bromley, B.C., 2010. Variations on debris disks. II. Icy planet formation as a function of the bulk properties and initial sizes of planetesimals. Astrophys. J. Suppl. 188, 242–279. https://doi.org/10.1088/0067-0049/188/1/242.

Kóspál, Á., Moór, A., Juhász, A., Ábrahám, P., Apai, D., Csengeri, T., Grady, C.A., Henning, T., Hughes, A.M., Kiss, C., Pascucci, I., Schmalzl, M., 2013. ALMA observations of the molecular gas in the debris disk of the 30 Myr old star HD 21997. Astrophys. J. 776, 77. https://doi.org/10.1088/0004-637X/776/2/77.

Kral, Q., Latter, H., 2016. The magnetorotational instability in debris-disc gas. Mon. Not. R. Astron. Soc. 461, 1614–1620. https://doi.org/10.1093/mnras/stw1429.

Kral, Q., Wyatt, M., Carswell, R.F., Pringle, J.E., Matrà, L., Juhász, A., 2016. A self-consistent model for the evolution of the gas produced in the debris disc of β Pictoris. Mon. Not. R. Astron. Soc. 461, 845–858. https://doi.org/10.1093/mnras/stw1361.

Kral, Q., Matrà, L., Wyatt, M.C., Kennedy, G.M., 2017. Predictions for the secondary CO, C and O gas content of debris discs from the destruction of volatile-rich planetesimals. Mon. Not. R. Astron. Soc. 469, 521–550. https://doi.org/10.1093/mnras/stx730.

Kral, Q., Marino, S., Wyatt, M.C., Kama, M., Matra, L., 2019. Imaging [CI] around HD 131835: reinterpreting young debris discs with protoplanetary disc levels of CO gas as shielded secondary discs. Mon. Not. R. Astron. Soc. 489, 3670–3691. https://doi.org/10.1093/mnras/sty2923.

Krivov, A.V., Booth, M., 2018. Self-stirring of debris discs by planetesimals formed by pebble concentration. Mon. Not. R. Astron. Soc. 479, 3300–3307. https://doi.org/10.1093/mnras/sty1607.

Krivov, A.V., Eiroa, C., Löhne, T., Marshall, J.P., Montesinos, B., del Burgo, C., Absil, O., Ardila, D., Augereau, J.-C., Bayo, A., Bryden, G., Danchi, W., Ertel, S., Lebreton, J., Liseau, R., Mora, A., Mustill, A.J., Mutschke, H.,

Neuhäuser, R., Pilbratt, G.L., Roberge, A., Schmidt, T.O.B., Stapelfeldt, K.R., Thébault, P., Vitense, C., White, G.J., Wolf, S., 2013. Herschel's "cold debris disks": background galaxies or quiescent rims of planetary systems? Astrophys. J. 772, 32. https://doi.org/10.1088/0004-637X/772/1/32.

Krivov, A.V., Ide, A., Löhne, T., Johansen, A., Blum, J., 2018. Debris disc constraints on planetesimal formation. Mon. Not. R. Astron. Soc. 474, 2564–2575. https://doi.org/10.1093/mnras/stx2932.

Lagrange, A.-M., Bonnefoy, M., Chauvin, G., Apai, D., Ehrenreich, D., Boccaletti, A., Gratadour, D., Rouan, D., Mouillet, D., Lacour, S., Kasper, M., 2010. A giant planet imaged in the disk of the young star β Pictoris. Science 329, 57. https://doi.org/10.1126/science.1187187.

Lawler, S.M., 2014. The debiased Kuiper belt: our solar system as a debris disk, In: Booth, M., Matthews, B.C., Graham, J.R. (Eds.) Exploring the Formation and Evolution of Planetary Systems, IAU Symposium, vol. 299, pp. 232–236.

Lawler, S.M., Greenstreet, S., Gladman, B., 2015. Fomalhaut b as a dust cloud: frequent collisions within the Fomalhaut disk. Astrophys. J. 802, L20. https://doi.org/10.1088/2041-8205/802/2/L20.

Lebreton, J., Augereau, J.-C., Thi, W.-F., Roberge, A., Donaldson, J., Schneider, G., Maddison, S.T., Ménard, F., Riviere-Marichalar, P., Mathews, G.S., Kamp, I., Pinte, C., Dent, W.R.F., Barrado, D., Duchêne, G., Gonzalez, J.-F., Grady, C.A., Meeus, G., Pantin, E., Williams, J.P., Woitke, P., 2012. An icy Kuiper belt around the young solar-type star HD 181327. Astron. Astrophys. 539, A17. https://doi.org/10.1051/0004-6361/201117714.

Liou, J.-C., Zook, H.A., 1999. Signatures of the giant planets imprinted on the Edgeworth-Kuiper belt dust disk. Astron. J. 118, 580–590. https://doi.org/10.1086/300938.

Lisse, C.M., Chen, C.H., Wyatt, M.C., Morlok, A., Song, I., Bryden, G., Sheehan, P., 2009. Abundant circumstellar silica dust and SiO gas created by a giant hypervelocity collision in the 12 Myr HD172555 system. Astrophys. J. 701, 2019–2032. https://doi.org/10.1088/0004-637X/701/2/2019.

Lisse, C.M., Wyatt, M.C., Chen, C.H., Morlok, A., Watson, D.M., Manoj, P., Sheehan, P., Currie, T.M., Thebault, P., Sitko, M.L., 2012. Spitzer evidence for a late-heavy bombardment and the formation of ureilites in η Corvi at 1 Gyr. Astrophys. J. 747, 93. https://doi.org/10.1088/0004-637X/747/2/93.

Löhne, T., Krivov, A.V., Rodmann, J., 2008. Long-term collisional evolution of debris disks. Astrophys. J. 673, 1123–1137. https://doi.org/10.1086/524840.

Löhne, T., Krivov, A.V., Kirchschlager, F., Sende, J.A., Wolf, S., 2017. Collisions and drag in debris discs with eccentric parent belts. Astron. Astrophys. 605, A7. https://doi.org/10.1051/0004-6361/201630297.

MacGregor, M.A., Lawler, S.M., Wilner, D.J., Matthews, B.C., Kennedy, G.M., Booth, M., Di Francesco, J., 2016. ALMA observations of the debris disk of solar analog τ Ceti. Astrophys. J. 828, 113. https://doi.org/10.3847/0004-637X/828/2/113.

MacGregor, M.A., Matrà, L., Kalas, P., Wilner, D.J., Pan, M., Kennedy, G.M., Wyatt, M.C., Duchene, G., Hughes, A.M., Rieke, G.H., Clampin, M., Fitzgerald, M.P., Graham, J.R., Holland, W.S., Panić, O., Shannon, A., Su, K., 2017. A complete ALMA Map of the Fomalhaut debris disk. Astrophys. J. 842, 8. https://doi.org/10.3847/1538-4357/aa71ae.

MacGregor, M.A., Weinberger, A.J., Hughes, A. M., Wilner, D.J., Currie, T., Debes, J.H., Donaldson, J.K., Redfield, S., Roberge, A., Schneider, G., 2018. ALMA detection of extended millimeter halos in the HD 32297 and HD 61005 Debris disks. Astrophys. J. 869, 75. https://doi.org/10.3847/1538-4357/aaec71.

Marino, S., Matrà, L., Stark, C., Wyatt, M.C., Casassus, S., Kennedy, G., Rodriguez, D., Zuckerman, B., Perez, S., Dent, W.R.F., Kuchner, M., Hughes, A.M., Schneider, G., Steele, A., Roberge, A., Donaldson, J., Nesvold, E., 2016. Exocometary gas in the HD 181327 debris ring. Mon. Not. R. Astron. Soc. 460, 2933–2944. https://doi.org/10.1093/mnras/stw1216.

Marino, S., Wyatt, M.C., Kennedy, G.M., Holland, W., Matrà, L., Shannon, A., Ivison, R.J., 2017a. ALMA observations of the multiplanet system 61 Vir: what lies outside super-Earth systems? Mon. Not. R. Astron. Soc. 469, 3518–3531. https://doi.org/10.1093/mnras/stx1102.

Marino, S., Wyatt, M.C., Panić, O., Matrà, L., Kennedy, G.M., Bonsor, A., Kral, Q., Dent, W.R.F., Duchene, G., Wilner, D., Lisse, C.M., Lestrade, J.-F., Matthews, B., 2017b. ALMA observations of the η Corvi debris disc: inward scattering of CO-rich exocomets by a chain of 3-30 $M_\oplus$ planets? Mon. Not. R. Astron. Soc. 465, 2595–2615. https://doi.org/10.1093/mnras/stw2867.

Marino, S., Bonsor, A., Wyatt, M.C., Kral, Q., 2018a. Scattering of exocomets by a planet chain: exozodi levels and the delivery of cometary material to inner planets. Mon. Not. R. Astron. Soc. 479, 1651–1671. https://doi.org/10.1093/mnras/sty1475.

Marino, S., Carpenter, J., Wyatt, M.C., Booth, M., Casassus, S., Faramaz, V., Guzman, V., Hughes, A.M., Isella, A., Kennedy, G.M., Matrà, L., Ricci, L., Corder, S., 2018b. A gap in the planetesimal disc around HD

107146 and asymmetric warm dust emission revealed by ALMA. Mon. Not. R. Astron. Soc. 479, 5423–5439. https://doi.org/10.1093/mnras/sty1790.

Marino, S., Yelverton, B., Booth, M., Faramaz, V., Kennedy, G.M., Matrà, L., Wyatt, M.C., 2019. A gap in HD 92945's broad planetesimal disc revealed by ALMA. Mon. Not. R. Astron. Soc. 484, 1257–1269. https://doi.org/10.1093/mnras/stz049.

Marois, C., Zuckerman, B., Konopacky, Q.M., Macintosh, B., Barman, T., 2010. Images of a fourth planet orbiting HR 8799. Nature 468, 1080–1083. https://doi.org/10.1038/nature09684.

Marsh, K.A., Dowell, C.D., Velusamy, T., Grogan, K., Beichman, C.A., 2006. Images of the Vega dust ring at 350 and 450 μm: new clues to the trapping of multiple-sized dust particles in planetary resonances. Astrophys. J. 646, L77–L80. https://doi.org/10.1086/506520.

Matrà, L., Panić, O., Wyatt, M.C., Dent, W.R.F., 2015. CO mass upper limits in the Fomalhaut ring—the importance of NLTE excitation in debris discs and future prospects with ALMA. Mon. Not. R. Astron. Soc. 447, 3936–3947. https://doi.org/10.1093/mnras/stu2619.

Matrà, L., Dent, W.R.F., Wyatt, M.C., Kral, Q., Wilner, D.J., Panić, O., Hughes, A.M., de Gregorio-Monsalvo, I., Hales, A., Augereau, J.-C., Greaves, J., Roberge, A., 2017a. Exocometary gas structure, origin and physical properties around β Pictoris through ALMA CO multitransition observations. Mon. Not. R. Astron. Soc. 464, 1415–1433. https://doi.org/10.1093/mnras/stw2415.

Matrà, L., MacGregor, M.A., Kalas, P., Wyatt, M.C., Kennedy, G.M., Wilner, D.J., Duchene, G., Hughes, A.M., Pan, M., Shannon, A., Clampin, M., Fitzgerald, M.P., Graham, J.R., Holland, W.S., Panić, O., Su, K.Y.L., 2017b. Detection of exocometary CO within the 440 Myr old Fomalhaut belt: a similar CO+CO_2 Ice Abundance in exocomets and solar system comets. Astrophys. J. 842, 9. https://doi.org/10.3847/1538-4357/aa71b4.

Matrà, L., Marino, S., Kennedy, G.M., Wyatt, M.C., Öberg, K.I., Wilner, D.J., 2018a. An empirical planetesimal belt radius-stellar luminosity relation. Astrophys. J. 859, 72. https://doi.org/10.3847/1538-4357/aabcc4.

Matrà, L., Wilner, D.J., Öberg, K.I., Andrews, S.M., Loomis, R.A., Wyatt, M.C., Dent, W.R.F., 2018b. Molecular reconnaissance of the β Pictoris gas disk with the SMA: a low HCN/(CO+CO_2) outgassing ratio and predictions for future surveys. Astrophys. J. 853, 147. https://doi.org/10.3847/1538-4357/aaa42a.

Matrà, L., Öberg, K.I., Wilner, D.J., Olofsson, J., Bayo, A., 2019a. On the ubiquity and stellar luminosity dependence of exocometary CO gas: detection around M dwarf TWA 7. Astron. J. 157, 117. https://doi.org/10.3847/1538-3881/aaff5b.

Matrà, L., Wyatt, M.C., Wilner, D.J., Dent, W.R.F., Marino, S., Kennedy, G.M., Milli, J., 2019b. Kuiper belt-like hot and cold populations of planetesimal inclinations in the β Pictoris belt revealed by ALMA. Astron. J. 157, 135. https://doi.org/10.3847/1538-3881/ab06c0.

Matthews, B.C., Kavelaars, J., 2016. Insights into planet formation from debris disks: I. The solar system as an archetype for planetesimal evolution. Space Sci. Rev. 205, 213–230. https://doi.org/10.1007/s11214-016-0249-0.

Matthews, B.C., Krivov, A.V., Wyatt, M.C., Bryden, G., Eiroa, C., 2014. Observations, modeling, and theory of debris disks. In: Protostars and Planets VI, pp. 521–544.

Mennesson, B., Millan-Gabet, R., Serabyn, E., Colavita, M.M., Absil, O., Bryden, G., Wyatt, M., Danchi, W., Defrère, D., Doré, O., Hinz, P., Kuchner, M., Ragland, S., Scott, N., Stapelfeldt, K., Traub, W., Woillez, J., 2014. Constraining the exozodiacal luminosity function of main-sequence stars: complete results from the keck nuller mid-infrared surveys. Astrophys. J. 797, 119. https://doi.org/10.1088/0004-637X/797/2/119.

Milli, J., Vigan, A., Mouillet, D., Lagrange, A.-M., Augereau, J.-C., Pinte, C., Mawet, D., Schmid, H.M., Boccaletti, A., Matrà, L., Kral, Q., Ertel, S., Chauvin, G., Bazzon, A., Ménard, F., Beuzit, J.-L., Thalmann, C., Dominik, C., Feldt, M., Henning, T., Min, M., Girard, J.H., Galicher, R., Bonnefoy, M., Fusco, T., de Boer, J., Janson, M., Maire, A.-L., Mesa, D., Schlieder, J.E., SPHERE Consortium, 2017. Near-infrared scattered light properties of the HR 4796 A dust ring. A measured scattering phase function from 13.6° to 166.6°. Astron. Astrophys. 599, A108. https://doi.org/10.1051/0004-6361/201527838.

Moór, A., Curé, M., Kóspál, Á., Ábrahám, P., Csengeri, T., Eiroa, C., Gunawan, D., Henning, T., Hughes, A.M., Juhász, A., Pawellek, N., Wyatt, M., 2017. Molecular gas in debris disks around young A-type stars. Astrophys. J. 849, 123. https://doi.org/10.3847/1538-4357/aa8e4e.

Morbidelli, A., Nesvorny, D., Laurenz, V., Marchi, S., Rubie, D.C., Elkins-Tanton, L., Wieczorek, M., Jacobson, S., 2018. The timeline of the lunar bombardment: revisited. Icarus 305, 262–276. https://doi.org/10.1016/j.icarus.2017.12.046.

Morey, É., Lestrade, J.-F., 2014. On the steady state collisional evolution of debris disks around M dwarfs. Astron. Astrophys. 565, A58. https://doi.org/10.1051/0004-6361/201322567.

Moro-Martín, A., Malhotra, R., 2002. A study of the dynamics of dust from the Kuiper belt: spatial distribution and spectral energy distribution. Astron. J. 124, 2305–2321. https://doi.org/10.1086/342849.

Moro-Martín, A., Wyatt, M.C., Malhotra, R., Trilling, D.E., 2008. Extrasolar Kuiper belt dust disks. In: Barucci, M.A., Boehnhardt, H., Cruikshank, D.P., Morbidelli, A. (Eds.) The Solar System Beyond Neptune, University of Arizona Press, Tucson, pp. 465–480.

Mouillet, D., Larwood, J.D., Papaloizou, J.C.B., Lagrange, A.M., 1997. A planet on an inclined orbit as an explanation of the warp in the beta Pictoris disc. Mon. Not. R. Astron. Soc. 292, 896. https://doi.org/10.1093/mnras/292.4.896.

Mustill, A.J., Wyatt, M.C., 2009. Debris disc stirring by secular perturbations from giant planets. Mon. Not. R. Astron. Soc. 399, 1403–1414. https://doi.org/10.1111/j.1365-2966.2009.15360.x.

Nesvorný, D., Vokrouhlický, D., 2016. Neptune's orbital migration was grainy, not smooth. Astrophys. J. 825, 94. https://doi.org/10.3847/0004-637X/825/2/94.

Nesvorný, D., Jenniskens, P., Levison, H.F., Bottke, W.F., Vokrouhlický, D., Gounelle, M., 2010. Cometary origin of the zodiacal cloud and carbonaceous micrometeorites. Implications for hot debris disks. Astrophys. J. 713, 816–836. https://doi.org/10.1088/0004-637X/713/2/816.

Olofsson, J., Juhász, A., Henning, T., Mutschke, H., Tamanai, A., Moór, A., Ábrahám, P., 2012. Transient dust in warm debris disks. Detection of Fe-rich olivine grains. Astron. Astrophys. 542, A90. https://doi.org/10.1051/0004-6361/201118735.

Pan, M., Nesvold, E.R., Kuchner, M.J., 2016. Apocenter glow in eccentric debris disks: implications for Fomalhaut and ϵ Eridani. Astrophys. J. 832, 81. https://doi.org/10.3847/0004-637X/832/1/81.

Pawellek, N., Krivov, A.V., Marshall, J.P., Montesinos, B., Ábrahám, P., Moór, A., Bryden, G., Eiroa, C., 2014. Disk radii and grain sizes in Herschel-resolved debris disks. Astrophys. J. 792, 65. https://doi.org/10.1088/0004-637X/792/1/65.

Phillips, N.M., Greaves, J.S., Dent, W.R.F., Matthews, B.C., Holland, W.S., Wyatt, M.C., Sibthorpe, B., 2010. Target selection for the SUNS and DEBRIS surveys for debris discs in the solar neighbourhood. Mon. Not. R. Astron. Soc. 403, 1089–1101. https://doi.org/10.1111/j.1365-2966.2009.15641.x.

Piquette, M., Poppe, A.R., Bernardoni, E., Szalay, J.R., James, D., Horányi, M., Stern, S.A., Weaver, H., Spencer, J., Olkin, C., New Horizons P&P Team, 2019. Student dust counter: status report at 38 AU. Icarus 321, 116–125. https://doi.org/10.1016/j.icarus.2018.11.012.

Quillen, A.C., Faber, P., 2006. Chaotic zone boundary for low free eccentricity particles near an eccentric planet. Mon. Not. R. Astron. Soc. 373, 1245–1250. https://doi.org/10.1111/j.1365-2966.2006.11122.x.

Reche, R., Beust, H., Augereau, J.-C., Absil, O., 2008. On the observability of resonant structures in planetesimal disks due to planetary migration. Astron. Astrophys. 480, 551–561. https://doi.org/10.1051/0004-6361:20077934.

Rieke, G.H., Su, K.Y.L., Stansberry, J.A., Trilling, D., Bryden, G., Muzerolle, J., White, B., Gorlova, N., Young, E.T., Beichman, C.A., Stapelfeldt, K.R., Hines, D.C., 2005. Decay of planetary debris disks. Astrophys. J. 620, 1010–1026. https://doi.org/10.1086/426937.

Riviere-Marichalar, P., Barrado, D., Montesinos, B., Duchêne, G., Bouy, H., Pinte, C., Menard, F., Donaldson, J., Eiroa, C., Krivov, A.V., Kamp, I., Mendigutía, I., Dent, W.R.F., Lillo-Box, J., 2014. Gas and dust in the beta Pictoris moving group as seen by the Herschel Space Observatory. Astron. Astrophys. 565, A68. https://doi.org/10.1051/0004-6361/201322901.

Schneider, G., Grady, C.A., Hines, D.C., Stark, C.C., Debes, J.H., Carson, J., Kuchner, M.J., Perrin, M.D., Weinberger, A.J., Wisniewski, J.P., Silverstone, M.D., Jang-Condell, H., Henning, T., Woodgate, B.E., Serabyn, E., Moro-Martin, A., Tamura, M., Hinz, P.M., Rodigas, T.J., 2014. Probing for exoplanets hiding in dusty debris disks: disk imaging, characterization, and exploration with HST/STIS multi-roll coronagraphy. Astron. J. 148, 59. https://doi.org/10.1088/0004-6256/148/4/59.

Sibthorpe, B., Kennedy, G.M., Wyatt, M.C., Lestrade, J.-F., Greaves, J.S., Matthews, B.C., Duchêne, G., 2018. Analysis of the Herschel DEBRIS Sun-like star sample. Mon. Not. R. Astron. Soc. 475, 3046–3064. https://doi.org/10.1093/mnras/stx3188.

Singer, K.N., McKinnon, W.B., Gladman, B., Greenstreet, S., Bierhaus, E.B., Stern, S.A., Parker, A.H., Robbins, S.J., Schenk, P.M., Grundy, W.M., Bray, V.J., Beyer, R.A., Binzel, R.P., Weaver, H.A., Young, L.A., Spencer, J.R., Kavelaars, J.J., Moore, J.M., Zangari, A.M., Olkin, C.B., Lauer, T.R., Lisse, C.M., Ennico, K., New Horizons Geology, Geophysics and Imaging Science Theme Team, New Horizons Surface Composition Science Theme Team, New Horizons Ralph and LORRI Teams, 2019. Impact craters on Pluto and Charon indicate a deficit of small Kuiper belt objects. Science 363, 955–959. https://doi.org/10.1126/science.aap8628.

Smith, B.A., Terrile, R.J., 1984. A circumstellar disk around beta Pictoris. Science 226, 1421–1424. https://doi.org/10.1126/science.226.4681.1421.

Smith, R., Wyatt, M.C., Haniff, C.A., 2009. Resolving the hot dust around HD69830 and η Corvi with MIDI and VISIR. Astron. Astrophys. 503, 265–279. https://doi.org/10.1051/0004-6361/200911626.

Strubbe, L.E., Chiang, E.I., 2006. Dust dynamics, surface brightness profiles, and thermal spectra of debris disks: the case of AU microscopii. Astrophys. J. 648, 652–665. https://doi.org/10.1086/505736.

Su, K.Y.L., Rieke, G.H., Stansberry, J.A., Bryden, G., Stapelfeldt, K.R., Trilling, D.E., Muzerolle, J., Beichman, C.A., Moro-Martin, A., Hines, D.C., Werner, M.W., 2006. Debris disk evolution around A stars. Astrophys. J. 653, 675–689. https://doi.org/10.1086/508649.

Su, K.Y.L., Rieke, G.H., Malhotra, R., Stapelfeldt, K.R., Hughes, A.M., Bonsor, A., Wilner, D.J., Balog, Z., Watson, D.M., Werner, M.W., Misselt, K.A., 2013. Asteroid belts in debris disk twins: Vega and Fomalhaut. Astrophys. J. 763, 118. https://doi.org/10.1088/0004-637X/763/2/118.

Tamayo, D., 2014. Consequences of an eccentric orbit for Fomalhaut b. Mon. Not. R. Astron. Soc. 438, 3577–3586. https://doi.org/10.1093/mnras/stt2473.

Telesco, C.M., Fisher, R.S., Pi na, R. K., Knacke, R. F., Dermott, S. F., Wyatt, M. C., Grogan, K., Holmes, E. K., Ghez, A. M., Prato, L., Hartmann, L. W., Jayawardhana, R., 2000. Deep 10 and 18 micron imaging of the HR 4796A circumstellar disk: transient dust particles and tentative evidence for a brightness asymmetry. Astrophys. J. 530, 329–341. https://doi.org/10.1086/308332.

Telesco, C.M., Fisher, R.S., Wyatt, M.C., Dermott, S.F., Kehoe, T.J.J., Novotny, S., Mari nas, N., Radomski, J.T., Packham, C., De Buizer, J., Hayward, T.L., 2005. Mid-infrared images of β Pictoris and the possible role of planetesimal collisions in the central disk. Nature 433, 133–136. https://doi.org/10.1038/nature03255.

Thébault, P., Wu, Y., 2008. Outer edges of debris discs. How sharp is sharp? Astron. Astrophys. 481, 713–724. https://doi.org/10.1051/0004-6361:20079133.

Thureau, N.D., Greaves, J.S., Matthews, B.C., Kennedy, G., Phillips, N., Booth, M., Duchêne, G., Horner, J., Rodriguez, D.R., Sibthorpe, B., Wyatt, M.C., 2014. An unbiased study of debris discs around A-type stars with Herschel. Mon. Not. R. Astron. Soc. 445, 2558–2573. https://doi.org/10.1093/mnras/stu1864.

Vitense, C., Krivov, A.V., Kobayashi, H., Löhne, T., 2012. An improved model of the Edgeworth-Kuiper debris disk. Astron. Astrophys. 540, A30. https://doi.org/10.1051/0004-6361/201118551.

Vitense, C., Krivov, A.V., Löhne, T., 2014. Will New Horizons see dust clumps in the Edgeworth-Kuiper belt? Astron. J. 147, 154. https://doi.org/10.1088/0004-6256/147/6/154.

Winn, J.N., Fabrycky, D.C., 2015. The occurrence and architecture of exoplanetary systems. Ann. Rev. Astron. Astrophys. 53, 409–447. https://doi.org/10.1146/annurev-astro-082214-122246.

Wyatt, M.C., 2003. Resonant trapping of planetesimals by planet migration: debris disk clumps and Vega's similarity to the solar system. Astrophys. J. 598, 1321–1340. https://doi.org/10.1086/379064.

Wyatt, M.C., 2005. The insignificance of P-R drag in detectable extrasolar planetesimal belts. Astron. Astrophys. 433, 1007–1012. https://doi.org/10.1051/0004-6361:20042073.

Wyatt, M.C., 2008. Evolution of debris disks. Ann. Rev. Astron. Astrophys. 46, 339–383.

Wyatt, M.C., Dent, W.R.F., 2002. Collisional processes in extrasolar planetesimal discs - dust clumps in Fomalhaut's debris disc. Mon. Not. R. Astron. Soc. 334, 589–607. https://doi.org/10.1046/j.1365-8711.2002.05533.x.

Wyatt, M.C., Dermott, S.F., Telesco, C.M., Fisher, R.S., Grogan, K., Holmes, E.K., Piña, R.K., 1999. How observations of circumstellar disk asymmetries can reveal hidden planets: pericenter glow and its application to the HR 4796 Disk. Astrophys. J. 527, 918–944. https://doi.org/10.1086/308093.

Wyatt, M.C., Holland, W.S., Greaves, J.S., Dent, W.R.F., 2003. Extrasolar analogues to the Kuiper belt. Earth Moon Planets 92, 423–434. https://doi.org/10.1023/B:MOON.0000031956.05551.b8.

Wyatt, M.C., Greaves, J.S., Dent, W.R.F., Coulson, I.M., 2005. Submillimeter images of a dusty Kuiper belt around η Corvi. Astrophys. J. 620, 492–500. https://doi.org/10.1086/426929.

Wyatt, M.C., Smith, R., Greaves, J.S., Beichman, C.A., Bryden, G., Lisse, C.M., 2007a. Transience of hot dust around Sun-like stars. Astrophys. J. 658, 569–583. https://doi.org/10.1086/510999.

Wyatt, M.C., Smith, R., Su, K.Y.L., Rieke, G.H., Greaves, J.S., Beichman, C.A., Bryden, G., 2007b. Steady state evolution of debris disks around a stars. Astrophys. J. 663, 365–382. https://doi.org/10.1086/518404.

Wyatt, M.C., Clarke, C.J., Booth, M., 2011. Debris disk size distributions: steady state collisional evolution with Poynting-Robertson drag and other loss processes. Celest. Mech. Dyn. Astron. 111, 1–28. https://doi.org/10.1007/s10569-011-9345-3.

Wyatt, M.C., Kennedy, G., Sibthorpe, B., Moro-Martín, A., Lestrade, J.-F., Ivison, R.J., Matthews, B., Udry, S., Greaves, J.S., Kalas, P., Lawler, S., Su, K.Y.L., Rieke, G.H., Booth, M., Bryden, G., Horner, J., Kavelaars, J.J., Wilner, D., 2012. Herschel imaging of 61 Vir: implications for the prevalence of debris in low-mass planetary systems. Mon. Not. R. Astron. Soc. 424, 1206–1223. https://doi.org/10.1111/j.1365-2966.2012.21298.x.

Wyatt, M.C., Panić, O., Kennedy, G.M., Matrà, L., 2015. Five steps in the evolution from protoplanetary to debris disk. Astrophysics and Space Science 357, 103. https://doi.org/10.1007/s10509-015-2315-6.

Wyatt, M. C., Bonsor, A., Jackson, A.P., Marino, S., Shannon, A., 2017. How to design a planetary system for different scattering outcomes: giant impact sweet spot, maximizing exocomets, scattered discs. Mon. Not. R. Astron. Soc. 464, 3385–3407. https://doi.org/10.1093/mnras/stw2633.

Yelverton, B., Kennedy, G.M., 2018. Empty gaps? Depleting annular regions in debris discs by secular resonance with a two-planet system. Mon. Not. R. Astron. Soc. 479, 2673–2691. https://doi.org/10.1093/mnras/sty1678.

PART V

Prospects for the future

CHAPTER

17

Plans for and initial results from the exploration of the Kuiper belt by New Horizons

S. Alan Stern[a], *John R. Spencer*[a], *Anne Verbiscer*[b], *Heather E. Elliott*[c], *Simon P. Porter*[a]

[a]Southwest Research Institute, Boulder, CO, United States [b]University of Virginia, Charlottesville, VA, United States [c]Southwest Research Institute, San Antonio, TX, United States

17.1 New Horizons Kuiper belt mission background

New Horizons is NASA's mission to explore the Pluto system and the Kuiper belt (KB). It was launched in 2006 as the first mission in NASA's New Frontiers program and conducted a successful Jupiter system gravity assist and science flyby in 2007, and the first exploration of the Pluto system in mid-2015 (Stern et al., 2015).

New Horizons has a payload consisting of seven scientific instruments. As described in Weaver et al. (2008), they are: Ralph—a visible/IR remote sensing suite comprised of panchromatic and color imagers and an IR mapping spectrograph; Alice—an ultraviolet mapping spectrograph; LORRI (LOng Range Reconnaissance Imager)—a long focal-length panchromatic visible imager; REX (Radio Experiment)—a radio science receiver; SWAP (Solar Wind Around Pluto)—a low-energy (~25–7500 eV) particle spectrometer; PEPSSI (Pluto Energetic Particle Spectrometer Science Investigation)—a high-energy (1–1000 keV) particle spectrometer; and Venetia Burney SDC (Student Dust Counter)—a student-built dust impact counter. Together this instrument suite provides these primary capabilities: (i) medium- and high-resolution panchromatic visible wavelength mapping, (ii) medium-resolution visible wavelength color mapping, (iii) IR surface composition mapping, (iv) stereo imaging for terrain height mapping, (v) ultraviolet spectroscopy for atmospheric composition and vertical

The Trans-Neptunian Solar System.
https://doi.org/10.1016/B978-0-12-816490-7.00017-5

structure studies, (vi) plasma spectroscopy to measure the atmospheric escape rate and the composition of ionized gases escaping from Pluto's atmosphere and the interaction of the atmosphere with the solar wind, (vii) radio science to measure the brightness temperature of Pluto's surface, to determine the vertical temperature-pressure profile of Pluto's lower atmosphere, and to make bistatic radar measurements at 4.2 cm wavelengths, and (viii) a dust detector to search for particulates in orbit around Pluto.

Following its exploration of the Pluto system, New Horizons was approved to conduct an extended mission from 2016 to 2021 to study the KB and KB objects (KBOs) out to a heliocentric distance of 50 AU (Stern et al., 2018). No other operating or planned space mission or ground-based work is capable of performing this science.

The KB is a rich scientific and intellectual frontier. Its exploration has important implications for better understanding comets, the origin of small planets, the solar system as a whole, the solar nebula, and extrasolar disks, as well as for studying thermally primitive material from the planet formation era. This exploration will transform KB and KBO science from a purely astronomical regime, as it is today, to a geological and geophysical regime, which radically changed paradigms when the same happened to asteroids and comets in past decades.

The centerpiece of the New Horizons KB Extended Mission (KEM) is the close flyby of the "cold classical" KBO 2014 MU_{69} (hereafter MU_{69}). This flyby will occur on January 1, 2019. The planned flyby will approach MU_{69} to ~3500 km, almost four times closer than New Horizons flew past Pluto; consequently, imaging and compositional mapping spectroscopy resolutions will be significantly enhanced. MU_{69} is a ~25-km-diameter KBO that is ~10^3 times more massive than comet 67P that the European Space Agency (ESA) Rosetta mission orbited, but ~5×10^5 times less massive than Pluto. This places MU_{69} in a key intermediate size regime to better understand the accretion processes that operated in the KB. Given its undisturbed orbit, and 4+ Gyr existence in storage at ~35 K, MU_{69} will be the most primitive body ever studied by spacecraft. The flyby of MU_{69} will obtain the first and only planned high-resolution geological and compositional studies of a small KBO and will make sensitive searches for coma, activity, and satellites/rings at MU_{69}.

However, KEM involves more than just the close flyby of MU_{69}. As described below, it also exploits the unique resource of New Horizons to act as an observatory in the KB, in order to: (i) study numerous other KBOs during 2016–2020 in multiple and otherwise unachievable ways; (ii) make measurements of the KB/heliospheric dust and plasma environment across the KB to 50 AU; and (iii) study the frequency of KBO dust rings using high-phase angle photometry that cannot otherwise be practically achieved.

17.2 Kuiper belt mission detailed objectives

As described above, the scientific investigations of the first New Horizons KEM fall into three distinct categories. Here we provide additional details on each, in turn.

17.2.1 Science objectives of the close flyby cold classical KBO target 2014 MU_{69}

As noted above, the centerpiece of the KEM is the January 1, 2019, close flyby of the cold classical KBO 2014 MU_{69}. Approach observations begin with LORRI images that are used to provide optical navigation data and but also to search for satellites and rings. Detected satellites will be tracked to determine their orbits. Throughout the 4-month long approach to MU_{69} spanning over 1 AU of approach distance, the SWAP, PEPSSI, and SDC instruments will observe the heliospheric plasma (solar wind, pick-up ions, and energetic particles) and the dust environment in the KB near MU_{69}, and the LORRI imager will obtain a rotational light curve for MU_{69}.

MU_{69} will only reach a resolved scale of 2 LORRI pixels some 2–3 days before closest approach. From this point inward New Horizons will study MU_{69}'s rotationally resolved photometric, color, and geological properties. Starting about 1 day before closest approach, Ralph instrument color images, Ralph near-IR composition spectroscopy, and Alice ultraviolet spectroscopy will be included with the LORRI imaging to obtain multiwavelength rotational coverage. Deep searches for satellites and rings will also intensify, allowing the detection of satellites as small as 0.1 km diameter for an assumed visible albedo of 0.1. During the final hours of approach these observations will be supplemented by Alice UV spectrograph integrations to search for airglow emissions from any gaseous coma surrounding the KBO. Color imaging, panchromatic imaging, and near-IR spectroscopy will be obtained through closest approach, with the best resolution being 300 m/pixel, 140 m/pixel, and 1.8 km/pixel, respectively. LORRI will obtain images with the best resolution at 35 m/pixel. Imaging at various phase angles up to 165 degree will be used to characterize surface photometric properties and to obtain digital elevation maps using stereo techniques. The flyby will also obtain disk-integrated observations of the day- and night-side microwave thermal emission and UV surface reflectance measurements. New Horizons will also attempt an uplink X-band bistatic radar detection reflected from MU_{69}'s surface. Plasma and dust observations with SWAP, PEPSSI, and SDC will be conducted during the close approach to search for MU_{69} dust and plasma interactions. After closest approach, the Alice ultraviolet spectrograph will observe the Sun and a star to search for absorption due to possible coma gases. Departure imaging observations will conduct deep searches for forward-scattering rings around MU_{69}. Table 17.1 more completely describes the complete set of scientific objectives for this flyby.

TABLE 17.1 New Horizons first extended mission primary science and measurement objectives.

Science objective	Measurement objective	Best achieved at MU_{69} in prime sequence (green: better than Pluto; blue: comparable to Pluto; red: not as good as Pluto)
	Group 1	
Characterize the global geology and morphology	Panchromatic full-disk close approach imaging	Full disk 0.13 km/pixel, partial coverage at 0.035 km/pix
	Panchromatic rotational coverage imaging	1.5 km/pix or better, all longitudes[a]
	Topography, digital elevation models (DEM)	0.14 km/pixel

Continued

TABLE 17.1 New Horizons first extended mission primary science and measurement objectives —cont'd

Science objective	Measurement objective	Best achieved at MU_{69} in prime sequence (green: better than Pluto; blue: comparable to Pluto; red: not as good as Pluto)
Map surface composition	Close approach full-disk IR maps	1.8 km/pix
	IR compositional spectroscopic rotational coverage	19 km/pix or better, all longitudes[a]
	Color full-disk close approach imaging	0.33 km/pix
	Color rotational coverage imaging	6 km/pix or better, ail longitudes[a]
	UV reflectance spectroscopy	Disk integrated
Search for satellites and rings	Deep high and low phase imaging for satellite and ring searches	Nested observations starting with full Hill sphere; most sensitive satellite detection threshold ~0.2 km diam to 5000 km distance (~5% of Hill sphere), ring detection threshold I/F ~5e−7
Group 2		
Characterize composition and magnitude of any volatile or dust escape	UV stellar occultation coma search	Alice stellar appulse
	UV solar occultation coma search	Alice solar appulse (0.5 h integration)
	UV coma airglow search	Several Alice scans of potential coma, high and low phase
	Heliospheric Ly-alpha coma absorption search	Alice scans near C/A
	High-phase imaging dust coma search	Multiple high-phase coma dust searches
	volatile escape detection via plasma interaction	Near-continuous SWAP and PEPSSI measurements near MU_{69}, including plasma rolls
Surface properties of MU_{69}	4 cm day and night brightness temperature	Both day and night, hemispheric unresolved
	Near-IR spectroscopic temperature measurements via band shifts	1.8 km/pix
	Range of phase angles for MU_{69} to determine photometric properties	Global color and pan imaging at phase angles 1.5–169 degrees
	Photometric properties of any satellites and rings	Global color and pan imaging at phase angles 4.5–169 degrees

Continued

TABLE 17.1 New Horizons first extended mission primary science and measurement objectives —cont'd

Science objective	Measurement objective	Best achieved at MU_{69} in prime sequence (green: better than Pluto; blue: comparable to Pluto; red: not as good as Pluto)
Crater size/frequency distributions	High-resolution imaging	Full disk 0.13 km/pix, partial coverage at 0.035 km/pix
Characterize any satellites and rings	Sizes, shapes, rotation periods of any satellites	0.6 km/plx to 1000 km distance, dozens of visits for lightcurves
	Geology of any satellites	0.6 km/pix to 1000 km distance
	Color of any satellites and rings	2.5 km/pix to 10,000 km distance
	Surface composition of any satellites and rings	6 km/pix to 1200 km distance

[a] *For typical rotation period of 5 h.*
Note: These quantitative measurements refer to the prime trajectory and flyby sequence.

17.2.2 Science objectives for other KBO studies

A major aspect of the KEM mission is that it uses the LORRI telescope/imager, which can make observations down to an apparent visual magnitude of $V \sim 21$, to observe ~25 KBOs, Centaurs, and dwarf planets (DPs) to make the following types of survey measurements: (1) light curves at multiple aspect (i.e., viewing) angles to determine shapes, rotation rates, and pole positions that cannot practically be obtained from the Earth; (2) high-phase photometric searches for ring and dust material; (3) high-parallax astrometry to refine their orbits; (4) photometry at moderate and high-phase angles to determine phase functions and photometric/regolith microphysical properties that cannot be obtained from the Earth; and (5) searches for both smaller satellites and closer binaries than can be achieved using ground-based systems, Hubble Space Telescope (HST), or James Webb Space Telescope (JWST). From its measurements of small KBOs, New Horizons will yield a statistical sample of their surface properties, shapes, and binarity to place its flyby target, 2014 MU_{69}, in context. From its measurements of DPs, New Horizons will better place Pluto and Charon in context with similar-sized bodies in the KB.

17.2.3 Science objectives for heliospheric studies

The primary goal of the KEM heliospheric observations is to create a set of transect measurements of the KB plasma, dust, and neutral gas environment from 33 to 50 AU. The 50 AU distance was chosen for KEM to correspond to the aphelion of Pluto, so that these heliospheric observations also inform the space-weathering environment of Pluto and KBOs in similar orbits. Throughout the KEM, PEPSSI, SWAP, and SDC combine to create a transect record of plasma and dust across the inner and classical KB out to 50 AU. Additionally, the REX radio science instrument makes integral, total electron content measurements along the path back to Earth approximately monthly, and the Alice UV spectrometer makes a series of sets of 6 great circle neutral H brightness maps approximately twice per year to tomograph the neutral H density across the KB.

17.3 Sample results to date

17.3.1 Distant KBO results

From its unique vantage point in the outer solar system, New Horizons has observed Kuiper belt objects (KBOs) from distances ranging between 0.1 and 70 AU and at solar phase angles (α) far larger than those attainable from Earth. While the size of the Earth's orbit limits Earth-based KBO observations to phase angles $\alpha < 2$ degree, New Horizons' LORRI can observe KBOs at nearly the full range of solar phase angles, with its viewing geometry limited only by flight rules that prohibit pointing LORRI close to the Sun at $\alpha > 165$ degrees. The sensitivity of LORRI is such that distant KBOs (DKBOs) must also have apparent magnitude $V < 21$ in order to be detected.

The first DKBO to be observed by New Horizons' LORRI at phase angles $\alpha > 10$ degrees was the Plutino (15810) Arawn (1994 JR1). By combining low-phase, Earth-based KBO observations from sources including the HST and the University of Hawaii 2.2-m telescope with those obtained at higher phase angles by LORRI, Porter et al. (2016) produced the first rotational lightcurve (Fig. 17.1) and solar phase curve (Fig. 17.2) for this target. Fitting Arawn's disk-integrated solar phase curve to the Hapke (2012) photometric model makes it possible to determine physical characteristics, such as macroscopic roughness and directional scattering properties, and compare those characteristics to other outer solar system bodies. Arawn has a rough surface with a $37° \pm 5°$ mean topographic slope and is strongly backscattering. Its rotation period is a relatively rapid 5.47 ± 0.33 h. These New Horizons observations of Arawn laid the groundwork for subsequent observations of nearly two-dozen distant KBOs by LORRI during the New Horizons KEM.

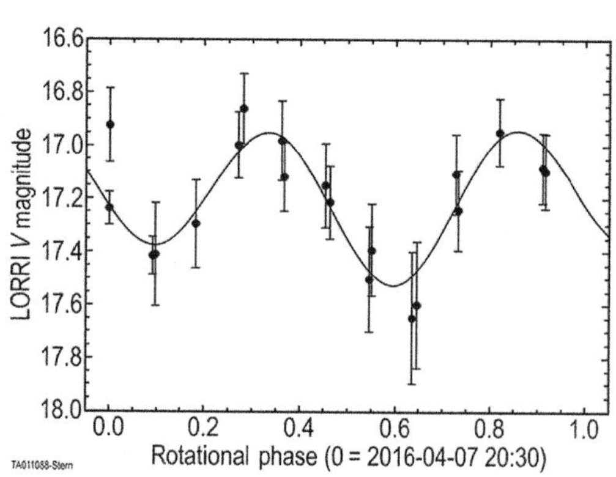

FIG. 17.1 From Porter et al. (2016), the first rotational lightcurve from a distant Kuiper belt object, (15810) Arawn (1994 JR1), seen from New Horizons LORRI at an average phase angle of 58.3 degree. The median V magnitude is 16.9 and the peak-to-peak amplitude is 0.8 magnitudes. This lightcurve demonstrates that Arawn rotates once every 5.47 h.

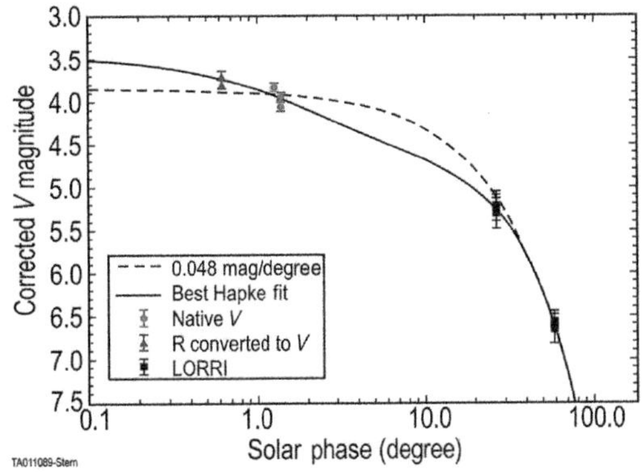

FIG. 17.2 From Porter et al. (2016), the first solar phase curve of a distant Kuiper belt object using observations acquired by New Horizons LORRI at phase angles unattainable from Earth. *Solid line* is the best fit to the Hapke (2012) photometric model, and the *dashed line* is simply a constant phase coefficient *(slope)* of $\beta = 0.048$ mag/degree. The low phase angle data are supplied by Green et al. (1997) and Benecchi et al. (2011) and D. Tholen, personal communication. R-band photometry was converted to V with V-R = 0.76 (Porter et al., 2016).

New Horizons' distant KBO targets successfully observed to date include DPs Haumea, Makemake, Quaoar, and 2002 MS_4; cold classical KBO 2011 HJ_{103} and hot classical 2012 HZ_{84}; resonant KBO 2012 HE_{85}; and Plutino Arawn. In order to combine these high-phase observations with those acquired from Earth at smaller phase angles to produce complete phase curves, accurate transformations between different photometric systems must be performed. Only following such transformations will it be possible to construct the first full KBO solar phase curves with substantial phase angle coverage using multiple observations at phase angles ranging from $\alpha = 0.06$ to 74 degrees. The production of a disk-integrated phase curve for each distant KBO enables the evaluation of the phase integral q (Verbiscer and Veverka, 1988). The spherical, or Bond albedo is then $A = pq$, the ratio between the total flux radiated in all directions to the total incident solar flux, a measure of the energy balance on the surface of the KBO as a whole. The full characterization of complete phase curves for all New Horizons distant KBO targets is beyond the scope of this chapter; however, we can compare the phase curve contributions from LORRI alone by evaluating observations of all distant KBOs observed to date at multiple phase angles and compare their phase coefficients β or slopes, in magnitudes/degree.

Fig. 17.3 compares the phase functions of distant KBOs from LORRI alone for all objects observed to date by New Horizons at multiple phase angles (Table 17.2). Evaluating the phase coefficients (slopes) over these ranges of phase angles illustrates clearly that the scattering properties of these distant KBOs are distinct. Arawn has the steepest slope while resonant KBO 2012 HE_{85} and hot classical KBO 2012 HZ_{84} have intermediate slopes between that of Arawn and the DPs. Given the Hapke photometric model derived from the disk-integrated analysis of Arawn observations, surface particles on the hot classical and resonant (nonplutino) DKBOs observed so far are more opaque than those on Arawn and they are more backscattering. The DPs exhibit shallow phase coefficients, meaning their surfaces are less backscattering and not as rough as that observed on Arawn. The full analysis of the distant KBO phase curves will

TABLE 17.2 Selected phase coefficients measured from New Horizons LORRI images.

KBO	Class	Phase angle range (degrees)	Phase coefficient β (mag/degree)
Makemake	Dwarf planet	8.4–32.0	0.0152
Quaoar	Dwarf planet	51.2–65.8	0.0199
Haumea	Dwarf planet	8.3–39.4	0.0243
2002 MS_4	Dwarf planet	37.8–51.4	0.0270
Arawn	Plutino	26.7–58.3	0.0425
2012 HE_{85}	Resonant	19.7–64.5	0.0328
2012 HZ_{84}	Hot class.	28.8–73.1	0.0326

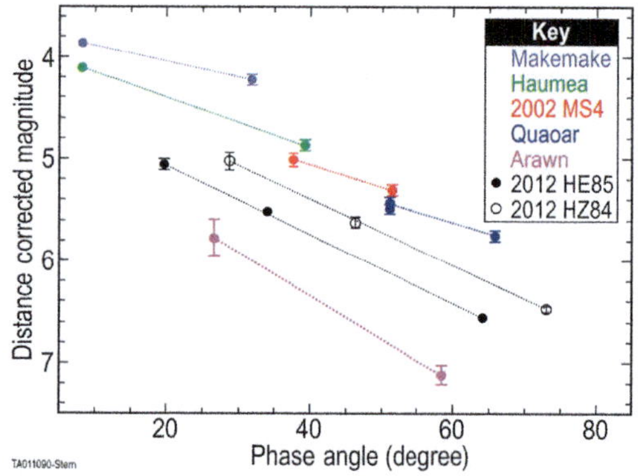

FIG. 17.3 Solar phase functions for all distant KBOs observed to date by New Horizons LORRI at multiple phase angles. All have been distance corrected to the magnitude at 1 AU from the observer and the Sun and displaced vertically as needed for clarity. Observations of Plutino Arawn *(magenta)* and all dwarf planets have been corrected for rotational lightcurve variation except 2002 MS_4. Dwarf planets Haumea *(green)*, Quaoar *(blue)* and 2002 MS_4 *(red)* have similar phase coefficients *(slopes)*. Dwarf planet Makemake *(cyan)* has the shallowest phase coefficient among the KBOs shown here. Plutino Arawn has the steepest slope (highest phase coefficient). The resonant KBO 2012 HE_{85} *(solid black circles)* and hot classical KBO 2012 HZ_{84} *(open black circles)* observations are averages of all LORRI observations at a given phase angle. They have similar phase coefficients intermediate between the Plutino Arawn and the dwarf planets.

only be possible following transformation from LORRI to each of the photometric systems at which these KBOs were observed from Earth at smaller phase angles.

17.3.2 Heliospheric results

The KEM New Horizons particle and UV observations are highly exploratory, extend the particle and UV observations of the successful Pioneer and Voyager missions, and provide new information for better understanding KB dust environment and the space weathering of objects.

Few missions have explored the outer heliosphere beyond 30 AU, and some key charged particle populations have been inaccessible owing to gaps in the energy coverage (e.g., Belcher et al., 1980) and sensitivity limitations. SWAP and PEPSSI data provide this missing, crucial energy coverage with much higher sensitivities. The only other missions to explore the solar wind and particle environment beyond 30 AU are the Pioneer and Voyager missions. The Pioneer 10 and 11 observations do not extend beyond 63.0 and 35.6 AU respectively, and the solar wind plasma instrument on Voyager 1 stopped working near Saturn at ~9.74 AU. Voyager 2 has the most extensive observations of the solar wind observations in the outer heliosphere, and the undisturbed solar wind observations extend to ~82.94 AU when Voyager 2 began to observe the foreshock of the termination shock and eventually crossed the termination shock at ~83.65 AU (Richardson and Stone, 2009). As the solar wind moves away from the Sun, it encounters interstellar neutral material. The interstellar material is ionized through charge exchange and photoionization, and these ions begin to spiral around the Interplanetary Magnetic Field (IMF) carried away from the Sun by the solar wind. Consequently, the ionized interstellar material is referred to as interstellar "pickup ions" (PUIs). The pickup process mass loads the solar wind causing the solar wind to slow and eventually become subsonic at the termination shock. The SWAP instrument simultaneously measures the slowing and heating of the solar wind owing to the interaction with the interstellar material, and unlike other missions also measures the interstellar pickup ions. The necessary analysis techniques to determine the solar wind (Elliott et al., 2016) and the interstellar H^+ PUIs (Randol et al., 2012, 2013; McComas et al., 2017) properties are already in place. Fig. 17.4 shows an example distribution with distinct H^+ and He^{++} solar wind peaks, a clear H^+ interstellar pickup ion cutoff and a portion of the He^+ interstellar pickup distribution (McComas et al., 2017). McComas et al. (2017) also found that by ~20 AU, the solar wind internal pressure is dominated by the interstellar pickup ions. Elliott et al. (2016) concluded that many solar wind structures have been worn down and merged beyond 20 AU, and the average solar wind speed at New Horizons was starting to

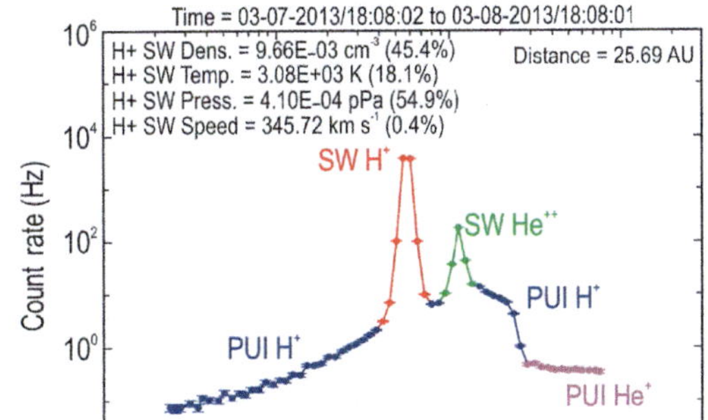

FIG. 17.4 A count rate versus energy per charge distribution averaged over 1-day with error bars taken at ~25.7 AU. Labels indicate the: solar wind (SW) or interstellar pickup ions (PUI) populations. The average solar wind conditions are shown in the upper left corner. Percentages in brackets on the solar wind parameters are normalized root mean square (RMS) variations of hourly values. *From McComas, D.J., Zirnstein, E.J., Bzowski, M., Elliott, H.A., Randol, B., Schwadron, N.A., 2017. Interstellar pickup ion observations to 38 AU. Astrophys. J. Supp. Series 233 (1), 8–12.*

decrease owing to the interaction with the interstellar material. More recent New Horizons observations indicate the speed decrease has become distinct (~6%) between 30 and 40 AU. Further, Zirnstein et al. (2018) analyzed the solar wind and interstellar pickup ions using SWAP data and found that an interplanetary shock at 34 AU is modified by the presence of the interstellar PUIs. Upstream of the shock, the solar wind is slowed and heated and the pickup ions downstream of the shock have six times the energy flux of the solar wind ions. Elliott et al. (2018) then found that many Interplanetary Coronal Mass Ejections (ICMEs), large releases of mass sporadically emitted from the Sun, can reliably be identified beyond 30 AU in the NH solar wind observations using an estimate of the solar wind alpha particle (He^{++}) to proton density ratio derived from the coarse energy sweep. Understanding the particle environment in the outer heliosphere has already proven useful for understanding how space weathering could affect the coloring of Charon Grundy et al. (2016).

The Alice UV spectrograph provides the first Lyman-α (Lyα) observations in the outer solar system since Voyager. These Lyα observations agree well with the prior Voyager results when the Voyager brightness is reduced by a factor of 2.4 as Quémerais et al. (2013) recommends in their reanalysis of the Voyager data (Gladstone et al., 2018). Fig. 17.5 shows the Lyα brightness in the upwind direction from Voyager and New Horizons. All the observations are scaled by a factor of (3×10^{11} photons/cm^2/s)/πF_{Sun}, where πF_{Sun} is the estimated subspacecraft 6-day average solar Lyα flux at 1 AU to remove the contribution from the solar flux. The fits have a $1/r$ functional form, but an additional 40 Rayleighs (R) is added to reproduce the brightness beyond 25 AU. Gladstone et al. (2018) concluded this additional brightness is consistent with Lyα emissions from the hydrogen wall. As the interstellar wind encounters the heliosphere, a bow shock is formed upstream in interstellar space. The hydrogen wall is a region of hot, interstellar material consisting mostly of hydrogen upstream, immediately outside the outer boundary of the heliosphere called the heliopause, but inside the bow shock in interstellar space.

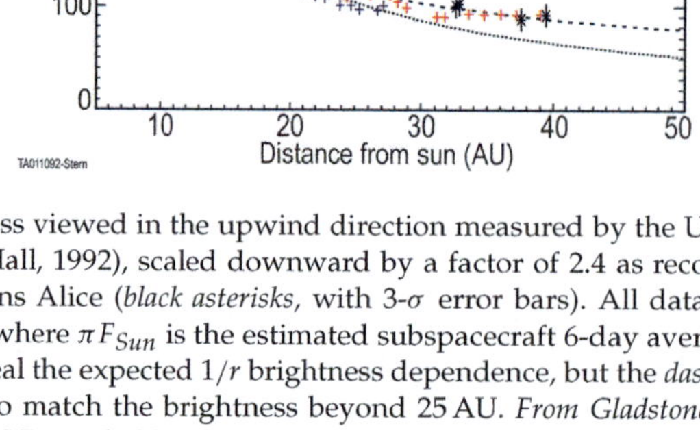

FIG. 17.5 The Lyα brightness viewed in the upwind direction measured by the UVS from Voyager 1 *(red crosses)* and Voyager 2 *(blue crosses)* (Hall, 1992), scaled downward by a factor of 2.4 as recommended by Quémerais et al. (2013), and from New Horizons Alice (*black asterisks*, with 3-σ error bars). All data are scaled by a factor of (3×10^{11}photons/cm^2/s)/πF_{Sun}, where πF_{Sun} is the estimated subspacecraft 6-day average solar Lyα flux at 1 AU. The brightness data are fit and reveal the expected $1/r$ brightness dependence, but the *dashed curve* has additional distant upstream brightness of 40 R to match the brightness beyond 25 AU. *From Gladstone, G.R., Pryor, W.R., Stern, S.A., Ennico, K., Olkin, C.B., Spencer, J.R., et al., 2018. The Lyman-α sky background as observed by New Horizons. Geophys. Res. Lett. (in press) https://arxiv.org/abs/1808.00400.*

The Student Dust Counter (SDC) is the first instrument dedicated to the study of dust in the KB. These interplanetary dust observations provide clues about the overall formation and evolution of the solar system. The only other dust observations beyond 18 AU are indirect dust observations from the Voyagers where they inferred the dust flux from plasma clouds created as dust grains impacted the spacecraft and caused a voltage pulse in the antenna Gurnett et al. (1997). Those observations indicate a nearly constant flux of about 4–5 detections per hour (Gurnett et al., 2005); however, estimating the true flux from such an indirect measure is challenging. SDC directly measures dust impacts for masses between 10^{-12} and 10^{-9} g and can determine the dust flux (Horanyi et al., 2008; Piquette et al., 2018a,b). The KB has been suggested to be a primary source of Interplanetary Dust Particles (IDPs; Landgraf et al., 2002), which are produced by collisions between KBOs (e.g., Stern, 1996) and by the bombardment of KBOs by IDPs and interstellar dust particles (ISDs; e.g., Yamamoto and Mukai, 1998). Recent models of KB dust production and loss use Pioneer 10/11 and NH observations to estimate the strengths of the IDP/ISD sources and sinks, and follow dust transport in the outer solar system Vitense et al. (2014); Poppe (2016). However, the two models predicted dust fluxes which dramatically diverge beyond ~35 AU (Fig. 17.6) (Poppe, 2016; Piquette et al., 2018a,b). Both models have similar parent distributions for the classical, scattered, and resonant populations of KBOs, but Poppe (2016) included an additional, currently undetectable population of "outer/detached" KBOs as described by Petit et al. (2011), yielding significant dust fluxes beyond 40 AU. By measuring the dust environment at distances beyond 40 AU, SDC will determine if such an outer source of extended dust exists, and in the process helps to constrain the collisional environment of the KB.

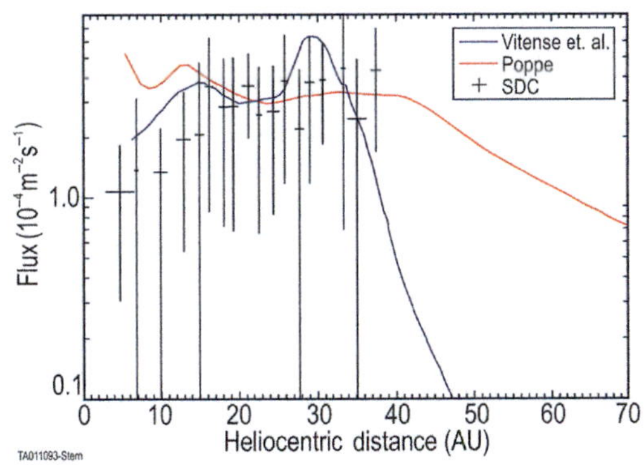

FIG. 17.6 Comparison of SDC observed dust fluxes and recent models, which dramatically diverge beyond ~35 AU with interplanetary dust flux onto SDC for grains with radii >0.6 μm. The *blue* (Vitense et al., 2014) and *red* (Poppe, 2016) curves represent two different models of the predicted dust flux. New Horizons measurements by SDC will resolve which model is correct as the spacecraft flies outward; it will reach late 70 AU in the late 2020s. *From Piquette, M., Poppe, A.R., Bernardoni, E., Szalay, J.R., James, D., Horányi, M., et al., 2018a. Student dust counter: status report at 38 AU. In: Lunar and Planet. Sci. Conf., vol. 49; Piquette, M., Poppe, A.R., Bernardoni, E., Szalay, J.R., James, D., Horányi, M., et al., 2018b. Student dust counter: status report at 38 AU. Icarus (submitted for publication).*

17.4 Anticipated future results

17.4.1 The MU_{69} flyby

New Horizons will fly closest to MU_{69} at 05:33 UT on January 1, 2019. The spacecraft will fly to the north (in ecliptic coordinates) of MU_{69}, with a nominal close approach distance of 3500 km, though an alternative trajectory and flyby sequence, with a close approach distance of 10,000 km, has been developed for use in contingency situations. Planned observations of MU_{69} extend from the first detection with LORRI on August 16, 2018 through January 3, 2019. Approach LORRI images will be used to provide optical navigation data and to search for both satellites and rings at resolutions and sensitivities not achievable from Earth or Earth orbit. An intensive ring and satellite search campaign between 32 and 19 days before encounter will return deep (limiting visual magnitude ~21.5) images to search for potentially hazardous dust structures, or satellites capable of generating hazardous material. These data will be used to inform the decision on whether to divert to the alternate trajectory and flyby sequence: this divert maneuver can occur as late as 10 days before closest approach. Detected satellites will be tracked to determine or constrain their orbits, both for more detailed imaging, and to determine the orbit of the MU_{69} primary around the system barycenter to better target close approach observations of the primary itself. Throughout the approach to MU_{69}, SWAP, PEPSSI, and SDC will observe the heliospheric plasma (solar wind, pickup ions, and energetic particles) and the dust environment near MU_{69}. Intense rotational coverage imaging will start approximately 2.5 days before the closest approach when MU_{69} will subtend about 2 LORRI pixels. Starting about 1 day before closest approach, Ralph color images, Ralph near-IR composition spectroscopy, and Alice ultraviolet spectroscopy will be interleaved with the panchromatic LORRI images to provide multiwavelength rotational coverage. Deep searches for satellites and rings will be able to detect moons as small as ~100 m in diameter.

About 3 h before the closest approach, New Horizons will obtain images and spectra blanketing a region of about 1000 km radius around MU_{69}, to characterize the shapes and compositions of any satellites or rings in this region.

In the final hour before the closest approach, assuming the prime aim point 3500 km from MU_{69}, color and panchromatic imaging with the MVIC camera inside Ralph and near-IR spectroscopy with the LEISA IR imaging spectrometer, also inside Ralph, will be obtained with the best resolution 300 m/pixel, 140 m/pixel, and 1.8 km/pixel, respectively. LORRI, also operating during these observations, will obtain images with the best resolution 35 m/pixel, though most LORRI images will have greater smear and lower SNR than the MVIC images. Images from multiple viewing directions, with phase angles up to 165 degree, will be used to characterize MU_{69}'s solar phase curve and textural surface properties and to obtain digital elevation maps using stereo techniques. New Horizons will also make disk–integrated observations of day and night side microwave thermal emission using REX, and attempt to measure UV surface reflectance using Alice. REX will also attempt an uplink bistatic radar experiment, looking for a signal sent from the Deep Space Network (DSN) reflected from MU_{69}'s surface. Plasma and dust observations with SWAP, PEPSSI, and SDC will be conducted during the close approach to search for associated MU_{69} dust and its plasma interactions.

Shortly after departure, Alice will observed both the Sun and a star to search for any UV absorption due to possible coma gases. Subsequent departure observations using the imagers will conduct deep searches for forward-scattering rings around MU_{69}.

The alternate flyby sequence will obtain similar observations, but best resolutions will be about a factor of 2 lower than for the prime trajectory.

Approximately 50 gigabits of data will be collected during the MU_{69} flyby. Owing to low data rates from the KB and the shared nature of the DSN, these data will require at least 20 months to be downlinked to Earth. The resulting dataset will provide the first definitive portrait of a small Kuiper belt object.

Note added in proof: Because the hazard environment around MU_{69} was determined to be benign, New Horizons flew its nominal trajectory to a closest approach distance near 3500 km.

17.4.2 Kuiper belt object studies

Through the remainder of KEM, more Kuiper belt objects will be observed by LORRI. We expect to obtain phase curves of these objects (including photometric searches for forward-scattering ring material), rotational lightcurves, and imaging searches for companions at better than HST resolution, down to 100 km/pixel for almost all of them.

17.4.3 Heliospheric studies

New Horizons can obtain unprecedented detail on the structure and time variability of the heliospheric plasma environment as the solar wind moves away from the Sun and picks up additional interstellar material. Based on Voyager and prior New Horizons experience, we expect increased interplanetary shock modification (Zirnstein et al., 2018) and continued slowing and heating of the solar wind (Richardson and Wang, 2003; Elliott et al., 2016; Richardson and Smith, 2003; McComas et al., 2017). ICMEs can be identified using enhancements in the alpha (He^{++}) to proton (H^{+}) density ratio ($n(He^{++})/n(H^{+})$) in the outer heliosphere (Elliott et al., 2018). It is estimated that New Horizons will be able to operate its particle instruments out to 90–100 AU when power decreases may terminate the mission (Stern et al., 2018); this is likely long enough to observe the termination shock and its foreshock, which were crossed by Voyager 1 (and 2) at ~94 AU (and 84 AU) (Richardson and Stone, 2009). New Horizons is headed along nearly the same longitude as Voyager 2, but unlike either Voyager it will remain in the ecliptic, heading toward the Energetic Neural Atom ribbon providing valuable constraints for simulations of the Interstellar Boundary Explorer (IBEX) mission Energetic Neutral Atom (ENA) maps.

Future Alice Lyα observations will be able to determine if the enhanced brightness of Lyα in the upstream direction is from the hydrogen wall associated with the interstellar boundary (Gladstone et al., 2018).

Future measurements of heliospheric dust at greater and greater distances will be able to test models of the distribution of dust parent bodies in the KB (e.g., Piquette et al., 2018a,b), including whether there is an extended dust source from KBOs in the 48 AU to several hundred AU region (Poppe, 2016; Petit et al., 2011; Vitense et al., 2014). These dust observations can thus provide key information on the nature of the outer solar system and are valuable for comparing dust in the KB to dust disks around other stars (Piquette et al., 2018a,b).

17.4.4 Potential astrophysical studies

While there are no plans for astrophysical (nonsolar-system) observations for the duration of the current extended mission through early 2021, the spacecraft's unique vantage point makes it potentially capable of unique astrophysical observations in future extended missions once outside the KB at 70 AU and beyond (Zemcov et al., 2018). For example, the location of New Horizons outside almost all our solar system's zodiacal light may allow sensitive studies of galactic and extragalactic background light. Any such work would, however, take fuel and so would limit future KB science, including flybys and studies of distant objects from New Horizons.

17.5 Future Kuiper belt exploration after New Horizons

At the conclusion of the current New Horizons extended mission, the spacecraft will be at 50 AU from the Sun. Power and fuel aboard are expected to be sufficient to operate the spacecraft until the mid-2030s, perhaps longer—which will take the spacecraft out to about 90 AU or more.

The New Horizons team plans to propose a second KB extended mission to observe more distant KBOs and to attempt another close flyby of a KBO—likely to be a still smaller one than MU_{69}, more akin to a pristine comet nucleus, should this prove feasible. The extended mission will also involve continued heliospheric plasma, dust, and neutral gas observations. As such, the first future exploration of the KB is likely to be carried out with New Horizons itself. Once the spacecraft is beyond the KB, it may be used to conduct heliospheric and astronomical observations until its fuel or power declines and terminates the mission.

However, in addition to the landmark science that New Horizons is performing in the KB, calls have been made for various new types of missions, both to explore the Pluto system in more detail and to sample the diversity of DPs and smaller bodies in the KB. The primary mission types being explored now in studies for future proposals to US Decadal Surveys and the NASA New Frontiers program include:

- *A Pluto orbiter*. This mission would study Pluto and its system of satellites at higher resolution over time to study temporal changes and with new techniques that New Horizons could not bring to bear such as thermal mappers, radars to sound the subsurface, and mass spectrometry to study atmospheric composition in more detail.
- *Flybys of other large DPs*. Such missions, which can also make additional studies of small (MU_{69}-like) KBOs (and which can be combined with ice giant flybys) would allow the known, exciting degree of diversity of the KB DP population to be studied at high resolution for the first time.
- *Missions to study a suite of Centaurs as KBOs*. This opportunity allows multiple KBO escapees to be studied in a single mission, which importantly can include mid-sized targets like 2060 Chiron (which also shows activity (e.g., Bus et al., 2001) and may have rings Ortiz et al., 2015); such mid-sized targets are too dilute a population to be likely targets for an orbiter on its way to KBO DPs or to Pluto.

No one can say when the next KB mission will be launched, but there are clearly scientifically exciting opportunities ahead for such exploration.

Acknowledgments

This work was funded by NASA's New Horizons project; we thank NASA for this support. We also thank both Ms. C. Conrad and Ms. R. Tedford for editorial support.

References

Belcher, J.W., et al., 1980. The low energy plasma in the Jovian magnetosphere. Geophys. Res. Lett. 7, 17–20.

Benecchi, S.D., Noll, K.S., Stephens, D.C., Grundy, W.M., Rawlins, J., 2011. Optical and infrared colors of transneptunian objects observed with HST. Icarus 213, 693–709.

Bus, S.J., A'Hearn, M.F., Bowell, E., Stern, S.A., 2001. Chiron: evidence for activity near Aphelion. Icarus 150, 94–103.

Elliott, H.A., McComas, D.J., Valek, P., Nicolaou, G., Weidner, S., Livadiotis, G., 2016. The new Horizons solar wind around Pluto (SWAP) observations of the solar wind from 11-33 AU. Astrophys. J. Supp. Series 223 (2), 19–22.

Elliott, H.A., Valek, P., McComas, D.J., Delamere, P.A., Bagenal, F., Gladstone, G.R., et al., 2018. Determining the alpha to proton density ratio for the New Horizons solar wind observations. Astrophys. J. 866, 2.

Gladstone, G.R., Pryor, W.R., Stern, S.A., Ennico, K., Olkin, C.B., Spencer, J.R., et al., 2018. The Lyman-α sky background as observed by New Horizons. Geophys. Res. Lett. 45, 8022–8028.

Green, S.F., McBride, N., O'Ceallaigh, D.P., et al., 1997. Surface reflectance properties of distant solar system bodies. MNRAS 290, 186–192.

Grundy, W.M., Cruikshank, D.P., Gladstone, G.R., Howett, C.J.A., Lauer, T.R., Spencer, J.R., et al., 2016. The formation of Charon's red poles from seasonally cold-trapped volatiles. Nature 539 (7), 65–68.

Gurnett, D.A., Ansher, J.A., Kurth, W.S., Granroth, L.J., 1997. Micron-sized dust particles detected in the outer solar system by the voyager 1 and 2 plasma wave instruments. Geophys. Res. Lett. 24, 3125–3128.

Gurnett, D.A., Wang, Z.Z., Persoon, A.M., Kurth, W.S., 2005. In: Dust particles detected in the outer solar system by Voyager 1 and 2, vol. 1280, Dust in Planetary Systems Workshop 2005, 1280, pp. 63.

Hall, D.T., 1992. Ultraviolet Resonance Line Radiation and the Structure of the Heliosphere (Doctoral Dissertation). The University of Arizona, Tucson.

Hapke, B., 2012. Theory of Reflectance and Emittance Spectroscopy, second ed. Cambridge Univ. Press. 528. https://doi.org/10.1017/CBO9781139025683.

Horanyi, M., Hoxie, V., James, D., Poppe, A., Bryant, C., Grogan, et al., 2008. The student dust counter on the New Horizons mission. Space Sci. Rev. 140, 387–402.

Landgraf, M., et al., 2002. Origins of solar system dust beyond Jupiter. Astron. J. 123, 2857–2861.

McComas, D.J., Zirnstein, E.J., Bzowski, M., Elliott, H.A., Randol, B., Schwadron, N.A., et al., 2017. Interstellar pickup ion observations to 38 AU. Astrophys. J. Supp. Series 233 (1), 8–12.

Ortiz, J.L., et al., 2015. Possible ring material around centaur (2060) Chiron. Astron. Astrophys. 576, A18.

Petit, J.M., Kavelaars, J.J., Gladman, B.J., Jones, R.L., Parker, J.W., Laerhoven, C.V., et al., 2011. The Canada-France ecliptic plane survey—full data release: the orbital structure of the Kuiper belt. Astron. J. 142, 131–136.

Piquette, M., Poppe, A.R., Bernardoni, E., Szalay, J.R., James, D., Horányi, M., et al., 2018a. Student dust counter: status report at 38 AU. In: Lunar and Planet. Sci. Conf., vol. 49.

Piquette, M., Poppe, A.R., Bernardoni, E., Szalay, J.R., James, D., Horányi, M., et al., 2018b. Student dust counter: status report at 38 AU Icarus 321, 116–125.

Poppe, A.R., 2016. An improved model for interplanetary dust fluxes in the outer solar system. Icarus 264, 369–86.

Porter, S.B., et al., 2016. The first high-phase observations of a KBO: New Horizons imaging of (15810) 1994 JR1 from the Kuiper belt. Astrophys. J. Lett. 828, L15–18.

Quémerais, E., Sandel, B.R., Izmodenov, V.V., Gladstone, G.R., 2013. Thirty years of interplanetary background data: a global view, cross-calibration of far-UV spectra of solar system objects and the heliosphere, ISSI Sci. Rep. Series. Berne, Switzerland, vol. 13. 141–162.

Randol, B.M., et al., 2012. Observations of isotropic interstellar pick-up ions at 11 and 17 AU from New Horizons. Astrophys. J. 755, 75–80.

Randol, B.M., McComas, D.J., Schwadron, N.A., 2013. Interstellar pick-up ions observed between 11 and 22 AU by new horizons. Astrophys. J. 768, 120.

Richardson, J.D., Smith, C.W., 2003. The radial temperature profile of the solar wind. Geophys. Res. Lett. 30 (5), 1206–1208.
Richardson, J.D., Stone, E.C., 2009. The solar wind in the outer heliosphere. Space Sci. Rev. 143, 7–20.
Richardson, J.D., Wang, C., 2003. The solar wind in the outer heliosphere at solar maximum. In: SOLAR WIND TEN: Proceedings of the Tenth International Solar Wind Conference. AIP Conference Proceedings. vol. 679. pp. 71–74.
Stern, S.A., 1996. Signatures of collisions in the Kuiper disk. Astron. Astrophys. 310, 999–1010.
Stern, S.A., et al., 2015. The Pluto system: initial results from its exploration by new horizons. Science 350, 292.
Stern, S.A., Weaver, H.A., Spencer, J.R., Elliott, H.A., 2018. The New Horizons Kuiper belt extended mission. Space Sci. Rev. 214 (4), 77–92.
Verbiscer, A., Veverka, J., 1988. Estimating phase integrals: a generalization of Russell's law. Icarus 73, 324–329.
Vitense, C., Krivov, A.V., Löhne, T., 2014. Will New Horizons see dust clumps in the Edgeworth-Kuiper belt? Astrophys. J. 147, 154–158.
Weaver, H.A., et al., 2008. Overview of the New Horizons payload. Space Sci. Rev. 140 (1), 75–82.
Yamamoto, S., Mukai, T., 1998. Dust production by impacts of interstellar dust on Edgeworth-Kuiper belt objects. Astron. Astrophys. 329, 785–791.
Zemcov, M., Arcavi, I., Arendt, R., et al., 2018. Astrophysics with New Horizons: making the most of a generational opportunity. Proc. Astron. Soc. Pacific 130, 993.
Zirnstein, E.J., McComas, D.J., Kumar, R., Elliott, H.A., Szalay, J.R., Olkin, C.B., et al., 2018. In situ observations of preferential pickup ion heating at an interplanetary shock. Phys. Rev. Lett. 121, 075102.

CHAPTER

18

Surface properties of large TNOs: Expanding the study to longer wavelengths with the James Webb Space Telescope

Noemí Pinilla-Alonso[a,b], *John A. Stansberry*[c], *Bryan J. Holler*[c]

[a]Florida Space Institute, UCF, Orlando, FL, United States [b]Arecibo Observatory, Arecibo, Puerto Rico [c]Space Telescope Science Institute, Baltimore, MD, United States

18.1 Introduction

The year 2005 saw the unprecedented discovery of three of the largest known Trans-Neptunian objects (TNOs): (136108) Haumea, (136472) Makemake, and (136199) Eris (Brown et al., 2005a; Santos-Sanz et al., 2005). One year later, the International Astronomical Union (IAU) revisited the definition of a planet and introduced a new category of objects in the solar system, the "dwarf planets" (DPs). According to this definition, DPs are small bodies that orbit the Sun and are large enough to be in hydrostatic equilibrium but not large enough to have cleared their orbit of other minor bodies (see Resolution B5 and B6 of the XXVI General Assembly of the IAU at https://www.iau.org/static/resolutions/Resolution_GA26-5-6.pdf). Hence, the fact that DPs and planets fall under different categories is a consequence of different evolutionary processes and a reflection of different physical and dynamical characteristics. In subsequent years, dedicated surveys (e.g., Brown et al., 2015) were used to search for additional TNOs with sizes similar to those of the known DPs. However, these searches, which proved to be 100% effective in detecting bright objects ($V \leq 19$) beyond 25 AU, did not identify any new dwarf planet. Consequently, the probability of finding one or two of these objects in the galactic plane, where the stellar background makes detections more difficult, is below 35%.

The Trans-Neptunian Solar System.
https://doi.org/10.1016/B978-0-12-816490-7.00018-7

The list of DPs recognized by the IAU has not changed since 2006, even though a handful of large TNOs are unofficially considered to be in this category.

In this work, we review the physical properties of these large TNOs to search for distinctive characteristics that could help to identify new DPs among the larger population. For that, we use the most recent results, which were not available in 2006 when the definition was published. We also discuss how new instrumentation, such as that on board the soon-to-be-launched *James Webb Space Telescope* (*JWST*), can expand our knowledge of the surface properties of DPs and candidates, as well as other TNOs.

18.2 Dwarf planets and candidate dwarf planets

According to the IAU definition, a large TNO must be in hydrostatic equilibrium to be considered a DP. It is difficult to determine if an object is in hydrostatic equilibrium using modern instrumentation; however, this property is related, among other factors, to the shape of the body. A combination of sufficient internal pressure caused by the body's gravitation, and sufficient plasticity to allow gravitational relaxation would cause bodies larger than a certain size to overcome their internal strength and adopt a shape characteristic of a body in hydrostatic equilibrium, typically a triaxial ellipsoid (e.g., Chandrasekhar, 1987). On the other end of the size spectrum, smaller TNOs are not dominated by gravitational forces and thus may adopt irregular shapes.

Tancredi and Favre (2008) review the geophysical criteria to separate a DP from a regular TNO. Assuming values for the density appropriate for icy bodies ($\rho \sim$ 1–2 g cm^{-3}), they find that the critical diameter is in the range of $D \sim$ 200–900 km, or 450 km for $\rho =$ 1.3 g cm^{-3}, typical of the comparably sized Uranian and Neptunian satellites (see the JPL "Planetary Satellite Physical Parameters" page at https://ssd.jpl.nasa.gov/?sat_phys_par) and compatible with the small sample of measured TNO densities (e.g., Bannister et al., 2019). The density criterion, although it is a rough estimation, provides enough information to construct a list of candidate dwarf planets (CDPs). Tancredi and Favre (2008) obtain a list of 46 candidates containing well-known large TNOs such as Orcus, Sedna, Quaoar, Varuna, Ixion, 2003 AZ_{84}, 2004 GV_9, and 2002 AW_{197}. Publication of the partial results of the *Herschel* Key Program "TNOs Are Cool," the largest database of TNO size estimates, motivated a revision of this list in Pinilla-Alonso (2016). That list contained 21 CDPs of TNOs with known sizes (25 if considering the 4 already accepted by the IAU).

In this work, we adopt the Tancredi and Favre (2008) diameter criterion (step 1 in their decision tree) for classifying TNO DPs. We also use the final list of sizes measured by "TNOs Are Cool," (Müller et al., 2019). Building on the preliminary work presented in Pinilla-Alonso (2016), we consider the error bars (σ_D) associated with the measured diameters. We include all the objects that have $D >$ 450 km, even when considering the error bar ($D <$ 450 km and $D + \sigma_D >$ 450 km or $D >$ 450 km and $D - \sigma_D <$ 450 km)

The list of CDPs is presented in Table 18.1, with 40 objects (including the 4 TNOs DPs currently defined by the IAU). This list is not meant to be a definition of which objects are or are not dwarf planets—it is only a starting point to study the physical properties of TNOs large enough to be considered as such. There are differences between our list and the list of Tancredi and Favre (2008) for two reasons: First, when Tancredi and Favre (2008) was published, knowledge of the size of TNOs was sparse and most of the diameter estimates were based on absolute magnitude (e.g., Harris, 1998) rather than thermal measurements. Second, we do not follow the full decision tree presented in Tancredi and Favre (2008) as they generally require broad assumptions about density, surface roughness, and axis ratios, all of which are poorly constrained for a majority of TNOs.

TABLE 18.1 List of TNO candidate dwarf planets (CDPs).

DP	CDP (36 TNOs)	
IAU accepted	$D > 900$ km	$450 < D < 900$ km
Pluto, Eris, Makemake, Haumea	2007 OR_{10}, Quaoar, Orcus, 2002 MS_4, Sedna, Salacia,	Varda, 2002 AW_{197}, 2003 AZ_{84}, 2002 UX_{25}, 2004 GV_9, 2005 RN_{43}, 2003 UZ_{413}, Varuna, Ixion, Chaos, 2007 UK_{126}, 2002 TC_{302}, 2002 XW_{93}, 2002 XV_{93}, 2003 VS_2, 2004 TY_{364}, Dziewanna, 2005 QU_{182}, 2005 UQ_{513}, 2005 TB_{190}, 2004 NT_{33}, 2004 PF_{115}, 2004 XA_{192}, 2003 FY_{128}, 2003 QW_{90}, 2002 GJ_{32}, 2002 KX_{14}, 2002 VR_{128}, 2001 KA_{77} Huya
4	**6**	**30**

It is worth mentioning that this list of CDPs could change with additional estimations of the size of TNOs. Unfortunately, a large increase on the TNOs with well-known size is unlikely to happen in the next decade based on the current timeline of NASA and ESA missions, as there are no confirmed plans for a successor to the *Herschel Space Observatory*. *Herschel* operated in the far-infrared (55–672 μm), which was ideal for measuring peak thermal emission of TNOs between ~70 and 100 μm (e.g., Stansberry et al., 2008). Thus, we must instead rely on TNO stellar occultations (Ortiz et al., 2019) for the foreseeable future to estimate diameters for other TNOs. However, these studies have been limited due to the uncertain knowledge of TNO orbits and the large amount of resources required to carry out these observations.

18.3 Surface compositions of dwarf planets and candidate dwarf planets

In terms of composition, extensive studies of the primitive small bodies in the solar system (asteroids, comets, and TNOs), interstellar particles, and stellar formation regions support the idea that the primordial disk that gave birth to the planets and small bodies was formed of volatile ices, macromolecular carbonaceous species, and refractory rock (e.g., McKinnon et al., 2017). TNOs are a compositionally diverse population that includes some of the most primitive bodies in the solar system, preserving a record of the composition of the planetary disk in these distant regions (e.g., de León et al., 2017). It is therefore not surprising that the surface compositions of TNOs, studied by means of models of the reflectance from their surfaces, have shown that they are a mélange of amorphous silicates, complex and red carbonaceous materials, and ultra processed amorphous carbon, with some presence of water ice and hints of other ices such as methanol (e.g., Cruikshank, 2005; Barucci et al., 2011; Brown, 2012; Chapter 5).

The size of a TNO and its surface temperature are the primary factors impacting an object's surface composition. Models of volatile retention on TNOs take these properties into consideration when used to evaluate whether or not a TNO should retain its initial volatile inventory over the age of the solar system (Schaller and Brown, 2007a; Johnson et al., 2015). It is clear from these models that the dominant factor for volatile retention is diameter, which is a stand-in for surface gravity, assuming comparable densities across the TNO population. These models predict that only the four currently recognized DPs, along with a small handful of the largest CDPs, are capable of retaining a detectable amount of their original inventory of volatile ices, including CH_4 (methane), N_2, and CO. This is confirmed by the abundant number of near-infrared (NIR) spectroscopic data (0.7–2.2 μm) collected in the last ~15 years (e.g., Barkume et al., 2008; Guilbert et al., 2009; Barucci et al., 2011; Brown et al., 2012)

The spectra of Pluto, Eris, and Makemake all exhibit strong CH_4 absorption features across the NIR spectral region (e.g., Cruikshank et al., 1976; Brown et al., 2005b; Licandro et al., 2006a; Lorenzi et al., 2015). N_2 ice is definitively detected on Pluto (e.g., Owen et al., 1993) and inferred on Eris and Makemake through measurement of CH_4 band shifts (Licandro et al., 2006b; Tegler et al., 2008, 2010; Alvarez-Candal et al., 2011). CO has only been detected on Pluto to this point (e.g., Owen et al., 1993), due to strong overlapping CH_4 absorption features. The presence of an atmosphere around Pluto is firmly established (e.g., Elliot et al., 1989), and the presence of volatile CH_4 and N_2 on Eris and Makemake make them good candidates to support an atmosphere, at least during part of their orbits (Young and McKinnon, 2013; Hofgartner et al., 2018). The lack of detectable atmospheres in the present-day combined with these objects' high albedos provides very strong evidence for ongoing resurfacing on both of these DPs (Sicardy et al., 2011; Ortiz et al., 2012).

The fourth Trans-Neptunian DP, Haumea, is quite peculiar in terms of surface composition. Haumea's NIR spectrum shows that its surface is composed of almost pure water ice, with a mixture of both, the crystalline and amorphous phases (Brown et al., 2007; Pinilla-Alonso et al., 2009). The presence of water ice on TNOs, independently of their sizes, is not exceptional, having also been recently detected on Pluto by New Horizons (Grundy et al., 2016), but the fact that the presence of other species is limited to less than 8% of the surface composition is unique (Pinilla-Alonso et al., 2009). This composition is shared only with a group of smaller objects with similar orbital parameters to Haumea (Pinilla-Alonso et al., 2007, 2008), which led to the conclusion that these TNOs make up a collisional family (Brown et al., 2007). No other members of the Haumea collisional family are considered CDPs by our criteria. The lack of detected volatile ices on Haumea may be due to the collision that formed the family, and the subsequent thermal heating that Haumea underwent.

Volatile retention models indicate that a limited number of CDPs, specifically 2007 OR_{10}, Quaoar, and Sedna may retain some volatile ices on their surfaces (Schaller and Brown, 2007a; Brown, 2012). Indeed, CH_4, the least volatile of the volatile ices (e.g., Fray and Schmitt, 2009), has been suggested on all three of these objects, with N_2 also posited on Sedna (Barucci et al., 2005; Schaller and Brown, 2007b; Emery et al., 2007). As shown in Fig. 18.1, the visible spectral slopes of these CDPs indicate their surfaces are significantly redder than those of the DPs, possibly due to the widespread presence of tholins. This implies that the equilibrium between the mechanisms altering the surface ices (Gil-Hutton, 2002) should be different to what is inferred on the volatile-dominated DPs. With a size between Sedna and Quaoar, Orcus makes a special case in the group of CDPs with $D > 900$ km. Orcus has neutral visible colors and a moderately high albedo. Models of the retention of volatiles suggest that it is too hot to retain volatiles, however, spectroscopic observations show that the water dominated surface of Orcus may also show evidence of CH_4, NH_3, and their irradiation products (Delsanti et al., 2010; Carry et al., 2011), though this is not confirmed. In this same size tier of CDPs (all with $D > 900$ km) are 2002 MS_4 and Salacia. The visible colors of these objects are neutral and their geometric albedos are remarkably low (Fig. 18.1), possibly indicating surfaces dominated by amorphous carbon and lacking volatile ices, including no traces of water ice from visible or near-infrared observations.

If we extend the study of albedo to the rest of the CDPs, this group coincides with what is referred to in Bannister et al. (2019) as "mid-sized TNOs." The distribution of albedo vs diameter for this group is more similar to that of the TNO population as a whole (Fig. 18.1, left panel), which is consistent with volatile-free surfaces. For these smaller objects, according to VIS/NIR spectroscopy, water ice is the most abundant ice, though it never dominates the surface. On the contrary, it appears mixed with other materials, typically featureless in the VIS/NIR, such as silicates and complex organics (tholins). It is therefore surprising that the smaller TNOs ($D \lesssim 450$ km) have a range of albedos from 2% to 30%, independent of their size. In the case of the CDPs smaller than 900 km in size, the albedo tends to decrease with an increase in size.

The comparative study of the visible colors of TNOs and CDPs shows that the majority of CDPs have red surfaces ($S' > 10\%/1000$ Å; Fig. 18.1, right panel). There is a notable group of neutral CDPs, including Orcus, but this group is smaller than the red group. An important feature of the right panel of Fig. 18.1 is the near lack of CDPs with spectral slopes between 6.6%/1000 Å (2003 UZ_{413}, Peixinho et al., 2015) and 17.1%/1000 Å (Varda, Peixinho et al., 2015); only Salacia, with a spectral slope of 12.6%/1000 Å, occupies this region (Pinilla-Alonso et al., 2008).

As shown in Fig. 18.1, the albedo and colors of the CDPs have some peculiarities that may be associated with the properties of their surface material. Lying in diameter between the geologically active DPs and the collisionally evolved small bodies, the CDPs represent the best tracers for the composition of the solar nebula. Improving compositional information about these objects will help to establish clearer connections between their surface compositions and their physical parameters (e.g., size, albedo, and dynamical class). These connections will, in turn, better constrain models for the formation and evolution of the solar system, and act as a model for planetary systems discovered around other stars. The biggest potential to extend this knowledge in the near future comes from gathering data at wavelengths longer than 2.2 μm, where ices such as CH_4, N_2, CO, H_2O, and CO_2 have their fundamental absorptions. This is also a wavelength regime where nonmethane hydrocarbons (ethane, ethylene, acetylene, propane, etc.) show very distinctive absorptions. Outstanding questions pertaining to DPs and CDPs include:

- Can we obtain definitive proof of the presence of the volatile ices N_2 and CO on Makemake and Eris, and if so, what does this mean for the possibility of atmospheres around these DPs?

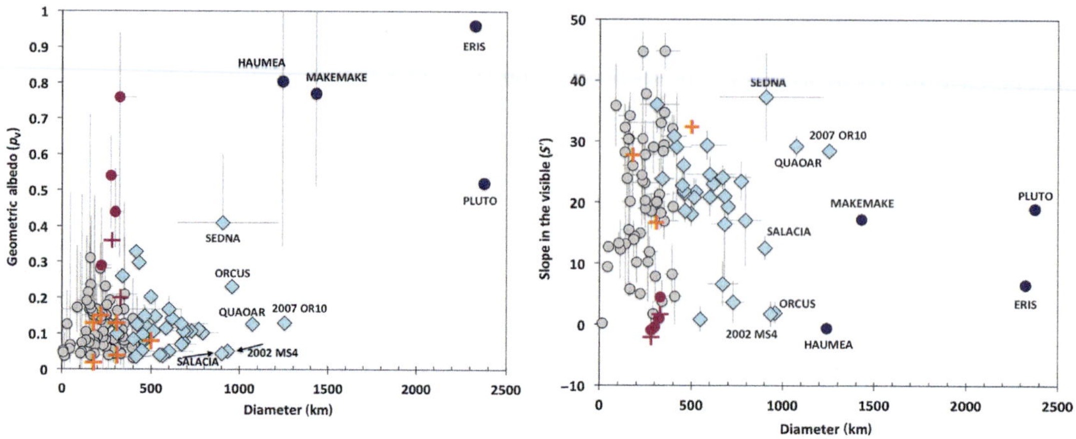

FIG. 18.1 *Left*: Geometric albedo vs diameter for a wide range of TNOs. *Blue dots* represent the DPs as defined by the IAU at this moment, *light blue diamonds* represent the CDPs, and *purple dots* represent the members of the Haumea family. Crosses represent TNOs with an estimation for the upper limit on diameter (lower limit in geometric albedo). *Purple crosses* represent members of the Haumea family with an estimation of the upper limit on diameter. *Gray dots* represent the TNOs with a good determination of the diameter and albedo that are not CDPs. Some well-known TNOs ($D > 900$ m) are labeled. The geometric albedo of the CDPs is more similar to that of medium and small TNOs than to that of the DPs. Only Sedna and some members of the Haumea family have a geometric albedo above 40%, which suggests a large amount of ice on their surfaces. *Right*: Visible spectral slope (%/1000 Å) vs diameter for the same collection of TNOs. Note that, on average, the CDPs are redder than the DPs, potentially due to ongoing resurfacing on the DPs. Note also that lack of CDPs with intermediate slopes ($6.6 < S' < 17.1$) (Spectral slope data from Pinilla-Alonso et al., 2007, 2008, 2009; Lorenzi et al., 2016; Hainaut et al., 2012; Bauer et al., 2013; Peixinho et al., 2015; Szabó et al., 2018.) Diameters and geometric albedos from "TNOs are Cool," http://public-tnosarecool.lesia.obspm.fr/Published-results.html.

- What is the inventory and nature of nonmethane hydrocarbons and complex organics (tholins) on DPs and CDPs, and what can a comparison of the two populations reveal about the chemical evolution of TNO surfaces?
- What, if any, volatiles are present on the largest CDPs?
- What ice species, if any, are present on the smallest TNOs, and what does this reveal about their formation environment?

The remainder of this paper is dedicated to a description of NASA's next great observatory, the *JWST*, which will help address the above questions with its large aperture, sensitive instrumentation, and expanded wavelength coverage.

18.4 Potential of the *James Webb Space Telescope*

The *JWST* offers significant promise for characterizing the compositions of DPs and other TNOs in detail, and for a significant sample, for the first time. Its wavelength coverage (0.6–28 μm), exceptional sensitivity, unprecedented spatial resolution, and comprehensive suite of modest resolution spectroscopic instrumentation could enable breakthroughs in linking composition to dynamical classification. This is particularly true in the critical 1–5 μm region, where many molecules of interest have numerous absorption bands. *JWST* will also have the spectral sensitivity and resolving power in the 5–10 μm range to characterize organic molecules at wavelengths that have previously been impossible, and will offer improvements in imaging sensitivity near 25 μm that could prove powerful for characterizing TNO albedos, temperatures, and sizes, particularly when combined with longer wavelength observations from Herschel (70–500 μm) and ALMA (>800 μm).

JWST is primarily designed as an astrophysical observatory covering wavelengths from 0.6 to 28 μm, but will also provide ground-breaking capabilities for solar system science, particularly for studies of TNOs. These capabilities result from the large (6.5-m effective diameter) primary mirror which, like the NIR instruments onboard, is passively cooled to around 40 K, as well as the significantly more complex and modern instrumentation when compared, for example, to those on *Spitzer*. There is no overlap in capabilities with *Herschel*. The observatory and science goals are described in Gardner et al. (2006). Solar system science capabilities are described in considerable detail in Milam et al. (2016). Ten companion papers are in the same volume of *Publications of the Astronomical Society of the Pacific*, including Parker et al. (2016), which focuses on TNO observations. Here we provide a brief summary, focused on TNO science applications. *JWST* is currently in the late stages of integration, with the instruments and telescope already assembled into a single subsystem and the spacecraft subsystem in the final stages of the test. The launch is now expected in March 2021. Science operations will be supported by the Space Telescope Science Institute (STScI), which also operates the *Hubble Space Telescope*.

18.4.1 Observatory and ground system capabilities

18.4.1.1 Orbit, field of regard, and moving-target tracking

JWST, like the *Spitzer* and *Herschel* observatories, will operate at the Earth-Sun L2 point, approximately 0.01 AU, external to Earth's solar orbit. The observatory telescope and instruments are passively cooled by means of a large sunshade; in order to keep those components shaded, the observatory is restricted to point between 85 and 135 degrees in solar elongation angle (Sun-*JWST*-Target angle). Note that observations at, or even near, opposition cannot be made. That limitation, combined with the fact that the observatory can be pointed to any azimuth around the Sun-observatory vector, defines the instantaneous "field of regard" (FOR). The FOR is thus an annulus of one celestial sphere, with two 50 degrees-wide regions centered on the ecliptic plane. The range of roll angles about the boresight is only ±5 degrees; for observations near the ecliptic, the available on-sky orientation is thus also limited to the same range (at higher ecliptic latitudes wider ranges of orientation can be accessed, depending on the epoch of the observation).

The observatory is required to be able to track moving targets at rates up to 108″/h (30 milliarcseconds/second, the maximum apparent rate of Mars as seen from L2), so it is more than adequate for observations of all TNOs and Centaurs (Milam et al., 2016). The length of science exposures is limited by the time a guide star remains in the field of view (FOV) of the guider. For moving-target observations, the effective FOV of the guider is 2.0′ × 2.0′. At the maximum track rate, the resulting limit on exposure time would thus be about 2000 s. For observations of Centaurs and TNOs, this guide-star limit on exposure time is superseded by the requirement that individual exposures be $\leq$10,000 s. Beyond that, observers must specify multiple exposures if additional time is needed.

18.4.1.2 Observation planning and documentation

Extensive online documentation for the *JWST* observatory, instrumentation, and planning tools is easily accessible at the STScI website: https://jwst-docs.stsci.edu/. High-level topics covered there are: calls for proposals (including science policies and a description of the guaranteed-time observation programs), proposal planning (including observatory performance and constraints), proposing tools (Exposure Time Calculator (ETC), Astronomer's Proposal Tool, and visibility tools), instrumentation, and data products. Links to the planning tools themselves are also available at the same website. The documentation currently includes about 15 separate pages giving specific guidance for planning observations of moving targets, including details of how to use the Astronomer's Proposal Tool (APT), ETC, and Moving-target Visibility Tool (MTVT).

JWST observations are defined using observing templates in APT, the same tool used to define observations using Hubble. The *JWST* templates are similar in function to the templates used for the Spitzer and Herschel observatories. They provide users with a fairly intuitive workflow for defining an observation, while helping to avoid making choices for instrument parameters that might negatively impact data quality. All templates for all instruments support observations of moving targets, with only a few minor restrictions relative to capabilities for fixed targets. An example of such a restriction is that target acquisition (TA) must be performed on the moving target itself, while for fixed targets acquisition on an offset target is supported.

Signal-to-noise calculations are performed using the *JWST* ETC (jwst.etc.stsci.edu). At present, the ETC is not well-suited for defining a solar system object and computing the SNR that would result from an observation of it on a date or over a range of dates. However, users can create a model spectrum using an appropriately normalized "solar" (G2V) stellar spectrum, and, if needed, combine that with a similarly normalized blackbody spectrum. For TNOs, whose observing circumstances change very little over a year-long observing cycle, this approach is adequate. Users can also compute a model spectrum separately and upload that into the ETC. The left panel of Fig. 18.2 shows an example ETC calculation using a 1000-s exposure with the NIRSpec integral field unit (IFU) and the low-resolution prism to observe TNO (55565) 2002 AW_{197}, where the spectrum was modeled as described above.

Visibility of targets from *JWST* can be calculated using the JPL Horizons system or the Python package jwst_mtvt (documentation and code available via the URL above). In Horizons, specify "@jwst" as the Observer Location and in the Table Settings limit the solar elongation angle to 85–135 degrees. The jwst_mtvt package generates tabular and graphical output giving the dates when a desired target is within that elongation range and provides the range of on-sky orientation angles of the *JWST* focal plane within those observability windows. (The orientation angle is generally not of great interest for observations of isolated sources such as TNOs, but can be critical for planetary satellites.) The right panel of Fig. 18.2 shows the graphical observability summary for (55565) 2002 AW_{197} between January 2021 and July 2022.

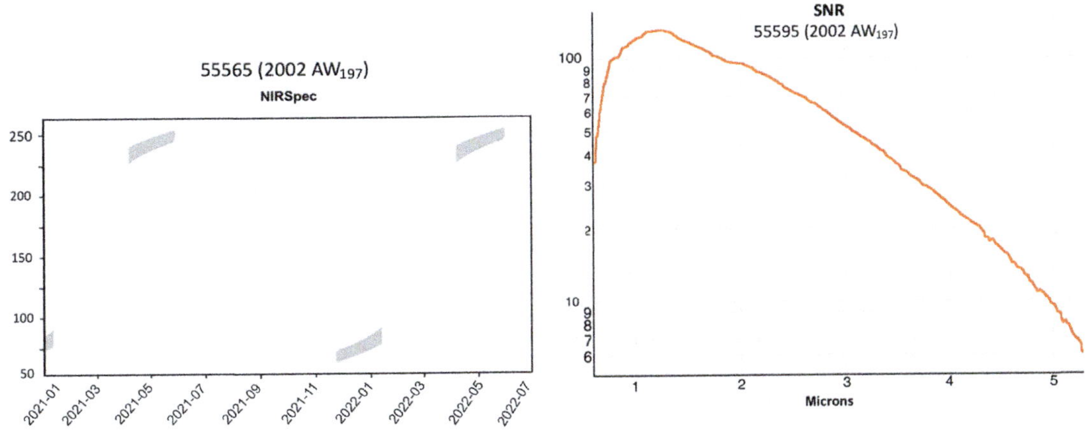

FIG. 18.2 *Right*: Example *JWST* ETC calculation of signal to noise for a 1000-s NIRSpec exposure on target (55565) 2002 AW_{197} using the integral field unit (IFU) and the PRISM. *Left*: Example output from the visibility tool, jwst_mtvt, for the same target.

18.4.1.3 *Pointing accuracy and target ephemerides*

The pointing accuracy of *JWST* is expected to be between 0.3 and 0.45″ (1-σ), depending on the distance between the science aperture and the guide star in the Fine Guidance Sensor (FGS). The pointing stability for moving targets over a 1000-s exposure is estimated to be between 6.2 and 6.7 mas (1-σ). The pointing accuracy rules out blind pointing for placement of targets in the NIRSpec fixed slits (Section 18.4.2.2) and the MIRI Low-Resolution Spectrometer (LRS) (Section 18.4.2.3), regardless of the quality of the target ephemeris. TA will be required to accurately place targets in slits. For targets with more uncertain ephemerides (~1″), TA may be required to place targets in the MIRI Medium Resolution Spectrometer and NIRSpec IFU apertures. Targets with ephemeris uncertainties of a few arcseconds or more are less likely to be targetable with *JWST*, given that they must first be blindly placed in the TA aperture or region of interest. Reporting additional astrometry of these targets to the Minor Planet Center (MPC) is recommended prior to proposing for spectroscopic observations with *JWST*.

18.4.2 Instrumentation

JWST has four science instruments providing imaging (NIRCam, NIRISS, and MIRI) and spectroscopy (the previous three and NIRSpec) covering wavelengths from 0.6 to 28 μm. The performance of the instruments is described in more detail below. The key modes for TNO science are likely to be NIRSpec IFU spectroscopy (0.7–5 μm), NIRCam imaging (0.7–5 μm), and MIRI imaging and spectroscopy (5–28 μm). The NIRSpec slitted spectroscopy mode could become important, but due to the small slit widths may require additional time to fully commission for moving targets.

All of the instruments utilize the "sample-up-the-ramp" approach for reading out the detectors. The 3 NIR instruments use 2048 × 2048 HgCdTe detector arrays while MIRI uses 1024 × 1024 Si:As arrays. By sampling the signal repeatedly and nondestructively as charge collects during an exposure, cosmic ray strikes can be identified during the fitting process that converts the data ramps to slope images. Additional benefits include better calibration of nonlinear effects and saturation of the detectors, as well as improved sensitivity. While cosmic rays will be detected and corrected at the individual integration ramp level, dithering will still be necessary to allow rejection of the fainter cosmic rays and for bad-pixel replacement in the final data products.

18.4.2.1 NIRCam

The Near-IR Camera (NIRCam) is the primary imager for *JWST*, with a total FOV of 9.5 square-arcminutes and providing simultaneous 2-filter imaging in separate 0.7–2.3 μm and 2.4–4.8 μm channels. The instrument consists of two fully redundant modules each with one short-wavelength (SW) and one long-wavelength (LW) channel. The four focal planes are composed of 10, 2048 × 2048 HgCdTe detector arrays, four in the shorter wavelength channels and two in the longer wavelength channels of both modules. For observations of specific TNOs, observers may prefer to use a single module (with a $2.2' \times 2.2'$ FOV). For surveys, both modules can be used simultaneously.

NIRCam has 13 SW and 16 LW filters. Their spectral widths fall into "wide," "medium," and "narrow" categories, with approximate fractional bandpasses of 25%, 10%, and 1%, respectively. Two ultrawide filters at 1.5 and 3.22 μm will be particularly useful for detecting the faintest TNOs, for example, in surveys.

The *JWST* telescope will provide diffraction-limited performance at wavelengths >2 μm, but image quality in terms of the full-width at half maximum (FWHM) of the point spread function (PSF) is nearly diffraction-limited through even the shortest filter at 0.7 μm. However, the NIRCam pixelscale in the SW and LW channels are 31 and 63 mas, set to provide Nyquist spatial sampling of the PSF at 2 and 4 μm, respectively. In order to take advantage of the full spatial resolution at shorter wavelengths, observers will need to include dithers in their observations. It is worth noting that, at the shortest wavelengths, the spatial resolution available with NIRCam will be approximately two times better than the best available with the Hubble Space Telescope and its WFC3 instrument.

With *JWST's* large aperture and modern detectors, NIRCam will provide significantly higher sensitivity than *Hubble* from 0.9 to 1.8 μm. Due to the IR-optimized throughput of *JWST*, however, the sensitivity at 0.7 μm is nearly the same as for *Hubble*. Between 1.8 and 5 μm *Hubble* has no capability, and observatories such as *Spitzer* and *WISE* are orders of magnitude less sensitive. Fig. 18.3 compares the 10-σ sensitivity of NIRCam imaging with 1000-s exposures to hypothetical spectral energy distributions of TNOs with a range of compositions (see Section 4.2.2 for more details of the spectral models). Where the spectra lie above the NIRCam sensitivity values signal to noise (SNR) will exceed 10; where the spectra fall below the sensitivity value SNR will be correspondingly lower. The figure illustrates that near-IR color photometry of small (~50 km diameter) TNOs in the cold-classical population will be possible in a modest amount of observing time, as will L-band (and shorter) characterization of Centaurs at the distance of Neptune. For brighter objects, observers may consider acquiring spectra with NIRSpec (see below) rather than obtaining NIRCam colors.

NIRCam has also great potential for performing surveys for faint TNOs. In particular, through the F150W2 filter it should be possible to detect objects at $m_V = 27$ (equivalent to a 35-km diameter object at 45 AU, comparable to (486958) 2014 MU_{69}) at an SNR of five in a 100-s exposure (short enough that typical TNOs will not be appreciably trailed in a fixed-pointing image). Using digital tracking techniques, it should be possible to push significantly fainter than that.

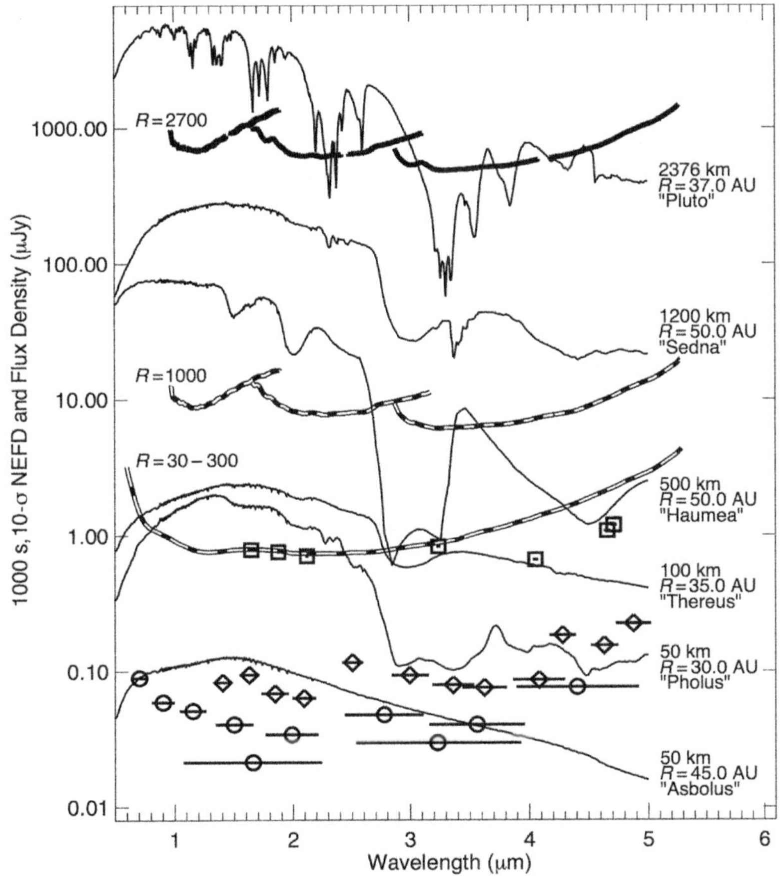

FIG. 18.3 NIRCam and NIRSpec 10-σ Noise Equivalent Flux Density (NEFD, aka sensitivity) compared to spectral energy distributions of hypothetical TNOs with albedo spectra based on those of (bottom to top) Asbolus, Pholus, Thereus, Haumea, Sedna, and Pluto. Albedo spectra are models fit to available VIS to NIR data for those targets. Diameters and distances of the hypothetical objects are given in the labels. NIRCam sensitivities are represented by *horizontal bars* indicating the filter bandpasses, with the wide, medium, and narrow filters shown as dots, diamonds, and squares, respectively. NIRSpec sensitivity for the IFU is shown by *dashed lines* labeled as $R = 30–300$ (PRISM, $30 < R < 300$), *dashed line* labeled as $R = 1000$ and *black thick line*, $R = 2700$ gratings Note that the resolving power of the gratings is not constant but varies much less than for the PRISM. Spectra for the high-resolution gratings are dispersed across two detectors, resulting in small gaps in spectral coverage (clearly visible for $R = 2700$).

18.4.2.2 NIRSpec

The Near-IR Spectrometer (NIRSpec) is the 0.6–5 μm spectrometer for *JWST* and offers imaging spectroscopy (via an integral field unit, or IFU), and fixed-slit and multiobject spectroscopy. Due to the small widths of the fixed slits (200 mas), the IFU with its $3'' \times 3''$ FOV will be more forgiving for observations of targets with any appreciable ephemeris uncertainty (see Section 18.4.1.3). This is especially true because TA for moving targets must be done using a small $1.6'' \times 1.6''$ aperture. This limitation of NIRSpec will require that target ephemerides be exceptionally well-known prior to scheduling.

NIRSpec utilizes a pair of the same H2RG detectors as NIRCam. Dispersers can be used in combination with the IFU or slits, and span wavelengths from 0.6 to 5.3 μm. The dispersers consist of a prism requiring a single observation to cover all wavelengths (resolving power $30 < R < 300$), three medium resolution ($R \approx 1000$), and three high-resolution ($R \approx 2700$) gratings, with the gratings requiring three separate observations to span the full wavelength range. When the high-resolution gratings are used with the IFU or fixed slits, small gaps in spectral coverage result from the physical gap between the two detectors; for the PRISM and medium resolution gratings, the spectra fall entirely on a single detector.

Fig. 18.3 compares the 10-σ sensitivity of NIRSpec IFU spectroscopy using 1000-s exposures to hypothetical spectral energy distributions of TNOs with a range of compositions. Sensitivity for the PRISM and medium- and high-resolution gratings are shown. While spectroscopy using the slits will be more sensitive, here we focus just on the IFU due to possible difficulties with placing moving targets into those slits, as mentioned earlier. The spectral models are based on albedo spectra for the objects named in the figure legend. For each object Hapke models (Hapke, 2012) were used to fit available data in the visible—near-IR range, and extended to 5 μm. The fitting included matching the known geometric albedo of each object. Those albedo spectra were then used to predict spectral energy distributions for hypothetical objects with sizes and distances not necessarily representative of the original objects. These hypothetical objects allow us to show the range of spectral features seen for TNOs, the dynamic range within the spectrum for single objects with different compositions, and the range of overall brightness to be expected for TNOs with a range of sizes and distances. Pluto is somewhat of a special case, as it is shown for its actual size and the distance appropriate for the end of the *JWST* mission, c. 2032. Additional details on NIRSpec observations of TNOs can be found in Métayer et al. (2019)

The NIRSpec IFU pixel scale is 100 mas, so the PSF (comparable to that of NIRCam) is undersampled at all wavelengths. Fixed slits are $\approx 3.5''$ long and either $0.2''$ or $0.4''$ wide. A larger $1.6'' \times 1.6''$ aperture is used for acquiring single targets and can also be used for spectroscopy. As mentioned earlier, the small sizes of these slits place tight requirements on ephemeris accuracy, and observers will be required to show that ephemerides for their targets are sufficient prior to observations being scheduled. Spectra from the slits fall on a region of the detectors that are behind an opaque mask, minimizing background light and offering higher sensitivity than is possible with the IFU.

The NIRSpec IFU offers imaging spectroscopy with a $3'' \times 3''$ FOV. Spaxels are $0.1''$ square, matched to the pixel scale. Dithers can be used to improve spatial sampling of the PSF if desired. Due to its larger aperture, observations with the IFU are less sensitive to target ephemeris uncertainties, and the location of the target within the scene can be determined after the fact to enable optimal spectral extraction. The IFU also simultaneously characterizes the background around the target in two dimensions, which is an advantage compared to the slits. However, spectra from the IFU fall on portions of the detectors that are masked by the microshutter array (MSA). The MSA is not perfectly opaque even for closed shutters, and there are a number of failed-open shutters that will allow dispersed light from objects and the zodi to fall on the detectors during IFU exposures. For these reasons the IFU is somewhat less sensitive than the slits, and dithers are strongly encouraged in order to help remove the effects of any sources that may fall on open shutters.

18.4.2.3 MIRI

The Mid-IR instrument (MIRI) offers both imaging and spectroscopy in the 5–28 μm wavelength range. The imager uses a single 1024 × 1024 pixel detector with a pixel scale of 0.11″ and a FOV of 74″ × 113″ (a portion of the detector is dedicated to coronagraphic imaging). The PSF is diffraction-limited at all wavelengths and is slightly under-sampled by the pixels at the shorter wavelengths (FWHM = 0.18″ at 5.6 μm) and highly oversampled at the longest (FWHM = 0.82″ at 25.5 μm). Fig. 18.4 shows the total system response through the nine MIRI filters. Images are acquired through a single filter at a time (in contrast to NIRCam, in which data is collected through two filters simultaneously).

MIRI also offers medium resolving power imaging spectroscopy from 4.9–28.8 μm via the medium resolution spectrometer (MRS), which is an IFU similar to the one in NIRSpec. Light is dispersed via gratings, resulting in spectral resolving power ranging from approximately 1400 at the longest wavelengths to about 3500 at the shortest wavelengths. In a single exposure, spectra are acquired in each of the four wavelength channels, two spectral channels being dispersed onto separate parts of each of the two MRS detectors. The four channels span wavelengths of 4.9–7.7 μm, 7.5–11.7 μm, 11.5–18.1 μm, and 17.7–28.8 μm, respectively. For a single grating setting the spectrum spans a wavelength band that is about one-third of the full wavelength range covered by each of the 4 channels. Continuous spectral coverage requires taking three exposures, each using a different grating setting. Fig. 18.4 shows the 12 MRS spectral sections (four channels, and three bands within each channel). The four MRS wavelength channels have fields of view that overlap on-sky, with footprints ranging from 3.3″ × 3.7″ in the shortest wavelength channel to 7.2″ × 7.9″ in the longest wavelength channel.

Finally, the MIRI LRS provides spectra spanning 5 μm to slightly beyond 10 μm in a single exposure (see Fig. 18.4). The light is dispersed by a prism and the resolving power ranges from 40 to 160 over that wavelength range. The spectrum is dispersed onto the same detector used for MIRI imaging (separate from the two MRS detectors) and can be taken through a 0.5″ × 4.7″ slit, or slitless (the latter intended primarily for observations of exoplanet transits).

Fig. 18.4 illustrates the sensitivity of MIRI in both imaging and spectroscopic modes. Generally, MIRI will provide much higher sensitivity than the Spitzer infrared spectrograph (IRS) instrument, hardly surprising given the huge disparity between the *JWST* and *Spitzer* apertures. However, beyond about 20 μm MIRI sensitivity is limited by thermal emission from the much warmer telescope (*JWST* operates at around 40 K while *Spitzer* was typically at around 15 K), and the sensitivity gain of *JWST*/MIRI over Spitzer is more modest. The much smaller PSF of *JWST* does, however, significantly reduce effective noise caused by background sources (referred to as confusion noise) in the scene relative to *Spitzer*. Confusion is primarily caused by thermal emission from dust in distant, un-resolved galaxies, and is more important at MIRI wavelengths than in the VIS and NIR. In the context of Fig. 18.4, Pluto provides a concrete example for which the *JWST*/MIRI sensitivity can be compared to that of *Spitzer*. MIRI imaging at 25.5 μm should detect Pluto with an SNR of ~50 in a 500 s exposure; *Spitzer* 500 s observations at 24 μm gave an SNR of 25 (Lellouch et al., 2011). For MIRI $R \approx 1500$

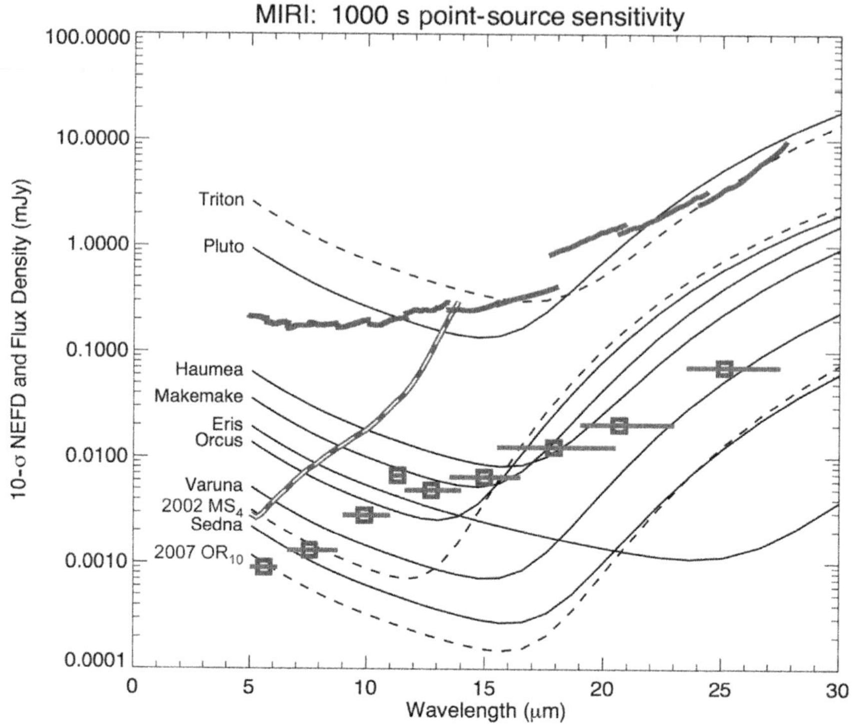

FIG. 18.4 MIRI 10-σ Noise Equivalent Flux Density (NEFD, aka sensitivity) compared to spectral energy distributions of various TNOs. Measured geometric albedos and diameters are used to predict the reflected component (assuming constant albedo), which dominates shortward of ~15 μm, and thermal emission using an assumed phase integral of 0.39. *Horizontal lines* with *square symbols* show the MIRI sensitivity in the imaging bandpasses, while the *thick short curves* at the top give the MRS sensitivity. The *dashed-solid curve* from 5 to 13 μm gives the LRS sensitivity.

spectroscopy of Pluto only (Fig. 18.4 does not include the contribution of Charon) using 1000 s exposures should give SNR ≈ 2 per spectral element, over wavelengths of 15–28 μm if the data are un-binned. If the data are binned by 15 spectral elements (giving $R \approx 100$), the resulting spectrum would have SNR ≈ 8. *Spitzer* 720 s spectra of Pluto (including Charon) with $R \approx 90$ also gave SNR ≈ 8 (Lellouch et al., 2011), but only over wavelengths of about 22–36 μm.

The 6.5 m primary mirror of *JWST* will enable much higher spatial resolution than any previous space observatories with instruments covering similar wavelengths, which all had ≤1 m primary mirrors (e.g., *IRAS ISO WISE Spitzer, AKARI*). The FWHM of the MIRI PSF ranges from 0.18″ at 5.6 μm to 0.81″ at 28.5 μm. In imaging mode, the pixel scale (0.11″) is such that the PSF is Nyquist sampled at the short wavelengths and highly oversampled at the longer wavelengths. For the medium resolution imaging spectrometer (MRS) the pixel scales in the 4 spectral channels mentioned above are 0.176″, 0.277″, 0.387″, and 0.645″, respectively, with the result that the spectral PSF is somewhat under-sampled by the pixels at all MRS wavelengths. This significant leap in spatial resolution for a space-based MIRI will be particularly important for Jupiter, Saturn, and their large satellites. In the Kuiper belt, however, the primary benefit may be in better detection and removal of flux from background sources. A few of the most well-separated binary systems (e.g., Pluto/Charon) can be resolved by MIRI at the shorter wavelengths where reflected light tends to dominate TNO spectra (see Fig. 18.4); at the longer wavelengths, where the thermal emission is brightest, even Pluto and Charon will be blended.

18.4.2.4 NIRISS

The Near-IR Imager and Slitless Spectrograph (NIRISS) offers a unique interferometric imaging mode (aperture masking interferometry, or AMI) in four filters spanning 2.5–5 μm. AMI delivers spatial resolution roughly 2.4× better (i.e., the PSF is that much sharper) than NIRCam imaging at those wavelengths. This capability relies on a pupil mask with seven small subapertures within the extent of the *JWST* primary mirror, where each pair-wise baseline vector is unique. Even though the pupil mask greatly reduces the throughput, the NIRISS optics are otherwise very efficient and in AMI mode NIRISS is only a factor of five less sensitive than NIRCam imaging. For $m_V = 23$ TNO with a neutral spectrum, NIRISS/AMI can achieve an SNR of around 200 in a 100 s exposure at 2.8 μm and would be able to resolve satellites at separations $\geq 0.09''$. AMI could be used to measure colors of TNO binaries in the L- and M-band region for the first time.

18.4.3 Data products and archive

Plans are in place to provide better support, relative to what has been provided for *Hubble*, for *JWST* moving-target observations, both in terms of the data processing and the ability to find data in the archive. Users are now strongly encouraged to use standard designations for their moving targets when submitting their observations. The APT makes it easy to do so because it now includes the capability to resolve target names and retrieve orbital elements from JPL Horizons. A secondary naming field can be used to specify details in addition to target names if desired (e.g., "west elongation" or "longitude 90"). This means that archive searches will reliably return data for users who search on normal target names.

The *JWST* data pipeline is also implementing the capability to co-add multiple exposures of moving targets in the frame of the target. This is critical, for example, for improving SNR and PSF sampling by combining dithered exposures into a final image or spectrum, or producing maps of extended objects such as comets. Such data products have been available for recent missions such as *Spitzer* and *Herschel* but have not been produced for *Hubble* observations.

18.4.4 Guaranteed-time observations (GTOs)

JWST includes a predetermined allocation of time for scientists who have contributed directly to the development of the observatory, for example, the principal investigators for each of the science instruments. Four guaranteed-time observer programs totaling about 70 h will be focused on the TNOs and Centaurs, primarily on the DPs including Pluto, Eris, Makemake, and Haumea. NIR spectroscopy with NIRSpec makes up the bulk of the time, with additional MIRI spectroscopy and MIRI imaging observations. Details regarding these programs can be found at http://www.stsci.edu/jwst/observing-programs/approved-gto-programs, including the observation specifications themselves. These programs will be finalized in March 2020. General observers wishing to make similar observations of the same targets will be required to provide a justification for the duplication, for example, the need to study time-variable phenomena or obtain significantly higher SNR.

18.5 Summary

Surface compositions of TNOs appear to be correlated with size, with the largest TNOs, the DPs, exhibiting dynamic, and volatile-dominated surfaces. We refer to the next-lowest size tier as CDPs. These objects appear to be vastly different from the dwarf planets in terms of color, albedo, and surface composition, even though they are closer in size to the DPs than the small TNOs. Fundamental questions remain about the connections between the DPs, CDPs, and small TNOs, many of which cannot be answered with current ground-based facilities and instrumentation. The *JWST* will provide a higher sensitivity and extended wavelength range needed to address these outstanding issues. *JWST* has a 6.5-m diameter primary mirror and is equipped with four science instruments with imaging and spectroscopic modes that cover 0.6–28 μm. All TNOs can be tracked with *JWST* and plans are already in place to process moving-target observations. Set to launch no later than March 2021, *JWST* is poised to revolutionize the study of TNO surface compositions in the coming decade.

Acknowledgments

The authors thank the conveners of the scientific workshop, "The Trans-Neptunian Solar System", held in the University in Coimbra, Portugal, in March 2018. The authors would also like to recognize the work of Michael S.P. Kelley on the jwst_mtvt and Michael Mommert on the astroquery adaptation for JPL Horizons.

References

Alvarez-Candal, A., et al., 2011. The spectrum of (1361999) Eris between 350 and 2350 nm: results with X-Shooter. Astron. Astrophys. 532, A130.

Bannister, M.T., et al., 2019. Expanding Horizons: the need for direct exploration of the diverse Trans-Neptunian solar system. Space Sci. Rev. (submitted for publication).

Barkume, K.M., Brown, M.E., Schaller, E.L., 2008. Near-infrared spectra of centaurs and Kuiper belt objects. Astron. J. 135, 55–67.

Barucci, M.A., et al., 2005. Is Sedna another Triton? Astron. Astrophys. 439, L1–L4.

Barucci, M.A., et al., 2011. New insights on ices in Centaur and Transneptunian populations. Icarus 214, 297–307.

Bauer, J.M., et al., 2013. Centaurs and scattered disk objects in the thermal infrared: analysis of WISE/NEOWISE observations. Astrophys. J. 773 (1). Id. 22.

Brown, M.E., et al., 2007. A collisional family of icy objects in the Kuiper belt. Nature 446, 294–296.

Brown, M.E., 2012. The compositions of Kuiper belt objects. Ann. Rev. Earth Plan. Sci. 40, 467–494.

Brown, M.E., et al., 2015. A serendipitous all sky survey for bright objects in the outer solar system. Astron. J. 149, 69.

Brown, M.E., Trujillo, C.A., Rabinowitz, D., 2005a. 2003 el_61, 2003 ub_313, and 2005 fy_9. iauc 8577, #1.

Brown, M.E., Trujillo, C.A., Rabinowitz, D.L., 2005b. Discovery of a planetary-sized object in the scattered Kuiper belt. Astrophys. J. 635, L97–L100.

Brown, M.E., Schaller, E.L., Fraser, W.C., 2012. Water Ice in the Kuiper belt. Astron. J. 143, 146.
Carry, B., et al., 2011. Integral-field spectroscopy of (90482) Orcus-Vanth. Astron. Astrophys. 534, Article ID: A115.
Chandrasekhar, S., 1987. Ellipsoidal Figures of Equilibrium. Dover Publications, New York.
Cruikshank, D.P., 2005. Triton, Pluto, Centaurs, and Trans-Neptunian bodies. Space Sci. Rev. 116, 421–439.
Cruikshank, D.P., Pilcher, C.B., Morrison, D., 1976. Pluto–evidence for methane frost. Science 194, 835–837.
de León, J., Licandro, J., Pinilla-Alonso, N., 2017. The diverse population of small bodies of the solar system. In: Deeg, H., Belmonte, J. (Eds.), Handbook of Exoplanets. Springer, Cham., pp. 55.
Delsanti, A., et al., 2010. Methane, ammonia, and their irradiation products at the surface of an intermediate-size KBO? A portrait of Plutino (90482) Orcus. Astron. Astrophys. 520, A40.
Elliot, J.L., et al., 1989. Pluto's atmosphere. Icarus 77, 148–170.
Emery, J.P., et al., 2007. Ices on (90377) Sedna: confirmation and compositional constrains. Astron. Astrophys. 466, 395–398.
Fray, N., Schmitt, B., 2009. Sublimation of ices of astrophysical interest: a bibliographic review. Planet. Space Sci. 57, 2053–2080.
Gardner, J.P., et al., 2006. The James Webb Space Telescope. Space Sci. Rev. 123, 485–606.
Gil-Hutton, R., 2002. Color diversity among Kuiper belt objects: the collisional resurfacing model revisited. Planet. Space Sci. 50, 57–62.
Grundy, W.M., et al., 2016. Surface compositions across Pluto and Charon. Science 351, aad9189.
Guilbert, A., et al., 2009. ESO-large program on TNOs: near-infrared spectroscopy with SINFONI. Icarus 201, 272–283.
Hainaut, O.R., Boehnhardt, H., Protopapa, S., 2012. Colours of minor bodies in the outer solar system II—a statistical analysis, revisited. Astron. Astrophys. 546, L115.
Hapke, B., 2012. Theory of Reflectance and Emittance Spectroscopy, second ed. Cambridge University Press, Cambridge, UK.
Harris, A.W., 1998. A thermal model for near-earth asteroids. Icarus 131, 291–301.
Hofgartner, J.D., et al., 2018. Ongoing resurfacing of KBO Eris by volatile transport in local, collisional, sublimation atmosphere regime. Icarus (in press). https://doi.org/10.1016/j/icarus.2018.10.028.
Johnson, R.E., et al., 2015. Volatile loss and classification of Kuiper belt objects. Astrophys. J. 809, 43.
Lellouch, E., et al., 2011. Thermal properties of Pluto's and Charon's surfaces from spitzer observations. Icarus 214, 701–716.
Licandro, J., et al., 2006a. The methane ice rich surface of large TNO 2005 FY_9: a Pluto-twin in the Trans-Neptunian belt? Astron. Astrophys. 445, L35–L38.
Licandro, J., et al., 2006b. Visible spectroscopy of 2003 UB_{313}: evidence for N2 ice on the surface of the largest TNO? Astron. Astrophys. 458, L5–L8.
Lorenzi, V., Pinilla-Alonso, N., Licandro, J., 2015. Rotationally resolved spectroscopy of dwarf planet (136472) Makemake. Astron. Astrophys. 577, id. A86.
Lorenzi, V., et al., 2016. The spectrum of Pluto, 0.40–0.93 microns I. Secular and longitudinal distribution of ices and complex organics. Astron. Astrophys. 585, A131.
McKinnon, W.B., et al., 2017. Origin of the Pluto-Charon system: constraints from the New Horizons flyby. Icarus 287, 2–11.
Métayer, R., et al., 2019. JWST/NIRSpec prospects on Transneptunian objects. Front. Astron. Space Sci. 6, 8.
Milam, S.N., et al., 2016. The James Webb Space Telescope's plan for operations and instrument capabilities for observations in the solar system. Publ. ASP 128, 018001.
Müller, T., Lellouch, E., Fornasier, S., 2019. Trans-Neptunian objects and Centaurs at thermal wavelengths. In: Prialnik, D., Barucci, M.A., Young L.A. (Eds.), The Transneptunian Solar System. Elsevier.
Ortiz, J.L., et al., 2012. Albedo and atmospheric constraints of dwarf planet Makemake from a stellar occultation. Nature 491, 566–569.
Ortiz, J.L., et al., 2019. Stellar occultation by TNOs: from predictions to observations. In: Prialnik, D., Barucci, M.A., Young L.A. (Eds.), The Transneptunian Solar System. Elsevier.
Owen, T.C., et al., 1993. Surface ices and the atmospheric composition of Pluto. Science 261, 745–748.
Parker, A., et al., 2016. Physical characterization of TNOs with the James Webb Space Telescope. Publ. ASP 128.
Peixinho, N., et al., 2015. Reanalyzing the visible colors of Centaurs and KBOs: What is there and what we might be missing. Astron. Astrophys. 577, A35.
Pinilla-Alonso, N., 2016. Icy dwarf planets: colored popsicles in the outer solar system. Proc. IAU 29A, 241–246.

Pinilla-Alonso, N., et al., 2007. The water Ice rich surface of (145453) 2005 RR43: a case for a carbon-depleted population of TNOs? Astron. Astrophys. 468, L25–L28.
Pinilla-Alonso, N., et al., 2008. Visible spectroscopy in the neighborhood of 2003 EL_{61}. Astron. Astrophys. 489, 455–458.
Pinilla-Alonso, N., et al., 2009. The surface of (136108) Haumea (2003 EL_{61}), the largest carbon-depleted object in the Trans-Neptunian belt. Astron. Astrophys. 496, 547–556.
Santos-Sanz, P., et al., 2005. 2003 el_61, 2003 ub_313, and 2005 fy_9. iauc 8577, #2.
Schaller, E.L., Brown, M.E., 2007a. Volatile loss and retention on Kuiper belt objects. Astrophys. J. 659, L61–L64.
Schaller, E.L., Brown, M.E., 2007b. Detection of methane on Kuiper belt object (50000) Quaoar. Astrophys. J. 670, L49–L51.
Sicardy, B., et al., 2011. A Pluto-like radius and a high albedo for the dwarf planet Eris from an occultation. Nature 478, 493–496.
Stansberry, J., et al., 2008. Physical properties of Kuiper belt and Centaur objects: constraints from the Spitzer space telescope. In: Barucci, M.A., Boehnhardt, H., Cruikshank, D.P., Morbidelli, A. (Eds.) The Solar System Beyond Neptune. University of Arizona Press, Tucson, pp. 161–179.
Szabó, G.M., et al., 2018. Surface ice and tholins on the extreme Centaur 2012 DR_{30}. Astron. J. 155, 170.
Tancredi, G., Favre, S., 2008. Which are the dwarfs in the solar system? Icarus 195, 851–862.
Tegler, S.C., et al., 2008. Evidence of N_2-ice on the surface of the icy dwarf planet 136472 (2005 FY9). Icarus 195, 844–850.
Tegler, S.C., et al., 2010. Methane and nitrogen abundances on Pluto and Eris. Astrophys. J. 725, 1296–1305.
Young, L., McKinnon, W.B., 2013. Atmospheres on volatile-bearing Kuiper belt objects. AAS/DPS 45, id. 507.02.

CHAPTER

19

Stellar occultations by Trans-Neptunian objects: From predictions to observations and prospects for the future

José L. Ortiz[a], *Bruno Sicardy*[b], *Julio I.B. Camargo*[c,d], *Pablo Santos-Sanz*[a], *Felipe Braga-Ribas*[c,d,e]

[a]Instituto de Astrofisica de Andalucia (CSIC), Granada, Spain [b]Paris Observatory, PSL Research University, CNRS, Sorbonne University, Univ. Paris Diderot, Sorbonne Paris City, LESIA, Meudon, France [c]Observatório Nacional/MCTIC, Rio de Janeiro, Brazil [d]LIneA, Rio de Janeiro, Brazil [e]Federal University of Technology, Paraná (UTFPR), Curitiba, PR, Brazil

19.1 Introduction

The study of stellar occultations by Trans-Neptunian Objects (TNOs) is a relatively new field because the first detection of a stellar occultation by a TNO was made only 9 years ago (Elliot et al., 2010). Although young, this field of work is quickly evolving and it is an essential one if we want to adequately characterize and understand the physical properties of the TNO population as a whole. This is because stellar occultations can provide extremely relevant information such as very accurate sizes and shapes, better than any other technique can do (except for spacecraft visits, which are extremely expensive, complex, and require a lot of time). Size and shape are basic physical parameters that must be accurately known if we want to fully characterize a body, and are the first step toward deriving accurate densities. We need to know densities if we aim to determine internal compositions and make first guesses on the internal structure of solar system bodies (Carry, 2012).

The Trans-Neptunian Solar System.
https://doi.org/10.1016/B978-0-12-816490-7.00019-9

On the other hand, once size and shape are correctly determined for a solar system body, combining this information with accurate brightness measurements allows deriving accurate geometric albedos, which are also an important piece of information to properly interpret many of the visible light observations of the TNOs, particularly their spectra (in order to derive surface composition, as surface reflectance composition models are highly degenerate unless the albedo is known, e.g., Chapters 6 and 20). And geometric albedo is also fundamental to analyze the temperatures and thermal behavior of the TNOs (see, e.g., Chapter 7). Even though measurements of the thermal output of solar system bodies can be used to obtain sizes and albedos (combining the thermal fluxes with visible light measurements), these so-called radiometric techniques have considerable limitations. For TNOs, sizes and albedos determined from radiometric techniques are only accurate to 10%–20% level (Mueller et al., 2010), and information on shape is not directly derived from the models. On the other hand, thermal measurements are extremely difficult to obtain for the TNOs because their peak of thermal emission is in the submillimeter range, around 100 μm, where our atmosphere blocks the electromagnetic radiation. Therefore, good radiometric studies in the past required the use of sophisticated space observatories with very limited lifetime such as Spitzer and Herschel. Because of that, only a sample of around 130 TNOs have radiometrically determined sizes and albedos (Mueller et al., 2009). And unfortunately the accuracy of the resulting products is not high. Stellar occultations do not suffer those problems and by comparing the occultation-based with the radiometric-based results for a given body, the radiometric models or thermophysical models can be tuned to derive other thermal parameters that play a role in the thermal measured flux, such as thermal inertia, but also information on spin axis orientation, spin rate, and other parameters can be retrieved or at least constrained using the combination of thermal data with occultation results (Mueller et al., 2018).

Besides, shapes derived from the occultations can be used to determine densities under the assumption of hydrostatic equilibrium of a homogeneous body, provided that the spin rate is known and some constraint on the spin axis orientation is used (Sicardy et al., 2011; Ortiz et al., 2012; Braga-Ribas et al., 2013; Benedetti-Rossi et al., 2016; Schindler et al., 2017; Dias-Oliveira et al., 2017). Nevertheless, at least for the case of Haumea this approach fails to derive the actual density (Ortiz et al., 2017). This implies the interesting consequence that the object is not in hydrostatic equilibrium and/or is not homogeneous. Apart from all the above, stellar occultations provide a powerful means to determine the presence of atmospheres through the gradual immersion and emersion of the stars (during the disappearance and reappearance of the star, respectively) or through the study of central flashes. The presence or absence of atmospheres down to a small fraction of the microbar level of surface pressure is easily determined nowadays and this is an important topic because highly volatile ices can potentially generate partial or even full atmospheres in TNOs (Stern and Trafton, 2008 and Chapter 9). So far, no clear evidence for global atmospheres on a TNO other than Pluto have been found, although a local atmosphere might be envisaged (Ortiz et al., 2012 and Chapter 9).

Another key aspect of stellar occultations is the possibility to detect narrow and dense dust structures such as rings around the occulting bodies through the dimmings that they can cause in the stellar flux. The occultation detection of rings around the centaur Chariklo (Braga-Ribas et al., 2014) as well as the detection of a dust structure in Chiron with many similarities to Chariklo's rings (see Ortiz et al., 2015, although interpretation in terms dust shells is possible Ruprecht et al., 2015) represented an important breakthrough in planetary science, and the subsequent discovery of a ring around Haumea, one of the dwarf planets in the Trans-Neptunian belt, has also been a major discovery with plenty of implications (Ortiz et al., 2017), which opens a new field and raises many questions, including how frequent these structures are in the current TNO population (see Chapter 12). All this calls for more observations of stellar occultations.

As a summary, it is clear that the interest in stellar occultations by TNOs is extremely high and the topic deserves particular attention in a review paper. In this chapter, we deal with general aspects of occultations by known TNOs and also include the Centaurs, which are not TNOs strictly speaking, but since they are closely related to them, it makes sense to address their occultations here too.

Serendipitous occultations by small unknown TNOs also provide a very powerful means of studying the Trans-Neptunian region, because the technique has the potential to derive the number density of bodies of a given size range and their size distribution (Roques et al., 2006; Bickerton et al., 2008; Bianco et al., 2009; Wang et al., 2010; Schlichting et al., 2012; Chang et al., 2013; Zhang et al., 2013; Liu et al., 2015; Pass et al., 2018). However, this topic is not covered here.

19.2 General results from stellar occultations thus far and lessons learned

19.2.1 Difficulties of predicting and observing occultations by TNOs

Although the power of stellar occultations is well recognized in the planetary science community and has been widely exploited to characterize the asteroid belt, it was clear that succeeding in this endeavor would be challenging for the Trans-Neptunian region, in contrast to stellar occultations by asteroids[1] (which are now relatively easy to predict and observe). Stellar occultations by TNOs have intrinsic problems; In order to predict if a TNO will occult a star, and given that the typical angular diameter of a 300-km TNO at 40 AU is 10 milliarcseconds (mas), one must know the positions of the stars in catalogs to that level of accuracy at least, and the ephemeris of the TNO must also be accurate to that level. Otherwise, predictions are extremely uncertain and thus useless. A TNO of that size at 40 AU has a brightness of $m_V = 22$ for an average geometric albedo of 0.08. For the stars, the positional accuracy is not a problem anymore thanks to the submas accurate astrometry from the Gaia space mission (Prusti et al., 2016; Lindegren et al., 2018).

Unfortunately, typical positional uncertainties for the TNOs from orbits in the JPL Horizons database are in the order of 300 mas, and that is true for other sources of orbits such as the

[1] In this chapter, we do not include the TNOs within the "asteroid" category, as asteroids have rocky compositions and are confined in the region inside 5.2 AU, whereas the TNOs are the progenitors of the short-period comets so their physical nature and characteristics are entirely different, and they reside much further away from the Sun. The IAU 2006 General assembly coined the term "small solar system body"' to refer to all the bodies that are not planets or dwarf planets. Some professional and mostly amateur astronomers use the term asteroid to collectively refer to all the bodies that are not planets, or comets (which display comas), and in that context TNOs could be asteroids, but we stick to the IAU terminology.

Minor Planet Center, Astorb or Astdys (see Fig. 19.1). The number of astrometric observations of a given distant small body, as well as the arclength (years) over which these observations are distributed, influences the accuracy of its ephemeris (Fig. 19.1).

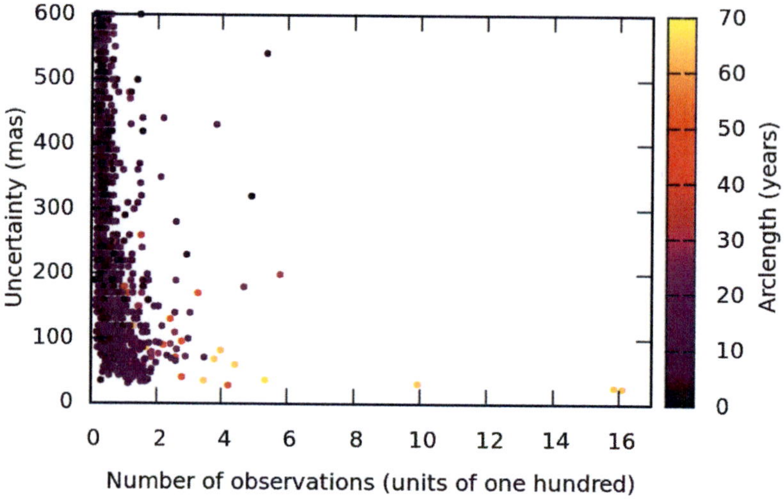

FIG. 19.1 Uncertainties (Current 1σ Ephemeris Uncertainty—CEU) in ephemerides for objects with semimajor axes greater than 15 AU as a function of the number of observations and of the arclength over which these observations are distributed. Only objects with an arclength greater than 1 year and more than five observations were considered. Uncertainties are calculated according to the Muinonen and Bowell (1993) formalism.*Source of data: astorb.dat (as obtained from ftp.lowell.edu, see also ftp://cdsarc.u-strasbg.fr/pub/cats/B/astorb/astorb.html.), as obtained on October 1, 2018.*

Our own statistics of monitoring 40 of the largest TNOs gives an rms deviation of the measurements with respect to the JPL orbits of 511 mas in right ascension and 273 mas in declination. Hence, we believe that 300 mas is a representative estimate of the uncertainties in the ephemeris, although obviously some objects have poorer orbits than others as shown in Fig. 19.1.

Given these overall numbers, it would appear nearly impossible to predict and observe occultations with a good success rate and not by mere chance after 30 or more "blind" attempts. However, there are ways to tackle this problem. The key issue is that even though the uncertainties in the absolute positions of the TNOs are as high as mentioned above, relative positions between the star to be occulted and the TNO can be obtained to the required accuracy of ~10 mas through specific observations. And once this is done, good enough predictions can be made. The approaches to make this possible are dealt with in Section 19.3.

However, once an accurate prediction is made, we are still left with the problem of observing it. There are many difficulties at the observing stage. The region of the Earth where the occultation shadow path falls, weather, the brightness of the star and its elevation above the horizon at the time of the occultation, the lunar phase, organization logistics, and contacts with the observers, training of the local observers (which are often newcomers in the occultation field), and many other complications play important roles. It is obvious that shadow paths on remote areas of the world, or mostly on oceans, result in observations not being practical.

Also, if the occultations take place in areas of the Earth that are in sunlight, they will not be observable either. These aspects are probably the most evident difficulties, but there are lots of logistic aspects that can ruin observations. Some specific aspects on equipment are dealt with in Sections 19.2.3, 19.4.1, 19.5.5, and 19.5.6.

19.2.2 General results

The evolution of the number of occultations that we are aware of (our own and those reported in amateur asteroid occultation networks, as well as in the scientific literature and conferences) observed since 2009, is shown in Fig. 19.2. Note that we did not include Pluto-Charon. Table 19.1 lists the occultations in our database. A version of this table that is continuously being updated can be found in http://occresults.ga/results/

It should be pointed out that the majority of the occultations detected thus far have been single chord (observed from just one site). This situation does not allow us to determine sizes, but some information can still be derived (Section 19.5). The number of multichord events is also shown in Fig. 19.2.

Since 2009, there has been a general increase in the number of occultation events observed per year except for a fluctuation in 2015 and 2016. Note that 2019 has not finished at the time of the revision of this paper, so it can be expected that the final number of occultations in 2019 will increase. The number of multichord occultations per year has been steadier, with a noticeable

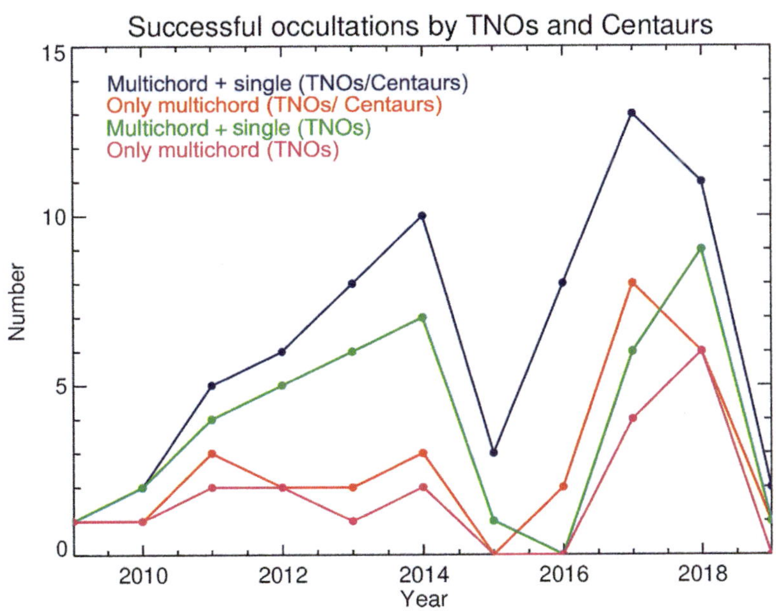

FIG. 19.2 Number of detected occultations per year (TNOs plus Centaurs) since 2009, Pluto excluded. In *blue*, all the events involving both TNOs and Centaurs. In *green*, only TNO events. In *red*, all the multichord occultations. In *purple*, multichord occultations involving only TNOs. Note that the data for 2019 contain only the first 2 months of the year, corresponding to the time of the revision of this paper. Hence the small number of detections in 2019.

TABLE 19.1 List of observed stellar occultations by TNOs and Centaurs since the first TNO occultation in 2009 (not counting Pluto's system).

Date	Object	Lucky star-related	Reference
9 Oct 2009	2002 TX_{300}	No	Elliot et al. (2010)
19 Feb 2010	Varuna	Yes	Sicardy et al. (2010)
6 Nov 2010	Eris	Yes	Sicardy et al. (2011)
8 Jan 2011	2003 AZ_{84}	Yes	Dias-Oliveira et al. (2017)
11 Feb 2011	Quaoar	Yes	Person et al. (2011)
23 Apr 2011	Makemake	Yes	Ortiz et al. (2012)
04 May 2011	Quaoar	Yes	Braga-Ribas et al. (2013)
29 Nov 2011	Chiron	No	Ruprecht et al. (2015)
3 Feb 2012	2003 AZ_{84}	Yes	Dias-Oliveira et al. (2017)
17 Feb 2012	Quaoar	Yes	Braga-Ribas et al. (2013)
26 Apr 2012	2002 KX_{14}	Yes	Alvarez-Candal et al. (2014)
25 Jun 2012	Echeclus	No	
15 Oct 2012	Quaoar	Yes	Braga-Ribas et al. (2013)
13 Nov 2012	2005 TV_{189}	No	
08 Jan 2013	Varuna	Yes	
13 Jan 2013	Sedna	Yes	
3 Jun 2013	Chariklo	Yes	Braga-Ribas et al. (2014)
9 Jul 2013	Quaoar	Yes	
29 Aug 2013	Eris	Yes	
24 Nov 2013	Asbolus	Yes	
2 Dec 2013	2003 AZ_{84}	Yes	Dias-Oliveira et al. (2017)
12 Dec 2013	2003 VS_2	Yes	
11 Feb 2014	Varuna	Yes	
16 Feb 2014	Chariklo	Yes	Berard et al. (2017)
1 Mar 2014	Orcus/Vanth	Yes	
4 Mar 2014	2003 VS_2	Yes	
16 Mar 2014	Chariklo	Yes	
29 Apr 2014	Chariklo	Yes	Leiva et al. (2017)
24 Jun 2014	Ixion	Yes	
28 Jun 2014	Chariklo	Yes	Leiva et al. (2017)
07 Nov 2014	2003 VS_2	Yes	
15 Nov 2014	2007 UK_{126}	Yes	Benedetti-Rossi et al. (2016)

TABLE 19.1 List of observed stellar occultations by TNOs and Centaurs since the first TNO occultation in 2009 (not counting Pluto's system)—cont'd

Date	Object	Lucky star-related	Reference
15 Nov 2014	2003 AZ_{84}	Yes	Dias-Oliveira et al. (2017)
26 Apr 2015	Chariklo	Yes	
12 May 2015	Chariklo	Yes	Berard et al. (2017)
3 Dec 2015	2002 VE_{95}	Yes	
12 Jun 2016	Chariklo	Yes	
25 Jul 2016	Chariklo	Yes	Berard et al. (2017)
08 Aug 2016	Chariklo	Yes	Berard et al. (2017)
10 Aug 2016	Chariklo	Yes	Berard et al. (2017)
10 Aug 2016	Chariklo	Yes	Berard et al. (2017)
15 Aug 2016	Chariklo	Yes	Berard et al. (2017)
20 Aug 2016	Chariko	Yes	
1 Oct 2016	Chariko	Yes	Leiva et al. (2017)
21 Jan 2017	Haumea	Yes	Ortiz et al. (2017)
08 Feb 2017	Chariklo	Yes	
7 Mar 2017	Orcus/Vanth	No	Sickafoose et al. (2018)
9 Apr 2017	Chariklo	Yes	
20 May 2017	2002 GZ_{32}	Yes	
24 May 2017	2003 FF_{128}	No	
22 Jun 2017	Chariklo	Yes	
10 Jul 2017	2014 MU_{69}	No	Zangari et al. (2017)
17 Jul 2017	2014 MU_{69}	No	Zangari et al. (2017)
23 Jul 2017	Chariklo	Yes	
24 Aug 2017	Chariklo	Yes	
17 Nov 2017	2004 NT_{33}	Yes	
29 Dec 2017	Bienor	Yes	
28 Jan 2018	2002 TC_{302}	Yes	
2 Apr 2018	Bienor	Yes	
15 Jul 2018	2010 EK_{139}	Yes	
26 Jul 2018	Quaoar	Yes	
4 Aug 2018	2014 MU_{69}	No	
2 Sep 2018	Quaoar	Yes	
10 Sep 2018	Varda	Yes	

Continued

TABLE 19.1 List of observed stellar occultations by TNOs and Centaurs since the first TNO occultation in 2009 (not counting Pluto's system)—cont'd

Date	Object	Lucky star-related	Reference
19 Sep 2018	2002 KX_{14}	Yes	
28 Sep 2018	2004 PF_{115}	Yes	
20 Oct 2018	2015 TG_{387}	No	
28 Nov 2018	Chiron	Yes	
24 Dec 2018	2005 RM_{43}	Yes	
30 Dec 2018	2002 WC_{19}	Yes	
11 Jan 2019	Bienor	Yes	
4 Feb 2019	2005 RM_{43}	Yes	

The date of the events is in the first column. The second column indicates the name or the provisional designation of the TNO or Centaur. The third column indicates whether the event had or not involvement of the Paris Granada and Rio teams' collaboration (currently in the frame of the ERC Lucky Star project). The fourth column indicates if there is a scientific publication for the event.

increase in 2017 and 2018, possibly as a result of better astrometric updates thanks to the Gaia Data Release 2 (DR2) (see Section 19.3).

There has been a wide variety of scientific results emerging from the occultations. For example, it is worth mentioning that a wide range of geometric albedos have been obtained. Also, a variety of shapes and sizes have been determined. Many of the results are still unpublished, partly because most of the events were single chord. In Table 19.2 some results from the already published events are summarized together with revisions of some values as explained in Section 19.5.

19.2.3 Some lessons learned

An important aspect is covering the shadow path with a sufficient number of observing stations. The larger the coverage with observers in the perpendicular direction to the shadow path, the higher the chances to succeed. Shadow paths on Earth are more frequently in the East-West direction than in the North-South direction because the apparent sky motion of the TNOs is mainly in this direction, but sometimes, near quadrature, where the TNO can move more in Declination than in Right Ascension we get shadow paths that move mainly North-South on Earth.

Note that the sky motion of the TNOs seen from Earth is basically due to parallax motion as the Earth travels in its orbit, because the orbital motions of the TNOs are very slow. As a result, typical speeds seen from Earth are in the range of 15–20 km/s (1–3 arcsecond/h in angular speed). For TNOs with sizes in the range of 200–2000 km, this means that occultations can last typically from 10 to 100 s. If one can achieve a time accuracy of a 10th of a second in determining the disappearance and reappearance times, typical accuracies in the chord lengths

TABLE 19.2 List of some physical parameters from the published stellar occultations and revised values with newer information on absolute magnitudes as described in the text.

Object	Effective diameter (km)	Geometric albedo p_V	References	Revised effective diameter (km)	Revised geometric albedo p_V
2002 TX_{300}	248 ± 10	0.88 ± 0.05	1	**320 ± 50**	**0.65 ± 0.15**
Eris	2326 ± 12	0.96 ± 0.03	2	2326 ± 12	**0.94 ± 0.03**
Makemake	1465 ± 47	0.77 ± 0.03	3	1465 ± 47	0.77 ± 0.03
Quaoar	1110 ± 5	0.109 ± 0.007	4	1110 ± 5	**0.112 ± 0.01**
2003 AZ_{84}[a]	764 ± 6	0.097 ± 0.009	5	764 ± 6	**0.094 ± 0.02**
2007 UK_{126}	638 ± 28	0.156 ± 0.013	6,7	**642 ± 28**	**0.142 ± 0.015**
Chariklo[b]	268 ± 12	Assumed	8	**268 ± 12**	**0.038 ± 0.008**
Haumea[c]	1392 ± 26	0.51 ± 0.02	9	1392 ± 26	0.51 ± 0.02

[a] *The values given here are for the 2014 occultation*
[b] *The values given here are for the oblate spheroid model in the 2014–2016 time frame.*
[c] *The effective diameter given is at the minimum of the rotational phase. The mean effective diameter (in equivalent area) is estimated at 1614 km.*
Results that differ with the initial findings are shown in bold.
1 (Elliot et al., 2010), 2 (Sicardy et al., 2011), 3 Ortiz et al. (2012), 4 Braga-Ribas et al., 2013, 5 (Dias-Oliveira et al., 2017), 6 (Benedetti-Rossi et al., 2016), 7 (Schindler et al., 2017), 8 Leiva et al. (2017), 9 Ortiz et al. (2017).
For 2002 TX_{300} the latest H_V magnitude from Alvarez-Candal et al. (2016) has been used.
For Eris the latest H_V magnitude from Alvarez-Candal et al. (2016) has been used. The effect of the satellite Dysnomia is negligible.
For Makemake, the effect of the discovered satellite is negligible.
For Quaoar the latest H_V magnitude from Alvarez-Candal et al. (2016) has been used.
For 2003 AZ_{84} the latest H_V magnitude from Alvarez-Candal et al. (2016) has been used. Also the effect of the satellite has been taken into account by adding 0.01 mag in H.
For 2007 UK_{126} the effect of the satellite has been taken into account by adding 0.032 mag in H.
For Haumea, no newer information is available.
For Chariklo we have used the minimum absolute magnitude of 7.35 in the historical time series when no ring contribution was present due to edge-on configuration (Duffard et al., 2014).
The effective diameter shown here is the diameter of a sphere with equal projected area. The values given are at the time of the occultation as published in the references.

are in the order of 1–3 km. Slow occultations near quadrature allow an even higher accuracy in the measured chords lengths, at least in theory. In practice, the final uncertainty in the timing of disappearance and reappearance of the star depends not only on clock time accuracy, but also (and primarily, in most of the cases) on the uncertainty of the photometry, which is the main parameter that plays a role in the square well models that must be fit to the occultation profiles (and from which the disappearance and appearance times are retrieved). This controls the final errors in the sizes of the chords and on the derived size for the TNO.

It is obvious that the fainter the stars, the more numerous they are, so the numbers of predicted occultations of faint stars are higher than those of brighter stars. For example, for a given TNO, there are 7 times more occultations of stars at magnitude 18 or brighter than at magnitude 15 or brighter (using the Gaia DR2 statistics of sources of 15 and 17.8 mag). Therefore, most of the occultation predictions will be for faint stars. However, detecting occultations of ~18-mag stars with widely available telescopes of diameter below ~40 cm become difficult. A typical limit for the brightness of a star to observe occultations with a large enough number of telescopes is around 17.5–18. Fainter than that, only "large" telescopes with a diameter around 60 cm or larger can contribute. Since the abundance of those telescopes on the surface of the Earth is much lower (or a fast deployment of telescopes in this size range is difficult), magnitude ~18 is currently a practical limit to pursue occultation observations in general, although in the future this might change (see the section on prospects for the future).

Even with a good prediction, on average, the success rate in our occultation program is typically 1 every 5.5 attempts. By success we mean a detection of the occultation from at least one site. Weather is the main reason ruining the campaigns, but also technical problems and observing mistakes are often present. Moreover, the number of participating stations is critical. Large campaigns involving at least 15–20 observing sites are typically needed to achieve multichord observations. From our own experience and from all the multichord stellar occultations published thus far, the campaigns that resulted in multichord detections typically involved >15 participating stations. The minimum to achieve a multichord detection was 15 stations, for Quaoar (Braga-Ribas et al., 2013). Hence, a number around 15 is probably the magical number of observing sites that need to be involved to guarantee success against weather, technical problems as well as other logistic issues and also give enough margin north and south (or east and west) of the nominal shadow path to accommodate changes in the shadow path within the typical uncertainties of the predictions. If predictions can be improved further than what we typically achieve, the number of participating stations may be decreased, but probably, a minimum number around 10 will still be needed. On average, attempts that do not reach this critical number of observing stations either fail or result in only single-chord or two-chord detections.

Most of the successes thus far have involved networks of telescopes from professional sites or specific networks established for TNO occultations, with some important contributions from amateur astronomers with access to high-performance equipment, as most of the TNO events are not reachable with the broadly used setups by the amateur community.

From Table 19.1, we can also see that most of the successful occultations were achieved thanks to the international collaboration of the groups at Paris, Granada, and Rio. For the last few years this is under the umbrella of the Lucky Star project, a European Research Council (ERC) advanced grant of the H2020 program. Hence, international collaboration and coordinated efforts are of paramount importance at that point.

19.3 The future of the predictions

19.3.1 Techniques to make accurate predictions

As already mentioned, for many years prior to the existence of the Gaia DR2 catalog, the positions of the stars in the best astrometric catalogs were known at no better than ~50 mas accuracy for the most favorable cases (Assafin et al., 2012). To make matters worse, the average uncertainty in TNO ephemeris is of the order of 300 mas. Hence, it appears that accurately predicting stellar occultations would be impossible.

The approach thus far has been to focus on those TNOs that have the best orbits, which also usually happen to be the largest ones so their angular diameters are also the largest, and try to improve the orbits with accurate astrometry. But since the star catalogs to which the astrometric measurements had to be referred also had large errors in the past, in practice it was difficult to improve the orbits substantially. However, this strategy, in connection with case-to-case analysis and appropriate weighting of reported measurements to the Minor Planet Center and our own measurements can allow us to generate our own orbits, such as in the Numerical Integration of the Motion of an Asteroid (NIMA) scheme, see Desmars et al. (2015).

For the TNOs with the best orbits and the largest sizes, the chances of success are the highest so we often focused our predictions on them. Most of the results shown in Table 19.1 and Fig. 19.2 come from astrometric efforts concentrated on around 50 TNOs/Centaurs. Once potentially good events are identified (with the best possible orbits), the predictions must still be refined. This requires observing the TNO close in time to the occultation date from weeks to days prior to the occultation and getting astrometry with respect to the same reference frame as the star to be occulted, either with relative astrometry or using a high accuracy intermediate star catalog. Hence, good astrometry of the TNO with respect to the occulted star allows us to decrease the large positional uncertainties and we often achieve the typical 10 mas level needed.

An example of predicted shadow path after an astrometric update and the real shadow path on Earth is shown in Fig. 19.3. The uncertainty in telescopic CCD astrometric observations has two main components, one due to accuracy in determining the TNO centroid (σ_C) and another one due to the astrometric solution of the plates using reference stars in the field of view, σ_S. The two uncertainties add quadratically so that $\sigma_t^2 = \sigma_c^2 + \sigma_s^2$.

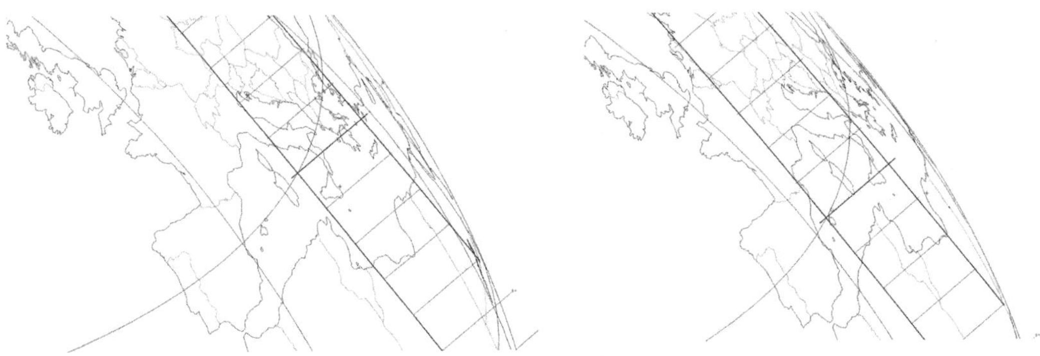

FIG. 19.3 *Left*: Prediction map of the stellar occultation by 2002 TC_{302} on January 28, 2018, from an astrometric update several days prior to the event using images acquired with the 1.5 m telescope at Sierra Nevada Observatory. The map shows the different countries as well as two meridians and a parallel. The *straight lines* delineate the star show path and the *perpendicular lines* to the shadow path indicate minutes. *Right*: The postoccultation reconstructed shadow path from a preliminary fit to the chords. Comparing the two figures and given that the shadow path has a width of 17 mas, it turns out that the prediction was good to the 10 mas level, with the prediction somewhat to the east compared with the reality. In this plot the shadow moves from south to north. Usually, most of the shadow paths of occultations by TNOs are in the east-west direction.

From our empirical tests on main belt asteroids with accurate orbits, the uncertainty of the TNO centroid determination is $\sigma_C \sim 1/2$ FWHM/SNR where FWHM is the full-width at half-maximum of the stellar images and SNR is the signal-to-noise ratio (SNR) achieved for the TNO, in fact, it can be shown that σ_C = FWHM/(2.355*SNR) for well sampled, signal dominated sources (see, e.g., Mighell, 2005 for a detailed discussion). For $m_V \sim 22$ TNOs under typical seeing (FWHM = 1 arcsecond), and a SNR = 100, the uncertainty is 0.005″ (5 mas). But apart from this error, as mentioned before, one has to add quadratically the error of the astrometric solution for the images σ_S, which typically goes as the square root of the number of stars used in the solution multiplied by the mean uncertainty in their positions. Thanks to Gaia DR2, we now have large enough number of stars even for small fields of view of the 2 m-diameter or larger telescopes that can be used to derive astrometry. Therefore, for images with a sufficient number of stars, the required uncertainty of 0.01″ can be obtained, provided that telescopes in the ~2 m size range are used. Note that in order to obtain SNR = 100 in typical ~1000 s exposures, telescopes in the 2 m size range are needed for TNOs with $m_V \sim 22$.

Differential chromatic refraction in the atmosphere can also give rise to systematic errors. A strategy to minimize this is to make observations near transit and use filters, ideally red filters, or even better, near IR filters. As given in Stone (2002), these errors are below the few mas level at zenith angles of ~20 degree, but depend on the color of the target and filter. Nevertheless, specific computations can be done to correct for this effect (Stone, 2002). Another solution involves the knowledge of spectra for the TNO and the reference stars. There are other sources of errors for the TNO centroid, such as contamination from faint background stars, which can shift the centroid. Observations on several days are recommended in order to minimize the problem, which may be very important in crowded fields and at deep magnitudes. In these cases, PSF fitting is usually superior to regular centroid determination. Also, techniques of deblending and/or image differencing are possible solutions.

Binary TNOs or TNOs with satellites also require especial treatment because the centroid will not lie in general on the center of the primary but in between the primary and satellite, and will give rise to a systematic error. This has an important effect on Haumea, which had to be carefully addressed (Ortiz et al., 2017) and was also seen in Orcus (Ortiz et al., 2011). This is also a well-known effect on Pluto due to its satellite Charon and can be modeled efficiently (Benedetti-Rossi et al., 2014). In many cases, we do not know whether the TNO could have an unknown satellite large enough to bias the centroid determination. Another complication is the possibility that the occultation star could be a close binary so that the catalog coordinates do not really correspond to any of the stars. Sometimes this can be anticipated by analyzing the colors or through spectroscopic observations of the star. The star duplicity aspects will be covered in future Gaia data releases but not nowadays.

Since the advent of Gaia, the astrometric updates are much more easily done and have good accuracy provided that the aspects above are taken into account. This is probably increasing our success rate of the multichord events. This can be seen in Fig. 19.2.

Before Gaia, the use of own star catalogues specifically tailored for the purpose of improving predictions of stellar occultations of several main TNO targets (Assafin et al., 2010, 2012; Camargo et al., 2014; Young et al., 2008; Porter et al., 2018), or small "subcatalogs" around pres-elected occultation events have usually been the main approach to make accurate updates. For specific cases, this may still be necessary nowadays (e.g., to make predictions of occultations by very faint stars not in the Gaia catalog, to be observed with very large telescopes only).

TNOs that move in dense stellar fields also cause more occultations. We can take advantage of that, but if the stellar fields are too crowded and the TNO is faint, good prediction refinement is difficult because good astrometry cannot be easily carried out from the ground due to the contamination by background stars, unless long campaigns are devoted to the TNO, with plenty of telescope time.

19.3.2 The role of big telescopic surveys

As already stated in Section 19.2.1, an accurate prediction of stellar occultations by small distant solar system Bodies relies on accurate stellar positions and ephemerides. Although the accuracy of stellar positions is no longer a problem, thanks to the submas accurate astrometry from the Gaia space mission, distant small bodies still need improvement of their orbits, and this can be achieved to some degree by frequent observations with powerful enough telescopes.

One important point to be considered here is how fast the uncertainties in the ephemerides of these bodies increase, as a function of time, after the most recent observation used in the determination of their orbits. As an example (Fig. 19.4) we can use the Centaur (55576) Amycus. After a time span of 7 years without observations, the uncertainty in the position grows considerably.

The NIMA scheme mentioned earlier is a numerical integration procedure that has been successfully used to compute accurate orbits in order to predict stellar occultations (Desmars et al., 2015). It uses a specific weighting scheme for the input astrometry data that considers the individual precision of the observation, the number of observations obtained during one night by the same observatory, and the possible presence of systematic errors in positions. It is easily seen from Fig. 19.4 that the ephemeris uncertainty ranges from few tens of mas in those dates close to the most recent observations (around 2013.4) to hundreds of mas a few years later. Although the increase rate of the uncertainty depends on the observational history of each object and its orbit; it is clear that regular observations of the bodies are important to achieve and keep accurate orbits, provided that the input astrometry is accurate. The big large-area astronomical surveys, like the Dark Energy Survey (Flaugher, 2005) and mainly the Large Synoptic Survey Telescope (LSST Science Book, Version 2.0, Abell et al., 2009), will play important roles in this regard (by providing their astrometric measurements of small solar system bodies to the Minor Planet Center).

As an example, the LSST, in particular, will record the entire sky visible from Cerro Pachón (Chile) twice a week and expects to observe around 40,000 TNOs over its 10 years of operations, with more than 100 observations for many of them. Single epoch (i.e., noncoadded) images should detect objects as faint as $m_r \sim 23.5$. Positions as accurate as 10 mas should be obtained under suitable seeing and for the expected SNR. The typical astrometric accuracy per coordinate per visit will range from 11 mas ($m_r = 21$) to 74 mas ($m_r = 24$).

Although it will still take one or two decades, the combination of many thousands of accurate positions of TNOs and stellar positions from Gaia will certainly move us from a context where observational campaigns for occultations are launched whenever an accurate prediction is obtained to another scenario where we will probably have to select events of interest and/or set up observational networks also allowing a data-driven approach in the

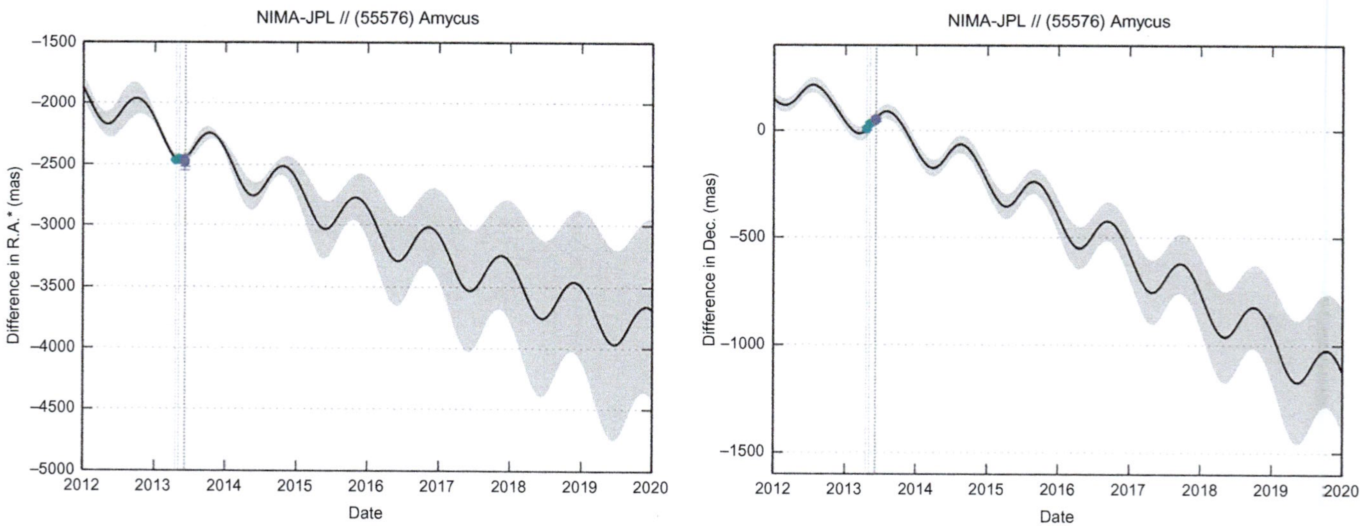

FIG. 19.4 Differences *(black wavy lines)* in right ascension *(left panel)* and declination *(right panel)* between the positions of (55576) Amycus as obtained from the NIMA and JPL (version 7) ephemerides. The sense of the differences is NIMA minus JPL. The gray area is the uncertainty of NIMA at 1σ level. The green dot represents a Gaia DR2-based position of Amycus. The blue dot represents a Gaia DR1-based position of Amycus. Positions of this Centaur, from 1987, were taken from the Minor Planet Center and also used in the orbit determination.

study of these distant objects. Whatever the case, the participation of large communities of observers or robotic, dedicated telescopes will be essential.

19.3.3 Expectations of accuracies and number of predictions

The number of predictions varies as a function of the angular diameter of the TNO and mainly of the stellar field that backgrounds the apparent sky path of the TNO. This said, the number of occultation events per TNO is typically 10 per year for a $m_V = 22$ target and for stars as faint as $m_V \sim 20$, but this depends on the position of the occulting body.

From Fig. 19.1, it looks safe (and conservative) to say that, after the 10 year LSST survey, orbits can be accurate to 50 mas or better (at least until few months after the latest observation of a given ephemeris) for at least several thousand TNOs. For an average of ~10 stellar occultations per year for stars as faint as 20 mag, there would be several tens of thousands of predictions. The bottleneck would be the needed updates from 50 to 10 mas accuracy. Setting a limit of $m_V = 18$ for the stars, the number of reasonably accurate predictions drops to a more tractable value of around a few thousand.

Within a decade or two from now, a first consequence from the big surveys and Gaia could therefore be a significant increase in the number of successful observed occultations, provided that the observational efforts are organized in an effective way, in a similar fashion as asteroidal occultations are being currently organized and observed within amateur networks, but some means of giving priority to specific events should be devised. Scientifically driven priorities should be established in some way.

19.4 The future of the observations

In this section we deal with observing possibilities and interesting opportunities in the future.

19.4.1 Telescope networks

The best performance in the TNO occultation field is achieved with telescope networks that are specifically designed for TNO occultations. If they can be arranged in a large north-south stretch of a good-weather area of the world, the success rate can be good. This is the reason that explains the successes in South America, which represent almost 50% of the published occultations thus far (see Tables 19.1 and 19.2).

The RECON Research and Education Collaborative Occultation Network (Buie and Keller, 2016), is a good example of a specific network created with the main goal of detecting occultations by TNOs. The greatest success of RECON thus far has been the detection of the occultation by 2007 UK126, within the context of a broad international collaboration (Benedetti-Rossi et al., 2016). The main current limitation of RECON is the fact that the telescopes are somewhat small and the magnitude limit for the setups is somewhat below 17, which is brighter than the stars in most of our predictions and observations.

Future robotic networks could afford to systematically observe predictions with low probability and given the expected overall future decrease of orbital uncertainties to 50 mas, the probabilities would not be too bad to achieve good results by mere serendipity. On the other hand, contributions have been made by large networks of amateur asteroid occultation observations. Networks exist in USA, Europe, Japan, Australia, and New Zealand. However, they are devoted to asteroid occultations, most of them using video cameras and telescopes smaller than 30 cm, which limits their observations to a few events involving relatively bright stars (m_V<14). New technologies and inexpensive, fast sensitive cameras, with nearly no deadtimes, and capable of integrating, are appearing in the market. Also, larger telescopes may be available to the amateurs in the future.

All this, in combination with portable, deployable networks, can give rise to an increase in the detection rate of occultations. The extremely remarkable success of the occultation by 2014 MU_{69}, the target of the New Horizons space mission, on July 17, 2017 is an example of the state of the art in the prediction and observation of stellar occultations by TNOs with the use of deployable networks (Zangari et al., 2017).

19.4.2 Large telescopes on Earth

The two largest optical telescopes planned to be built on Earth are the Extremely Large Telescope (ELT) and the Thirty Meter Telescope (TMT). They may allow recording occultations of extremely faint stars, which implies that many stellar occultations might be detectable per year, but currently there are no stellar catalogs reaching the needed depth and the detections would be single chord. If there are occultations of brighter stars near their sites, extremely high signal to noise may be obtained, which is an aspect that can improve the scientific output especially for searches for rings or an atmosphere (see Section 19.5)

19.4.3 James Webb Space Telescope

The James Webb Space Telescope (JWST) has already specific programs to observe stellar occultations in its GTO programs. The feasibility of observing occultations with JWST has been analyzed in (Santos-Sanz et al., 2016). Although the detections will be single chord, there will potentially be good science emerging from the infrared capabilities, particularly in atmospheric science.

19.4.4 Stratospheric balloon observatories

Small-to-medium light weight telescopes on board stratospheric balloons are being studied as potential future facilities for astronomical use (Maier et al., 2018). If long-lasting flights of around a month or more, launched from several sites on Earth, are possible, occultation events may be monitored even in day light, and the improved seeing at the stratosphere would allow reaching fainter stars than usual. Maybe the highest potential of these instruments is related to the possibility of being deployed within the shadow paths of predicted occultations, as this would be a guarantee against clouds or bad weather ruining the observations, for high scientific value occultations.

19.4.5 SOFIA

The SOFIA airborne observatory has also similar advantages to those mentioned in the previous paragraph. The main limitations are star brightness that can be accessible (because of high turbulence induced by the plane door) and mainly the deployment options and range, as well as its cost. For particularly interesting events, SOFIA can play an important role.

19.4.6 Cherenkov Telescopes

Large telescopes designed to detect the Cherenkov radiation from the avalanche of particles as a result of gamma rays hitting the atmosphere are the so-called Cherenkov Telescopes. They usually have very large collecting areas and cover large fields of views but have poor spatial resolution and cannot observe faint stars. However, they are capable of achieving extremely high time resolution measurements of bright targets. The use of this type of facilities for the study of serendipitous occultations has been assessed by Lacki (2014), but their use in predicted occultations has not been analyzed in detail yet. Currently, the central detector of telescopes such as the MAGIC telescope at La Palma can observe stars as faint as 13 mag with extremely high time resolution (J. Cortina, personal communication). In the future, this may be improved.

19.4.7 Radiotelescopes and radio occultations

Asteroidal occultations in the radio domain have recently been obtained for the first time (Lehtinen et al., 2016; Harju et al., 2018). Because the profiles are dominated by diffraction effects at radio wavelengths, the size of the asteroid can be retrieved from a single site. Therefore, the technique may prove useful for TNOs as well. Unfortunately, the number of radio sources in the radio catalogs is still very low compared to the number of stars, but this can change considerably after the square kilometer array (SKA) or its precursors enter operation. Also, the number of powerful enough radiotelescopes on Earth is not comparable to that of optical telescopes, but in the long term the technique may be fruitful for TNOs.

19.5 Aspects needed to improve the scientific output

Even though we have shown the great interest of stellar occultations by TNOs, often the scientific results from the detected events are not the best possible. In this section, we deal with aspects that can help to get more and better scientific results.

19.5.1 Efforts for multiple-chord vs single-chord occultations

Single-chord occultations do not allow us to determine the size of a TNO, but just a lower limit to it (assuming spherical shape). Note that such occultations can also be useful to detect or put stringent limits of an atmosphere. Two-chord occultations can also provide limited results, because we know that, in general, the projected shape of a TNO should be elliptical, not circular. The most likely shape of a TNO is an oblate spheroid, followed, with less probability by triaxial ellipsoids (Duffard et al., 2009) all of which give elliptical projections. A circle can be uniquely determined with four points (two chords extremities), but not an ellipse. For a unique elliptical solution at least three chords are needed. Therefore, efforts to achieve multichord events are important. Typically, this requires ~15 participating stations as discussed in Section 19.2.3.

19.5.2 Taking advantage of single-chord and two-chord occultations

Single-chord occultations at least provide an accurate position of the TNO to a level below the angular size of the TNO, which helps to improve the orbit of the TNO. But single-chord occultations combined with accurate astrometry of the TNO at the moment of the occultation to pinpoint the centroid position with respect to the chord, can result in at least estimates of the size, as was done in Alvarez-Candal et al. (2014). In order to apply this technique, the telescope used for the occultation observations must be powerful enough to record the occulting target with high SNR and the images must have several stars in the field to allow for relative astrometry to be performed. Since the occulting body is usually several magnitudes fainter than the occulted star this requires telescopes of the 2 m class and above in practice. Also, regular astrometry, if it can be accurate to the few mas level, can also be used to place the centroid relative to the observed chord so that a size can be estimated assuming a circular projection or an elliptical projection, in a similar way as mentioned above.

19.5.3 Deriving accurate geometric albedos

The geometric albedo of a TNO can be derived once a good projected area has been obtained from an occultation, using this simple equation:

$$p_v = \frac{10^{0.4(H_{\text{Sun}}-H)}}{(A/\pi)}$$

where H is the absolute magnitude of the TNO in a specific photometric band, H_{Sun} is the magnitude of the Sun in the same band, and A is the projected area in square astronomical units. Since the absolute magnitude H enters in the computation of geometric albedo and H is often not well known from data in the scientific literature, efforts to derive the correct H value are important. Also, it is necessary to discount the contributions of satellites and rings to H. If the satellites are large, this can make a difference of up to several percent. Rings or dust around the body can also make a contribution in reflected light, which must be discounted from the H magnitude. In the case of Haumea this contribution appears to be very small, of the order of 1% only, but on Chariklo and Chiron, their contribution is significant (Duffard et al., 2014; Ortiz et al., 2015).

The tiny contribution of the recently discovered small satellite of Makemake (Parker et al., 2016) to its absolute magnitude does not have an important effect on the albedo, but if there is a yet-undiscovered close satellite to Makemake, the albedo of Makemake could be lower than what has been derived thus far. The two-terrain thermal model (Lim et al., 2010) needing 10% of total Makemake's area in dark terrain might be an indication for such a large satellite.

Also, rotational variability in the value of H must be accounted for because the H value used must be the one at the moment of the occultation. This is especially important for objects with large amplitude of variability. The phase angle law is needed and it is often lacking. Observations of the TNO at different solar phase angles to derive *H* are usually important. Also, *H* is a function of epoch, given that the aspect angle changes somewhat in decades, especially for Centaurs. This must be modeled with a shape model if data separated by several years is used. For Haumea, all this was taken into account (Ortiz et al., 2017), but this has not often been possible for all the occultations.

A reassessment of all the published geometric albedos of TNOs from stellar occultations is made in Table 19.2, taking all the above effects into consideration. An example of reassessment for 2007 UK_{126}, with the inclusion of a new elliptical fit (combining the data from two different publications) is shown in Fig. 19.5.

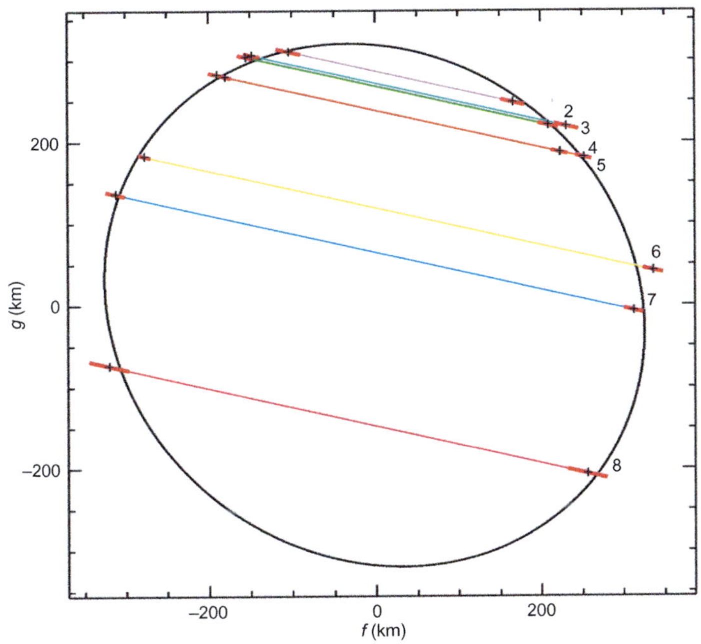

FIG. 19.5 Chords and new elliptical fit for 2007 UK126, combining the seven chords in Benedetti-Rossi et al. (2016) (marked here as (1) Reno, (2) Carson City S, (3) Carson City B, (4) Yerington, (5) Gardnerville Nevada, (6) Tonopah, (8) Urbana Illinois) with a nonredundant chord of the three-chord event reported in Schindler et al. (2017) for the same occultation. This chord from Sierra Remote Observatories is labeled with 7 here. The *red segments* indicate the uncertainties in the chords. North is up, East to the left. The new fit ($a = 678.0 \pm 11$ km, $b = 611.4 \pm 18$ km), with the new information on absolute magnitude (and taking into account a contribution of the known satellite of 0.036), gives rise to a new geometric albedo of 0.142.

19.5.4 Deriving three-dimensional shapes

For bodies without rotational symmetry, occultations take just a snapshot of the complete shape, at a particular rotational phase. More information from other techniques is needed in order to derive the full three-dimensional shape, or several occultations are needed together with a model shape to determine the 3D shape of the body.

Lightcurve observations of the TNO a few days around the occultation date are often needed so that the extrapolation of the rotation period does not give a large uncertainty in the rotational phase at which the occultation took place. Besides, this kind of observation is also needed to use the proper H value at the time of the occultation for a good geometric albedo estimate, as mentioned above.

19.5.5 Timing accuracy

Unfortunately, telescope equipment that is not specific for occultations can produce inaccurate absolute timings for a variety of reasons. Although efforts are made to avoid this problem, it is sometimes present in the data sets. Then, shifts in time must be applied to the chords of one or several sites. However, this must be done with a suitable methodology. A reasonable method was outlined in Braga-Ribas et al. (2013) to correct for small timing problems, by looking for the best time shifts that aligned the centers of the chords in a straight (but tilted) line.

The case of the stellar occultation by 2002 TX_{300} is a clear example in which a large timing problem of almost 32 s was present (Elliot et al., 2010). And unfortunately, only two chords were observed. By shifting the problematic chord as in Elliot et al. (2010) to make their centers coincide the authors were assuming that 2002 TX_{300} would be spherical, whereas we know that this is not the case and the most likely shape of a TNO is an oblate spheroid (Duffard et al., 2009). As a result, the albedo determined by Elliot et al. was too large. An example of an elliptical fit, compatible with the observations and an axial ratio 1.28 is shown in Fig. 19.6., but implies a larger effective diameter (320 km) than that in Elliot et al. If 2002 TX_{300} has a hydrostatic equilibrium shape, for the rotation rate of ~8 h, the minimum axial ratio a/c would be 1.28 for a density around 1000 kg/m^3. For a more plausible density of 800 kg/m^3 the axial ratio would be 1.39. Alternatively, the 32 s difference might be real and could be reflecting a binary body with a configuration as sketched in Fig. 19.6. In this case, a lower limit to the effective diameter would be also ~320 km. This implies a geometric albedo of 0.76 if we use

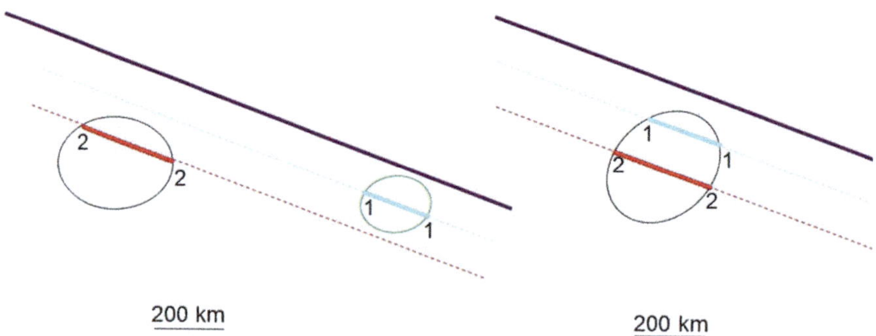

FIG. 19.6 Sketch of two possibilities for fits to the chords of the 2002 TX_{300} event of October 9, 2009. The continuous *red* and *blue lines* correspond to the chords in Mauna Kea and Faulkes telescope chords respectively (labeled 2 and 1, respectively). The *purple line* marks the negative chord at Leeward Community College. The *dashed lines* on the left panel indicate that just single events would be detected in sites 1 and 2 for the two proposed nuclei. The minimum size for the body is ~320 km. North is up, East to the left.

the H_V value of 2002 TX_{300} from Rabinowitz et al. (2008) or 0.68 if we use the same H_V used in Elliot et al. With the latest $H_V = 3.574$ determination by Alvarez-Candal et al. (2016), the resulting geometric albedo would be 0.65. Either in the elliptical case or in the binary scenario this means that 2002 TX_{300} would have an albedo more in accord with the recently determined value for Haumea (0.51), which makes sense given that 2002 TX_{300} is in the family of bodies orbitally related to Haumea.

Also, it is worth mentioning that topographic features can be determined with good accuracy provided that the timings are accurate. Otherwise, some of the deficiencies in the fits translate into residuals that can be wrongly interpreted as topographic features.

19.5.6 Detection of satellites and rings

In order to detect rings, given that they could be as narrow as 1 km, or even less, good temporal subsecond resolutions are needed, with good enough SNRs. At the typical occultation speed of 20 km s^{-1}, a 0.1 s integration corresponds to 2 km and depending on the opacity of the ring, the brightness drop may reach less than 50%. Hence, in order to detect rings as thin as 1 km, time resolutions of <0.1 s and SNRs around 10 are needed to detect the events at 5 sigma. This requires fast readout cameras, ideally with no deadtimes and very low readout noise. EMCCD cameras fall in this category but they are usually expensive for large telescope networks. Currently, cameras based on CMOS sensors with extremely low noise levels and fast readout are becoming available at low cost, which will have a positive impact. However, the telescope diameters needed are typically beyond those in the amateur level in order to reach the needed signal to noise. But for bright ~15-mag stars this can be possible with amateur equipment. Detection of satellites around the main occulting bodies is possible too, although of low probability. It may be an important scientific output, but in practice very dense coverage in terms of chords would be needed, which is still very difficult nowadays. In the future, with large and dense networks, this may become possible. This possibility is especially interesting for satellites that are too small and too close to the body to be detectable through imaging even with the JWST.

19.5.7 Atmospheres

Global atmospheres are easy to identify through occultations, as it is the case for Pluto (a topic not covered here). However, local atmospheres are not, because it is not likely that one of the chords will sample, by serendipity, the specific part where the atmospheric dome could be present. Detection is made even harder for an atmospheric dome, since the maximum pressure would be close to the subsolar point (depending on the thermal inertia, rotation rate, and pole position), and the pressure can be much less at the dawn and dusk limbs. This is quantified in (for example, Young, 2017). Ground-based stellar occultations of TNOs essentially only probe the dawn and dusk limbs.

Dense, multichord observations are needed for this. Alternatively, or complementary, central flashes from local atmospheres can potentially be detected provided that one or several of the chords sample a region near the center of the body. From ray-tracing simulations of local atmospheres on nonspherical bodies, one can conclude that some rays can reach the observers near the centrality, and if the object departs considerably from spherical, the regions of the body where anomalous flux can be measured can be larger than just a central spot (Ortiz et al., 2012, supplementary material). Therefore, achieving very high SNR at the bottom of the occultation profiles may allow detecting anomalous flux that could arise from such atmospheres. Features of that type were tentatively seen in the Makemake stellar occultation in 2011 although a cosmic-ray hit cannot be discarded as a possible source for the main spike (Ortiz et al., 2012). Quaoar may have also exhibited something similar near the centrality in its 2011 occultation, but the SNR was not high enough (Braga-Ribas et al., 2013). The largest TNOs are the best candidates to look for such features, if they exist. Spectrally resolved occultations with high-cadence spectrographs would also allow detecting absorptions at specific wavelengths, in order to determine gas composition and dust properties.

Note added in proof. Since the writing of this chapter, nineteen more occultations have been detected in the following dates for the indicated TNOs: 20 Oct 2018, 2015 TG_{387}, 28 Nov 2018, Chiron, 24 Dec 2018 2005 RM_{43}, 30 Dec 2018 2002 WC_{19}, 11 Jan 2019 Bienor, 3 Feb 2019 2005 RM_{43}, 25 Feb 2019, 2015 EZ_{51}, 18 Mar 2019 Huya, 27 Mar 2019 Quaoar, 28 Apr 2019 Quaoar, 28 May 2019 Quaoar, 5 Jun 2019 Quaoar, 9 Jul 2019 2002 MS_4, 11 Jul 2019 2007 JJ_{43}, 26 Jul 2017 2002MS_4, 7 Aug 2019 2007 JJ_{43}, 19 Aug 2019 2007 JJ_{43}, 28 Aug 2019 2017 OF_{69}, and 8 Sep 2019 Chiron.

Acknowledgments

J.L.O. acknowledges support from grants AYA2017-89637-R and FEDER funds. J.L.O. and P.S.S. acknowledge financial support from the State Agency for Research of the Spanish MCIU through the "Center of Excellence Severo Ochoa" award for the Instituto de Astrofísica de Andalucía (SEV-2017-0709). B.S. acknowledges support from the French grants "Beyond Neptune" ANR-08-BLAN-0177 and "Beyond Neptune II" ANR-11-IS56-0002. Part of the research leading to these results has received funding from the European Research Council under the European Community's H2020 (2014-2020/ERC grant agreement no. 669416 "Lucky Star"). P.S.S. acknowledges financial support by the European Union's Horizon 2020 Research and Innovation Programme, under Grant Agreement No. 687378. J.I.B.C. acknowledges CNPq grant 308150/2016-3. F.B.R. acknowledges CNPq grant 309578/2017-5. We acknowledge the use of Occult software by D. Herald.

References

Abell, P.A., Allison, J., Anderson, S.F., et al., 2009. LSST Science Book. Version 2.0. 2009arXiv0912.0201L.

Alvarez-Candal, A., Ortiz, J.L., Morales, N., et al., 2014. Stellar occultation by (119951) 2002 KX_{14} on April 26, 2012. Astron. Astrophys. 571, 8. Id.A48.

Alvarez-Candal, A., Pinilla-Alonso, N., Ortiz, J.L., et al., 2016. Absolute magnitudes and phase coefficients of Trans-Neptunian objects. Astron. Astrophys. 586, 33. Id.A155.

Assafin, M., Camargo, J.I.B., Vieira Martins, R., et al., 2010. Precise predictions of stellar occultations by Pluto, Charon, Nix, and Hydra for 2008–2015. Astron. Astrophys. 515, 14. Id.A32.

Assafin, M., Camargo, J.I.B., Vieira Martins, R., et al., 2012. Candidate stellar occultations by large Trans-Neptunian objects up to 201. Astron. Astrophys. 541, 10. Id.A142.

Benedetti-Rossi, G., Vieira Martins, R., Camargo, J.I.B., et al., 2014. Pluto: improved astrometry from 19 years of observations. Astron. Astrophys. 570, 12. Id. A86.

Benedetti-Rossi, G., Sicardy, B., Buie, M.W., 2016. Results from the 2014 November 15th multi-chord stellar occultation by the TNO (229762) 2007 UK126. Astron. J. 152, 11. Article id. 156.

Berard, D., Sicardy, B., Camargo, J.I.B., et al., 2017. The structure of Chariklo's rings from stellar occultations. Astron. J. 154, 21. Article id. 144.

Bianco, F.B., Protopapas, P., McLeod, B.A., et al., 2009. A search for occultations of bright stars by small Kuiper belt objects using Megacam on the MMT. Astron. J. 138, 568–578.

Bickerton, S.J., Kavelaars, J.J., Welch, D.L., 2008. A search for sub-km Kuiper belt objects with the method of serendipitous stellar occultations. Astron. J. 135, 1039–1049.

Braga-Ribas, F., Sicardy, B., Ortiz, J.L., et al., 2013. The size, shape, Albedo, density, and atmospheric limit of transneptunian object (50000) Quaoar from multi-chord stellar occultations. Astrophys. J. 773, 13. Article id. 26.

Braga-Ribas, F., Sicardy, B., Ortiz, J.L., et al., 2014. A ring system detected around the centaur (10199) Chariklo. Nature 508, 72–75.

Buie, M.W., Keller, J.M., 2016. The research and education collaborative occultation network: a system for coordinated TNO occultation observations. Astron. J. 151, 13. Article id. 73.

Camargo, J.I.B., Vieira-Martins, R., Assafin, M., et al., 2014. Candidate stellar occultations by centaurs and Trans-Neptunian objects up to 2014. Astron. Astrophys. 561, 10. Id.A37.

Carry, B., 2012. Density of asteroids. Planet. Space Sci. 73, 98–118.

Chang, H.-K., Liu, C.-Y., Chen, K.-T., 2013. Search for serendipitous Trans-Neptunian object occultation in X-rays. Mon. Not. R. Astron. Soc. 429, 1626–1632.

Desmars, J., Camargo, J.I.B., Braga-Ribas, F., et al., 2015. Orbit determination of Trans-Neptunian objects and centaurs for the prediction of stellar occultations. Astron. Astrophys. 584, 11. Id.A96.

Dias-Oliveira, A., Sicardy, B., Ortiz, J.L., et al., 2017. Study of the Plutino object (208996) 2003 AZ_{84} from stellar occultations: size, shape, and topographic features. Astron. J. 154, 13. Article id. 22.

Duffard, R., Ortiz, J.L., Thirouin, A., et al., 2009. Transneptunian objects and centaurs from light curves. Astron. Astrophys. 505, 1283–1295.

Duffard, R., Pinilla-Alonso, N., Ortiz, J.L., et al., 2014. Photometric and spectroscopic evidence for a dense ring system around Centaur Chariklo. Astron. Astrophys. 568, 9. Id.A79.

Elliot, J.L., Person, M.J., Zuluaga, C.A., et al., 2010. Size and Albedo of Kuiper belt object 55636 from a stellar occultation. Nature 465, 897–900.

Flaugher, B., 2005. The dark energy survey. Int. J. Modern Phys. A 20, 3121–3123.

Harju, J., Lehtinen, K., Romney, J., et al., 2018. Radio interferometric observation of an asteroid occultation. Astron. J. 156, 10. Article id. 155.

Lacki, B.C., 2014. On the use of Cherenkov Telescopes for outer solar system body occultations. Mon. Not. R. Astron. Soc. 445, 1858–1877.

Lehtinen, K., Bach, U., Muinonen, K., et al., 2016. Asteroid sizing by radiogalaxy occultation at 5 GHz. Astrophys. J. Lett. 822, 5. Article id. L21.

Leiva, R., Sicardy, B., Camargo, J.I.B., et al., 2017. Size and shape of Chariklo from multi-epoch stellar occultations. Astron. J. 154. Article id. 159.

Lim, T. L., Stansberry, J., Müller, T. G., Mueller, M., Lellouch, E., Kiss, C., Santos-Sanz, P., Vilenius, E., Protopapa, S., Moreno, R., Delsanti, A., Duffard, R., Fornasier, S., Groussin, O., Harris, A.W., Henry, F., Horner, J., Lacerda, P., Mommert, M., Ortiz, J.L., Rengel, M., Thirouin, A., Trilling, D., Barucci, A., Crovisier, J., Doressoundiram, A., Dotto, E., Gutiérrez Buenestado, P.J., Hainaut, O., Hartogh, P., Hestroffer, D., Kidger, M., Lara, L., Swinyard, B., Thomas, N., "TNOs are cool": a survey of the trans-Neptunian region. III. Thermophysical properties of 90482 Orcus and 136472 Makemake. Astron. Astrophys. 518, Article ID: L148, 5 pp.

Lindegren, L., Hernández, J., Bombrun, A., Klioner, S., et al., 2018. Gaia data release 2. The astrometric solution. Astron. Astrophys. 616, 25. Id.A2.

Liu, C.-Y., Doressoundiram, A., Roques, F., et al., 2015. Search for sub-kilometre Trans-Neptunian objects using CoRoT asteroseismology data. Mon. Not. R. Astron. Soc. 446, 932–940.

Maier, P., Wolf, J., Keilig, T., et al., 2018. Towards a European stratospheric balloon observatory: the ESBO design study. Proc. SPIE 10700, 12. Id. 107004M.

Mighell, K.J., 2005. Stellar photometry and astrometry with discrete point spread functions. Mon. Not. R. Astron. Soc. 361, 861–878.

Mueller, T.G., Lellouch, E., Böhnhardt, H., et al., 2009. TNOs are cool: a survey of the transneptunian region. Earth Moon Planets 105, 209–219.

Mueller, T.G., Lellouch, E., Stansberry, J., et al., 2010. "TNOs are cool": a survey of the Trans-Neptunian region. I. results from the Herschel science demonstration phase (SDP). Astron. Astrophys. 518, 5. Id. L146.

Mueller, T.G., Marciniak, A., Kiss, C.S., et al., 2018. Small bodies near and far (SBNAF): a benchmark study on physical and thermal properties of small bodies in the solar system. Adv. Space Res. 62, 2326–2341.
Muinonen, K., Bowell, E., 1993. Asteroid orbit determination using Bayesian probabilities. Icarus 104, 255–279.
Ortiz, J.L., Cikota, A., Cikota, S., et al., 2011. A mid-term astrometric and photometric study of Trans-Neptunian object (90482) Orcus. Astron. Astrophys. 525, 12. Id.A31.
Ortiz, J.L., Sicardy, B., Braga-Ribas, F., et al., 2012. Albedo and atmospheric constraints of dwarf planet Makemake from a stellar occultation. Nature 491, 566–569.
Ortiz, J.L., Duffard, R., Pinilla-Alonso, et al., 2015. Possible ring material around centaur (2060) Chiron. Astron. Astrophys. 576, 12. Id.A18.
Ortiz, J.L., Santos-Sanz, P., Sicardy, B., et al., 2017. The size, shape, density and ring of the dwarf planet Haumea from a stellar occultation. Nature 550, 219–223.
Parker, A.H., Buie, M.W., Grundy, W.M., Noll, K.S., et al., 2016. Discovery of a Makemakean moon. Astrophys. J. Lett. 825, 5. Article id. L9.
Pass, E., Metchev, S., Brown, P., et al., 2018. Pipeline for the detection of serendipitous stellar occultations by Kuiper belt objects with the Colibri fast-photometry array. Publ. Astron. Soc. Pac. 130 (983).
Person, M.J., Elliot, J.L., Bosh, A.S., et al., 2011. Constraints on the size of KBO (50000) Quaoar from a single-chord stellar occultation. In: American Astronomical Society, AAS Meeting #218, id.224.12, Bulletin of the American Astronomical Society, vol. 43, 2011.
Porter, S.B., Buie, M.W., Parker, A.H., et al., 2018. High-precision orbit fitting and uncertainty analysis of (486958) 2014MU$_{69}$. Astron. J. 156, 7. Article id. 20.
Prusti, T., de Bruijne, J.H.J., Brown, A.G.A., et al., 2016. The Gaia mission. Astron. Astrophys. 595, 36. Id. A1.
Rabinowitz, D.L., Schaefer, B.E., Schaefer, M., Tourtellotte, S.W., 2008. The youthful appearance of the 2003 EL$_{61}$ collisional family. Astron. J. 136, 1502–1509.
Roques, F., Doressoundiram, A., Dhillon, V., et al., 2006. Exploration of the Kuiper belt by high-precision photometric stellar occultations: first results. Astron. J. 132, 819–822.
Ruprecht, J.D., Bosh, A.S., Person, M.J., et al., 2015. 29 November 2011 stellar occultation by 2060 Chiron: symmetric jet-like features. Icarus 252, 271–276.
Santos-Sanz, P., French, R.G., Pinilla-Alonso, N., et al., 2016. James Webb Space Telescope observations of stellar occultations by solar system bodies and rings. Publ. Astron. Soc. Pac. 128 (959).
Schindler, K., Wolf, J., Bardecker, J., et al., 2017. Results from a triple chord stellar occultation and far-infrared photometry of the Trans-Neptunian object (229762) 2007 UK126. Astron. Astrophys. 600, 16. Id.A12.
Schlichting, H.E., Ofek, E.O., Sari, R., et al., 2012. Measuring the abundance of sub-kilometer-sized Kuiper belt objects using stellar occultations. Astrophys. J. 761, 10. Article id. 150.
Sicardy, B., Colas, F., Maquet, L., et al., 2010. The 2010. February 19 stellar occultation by Varuna, In: American Astronomical Society, DPS meeting #42, id. 23.11, Bulletin of the American Astronomical Society, vol. 42, pp. 993.
Sicardy, B., Ortiz, J.L., Assafin, M., et al., 2011. A Pluto-like radius and a high Albedo for the dwarf planet Eris from an occultation. Nature 478, 493–496.
Sickafoose, A.A., Bosh, A.S., Levine, S., et al., 2019. A stellar occultation by Vanth, a satellite of (90482) Orcus. Icarus 319, 657–668.
Stern, S.A., Trafton, L.M., 2008. On the atmospheres of objects in the Kuiper belt. In: Barucci, M.A., Boehnhardt, H., Cruikshank D.P., Morbidelli, A. (Eds.) The Solar System Beyond Neptune. University of Arizona Press, Tucson, pp. 365–380.
Stone, R.C., 2002. A new method for computing differential color refraction. Publ. Astron. Soc. Pacific 114, 1070–1086.
Wang, J.-H., Protopapas, P., Chen, W.-P., et al., 2010. Searching for sub-kilometer Trans-Neptunian objects using Pan-STARRS video mode light curves: preliminary study and evaluation using engineering data. Astron. J. 139, 2003–2013.
Young, L.A., 2017. Volatile transport on inhomogeneous surfaces: II. Numerical calculations (VT3D). Icarus 284, 443–476.
Young, E.L., French, R.G., Young, L.A., et al., 2008. Vertical structure in Pluto's atmosphere from the 2006 June 12 stellar occultation. Astron. J. 136, 1757–1769.
Zangari, A., Buie, M.W., Stern, S.A., et al., 2017. A stellar occultation by (486958) 2014 MU$_{69}$: results from the 2017 July 17 portable telescope campaign. In: American Astronomical Society, DPS Meeting #49, id.504.03.
Zhang, Z.-W., Lehner, M.J., Wang, J.-H., et al., 2013. The TAOs project: results from seven years of survey data. Astron. J. 146, 10. Article id. 14.

Further reading

Bosh, A.S, Zuluaga, C., Levine, S., et al., 2017. Astrometry of the Orcus/Vanth occultation on UT 7 March 2017. In: American Astronomical Society, DPS meeting #49, id.216.01.
Braga-Ribas, F., Vachier, F., Camargo, et al., 2017. Stellar Occultations by TNOS: Probing rings, Surface, and Satellites, Asteroids, Comets, Meteors (ACM2017). Montevideo, Uruguay.

CHAPTER

20

A darkness full of worlds: Prospects for discovery surveys in the outer solar system

Michele T. Bannister

Astrophysics Research Centre, School of Mathematics and Physics, Queen's University Belfast, Belfast, United Kingdom

20.1 Introduction

> Is it not likely that in Pluto there has come to light the first of a series of ultra-Neptunian bodies, the remaining members of which still await discovery but which are destined eventually to be detected? ***Frederick C. Leonard (1930)***

Exploring the population and structure of the outer solar system requires us to examine the sky. The improving sophistication of astronomical techniques have brought us from painstaking naked-eye examination of photographic plates to identify moving sources, to computerized techniques that can pull moving sources from the crowded depths of the Galactic plane. Such advances in our ability to discover new objects have allowed us to build an understanding of the solar system's distant small-body populations.

The discovery of objects in the region beyond Neptune has always been driven by the capacity of telescope apertures and detection technology. This naturally led to the division of surveying for new discoveries into two paths: the wide-aperture, small field-of-view "pencil-beam" surveys, and the small-to medium-aperture, wide field-of-view "large-area" surveys.

These surveys have been productive. As many discoveries were lost (sufficiently uncertain in on-sky position as to be unrecoverable[1]), it is most useful to compare orbits measured with

[1] See Joel Parker's recovery statistics at https://www.boulder.swri.edu/ekonews/objects/recov_stats.jpg. As of May 2019, some 20% of all known TNOs/Centaurs (690 objects) were either endangered or lost.

The Trans-Neptunian Solar System.
https://doi.org/10.1016/B978-0-12-816490-7.00020-5

sufficient observational arc to either be dynamically classifiable or, more generally, seen over multiple oppositions. At the publication of the major field review of *The Solar System Beyond Neptune* (Barucci et al., 2008), there were 584 distant orbits that Gladman et al. (2008) could classify as "reasonable-quality," as of May 2006. The Minor Planet Center records 2727 distant objects with multiopposition orbits, as of mid-April 2019.

20.2 Discovery surveys

Table 20.1 updates and expands on the collation of surveys up to 2008 by Kavelaars et al. (2008) and Brown and Webster (1998). I include optical discovery surveys that could at least in principle and with sufficient follow-up time have led to secure orbits for their discoveries; this choice retains surveys with Hubble Space Telescope (HST), but does exclude occultation surveys, whose discoveries are inherently irretrievable. In both tables, facility refers to telescope by aperture/instrument, where ambiguous. Survey limiting magnitude is normally given at m_{50}, the magnitude at which the detection efficiency of TNOs drops to 50%; some surveys instead report m_{40}. The degree to which this is accurately characterized improves over time, and becomes a priority of more surveys in the mid-2000s.

Most of the early surveys up to 2000 (Table 20.1) have been discussed at length; see Davies et al. (2008) for such a review. The magnitude and spatial completeness of the painstaking handwork of Tombaugh (1946, 1961) meant that few further surveys were undertaken until the late 1980s. Luyten's Palomar Sky Survey of the 1950s used a plate[2] measuring machine that used color illumination of the plates to show unmatched objects statically, rather than through blinking. Sadly, this device was not used for planetary astronomy, though it was used by Luyten to successfully find many nearby stars. Tombaugh (1961) looked ahead hopefully to a search "to the twentieth magnitude" with the new, optically faster design of Schmidt telescope, and suggested that it be tried with the 48″ Schmidt telescope at Palomar.

Surveying techniques changed with the development of CCD cameras in the late 1980s (Table 20.1). These very efficient detectors could only be manufactured with painfully small surface areas, which limited their utility for wide-field surveys for nearly another decade. A number of researchers made pilot digital surveys with a mix of hand and computer-blinked analysis techniques in the late 1980s and early 1990s, though all were unsuccessful (Luu and Jewitt, 1988; Kowal, 1989; Levison and Duncan, 1990; Cochran et al., 1991; Tyson et al., 1992).

The perseverance of Jewitt and Luu (1993) in a subsequent larger survey brought their discovery of (15760) 1992 QB_1. Nearly a hundred more Kuiper belt objects followed in the next 6 years (Fitzsimmons, 2000), predominantly from a scattering of small but quite deep algorithmically searched surveys in the Northern Hemisphere. These surveys (Table 20.1) were hampered in exploring Trans-Neptunian orbital parameter space by their small area; though the total area of the sky is 41,253 degree^2, Fitzsimmons (2000) noted the 52 degree^2 of Jewitt et al. (1998) was "huge."

[2] Note that the limiting magnitudes on all the plate surveys should be used with caution, in part due to variations between emulsions.

TABLE 20.1 Optical TNO surveys up to April 2019.

Survey[a] (if named)	Facility	Area (deg^2)	m_{50}	Bandpass	Reference
OSSOS	CFHT/MegaCam	155.3	24.09–25.15	r	Bannister et al. (2016); Bannister et al. (2018)
DES SN*	CTIO/DECam	250.0	23.5–23.8	griz	Gerdes et al. (2017); Becker et al. (2018); Lin et al. (2019)
KiDS DR3*	VLT/OmegaCam	346.0	22.9–25.4	ugri	Mahlke et al. (2018)
HSC-SSP DR1*	Subaru/HSC	221.0	25.4	HSC-g,r,i,r2,i2	Chen et al. (2018)
—	CTIO/DECam; Subaru HSC	~2130	25.0–25.5	VR ; r	Trujillo and Sheppard (2014); Sheppard et al. (2016); Sheppard and Trujillo (2016); Sheppard et al. (2019)
CFEPS-HiLat	CFHT/MegaCam	701.0	22.4–24.8	g′	Petit et al. (2017)
PS1 2010–15*	Pan-STARRS1	~20,000	22.5	griw	Weryk et al. (2016)
PS OSS 2010–14*	Pan-STARRS1	~20,000	22.5	griw	Chen et al. (2016); Lin et al. (2016); Holman et al. (2018b)
—	CFHT/MegaCam	32.0	24.6	r	Alexandersen et al. (2016)
—*	CSS 0.7; Uppsala 0.5	~29,700	19.4; 18.9	R; clear	Brown et al. (2015)
Siding Spring*	Uppsala 0.5	6275.0	19.0	Clear	Bannister (2014)
NGVCS *	CFHT/MegaCam	76.0	25.5	g′	Chen et al. (2013)
LSQKBO	La Silla 1.0	~7500	21.5	Q_{st}	Rabinowitz et al. (2012)
OCKS	Las Campanas 1.3	~2500	21.6	V, I, R	Sheppard et al. (2011)
—	Subaru/Suprime	8.93	24.7	r′	Fraser et al. (2010)
—*	HST/ACS	1.83	26.43	8; see paper	Fuentes et al. (2011), 5–20° off ecliptic
—*	HST/ACS	0.45	26.1	8; see paper	Fuentes et al. (2010), ecliptic
CFEPS	CFHT/MegaCam	321.0	23.5–24.4	g′	Jones et al. (2006); Kavelaars et al. (2009); Petit et al. (2011)
PDSSS	Palomar 1.2	11,786.0	21.3	RG610	Schwamb et al. (2009, 2010)
—*	Subaru/Suprime	0.26	26.8	R	Fuentes et al. (2009)
SDSS Stripe 82 *	Sloan	300.0	22.6	r	Unpublished; cf. Kaib et al. (2009)

Continued

TABLE 20.1 Optical TNO surveys up to April 2019—cont'd

Survey[a] (if named)	Facility	Area (deg^2)	m_{50}	Bandpass	Reference
—	Subaru/Suprime	0.3	26.8	r	Fraser and Kavelaars (2009)
—*	Subaru/Suprime	2.8	25.7	R	Fuentes and Holman (2008)
—*	CFHT; Blanco	3.0	26.4	R, g′, VR	Fraser et al. (2008)
ESSENCE*	Blanco	11.52	23.1–23.7	I, R	Becker et al. (2008)
Spacewatch	KPNO 0.9	7790.6	21.0	OG-515	Larsen et al. (2007)
Irregular Satellites	CFHT/12K	11.85	24.0	R	Petit et al. (2006)
DSS NGS-POSS I*	Palomar 1.2	4992.0	20–21, ~20	O, E	Rhoads (2005)
Deep Ecliptic Survey	KPNO; Blanco	550.0	22.0	VR	Millis et al. (2002); Elliot et al. (2005)
—	HST/ACS	0.02	29.2	F606W	Bernstein et al. (2004, 2006)
Caltech Wide Area	Palomar 1.2	19,389.0	20.5	Clear	Trujillo and Brown (2003); Brown (2008)
SEKS	Mt Stromlo 1.3	2900.0	19.5	Red, blue	Moody et al. (2003); Moody (2004)
—	La Silla 2.2	18.5	23.0?	R	Delsanti (2003)
—	KPNO	2.3	24.9–25.4	VR	Allen et al. (2002)
—	CTIO	1.5	24.9–25.5	R, VR	Allen et al. (2001)
SDSS*	Sloan	500	21.5	ugriz	Ivezic et al. (2001)
KPNO-Large	KPNO 0.9/Mosaic	164.0	21.1	R	Trujillo et al. (2001b)
—	CFHT; UT1	0.31	25.93	R, V; R(Bessel)	Gladman et al. (2001)
CFHT	CFHT	73.0	23.7	R	Trujillo et al. (2001a)
Spacewatch*	KPNO 0.9	1483.5	21.5	OG-515	Larsen et al. (2001)
Baker-Nunn	APT 0.5	1428.0	18.8	R	Sheppard et al. (2000)
EROS-2*	La Silla 1.5	30.0	~19.0	I	P. Tisserand (pers. comm.)
QUEST	Llano del Hato 1.0	66.8	20.0	BVRI	Rabinowitz and the Quasar Equatorial Survey Team (2000)
sKBO	CFHT/12K	20.2	23.7	R	Trujillo et al. (2000)
sKBO	UH 2.2/8K	51.5	22.5	VR	Trujillo et al. (2000)
—	KECK/LRIS	0.01	27.9	V	Chiang and Brown (1999)

—	SSO 40″	12.0	21.0	R	Brown and Webster (1998)
JL Deep	Keck	0.3	26.3	R	Luu and Jewitt (1998)
—	CFHT/UH8k	51.5	23.4	R	Gladman et al. (1998)
—	Palomar 5	0.049	25.6	Gunn r	Gladman et al. (1998)
—	Palomar 5	0.075	25.0	Gunn r	Gladman et al. (1998)
Maunakea 8kCCD	CFHT/UH8k	51.5	22.5	VR	Jewitt et al. (1998)
Pluto-Express	CFHT/12k	2.22	23.1	VR	Trujillo and Jewitt (1998)
MKCT	UH 2.2	3.9	24.2	VR	Jewitt et al. (1996)
MKCT	CTIO 1.5	4.4	23.2	R	Jewitt et al. (1996)
—	HST/WFPC2	0.001	27.9	Wide V	Cochran et al. (1995); see also Brown et al. (1997); Cochran et al. (1998)
—	WHT	0.7	23.5	R	Irwin et al. (1995)
—	UH 2.2	1.2	24.8	Mould R	Jewitt and Luu (1993, 1995)
—	INT 2.5	0.7	22.5	KPNO R	Williams (1995)
Unpublished	McDonald 2.7	Unknown	Unknown	V	Cochran et al. (1991)
—	Flagstaff 40″	4.9	21.0	V	Levison and Duncan (1990); limit per Brown and Webster (1998)
—	KPNO 1.3	0.338	24.0	R	Luu and Jewitt (1988)
—	KPNO 0.6/CTIO 0.9/plate	297.0	19.5	V	Luu and Jewitt (1988)
Unpublished	unreported	0.01	24.9	R	Tyson et al. (1992)
—	Palomar 48″/plate	6400	18.5	Wratten 2c	Kowal (1989)
—	Lowell 13″/plate	1530	18.0?	Clear	Tombaugh (1961)
—	Lowell 13″/plate	30,000	16.8?	Clear	Tombaugh (1946, 1961)
—	26″ USNO refractor/eye	100	10.0?	Clear	Todd (1880)

[a] *Surveys that have been conducted by data mining of images primarily acquired for a purpose other than TNO surveying are indicated with an asterisk (*).*
For Spacewatch, CFEPS and OSSOS, the magnitude limit is sky motion-rate dependent.
For POSS I, the photographic O-band is intermediate between Johnson B and U, E is approximately R (~6500 Å).
While Gladman et al. (1998) observed in Gunn r, they reported their Palomar mags in R.
For Kowal (1989), the plate-filter combination places the magnitude between B and V.

Around the turn of the 21st century, the technology of CCDs had matured into mosaic arrays, and a focus on wide-field surveys developed in the early 2000s; several meter-aperture and smaller telescopes were dedicated in large part to Trans-Neptunian surveys (Table 20.1), including several digital surveys with Tombaugh's hoped-for 48″ Schmidt, noted for their dwarf planet discoveries (Trujillo and Brown, 2003; Brown, 2008; Schwamb et al., 2009, 2010).

Increased access to larger apertures, and massive efforts including the Deep Ecliptic Survey and the Canada-France Ecliptic Plane Survey for mid-aperture, multihundred square degree surveys have been the hallmark of the 2000s and 2010s (Table 20.1). This decade saw a threefold increase in the number of known TNOs.

Some of the most recent surveys in Table 20.1 are ongoing. An online version of this table will be updated as future surveys are published.[3]

20.3 Current and future surveys

The continued effort to survey the sky for new discoveries has explored the phase space of much of the TNO size distribution. At the largest end, from wide-field surveys with small-to-medium optical telescopes in both Northern and Southern Hemispheres, the brightest (and correspondingly largest) objects are now effectively complete to $m \sim 19.5$ (Schwamb et al., 2014; Brown et al., 2015) and have substantial sky completion to $m \sim 22$ (Weryk et al., 2016; Holman et al., 2018a; Gerdes et al., 2017), with growing patches complete to greater depth. Continuing moderate-aperture optical surveys such as the 1.8 m PanSTARRS PS1/PS2 (R. Weryk et al., in prep) will ensure sky coverage is complete in the near term to $m \sim 22$. Surveys have now been running long enough that the closer part (e.g., plutinos) of the population are now beginning to clear the Galactic plane and be discovered. The Galactic plane has been systematically avoided by surveys, due to the difficulty of dealing with stellar crowding.

The need for abundant new discoveries, both to smaller sizes and in specific populations, including the remaining distant system, has been well recognized by the community. For understanding small objects, occultation surveys offer the promise of exquisite constraints on the size distribution (see next chapter). Additionally, the 194 orbits of HST that were allocated in support of the New Horizons mission search for a post-Pluto encounter target, which surveyed in the Galactic plane and succeeded in the discovery of 2014 MU_{69} for flyby in January 2019 (e.g., Porter et al., 2018), will provide a valuable constraint on the size distribution of the cold classical KBOs. This search (M. Buie et al., in prep) expands on the previous HST searches of Bernstein et al. (2004, 2006) (125 orbits) and the serendipitous HST archive searches of Fuentes et al. (2010, 2011).

In the infrared, survey missions have only had sensitivity within the Kuiper belt to the thermal brightness of the larger TNOs. The survey data of the WISE/NEOWISE spacecraft mission have provided serendipitous thermal measurements of sufficiently bright known Centaurs (Bauer et al., 2013), and have been analyzed for new TNOs, without discoveries (e.g., Perdelwitz et al., 2018). At greater distances, WISE/NEOWISE data offered further limits on

[3] Updated table at http://github.com/mtbannister/tnosurveys.

planets large and warm enough to be visible at shorter WISE wavelengths under a range of assumptions about their atmospheres, in addition to placing limits that excluded Saturn-mass planets out to 28,000 AU and Jupiter-mass planets out to 82,000 AU (Luhman, 2014; Meisner et al., 2018).

Two factors have led to an increase of surveying with 4+ meter optical telescopes: significant new wide-field cameras on these facilities have recently come online, and as existing wide-field facilities mature, time has become more accessible to conduct large dedicated TNO surveys. Primary survey roles are with the Dark Energy Camera (DECam) (DePoy et al., 2008) on CTIO, HyperSuprimeCam (Miyazaki et al., 2012) on Subaru, MegaPrime on CFHT, and MegaPrime on the Magellan 6.5 m. Major efforts with these facilities include work to add new TNOs on orbits like that of Sedna to map out the most distant regions (Trujillo and Sheppard, 2014, and related papers), to obtain precision estimates of the resonant populations in the Kuiper belt (e.g., CFEPS, OSSOS; Petit et al. (2011); Bannister et al. (2016); Bannister et al. (2018)), and to search for a new and massive planet in the outer solar system (Brown and Batygin, 2016; Batygin et al., 2019). Dedicated surveys collecting data specifically for the purpose of TNO observation have been the primary tool of advancement in these first two decades of rapid Kuiper belt discovery.

The coming decades will also see a surge in Trans-Neptunian discovery by data mining. Surveys that have been conducted by data mining of images primarily acquired for a purpose other than TNO surveying are indicated with an asterisk (*) in Table 20.1. The multinight detection surveying of Larsen et al. (2007), foreshadowed by SDSS commissioning trials (Ivezic et al., 2001) and echoed in the exploitation of supernovae data of Becker et al. (2008) and Kaib et al. (2009), has and will continue to increase: Table 20.1 demonstrates the uptick of using existing data sets for serendipitous discovery in the last decade, which can be considered together with the NEOWISE analyses. Dedicated surveys will remain highly important, for targeting specific aspects of known populations, for example, deep pencil-beam surveys will be possible with the coming suite of 30 m-class telescopes—but surveying by serendipity will become a prime, complementary discovery technique. Holman et al. (2018a) were able to recover otherwise lost TNOs among some 14 million observations in the MPC's Isolated Tracklet File. The ever-growing archive of imaging from large ground-based surveys such as KiDS (Kuijken et al., 2019), CFIS, DES, SkyMapper, and VISTA (though less successfully in the infrared) can be mined for additional astrometry and photometry of known TNOs (e.g., Popescu et al., 2016; Banda-Huarca et al., 2019; Mahlke et al., 2018), with tools like SSOIS (Gwyn et al., 2012) and EURONEAR (Vaduvescu et al., 2019). Space-based surveys like Gaia and in the near-infrared Euclid and WFIRST will complement such efforts, together with their own substantial discoveries (Carry, 2018; Holler et al., 2018).

With the extra complexities of discovery in datasets not designed for a TNO survey, the characterization of such surveys, though difficult, will become even more important, so that flux bias, pointing bias, ephemeris bias, and detection bias (Kavelaars et al., 2008) are well understood. The highly repetitious cadence of supernovae and asteroid surveys does offer reinforcement of discovery efficiencies against ephemeris bias (e.g., Bannister, 2014; Brown et al., 2015). The substantial effort required to characterize a survey means it needs to be considered as early as possible in the design of new surveys. The use of survey simulators is now a community-standard technique for evaluating the intrinsic population, given the survey biases inherent in obtaining each detected sample (e.g., Lawler et al., 2018).

Techniques for data analysis will also shift toward ever-greater automation. Analysis for discovery, whether by sophisticated variants of blink comparison or by shift-and-stack, has required an integral component of final vetting of the TNO candidates by people. While this can be characterized, as for instance in OSSOS (Bannister et al., 2016), it is exhausting for all astronomers concerned as the data volume increases. Machine learning is becoming a prime technique for candidate recognition in sidereally focused sky surveying, often with supervised learning algorithms (e.g., Haiman, 2018; Schanche et al., 2019; Baron, 2019). Use in solar system discovery surveying has been limited, but is now growing; for example, the use of convolutional neural nets, implemented with transfer learning for Euclid (Lieu et al., 2019) and with deep learning for ZTF (Duev et al., 2019). Its use has now been tried in TNO studies; Lin et al. (2018) tried both a supervised random forest and an unsupervised isolation forest implementation for HSC data and found the supervised learning more effective.

The community should recognize the value of the large sets of data checked by people that exist from past TNO surveys, and try to ensure that these classifications themselves are publicly archived and made accessible. These sets of many hundreds of thousands of classified candidates (the combined photometry, astrometry and decision of real object vs artifact) can be used as training sets for future algorithms—retaining, naturally, a detection efficiency bias, but this is satisfactory as long as the bias remains quantified. The increasing use of machine learning will require two areas of caution: ensuring a solid understanding of the physical features that neural-net-based architectures use to provide their classification, and ensuring that the truly unexpected, which is most likely absent from the training sets, can be successfully retained and flagged. Understanding the limitations of one's training data is very important, and a bias that should also be considered.

The future holds the promise from the Large Synoptic Survey Telescope (LSST) to reach $m_r = 24.7$ (5σ single visit) on the order of every three nights across at least 18,000 square degrees of the whole southern sky (Ivezić et al., 2019). The survey will go deeper in small targeted areas, or by shift-and-stack reduction of the full decade-long survey (Whidden et al., 2019). These observations will yield some 40,000 TNOs, observed 200–350 times in r for the brighter objects, which will provide insight into a wide range of open questions in TNO science (Schwamb et al., 2018). LSST's vast new dataset will challenge current approaches to both interacting with and assessing TNO datasets—once its initial round of observational arcs are well constrained after the first 2–3 years of operation and its quantified survey biases are made available with a survey simulator. It will also retrieve effectively all TNOs listed at the Minor Planet Center that are currently lost. The situation for understanding the TNO populations will become much more like that experienced by the asteroid community, and substantial amounts of software will need to be built by the TNO community (Schwamb et al., 2019). Eventually, LSST will become the epitome of data-mining surveys.

20.4 Conclusion

More than 60 sky surveys of the Trans-Neptunian region have now been conducted, the vast majority in the last 27 years. The last decade alone has more than doubled the number of known objects. I provide this compilation of the published and unpublished efforts of the community for easier recall. While targeted surveys have dominated the discovery strategy to date, there is a clear trend to mine discoveries from existing survey data sets, and with the advent of further medium-aperture and large-aperture wide-field imagers on more facilities, this surveying by serendipity is set to continue. With the advent of LSST, all that has gone before will shortly be extended ten-fold. In the early 2020s, we will truly begin mapping the deep.

Acknowledgments

M.T.B. appreciates support from UK STFC grant ST/P0003094/1. Thanks to Jeff Larsen for historical detail, and to an anonymous referee, Sam Lawler, JJ Kavelaars, and Paul Francis for helpful comments on various versions of this chapter.

References

Alexandersen, M., Gladman, B., Kavelaars, J.J., Petit, J.-M., Gwyn, S.D.J., Shankman, C.J., Pike, R.E., 2016. A carefully characterized and tracked Trans-Neptunian survey: the size distribution of the plutinos and the number of Neptune trojans. Astron. J. 152 (5), 1–25.

Allen, R.L., Bernstein, G.M., Malhotra, R., 2001. The edge of the solar system. Astrophys. J. 549 (2), L241–L244.

Allen, R.L., Bernstein, G.M., Malhotra, R., 2002. Observational limits on a distant cold Kuiper belt. Astron. J. 124 (5), 2949–2954.

Banda-Huarca, M.V., Camargo, J.I.B., Desmars, J., Ogando, R.L.C., Vieira-Martins, R., Assafin, M., da Costa, L.N., Bernstein, G.M., Carrasco Kind, M., Drlica-Wagner, A., Gomes, R., Gysi, M.M., Braga-Ribas, F., Maia, M.A.G., Gerdes, D.W., Hamilton, S., Wester, W., Abbott, T.M.C., Abdalla, F.B., Allam, S., Avila, S., Bertin, E., Brooks, D., Buckley-Geer, E., Burke, D.L., Carnero Rosell, A., Carretero, J., Cunha, C.E., Davis, C., De Vicente, J., Diehl, H.T., Doel, P., Fosalba, P., Frieman, J., García-Bellido, J., Gaztanaga, E., Gruen, D., Gruendl, R.A., Gschwend, J., Gutierrez, G., Hartley, W.G., Hollowood, D.L., Honscheid, K., James, D.J., Kuehn, K., Kuropatkin, N., Menanteau, F., Miller, C.J., Miquel, R., Plazas, A.A., Romer, A.K., Sanchez, E., Scarpine, V., Schubnell, M., Serrano, S., Sevilla-Noarbe, I., Smith, M., Soares-Santos, M., Sobreira, F., Suchyta, E., Swanson, M.E.C., Tarle, G., DES Collaboration, 2019. Astrometry and occultation predictions to Trans-Neptunian and centaur objects observed within the Dark Energy Survey. Astron. J. 157 (3), 120.

Bannister, M.T., 2014. Bright Trans-Neptunian Objects in the Southern Sky (Ph.D. thesis). RSAA, the Australian National University, Canberra.

Bannister, M.T., Kavelaars, J.J., Petit, J.-M., Gladman, B.J., Gwyn, S.D.J., Chen, Y.-T., Volk, K., Alexandersen, M., Benecchi, S.D., Delsanti, A., Fraser, W.C., Granvik, M., Grundy, W.M., Guilbert-Lepoutre, A., Hestroffer, D., Ip, W.-H., Jakubik, M., Jones, R.L., Kaib, N., Kavelaars, C.F., Lacerda, P., Lawler, S., Lehner, M.J., Lin, H.W., Lister, T., Lykawka, P.S., Monty, S., Marsset, M., Murray-Clay, R., Noll, K.S., Parker, A., Pike, R.E., Rousselot, P., Rusk, D., Schwamb, M.E., Shankman, C., Sicardy, B., Vernazza, P., Wang, S.-Y., 2016. The Outer Solar System Origins Survey. I: design and first-quarter discoveries. Astron. J. 152 (3), 1–25.

Bannister, M.T., Gladman, B.J., Kavelaars, J.J., Petit, J.-M., Volk, K., Chen, Y.-T., Alexandersen, M., Gwyn, S.D.J., Schwamb, M.E., Ashton, E., Benecchi, S.D., Cabral, N., Dawson, R.I., Delsanti, A., Fraser, W.C., Granvik, M., Greenstreet, S., Guilbert-Lepoutre, A., Ip, W.-H., Jakubik, M., Jones, R.L., Kaib, N.A., Lacerda, P., Van Laerhoven, C., Lawler, S., Lehner, M.J., Lin, H.W., Lykawka, P.S., Marsset, M., Murray-Clay, R., Pike, R.E., Rousselot, P., Shankman, C., Thirouin, A., Vernazza, P., Wang, S.-Y., 2018. OSSOS. VII. 800+ Trans-Neptunian objects—the complete data release. Astrophys. J. Suppl. 236, 18.

Baron, D., 2019. Machine learning in astronomy: a practical overview. arXiv e-prints arXiv:1904.07248.

Barucci, M.A., Boehnhardt, H., Cruikshank, D.P., Morbidelli, A. (Eds.), 2008. The Solar System Beyond Neptune. The University of Arizona Press, Tucson.

Batygin, K., Adams, F.C., Brown, M.E., Becker, J.C., 2019. The planet nine hypothesis. Phys. Rep. 805, 1–53.

Bauer, J.M., Grav, T., Blauvelt, E., Mainzer, A.K., Masiero, J.R., Stevenson, R., Kramer, E., Fernández, Y.R., Lisse, C.M., Cutri, R.M., Weissman, P.R., Dailey, J.W., Masci, F.J., Walker, R., Waszczak, A., Nugent, C.R., Meech, K.J., Lucas, A., Pearman, G., Wilkins, A., Watkins, J., Kulkarni, S., Wright, E.L., WISE Team, PTF Team, 2013. Centaurs and scattered disk objects in the thermal infrared: analysis of WISE/NEOWISE observations. Astrophys. J. 773 (1), 22.

Becker, A.C., Arraki, K., Kaib, N.A., Wood-Vasey, W.M., Aguilera, C., Blackman, J.W., Blondin, S., Challis, P., Clocchiatti, A., Covarrubias, R., Damke, G., Davis, T.M., Filippenko, A.V., Foley, R.J., Garg, A., Garnavich, P.M., Hicken, M., Jha, S., Kirshner, R.P., Krisciunas, K., Leibundgut, B., Li, W., Matheson, T., Miceli, A., Miknaitis, G., Narayan, G., Pignata, G., Prieto, J.L., Rest, A., Riess, A.G., Salvo, M.E., Schmidt, B.P., Smith, R.C., Sollerman, J., Spyromilio, J., Stubbs, C.W., Suntzeff, N.B., Tonry, J.L., Zenteno, A., 2008. Exploring the outer solar system with the ESSENCE supernova survey. Astrophys. J. 682 (1), L53–L56.

Becker, J.C., Khain, T., Hamilton, S.J., Adams, F.C., Gerdes, D.W., Zullo, L., Franson, K., Millholland, S., Bernstein, G.M., Sako, M., Bernardinelli, P., Napier, K., Markwardt, L., Lin, H.W., Wester, W., Abdalla, F.B., Allam, S., Annis, J., Avila, S., Bertin, E., Brooks, D., Carnero Rosell, A., Carrasco Kind, M., Carretero, J., Cunha, C.E., D'Andrea, C.B., da Costa, L.N., Davis, C., De Vicente, J., Diehl, H.T., Doel, P., Eifler, T.F., Flaugher, B., Fosalba, P., Frieman, J., García-Bellido, J., Gaztanaga, E., Gruen, D., Gruendl, R.A., Gschwend, J., Gutierrez, G., Hartley, W.G., Hollowood, D.L., Honscheid, K., James, D.J., Kuehn, K., Kuropatkin, N., Maia, M.A.G., March, M., Marshall, J.L., Menanteau, F., Miquel, R., Ogando, R.L.C., Plazas, A.A., Sanchez, E., Scarpine, V., Schindler, R., Sevilla-Noarbe, I., Smith, M., Smith, R.C., Soares-Santos, M., Sobreira, F., Suchyta, E., Swanson, M.E.C., Walker, A.R., DES Collaboration, 2018. Discovery and dynamical analysis of an extreme Trans-Neptunian object with a high orbital inclination. Astron. J. 156, 81.

Bernstein, G.M., Trilling, D.E., Allen, R.L., Brown, M.E., Holman, M., Malhotra, R., 2004. The size distribution of Trans-Neptunian bodies. Astron. J. 128, 1364.

Bernstein, G.M., Trilling, D.E., Allen, R.L., Brown, M.E., Holman, M., Malhotra, R., 2006. Erratum: "The Size Distribution of Trans-Neptunian Bodies" (AJ, 128, 1364 [2004]). Astron. J. 131, 2364.

Brown, M.E., 2008. The largest Kuiper belt objects. In: Barucci, M.A., Boehnhardt, H., Cruikshank, D.P., Morbidelli, A., Dotson, R. (Eds.) The solar system Beyond Neptune. University of Arizona Press, Tucson, pp. 335–344.

Brown, M.E., Batygin, K., 2016. Observational constraints on the orbit and location of planet nine in the outer Solar System. Astrophys. J. 824 (2), 1–9.

Brown, M.J.I., Webster, R.L., 1998. A search for bright Kuiper belt objects. Publ. Astron. Soc. Aust. 15, 176.

Brown, M.E., Kulkarni, S.R., Liggett, T.J., 1997. An analysis of the statistics of the Hubble Space Telescope Kuiper belt object search. Astrophys. J. 490 (1), L119–L122.

Brown, M.E., Bannister, M.T., Schmidt, B.P., Drake, A.J., Djorgovski, S.G., Graham, M.J., Mahabal, A., Donalek, C., Larson, S., Christensen, E., Beshore, E., McNaught, R., 2015. A serendipitous all sky survey for bright objects in the outer solar system. Astron. J. 149 (2), Article ID: 69, 6 pp.

Carry, B., 2018. Solar system science with ESA Euclid. Astron. Astrophys. 609, A113.

Chen, Y.-T., Kavelaars, J.J., Gwyn, S., Ferrarese, L., Côté, P., Jordán, A., Suc, V., Cuillandre, J.-C., Ip, W.-H., 2013. Discovery of a new member of the inner Oort cloud from the Next Generation Virgo Cluster Survey. Astrophys. J. Lett. 775 (1), Article ID: L8, 5 pp.

Chen, Y.-T., Lin, H.W., Holman, M.J., Payne, M.J., Fraser, W.C., Lacerda, P., Ip, W.-H., Chen, W.-P., Kudritzki, R.-P., Jedicke, R., Wainscoat, R.J., Tonry, J.L., Magnier, E.A., Waters, C., Kaiser, N., Wang, S.-Y., Lehner, M., 2016. Discovery of a new retrograde Trans-Neptunian object: hint of a common orbital plane for low semimajor axis, high-inclination TNOs and centaurs. Astrophys. J. 827 (2), L24.

Chen, Y.-T., Lin, H.-W., Alexandersen, M., Lehner, M.J., Wang, S.-Y., Wang, J.-H., Yoshida, F., Komiyama, Y., Miyazaki, S., 2018. Searching for moving objects in HSC-SSP: pipeline and preliminary results. Publ. Astron. Soc. Jpn. 70, S38.

Chiang, E.I., Brown, M.E., 1999. Keck pencil-beam survey for faint Kuiper belt objects. Astron. J. 1411–1422.

Cochran, A.L., Cochran, W.D., Torbett, M.V., 1991. A deep imaging search for the Kuiper disk of comets. Bull. Am. Astron. Soc. 23, 1314.

Cochran, A.L., Levison, H.F., Stern, S.A., Duncan, M.J., 1995. The discovery of Halley-sized Kuiper belt objects using the Hubble Space Telescope. Astrophys. J. 455, 342.

Cochran, A.L., Levison, H.F., Tamblyn, P., Stern, S.A., Duncan, M.J., 1998. The calibration of the Hubble Space Telescope Kuiper belt object search: setting the record straight. Astrophys. J. 503 (1), L89–L93.

Davies, J.K., McFarland, J., Bailey, M.E., Marsden, B.G., Ip, W.H., 2008. The early development of ideas concerning the transneptunian region. In: Barucci, M.A., Boehnhardt, H., Cruikshank, D.P., Morbidelli, A., Dotson, R. (Eds.) The Solar System Beyond Neptune. University of Arizona Press, Tucson, pp. 11–23.

Delsanti, A., 2003, Dec. Les objets Trans-Neptuniens : Détection et Caractérisation Physique (Ph.D. thesis). Université de Paris VI, Pierre et Marie Curie, Paris.

DePoy, D.L., Abbott, T., Annis, J., Antonik, M., Barceló, M., Bernstein, R., Bigelow, B., Brooks, D., Buckley-Geer, E., Campa, J., Cardiel, L., Castander, F., Castilla, J., Cease, H., Chappa, S., Dede, E., Derylo, G., Diehl, H.T., Doel, P., DeVicente, J., Estrada, J., Finley, D., Flaugher, B., Gaztanaga, E., Gerdes, D., Gladders, M., Guarino, V., Gutierrez, G., Hamilton, J., Haney, M., Holland, S., Honscheid, K., Huffman, D., Karliner, I., Kau, D., Kent, S., Kozlovsky, M., Kubik, D., Kuehn, K., Kuhlmann, S., Kuk, K., Leger, F., Lin, H., Martinez, G., Martinez, M., Merritt, W., Mohr, J., Moore, P., Moore, T., Nord, B., Ogando, R., Olsen, J., Onal, B., Peoples, J., Qian, T., Roe, N., Sanchez, E., Scarpine, V., Schmidt, R., Schmitt, R., Schubnell, M., Schultz, K., Selen, M., Shaw, T., Simaitis, V., Slaughter, J., Smith, C., Spinka, H., Stefanik, A., Stuermer, W., Talaga, R., Tarle, G., Thaler, J., Tucker, D., Walker, A., Worswick, S., Zhao, A., 2008. The Dark Energy Camera (DECam). In: McLean, Ian S., Casali, Mark M. (Eds.) Astronomical Telescopes and Instrumentation: Synergies Between Ground and Space. SPIE, pp. 70140E–70140E-9.

Duev, D.A., Mahabal, A., Ye, Q., Tirumala, K., Belicki, J., Dekany, R., Frederick, S., Graham, M.J., Laher, R.R., Masci, F.J., Prince, T.A., Riddle, R., Rosnet, P., Soumagnac, M.T., 2019. DeepStreaks: identifying fast-moving objects in the Zwicky transient facility data with deep learning. arXiv e-prints arXiv:1904.05920.

Elliot, J.L., Kern, S.D., Clancy, K.B., Gulbis, A.A.S., Millis, R.L., Buie, M.W., Wasserman, L.H., Chiang, E.I., Jordan, A.B., Trilling, D.E., Meech, K.J., 2005. The Deep Ecliptic Survey: a search for Kuiper belt objects and centaurs. II. Dynamical classification, the Kuiper belt plane, and the core population. Astron. J. 129 (2), 1117–1162.

Fitzsimmons, A., 2000. Surveys of the distant solar system, In: Minor Bodies in the Outer Solar System: Proceedings of the ESO Workshop held at Garching, pp. 87.

Fraser, W.C., Kavelaars, J.J., 2009. The size distribution of Kuiper belt objects for $D_g \sim 10$ km. Astron. J. 137, 72.

Fraser, W.C., Kavelaars, J.J., Holman, M.J., Pritchet, C.J., Gladman, B.J., Grav, T., Jones, R.L., MacWilliams, J., Petit, J.M., 2008. The Kuiper belt luminosity function from $m_r = 21$ to 26. Icarus 195 (2), 827–843.

Fraser, W.C., Brown, M.E., Schwamb, M.E., 2010. The luminosity function of the hot and cold Kuiper belt populations. Icarus 210 (2), 944–955.

Fuentes, C.I., Holman, M.J., 2008. A Subaru archival search for faint Trans-Neptunian objects. Astron. J. 136 (1), 83–97.

Fuentes, C.I., George, M.R., Holman, M.J., 2009. A Subaru pencil-beam search for $m_r \sim 27$ Trans-Neptunian bodies. Astrophys. J. 696 (1), 91–95.

Fuentes, C.I., Holman, M.J., Trilling, D.E., Protopapas, P., 2010. Trans-Neptunian objects with Hubble Space Telescope ACS/WFC. Astrophys. J. 722 (2), 1290–1302.

Fuentes, C.I., Trilling, D.E., Holman, M.J., 2011. Dynamically excited outer solar system objects in the Hubble Space Telescope archive. Astrophys. J. 742 (2), 118.

Gerdes, D.W., Sako, M., Hamilton, S., Zhang, K., Khain, T., Becker, J.C., Annis, J., Wester, W., Bernstein, G.M., Scheibner, C., Zullo, L., Adams, F., Bergin, E., Walker, A.R., Mueller, J.H., Abbott, T.M.C., Abdalla, F.B., Allam, S., Bechtol, K., Benoit-Lévy, A., Bertin, E., Brooks, D., Burke, D.L., Carnero Rosell, A., Carrasco Kind, M., Carretero, J., Cunha, C.E., da Costa, L.N., Desai, S., Diehl, H.T., Eifler, T.F., Flaugher, B., Frieman, J., García-Bellido, J., Gaztanaga, E., Goldstein, D.A., Gruen, D., Gschwend, J., Gutierrez, G., Honscheid, K., James, D.J., Kent, S., Krause, E., Kuehn, K., Kuropatkin, N., Lahav, O., Li, T.S., Maia, M.A.G., March, M., Marshall, J.L., Martini, P., Menanteau, F., Miquel, R., Nichol, R.C., Plazas, A.A., Romer, A.K., Roodman, A., Sanchez, E., Sevilla-Noarbe, I., Smith, M., Smith, R.C., Soares-Santos, M., Sobreira, F., Suchyta, E., Swanson, M.E.C., Tarle, G., Tucker, D.L., Zhang, Y., DES Collaboration, 2017. Discovery and physical characterization of a large scattered disk object at 92 au. Astrophys. J. Lett. 839, L15.

Gladman, B., Kavelaars, J.J., Nicholson, P.D., Loredo, T.J., Burns, J.A., 1998. Pencil-beam surveys for faint Trans-Neptunian objects. Astron. J. 116, 2042.

Gladman, B., Kavelaars, J.J., Petit, J.-M., Morbidelli, A., Holman, M.J., Loredo, T., 2001. The structure of the Kuiper belt: size distribution and radial extent. Astron. J. 122, 1051.

Gladman, B., Marsden, B.G., Vanlaerhoven, C., 2008. Nomenclature in the outer solar system. In: Barucci, M.A., Boehnhardt, H., Cruikshank, D.P., Morbidelli, A., Dotson, R. (Eds.) The Solar System Beyond Neptune. The University of Arizona Press, pp. 43–57.

Gwyn, S.D.J., Hill, N., Kavelaars, J.J., 2012. SSOS: a moving-object image search tool for asteroid precovery. Publ. Astron. Soc. Pac. 124 (916), 579–585.

Haiman, Z., 2018. Machine learning uncovers new cosmological information. Nat. Astron. arXiv:1810.12220.

Holler, B.J., Milam, S.N., Bauer, J.M., Alcock, C., Bannister, M.T., Bjoraker, G.L., Bodewits, D., Bosh, A.S., Buie, M.W., Farnham, T.L., Haghighipour, N., Hardersen, P.S., Harris, A.W., Hirata, C.M., Hsieh, H.H., Kelley, M.S.P., Knight, M.M., Kramer, E.A., Longobardo, A., Nixon, C.A., Palomba, E., Protopapa, S., Quick, L.C., Ragozzine, D.,

Reddy, V., Rhodes, J.D., Rivkin, A.S., Sarid, G., Sickafoose, A.A., Simon, A.A., Thomas, C.A., Trilling, D.E., West, R.A., 2018. Solar system science with the Wide-field Infrared Survey Telescope. J. Astron. Telescopes Instrum. Syst. 4, 034003.

Holman, M.J., Payne, M.J., Blankley, P., Janssen, R., Kuindersma, S., 2018a. HelioLinC: a novel approach to the minor planet linking problem. Astron. J. 156 (3), Article ID: 135, 15 pp.

Holman, M.J., Payne, M.J., Fraser, W., Lacerda, P., Bannister, M.T., Lackner, M., Chen, Y.-T., Lin, H.W., Smith, K.W., Kokotanekova, R., Young, D., Chambers, K., Chastel, S., Denneau, L., Fitzsimmons, A., Flewelling, H., Grav, T., Huber, M., Induni, N., Kudritzki, R.-P., Krolewski, A., Jedicke, R., Kaiser, N., Lilly, E., Magnier, E., Mark, Z., Meech, K.J., Micheli, M., Murray, D., Parker, A., Protopapas, P., Ragozzine, D., Veres, P., Wainscoat, R., Waters, C., Weryk, R., 2018.b, A dwarf planet class object in the 21:5 resonance with Neptune. Astrophys. J. Lett. 855 (1), Article ID: L6, 9 pp.

Irwin, M., Tremaine, S., Zytkow, A.N., 1995. A search for slow-moving objects and the luminosity function of the Kuiper belt. Astron, J. 110, 3082.

Ivezic, Z., Tabachnik, S., Rafikov, R., Lupton, R.H., Quinn, T., Hammergren, M., Eyer, L., Chu, J., Armstrong, J.C., Fan, X., Finlator, K., Geballe, T.R., Gunn, J.E., Hennessy, G.S., Knapp, G.R., Leggett, S.K., Munn, J.A., Pier, J.R., Rockosi, C.M., Schneider, D.P., Strauss, M.A., Yanny, B., Brinkmann, J., Csabai, I., Hindsley, R.B., Kent, S., Lamb, D.Q., Margon, B., McKay, T.A., Smith, J.A., Waddell, P., York, D.G., 2001. Solar system objects observed in the Sloan Digital Sky Survey commissioning data. Astron. J. 122 (5), 2749–2784.

Ivezić, Ž., Kahn, S.M., Tyson, J.A., et al., 2019. LSST: from science drivers to reference design and anticipated data products. Astrophys. J. 873 (2), 111.

Jewitt, D., Luu, J., 1993. Discovery of the candidate Kuiper belt object 1992 QB_1. Nature 362, 730 (ISSN 0028-0836).

Jewitt, D.C., Luu, J.X., 1995. The solar system beyond Neptune. Astron. J. 109, 1867–1876 (ISSN 0004-6256).

Jewitt, D., Luu, J., Chen, J., 1996. The Mauna Kea-Cerro-Tololo (MKCT) Kuiper belt and Centaur survey. Astron. J. 112, 1225.

Jewitt, D., Luu, J., Trujillo, C., 1998. Large Kuiper belt objects: the Mauna Kea 8K CCD survey. Astron. J. 115, 2125.

Jones, R.L., Gladman, B., Petit, J.M., Rousselot, P., Mousis, O., Kavelaars, J.J., Campo Bagatin, A., Bernabeu, G., Benavidez, P., Parker, J.W., Nicholson, P., Holman, M., Grav, T., Doressoundiram, A., Veillet, C., Scholl, H., Mars, G., 2006. The CFEPS Kuiper belt survey: strategy and presurvey results. Icarus 185 (2), 508–522.

Kaib, N.A., Becker, A.C., Jones, R.L., Puckett, A.W., Bizyaev, D., Dilday, B., Frieman, J.A., Oravetz, D.J., Pan, K., Quinn, T., Schneider, D.P., Watters, S., 2009. 2006 SQ 372: a likely long-period comet from the inner Oort cloud. Astrophys. J. 695 (1), 268–275.

Kavelaars, J.J., Jones, L., Gladman, B., Parker, J.W., Petit, J.M., 2008. The orbital and spatial distribution of the Kuiper belt. In: Barucci, M.A., Boehnhardt, H., Cruikshank, D.P., Morbidelli, A., Dotson, R. (Eds.) The Solar System Beyond Neptune. University of Arizona Press, Tucson, pp. 59–69.

Kavelaars, J.J., Jones, R.L., Gladman, B.J., Petit, J.-M., Parker, J.W., Van Laerhoven, C., Nicholson, P., Rousselot, P., Scholl, H., Mousis, O., Marsden, B., Benavidez, P., Bieryla, A., Campo Bagatin, A., Doressoundiram, A., Margot, J.L., Murray, I., Veillet, C., 2009. The Canada-France Ecliptic Plane Survey—L3 data release: the orbital structure of the Kuiper belt. Astron. J. 137 (6), 4917–4935.

Kowal, C.T., 1989. A solar system survey. Icarus 77 (1), 118–123.

Kuijken, K., Heymans, C., Dvornik, A., Hildebrandt, H., de Jong, J.T.A., Wright, A.H., Erben, T., Bilicki, M., Giblin, B., Shan, H. -Y., Getman, F., Grado, A., Hoekstra, H., Miller, L., Napolitano, N., Paolilo, M., Radovich, M., Schneider, P., Sutherland , W., Tewes, M., Tortora, C., Valentijn, E.A., Verdoes Kleijn, G.A., 2019. The fourth data release of the Kilo-Degree Survey: ugri imaging and nine-band optical-IR photometry over 1000 square degrees. arXiv e-prints Astron. Astrophys. 625, Article ID: A2, 25 pp.

Larsen, J.A., Gleason, A.E., Danzl, N.M., Descour, A.S., McMillan, R.S., Gehrels, T., Jedicke, R., Montani, J.L., Scotti, J.V., 2001. The Spacewatch wide-area survey for bright centaurs and Trans-Neptunian objects. Space Sci. Rev. 121 (1), 562–579.

Larsen, J.A., Roe, E.S., Albert, C.E., Descour, A.S., McMillan, R.S., Gleason, A.E., Jedicke, R., Block, M., Bressi, T.H., Cochran, K.C., Gehrels, T., Montani, J.L., Perry, M.L., Read, M.T., Scotti, J.V., Tubbiolo, A.F., 2007. The search for distant objects in the solar system using Spacewatch. Astron. J. 133 (4), 1247–1270.

Lawler, S.M., Kavelaars, J.J., Alexandersen, M., Bannister, M.T., Gladman, B., Petit, J.-M., Shankman, C., 2018. OSSOS: X. How to use a survey simulator: statistical testing of dynamical models against the real Kuiper belt. Front. Astron. Space Sci 5, 14.

Leonard, F.C., 1930. The new planet Pluto. Astron. Soc. Pac. Leaflets 1, 121.

Levison, H.F., Duncan, M.J., 1990. A search for proto-comets in the outer regions of the solar system. Astron. J. 100, 1669 (ISSN 0004-6256).

Lieu, M., Conversi, L., Altieri, B., Carry, B., 2019. Detecting solar system objects with convolutional neural networks. Mon. Not. R. Astron. Soc. 485 (4), 5831–5842.

Lin, H.W., Chen, Y.-T., Holman, M.J., Ip, W.-H., Payne, M.J., Lacerda, P., Fraser, W.C., Gerdes, D.W., Bieryla, A., Sie, Z.F., Chen, W.P., Burgett, W.S., Denneau, L., Jedicke, R., Kaiser, N., Magnier, E.A., Tonry, J.L., Wainscoat, R.J., Waters, C., 2016. The Pan-STARRS1 discoveries of five new Neptune Trojans. Astron. J. 152 (5), 1–8.

Lin, H.-W., Chen, Y.-T., Wang, J.-H., Wang, S.-Y., Yoshida, F., Ip, W.-H., Miyazaki, S., Terai, T., 2018. Machine-learning-based real-bogus system for the HSC-SSP moving object detection pipeline. Publ. Astron. Soc. Jpn. 70, S39.

Lin, H.W., Gerdes, D.W., Hamilton, S.J., Adams, F.C., Bernstein, G.M., Sako, M., Bernadinelli, P., Tucker, D., Allam, S., Becker, J.C., Khain, T., Markwardt, L., Franson, K., Abbott, T.M.C., Annis, J., Avila, S., Brooks, D., Carnero Rosell, A., Carrasco Kind, M., Cunha, C.E., D'Andrea, C.B., da Costa, L.N., De Vicente, J., Doel, P., Eifler, T.F., Flaugher, B., García-Bellido, J., Hollowood, D.L., Honscheid, K., James, D.J., Kuehn, K., Kuropatkin, N., Maia, M.A.G., Marshall, J.L., Miquel, R., Plazas, A.A., Romer, A.K., Sanchez, E., Scarpine, V., Sevilla-Noarbe, I., Smith, M., Smith, R.C., Soares-Santos, M., Sobreira, F., Suchyta, E., Tarle, G., Walker, A.R., Wester, W., 2019. Evidence for color dichotomy in the primordial Neptunian Trojan population. Icarus 321, 426–435.

Luhman, K.L., 2014. A search for a distant companion to the Sun with the Wide-Field Infrared Survey Explorer. Astrophys. J. 781 (1), 4.

Luu, J.X., Jewitt, D., 1988. A two-part search for slow-moving objects. Astron. J. 95, 1256 (ISSN 0004-6256).

Luu, J.X., Jewitt, D.C., 1998. Deep imaging of the Kuiper belt with the Keck 10 meter telescope. Astrophys. J. Lett. 502, L91.

Mahlke, M., Bouy, H., Altieri, B., Verdoes Kleijn, G., Carry, B., Bertin, E., de Jong, J.T.A., Kuijken, K., McFarland, J., Valentijn, E., 2018. Mining the Kilo-degree survey for solar system objects. Astron. Astrophys. 610, A21.

Meisner, A.M., Bromley, B.C., Kenyon, S.J., Anderson, T.E., 2018. A 3π search for planet nine at 3.4 μm with WISE and NEOWISE. Astron. J. 155 (4), Article ID: 166, 10 pp.

Millis, R.L., Buie, M.W., Wasserman, L.H., Elliot, J.L., Kern, S.D., Wagner, R.M., 2002. The Deep Ecliptic Survey: a search for Kuiper Belt objects and centaurs. I. Description of methods and initial results. Astron. J. 123 (4), 2083–2109.

Miyazaki, S., Komiyama, Y., Nakaya, H., Kamata, Y., Doi, Y., Hamana, T., Karoji, H., Furusawa, H., Kawanomoto, S., Morokuma, T., Ishizuka, Y., Nariai, K., Tanaka, Y., Uraguchi, F., Utsumi, Y., Obuchi, Y., Okura, Y., Oguri, M., Takata, T., Tomono, D., Kurakami, T., Namikawa, K., Usuda, T., Yamanoi, H., Terai, T., Uekiyo, H., Yamada, Y., Koike, M., Aihara, H., Fujimori, Y., Mineo, S., Miyatake, H., Yasuda, N., Nishizawa, J., Saito, T., Tanaka, M., Uchida, T., Katayama, N., Wang, S.-Y., Chen, H.-Y., Lupton, R., Loomis, C., Bickerton, S., Price, P., Gunn, J., Suzuki, H., Miyazaki, Y., Muramatsu, M., Yamamoto, K., Endo, M., Ezaki, Y., Itoh, N., Miwa, Y., Yokota, H., Matsuda, T., Ebinuma, R., Takeshi, K., 2012. Hyper Suprime-Cam. In: McLean, I.S., Ramsay, S.K., Takami, H. (Eds.) SPIE Astronomical Telescopes + Instrumentation. SPIE, pp. 84460Z–84460Z-9.

Moody, R., 2004. The Southern Edgeworth-Kuiper Belt Survey (Ph.D. thesis). RSAA, the Australian National University.

Moody, R., Schmidt, B., Alcock, C., Goldader, J., Axelrod, T., Cook, K.H., Marshall, S., 2003. Initial results from the Southern Edgeworth-Kuiper belt survey. Earth 92, 125.

Perdelwitz, V., Völschow, M., Müller, H.M., 2018. A new approach to distant solar system object detection in large survey data sets. Astron. Astrophys. 615, A159.

Petit, J.M., Holman, M.J., Gladman, B.J., Kavelaars, J.J., Scholl, H., Loredo, T.J., 2006. The Kuiper belt luminosity function from $m_r = 22$ to 25. Mon. Not. R. Astron. Soc. 365 (2), 429–438.

Petit, J.M., Kavelaars, J.J., Gladman, B.J., Jones, R.L., Parker, J.W., Van Laerhoven, C., Nicholson, P., Mars, G., Rousselot, P., Mousis, O., Marsden, B., Bieryla, A., Taylor, M., Ashby, M.L.N., Benavidez, P., Campo Bagatin, A., Bernabeu, G., 2011. The Canada-France Ecliptic Plane Survey—full data release: the orbital structure of the Kuiper belt. Astron. J. 142 (4), 131.

Petit, J.-M., Kavelaars, J.J., Gladman, B.J., Jones, R.L., Parker, J.W., Bieryla, A., Van Laerhoven, C., Pike, R.E., Nicholson, P., Ashby, M.L.N., Lawler, S.M., 2017. The Canada-France Ecliptic Plane Survey (CFEPS)—high-latitude component. Astron. J. 153 (5), Article ID: 236, 11 pp.

Popescu, M., Licandro, J., Morate, D., de León, J., Nedelcu, D.A., Rebolo, R., McMahon, R.G., Gonzalez-Solares, E., Irwin, M., 2016. Near-infrared colors of minor planets recovered from VISTA-VHS survey (MOVIS). Astron. Astrophys. 591, A115.

Porter, S.B., Buie, M.W., Parker, A.H., Spencer, J.R., Benecchi, S., Tanga, P., Verbiscer, A., Kavelaars, J.J., Gwyn, S.D.J., Young, E.F., 2018. High-precision orbit fitting and uncertainty analysis of (486958) 2014 MU_{69}. Astron. J. 156 (1), 20. https://doi.org/10.3847/1538-3881/aac2e1.

Rabinowitz, D., The Quasar Equatorial Survey Team, 2000. Discovery circumstances and photometric properties of 2000 EB173, the brightest known Trans-Neptunian object. Am. Astron. Soc. 32, 1028.

Rabinowitz, D., Schwamb, M.E., Hadjiyska, E., Tourtellotte, S., 2012. The La Silla-QUEST Kuiper belt survey. Astron. J. 144 (5), 140.

Rhoads, E., 2005. A Survey for TNOs Using the APS POSS I Database (Ph.D. thesis). University of Minnesota. Publication Number: AAT 3167711. DAI-B 66/03, p. 1511.

Schanche, N., Collier Cameron, A., Hébrard, G., Nielsen, L., Triaud, A.H.M.J., Almenara, J.M., Alsubai, K.A., Anderson, D.R., Armstrong, D.J., Barros, S.C.C., Bouchy, F., Boumis, P., Brown, D.J.A., Faedi, F., Hay, K., Hebb, L., Kiefer, F., Mancini, L., Maxted, P.F.L., Palle, E., Pollacco, D.L., Queloz, D., Smalley, B., Udry, S., West, R., Wheatley, P.J., 2019. Machine-learning approaches to exoplanet transit detection and candidate validation in wide-field ground-based surveys. Mon. Not. R. Astron. Soc. 483 (4), 5534–5547.

Schwamb, M.E., Brown, M.E., Rabinowitz, D.L., 2009. A search for distant solar system bodies in the region of Sedna. Astrophys. J. Lett. 694, L45.

Schwamb, M.E., Brown, M.E., Rabinowitz, D.L., Ragozzine, D., 2010. Properties of the distant Kuiper belt: results from the Palomar distant solar system survey. Astrophys. J. 720, 1691.

Schwamb, M.E., Brown, M.E., Fraser, W.C., 2014. The small numbers of large Kuiper belt objects. Astron. J. 147 (1), 6.

Schwamb, M.E., Jones, R.L., Chesley, S.R., Fitzsimmons, A., Fraser, W.C., Holman, M.J., Hsieh, H., Ragozzine, D., Thomas, C.A., Trilling, D.E., Brown, M.E., Bannister, M.T., Bodewits, D., de Val-Borro, M., Gerdes, D., Granvik, M., Kelley, M.S.P., Knight, M.M., Seaman, R.L., Ye, Q.-Z., Young, L.A., 2018. Large Synoptic Survey Telescope solar system science roadmap. arXiv e-prints arXiv:1802.01783.

Schwamb, M.E., Hsieh, H., Bannister, M.T., Bodewits, D., Chesley, S.R., Fraser, W.C., Granvik, M., Jones, R.L., Jurić, M., Kelley, M.S.P., Ragozzine, D., Trilling, D.E., and, K.V., 2019. A software roadmap for solar system science with the Large Synoptic Survey Telescope. Res. Not. AAS 3 (3), 51. https://doi.org/10.3847%2F2515-5172%2Fab0e10.

Sheppard, S.S., Trujillo, C., 2016. New extreme Trans-Neptunian objects: toward a super-Earth in the outer solar system. Astron. J. 152, 221.

Sheppard, S.S., Jewitt, D.C., Trujillo, C.A., Brown, M.J.I., Ashley, M.C.B., 2000. A wide-field CCD survey for centaurs and Kuiper belt objects. Astron. J. 120, 2687.

Sheppard, S.S., Udalski, A., Trujillo, C., Kubiak, M., Pietrzynski, G., Poleski, R., Soszynski, I., Szymański, M.K., Ulaczyk, K., 2011. A southern sky and galactic plane survey for bright Kuiper belt objects. Astron. J. 142 (4), 98.

Sheppard, S.S., Trujillo, C., Tholen, D.J., 2016. Beyond the Kuiper belt edge: new high perihelion Trans-Neptunian objects with moderate semimajor axes and eccentricities. Astrophys. J. 825, L13.

Sheppard, S.S., Trujillo, C.A., Tholen, D.J., Kaib, N., 2019. A new high perihelion Trans-Plutonian inner Oort cloud object: 2015 TG_{387}. Astron. J. 157, 139.

Todd, D.P., 1880. Preliminary account of a speculative and practical search for a transneptunian planet. Space Sci. Rev. 1, 167–169.

Tombaugh, C.W., 1946. The search for the ninth planet, Pluto. Astron. Soc. Pac. Leaflets 5, 73.

Tombaugh, C.W., 1961. The Trans-Neptunian planet search. Planets Satell. 12.

Trujillo, C.A., Brown, M.E., 2003. The Caltech wide area sky survey. Earth 92, 99.

Trujillo, C., Jewitt, D., 1998. A semiautomated sky survey for slow-moving objects suitable for a Pluto express mission encounter. Astron. J. 115, 1680.

Trujillo, C.A., Sheppard, S.S., 2014. A Sedna-like body with a perihelion of 80 astronomical units. Nature 507 (7493), 471–474.

Trujillo, C.A., Jewitt, D.C., Luu, J.X., 2000. Population of the scattered Kuiper belt. Astrophys. J. 529, L103.

Trujillo, C.A., Jewitt, D.C., Luu, J.X., 2001a. Properties of the Trans-Neptunian belt: statistics from the Canada-France-Hawaii Telescope survey. Astron. J. 122, 457.

Trujillo, C.A., Luu, J.X., Bosh, A.S., Elliot, J.L., 2001b. Large bodies in the Kuiper belt. Astron. J. 122, 2740.

Tyson, J.A., Guhathakurta, P., Bernstein, G.M., Hut, P., 1992. Limits on the surface density of faint Kuiper belt objects. Bull. Am. Astron. Soc. 24, 1127.

Vaduvescu, O., Curelaru, L., Popescu, M., 2019. Mega-archive and the EURONEAR tools for datamining world astronomical images. Astron. Comput. arXiv:1905.08847 (submitted for publication).
Weryk, R.J., Lilly, E., Chastel, S., Denneau, L., Jedicke, R., Magnier, E., Wainscoat, R.J., Chambers, K., Flewelling, H., Huber, M.E., Waters, C., The PS1 Builders, 2016. Distant Solar System objects identified in the Pan-STARRS1 survey. arXiv.org arXiv:1607.04895.
Whidden, P.J., Bryce Kalmbach, J., Connolly, A.J., Jones, R.L., Smotherman, H., Bektesevic, D., Slater, C., Becker, A.C., Ivezić, Ž., Jurić, M., Bolin, B., Moeyens, J., Förster, F., Golkhou, V.Z., 2019. Fast algorithms for slow moving asteroids: constraints on the distribution of Kuiper belt objects. Astron. J. 157 (3), 119. 1901.02492.
Williams, I., 1995. The slow-moving objects 1993 SB and 1993 SC. Icarus 116 (1), 180–185.

Epilogue

As a practical matter as well as a subject of meaningful scientific research, the Kuiper belt is about as old as the graduate students who study it. In a rather short time, our understanding has jumped, not just regarding the contents of Trans-Neptunian space, but also concerning the origin and evolution of the solar system and its relationship to the dust disks of other stars. To say that the Kuiper belt has changed our perception of the solar system sounds like hyperbole, but it is not, because it has.

In a strange way, the most basic advances are the easiest to forget. First was the realization that the region beyond the known planets is not empty, quickly followed by the discovery of the different dynamical classes (the classical, resonant, scattered, and detached populations), the unexpected finding that the Kuiper belt is mostly dynamically excited ("hot") and is of low mass ($\lesssim 0.1 M_{\oplus}$), that it contains a uniquely wide range of surface compositions and that binaries are abundant. Short-period comets, previously thought by many to be derived somehow from the Oort cloud, suddenly found a source in the Neptune-scattered population, with the Centaurs being an intermediate stage (Chapters 7 and 14). The Kuiper belt might also have fed other populations, including the irregular satellites, the Trojans, and possibly even Saturn's rings, but further work is warranted. The overabundance of resonant objects provided proof of planetary migration, now incorporated as a feature of all respectable (and some not so respectable) models of dynamical evolution (Chapter 2). Migration opened the door to a slew of chaotic evolution models that can match many of the known properties of the planetary system, albeit by invoking a large number of parameters, many of them free. All these were established in the first decade of the Kuiper belt but set the context for work undertaken since.

More recently, astonishing images from the New Horizons mission have transformed Pluto from an astronomical body to a geological world and, separately, revealed the intriguing shape of 2014 MU_{69} (Chapter 12). Ground-based observations continue to refine Kuiper belt architecture, principally through a series of large and impressive surveys (Chapter 3). A separate survey has revealed tantalizing evidence for an alignment of the orbits of high-perihelion objects which, if real (the reported significance remains $<3\sigma$), suggests to some extent the action of an undetected planet (Chapter 4).

What will the future bring? It is impossible to know, of course, but it is permissible to offer a few guesses by extrapolating from what has been done so far. Deeper surveys will probe the outer solar system to larger distances, either confirming or refuting the suspected alignment of detached objects and perhaps detecting the hoped-for planet (Chapter 20). A few reports of chance occultations by Kuiper belt objects have already been published, although we do not yet know which, if any, can be trusted (Chapter 19). Nevertheless, the statistical occultation technique is being perfected (e.g., TAOS 2) and should soon be an important source of data, particularly on objects too small to be studied in scattered light. Targeted occultations are also becoming more feasible with the aid of accurate stellar astrometry from Gaia.

The physical characterization of Kuiper belt objects has largely stagnated, owing to their faintness and general lack of detectable, compositionally diagnostic molecular bonds (Chapters 5 and 8). Perhaps the James Webb Space Telescope will reinvigorate this subject by providing orders of magnitude greater sensitivity in the thermal infrared region (Chapter 18). So far, our museums contain many meteorites from the asteroid belt, but no known samples of Kuiper belt material. With improved laboratory techniques and a deeper understanding of the transport mechanisms (Chapter 16), I see the possibility that such an identification might soon be made, either in material delivered to Earth by the impact of a comet from the belt, or as collisionally generated dust pulled in by Poynting-Robertson drag. Once we have a sample, we can begin to glean isotopic and compositional clues that relate to the accumulation and radioisotope decay histories of Kuiper belt objects, the larger of which might sustain internal oceans (Chapter 8) and even life.

Few features of the Kuiper belt were predicted or even imagined before they were discovered and measured. This is quite amazing, as is the incredible diagnostic power the region holds for understanding the history of the solar system. The exploration of Trans-Neptunian space has been a deep dive into the abyss. Long may it continue to be so.

David Jewitt
Department of Earth, Planetary and Space Sciences
UCLA

Index

Note: Page numbers followed by *f* indicate figures and *t* indicate tables.

O

P

Q

R

S

T

CPI Antony Rowe
Eastbourne, UK
November 26, 2019

Hydrobiologia (2007) 581:3–14
DOI 10.1007/s10750-006-0521-5

EUTROPHICATION IN LAKES

Environmental issues of Lake Taihu, China

Boqiang Qin · Pengzhu Xu · Qinglong Wu · Liancong Luo · Yunlin Zhang

Abstract Lake Taihu is characterized by its shallowness (mean depth = 1.9 m) and large surface area (2,338 km^2). Runoff sources are mostly from the mountainous west and southwest, and outflows are located throughout East Taihu. This causes shorter retention times in the south. In contrast, urban pollutants discharge into northern Taihu and result in poor water quality. Non-point pollution from rural areas and sewage wastewater is the primary pollution source. Water current velocity ranges from 10–30 cm s^{-1}, and surface currents normally follow wind direction. Bottom currents appear to be a compensation flow. Most wave heights are less than 40 cm, and underwater irradiance correlates to seston in the water column. Lacustrine sediment is distributed in littoral zones, mostly along the western shoreline, with almost no accumulation in the lake center. Intensive aquaculture in East Taihu caused eutrophication and hampered water supply in surrounding areas. In addition, development of marshiness in the eastern littoral zones and East Taihu has occurred. The function of flood discharging of East Taihu has been limited by flourishing macrophytes. The problems facing in Lake Taihu will be alleviated by improving the management of nutrient sources into the lake.

Guest editors: B. Qin, Z. Liu & K. Havens
Eutrophication of shallow lakes with special reference to Lake Taihu, China

B. Qin (✉) · P. Xu · Q. Wu · L. Luo · Y. Zhang
Nanjing Institute of Geography & Limnology, Chinese Academy of Sciences, 73 East Beijing Road, 210008 Nanjing, P.R. China
e-mail: qinbq@niglas.ac.cn

Keywords Lake Taihu · Hydrography · Water quality · Eutrophication · Aquaculture

Introduction

Lake Taihu is situated in the Changjiang (Yangtze) delta, the most industrialized area in China with high population density, urbanization, and economic development. Although Taihu Basin occupies only 0.4% of China's territory and 3% of China's population, it contributes about 10% of Gross National Product (GNP) and 15% of province revenues (NSB, 2000; SBSC, 2000; SBZP, 2000; SBJP, 2000). Lake Taihu is important for water supply, flood control, tourism and recreation, shipping and aquaculture. It is the drinking water source for several cities, such as Shanghai, Suzhou, Wuxi, and Huzhou. Since 1980's, however, economic development has resulted in pollutants being produced and discharged into rivers and the lake. With the deterioration of water quality, eutrophication and algal blooms (*Microcystis* spp.) have occurred. Recently, the algal bloom has extended its coverage and persists throughout the summer, which affects the function of the lake as a drinking water supply. In 1996, the Chinese government targeted pollution and eutrophication control before 2000 and 2010. Factories within the watershed now must treat wastewater before it is discharged to rivers. Also, rivers flowing into Lake Taihu should meet grade III standards (National Surface Water Quality Standard GB3838-2002) before 2000, and lake water should be clarified before 2010. However, lake water quality has not improved to date, and, in 2001, pollution control work was reevaluated and redirected. In fact, problems facing Lake Taihu are similar to most freshwater lakes in eastern China. Based on field investigations and literature review, this paper attempts to review and address main eco-environmental issues in Lake Taihu.

General description of the lake

Lake Taihu is the third largest freshwater lake in China and is located between 30°55′40″–31°32′58″ N and 119°52′32″–120°36′10″ E. Lake length (from north to south) is 68.5 km and width (from east to west) is 56 km. Mean depth is 1.9 m, and maximum depth is 2.6 m corresponding to an elevation of 3.0 m above sea level. The lake bottom features flat terrain with an average topographic gradient of 0°0′19.66″ and elevation of 1.1 m above sea level. Shallow-water area with mean depth <1.5 m is about 452 km^2, mostly in East Taihu, accounting for 19.3% of the total surface area. Deepest areas (>2.5 m) are in the north and west occupying 197 km^2 or 8.4% of the total lake area.

Origin of the lake

Many researchers have discussed the origin and history of Lake Taihu (Wang & Ding, 1936; Wissman, 1941; Chen et al., 1959; NIGLAS, 1965; Yang et al., 1985; Yan & Xu, 1987; Sun et al., 1987; Sun & Huang, 1993; Huang, 2000; Sun, 2004). In general, Lake Taihu probably was an ancient lagoon, as suggested by its location in the Changjiang delta where solid particles have been transported and deposited in the transitional estuarine zone. Evidence of marine sediment materials found in drill cores supported this assumption (Chen et al., 1959). Continental shelf extension resulted in enclosing of the lagoon and it eventually became a freshwater lake (Chen et al., 1959; Yang et al., 1985; Yan & Xu, 1987). Recently, new evidence from stratigraphy and archaeological sites suggest that Taihu originated from extreme flooding. Flood water discharged by man-made channels was retained in the depression (Sun & Huang, 1993; Sun, 2004).

Hydrology of the lake

Lake Taihu has a complicated river and channel network. There are 172 rivers or channels connecting to the lake (Xu & Qin, 2005). The total length of rivers in Taihu basin is ca. 12,000 km, i.e., about 3.24 km km^{-2}. During flooding season (spring and summer), flooding runoff from the west or southwest goes through lake to the east and empties into the East China Sea (Fig. 1). In the dry season (autumn and winter), however, water flow can reverse, especially in rivers in the east. Generally, rivers in the west of the lake are defined as upstream and vice versa. Runoff is defined as positive or negative. Table 1 shows the runoff of main rivers flowing into or out of the lake from May 2001 to April 2002. The greatest inflow rivers are Chendonggang, Xitiaoxi, and Yincungang. Three are located in the west or southwest watershed. The main outflow rivers are Taipu, Xinyunhe, and Xijiang located in the southeast (Table 1). Table 2 shows the monthly distribution of runoff from May 2001 to April 2002. Some rivers flow into the lake year-round, some flow out all year, and some reverse flow direction in different seasons (Table 2). There is no definitive estimate of water balance yet because of the complicated hydraulic connections between rivers and the lake. Table 3 is a rough estimate of water balance based on surveys of 115 rivers from May 2001 to April 2002. Flowing water from these rivers account for 90% of total runoff input and output, but the measuring frequency was insufficient because measurements were taken once in normal months and twice in

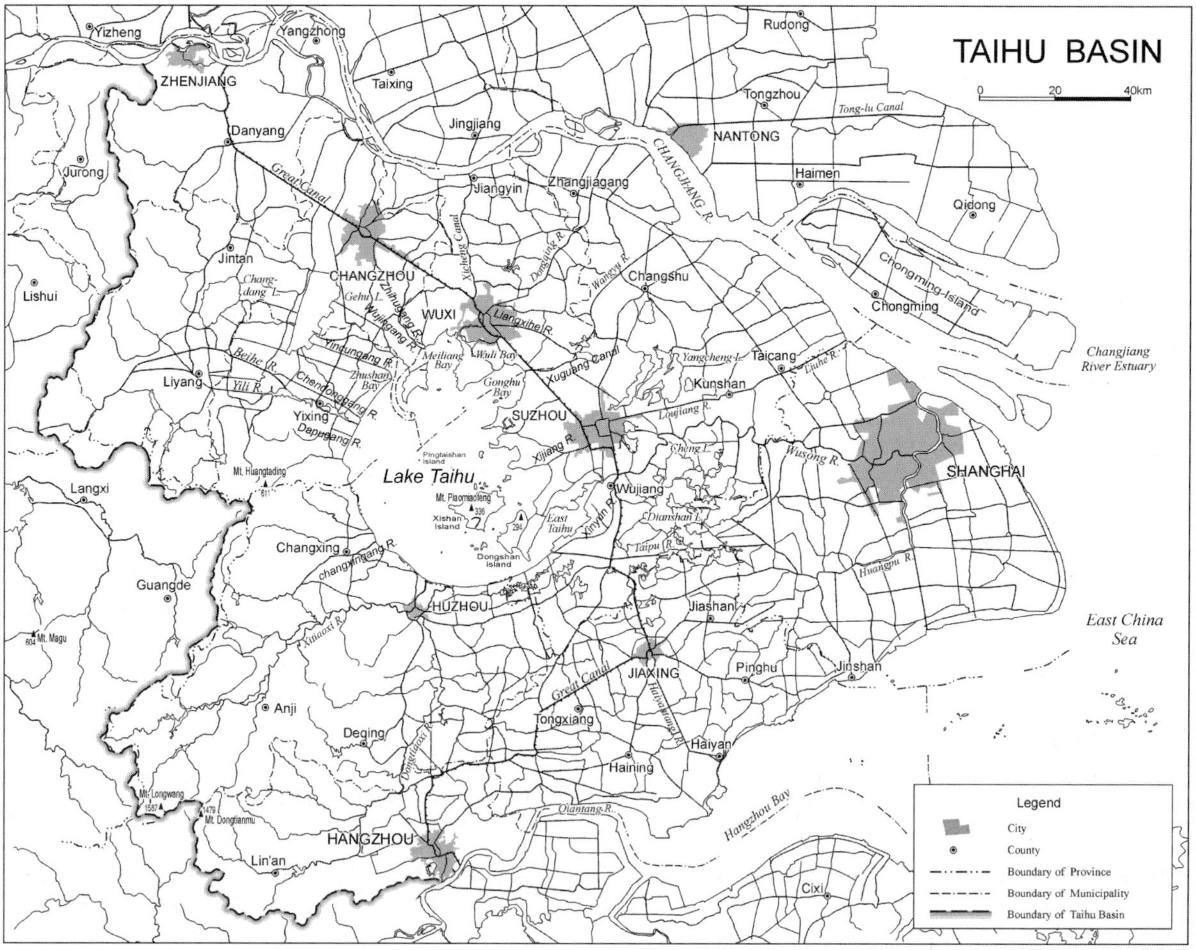

Fig. 1 Sketch map of Taihu Basin

flooding months. Annual runoff input, output, and water balance from May 2001 to April 2002 is estimated based on the measurement (Table 3). This water balance represents more flooding years versus the multi-year averaged water balance (Mao et al., 2004). Water retention time is about 5 months, but it is shorter in the south since most runoff water is discharged via the Taipu River, in the southeast. Water quality, therefore, is better in the south than in the north.

Hydrodynamics

There still is not a complete understanding of water currents in Lake Taihu. Earliest measurements were made in 1960, when two boats cruised around the lake at the same speed (NIGLAS, 1965). Other measurements were made in 1986–1987 (Sun & Huang, 1993) and summer in 1997 (Qin et al., 2000). These measurements were incomplete and incomparable since records were collected on different dates or at different sites. In summer, prevailing winds from the southeast or southwest generate a counter-clockwise water current in western Taihu centered near Pingtaishan Island (Fig. 2). Current velocity ranged from 10 cm s^{-1} to 30 cm s^{-1} (Qin et al., 2000). There was a stable southward current along the western coast and a weak and unstable current in Meiliang Bay (Fig. 2). This current pattern was reproduced by a numerical simulation under the same conditions (Luo & Qin, 2004; Hu et al., 2004). Regardless of season, there is a weak counter-clockwise current in northern Meiliang Bay (Qin et al., 2000; Luo et al., 2004a, b), which may explain

Table 1 Inflow and outflow of main rivers around Lake Taihu during May 2001 to April 2002 (if the sum of number of days of inflow and outflow less than 365, the remainder is days of no flow)

River	Inflow				Outflow				Annual runoff (10^8 m^3)
	Days	Mean daily inflow rate (m^3 s^{-1})	Mean inflow rate (m^3 s^{-1})	Total inflow (10^8 m^3)	Days	Mean daily outflow rate (m^3 s^{-1})	Mean outflow (m^3 s^{-1})	Total outflow (10^8 m^3)	
Xitiaoxi R.	234	273.0	34.9	7.05	125	–57.8	–12.79	–1.38	5.67
Dongtiaoxi R.	118	400.0	71.5	7.29	247	–122.0	–45.37	–9.68	–2.39
Changxingang R.	301	61.0	6.8	1.76	64	–10.4	–2.21	–0.12	1.64
Chendonggang R.	338	147.0	34.2	9.98	26	–38.7	–10.07	–0.23	9.76
Yincungang R.	347	72.9	18.5	5.56	18	–14.8	–5.47	–0.09	5.47
Wujingang R.	339	49.8	6.6	1.95	26	–7.1	–2.89	–0.06	1.88
Zhihugang R.	244	78.1	8.3	1.75	27	–14.1	–5.56	–0.13	1.62
Liangxihe R.	174	60.7	11.8	1.78	60	–23.3	–9.58	–0.50	1.28
Wangyu R.	116	240.0	82.8	8.30	249	–177.0	–44.10	–9.49	–1.19
Xinyun River	32	33.1	15.4	0.43	333	–183.0	–31.47	–9.05	–8.63
Dapugang R.	14	4.2	2.7	0.03	348	–14.2	–5.78	–1.74	–1.71
Xijiang R.	7	16.5	8.7	0.05	358	–44.3	–18.72	–5.79	–5.74
Xuguang Canal	38	7.7	3.3	0.11	290	–27.3	–8.03	–2.01	–1.91
Taipu River	8	73.2	27.5	0.19	357	–296.0	–119.47	–36.85	–36.66

Table 2 Monthly runoff rate of rivers around Lake Taihu during May 2001 to April 2002 (m^3 s^{-1}) (inflow is defined as positive and outflow is defined negative)

River	2001								2002				Yearly Average
	May	Jun	Jul	Aug	Sep	Oct	Nov	Dec	Jan	Feb	Mar	Apr	
Xitiaoxi R.	8.8	49.3	22.0	57.0	–9.4	–1.7	–6.3	6.8	7.9	–4.1	28.5	55.6	17.9
Dongtiaoxi R.	–2.2	80.6	–5.9	8.2	–76.7	–45.6	–54.5	–20.3	–15.1	–31.0	9.6	61.1	–7.7
Changxingang R.	3.7	10.7	6.6	10.3	1.1	0.1	1.0	3.6	3.1	1.2	9.8	11.0	5.2
Chendonggang R.	15.7	44.8	33.1	67.1	31.1	26.4	24.7	32.0	22.2	12.7	37.4	22.3	30.8
Yincungang R.	14.8	28.8	11.9	37.2	18.9	15.4	14.0	19.3	10.0	7.1	18.4	11.6	17.3
Wujingang R.	4.5	11.2	4.9	11.7	5.5	4.9	5.8	4.2	3.1	4.2	7.0	4.6	6.0
Zhihugang R.	5.7	13.4	0.1	12.9	5.4	3.9	2.9	2.1	0.9	2.9	9.0	2.4	5.1
Liangxihe R.	5.9	14.1	–5.8	2.6	3.6	10.3	5.8	10.4	1.6	0	0	0	4.0
Wangyu R.	12.0	–12.4	–87.4	–48.5	–41.6	15.0	–12.8	–18.0	–37.2	115.0	116.0	–36.5	–3.0
Xinyun R.	–23.4	–47.4	–34.3	–24.8	–22.0	–31.3	–26.6	–28.7	–30.9	–21.8	–15.8	–21.1	–27.3
Dapugang R.	–2.7	–2.0	–5.9	–5.2	–5.5	–5.3	–6.5	–8.4	–9.5	–5.5	–5.8	–2.4	–5.4
Xijiang R.	–10.7	–9.3	–27.4	–27.9	–25.2	–17.6	–17.0	–15.3	–17.4	–14.0	–18.5	–17.6	–18.2
Xuguang Canal	–1.6	0.2	–10.5	–12.4	–10.2	–4.7	–6.1	–3.9	–7.3	–2.2	–7.7	–5.6	–6.0
Taipu R.	–49.6	–84.9	–132.0	–135.0	–125.0	–85.4	–107.0	–127.0	–118.0	–116.0	–160.0	–156.0	–116.3

Table 3 The water balance of Lake Taihu during May 2001 to April 2002

Input runoff (10^8 m^3)	Output runoff (10^8 m^3)	Precipitation (10^8 m^3)	Evaporation (10^8 m^3)	Change in capacity (10^8 m^3)	Inflow from uncontrolled areas and river net (10^8 m^3)
80.11	–96.67	27.19	–17.71	–5.38	12.46

why this area is favorable for concentrating algal blooms (Qin et al., 2000).

In July 2003, an Acoustic Doppler Current Profiler was deployed at the center of Meiliang Bay, and 10-min interval current velocity and direction data from 11 layers were collected. Surface currents followed the direction of wind forcing (Fig. 3a), but current velocity attenuated

Fig. 2 The prevailing current pattern during the summer, the velocity of current ranged from 10 cm s^{-1} to 30 cm s^{-1}

with depth and reversed direction at the bottom (Fig. 3b). This indicates that there is a compensation current at the bottom opposite wind forcing on the surface (Fig. 3b).

Wave action is a key factor in interactions between the sediment and water (Qin et al., 2004).

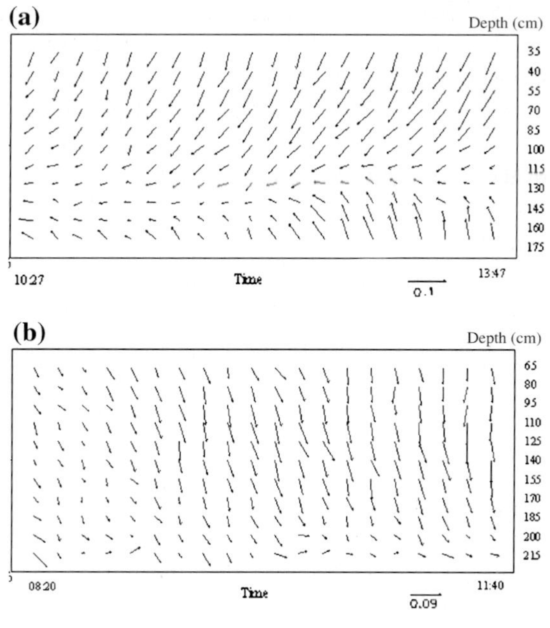

Fig. 3 (**a**) Current vectors at 35, 40, 55, 70, 85, 100, 115, 130, 145, 160, 175 cm below surface (water depth is 2.4 m) at 10-min intervals in Meiliang Bay during 10:27–13:47 on 14 July 2003. (**b**) Current vectors at 65, 80, 95, 110, 125, 140, 155, 170, 185, 200, 215 cm below surface (water depth is 2.4 m) at 10-min intervals in Meilaing Bay during 8:20–11:40 on 17 July 2003

In August 1986, a continuous observation of waves was made in the northwest (Wu et al., 1987). Statistical analyses with cumulative frequency of one-third highest wave height ($H_{1/3}$) and maximum wave height (H_{max}) showed that most waves were low in height with 80% of $H_{1/3} < 8$ cm, 98% of $H_{1/3} < 20$ cm, and 5% > 40 cm (Wu et al., 1987). From field observations, the empirical relationship among average wave height ($\overline{H}$), wind speed (W), wind fetch (F), and period (T) was derived as follows (Hu et al., 2004):

$$\frac{g\overline{H}}{W^2} = 0.00168\left(\frac{gF}{W^2}\right)^{0.46}$$

If water depth was assumed not to change, then wave height was proportional to wind speed (Hu et al., 2004):

$$\overline{H} = 5.497W^{0.5} - 4.516$$

Similarly, the relationship between maximum wave height (H_{max}) and wind speed can be expressed as (Wu et al., 1987):

$$\frac{gH_{max}}{W^2} = 0.1094\left(\frac{gD}{W^2}\right)^{0.7508}$$

$$H_{max} = 5.237 \times 10^{-2}\ W$$

Underwater light climate

Lake Taihu is a shallow and eutrophic lake, and Secchi depths are normally low, i.e., about 30–40 cm in the north and 40–50 cm in the south (Qin et al., 2004; Zhang et al., 2004a). Based on water quality monitoring conducted by Taihu Laboratory for Lake Ecosystem Research (TLLER), transparency is related to seston abundance (Fig. 4):

$$S^{1/4} = 8.103 - 5.847 \ln ST$$

$$(R = 0.87,\ SD = 0.32,\ N = 135,\ P < 0.0001)$$

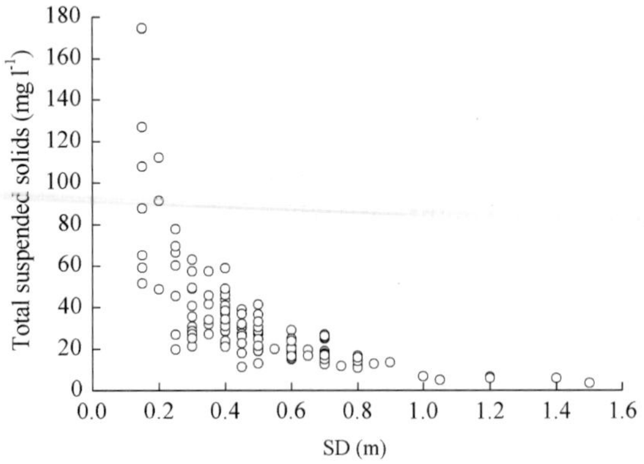

Fig. 4 The relationship between the solid suspension and secchi depth (SD) in Lake Taihu

where *S* is seston abundance (mg l^{-1}), *ST* is secchi depth (m). Correlation and regression analysis based on measurements of light attenuation and seston, including inorganic and organic seston, and chlorophyll-*a* showed that inorganic solids were important in autumn, whereas organic solids were important in summer, especially in the north (Zhang et al., 2004b). Correlation analysis also indicated that chlorophyll was not a large factor on light attenuation (Zhang et al., 2004b).

Underwater irradiance measurements show that daily light distribution at specific depth followed solar radiation, i.e., increase in the morning and decrease in the afternoon (Fig. 5) (Zhang et al., 2004b). Irradiance at 1 m was about 10% of that at the surface (Fig. 5). Ultraviolet (UV) radiation, which is harmful for phytoplankton growth, was attenuated rapidly, i.e., UV-B at 20 cm was about 1% of surface levels (Zhang et al., 2004b). Short-wave irradiance, therefore, is attenuated faster than long-wave irradiance in this lake.

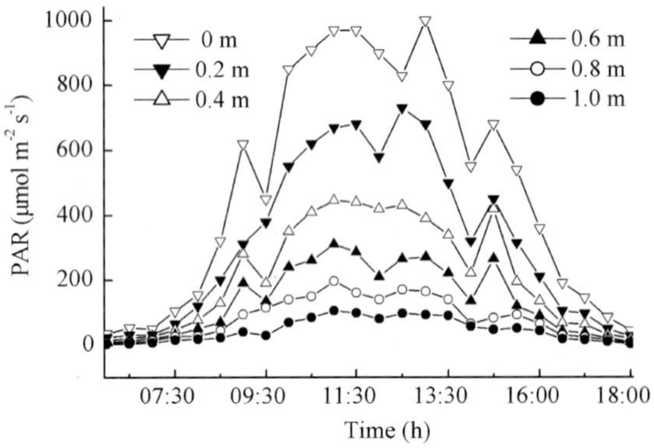

Fig. 5 Daily variations of underwater irradiance at different depth (0.0, 0.2, 0.4, 0.6, 0.8, 1.0 m under the surface) at the site nearby Taihu Laboratory for Lake Ecosystem Research (TLLER) on May 17, 1998

Sediment in Lake Taihu

Spatial and vertical distribution of lacustrine sediment was investigated at 723 sites in Lake Taihu. Successive correction schemes and geostatistical methods were used to determine sediment thickness in the 69 × 69 whole lake grids. Spatial statistical analysis revealed that sediment covers about 1,100 km^2 of lake bottom, which accounts for about 47.5% of lake area (Fig. 6) (Luo et al., 2004a, b). Total volume of lacustrine sediment is estimated to be 1.86×10^9 m^3 (Luo et al., 2004a, b), and, in most regions, lacustrine sediment depth was 0.5–2.0 m (Luo et al., 2004a, b). Most lacustrine sediments are distributed along the western shoreline from Meiliang Bay to the south. There is little lacustrine sediment in the center of the lake (Fig. 6).

Dating sediment at different depths from drill cores showed that sedimentation rates range from 0.6 mm $year^{-1}$ to 3.6 mm $year^{-1}$ with an average of 2 mm $year^{-1}$ (Fan et al., 2004). Bulk density of surface sediment is 1.2 g cm^{-3} with lowest bulk density from East Taihu and highest from the western lake (Fan et al., 2004). Organic matter content ranges from 0.13% to 16.6% with maxima from East Taihu and minima from the center and west (Fan et al., 2004). TN and TP in sediments, range from 0.01% to 0.66% and 0.01% to 0.015%, respectively (Fan et al., 2004).

Pollution and eutrophication

In 1960, Lake Taihu was categorized as oligotrophic because total inorganic nitrogen (TIN), total inorganic phosphorus (PO_4^{3-}-P), and chemical oxygen demanded (COD_{Mn}) were low (Table 4) (NIGLAS, 1965). TIN increased dramatically until 1981 to about 18 times greater than that in 1960, COD_{Mn} increased by 49%, and PO_4^{3-}-P remained stable (Table 4) (Sun & Huang,

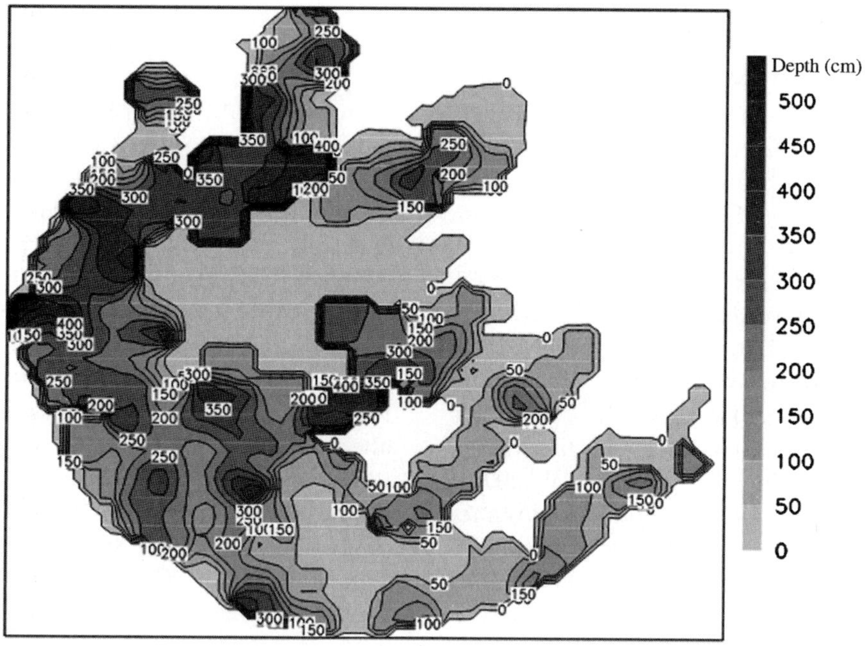

Fig. 6 Distribution of lacustrine sediment in Lake Taihu

Table 4 Change of water quality indexes since 1960's

Year	TIN (mg l^{-1})	TN (mg l^{-1})	PO_4^{3-}-P (mg l^{-1})	TP (mg l^{-1})	COD_{Mn} (mg l^{-1})	Sources
1960	0.05		0.02		1.90	NIGLAS (1965)
1981	0.894	0.9	0.014		2.83	Sun & Huang (1993)
1988	1.115	1.84	0.012	0.032	3.3	Sun & Huang (1993)
1994	1.135	2.05	0.010	0.086	5.77	SEPA (2000)
1995	1.157	3.14	0.011	0.111	5.53	SEPA (2000)
1998[a]	1.582	2.34	0.007	0.085	5.03	TLLER
1999[a]	1.79	2.57	0.004	0.105	4.99	TLLER

[a] Data provided by TLLER

1993). From 1981 to 1998, TP and COD_{Mn} increased by 2.7 and 1.5 times, respectively (Table 4). Total nitrogen (TN), total phosphorus (TP), and chemical oxygen demand (COD_{Mn}) increased from 1.8, 0.03, and 3.3 mg l^{-1} in 1988 to 2.3, 0.085, and 5.03 mg l^{-1} in 1998, respectively (Table 4).

Table 5 shows quantities of pollutants which rivers transport into the lake. Most pollutant inflow rivers are located in the west or northwest, whereas pollutant outflow rivers are located in the south or southeast. Based on studies in 2002 and 2003, about 30–40% of nitrogen and phosphorus is retained inside the lake (Table 5).

Table 5 Nutrient flux of inflow and outflow of Lake Taihu during 2001–2002

	COD_{Mn} (ton)	NH_4^+-N (ton)	NO_2^--N (ton)	NO_3^--N (ton)	TN (ton)	TP (ton)
Inflow	37,571.06	12,431.70	1,007.86	10,469.12	28,657.97	1,029.37
Outflow	–35,431.36	–4,864.99	–379.77	–5,862.28	–14,599.90	–668.27
Inflow–outflow	2,139.70	7,566.71	628.09	4,606.83	14,058.07	361.10
Retention rate	5.7%	60.9%	62.3%	44.0%	49.1%	35.1%

Wastewater produced within the basin amounted to 2.25×10^5 t year^{-1} of COD_{Cr}, 4.13×10^4 t year^{-1} of TN and 3.8×10^3 t year^{-1} of TP in the 1980's (Sun & Huang, 1993), increasing to 2.82×10^5 t year^{-1} of COD_{Cr} and 8.0×10^4 t year^{-1} of TN and 5.66×10^3 t year^{-1} of TP in the early 1990's (SEPA, 2000), and increasing further to 7.46×10^5 t year^{-1} of COD_{Cr}, 1.09×10^5 t year^{-1} of TN and 1.56×10^4 t year^{-1} of TP in the late 1990's (Huang, 2004). However, pollutants discharged into the lake were less than the pollutants produced within the basin. About 1.06×10^5 t year^{-1} of COD_{Cr}, 2.0×10^4 t year^{-1} of TN and 1.55×10^3 t year^{-1} of TP in the 1980's were discharged into the lake (Huang et al., 2001). During the early 1990's, total pollutants discharged increased to 1.31×10^5 t year^{-1} of COD_{Cr}, 3.1×10^4 t year^{-1} of TN and 1.75×10^3 t year^{-1} of TP (SEPA, 2000). In the late 1990's, the COD_{Cr} empting into the lake increased to 1.71×10^5 t year^{-1}, but TN and TP decreased to 2.5×10^4 t year^{-1} and 1.3×10^3 t year^{-1}, respectively (Huang, 2004). During the early 2000's, inputs of COD_{Mn}, TN, and TP were 3.76×10^4 t year^{-1}, 2.87×10^4 t year^{-1} and 1.03×10^3 t year^{-1}, respectively (Table 5) (Xu & Qin, 2004), which are much lower than earlier estimates. These results varied among different investigators, but pollutants empting into the lake probably increased in the early 1990's and decreased in the late 1990's. This change may relate to increased wastewater treatment in surrounding areas.

Analysis of pollutant composition in wastewater in 1994 showed that 16% of total nitrogen (TN) was from industrial sources, 25% from domestic sewage, and 38% from non-point sources (agriculture production) (SEPA, 2000). For total phosphorus (TP), 10% was from industry, 60% from sewage, and 15% from non-point sources (SEPA, 2000). For COD, 39% was from industry, 42% from sewage, and 10% from non-point sources (SEPA, 2000) (Table 6). In 1998, industrial pollution to the lake, decreased dramatically either in total COD, or TN and TP (Table 6), whereas the contribution from agriculture to COD and TN increased significantly and sewage pollution contributed 70% of TP (Table 6). This change in pollution composition reflects rapid economic growth and land use alteration.

Based on data from 1998, pollutants produced by industry were mainly from the east or southeast of the lake basin, such as Suzhou, Wuxi, and Jiaxing (Huang, 2004). Pollutants from agriculture (cropping, rice growing, etc.) accounted for 37% of COD, 49.5% of TN and 48% of TP produced in the basin (Huang, 2004). These non-point pollutants came mostly from the west. In addition, poultry culture scattered in the rural region accounted for 5% of COD, 6% of TN, and 18% of TP (Huang, 2004). Fish culture made up 5.5% of COD, 5.2% of TN, and 4.1% of TP. With strengthening control of point sources, the contribution from non-point sources, i.e., agriculture and animal culture, likely will increase with time.

Aquaculture in the lake

The main form of aquaculture in Lake Taihu is pen-fish-culture. Aquaculture has been limited to East Taihu, a macrophyte-dominated bay in the southeast part of the lake with an area of

Table 6 Comparison of composition of different pollution sources in COD, TN, and TP in 1994 and 1998

Year	Pollution source	COD_{Cr}		TN		TP	
		Amount (ton)	Percentage (%)	Amount (ton)	Percentage (%)	Amount (ton)	Percentage (%)
1994	Industry	111,061	39	12,544	16	591	10
	Domestic	119,029	42	19,948	25	3,394	60
	Agriculture	28,138	10	29,842	38	852	15
	Aquaculture			13,195	17	533	9
1998	Industry	93,822	26	2,686	4	191	2
	Domestic	96,782	27	16,232	25	6,510	70
	Agriculture	91,582	25	17,815	28	1,362	15
	Aquaculture	49,627	14	17,750	28	733	8

126 km^2. Although the documented area of pen-fish-culture was 2,830 ha (Chen & Wu, 2004), field observations reveal that almost the entire water surface of East Taihu is occupied by net pens. Fish culture in this area has raised some environmental problems. East Taihu is a shallow bay with 95% coverage by macrophytes. In 1993, annual macrophyte production in East Taihu was 1.12×10^6 t, which accounts for 3,920 tons of nitrogen and 496 tons of phosphorus (Wu et al., 1995; Yang & Li, 1996), corresponding to 58% and 95%, respectively, of total external loading to East Taihu. About 60×10^4 t of macrophytes were harvested for fish farming resulting in removal of 1,890 t of nitrogen and 296 t of phosphorus, corresponding to 28% and 57%, respectively, of total external loading to East Taihu (Wu et al., 1995; Yang & Li, 1996). Harvesting of macrophytes now is blocked due to the proliferation of net pens. Consequently, the lake lost an important pathway for nutrient export. This loss has promoted sedimentation and eutrophication in East Taihu (Table 7). Aquaculture activities also contribute to increases in nutrient loading and eutrophication. It is estimated that producing 1 ton of fish requires 141 kg nitrogen and 14 kg phosphorus (Yang & Li, 1996; Wu et al., 1995; Li, 1998), and about 65% of each is not assimilated and is subsequently released into the overlying water (Yang & Li, 1996). In pen-fish-culture areas, increased nutrient loading leads to rapid growth of phytoplankton, zooplankton, and bacteria. After one year of fish farming, the phytoplankton abundance was three times higher than in non-culture areas, and heterotrophic bacteria abundance increased 3- to 4-fold (Wu et al., 1995). Total organic carbon, total nitrogen, and total organic nitrogen in surface sediments increased by 141, 87.5, and 86%, respectively, after 2 years of fish culturing (Li, 2004). Recently, fish culturing has been replaced by the more profitable freshwater crab culturing, which will increase the input of feed, and further increase the deposit of organic materials from the remnants of feed. Because fish culture will make use of macrophytes for fish food, the lake will output the nutrients and retard the development of marsh in this way. In turn, the recent aquaculture aggravates the deterioration of water quality and degradation of ecosystem.

Table 7 Changes in water quality parameters in East Taihu in 1990's

Year	TN (mg l^{-1})	NH_4^+-N (mg l^{-1})	NO_3^--N (mg l^{-1})	TP (ug l^{-1})	COD_{Mn} (mg l^{-1})
1991	1.01	0.14	0.32	43	5.5
1993	1.04	0.20	0.25	75	5.6
1997	1.39	0.19	0.10	31	5.5
1999	1.65	0.475	0.23	40	6.6
2000	1.08	0.26	0.30	100	6.4

Marsh development

Marsh development mainly takes place in the eastern shoreline areas, i.e., East Taihu, Xikou Bay, and Gonghu Bay. Figure 7 shows the expansion of spatial macrophyte distribution (TBA & NIGLAS, 2000). From Fig. 7, macrophytes have extended from East Taihu in the 1960's (Fig. 7a) to Xikou Bay and Gonghu Bay in the 1980's (Fig. 7b). Marsh development in East Taihu is more serious in East Taihu because it also functions as a flood discharge area and drinking water supply for Suzhou and Shanghai. These functions, however, are hampered by marsh development. Generally, marsh development in lakes results from organic or inorganic material deposition, which elevates the lake bottom, decreases water capacity, and eventually leads to elimination of the lake (Li, 2004).

According to field investigations and TM satellite image data in 1997, 95% of East Taihu is covered by aquatic vegetation. Dominant species include *Potamogeton maackianus, Vallisneria natans*, and *Zizania latifolia*. Average macrophyte biomass in East Taihu is about 3.8 kg m^{-2} with a range of 2.0–5.6 kg m^{-2} (Li, 2004). Total standing biomass of aquatic plants in Lake Taihu is about 500,000 t, and *Potamogeton maackianus* accounted for 42%, *Zizania latifolia* 29.5% and *Vallisneria natans* 0.2% (Li, 2004). Floating-leaf and emergent vegetation, such as *Zizania latifolia*, indicate accelerated marsh development in East Taihu (Li, 2004).

Since East Taihu is the major outlet of Taihu Lake, particles are transported and deposited in

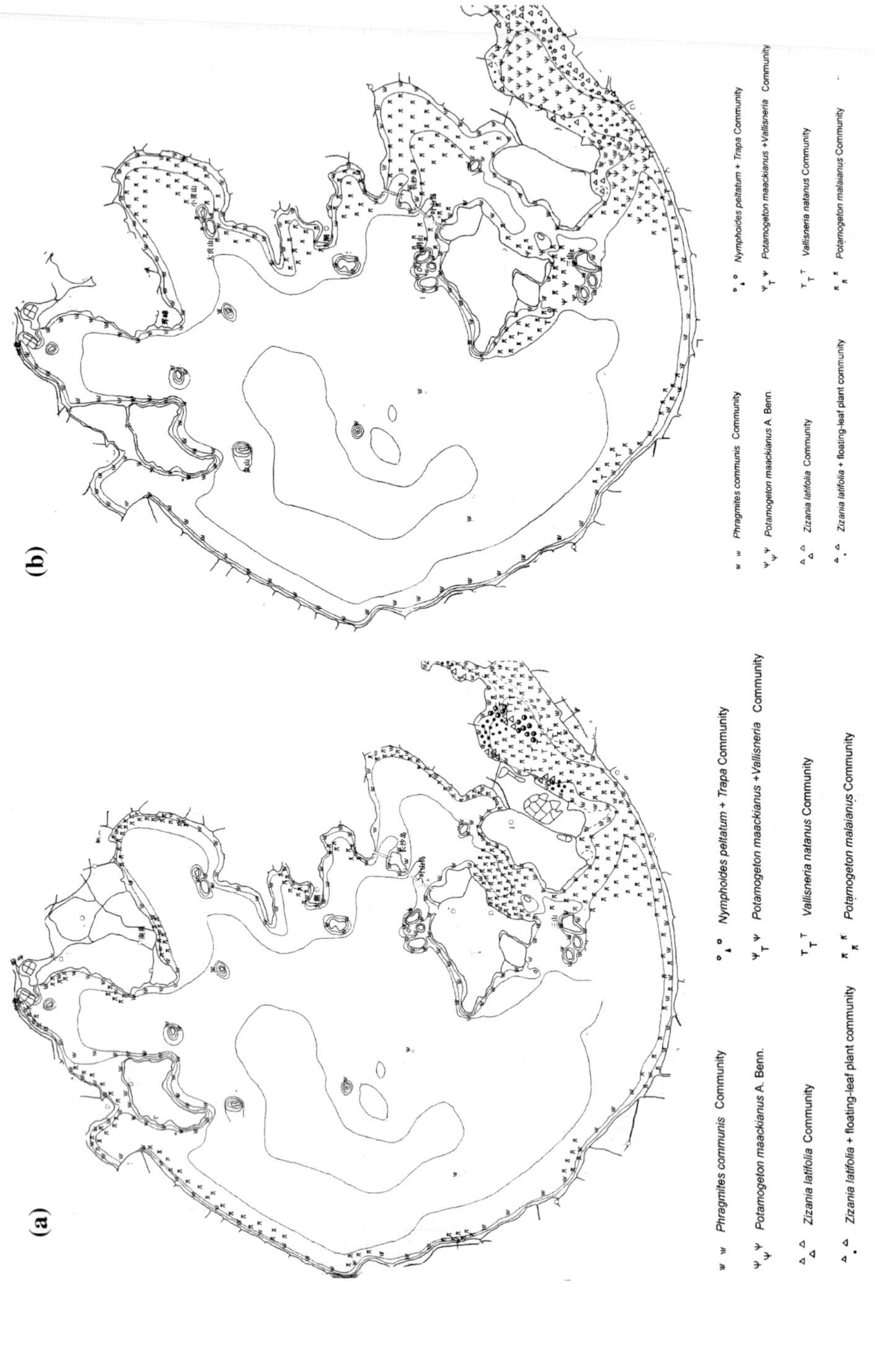

Fig. 7 Spatial distribution of macrophytes in 1960's **(a)** and in 1980's **(b)**

East Taihu. Macrophytes reduce water flow velocity and, hence, increase silting. Investigation of sediment distribution in East Taihu showed that the top 1 m of sediment was soft and composed of sand (98%) and organic materials (1.5%) (Li, 2004). Sediment composition showed that most deposit materials were allocthonous. Thus, increased sedimentation rates and marsh development may be attributed to pen-fish-culture, since it limited macrophyte harvest, promoted succession from submerged to emergent and floating-leaf vegetation, inhibited water movement, and augmented deposition of macrophyte detritus and other particles.

Management

Lake Taihu has many functions, including drinking water supply, flood water retention, fisheries, tourism and shipping. These varied functions have resulted in multi-purpose-oriented management without sufficient coordination. Over-exploration and over-utilization caused degradation of the lake ecosystem and deterioration of water quality, since some functions are contradictory, such as water supply and fish culture. Development of a centralized-management system, therefore, should be considered to decelerate eutrophication and marsh development, restore the ecosystem, and improve water quality. Regulations and administrative policies should be enacted, e.g., reclamation should be forbidden and fish-culture should be limited in spatial distribution. A long-term plan for pollution and eutrophication control should be based on accurate prediction of economic development and contaminant sources. Watershed management should be planned and implemented aiming at pollution source control. Scientific research should be emphasized and focus on internal loading from sediment, water and nutrient budgets, and ecological restoration in the littoral zone.

Conclusion

1. Lake Taihu has a complicated inflow and outflow system. Water retention time is shorter and water quality is better in the south or southeast, whereas retention time is longer and water quality is worse in the north or northwest.
2. Major pollutants are mostly from the western watershed. Most COD input is balanced by river output, and about 30–40% of nutrient input is retained in the lake. Most pollution is from non-point agriculture production and domestic sewage.
3. Light attenuation is associated with suspended solids, which is related to wave forcing. Most sediments are distributed in the littoral zone, especially along the western shoreline.
4. Aquaculture in East Taihu increases nutrient concentrations in water and sediments, which accelerates eutrophication and marsh development. Swamping in East Taihu results from macrophyte success, which increases sedimentation rate, reduces the ability of flooding water passing through the system and quickens the tempo of extinguishment of the lake.

Acknowledgements This work was funded by Chinese Academy of Sciences (No. KZCX1-SW-12), Ministry of Science and Technology of China (No. 2002AA601011), and National Natural Science Foundation of China (No. 40501078). We thank Prof. Karl Havens, Prof. Wayne Gardner, and Prof. Koren Nydick for their careful reviews and constructive suggestions. Thanks are also to Dr. Mark McCarthy for his help with English polishing.

References

Chen, J., Z. Yu & C. Yun, 1959. Morphology of Changjiang Delta. Acta Geographica Sinica 25: 201–220 (in Chinese).

Chen, W. & Q. Wu, 2004. Structure and evolution of ecosystem in Lake Taihu. In Qin, B., W. Hu & W. Chen (eds), Process and Mechanism of Environmental Changes of Lake Taihu. Science Press, Beijing: 289–297 (in Chinese).

Fan, C., G. Zhu, W. Li, S. Sun, C. Hu & B. Qin, 2004. Exchange between the water–sediment interface and its environmental effects. In Qin, B., W. Hu & W. Chen (eds), Process and Mechanism of Environmental Changes of Lake Taihu. Science Press, Beijing: 170–186 (in Chinese).

Hu, W., B. Qin & L. Luo, 2004. The hydrodynamic processes and the environmental effects in Lake Taihu. In Qin, B., W. Hu & W. Chen (eds), Process and Mechanism of Environmental Changes of Lake Taihu. Science Press, Beijing: 107–161 (in Chinese).

Huang, W., 2004. The pollutant budget in Lake Taihu. In Qin, B., W. Hu & W. Chen (eds), Process and Mechanism of Environmental Changes of Lake Taihu. Science Press, Beijing, 21–28 (in Chinese).

Huang, X., 2000. Layout and Integrated Management of Taihu Basin. Press of Water & Electricity, Beijing, 1–20 (in Chinese).

Huang, Y., C. Fan, P. Pu, J. Jiang & Q. Dai, 2001. Aquatic Environment and Pollution Control of Lake Taihu. Science Press, Beijing, 187–206 (in Chinese).

Li, W., 2004. Tendency and reason for evolution toward mash in the East Taihu. In Qin, B., W. Hu & W. Chen (eds), Process and Mechanism of Environmental Changes of Lake Taihu. Science Press, Beijing, 33–51 (in Chinese).

Li, W., 1998. Utilization model of aquatic macrophyte for fishery in Chinese shallow lakes. Ecological Engineering 11: 61–72.

Luo, L. & B. Qin, 2004. Numerical simulation based on a three-dimensional shallow-water hydrodynamic model-current circulations in Lake Taihu with prevailing wind-forcing. Journal of Hydrodynamics (Serial B) 16(3): 341–349.

Luo, L., B. Qin, G. Zhu & Y. Zhang, 2004a. The current pattern of Meiliang Bay in the winter time. Journal of Lake Sciences 16(1): 73–76 (in Chinese).

Luo, L., B. Qin & G. Zhu, 2004b. Sediment distribution pattern maped from the combination of objective analysis and geostatistics in the large shallow Lake Taihu, China. Journal of Environmental Sciences 16(6): 908–911.

Mao, R., P. Xu & J. Gao, 2004. Hydrography and water balance of Lake Taihu. In Qin, B., W. Hu & W. Chen (eds), Process and Mechanism of Environmental Changes of Lake Taihu. Science Press, Beijing, 9–18 (in Chinese).

Nanjing Institute of Geography, Academia Sinica (NIGLAS), 1965. Investigation of Lake Taihu. Science Press, Beijing, 1–84 (in Chinese).

National Statistics Bureau (NSB), 2000. Statistics Almanac of China. China Statistics Press, Beijing (in Chinese).

Qin, B., W. Hu, W. Chen, J. Ji, C. Fan, Y. Chen, X. Gao, L. Yang, G. Gao, W. Huang, J. Jiang, S. Zhang, Y. Liu & Z. Zhou, 2000. Studies on the hydrodynamic processes and related factors in Meiliang Bay, Northern Taihu. Journal of Lake Sciences 12(4): 327–334 (in Chinese).

Qin, B., W. Hu, G. Gao, L. Luo & J. Zhang, 2004. The dynamics of resuspension and conceptual mode of nutrient releases from sediments in large shallow Lake Taihu, China. Chinese Sciences Bulletin 49(1): 54–64.

State Environmental Protection Administration (SEPA), 2000. Water Protection in the Three Rivers and Three Lakes of China. Chinese Environmental Press, Beijing: 151–175 (in Chinese).

Statistics Bureau of Jiangsu Province (SBJP), 2000. Jiangsu Statistics Almanac. China Statistics Press, Beijing (in Chinese).

Statistics Bureau of Shanghai City (SBSC), 2000. New Shanghai Statistics Almanac of National Economic and Social History in Past 50 Years. Statistics Bureau of Shanghai City (in Chinese).

Statistics Bureau of Zhejiang Province (SBZP), 2000. New Zhejiang Statistics Almanac. China Statistics Press, Beijing (in Chinese).

Sun, S., 2004. Topographic features and origin of Lake Taihu and its catchment. In Qin, B., W. Hu & W. Chen (eds), Process and Mechanism of Environmental Changes of Lake Taihu. Science Press, Beijing, 1–9 (in Chinese).

Sun, S. & Y. Huang, 1993. Lake Taihu. China Ocean Press, Beijing, 23–89 (in Chinese).

Sun, S., Y. Wu & B. Dong, 1987. The Bottom Configuration and Recent Deposition of Lake Taihu. Memoirs of Nanjing Institute of Geography and Limnology, Academia Sinica, No. 4. Science Press, Beijing, 1–14 (in Chinese).

Taihu Basin Authority (TBA) & Nanjing Institute of Geography & Limnology (NIGLAS), 2000. Atlas of Eco-environment of Lake Taihu. Science Press, Beijing, 1–118.

Wang, H. & W. Ding, 1936. Formation and shrinkage of Lake Taihu. Monthly of Water Resources 11(6): 8–13 (in Chinese).

Wissman, F., 1941. Geographical problem in the lower delta of Changjiang River (translated by Wang, D.). Geography 1(1): 74–77 (in Chinese).

Wu, J., X. Han & J. Jiang, 1987. A study of the wind waves at the platform and resuspension of the sediments in Taihu Lake. In ISTB'87 – International Symposium on Lake Taihu and Lake Biwa, Nanjing, 8/1–8/6.

Wu, Q., K. Chen & G. Gao, 1995. Effects on aquatic environment of large-scale inning for aquaculture. Journal of Aquaculture 19: 343–349 (in Chinese).

Xu, P. & B. Qin, 2005. Water quantity and pollutant fluxes of the surrounding rivers of Lake Taihu during the hydrological year of 2001–2002. Journal of Lake Sciences 17(3): 213–218 (in Chinese).

Yan, Q. & S. Xu, 1987. Recent Yangtze Delta Deposits. East China Normal University Press, Shanghai, 68–75 (in Chinese).

Yang, H., Z. Xie & D. Yang, 1985. Sea-level change in holocene and evolution of Lake Taihu. In Yang, H. (ed.), Contribution to the Quaternary Glaciology and Quaternary Geology. Geological Publishing House, Beijing, 49–64 (in Chinese).

Yang, Q. & W. Li, 1996. Fishery impacts on aquatic plants and possible strategy. Journal of Applied Ecosystem 7: 83–88 (in Chinese).

Zhang, Y., B. Qin, W. Chen & L. Luo, 2004a. A study on total suspended matter in Lake Taihu. Resources and Environment in the Changjiang River 13(3): 266–271 (in Chinese).

Zhang, Y., Q. Cai & P. Yang, 2004b. Light transmission in the water. In Qin, B., W. Hu & W. Chen (eds), Process and Mechanism of Environmental Changes of Lake Taihu. Science Press, Beijing, 53–96 (in Chinese).

Hydrobiologia (2007) 581:15–24
DOI 10.1007/s10750-006-0524-2

The art and science of lake restoration

Brian Moss

Abstract Science and art encompass understanding, through human hand and mind; both are applied in the form of creation by human hand and mind. Like Janus, the Roman god of doorways and bridges, they are two faces of the same thing. Great art requires great understanding and great understanding is wasted unless it is applied to create meaningful art, whether this is in the form of written works or paintings. These principles apply not least to the restoration of lakes and whole landscapes from the human-induced damage that is the antithesis of significant art and fundamental science. Because they lie at the lowest points of hydrologic basins, lakes reflect the balance of human activities in their catchments. If these activities are carried out well, they contribute greatly to the landscape and to the welfare of the human population. If carried out badly in the sole interests of exploitation of resources, both human and otherwise, they may become problems reflected, for example, in eutrophication, algal blooms and fish kills. Lake restoration involves first an understanding in a reductionist (scientific) way of the processes that drive lake ecosystems, but then of the ultimate reasons that create these proximate problems. Ultimate reasons lie in a much wider sphere of human nature and the organisation of society. Lake restoration, in its most trivial form, may be simply a form of gardening to allay the symptoms of problems and create the illusion of a solution. Lake restoration in its most profound form involves an understanding of cultural significance and the workings of human societies and forms an epitome for the solution of much greater, global problems. Only in this form does it become truly creative. Approval must come from both of the faces of Janus.

Keywords Eutrophication · Shallow lakes · Culture · China

Guest editors: B. Qin, Z. Liu & K. Havens
Eutrophication of shallow lakes with special reference to Lake Taihu, China

B. Moss (✉)
School of Biological Sciences, Biosciences Building, University of Liverpool, Liverpool L69 3BX, UK
e-mail: brmoss@liverpool.ac.uk

Introduction

There has arisen a divide in the way that knowledge is perceived. It has been called the two cultures by the English novelist and physicist. Snow (1959) and it concerns the gulf between scientific perception and perception by artists, linguists, historians, social scientists and all those loosely grouped into the subject areas of the humanities. Science is perceived as objective, collects facts from survey and experiments, and interprets them in a way that is alleged to be value-free and objective. Scholars in the humanities also collect facts. The methodology of a historian is no different from that of an ecologist describing past and present ecosystems, but controlled experiments are not conceivable in the arts and thus determination of mechanism has to be left to the perception of the observer and therefore is regarded as subjective rather than objective. This simplified overview disguises the fact that there can be a great deal of subjectivity in the interpretation of information collected by scientists and that there are insights available to other scholars that are too readily rejected by some scientists because they are not amenable to controlled experiment. A better way of dividing approaches to understanding is to distinguish the reductionist from the holistic. Scientists and those in the humanities use both approaches, but current practice is for scientists to be more reductionist and humanities' scholars more holistic.

Where ecological restoration is concerned, there can be a harmony between the two approaches and the lessons of the humanities, indeed of fine art, can be applied to guide the process. Ecological restoration, for example, involves several stages. First there is a problem. A habitat has to be restored because it has been degraded to some extent. The reasons for the damage need first to be understood or otherwise restoration may be misguided. Secondly, the extent of the damage must be assessed and thirdly a target towards which restoration should be directed must be decided. Then, fourthly, the scientific understanding of the system is brought into play to determine how to achieve the target and lastly it is likely that maintenance of this new status will require management of some sort. The lessons of the humanities are applicable at all but the fourth stage of technical restoration.

The parallels with the conception and creation of fine art are considerable. Art exists to interpret society as much as to decorate it. Science likewise diagnoses problems as well as helps create the artefacts of comfortable life styles. The power with which artworks are expressed mirrors the degree of the problem just as scientific approaches assess the degree of damage. Restoration involves design and the parallels are obvious here and management may be both a scientific and aesthetic process if the results of the restoration are to be permanent and significant.

The existence of the problems

Lakes lie at the bottoms of catchment basins. Simply because of the operation of gravity, they will reflect activities in their catchments, manifested in changed water quality, but also in alterations to watercourses for the purposes of water storage, flood control, pollutant disposal and farming. A catchment covered by pristine ecosystems will pose no problems for the freshwater systems that drain it and indeed there is a seamless connectance between the terrestrial and freshwater systems. Intact ecosystems are parsimonious with nutrients and very little available nutrient leaches from the land to the water. Most key nutrients are held and recycled within the terrestrial biomass. Pristine freshwater systems thus have extremely low nitrogen and phosphorus concentrations, even when the catchment soils are made fertile from easily weathered rocks.

Pristine freshwater systems will also have considerable structure. They are not just masses of mixed water. Rivers have structures created by geomorphological processes (sand bars, riffles, floodplains, levees) and by woody debris entering from the immediate catchment. Lakes have littoral zones with a rich diversity of emergent and submerged plants. Maintenance of this structure may depend on the intactness of the diversity of the ecosystem. Top predators, wolves and bears, salmonids and other piscivorous fish and birds may exert considerable top-down effects (Ripple & Beschta, 2004a, b) that maintain both this structure and the efficiency of cycling of nutrients in a system where nutrients are naturally extremely scarce. There will also be a great connectivity between systems. A lake does not exist as an isolated system. It depends on organic matter washed in from the catchment through the inflow rivers for part of its productivity (Hanson et al., 2004). The surrounding wetlands may be crucial for fish and invertebrate life histories and in the interesting case of the functioning of northern rivers, nutrients brought from the ocean in the bodies of salmonid fish are recycled through the faeces of their brown bear predators into the riparian woodland. In turn tree debris retains carcases after spawning to provide the nutrients needed by the periphyton and micro invertebrates that sustain the hatched fish once they have used up their yolk sacs. Movements of pike (*Esox lucius*) between lakes and wetlands perform parallel roles (Klinge et al., 1995).

Parsimony of nutrients, preservation of structure, including appropriate biodiversity, and maintenance of connectivity are the three key elements that constitute the 'gold standard' for any ecosystem restoration. A system with these three characteristics intact is self-maintaining. Scientists comment on deviation from this state by recording the impacts on freshwater systems that damage these characteristics. Nutrient parsimony is eroded by eutrophication, which may also result in a loss of structure. Catchment management for efficient forestry may deny to rivers the structure provided by forest debris (Harmon et al., 1986) and connectance is easily lost through the construction of dams and the channelisation and embankment of lowland rivers so that they lose hydrological connection with their floodplains. Essentially, a long catalogue of impacts that damage freshwater systems ultimately boils down to damage to the three key elements.

The problems of freshwaters are ultimately the problems of how landscapes are managed and therefore of the way that human societies are controlled and managed. In the developed world the current philosophy is of maximising resource use in the interests of profit with rather little consideration of the sustainability of the resource. Early ecology contented itself with description of the workings of intact systems; now it is equally concerned with documenting the damage. Likewise fine art once celebrated the contemplative delights of pristine or lightly used landscapes. The traditions of Chinese scroll painting very much manifest this. But currently much of it is dedicated to protest and comment. Pablo Picasso's painting of *Guernica*, a horrific view of the bombing of a Spanish town in support of extreme right wing policies is a famous example. Peter Scott's painting. 'The Natural Environment of Man' (Fig. 1) uses art as a means of expressing environmental controversy.

Fig. 1 The Natural World of Man. Peter Scott, 1961. The painting is a comment on the dilemma faced by humans in contemplating the advances of technology, to the right of the picture and the consequent losses to ecosystems, represented by a selection of endangered mammals and birds to the left. The figure stands apart from the scene, representing philosophies that regard the natural world as something to be controlled by humans in their own interests but is also transparent and melds into the natural world. Humans are as dependent on the biosphere as all other creatures. The figure also holds his or her arms in a gesture of uncertainty, not knowing quite what to do about the problems being caused by human activities. By permission of Lady Scott, Wildfowl and Wetlands Trust, Slimbridge

The six artistic principles of Xi He and their equivalents in lake restoration

It is in the design of targets for restoration, however that the parallel approaches between art and science become most interesting. In the sixth century (Northern and Southern Dynasty) Xie He (Hsieh Ho), in his *Ku-hua p'inlu* (Acker, 1954; Tregear, 1997) posited, for the judgement of Chinese paintings, six principles, which have been so influential that they influenced Chinese painting into the twentieth century and may still influence it. These were:

Q'i yun sheng tung: The painter should endow his work with life and movement through harmony with the spirit of nature

Ku fa yung pi: Refers to structural power and tension of the brush stroke
Ying wu hsiang hsing: Fidelity to the object in portraying forms
Sui lei fu ts'ai: Conforming to kind in applying colours
Ching ying wei chih: Plan and design, place and composition
Ch'uan I mu hsieh: Transmission of ancient models by copying

Gombrich (1995) talks of a wonderful restraint in Chinese painting (Fig. 2). There is an economy of line, a sense of contemplation and inner feeling for the subject rather than a precise elaboration of details. Technique in brushwork is economical and firm and painted scrolls are usually adorned with philosophic comment, also elegantly executed in text characters. The long tradition of Chinese painting, adopted also by Japanese painters, has its dangers in that rigid tradition inhibits individual inspiration and innovation, but a balance between tradition and innovation often gives the best of both worlds and in such a balance lies the relevance of art to lake restoration.

Fig. 2 A characteristic Chinese painting by Sheng Mou, 1330–1369. Waiting for the Ferry on an autumn river (1351). Hanging scroll on paper. Palace Museum, Beijing

With some contraction of the third and fourth principles, which concern fidelity of form and colour, there are equivalents to Xie Hi's principles in the guidance that might be given for effective lake restoration, especially in the complex case of shallow lakes. There should be skill in carrying out the work and economy in resources and effort; there should be fidelity to the structures and nature of the system; there should be taste in reflecting the nature of the landscape; there should be wisdom in understanding the essence of the relationships by reference to pristine examples; and there should be harmony with the spirit of wise and sustainable use. It is not known whether there was particular significance in the order of Xie Hi's strictures on painting, except that the first was the most important. The logical parallels for lake restoration however might re-order the equivalents in terms of: Wisdom, Skill, Fidelity, Taste and Harmony.

Wisdom in understanding the essence of the relationships

Understanding of the functioning of shallow lake systems has advanced greatly since 1970. The natural state of all lakes is to have very low available nutrient concentrations, as discussed above. If the natural catchment vegetation is destroyed for urbanisation or agriculture of any kind, these mechanisms are destroyed and leaching losses increase greatly. Such eutrophication has been associated with the loss of clear water and submerged plant communities from shallow lakes and their replacement with filamentous algae and phytoplankton.

The idea that this was a process entirely driven by nutrient increase is still held by some workers, but was challenged by Balls et al. (1989) and Irvine et al. (1989) who found that adding large amounts of nutrients to such plant communities in experimental ponds did not displace the plants. Indeed they grew better. A switch to phytoplankton only occurred if the plants were independently damaged by mechanical removal. In subsequent years a range of switch mechanisms has been recognised, each of which, especially but not necessarily, in tandem with nutrient enrichment may displace the plants (Moss et al., 1996). These include herbicide application, grazing by introduced birds and fish like common carp (*Cyprinus carpio* L.), pesticide leachate and increasing salinity (both of which disturb the zooplankton grazer interactions that maintain clear water), rise in water level and violent storms. Thus over a large range of nutrient loadings, dependent on whether such a switch has occurred, shallow lakes may have alternative states of clear water and plant dominance and turbid water with phytoplankton dominance (Fig. 3). Sometimes, turbidity may be created through suspended sediment raised by wind or benthic fish in plant-devoid waters. This understanding has come from a range of approaches, including analysis of very large data sets (Jeppesen et al., 1999), mesocosm (Moss et al., 2004) and pond experiments (Balls et al., 1989; McKee et al., 2003), whole-lake experiments (Meijer et al., 1999) and theoretical treatments (Scheffer et al., 1993). The general model applies to small and moderate sized lakes (perhaps up to the order of 10^2–10^3 km^2 but is relevant also to very large shallow lakes though with very large wind fetches, the more exposed portions of these may behave differently.

Part of the understanding has also come from attempts to restore the plant communities to

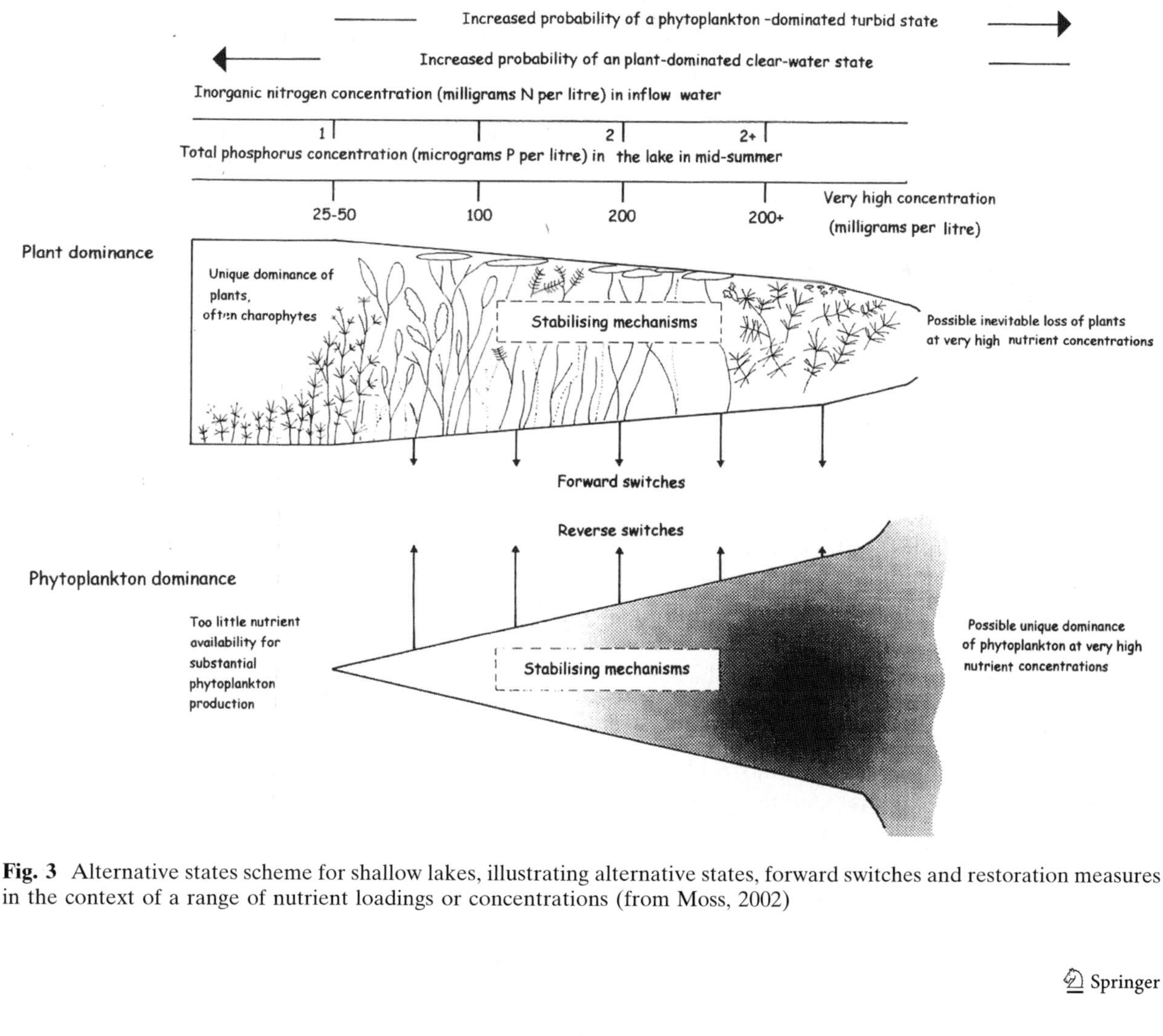

Fig. 3 Alternative states scheme for shallow lakes, illustrating alternative states, forward switches and restoration measures in the context of a range of nutrient loadings or concentrations (from Moss, 2002)

turbid lakes. Some cultures, not least those that have large populations to feed, may prize the turbid state for its ease of fishing and productivity of fish for food. The general presumption, however, is that the more desirable state of a shallow lake is to be plant-dominated, because this state is associated with the greatest biodiversity and aesthetic value and the least problems of smell, fish kills and toxicity of algal blooms. Plant communities use a number of mechanisms to maintain clear waters, including competition for nitrogen with algae, allelopathy, and reduction of eddy disturbance necessary for phytoplankton to remain suspended. Indirect, but equally important, mechanisms include provision of refuges against fish predators for zooplankton grazers on phytoplankton or macroinvertebrate grazers on periphyton (Moss et al., 1996; Scheffer, 1998).

Restoration attempts initially centred on reinstating the role of grazers by temporary removal of their fish predators, a process called biomanipulation. This resulted in clear water in which plants could grow, but protection against damage from grazing birds was sometimes necessary to establish self-sustaining communities. The communities were, however, usually of low diversity and vulnerable to slight disturbance that led to return of turbid conditions. Few restorations have been entirely successful in restoring diverse, stable plant communities (Moss et al., 1996; Meijer et al., 1999).

The reasons have perhaps been because the scale of restoration attempts has been too limited, that nutrients have not been adequately controlled or that only one of two important nutrients has been controlled. Shallow lakes have strong links in their natural states with surrounding wetlands that provide habitat for piscivorous fish that control zooplanktivorous fish and hence result in large zooplankton populations (Klinge et al., 1995). Extensive restoration of whole floodplains rather than just isolated lake basins is needed. Although alternative states exist over a wide range of nutrient loadings, it appears that the threshold for switches to act to remove plants is lower when nutrients are high and that biomanipulation is more effective if nutrients are reduced and that perhaps nutrient control should involve both phosphorus and nitrogen. Phosphorus may control algal production but the diversity of the plant community and hence its stability seems to be determined by nitrogen (James et al., 2005). Nutrient control is difficult especially when diffuse sources dominate and generally only phosphorus is controlled because point sources (for example, waste water treatment works) tend to be prominent and phosphorus can easily be precipitated at them.

Moss et al. (1996) developed a technical strategy for restoration, which forms an idealised basis for restoration attempts. First the target for restoration should be determined. The nature of the plant community is linked to the nutrient loading and generally diverse charophyte or isoetid communities require very low nutrient concentrations; mixed pondweed communities will persist over a middle range; and nymphaeid communities will still grow at very high concentrations and even when conditions are switching to turbid conditions. Once the target is decided, any existing or potential forward switches should be removed. There can be no success if this is not done. Thus, damaging fish or unusually high numbers of grazing birds need to be removed. Usually these are of introduced species and have built up to large numbers in the absence of natural predators. Risks from pesticide and herbicide contamination must be reduced and any damaging management such as mechanical clearance eliminated.

Nutrients should then be reduced, including both phosphorus and nitrogen. A range of methods exists for controlling point and diffuse sources but generally it is difficult to reduce concentrations to those that are likely to favour successful restoration. Restorations are most likely to be successful if total phosphorus concentration is reduced to winter values of less than 0.05 mg l^{-1} and total nitrogen to less than 1 mg l^{-1}. These suggestions are subject to much variability dependent on local circumstances however and emulation of the pristine condition of very low available nutrients should be the guiding principle. The fourth step is that of biomanipulation and a common source of failure is for too few fish to be removed. Successful restoration will be more likely if stocks are reduced to lower than 2 g m^{-2} and if some piscivores are left in the lake. This will allow daphnids and other zooplankters to increase and will clear the water, although if retention times are long and favourable to Cyanobacteria, there may be difficulties (Moss et al., 1991). Cyanobacteria can be controlled by grazing, but generally only if the proportion of such algae is low. Thick, surface blooms generally cannot. In the clear water, plants may spontaneously return from residual fragments but may have to be reintroduced. New plantings are vulnerable to bird damage and may need to be protected by caging. Thereafter, once plants have established vigorous beds, perhaps after two years, a fish community can be reintroduced, but should reflect the native community for the lake, should not include exotics and should always include piscivorous fish. Finally restoration projects should always be carefully monitored for several years afterwards. Many are not and valuable lessons as to why they succeed or fail will be lost. And in all things, patience is a virtue. It is not possible to replace over a year what may have been lost over a century.

Beyond this catalogue of necessary technicalities, however, is a philosophical basis for restoration and it is in this that the parallels with the principles of Xie He for painting become very relevant. Wisdom, skill, fidelity, taste and harmony are further contracted here into a hierarchy of skill, fidelity and taste, harmony and wisdom.

Skill in carrying out the work with economy of resources and effort

Lake restoration is often carried out by contractors, for whom it is simply a means of making profit. The results can then fall far short of the exquisiteness that Xie He would have expected of his painters and that sensitive ecologists would expect in the final result of a restoration. Mass-produced art is rarely prized. Skill comes in appreciating details and in minimising costs without compromise on the final result.

In one restoration attempt in which I was involved, at a lake in Wales (Moss et al., 2002), a detailed specification of what should be done was drawn up, including a strong proviso that certain plant species, like *Ceratophyllum demersum* L. should not be used, for they were likely to create dense beds that would interfere with angling. The use of plant stock bought from garden suppliers, which might include exotics and cultivars, was also precluded. In the event, the work was carried out by contractors who took the cheapest options of putting in any plant they could find, which included the very common *Ceratophyllum*, and at least effort, which meant they bought water lily stock from garden suppliers. The result was a lake choked by *Ceratophyllum* and fringed with an inappropriate, for the site, and probably non wild-type nymphaeid, *Nymphoides peltata* L. There was public protest at the *Ceratophyllum* choking and the result was that the local authority treated the lake with herbicide and effectively reversed the entire restoration after spending of a considerable sum of money ultimately to achieve nothing.

This however, is a small example of this particular precept. Skill involves choosing the right level of approach, and economy has to be considered in the long, not the short term. There are three levels of approach to restoration that correspond to different levels of cause of the problem. These are immediate, proximate and ultimate. For a shallow lake that has switched to phytoplankton dominance with troublesome, perhaps toxic cyanobacterial blooms, the immediate cause of the problem is the bloom. The immediate treatment is to use mixing methods or algicide. Both are cheap, temporary and ultimately ineffective in solving the problem for they treat only symptoms and the money spent on them is ultimately wasted.

The next level of treatment is to carry through the technical strategy outlined above. This deals with the proximate causes such as increased nutrient loading in the catchment, introduction of exotic fish or profligate use of herbicides and pesticides. It will cost much more but will be effective for a longer term. It may not be ultimately effective, however, for continuous control of nutrients will be needed and should funds become unavailable or should inappropriate fish be reintroduced for some new commercial intention, the money will have been wasted. What is really needed in this, as in all environmental problems, is a profound enquiry into the ultimate causes, which will be political, social and cultural. Treatment of the ultimate cause of the problem may have to be a change in agricultural policy so that large areas of the catchment are used as buffer zones to allow intensive agriculture to continue in the remainder. Or they may mean an enquiry into the relative rights of some parts of society to exploit the resources of the lake (for a fishery for introduced, very productive species for example), and the rights of other parts of society not to tolerate this in the interests of an environment that preserves biodiversity. Skill may come in realising these levels of approach and in aiming to tackle the problem as high up the hierarchy of causation as is possible.

Fidelity and taste in reflecting the nature of the landscape

Lakes that have been damaged are frequently incapable of being restored to anything remotely like their pristine equivalents. Shorelines often have been developed and damaged and may support building development that cannot be removed. Restoration may have to take on board such changes. A parallel aspect of restoration is re-creation of new lakes from quarries, gravel excavations and other such engineering activities. Pond creation is a widespread activity in Europe (Andrews & Kinsman, 1990; Williams, 1999; Moore, 2002) where former small ponds have often been filled in as a result of agricultural intensification to ease the use of large agricultural machines. Taste and design are very important in both these activities.

The guidance that Xie He would doubtless have given to lake restorers would have been that if natural structures such as emergent wetland margins cannot be restored, and hard edges have to be maintained, these should be of natural materials, appropriate to the area, especially its local geology, and not concrete or building stone of colours and textures that do not fit into the surrounding landscape. He would have spurned the artificiality of structures like fountains and artificial rafts for bird roosting. Fountains are particularly favoured by local government authorities, partly because the idea is rife that water needs continually oxygenating and partly by a perception that moving waters are tourist attractions. Water does not need artificially oxygenating, unless there has been such severe organic pollution that it is in danger of becoming permanently anaerobic and fountains are not good oxygenating devices anyway. Sub-surface circulation using powerful pumps is needed in such situations. Fountains may be appropriate to man-made pools in urban situations but not to natural or large lakes.

In designing new lakes, there is a tendency also to ignore the tenets of art in favour of the conveniences of engineering. Newly excavated lakes tend to have either geometric shapes, square or oblong, or exaggerated sinuosity in an attempt to avoid geometricity. Both are equally offensive to the eye. In natural lake basins the shape and outline are determined by the local geology and geomorphology. No general prescription can be given except that design should take account of the subtleties of the landscape and that this takes different talents from those needed for the engineering of earth movement and water retention. An appreciation of the nature of unmodified local lake basins and what gives them their particular characteristics is crucial to this. In an aerial photograph the ultimate results of lake creation or restoration should blend with the landscape and not stand out prominently as clearly different. One group of man-made lakes, the Norfolk Broads (Moss, 2002) formed in the 14th century in the United Kingdom, fits so well into the landscape that for many years their anthropogenic origin was not even realized.

Harmony with the spirit of wise and sustainable use

Xie He's first and most important stricture for painters is ultimately the most important one in lake restoration also. Problems in lakes reflect problems in society. An algal bloom is a symptom of a society that has not satisfactorily dealt with its immediate problems of waste disposal and food production in the best possible way. It reflects a dominance of some interests over others and generally these will have been the richest and most powerful interests. The current dominant philosophy of the human population of this planet is one of short-term exploitation of resources with rather little thought for ultimate consequences. Were this not the case we would not be facing severe climate change, heavily overfished seas, severe water shortages and polluted lakes.

There is acknowledgement of the problems in many political quarters but a policy of acknowledging problems but doing little and hoping that solutions will emerge spontaneously is a dangerous one. High quality lakes are set in high quality landscapes. High quality landscapes reflect sustainable values in society. Xie He taught that great art reflects great understanding, fundamental values and worthy objectives. Great lake restoration likewise reflects understanding and objectives worthy of a society that can be sustained, that reflects the needs of all people and that is not just a commercial activity. And if a Chinese art critic can be seen to offer guidance in lake restoration, perhaps ancient Chinese philosophy (Lao Tzu, approx. 500BC) can provide a solid basis for creating the sort of society that will sustain itself and its associated natural world, not least its freshwater resources. It defies all pleas above for connectedness, and a holistic and profound approach to quote a single passage to illustrate an entire philosophy, but Verse 67 will have to suffice:

Some say that my teaching is nonsense.
Others call it lofty but impractical,
But to those who have looked inside themselves,
This nonsense makes perfect sense.
And to those who put it into practice,
This loftiness has roots that go deep.

I have just three things to teach:
Simplicity, patience, compassion.
These three are your greatest treasures.
Simple in actions and in thoughts,
You return to the source of being.
Patient with both friends and enemies,
You accord with the way things are.
Compassionate towards yourself,
You reconcile all beings in the world.
(Translated by S. Mitchell, 1999)

References

Acker, W. B. R., 1954. Some T'ang and pre T'ang texts on Chinese paintings. Leiden University press, Leiden.

Andrews, J. & D. Kinsman, 1990. Gravel pit restoration for wildlife. Royal Society for the Protection of Birds, Sandy.

Balls, H., B. Moss & K. Irvine, 1989. The loss of submerged plants with eutrophication. I. Experimental design, water chemistry, aquatic plant and phytoplankton biomass in experiments carried out in ponds in the Norfolk Broadland. Freshwater Biology 22: 71–87.

Gombrich, E. H., 1995. The Story of art. 16 edn. Phaidon Press, London.

Hanson, P. C., A. I. Pollard, D. L. Bade, K. Predick, S. R. Carpenter & J. A. Foley, 2004. A model of carbon evasion and sedimentation in temperate lakes. Global Change Biology 10: 1285–1298.

Harmon, M. E., J. F. Franklin, F. J. Swanson, P. Sollins, S. V. Gregory, J. D. Lattin, N. H. Anderson, S. P. Cline, N. G. Aumen, J. R. Sedell, G. W. Lienkaemper, J. Cromak & K. W. Cummins, 1986. Ecology of coarse woody debris in temperate ecosystems. Advances in Ecological Research 15: 133–302.

Irvine, K., B. Moss & H. R. Balls, 1989. The loss of submerged plants with eutrophication II Relationships between fish and zooplankton in a set of experimental ponds, and conclusions. Freshwater Biology 22: 89–107.

James, C. S., J. Fisher, V. Russell, S. Collings & B. Moss, 2005. Nitrate availability and hydrophyte species richness in shallow lakes. Freshwater Biology 50:1049–1063.

Jeppesen, E., M. Sondergaard, B. Kronvang, J. P. Jensen, L. M. Svendsen & T. L. Lauridsen, 1999. Lake and catchment management in Denmark. Hydrobiologia 395/396: 419–432.

Klinge, M., M. P. Grimm & S. H. Hosper, 1995: Eutrophication and ecological rehabilitation of Dutch lakes: presentation of a new conceptual framework. Water Science, Technology 31: 207–218.

Lao Tzu C5 BC approx. Tao Te Ching, 1999. An illustrated journey (translated by Mitchell, S. Tao Te Ching). Frances Lincoln, Ltd, London.

McKee, D., D. Atkinson, S. E. Collings, J. W. Eaton, A. B. Gill, I. Harvey, K. Hatton, T. Heyes, D. Wilson & B. Moss, 2003. Response of freshwater microcosm communities to nutrients, fish, and elevated temperature during winter and summer. Limnology & Oceanography 48: 707–722.

Meijer, M.-L., I. de Boois, M. Scheffer, R. Portielje & H. Hosper, 1999. Biomanipulation in shallow lakes in the Netherlands: an evaluation of 18 case studies. Hydrobiologia 408/409: 13–30.

Moore, N. W., 2002. Oaks, dragonflies and people. Harley Books, Colchester.

Moss, B., 2002. The Broads: the people's wetland. Harper Collins, London.

Moss, B., J. Madgwick & G. L. Phillips, 1996. A guide to the restoration of nutrient enriched shallow lakes.

Environment Agency, Broads Authority & European Union Life Programme, Norwich.
Moss, B., D. Stephen, D. M. Balayla, E. Becares, S. E. Collings, C. Fernandez-Alaez, M. Fernandez-Alaez, C. Ferriol, P. Garcia, J. Goma, M. Gyllstrom, L. A. Hansson, J. Hietala, T. Kairesalo, M. R. Miracle, S. Romo, J. Rueda, V. Russell, A. Stahl-Delbanco, M. Svensson, K. Vakkilainen, M. Valentin, W. J. Van de Bund, E. Van Donk, E. Vicente & E. G. Villena, 2004. Continental-scale patterns of nutrient and fish effects on shallow lakes: synthesis of a pan-European mesocosm experiment. Freshwater Biology 49: 1633–1649.
Moss, B., J. Stansfield & K. Irvine, 1991. Development of daphnid communities in diatom- and cyanophyte-dominated lakes and their relevance to lake restoration by biomanipulation. Journal of Applied Ecology 28: 586–602.
Moss, B., L. Carvalho & J. Plewes, 2002. The Lake at Llandrindod Wells – a restoration comedy ? Aquatic Conservation: Marine & Freshwater 12: 229–245.
Ripple, W. J. & R. I. Beschta, 2004a. Wolves, elk, willows, and tropic cascades in the upper Gallantin Range of South-western Montana, USA. Forest Ecology and management 200: 161–181.
Ripple, W. J. & R. L. Beschta, 2004b. Wolves and the ecology of fear: can predation risk structure ecosystems? BioScience 54: 755–766.
Scheffer, M., S. H. Hosper, M. L. Meijer, B. Moss & E. Jeppesen, 1993. Alternative equilibria in shallow lakes. Trends in Ecology & Evolution 8: 275–279.
Scheffer, M., 1998. Ecology of shallow lakes. Chapman & Hall, London.
Snow, C. P., 1959. The two cultures. Cambridge University Press, Cambridge.
Tregear, M., 1997. Chinese art. Thames & Hudson, London.
Williams, P., 1999. The pond book: a guide to the management and creation of ponds. The ponds conservation trust, Oxford.

Hydrobiologia (2007) 581:25–42
DOI 10.1007/s10750-006-0502-8

EUTROPHICATION IN LAKES

Phosphorus dynamics at multiple time scales in the pelagic zone of a large shallow lake in Florida, USA

Karl E. Havens · Kang-Ren Jin · Nenad Iricanin · R. Thomas James

Abstract Phosphorus (P) dynamics in large shallow lakes are greatly influenced by physical processes such as wind-driven sediment resuspension, at times scales from hours to years. Results from long-term (30 year) research on Lake Okeechobee, Florida (area 1,730 km^2, mean depth 2.7 m) illustrate key features of these P dynamics. Variations in wind velocity result in changes in water column transparency, suspended solids, and total P (TP). In summer there are diurnal changes in TP associated with afternoon winds, and in winter, when strong winds occur for multiple days, monthly average TP remains high compared to summer. The magnitude of daily and seasonal TP changes can exceed 100 μg l^{-1}. Hurricanes and tropical storms also cause extreme changes in TP that are superimposed on seasonal dynamics. When a hurricane passed 80 km south of the lake in October 1999, mean pelagic TP increased from 88 to 222 μg l^{-1}. During large resuspension events, light attenuation is substantially increased, and this influences the biomass and spatial extent of submerged plants, as well as water column TP. In Lake Okeechobee, TP concentrations typically are ~20 μg l^{-1} when submerged plants are dense, and soluble reactive P concentrations are reduced below detection, perhaps by the periphyton and plant uptake and by precipitation with calcium at high pH. In contrast, TP exceeds 50 μg l^{-1} when submerged plants and periphyton are absent due to prolonged deep water, and phytoplankton biomass and algal bloom frequency both are increased. In Lake Okeechobee and other large shallow lakes, complex models that explicitly consider wind-wave energy, hydrodynamics, and sediment resuspension, transport, and key biological processes are needed to accurately predict how lake water TP will respond to different management options.

Keywords Phosphorus · Shallow lakes · Sediment resuspension · Wind effects · Physical processes · Diffusion · Eutrophication · Long-term studies

Guest editors: B. Qin, Z. Liu & K. Havens
Eutrophication of shallow lakes with special reference to Lake Taihu, China

K. E. Havens (✉)
Department of Fisheries and Aquatic Sciences, University of Florida, Gainesville, FL 33458, USA
e-mail: khavens@ufl.edu

K.-R. Jin · N. Iricanin · R. T. James
South Florida Water Management District, West Palm Beach, FL 33406, USA

Introduction

The seasonal and year-to-year dynamics of TP in deep dimictic lakes have been studied for decades, and they follow generally predictable patterns (e.g., Wetzel, 2001). During summer stratification, dissolved P is incorporated into the biomass of plankton, which subsequently settles from the epilimnion, or is incorporated into higher trophic levels of the food web. With the exception of buoyant cyanobacteria or other algae that can actively maintain their position in the upper water (Reynolds, 1994), there is a net downward movement of P from the epilimnion to the hypolimnion and sediments. Once particles have settled, the only time they return to surface waters in a deep stratified lake may be at fall overturn. Various authors have examined and summarized these patterns and their underlying processes (e.g., Hutchinson, 1975; Wetzel, 2001) that are largely based on work done decades ago by Mortimer (1941), Rigler (1956, 1964, 1973), and Lean (1973a, b).

In contrast to their stratified counterparts, shallow lakes may display highly irregular seasonal and year-to-year variations in TP, largely driven by stochastic processes that are not as readily predicted. These lakes also are considerably less studied, although in recent years that tendency has changed (Scheffer, 1998). At any given time, the physical transport of particulate and interstitial P into the water column of a shallow lake by wind energy may be orders of magnitude greater than other inputs of P, including diffusion, bioturbation, and external loading. As long as a concentration gradient exists between sediment-bound P and dissolved P in the water ($P_{water} > P_{sediment}$), downward transport of P with sediment particles may greatly exceed other loss processes, such as algal settling (e.g., James et al., 1997). In shallow lakes where wind strongly influences sediments, physical resuspension by wind-waves dominates the upward transport of P, and settling of particles dominates the downward flux (Fig. 1). In lakes not as heavily influenced by wind, other processes in the P cycle, including bioturbation, uptake of allochthonous P by plankton, plants, and periphyton may represent downward fluxes of greater magnitude.

There are several ways to predict whether P dynamics of a lake will be strongly driven by wind-waves. These predictions range from complex hydrodynamic models dealing with conservation of momentum and mass (e.g., Jin et al., 2000, 2002) to calculations of intermediate complexity based on wavelength and water depth (Caper & Bachmann, 1984) to simple index values. An example of an index is the dynamic ratio of Hakanson (1982), calculated as the square root of lake surface area (km^2) divided by mean depth (m). This index was developed to identify the relative portion of a lake bottom likely to be influenced by wind-driven sediment resuspension. Using data from 36 Florida lakes, Bachmann et al. (2000) showed that when the dynamic ratio is above 0.8, 100% of the lake bottom is subject to resuspension, whereas at ratios below 0.8, the risk of resuspension rapidly declines in a linear manner. This method can be used to characterize the influence of wind-generated waves on large lakes of the world (Fig. 2). Lakes are distributed according to their mean depth and area, with the diagonal line indicating the critical depth for

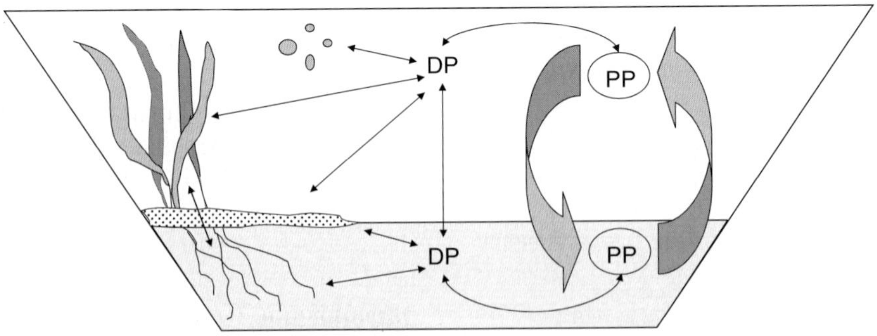

Fig. 1 Conceptual diagram of phosphorus (P) dynamics in a shallow lake that is heavily influenced by wind-induced waves. The arrows indicate fluxes of P between plants and periphyton, benthic algal mats, plankton, sediments, and water/sediment dissolved (DP) and particulate (PP) forms. In this example, resuspension and settling of PP dominates the lake's P dynamics

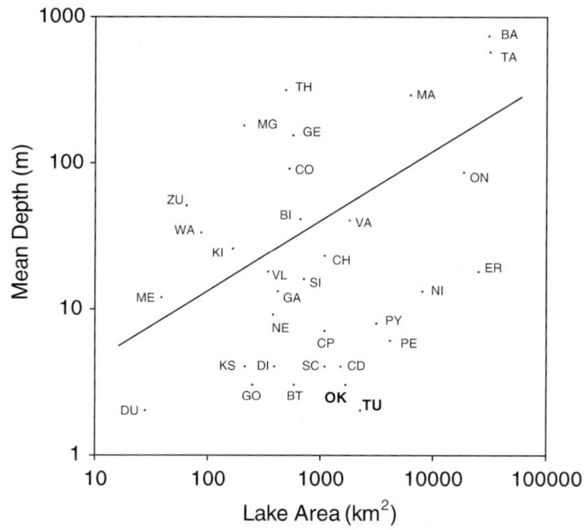

Fig. 2 Large lakes of the world, plotted as a function of their mean depth and surface area. The diagonal line indicates the mean depth for any particular surface area where the dynamic ratio (Hakanson, 1982) equals 0.8. Lakes lying below this line have higher dynamic ratios, and as such, are expected to be increasingly influenced by wind-induced waves. Symbols: Balaton (BT), Constance (CO), Biwa (BI), Chad (CD), Champlain (CH), Chapala (CP), Dianchi (DI), Donghu, (DU), Gatiun (GA), George (GO), Kasumigaura (KA), Kinneret (KI), Malawi (MA), Maggiore (MG), Neagh (NE), Peipsi (PE), Poyang (PY), Saint Clair (SC), Simcoe (SI), Tahoe (TH), Taupo (TZ), Valencia (VL), Vattern (VA), Baikal (BA), Erie (ER), Mendota (ME), Nicaragua (NI), Okeechobee (**OK**), Ontario (ON), Taihu (**TU**), Tanganyika (TA), Washington (WA), and Zurich (ZU). Data were obtained from the World Lake Database of the International Lake Environment Committee (www.ilec.or.jp)

any given area where the dynamic ratio is exactly 0.8. Lakes lying far below the 0.8 line are expected to be more influenced by wind-induced waves and sediment resuspension. Examples of these types of lakes are Okeechobee (OK), Taihu (TU), Dianchi (DI), Balaton (BT), George (GO), Kasumigaura (KS), Saint Clair (SC), Chad (CD), Chapala (CP), and Piepsi (PE). These lakes also fall into the range of large size and low maximal depth that indicates a predominance of random redistribution of sediments by wind-induced waves (Hilton, 1985; Douglas et al., 2002). Results presented in this paper can be most readily generalized to these other large lakes, as well as to smaller lakes with shallower depth.

Qin et al. (2006) describe the processes involved in P flux from sediments to water in Lake Taihu, P.R. China. Lake Taihu is very similar in size and depth to a lake in the USA – Lake Okeechobee, Florida. In this paper, we examine the P dynamics of Lake Okeechobee as another representative case study for large shallow lakes, and focus in particular on how P dynamics vary at different temporal scales in these highly dynamic ecosystems.

Materials and methods

Characteristics of the study site

The physical, chemical, and biological characteristics of Lake Okeechobee have been extensively documented (e.g., Aumen, 1995; James et al., 1995a, b; Havens et al., 1996; Steinman et al., 2002a). Therefore, only a brief overview is provided here except for attributes specifically related to the topic of this paper. The lake has a nutrient residence time of 3.4 years (James et al., 1995a) and a high rate of external P loading (>0.3 g m^{-2} y^{-1}). That loading is substantially higher than a background load of 0.08 g m^{-2} y^{-1} estimated from sediment accumulation rates (Brezonik & Engstrom, 1998). Pelagic TP concentration now averages 110 μg l^{-1}, the ratio of total nitrogen to total P (TN:TP ratio, by mass) is 12:1, the ratio of dissolved inorganic N to soluble reactive P (DIN:SRP ratio, by mass) is 6:1, and cyanobacteria dominate the phytoplankton (Cichra et al., 1995; Havens et al., 2003). The lake is N-limited with surplus SRP in the water (Aldridge et al., 1995; Phlips et al., 1997). Despite N limitation and low N:P ratios, non N-fixing cyanobacteria (*Oscillatoria* and *Lyngbya*) are most common in the plankton because low underwater irradiance restricts the occurrence of N-fixing taxa (*Anabaena*, *Aphanizomenon*) to infrequent calm summer days (Havens et al., 1998, 2003).

The lake bottom is relatively flat and is 60–90 cm above sea level (Fig. 3a). An underwater reef of limestone near the southern end separates the central pelagic zone from shallower south and southwestern regions. On the western side of the lake, there is a relatively steep bottom slope at the edge of a low-gradient littoral marsh that supports

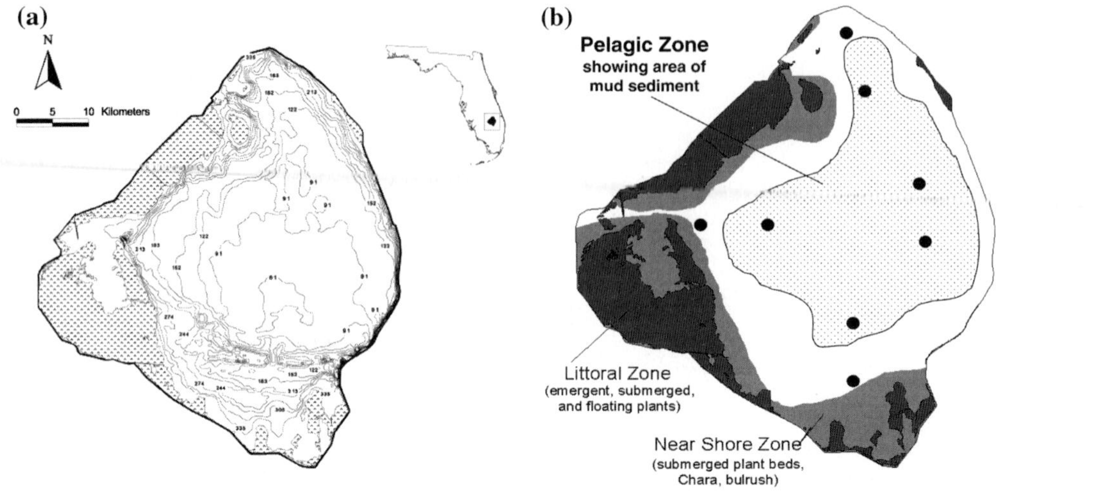

Fig. 3 (**a**) Bathymetric map of Lake Okeechobee, showing bottom elevations in cm above NGVD (National Geodetic Vertical Datum, formerly mean sea level). The lake's littoral marsh, a region of emergent vegetation, is indicated by the stippled area. The inset map indicates the lake's location in Florida, USA. (**b**) Map of Lake Okeechobee, showing three distinct habitats (pelagic, littoral, and near-shore), as well as location of fine mud sediments on the lake bottom and the eight long-term water quality monitoring stations of the South Florida Water Management District (black circles)

emergent vegetation over an area of 400 km^2. The entire lake, including the littoral zone, is enclosed by a dike constructed in the mid-1900s by the United States Army Corps (USACE) of Engineers.

As a result of two factors—water depth and sediment type—there also are three distinct zones (Fig. 4) in the lake that support different communities of primary producers and have markedly different water chemistry. A central pelagic zone is the deepest region of the lake. It is underlain by fine-grained mud sediments (Fisher et al., 2001) that are readily resuspended by wind-induced waves (Jin & Ji, 2001) as in other large shallow lakes (Otsubo & Muraoka, 1987; Luettich et al., 1990; Kristensen et al., 1992; Qin et al., 2006). The water column in this zone has high concentrations of P (Phlips et al., 1995), frequent light limitation of phytoplankton (Phlips et al., 1997), and it does not support vascular plants or benthic algae. Phytoplankton chlorophyll *a* concentrations do not correlate with TP (Havens et al., 1999) because most of the P is associated with resuspended sediment particles (details below). In sharp contrast, TP concentrations in the interior littoral zone are frequently below 10 μg l^{-1} (Hwang et al., 1998) because the main source of water input for this zone is rainfall. Little transport of nutrients occurs between the pelagic and littoral zones except when water level is very high (Havens, 1997). Lying between the pelagic and littoral zones is an area of intermediate depth, turbidity, and nutrient enrichment (Hwang et al., 1998; Havens & Walker, 2002). This region is referred to as the near-shore zone (Havens, 2002). When the lake experiences prolonged high water, as in the late 1990s, the near-shore zone becomes dominated by phytoplankton (Havens et al., 2001a). However, under shallow conditions it develops a dense community of submerged plants and clear water (Havens, 2003). The relationship between chlorophyll *a* and TP in the near-shore zone is similar to what is found in deeper Florida and North America lakes (Mazumder & Havens, 1998).

Data sources

This paper synthesizes prior work dealing with P dynamics in Lake Okeechobee, and aims to bring this information together in a manner that allows for comparisons of processes occurring at different temporal scales. Citations are provided to the original published sources of the information, where details can be found for field sampling,

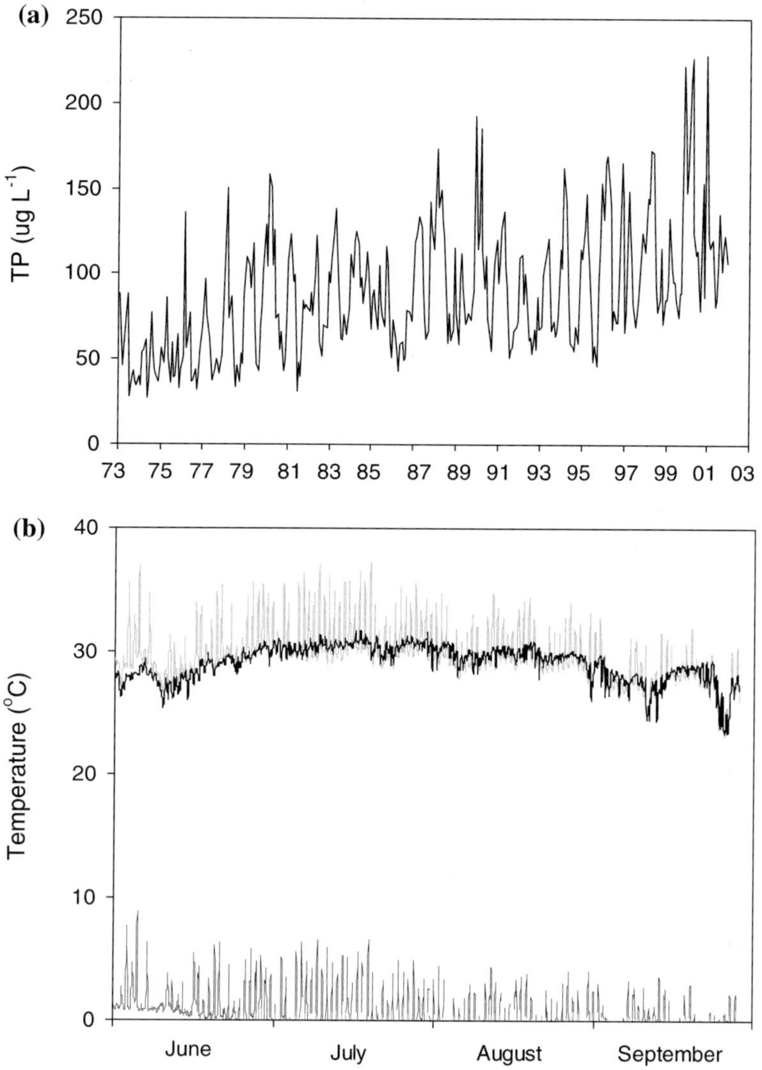

Fig. 4 (**a**) Historical record (1973–2002) of pelagic total phosphorus (TP) concentrations in Lake Okeechobee, based on monthly data collected at eight stations shown in Fig. 3. (**b**) Representative data regarding water temperatures near the surface (upper grey trace) and at the bottom (black trace) of the mid-lake water column in Lake Okeechobee. These are hourly data collected during summer 1999. The lower data set represents the difference between top and bottom temperature

laboratory methods, and experimental protocols. Only abbreviated information is presented here, coincident with presentation of the particular results and conclusions of each study. Most of the data were collected in the pelagic zone where the South Florida Water Management District (SFWMD) conducts monthly or semi-monthly water quality assessment at eight stations, which have existed since 1973 (Fig. 3b). Data from the near-shore zone, which has been sampled since the mid-1980s, also are presented, along with results from studies done by the University of Florida, under contract with the SFWMD. The littoral zone is not addressed here because it typifies a wetland more than a lake ecosystem. All of the water quality sampling and analyses carried out by the South Florida Water Management District (SFWMD), and the agencies and universities working under contract to the SFWMD, follows standard quality control standards that have been accepted by the State of Florida (SFWMD, 1998). Unless otherwise indicated, statistical significance is defined in this paper at $P = 0.05$.

Results

Greater than 30 years of data indicate that temporal dynamics in pelagic TP in Lake Okeechobee occur at seasonal, yearly, multi-year, and multi-decadal time scales (Fig. 4a). Considering first the short-term dynamics, we examined the hypothesis of Moore & Reddy (1994) that large fluxes of P from lake sediments might occur during brief periods of anoxia associated with transitory thermal stratification. During 1997–1999 we measured water temperature and dissolved oxygen (DO) at surface and bottom depths every 30 min with a YSI multi-parameter sonde. Representative data (Fig. 4b) for temperature indicate that in summer daily thermal stratification does occur, forming in morning and breaking up in the afternoon. However, anoxia never was observed (data not shown), with 95% of DO measurements exceeding 5 mg L^{-1} in the study period. These data do not support a major role for anoxia in stimulating sediment P release in this lake.

At the seasonal time scale, there is a regular pattern in water column TP (Fig. 5) that closely tracks seasonality of wind velocity. Concentrations of TP are highest in winter, when prevailing winds are strongest and lowest in summer when winds are relatively weak. The exception to this pattern is when hurricanes pass close to the lake, as occurred in October 1999 when a Category I storm crossed the Everglades (Fig. 6) and produced maximal sustained winds of 90 km h^{-1} over the center of Lake Okeechobee.

Results from the Lake Okeechobee Ecosystem Model (LOEM, Jin et al., 2000, 2002) indicate that the hurricane generated currents as strong as 50 cm s^{-1} over a large portion of the lake, and that large amounts of mud sediment were suspended from the center of the lake and transported to shoreline areas (Fig. 7). Field observations confirmed that fine mud sediment was deposited into bays in the south and west

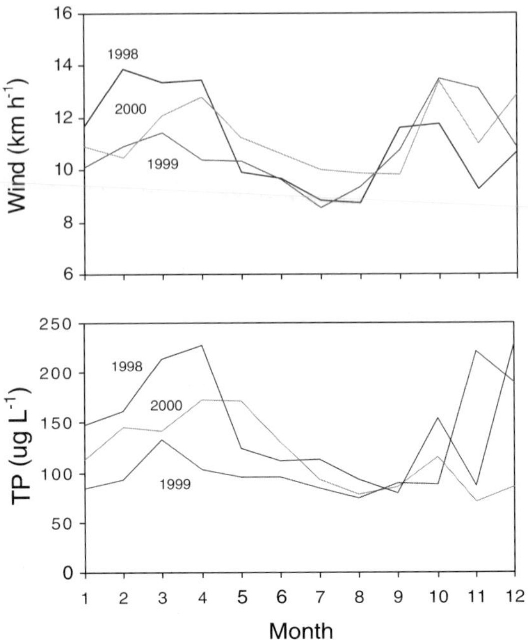

Fig. 5 Representative data regarding wind velocity at mid-lake and pelagic mean total phosphorus (TP) from the eight stations shown in Fig. 3. The three selected years are 1999, 2000, and 2001

regions of the lake where prior to the storm the sediments were comprised of peat and sand, respectively. Water quality monitoring data indicate that the hurricane resulted in lake-wide increases of TP from an average of 88 $\mu g\ l^{-1}$ before the storm to 222 $\mu g\ l^{-1}$ afterwards (Fig. 8).

Year-to-year variation in TP in Lake Okeechobee is quite pronounced and some of the variation can be explained by changes in water level (Fig. 9). A long-term increasing trend in water column TP (Fig. 10) coincides with increasing inflow TP from 1973 to 1985, but then contrasts with decreasing inflow TP afterwards.

Discussion

Pelagic phosphorus dynamics at hourly time scales – effects of wind and diffusion

In lakes where wind-generated waves and circulation are major factors controlling pelagic P dynamics, large changes in TP concentration can occur in short time intervals. Kristensen et al.

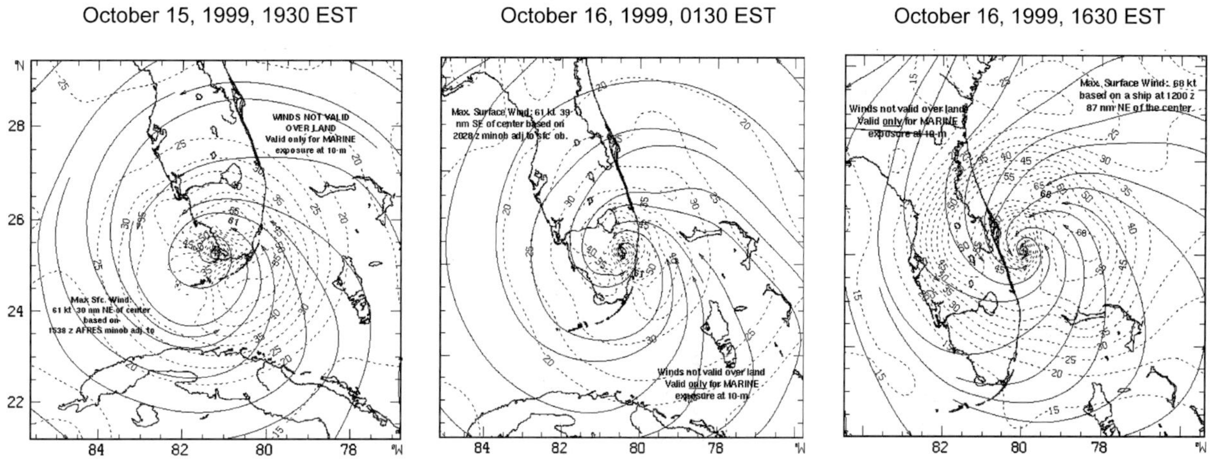

Fig. 6 Graphics from the National Hurricane Center, showing the movement of Hurricane Irene across the Florida peninsula, from 1930 h on October 15 to 1630 h on October 16, 1999. These figures are presented to illustrate the location of the eye of the storm as it moved south of Lake Okeechobee, as well as the wind field patterns. As indicated on the panels, the specific wind velocities are not valid over land

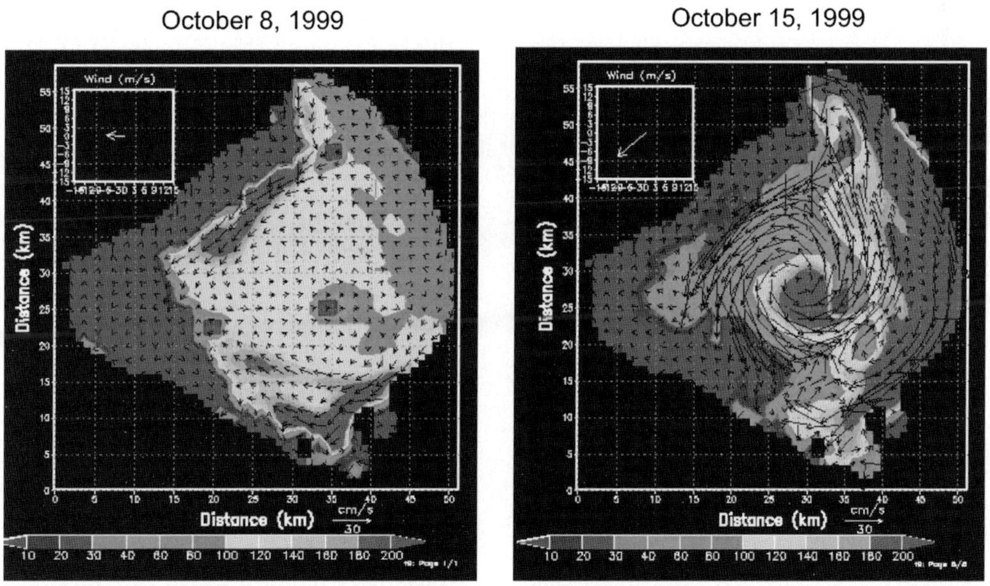

Fig. 7 Results of the Lake Okeechobee hydrodynamic and sediment transport model (Jin et al., 2000, 2001; Jin & Ji, 2003), showing surface current velocity vectors (black arrows) and resuspended sediment concentrations (grey scale) on October 8, before Hurricane Irene, and October 15, 1999, during the onset of the storm. The inset panel indicates wind velocity and direction recorded at the meteorological station at mid-lake

(1992) evaluated changes in TP in Lake Arreso, Denmark (mean depth 2.9 m) and documented increases of 100–400 μg P l^{-1} over time periods less than 12 h. Large increases in TP followed short-lasting periods of increased wind velocity with rapid declines in TP concentration after the winds had subsided. Similarly, Hamilton & Mitchell (1997) examined TP concentration on calm vs. windy days in seven lakes in New Zealand (mean depth < 1.5 m) and estimated that daily variations of TP were as great as 700 μg P l^{-1}. Sub-daily data are not available for Lake Okeechobee.

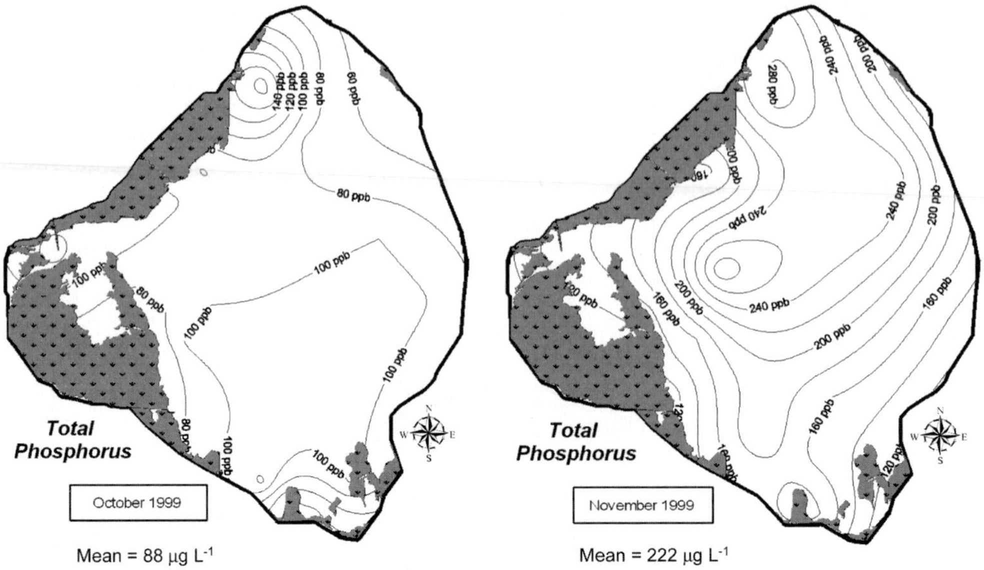

Fig. 8 Total phosphorus (TP) concentrations in the open water regions (pelagic and near-shore zones) of Lake Okeechobee, during October 1999, two weeks before Hurricane Irene, and November 1999, two weeks after the storm. Contours of TP concentration were interpolated from point data using Surfer® Software.

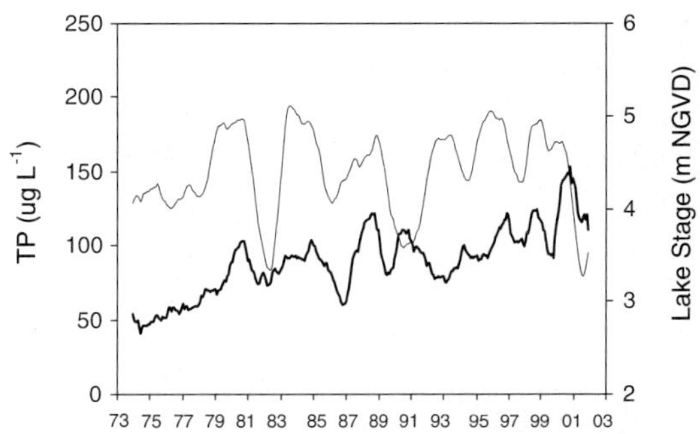

Fig. 9 Pelagic total phosphorus (TP) concentrations and stage in Lake Okeechobee, shown as 12-month rolling averages from 1973 to 2002

However, correlations exist between total suspended solids, vertical attenuation coefficient (K_o) and TP (Reddy et al., 1995; Hanlon et al., 1999; Sheng, 1999), allowing for an estimated occurrence of wind-induced high TP events. During summer, wind velocities are maximal in afternoon (coinciding with thunderstorm activity) and again in the evening due to winds off the Atlantic Ocean. Hanlon et al. (1999) found large increases in K_o (from below 2 up to as high as 9 m^{-1} and lasting for just 1–2 h) within 2 h of maximal wind occurrence. Output from the LOEM indicate a similar time lag in Lake Okeechobee (Jin et al., 2002). The predicted time lag does not appear to be associated with energy transfer to the lake bottom, which occurs within 30 minutes or less. Rather, it seems to be associated with vertical transport of fine sediment particles to the upper water column. Sheng (1993) estimated that wind events of 30 km h^{-1} could increase water column TP by as much as 180 μg P l^{-1} within approximately the same period of time as estimated by the LOEM.

Wind-induced sediment resuspension is estimated to transport 6–18 times the amount of P to the water column as diffusive flux in Lake Okeechobee (Sheng, 1993; Moore & Reddy, 1994), and up to 6 times the amount of P found in external loads, when the data are compared on a yearly basis. This compares with the estimate by Qin et al. (2006) that TP loading due to resuspension in Lake Taihu is 5–10 times the external load. However, Reddy et al. (1995) noted that the amount of SRP released to the water column in a typical diurnal event is only a small fraction of the TP that is resuspended in a shallow lake. They concluded that the increase in SRP is limited by the high Fe content of this lake's sediments, and their low equilibrium P concentration, which

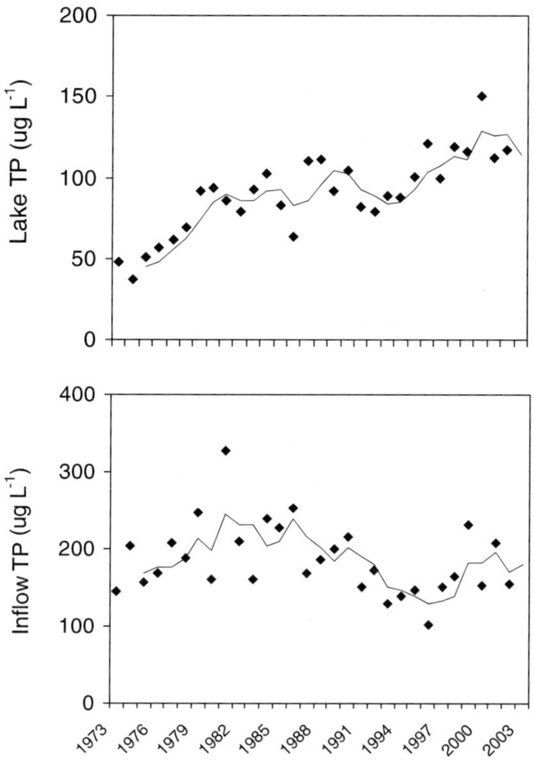

Fig. 10 Monthly mean concentrations of total phosphorus (TP) for Lake Okeechobee pelagic zone and inflows, from 1973 to 2002. The line indicates a long-term trend in the data for pelagic TP. The inflow data do not display a linear trend

favors sorption over desorption on sediment particles in the water column. Not all lakes may behave in this manner. Sediment resuspension may be a source, a sink, or have no effect on water column SRP depending on sediment geochemistry and other factors (e.g., Andersen, 1974; Peters & Cattaneo, 1984; Sondergaard et al., 1992). An opposite extreme from Lakes Okeechobee and Taihu is Lake Arreso, Denmark, where Sondergaard et al. (1992) documented a large net transport of SRP into the water column during resuspension events (20–30 times the SRP release from undisturbed sediments). They concluded that this massive SRP release occurred because (1) a large portion of sediment P in Lake Arreso occurred in a loosely-adsorbed organic form; (2) the sediments had a low ratio of Fe to P; and (3) there was a high equilibrium P concentration (50 to 180 μg l^{-1}). In contrast, Lakes Okeechobee and Taihu have high Fe:P ratios in their sediments, so that desorption of SRP during resuspension is not favored (Moore & Reddy, 1994; Qin et al., 2006).

Sondergaard et al. (1992) described a process whereby the combination of SRP desorption from resuspended sediment particles and rapid uptake of SRP by phytoplankton acts like a "pump" for transferring P into the water column of Lake Arreso. In that lake, light availability in the mixed layer is sufficient to allow a high biomass and productivity of phytoplankton to occur. In Lake Okeechobee, phytoplankton is severely light limited (Phlips et al., 1997) in the pelagic zone, has a low biomass, and is less likely to have this impact on P dynamics. Hence, while generalizations may be made about effects of wind resuspension on TP in large shallow lakes, the effects on SRP cannot be predicted without information regarding such factors as sediment biogeochemistry and mean irradiance in the water column.

Hourly variations of TP concentration in Lake Okeechobee are most pronounced in summer when there are large-scale changes in wind velocity between late morning (low wind) and afternoon-evening (higher wind, due to thunderstorms). During winter, when convective thunderstorms do not form around the lake and daily variations in wind from the ocean are less pronounced, regional low versus high pressure systems generally determine wind velocities over the lake. In that case it is common to have periods lasting for two or more days with strong sustained winds from a particular direction. As a result, the concentration of TP is more likely to remain elevated over multiple days, with less intra-daily variation.

Passive diffusion of P from lake sediments to the water column also can be important in shallow eutrophic lakes, as in deeper systems. However, when Fe chemistry regulates P release, the flux will depend on the occurrence of calm conditions of a sufficient duration to allow thermal stratification and bottom water anoxia. In profiles of intact sediments from Lake Okeechobee, soluble reactive P increased from near 0.1 mg l^{-1} at the sediment-water interface to 1.1 mg l^{-1} at sediment depths below 10 cm (Moore & Reddy, 1994). The authors concluded that large fluxes of P from the sediment could occur if the water column were to experience low dissolved oxygen levels, but as shown here that is unlikely to occur on a regular basis. Thus a relatively low rate of P release under oxidized conditions dominates the diffusive release of P to the water column in this lake. Similar results were obtained for Lake Taihu, where Qin et al. (2006) note that "constant dynamic disturbance frequently transports oxygen to the sediment-water interface" such that variations in dissolved oxygen and Eh are quite small.

Although diffusive P release may be small compared to the P transport associated with sediment resuspension, it still could play an important role in providing readily available P to the water column (Moore & Reddy, 1994). The effects on water quality then will depend on whether there are adequate light conditions for net phytoplankton growth during or subsequent to the resuspension event.

The results regarding P dynamics at hourly to daily time scales in Lake Okeechobee also are consistent with the findings of Kristensen et al. (1992), who concluded that in the shallow Danish lake (Lake Aresso, area 41 km^2, mean depth 3 m), the input of P to the lake water is dominated by resuspension. The same result is found in Lake Kasumigaura, Japan, where despite high rates of P release by diffusion (Ishikawa & Tanaka, 1993), the ratio of TP supply to the water column from resuspension vs. diffusion is 10:1 (Havens et al., 2001b). This is similar to the ratio estimated for Lake Okeechobee.

Seasonal dynamics of pelagic phosphorus

As a result of seasonal variation in wind velocity, and its influence on duration and magnitude of sediment resuspension (Jin & Ji, 2001), a predictable seasonal pattern exists for TP in Lake Okeechobee. Phlips et al. (1995) identified this seasonal pattern using contour maps of winter vs. summer TP content based on four years of integrated water samples from 84 locations in the lake. In winter, TP ranged from 50 to 110 $\mu g\ l^{-1}$, while in summer the range was between 30 and 80 $\mu g\ l^{-1}$. We found that in any given year, over 70% of the monthly variation in pelagic mean TP could be explained by wind velocity.

The response of phytoplankton to the winter increase in pelagic TP is minimal because most of that P is associated with abiotic seston particles. Those particles attenuate light to the extent that the algae are light limited during most of the winter months, when there is complete mixing of the water column (Aldridge et al., 1995; Phlips et al., 1995). One might consider that lower water temperatures and shorter photoperiod in winter also could limit phytoplankton biomass. However, this is unlikely because chlorophyll *a* concentrations in near-shore regions not influenced by sediment resuspension often are maximal during winter. That differing seasonality has been attributed to a reduced competition with plants and periphyton for water column phosphorus during winter months (Phlips et al., 1993).

Hurricane effects on pelagic phosphorus

In regions of the world that are influenced by tropical storms (hurricanes and cyclones), these extreme and unpredictable events may have strong impacts on P dynamics of lakes, yet those impacts are little studied. A major limitation is lack of substantive data prior to and after the event, which reflects the unpredictable timing and location of these severe storms.

There is evidence that a tropical storm contributed to a large decline in submerged plants in Chesapeake Bay in the 1970s and accompanying changes in water quality (Bayley et al., 1978), and it has been suggested that a hurricane caused the large-scale loss of plants and development of dense phytoplankton blooms in Lake Apopka, Florida in the 1940s (Clugston, 1963; Bachmann et al., 1999). We documented major impacts of a Category I hurricane in 1999 on sediment suspension and transport and water column TP in Lake Okeechobee.

When wind velocities decline abruptly, as they did after the 1999 hurricane passed beyond Florida, strong currents persist in the lake as residual circulation from the previous wind forcing event (Jin & Ji, 2001). Hence, sediment particles may be transported long distances from where they are resuspended, in particular if they are very fine-grained like the mud sediments that predominate in the center of Lake Okeechobee.

The modeling scenario illustrated here (Fig. 7) indicated that over 10,000 metric tons of fine-grained mud sediment was uplifted during the storm. Concentrations of TP remained high in the lake for two months after the hurricane, and displayed a second large increase (to above 230 $\mu g\ l^{-1}$) when another wind storm passed over the lake in late winter 2001.

During the last century, the Florida peninsula experienced landfall by a named tropical storm, on average, once per year, and landfall by a hurricane (wind velocity >120 km h^{-1}) every 2–3 years. A major hurricane (wind velocity >180 km h^{-1}) impacted the peninsula every 5–6 years. Wind velocities decline as a tropical storm moves over land, so sustained winds over the lake in any given storm might be considerably lower than at the time of landfall. Nevertheless, Lake Okeechobee is located in a position that has a high probability for tropical storm impacts. This also may be true for some large shallow lakes in lowland areas of Asia. Given that Hurricane Irene had low wind velocities for a tropical storm and did not pass directly over Lake Okeechobee, the impacts summarized here may be in the lower range of what can occur when a tropical storm passes over a large shallow lake. In August and September 2004, for example, three major hurricanes passed over the Florida peninsula, all causing substantial storm surges and resuspension events in Lake Okeechobee. We are in the process of analyzing the impacts, and will report those results in a future publication.

Of additional interest is the expected influence of long-term trends in ocean currents on future hurricane frequencies. During the period from 1995 to 2000, the Atlantic Ocean experienced the highest activity of hurricanes ever measured, and this has been linked with a warm phase of the Atlantic Ocean Thermohaline Current (AOTC), which is projected to last 20–40 years (Landsea, 2001). Thus the P dynamics of a lake impacted by hurricanes may be influenced by environmental changes occurring at a much greater spatial scale than the lake's own watershed. While limnologists have become attuned to the need to manage lakes in a watershed context, an even broader context may sometimes be required.

Inter-annual variations in pelagic phosphorus

Canfield & Hoyer (1988) suggested that the variation in yearly mean TP of the lake's pelagic zone was due to variations in water level ($r = 0.70$, 1973–1984). Havens (1997) confirmed that a statistical relationship existed for the years 1973–1995, albeit with a lower correlation ($r = 0.50$) than in Canfield & Hoyer's data set, and he summarized the various mechanisms that might explain the correspondence between lake stage and TP. Four hypotheses have been proposed: (1) littoral flooding occurs at high stage, and is followed by P release from dead plants (Canfield & Hoyer, 1988); (2) increased horizontal transport of resuspended sediment occurs when water levels are high (Maceina, 1993); (3) greater uptake of P by submerged plants and periphyton occurs when water levels are low (Phlips et al., 1993); and (4) co-variation of wind and water level occurs on a yearly basis, but wind actually is the driving factor (Maceina & Soballe, 1990) and water level only is a covariate, not a causal factor.

The first hypothesis is rejected based on several lines of evidence (Havens, 1997). The lake's littoral zone is nutrient-poor compared to the pelagic, and taken together, the biomass of all submerged and emergent plants and periphyton in the littoral and near-shore zones contain just 0.8% of the lake's P, compared to 97.1% in the sediments (Steinman et al., 1999). Dierberg (1992) and Harris et al. (1995) carried out decomposition experiments to quantify P release from submerged and emergent vegetation, respectively, in this lake. They concluded that the potential for increased TP in the pelagic zone due to P release from dead plants was on the order of 3–11 $\mu g\ l^{-1}$ much lower than the year-to-year variation in TP that has been observed with rising and falling lake water levels. Therefore, we focus here on the other three hypotheses, which seem more likely to be responsible for the TP variability at an intermediate time scale.

Water level, phosphorus, and horizontal mixing

Maceina (1993) suggested that the bottom topography of Lake Okeechobee, in particular the limestone reef near the southern and south-western area, influences horizontal transport of sediment particles that are resuspended by wind-induced waves at mid-lake. He further suggested that at high lake stage (surface elevation > 4.9 m mean sea level) sediments could readily be transported over the top of the reef, whereas at low water levels (surface elevation < 4.6 m) this transport would be partially blocked. As a result, the lake-wide average TP, which is based on eight stations, two of which occur in the near-shore zone, would be higher at higher lake stages. To support this proposed mechanism, Maceina (1993) showed that the difference between pelagic and near-shore TP concentration declined with increasing lake water level, i.e., the lake became more homogenized. We examined this relationship using recent data from the lake, including data from the eight long-term stations, plus data from 30 stations located in the near-shore zone that have been sampled since 1986. The result was consistent with what Maceina (1993) observed. In the pelagic zone, TP increased only slightly with increasing lake stage, whereas in the near-shore zone, TP increased much more rapidly, from TP near 40 $\mu g\ l^{-1}$ at stage <4 m to TP near 80 $\mu g\ l^{-1}$ at stage ~5 m. As a result the difference in TP concentration between pelagic and near-shore zones dropped from near 60 $\mu g\ l^{-1}$ at low stage to near 20 $\mu g\ l^{-1}$ at high stage. The two pelagic stations located furthest south and west (Fig. 3b) respond to stage in a similar manner to the near-shore stations, and are largely responsible for the slight decline in TP at lower stage when the eight-station pelagic network is averaged.

In addition to the empirical relationships, hydrodynamic modeling (Sheng, 1993) supports the concept of greater horizontal mixing of the lake at high stage. Sheng concluded that "during low lake stage, wind-driven circulation gyres are generally confined to the open water (pelagic) zone, with little mixing into the vegetation (near-shore) zone." The LOEM gave a similar result when it is run under low vs. high stage conditions, with all other factors (e.g., wind velocity) being held constant.

In this particular lake it is important to understand why the near-shore zone displays a strong correspondence between TP and lake stage. That zone includes drinking water intakes for several municipalities, it is the main location of recreational fishing, and it provides critical vegetative habitat for fish, wading birds, and other wildlife (Havens, 2002). Hence, changes in TP concentration of the near-shore zone have a greater potential to impact ecological and societal values of the lake than changes in TP of the pelagic zone. This situation likely exists in other large shallow lakes.

Water level, phosphorus, and submerged plants

Another explanation for the reduced lake-wide TP at lower stage is that submerged plants and algae sequester large amounts of P from the water column, either by direct uptake (Burkholder et al., 1990; Hansson, 1990) or other processes associated with dense plant beds, including increased particle settling (Kufel & Kufel, 2002), sediment stabilization (Vermaat et al., 2000), and co-precipitation of P with calcium at high pH (Murphy et al., 1983). In Lake Okeechobee, it has been established that the biomass and spatial extent of submerged plants and algae is controlled by water level (Steinman et al., 2002b; Havens, 2003). In the late 1980s, following a drought and low lake stage, submerged plants covered between 18,000 and 21,000 hectares of the near-shore zone (Zimba et al., 1995), and nutrient-poor water and low biomass of phytoplankton coincided with seasonal maxima of submerged plants (Phlips et al., 1993). As noted above, this generally happens in winter. From 1995 to 1999, water levels in the lake were high, and the spatial extent of submerged plants declined to below 2,000 ha. Water in the near-shore zone displayed high TP concentrations and was extremely turbid (Havens et al., 2001a). During 2000–2001, another drought resulted in record low water levels in the lake, and the spatial extent of submerged plants again increased, to over 17,000 hectares (Havens et al., 2002). Havens (2003) found that the average biomass of submerged plants in the near-shore zone could be predicted from water depth and non-volatile suspended solids.

These results demonstrate correlation between plants and clear water, but cannot tell us whether plants are responsible for the improved water quality, or were simply responding to clear water caused by some other processes. Even if plants have local impacts on water quality, it is important to determine the spatial scale at which they influence TP in such a large lake, where much of the bottom (due to soft mud sediments) is not suitable for their colonization. At their maximal spatial extent in 2002, submerged plants covered just 13% of the combined surface area of the pelagic and littoral zones. It therefore is unlikely that they would substantially control TP concentration at a lake-wide scale. The vegetation might influence the 8-station average calculated in the long-term monitoring program, because 2 of those stations occur in areas that can support plants, but this effect on 25% of the stations should be limited. Only if there are strong currents passing through the near-shore zone would submerged plants be expected to influence lake-wide TP, and as discussed, water and nutrient transport to the near-shore zone is reduced when lake stage is low and plants are at their greatest density (Sheng, 1993). Submerged plants in Lake Okeechobee likely have a localized effect that somewhat dampens seasonal and year-to-year variations of TP, but physical processes are of considerably greater importance. Considering data from a seasonal survey program for submerged plants in Lake Okeechobee (Havens, 2003), there is a synchronized rise and fall in water depth, non-volatile solids (indicative of inorganic sediments), and phytoplankton chlorophyll *a* between 1999 and 2002 (Havens et al., 2004). This pattern occurs both at sites with and without submerged plants.

Wind and phosphorus at a yearly time scale

Wind has been identified as the main driving force behind P dynamics at hourly, daily, and seasonal time scales, and it also may be the main factor influencing year-to-year variations in pelagic TP. Maceina & Soballe (1990) examined this relationship, using pelagic TP data from 1975 to 1988. They found that yearly-averaged TP and stage were correlated ($r = 0.61$), but noted that the correlation between TP and wind velocity was stronger ($r = 0.78$). Havens (1997) examined a longer dataset (1973–1995) and obtained similar results (TP vs. stage, $r = 0.50$; TP vs. wind, $r = 0.64$).

We cannot identify the causal mechanisms behind year-to-year variations in TP based on field observations and correlations, and it is not possible to perform controlled experiments at a scale that is relevant to what occurs in the >1,000 km^2 pelagic zone of this lake. Furthermore, simple empirical models cannot effectively predict temporal changes in water column TP, chlorophyll *a*, and other attributes. In the case of Lake Okeechobee, our approach to understanding the driving factors for variations in TP and reasonably predicting future trends has been to develop the LOEM. Similar tools are used for management of large shallow coastal systems, such as the Chesapeake Bay (Cerco & Cole, 1994) and Tampa Bay, Florida (Yassuda & Sheng, 1997), and in other large shallow lakes including Lake Taihu.

Phosphorus dynamics over multiple decades

Underlying the seasonal and year-to-year variation in pelagic TP of Lake Okeechobee is a 30-year trend of increasing concentrations. This increase is approximately 20 μg l^{-1} per decade ($P < 0.01$, seasonal Kendall's Tau test). The minima, maxima, and central tendency of the data all display a systematic increase, and a non-linear pattern can be seen in the smooth curve fit to these data. Paleolimnological data (Brezonik & Engstrom, 1998) indicate a trend of increasing P deposition rate in the lake that extends back as far as 1910.

The trend in pelagic TP documented during the last 30 years may be a result of several factors, including: (1) substantial changes in the TP concentration of water flowing into the lake; (2) a delayed response of the lake to recent reductions in TP input due to P recycling from sediments; and (3) reduced P assimilation by the ecosystem due to saturation effects. The concentration of TP in water flowing into the lake was near 150 μg l^{-1} in the early 1970s, increased to over 200 μg l^{-1} in the early 1980s, and then declined back to near 150 $\mu g\ l^{-1}$ in the early 1990s. Two recent years (1999 and 2001) are outliers, perhaps reflecting effects of (1) intense runoff during and after the hurricane in 1999, and (2) flushing of the watershed after a 2-year drought in 2001. Otherwise, there has not been any upward or downward trend in inflow TP during the last decade.

Declining inflow TP concentration after the 1980s has been attributed to an array of P management programs that were implemented in the watershed (Flaig & Havens, 1995; Havens et al., 1996). Coincident increases in lake water TP from the 1970s to 1980s, and a plateau of lake water TP from the 1980s to 1990s, are typical response patterns for shallow eutrophic lakes with high internal recycling of P (Sas, 1989). Concentrations of TP in the water column increase with increasing external inputs, but then remain high because the surface sediments (the source of internal P loading) reflect the past loading conditions until they are buried by new sediment material with a lower concentration of P. The sediment accrual rate in Lake Okeechobee is ~0.1 cm y^{-1} with the upper 5 cm of sediments interacting with the overlying water column through diffusion and resuspension (Reddy et al., 1995). As a result, the lake response time to changes in external P inputs is expected to be on the order of decades. For predicting response time of other shallow lakes to P load reductions, deposition rate and depth of active sediment layer are of particular importance, yet they are difficult to determine. Estimating net sediment deposition rate, for example, requires paleolimnological analysis and undisturbed sediment cores (Brezonik & Engstrom, 1998). In large shallow lakes that are heavily influenced by wind, it is important to confirm that sediments preserve an undisturbed historic record before using profiles to estimate process rates or to infer past conditions. In some lakes, this may not be the case. Depth of the active layer also can be influenced by bioturbation of macroinvertebrates (Van Rees et al., 1996), and in some lakes with reduced P loading, SRP migrates from deeper sediment layers as surface sediment P content is depleted through diffusion (Sondergaard et al., 2001).

There also is evidence that the sediments of Lake Okeechobee have become saturated with P over recent decades, and that this may contribute in part to the lake's poor response to reduced P inputs. Fisher et al. (2001), for example, documented that porewater SRP concentration in the lake's surface sediments increased significantly between 1988 and 1998, suggesting a saturation of binding sites. Further, the P mass balance for Lake Okeechobee indicates that the net P sedimentation rate declined four-fold from the 1970s to the 1990s (Havens & Schelske, 2001). This apparent P saturation led resource managers to consider sediment treatment, with calcium or alum, as a way to help speed recovery of the lake. However, due to the large size of Lake Okeechobee, its extreme shallow depth, and a variety of other engineering, ecological, and economic issues, this active management was determined to be non-feasible. Smaller lakes may be more suitable for this type of management (e.g., Cooke et al., 1993).

Conclusions

A substantial portion of the global storage of surface freshwater occurs in large shallow lakes (Herdendorf, 1984). These lakes typically are located in lowland areas (Fig. 11) that are readily used for agriculture and other human activities. As such, large shallow lakes are likely to receive substantial inputs of non-point source agricultural pollution, as well as solids from soil erosion in upstream watersheds. Unless they are rapidly flushed, their sediments become natural sinks for accumulation of P, nutrients, and other pollutants (Havens & Steinman, 1995). Based on the information presented in this case study of Lake Okeechobee, the following conclusions are arrived at regarding large shallow lakes, with emphasis on lakes that have a dynamic ratio substantially greater than 0.8 and occur in coastal areas influenced by sea breeze winds and/or tropical storms.

(1) Physical processes of sediment resuspension and horizontal circulation dominate ecosystem P dynamics.

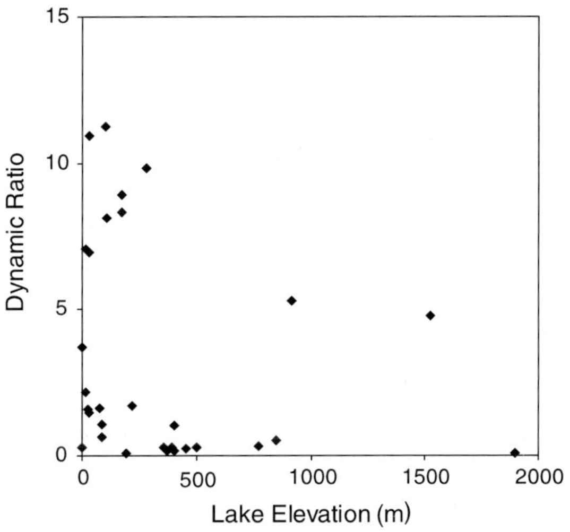

Fig. 11 Dynamic ratio of large lakes (>500 km^2) plotted as a function of their elevation above sea level. The dynamic ratio, calculated as the square root of surface area divided by mean depth, indicates a lake's susceptibility to wind effects. Ratios above 0.8 indicate high susceptibility (Bachmann et al., 2000). This figure illustrates that large lakes with high dynamic ratios occur in lowland regions of the world. The lakes plotted here are from the same dataset used to generate Fig. 2

(2) Rapid increases in P concentration can occur at time scales from days to weeks, and at a magnitude that is considerably greater than at longer time scales.
(3) The potential to modify trophic state by manipulating the food web is minimal because physics overwhelms biology in terms of controlling P dynamics, light availability, and other key attributes in this ecosystem.
(4) One exception is that by controlling water level, there might be potential to influence TP by promoting growth of submerged aquatic vegetation. However, depending on sediment type and bottom profile, this has a variable influence on lake-wide TP.
(5) Responses of lake water TP concentration to external P load reduction are significantly delayed by internal recycling, as indicated by Sas (1989), and this situation may be exacerbated where lakes are influenced periodically by tropical storms.
(6) Simple empirical models are not effective as tools for eutrophication management in lakes of this type. Resource managers will require more complex models that take into consideration lake hydrodynamics, with wind as a major external driver.

The emphasis in limnology clearly has shifted from traditional studies of temperate dimictic lakes to a more holistic consideration of the variety of lake types in the world (e.g., see Wetzel, 2001). Large shallow lakes, in particular, have received attention in special symposia in Europe (e.g., Van Liere & Gulati, 1992), and now in Asia (this special issue). Such lakes contain a large percentage of the world's fresh water, provide ecosystem services to hundreds of millions of people, and are located in areas prone to anthropogenic and climactic impacts. An understanding of their main driving factors, especially those related to limiting nutrients such as P, is critical to their protection. A challenge for the future is how to most effectively generalize results of research on lakes such as Okeechobee and Taihu, where funding for research and numerical model development has been favorable, to help address the water quality issues in similar ecosystems at locations where that level of investigation has not been possible.

Acknowledgements The authors are grateful to Peter Doering, Susan Gray, Garth Redfield, and two anonymous reviewers for providing comments on an earlier version of this paper.

References

Aldridge, F. J., E. J. Phlips & C. L. Schelske, 1995. The use of nutrient enrichment bioassays to test for spatial and temporal distribution of limiting factors affecting phytoplankton dynamics in Lake Okeechobee, Florida. Archiv für Hydrobiologie, Advances in Limnology 45: 177–190.

Andersen, J. M., 1974. Nitrogen and phosphorus budgets and the role of sediments in six shallow Danish lakes. Hydrobiologia 74: 528–550.

Aumen, N. G., 1995. The history of human impacts, lake management, and limnological research on Lake Okeechobee, Florida (USA). Archiv für Hydrobiologie, Advances in Limnology 45: 1–16.

Bachmann, R. W., M. V. Hoyer & D. E. Canfield Jr., 1999. The restoration of Lake Apopka in relation to alternative stable states. Hydrobiologia 394: 219–232.

Bachmann, R. W., M. V. Hoyer & D. E. Canfield Jr., 2000. The potential for wave disturbance in shallow Florida lakes. Lake and Reservoir Management 16: 281–291.

Bayley, S., V. Scotts, P. F. Springer & J. Stennis, 1978. Changes in submerged aquatic macrophyte populations at the head of Chesapeake Bay, 1958–1975. Estuaries 1: 171–182.

Brezonik, P. L. & D. R. Engstrom, 1998. Modern and historic accumulation rates of phosphorus in Lake Okeechobee, Florida. Journal of Paleolimnology 20: 31–46.

Burkholder, J. M., R. G. Wetzel & K. L. Klomparens, 1990. Direct comparison of phosphate uptake by adnate and loosely attached microalgae within an intact biofilm matrix. Applied and Environmental Microbiology 56: 2882–2890.

Canfield, D. E. Jr. & M. V. Hoyer, 1988. The eutrophication of Lake Okeechobee. Lake and Reservoir Management 4: 91–99.

Cerco C. F. & T. Cole, 1994. User's guide to the CE-QUAL-ICM three-dimensional eutrophication model. Technical Report EL-95-15, Waterways Experiment Station, US Army Corps of Engineers, Vicksburg, Mississippi, USA.

Caper, G. L. & R. W. Bachmann, 1984. Wind resuspension of sediments in a prairie lake. Canadian Journal of Fisheries and Aquatic Sciences 41: 1763–1767.

Cichra, M. F., S. Badylak, N. Henderson, B. H. Reuter & E. J. Phlips, 1995. Phytoplankton community structure in the open water zone of a shallow subtropical lake (Lake Okeechobee, Florida, USA). Archiv für Hydrobiologie, Advances in Limnology 45: 157–175.

Clugston, J. P., 1963. Lake Apopka, Florida, a changing lake and its vegetation. Florida Academy of Science 26: 168–174.

Cooke, G. D., E. B. Welch, S. A. Peterson & P. R. Newroth, 1993. Restoration and management of lakes and reservoirs. Lewis Publishers, Florida, USA.

Dierberg, F. E., 1992. The littoral zone of Lake Okeechobee as a source of phosphorus after drawdown. Environmental Management 6: 371–380.

Douglas, R. W., B. Rippey & C. E. Gibson, 2002. Interpreting sediment trap data in relation to the dominant redistribution process in a lake. Archiv für Hydrobiologie 155: 529–539.

Fisher, M. M., K. R. Reddy & R. T. James, 2001. Long-term changes in the sediment chemistry of a large shallow subtropical lake. Lake and Reservoir Management 17: 217–232.

Flaig, E. J. & K. E. Havens, 1995. Historical trends in the Lake Okeechobee ecosystem I. Land use and nutrient loading. Archiv für Hydrobiologie, Supplement 107: 1–24.

Hakanson, L., 1982. Lake bottom dynamics and morphometry: the dynamic ratio. Water Resources Research 18: 1444–1450.

Hamilton, D. P. & S. F. Mitchell, 1997. Wave-induced shear stresses, plant nutrients, and chlorophyll in seven shallow lakes. Freshwater Biology 38: 159–168.

Hanlon, C. G., R. L. Miller & B. F. McPherson, 1999. Relationships between wind velocity and underwater irradiance in a shallow lake (Lake Okeechobee, Florida, USA). Journal of the American Water Resources Association 34: 951–961.

Hansson, L. A., 1990. Quantifying the impact of periphytic algae on nutrient availability for phytoplankton. Freshwater Biology 24: 265–273.

Harris, T. T., K. A. Williges & P. V. Zimba, 1995. Primary productivity and decomposition of five emergent macrophyte communities in the Lake Okeechobee marsh ecosystem. Archiv für Hydrobiologie, Advances in Limnology 45: 63–78.

Havens, K. E., 1997. Water levels and total phosphorus in Lake Okeechobee. Lake and Reservoir Management 13: 16–25.

Havens, K. E., 2002. Development and application of hydrologic restoration goals for a large subtropical lake. Lake and Reservoir Management 18: 285–292.

Havens, K. E., 2003. Submerged aquatic vegetation correlations with depth and light attenuating materials in a shallow subtropical lake. Hydrobiologia 493: 173–186.

Havens, K. E., N. G. Aumen, R. T. James & V. H. Smith, 1996. Rapid ecological changes in a large subtropical lake undergoing cultural eutrophication. AMBIO 25: 150–155.

Havens, K. E., H. J. Carrick, E. F. Lowe & M. F. Coveney, 1999. Contrasting relationships between nutrients, chlorophyll a, and Secchi transparency in two shallow subtropical lakes: Lakes Okeechobee and Apopka (Florida, USA). Lake and Reservoir Management 15: 298–309.

Havens, K. E., T. Fukushima, P. Xie, T. Iwakuma, R. T. James, N. Takamura, T. Hanazato & T. Yamamoto, 2001b. Nutrient dynamics and the eutrophication of shallow lakes Kasumigaura (Japan), Donghu (P.R. China), and Okeechobee (USA). Environmental Pollution 111: 263–272.

Havens, K. E., R. T. James, T. L. East & V. H. Smith, 2003. N:P ratios, light limitation, and cyanobacterial dominance in a subtropical lake impacted by non-point source nutrient pollution. Environmental Pollution 122: 379–390.

Havens, K. E., M. C. Harewll, M. A. Brady, B. Sharfstein, T. L. East, A. J. Rodusky, D. Anson & R. P. Maki, 2002. Large-scale mapping and predictive modeling of submerged aquatic vegetation in a shallow eutrophic lake. The Scientific World Journal 2: 949–965.

Havens, K. E., K.-R. Jin, A. J. Rodusky, B. Sharfstein, M. A. Brady, T. L. East, N. Iricanin, R. T. James, M. C. Harwell & A. D. Steinman, 2001a. Hurricane effects on a shallow lake ecosystem and its response to a controlled manipulation of water level. The Scientific World Journal 1: 44–70.

Havens, K. E., E. J. Phlips, M. F. Cichra & B. L. Li, 1998. Light availability as a possible regulator of cyanobacteria species composition in a shallow subtropical lake. Freshwater Biology 39: 547–556.

Havens, K. E. & C. L. Schelske, 2001. The importance of considering biological processes when setting total maximum daily loads (TMDL) for phosphorus in shallow lakes and reservoirs. Environmental Pollution 113: 1–9.

Havens, K. E., B. Sharfstein, M. A. Brady, T. L. East, M. C. Harwell, R. P. Maki & A. J. Rodusky, 2004. Recovery of submerged plants from high water stress in a large subtropical lake in Florida, USA. Aquatic Botany 78: 67–82.

Havens, K. E. & A. D. Steinman, 1995. Aquatic systems. In Rechcigl, J. E. (ed.), Soil Amendments—Impacts on Biotic Systems. Lewis Publishers, Florida, USA.

Havens, K. E. & W. W. Walker Jr., 2002. Development of a total phosphorus concentration goal in the TMDL process for Lake Okeechobee, Florida (USA). Lake and Reservoir Management 18: 227–238.

Herdendorf, C. E., 1984. Inventory of the morphometric and limnological characteristics of the large lakes of the world. Tech. Bull. Ohio Sea Grant, Columbus, Ohio, USA.

Hilton, J., 1985. A conceptual framework for predicting the occurrence of sediment focusing and sediment redistribution in small lakes. Limnology and Oceanography 30: 1131–1143.

Hutchinson, G. E., 1975. A Treatise on Limnology, volume 1, part 2—Chemistry of Lakes. John Wiley and Sons, New York, USA.

Hwang, S.-J., K. E. Havens & A. D. Steinman, 1998. Phosphorus kinetics of planktonic and benthic assemblages in a shallow subtropical lake. Freshwater Biology 40: 729–745.

Ishikawa, T. & M. Tanaka, 1993. Diurnal stratification and its effects on wind-induced currents and water qualities in Lake Kasumigaura, Japan. Journal of Hydraulic Research 31: 307–322.

James, R. T., B. L. Jones & V. H. Smith, 1995a. Historical trends in the Lake Okeechobee ecosystem II. Nutrient budgets. Archiv für Hydrobiologie, Supplement 107: 25–47.

James, R. T., J. Martin, T. Wool & P. F. Wang, 1997. A sediment resuspension and water quality model of Lake Okeechobee. Journal of the American Water Resources Association 33: 661–680.

James, R. T., V. H. Smith & B. L. Jones, 1995b. Historical trends in the Lake Okeechobee ecosystem III. Water quality. Archiv für Hydrobiologie, Supplement 107: 49–69.

Jin, K.-R. & Z.-G. Ji, 2001. Calibration and verification of a spectral wind-wave model for Lake Okeechobee. Ocean Engineering 28: 571–584.

Jin, K.-R., Z.-G. Ji & J. H. Hamrick, 2002. Modeling winter circulation in Lake Okeechobee, Florida. Journal of Waterway, Port, Coastal and Ocean Engineering 27: 114–125.

Jin, K.-R., J. H. Hamrick & T. Tisdale, 2000. Application of three-dimensional hydrodynamic model for Lake Okeechobee. Journal of Hydraulic Engineering 126: 758–771.

Kristensen, P., M. Sondergaard & E. Jeppesen, 1992. Resuspension in a shallow eutrophic lake. Hydrobiologia 228: 101–109.

Kufel, L. & I. Kufel, 2002. *Chara* beds acting as nutrient sinks in shallow lakes—a review. Aquatic Botany 72: 249–260.

Landsea, C., 2001. Testimony before the Committee on Science, Environment, Technology, and Standards Sub-Committee. United States House of Representatives, Washington, D.C.

Lean, D. R. S., 1973a. Phosphorus dynamics in lake water. Science 179: 678–680.

Lean, D. R. S., 1973b. Movements of phosphorus between its biologically important forms in lake water. Journal of the Fisheries Research Board of Canada 30: 1525–1536.

Luettich, R. A. Jr., D. R. F. Harleman & L. Somlyudy, 1990. Dynamic behavior of suspended sediment concentrations in a shallow lake perturbed by episodic wind events. Limnology and Oceanography 35: 1050–1067.

Maceina, M. J., 1993. Summer fluctuations in planktonic chlorophyll *a* concentrations in Lake Okeechobee, Florida: the influence of lake levels. Lake and Reservoir Management 8: 1–11.

Maceina, M. J. & D. M. Soballe, 1990. Wind-related limnological variation in Lake Okeechobee. Lake and Reservoir Management 6: 93–100.

Mazumder, A. & K. E. Havens, 1998. Nutrient-chlorophyll relationships in temperate vs. subtropical lakes. Canadian Journal of Fisheries and Aquatic Sciences 55: 1652–1662.

Moore, P. A. Jr. & K. R. Reddy, 1994. Role of Eh and pH on phosphorus biogeochemistry in sediments of Lake Okeechobee, Florida. Journal of Environmental Quality 23: 955–964.

Mortimer, C. H., 1941. The exchange of dissolved substances between mud and water in lakes (parts I and II). Journal of Ecology 29: 280–329.

Murphy, T., K. Hall & I. Yesaki, 1983. Co-precipitation of phosphate and calcite in a naturally eutrophic lake. Limnology and Oceanography 28: 58–67.

Otsubo, K. & K. Muraoka, 1987. Field studies on physical properties of sediment and sediment resuspension in Lake Kasumigaura. Japanese Journal of Limnology 48: 131–138.

Peters, R. H. & A. Cattaneo, 1984. The effects of turbulence on phosphorus supply in a shallow bay of Lake Memphremagog. Verhandlungen Internationale Vereinigung für Theoretische und Angewandte Limnologie 22: 185–189.

Phlips, E. J., M. Cichra, K. E. Havens, C. Hanlon, S. Badylak, B. Rueter, M. Randall & P. Hansen, 1997. Relationships between phytoplankton dynamics and the availability of light and nutrients in a shallow subtropical lake. Journal of Plankton Research 19: 319–342.

Phlips, E. J., F. J. Aldridge & P. Hansen, 1995. Patterns of water chemistry, physical and biological parameters in a shallow subtropical lake (Lake Okeechobee, Florida, USA). Archiv für Hydrobiologie, Advances in Limnology 45: 117–135.

Phlips, E. J., P. V. Zimba, M. S. Hopson & T. L. Crisman, 1993. Dynamics of the plankton community in submerged plant dominated regions of Lake Okeechobee, Florida, USA. Verhandlungen Internationale Vereinigung für Theoretische und Angewandte Limnologie 25: 423–426.

Qin, B., G. Zhu, L. Zhang, L. Luo, G. Gao, W. Sumin & B. Gu, 2006. Models and estimation methods of nutrient release from sediments of large shallow lakes:

a case study of Lake Taihu. Science in China (in press).

Reddy, K. R., 1993. Lake Okeechobee Phosphorus Dynamics Study. volume III - biogeochemical processes in the sediments. Technical Report to the South Florida Water Management District, West Palm Beach, Florida, USA.

Reddy, K. R., Y. P. Sheng & B. L. Jones, 1995. Lake Okeechobee Phosphorus Dynamics Study. volume I - summary. Technical Report to the South Florida Water Management District, West Palm Beach, Florida, USA.

Rigler, F. H., 1956. A tracer study of the phosphorus cycle in lake water. Ecology 37: 675–683.

Rigler, F. H., 1964. The phosphorus fractions and the turnover time of inorganic phosphorus in the epilimnion of lakes. Canadian Fish Culturist 32: 3–9.

Rigler, G. H., 1973. A dynamic view of the phosphorus cycle in lakes. In Griffith, E. J., A. Beeton, J. M. Spencer & D. T. Mitchell (eds), Environmental Phosphorus Handbook. John Wiley and Sons, New York, USA.

Reynolds C. S., 1994. The Ecology of Freshwater Phytoplankton. Cambridge Univiversity Press.

Sas, H., 1989. Lake restoration by reduction of nutrient loadings: expectations, experiences, extrapolations. Academia Verlag Richarz, Germany.

Scheffer, M., 1998. Ecology of Shallow Lakes. Chapman and Hall, New York, USA.

Sheng, Y. P., 1993. Hydrodynamics, sediment transport and their effects on phosphorus dynamics in Lake Okeechobee. In Mehta, A. J. (ed.), Coastal and Estuarine Studies 42: Near-shore and Estuarine Cohesive Sediment Transport. American Geophysical Union, Washington, D.C.

Sheng, Y. P., 1999. Effects of hydrodynamic processes on phosphorus distribution in aquatic ecosystems. In Reddy, K. R., G. A. O'Conner & C. L. Schelske (eds), Phosphorus Biogeochemistry in Subtropical Ecosystems. Lewis Publishers, Florida, USA.

Sondergaard, M., J. P. Jensen & E. Jeppesen, 2001. Retention and internal loading of phosphorus in shallow, eutrophic lakes. The ScientificWorld Journal 1: 427–442.

Sondergaard, M., P. Kristensen & E. Jeppesen, 1992. Phosphorus release from resuspended sediment in the shallow and wind-exposed Lake Arreso, Denmark. Hydrobiologia 228: 91–99.

Steinman, A. D., K. E. Havens, N. G. Aumen, R. T. James, K-R. Jin, J. Zhang & B. H. Rosen, 1999. Phosphorus in Lake Okeechobee: sources, sinks, and strategies. In Reddy, K. R., G. A. O'Conner & C. L. Schelske (eds), Phosphorus Biogeochemistry in Subtropical Ecosystems. Lewis Publishers, Florida, USA.

Steinman, A. D., K. E. Havens, H. J. Carrick & R. VanZee, 2002a. The past, present, and future hydrology and ecology of Lake Okeechobee ant its watersheds. In Porter, J. W. & K. G. Porter (eds), The Everglades, Florida Bay, and Coastal Reefs of the Florida Keys, an Ecosystem Restoration Sourcebook. CRC Press, Florida, USA.

Steinman, A. D., K. E. Havens, A. J. Rodusky, B. Sharfstein, R. T. James & M. C. Harwell, 2002b. The influence of environmental variables and a managed water recession on the growth of charophytes in a large, subtropical lake. Aquatic Botany 72: 297–313.

Van Liere, L. & R. D. Gulati, 1992. Restoration and Recovery of Shallow Eutrophic Lake Ecosystems in the Netherlands. Kluwer Academic Publishers, The Netherlands.

Van Rees, K. C. J., K. R. R. Reddy & P. S. C. Rao, 1996. Influence of benthic organisms on solute transport in lake sediments. Hydrobiologia 317: 31–40.

Vermaat, J. E., L. Santamaria & P. J. Roos, 2000. Water flow across and sediment trapping in submerged macrophyte beds of contrasting growth form. Arch. Hydrobiologia 148: 549–562.

Wetzel, R. G., 2001. Limnology—Lake and River Ecosystems, 3rd edn. Academic Press, New York, USA.

Yassuda, E. A. & Y. P. Sheng, 1997. Integrated modeling of Tampa Bay estuarine system. Technical Report, Coastal and Ocean Engineering Department, University of Florida, Gainesville, Florida, USA.

Zimba, P. V., M. S. Hopson, J. P. Smith, D. E. Colle & J. V. Shireman, 1995. Chemical composition and distribution of submersed aquatic vegetation in Lake Okeechobee, Florida (1989–1991). Archiv für Hydrobiologie, Advances in Limnology 45: 241–246.

Hydrobiologia (2007) 581:43–52
DOI 10.1007/s10750-006-0520-6

EUTROPHICATION IN LAKES

Chromophoric dissolved organic matter (CDOM) absorption characteristics in relation to fluorescence in Lake Taihu, China, a large shallow subtropical lake

Yunlin Zhang · Boqiang Qin · Guangwei Zhu · Lu Zhang · Longyuan Yang

Abstract Absorption measurements from chromophoric dissolved organic matter (CDOM) and their relationships with dissolved organic carbon (DOC) and fluorescence were studied in Lake Taihu, a large, shallow, subtropical lake in China. Absorption spectra of lake water samples were measured from 240 nm to 800 nm. Highest values of $a(\lambda)$, DOC and $F_n(355)$ occurred near the river inflow to Meiliang Bay and decreased towards the central lake basin. A significant spatial difference was found between Meiliang Bay and the central lake basin in absorption coefficient, DOC-specific absorption coefficient, exponential slope coefficient, DOC concentration and fluorescence value. The spatial distribution of CDOM suggested that a major part of CDOM in the lake was from river input. CDOM absorption coefficients were correlated with DOC over the wavelength range 280–500 nm, and $a(355)$ was also correlated with $F_n(355)$, which showed that CDOM absorption could be inferred from DOC and fluorescence measurement. The coefficient of variation between $a(\lambda)$ and DOC concentration decreased with increase in wavelength from 240 nm to 800 nm. Furthermore, a significant negative linear relationship was recorded between S value and CDOM absorption coefficient, as well as DOC-specific absorption coefficient. S value and DOC-specific absorption coefficient were used as a proxy for CDOM composition and source. Accurate CDOM absorption measurements are very useful in explaining UV attenuation and in developing, validating remote sensing model of water quality in Lake Taihu.

Keywords Chromophoric dissolved organic matter (CDOM) · Fluorescence · Dissolved organic carbon (DOC) · DOC-specific absorption coefficient · Spectral slope coefficient · Lake Taihu

Guest editors: B. Qin, Z. Liu & K. Havens
Eutrophication of shallow lakes with special reference to Lake Taihu, China

Y. Zhang · B. Qin (✉) · G. Zhu · L. Zhang · L. Yang
Nanjing Institute of Geography & Limnology, Chinese Academy of Sciences, Nanjing 210008, P. R. China
e-mail: qinbq@niglas.ac.cn

Y. Zhang · L. Zhang
Graduate school of Chinese Academy of Sciences, Beijing 100039, P. R. China

Introduction

Chromophoric dissolved organic matter (CDOM) is the light absorbing (ultraviolet and visible) fraction of dissolved organic carbon (DOC) (Kirk, 1994). Spatial and temporal distribution of CDOM in aquatic ecosystems can affect ecosystem productivity; negatively impacting primary productivity since CDOM absorbs light, while positively impacting secondary productivity by providing a substitute for microbial respiration via photo-degraded CDOM (Benner & Biddanda, 1998; Reche et al., 1998; Mazzuoli et al., 2003). Overlaps of pigment absorption spectra with CDOM absorption at blue wavelengths complicate the use of chlorophyll *a* retrieval algorithms based on remotely sensed ocean color (Carder et al., 1991) and lead to chlorophyll *a* concentration overestimation (Nelson et al., 1998; Rochelle-Newall & Fisher, 2002a). In addition, high CDOM concentrations can act as a photoprotectant against UV damage to aquatic organisms (Williamson et al., 1996). Several studies have identified CDOM as the most important factor controlling UV attenuation in natural freshwaters (Morris et al., 1995; Laurion et al., 2000). However, protection of aquatic biota against UV radiation provided by dissolved humic material may be diminished with CDOM photodegradation by UV radiation (Morris & Hargreaves, 1997) and acidification (Schindler et al., 1996; Yan et al., 1996).

In shallow, inland lakes and coastal waters, absorption by non-chlorophyllous substances often exceeds that by phytoplankton (Carder et al., 1991; Frenette et al., 2003). CDOM originates from phytoplankton degradation and river inputs and land drainage, especially in inland waters. Although CDOM influences lake optical properties, remotely sensed spectra, and biogeochemical processes, the CDOM dynamics in diverse aquatic environments are not well known. A thorough inventory of regional CDOM distribution and its light absorption in wavebands covering UV and visible radiation spectra have not been conducted in Lake Taihu. This information is needed to interpret satellite and airplane remote sensing imagery to be used to record lake response to changes in environmental conditions, such as water pollution, eutrophication and cyanobacterial blooms. In this paper, we describe the absorption spectrum of CDOM in lake water samples from north Lake Taihu. To examine CDOM, the following aspects are addressed in this paper:

(1) Spatial distribution of CDOM in different lake regions interpreted by absorption spectra;
(2) The relationship between CDOM absorption coefficients and DOC concentrations;
(3) Variations in the absorption spectra with wavelength and fluorescence.

Materials and methods

Study area

Forty sampling sites were chosen in the northern lake. Sampling site 24 was located at the boundary

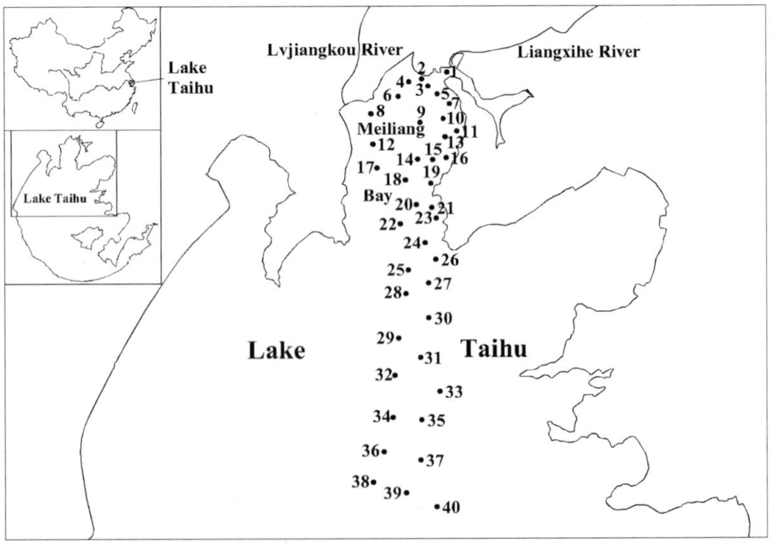

Fig. 1 Position of sampling sites during the cruise

of Meiliang Bay and the central lake. Sites 1–24 were distributed in Meiliang Bay, and other sites were in the central lake (Fig. 1). Meiliang Bay is eutrophic with an area of 132 km^2 and mean depth of 2.0 m. Lujiangkou and Liangxihe Rivers flow into the bay (Fig. 1).

Absorption of CDOM and DOC concentration

Water from 0 m to 0.5 m depth for CDOM absorption and DOC analysis was collected in 250 ml acid-cleaned plastic bottles and held on ice while in the field. In the laboratory, these samples were kept at 4°C until analysis within two days. All samples were filtered at low pressure through pre-combusted Whatman GF/F filters (0.7 μm) into glass bottles pre-combusted at 550°C for 6 h (Del Castillo et al., 1999; Del Castillo & Coble, 2000; Rochelle-Newall & Fisher, 2002b; Yacobi et al., 2003; Frenette et al., 2003; Callahan et al., 2004; Chen et al., 2004). Absorption spectra were obtained between 240 nm and 800 nm at 1-nm intervals using a Shimadzu UV-2401PC UV-Vis recording spectrophotometer with matching 4 cm quartz cells. Milli-Q water was used in the reference cell. Absorption coefficients were obtained by using following expression (Kirk, 1994):

$$a(\lambda') = 2.303D(\lambda)/r \quad (1)$$

where $a(\lambda')$ is uncorrected CDOM absorption coefficient at wavelength λ, $D(\lambda)$ is the optical density at wavelength λ, and r is the cuvette path length in m. Absorption coefficients were corrected for backscattering of small particles and colloids, which pass through filters, using Eq. 2 (Bricaud et al., 1981; Green & Blough, 1994).

$$a(\lambda) = a(\lambda') - a(750) \times \lambda/750 \quad (2)$$

where $a(\lambda)$ = absorption coefficient at wavelength (λ) corrected for scattering, and $a(750)$ = measured absorption coefficient at 750 nm corrected for scattering. Because of the chemical complexity of CDOM, concentration is expressed using the absorption coefficient at some reference wavelength, typically 355 nm, 375 nm or 440 nm (Kirk, 1994; Del Castillo & Coble, 2000; Stedmon et al., 2000). Here, absorption at 355 nm was used to analyze the relationship of CDOM and fluorescence. Molecular size of humic molecules was estimated from the ratio of absorption coefficients at 250 and 365 nm [$a(250)/a(365)$] (De Haan, 1993; Peuravuori & Pihlaja, 1997). The apparent DOC-specific absorption coefficient $a^*(\lambda)$ was calculated by the relationship below (Seritti et al., 1998):

$$a*(\lambda) = a(\lambda)/\mathrm{DOC} \quad (3)$$

where DOC is expressed in mg l^{-1}. DOC in the filtrates was measured on a 1020 TOC analyzer.

Exponential slope coefficient calculation

Exponential slope coefficient was calculated from the absorption spectrum between 280 and 500 nm ($S1$), 280 and 360 nm ($S2$), and 360 and 440 nm ($S3$) using the non-linear regression of Equation (4) (Bricaud et al., 1981). The coefficient of determination (r^2) for the fit of S was consistently higher than 0.99.

$$a(\lambda) = a(\lambda_0)\exp[S(\lambda_0 - \lambda)] \quad (4)$$

where $a(\lambda)$, $a(\lambda_0)$ are the absorption coefficients at wavelengths λ and λ_0, respectively, and S is the exponential slope coefficient.

Fluorescence measurements

Fluorescence was measured using a 1-cm quartz cell in a Shimadzu 5301 spectrofluorometer with an excitation wavelength of 355 nm and an emission wavelength of 450 nm. A Milli-Q water blank was used for comparison (Seritti et al., 1998; Rochelle-Newall & Fisher, 2002b).

Data analysis

Statistical analyses were performed with SPSS 11.0 software (Statistical Program for Social Sciences).

Results and discussion

Spectral character and spatial distribution of CDOM

Figure 2a, b shows representative absorption spectra of samples from northern Meiliang Bay (3), the Bay inlet (24), and the central lake basin (31, 40), which represent a range of CDOM concentrations. Spectral curves for CDOM generally showed near zero absorption at the red end of the visible spectrum (700 nm) and exponentially increase through the ultraviolet (UV) wavelength regimes (280–400 nm). Mean corrected CDOM absorption coefficients at 280 nm [$a(280)$] and 355 nm [$a(355)$] were 18.3 ± 5.6 and 4.6 ± 1.8 m^{-1}, respectively, for the 40 sites. CDOM absorption coefficients at 280, 355 and 440 nm for each site are shown in Fig. 3. The highest value of $a(355)$ was 8.6 m^{-1} at site 15 close to shore in Meiliang Bay, and the lowest value was 2.2 m^{-1} at site 40 near the center of the lake (Figs. 1, 3). In general, $a(355)$ decreased away from the river inflow to Meiliang Bay towards the lake center (Figs. 1–3). CDOM absorption in the Meiliang Bay inlet were around twice those in the central basin. All sites were divided into two groups, one representing Meiliang Bay (sites 1–24) and one representing the central basin (sites 25–40). Significant differences were found between Meiliang Bay and the central basin arising from spatial differences in absorption coefficients, DOC-specific absorption coefficients, exponential slope coefficients, DOC concentrations and fluorescence values using ANOVA ($P \leq 0.005$). Values of $a(280)$ and $a(355)$ in Lake Taihu are higher than those in estuarine and coastal waters (Seritti et al., 1998; Chen et al., 2004), but CDOM absorption values are close to those in large, shallow, fluvial Lake Saint-Pierre (Frenette et al., 2003) and others (Kirk, 1994; Morris et al., 1995). The ratio of $a(250)$ to $a(365)$ ranged from 5.28 to 9.24 with a mean value of 7.10 ± 0.94 (standard deviation), which was close to results from other studies (De Haan, 1993; Huovinen et al., 2003). Higher ratios corresponding to smaller molecules were found in the lake center in agreement with findings from Huovinen et al. (2003) and Yacobi et al. (2003).

Relationship between CDOM absorption and DOC

DOC concentrations ranged from a maximum of 10.9 mg l^{-1} at the Liangxihe mouth (site 1) to a minimum of 6.0 mg l^{-1} in the central basin (Site 37), with a mean value of 8.17 ± 1.38 mg l^{-1}. A significant positive correlation was found between DOC and CDOM absorption coefficients in the ultraviolet and blue band (280–500 nm), especially at shorter wavelengths (Fig. 4a), in agreement with other studies where DOC was dominated by CDOM (Seritti et al., 1998; Stedmon et al., 2000; Laurion et al., 2000). Figure 4b shows that the coefficients of determination (r^2) between CDOM absorption and DOC decrease with increasing wavelength from 280 nm to 500 nm. Similar results were found in shallow lakes in the middle and lower reaches of the Changjiang (Yangtze) River (Zhang et al., 2005).

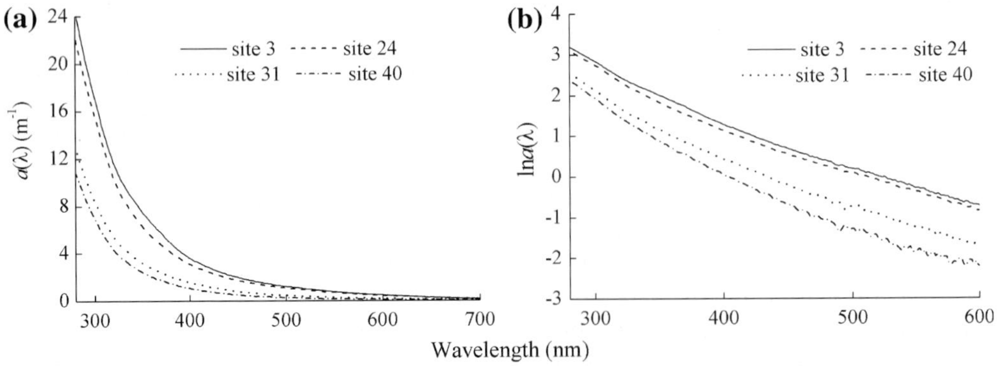

Fig. 2 CDOM absorption spectra of representative sites: (**a**) Wavelength versus $a(\lambda)$ and (**b**) wavelength versus $\ln[a(\lambda)]$

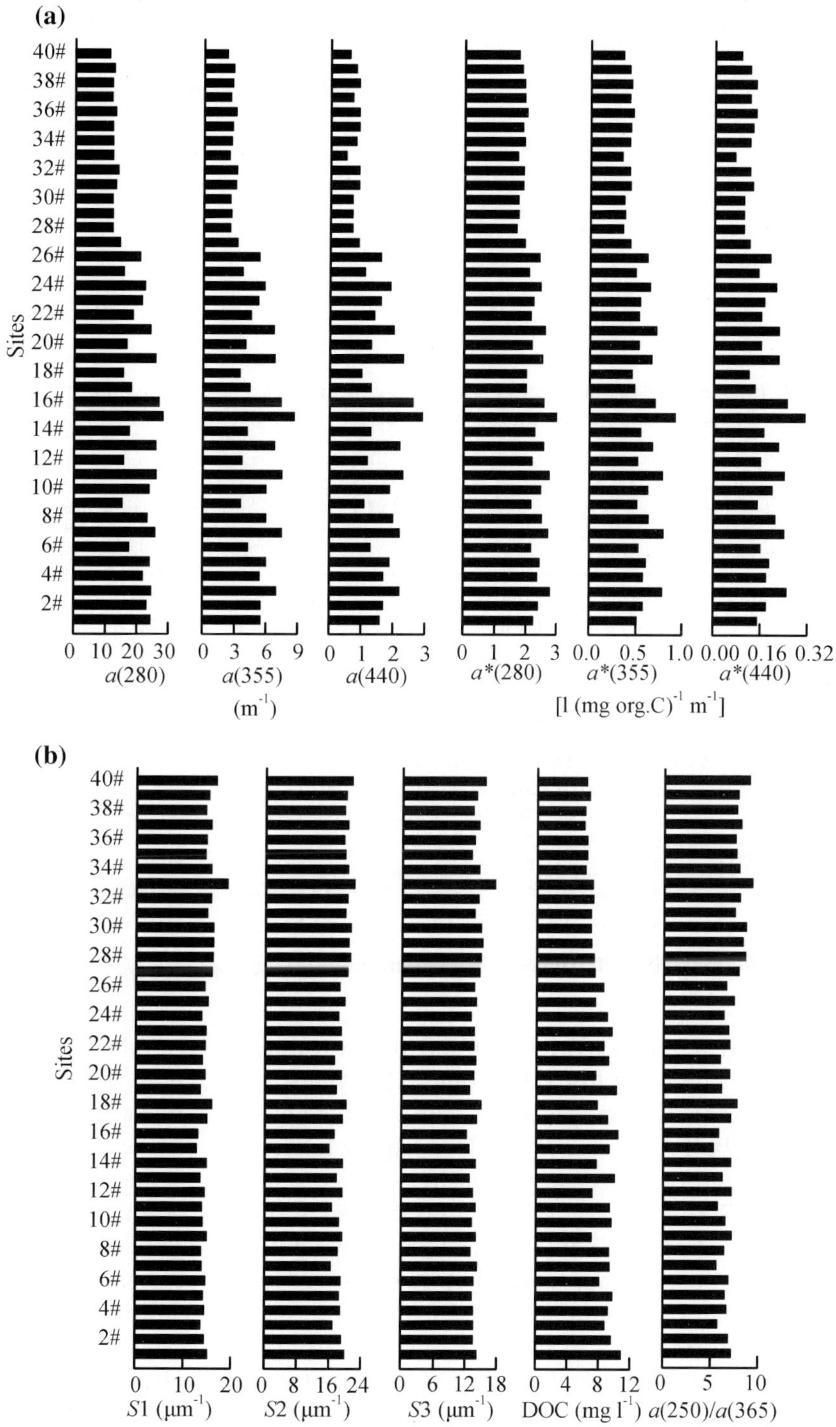

Fig. 3 Absorption coefficients $a(280)$, $a(355)$, $a(440)$, DOC-specific absorption coefficients, $a^*(280)$, $a^*(355)$, $a^*(440)$, spectral slope $S1$ (280–500 nm), $S2$ (280–360 nm), $S3$ (360–440 nm), DOC, fluorescence and ratio of $a(250)$ to $a(365)$

Relationships between DOC and CDOM absorption from this study and other studies are presented in Table 1. Published relationships of DOC and $a(355)$ are valuable tools to model DOC concentration (Seritti et al., 1998; Rochelle-Newall & Fisher, 2002b). However, differences between models suggest regional and lake-to-lake variations in bio-optical characteristics.

Calculated DOC-specific absorption provides an additional means for distinguishing bio-optical

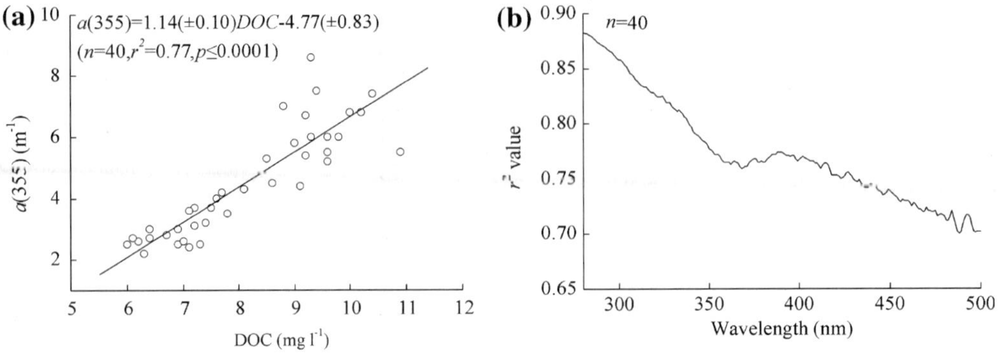

Fig. 4 Plots of CDOM absorption and DOC concentration (**a**) CDOM absorption $a(355)$ versus DOC concentration and (**b**) wavelength versus determination coefficients r^2 value between $a(\lambda)$ and DOC concentration from 280 nm to 500 nm

Table 1 Linear models for the relationship between DOC concentrations and CDOM absorption coefficients

Equations		r^2	n	P	DOC (mg l^{-1})	References
DOC, $a(\lambda)$	$a(355) = 1.465\text{DOC} - 1.315$	0.98	31	≤0.0001	0.89–4.82	Seritti et al. (1998)
	$a(440) = 0.280\text{DOC}$	0.91	6	≤0.0007	4.9–40.7	Yacobi et al. (2003)
	$a(440) = 0.235\text{DOC}$	0.92	6	≤0.0005	7.1–40.5	Yacobi et al. (2003)
	$a(440) = 0.429\text{DOC} - 0.744$	0.73	64	≤0.0001	0.24–23.5	Morris et al. (1995)[a]
	$a(300) = 3.584\text{DOC} - 2.690$	0.71	26	≤0.0001	0.77–3.31	Del Castillo et al. (1999)[a]
	$a(440) = 0.131\text{DOC} + 0.020$	0.58	39	≤0.0001	0.21–3.5	Laurion et al. (2000)[a]
	$a(440) = 0.552\text{DOC} - 1.477$	0.77	12	≤0.0005	4.9–14.9	Huovinen et al. (2003)[a]
	$a(355) = 0.015\text{DOC} - 5.5$	0.90	92	≤0.0001	–	Ferrari et al. (1996)[b]
	$a(355) = 0.00036\text{DOC} + 0.028$	0.66	43	≤0.0001	–	Ferrari (2000)[b]
	$a(355) = 0.744\text{DOC} - 2.176$	0.58	22	≤0.001	2.7–10.1	Zhang et al. (2005)
	$a(355) = 1.14\text{DOC} - 4.77$	0.77	40	≤0.0001	6.0–10.9	Present study
	$a(375) = 0.87\text{DOC} - 3.74$	0.77	40	≤0.0001	6.0–10.9	Present study
	$a(440) = 0.38\text{DOC} - 1.73$	0.73	40	≤0.0001	6.0–10.9	Present study

[a] Fitted with the data in papers ; [b] DOC unit: μM

characteristics of CDOM. Low CDOM absorption was associated with low values of $a^*(\lambda)$. Values of $a^*(355)$ ranged from 0.34 to 0.92 with an average 0.54 ± 0.14 [l (mg org C)$^{-1}$ m^{-1}], which are higher than those observed in estuarine and coastal waters (Seritti et al., 1998). Lowest values of $a^*(320)$ and $a^*(355)$ corresponded with low DOC concentrations and approached those measured in coastal water and mountain lakes (Seritti et al., 1998; Laurion et al., 2000). DOC-specific absorption coefficient represents CDOM composition, which is used to establish relationships with other parameters. Specific absorption was correlated with pH and DOC concentration in three acid lakes (Gallie, 1997). In our study, specific absorption was correlated positively with DOC concentration. The linear regression equation is $a^*(355) = 0.075\text{DOC} - 0.070$ ($r^2 = 0.53$, $n = 40$, $P \le 0.0001$). The linear correlation between DOC concentration and specific absorption is known from other studies (Yacobi et al., 2003). Furthermore, the higher average $a^*(440)$ corresponded to the higher proportion of large and medium fractions in CDOM (Yacobi et al., 2003). In this study, higher $a^*(355)$ and $a^*(440)$ values corresponded to lower $a(250)/a(365)$ values towards increasing molecular size. The regression equation between $a^*(355)$ and $a(250)/a(365)$ values is: $a^*(355) = -0.145a(250)/a(365) + 1.570$ ($r^2 = 0.92$, $n = 40$, $P \le 0.0001$).

Estimation of the spectral slope S

It is widely accepted that spectral slope S can be used as a proxy for CDOM composition (Kowalczuk et al., 2003). The values of the exponential slope coefficient (S) ranged from 12.7 μm^{-1} to 19.0 μm^{-1} with an average value of 14.74 ± 1.13 μm^{-1}, from 15.9 μm^{-1} to 22.0 μm^{-1} with an average value 19.11 ± 1.42 μm^{-1}, and from 12.2 to 17.4 μm^{-1} with an average value 13.86 ± 0.91 μm^{-1} for the wavelength domain from 280 nm to 500 nm ($S1$), 280–360 nm ($S2$) and 360–440 nm ($S3$), respectively. The coefficient of determination (r^2) for the fit of S was higher than 0.99. Although slope values in the present investigation fell within the range of previously reported values (Bricaud et al., 1981; Davies-Colley, 1983; Seritti et al., 1998; Markager & Vincent, 2000), variability of the spectral slope in different wavelength ranges is significant, which illustrates the importance and sensitivity of the wavelength range chosen to calculate the slope coefficient (S). S generally increases with decrease in the wavelength range and may be influenced by CDOM composition (i.e. humic and fulvic acids affecting spectral properties). Samples collected from the Mississippi River plume in the Gulf of Mexico, where CDOM was comprised predominantly from fulvic acids, had a steeper slope (S = 0.0194) than marine samples, where CDOM was mostly humic acids (S = 0.010) (Carder et al., 1989). S values in Lake Taihu were higher than those for oceanic and coastal waters (Markager & Vincent, 2000).

In our study, there are significant negative correlations between spectral slope S and $a(355)$ (Fig. 5a) and $a^*(355)$ (Fig. 5b, Table 2). Determination coefficients between S and $a^*(\lambda)$ were higher than those between S and $a(\lambda)$. It is also interesting that determination coefficients are the highest between $S2$ (280–306 nm) and $a(\lambda)$, $a^*(\lambda)$. Furthermore, significant negative correlations have been obtained between S values and $a^*(440)$ (Yacobi et al., 2003), $a(355)$ (Rochelle-Newall et al., 2004) and $a(375)$ (Del Castillo &

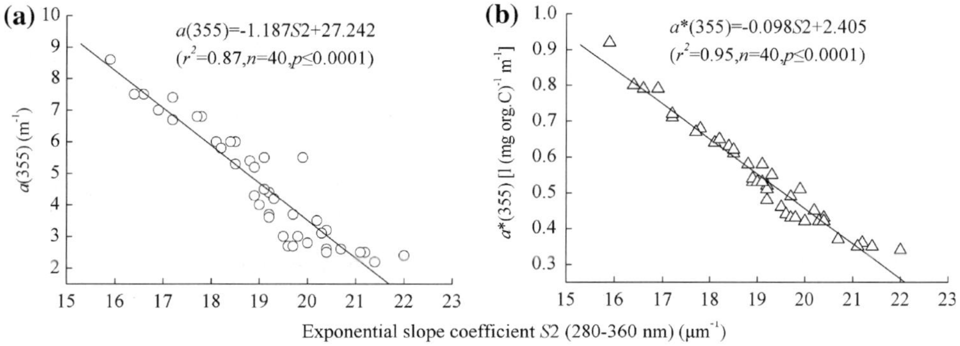

Fig. 5 Plots of exponential slope coefficient versus (**a**) absorption coefficient $a(355)$ and (**b**) specific absorption coefficient $a^*(355)$

Table 2 Linear models for the relationship between $S1$ (280–500 nm), $S2$ (280–360 nm), $S3$ (360–440 nm) and absorption coefficient, DOC-specific absorption coefficient of CDOM (n = 40, $P \le 0.0001$)

Parameters		Equations	r^2
$a(355)$, S	$S1$ (280–500 nm)	$a(355) = -1.254(\pm0.160)S1 + 23.045(\pm2.357)$	0.62
	$S2$ (280–360 nm)	$a(355) = -1.187(\pm0.075)S2 + 27.242(\pm1.444)$	0.87
	$S3$ (360–440 nm)	$a(355) = -1.158(\pm0.262)S3 + 20.613(\pm3.646)$	0.34
$a(375)$, S	$S1$ (280–500 nm)	$a(375) = -0.976(\pm0.119)S1 + 17.772(\pm1.751)$	0.64
	$S2$ (280–360 nm)	$a(375) = -0.910(\pm0.057)S2 + 20.779(\pm1.087)$	0.87
	$S3$ (360–440 nm)	$a(375) = -0.919(\pm0.197)S3 + 16.133(\pm2.733)$	0.37
$a^*(355)$, S	$S1$ (280–500 nm)	$a^*(355) = -0.103(\pm0.011)S1 + 2.064(\pm0.168)$	0.68
	$S2$ (280–360 nm)	$a^*(355) = -0.098(\pm0.004)S2 + 2.405(\pm0.067)$	0.95
	$S3$ (360–440 nm)	$a^*(355) = -0.095(\pm0.020)S3 + 1.858(\pm0.279)$	0.37
$a^*(375)$, S	$S1$ (280–500 nm)	$a^*(375) = -0.083(\pm0.008)S1 + 1.627(\pm0.124)$	0.72
	$S2$ (280–360 nm)	$a^*(375) = -0.077(\pm0.003)S2 + 1.868(\pm0.049)$	0.96
	$S3$ (360–440 nm)	$a^*(375) = -0.079(\pm0.015)S3 + 1.495(\pm0.211)$	0.42

Coble, 2000; Stedmon et al., 2000). Kowalczuk et al. (2003) presented a trend of increased slope coefficient with a decrease in absorption level.

Relationship between CDOM absorption and fluorescence

CDOM absorption may be measured directly using a spectrophotometer or inferred from fluorescence, which can be measured more rapidly and with greater sensitivity than absorption. In previous studies, a strong relationship was established between CDOM absorption and fluorescence intensity (Vodacek et al., 1995; Seritti et al., 1998; Ferrari & Dowell, 1998; Hoge et al., 1993; Kowalczuk et al., 2003) despite numerous sources of CDOM and different chemical characteristics. CDOM fluorescence $F_n(355)$ and absorption coefficients $a(355)$ were correlated positively in Lake Taihu in this study (Fig. 6). The correlation may enable the use of fluorescence as a proxy for CDOM absorption despite different sources from river inputs and autochthonous humic substances and locations within the lake.

Conclusions

(1) Higher CDOM absorption coefficients were found in the Meiliang Bay inlet of Lake Taihu in association with increased DOC concentrations and fluorescence intensity. There were significant differences between Meiliang Bay and the central lake basin in absorption coefficients, DOC-specific absorption coefficient $a^*(355)$, exponential slope coefficient (S), DOC concentrations and fluorescence values.

(2) There is a significant linear relationship between CDOM absorption and DOC concentrations, especially at shorter wavelengths in Lake Taihu samples.

(3) Exponential slope coefficients fall in the range reported in the literatures. There is a significant negative linear relationship between S values and CDOM absorption coefficients, as well as DOC-specific absorption coefficients.

(4) There is a linear relationship ($r = 0.66$, $P \leq 0.0001$) between CDOM absorption and CDOM fluorescence intensity: $a(355) = 2.12(\pm0.33) + 3.72(\pm0.43)F_n(355)$.

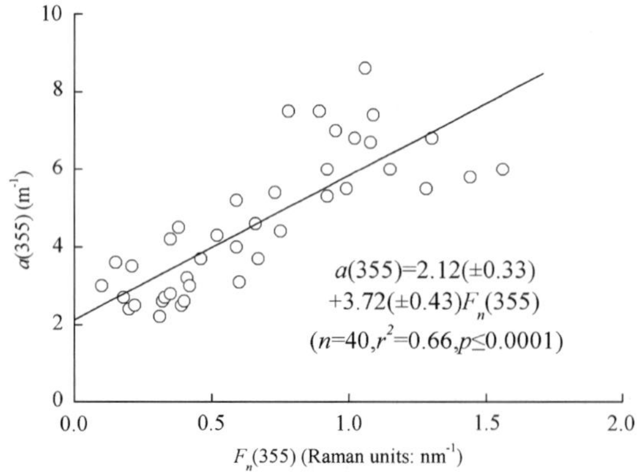

Fig. 6 Plot of $a(355)$ versus $F_n(355)$

Acknowledgements This work was supported by Chinese Academy of Sciences (No. KZCX1-SW-12) and National Natural Science Foundation of China (Nos. 40601099, 40501078), and Director Foundation of Nanjing Institute of Geography and Limnology, CAS (No. NIGLA-S220445). We specially thank Dr. Mark J. McCarthy for his linguistic improvement.

References

Benner, R. & B. Biddanda, 1998. Photochemical transformations of surface and deep marine dissolved organic matter: Effects on bacterial growth. Limnology and Oceanography 43: 1373–1378.

Bricaud, A., A. Morel & L. Prieur, 1981. Absorption by dissolved organic matter of the sea (yellow substance) in the UV and visible domain. Limnology and Oceanography 26: 43–53.

Callahan, J., M. H. Dai, R. F. Chen, X. L. Li, Z. M. Lu & W. Huang, 2004. Distribution of dissolved organic matter in the Pearl River Estuary, China. Marine Chemistry 89: 211–224.

Carder, K. L., S. K. Hawes, R. C. Smith, R. G. Steward & B. G. Mitchell, 1991. Reflectance model for quantifying chlorophyll *a* in the presence of productivity degradation products. Journal of Geophysical Research 9: 20599–20611.

Carder, K. L., R. G. Steward, G. R. Harvey & P. B. Ortner, 1989. Marine humic and fulvic acids: their effects on remote sensing of ocean chlorophyll. Limnology and Oceanography 34: 68–81.

Chen, Z. Q., Y. Li & J. M. Pan, 2004. Distributions of colored dissolved organic matter and dissolved organic carbon in the Pearl River Estuary, China. Continental Shelf Research 24: 1845–1856.

Davies-Colley, R. J., 1983. Absorption of light by yellow substance in freshwater lakes. Limnology and Oceanography 32: 416–425.

De Haan, H., 1993. Solar UV-light penetration and photodegradation of humic substances in peaty lake water. Limnology and Oceanography 38: 1072–1076.

Del Castillo, C. E., P. G. Coble, J. M. Morell, J. M. López & J. E. Corredor, 1999. Analysis of the optical properties of the Orinoco River plume by absorption and fluorescence spectroscopy. Marine Chemistry 66: 35–51.

Del Castillo, C. E. & P. G. Coble, 2000. Seasonal variability of the colored dissolved organic matter during the 1994–95 NE and SW Monsoons in the Arabian Sea. Deep–Sea Research II 47: 1563–1579.

Ferrari, G. M., 2000. The relationship between chromophoric dissolved organic matter and dissolved organic carbon in the European Atlantic coastal area and in the West Mediterranean Sea Gulf of Lions. Marine Chemistry 70: 339–357.

Ferrari, G. M. & M. D. Dowell, 1998. CDOM absorption characteristics with relation to fluorescence and salinity in coastal areas of the Southern Baltic Sea. Estuarine, Coastal and Shelf Science 47: 91–105.

Ferrari, G. M., M. D. Dowell, S. Grossi & C. Targa, 1996. Relationship between the optical properties of chromophoric dissolved organic matter and total concentration of dissolved organic carbon in the southern Baltic Sea region. Marine Chemistry 55: 299–316.

Frenette, J. J., M. T. Arts & J. Morin, 2003. Spectral gradients of downwelling light in a fluvial lake (Lake Saint-Pierre, St-Lawrence River). Aquatic Ecology 37: 77–85.

Gallie, E. A., 1997. Variation in the specific absorption of dissolved organic carbon in Northern Ontario lakes. In Steven, G. A. & F. Robert (eds), Ocean Optics XIII, Proc. SPIE 2963, 417–422.

Green, S. & N. Blough, 1994. Optical absorption and fluorescence properties of chomophoric dissolved organic matter in natural waters. Limnology and Oceanography 39: 1903–1916.

Hoge, F. E., A. Vodacek & N. V. Blough, 1993. Inherent optical properties of the ocean: retrieval of the absorption coefficient of chromophoric dissolved organic matter from fluorescence measurements. Limnology and Oceanography 38: 1394–1402.

Huovinen, P. S., H. Penttolä & M. R. Soimasuo, 2003. Spectral attenuation of solar ultraviolet radiation in humic lakes in Central Finland. Chemosphere 51: 205–214.

Kirk, J. T. O., 1994. Light and Photosynthesis in Aquatic Ecosystem. Cambridge University Press, Cambridge, Britain, 57–71.

Kowalczuk, P., W. J. Cooper, R. F. Whitehead, M. J. Durako & W. Sheldon, 2003. Characterization of CDOM in an organic-rich river and surrounding coastal ocean in the South Atlantic Bight. Aquatic Science 65: 384–401.

Laurion, I., M. Ventura, J. Catalan, R. Psenner & R. Sommaruga, 2000. Attenuation of ultraviolet radiation in mountain lakes: factors controlling the among- and within-lake variability. Limnology and Oceanography 45: 1274–1288.

Markager, W. & W. F. Vincent, 2000. Spectral light attenuation and absorption of UV and blue light in natural waters. Limnology and Oceanography 45: 642–650.

Mazzuoli, S., S. Loiselle, V. Hull, L. Bracchini & C. Rossi, 2003. The analysis of the seasonal, spatial, and compositional distribution of humic substances in a subtropical shallow lake. Acta Hydrochimica et Hydrobiologia 31: 461–468.

Morris, D. P. & B. P. Hargreaves, 1997. The role of photochemical degradation of dissolved organic carbon in regulating the UV transparency of three lakes on the Pocono Plateau. Limnology and Oceanography 42: 239–249.

Morris, D. P., H. Zagarese, C. E. Williamson, E. G. Balseiro, B. R. Hargreaves, B. Modenutti, R. Moeller & C. Queimalinos, 1995. The attenuation of solar UV radiation in lakes and the role of dissolved organic carbon. Limnology and Oceanography 40: 1381–1391.

Nelson, N. B., D. A. Siegel & A. F. Michaels, 1998. Seasonal dynamics of colored dissolved organic material in the Sargasso Sea. Deep–Sea Research I 45: 931–957.

Peuravuori, J. & K. Pihlaja, 1997. Molecular size distribution and spectroscopic properties of aquatic humic substances. Analytica Chimica Acta 337: 133–149.

Reche, I., M. L. Pace & J. J. Cole, 1998. Interactions of photobleaching and inorganic nutrients in determining bacterial growth on colored dissolved organic carbon. Microbial Ecology 6: 270–280.

Rochelle-Newall, E. J. & T. R. Fisher, 2002a. Production of chromophoric dissolved organic matter fluorescence in marine and estuarine environments: an investigation into the role of phytoplankton. Marine Chemistry 77: 7–21.

Rochelle-Newall, E. J. & T. R. Fisher, 2002b. Chromophoric dissolved organic matter and dissolved organic carbon in Chesapeake Bay. Marine Chemistry 77: 23–41.

Rochelle-Newall, E., B. Delille, M. Frankignoulle, J. P. Gattusol, S. Jacquet, U. Riebesell & A. Terbruggen, 2004. Chromophoric dissolved organic matter in experimental mesocosms maintained under different pCO_2 levels. Marine Ecology Progress Series 272: 25–311.

Schindler, D. W., P. J. Curtis, B. R. Parker & M. P. Stainton, 1996. Consequences of climate warming and lake acidification for UV-B penetration in North American boreal lakes. Nature 379: 705–708.

Seritti, A., D. Russo, L. Nannicini & R. Del Vecchio, 1998. DOC, absorption and fluorescence properties of estuarine and coastal waters of the Northern Tyrrhenian Sea. Chemical Speciation and Bioavailability 10: 95–106.

Stedmon, C. A., S. Markager & H. Kaas, 2000. Optical properties and signatures of chromophoric dissolved organic matter (CDOM) in Danish coastal waters. Estuarine, Coastal and Shelf Science 51: 67–278.

Vodacek, A., F. E. Hoge, R. N. Swift, J. K. Yungel, E. T. Peltzer & N. V. Blough, 1995. The use of in situ and airborne fluorescence measurements to determine UV absorption coefficients and DOC concentrations in surface waters. Limnology and Oceanography 40: 411–415.

Williamson, C. E., R. S. Stemberger, D. P. Morris, T. M. Frost & S. G. Paulsen, 1996. Ultraviolet radiation in North American lakes: attenuation estimates from DOC measurements and implications for plankton communities. Limnology and Oceanography 41: 1024–1034.

Yacobi, Y. Z., J. J. Alberts, M. Takács & M. Mcelvaine, 2003. Absorption spectroscopy of colored dissolved organic carbon in Georgia (USA) rivers: the impact of molecular size distribution. Journal of Limnology 62: 41–46.

Yan, N. D., W. Keller, N. M. Scully, D. R. S. Lean & P. J. Dillon, 1996. Increased UV-B penetration in a lake owing to drought-induced acidification. Nature 381: 141–143.

Zhang, Y. L., B. Q. Qin, L. Zhang, G. W. Zhu & W. M. Chen, 2005. Optical properties and fluorescence of chromophoric dissolved organic matter (CDOM) of shallow lakes in the middle and lower reaches of the Yangtze River. Journal of Freshwater Ecology 20(3): 451–459.

Hydrobiologia (2007) 581:53–61
DOI 10.1007/s10750-006-0519-z

EUTROPHICATION IN LAKES

Effects of hydrodynamics on phosphorus concentrations in water of Lake Taihu, a large, shallow, eutrophic lake of China

Guangwei Zhu · Boqiang Qin · Guang Gao · Lu Zhang · Liancong Luo · Yunlin Zhang

Abstract To understand the effect of hydrodynamical process on water phosphorus concentration, wind, wave, and several water quality indices were observed in Meiliang Bay, a shallow and eutrophic bay locates in north of Lake Taihu. During the 7 day observation period, wind speed and significant wave height were recorded more than 3 h per day, and water samples were collected in five water-depth layers once a day. Hydrodynamical disturbance had no significant correlationship with the water quality at the top layer when the significant wave height was smaller than 30 cm, but it significantly increased suspended solids (SS) concentration of the bottom water layer. Concentrations of nutrients showed no positive correlationship with SS concentration in the water body. Intensive sediment resuspension may not have occurred when the hydrodynamic stress on sediment was only a little higher than the critical stress for sediment resuspension. A new method for confirming the critical stress for intensive sediment resuspension and nutrient release still needs to be developed. The range of the water quality indices was quite high during the seven days of observation. High variation seems to be a common character of large shallow lakes like Taihu.

Keywords Shallow lake · Hydrodynamics · Eutrophication · Phosphorus · Algae bloom · Lake Taihu

Guest editors: B. Qin, Z. Liu & K. Havens
Eutrophication of shallow lakes with special reference to Lake Taihu, China

G. Zhu (✉) · B. Qin · G. Gao · L. Zhang · L. Luo · Y. Zhang
Nanjing Institute of Geography and Limnology, Chinese Academy of Sciences, Nanjing 210008, China
e-mail: gwzhu@niglas.ac.cn

Introduction

In shallow lakes, wind-induced waves frequently disturb the water-sediment interface, and often cause intensive sediment resuspension. Strong winds may cause significant increases of suspended solids (SS) concentration in overlying water (Aalderink et al., 1984; Carper & Bachmann, 1984; Luettich et al., 1990; Qin et al., 2000; Zhu et al., 2004; Zhu et al., 2005b). The intensive sediment resuspension may increase the internal release of nutrients from sediment (Søondergåard et al., 1992), but sometimes it may also help deposit soluble nutrients into the bottom sediments by sorption and precipitation (Gunatilaka, 1982). Simulated experiments also indicated that hydrodynamics may have both positive and negative effects on sediment nutrient release (Kristensen et al., 1992; Fan et al., 2001). Therefore, effects of hydrodynamical processes on the exchange of nutrients between sediment and water are quite complicated.

Lake Taihu is the third largest freshwater lake in China (Qin et al., 2007). Water quality of Lake Taihu has been deteriorated quickly since the 1960's due to the drastic effects of anthropogenic activities (Dokulil et al., 2000). The nutrition status of Meiliang Bay (a bay in the northern Lake Taihu) changed from oligo-mesotrophic stage with low algal biomass before 1981 to a eutrophic condition with blooms of *Microcystis* during 1988–1995, and a hyper-eutrophic conditions with the dominance of *Planctonema* and total phosphorus up to 200 mg m^{-3} during 1996–1997 (Chen et al., 2003). Sediment investigations also showed a strong effect of municipal drainage on the water quality of northern bay of Lake Taihu (Qu et al., 2001). At present, the dominant phytoplankton species of Lake Taihu are *Microcystis, Anabaena, Melosira, Cyclotella and Cryptomonas*. *Microcystis* spp. occupies 85% of algae biomass and forms an algal bloom each summer (Chen et al., 1997). Zooplankton grazing is inadequate to control the algae bloom of *Microcystis* (Chen & Nauwerck, 1996; McNaught et al., 1997). Pen fish culture and reconstruction of floating, leave-floating, and submerged plants may benefit water quality improvement (Li & Yang, 1995; Hu et al., 1998; Pu et al., 1998), but are difficult to carry out in a large area like Meiliang Bay.

Wave processes occur frequently in Lake Taihu due to its shallowness and special shape. For example, from 1997 to 1999, daily maximum wind speeds (calculated as hourly average wind speed) higher than 5 ms^{-1} occurred 89.5% of the time, and wind speeds higher than 8 ms^{-1} occurred 34.2% of the time (Fan et al., 2004). A simulation study on Lake Taihu sediment showed that water disturbance caused soluble reactive phosphorus (SRP) release from sediment which was 8–10 times higher than release under undisturbed conditions (Fan et al., 2001). Wave flume experiments showed that the critical shear stress (0.03–0.04 N m^{-2}) that leads to extensive sediment resuspension in Lake Taihu could be induced by a 4–6 ms^{-1} southeast wind event (Qin et al., 2004). Nutrients released from sediment resuspension caused by wind-induced wave may be the main type of internal load of nutrients in Lake Taihu (Qin et al., 2004), but there is little knowledge about the amounts of suspended sediment and released nutrients and also the form of released nutrients in the lake due to the great difficulty in making field observations.

The objectives of this paper are to present field observations on hydrodynamics, wind, water chemistry and algae biomass in Lake Taihu, and to discuss the effect of hydrodynamics on nutrient release from sediments.

Materials and methods

Field observation description

Meiliang Bay locates in the north of Lake Taihu, with an area of 130 km^2. The observation site was selected in the centre of Meiliang Bay to minimize the effect of lake shore on hydrodynamics. Near the observation site, is a regular water quality observation site of Taihu Laboratory for Lake Ecosystem Research (TLLER), which has been monitored every month since 1991. The water depth was 2.4 m at the observation site and did not change significantly during the observation period. On 10th July 2003, a steel-tube platform was set up in the observation site. A wave-meter with three sensors was fixed on the observation platform for wave measurement. An Acoustic Doppler Profile (ADP) was fixed on the platform for lake current measurement. And a Davis meteorological station was fixed on the platform for wind-speed and direction measurement. All of the data were downloaded to computer regularly. A 20 cm long sediment core was collected and sliced to sub-sample each 2 cm depth. Water content (WC, %), bulk density (g cm^{-3}), porosity (%) and loss on ignition (LOI, %) of each sub-sample were measured within 3 h. At the same time, a polymethyl-methacrylate peeper interstitial water sampler with 36 cm long, 1 cm thickness and 0.45 μm membrane was laid at the observation site. After 7 days, the peeper sampler was pulled out and the interstitial water was sampled immediately in each centimeter for Fe^{2+}, PO_4^{3-} and NH_4^+ analysis. Redox potential (Eh) of the interstitial water was immediately measured, too. The meteorological station did not work well the first day, resulting in the loss of some data, so data of wind-speed and direction per hour from the TLLER standard weather station were used for that period.

From 11 July to 17 July, lake water was sampled at 13:00 each day at the five layers (20, 50, 100, 200, and 230 cm). For each water sample, about 100 ml was filtered in the field by 0.45 μm aperture GF/F filter membrane for measuring dissolved total phosphorus (DTP), soluble reactive phosphorus (SRP), ammonia nitrogen (NH_4^+-N), and dissolved organic carbon (DOC). A 250 ml water sample was fixed by alkaline potassium iodide (KI) and manganese sulfate ($MnSO_4$) for dissolved oxygen (DO) determination. Another 1000 ml unfiltered water sample was preserved for total phosphorus (TP), suspended solids (SS) and chlorophyll a (Chl-*a*) analysis. All of the water samples were sent to TLLER for chemical analysis within 2 h. Organic matter in SS was evaluated by loss on ignition (LOI) on the GF/C filter membrane.

Chemical analysis methods

For TP and DTP analysis, water sample was digested by alkaline potassium persulphate in high pressure sterilization pot at 120°C. Then phosphate concentration was determined by molybdenum–antimony–ascorbic acid colorimetry with a Shimadzu UV 2401 spectrophotometer. SRP and NH_4^+-N were determined by colorimetry with Skalar flow-injection analyzer (detection limit for phosphate is 0.001 mg P l^{-1}). DO was determined by sodium hyposulfite titration. DOC was analyzed by I/O 1020A TOC Analyzer. SS was determined by a gravimetric method with Whatman GF/C filter membrane filtration and oven-dried the filtrating residua over 4 h at 105°C. LOI of SS was also determined by gravimetric method with 550°C ignite the SS contained GF/C membrane over 5 h. To determine Chl-*a* concentration, 250 ml water sample was filtered through 0.45 μm pore size synthetic fabric membrane. Chl-*a* concentration was determined by colorimetry after the residue on the membrane was frozen, grinded and extracted by ethanol.

Statistic methods

One way ANOVA analyses of water quality among different wave conditions with Scheffe's method and Tukey's method were carried out by SPSS 11.0 statistical software. Correlation analysis among observed indices was also done by SPSS 11.0.

Results

Physicochemical characteristics of sediment and interstitial water

Eh of interstitial water showed a steep decline in the top 7 cm layer of sediment (Fig. 1), and remained constant in the layer deeper than 7 cm. Concentration of Fe^{2+} showed a continuously increasing trend with depth (Fig. 1), suggesting there is a sharp concentration gradient of Fe^{2+}. This may owe to the release of Fe^{2+} from interstitial water to more oxidation overlying water with increased Eh. Concentration of PO_4^{3-} also showed an increasing trend from overlying water to sediment depth. A peak value at 5–7 cm depth may indicate that internal release of P from sediment by diffusion occurred mainly in the surface 7 cm. The alkaline phosphate activity (APA) and organic phosphorus decomposing bacteria (OPDB) also showed a peak value at about 7 cm and declined to background value at sediment depths over 11 cm in the sediment of Lake Taihu (Wang, 2004). The sediment properties and chemical forms of phosphorus changed drastically in the depth of 5–15 cm in an investigation of vertical distribution of physicochemical characteristics and phosphorus chemical forms of sediment in four sub-lake areas of Lake Taihu (Zhu et al., 2003). Therefore, 7 cm depth of sediment in Lake Taihu may be the peak value of activity of anaerobic bacteria for organic matter and phosphorus regeneration from organic debris. However, no significant increase of NH_4^+ concentration occurred in interstitial water of the top 15 cm except in the surface 1 cm.

Water content, LOI percentage, porocity and bulk density of sediment also indicated that the surface 7 cm sediment is quite different from the

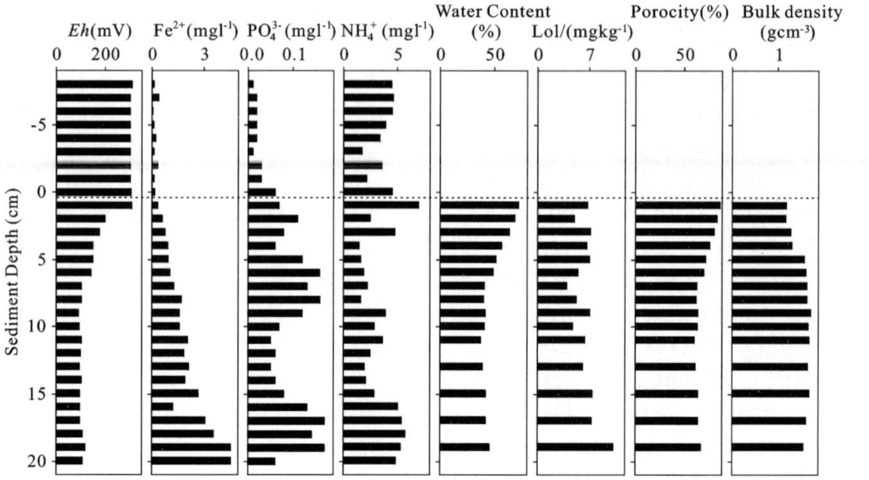

Fig. 1 Vertical distribution of Eh, Fe^{2+} concentration, PO_4^{3-} concentration, NH_4^+ concentration in interstitial water and physicochemical characteristics of sediment of observed site

deeper sediment. Qin et al. (2004) investigated the vertical distribution of physiochemical characteristics of sediment in the whole Lake Taihu and found that the surface 5–15 cm sediment was significantly different from the deeper layer. They concluded that the top 5–15 cm sediment was the active layer participating in the material cycling of Lake Taihu. Thereby, hydrodynamic processes in Lake Taihu may mainly affect the top 5–15 cm sediment that affects nutrient exchange with lake water.

Wind and wave conditions during field observation

During the observation, wind speed varied from 0 ms^{-1} to 7.0 ms^{-1}. Significant wave height varied from 5.85 cm to 28.25 cm. No strong wind-wave process occurred during the observation. The mean wind speed and direction between 12:30–13:30, mean significant wave height per hour of 10:00–11:00, 11:00–12:00, 12:00–13:00 per day, and the weighted mean value of the 3 h are presented in Table 1. Generally, the hydrodynamical disturbance intensity was divided into three kinds of situation according to significant wave height: Over 25 cm at 13th July, 12–20 cm on 16th July, 12th July, and 15th July, and lower than 12 cm on 17th July, 14th July, and 11th July.

The intensity of wind-induced waves was affected significantly by wind direction, though the observation site was located in the central of Meiliang Bay. The wave heights were significantly lower on 12th July, 15th July and 16th July than on other dates because of northeast and southeast winds (Fig. 2). A small hill near the east shore of Meiliang Bay and a relatively short wind-fetch-length explain this observation.

Suspended solids concentration and its LOI percentage

No significant differences of SS concentrations were observed among the 7 days when the SS concentrations of the top four layers were used to denote the condition of water body. However, SS concentrations in the bottom layer were quite different among the different wave conditions. The SS concentration of the bottom layer water was the highest value (154 mg l^{-1}) in 13th July. It is consistent with the strongest wind-induced wave during sampling than the other days (Fig. 2). This result shows that hydrodynamic disturbances did not have significant effects on the top water in Lake Taihu because the significant wave height was lower than 20 cm. This wave intensity may be too common and frequent in Lake Taihu to cause distinguished differences among different wind-induced wave processes.

Concentration of LOI in SS showed a similar trend to that of SS concentration. Generally, there were no significant differences of LOI

Table 1 Wind speed and H_s during the field observation

Date	Wind speed/direction/(ms^{-1})			H_s /(cm)			
	10:00–11:00	11:00–12:00	12:00–13:00	10:00–11:00	11:00–12:00	12:00–13:00	Weighted mean value[a]
11-July	4.1/307	2.1/55	1.2/7	15.14	10.63	6.95	9.54
12-July	3.4/132	2.7/134	2.0/129	19.33	18.52	12.91	15.85
13-July	2.3/235	1.2/239	6.1/278	20.35	26.46	27.05	25.74
14-July	3.3/119	2.1/106	1.8/114	13.82	10.82	8.83	10.32
15-July	3.5/302	4.0/34	4.6/14	12.18	12.57	11.90	12.17
16-July	4.8/71	6.2/60	6.3/70	15.98	17.10	20.28	18.50
17-July	2.3/246	1.9/233	1.1/246	13.48	10.58	8.88	10.21

[a] Weighted mean value of H_s was calculated as following: H_s (mean value) = (H_s (10:00–11:00) + 2 × H_s (11:00–12:00) + 3 × H_s (12:00–13:00))/6

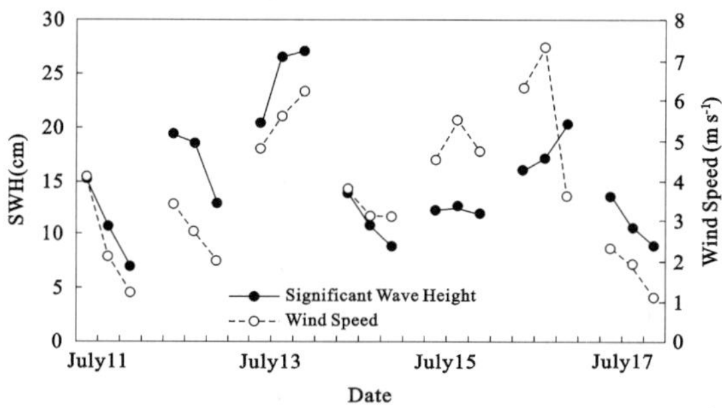

Fig. 2 Wind speed and significant wave height three hours before each days water sampling

concentration in the top four layers among the first 6 days. However, LOI concentration in the top four layers was significantly higher during the last days. This result could be caused by the algae bloom that occurred during the last day. The concentration of Chl-*a* also confirmed this presumption.

Nutrients, DO, and Chl-*a* concentrations

Vertical distributions of TP, DTP and SRP concentrations are presented in Fig. 3. For TP, there was no difference among different days except for 11 July, which was quite low compared to the next 6 days. For DTP, there was no difference among the 7 days. SRP was remarkably high on 15th July. And for NH_4^+, the last 3 days were significantly higher than the first 3 days. For Chl-*a* and DO, the last day is markedly higher than the previous days.

Correlation analysis indicated that there was not any significantly correlation between hydrodynamical disturbance intensive and water quality indices in the top four layers of water. DTP concentration correlated with SRP concentration in the top four layers ($r = 0.83$, $n = 7$), and in the bottom layer ($r = 0.79$, $n = 7$). LOI concentration correlated with Chl-*a* concentration and DO concentration in the top four layers, which means the water DO and organic suspended solids are determined by algae biomass.

SS and LOI concentrations in bottom layer were significantly affected by the intensive hydrodynamical disturbance ($r = 0.83$, $n = 7$). However, there was no significant relationship between SS and the intensive hydrodynamic disturbance in the top four layers. It suggested that wind-induced wave may have caused surface sediment resuspension, but the effect was limited in the bottom layer.

Discussion

Effects of wind-induced wave disturbance on phosphorus release from sediment in shallow lake

Do wind-induced waves cause phosphorus release from sediment to overlying water in shallow lakes? Our observations in Meiliang Bay suggest that they do not. However, we can not conclude that sediment resuspension will not affect nutrient concentrations in overlying water. In fact, field

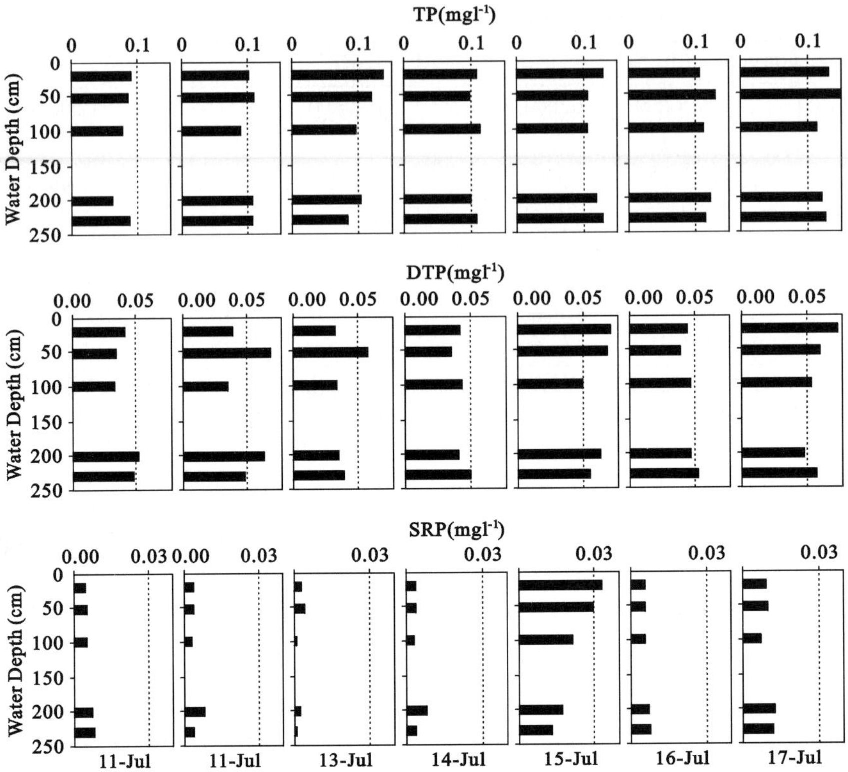

Fig. 3 Vertical distribution of TP, DTP and SRP concentrations in overlying water

observations have indicated that strong wind could increase TP, DTP and SRP concentrations in lake water as the wind speed was over 12 ms^{-1} in Lake Taihu (Zhu et al., 2004, 2005b). However, no significant release of phosphorus from sediment was found in this field observation, perhaps due to the relative low wind speed. In the wave flume experiments, Qin et al. (2004) found that the critical shear stress that leads to extensive sediment resuspension in Lake Taihu was about 0.03–0.04 N m^{-2}, equivalent to a wind speed in situ up to 4 ms^{-1}. The stress of wind-induced wave calculated minute by minute using observed data varied between 0.01 N m^{-2} and 0.11 N m^{-2}, and its peak value occurred between 18:00–18:30, 16th July. But intensive sediment resuspension still was not observed. This result suggests that the critical shear stress of 0.03–0.04 N m^{-2} isn't the correct stress for extensive sediment resuspension, but is simply the stress needed for surface sediment particles to begin to move up. According to hydraulic judging criteria, critical stress for sediment resuspension means the surface sediment particles begin to gradually float upward. In another wave flume experiments, Zhu et al. (2005a) found that SS concentration increased from 2.5 mg l^{-1} without any waves to 13.6 mg l^{-1} when wave height just exceeded the critical wave height (5.93 cm). But SS concentration increased to 224 mg l^{-1} as wave height increased to 13.29 cm. This means that sediment won't suspend extensively as the wave height become a little higher than the critical wave height. The wave height that will cause intensive sediment resuspension by field observation and simulated flume experiment is still not known.

In other field observations of hydrodynamical processes in Meiliang Bay, SS concentration in overlying water increased significantly when the wind-speed over 6.5 ms^{-1} in Meiliang Bay (Qin et al., 2000). SS concentration was about 120 mg l^{-1} with wind-speed over 6.5 ms^{-1}. The SS concentration was over 400 mg l^{-1} in top layer water with wind speed over 12 ms^{-1} (Zhu et al.,

2004, 2005b). At the same time, TP, DTP, SRP concentration in the top layer water increased significantly, too. These results indicate that the wind-speed needs to be between 7 and 12 ms^{-1} for intensive sediment resuspension. However, wind speed seldom exceeded 6.5 ms^{-1} in this field observations. Thus, SS and nutrients concentration in overlying water had not significantly varied with the intensity of hydrodynamical disturbance perhaps due to lower wave height than the critical wave height for intensive sediment resuspension.

Sediment resuspension does not always increase phosphorus concentration in overlying water in shallow lakes. Correlative analysis indicates that concentrations of DTP correlate inversely with SS concentration ($r = 0.83$, $p < 0.05$). The reason is that surface sediments are often oxidized in large shallow lakes. The surface floc is in constant equilibrium with the water column. When the surface floc suspended to overlying water, it absorbs dissolved phosphorus from water phase. This phenomenon was also found in shallow Lake Yangebup (Linge & Oldham, 2004).

Although a positive correlationship between water SS concentration and intensity of hydrodynamic disturbance was not observed in this field observation, it should be noted that a relative strong wave occurred in the third day, with subsequent nutrient concentration increases in the fifth day followed by an algae bloom on the seventh day. However, cause and effect relationships among the three phenomenons are still unclear.

Variation of water quality indices in large shallow lake

Water quality indices of shallow lake are highly variable because of the spatial and vertical inhomogeneity of water quality, as well as the violent influence of hydrodynamical disturbance. Most of the observed water quality indices varied drastically in the 7 days of observation (Fig. 4). For example, DTP concentration in the surface 20 cm layer water varied between 0.34 mg l^{-1} and 0.77 mg l^{-1}, SRP varied between 0.002 mg l^{-1} and 0.033 mg l^{-1}, NH_4^+–N varied between 0.056 mg l^{-1} and 0.832 mg l^{-1}, and Chl-*a* varied between 33 mg l^{-1} and 117 mg l^{-1} during the seven observation days.

Compared to the monthly monitored data from 2000 to 2003, considerable variation of the water quality indices occurred among the observed 7 days (Fig. 4). This observation suggests that monthly monitored data may be inadequate to reflect the water quality variation in large shallow lakes like Lake Taihu. For example, the cycle time of phytoplankton colony is generally a week. Thus, Chl-*a* concentration must be quite different between water sampling events depending on the

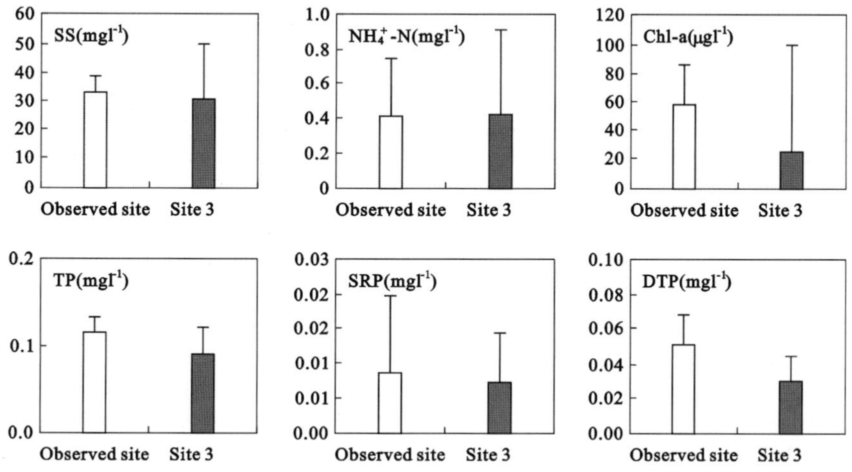

Fig. 4 Variation of the observed surface 20 cm water quality indices during the 7 days and that of the monthly monitored data at No. 3 regular monitoring site of TLLER

presence or absence of an algae bloom. Additionally, wind speed and direction during water sampling can affect water quality indices, such as Chl-*a* concentration, phytoplankton structure (Chen et al., 2003). Therefore, one must be careful in using monthly monitored data to explain water quality change in large shallow lakes.

Acknowledgements This work was supported by The Key Project of Chinese Academy of Sciences (KZCX1-SW-12) and Natural Science Foundation of China (No. 40673078, 40573062). We also would like to thank Chen Weimin, Hu Weiping, Zhang Fabing, Ji Jiang, Qian Rongshu, Song Yuzhi, Yang Liuyan, Zhang Xiaofeng, Yang Longyuan for their help on field observation, sampling and chemical analysis.

References

Aalderink, R. H., L. Lijklema, J. Breukelman, W. Raaphorts & A. G. Brinkman, 1984. Quantification of wind-induced resuspension in a shallow lake. Water Science & Technology 17: 903–914.

Carper, G. L. & R. W. Bachmann, 1984. Wind Resuspension of Sediments in a Prairie Lake. Canadian Journal of Fisheries and Aquatic Sciences 41: 1763–1767.

Chen, W. M. & A. Nauwerck, 1996. A note on composition and feeding of the crustacean zooplankton of Lake Taihu, Jiangsu Province, China. Limnologia 26: 275–279.

Chen, W. M., Y. W. Chen, X. Y. Gao & I. Yoshida, 1997. Eutrophication of Lake Taihu and its control. Agricultural Engineering Journal 6(2): 109–120.

Chen, Y. W., B. Q. Qin, K. Teubner & M. Dokulil, 2003. Long-term dynamics of phytoplankton assemblages: Microcystis- domination in a large shallow lake in China. Journal of Plankton Research 25: 445–453.

Dokulil, M., W. Chen & Q. Cai, 2000. Anthropogenic impacts to large lakes in China: the Tai Hu example. Aquatic Ecosystem Health and Management 3: 81–94.

Fan, C. X., L. Zhang & W. C. Qu, 2001. Lake sediment resuspension and caused phosphate release - a simulation study. Journal of Environmental Sciences 13: 406–410.

Fan, C. X., L. Zhang, B. Q. Qin, S. M. Wang, W. P. Hu & C. Zhang, 2004. Estimation on dynamic release of phosphorus from wind-induced suspended particulate matter in Lake Taihu. Science in China Series D 47: 710–719.

Gunatilaka, A., 1982. Phosphate adsorption kinetics of resuspended sediments in a shallow lake, Neusiedlersee, Austria. Hydrobiologia 91: 293–298.

Hu, W. P, J. Salomonsen, F. L. Xu & P. M. Pu, 1998. A model for the effects of water hyacinths on water quality in an experiment of physico-biological engineering in Lake Taihu, China. Ecological Modelling 107: 171–188.

Kristensen, P. M., M. Søondergåard & E. Jeppesen, 1992. Resuspension in a shallow lake. Hydrobiologia 228: 101–109.

Li, W. C. & Q. X. Yang, 1995. Wetland utilization in Lake Taihu for fish faming and improvement of lake water quality. Ecological Engineering 5: 107–121.

Linge, K. L. & C. E. Oldham, 2004. Control mechanisms for dissolved phosphorus and arsenic in a shallow lake. Applied Geochemistry 19: 1377–1389.

Luettich, R. A., D. R. F. Harleman & L. Somlódy, 1990. Dynamic behaviour of suspended sediment concentrations in a shallow lake perturbed by episodic wind events. Limnology & Oceanography 35: 1050–1067.

McNaught, D. C., W. M. Chen & H. J. Cai, 1997. Grazing on a Microcystis bloom by cladocerans and copepods in Tai Lake. Verhandlungen Internationale Vereinigung fur Theoretische und Angewandte Limnologie 26: 550–552.

Pu, P. M., W. P. Hu, J. S. Yan, G. X. Wang & C. H. Hu, 1998. A physico-ecological engineering experiment for water treatment in a hypertrophic lake in China. Ecological Engineering 10: 179–190.

Qin, B. Q., W. P. Hu, W. M. Chen, C. X. Fan, J. Ji, Y. W. Chen, X. Y. Gao, L. Y. Yang, G. Gao, W. Y. Huang, J. H. Jiang, C. Zhang, Y. B. Liu & Z. Y. Zhou, 2000. Studies on the hydrodynamic processes and related factors in Meiliang Bay, Northern Taihu Lake, China. Journal of Lake Sciences 12: 327–334 (in Chinese).

Qin, B. Q., W. P. Hu, G. Gao, L. C. Luo & J. S. Zhang, 2004. Dynamics of sediment resuspension and the conceptual schema of nutrient release in the large shallow Lake Taihu, China. Chinese Science Bulletin 49: 54–64.

Qin, B. Q., P. Z. Xu, Q. L. Wu & L. C. Luo, 2007. Environmental issues of Lake Taihu, China. Hydrobiologia 581: 3–14.

Qu, W. C., M. Dickman & S. M. Wang, 2001. Multivariate analysis of heavy metal and nutrient concentrations in sediments of Taihu Lake, China. Hydrobiologia 450: 83–89.

Søondergåard, M., P. Kristensen & E. Jeppesen, 1992. Phosphorus release from resuspended sediment in the shallow and wind exposed Lake Arresø, Denmark. Hydrobiologia 228: 91–99.

Wang, J. J., 2004. Effect of biogeochemistry processes of new born surface caused by sludge dredging on the internal loading recovery in a lake. Master thesis of Nanjing Institute of Geography and Limnology, Chinese Academy of Sciences.

Zhu, G. W., G. Gao, B. Q. Qin, L. Zhang & L. C. Luo, 2003. Geochemical characteristics of phosphorus in sediments of a large shallow lake. Advances in Water Science 14: 714–719 (in Chinese).

Zhu, G. W., B. Q. Qin & G. Gao, 2004. Vertical distribution of the concentrations of phosphorus and suspended solid in Lake Taihu affected by wind-induced wave. Advances in Water Science 15: 775–780 (in Chinese).

Zhu, G. W., B. Q. Qin, L. Zhang, L. C. Luo, X. J. Sun, D. L. Hong, Y. J. Gao & R. Xie, 2005a. Study on the effects of wave on nutrients releasing of sediments from Lake Taihu by flume experiments. Journal of Lake Sciences, 17(1): 61–68 (in Chinese).

Zhu, G. W., B. Q. Qin & G. Gao, 2005b. Direct evidence of phosphorus outbreak release from sediment to overlying water in a large shallow lake caused by strong wind. Chinese Science Bulletin 50: 577–582.

Hydrobiologia (2007) 581:63–70
DOI 10.1007/s10750-006-0515-3

EUTROPHICATION IN LAKES

Total inputs of phosphorus and nitrogen by wet deposition into Lake Taihu, China

Liancong Luo · Boqiang Qin · Longyuan Yang · Yuzhi Song

Abstract Lake Taihu suffers from eutrophication caused by riverine nutrient inputs and air deposition. To characterize wet deposition of phosphorus (P) and nitrogen (N) to the lake, precipitation collection and measurements of total phosphorus (TP) and total nitrogen (TN) and other components at five cities around Lake Taihu were made from July 2002 to June 2003. TP and TN concentrations and deposition rates exhibited strong spatial variation in the whole catchment. An inverse correlation between station-averaged TP and TN concentrations and precipitation amount was found. Maximal TP concentration in rainfall was found in Suzhou, and maximal TN in Wuxi. However, highest wet deposition rates of TP and TN were found in Suzhou, which suggests that atmospheric nutrients are mostly from the east and northwest area of Lake Taihu. Mean TP and TN deposition rates were 0.03 and 2.0 t km^{-2} year^{-1} respectively in Lake Taihu, which are greater than reported values in other areas by comparision. Total N and P contributed to the lake by wet deposition were 75 and 4720 t per year, respectively, which represent about 7.3% and 16.5% of total annual N and P inputs via inflow rivers. Wet deposition, especially N, could have significant effects on eutrophication in the lake, which shows that air deposition should be taken into account while reducing the external nutrients in the lake.

Keywords Nitrogen · Phosphorus · Wet deposition · Eutrophication · Riverine inputs · Lake Taihu

Guest editors: B. Qin, Z. Liu & K. Havens
Eutrophication of shallow lakes with special reference to Lake Taihu, China

L. Luo · B. Qin (✉) · L. Yang · Y. Song
Nanjing Institute of Geography & Limnology, Chinese Academy of Sciences, Nanjing 210008, China
e-mail: qinbq@niglas.ac.cn

Introduction

Wet deposition may impact freshwater ecosystems, terrestrial environments (Buijsman & Erisman, 1988; Likens, 1989; Khemani et al., 1989; Lovett & Kinsman, 1990; Owens et al., 1992; Spokes et al., 1993; Helmers & Schrems, 1995; Raper & Lee, 1996) and coastal seas (De Leeuw et al., 2003). Input of nutrients by rainfall (e.g. P and N) and acidic rain characterized by low pH (<5.6) can affect the chemistry of lakes, soils, forests and wetlands (Gorham, 1998) and make ecosystems nutrient-limited or susceptible to acid deposition. For instance, it was noted (Gordon & Gorham, 1963) that acid deposition close to the iron-sintering plant at Lakes Wawa and Ontario was causing substantial leaching of calcium from local soils into adjacent ponds and lakes. It then was observed that acid deposition led to dissolution of aluminum in soil minerals and strong enrichment of aluminum ions in lake water (Wright et al., 1976). Many other researchers have addressed acidic rain in different areas (Cowling & Nilsson, 1995; Lin et al., 1998; Okochi et al., 2000; Peart, 2000; Vasjari et al., 2000).

Polluted rain also may deposit nutrients and trace elements, resulting in eutrophication and iron-concentration changes. Input of nutrients by wet deposition can contribute to total nutrients in water, especially in seepage lakes, regulate primary productivity and affect algal community structure (Pollman et al., 2002). Atmospheric N can be available for algae growth, which makes this contribution more significant because only a fraction of river runoff contributes N that is incorporated into biological material (De Leeuw et al., 2003). Heavy metals from rainfall (e.g. Zn, Cu, Fe, Ca and Mg) can affect surface water production (Morel et al., 1991) and influence biological activity in marine waters (Coale & Bruland, 1990; Sunda & Huntsman, 1992).

Lake Taihu is located in the lower Changjiang (Yangtze) River delta and is important for water quantity regulation, water supply, cultivation, irrigation, navigation, culture fisheries and tourism. In the 1950s, lake water was oligotrophic (Chang, 1995). With rapid industrial and agricultural development and population growth, increasing amounts of nutrients, heavy metals (Shen et al., 2001) and other pollutants (Feng et al., 2003) have been delivered to the lake via river runoff (Zhou & Zhu, 2003; Chen et al., 2003; Qu et al., 2001) or bulk deposition (Qu et al., 2002; Pu et al., 1998). Pollutants can be absorbed onto or associated with fine-grained sediment particles, resulting in limnic ecosystem recovery prevention and long-term eutrophication problems by sediment-water exchange. The lake is hypertrophic in three northern bays, with seasonal algae blooms increasing the cost of water purification. Consequently, reasons for deterioration and improvement of lake water-quality have received much attention in recent years (Wang et al., 1998; Qin, 1998; Yang et al., 2001; Chen, 2002; Yang et al., 2003; Qin et al., 2004). The largest project, "Zero-point Action", initiated by the State Environmental Protection Administration (SEPA) in 1998, required "Standard Discharge" of all industrial units. However, water quality has not improved remarkably despite reduction of external loading from rivers, which raises the possibility that net air-deposition and sediment resuspension may have considerable impacts.

Many reports have focused on effects of sediment and river inputs (Zheng et al., 2001) on water quality degradation in Lake Taihu, including acid rain impacts (Chen et al., 2000, 2001; Yang et al., 2001; Yang, 2001). Little is known about amounts of N & P contributed by wet deposition. The aim of this study is to estimate wet-deposition of P and N into the lake based on data from samples collected from July 2002 to June 2003.

Methodology

Site description

Fifteen sampling sites, located in five cities (Wuxi, Suzhou, Huzhou, Jintan and Changzhou; Fig. 1) around Lake Taihu, were selected for rain collection. Four were at Changzhou situated northwest of the lake, four in Suzhou in the east, three in Huzhou in the south, two in Wuxi in the north and two in Jintan in the northwest (Table 1). Several sites in each city were sampled

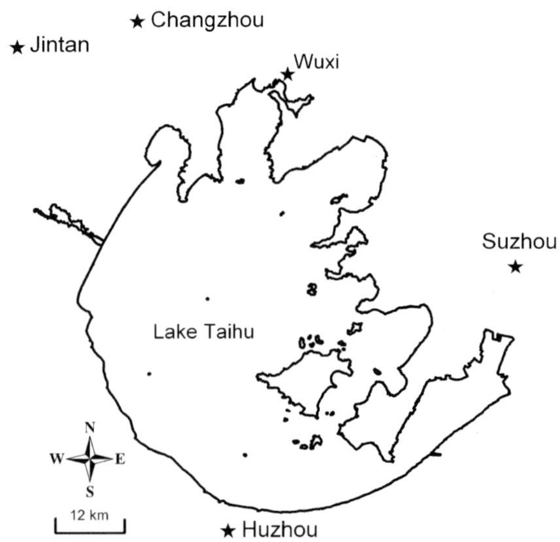

Fig. 1 Map of the study area

Table 1 Sample stations and their locations in Taihu Basin

Station	Longitude (°E)	Latitude (°N)	Site number	Location
Wuxi	120.25	31.58	2	North
Suzhou	120.61	31.30	4	East
Changzhou	119.97	31.77	4	Northwest
Jintan	119.58	31.73	2	Northwest
Huzhou	120.10	30.89	3	South

to evaluate different land-use types (e.g. rural, urban, industrial, forest, etc.).

Sample collection and analysis

Precipitation samples were collected automatically by a tipping-bucket rain gauge at all sites from July 2002–June 2003. Analytical parameters included pH, conductivity, K^+, Na^+, Ca^{2+}, Mg^{2+}, Cl^-, SO_4^{2-}, NO_3^-, NH_4^+, TP and TN (no TP & TN in Huzhou due to technical limits). All chemical components were analyzed monthly in the local Environmental Monitoring Center (EMC). pH was measured immediately without stirring by a digital pH meter, and conductivity was measured using an ion chromatography detector. All ions were determined using suppressed ion chromatography (Chen et al., 2001). All data from different EMCs were assumed to have similar accuracy since ion concentration measurements were done using standard SEPA methods.

Calculation of annual amount of deposited nutrients

Annual amounts (in: t $year^{-1}$) of TP and TN can be calculated from the following equation,

$$AA = \frac{1}{m}\sum_{j=1}^{m}\sum_{i=1}^{n} S \times RR_{ij} \times C_{ij} \times 10^{-3} \qquad (1)$$

where S is the lake surface area (km^2), RR_{ij} (i = 1,2,...12; j = 1,2,3,4) is the precipitation rate (mm $month^{-1}$) at station j and month i, C_{ij} is the concentration of a substance in precipitation (mg l^{-1}) with the same subscript as RR_{ij}, m is the number of stations (4 in this case due to lack of data in Huzhou). The wet deposition rate (in: t km^{-2} $year^{-1}$) of chemical components can be derived from Eq. 1 without consideration of S.

Results

Seasonality

Monthly precipitation (RR) and TN and TP concentrations are shown in Fig. 2. RR ranged from 24–130 mm per month. Minimum RR was in January, with a value of 24.3 mm, and November, with a value of 50.7 mm. There was more rainfall in summer (128 mm in August and 117 mm in July). RR exhibited a summer (June–August) maximum and a fall (September–November) minimum, demonstrating strong seasonality.

Fig. 2 Monthly station-averaged precipitation amount (RR, in mm), precipitation-weighted concentrations of TP (in: 100 mg l^{-1}) and TN (in: mg l^{-1})

TP concentration varied from 0.01 to 0.11 mg l^{-1} with the peak value in January, which is, in magnitude, similar with those reported by Brown et al. (1984, 0.015–0.15 mg l^{-1}) and Migon and Sandroni (1999, 0.002–0.14 mg l^{-1},), but significantly greater than those reported by Chen et al. (1985) and Filoso et al. (1999). Although no significant seasonal variation in TN was found (Fig. 2), there was a seasonality of TP and TN despite slight differences among the 12 months. TN and TP followed an inverse pattern to RR.

TN contributed by rain followed a similar seasonal pattern to RR and peaked in summer and fall, with minima in winter and spring. However, TP had a maximum in fall and winter and minimum in spring and summer (Fig. 3). These seasonal differences varied from those of TP and TN concentrations throughout the year due to differences in precipitation amounts.

Total annual TP and TN deposition rates

Deposition rates of NO_3^-, NH_4^+, TP and TN were estimated from Eq. (1) with lake surface area excluded. In each city, monthly precipitation amounts were used to calculate average precipitation rates. Table 2 shows month-averaged wet deposition rates of the five cities. There was no data on TN and TP in Huzhou, but concentrations of NO_3^- and NH_4^+ suggest a moderate contribution. Spatial differences suggest that the east region of Lake Taihu (Suzhou), and then the northwest (Changzhou and Jintan) have the largest N and P contribution from rainfall. Wuxi and Huzhou have moderate influences. Wet deposition of TP within the lake catchment was 0.03 t km^{-2} $year^{-1}$, TN was 2.0 t km^{-2} $year^{-1}$, NO_3^- was 3.70 t km^{-2} $year^{-1}$, and NH_4^+ was 1.56 t km^{-2} $year^{-1}$. Based on individual deposition rates, total annual contribution from precipitation was 75 and 4720 t per year for TP and TN, respectively.

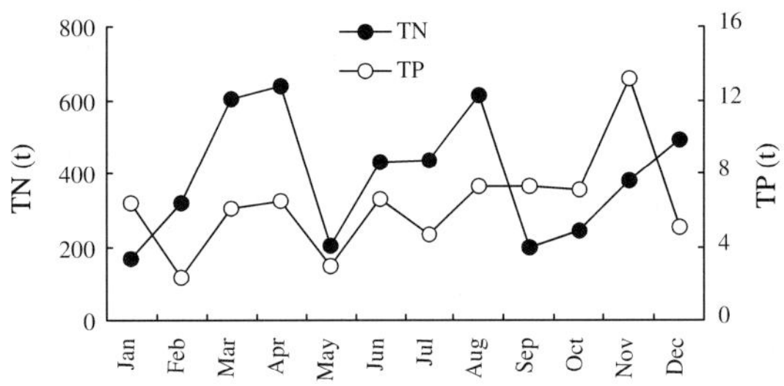

Fig. 3 Monthly station-averaged amounts of TP & TN (in: t)

Discussion

The simple model based on Eq.1 may overestimate total input of N & P into Lake Taihu because precipitation chemistry over the lake may differ from cities. Raindrops are formed when small water droplets adhere to particles, including chemical components in air, and, consequently, depend on local air movement and concentrations of vapor and particles in the atmosphere. Due to low nutrient concentrations in air over the lake, especially the center, precipitation inputs of P and N in these regions may be lower than those in the shore areas. However, this uncertainty for regional differences is difficult to quantify unless many observation sites are established over the water surface of Lake Taihu.

Despite the spatial variability of nutrient concentrations in precipitation, results obtained from observation data and calculations should indicate

Table 2 Wet-deposition rates (t km^{-2} $year^{-1}$) and concentrations (mg l^{-1}) of TP and TN at different stations

City	Wet deposition rate (t km^{-2} $year^{-1}$)				Averaged concentration (mg l^{-1})	
	TP	TN	NO_3^-	NH_4^+	TP	TN
Wuxi	0.016	2.25	2.76	1.44	0.022	3.28
Suzhou	0.06	2.81	4.36	2.26	0.075	2.58
Changzhou	0.018	1.81	0.82	1.52	0.023	2.22
Jintan	0.033	1.20	6.70	0.95	0.057	1.33
Huzhou			3.87	1.65		
Average	0.03	2.0	3.70	1.56	0.044	2.35

effects of wet deposition on lake eutrophication. In Lake Taihu, the primary source of N and P is from Suzhou, where the highest TP concentration and deposition rates were found. Changzhou and Jintan had the second greatest contribution of TN and TP. Thus, the main N and P inputs via rainfall are from the east and northwest. The smallest TP deposition rate and the greatest TN concentration were found in Wuxi. Therefore, Wuxi, with a greater annual precipitation, may make the greatest contribution to total N input into the lake.

Deposition rates of NO_3^- and NH_4^+ are high relative to TN inputs, with values of 3.70 and 1.56 t km^{-2} year^{-1}, respectively, which are 2 and 5 times higher than those for Barnegat Bay (1.81 and 0.34 t km^{-2} year^{-1} for NO_3^- and NH_4^+; Gao, 2002). Estimated TP deposition rate (0.03 t km^{-2} year^{-1}) near Lake Taihu also is high compared to that in the Amazon (0.004 t km^{-2} year^{-1}; Swap et al., 1992), Connecticut (0.004 t km^{-2} year^{-1}; Yang et al., 1996), New Zealand (0.01 t km^{-2} year^{-1}; Chen et al., 1985), Corsica (0.01–0.02 t km^{-2} year^{-1}; Bergametti et al., 1992), Apalachicola, FL (0.012 t km^{-2} year^{-1}; Fu & Winchester, 1994) and Ireland (0.021 t km^{-2} year^{-1}; Gibson et al., 1995). Also, TN deposition rate is about 10 times that in Colorado and southern Wyoming (<0.2 t km^{-2} year^{-1}; Burn, 2003). High N and P deposition rates result from livestock and fertilization of croplands (Vitousek et al., 1997), crop senescence, fossil fuel combustion, automobile emissions and factories. In Taihu Basin, agriculture and coal combustion are the main contributors to high rates of N and P deposition.

Wet deposition is greater than dry deposition, especially in ocean areas (Owens et al., 1992; Hertel et al., 2002), the Mississippi River Basin (Lawrence et al., 2000), and Barnegat Bay (Gao, 2002). With this hypothesis, annual net atmospheric inputs of TP and TN into Lake Taihu should be slightly higher than estimated values. However, deposition effects on lake eutrophication are significant even excluding dry deposition. Based on nutrient concentrations observed at inflow rivers connected to Lake Taihu, annual riverine input of TP was estimated to be 1,030 t and TN about 28,650 t (Xu & Qin, 2005). Therefore, atmospheric inputs of TP and TN represent about 7.3% and 16.5% of annual total inputs of P and N via inflow rivers, which suggests that air deposition should be considered in the evaluation nutrient inputs to Lake Taihu. Nitrogen from air deposition may promote primary productivity. For example, 5.3 mmol C m^{-2} day^{-1} could result from average deposition of 0.8 mmol N m^{-2} day^{-1} (De Leeuw et al., 2003). Also, atmospheric deposition can affect nitrogen-saturated forests and acidic freshwater lakes and streams (Stoddard, 1994; Aber et al., 1995; Fenn et al., 1998). However, Lake Taihu generally is P-limited and inherently vulnerable to N deposition (Yin et al., 1996). As a result, deposited P could promote primary productivity. Moreover, some deposited N and P may be absorbed onto or associated with sediment particles, which could make bottom sediment a source of nutrients. In shallow lakes, perturbation has substantial effects on sediment-water interactions (Sheng & Lick, 1979; Furumai et al., 1989; Shrestha, 1996; James et al., 1997; Bailey and Hamilton, 1997; Blom & Winkels, 1998; Poelma & Ooms, 2002), and therefore, waves may resuspend sediment and release nutrients resulting in increased N & P. For example, 1.64×10^4 t P can be suspended from sediment into water of Lake Taihu with a 15.0 m s^{-1} SE wind event, resulting in an average TP increase of 4.70 mg l^{-1} (Luo, 2004). Some suspended P may be available for aquatic plants, and some may be redeposited.

Atmospheric N originates mainly from ammonium nitrate aerosol (NH_4NO_3), ammonia gas (NH_3) and nitric acid gas (HNO_3) (Lindberg et al., 1990). NO_x exists primarily as NH_4NO_3 and HNO_3, and can be transformed to HNO_3 after photochemical processes. HNO_3 and NH_3 (primary pollutant) can form an equilibrium with NH_4NO_3 (Tarnay et al., 2001). These compounds are mainly derived from fossil fuel combustion and agricultural production (e.g. fertilizer and animal waste). The origin of atmospheric deposition can reflect nutrient-source patterns of riverine inputs. Consequently, external N loading control in Lake Taihu Basin should include production of agricultural and fuel-combustion emissions. The "Zero-point Action" initiated by SEPA in 1998 has resulted in significant reduction of TP and TN inputs in Lake Taihu, but agricultural inputs of TP and TN still constitute a large proportion. Thus, control of only industrial discharge is not an effective method to control eutrophication. Municipal policy should include control of agricultural emission and coal combustion inputs in Lake Taihu Basin.

Conclusion

A study of wet deposition in Lake Taihu Basin examined the effects of N and P contributed by rainfall on lake eutrophication. Fifteen sites were selected in five cities around the lake. Results lead to the following conclusions:

(1) Deposition rates and concentrations of TP and TN in precipitation over Lake Taihu exhibited strong spatial variation. N and P in precipitation come mainly from the east (Suzhou) and northwest (Changzhou and Jintan). Wuxi and Huzhou make only moderate contributions to atmospheric inputs into Lake Taihu.

(2) Precipitation had a summer maximum and a fall minimum and followed an inverse seasonal cycle compared to TP and TN concentrations.

(3) Deposition rates of NO_3^- and NH_4^+ were 3.70 and 1.56 t km^{-2} year^{-1}, respectively. TP and TN deposition rates were 0.03 and 2.0 t km^{-2} year^{-1}, respectively.

(4) Total amounts of N and P contributed to the lake by wet deposition were 75 and 4720 t per year, respectively, which may contribute to lake eutrophication.

Acknowledgements We would like to gratefully acknowledge financial support from Chinese Academy of Sciences (No. KZCX1–SW–12), National Natural Science Foundation of China (No. 40501078, 40131160734), Ministry of Science and Technology of China (No. 2002AA601011) and Director Foundation of Nanjing Institute of Geography & Limnology, Chinese Academy of Sciences. Thanks to the Environmental Monitoring Centers of Wuxi, Suzhou, Changzhou, Jintan and Huzhou for their sample collection and analysis. The authors are indebted to Professor Guang Gao, Dr. Guangwei Zhu, Dr. Xiaoli Shi and Dr. Lu Zhang for their help with mathematical analysis on chemical data. This manuscript benefited from the comments of two anonymous reviewers.

References

Aber, J. D., A. Magill, S. G. McNulty, R. D. Boone, K. J. Nadelhoffer, M. Downs & R. Hallett, 1995. Forest biogeochemistry and primary production altered by nitrogen saturation. Water, Air and Soil Pollution 85: 1665–1670.

Bailey, M. C. & D. P. Hamilton, 1997. Wind induced sediment resuspension: a lake-wide model. Ecological Modeling 99: 217–228.

Bergametti, G., E. Remoudaki, R. Losno & E. Steiner, 1992. Source, transport and deposition of atmospheric phosphorus over the northwestern Mediterranean. Journal of Atmospheric Chemistry 14: 501–513.

Blom, G. & H. J. Winkels, 1998. Modeling sediment accumulation and dispersion of contaminants in Lake Ijsselmeer (the Netherlands). Journal of Environmental Management 37: 17–24.

Brown, G., D. T. Mitchell & W. D. Stock, 1984. Atmospheric deposition of phosphorus in a Coastal Fynbos Ecosystem of the southwestern cape, South-Africa. Journal of Ecology 72(2): 547–551.

Buijsman, E. & J. W. Erisman, 1988. Wet deposition of ammonium in Europe. Journal of Atmospheric Chemistry 6: 265–280.

Burn, D. A., 2003. Atmospheric nitrogen deposition in the Rocky Mountains of Colorado and southern Wyoming – a review and new analysis of past study results. Atmospheric Environment 37: 921–932.

Chang, Y. B., 1995. Major environmental changes since 1950 and the onset of accelerated eutrophication in Taihu Lake, China. Acta Palaeontologica Sinica 35: 155–174 (in Chinese).

Chen, H. S., 2002. Analysis of water environmental changes in Taihu Basin. Journal of Lake Sciences 14: 111–116 (in Chinese).

Chen, J. M., G. X. Pan, L. Cang, J. J. Yang & J. H. Wang, 2000. Effect of simulated acid rain on plant growth and paddy soils pH in Taihu area. Journal of Nanjing Agricultural University 23: 116–118 (in Chinese).

Chen, J. M., G. X. Pan & J. W. Zheng, 2001. Effect of simulated acid rain on adsorption and desorption of copper by paddy soils in Lake Taihu area. Acta Pedologica Sinica 38: 333–340 (in Chinese).

Chen, L., R. Arimoto & R. A. Duce, 1985. The sources and forms of phosphorus in marine aerosol – particles and rain from Northern New Zealand. Atmospheric Environment 19(5): 779–787.

Chen, X. M., Q. R. Shen & G. X. Pan, 2003. Characteristics of nitrate horizontal transport in a paddy field of the Tai Lake region, China. Chemosphere 50: 703–706.

Coale, K. H. & K. W. Bruland, 1990. Spatial and temporal variability in copper complexation in the North Pacific. Deep-sea Research 34: 317–336.

Cowling, E. B. & J. Nilsson, 1995. Acidification research: lessons from history and visions of environmental futures. Water, Air and Soil Pollution 85: 279–292.

De Leeuw, G., L. Spokes, T. Jickells, C. A. Skjoth, O. Hertel, E. Vigniti, S. Tamm, M. Schulz, L. L. Sorensen, B. Pedersen, L. Klein & K. H. Schlunzen,

2003. Atmospheric nitrogen inputs into the North Sea: effect on productivity. Continental Shelf Research 23: 1743–1755.

Feng, K., B. Y. Yu, D. M. Ge, M. H. Wong, X. C. Wang & Z. H. Cao, 2003. Organo-chlorine pesticide (DDT and HCH) residues in the Taihu Lake Region and its movement in soil – water system I. Field survey of DDT and HCH residues in ecosystem of the region. Chemosphere 50: 683–687.

Fenn, M. E., M. A. Poth, J. D. Aber, J. S. Baron, B. T. Bormann, D. W. Johnson, A. D. Lemly, S. D. McNulty, D. F. Ryan & R. Stottlemyer, 1998. Nitrogen excess in North American ecosystems: a review of predisposing factors, geographic extent, ecosystem responses and management strategies. Ecological Applications 8: 706–733.

Filoso, S., M. R. Williams & J. M. Melack, 1999. Composition and deposition of throughfall in a flooded forest archipelago (Negro River, Brazil). Biogeochemistry 45(2): 169–195.

Fu, J. M. & J. W. Winchester, 1994. Sources of nitrogen in 3 watersheds of Northern Florida, USA – mainly atmospheric deposition. Geochimica et Cosmochimica Acta 58: 1581–1590.

Furumai, H., T. Kondo & S. Ohgaki, 1989. Phosphorus exchange kinetics and exchangeable phosphorus forms in sediments. Water Research 23: 685–691.

Gao, Y., 2002. Atmospheric nitrogen deposition to Barnegat Bay. Atmospheric Environment 36: 5783–5794.

Gibson, C. E., Y. Wu & D. Pinkerton, 1995. Substance budgets of an upland catchment-the signi.cance of atmospheric phosphorus inputs. Freshwater Biology 33: 385–392.

Gordon, A. G. & E. Gorham, 1963. Ecological aspects of air pollution from an iron-sintering plant at Wawa, Ontario. Canadian Journal of Botany 41: 1063–1078.

Gorham, E., 1998. Acid deposition and its ecological effects: a brief history of research. Environmental Science & Policy 1: 153–166.

Helmers, E. & O. Schrems, 1995. Wet deposition of metals to the tropical north and the south Atlantic Ocean. Atmospheric Environment 29: 2475–2484.

Hertel, O., C. Ambelas Skjoth, L. M. Frohn, E. Vignati, J. Frydendall, G. De Leeuw, U. Schwarz & S. Reis, 2002. Assessment of the atmospheric nitrogen inputs into the North Sea using a Lagrangian model. Physics and Chemistry of the Earth 27: 1507–1515.

James, R. T., J. Martin, T. Wool & P. F. Wang, 1997. A sediment resuspension and water quality model of Lake Okeechobee. Journal of the American Water Resources Association 33: 661–680.

Khemani, L. T., G. A. Momin, P. S. P. Rao, P. D. Safai, G. Singh & R. K. Kapoor, 1989. Spread of acid rain over India. Atmospheric Environment 23: 757–762.

Lawrence, G. B., D. A. Goolsby, W. A. Battaglin & G. J. Stensland, 2000. Atmospheric nitrogen in the Mississippi River Basin – emissions, deposition and transport. The Science of the Total Environment 248: 87–99.

Likens, G. E., 1989. Some aspects of air pollutant effects on terrestrial ecosystems and prospects for the future. AMBIO 14: 9–14.

Lin, C., M. D. Melville, M. M. Islam, B. P. Wilson, X. Yang & P. Van Oploo, 1998. Chemical controls on acid discharges from acid sulfate soils under sugarcane cropping in an eastern Australian estuarine floodplain. Environmental Pollution 103: 269–276.

Lindberg, S. E., M. Bredemeier, D. A. Schafer & L. Qi, 1990. Atmospheric concentrations and deposition of nitrogen and major ions in conifer forests in the United States and Federal Republic of Germany. Atmospheric Environment 24: 2207–2220.

Lovett, G. M. & J. D. Kinsman, 1990. Atmospheric pollutant deposition to high-elevation ecosystems. Atmospheric Environment 24: 2767–2786.

Luo, L. C., 2004. Hydrodynamics and its effects on aquatic environment in Lake Taihu, Ph.D thesis, Nanjing Institute of Geography and Limnology, Chinese Academy of Sciences.

Migon, C. & V. Sandroni, 1999. Phosphorus in rainwater: partitioning inputs and impact on the surface coastal ocean. Limnology and Oceanography 44(4): 1160–1165.

Morel, F. M. M., R. J. M. Hudson & N. M. Price, 1991. Limitation of productivity by trace metals in the sea. Limnology & Oceanography. 36: 1742–1755.

Okochi, H., H. Kameda, S. Hasegawa, N. Saito, K. Kubota & M. Igawa, 2000. Deterioration of concrete structures by acid deposition an assessment of the role of rainwater on deterioration by laboratory and field exposure experiments using mortar specimens. Atmospheric Environment 34: 2937–2945.

Owens, N. J. P., N. J. Galloway & R. A. Duce, 1992. Episodic atmospheric nitrogen deposition on oligotrophic oceans. Nature 357: 397–399.

Peart, M. R., 2000. Acid rain, storm period chemistry and their potential impact on stream communities in Hong Kong. Chemosphere 41: 25–31.

Poelma, C. & G. Ooms, 2002. Influence of hydrodynamic interactions between particles on the turbulent flow in a suspension. Experimental Thermal and Fluid Sciences 26: 653–659.

Pollman, C. D., W. M. Landing, J. J. Perry, Jr & T. Fitzpatrick, 2002. Wet deposition of phosphorus in Florida. Atmospheric Environment 36: 2309–2318.

Pu, P. M., W. P. Hu, J. S. Yan, G. X. Wang & C. H. Hu, 1998. A physico-ecological engineering experiment for water treatment in a hypertrophic lake in China. Ecological Engineering 10: 179–190.

Qin, B. Q., W. P. Hu, G. Gao, L. C. Luo & J. S. Zhang, 2004. The Dynamics of resuspension and conceptual mode of nutrient releases from sediments in large shallow Lake Taihu, China. In Chinese Sciences Bulletin 49: 54–64.

Qin, B. Q., 1998. A review and prospect about the aquatic environment studies in Lake Taihu. Journal of Lake Sciences 10: 1–9 (in Chinese).

Qu, W. C., M. Dickman, C. X. Fan, S. M. Wang, W. C. Su, L. Zhang & H. X. Zou, 2002. Distribution, sources and potential toxicological significance of polycyclic aromatic hydrocarbons (PAHs) in Lake Taihu sediments, China. Hydrobiologia 485: 163–171.

Qu, W. C., M. Dickman & S. M. Wang, 2001. Multivariate analysis of heavy metal and nutrient concentrations in sediments of Lake Taihu, China. Hydrobiologia 450: 83–89.

Raper, D. W. & D. S. Lee, 1996. Wet deposition at the sub–20 km scale in a rural upland area of England. Atmospheric Environment 30: 1193–1207.

Shen, J. H., B. Gutendorf, H. H. Vahl, L. Shen & J. Westendorf, 2001. Toxicological profile of pollutants in surface water from an area in Taihu Lake, Yangtze Delta. Toxicology 166: 71–78.

Sheng, Y. P. & W. Lick, 1979. The transport and resuspension of sediments in a shallow lake. Journal of Geophysical Research 84: 1809–1826.

Shrestha, P. L., 1996. An integrated model suite for sediment and pollutant transport in shallow lake. Advances in Engineering Software 27: 201–212.

Spokes, L., T. Jickells, A. Rendell, M. Schulz, A. Rebers, W. Dannecker, O. Kruger, M. Leermakers & W. Baeyens, 1993. High atmospheric nitrogen deposition events over the North Sea. Marine Pollution Bulletin 26: 689–703.

Stoddard, J. L., 1994. Long-term changes in watershed retention of nitrogen. In Baker, L. A. (ed.), Environmental Chemistry of Lakes and Reservoirs. American Chemical Society Advances in Chemistry 227: 223–284.

Sunda, W. G. & S. A. Huntsman, 1992. Feedback interactions between zinc and phytoplankton in seawater. Limnology & Oceanography 37: 9–40.

Swap, R., M. Garstang, S. Greco, R. Talbot & P. Kallberg, 1992. Saharan dust in the Amazon Basin. Tellus, Series B. Chemical and Physical Meteorology 44: 133–149.

Tarnay, L., A. W. Gertler, R. R. Blank & G. E. Taylor, Jr, 2001. Preliminary measurements of summer nitric acid and ammonia concentrations in the Lake Tahoe Basin air-shed: implications for dry deposition of atmospheric nitrogen. Environmental Pollution 113: 145–153.

Vasjari, M., A. Merkoci & S. Alegret, 2000. Potentiometric characterisation of acid rains using corrected linear plots. Analytica Chimica Acta 405: 173–178.

Vitousek, P. M., J. D. Aber, R. W. Howarth, G. E. Likens, P. A. Matson, D. W. Schindler, W. H. Schlesinger & D. G. Tilman, 1997. Human alternation of the global nitrogen cycle: sources and consequences. Ecological Applications 7: 737–750.

Wang, G. X., P. M. Pu, S. Z. Zhang, W. C. Li, W. P. Hu & C. H. Hu, 1998. The purification of artificial complex ecosystem for local water in Lake Taihu. China Environmental Science 18: 410–414 (in Chinese).

Wright, R. F., T. Dahl, E. T. Gjessing, G. R. Hendrey, A. Henriksen, M. Johannessen & I. P. Muniz, 1976. Impact of acid precipitation on freshwater ecosystems in Norway. Water, Air and Soil Pollution 6: 483–499.

Xu, P. Z & B. Q. Qin, 2005. Water quantity and pollutant fluxes of the surrounding rivers of Lake Taihu during the hydrological year of 2001–2002. Journal of Lake Sciences 17(3): 213–218 (in Chinese).

Yang, L. Y., B. Q. Qin & R. J. Wu, 2001. Preliminary study for potential impacts on the aquatic environment of Lake Taihu by acid rain. Journal of Lake Sciences, 13: 135–142 (in Chinese).

Yang, X. S., D. R. Miller & X. Xu, 1996. Spatial and temporal variations of atmospheric deposition in interior and coastal Connecticut. Atmospheric Environment 30: 3801–3810.

Yang, X. H., 2001. Acid rain and countermeasures in southern Lake Taihu. Journal of Huzhou Teachers College 23: 68–72 (in Chinese).

Yang, Z. F., W. G. Shi, L. Q. Chen, Y. Chen & Z. L. Zhou, 2003. Ecological environmental succession and countermeasure of East Taihu. China Environmental Science 23: 64–68 (in Chinese).

Yin, D. Q., Z. H. Wu, X. R. Wang & S. Yan, 1996. Algal growth potential as affected by lake water and phosphorus released from sediments in Lake Taihu. Journal of Nanjing University (Natural Sciences) 32(2): 253–260 (in Chinese).

Zheng, Y., X. J. Wang & Y. C. Jiang, 2001. Water quality of the inflow rivers connected to Lake Taihu and totally inputted nutrients. Geography and Territorial Research 17: 40–44 (in Chinese).

Zhou, Q. X. & Y. M. Zhu, 2003. Potential pollution and recommended critical levels of phosphorus in paddy soils of the southern Lake Tai area, China. Geoderma 115: 45–54.

Hydrobiologia (2007) 581:71–77
DOI 10.1007/s10750-006-0499-z

EUTROPHICATION IN LAKES

The effects of different electron donors on anaerobic nitrogen transformations and denitrification processes in Lake Taihu sediments

Donghong Wang · Qinghui Huang · Chunxia Wang · Mei Ma · Zijian Wang

Abstract Nitrogen transformations in anaerobic sediments and leachate in Lake Taihu were simulated in the laboratory. Ammonium, nitrate and nitrite were analyzed after incubation under anaerobic conditions. Different reductive states and pH values were obtained by using different electron donors, such as glucose, sucrose, potato starch and sodium acetate. Chemical nitrogen transformation mechanisms were discussed relative to physico-chemical properties of lake sediment. Results demonstrated that nitrogen transformations in anaerobic conditions supplemented with different electron donors varied, and supplementation with certain electron donors may enhance nitrogen removal from anaerobic sediments. Among the four electron donors studied, higher nitrogen removal efficiencies were observed with acetate and starch. Saccharides, such as glucose, sucrose and starch, stimulate nitrate reduction to nitrite, while acetate stimulates nitrate reduction to ammonium.

Guest editors: B. Qin, Z. Liu & K. Havens
Eutrophication of shallow lakes with special reference to Lake Taihu, China

D. Wang · Q. Huang · C. Wang · M. Ma · Z. Wang (✉)
State Key Laboratory of Environmental Aquatic Chemistry, Research Center for Eco-Environmental Sciences, Chinese Academy of Sciences, Shuangqing Road, Haidian District, Beijing 100085, China
e-mail: zjwang@mail.rcees.ac.cn

Keywords Electron donor · Anaerobic · Nitrogen · Transformation · Sediment · Lake Taihu

Introduction

Denitrification is a microbial process that converts nitrate to nitrogen gas via several intermediates, including nitrite and nitrous oxide (Tiedje, 1982). It is recognized as the most important process removing fixed nitrogen in most natural environments (Boicourt et al., 1996). Transformations involve nitrogen reduction and therefore, require electron donors (Devlin et al., 2000). Denitrification is a respiratory process which also requires an electron donor for energy (Islas-Limaa et al., 2004). Denitrifying bacteria can use nitrate and/or nitrite as an alternative electron acceptor for metabolic activities when molecular oxygen is unavailable (Bae et al., 2004). However, in natural anaerobic systems including sediments, denitrification often is limited by readily biodegradable electron donors.

Studies have demonstrated enhancement of nitro-reduction rates following addition of organic electron donors, such as glucose, pyruvate, formate, starch, acetate, hydrogen gas, elemental sulphur, thiosulphate, aqueous ferrous iron and pyrite (Preuss et al., 1993; Boopathy, 1994; Roberts et al., 1995; Devlin et al., 2000).

Saccharide is an abundant biomolecule in the natural environment. Glucose is a monosaccharide, and sucrose is a disaccharide, which yields two monosaccharides upon hydrolysis. Starch contains two polysaccharides: amylose and amylopectin (Yu et al., 2004). Acetate is an organic acid and has been used in waste water treatment for nitrogen removal (Graaf et al., 1996; Cervantes et al., 1999).

Addition of electron donors to enhance reduction processes is, therefore, a potential method for in-place treatment of contaminated sediments (Gerlach et al., 1999). This may be an important treatment for eutrophic lakes, such as Lake Taihu, which is the third largest freshwater lake in China (Qin et al., 2007). Water quality in Lake Taihu has deteriorated due to rapid economic development and intensive use of water resources in recent years. As a result, cyanobacteria blooms occur over large areas in summer, and these blooms impact the aquatic ecosystem detrimentally (Pu et al., 1998a, b).

Most studies have focused on nitrate and ammonium removal in waste water systems (Cervantes et al., 2001; Schmidt et al., 2003), and comparative studies on denitrification and electron donor supplements in fresh water sediments are scarce.

Objectives of the present study were to compare total nitrogen (TN) reduction effects in Lake Taihu sediments stimulated by saccharide electron donors, such as glucose, sucrose, potato starch and organic acid acetate; and to evaluate effects of nitrogen species transformation under different electron donor additions.

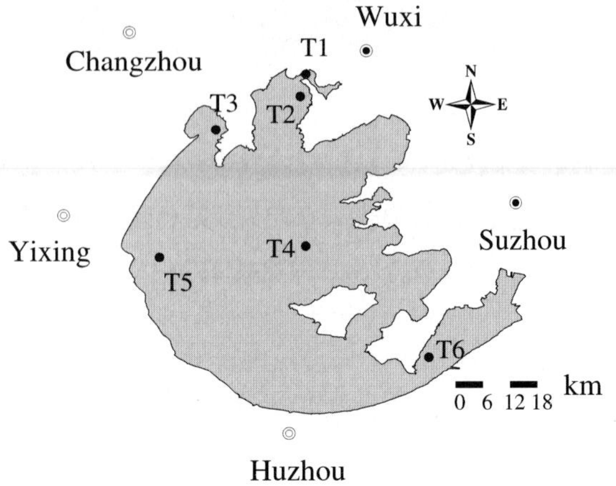

Fig. 1 Geographic locations of sampling sites

Materials and methods

Sediment samples collection

Six surface sediment samples were collected using a Peterson grab bucket on 13–15 July 2003 from Lake Taihu. Locations of sampling sites are shown in Fig. 1. Samples were stored in a cooler until return to the laboratory, freeze-dried, sieved to ~0.28 mm aggregate size, mixed together, homogenized by gentle shaking, and stored at 4°C in a sealed polyethylene bag.

Anaerobic incubation experiment

Microcosms were constructed using 20 g of sediment and 150 ml of artificial freshwater (1 mg l^{-1} ammonium, 1 mg l^{-1} nitrate and 0.1 mg l^{-1} nitrite), which was purged with high purity nitrogen (99.99% N_2) for 10 min to remove oxygen before use in 250-ml serum bottles. Microcosms were prepared in triplicate, and electron donors glucose, sucrose, potato starch (Sigma Co.) and sodium acetate added 2 g in stoichiometric excess. Two electron donor-free control groups, one sterilized with 0.1 g $HgCl_2$, were established. Table 1 lists the treatment conditions and microcosm abbreviations. Test bottles were aerated with high purity nitrogen for 3–5 additional minutes, sealed with butyl rubber stoppers and continuously mixed with a planar shaker rotating at 200 rpm at 25°C.

Table 1 Treatment conditions and abbreviations of microcosms

Abbreviation	Treatment
A	Sterilized control
B	Electron donor free control
C	Microcosm of glucose as electron donor
D	Microcosm of sucrose as electron donor
E	Microcosm of potato starch as electron donor
F	Microcosm of sodium acetate as electron donor

Sampling was done in a glove bag (I^2R Glove Bag™, Model X-27-27, USA) purged with nitrogen before use. Bottles were vigorously shaken; pH and Eh were determined using a portable meter (Thermo Orion Model 250, USA). Ten milliliter of sediment/water slurry was poured into a 10 ml centrifuge tube, and the same quantity was added (10 ml artificial freshwater contained 1.33 g sediment). After sampling, sediment samples were centrifuged at 4000 rpm for 20 min and filtered through a 0.45 μm pore-size cellulose acetate filter to separate sediment and leachate at once.

Chemical analysis

For leachate samples, ammonium, nitrate and nitrite were determined within 24 h after sampling. Nitrate was determined by ion chromatography (Dionex 4500i using an AS-14 4-mm anion column), nitrite with *N*-(1-naphthyl) ethylene diamine dihydrochloride (APHA, 1998a), and ammonium with salicylic-hypochlorite colorimetry (Standard method of People's Republic of China, 1987).

Total nitrogen in sediments was analyzed using persulfate according to standard methods for examination of water and wastewater (APHA, 1998b). Weights of sediments reported are on an oven-dry basis.

Results and discussion

Formal potentials and pH conditions in the microcosms

Formal potential consists of measurement of the electromotive force of an electrochemical cell in which, under the specified conditions, the analytical concentration of the two oxidation states is varied. The formal potential includes correction factors for activity coefficients, acid-base phenomena, complex formation, and the liquid junction potential used between the reference electrode and the half-cell in question. It may often leads to better predictions than standard potential because it represents quantities subject to direct experimental measurement. In this paper, it is expressed by pε, which is calculated from the determined Eh values using Nernst Equation. The pε can be defined as the negative logarithm of electron activity in a solution. It is a parameter for the redox intensity gives the electron activity at equilibrium and measures the relative tendency of a solution to accept or transfer electrons (Stumm & Morgan, 1981).

Values of pH and pε are shown in Fig. 2. In all microcosms, pH declined with time. Trends are divided into three groups. Microcosms treated with glucose and sucrose had similar trends. pH declined sharply in the first 96 h and dropped to 4.3 by completion. pH in starch additions also declined, but more slowly than glucose and sucrose, dropping to 4.8 at the end of the incubation. pH in the sterilized control, electron donor-free control and sodium acetate addition declined in the first 48 h, then remained around 7.4 to 7.5.

Formal potentials in electron donor systems reached a strong reductive environment after 2 or 3 days. pε dropped in the first 48 h in glucose and sucrose microcosms and in the first 96 h in starch and sodium acetate microcosms. Afterwards, the

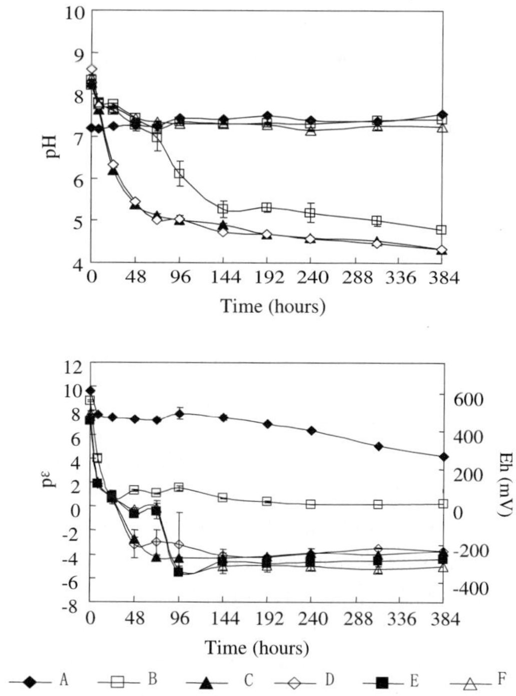

Fig. 2 pH and pε values of microcosms

reductive environment in these microcosms reached steady-state and remained constant at –4 to –6. The sterilized control did not attain reductive conditions, and the electron donor-free control barely reached a mild reductive environment.

Systems supplemented with electron donors achieved more reduction potential than controls. It is believed that Eh values is featured by reduction when they below 200 mV. (Song et al., 1990) Sediments from Lake Taihu in our experiment were strongly anaerobic except two control groups, and additional electronic donors could accelerate reducing conditions in microcosms.

pH and reductive potential are important factors controlling the fate of pollutants, especially those at the soil-water interface (Cao et al., 2001). pH and reductive potential also affect nitrogen composition (Stumm & Morgan, 1981). pH in all microcosms treated with electron donors declined, except the system treated with sodium acetate, likely due to acetate hydrolysis. In our experiment, the acetate microcosm may have maintained pH of 7.2–7.4 via its strong buffering capacity (Terra & Regel, 1995). However, in other three electron donor microcosms, pH values declined and remained below 5. At this pH, acid producing microorganisms may not survive, leading to stable pH.

Transformation of nitrogen species

Nitrogen transformations in control and electron donor microcosms are shown in Figs. 3 and 4. Results show that, in all microcosms, nitrate in leachate increased early in the incubation, then decreased. Nitrate, which is formed during organic matter degradation under oxic conditions (ammonification/nitrification), may be transported (mixing, currents, upwelling etc.) to anoxic environments where it can undergo denitrification (Boyd, 2001). In the present study, early nitrate increases may be caused by degradation of organic matter using residual oxygen in the sediment. For in the first 24 h, formal potential in all microcosms was positive, and pH was about 7, indicating characteristics of sediments were weak oxic or weak reduction in this condition.

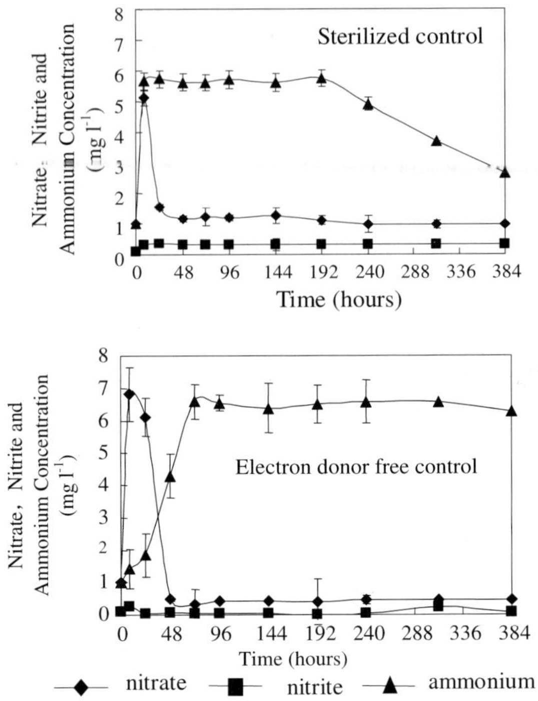

Fig. 3 Transformation of nitrogen species in two control microcosms

Nitrogen transformations in glucose and sucrose microcosms were similar. Nitrate in glucose and sucrose systems were lower than controls. Highest nitrate concentrations in glucose and sucrose microcosms were 5.2 and 4.7 mg l^{-1}, respectively, in the first 8 h and then decreased sharply. Highest nitrite concentrations in leachate occurred simultaneously with nitrate maxima. Ammonium decreased to below detection limits in 48 h, which indicates that monosaccharide and disaccharide electron donors may stimulate anerobic nitrogen transformations.

Nitrate concentration in the starch addition reached 7.4 mg l^{-1} in 24 h, and ammonium concentration increased to a similar value as the controls, then decreased to below detection limits after 96 h.

In anoxic, nitrate-containing environments, two pathways of dissimilatory nitrate reduction have been identified: denitrification, by which nitrogen oxides (mainly NO_3^-) are reduced to nitrite (NO_2^-) and then to dinitrogen gases (N_2O and N_2), and dissimilatory nitrate reduction to ammonia (DNRA), producing ammonia by

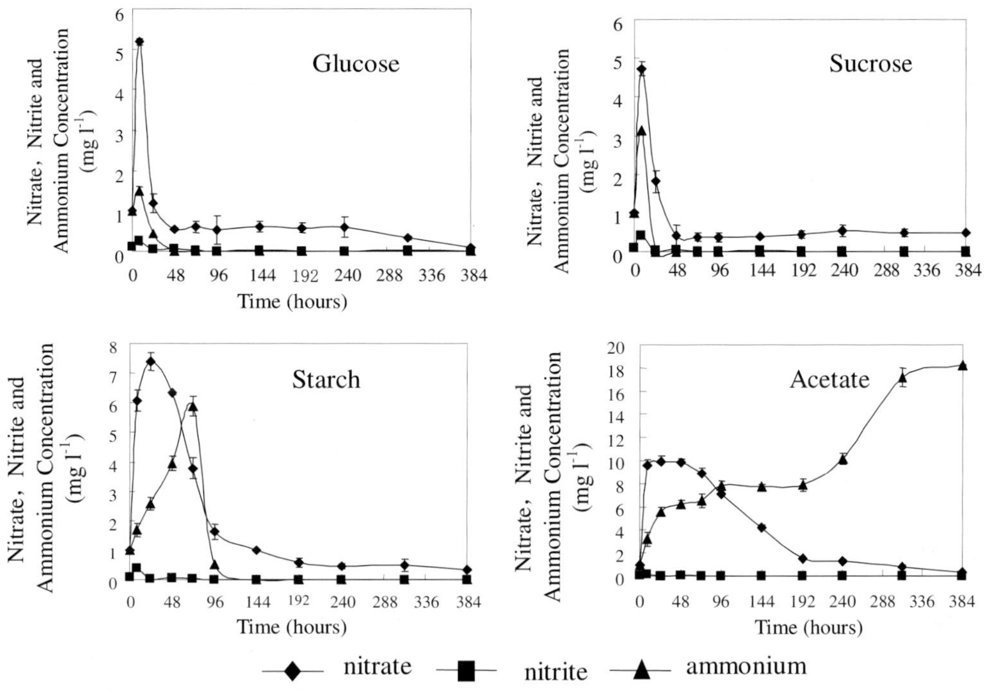

Fig. 4 Transformation of nitrogen species in different electron donor treated microcosms

reduction of nitrate and/or nitrite (Brunet & Garcia-Gil, 1996).

Apparently, nitrate reduced in microcosms supplemented with glucose, sucrose and starch as electronic donors were denitrified. Ammonium could not be detected after 48 h in glucose and sucrose systems and 96 h in the starch system. This indicates that saccharide, including monosaccharide (glucose), disaccharide (sucrose) and polysaccharides (starch), may stimulate nitrate reduction to nitrogen gas, although the dinitrogen gases (N_2O and N_2) were not determined in our experiment.

Nitrate and nitrite in the sodium acetate system was different from other electron donor systems. Nitrate in leachate reached 10 mg l^{-1} and maintained this value for 96 h. Nitrite concentration was half that of glucose and sucrose microcosms. However, ammonium increased to about 8 mg l^{-1} within 96 h and remained steady to 240 h. Afterwards, ammonium concentration increased to 18.3 mg l^{-1} by the end of incubation. Ammonium increased early maybe caused by ammonification of organic matter, for nitrate did not decrease at the first 96 h. But latter the high ammonium value indicated that nitrate was reduced to ammonium in this microcosm. The removal of eight electrons during NO_3^- reduction to NH_4^+ makes this process potentially advantageous in reduced environments (Tiedje et al., 1982). DNRA may be quantitatively important nitrogen cycling of marine sediments (Sorensen, 1978).

Ammonium concentrations in the control systems were higher than electron donor addition systems except for acetate. Ammonium in the sterilized control increased for 8 h, then remained constant at approximately 6 mg l^{-1} from 8 h to 192 h. In the electron donor-free control system, ammonium increased slowly for 192 h. These systems may be controlled by pH and redox potential. According to pε~pH diagrams (Wang, 2001), at pH about 7.5 and pε about 0–8, ammonium is the predominant nitrogen species. This may explain why ammonium in the starch microcosm increased for 96 h and then decreased to detection limits. pH and pε conditions in the starch microcosm up to 96 h were within the ammonium predominance area. And, nitrate reduction seems follow denitrification for 96 h in the starch system since pH was lower than 6 and pε decreased sharply to about –6. This condition overstepped the ammonium predominance area. So, nitrate reduction to ammonium stimulated by sodium acetate may be due to comparatively high pH caused by acetate hydrolysis. In soil, similar results were reported that DNRA was favored at high pH (Stevens et al., 1998).

TN removal by denitrification

The ability to enhance sediment denitrification via addition of various electron donors was tested in sediment–water microcosms using TN reduction as a gross indicator of denitrification activity. TN removal from sediments in microcosms over 384 h are shown in Fig. 5. TN removal was observed in all systems but was not significant in the sterilized control, which increased for 48 h, then stayed steady for 384 h. Denitrification in the electron donor-free control increased for 144 h, and TN reduction percentage was 21.8%. Compared to control microcosms, potato starch and sodium acetate systems obtained higher reductive efficiency (45.5% and 42.5%, respectively). At the beginning of the incubation, denitrification in glucose and sucrose additions was lower than controls. However, reductive efficiency increased sharply at the end of the incubation (31.4% and 34.7%, respectively).

These results demonstrate that additional electronic donors may accelerate. Among the four electron donors, starch and acetate have largest effects. TN reduction ratios of starch and acetate were 10% higher than glucose and sucrose. Since the starch used had low solubility, starch may have been more associated with sediments than the other electron donors (Gerlach et al., 1999) However, TN reduction in the sterilized control increased slightly throughout the experiment, indicating that $HgCl_2$ addition may not have been sufficient to eliminate microbial activity; but the increase in TN reduction was significantly less than the non-inhibited microcosms. However, this TN reduction was not observed in the electronic donor-free controls, indicating that either the sterilization agent affected abiotic redox activity, or microorganisms could reduce TN without electron donor addition.

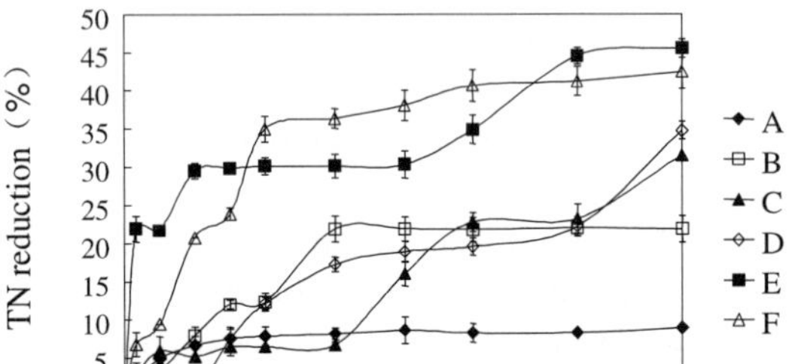

Fig. 5 TN removal ratio of sediments in microcosms

Conclusions

The results presented in this paper show certain electron donors may be used to enhance TN removal in anaerobic sediments. Higher nitrogen removal efficiencies were observed when acetate and starch were added. However, nitrogen transformations in anaerobic conditions supplemented with different electron donors varied. Saccharides, such as glucose, sucrose and starch, stimulated nitrate reduction to nitrite, and acetate stimulated nitrate reduction to ammonium. Nitrogen transformations in these microcosms were controlled by pH and formal potential.

Acknowledgements This work was supported by the Key Project of Chinese Academy of Sciences, (Grant No.KZCX1-SW-12-II-32) and National Key Basic Research Support Foundation of China (Grant No.2002CB412304).

References

APHA, AWWA & WEF, 1998a. 4500- NO_2^--B Colorimetric Method. Standard Methods for the Examination of Water and Wastewater, 20th edn.

APHA, AWWA & WEF, 1998b. Persulfate Method. Standard Methods for the Examination of Water and Wastewater, 20th edn, 4500-N C: 4–102.

Bae, H.-S., T. Yamagishi & Y. Suwa, 2004. An anaerobic continuous-flow fixed-bed reactor sustaining a 3-chlorobenzoate-degrading denitrifying population utilizing versatile electron donors and acceptors. Chemosphere 55: 93–100.

Boicourt, W. C., W. R. Boynton, T. M. Church, D. M. Ditoro, R. Elmgren, J. H. Garber, A. E. Giblin, R. A. Jahnke, N. J. P. Owens, M. E. Q. Pilson & S. P Seitzinger, 1996. The fate of nitrogen and phosphorus at the land-sea margin of the North Atlantic Ocean. Biogeochemistry 35: 141–180.

Boopathy, R., 1994. Transformation of nitroaromatic compounds by a methanogenic bacterium, *Methanococcus* sp. strain B. Archives of Microbiology 162: 167–172.

Boyd, S. R., 2001. Nitrogen in future biosphere studies. Chemical Geology 176: 1–30.

Brunet, R. C. & L. J. Garcia-Gil, 1996. Sulfide-induced dissimilatory nitrate reduction to ammonia in anaerobic freshwater sediments. FEMS Microbiology Ecology 21: 131–138.

Cao, X., Y. Chen, X. R. Wang & X. H. Deng, 2001. Effects of redox potential and pH value on the release of rare earth elements from soil. Chemosphere 44: 655–661.

Cervantes, F., O. Monroy & J. Gómez, 1999. Influence of ammonium on the performance of a denitrifying culture under heterotrophic conditions. Applied Biochemistry and Biotechnology 81: 13–23.

Cervantes, F. J., A. David De la Rose & J. Gómez, 2001. Nitrogen removal from wastewater at low C/N ratios with ammonium and acetate as electron donors. Bioresource Technology 79: 165–170.

Devlin, J. F., R. Eedy & B. J. Butler, 2000. The effects of electron donor and granular iron on nitrate transformation rates in sediments from a municipal water supply aquifer. Journal of Contaminant Hydrology 46: 81–97.

Gerlach, R., M. Steiof, C. L. Zhang & J. B. Hughes, 1999. Low aqueous solubility electron donors for the reduction of nitroaromatics in anaerobic sediments. Journal of Contaminant Hydrology 36: 91–104.

Graaf, A., P. Bruijin, L. A. Robertson, M. S. M. Jetten & J. G. Kuenen, 1996. Autotrophic growth of anaerobic, ammonium-oxidizing microorganisms in a fluidized bed reactor. Microbiology 142: 2187–2196.

Islas-Limaa, S., F. Thalassoa & J. Gómez Hernandezb, 2004. Evidence of anoxic methane oxidation coupled to denitrification. Water Research 38: 13–16.

Pu, P. M., W. P. Hu, G. X. Wang, S. Z. Zhang, C. G. Hu & J. S. Yan, 1998a. The new strategy for improving the aqua-ecological environmental in Taihu Lake Basin, China. How can we solve the problem of lack of qualified water and deterioration of environment and natural resources in Taihu Lake basin. Journal of Lake Science 10: 47–58 [Suppl].

Pu, P. M., W. P. Hu, J. S. Yan, G. X. Wang & C. G. Hu, 1998b. A physico-ecological engineering experiment for water treatment in a hypertrophic lake in China. Ecologied Engineering 10: 179–190.

Preuss, A., J. Fimpel & G. Diekert, 1993. Anaerobic transformation of 2,4,6-trinitrotoluene TNT. Archives of Microbiology 159: 345–353.

Qin, B. Q, P. Z. Xu, Q. L. Wu, L. C. Luo & Y. L. Zhang, 2007. Environmental Issues of Lake Taihu, China. Hydrobiologia 581: 3–14.

Roberts, D. J., S. Pendharkar & F. Ahmad, 1995. Factors Affecting TNT Degradation by Anaerobic Consortia. Platform-Presentation, 3rd International Symposium on In Situ and On-site Bioreclamation. San Diego, CA, April 24–27.

Schmidt, I., O. Sliekers, M. Schmid, E. Bock, J. Fuerst, J. G. Kuenen, S. M. Jetten & M. Strous, 2003. New concepts of microbial treatment processes for the nitrogen removal in wastewater. FEMS Microbiology Reviews 27: 481–492.

Song, J. M., Y. Li & Z. Zhu, 1990. Relationship between Eh value and redox environment in marine sediments. Marine Science Bulletin 9(4): 33–39.

Sorensen, J., 1978. Capacity for denitrification and reduction of nitrate to ammonia in a coastal marine sediment. Applied & Environmental Microbiology 35: 301–305.

Standard method of People's Republic of China, 1987. Water quality-determination of ammonium-spectrophotometric method with salicylic acid, GB 7481–87:211.

Stevens, R. J., R. J. Laughlin & J. P. Malone, 1998. Soil pH affects the processes reducing nitrate to nitrous oxide and di-nitrogen. Soil Biology and Biochemistry 30: 1119–1126.

Stumm, W., J. J. Morgan, 1981. Aquatic Chemistry: An Introduction Emphasizing Chemical Equilibria in Natural Waters, 2nd edn. John Wiley, Sons, Inc.

Terra, W. R. & R. Regel, 1995. pH buffering in *Musca domestica* midguts, comparative biochemistry and physiology Part A. Physiology 112: 559–564.

Tiedje, J. M., 1982. Denitrification. In Page, A. L. (ed.), Methods of Soil Analysis: Part 2. Chemical and Microbiological Properties. 2nd edn. ASA-ASSA, Madison, WI, 1011–1026.

Tiedje, J. M., A. J. Sexstone, D. D. Myrold & J. A. Robinson, 1982. Denitrification: ecological niches, competition and survival. Antonie Van Leeuwenhoek Journal of Microbiology 48: 569–583.

Wang, K. X., 2001. Aquatic chemistry, Beijing: Chemical Industry Press, Ch.6, Oxidation and Reduction Chemistry, 207 pp.

Yu, H., M. Amann, T. Hansson, J. Köhler, G. Wich & W. F. van Gunsterena, 2004. Effect of methylation on the stability and solvation free energy of amylose and cellulose fragments: a molecular dynamics study. Carbohydrate Research 339: 1697–1709.

Hydrobiologia (2007) 581:79–87
DOI 10.1007/s10750-006-0514-4

EUTROPHICATION IN LAKES

A 200-year historical modeling of catchment nutrient changes in Taihu basin, China

Ge Yu · Bin Xue · Geying Lai · Feng Gui · Xiaomei Liu

Abstract Sedimentary records provide important information for understanding changes in the history of eutrophication in Lake Taihu. In addition, the catchment nutrient model SWAT provides a powerful tool to examine eutrophic changes in a long-term context. Since it is difficult to evaluate impacts of natural eutrophic development and anthropogenic changes in catchment discharge and land use, simulation of past changes provides a mirror on processes and dynamics. Boundaries in the simulations are set to a pre-industrial time to evaluate natural-agricultural nutrient changes in Taihu Basin a 100 years ago. Total nitrogen (TN) and total phosphorus (TP) in the main channel flowing into the lake are simulated in four sub-basins for 200 model years. Results show that modeling can capture basic features of basin nutrient development, where mean TN concentration (0.12 mg l^{-1}) can be compared in broad scale to mean TN concentration (0.17 mg kg^{-1}) from Lake Taihu sedimentary cores dating back about 100 years. Spatial nutrient simulations suggest that the two major nutrient sources are from the southwestern sub-basin (48% TN and 68% TP of the basin total) and the northwestern sub-basin (18% TN and 17% TP). There are differences of $+7.3 \times 10^4$ kg TN and $+2.0 \times 10^5$ kg TP between total input and output values, simulating mean annual amounts of nutrient deposited into the lake. TN and TP concentration differences between input and output sub-basins become smaller in the second 100 years than the first 100 years, suggesting a 100 year period to reach a balance of net nutrients. Catchment nutrient modeling provides a basis to evaluate how nutrient production and balance responded to environmental changes over 200 years in Taihu Basin.

Keywords Natural-agricultural environment · TN and TP simulations · Nutrient change · Decadal-centurial time scale · Taihu Basin

Introduction

Eutrophication in Lake Taihu became a problem in the last 20 years, and anthropogenic change is a major factor (Jin et al., 1990; Huang et al., 2001; Qin et al., 2004). Nutrient release from lake bottom sediments (Qin et al., 2004) and nutrient discharge from the catchment (Xia & Yang, 2003) are two important sources leading to eutrophication, whereas nutrient source changes from

Guest editors: B. Qin, Z. Liu & K. Havens
Eutrophication of shallow lakes with special reference to Lake Taihu, China

G. Yu (✉) · B. Xue · G. Lai · F. Gui · X. Liu
Nanjing Institute of Geography & Limnology,
Chinese Academy of Sciences, Nanjing 210008, China
e-mail: geyu@niglas.ac.cn

bottom sediments or the catchment are not only controlled by human activities on an annual-interannual timescale, but also are related to natural system changes in a decadal-century time scale (Last & Smol, 2001; Smol, 2002). Paleolimnological techniques can be used to reconstruct past water quality changes on a number of time scales using information preserved in sedimentary profiles (e.g. Ramstack et al., 2003; Rose et al., 2004; Bradbury et al., 2004). When tracking eutrophic changes using lake sediments, a number of rich-organic layers with high nitrogen content have been found in lacustrine sediments from Lake Taihu. At depths of 22, 30, 35, and 42 cm in cores from Lake Taihu (Liu et al., 2004a; b), total nitrogen contents are similar to or higher than values at 1–5 cm (1990s; Fig. 1). While examining longer sediment cores from Lake Taihu (Xue et al., 1998), nitrogen contents at two layers ^{14}C-dated to 5,930 year B.P. (i.e. Before Present) and 7,899 year B.P. were higher than those in surface sediments. The lake may have experienced natural eutrophication before industrial and urban modernization in Taihu Basin. Examining and modeling long-term changes in lake eutrophication will provide a better understanding of the trajectories of lake environmental degradation and help set realistic mitigation goals for recovery.

Taihu Basin (Fig. 2) is located in a lowland area of the Changjiang River Plains in China approximately between 29–32° N and 118–121° E and about 50–100 km from the East China Sea (Sun & Huang, 1993). Climate is warm and wet, typical of subtropical monsoon climate. The basin is covered naturally by subtropical evergreen broadleaf forest, and the lake has a high aquatic production (Huang et al., 2001). Natural soil horizons developed on parent rock are composed of loose and organic-rich fluvial-lacustrine sediments deposited in the Holocene (Sun & Huang, 1993; Wang et al., 1996). The basin is in a shallow depression in the Changjiang River Plains, and there is a seasonal water exchange between the Changjiang River and Lake Taihu. Thus these natural conditions are potentials to develop the lake eutrophication.

To understand eutrophic changes and factors controlling nutrients in Taihu Basin (e.g. Zhang et al., 2002; Qin, 2002), it must be determined if natural nutrient sources exist that can result in eutrophication. If so, what are the rates and trends of these natural changes? Long-term catchment nutrient modeling of a natural system is needed to address this question. Other issues, such as how these natural changes were different from anthropogenic effects, need to be addressed. Hydrological-based catchment models provide a powerful tool to understand these issues in processes and dynamics (Schulze, 2000; Wang et al., 2003).

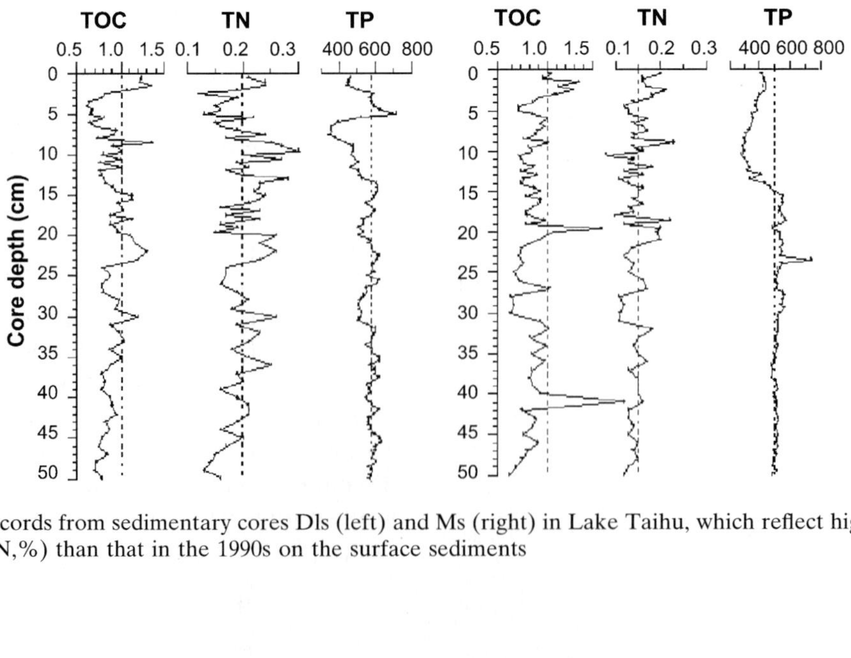

Fig. 1 Nutrient records from sedimentary cores Dls (left) and Ms (right) in Lake Taihu, which reflect higher total nitrogen concentrations (TN,%) than that in the 1990s on the surface sediments

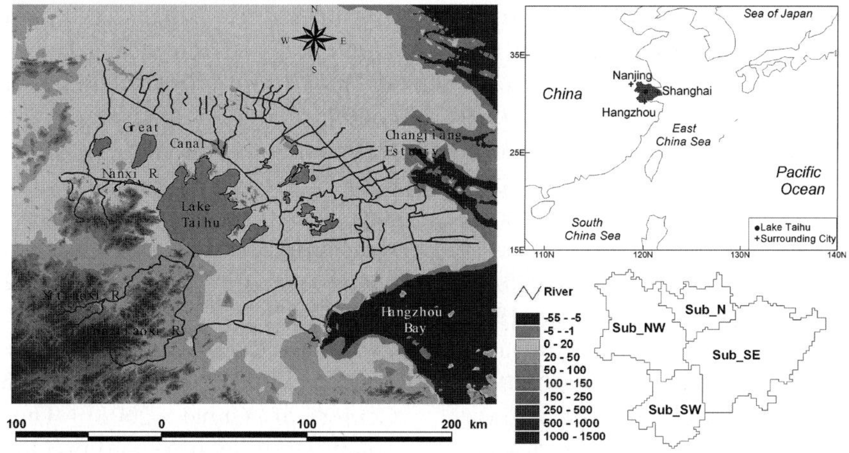

Fig. 2 Geographical location (Right upper), physical and hydrological settings (Left) and four sub-basin divisions (Right lower) of Taihu Basin

The SWAT (Soil and Water Assessment Tool) model is a river basin-scale model developed to predict the impact of land management practices on water, sediment, and agricultural chemical yields in large, complex basins with varying soils, land use, and management conditions over long periods of time (Arnold et al., 1998; Di Luzio et al., 2002a). The model is physically based and computationally efficient, uses readily available inputs, and enables users to study long-term impacts (Di Luzio et al., 2002b). SWAT is being used extensively to assess the impact of global climate on water supply and quality (e.g. Rosenberg et al., 1999). In this study, SWAT was applied to simulate nutrient changes in Taihu Basin dating back to pre-industrial time. The model highlights the whole basin and can simulate catchment nutrients as sources and feedback. Second, climatic and geographic conditions are major controls in natural nutrient development, and the model can be driven by climate-based hydrology, topographical-based Hydrological Response Units (HRUs), and nutrients sourced from vegetation and soil coverage. Finally, natural eutrophication in a lake is a long-term process on a decade and century timescale (Smol, 2002), and SWAT is a continuous integration model capable of simulation 10^1–10^2 year timescales.

This study examines major nutrient changes in a pre-industrial time mode for Taihu Basin to identify nutrient sources from the sub-basins, estimate nutrient balance in the lake, and assess trends on a timescale of 10^1–10^2 years.

Model and experimental design

The model used in this study is the AVSWAT-2000 model (Di et al., 2002a, b), which is an ArcView graphical user interface for the SWAT model (Arnold et al., 1998). Basin topography is a basic boundary in the SWAT model. Vector-based 1:250,000 topographical maps in electronic files (National GIS Center of China, 1999) are converted using ARCINFO into a grid-based Digital Elevation Model (DEM). Grids within 28–3° N and 118–121° E were taken as the boundary of Taihu Basin, and total area was 30,246 km^2 with a maximum elevation difference of 1,266 m from the southwestern mountains to the eastern coastal plains. Four river systems in the basin constitute four sub-basins (Fig. 2). The Nanxi River and tributaries flows to the lake from the northwestern part of the basin (Sub_NW). The Tiaoxi River, with two eastern and western tributaries, discharges to the lake from the southwestern part of the basin (Sub_SW). The southeastern part of the basin is a watershed area from Lake Taihu towards the Changjiang Estuary – Hangzhou Bay (Sub_SE). The northern sub-basin connects Changjiang River and Lake Taihu by the Great Canal of China system (Sub_N).

Thus, Lake Taihu receives stream flows from Sub_SW and Sub_NW and drains to Sub_SE. Different from the other subbasins, there is seasonal water exchange between the Changjiang River and Lake Taihu in Sub_N; i.e. during Changjiang flooding, discharge is orientated from the River to the Lake, while in other seasons, water flows toward the River. Based on observations in extreme years during the 1960–1980s at Liangxihe River Station (Sun and Huang, 1993; Huang et al., 2001), the ratio of river discharge volumes to lake discharge volumes was 0.55:0.45, and the ratio of river discharge days to lake discharge days was 0.46:0.54. Thus, the ratios are approximately 1:1. Based on the estimates above, water input is from the 2.5 sub-basins, and outflow is from 1.5 sub-basins (Table 1).

Land-use in the SWAT contains vegetation coverage and agricultural, industrial, and urban land. In this study, natural and agricultural plants were set at a pre-industrial time mode. Vegetation data were sourced from a vector-based 1:1,000,000 China Vegetation Map (Hou et al., 1982), which summarizes Chinese vegetation investigations since the early 1960s. A digital map from Hou et al. (1982) at a resolution of 10 × 10 min was interpolated into the same resolution as the DEM, consistent with applications in SWAT modeling. According to the dominant plants and relative physical and biological properties (Hou et al., 1982; China Academy of Sciences, 1988) and the agricultural crop atlas (Zui et al., 1984; Bi, 1995), vegetation types in Taihu Basin consistent with the land-use types in the SWAT model setting were assigned (Table 2).

Soil data were obtained from Chinese soil investigations and mapping post-1980s (Institute of Soil Science, 1986; National Soil Survey Office of China, 1993–1997). Vector-based soil data at a resolution of 1:1,000,000 (Unpublished data; ref. to Shi et al., 2002, 2003) were used in the simulation boundary. The data contain soil types and major properties, including sand, silt, and clay contents (%), organic matter content (%), pH, total nitrogen (TN mg l^{-1}), total phosphorus (TP mg l^{-1}), available phosphorus (AP mg kg^{-1}), and bulk density (g cm^{-3}). Additional soil parameters were based on documentation of regional soils in China (Institute of Soil Science, 1986; Zhang, 1998).

Climate data were sourced from observations made from 1960 to 1990 (National Meteorology Center of China, 2001). The 30-year mean monthly temperature and precipitation data were the two major climate variables for the long-term simulation. Relative climate parameters were based on the Atlas of China Climate (Lu & Gao, 1984).

The pre-industrial model was designed to simulate natural-agricultural impacts on water, sediments, and nutrients in Taihu Basin. Model results were compared to geological records from lake cores, which preserve information on decadal-centurial time scales.

Results and analysis

The experiment runs 200 model years in four sub-basins and simulates the flow of major nutrients through the main channel in the catchment, nutrients transported with water during each time step in concentrations in HRUs, and production in each sub-basin. The simulation outputs contain organic nitrogen, NO_3^-, NO_2^- and NH_4^+, organic phosphorus and mineral phosphorus. Total nitrogen (TN) is calculated as a sum of organic nitrogen, NO_3^-, NO_2^- and NH_4^+ and total phosphorus (TP) as the sum of mineral phosphorus and organic phosphorus.

Table 1 Sub-basin topography and discharge features

Subbasin	Sub_SW	Sub_NW	Sub_N	Sub_SE	Total
Area (km^2)	5254	8298	3710	12984	30246
Elevation range (m a s l.)	21–1266	2–453	1–283	0–178	–
Mean elevation (m a s l.)	180	31	11	7.8	–
Input basin	1	1	0.5	0	2.5
Output basin	0	0	0.5	1	1.5

Table 2 Dominant plants in Taihu Basin and assigned land use types in the SWAT modeling

Land use type in SWAT model	Dominant plants in Taihu Basin
Forest - Mixed	Pinus massoniana forest: Shrubby layer with *Rhododendron simsii*, *Vaccinium bracteatum*, *Loropetalum chinense* and *Eurya nitida*; *Cunninghamia lanceolata* forest
Forest - Deciduous	Deciduous oak forest: *Quercus Mongolic*, *Quercus variabilis*, *Q. Acutedentata* and *Q. glandulifera*
Forest - Evergreen	Mixed forest containing *Cyclobalanopsis*, *Castanopsis* and *Fagus*; *Phyllostachys pubescens* forest
Wetlands - Mixed	Halophytic grass and forb meadow *Aeluropus littoralis* var. *sinensis* and *Suaeda salsa*
Winter Wheat	Winter wheat, coarse grains (corn, millet, sweet potatoes), two crops annually
Rice	Summer rice, winter wheat (rapeseed), two crops annually (double cropping rice locally); Double-cropping rice followed by winter wheat (rapeseed) or green manure; Single or double-cropping rice followed by winter wheat, rapeseed, green manure annually
Potato	Potatoes, sweet potatoes, grains, soybean green upland crops annually
Water	Water

Simulation of TN and TP production from four sub-basins

The model simulates 200-year processes for nitrogen and phosphorus production for each sub-basin (Fig. 3). When annual means for each sub-basin (column diagrams in Fig. 4) are divided by sub-basin areas (pie diagrams in Fig. 4), the largest portions of TN and TP are produced in the southwestern mountain areas of Sub_SW, in which TN is 48% and TP is 68% of total production in the basin. Smaller portions of nutrients are produced in Sub_NW (18% TN and 17% TP) and Sub_N (18% TN and 10% TP). The smallest portion is generated in Sub_SE (16% TN and 5% TP).

About half of the nitrogen production comes from Sub_SW, which relates to high biome productions of evergreen forests over the areas (Huang et al., 2001). The model also simulated a

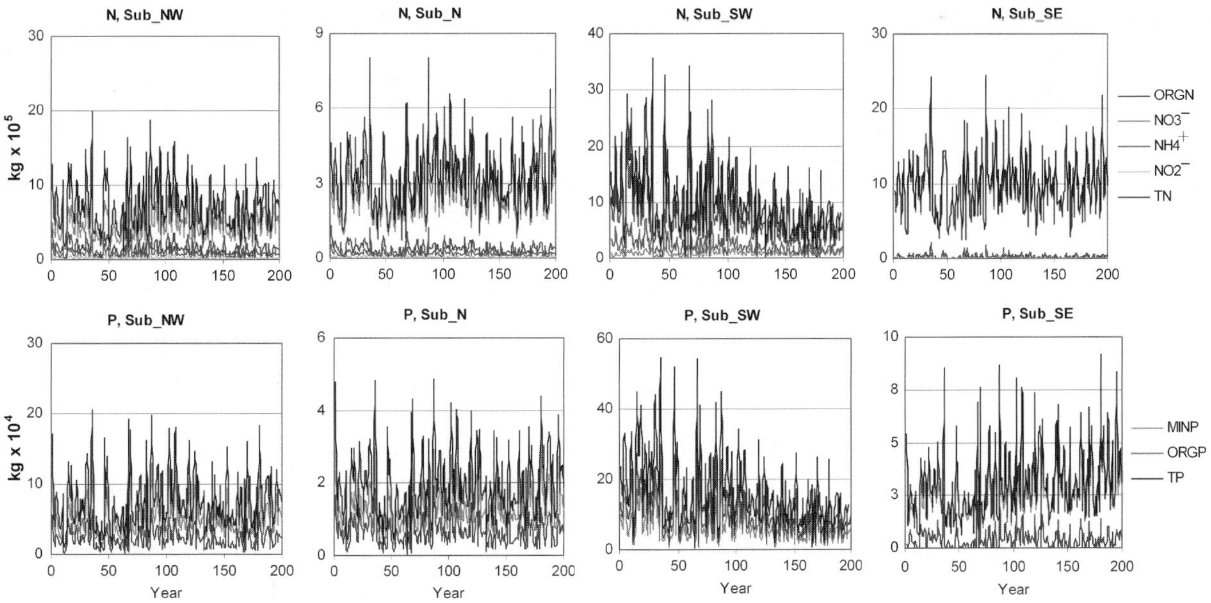

Fig. 3 Annual nitrogen and phosphorus production (kg) simulated from four sub-basins in Taihu Basin, where total nitrogen (TN) as a sum of organic nitrogen (ORGN), NO_3^-, NO_2^- and NH_4^+, total phosphorus (TP) a sum of mineral phosphorus (MINP) and organic phosphorus (ORGP)

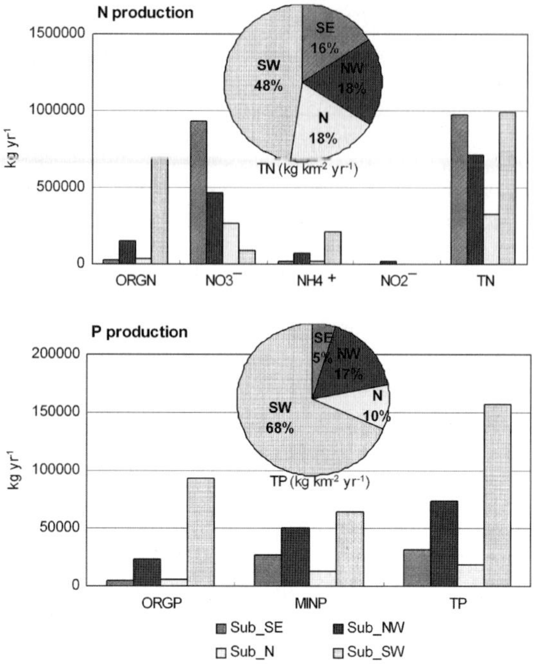

Fig. 4 Distribution simulations of TN and TP in the Lake Taihu sub-basins. Nutrient codes as in Fig. 3

high phosphorus production (~67%) for this area, reflecting strong weathering and erosion of limestone and metamorphic bedrock/outcroppings in the sub-basin. Additionally, relatively steep slopes in this mountain area (mean 2.0%) result in a high capability of discharge and transportation. Lower nutrient production was simulated in Sub_SE, which is an output area for Lake Taihu. Although nutrients discharge to coastal areas, there is small slope (mean < 0.1%), which often slacks water flows and favors nutrient deposition into the lake.

Annual and areal means of TN and TP production are plotted in Fig. 5. Total TN input from the 2.5 sub-basins to the lake is 1.86×10^5 kg, and total TN from 1.5 output sub-basins is 1.14×10^5 kg. This indicates that there is net deposition of 7.3×10^4 kg TN to the lake. Annual TP mean is 2.39×10^5 kg in 2.5 input sub-basins and 0.41×10^5 kg in 1.5 output sub-basins. TP difference between input and output is 1.99×10^5 kg, suggesting that there is TP deposition in the lake each year.

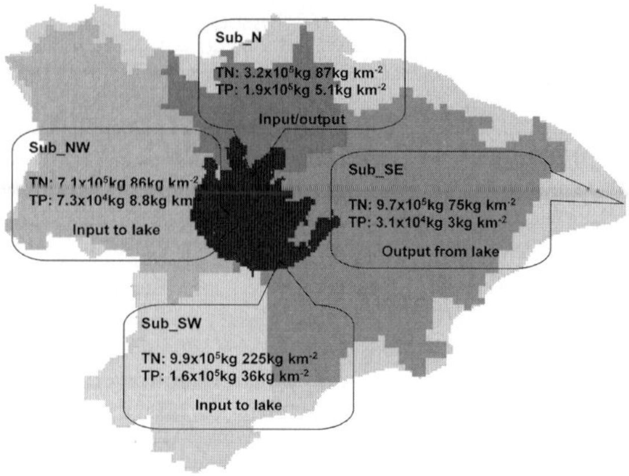

Fig. 5 Simulations of TN and TP annual means (kg) and the annual area means (kg km^{-2}) in four sub-basins

Analysis of catchment-sourced lake eutrophication

To examine lake eutrophication changes, the nutrient balance was evaluated by calculating nutrient differences between input sub-basins and output sub-basins. TN and TP concentrations used are weighted averages of all HRUs within the sub-basins.

The 200-year simulations show that annual TN concentration changes are characteristic of 12-month cycles (Fig. 6a). In the first 100 years, mean TN concentration difference was 0.044 mg l^{-1}, suggesting a positive deposition in the lake. In the second 100 years, the mean difference decreases to 0.017 mg l^{-1}, reflecting positive nitrogen deposition but at a lower rate than the first 100 years. Slopes for the trends are -3.3×10^{-5} in the first 100 years and -1.4×10^{-5} in the second 100 years (Fig. 6a), indicating that TN change stabilized after the first 100 years.

TP concentration simulations show phosphorus deposition in the lake (Fig. 6b). The TP difference in the first 100 years (+0.012 mg l^{-1}) is higher than the second 100 years (+0.008 mg l^{-1}), indicating a declining trend in phosphorus deposition. The slope for the first 100 years (-5.0×10^{-6}) was steeper than the second 100 years (slope -2.4×10^{-6}), suggesting that phosphorus concentration changes are small in the lake.

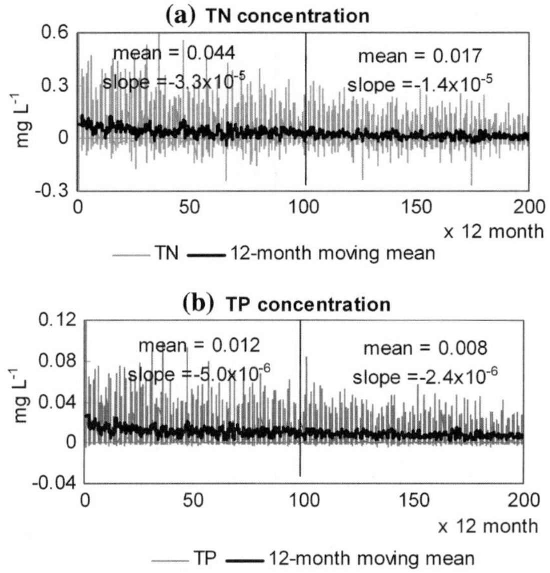

Fig. 6 Monthly differences of TN and TP concentration with the 12-month moving means between input and output sub-basins in Taihu Basin

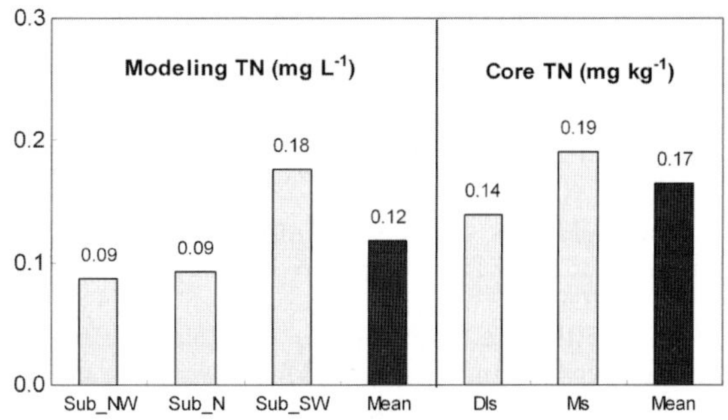

Fig. 7 Modeling TN means for a 100 year ago and measured TN from sediment records in Lake Taihu. Modeling TN concentrations (mg l^{-1}) for the lake water are tentatively compared to concentrations from sediment deposits (mg kg^{-1})

Discussion and conclusions

Comparison with nutrient records from lake sediment cores allows evaluation of the simulations when the area was agricultural before modern industrial development in Taihu Basin. Two sediment cores provided TN concentrations dating back 100 years (Liu et al., 2004a; Fig. 1). Core Dls was taken from the southern part and Core Ms from the northern part of Lake Taihu. TN concentrations at core depth of 20–40 cm are 0.14–0.19 mg kg^{-1} (Liu et al., 2004b), where sediment samples were ^{210}Pb- and ^{137}Cs-dated to before 1900 AD (Liu et al., 2004a). 200-year means show TN concentrations between 0.09 and 0.18 mg l^{-1} (Fig. 7). Geochemical nitrogen values from lake sediments may be subjected to post-depositional mobility (Smol, 2000). Diatom-inferred nutrients, together with geochemical data, may be used for historical eutrophication studies in the future.

The simulation mean (0.12 mg l^{-1}) and sub-basin means (0.09–0.18 mg l^{-1}) are significantly lower than the data mean (0.17 mg kg^{-1}) and two cores (0.14–0.19 mg kg^{-1}) with 95% confidence interval in T-test. Nitrogen from sediment cores would respond to both catchment sources and lake water production. However, the present modeling is focused on nutrient production in the catchment and transport to the lake. Thus, the nitrogen concentration simulation should be lower than in the cores. Modeling may adequately simulate basin nutrients for pre-industrial times.

In summary, the present simulation may capture major factors of lake eutrophication over a 200-year history. Analysis of spatial nutrient distributions suggests that the major source (48% TN and 68% TP) comes from the southwestern sub-basin. Due to a gentle slope in the eastern plains of Taihu Basin, the smallest portion of nutrient input (16% TN and 5% TP), from the southeastern sub-basin, cannot be fully discharged to the lake. There is a net annual nutrient deposition of 7.3×10^4 kg TN and 1.99×10^5 kg TP to the lake. Time series analysis for the 200-year simulation suggests that nitrogen and phosphorus deposition has decreased over the last 100 years.

Catchment nutrient modeling provides a basis to evaluate nutrient production and balance response to environmental changes for the last 200 years in Taihu Basin. There is considerable uncertainty about the direction of future nutrient changes in Lake Taihu due to anthropogenic changes. The accuracy of model-based predictions of future eutrophication cannot be evaluated, but the models used to generate such predictions can be tested by simulating the past and comparing these simulations with sediment records. This effort, along with future investigations, e.g. using regional climate modeling (Zheng et al., 2003) and fossil pollen-based vegetation reconstructions (Yu et al., 2000), would provide more credibility for modeling the future.

Acknowledgements Financial support was provided by Chinese Academy of Sciences (No. KZCX1-SW-12-1) and National Basic Research Program of China (No. 2002CB412301). We thank Jinfang Dai (Nanjing Institute of Geography & Limnology, CAS) for providing land use data, Haiyu Li (Department of Geography, Nanjing University) for assistance in DEM transforming, and Xuesheng Shi (Institute of Soil Sciences, CAS) for providing soil data. We thank Sumin Wang for constructive suggestions on the manuscript.

References

Arnold, J. G., R. Srinivasan, R. S. Muttiah & J. R. Williams, 1998. Large area hydrologic modeling and assessment part I: model development. Journal of American Water Resources Association 34:73–89.

Bi, Y. Y., 1995. Farmland in China. Agricultural Science and Technology Press, Beijing, 231 (in Chinese).

Bradbury, J. P., S. M. Colman & R. L. Reynolds, 2004. The history of recent limnologic changes and human impact on Upper Klamath Lake, Oregon. Journal of Paleolimnology 31: 151–165.

Chinese Academy of Sciences, 1988. Physical Geography of China. Biogeography Science Press, Beijing, 318 (in Chinese).

Di Luzio, M., R. Srinivasan, J. G. Arnold & S. L. Neitsch, 2002a. Soil and Water Assessment Tool. ArcView GIS Interface Manual: Version 2000. GSWRL Report 02–03, BRC Report 02–07, Published by Texas Water Resources Institute TR-193, College Station, TX, 346.

DiLuzio, M., R. Srinivasan & J. G. Arnold, 2002b. Integration of watershed tools and SWAT model into basins. Journal of the American Water Resources Association 38: 1127–1141.

Hou, X. Y., S. Z. Sun, J. W. Zhang, M. G. He, Y. F. Wang, D. Z. Kong & S. Q. Wang, 1982. Vegetation Map of the People's Republic of China. China Cartography, Beijing (in Chinese).

Huang, Y. P., C. F. Fan, W. M. Pu, J. H. Jiang & D. Y. Dai, 2001. Water Environments of Lake Taihu and Controlling the Pollution. Science Press, Beijing, 298 (in Chinese).

Institute of Soil Science of Chinese Academy of Sciences, 1986. The Soil Atlas of China (1:14,000,000). Cartographic Publishing House, Beijing (in Chinese).

Jin, X. C., H. L. Liu, Q. Y. Tu, Z. S. Zhang & X. Zhu, 1990. Lake Eutrophications of China. Chinese Environment Press, Beijing, 614 (in Chinese).

Last, W. M. & J. P. Smol, 2001. Tracking Environmental Change Using Lake Sediments, Volume 1, Basin Analysis, Coring, and Chronological Techniques. Kluwer Academic Publishers, Dordrecht, 548.

Liu, E. F., J. Shen, S. C. Yao & B. Xue, 2004a. Analysis of heavy metal element sources in surface sediment of Lake Taihu. Journal of Lake Sciences 16: 113–119 (in Chinese).

Liu, E. F., J. Shen, Y. X. Zhu, W. L. Xia, H. X. Pan & Z. D. Jin, 2004b. Heavy metals and nutrients pollution in sediments of Taihu Lake. Acta Sedimentologica Sinica 22: 507–512 (in Chinese).

Lu, Y. R. & G. D. Gao, 1984. Atlas of Atmosphere-Hydrology Climate of China. Meteorology Press, Beijing, 183 (in Chinese).

National GIS Center of China, 1999. Data Base of 1:2,500,000 topography of China, Beijing.

National Meteorology Center of China, 2001. Annual Meteorological Observation Report in land surface of China, Beijing (in Chinese).

National Soil Survey Office of China, 1993–1997. Soils in China. Agricultural Publishing House: Beijing, 1–6 (in Chinese).

Qin, B. Q, 2002. Eutrophication of shallow water lakes in the mid-lower reaches of Changjiang and their controlling strategy. Journal of Lake Sciences 14: 193–202 (in Chinese).

Qin, B. Q., W. P. Hu & G. Gao, 2004. Dynamics of sediment resuspension and the conceptual schema of nutrient release in the large shallow Lake Taihu, China. Chinese Science Bulletin 49: 54–64.

Ramstack, J. M., S. C. Fritz, D. R. Engstrom & S. A. Heiskary, 2003. The application of a diatom -based transfer function to evaluate regional water-quality trends in Minnesota since 1970. Journal of Paleolimnology 29: 79–94.

Rose, N. L., J. F. Boyle, Y. Du, C. Yi, X. Dai, P. G. Appleby, H. Bennion, S. Cai & L. Yu, 2004. Sedimentary evidence for changes in the pollution status of Taihu in the Jiangsu region of eastern China. Journal of Paleolimnology 32: 41–51.

Rosenberg, N. J., D. L. Epstein, D. Wang, L. Vail, R. Srinivasan & J. G. Arnold, 1999. Possible impacts of global warming on the hydrology of the Ogallala aquifer region. Journal of Climate 42: 677–692.

Schulze, R. E., 2000. Modelling hydrological responses to land use and climate change: a southern african perspective. AMBIO 29: 12–22.

Shi, X. Z. & D. S. Yu, 2002. A framework for the 1:1,000,000 soil database of China. In Warner, E. D. & G. W. Peterson (eds), Proceedings of the 17th World Congress of Soil Science, Bangkok, 1757: 1–5.

Shi, X. Z., D. S. Yu, X. Z. Pan, W. X. Sun, H. J. Wang & Z. T. Gong, 2003. The 1:1,000,000 China Soil Database and the Application. The Proceedings of China Soil Science Congress (in Chinese).

Smol, J. P., 2002. Pollution of Lakes and Rivers: A Paleoenvironmental Perspective. London, Arnold Publishers, New York, 280.

Sun, S. C. & Y. P. Huang, 1993. Lake Taihu. Ocean Press, Beijing, 273 (in Chinese).

Wang, J., Y. J. Wang & J. L. Liu, 1996. Evolution of sedimentary environment in Lake Taihu during the last 16,000 years. Acta Palaeontologica Sinica 35: 213–223 (in Chinese).

Wang, Z. G., C. M Liu & X. F. Wu, 2003. A perspective for the DEM-based hydrological catchment model. Journal of Natural Resources 18: 168–173 (in Chinese).

Xia, L. Z. & L. Z. Yang, 2003. Studying and controlling for catchment – sourced pollutions in Lake Taihu. Resources and Environments in the Changjiang Catchment 12: 45–49 (in Chinese).

Xue, B., W. C. Qu, Y. H. Wu, S. M. Wang & R. J. Wu, 1998. Sedimentological record of paleoenvironment of Lake Taihu in Late-Glacial to Holocene. Journal of Lake Sciences 10: 30–36 (in Chinese).

Yu, G., X. Chen, J. Ni, R. Cheddadi, J. Guiot, H. Han, S.P. Harrison, C. Huang, M. Ke, Z. Kong, S. Li, W. Li, P. Liew, G. Liu, J. Liu, Q. Liu, K. B. Liu, I. C. Prentice, W. Qui, G. Ren, C. Song, S. Sugita, X. Sun, L. Tang, E. Van Campo, Y. Xia, Q. Xu, S. Yan, X. Yang, J. Zhao & Z. Zheng, 2000. Palaeovegetation of China: a pollen data-based synthesis for the mid-Holocene and last glacial maximum. Journal of Biogeography 27: 635–664.

Zhang, W. R., 1998. Soil Conditions of the Dominant Plants in the Forest of China. China Science and Technique Press, Beijing, 432 (in Chinese).

Zhang, Y. L., B. Q. Qin & F. Q. Huang, 2002. Evolution and zonal analysis on lake eutrophication in East Plain Region. Shanghai Environmental Sciences 21: 549–553 (in Chinese).

Zheng, Y. Q., G. Yu & S. M. Wang, 2003. Numerical climate modeling since the last 140 years by RegMC2 for East Asia. Progresses of Natural Sciences 13: 951–956 (in Chinese).

Zui, D. C., H. S. Liu, J. R. Min & J. M. He, 1984. Atlas of Climatic Resources for Major Crops in China (1:17,000,000). Meteorological Press, Beijing (in Chinese).

Hydrobiologia (2007) 581:89–95
DOI 10.1007/s10750-006-0510-8

EUTROPHICATION IN LAKES

Evaluation of biogenic and anthropogenic inputs of aliphatic hydrocarbons to Lake Taihu sediments using biomarkers

Wenchuan Qu · Bin Xue · Chenwei Su · Sumin Wang

Abstract Surficial sediments from 13 sites throughout Lake Taihu, one of the largest urbanized freshwater lake systems in China, were analyzed for biomarkers (e.g., *n*-alkanes and hopanes) to track the origin of organic inputs (biogenic or anthropogenic), and, thus, to identify any 'hot spots' of hydrocarbon contamination. A distinct spatial distribution of aliphatic hydrocarbons in sediments was observed in Lake Taihu. At the northern tip of the lake (i.e., Meiliang and Wuli Bays), the highest mean aliphatic hydrocarbon concentration, with a significant contribution of an unresolved complex mixture (UCM), was found, indicating possible anthropogenic petroleum contamination (mainly from untreated and partially treated industrial and domestic sewage from Wuxi, Changzhou and other cities). This was supported by the n-alkane indices (e.g., small Carbon Preference Index and odd-to-even values) and a high degree of maturity of the hopane biomarkers. However, hydrocarbons from East Taihu were mainly biogenic, with the lowest mean concentrations, negligible or no contribution of UCM, abundance of vascular plant C_{23}–C_{33} *n*-alkanes with a high odd-to-even predominance, and the presence of biogenic hopanes (e.g., 17β(H), 21β(H)-hopanes and hopenes). In the other areas of the lake, however, the predominance of biogenic in combination with petrogenic hydrocarbons was indicated by the biomarkers.

Keywords Lake Taihu · Sediment · Biomarkers · Sources · Biogenic · Anthropogenic

Guest editors: B. Qin, Z. Liu & K. Havens
Eutrophication of shallow lakes with special reference to Lake Taihu, China

W. Qu (✉) · B. Xue · S. Wang
Key Laboratory of Lake Sediment & Environment, Chinese Academy of Sciences, Nanjing 210008, China
e-mail: wenchuan.qu@gmail.com

W. Qu
Australian Bureau of Statistics, Canberra, ACT 2615, Australia

C. Su
Research School of Biological Sciences, The Australian National University, Canberra, ACT 0200, Australia

Introduction

Aliphatic hydrocarbons (AHCs) are ubiquitous in aquatic sediments, and they derive from anthropogenic and natural sources (Zanardi et al., 1999). Potential long-term deleterious effects of hydrocarbon pollutants to organisms and human health, through trophic transfer, has driven intensive research to study hydrocarbon pollutant sourcing in aquatic environments using biomarkers and related indices (Albaiges & Albrecht,

1979; Qu et al., 1998; Commendatore et al., 2000; Barakat et al., 2002).

Lake Taihu is the third largest freshwater lake in China (Sun & Huang, 1993; Qu et al., 1999; Qin et al., 2007). Pollution is one of the key environmental problems in Lake Taihu due to recent urbanization and industrial development, with associated increases in urban, industrial and rural pollution (TBA, 1998). For example, Meiliang Bay (at the northern tip) receives large amounts of untreated effluent from industries in Changzhou and Wujin, mainly through Zhihugang. Just beside Meiliang Bay, there is a small lagoon (area ~10 km^2) named Wuli Bay (Fig. 1), which receives millions of tons of domestic and industrial sewage each day, mainly via Liangxihe from Wuxi, a large city located ~2 km northeast of Wuli Bay (Zou et al., 1996). The southeast part of Lake Taihu (East Taihu; area 131 km^2) has macrophytes covering ~97% of its total area and is referred to as a macrophyte-type basin.

In this article, surficial sediments from 13 sites in Lake Taihu were analyzed for biomarkers (e.g., *n*-alkanes and hopanes) to track origins of organic inputs (biogenic or anthropogenic) and identify 'hot spots' of hydrocarbon contamination.

Materials and methods

Sample collection

Surface sediments were collected using a metallic bucket grab sampler in February 2000 at stations shown in Fig. 1. Sites 1, 2 and 3 are located in Wuli Bay. Site 1 is close to the mouth of Liangxihe. Site 4 is located in west Meiliang Bay near the mouth of Zhihugang. Site 5 is located in central Meiliang Bay. Sites 11, 12, and 13 are located in East Taihu, and the other sites (6–10) are distributed in the West Basin. Sediment samples were returned to the laboratory and kept frozen until analyzed.

Extraction and fractionation

Sediment samples were dried in the air and sieved (100 mesh or 149 μm) to remove large particles of plant debris. About 50–100 g of sediment were extracted in a Soxhlet (500 ml; (2:1) CH_2Cl_2/MeOH) solution for 72 h. Sulphur was removed by activated copper. Extracts were fractionated into aliphatic hydrocarbons and aromatic hydrocarbons on an activated silica/alumina column using hexane and benzene as eluents, respectively.

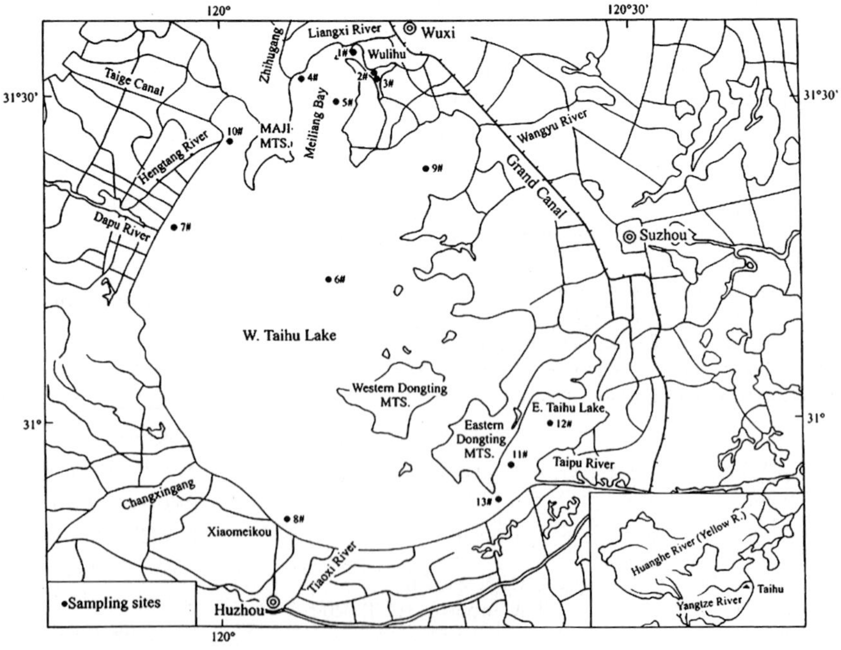

Fig. 1 Sampling sites in Lake Taihu

Gas chromatography and mass spectrometry

Aliphatic hydrocarbons were analyzed on a gas chromatography-mass spectrometry (GC-MS) instrument (HP5890 capillary GC-HP5989A quadrupole MS). Operating conditions were: a HP-5 fused silica capillary column (30 m × 0.25 mm), high-purity He carrier gas (>4.6); gradient temperature: 70 → 300°C, at 4°C/60 s; ion source temperature: 250°C and electron energy: 1.12×10^{-17} J (Qu et al., 1998). Biomarkers were analyzed by monitoring m/z 85 (alkanes) and m/z 191 (hopanes) using the GC/MS in Selected Ion Monitoring (SIM) mode (Qu et al., 1998).

Total organic carbon analysis

Total organic carbon (Org-C) was determined on dried sediments by $K_2Cr_2O_7$ volumetry (Marr & Cresser, 1983).

Results and discussion

Gas Chromatography—total aliphatic hydrocarbon fraction

Concentration of total aliphatic hydrocarbons (AHCs, defined as the sum of resolved peaks plus the unresolved complex mixture, UCM) and organic carbon for 13 surface sediments in Lake Taihu are summarized in Table 1.

A distinct spatial distribution of aliphatic hydrocarbons in surface sediments was observed. High AHC concentrations (>300 μg g^{-1} dw) were found in sediments from Wuli Bay (Sites 1–3), with the highest concentration (1,630 μg g^{-1} dw) occurring at Site 1 (Table 1). This may be because Site 1 is close to the mouth of the Liangxihe, which carries large amounts of untreated domestic and industrial sewage from Wuxi and other small cities near Wuli Bay (Fig. 1). AHC concentration at Site 4 also was high (342 μg g^{-1} dw). This may be explained by its proximity to the mouth of Zhihugang, which carries about 20 million tons of industrial effluents annually from industrialized cities and towns (e.g., Changzhou and Wujin) (Sun & Huang, 1993). AHC concentrations in sediments from other sites were lower (<100 μg g^{-1} dw), with the lowest occurring in East Taihu (36, 33 and 54 μg g^{-1} dw for Sites 11, 12 and 13, respectively) (Table 1), which is likely due to the greater distance from major urban and industrial areas and, thus, anthropogenic hydrocarbon sources. The same spatial distribution pattern was found for other micropollutants, such as polycyclic aromatic hydrocarbons (PAHs) (Qu et al., 2002).

To reduce variability from organic matter dilution by inorganics and, thus, better evaluate sediment hydrocarbon pollution, AHC concentrations (in μg g^{-1}) were normalized to organic carbon content (Org-C%). The ratio of AHC concentration to Org-C% has been used as a diagnostic tool, proposing that hydrocarbon (petroleum) pollution is present when the ratio is >50 (Boehm, 1984; Hong et al., 1995). In this study, Sites 1–4 showed the highest Org-C-normalized AHC concentrations (ranging from 144 to 426), indicating the presence of substantial hydrocarbon contamination at the northern tip of the lake. Sediments from East Taihu (Sites 11,12 and 13) had the lowest ratios (ranging from 14 to 52) and were considered unpolluted. AHCs in the remaining sediment samples (mainly from the West Basin) were intermediate relative to Org-C and appeared to be slightly contaminated

GC/MS SIM analysis of alkane biomarkers for AHC source tracing

To distinguish probable sources of AHCs (anthropogenic or biogenic) in surface sediments, different n-alkane evaluation parameters, such as UCM index, carbon preference index (CPI) and even-to-odd carbon number predominance (OEP) ratios, were analysed in addition to absolute concentrations.

UCM is a mixture of chromatographically unresolved components (e.g., many structurally complex isomers and homologues of branched and cyclic hydrocarbons) seen as a 'hump' in the chromatogram (Hostettler et al., 1989; Hong et al., 1995). UCM magnitude is related to the degree of anthropogenic (petroleum) contribution (Mazurek & Simoneit, 1984; Commendatore et al., 2000). UCM is shown clearly in gas chromatograms of aliphatic hydrocarbons in heavily

Table 1 Concentrations of total aliphatic hydrocarbons (AHCs), organic carbon (Org-C), alkane and hopane biomarker parameters for 13 sediment samples in Lake Taihu

Locations	Org-C (%)	AHC $\mu g\ g^{-1}$ dw	UCM[a](%)	AHC/Org-C	Odd/Even (OEP)	CPI^{b}_{24-34}	%C_{30}-hopenes/total hopanes	Tm/Ts[c]	$C_{30}\beta\alpha/(\alpha\beta+\beta\alpha)$	$C_{30}\beta\beta/\alpha\beta$	$C_{31}S/(S+R)$
1	3.83	1630	80	426	1.6	1.8	1.1	0.90	0.08	0.00	0.53
2	2.05	337	60	164	1.9	2.4	4.9	0.91	0.10	0.00	0.54
3	2.16	312	65	144	2.0	2.0	4.7	0.92	0.10	0.00	0.55
4	1.86	342	60	184	1.6	2.1	1.6	0.80	0.06	0.00	0.58
5	1.06	93	45	87	2.1	2.6	6.0	0.94	0.18	0.00	0.53
6	0.87	66	15	76	2.3	3.7	6.9	1.50	0.18	0.00	0.48
7	0.82	73	40	89	2.8	3.7	6.3	1.00	0.13	0.00	0.52
8	0.81	74	35	91	2.1	2.8	5.1	1.00	0.11	0.00	0.53
9	0.94	52	15	55	2.3	3.3	9.6	1.70	0.15	0.00	0.52
10	0.81	112	45	138	2.5	3.8	9.7	1.00	0.11	0.00	0.50
11	1.99	36	6	18	3.1	5.8	36.1	2.00	0.20	0.50	0.41
12	2.31	33	8	14	3.4	8.0	13.1	3.28	0.19	0.12	0.46
13	1.04	54	7	52	2.9	4.4	10.0	1.80	0.18	0.10	0.48

[a] UCM = Unresolved complex mixture

[b] $CPI_{24-34} = 1/2([(C_{25}+C_{27}+C_{29}+C_{31}+C_{33})/(C_{24}+C_{26}+C_{28}+C_{30}+C_{32})]+[(C_{25}+C_{27}+C_{29}+C_{31}+C_{33})/(C_{26}+C_{28}+C_{30}+C_{32}+C_{34})])$

[c] Tm/Ts = 17α(H)-22,29,30-trisnorneohopane /18α(H)-22,29,30-trisnorneohopane (Please refer to the text for more explanations)

AHC polluted sediments (e.g., Site 4, Fig. 2a). Table 1 suggests that UCM accounted for >60% of total AHCs in sediments from Sites 1 to 4, indicating possible petroleum contamination. In contrast, sediments from East Taihu (Sites 11, 12 and 13) showed negligible UCM contribution (<10%) (Fig. 2b), indicating absence of petrogenic inputs. UCM in other samples contributed about 10–50% to total AHCs, suggesting predominance of biogenic in combination with petrogenic hydrocarbons.

The CPI and OEP indices also are used as diagnostic criteria of hydrocarbon sources (anthropogenic or biogenic) in sediments (Hostettler et al., 1989). In most plant waxes, odd-chain alkanes are 8–10 times more abundant than even-chain *n*-alkanes. Thus, vascular plants and unpolluted sediments usually show high OEP and CPI_{24-34} values (ranging from 3 to 6). However, petrogenic hydrocarbons often show a wide distribution range of *n*-alkanes, and the even/odd ratio shows no predominance, with CPI values around 1 (Colombo et al., 1989). As expected, samples from Wuli and Meiliang Bays displayed low CPI (ranging from 1.8 at Site 1 to 2.6 at Site 5) and OEP (ranging from 1.6 at Site 1 to 2.1 at Site 5) ratios (Table 1, Fig. 2c), which evidenced the presence of (riverine) petroleum residues in surface sediments at the northern tip of the lake. Lower CPI and OEP values occurred in samples from the West Basin (Sites 6–10), which indicates decreased contribution of petrogenic hydrocarbons. Samples from East Taihu (Sites 11, 12 and 13) showed the highest CPI (5.8, 8 and 4.4, respectively) and OEP ratios (3.1, 3.4 and 2.9, respectively) (Table 1, Fig. 2d), confirming that these sediments contain mainly contributions from biogenic sources (i.e., vascular plants).

GC/MS SIM analysis of hopane biomarkers for source tracing

Chromatograms of the hexane fraction (Fig. 2) also showed some hopane biomarker peaks. Therefore, GC/MS SIM analysis of hopanes (m/z 191) was conducted to provide more information on AHC sources and spatial patterns, since these biomarkers are more resistant to degradation than alkanes and isoprenoids and retain their composition and distribution during various processes (Barakat et al., 2002).

In the AHC polluted sediments (Sites 1–4), the thermodynamically stable 17α(H), 21β(H)-hopane (that is, αβ-hopane) series were more

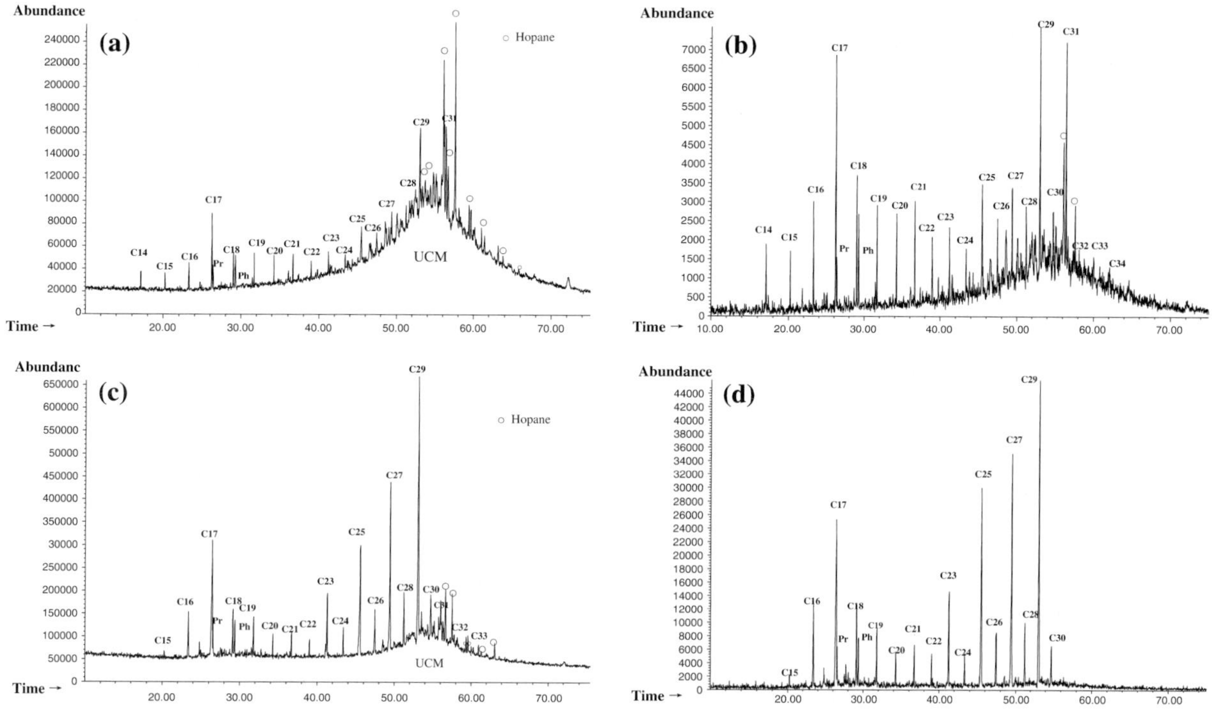

Fig. 2 Capillary column gas chromatograms of the hexane fraction of sediment samples from (**a**) Site 4 (representative of locations with mainly anthropogenic contributions) and (**b**) Site 13 (representative of locations with mainly biogenic contributions); the m/z 85 ion chromatograms of sediment samples from (**c**) Site 4 and (**d**) Site 13 in Lake Taihu. Peaks labeled are the *n*-alkanes from C_{14} to C_{33} (14–33), pristine (Pr), phytane (Ph), and hopanes (*o*)

abundant than the less stable $\beta\alpha$-hopane series, natural biogenic $\beta\beta$-hopane series or mono-unsaturated hopanes (e.g., C_{30}-hopenes) (Fig. 3a). These samples thus had a high degree of hopane biomarker maturity (e.g., lower ratios of $C_{30}\beta\alpha/(\alpha\beta + \beta\alpha)$, $C_{30}\beta\beta/\alpha\beta$ and % C_{30}-hopenes) (Table 1), suggesting a contribution of petrogenic hydrocarbons. Further confirmation of petrogenic pollution was obtained from other geochemical maturity and source indicators. Mature sediments often have low 17α(H)-/18α(H)-22,29,30-trisnorneohopane (Tm/Ts) ratios; extended $\alpha\beta$-hopanes (i.e., series with more than 30 carbon atoms) are resolved into two diastereomers (22S and R) with high S/(S+R) ratios near values for full maturity (equilibrium distribution) (e.g., approximately 0.6 for $C_{31}\alpha\beta$-hopanes) (Hostettler et al., 1999). This is the case in samples from Sites 1 to 4 (Fig. 3a, Table 1).

However, hopane in East Taihu sediments were biogenic (e.g., $\beta\beta$-hopane series and mono-unsaturated hopanes) (Fig. 3b). These samples had the highest ratios of $C_{30}\beta\alpha/(\alpha\beta + \beta\alpha)$ (ranging from 0.18 to 0.20), $C_{30}\beta\beta/\alpha\beta$ (0.10 to 0.50) and % C_{30}-hopenes (10.0–36.1%) (Table 1). Mass chromatograms (m/z 191) of these samples (e.g., Site 13, Fig. 3b) also showed extended $\alpha\beta$-hopane epimers and trisnorhopanes, Ts and Tm, indicating lower maturity (i.e., higher Tm/Ts and lower C_{31} S/(S+R) epimer ratios) (Table 1). All these features are indications of low geochemical maturity and biogenic hydrocarbon sources. In addition, the rest of the samples, mainly from the Western Basin, showed a mixture of mature and immature constituents (Table 1), supporting the conclusion that these sediments contain both biogenic (i.e., vascular plants) and anthropogenic (i.e., petrogenic) sources.

Conclusions

In summary, Lake Taihu surficial sediments showed different levels and sources of aliphatic

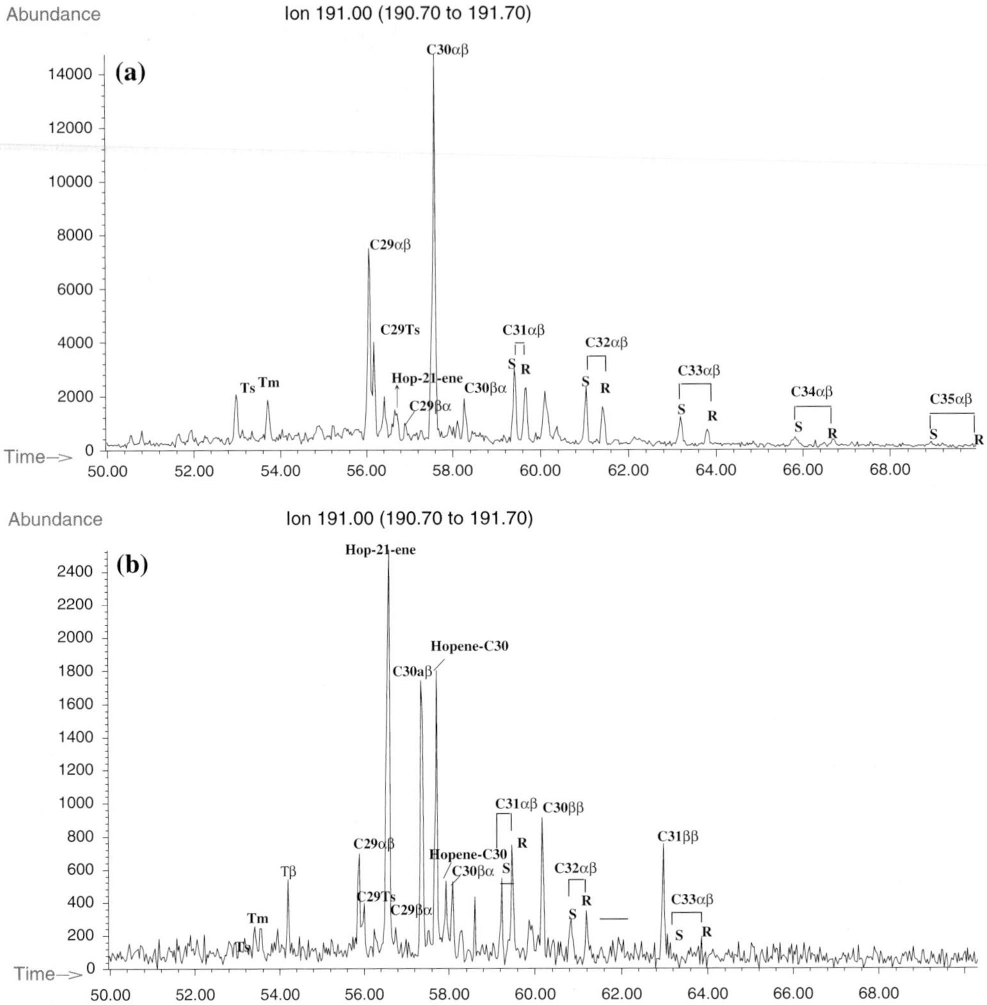

Fig. 3 Mass chromatograms of m/z 191 (hopanes) of surface sediment samples from (**a**) Site 4 and (**b**) Site 13 in Lake Taihu. Peaks identified are hopanes: Ts = 18α(H)-22,29,30-trisnorneohopane; Tm = 17α(H)-22,29,30-trisnorneohopane; T_β = 17β(H)-22,29,30-trisnorneohopane; $C_{29}\alpha\beta$ = 17α(H), 21β(H)-30-norhopane; 29Ts = 18α(H)-norneohopane; $C_{29}\beta\alpha$ = 17β(H), 21α(H)-30-normoretane; $C_{30}\alpha\beta$ = 17α(H), 21β(H)-hopane; $C_{30}\beta\alpha$ = 17β(H), 21α(H)-moretane; S- and R-$C_{31}\alpha\beta$ = S and R isomers of 17α(H), 21β(H)-homohopanes; $C_{30}\beta\beta$=17β(H), 21β(H)-hopane; S- and R- $C_{32}\alpha\beta$ = S and R isomers of 17α(H), 21β(H)-bishomohopanes; $C_{31}\beta\beta$=17β(H), 21β(H)-homohopane; S- and R- $C_{33}\alpha\beta$ = S and R isomers of 17α(H), 21β(H)-trishomohopanes

hydrocarbons, and biomarkers and related indices can provide information on anthropogenic contamination, especially related to petrogenic sources, as well as recent biogenic materials.

Near urban (e.g., Wuxi and Changzhou) and riverine (e.g., Zhihugang and Liangxihe) discharges, anthropogenic petrogenic AHC inputs were predominant. This was supported by: (1) high Org-C-normalized AHC concentrations; (2) significant contributions of UCM; (3) low n-alkane CPI and OEP; and (4) the high degree of maturity of hopane biomarkers. On the contrary, East Taihu sediments exhibited: (1) the lowest level of AHCs; (2) negligible or no contributions of UCM; (3) high n-alkane CPI and OEP; and (4) low degree of maturity of hopane biomarkers. These features suggested that AHCs in these sediments have a dominant biogenic (vascular plants) origin. Biomarker parameters for Western Basin sediment samples, however, indicated the predominance of biogenic in combination with petrogenic hydrocarbons.

Information on sources and spatial variations of AHCs, together with other micropollutants, such as PAHs and heavy metals, may be useful in designing future strategies for protection and management of Lake Taihu.

Acknowledgements The authors thank Jiang Ji of Nanjing Institute of Geography & Limnology, Chinese Academy of Sciences (CAS), for his help in the field. They are grateful to Minzhong Cui and Wanren Ding of Lanzhou Institute of Geology, CAS, for their assistance in the experimental analysis. Thanks are also due to the National Key Project for Basic Research on the Processes of Lake Eutrophication and the Mechanism of Cyanobacterial Blooming (No. 2002CB412301), Chinese Academy of Sciences; Knowledge Innovation Project "Environmental Effects Studies of Lake Sediment Dredging" (No. KZCX3-SW-348) and the 2005 Endeavour Australia Cheung Kong Awards for their financial aid.

References

Albaiges, J. & P. Albrecht, 1979. Fingerprinting marine pollutant hydrocarbons by computerized gas chromatography-mass spectrometry. International Journal of Environmental Analytical Chemistry 6: 171–190.

Barakat, A. O., A. Mostafa, Y. Qian & M. C. Kennicutt, 2002. Application of petroleum hydrocarbon chemical fingerprinting in oil spill investigation - Gulf of Suez, Egypt. Spill Science & Technology Bulletin 7: 229–239.

Boehm, P. D., 1984. Aspects of the saturated hydrocarbon geochemistry of recent sediments in the Georges Bank region. Organic Geochemistry 7: 11–23.

Colombo, J. C., E. Pelletier, C. Brochu & M. Khalil, 1989. Determination of hydrocarbons sources using n-Alkane and polyaromatic hydrocarbon distribution indexes. Case study: Rio de La Plata Estuary, Argentina. Environmental Science & Technology 23: 888–894.

Commendatore, M. G., J. L. Esteves & J. C. Colombo, 2000. Hydrocarbons in coastal sediments of Patagonia, Argentina: levels and probable sources. Marine Pollution Bulletin 40(11): 989–998.

Hong, H., L. Xu, L. Zhang, J. C. Chen, Y. S. Wong & T. S. M. Wan, 1995. Environmental fate and chemistry of organic pollutants in the sediment of Xiamen and Victoria Harbours. Marine Pollution Bulletin 31: 229–236.

Hostettler, F. D., W. E. Pereira, K. A. Kvenvolden, A. Geen, S. N. Luoma, C. C. Fuller & R. Anima, 1999. A record of hydrocarbon input to San Francisco Bay as traced by biomarker profiles in surface sediment and sediment cores. Marine Chemistry 64: 115–127.

Hostettler, F. D., J. B. Rapp, K. A. Kvenvolden & S. N. Luoma, 1989. Organic markers as source discriminants and sediment transport indicators in south San Francisco Bay, California. Geochimica et Cosmochimica Acta 53: 1563–1576.

Marr, I. L. & M. S. Cresser, 1983. Environmental Chemical Analysis. Blackie & Son Ltd, London: 258.

Mazurek, M. A. & B. R. T. Simoneit, 1984. Characterization of biogenic and petroleum-derived organic matter in aerosols over remote, rural and urban areas. Ann Arbor Science. Boston: 353–378.

Qin, B. Q, P. Z. Xu, Q. L. Wu, L. C. Luo & Y. L. Zhang, 2007. Environmental issues of Lake Taihu, China. Hydrobiologia 581: 3–14.

Qu, W. C., M. Dickman, C. X. Fan, S. M. Wang, C. W. Su, L. Zhang & H. X. Zou, 2002. Distribution, sources and potential toxicological significance of polycyclic aromatic hydrocarbons (PAHs) in Lake Taihu sediments, China. Hydrobiologia 485: 163–171.

Qu, W. C., M. Dickman, S. M. Wang & R. J. Wu, 1999. Evidence for an aquatic origin of ketones found in Lake Taihu sediments. Hydrobiologia 397: 149–154.

Qu, W. C., M. Dickman, S. M. Wang, R. J. Wu, P. Z. Zhang, J. F. Chen & H. J. He, 1998. Lake typology based on the use of sediment alkanes in the east and west basins of Lake Taihu, China. Hydrobiologia 364: 219–223.

Sun, S. C. & Y. P. Huang, 1993. Taihu. The Ocean Press, Beijing: 271.

TBA, 1998. Eutrophication Assessment of Lake Taihu - Report of China-Japan Co-operation Project in Lake Taihu. Taihu Basin Authority, Shanghai, 200 pp.

Zanardi, E., M. C. Bicego, L. B. Miranda & R. R. Weber, 1999. Distribution and origin of hydrocarbons in water and sediment in Sao Sebastiao, SP, Brazil. Marine Pollution Bulletin 38(4): 261–267.

Zou, H. X., G. Y. Sheng, C. Sun & O. Y. Xu, 1996. Distribution of organic contaminants in Lake Taihu. Water Research 30: 2003–2008.

Hydrobiologia (2007) 581:97–108
DOI 10.1007/s10750-006-0494-4

EUTROPHICATION IN LAKES

Spatio-temporal distribution of nitrogen in the undulating littoral zone of Lake Taihu, China

Hongjun Wang · Jinwei Lu · Weidong Wang · Peisheng Huang · Chengqing Yin

Abstract Spatio-temporal distribution of nitrogen was examined along the gradient from open water to lakeshore in Lake Taihu, China. Two types of undulating littoral zones were selected: natural reed belt and bare lakeshore. The reed belt affected nitrogen transformation and was sink for internal-lake nitrogen, whereas the bare lakeshore showed little effect. During the growing season, NO_3^--N concentration increased by up to 3–5 times from open water to reed belt, while NH_4^+-N concentration decreased. It suggested that nitrification was the main nitrogen process in reed belt. Total dissolved nitrogen (TDN) showed little spatial variation, indicating that most of nitrogen released from sediment did not move into open water again. Significant temporal variation of dissolved nitrogen occurred and was similar in both the littoral zone and the open water. Maximum TDN and NH_4^+-N concentrations occurred in January, and NO_3^--N in March. Minimum NH_4^+-N and NO_3^--N concentrations occurred in July and August, respectively. An increasing total soil nitrogen was found in the surface sediments from reed belt to open water. This further suggested that the reed-covered littoral zone had strong nitrogen transformation potential.

Keywords Nitrogen · Distribution · Spatial · Temporal · Littoral zone

Guest editors: B. Qin, Z. Liu & K. Havens
Eutrophication of shallow lakes with special reference to Lake Taihu, China

H. Wang · J. Lu · W. Wang · C. Yin (✉)
SKLEAC, Research Center for Eco-Environmental Sciences, CAS, Beijing 100085, P.R. China
e-mail: cqyin@263.net

H. Wang · J. Lu
Graduate University of the Chinese Academy of Sciences, Beijing 100039, P.R. China

P. Huang
Institute of Hydrobiology, Ji'nan University, Guangzhou 510632, P.R. China

Introduction

Riparian zone, the transitional boundary between terrestrial and aquatic ecosystems, influences water movement and waterborne contaminants (Fennessy & Cronk, 1997) and is the principal route for transport of allochthonous matter from terrestrial to aquatic systems (Naiman et al., 1989). Many studies have shown that riparian buffer strips with small dimensions but playing a disproportionately large role, can attenuate terrestrial non-point source pollutants, including sediment, phosphorus and nitrogen (Cooper et al., 1987; Yin & Lan, 1995; Bratli et al., 1999). Recent industrial growth and intensive agriculture have resulted in many riparian zones being disturbed or destroyed (Mitsch, 1995). If 70–95% of natural riparian systems are disturbed or destroyed, the landscape would lose much of its water purification and protection capacity (Fennessy & Cronk, 1997). In China, many lakes, such as Lake Taihu (Hu et al., 1998) and Lake Chaohu (Xu et al., 2003), have deteriorated with nutrient loading sufficiently high to maintain a eutrophic state, even after external nutrient source reduction (Xu et al., 2003). Removing internal nutrients from aquatic systems may help control eutrophication. Dredging sediment from lakes is a direct, physical method (Xu et al., 2003), but it is expensive and difficult to apply to the whole lake. Instead, it is urgent to find alternative biological techniques based on natural processes.

Lake Taihu is an important drinking water source and supports local agriculture, tourism and fisheries. Since the 1980s, rapid industrial development, heavy application of chemical fertilizers and intensive use of water resources have resulted in nutrient-rich water discharge into the lake, causing hyper-eutrophication. According to the Chinese environmental quality standards for surface water (GB 3838-2002), Lake Taihu water quality has deteriorated from grade I–II to grade IV–V, affecting life and economic development. Thus, studies on improving water quality, via non-point and point pollutant attenuation schemes, have been carried out in Lake Taihu (Dou et al., 1995; Pu & Hu, 1996). While internal pollution and sediment/water interactions are stronger in shallow versus deep lakes (Threlkeld, 1994), internal loading may result from accumulation of pore water nutrients and subsequent release across the sediment–water interface in shallow lakes (Kleebery & Kozerski, 1997). Internal nutrient loading from sediments may have an equal or greater contribution than external loading from the watershed (Detenbeck et al., 1993). Reducing internal nutrient loading is difficult, especially with respect to nitrogen, which is comparatively active.

Riparian zones can decrease the energy of flowing water, often resulting in increased sedimentation (Naiman et al., 1989). Finer suspended or resuspended sediment with abundant labile nutrient and absorbed pollutants (Naiman & Décamps, 1997) carried by flowing water enrich riparian sediments proportionally (Johnston et al., 1984). Whether riparian zones will be a sink or source to open water is not clear. A review of nutrient budgets for lacustrine wetlands revealed that the lakeshore could be a net sink for nitrogen and phosphorus either seasonally or annually (Van der Valk et al., 1978). Prentki et al. (1978) suggested that wetland plants and macrophytes act as nutrient pumps from the sediment, while Sager et al. (1985) implied that lakeshore marshes act as nutrient transformers rather than nutrient sinks. If plant-covered littoral zones can act as nutrient sinks, with nutrients mostly converted to inactive forms with low mobility and bioavailability or removed completely, then the riparian zones could mitigate internal pollution. However, little information is available, since most researchers have focused on non-point source or external pollutant removal (Mitsch, 1995). To explore this problem, two types of littoral zones with different plant coverage and lacking terrestrial inputs were selected. In the first year, we mainly developed some fundamental work for our further long-term research in the two littoral zones, involving water, sediment, plant and phytoplankton, and focused on nitrogen in this paper.

Riparian zones are biogeochemical hot-spots for nitrogen (Cirmo & McDonnell, 1997), where complicated nitrogen transformations take place. Nitrogen distribution is a function of physical, chemical and biological processes in the water column and sediments (Watts et al., 1998). Spatio-temporal distribution of nitrogen in mesocosmic environments may be representative of such processes. This paper presents results of a 13-month field study on spatio-temporal nitrogen distribution in the littoral zone of Meiliang Bay, Lake Taihu.

Materials and methods

Lake Taihu is a shallow, eutrophic lake in Jiangsu Province, East China. Monsoon climate dominates this region with annual mean air temperature of 14.9–16.2°C and a growing season lasting from April to November. Meiliang Bay, in the

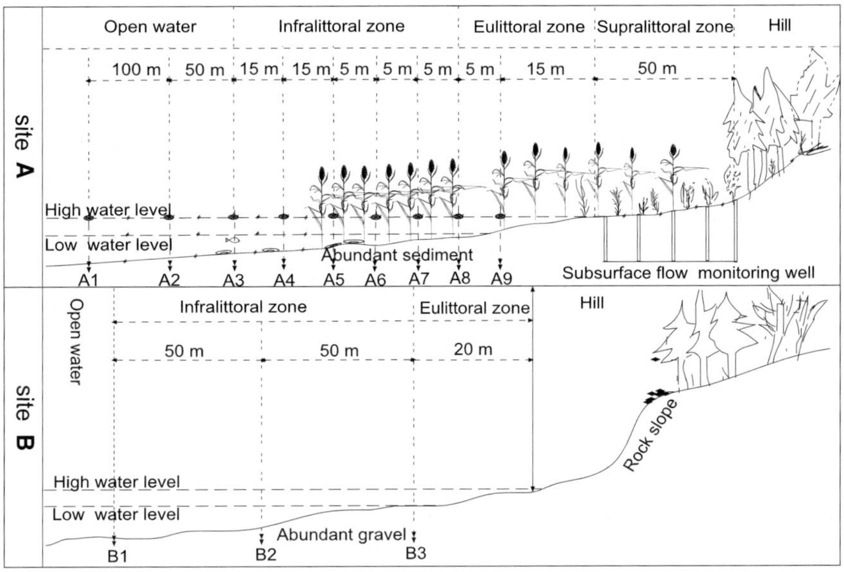

Fig. 1 Sampling station distribution in Sites A and B

northern part of the lake, is hyper-eutrophic. Two types of littoral zones with differing plant coverage and one open water site (control) were chosen in Meiliang Bay (Fig. 1). Site A is a reed-covered littoral zone with an above-ground dry biomass density of 3–4 kg m^{-2}, and Site B is a bare zone without plants. Site A is covered by a well-developed reed population with few other macrophytes. From open water, the first 25 m is inundated continuously, and the next 20 m is inundated intermittently. Toward land, the first 50 m are dominated by a reed (*Phragmites australis*) community and *Gramineae*, then a change from herbs to shrubs toward hills. All shrub stands are at least 30 years old, dominated by *Myrica rubra*, *Symplocos setchensis*, *Loropetalum chinense* and *Eurya murica*, and without visual evidence of recent natural or human disturbance, such as fire, wind-throw, logging or widespread insect infestation. In the lake-ward zone, there are sparse macrophytes with *Myriophyllum verticillatum* predominant in spring. The lake bottom is overlain by 30–40 cm soft sediment. Site B is a bare zone with little bottom sediment, and the interface between water and hills is cliff rock about 10 m wide. Shore vegetation resembles Site A with abundant shrubs. The open water site (C) is regarded as a control point. Initially, Site C was about 500 m from shore. After three preliminary surveys, due to little variation from 100 to 500 m, another station about 200 m from Site A was chosen as the control point.

To collect water and sediment and reduce sampling disturbance, a sampling bridge (50 cm width, 60 m long) was built from dry shore to open water at Site A. To investigate subsurface water in the supralittoral zone, a soil auger (Ø = 11.5 cm) was used to dig 5 vertical holes about 2 m deep as monitoring wells, and polypropylene pipes (Ø = 11 cm) equipped with nylon netting (< 300 μm) at the bottom were inserted into the wells.

Nine sampling stations at Site A and three at Site B were established for monthly sampling along open water to lakeshore (Fig. 1), named A1–A9 and B1–B3, respectively. The interval distances of stations A1–A9 were 100, 50, 15, 15, 5, 5, 5 and 5 m, respectively, with A9 10–15 m from the lake edge at high water, A5 at the fringe of the reed belt, and A1–A4 in the water without macrophytes. On occasion, low water levels left A9 out of water. At Site B, the interval was 50 m. Sampling stations were selected after an extensive survey of hydrochemistry reflecting the fringe effect of reed belt.

Monthly water sampling was carried out from Sep. 2003 to Sep. 2004. At each sampling, after 4–5 days of calm water, triplicate samples were taken using a 3-m long tube providing a vertical sample from the whole water column (Présing et al., 2001). Samples were placed in acid-washed polyethylene bottles and stored in a heat-preservation box on site. Dissolved oxygen (DO), pH, and Eh were measured in situ using an oxygen meter (YSI Model 57) and a pH/Eh meter (HI 8424 Microcomputer HANNA), respectively. In the laboratory, 250 ml from each water sample was filtered within 24 h for analyses of nutrient and chlorophyll-*a* concentrations. Ammonium nitrogen (NH_4^+-N), nitrate nitrogen (NO_3^--N), nitrite nitrogen (NO_2^--N), and chlorophyll-*a* (chl-*a*) were determined using APHA standard methods (APHA, 1998). Total nitrogen (TN), total phosphorus (TP), total dissolved nitrogen (TDN), and total dissolved phosphorus (TDP) were determined by peroxodisulfate oxidation (Ebina et al., 1983) in original water samples and filtrates, respectively. Total particulate nitrogen (TPN) and total particulate phosphorus (TPP) were calculated as the difference between TN and TDN, and TP and TDP, respectively.

About 300–400 g wet weight of mixed sediment (0–10 cm) and surface sediment (0–1 cm) was collected in duplicate using an Ekman-bottom sampler (Hydro-Bios, Kiel, Germany). At Site B, no sediment was present or collected. Sediment samples were stored in polyethylene bags and air-dried in the laboratory. Dried sediments were sieved (≤150 μm) for analyses of soil organic matter (OM) and total soil nitrogen (TSN) contents. OM was obtained from the loss on ignition (500°C, 2 h; Zhu & Carreiro, 2004), and TSN was determined by the semi-microdistillation method using a K_2SO_4–$CuSO_4$–Se catalyst (Lao, 1996).

From Apr. to Sep. 2004, particle sedimentation rates were measured with three sediment traps (Wahbah & Zughul, 2001) placed in different locations in the two littoral zones. Each trap consisted of a cylindrical plastic jar (19 cm height, 7 cm diameter) attached to a rope stretched by an anchor and subsurface buoy marked by a floating buoy. Traps were retrieved monthly, sampled and then redeployed. Material collected in the trap was oven-dried at about 60°C for measuring total weight, and TSN and OM contents were measured as described above.

For subsurface water, samples were collected twice in the growing (May and June 2004) and non-growing season (Jan. and Feb. 2004). Two days prior to sampling, stagnant water in each well was removed three times, allowing the water level to recover to its original level on the sampling day. Nutrient analyses were carried out as above.

Storm runoff may be an important nutrient source in littoral zone. During the study, storms with duration more than 1 h were few. Surface runoff samples in the littoral zone could not be collected during any of the storms, so overland runoff effects remain untested.

One-way ANOVA was used to test temporal and spatial differences by means of parameters. Unless otherwise noted, all significant differences were at a probability of 0.05 or less. All analyses were performed using SAS 8.01 software (SAS, Inc., Cary, NC, USA).

Results

Spatial distribution of inorganic and total nitrogen

At Site A, water column NH_4^+-N and NO_3^--N concentrations exhibited opposite trends from open water to reed belt in the growing season (Fig. 2). NH_4^+-N concentrations decreased from 3.3 to 1.5 mg l^{-1} in April and 1.5 to 0.7 mg l^{-1} in August, while NO_3^--N increased from 2.4 to 4.4 mg l^{-1} in April and 0.4 to 2.3 mg l^{-1} in August. Moreover, spatial variation was highly significant ($P < 0.01$) between the sampling points in reed belt. Figure 2 shows that reed belt fringe effects on open water extend to 20 m, while between 50 and 100 m offshore, inorganic nitrogen concentrations change little. In the non-growing season, the spatial variation trend was similar to that in the growing season, but with a gentler gradient. From surface water in reed belt to subsurface water in supralittoral zone, each nitrogen form decreases in concentrations, except NH_4^+-N in the growing season (Table 1). One-way ANOVA showed significant difference between reed belt and open water in the growing season ($P < 0.01$) but not in the non-growing season. At Site B, inorganic nitrogen concentrations in the shore were similar to those in open water and no significant variation was detected.

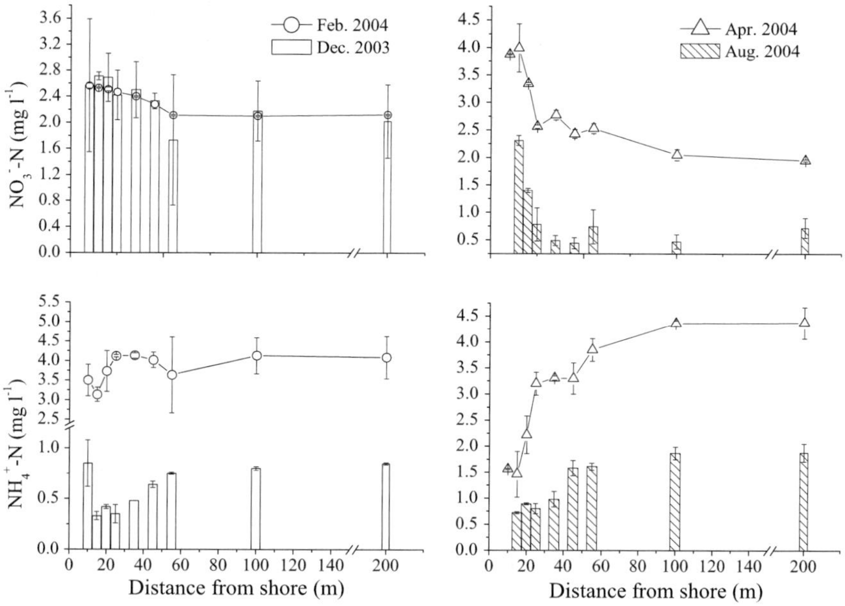

Fig. 2 Spatial variation of nitrate (NO_3^--N, top) and ammonium nitrogen (NH_4^+-N, bottom) concentrations (mean ± SD) in the reed-covered littoral zone (A) on two occasions

Table 1 The nitrogen forms in surface water of reed belt and subsurface water of supralittoral zone, n = 36

Parameters (mg l^{-1})	Growing season (mean ± SD)		Non-growing season (mean ± SD)	
	Surface water	Subsurface water	Surface water	Subsurface water
NH_4^+-N	0.37 ± 0.26	0.40 ± 0.24	3.37 ± 0.37	1.23 ± 0.81
NO_3^--N	2.32 ± 0.24	0.52 ± 0.59	2.83 ± 0.21	1.28 ± 0.59
NO_2^--N	0.12 ± 0.14	0.04 ± 0.03	0.22 ± 0.13	0.11 ± 0.15
TN	4.17 ± 1.15	2.62 ± 1.97	17.72 ± 7.82	5.81 ± 2.24

Spatial distribution of TN and TDN at Site A is shown in Fig. 3. TDN in reed belt varied little, and concentrations were similar to those in open water. Only at Station A9 at 10 cm water depth, TN concentrations increased from 12 to 28 mg l^{-1} and 4 to 54 mg l^{-1} in February and August, respectively. TN and TDN in the shore and open water was homogeneous at Site B.

Temporal variation of inorganic and total nitrogen

Temporal variations of NH_4^+-N, NO_3^--N and NO_2^--N in various zones are given in Fig. 4. Every form of nitrogen showed noticeable variation trend. NH_4^+-N concentration maxima (3.2–5.2 mg l^{-1}) occurred in January, with temperature rise and increasing biological activities and NH_4^+-N concentrations decreased to 0.3–0.9 mg l^{-1} in September. NO_3^--N concentrations reached highest values (2.9–4.8 mg l^{-1}) in March, with a "lag" of about 2 months, and the minimum value occurred in August. Temporal variation of NO_2^--N concentrations showed peak values (1.4 and 0.7 mg l^{-1}) in May and September, respectively.

TN can be categorized as two components, TDN and TPN. There was a similar temporal trend between open water and bare littoral zones (Fig. 5). Maxima for TN (14.9–18.8 mg l^{-1}) and TDN (9.8–11.6 mg l^{-1}) occurred in January while minima in summer. TPN was highest (4.4–11.1 mg l^{-1}) in winter and lowest (0.5–1.6 mg l^{-1}) in spring. However, irregular TN and TPN patterns were observed in the reed-covered littoral

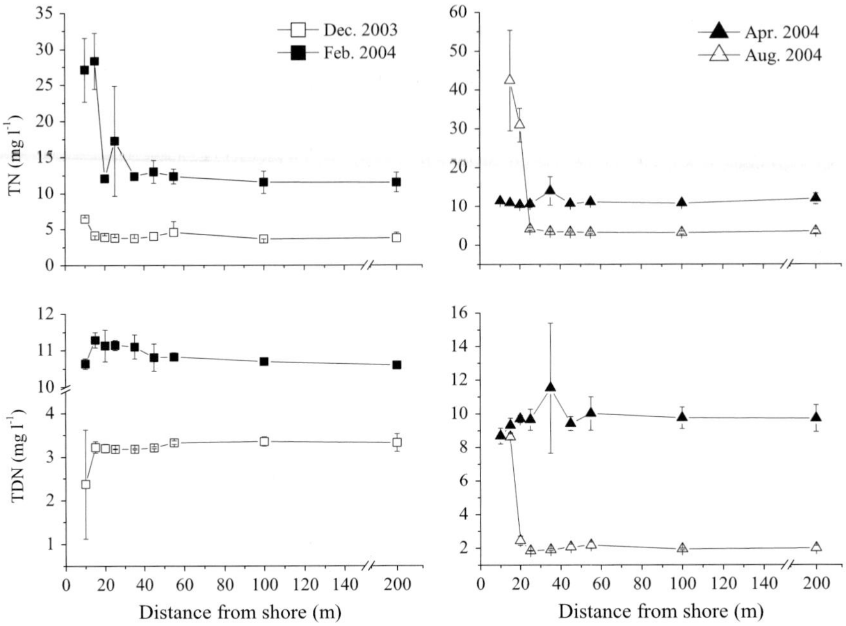

Fig. 3 Spatial distribution of total nitrogen (TN, top) and total dissolved nitrogen (TDN, bottom) concentrations (mean ± SD) in the reed-covered littoral zone (A)

zone, with peaks in summer and winter, respectively. From July to August, an algal bloom developed, and algae accumulated in reed belt (100,000–300,000 μg l^{-1} chl-*a*), so there was an extraordinary peak value (about 30 mg l^{-1}) of TPN with a positive correlation between TN and chl-*a* in summer ($r^2 = 0.925$, $P < 0.0001$).

Spatial variation of TSN and OM in sediment

In the reed-covered littoral zone, there were plentiful soft sediments about 30–40 cm in thickness. Most were freshly settled from suspension in open water where mechanical energy is dampened by littoral reed stands. These sediments were dominated by decayed algae rich in nutrients, and TSN and OM contents in vertical mixed sediment (0–10 cm) were up to 0.5–0.6% and 7–8%, respectively. TSN and OM contents in vertical mixed sediment had little spatial variation, however, they decreased from 0.5–0.6% to 0.1–0.2% and 7–8% to 2–3% in surface sediment (0–1 cm), respectively from open water to reed belt (Fig. 6).

Discussion

External sources of nitrogen in the littoral zone

Littoral zone is the interface between terrestrial and aquatic systems, and its dynamics are influenced by both. External nitrogen sources mainly come from land and open water, and the contribution from wet precipitation is negligible (0.18 g m^{-2} a^{-1} in the case of Lake Taihu: Qin et al., 2004). Nitrogen enters littoral zone via surface runoff and subsurface flow from land. There were no conspicuous point sources, such as a factory or village, or non-point pollution sources in this study. Vegetation covered the littoral zone, and the rainy season coincided with vegetation density, where plants dampened the energy of flowing water and retained most waterborne materials with little subsequent erosion. However, there were few long duration rainstorms. Five surface runoff events were observed, but overland transport to littoral zone was not detected. Although some heavy rainstorms may

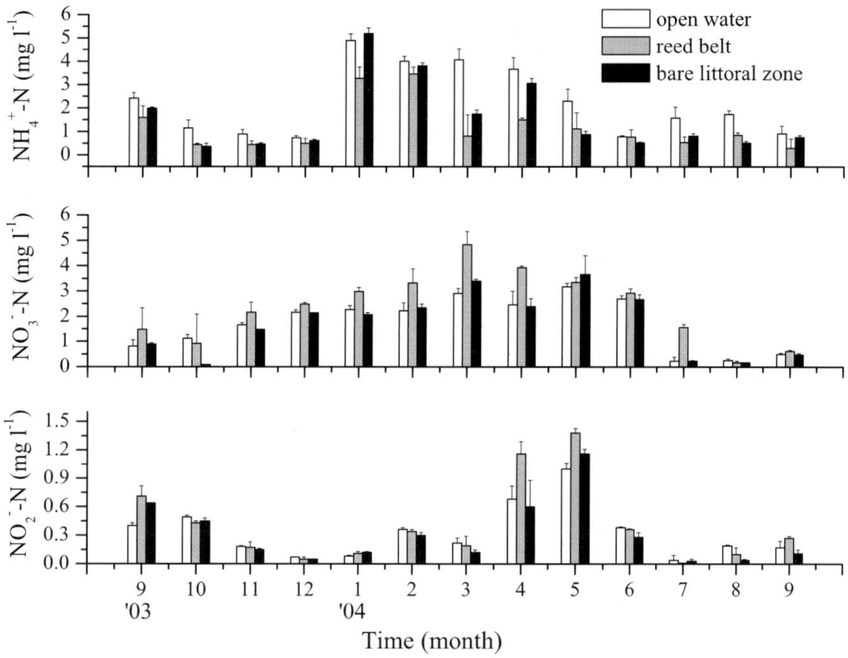

Fig. 4 Temporal variation of ammonium nitrogen (NH_4^+-N), nitrate nitrogen (NO_3^--N), and nitrite nitrogen (NO_2^--N) concentrations (mean ± SD) in open water, reed belt and bare littoral zone

have been missed, we believe they were few. In this study, overland runoff effects have been disregarded. Subsurface flow is mostly unidirectional toward aquatic systems across the terrestrial boundary (Naiman & Décamps, 1997), and our investigation suggested that few nutrients came from subsurface flow. TN and NO_3^--N concentrations in subsurface water were less than those in the reed belt water column in the growing and non-growing seasons. Moreover, NO_3^--N concentrations decreased from the lakeside to the hill and was lowest in the saturated area, which indicates (1) subsurface flow can carry little NO_3^--N to the inundated littoral zone; and (2) water column NO_3^--N may infiltrate into the supralittoral zone; and (3) the littoral reed belt may remove NO_3^--N efficiently from infiltrated lake water, as found in Lake Baiyangdian (Wang et al., 2002).

Material retention was high in ecotones or riparian zones with a sharp decrease in the kinetic energy of wind or water (Naiman et al., 1989). In a eutrophic lake, the retentate contained abundant nutrients, which would enrich the riparian zone (Johnston et al., 1984). In the littoral, especially where there is a macrophyte cover, sediments carried by lake currents also may settle. In our study, there was a distinct difference between average sedimentation rates at Sites A and B (250 and 75 g m^{-2} d^{-1}, respectively). At Site A, 30–40 cm thick soft sediments contained 0.5–0.6% TSN and 7–8% OM, while at Site B, there was almost no sediment and bottom gravels were exposed.

These preliminary investigations indicate that particulates from open water were the main nitrogen source to the reed-covered littoral zone, perhaps over 90% on a rough estimate. Reed-covered littoral zones may be a sink for internal TPN sources in the lake if these enriched areas do not release nitrogen back into open water.

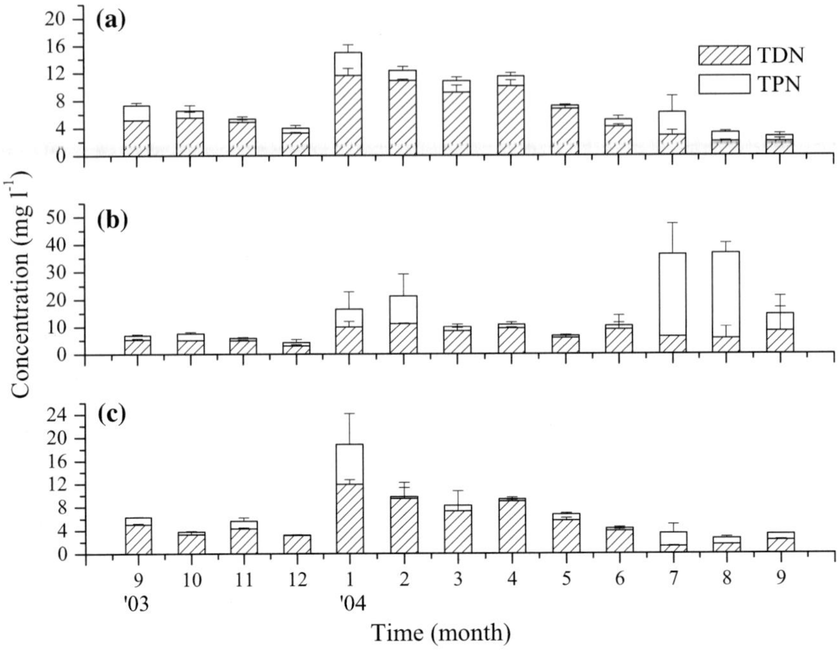

Fig. 5 Temporal variation of total dissolved nitrogen (TDN) and total particle nitrogen (TPN) in open water (**a**), reed belt (**b**), and bare littoral zone (**c**)

Spatial distribution of several forms of nitrogen

In the aquatic nitrogen cycle, energy utilization is more efficient and turnover is more rapid than in terrestrial systems (Heathwaite, 1993). Riparian zones may provide ideal conditions for nitrogen transformations, including ammonification, nitrification and denitrification (Fennessy & Cronk, 1997). Continuous deposition of plant and algae detritus provides rich carbon sources. Many studies have shown that ammonification in vegetated sediments is higher than in bare sediments (Herbert, 1999: 570, Table 5). Nitrification and denitrification in vegetated sediments also may be high (Christensen & Sørensen, 1986; Ottosen et al., 1999; Saunders & Kalff, 2001; Matheson et al., 2002). There is enough oxygen for nitrification in surface sediment of shallow vegetated zones, and denitrification may occur at greater depths. However, wetland vegetation transports atmospheric oxygen through the stem, to the root, and into adjacent soil, which may oxidize the reduced microzone around the root surface (Armstrong, 1964; Brix & Schierup, 1990). Two distinct layers are formed around a root surface favoring nitrification–denitrification (Reddy & Patrick, 1984). Trapped sediments provide ammonification sites; nitrification can utilize ammonification product in the aerobic water, surface sediments and root surface; and denitrification occurs in the underlying sediment, releasing dinitrogen or nitrous oxide. These processes occur simultaneously and are controlled, in part, by NH_4^+-N and NO_3^--N in the reed-covered littoral zone.

In the present study, there were opposite trends in NH_4^+-N and NO_3^--N dynamics in the reed covered littoral zone. NH_4^+-N concentrations decreased and NO_3^--N increased. In addition, other biotic factors influenced inorganic nitrogen concentrations, such as algal uptake and zooplankton and zoobenthos excretion. Moreover, NH_4^+-N is taken up by cyanobacteria more rapidly than NO_3^--N or NO_2^--N. Algae biomass in reed belt was not always higher than in open water, while the NH_4^+-N concentration gradient existed at all times, especially in the growing season. These results implied that nitrification was more intensive than ammonification, and nitrification

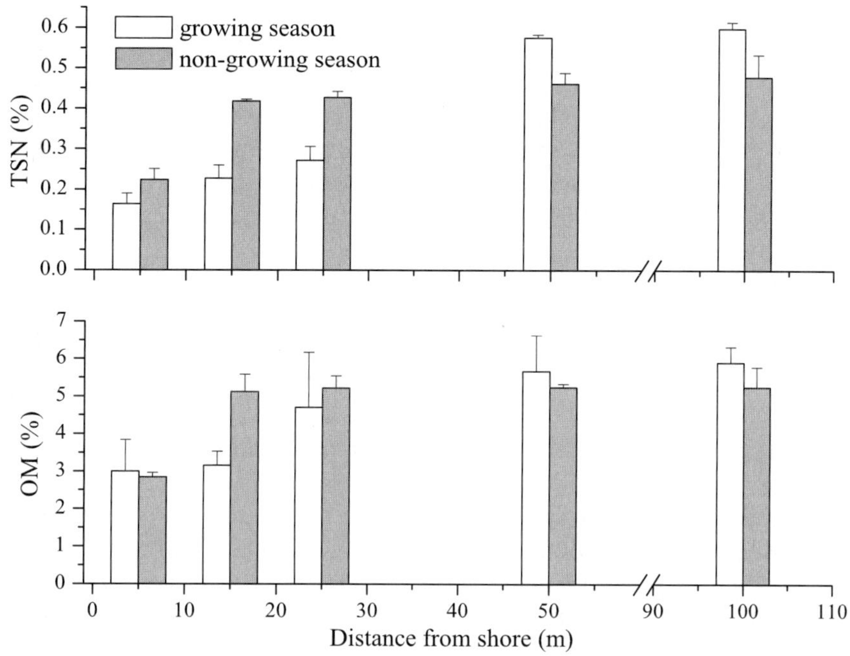

Fig. 6 Spatial variation of total soil nitrogen (TSN) and organic matter (OM) contents (mean ± SD) in surface soft sediment in the reed-covered littoral zone (A)

exceeded denitrification. Our data are inadequate to explain the transformations, and further biotic research is needed.

Inorganic nitrogen forms variations showed that the plant-covered littoral zone is a strong transformer of nutrients. TN and TDN in water column, combined with TSN and OM in surface sediment, provide clues about their fates. Spatial variation of TN concentrations in reed belt varied but generally was higher than, and sometimes equal to, that in open water, while TDN concentrations showed little spatial variation. The difference between TN and TDN suggested that particulate nitrogen dominated the spatial variation of TN. In summer, chl-*a* concentrations (100,000–300,000 μg l^{-1}) in reed belt were 500–1,000 times higher than those in open water. Second, there was no or inconspicuous secondary nitrogen release into open water from the littoral zone. Furthermore, periodic high concentrations of particulate nitrogen indicate that these particles mainly come from open water by intermittent wind forcing.

Denitrification is a key process of nitrogen removal from aquatic systems (Fennessy & Cronk, 1997; Martina et al., 1999). Spatial variation of inorganic nitrogen suggested that it was not intensive in the present study. To verify this, we investigated surface sediment (0–1 cm) and found an interesting phenomenon. TSN and OM contents in surface sediment showed synchronous spatial variation with both decreasing from open water to reed belt. The decrease indicates that nitrogen in settling sediment is lost in reed belt via release or denitrification. Because the dissolved nitrogen bulk in littoral zone was similar to that in open water, we deduced that nitrogen decrease in surface sediment was removed by denitrification. The sediment results further demonstrated that the reed-covered littoral zone could act as nitrogen sink of the lake, which played a vital role in removing internal nutrients.

Compared with the reed-covered littoral zone, spatial variation of nitrogen in bare zone was similar to that in open water. Lacking macrophytes, littoral zone erodes and particulates do not settle, so nutrients are not stored, and many biotic activities and biogeochemical cycles are limited. Altered physical conditions, when riparian zone is damaged, may result in loss of function (Fennessy & Cronk, 1997).

Temporal variation of inorganic nitrogen and TN

Littoral zone is often a small part of the aquatic system, and temporal variation of dissolved nitrogen fractions are controlled by variations in open water. TN and NH_4^+-N concentrations show winter maxima. These winter maxima possibly arise from (1) around lake, surface and subsurface catchment inputs are high in the non-growing season; (2) nutrients are concentrated in lower water levels; and (3) microbial activity is low due to low temperatures. NO_3^--N produced by nitrification, reaches peak value in spring often after a lag, since nitrification is sensitive to temperature. Below 15°C, nitrification rates decrease (Focht & Verstraete, 1977). In our study, TPN mainly came from suspended sediment and algae, and there were two maxima in reed belt (algae in summer, resuspended sediment in winter). While, in open water and bare zones, only one maxima in winter from suspended sediment was observed (Zhang et al., 2004).

Conclusions

Spatio-temporal distributions of different nitrogen forms were interpreted along gradients from open water to lakeshore in undulating littoral zones. In our two study zones, direct terrestrial nutrients accounted for little of the input, whereas more than 90% came from open water as resuspended sediments and algae. Between bare zone and reed-covered littoral zone, there were significant differences in hydrochemistry and physical conditions. The bare zone without soft sediment was almost homogeneous with open water, while the reed-covered littoral zone had abundant soft sediments and displayed spatial variations of inorganic nitrogen concentrations in water column. By comparison of spatial distribution of nitrogen in reed belt, the results showed that (1) nitrification was a key process controlling the nitrogen cycle in the reed-covered littoral zone, and most ammonification production was transformed in to nitrate; (2) little of the nitrogen enrichment to reed belt was released back into open water; (3) nitrification provided adequate nitrogen for denitrification; and (4) the reed-covered littoral zone was an important nitrogen sink. TSN and OM contents in surface sediments decreased from open water to reed belt. Denitrification appears to be significant in reed belt and most released nitrogen from sediments are removed, which further suggests that the reed covered littoral zone is a nitrogen sink and a self-balancing system.

Acknowledgements This research was funded by Key Project of Knowledge Innovation Program of CAS (KZCX1-SW-12), National Hi-Tech Research Program (2002AA601011-05), and National Basic Research Program of China (2006CB403306).

References

APHA, 1998. Standard Methods for the Examination of Water and Wastewater, 20th ed. American Public Health Association, NW, Washington DC.

Armstrong, W., 1964. Oxygen diffusion from the roots of some British bog plants. Nature (London) 204: 801–802.

Brix, H. & H. H. Schierup, 1990. Soil oxygenation in constructed reed beds: the role of macrophyte and soil atmosphere interface oxygen transport. In Cooper, P. & F. B. C. Findlater (eds), Constructed Wetlands in Water Pollution Control. Pergamon Press, London: 53–66.

Bratli , J. L., A. Skiple & M. Mjelde, 1999. Restoration of Lake Borrevannet – self-purification of nutrients and suspended matter through natural reed belts. Water Science and Technology 40: 325–332.

Christensen, P. B. & J. Sørensen, 1986. Temporal variation of denitrification activity in plant-covered littoral sediment from Lake Hampen, Denmark. Applied and Environmental Microbiology 51: 1174–1179.

Cirmo, C. P. & J. J. McDonnell, 1997. Linking the hydrologic and biogeochemical controls of nitrogen transport in near-stream zones of temperate-forested catchments: a review. Journal of Hydrology 199: 88–120.

Cooper, J. R., J. W. Gilliam, R. B. Daniels & W. P. Robarge, 1987. Riparian areas as filters for agricul-

tural sediment. Soil Science Society of America Journal 51: 416–420.
Detenbeck, N. E., C. A. Johnston & G. J. Niemi, 1993. Wetland effects on lake water quality in the Minneapolis/St. Paul metropolitan area. Landscape Ecology 8: 39–61.
Dou , H. S., P. M. Pu, S. Z. Zh, W. P. Hu & Y. Pang, 1995. An experimental study on culture of *Eichhornia crassipes* (Mart.) Solms on open area of Taihu Lake. Journal of Plant Resources and Environment 4: 54–60 (in Chinese).
Ebina, J., T. Tsutsui & T. Shirai, 1983. Simultaneous determination of total nitrogen and total phosphorus in water using peroxodisulfate oxidation. Water Research 17: 1721–1726.
Fennessy, M. S. & J. K. Cronk, 1997. The effectiveness and restoration potential of riparian ecotones for the management of nonpoint source pollution, particularly nitrate. Critical Review in Environmental Science and Technology 27: 285–317.
Focht, D. D. & W. Verstraete, 1977. Biochemical ecology of nitrification and denitrification. Advance in Microbial Ecology 1: 135–214.
Heathwaite, A. L., 1993. Nitrogen cycling in surface waters and lakes. In Burt, T. P., A. L. Heathwaite & S. T. Trudgill (eds), Nitrate: Processes, Patterns and Management. John Wiley & Sons Ltd., Chichester: 99–100.
Herbert, R. A., 1999. Nitrogen cycling in coastal marine ecosystems. FEMS Microbiology Reviews 23: 563–590.
Hu, W. P., J. Salomonsen, F. L. Xu & P. M. Pu, 1998. A model for the effects of water hyacinths on water quality in an experiment of physico-biological engineering in Tai Lake, China. Ecological Modelling 107: 171–188.
Johnston, C. A., G. D. Bubenzer, G. B. Lee, F. W. Madison & J. R. McHenry, 1984. Nutrient trapping by sediment deposition in a seasonally flooded lakeside wetland. Journal of Environmental Quality 13: 283–290.
Kleebery, A. & H. P. Kozerski, 1997. Phosphorus release in Lake Grosser Muggelsee and its implications for lake restoration. Hydrobiologia 342: 9–26.
Lao, J. C., 1996. Soil agrochemical analysis methods [M], 2nd ed. Agricultural Press, Beijing.
Martina, T. L., N. K. Kaushik, J. T. Trevors & H. R. Whiteley, 1999. Review: denitrification in temperate climate riparian zones. Water, Air, and Soil Pollution 111: 171–186.
Matheson, F. E., M. L. Nguyen, A. B. Cooper, T. P. Burt & D. C. Bull, 2002. Fate of ^{15}N-nitrate in unplanted, planted and harvested riparian wetland soil microcosms. Ecological Engineering 19: 249–264.
Mitsch, W. J., 1995. Restoration of our lakes and rivers with wetlands – an important application of ecological engineering. Water Science and Technology 31: 167–177.
Naiman, R. J. & H. Décamps, 1997. The ecology of interface: Riparian zones. Annual Review of Ecology and Systematics 28: 621–658.
Naiman, R. J., H. Décamps & F. Fournier (eds), 1989. The role of land/inland water ecotones in landscape management and restoration: a proposal for collaborative research. MAB Digest 4, UNESCO.
Ottosen, L. D. M., N. Risgaard-Petersen & L. P. Neilsen, 1999. Direct and indirect measurements of nitrification and denitrification in the rhizosphere of aquatic macrophytes. Aquatic Microbiol Ecology 19: 81–91.
Prentki, R. T., T. D. Gustafson & M. S. Adams, 1978. Nutrient movements in lakeshore marshes. In Good, R. E., D. F. Whigham & R. L. Simpson (eds), Freshwater Wetlands: Ecological Processes and Management Potential. Academic Press, New York: 169–194.
Présing, M., S. Herodek, T. Preston & L. Voros, 2001. Nitrogen uptake and the importance of internal nitrogen loading in Lake Balaton. Freshwater Biology 46: 125–139.
Pu, P. M. & W. P. Hu, 1996. Key strategies for controlling eutrophication of large shallow lakes in China. In 28th International Geographical Congress, Utrecht.
Qin, B. Q., W. P. Hu & W. M. Chen, 2004. Process and mechanism of environmental changes of the Taihu Lake. Science Press, Beijing.
Reddy, K. R. & W. H. Patrick, 1984. Nitrogen transformations and loss in flooded soils and sediments. CRC Critical Reviews in Environmental Control 13: 273–309.
Sager, P. E., S. Richman & H. J. Harris, 1985. Preliminary observations on the flux of carbon, nitrogen, and phosphorus in a Great Lakes coastal marsh. In Prince, H. H. & F. M. D'Itri (eds), Coastal Wetlands. Lewis Publishers, Inc., Chelsea, MI: 59–65.
Saunders, D. L. & J. Kalff, 2001. Denitrification rates in the sediments of Lake Memphremagog, Canada-USA. Water Research 35: 1897–1904.
Threlkeld, S. T., 1994. Benthic-pelagic interactions in shallow water columns: an experimentalist's perspective. Hydrobiologia 275/276: 293–300.
van der Valk, A. G., C. B. Davis, J. L. Baker & C. E. Beer, 1978. Natural fresh water wetlands as nitrogen and phosphorus traps for land runoff. In Greeson, P. E., J. R. Clark & J. E. Clark (eds), Wetland Functions and Values: The State of our Understanding. Proceeding of National Symposium on Wetlands, Lake Buena Vista, FL. American Water Resources Association, Minneapolis, MN: 457–467.
Wahbah, M. I. & M. B. Zughul, 2001. Temporal distribution of chlorophyll *a*, suspended matter, and the vertical flux of particles in Aqaba (Jordan). Hydrobiologia 459: 147–156.
Wang, W. D., D. L. Wang & C. Q. Yin, 2002. A field study on the hydrochemistry of land/inland water ecotones with reed domination. Acta Hydrochimica et Hydrobiologica 30: 117–127.
Watts, L. J., T. P. Rippeth & A. Edwards, 1998. The roles of hydrographic and biogeochemical processes in the distribution of dissolved inorganic nutrients in a cottish Sea-loch: consequences for the spring phytoplankton bloom. Estuarine, Coastal and Shelf Science 46: 39–50.

Xu, F. L., S. Tao, R. W. Dawson & Z. R. Xu, 2003. The distributions and effects of nutrients in the sediments of a shallow eutrophic Chinese lake. Hydrobiologia 429: 85–93.

Yin, C. Q. & Z. W. Lan, 1995. The nutrient retention by ecotone wetlands and their modification for Baiyangdian Lake restoration. Water Science and Technology 32: 159–167.

Zhang, Y. L., B. Q. Qin, W. M. Chen & L. C. Luo, 2004. A study on total suspended matter in Tai Lake. Resources and Environment in the Yangtze Basin (in Chinese) 13: 266–271.

Zhu, W. X. & M. M. Carreiro, 2004. Temporal and spatial variations in nitrogen transformations in deciduous forest ecosystems along an urban–rural gradient. Soil Biology & Biochemistry 36: 267–278.

Hydrobiologia (2007) 581:109–116
DOI 10.1007/s10750-006-0503-7

EUTROPHICATION IN LAKES

Variations in kinetics of alkaline phosphatase in sediments of eutrophic, shallow, Chinese lakes

Yiyong Zhou · Xiuyun Cao · Chunlei Song · Jianqiu Li · Guoyuan Chen · Liang Peng

Abstract Kinetics of alkaline phosphatase in sediments of a shallow Chinese freshwater lake (Lake Donghu) were investigated. Spatially, among 20 sites sampled, V_{max} and K_m values of alkaline phosphatase in surface sediments were higher in the zone adjacent to sites with the highest chlorophyll *a* concentrations. Vertically, there was a peak in V_{max} at intermediate sediment depths in addition to the expected maximum at the surface. Some inhibitors, such as $CuSO_4$, $ZnSO_4$ and Na_2WO_4, showed significantly different effects on kinetics of alkaline phosphatase in interstitial water and sediments. Moreover, alkaline phosphatase in interstitial water and sediments responded to Na_2WO_4 in different ways in Lake Taihu. These observations imply that the enzyme is immobilized in sediments, which became more stable with accelerated eutrophication, as suggested by highest alkaline phosphatase activity (APA) in sediments corresponding with highest water column chlorophyll *a* concentrations in Lake Donghu.

Keywords Alkaline phosphatase · Kinetics · Sediment · Eutrophication · Inhibitions · Immobilization

Guest editors: B. Qin, Z. Liu & K. Havens
Eutrophication of shallow lakes with special reference to Lake Taihu, China

Y. Zhou (✉) · X. Cao · C. Song · J. Li · G. Chen · L. Peng
Institute of Hydrobiology, The Chinese Academy of Sciences, Wuhan 430072, China
e-mail: zhouyy@ihb.ac.cn

X. Cao · C. Song · G. Chen · L. Peng
Graduate School of the Chinese Academy of Sciences, Beijing 100039, China

Introduction

Phosphorus often limits phytoplankton growth in freshwater systems, while organic phosphorus may account for larger part of phosphorus in lake sediments. For example, in a eutrophic lake (Lake Alserio, northern Italy), the results of the sedimentary phosphate fractionation showed that the most important P fraction was an organic fraction, and a digestion of the supernatant of the P-fraction bound to $CaCO_3$ allowed the detection of a large pool of org-P (Vicente et al., 2006). Therefore, regeneration of organic P in sediments is emphasized in aquatic ecology research. The phosphatase has been studied in relation to organic phosphorus decomposition in aquatic ecosystems (reviewed by Jansson et al., 1988). The relationship between alkaline phosphatase and substrate is of ecological interest because of implications for phosphorus cycling. However, most attempts to determine phosphatase in sediments have focused on activity. Kinetics of phosphatase in sediments were influenced by discharge rate, water quality, concentration of dissolved orthophosphate and substratum in stream sediments (Marxsen & Schmidt, 1993), submerged macrophytes (Zhou et al., 2000) and P status in lake sediments (Zhou et al., 2001, 2002). Data on spatial and vertical variations in kinetics and properties of alkaline phosphatase in sediments are limited.

In this paper, alkaline phosphatase kinetics and its response to various inhibitors in a shallow freshwater lake (Lake Donghu) in China were described. The aims of the work were to establish whether enzyme regimes in sediments (1) were unique for the whole lake and depth, (2) shared a similarity in kinetic behavior with interstitial water, and (3) were related to trophic status in the lake.

Materials and methods

The experimental Lakes (Lake Donghu and Lake Taihu) are described elsewhere (Zhou et al., 2002; Qin et al., 2006). To evaluate the trophic state in Lake Donghu on an annual basis, surface water was collected monthly from sites 1, 12 and 21 (Fig. 1), from February 1997 through January 1999. To determine spatial variation of alkaline phosphatase kinetics, 20 sites were sampled in the largest basin (Fig. 1) in May 1995. Three sites in Lake Taihu were sampled in December 2004 to assay alkaline phosphatase activity (APA) in surface sediments. In May 1997, vertical samples were taken at five sites in Lake Donghu with Site 1 at the center and others (1A, 1B, 1C and 1D) located approximately 100 m away (Fig. 1).

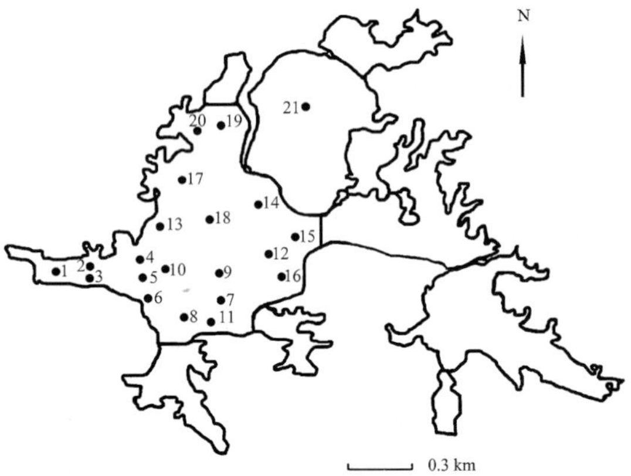

Fig. 1 The map of Lake Donghu, showing the location of the sampling sites

For spatial profiles, sediments were sampled by Peterson dredge. Sediment columns were obtained using a hand-driven stainless steel corer 50 cm long with an internal diameter of 3.5 cm. For depth profiles, columns were sliced at 4 cm intervals. Samples were transferred to the laboratory for analysis. Interstitial water was extracted by centrifugation at 3,000 rpm for 30 min.

Total chlorophyll *a* was measured by acetone extraction. Lake surface water samples (0.4–0.7 l) were filtered through a Whatman GF/C filter and absorbance measured at 663 and 750 nm in 1-cm path-length glass cuvettes after overnight extraction in acetone (90% v/v) (Golterman & Clymo, 1969).

APA assays used *p*-nitro-phenylphosphate (*p*NPP), which is hydrolyzed at 37°C by alkaline phosphatase to yield *p*-nitrophenol; with this system, enzyme activity is indicated by an increase in light absorbance (Sayler et al., 1979). Sediment samples were slurried in Tris buffer (pH 8.9). *p*NPP was added to slurries at eight final concentrations ranging from 0.05 mmol l^{-1} to 10 mmol l^{-1}. Samples were incubated at 37°C. After 1 h, 1.6 ml of slurry were centrifuged at 3,000 rpm. One ml of the supernatant was mixed with 4 ml 0.1 M NaOH to stop the reaction. Absorption of the final solution was measured at 400 nm. *p*NPP was added to reagent blanks after NaOH addition. APA was converted to absolute units using a standard curve based on enzymatically hydrolyzed *p*-nitrophenol. V_{max} and K_m values were estimated by fitting linearized Michaelis–Menten equations per the Lineweave–Burk plot. APA in surface sediments was determined with a final substrate concentration of 6.0 mmol l^{-1}.

Kinetics of total APA in interstitial water was determined using a method adapted from Berman (1970). Substrate was added at eight final concentrations ranging from 0.01 mmol l^{-1} to 1.8 mmol l^{-1}. APA in interstitial water was determined with a final substrate concentration of 0.3 mmol l^{-1}. All samples were run in triplicate.

Several inhibitors were added to sediments and interstitial water. Final concentrations of inhibitors are described in results tables. Fisher's least significant difference (LSD) test and independent-samples T test were used to make comparisons among treatment levels for the inhibitor addition, using the SPSS statistical package.

Results

Seasonally, chlorophyll *a* concentration in surface water was highest at Site 1 (Fig. 2). APA kinetics in sediments exhibited spatial heterogeneity (Table 1). Higher V_{max} values were found at Sites 2, 6, 9, 10 and 12, which are adjacent to Site 1. However, in the northeastern zone represented by Sites 14, 15 and 19, both V_{max} and K_m values were markedly lower.

There were variations in vertical distribution of APA kinetics at different sites (Fig. 3). At Sites 1B and 1D, V_{max} decreased with depth, while, at Sites 1A, 1, and 1C, there was a peak in the middle layers. At the same time, highest values for K_m were observed in the surface sediment at Sites 1B, 1C, and 1D, and in the middle layers at Sites 1A and 1.

Responses of APA kinetic parameters in sediments and interstitial water to various inhibitors were examined (Table 2). In sediments, V_{max} decreased significantly with Cu^{2+} concentration ($p < 0.01$), while in interstitial water, it significantly increased ($p < 0.01$). It also increased markedly ($p < 0.01$) with tungstate concentration in interstitial water. With Zn^{2+} added, V_{max} increased significantly in sediments ($p < 0.01$) but decreased at higher concentration in interstitial water ($p < 0.01$).

Responses of K_m values were also variable in both sediments and interstitial water. In sediments, it decreased significantly with lower Cu^{2+} concentration but increased with higher concen-

Table 1 The kinetic parameters of alkaline phosphatase in the sediments of Lake Donghu

Sampling sites	V_{max} (SD) (μmol G^{-1} H^{-1})	K_m (SD) (mmol l^{-1})
1	147.0 (2.0)	0.32 (0.02)
2	557.3 (50.6)	3.18 (0.09)
3	208.1 (1.0)	0.56 (0.04)
4	139.7 (8.1)	0.66 (0.02)
5	62.5 (3.1)	0.81 (0.04)
6	424.3 (24.3)	3.74 (0.67)
7	258.9 (49.5)	3.38 (0.27)
8	143.4 (21.4)	0.80 (0.12)
9	458.3 (11.64)	1.39 (0.04)
10	305.6 (30.0)	1.03 (0.23)
11	74.4 (1.8)	0.47 (0.03)
12	293.4 (27.7)	0.61 (0.07)
13	133.2 (12.8)	0.77 (0.16)
14	64.5 (17.6)	0.33 (0.08)
15	68.6 (6.6)	0.28 (0.06)
16	109.6 (3.8)	2.19 (0.16)
17	141.8 (31.8)	0.69 (0.09)
18	144.3 (21.8)	2.67 (0.56)
19	94.6 (8.4)	0.74 (0.05)
20	169.4 (17.0)	1.00 (0.12)

Fig. 2 Seasonal variation in chlorophyll *a* concentrations in surface water at three experimental stations in Lake Donghu

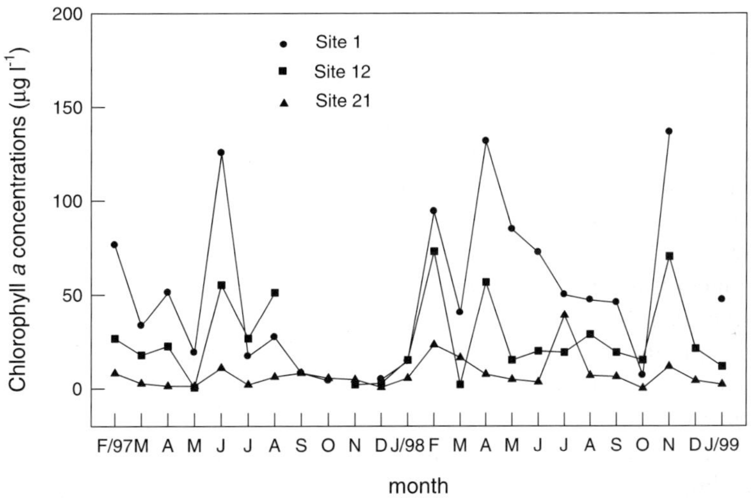

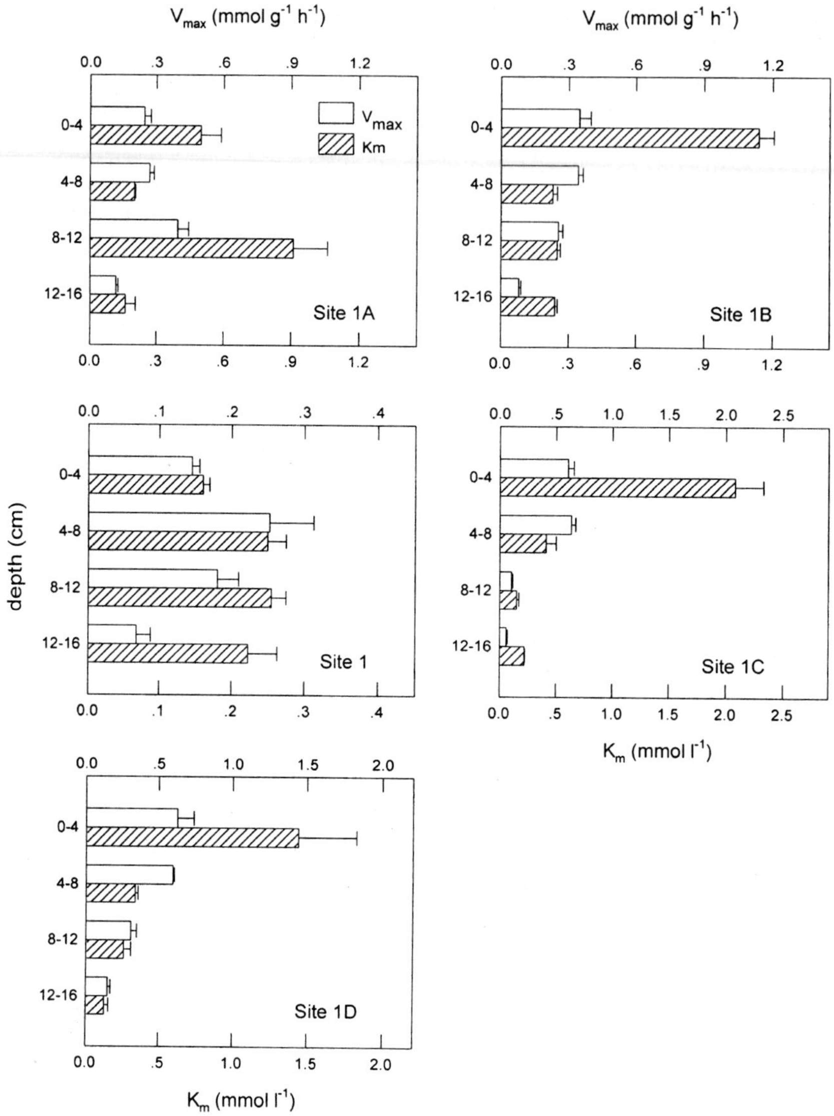

Fig. 3 Vertical distribution of V_{max} and K_m values of alkaline phosphatase activity (APA) in core sediments of Lake Donghu. Samples were taken on 12 May 1997

tration ($p < 0.01$). It also decreased significantly at lower tungstate concentration ($p < 0.05$) but increased significantly at higher concentration ($p < 0.01$) in sediments. While in interstitial water, it significantly decreased with tungstate concentrations ($p < 0.01$). With Zn^{2+} added, K_m significantly increased at lower concentration in sediments ($p < 0.05$), whereas in interstitial water, it greatly increased at higher concentration ($p < 0.01$).

In Lake Taihu, tungstate inhibited APA in sediment at Site T1 ($p < 0.01$) and interstitial water at Sites T1 and T3 ($p < 0.01$) at higher concentration. At Site T2, it stimulated APA in sediments at higher concentration, but inhibited APA in interstitial water at both concentrations ($p < 0.01$, Table 3).

Discussion

At Site 1, chlorophyll *a* in surface water was highest among experimental stations from 1997 to 1999 (Fig. 2), a situation commonly recorded in

Table 2 Effects of some inhibitors of alkaline phosphatase on kinetics of alkaline phosphatase activity (APA) in the surface sediments and the interstitial water in Lake Donghu

Sampling time	Inhibitors	Concentration (mmol l^{-1})	Surface sediments		Interstitial water	
			V_{max} (SD) (μmol G^{-1} H^{-1})	K_m (SD) (mmol l^{-1})	V_{max} (SD) (nmol l^{-1} min^{-1})	K_m (SD) (μmol l^{-1})
15 March, 1996	$CuSO_4$	0	157.5 (21.5)	3.25 (0.15)	51.9 (0.7)	4.11 (0.21)
		0.2	65.4 (6.6)	0.30 (0.06)	54.9 (0.5)	4.22 (0.29)
		2.0	52.5 (0.3)	4.78 (0.25)	105.5 (1.0)	1.65 (0.12)
		0	82.5 (0.8)	0.51 (0.02)	53.2 (0.9)	3.43 (0.23)
29 March, 1996	Na_2WO_4	0.05	90.8 (3.4)	0.40 (0.01)	67.9 (0.5)	1.81 (0.21)
		0.1	127.0 (37.7)	0.68 (0.02)	69 (0.5)	1.63 (0.11)
		0	143.7 (18.2)	0.54 (0.20)	46.7 (0.4)	1.59 (0.16)
8 February, 1996	$ZnSO_4$	0.2	291.9 (8.7)	1.07 (0.06)	44.6(0.6)	1.70 (0.10)
		2.0	287.4 (2.6)	0.48 (0.14)	38.6(0.2)	2.13 (0.08)

Table 3 Effects of tungsten on alkaline phosphatase activity (APA) in the surface sediments and interstitial water in Lake Taihu (Samples were taken on 14 December, 2004)

Sampling sites	Tungstate concentrations (mmol l^{-1})	APA in surface sediments (SD) (μmol G^{-1} H^{-1})	APA in interstitial water (SD) (nmol l^{-1} min^{-1})
T1	0	428.5 (6.0)	42.6 (1.5)
(31°32′4.6″ N, 120°13′22.6″ E)	0.16	157.0 (14.9)	35.2 (0.3)
T2	0	137.8 (16.3)	48.4 (1.0)
(31°32′1.7″ N,	0.08	144.3 (7.1)	39.8 (1.1)
120°13′18.7″ E)	0.16	309.2 (20.5)	35.9 (1.5)
T3	0	483.1 (53.5)	49.31(1.2)
(31°31′45.5″ N, 120°13′42.8″ E)	0.16	163.9 (13.4)	39.9 (0.6)

the lake (Zhou et al., 2002, 2004). Higher V_{max} and K_m values of APA in surface sediments were found at sites adjacent to Site 1, while, in the northeast, both were lower (Table 1). Significant inter-station variation in APA was observed in sediments of the Mandovi Estuary (Silva & Bhosle, 1990). Distinct areas of phosphatase activity also were seen in the Venice lagoon (Sabil et al., 1994). Variability of phosphatase activity in freshwater sediments may be related to heterogeneity within a site (Sayler et al., 1979). At sites along a eutrophication gradient in Nordruegensche Bodden (Baltic Sea, Germany), water column measurements revealed that turbidity, seston content, and chlorophyll *a* and inorganic nutrient concentrations increased from outer to inner parts of the Bodden. Sediment investigations confirmed this eutrophication gradient. Generally, hydrolytic enzyme activities increased with eutrophication (Koester et al., 1997). In Lake Donghu, the V_{max} values in sediment increased during the summer, in conjunction with lower K_m values in interstitial water that suggests a higher affinity for the substrate. The accumulation of organic matter in the sediment could be traced back to the breakdown of the algal bloom, which may stimulate APA with higher kinetic efficiency, by a combination of the higher V_{max} in sediments plus lower K_m values in interstitial water, in summer (Zhou et al., 2002). Our results highlight the connection of the kinetics of APA in sediments with phytoplankton abundance in surface water, indicating the induction of enzyme by organic matter.

The V_{max} of alkaline phosphatase was highest at the sediment surface (Sites 1B and 1D, Fig. 3). APA also was highest at the surface of marine (Kobori & Taga, 1979) and lake sediments (Sinke et al., 1991). In three lakes in Finland, the eleven hydrolytic enzyme activities were high into deep sediment layers indicating potential for turnover of organic matter in the permanently anoxic zones (Hakulinen et al., 2005). In the present study, there was a peak in the middle layers (Sites 1A, 1, and 1C, Fig. 3), showing an additional dimension for the degradation of organic matter mediated by alkaline phosphatase in lake sediments.

Spatial and vertical variations in V_{max} and K_m values of APA in sediments were similar (Table 1; Fig. 2). This may be interpreted with reference to organic matter and orthophosphate enrichment. At first, organic matter may stimulate V_{max} of APA in sediments. Surface sediment profiles were taken from the deepest part of six central Finnish lakes representing six different trophic states, and each profile was divided into three layers. Phosphatase activity and levels of organic substances were highest in the top layer of every sediment profile. Microbiological phosphate mineralization from organic substances may determine the internal phosphorus load in these lakes (Matinvesi & Heinonen-Tanski, 1992). In fish pond, inorganic phosphate added to the sediment scarcely restricted phosphatase activity. At the same time the enrichment with organic phosphorus compounds highly raised enzyme activity (Olah & Toth, 1978). In marsh sediments, the V_{max} of acid phosphatase followed the same trend as in situ activity. Sediment salinity and pH were negatively correlated with the enzyme activity, while soil organic matter content, clay content and sediment organic P were positively correlated (Huang & Morris, 2005). In tidal freshwater habitats adjacent to the Cooper River, acid phosphatase activity was highly correlated with the organic matter content of the sediment. The V_{max} of all phosphatases increased along the successional gradient. Trends in phosphatase activity and V_{max} correlated positively with plant biomass and negatively with concentrations of soluble reactive phosphorus in porewater, sediment extractable phosphorus, and total phosphorus (Huang & Morris, 2003); Secondly, organic matter may alter K_m of APA in sediments. Alkaline phosphatase kinetics in sediment associated with cage culture of *Oreochromis niloticus* was studied near Site 1 in Lake Donghu. Both V_{max} and K_m increased with the addition of fish feces (Zhou et al., 2001). A possible explanation is that humic substances would adsorb the substrates of enzymes. In the consecutive sapropel layers collected from bathyal sediments of the eastern Mediterranean Sea, the determination of exoenzyme activity with fluorescently labeled substrate analogues was impaired by the strong adsorption of up to 97% of the enzymatically liberated fluorophores to the sediment particles. High activities of aminopeptidase and alkaline phosphatase were detected even in a 124,000-year-old sapropel layer, whereas the activity of β-glucosidase was low in all layers. It had been assumed that the organic matter which constitutes the sapropels is highly refractory. Since a high adsorption capacity was determined not only for the low-molecular-weight compounds but also for DNA, the extraordinarily strong adsorption of structurally different substrates to the sapropel matrix appears to be the major reason for the long-term preservation of biodegradable carbon in this environment (Coolen & Overmann, 2000). This strong adsorption would weaken the affinity of substrates for enzymes, leading to increase in apparent K_m values. In addition, orthophosphate may be a competitive inhibitor for alkaline phosphatase, causing a significant increase in K_m values. Spatially, orthophosphate concentrations in interstitial water were highest at Site 1 among experimental stations in Lake Donghu in 1995 to 1996 (Zhou et al., 2002), it could act as a competitive inhibitor of extracellular phosphatase in lake (Chrost & Overbeck, 1987).

Alkaline phosphatase in sediments showed different responses to various inhibitors from those observed in interstitial water. For example, in interstitial water of lake Donghu, V_{max} increased with copper sulphate addition at both lower ($p < 0.05$) and higher ($p < 0.01$) concentrations (Table 2). Increased phosphatase activities were noted in green algae isolated from Lake Kinneret and pretreated with intermediary Cu^{2+} concentrations (Wynne & Pieterse, 2000). However, the enzyme was inhibited by zinc at higher concentration ($p < 0.01$, Table 2). Zhang et al. (2001) studied the kinetics of inactivation of alkaline phosphatase from green crab by zinc ions. The enzyme reversibly and quickly bound

Zn^{2+} and then underwent a slow, reversible inactivation and slow conformational change. In sediments, V_{max} and K_m decreased with Cu^{2+} at lower concentration ($p < 0.01$, Table 2). Accordingly, APA was affected by Cu^{2+} in soil polluted by heavy metals (Kuperman & Carriero, 1997; Kunito et al., 2001). The influence of Cu^{2+} and Zn on the activity and kinetics of acid phosphatase immobilized by two soil clays, kaolin or goethite, indicated that Cu decreased V_{max} of the enzymes, but increased affinity of the enzymes for the substrate (Huang & Shindo, 2001). In this context, alkaline phosphatase in sediment and interstitial water showed different kinetics with different inhibitors, implying enzyme immobilization in sediment. Concurrently, various inhibition patterns by Zn were observed for free and immobilized acid phosphatase at different pH values, likely related to the degree of deactivation by Zn (Huang & Shindo, 2000). In soils, kinetics of the immobilized enzyme conformed to Michaelis–Menten, and V_{max} was lower and K_m higher than those of the free enzyme (Rao & Gianfreda, 2000). Immobilized enzymes were more resistant to environmental changes compared to their soluble counterparts (Sabil et al., 1993). In Venice lagoon sediments within a shallow water area, phosphatase activity was prolonged by its insolubility (Sabil et al., 1994).

Shortly, there is a correlation between higher sediment APA in term of kinetics and chlorophyll *a* concentration in water of Lake Donghu. The kinetic properties of phosphatase and their distributions in sediments were linked to the process of lake eutrophication, which might become more stable upon immobilization in sediments with accelerated eutrophication.

Acknowledgements This work was supported by the grant (KZCX1-SW-12-II-02-02) from the Chinese Academy of Sciences (CAS), National Basic Research Program of China (2002CB412300) and funding from the National Science Foundation of China (39170165, 39670149 and 20177033). We also obliged to the funds (2002AA601013). The authors also gratefully acknowledge the support of the Donghu Lake Ecosystem Experimental Station of CAS for the help in sampling collection.

References

Berman, T., 1970. Alkaline phosphatase and phosphorus availability in Lake Kinneret. Limnology and Oceanography 15: 663–674.

Chrost, R. J. & J. Overbeck, 1987. Kinetics of alkaline phosphatase activity and phosphorus availability for phytoplankton and bacterioplankton in Lake Plusssee (North German eutrophic lake). Microbial Ecology, New York, NY 13(3): 229–248.

Coolen, M. J. & J. Overmann, 2000. Functional exoenzymes as indicators of metabolically active bacteria in 124,000-year-old sapropel layers of the eastern Mediterranean Sea. Applied and Environmental Microbiology 66(6): 2589–2598.

de Vicente, I., K. Cattaneo, L. Cruz-Pizarro, A. Brauer & P. Guilizzoni, 2006. Sedimentary phosphate fractions related to calcite precipitation in an eutrophic hardwater lake (Lake Alserio, northern Italy). Journal of Paleolimnology 35: 55–64.

Golterman, H. L. & R. S. Clymo, 1969. Method for chemical analysis of fresh water. In IBP Handbook No. 8. Blackwell Sci. Publ., Oxford, 1–188.

Hakulinen, R., M. A. Kahkonen & M. Salkinoja-Salonen, 2005. Vertical distribution of sediment enzyme activities involved in the cycling of carbon, nitrogen, phosphorus and sulphur in three boreal rural lakes. Water Research 39: 2319–2326.

Huang, Q. & H. Shindo, 2001. Comparison of the influence of Cu, Zn, and Cd on the activity and kinetics of free and immobilized acid phosphatase. Soil Science and Plant Nutrition 47: 767–772.

Huang, Q. & H. Shindo, 2000. Inhibition of free and immobilized acid phosphatase by zinc. Soil Science 165: 793–802.

Huang, X. & J. T. Morris, 2003. Trends in *phosphatase* activity along a successional gradient of tidal freshwater marshes on the cooper river, South Carolina. Estuaries 26(5): 1281–1290.

Huang, X. & J. T. Morris, 2005. Distribution of *phosphatase* activity in marsh *sediments* along an estuarine salinity gradient. Marine Ecology Progress Series 292: 75–83.

Jansson, M., H. Olsson & K. Pettersson, 1988. Phosphatases; origin, characteristics and function in lakes. Hydrobiologia 170: 157–175.

Kobori, H. & N. Taga, 1979. Occurrence and distribution of phosphatase in neritic and oceanic sediments. Deep-Sea Research 26A: 799–808.

Koester, M., S. Dahlke & L.-A. Meyer-Reil, 1997. Microbiological studies along a gradient of eutrophication in a shallow coastal inlet in the southern Baltic Sea (Nordruegensche Bodden). Marine Ecology Progress Series 152: 27–39.

Kunito, T., K. Saeki, S. Goto, H. Hayashi, H. Oyaizu & S. Matsumoto, 2001. Copper and zinc fractions affecting microorganisms in long-term sludge-amended soils. Bioresource Technology 79: 135–146.

Kuperman, R. G. & M. Carriero, 1997. Soil heavy metal concentrations, microbial biomass and enzyme activities in a contaminated grassland ecosystem. Soil Biology and Biochemistry 29: 179–190.

Marxsen, J. & H.-H. Schmidt, 1993. Extracellular phosphatase activity in sediments of the Breitenbach, a Central European mountain stream. Hydrobiologia 253: 207–216.

Matinvesi, J. & H. Heinonen-Tanski, 1992. Biodegradable substances in lake sediments and their relation to sediment microbiological activity and phosphorus recycling. Aqua Fennica 22: 193–200.

Olah, J. & E. O. Toth, 1978. The function of alkaline *phosphatase* enzyme in the phosphorus cycle of fertilized fishponds. Aquacult Hung 1: 15–23.

Qin, B., P. Xu, Q. Wu, L. Luo & Y. Zhang, 2006. Environmental issues of Lake Taihu, China. Hydrobiologia.

Rao, M. A. & L. Gianfreda, 2000. Properties of acid phosphatase-tannic acid complexes formed in the presence of Fe and Mn. Soil Biology and Biochemistry 32: 1921–1926.

Sabil, N., A. Cherri, D. Tagliapietra & M.-A. Coletti-previero, 1994. Immobilized enzymic activity in the Venice Lagoon sediment. Water Research 28: 77–84.

Sabil, N., D. Tagliapietra & M.-A. Coletti-Previero, 1993. Insoluble biodegradative potential of the Venice Lagoon. Environmental Technology 14: 1089–1095.

Sayler, G., S. M. Puziss & M. Silver, 1979. Alkaline phosphatase assay for freshwater sediments: Application to perturbed sediment system. Applied and Environmental Microbiology 38: 922–927.

Silva, C. D. & N. B. Bhosle, 1990. Phosphorus availability and phosphatase activity in the sediments of Mandovi estuary, Goa. Indian Journal of Marine Sciences 19: 143–144.

Sinke, A. J. C., A. A. Cornelese & T. E. Cappenberg, 1991. Phosphatase activity in sediments of the Loosdrecht lakes. Internationale Vereinigung für Theoretische und Angewandte Limnologie 24: 719–721.

Wynne, D. & A. J. H. Pieterse, 2000. The effect of copper on photosynthesis, nitrate reductase and phosphatase activities in Lake Kinneret phytoplankton. Limnology and Lake Management 55: 581–593.

Zhang, R. Q., Q. X. Chen, R. Xiao, L. P. Xie, X. G. Zeng & H. M. Zhou, 2001. Inhibition kinetics of green crab (*Scylla serrata*) alkaline phosphatase by zinc ions: A new type of complexing inhibition. Biochimica et Biophysica Acta 1545: 6–12.

Zhou, Y., J. Li & Y. Fu, 2000. Effects of submerged macrophytes on kinetics of alkaline phosphatase in Lake Donghu. I. Unfiltered water and sediments. Water Research 34: 3737–3742.

Zhou, Y., J. Li, Y. Fu & M. Zhang, 2001. Kinetics of alkaline phosphatase in lake sediment associated with cage-culture of *Oceochrolmis miloticus*. Aquaculture 203: 23–32.

Zhou, Y., J. Li, Y. Fu & M. Zhang, 2002. Temporal and spatial variations in kinetics of alkaline phosphatase in sediments of a shallow Chinese eutrophic lake (Lake Donghu). Water Research 36: 2084–2090.

Zhou, Y., J. Li, C. Song & X. Cao, 2004. Variations and possible source of potentially available phosphorus in a Chinese shallow eutrophic lake. Journal of Freshwater Ecology 19: 87–96.

Hydrobiologia (2007) 581:117–123
DOI 10.1007/s10750-006-0513-5

EUTROPHICATION IN LAKES

Environmental changes in Lake Taihu during the past century as recorded in sediment cores

Bin Xue · Shuchun Yao · Weilan Xia

Abstract The Lake Taihu drainage basin is an economically developed area with some of the highest population densities in China. The lake has deteriorated due to ecological destruction and eutrophication. Three short sediment cores from eastern, northeastern and southwestern Lake Taihu were collected. Total organic carbon (TOC), total nitrogen (TN), pigments, elements and particle size were analyzed for the purpose of understanding past trophic status and pollution levels. Sedimentation rates were based on ^{137}Cs or ^{210}Pb methods. Results indicated that sediment particle size became coarser since the 1920s, and the lake was contaminated by heavy metals, such as Cu and Zn, since the 1970s. A remarkable increase in eutrophication since the 1980s due to increased loading of untreated effluents from industry, agriculture and urbanization is reflected by total organic carbon, total nitrogen and pigments in the studied cores. However the onset times of eutrophication in different parts of Lake Taihu were not synchronous.

Guest editors: B. Qin, Z. Liu & K. Havens
Eutrophication of shallow lakes with special reference to Lake Taihu, China

B. Xue (✉) · S. Yao · W. Xia
Nanjing Institute of Geography and Limnology, Chinese Academy of Sciences (CAS), Nanjing 210008, China
e-mail: bxue@niglas.ac.cn

S. Yao
Graduate School of CAS, Beijing 100039, China

Keywords Lake Taihu · Sediment core · Eutrophication · Heavy metal

Introduction

Lake Taihu, situated in the southern Yangtze River Delta, is one of the five largest freshwater lakes in China (Shun & Huang, 1993). Anthropogenic activities in the basin can be traced to 7,000 B.P. (Chen & Wang, 1999; Chen et al., 1997). Human impacts have increased dramatically in the last two decades.

Lake sediments can serve as information archives of environmental changes through time (Mônica & Carlos, 2002), since each layer of buried sediments represents a record of past environmental conditions (Von Guten et al., 1997). Changing heavy metals and organic matter concentrations in dated sediment cores indicate the evolution of adjacent emission sources and terrestrial ecosystems. This evolution often is influenced by social changes, such as industrial development, deforestation, mining, population increase and energy consumption (Palanques et al., 1998). Pigments in lake sediments can also provide an effective means for reconstructing the paleoenvironmental history of lake ecosystems. As a result, sediment cores may provide a valuable record of anthropogenic and environmental changes.

The purpose of this study was to track trophic status and human impact in the past century as recorded in three short sediment cores taken from eastern, northeastern and southwestern Lake Taihu. By extending the record to the beginning of the last century, changes in water quality and the ecosystem may be identified.

Materials and methods

Cores were taken from two sites in Lake Taihu in 1995 using a gravity corer (Fig. 1). One core (W) was taken from western Lake Taihu; the other one (E) was taken from East Taihu Bay. In 2003, another core (THS) was taken from southern Lake Taihu. W station was near Wuxi, an industrial city located 2 km northeast of Wuli Bay. Wuli Bay receives industrial waste water and domestic sewage from Wuxi, and water flows from Wuli Bay to Lake Taihu in winter/spring (Zou et al., 1996). Site E receives a small input of waste water and has dense submerged macrophyte coverage, while THS receives only waste materials from small towns.

Cores W and E were sliced at 2 cm intervals for total organic carbon (TOC), total nitrogen (TN) and sediment pigments analysis. Core THS was sliced at 1 cm intervals for chemical analysis of sediments, physics and radioactivity dating.

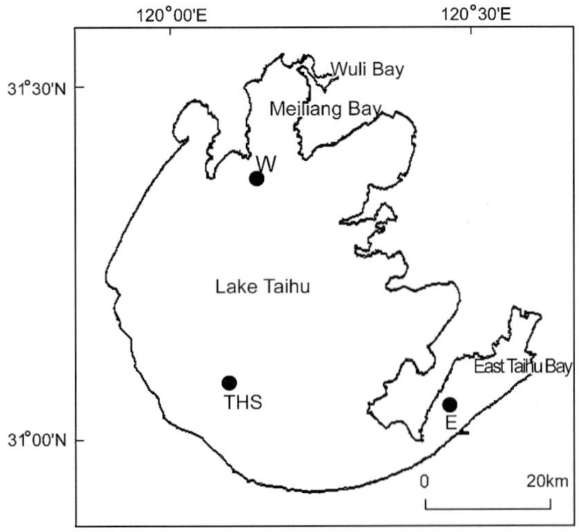

Fig. 1 Map of study area showing sampling sites

TOC was measured using concentrated sulfuric acid digestion with potassium dichromate, and TN was measured by the Kjeldahl technique (Institute of Soil Science, 1978; Professional Committee of agricultural chemistry of Chinese Society for Soil Sciences, 1983).

Samples were transferred quickly to air-tight, darkened containers for pigment analysis. Samples were stored at 4 °C before analysis. Pigments were determined using the method from Swain (1985). Samples were extracted using 90% acetone, and pigment concentrations were determined by measuring absorption at appropriate wavelengths with a spectrophotometer.

Samples for determination of heavy metals (Fe, Mn, Cu, Zn, Ni) were determined using EPA Method 3052 (HCl-HNO3-HF digestion) and ICP-AES (Leeman Labs Profile) (USEPA 1995). Accuracy was assessed with reference material from the National Institute of Standard and Technology (SRM 1646a).

Grain sizes of sediments were determined using a Malvern Mastersizer 2000. In addition, sediment age was estimated from available dates determined by ^{210}Pb or ^{137}Cs (Wan et al., 1987). ^{210}Pb and ^{137}Cs were determined using EG and G Ortec Gamma Spectrometry at Nanjing Institute of Limnology and Geography. ^{137}Cs was measured at 662 keV, while ^{210}Pb was determined via gamma emission at 46.5 keV and ^{226}Ra at 295 and 352 keV γ-rays emitted by its daughter isotope ^{214}Pb. A stable depositional rate was assumed.

Results

Sedimentation rate

Average sedimentaion rate for northern East Taihu Bay was 1.8 mm yr^{-1}, and average depositional flux was 0.13 g cm^{-2} yr^{-1}. Average depositional flux for southern East Taihu Bay was 0.24 g cm^{-2} yr^{-1}. Depositional flux was 0.08 g cm^{-2} yr^{-1} for western Lake Taihu (Shun & Huang, 1993). Average sedimentation rate for East Taihu Bay based on ^{210}Pb was 0.39 cm yr^{-1} according to Chang (1996), and our core E location was the same as Changs'. In 2002, a core from the same location as W had an average

Table 1 Sedimentation rates of W, E and THS

Core name	W		E		THS	
	^{210}Pb	^{137}Cs	^{210}Pb	^{137}Cs	^{210}Pb	^{137}Cs
Sedimentation rate (cm yr^{-1})	–	0.17 (Liu et al,2004)	0.39 (Chang, 1996)	–	0.34	0.41

sedimentation rate based on ^{137}Cs of 0.17 cm yr^{-1} (Liu et al., 2004). For core THS, the average sedimentation rate was 0.34 cm yr^{-1} based on ^{137}Cs, and 0.41 cm yr^{-1} using ^{210}Pb (Table 1).

Nutrients in the sediment

TOC, TN and C/N ratios for the three cores are shown in Fig. 2. Highest TOC and TN concentrations in core W occur in the top 4 cm, which represents about 20 years of deposits estimated from ^{137}Cs. Highest TOC and TN concentrations also occur in the top 4 cm in core E, representing about 10 years of deposits based on ^{210}Pb. In core THS, the TOC concentration maximum occur in the top 8 cm, representing about 20–24 years based on ^{210}Pb or ^{137}Cs. An abrupt increase in TN concentration in core THS occur in the top

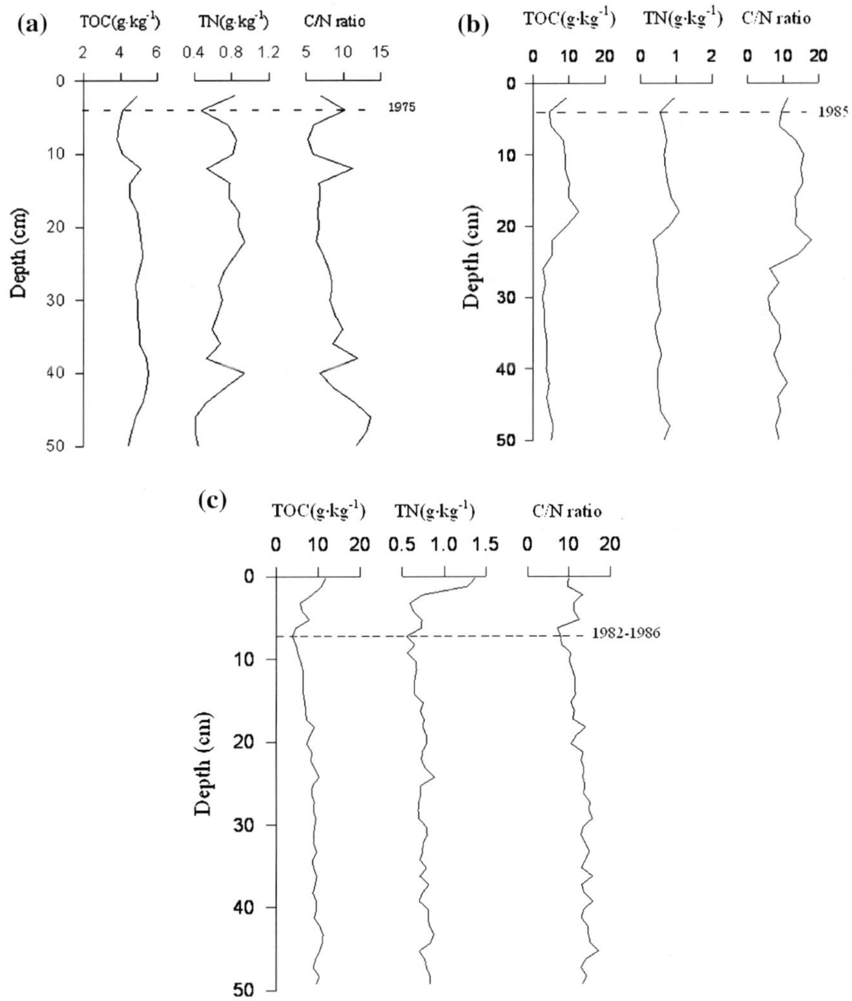

Fig. 2 Variation of TOC, TN and C/N in (**a**) W, (**b**) E and (**c**) THS core sediments

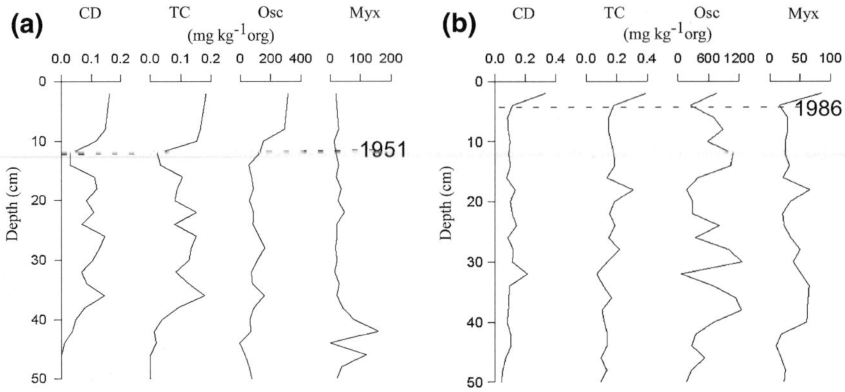

Fig. 3 Variation of pigments in (**a**) W and (**b**) E core sediments

3 cm, which represents about 7–9 years based on ^{210}Pb or ^{137}Cs.

In East Taihu Bay, C/N ratios range from 10 to 15 since the 1930s, while in core W C/N ratios are from 5 to 10 since 1975. C/N ratios vary little throughout the THS core. In general, lower C/N ratios are found in the top 2 cm of cores W and THS.

Pigments in the sediment

Pigments preserved in lake sediments contain detailed information about eutrophication history. An interpretation of major pigment types (e.g. chlorophyll and carotenoids) is often difficult and ambiguous. In contrast, pigments unique to blue-green algae are useful in tracing eutrophication development. Oscillaxanthin is a pigment unique to Oscillatoriaceae and has been found in only one genus other than Oscillatoria. However, myxoxanthophyll is found in most blue-green genera and all six families examined (Swain, 1985).

Results of pigment analysis in cores W and E are shown in Fig. 3. An increase in pigments is noted in the top 8 cm of core W. Based on ^{137}Cs, these sediments represent about 44 years. Distinct increases in pigments from core E are found in the top 4 cm. Based on ^{210}Pb, the top 4 cm represents about 9 years. The largest change in pigment concentration is in chlorophyll derivatives (CD) and total carotenoids (TC) in cores W and E. Differences exist between cores W and E. A large increase in myxoxanthophyll occurred in E core, and a dramatic rise in oscillaxanthin was found in W core.

Heavy metal pollution

Heavy metal concentrations and particle sizes in core THS are shown in Fig. 4. Mn, Cu, Zn, Ni and Pb concentrations decrease from bottom to depth

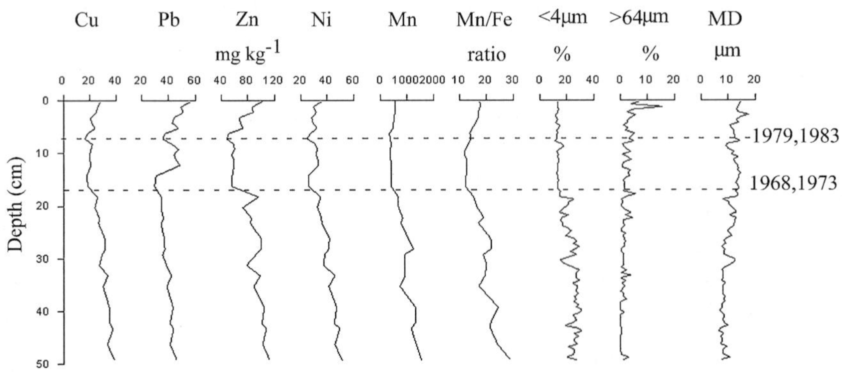

Fig. 4 Variation of metals in THS core sediments

17 cm, then become stable until depth 7 cm. Cu, Zn, Ni and Pb values shows the trend of increasing from depth 7 cm to 0 cm. Elevated organic carbon and Mn/Fe ratio also were observed, so heavy metals may be bound with organic matter or the Fe/Mn oxide fraction under oxydic conditions. However, coarse materials were found in these layers.

Others have reported that Hg originated partly from human pollution from the mid 1920s to 1970s at a site near W (Liu et al., 2004). Heavy metals, such as Cu, Zn, Pb, Hg, and As are indicative of human pollution. Source characteristics of heavy metals correspond to human activities and economic development in the Lake Taihu drainage basin (Liu et al., 2004).

Discussion

C/N ratios are useful for distinguishing between algal and vascular origins in organic sediments (Meyers et al., 1995). Dense submerged macrophytes may explain the high C/N ratios in East Taihu Bay. Lower C/N ratios in the top 2 cm of cores W and THS compared to the deep depth sediment indicate increasing algal origin.

Increasing values of total organic matter (TOM), TN and total phosphate (TP) concentrations were found in Lake Taihu surfical sediments dated between 1980 and the 1990s (Table 2; Hu, 2003). Increases in TOC and TN are also observed since the early 1980s in all the three cores. High TOC and TN in the three cores since the early 1980s agree with others concluding that accelerated eutrophication began in 1983/1984 in Lake Taihu (Chang, 1996). Rose et al. (2004) also got the same results from northern parts of Lake Taihu that nitrogen levels increase dramatically from the mid-1980s which is in agreement with N concentrations in lake water. However, estimates of nutrient enrichment onset may differ among cores if the stable depositional rate from ^{210}Pb or ^{137}Cs is accepted (i.e., about 1975 in core W, early 1980s in core THS and mid 1980s in core E).

The reason why Oscillaxanthin increase during 1940s and 1950s at site west while another variables do not increase until much later is unknown. Deep work should be done for this. Pigments are labile compounds, so the increases in the surfical sediment may indicate either better preservation via sediment anoxia or increased production. But changes of CD/TC profiles can help identify the cause for increased blue-green pigment concentrations, lower CD/TC ratios represent good preservation condition (Wu et al., 2004). Relatively lower CD/TC ratios in top sediments of cores W and E keep stable or show up-core increase trend. According Wu's method (Wu et al., 2004), increased production could be acceptable reason for the increase in blue-green algae.

Pigment enrichment in core W was caused by anthropogenic inputs due to increasing population and human activities in the watershed since the 1950s. However, the change was slow in the 1950s and 1970s, and inputs were primarily low amounts of agricultural and household wastes (Chang, 1996). Untreated effluents from the watershed have increased substantially in the last two decades. Total wastewater discharge in 1979 was 662 million tons, 1153 million tons in 1989, and 1387 million tons in 1999. Wastewater treatment facility capacity is only 328 million tons per year (Hu, 2003). This untreated effluent rapidly increased eutrophication in Lake Taihu. This

Table 2 Comparison of amount of total nitrogen, total phosphorus and total organic matter in the sediment of Lake Taihu over years

Year	TN(%)		TP(%)		TOM(%)	
	Range	Average	Range	Average	Range	Average
1960	–	0.067	–	0.044	0.54–6.23	0.68
1980	0.022–0.147	0.065	0.037–0.067	0.052	0.24–2.78	1.04
1990–1991	0.049–0.558	0.080	0.040–0.107	0.056	0.57–15.10	1.90
1995–1996	0.022–0.450	0.094	0.039–0.237	0.058	0.31–9.04	1.70
1997–1999	0.022–0.618	0.092	0.028–0.280	0.060	0.31–15.73	1.83

corresponds to increased pigment content in core E since 1986. Differences in onset time of sediment pigment enrichment may be due to the distance from river input locations. And Core W is located in northern Lake Taihu, where large amounts of untreated effluent are discharged, while E receives low waste water inputs.

Higher concentrations of Mn, Cu, Zn, Ni and Pb in the deepest layers may be related to differences in Mn/Fe ratio and fine-grained sediment. Largest concentration of heavy metals in finer sediments agrees with other observations (Salomons & Fostner, 1984). Rising heavy metal concentrations in core THS may indicate recent contamination. However, high Pb concentrations also were observed from 7–12 cm. Differences may indicate earlier discharges of Pb. And Pb, Cu, Ni and Zn were the main pollutants in some rivers investigated around Lake Taihu (Wang & Cao, 2002).

Conclusions

Combined with ^{210}Pb and ^{137}Cs dating and geochemical analyses, the sediment record of Lake Taihu demonstrates a complex history of environmental changes in the past century. Heavy metals concentration reveals contamination caused by human activities such as industry and urbanization. TOC and C/N ratios indicate changes in organic matter enrichment in sediment, such as enhanced algae contribution due to waste water and nutrient matter input to the lake. Changes in TN, TP and pigments from the cores confirm eutrophication of Lake Taihu. However the onset times of eutrophication in different parts of Lake Taihu were not synchronous. This means multi-core analyses would be better to understand eutrophication history of such a large shallow lake.

Acknowledgements The authors would like to thank colleagues in Lake Sedimentation and Environment Laboratory of NIGLAS for field and laboratory support. We are indebted to Dr Qu Wenchuan for reviewing of this manuscript. Thanks are due to Dr. Li Shiji for providing facilities. Financial support was from the CAS project (KZCX2-YW-319-2), NSFC (40472085) and a 973 project (2002CB412301).

References

Chang, W. Y. B., 1996. Major in environmental changes since 1950 and the onset of accelerated eutrophication in Lake Taihu, China. Acta Palaeontologica Sinica 35: 155–174 (in Chinese).

Chen, Z. Y., X. Q. Hong, S. Li, L. Wang & X. M. Shi, 1997. Archaeology-related environmental change: Yangtze Delta, China. Acta Geographic Sinica 52: 131–137 (in Chinese).

Chen, Z. Y. & Z. H. Wang, 1999. Yangtze Delta, China: Lake Taihu level variation since 1950's, response to sea level and human impact. Environmental Geology 37: 333–339.

Hu, B. B, 2003. Analysis on effect of water pollution in Lake Taihu basin on water quality of Lake Taihu, Shanghai Environmental Sciences 22: 1017–1021 (in Chinese).

Institute of Soil Science, Chinese academy of sciences, 1978. Physical and chemical analysis of soil. Shanghai scientific and Technical Publishers, Shanghai (in Chinese).

Liu, E. F., J. Shen, Y. X. Zhu, W. L. Xia & G. W. Zhu, 2004. Source analysis of heavy metals in surface sediments of Lake Taihu. Journal of Lake Sciences 16: 113–119 (in Chinese).

Meyers, P. A., M. J. Leenheer & R. A. Bourbonniere, 1995. Diagenesis of vascular plant organic matter components during burial in lake sediments. Aquatic Geochemistry 1: 35–52.

Mônica, A. L. S. & E. R. Carlos, 2002. Behavior of selected micro and trace elements and organic matter in sediments of a freshwater system in south-east Brazil. The Science of the Total Environment 292: 121–128.

Palanques, A., J. A. Sanchez-Cabeza & P. Masqué, 1998. Historical record of heavy metals in a highly contaminated Mediterranean deposit: the Besòs prodelta. Marine Chemistry 61: 209–217.

Professional Committee of agricultural chemistry of chinese society for soil sciences, 1983. Routine analysis method for soil agricultural chemistry, Beijing, 103 (in Chinese).

Rose, N. L., J. F. Boyle, Y. Du, C. Yi, X. Dai, P. G. Appleby, H. Bennionl, S. Cai & L. Yu, 2004. Sedimentary evidence for changes in the pollution status of Taihu in the Jiangsu region of eastern China. Journal of Paleolimnology 32: 41–51.

Salomons, W. & U. Fostner, 1984. Metals in the Hydrocycle. Springer-Verlag, Berlin: 349.

Shun, S. C. & Y. P. Huang, 1993. Taihu. Ocean publishing, Beijing, 271 (in Chinese).

Swain, E. B, 1985. Measurement and interpretation of sedimentary pigments. Journal of Freshwater Biology 15: 53–75.

USEPA 1995. Test methods for evaluating solid waste. Vol. IA: Laboratory manual physical/chemical methods (3rd edn.). U.S. Gov. Print. Office, Washington, DC, SW 846.

Von Guten, H. R., M. Sturm & R. N. Moser, 1997. 200-year record of metals in lake sediments and natural background concentrations. Environmental Science Technology 31: 2193–2197.

Wan, G. J., P. H. Santschi, M. Sturm, K. Farrenkothen, A. Lueck, E. Werth & Ch. Schuler, 1987. Natural (^{210}Pb, ^{7}Be) and fallout (^{137}Cs, $^{239,240}Pu$, ^{90}Sr) radionuclides as geochemical tracers of sedimentation in Greifensee, Switzerland. Chemical Geology 63: 181–196.

Wu, J. L., K. G. Michael, X. Z. Jiang, W. L. Xia & S. M. Wang, 2004. Sedimentary geochemical evidence for recent eutrophication of Lake Chenghai, Yunnan, China. Journal of Paleolimnology 32: 85–94.

Zou, H. X., G. Y. Sheng, C. Sun & O. Y. Xu, 1996. Distribution of organic contaminants in Lake Taihu. Water Research 30: 2003–2008.

Hydrobiologia (2007) 581:125–134
DOI 10.1007/s10750-006-0504-6

EUTROPHICATION IN LAKES

Role of climate and agricultural practice in determining matter discharge into large, shallow Lake Võrtsjärv, Estonia

Peeter Nõges · Marju Kägu · Tiina Nõges

Abstract This article addresses how seasonal and annual matter discharge into large, shallow Lake Võrtsjärv (Estonia) is determined by climate and changes in agricultural practices. Climate variability involved increasing winter air temperatures and large inter-annual temperature and precipitation variation. Agriculture practices have transformed from high fertilizer usage (i.e. swine slurry) in the 1970s and 1980s, leading to high phosphate, ammonium, and BOD_5 loadings, to low loadings after the collapse of soviet-type agriculture in the early 1990s. The 28-year monthly record on river flow and concentrations of nutrients (N, P) and dissolved organic matter (BOD_5, COD_{Mn}) from four main tributaries was analysed with seasonal air temperature and precipitation data. Long-term trends in nutrient and organic matter loadings to Lake Võrtsjärv resulted from agriculture and climate changes. The change could be traced as a linear trend in loadings and a highly inter-correlated cluster of the slurry-related pollutants. Coincidental trends in air temperature and fertiliser use caused strong correlations between air temperature and pollutant loadings to the rivers, which turned non-significant after removing trends showing that the relationships were not based on year-to-year differences. Residual analysis revealed significant positive correlations between precipitation and annual loadings of ammonium, phosphates, and COD_{Mn}. Both components forming the load (water discharge and concentrations of substances) increased in wet years. The effect of high winter North Atlantic Oscillation Index (NAO) was expressed as more intensive river flow during winter months and decreased flow during the flood peak.

Keywords Large and shallow lake · Climate impact · Agricultural practice · Discharge · Nitrogen · Phosphorus · Dissolved organic matter

Guest editors: B. Qin, Z. Liu & K. Havens
Eutrophication of shallow lakes with special reference to Lake Taihu, China.

P. Nõges (✉) · M. Kägu · T. Nõges
Limnological Centre, Institute of Agricultural and Environmental Sciences,
Estonian Agricultural University,
61101 Rannu, Tartumaa, Estonia
e-mail: peeter.noges@jrc.it

P. Nõges
Institute for Environment and Sustainability,
European Commission – DG Joint Research Centre,
via E. Fermi 1, 21020 Ispra (VA), Italy

M. Kägu
Institute of Zoology and Hydrobiology, University of Tartu, Vanemuise 46, 51014 Tartu, Estonia

Introduction

Studies on global climate have revealed that conditions in the Northern Hemisphere are warming, with wetter winters and drier summers (Magnuson et al., 1997). These changes are attributed to increasing content of green-house gases in the atmosphere, which may be reflected in the increasing tendency of the North Atlantic Oscillation Index (NAO) (Hurrell et al., 2001, 2003). The increase of NAO is most pronounced in winter during the last 30 years (Cook, 2003). The signal of the winter NAO can be traced in thermal (Gerten & Adrian, 2001), hydrological (Yoo & D'Odorico, 2002) and biological parameters (Straile et al., 2003) of lakes, where its persistence is modified mostly by lake morphology (Gerten & Adrian, 2001). Climate change affects the balance of heat, water and substances in lake catchments resulting in changes to the hydrological regime and loadings. Hydrological changes have the strongest impact on shallow lakes, where they cause large fluctuations in water volume and lake depth. In Lake Võrtsjärv (mean depth 2.8 m), mean annual amplitude of water level fluctuations is 1.4 m, and the maximum range reaches 3.2 m. The latter corresponds to a 1.4-times difference in lake area, 2.4-times difference in mean depth and 3-times difference in lake volume (Nõges & Nõges, 1999). Changes in lake depth affect average illumination of the mixed water column, resuspension intensity, nutrient release from sediments, and denitrification rate in non-stratified shallow lakes. These factors control growth and composition of phytoplankton, which is the first link of the pelagic food web. Thus, changing water level has been considered the leading factor controlling ecosystem dynamics of Lake Võrtsjärv, primarily through phytoplankton (Nõges et al., 2003).

The impact of climate change on discharge of substances into the lake also must be considered. To now, this aspect of climate change has been less studied compared to better-known impacts on water temperature, stratification and ice cover. Phosphorus, nitrogen and silica are the most important substances coming from the watershed and shaping lake ecosystems. In warm, dry years, nutrient loading usually decreases, while in-lake concentrations may increase due to increased evaporation and more intensive release from sediments at a low water stage (Magnuson et al., 1997). The direction of the changes caused by climatic alterations can differ between deep and shallow lakes as well as regions of different climatic conditions. For example, cold low-NAO winters in Sweden bring about lower nutrient input to lakes, while, in Great Britain, such winters are associated with increased nutrient loading (Monteich et al., 2000; Straile et al., 2003).

Recently, limnologists have studied connections between climate change and coloured dissolved organic matter (CDOM) loading. CDOM affects underwater light conditions (Arst, 2003) and is important for phytoplankton and food webs. This may be crucial in large, very shallow lakes, like Lake Võrtsjärv, where light conditions play a leading role among factors limiting phytoplankton development (Nõges et al., 1998). In addition, CDOM is a substrate for aquatic bacteria. Water in European lakes is getting darker, and this phenomenon has been connected with intensified CDOM leakage from peaty catchments in warmer conditions (Freeman et al., 2001). There is, however, a current debate regarding effects of global warming and climate change on terrestrial organic matter export to aquatic environments (Tranvik & Jansson, 2002) since, e.g., northern boreal lakes generally have high CDOM concentrations (Curtis, 1998). It is clear that knowledge about the nature of climate change and its impacts is far from complete. The response of lakes to climate changes via temperature regime and stratification is complex and unpredictable, and different lakes may react in individual ways. Eutrophication may either diminish or magnify effects of climate change.

Answering questions regarding climate change impacts on lakes often is hindered by the scarcity of reliable, long-term limnological data. Changes in agricultural practices also may affect loadings and either mask or magnify climate change effects. The nearly 30-year record of monthly meteorological, hydrological, and chemical data available for Lake Võrtsjärv and its main tributaries offers a unique opportunity to study seasonal and annual dynamics of loading of different substances into the lake. This paper addresses how seasonal and annual matter discharge is determined by climate and changes in agricultural practices in large, shallow Lake Võrtsjärv (Estonia).

Site description

Lake Võrtsjärv is a large, shallow lake (270 km^2, mean depth 2.8 m) located in central Estonia (58°05′–58°25′ N, 25°55′–26°10′ E). Its catchment area (3374 km^2; Fig. 1) belongs to the southern Boreal forest zone. Forests cover 30–40% of the southern catchment, reaching 50–60% in the Väike Emajõgi Valley. The largest wetland areas cover the eastern coastal region and the mouth the Väike Emajõgi and Õhne rivers. The proportion of arable land is the lowest (10–20%) in the northern part and highest (30–40%) in the western and eastern parts of the catchment (Järvet et al., 2004). Agricultural activities are the main anthropogenic impact on the lake. In the 1980s, large areas, especially in the R. Tänassilma sub-catchment, were fertilized with swine slurry causing increased nutrient loading (Järvet & Laanemets, 1989, 1990). Improvements in wastewater treatment since the 1990s has decreased point source loadings from Valga (pop. 14,000) and Viljandi (pop. 21,000), located on the watershed boundary, and other smaller settlements. Abandonment of several large cattle farms has contributed to this decrease. The lake is

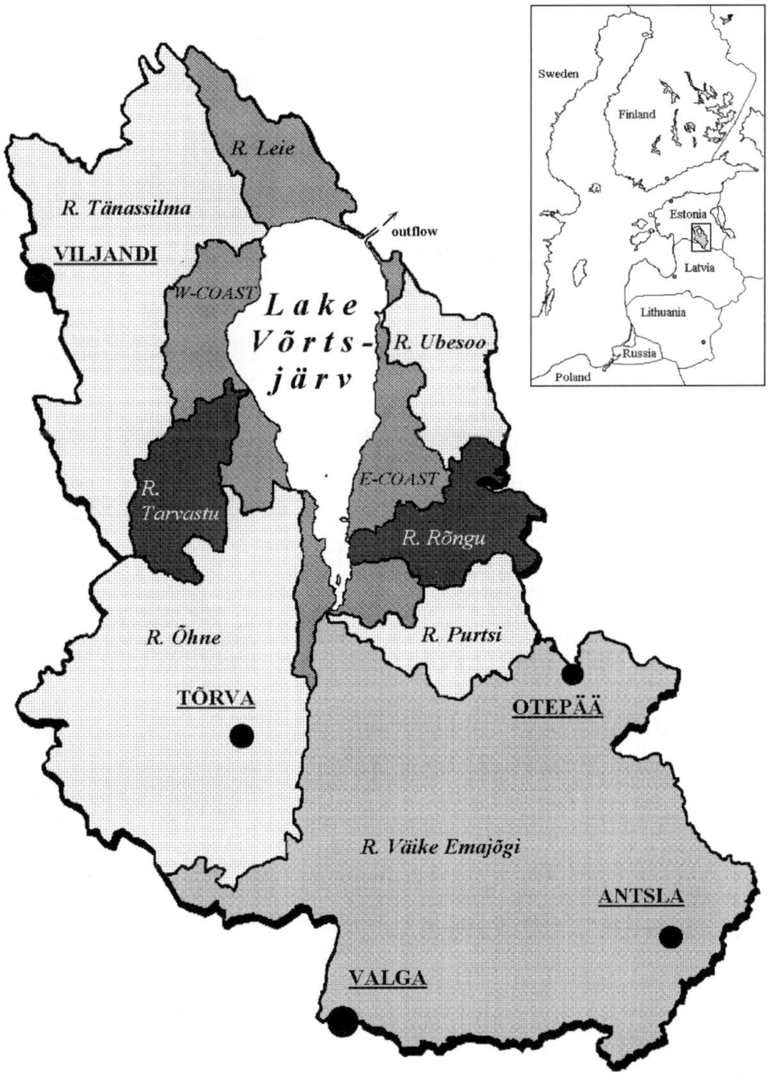

Fig. 1 Location map and division of Lake Võrtsjärv catchment area

ice-covered from mid-November to mid-April (average 135 days). The flow regime of inflowing rivers is natural, and inflow usually peaks in April with about one-quarter of its annual total inflow. Water retention time is about 1 year, and Lake Võrtsjärv is highly eutrophic, characterised by a mean total nitrogen concentration of 1.6 mg l^{-1}, total phosphorus of 54 μg l^{-1}, and chlorophyll *a* of 24 μg l^{-1} (Haberman et al., 1998). During the ice-free period, Secchi depth commonly is less than 1 m, mainly due to resuspended sediments.

Material and methods

Nutrient concentration data (different forms of N and P), biochemical oxygen demand (BOD_5), chemical oxygen demand by permanganate (COD_{Mn}), and water colour (Pt-Co°) in rivers were measured from 1975 to 2002 in the State Monitoring Programme, and data were obtained from the Information Centre of the Estonian Ministry of Environment. Yearly riverine loadings (*L*) were calculated based on daily water discharges and monthly concentrations of substances:

$$L = \sum_{i=1}^{12} W_{ki} \cdot C_{ki}$$

W_{ki}—volume of monthly runoff
C_{ki}—mean monthly concentration

NAO data from 1886 to 2002 were obtained from http://www.cru.uea.ac.uk/cru/data/nao.htm. Winter NAO (NAO_w), calculated as the December–February mean, was used in this analysis. Air temperature (1894–2002), precipitation (1866–2002), lake water level (1923–2002), and riverine discharges were measured routinely during meteorological and hydrological surveys by the Estonian Institute of Hydrology and Meteorology.

Statistica for Windows (v 6.0) was used for statistical analyses. Data were detrended linearly using the Time Series Analysis module. Spearman rank correlation was used throughout the analysis.

Results

Twenty-eight-years of data on annual nutrient loading were analyzed from four Lake Võrtsjärv sub-catchments: Tänassilma, Tarvastu, Õhne, and V. Emajõgi (Fig. 1) together with precipitation and air temperature data. Several peak loadings of pollutants visible in the long-term dynamics were related directly to extreme wet years (Fig. 2). For example, rainy years 1981 and 1998 were distinguished by increased runoff of COD_{Mn} and phosphates, while 1990 was characterized by that of COD_{Mn} and nitrates (Fig. 2). Besides rainy years, peaks of ammonium, phosphate and BOD_5 (latter not shown) in 1986 and 1987 coincided with cold winters (Fig. 2). Time-series of winter air temperature data (Jan.–Mar.) can be

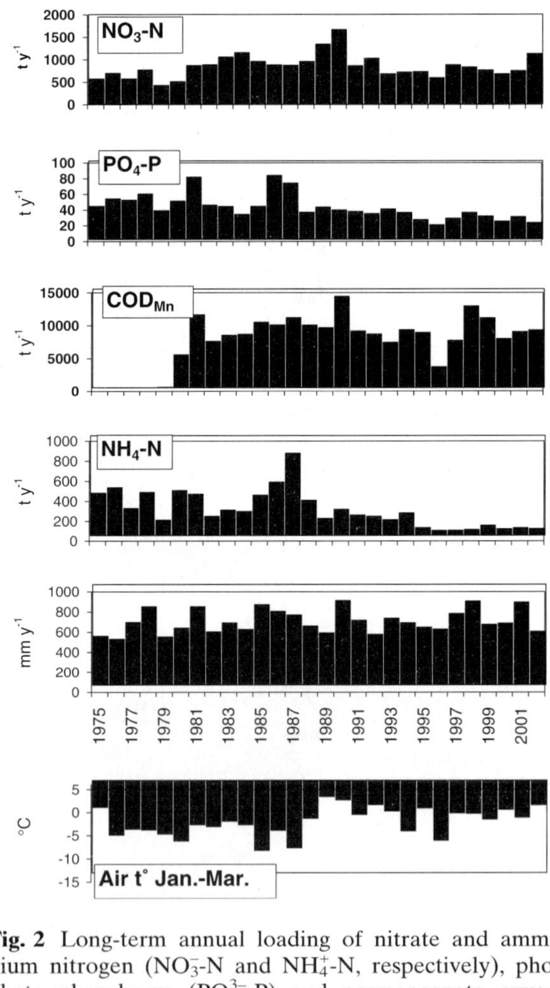

Fig. 2 Long-term annual loading of nitrate and ammonium nitrogen (NO_3^--N and NH_4^+-N, respectively), phosphate phosphorus (PO_4^{3-}-P) and permanganate oxygen demand (COD_{Mn}) from four sub-catchments of Lake Võrtsjärv (Tänassilma, Tarvastu, Õhne, and V. Emajõgi), annual amount of precipitation (Prec.) and winter air temperature measured at Tartu hydrometeorological station

roughly subdivided into two periods: 1975–1988 and 1989–2002 (Fig. 6). In the second period, temperature was notably higher. At the same time, loading of some substances, phosphates and ammonium in particular, showed a similar relationship: average loadings were different before and after the late 1980s. Presumably, this trend, which coincidentally was synchronized with the climate shift and caused, in part, by agricultural practice changes, may result in a negative relationship between temperature and substance loadings and bias other relationships. Correlation differences before and after trend removal were analyzed to distinguish between effects caused by year-to-year variations of (mostly climatic) indices and those caused by coinciding trends. Annual loading data was analyzed with annual and seasonal data on climatic variables: air temperature and precipitation. Trend removal caused several changes in the correlations (Table 1):

(1) all correlations between air temperature and loadings of substances became insignificant,
(2) correlations between phosphates, ammonium and biochemical oxygen demand weakened,
(3) statistical relationships between loadings and precipitation strengthened,
(4) correlations between loadings of chemical oxygen demand and other substances strengthened.

Second, emphasis was placed on two sides of the loading formation: river discharge and concentration of substances and their relationships with NAO. Relationships were analyzed between detrended NAO_w, discharge, and concentrations of CDOM and nutrients (N and P) in two major inflows to Lake Võrtsjärv (R. Väike Emajõgi and R. Õhne). These two rivers account for 60% or more of total discharge and nutrient loading to the lake. Figure 3a, b show the monthly inflow regime, and both rivers show peak flow in April. From January to March, discharge (Q) of both rivers was related positively to NAO_w, while the relationship was negative in April–May (Fig. 3c).

Water colour and COD_{Mn} were used as proxies for CDOM. These two parameters were related strongly to each other ($r = 0.57$ in the R. Õhne and $r = 0.66$ in the R. Väike Emajõgi; $p < 0.00001$). NAO_w showed strong correlations with both CDOM proxies. From January to March, correlations were mostly positive ($p < 0.05$ in February and March) and turned mostly negative for the rest of the year (Fig. 4a,

Table 1 Spearman correlations between climatic variables and loadings of substances to Lake Võrtsjärv (1975–2002) before and after removing linear trends from all data series

Direction of change	Between variables	Spearman r before detrending	Spearman r after detrending
Weakening	T_{1-3} - PO_4^{3-}	–0.438*	–0.177
	T_{1-3} - NH_4^+	–0.448*	–0.303
	T_{4-6} - BOD_5	–0.472*	0.027
	T_{4-6} - PO_4^{3-}	–0.507**	–0.301
	T_{4-6} - NH_4^+	–0.418*	–0.174
	T_{7-9} - BOD_5	–0.391*	–0.174
	T_{7-9} - PO_4^{3-}	–0.512**	–0.269
	T_{7-9} - NH_4^+	–0.377*	–0.033
	BOD_5 - PO_4^{3-}	0.875**	0.552**
	BOD_5 - NH_4^+	0.862**	0.432*
	NH_4 - PO_4^{3-}	0.857**	0.589**
Strengthening	P_{7-9} - NH_4^+	0.322	0.391*
	P_{7-9} - COD_{Mn}	0.374	0.420*
	P_{7-9} - PO_4^{3-}	0.383*	0.522**
	P_{1-12} - PO_4^{3-}	0.152	0.605**
	COD_{Mn} - NH_4^+	0.368	0.517**
	COD_{Mn} - PO_4^{3-}	0.263	0.517**
	COD_{Mn} - NO_3^-	0.437	0.629**
Unchanged	P_{1-12} - COD_{Mn}	0.570**	0.554**

T—temperature, P—amount of precipitation (numbers in the foot index show the period in months to which the data pertain); loadings: PO_4^{3-}—phosphates, NO_3—nitrates, NH_4^+—ammonium, BOD_5–biochemical oxygen demand, COD_{Mn}—permanganate oxygen demand

* $p < 0.05$, ** $p < 0.01$

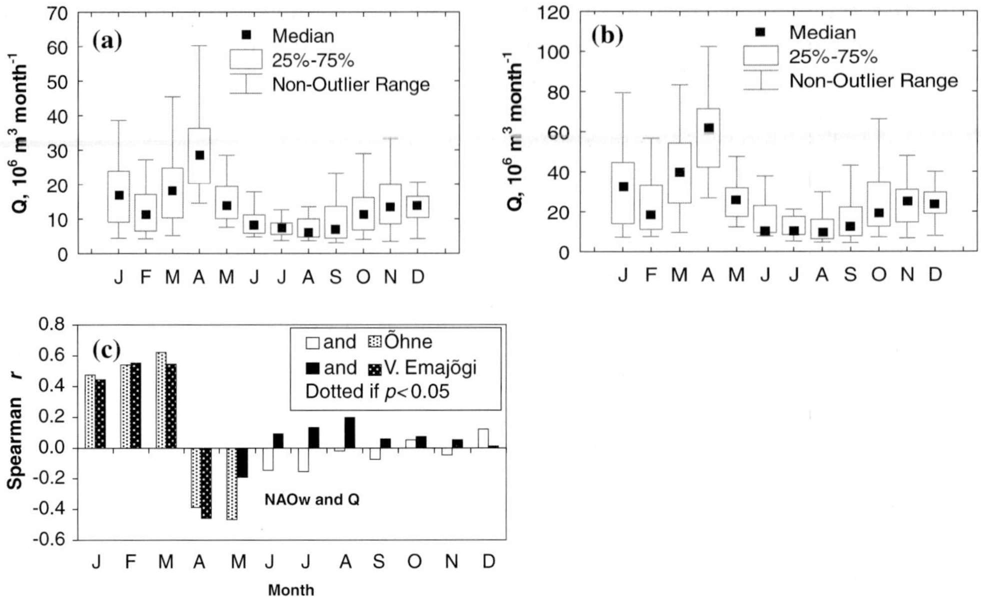

Fig. 3 Monthly discharges (Q) of rivers Õhne (**a**) and Väike Emajõgi (**b**) and their Spearman correlations with the NAO index for winter (**c**). All series detrended

b). Strongest negative correlations occurred in both rivers and for both parameters in July. In both rivers, water colour and COD_{Mn} were related positively to discharge throughout the year, and strong correlations occurred in March, June and August–November (Fig. 4c, d).

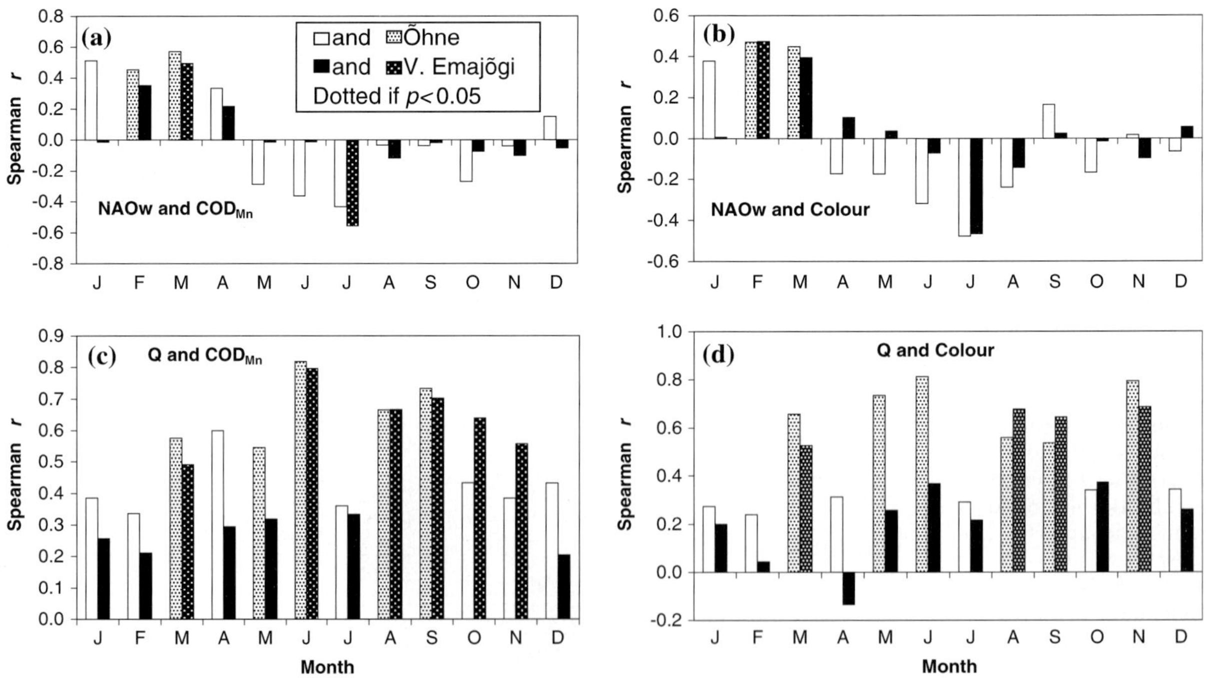

Fig. 4 Monthly Spearman correlations of permanganate oxygen demand (COD_{Mn}) and water colour with the winter NAO (NAO_w: **a** & **b**, respectively), and discharge (Q: **c** & **d**, respectively) in two major rivers in Lake Võrtsjärv catchment. All series detrended

Increased nutrient concentrations in January were characteristic of high-NAO_w years. For total nitrogen (TN) and, especially, nitrates, the positive relationship with NAO_w continued in February and March (Fig. 5a, b). For the rest of the year, the relationship with NAO_w turned negative for TN, while nitrates showed a positive relationship from July to September. For phosphates and total phosphorus (TP), the correlation with NAO_w remained mostly negative from February to July-August (strongest in April; Fig. 5c, d). In autumn, the correlation with NAO_w was mostly positive for TP and variable for phosphates. Increased discharge generally brought about higher nitrogen concentrations in rivers (Fig. 6a, b) with exceptions for the flood peak in April and the end of the year (November or December). Strong flushing of nitrates from the watershed by increased runoff in summer and autumn especially was pronounced in the R. Väike Emajõgi. The dependence of phosphorus concentration on river flow was more variable (Fig. 5c, d): in both rivers the correlation was negative in February and March, and strongest positive correlations occurred in August for phosphates and September–October for TP.

Discussion

Analysis showed that correlations between the original series of air temperature and substance loadings were caused by common trends in variables and not an inherent causal relationship. The trend in winter air temperatures, expressed as a step-wise change since the late 1980s, perhaps could be attributed to climate change or climate warming observed over large areas of the northern hemisphere (IPCC, 2001). Climate investigations in Estonia show that air temperature increased by 0.3–0.5°C from 1966 to 1998 compared to the earlier 90-year period, and the largest increase (>1°C) has taken place in March (Jaagus, 1999). Changes in winter temperatures may cause changes in catchment processes and affect loadings of substances to the lakes. Year-to-year variation in winter air temperature influences nitrate concentration in lakes in the English

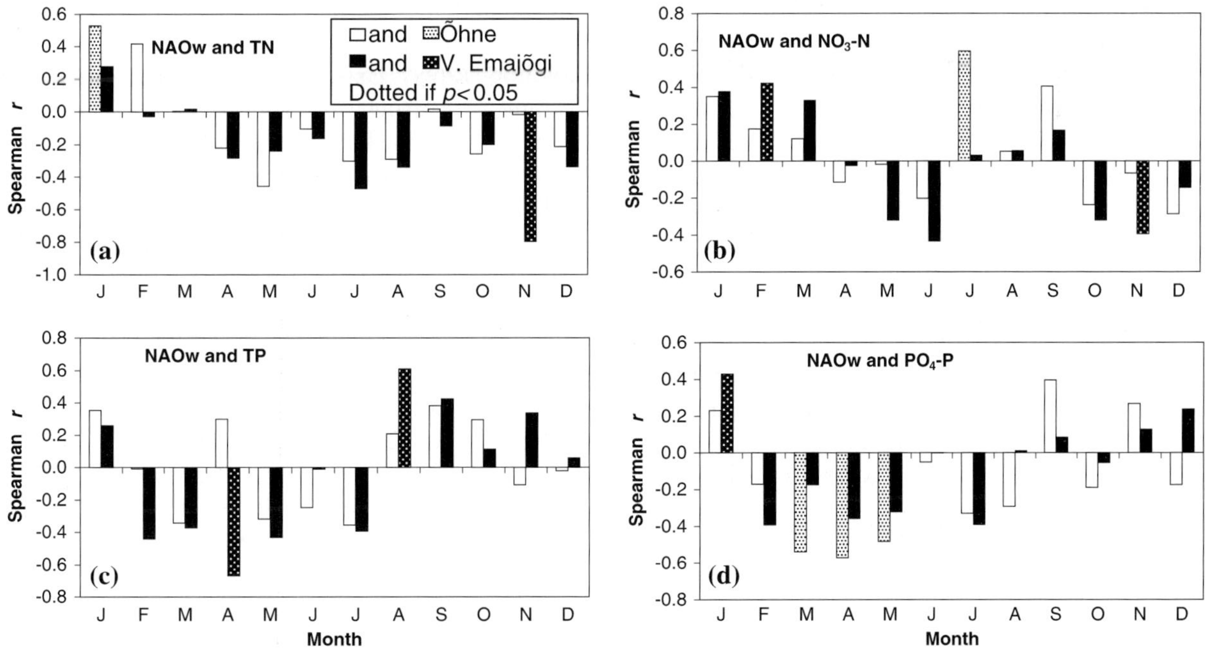

Fig. 5 Monthly Spearmen correlations of the winter NAO (NAO_w) with (**a**) total nitrogen (TN), (**b**) nitrate (NO_3^--N), (**c**) total phosphorus (TP), and (**d**) phosphate (PO_4^{3-}-P) concentrations in two major rivers in Lake Võrtsjärv catchment. All series detrended

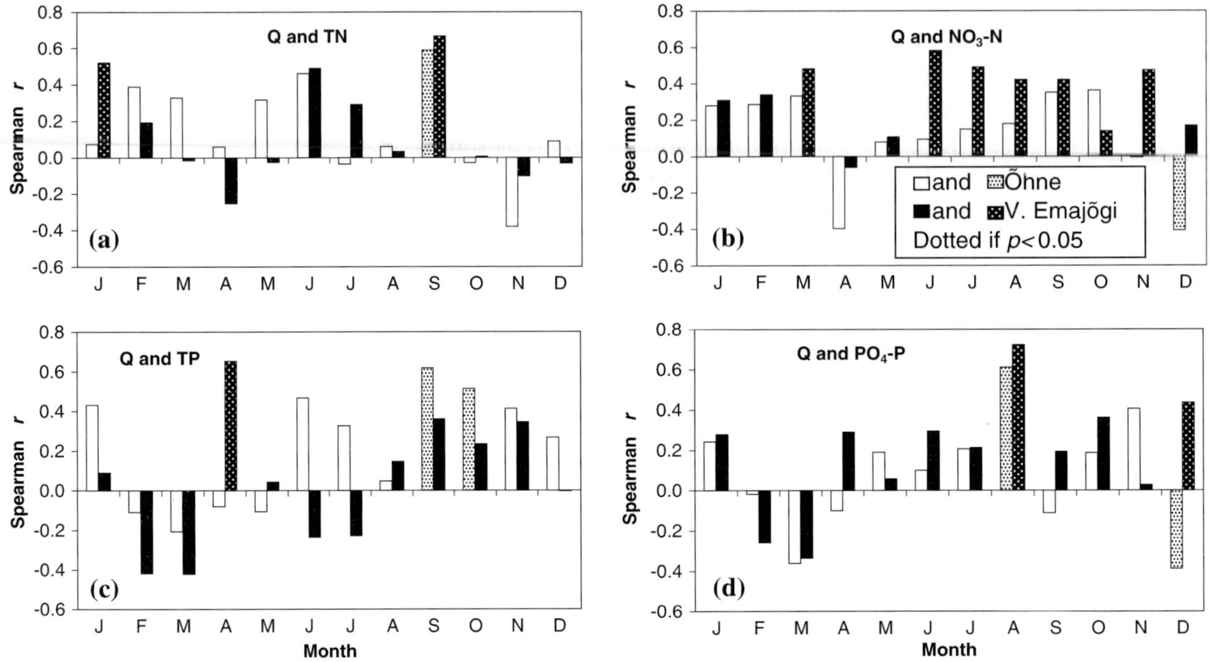

Fig. 6 Monthly Spearmen correlations of the discharge (Q) with (**a**) total nitrogen (TN), (**b**) nitrate (NO_3^--N), (**c**) total phosphorus (TP), and (**d**) phosphate (PO_4^{3-}-P) concentrations in two major rivers in Lake Võrtsjärv catchment. All series detrended

Lake District, since warm winters lead to a reduced nitrate concentration (George et al., 2004). As an extreme example, a 2° temperature increase melted the permafrost in the Lake Schwarzensee catchment, a remote high-alpine lake in Austria, and totally changed the chemical regime of the lake. Conductivity and silica concentration doubled, and pH increased dramatically (Psenner & Schmidt, 1992).

However, climate variability had a clear influence on loadings via precipitation changes. Despite several cases (1981, 1990, 1998) where effects of rainy years on loadings was visible in the time-series (Fig. 2), relationships remained insignificant without detrending the data. Residual analysis revealed significant positive correlations between precipitation and annual loadings of ammonium, phosphates, and COD_{Mn}. Ammonium and COD_{Mn} loadings showed strongest relationships with rainfall from July to September, and phosphate loading with annual precipitation.

After detrending, relationships between BOD_5, phosphates, and ammonium became weaker. Increased runoff of these pollutants is typical for agricultural catchments fertilised with slurry and, hence, changes in slurry application were expressed as a common trend in their loading rates. Experiments carried out on a sloping site using purpose-built facilities (Smith et al., 2001a, b) showed that application of cattle slurries to the silty clay loam soil increased loss of solids, ammonium nitrogen, and particulate and soluble phosphorus in surface water flow compared to control plots receiving inorganic fertiliser only, or no treatment, but had little effect on nitrate losses by this route. Rainfall events immediately after manure applications may cause large run-off of nutrients (Jarvis et al., 1987; Smith et al., 2001a), which may explain the synchronous increase of ammonium and phosphate loadings observed in the present study in some rainy years (e.g. in 1981).

In the original data, COD_{Mn} loading showed a different behaviour compared to other substances: it did not correlate with any others and had a significant positive relationship only with annual precipitation. After trend removal, the latter did not change, but correlations between loadings of all substances became highly significant. COD_{Mn} is represented in runoff waters from the Lake Võrtsjärv catchment mainly by humic substances, which give a brownish colour to the water. Dynamics of COD_{Mn} loadings reflected natural changes in catchment hydrology and was not affected by changes in agriculture, which caused different behaviour of other substances. After trend removal, dynamics of the slurry-related components became more natural, while the role of precipitation as the most important climatic parameter became a clearer manifestation.

NAO is one of the often-used proxy indicators allowing complex atmospheric circulation patterns to be expressed as a simple index. It combines effects of changes in temperature, affluence, and wind patterns. The relationship between NAO_w and river flow is established through precipitation in winter and spring but, first, by a shift in the timing and height of the flood peak. Milder winters (increased air temperature in February from –7.9 to –5.5°C during 1950–1997) and changed precipitation patterns have influenced mean annual water discharge, which has resulted in more intensive flow during colder seasons and decreased runoff in summer (Mander et al., 2000). During mild winters (high NAO_w), recurrent snowmelts keep water flowing throughout the winter, thus reducing peak discharges in spring (Fig. 3). Since temperature affects timing of the flood peak and, to a lesser extent, annual discharge (if differences in evaporation are considered), its effect on annual loading remained marginal and could not be revealed in our data. Concentrations of substances in rivers were related positively to river flow (Figs. 4c, d, 6). Thus, both components forming the load (water discharge and concentrations of substances) increased in wet years. As mentioned, summer and autumn rains (July–September) correlated most strongly with annual loadings of substances. Runoff of substances in this period, characterised usually by low flow in rivers (Fig. 3), was sensitive to precipitation, evidenced by strong correlations between water discharge and concentrations. Periods of negative correlations between water flow and concentrations occurred only for nutrients and were observed mostly in winter and spring. For phosphorus, the negative relationship in February and March may be attributed to the winter low flow period when most rivers switch to groundwater feeding (Järvet, 2001). Negative correlations between water flow and nitrogen concentration coincided with the April flood peak and may be caused by dilution.

Conclusions

(1) Long-term changes in loadings of nutrients and organic matter to Lake Võrtsjärv resulted from changes in agricultural practices as well as climate change.

(2) Agriculture changes involved the transition from high loadings of phosphates, ammonium, and BOD_5 in the 1970s and 1980s from fertilisers, in particular swine slurry, to low loadings after the collapse of soviet-type agriculture in the early 1990s. Changes are traced as a linear trend in loadings and an inter-correlated cluster of slurry-related pollutants.

(3) Climate variability consisted of an increasing trend in winter air temperatures and large inter-annual fluctuations of temperature and precipitation.

(4) Coinciding trends in air temperature and fertiliser use caused strong correlations between air temperature and pollutant loadings to rivers. All these relationships were insignificant after removing the trends, showing that the relationships were based on a coincidental common trend and not year-to-year differences.

(5) Residual analysis revealed significant positive correlations between precipitation and annual loadings of ammonium, phosphates, and COD_{Mn}. Both components forming the load (water discharge and concentrations of substances) increased in wet years.

(6) The effect of high winter NAO was expressed in more intensive river flow during winter months and decreased flow during the flood peak.

Acknowledgements Funding was provided by the Estonian Ministry of Education (No. 0362480s03), Estonian Science Foundation grants 5738 and 5425, and the EC project CLIME (No. EVK1-CT-2002-00121). Data collection in state monitoring program was supported by the Estonian Ministry of Environment.

References

Arst, H., 2003. Optical Properties and Remote Sensing of Multicomponental Water Bodies. Springer-Verlag, 238.

Cook, E., 2003. Multi-proxy reconstructions of the North Atlantic Oscillation (NAO) index: a critical review and a new well verified NAO index reconstruction back to AD 1400. In Hurrell, J. W., Y. Kushnir, G. Ottersen & M. Visbeck (eds), The North Atlantic Oscillation: Climatic Significance and Environmental Impact. American Geophysical Union, Washington DC, 63–79.

Curtis, P. J., 1998. Climatic and hydrologic control of DOM concentration and quality in Lakes. In Hessen, D. O. & L. J. Tranvik (eds), Aquatic Humic Substances, Vol. 133, Springer-Verlag, 93–104.

Freeman, C., C. D. Evans, D. T. Monteih, B. Reynolds & N. Fenner, 2001. Export of organic carbon from peat soils. Nature 412: 785.

George, D. G., S. C. Maberly & D. P. Hewitt, 2004. The influence of the North Atlantic Oscillation on the physical, chemical and biological characteristics of four lakes in the English Lake District. Freshwater Biology 49(6): 760–774.

Gerten, D. & R. Adrian, 2001. Differences in the persistency of the North Atlantic Oscillation signal among lakes. Limnology & Oceanography 46: 448–455.

Haberman, J., P. Nõges, E. Pihu, T. Nõges, K. Kangur & V. Kisand, 1998. Characterization of Lake Võrtsjärv. Limnologica 28(1): 3–11.

Hurrell, J. W., Y. Kushnir & M. Visbeck, 2001. The North Atlantic Oscillation. Science 291: 603–608.

Hurrell, J. W., Y. Kushnir, G. Ottersen, M. Visbeck, 2003. An overview of North Atlantic Oscillation. In Hurrell, J. W., Y. Kushnir, G. Ottersen & M. Visbeck (eds), The North Atlantic Oscillation: Climatic Significance and Environmental Impact. American Geophysical Union, Washington DC, 7–36.

IPCC, 2001. Summary for Policymakers, the third assessment report: 1–18.

Jaagus, J. 1999. Uusi andmeid Eesti kliimast. (New data about the climate in Estonia). In Jaagus, J. (ed), Uurimusi Eesti kliimast. Publicationes Instituti Geographici Universitatis Tartuensis 85: 28–40.

Järvet, A., 2001. Veekogude klimaatilised aastaajad. Publicationes Instituti Geographici Universitatis Tartuensis 90: 48–81.

Järvet, A., R. Karukäpp & I. Arold, 2004. Location and physico-geographical conditions of the catchment area. In Haberman, J., E. Pihu & A. Raukas (eds), Lake Võrtsjärv. Estonian Encyclopedia Publishers, Tallinn: 11–28.

Järvet, A. & A. Laanemets, 1989. Võrtsjärv on ohus. Eesti Loodus 5: 308–312.

Järvet, A. & A. Laanemets, 1990. Võrtsjärv kui looduslik puhastusseade. Eesti Loodus 11: 708–712.

Jarvis, S. C., M. Sherwood, J. H. A. M. Steenvoorden, 1987. Nitrogen losses from animal manures: from grazed pastures and from applied slurry. In: van der Meer, H. G. et al. (eds), Animal Manure On Grassland And Fodder Crops. Martinus Nijhoff Publishers, The Netherlands, 195–212.

Magnuson, J. J., K. E. Webster, R. A. Assel, C. J. Bowser, P. J. Dillon, J. D. Eaton, H. E Evans, E. J. Fee, R. I. Hall, L. R. Mortsch, D. W. Schindler & F. H. Quinn, 1997. Potential effects of climate changes on aquatic ecosystems: Laurentian Great Lakes and Precambrian Shield Region. In Cushing, C. E. (ed.), Freshwater Ecosystems and Climate Change in North America, A Regional Assessment. John Wiley & Sons, New York, 7–53.

Mander, Ü., A. Kull, V. Kuusemets & T. Tamm, 2000. Nutrient runoff dynamics in a rural catchment: Influence of land-use changes, climatic fluctuations and ecotechnological measures. Ecological Engineering 14: 405–417.

Monteich, D. C., C. D. Evans & B. Reynolds, 2000. Are temporal variations in the nitrate content of UK upland freshwaters linked to the North Atlantic Oscillation? Hydrological Processes 14: 1745–1749.

Nõges, T. & P. Nõges, 1999. The effect of extreme water level decrease on hydrochemistry and phytoplankton in a shallow eutrophic lake. Hydrobiologia 408/409: 277–283.

Nõges, T., V. Kisand, P. Nõges, A. Põllumäe, L. Tuvikene & P. Zingel, 1998. Plankton seasonal dynamics and its controlling factors in shallow polymictic eutrophic Lake Võrtsjärv, Estonia. International Review of Hydrobiology 83: 279–296.

Nõges, T., P. Nõges & R. Laugaste, 2003. Water level as the mediator between climate change and phytoplankton composition in a large shallow temperate lake. Hydrobiologia 506(1): 257–263.

Psenner, R. & R. Schmidt, 1992. Climate-driven pH control of remote alpine lakes and effects of acid deposition. Nature 356: 1–3.

Smith, K. A., D. R. Jackson & T. J. Pepper, 2001a. Nutrient losses by surface run-off following the application of organic manures to arable land. 1. Nitrogen. Environmental Pollution 112: 41–51.

Smith, K. A., D. R. Jackson & P. J. A. Withers, 2001b. Nutrient losses by surface run–off following the application of organic manures to arable land. 2. Phosphorus. Environmental Pollution 112: 53–60.

Straile, D., D. M. Livingstone, G. A. Weyhenmeyer & D. G. George, 2003. The response of freshwater ecosystems to climate variability associated with the North Atlantic Oscillation. In Hurrell J. W, Y. Kushnir, G. Ottersen & M. Visbeck (eds), The North Atlantic Oscillation: Climatic Significance and Environmental Impact. American Geophysical Union, Washington DC, 263–279.

Tranvik, L. J. & M. Jansson, 2002. Climate change and terrestrial export of organic carbon. Nature 415: 861–862.

Yoo, J. C. & P. D'Odorico, 2002. Trends and fluctuations in the dates of ice break-up of lakes and rivers in Northern Europe: the effect of the North Atlantic Oscillation. Journal of Hydrology 268: 100–112.

Hydrobiologia (2007) 581:135–140
DOI 10.1007/s10750-006-0505-5

EUTROPHICATION IN LAKES

Stable isotopic composition of nitrate in Lake Taihu, China, and major inflow rivers

Amy Townsend-Small · Mark J. McCarthy · Jay A. Brandes · Longyuan Yang · Lu Zhang · Wayne S. Gardner

Abstract Identification of sources of nutrient pollution is a first step towards remediation of eutrophication in aquatic ecosystems. The stable isotope nitrogen-15 (^{15}N) is a natural indicator of nitrogen (N) source and biogeochemistry. We sampled Lake Taihu, a hyper-eutrophic lake in eastern China, and major inflow rivers during winter and spring of 2004 to determine concentration and $\delta^{15}N$ of nitrate (NO_3^-). Nitrate concentrations in rivers and the lake were higher, in most cases, in spring than in winter. $\delta^{15}N$ of NO_3^- was not correlated with NO_3^- concentration, indicating that concentrations alone are insufficient to describe N sources. Results show that riverine N inputs in winter are influenced by discharge of human sewage into rivers and the lake. In spring, however, wastewater inputs to the lake appear to be balanced by fertilizers, atmospheric, and/or N_2 fixation sources. Rain NO_3^- concentrations were seasonally high and isotopically enriched compared to potential sources, indicating that rain may be a significant or even dominant source of N to the lake during the rainy season. $\delta^{15}N$ values show that urbanized areas of the lake have more sewage-derived N than those areas dominated by agriculture, aquaculture, or industry. This observation has important implications for human health, since Lake Taihu is a source of drinking and irrigation water as well as fish for human consumption.

Keywords Nitrate · Nitrogen-15 · Lake Taihu · China · Stable isotopes · Eutrophication

Guest editors: B. Qin, Z. Liu & K. Havens
Eutrophication of shallow lakes with special reference to Lake Taihu, China

A. Townsend-Small (✉) · M. J. McCarthy · J. A. Brandes · W. S. Gardner
Marine Science Institute, The University of Texas at Austin, 750 Channel View Drive, Port Aransas, TX 78373, USA
e-mail: amy@utmsi.utexas.edu

L. Yang · L. Zhang
Nanjing Institute of Limnology and Geography, Chinese Academy of Sciences, 73 East Beijing Road, Nanjing 21008, China

Introduction

Eutrophication is a serious environmental problem in many aquatic ecosystems around the world, including Lake Taihu, located in eastern China in the Changjiang River delta. Excess nutrient inputs into Lake Taihu have resulted in seasonally high population densities of phytoplankton and macrophytes with detrimental effects on water quality and fish populations (Pu & Yan, 1998; Qin et al., 2007). High nutrient inputs can be concurrent with loading of other toxins, such as heavy metals and harmful bacteria that can be associated with wastewater (Wu et al., 1999).

Nitrogen (N) has two naturally occurring stable isotopes, ^{14}N and ^{15}N, which are present as 99.64 and 0.36% of all N atoms, respectively (Faure, 1986). Nitrogen isotope ratios usually are expressed in delta notation with respect to the N isotope ratio of air; therefore, samples containing an N isotope ratio close to that of air will have $\delta^{15}N$ values near zero (Mariotti, 1984). For this reason, inorganic fertilizers derived from atmospheric N_2 and biomass of N_2–fixing plants and microorganisms have a $\delta^{15}N$ value around zero (Peterson & Fry, 1987). Most inorganic fertilizers have a $\delta^{15}N$ of nitrate (NO_3^-) between –1 and 2‰ (Macko & Ostrom, 1994), whereas organic fertilizers such as manure may produce NO_3^- with higher $\delta^{15}N$ values (Jin et al., 2004). N cycling processes such as denitrification and ammonia volatilization discriminate against heavier ^{15}N, leaving remaining substrate enriched in ^{15}N relative to initial composition and the end product. Dissolved N compounds derived from human waste can be enriched in ^{15}N as a result of the sewage treatment process and the high trophic status of humans. Nitrate derived from wastewater has a $\delta^{15}N$ from 10 to 20‰, whereas natural soil nitrate $\delta^{15}N$ values usually range from 2 to 8‰ (McClelland & Valiela, 1998). It is important to note that $\delta^{15}N$ of NO_3^- is not an unambiguous N source indicator in lakes (Bedard-Haughn et al., 2003), especially since fertilizers and natural soil NO_3^- may become isotopically enriched in ground waters and rivers before they reach the lake (DeSimone & Howes, 1996; Kellman & Hillaire-Marcel, 1998). However, isotopic differences of N sources and changes in ^{15}N-content of those sources can be used to identify seasonal N contributions to aquatic ecosystems, especially when considered in the context of other nutrient monitoring and flux studies.

Site description

Lake Taihu (Fig. 1) is a large (2338 km^2), shallow (average depth 1.9 m) lake located in the Changjiang River delta in southeastern China (Qin et al., 2007). The lake provides drinking water, irrigation water, and fish for the approximately 36 million people residing in the catchment (Qin et al., 2007). Lake Taihu suffers from eutrophication due to point and non-point nutrient sources (Guo et al., 2004). Meiliang Bay and

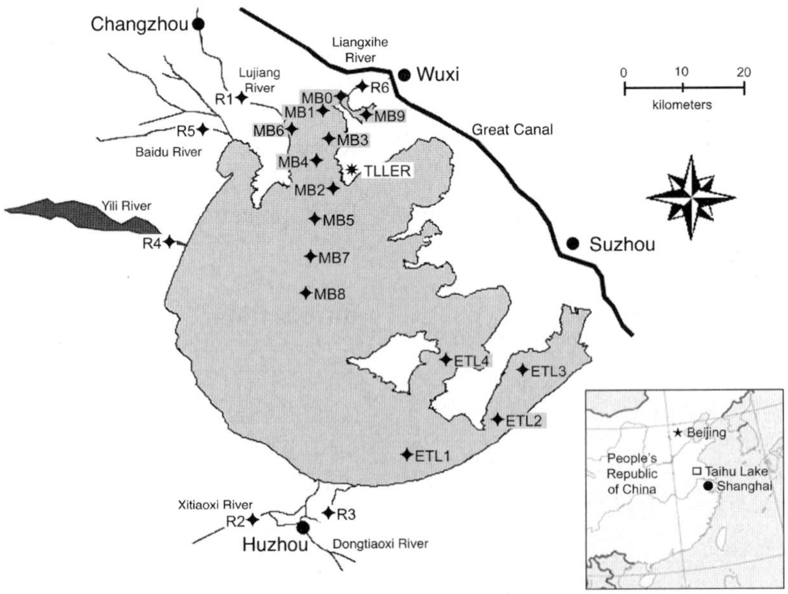

Fig. 1 Map of Lake Taihu and major inflow rivers (outflow rivers not shown) with inset showing location within China. Sampling locations within the lake and in major inflow rivers are shown. Also depicted are major cities in the area and TLLER, where rainwater was collected

Wuli Bay, located in the urbanized northern part of the lake, are hyper-eutrophic and characterized by annual cyanobacterial blooms (Chen et al., 2003). The northwest and southern portions of the lake are high-density urban areas. The landscape to the east and west of the lake is largely industrial and agricultural. Annual rainfall is about 1250 mm, 75% of which falls during the summer monsoon season from May to October (Pu & Yan, 1998). In 2004, rainfall rates were highest in the summer months, peaking at 181 mm for July, but only about 55% of total rain occurred during the monsoon season (L. Yang, personal communication). Annual river inflow is about 80×10^8 m^3 and the residence time of lake water is 5 months (Qin et al., 2007).

Materials and methods

Samples were taken from Lake Taihu and major inflow rivers in January and May of 2004 (Fig. 1). Most sampling sites were concentrated in Meiliang Bay (MB, Fig. 1). Some samples were taken from East Taihu (ETL), an area dominated by rooted macrophytes and aquaculture enclosures. Rainwater was sampled once during each field season at the Taihu Laboratory for Lake Ecosystem Research (TLLER; Fig. 1). Water samples were filtered in the field through 0.2 μm nylon syringe filters into glass vials and frozen. Samples not frozen immediately were filtered and acidified with 1 $\mu l\ ml^{-1}$ of H_2SO_4, then neutralized prior to isotopic analysis.

Nitrate concentrations were determined by cadmium reduction using a flow-injection autoanalyzer (Lachat QuikChem 8000) at The University of Texas Marine Science Institute (UTMSI). Water samples were analyzed in triplicate for $\delta^{15}N$ of NO_3^- using the "denitrifier" method (Sigman et al., 2001). Nitrous oxide produced via this method was analyzed on a Finnegan MAT DELTA^{plus} with a ThermoFinnegan Trace GC and a PreCon interface, also at UTMSI. The precision of this method for $\delta^{15}N$ analysis is 0.5‰ or greater.

Results

Dissolved NO_3^- plus nitrite (NO_2^-) concentrations and $\delta^{15}N$ of NO_3^- values determined at each site during each sampling season are shown in Table 1. $NO_2^- + NO_3^-$ concentrations were higher in spring than winter at all but four sites (Table 1). Concentrations were not correlated to $\delta^{15}N$ of NO_3^- ($r^2 = 0.0884$, $n = 40$).

$\delta^{15}N$ of NO_3^- also showed seasonal differences. In winter, $\delta^{15}N$ of NO_3^- values in rivers draining urban areas were generally higher than those from rivers draining agricultural or industrial areas. In January, the Liangxihe (R6), Lujiang (R1), and Dongtiaoxi Rivers (R3), which drain the urban watersheds of Wuxi, Changzhou, and Huzhou, respectively, all exhibited $\delta^{15}N$ values above 10‰ (Table 1). The other three rivers sampled in January had $\delta^{15}N$ values below 7.5‰. During spring, however, only the Dongtiaoxi River (R3) had a $\delta^{15}N$ of NO_3^- above 10‰.

In winter, lake sites located near river mouths reflected the $\delta^{15}N$ of river water. In general, sampling stations near river mouths had higher $\delta^{15}N$ of NO_3^- values, and isotopic signatures decreased with distance from rivers. Stations in central Lake Taihu had low $\delta^{15}N$ values, reflecting either an increased contribution of rain (4.4‰) in the central lake or dilution of enriched riverine N. In spring, NO_3^- in Meiliang Bay maintained high $\delta^{15}N$ levels. Stations near the middle of the lake had higher $\delta^{15}N$ in spring than winter.

Stations in East Taihu had consistent $\delta^{15}N$ values, between 6.3 and 7.6‰. Maximum $\delta^{15}N$ in East Taihu during January occurred near the Dongtiaoxi River mouth. In spring, two areas of East Taihu had NO_3^- levels too low to measure $\delta^{15}N$, but overall $\delta^{15}N$ of NO_3^- was comparable from season to season at this location.

Rainwater $\delta^{15}N$ of NO_3^- varied little with sampling season (Table 1), although rainwater NO_3^- concentrations were very high in the summer (145.8 μM, Table 1).

Discussion

In January, rivers on the northwest and southern parts of the lake that drain large urban areas had elevated $\delta^{15}N$ of NO_3^- values, indicating the predominance of sewage-derived N. In comparison, rivers draining agricultural or industrial

Table 1 Nitrate and nitrite concentrations and $\delta^{15}N$ of NO_3^- values for water samples collected at Lake Taihu during January and May 2004

	JANUARY 2004		MAY 2004	
	$NO_2^- + NO_3^-$ (μM)	$\delta^{15}N$ (‰)	$NO_2^- + NO_3^-$ (μM)	$\delta^{15}N$ (‰)
R1	55.1	19.4	93.0	5.7
R2	59.1	7.3	130.4	7.1
R3	74.7	11.1	212.9	13.8
R4	184.1	6.3	131.8	8.6
R5	28.1	5.9	56.7	6.7
R6	84.9	10.1	22.9	7.4
ETL1	49.8	7.6	59.0	3.9
ETL2	7.45	6.8	0.8	undetectable
ETL3	6.95	6.5	0.7	undetectable
ETL4	30.0	6.3	52.5	7.1
Rain water	17.6	4.4	145.8	6.0
MB0	127.0	9.7	96.1	7.1
MB1	84.1	4.0	169.7	8.3
MB2	115.7	6.1	218.0	27.3
MB3	109.7	3.0	228.5	10.0
MB4	111.8	16.2	193.0	11.7
MB5	110.9	6.5	201.0	6.7
MB6	118.2	9.7	89.0	1.6
MB7	85.4	6.0	186.8	8.4
MB8	84.7	3.9	193.5	9.9
MB9	68.9	5.4	329.0	5.9

areas, such as those on the southwest side of the lake, had lower isotopic signatures, presumably due the to dominance of fertilizer or plant-derived N. In spring, river $\delta^{15}N$ of NO_3^- was lower at all locations, with only one value above 10‰. This result may indicate the additional contribution of fertilizers or N_2 fixation to river NO_3^- in northwest or southern rivers during the growing season, especially when considered in the context of higher springtime riverine NO_3^- concentrations. Alternatively, the reduction in $\delta^{15}N$ and increase in NO_3^- concentration in rivers during spring may be due to an increased contribution of high concentration, low $\delta^{15}N$ NO_3^- rainwater to rivers, since there is an increase in rainfall at this time.

As expected in a monsoonal climate, river discharge in the Lake Taihu basin is higher in spring than winter (Qin et al., 2007). For January, concentration and discharge weighted average riverine $\delta^{15}N$ values were 93% successful at predicting the average $\delta^{15}N$ of NO_3^- in Meiliang Bay and the central lake (weighted riverine average $\delta^{15}N = 6.6‰$; lake average $\delta^{15}N = 7.1‰$), indicating that river discharge was the primary control on lake $\delta^{15}N$ in the winter months. Such integration of river values in lake water was not unexpected, since Taihu is a large lake with a long residence time. However, in May the same calculation was only 77% successful at predicting lake $\delta^{15}N$ values (weighted riverine average $\delta^{15}N = 7.5‰$; lake average $\delta^{15}N = 9.7‰$), indicating that biological processes that fractionate N isotopes are important controls on lake isotopic composition in spring and summer. Also, since there is a greater turnover of lake water in spring due to higher inflow, it is expected that springtime river and rainfall values become less effective predictors of lake conditions.

During both sampling seasons, some sites in Meiliang Bay had higher $\delta^{15}N$ values than river or rain sources, indicating fractionation during N cycling processes. Denitrification rates in Lake Taihu decrease with distance from inflow rivers (McCarthy et al., 2007), so high denitrification rates may explain locally high $\delta^{15}N$ values in Meiliang Bay. Samples from May were taken during initiation of the annual phytoplankton bloom, and isotopic discrimination during uptake may explain some of the high $\delta^{15}N$ values observed in spring in Meiliang Bay. However, primary production in Meiliang Bay appears to be limited by phosphorus, and inorganic N is plentiful (McCarthy et al., 2007), indicating minimal isotopic discrimination of the DIN pool during N uptake. On the other hand, $\delta^{15}N$ values from East Taihu were within the range of natural soil NO_3^- and rainwater. This area lacks major inflow rivers and is used for aquaculture (Qin et al., 2007). More sampling of end-members, especially aquaculture sources, is needed to determine the major N sources to this part of the lake, but $\delta^{15}N$ values from this study indicate that sewage is minimally present.

Rain could be a major source of N to Lake Taihu and the surrounding watershed, considering the high NO_3^- concentrations observed in spring. Atmospheric NO_3^- derived from volatilization or nitrification of fertilizers usually is depleted in ^{15}N, with reported values as low as –15‰ (Wada & Hattori, 1991, Macko & Ostrom, 1994). Lightning is another source of isotopically light fixed atmospheric N, with $\delta^{15}N$ of NO_3^- values reported from 0 to –15‰ (Macko & Ostrom, 1994). Seasonal variation in $\delta^{15}N$ of NO_3^- in rain has been attributed to changes in atmospheric NO_3^- sources, with the dominant sources in winter and summer being fossil fuel combustion ($\delta^{15}N$ around 0‰) and soil ($\delta^{15}N$ around –5‰) NO_3^-, respectively (Freyer, 1991; Hastings et al., 2003). The $\delta^{15}N$ of NO_3 in rain observed at Lake Taihu (4.4 to 6‰) is more enriched than these expected sources, implying either a greater influence of fossil fuel combustion than previously observed, or a progressive isotopic enrichment of atmospheric fixed N as rain clouds move east over the European and Asian land masses. Enriched rainwater N also complicates the use of stable N isotopes to identify N sources, since the observed rainwater $\delta^{15}N$ values fall within the range of natural soil N. Factors influencing the N concentration and isotopic composition of rain in the Lake Taihu area warrant further study, especially since nitrogen oxides contribute to the formation of acid rain and since it represents a non-point source of pollution that is difficult to attenuate.

This study indicates that seasonal fertilizer application in the Lake Taihu watershed and rain are the major sources of NO_3^- in the summer months. In the winter months, sewage is a major source of NO_3^- to urbanized areas of the lake. This finding raises important questions about water quality, since most drinking water in China is derived from rivers and lakes. Wastewater discharge into these water bodies has been linked to illnesses such as typhoid and other bacterial or parasitic infections in China (Wu et al., 1999). Future studies should include N loading estimates, quantification of sewage and fertilizer discharge, and greater spatial and temporal coverage of $\delta^{15}N$ of NO_3^- measurements, especially rainfall, in order to completely characterize the NO_3^- budget for this system. An annual record of the isotopic composition of NO_3^- in lake water combined with a complete water and N budget would be a useful next step in the characterization of N cycling and sources in this ecosystem.

Acknowledgements This project was supported by National Natural Science Foundation of China (No. 40371104) and National Science Foundation of the United States. We thank the many faculty members, staff, and students at TLLER who assisted with project coordination and sampling. Patricia Garlough of the University of Texas Marine Science Institute assisted with stable isotopic analyses. The manuscript benefited from the comments of three anonymous reviewers.

References

Bedard-Haughn, A., J. W. van Groenigen & C. van Kessel, 2003. Tracing ^{15}N through landscapes: potential uses and precautions. Journal of Hydrology 272: 175–190.

Chen, Y., B. Qin, K. Teubner & M. T. Dokulil, 2003. Long-term dynamics of phytoplankton assemblages: *Microcystis*-domination in Lake Taihu, a large shallow lake in China. Journal of Plankton Research 25: 445–453.

DeSimone, L. A. & B. L. Howes, 1996. Denitrification and nitrogen transport in a coastal aquifer receiving wastewater discharge. Environmental Science and Technology 30: 1152–1162.

Faure, G., 1986. Principles of Isotope Geology. Wiley & Sons Publishers, New York.

Freyer, H. D, 1991. Seasonal variation of $^{15}N/^{14}N$ ratios in atmospheric nitrate species. Tellus 43B: 30–44.

Guo, H. Y., X. R. Wang & J. G. Zhu, 2004. Quantification and index of non–point source pollution in Lake Taihu region with GIS. Environmental Geochemistry and Health 26: 147–156.

Hastings, M. G., D. M. Sigman & F. Lipschultz, 2003. Isotopic evidence for source changes of nitrate in rain at Bermuda. Journal of Geophysical Research-Atmospheres 108: doi: 10.1029/2003JD003789.

Jin, Z., C. Yingxu, F. Wang & N. Ogura, 2004. Detection of nitrate sources in urban groundwater by isotopic and chemical indicators, Hangzhou City, China. Environmental Geology 45: 1017–1024.

Kellman, L. & C. Hillaire-Marcel, 1998. Nitrate cycling in streams: using natural abundance of $NO_3^-\delta^{15}N$ to measure in-situ denitrification. Biogeochemistry 43: 273–292.

Macko, S. A. & N. E. Ostrom, 1994. Pollution studies using stable isotopes. In Lajtha, K. & R. H. Michener (eds), Stable isotopes in ecology and environmental science. Blackwell Scientific Publications, Oxford, 45–62.

Mariotti, A., 1984. Natural ^{15}N abundance measurements and atmospheric nitrogen standard calibration. Nature 311: 251–252.

McCarthy, M. J., P. J. Lavrentyev, L. Yang, Y. Chen, L. Zhang, B. Qin & W. S. Gardner, 2007. Nitrogen dynamics and microbial food web structure during a summer cyanobacterial bloom in a subtropical, shallow, well-mixed, eutrophic lake (Lake Taihu, China). Hydrobiologia 581: 195–207.

McClelland, J. W. & I. Valiela, 1998. Linking nitrogen in estuarine producers to land-derived sources. Limnology and Oceanography 43: 577–585.

Peterson, B. J. & B. Fry, 1987. Stable isotopes in ecosystem studies. Annual Review of Ecology and Systematics 18: 293–320.

Pu, P. & J. Yan, 1998. Lake Taihu - a large shallow lake in the East China Plain. Journal of Lake Sciences 10 (supplement): 1–12 (in Chinese).

Qin, B., P. Xu, Q. Wu & L. Luo, 2007. Environmental issues of Lake Taihu, China. Hydrobiologia 581: 3–14.

Sigman, D. M., K. L. Casciotti, M. Andreani, C. Barford, M. Galanter & J. K. Böhlke, 2001. A bacterial method for the nitrogen isotopic analysis of nitrate in seawater and freshwater. Analytical Chemistry 73: 4145–4153.

Wada, E. & A. Hattori, 1991. Nitrogen in the Sea: Forms, Abundances, and Rate Processes. CRC Press, Boca Raton.

Wu, C., C. Maurer, Y. Wang, S. Xue & D. L. Davis, 1999. Water pollution and human health in China. Environmental Health Perspectives 107: 251–256.

Hydrobiologia (2007) 581:141–150
DOI 10.1007/s10750-006-0523-3

EUTROPHICATION IN LAKES

Distribution and chemical fractionation of heavy metals in recent sediments from Lake Taihu, China

Ji Shen · Enfeng Liu · Yuxin Zhu · Shouyun Hu · Wenchuan Qu

Abstract Lake Taihu is one of the most contaminated lakes in China. Surface sediment data show that the northern area of the Lake has the worst heavy metals pollution, and high heavy metal concentrations were attributed to discharge of untreated and partially treated industrial waste water from cities to the north of the lake. To study geochemical features and pollution history of heavy metals, total content and chemical fractionations of Cu, Fe, Mn, Ni, Pb, and Zn were analyzed for core sediments from western Lake Taihu using the speciation extraction procedure, proposed by the Commission of the European Communities Bureau of Reference (BCR), together with grain size and organic carbon measurements. Results show that sediments are composed of organic-poor clayey-fine silts for Cores MS and DLS, and have similar geochemical features shown by heavy metals. Cu, Fe, Ni, and Zn mainly are associated with the residue fraction, Mn is concentrated in the exchangeable-carbonate and residue fractions, and Pb is concentrated in the Fe–Mn oxide fraction and organic-sulfide fraction. The fractions of Ni, Pb, and Zn bound to Fe–Mn oxide show significant correlations with Mn from the Fe–Mn oxide fraction, and the organic-sulfide fractions of Cu, Mn, Ni, Pb, and Zn are correlated with TOC. The increase of Cu, Mn, Ni, Pb and Zn content and percentage of extractable fractions in the upper layers of the sediments are correlated with anthropogenic input of heavy metals due to rapid industrial development. This coincides with rapid economic development in the Taihu basin since late 1970s. Heavy metals in the surface sediments have certain potential biological toxicity as shown by the higher SEM/AVS ratio.

Keywords Heavy metals · Sediments · Distribution · Fractionation · Lake Taihu

Guest editors: B. Qin, Z. Liu & K. Havens
Eutrophication of shallow lakes with special reference to Lake Taihu, China

J. Shen (✉) · E. Liu · Y. Zhu · S. Hu · W. Qu
Nanjing Institute of Geography and Limnology, Chinese Academy of Sciences, Nanjing 210008, China
e-mail: Jishen@niglas.ac.cn

Introduction

Lake Taihu, the third largest fresh water lake in China, is situated in the center of the Changjiang River Delta (Qin et al., 2007) (Fig. 1). With intensive development of agriculture and industry in this densely populated area since the 1950s, especially since 1978, untreated or partially treated wastewater from industry, agriculture and sewage increasingly has been discharged into Lake Taihu, which induced degeneration of lake ecology and water quality. The quality of Lake Taihu

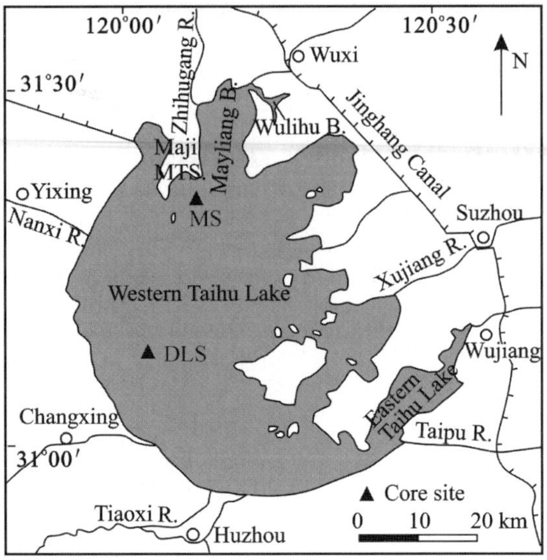

Fig. 1 Location of Lake Taihu and core site

water degraded 2–3 levels from the 1970s to 1990s (Lu, 1998), which caused about 1 billion RMB economic loss per year during the late 1980s, accounting for 3.12% of the local GNP (Ge, 1992).

As one of many environmental problems in Lake Taihu, heavy metal pollution has drawn great attention in recent years. A few studies have been conducted on heavy metal pollution of surface sediments (Dai & Sun, 2001; Qu et al., 2001; Yuan et al., 2002a), which mainly address heavy metal total concentration. However, work on chemical fractionation of heavy metals and pollution history of the sediments are rare (Wang et al., 2002; Rose et al., 2004). There is a lack of systemic investigation on distribution and chemical forms of heavy metals in sediment cores, which are an important record of historical pollution, biogeochemical recycling and toxicity for heavy metals (Luoma, 1983).

In the present study, two sediment cores in western Lake Taihu were obtained. Total concentration of heavy metals and their chemical forms deduced from the BCR method (proposed by the Community of the European Bureau of Reference) (Quevauviller et al., 1997) in the sediment cores are presented. The aim of this study is to understand geochemical characteristics of the heavy metals and pollution features in the sediments and their correlations with economic development and human activities in the Lake Taihu catchment.

Material and methods

Sample collection

According to sedimentary features and the degree of pollution in the sediments (Sun & Huang, 1993; Qu et al., 2001), two sediment cores (MS and DLS) were obtained in northern and southern Lake Taihu, respectively, in October 2002 using a gravity corer (Fig. 1). The sediment exhibited no bioturbation and can be used to reconstruct environmental evolution. Sediments were sliced into sections at 0.5 cm intervals. Samples were freeze dried using a LABCONCO freeze drier. Physical and chemical data from the upper 13 and 7 cm sediments for Cores MS and DLS, respectively, were used.

Analysis methods

TOC concentration in samples was determined using a CNHO-S elemental analyzer (Leeman Company). Particle size was measured with a Mastersizer–2000 laser diffraction particle size analyzer (Malvern Instruments Ltd.). Analysis errors for experiments were within ±5%.

For heavy metal analysis, the HCl–HNO_3–HF digestion method was applied. About 125 mg of sediment, ground in an agate mortar to pass 100-μm sieve, was digested with 6.0 ml HNO_3, 0.5 ml HCl, and 3.0 ml HF at 180 ± 5°C for 15 min in a microwave oven (Berghof MWS–3 Digester). Residue was dissolved with 0.5 ml $HClO_4$ by braising in a heating block at <200°C and diluted to 25 ml with double-distilled deionized water. The solution was analyzed for heavy metals and major elements using inductively coupled plasma-atomic emission spectrometry (ICP-AES).

Chemical fractionations of Cu, Fe, Mn, Ni, Pb and Zn were determined using the BCR sequential extraction procedure (Quevauviller et al., 1997; Tokalioğlu et al., 2000) (Fig. 2). Only the three extractable fractions (exchangeable-carbonate, Fe–Mn oxide, and organic-sulfide fraction) were analyzed. Heavy metals concentrations in the extracted solution were analyzed by ICP-AES.

Sediments were dated using the ^{137}Cs methods described by Rosales-Hoza et al. (2003).

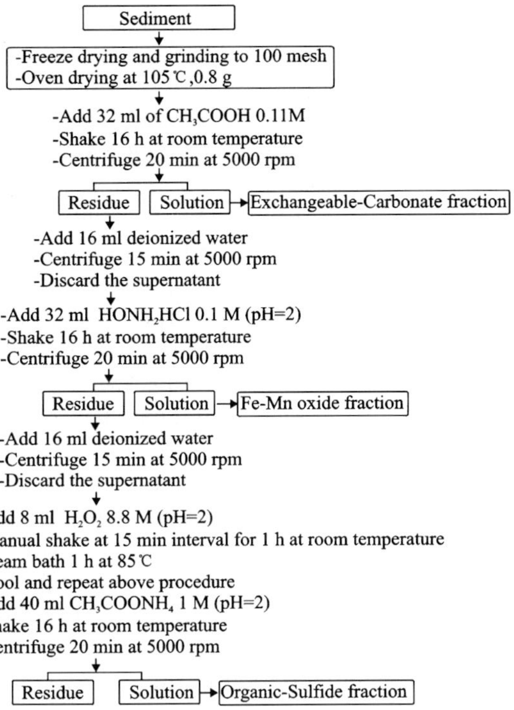

Fig. 2 Heavy metals sequence extraction procedure of BCR method

Accuracy of heavy metals analysis method

Four parallel analyses of standard sediment reference material (GSD–9 from Institute of Geophysical and Geochemical Exploration, Chinese Academy of Geological Sciences) were conducted to verify the accuracy of the heavy metal digestion method. The maximum error of all results was <3% of the standard value, showing that the HCl–HNO_3–HF digestion method and ICP-AES analysis have reasonable precision and reproducibility.

Three lacustrine samples from Lake Taihu were measured in parallel analyses to verify the accuracy of the BCR procedure due to the lack of suitable standard reference material. Blanks were also used for background correction. Results are shown in Table 1. Maximum errors of the results are within 10%, except 23.0 and 18.6% for the exchangeable-carbonate and organic-sulfide fractions of Pb in samples MS–5 and TH-2, respectively, and 24.4% for the Fe–Mn oxide fraction of Zn in sample MS-5. Errors for Pb and Zn may be related to contamination of certain procedure during the sequential extraction process, such as losing of the liquid during shaking etc (Tokalioğlu et al., 2000).

Table 1 Error analysis of BCR sequential extraction method for sedimentary samples of Lake Taihu

Elements		MS-5				TH-1				TH-2			
		AC	S.D.	CV	ME	AC	S.D.	CV	ME	AC	S.D.	CV	ME
Cu	F1	3.05	0.07	2.19	2.52	2.01	0.05	2.35	2.71	3.54	0.03	0.75	0.85
	F2	3.62	0.01	0.16	0.18	2.97	0.05	1.52	1.57	3.21	0.14	4.21	4.36
	F3	5.28	0.16	3.08	3.35	5.05	0.12	2.28	2.31	10.0	0.30	3.00	3.35
Fe	F1	250	10.4	4.17	4.79	595	14.2	2.38	2.45	1590	17.4	1.10	1.17
	F2	5070	41.2	0.81	0.90	5570	66.7	1.20	1.23	8390	246	2.93	3.38
	F3	989	53.4	5.40	5.81	1310	46.7	3.55	3.77	2190	175	7.99	8.52
Mn	F1	302	4.48	1.48	1.71	192	2.78	1.45	1.52	823	4.34	0.53	0.59
	F2	92.1	3.10	3.37	3.83	49.3	0.50	1.01	1.11	208	4.01	1.93	1.97
	F3	36.2	1.35	3.73	4.09	33.1	1.00	3.03	3.37	64.8	4.74	7.31	7.32
Ni	F1	3.39	0.06	1.62	1.67	2.10	0.07	3.45	4.00	2.96	0.01	0.39	0.45
	F2	4.00	0.06	1.38	1.58	2.86	0.06	1.92	2.21	3.49	0.19	5.48	6.11
	F3	4.59	0.11	2.30	2.40	3.46	0.04	1.17	1.35	4.12	0.06	1.38	1.54
Pb	F1	1.59	0.34	21.3	23.0	0.82	0.07	8.15	9.89	0.42	0.00	0.81	0.55
	F2	8.97	0.51	5.72	6.39	3.74	0.08	2.09	2.41	2.96	0.09	3.04	3.04
	F3	14.19	0.51	3.62	4.13	9.08	0.24	2.59	2.61	11.9	1.95	16.41	18.6
Zn	F1	12.3	0.56	4.56	5.19	3.27	0.07	2.04	2.21	4.12	0.15	3.52	3.56
	F2	16.8	3.59	21.4	24.4	4.48	0.12	2.60	2.83	5.49	0.20	3.70	4.25
	F3	9.10	0.48	5.30	5.86	3.68	0.33	8.89	9.96	6.40	0.21	3.21	3.28

AC, average content (mg kg^{-1}); S.D, standard deviation; CV, coefficient of variation; ME, maximum error

F1, exchangeable-carbonate fraction; F2, Fe–Mn oxide fraction; F3, organic-sulfide fraction

Due to good precision and reproducibility for the digestion and BCR methods, the residue fraction concentration of heavy metals can be represented by the difference of total concentration and extractable fractions of each heavy metal.

Results

Dating results

The chronological framework of Cores MS and DLS were determined by ^{137}Cs dating. Average sedimentation rates were 1.7 mm $year^{-1}$ and 0.9 mm $year^{-1}$, respectively, for the two cores (Liu et al., 2005). These results are similar to those from other researchers (Sun & Huang, 1993). The ages for Core MS and DLS are shown in Fig. 3.

Heavy metals, grain size and TOC in Cores MS and DLS

Grain size and TOC in the cores are shown in Fig. 3. Sediments were dominated by silt (4–64 μm) and clay (<4 μm). The grain size becomes finer gradually from 14 to 4 cm depth and coarser from 4 cm to the surface in Core MS; and it becomes finer gradually from 7 to 2.5 cm and coarser from 2.5 cm to the surface in Core DLS.

TOC content was <2% for most of the surface sediments (Sun & Huang, 1993). TOC was around 0.62–1.35% in Cores MS and DLS and increases near the surface (Fig. 3).

Concentrations of the major elements (Al, Mg, K and etc) increased from 14 to 4 cm and decreased from 4 cm to the surface in Core MS, having consistent variation with the finer fraction (clay and fine silt). For Core DLS, major elements increased from 7 to 2.5 cm and decreased from 2.5 cm to the surface, also having consistent variation with the finer fraction.

Heavy metals, such as Mn, Ni, Pb and Zn, also had similar variation patterns as major elements in the 14–4 cm interval for Core MS, and 7–2.5 cm for Core DLS, except for higher Mn and Fe in the middle part of the core. In contrast to major elements and the finer fraction, heavy metals generally increased in the top 4–0 cm and 2.5–0 cm for the two sediment cores.

Chemical forms of heavy metals in Cores MS and DLS

Chemical fractionation compositions of Cu, Fe, Mn, Ni, Pb and Zn determined by BCR are illustrated in Fig. 4. Cu, Fe, Ni and Zn were

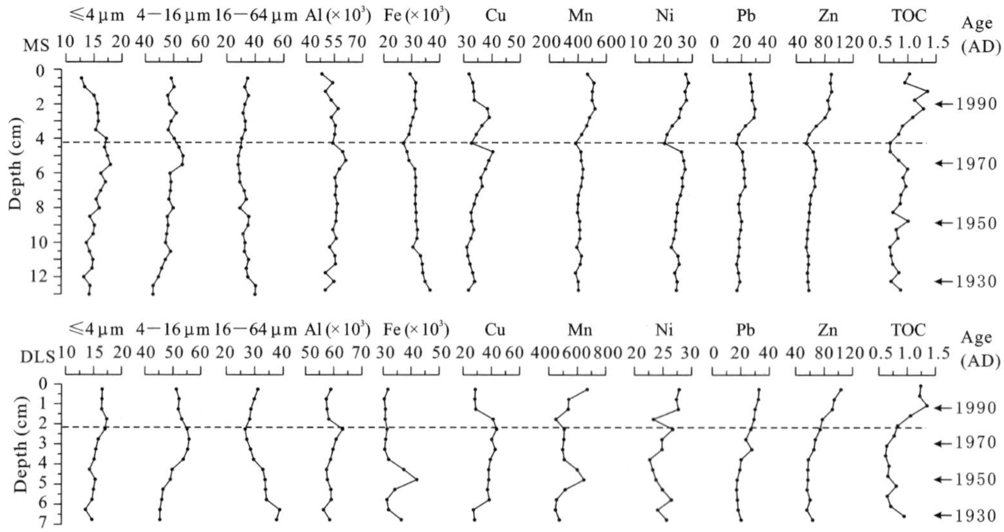

Fig. 3 The variation of grain size (%), heavy metals (mg kg^{-1}) and TOC (%) in Cores MS and DLS

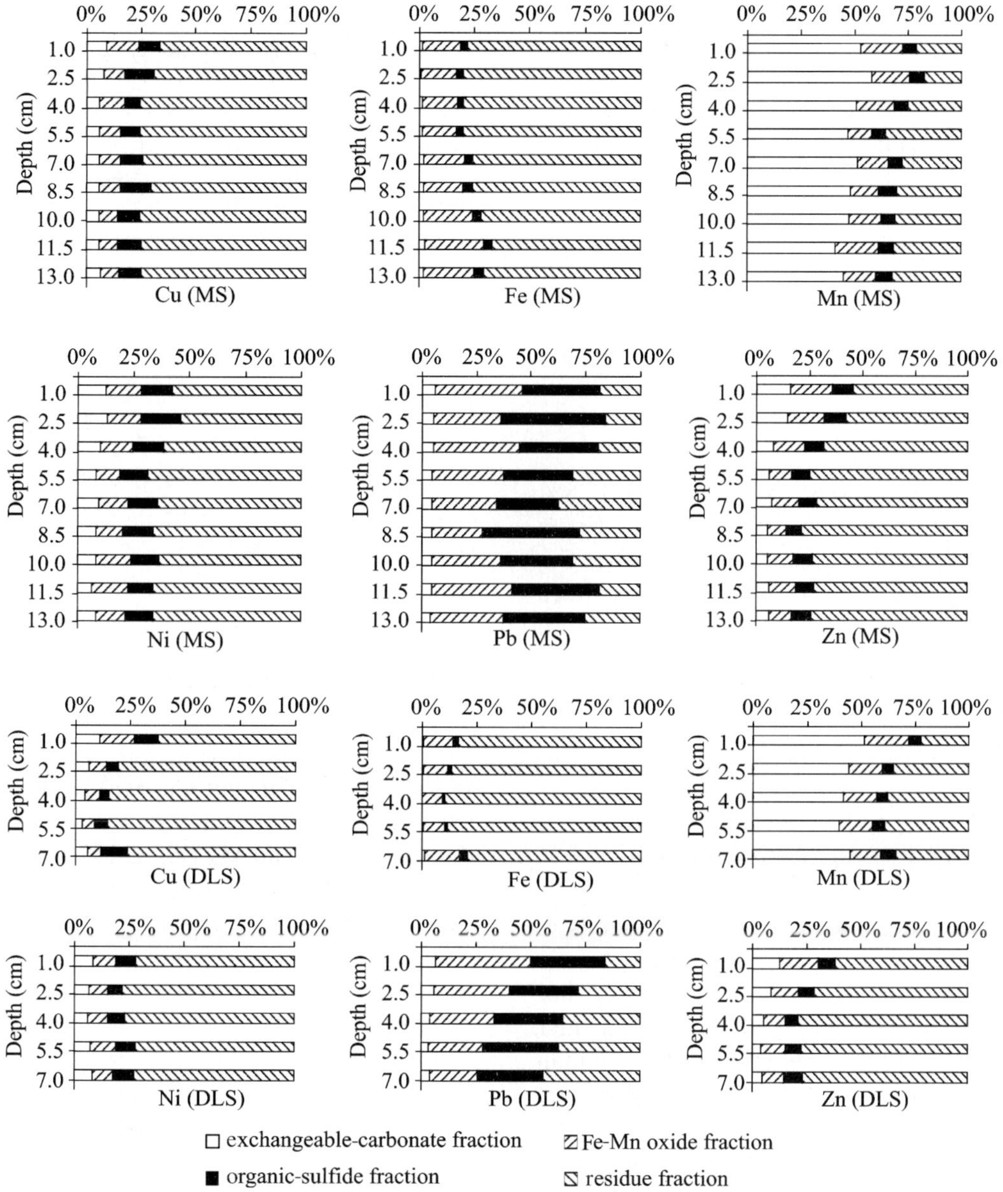

Fig. 4 The chemical fractionation percentages of Cu, Fe, Mn, Ni, Pb and Zn in the sediments of Cores MS and DLS (with 1.5 cm interval)

concentrated in the residue fraction, and average percentages were 72.9%, 75.6%, 62.9% and 69.7% in Core MS, and 77.8%, 85.4%, 72.9% and 73.3% in Core DLS, respectively. Mn was present in the exchangeable-carbonate fraction, and its percentages were 48.9 and 44.3% in Cores MS and DLS, respectively, whereas Pb was present in both the Fe–Mn oxide fraction and organic-sulfide fraction.

Percentages of extractable fractions of Cu, Mn, Ni, Pb and Zn, including exchangeable-carbonate, Fe–Mn oxide and organic-sulfide fractions, had a consistent variation in the vertical sequence for both cores, which were stable at 14–4 and 7–2.5 cm, but increased from 4 to 0 cm and from 2.5 to 0 cm for Cores MS and DLS, respectively (Fig. 4). The percentages of extractable fractions of Cu, Mn, Ni, Pb and Zn in the top 4 and 2.5 cm

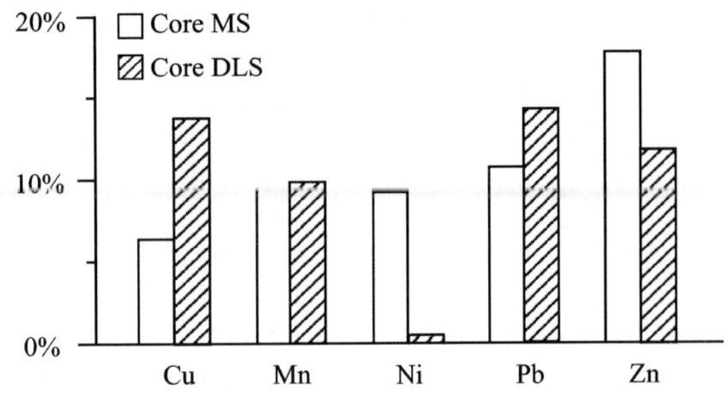

Fig. 5 The increased percentages of Cu, Mn, Ni, Pb and Zn in extractable fractions in the top 4 and 2.5 cm of Cores MS and DLS compared with that in the deeper layers

of Cores MS and DLS increased 0.5–17.8% comparing with that in the deeper layers (Fig. 5). The average percentage of Pb, Ni, Cu, Zn and Mn in the organic-sulfide fraction were 35.8%, 13.5%, 10.3%, 9.3% and 6.8% in Core MS, and 32.3%, 8.7%, 7.9%, 7.6% and 5.9% in Core DLS.

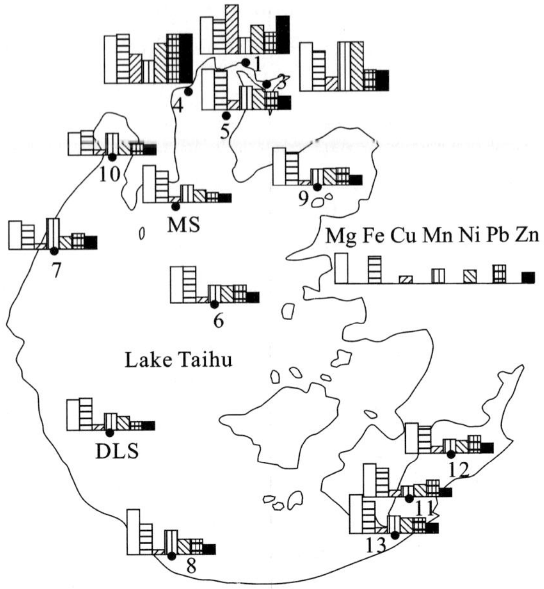

Fig. 6 The relative concentration of heavy metals in each site compared with the average of the 14 sites of Lake Taihu surface sediments (the data of site 1–13 from Qu et al., 2001)

Discussion

Pollution features of the surface sediment

There are two main sources for heavy metals in sediments: natural sources and anthropogenic sources. Heavy metals content commonly has good correlation with major elements and finer fractions in unpolluted sediments, but human pollution can alter this relationship (Abraham, 1998; Covelli & Fontolan, 1997). Therefore, variation patterns of major elements, heavy metals and their correlation with grain size composition can be used to evaluate heavy metals pollution (Rubio et al., 2000). Al, Fe, Mg and K mainly come from natural versus anthropogenic sources and can be used as reference elements for heavy metal pollution evaluation (Covelli & Fontolan, 1997; Soto-Jiménez & Páez-Osuna, 2001).

Heavy metals in surface sediments of Core MS and DLS and the other 12 sites from Qu et al. (2001) were analyzed for spatial distribution of heavy metals in Lake Taihu. Contents of Fe, Mg and other major elements in the sediments of the 14 sites are similar, but Cu, Mn, Ni, Pb and Zn in the sediments of sites 1–5 have higher concentrations versus Fe and Mg (Al and K not determined at sites 1–13) than other sites (Fig. 6). Concentrations of the three extractable fractions of Cu, Ni and Zn in surface sediments also were higher in the northern lake compared with other areas (Wang et al., 2002). These indicate that sediments in northern Lake Taihu, such as Meiliang Bay and Wuli Bay, are polluted with heavy metals. Pollution in the southern lake is much lighter (Qu et al., 2001; Wang et al., 2002). Larger amounts of waste water with higher concentrations of heavy metals from industries in Changzhou and Wuxi discharged into Wuli and Meiliang Bays via the Zhihugang and Liangxihe is the source of heavy metal pollution in the northern Lake Taihu (Huang et al., 2001; Qu et al., 2001).

Geochemical features of the core sediment

For total heavy metal concentrations, Cu, Mn, Ni, Pb and Zn have similar variation with major elements (Al, Mg and K) and finer fractions before the late 1970s. However, there is a sharp

Hydrobiologia (2007) 581:151–159
DOI 10.1007/s10750-006-0495-3

EUTROPHICATION IN LAKES

Vegetation and soil properties in restored wetlands near Lake Taihu, China

Jinwei Lu · Hongjun Wang · Weidong Wang · Chengqing Yin

Abstract Riparian wetlands are important components of the lake ecosystem, and they play essential roles in maintaining system health. Remediation of degraded lakeshore wetlands is an essential component of lake restoration. A study was conducted to investigate the restoration of lakeshore wetlands, which were converted to rice fields and then abandoned for 2, 5, 10 and 15 years, near Lake Taihu. Soil samples (0–20 cm and 20–40 cm) were taken and plant species were investigated. The carbon content in the soil had increased significantly, rising from 0.71% to 1.85% between 2 and 15 years. Organic matter accumulation improved soil texture, and water stable aggregate content (>0.25 mm) and soil porosity increased. Total nitrogen in the soil increased from 0.06% to 0.13%, and total Kjeldahl nitrogen increased from 124.4 mg kg^{-1} to 351.5 mg kg^{-1}. Total phosphorus in the soil increased from 0.045% to 0.071%, and the Olsen-*P* value increased from 5.13 mg kg^{-1} to 16.0 mg kg^{-1}. Results showed that phosphorous did not increase as much as nitrogen. In the vegetation restoration process, plant species composition moved towards a natural wetland community, and spatial heterogeneity and landscape diversity increased. The richness of plant biodiversity increased rapidly in the first 2 years, then more slowly in later restoration stages. The wetlands recovery process may be complicated by interactions of biota and soil and hydrological conditions.

Keywords Lakeshore · Wetlands · Restoration · Dynamics · Process · Vegetation · Soil nutrients · Lake Taihu

Guest editors: B. Qin, Z. Liu & K. Havens
Eutrophication of shallow lakes with special reference to Lake Taihu, China

J. Lu · H. Wang · W. Wang · C. Yin (✉)
SKLEAC, the Research Center for Eco-Environmental Sciences, Chinese Academy of Sciences, Beijing 100085, China
e-mail: cqyin@263.net

J. Lu · H. Wang
Graduate School of the Chinese Academy Sciences, Beijing 100039, China

C. Yin
Northeast Institute of Geography and Agricultural Ecology, CAS, Jinlin Province, Changchun 130000, China

Introduction

Eutrophication in lakes and reservoirs is a worldwide environmental problem (Vollenweider, 1992), particularly since the 1980s in China due to rapid economic development (Jin et al., 1990). Riparian wetlands are important components of the lake ecosystem, and they play essential roles for maintaining system health (Naiman & Décamps, 1997). These wetlands are effective buffers, retaining water and nutrients, improving water quality, and providing diverse and dynamic habitats for wildlife (Naiman et al., 1994; Yin & Lan, 1995; Tabacchi et al., 1998; Coveney et al., 2002). Therefore, remediation and restoration of lakeshore wetlands is an essential component of lake restoration.

Due to historical lack of recognition of the ecological value of riparian wetlands, many wetlands were destroyed in the last 50 years. Most of the destruction is a result of conversion to agriculture and other economic uses (National Bureau of Statistics of China, 2001; Chen & Chen, 2002). Approximately 160 km^2 of riparian areas are cultivated along Lake Taihu (Hu & Jiang, 2004; Xiong, 2004). Losing the buffering capacity of riparian zones enhanced eutrophication, and, in recent years, wetland restoration has been implemented extensively in China (Chen, 2001). Hydrology, soil, and vegetation are interconnected components of wetlands and undergo dynamic changes during restoration. This paper describes soil and vegetation changes in the riparian wetlands and discusses the effectiveness and feasibility of natural restoration. This information will be useful for future wetland restoration efforts.

Study area, materials and methods

Description of study area

The study area is located at the lakeshore of northeast Lake Taihu. The land in this region is flat with adjacent undulating steep hills. Natural riparian wetlands were changed into paddy fields and other agricultural uses from the 1960's to 1980's. After 1988, with a local policy for protecting the lake and an increase in labor costs, some paddy fields were abandoned. Vegetation recolonized these paddy fields, and the weir was destroyed, allowing water to move freely into the wetlands. In investigation years 2003 and 2004, the lake water table was low, so the wetlands were not inundated.

Field investigation and sampling

Experiment design

Five sections of paddy fields, with an area ~0.15 ha and an age series of 2-, 5-, 10-, and 15-years since abandonment, and a reference wetland were selected in the riparian areas of Lake Taihu in the Dafu Town of Wuxi (Fig. 1). Since it was difficult to find an undisturbed natural wetland as a reference site, an uncultivated and less disturbed 30-year old riparian wetland was chosen. All of the abandoned lands in the study areas were flat and subject to similar hydrological conditions. The age of abandoned sites were

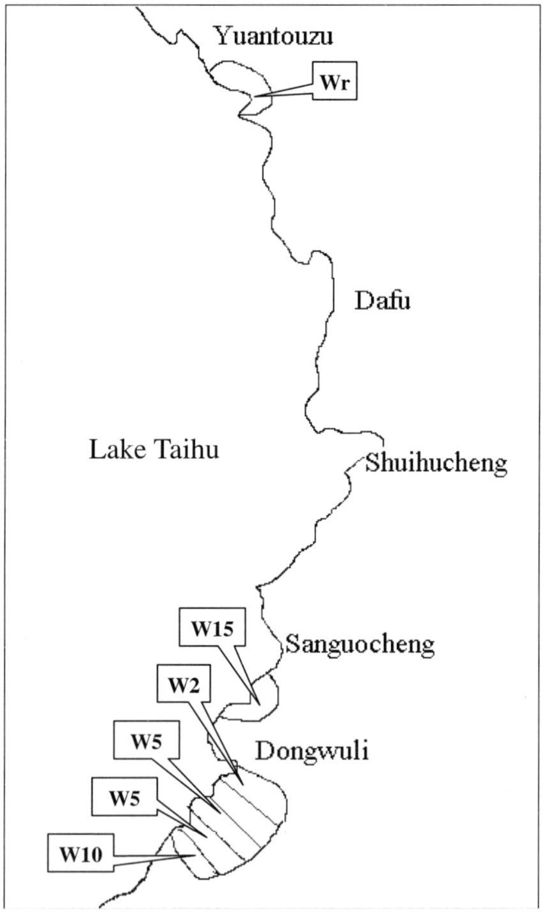

Fig. 1 Map of the distribution of restored wetlands in the studied area in Lake Taihu. Numbers refer to the numbers of years after they were abandoned as rice fields. W indicates wetland, and Wr indicates reference wetland

determined from field surveys and interviews with local landowners. Each wetland was numbered according to the time since abandonment as paddy fields (e.g. W5 indicates a wetland in its fifth year or growing season since abandonment as a paddy field). Wr means reference wetland.

Investigation and measurement of restored vegetation

Vegetation plays a critical role in wetland restoration and may indicate wetland value and function. In 2003 and 2004, four vegetation investigations were conducted in five different wetlands. Belts were identified according to vegetation characteristics. Six 1 × 1 m quadrats were chosen randomly in each wetland belt. Species, coverage, diversity and biomass were recorded in each quadrat every three months from March to October. Six 0.5 × 0.5 m quadrats also were chosen for ratio analysis of biomass between upland and low land parts in the same season. Height and density of the dominant vegetation were measured in three of these quadrats selected at random (Song, 2001). Data on hydrology, vegetation, soil and other conditions in the study area were collected as reference.

Soil analysis

Soil was collected from five wetlands in different stages of restoration. Three to five sample sites were selected in each vegetation belt. The soil profile was divided into two horizontal layers: 0–20 cm and 20–40 cm, and five samples were collected at each sampling site. Soil samples were air-dried and hand-sieved through a 20-mesh sieve screen to remove roots and debris. Sieved soil samples were sealed in a plastic bag for pH, biologically available N, and Olsen-P analysis. Remaining soil was crushed and sieved through a 100-mesh sieve screen, then stored for soil organic matter (SOM), total N (TN), and total P (TP) analysis. Soil bulk density and porosity were evaluated with a cutting ring. All soil samples were analyzed with standard methods for soil analysis (Soil Science Society of America and American Society of Agronomy, 1996).

Statistical analysis

Locating sites abandoned for different periods of time but similar characteristics is difficult. Therefore, data interpretation techniques for the four durations of abandonment were used. While the experiments were pseudo-replicates, the distance between sampling sites was sufficient to ensure adequate variation within each site.

Similarities among communities at each site and the reference natural wetland were determined using the Sorensen similarity coefficient (SC) (Song, 2001) calculated as $SC = 2\omega/(a + b)$, where ω is the number of species found in both the natural and a given abandoned site; a is the total number of species found in the reference natural wetland, and b is the total number of species found in a given abandoned site. SAS software was used to analyze experimental data, and all significant differences are at a probability of 0.05 or less, unless otherwise noted.

Results

C, N, and P changes in restored wetlands soil

An increasing trend was observed for TOC, TN, KN, TP, and Olsen-P content along the succession gradient defined by 2-, 5-, 10-, and 15-year-old paddy field abandonment sites and the reference wetland. TOC ranged from 0.73% to 2.76% in the 0–20 cm layer and 0.31–0.46% in the 20–40 cm layer (Fig. 2). There was lower

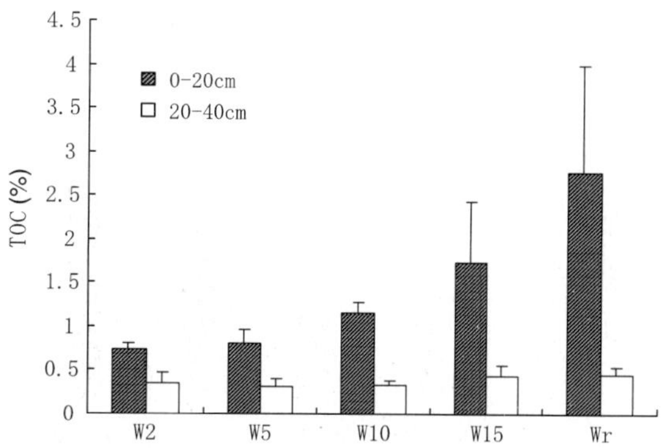

Fig. 2 Changes of soil Organic C in different restored wetlands

TOC content in the 2- to 5-year periods and higher content in the 10- to 15-year period and lay obvious accumulation of TOC at 0–20 cm layer. One-way ANOVA showed a significant difference in the 0–20 cm layer, but not in the 20–40 cm layer.

TN and KN ranged from 0.039% to 0.098% and 124.35 mg kg^{-1} to 351.48 mg kg^{-1}, respectively, in the 0–20 cm layer and 0.020–0.044% and 52.5–83.1 mg kg^{-1}, respectively, in the 20–40 cm layer (Fig. 3). There lay while an obvious decrease trend of TN content in the earlier restoration periods and an increase trend in the later restoration period. One-way ANOVA showed significant difference in KN in the 0–20 cm layer, and no significance difference in the 20–40 cm layer or TN in either layer.

TP and Olsen-P ranged from 0.046% to 0.070% and 5.13 mg kg^{-1} to 15.99 mg kg^{-1}, respectively, in the 0–20 cm layer and 0.031–0.037% and 1.87–5.57 mg kg^{-1}, respectively, in the 20–40 cm layer (Fig. 4). One-way ANOVA analysis showed no significant difference in Olsen-P and TP in both 0–20 cm layer 20–40 cm layer. The results showed that although the content of TP and Olsen-P indicate an increasing trend, there were no significant change of TP and Olsen-P in restored wetlands in all.

Soil nutrient (TN, TP, TOC, KN, and Olsen-P) concentrations were lower in the 20–40 cm layer than the 0–20 cm layer. This indicates that nutrients are stored primarily in recent sediment deposits.

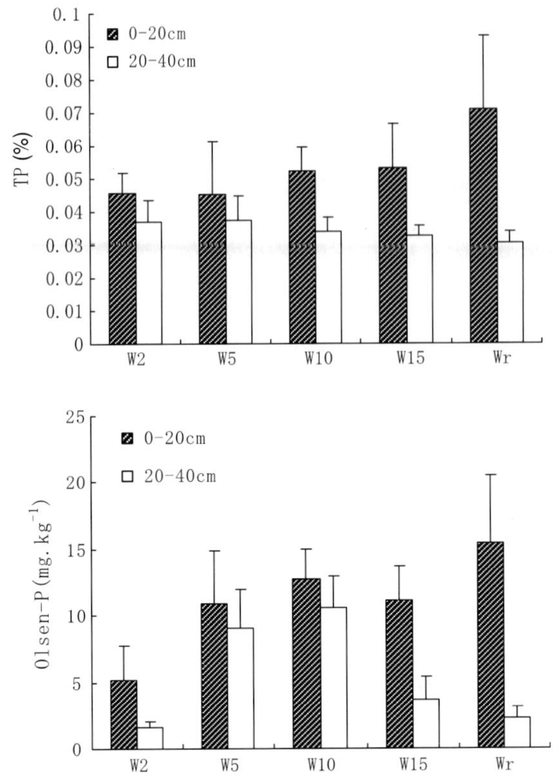

Fig. 4 Changes of total P and Olsen-P in different restored wetlands

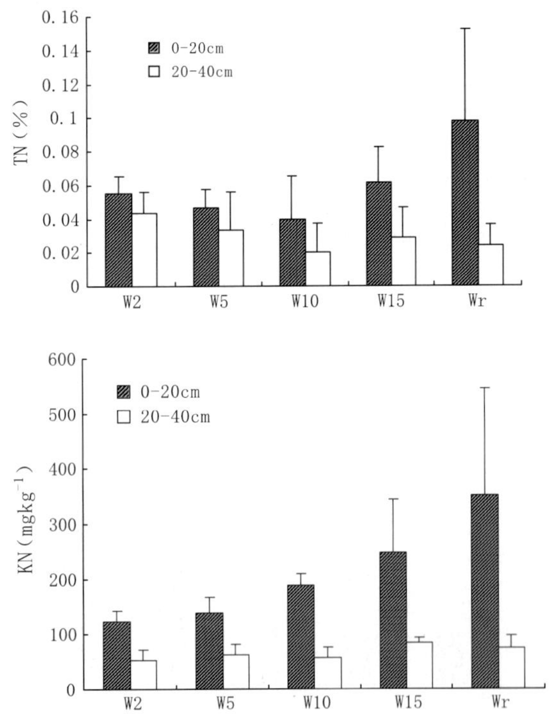

Fig. 3 Changes of soil total N and K N in different restored wetlands

Table 1 Changes of soil bulk density, porosity and water stable aggregate content in different restored wetlands in Lake Taihu, East China

Time abandoned	Soil bulk density	Soil porosity (%)	WSA (>0.25 mm)
2 years	1.06 g cm^{-3}	59.9	20.2%
5 years	0.87 g cm^{-3}	71.2	30.6%
10 years	0.74 g cm^{-3}	72.0	44.3%
15 years	0.74 g cm^{-3}	72.0	54.8%
Reference wetland	0.58 g cm^{-3}	77.9	57.1%

Table 2 The species colonized after rice fields abandonment

Species	Length of land abandonment (years)				Natural reference wetland
	2	5	10	15	
Phragmites communis	0	1	1	1	1
Miscanthus sacchariflorous	0	1	1	1	1
Zizania latifolia stapf L.	0	0	1	1	0
Cyperus iria L.	0	0	0	1	0
Eleocharis yakoscensis	0	0	1	0	0
Polygonum amphibium	0	0	0	1	0
Polygonum perfoliatum L.	0	0	1	1	1
Juncus setchueusis Bush.	1	1	0	0	0
Murdannia triquetra (Wall) Bruckn	0	1	0	0	0
Lycopus lucidus Turcz var. *hirtus* Regel	0	1	1	1	0
Poa annua	1	1	0	0	0
Poa faberi Rendle	1	1	0	0	0
Glgcine sojasieb et Zucc	1	0	0	0	0
Solanum nigrum	1	0	0	0	0
Setaria viridis (L.) Beauv	1	1	0	0	0
Lactuca indica. L.	1	0	0	0	0
Artemisia lavandulaefolia DC	1	1	0	0	0
Artemisia annua	0	1	0	0	0
Gynura crepidioides	0	1	0	0	0
Clematis	0	0	1	0	0
Humulus scandens	1	1	1	1	1
Bidens bipinata	0	0	1	1	1
Cocculus trilobus (Thrund.) DC.	0	0	0	0	1
Rosa cymosa Tratt	1	1	1	0	0
Ampelopsis Michx	0	0	0	1	1
Sapium P. Br	0	0	0	0	1
Chinese Fevervine scandens	0	0	1	1	1
Total number of species	10	13	11	10	9

Soil texture changes in restored wetlands

Table 1 shows soil bulk density, soil water content, and soil porosity along the succession gradient defined by paddy fields abandoned for 2, 5, 10, and 15 years and the natural reference wetland. Soil bulk density ranged from 0.58 g cm^{-3} to 1.06 g cm^{-3} with a decreasing progression with time abandonment. During the wet season, water content increased from 50% to 300% along the succession gradient. In the dry season, land that had been abandoned longer had more water than more recently abandoned paddy fields. There also was an increase in soil water stable aggregates (>0.25 mm) in abandoned paddy fields along the succession gradient, and the proportion of large water stable aggregate at later stages was higher than earlier stages of succession. These results indicate that restoration of abandoned paddy fields improves soil quality and texture.

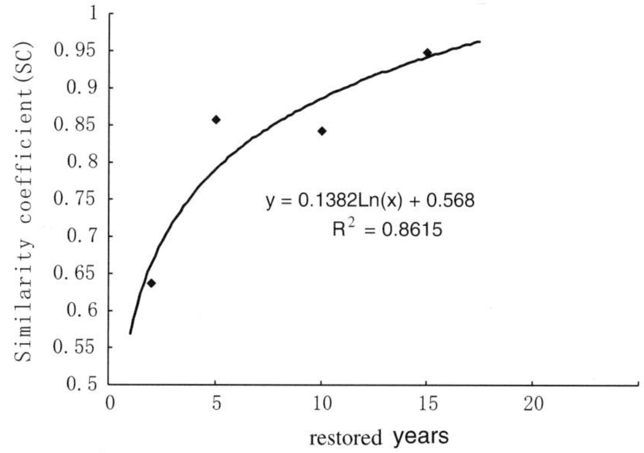

Fig. 5 Similarity coefficient (SC) of vegetation in different restored wetlands

Vegetation changes in restored wetlands

Dominant plant species changed in the first 10 years after the paddy fields were abandoned (Table 2). In the first 2 years, abandoned land became covered by a plant community dominated by *Poa annua, Humulus scandens, Gynura crepidioides*, and *Glgcine sojasieb*. This situation was

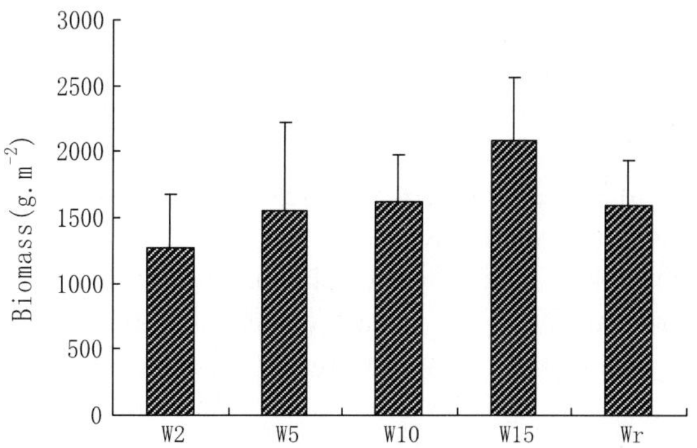

Fig. 6 Changes of vegetation biomass in different restored wetlands

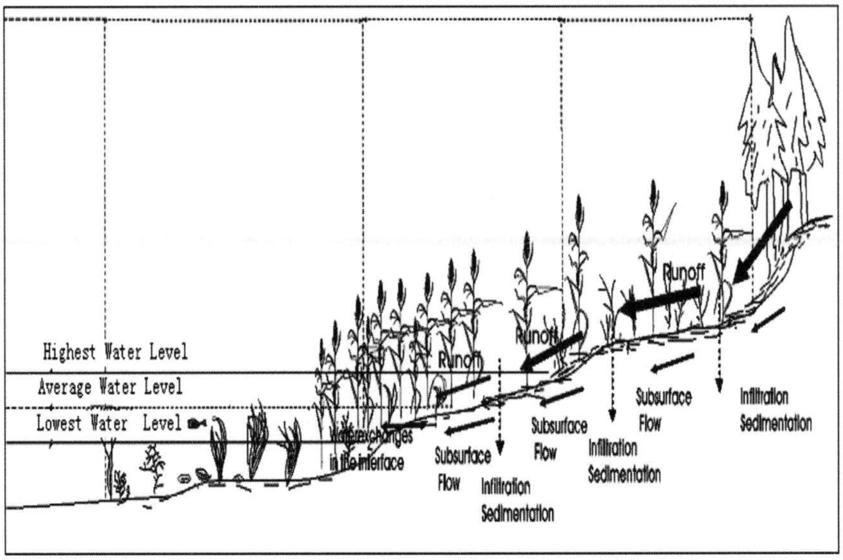

Fig. 7 The sketch map of the function of riparian wetlands

common in most abandoned paddy fields. During the early stage of abandonment (2–5 years), the number of plant species increased from 10 at the 2-year abandonment site to 13 at the 5-year site. Dominant plant species were *Humulus scandens, Miscanthus saccharifloous, Lycopus lucidus* Turcz, Var. *hirlus* Regel, *Poa annua*, and *Poa orinosae*. At the 5-year site, some wetland species, such as *Miscanthus saccharifloous, Zizania latifolia stapf*, and *Polygonum amphibium*, increased significantly, but some annual and biennial species and weeds still existed. After 5 years, some initial annual species, such as *Poa annua, Gynura crepidioides*, and *Glgcine sojasieb* declined. The number of species decreased from 13 at 5-years to 10 and 9 in the 10- and 15-year fields, respectively. Species composition after 15 years of abandonment was similar to the reference wetland, and *Phragmites communis, Miscanthus saccharifloous*, and *Zizania latifolia stapf* were dominant. SC was 0.64 at the 2-year site and 0.94 at the 15-year site compared with the reference wetland (Fig. 5). There was a significant relationship between SC and number of years since abandonment ($r^2 = 0.8615$). Dry biomass increased from 1270 g m^{-3} to 2100 g m^{-3} over the whole period (Fig. 6). One-way ANOVA showed a significant difference in dry biomass in wetland restoration, although there was no significance difference between W5 and W10. That showed there lay an obvious trend of dry biomass in the wetland restoration process. Plant community changes after land abandonment suggest that vegetation self-regenerates and species diversity recovers by returning paddy fields or lakeshore land to a more natural state.

Discussion

Riparian zones are critical in maintaining biological connections along environmental gradients (as indicated in Fig. 7). They reduce runoff from upland and increase infiltration; they are a physical buffer to trap sediments and sediment-bound pollutants; they are also a biological buffer to remove nutrients via plant uptake.

Vegetation is important in riparian ecosystems, and wetlands converted from abandoned paddy field showed secondary succession. Height and density of colonized plants increased along the succession series. With longer abandonment time, species composition increased in similarity and stability and reflected functional characteristics of a natural riparian wetland vegetation community. Wetland species, such as *Phragmites communis, Miscanthus saccharifloous, and Zizania latifolia stapf* dominated in abandoned paddy fields and colonized gradually over 15 years. Other plant species may be more competitive in other environments (Cooper et al., 1987). Changes in the similarity index and aboveground biomass indicated an increasing trend during restoration. Comparison of vegetation changes and related indices between restored wetlands showed a general trend towards a mature community or ecosystem, which has also been observed in terrestrial systems (Peng, 1995; Comin et al., 2001). In this study, we observed that groundwater table fluctuation might limit riparian wetland ecosystem succession. In some situations, only a high groundwater table allowed abandoned paddy fields or degraded wetlands to be restored to natural riparian wetlands (Andrew & Baldwin, 2002; Castelli et al., 2000).

Soil restoration is important to consider in evaluating wetland restoration. Results showed that soil quality improved with time due to accumulation of TOC, TN, TP, and improvement of soil texture. Similar results were reported in terrestrial ecosystems (Browman et al., 1990; Peng, 1995; Kosmas et al., 2000; Comin et al., 2001; Singh et al., 2001; Lee et al., 2002; Bruland & Richardson, 2004, 2005). Soil pH in riparian areas ranged from 4.90 to 5.28 in the 0–20 cm layer and 5.68 to 5.92 in the 20–40 cm layer in Fig. 5. In this study, TOC, TN, TP, KN, and Olsen-P increased with time and eventually reached similar concentrations to natural, undisturbed wetlands. Recovery of vegetation is key to successful soil restoration.

The results suggest a faster increase for soil TOC, KN, TP and Olsen-P in the early vegetation succession stage followed by a slower rate in later stages, while TN decreased in early stages. Nutrient accumulation probably resulted from decomposition of plant debris and retention of nutrients from upland areas. The recovery of vegetation and litter decomposition result in lower soil pH during restoration and create a suitable environment for soil C, N, P accumulation. Although nutrient accumulation in riparian restoration may result in leaching into lakes by subsurface flows, it is negligible compared with controlling non-point source pollution (Naiman & Décamps, 1997).

Low TOC accumulation in the early restoration period mainly was due to high soil TN concentration, which caused a higher decomposition rate of vegetation litter. The TN decrease early in the succession period may be due to growth of wetland vegetation, high nitrogen fertilizer consumption in the paddy soil before land abandonment, and ammonia volatilization and denitrification. Results also showed that soil TOC, TN, TP, KN, and Olsen-P increased towards levels found in natural wetlands, but restoration of degraded ecosystems likely would need a long time to complete the process. Vegetation restoration in Lake Taihu degraded riparian wetlands probably takes 15 years or more, and restoration of soil characteristics may take more time. In comparison to other ecosystems, restoration of degraded tidal wetlands needs about 6 years (Comín et al., 2001). Craft (2000) suggested that restoration of salt marsh vegetation requires 15 years, and restoration of soil C, N and bulk density needs 25 years. Restoration of degraded pasture may require 30 years (Aronson et al., 1993; Grantz et al., 1998; Li, 2003). Restoration of degraded forest ecosystems may take 50–100 years (Peng, 1995).

Restored vegetation not only protects surface soil from erosion and improves soil texture, but it also traps nutrient-rich material carried by runoff from uplands to riparian areas. Organic C, N and P accumulation in soil facilitated dominant wetland species in later succession stages. Vegetation restoration increased the roughness of land and coverage. Improvement in soil texture increases soil porosity and infiltration, facilitates accumulation of soil nutrients, and provides a suitable environment for microbe activity. Implications are that more non-point source pollutants may be treated in restored wetlands. It is expected that, when water level reaches inundation conditions, wetlands still remove pollutants, but is not a pollution source to the lake.

Conclusions

Changing land use in abandoned paddy fields may exert a profound influence on soil, vegetation, and the environment. With increasing time, vegetation in abandoned fields increased in the maturity of species composition and biomass. Results also showed that soil TOC, TN, TP, KN, and Olsen-P improved with time. Soil and vegetation facilitated one another in the restoration process. It may take 15 years or more for vegetation and soil characteristics to reach levels in natural wetlands.

Acknowledgments This research was supported by Chinese Academy of Sciences (No. KZCX1-SW-12) and the National Hi-Tech Research Program (No. 2002AA601011-05). We gratefully thank Yaohui Hu, Zexiang Lei, Nian'gen Du and Hongmei Hao for their assistance with sampling and analysis.

References

Andrew, E. C. & H. Baldwin, 2002. Responses of wetland plants to ammonia and water level. Ecological Engineering 18: 257–264.

Aronson, J., C. Floret, E. Le Floc'h, C. Ovalle & R. Pontanier, 1993. Restoration and rehabilitation of degraded ecosystems in arid and semiarid lands. I. A view from the south. Restoration Ecology 1: 8–17.

Browman, R. A., J. D. Reeder & R. W. Lober, 1990. Changes in soil properties in Central Plain rangeland soil after 3, 20, and 60 years of cultivation. Soil Science 150: 851–857.

Bruland, G. L. & C. J. Richardson, 2004. A spatially explicit investigation of phosphorus sorption and related soil properties in two riparian wetlands. Journal of Environmental Quality 33: 785–794.

Bruland, G. L. & C. J. Richardson, 2005. Spatial variability of soil properties in created, restored, and paired natural wetlands. Soil Science Society of America Journal 69: 273–284.

Castelli, R. M., J. C. Chambers & R. J. Tausch, 2000. Soil–water relations along a soil-water gradient in Great Basin riparian meadows. Wetlands 20: 251–266.

Chen, H. S., 2001. Restoration project of the ecosystem in Lake Taihu. Resources and Environment in the Changjiang River Basin 10(2): 173–178 (in Chinese).

Chen, J. Y. & S. L. Chen, 2002. Challenges to estuary and shoreline in China. Marine Geology 18(1): 1–5 (in Chinese).

Comin, F. A., J. A. Romero, H. Oliver & M. Menéndez, 2001. Restoration of wetlands from abandoned paddy fields for nutrient removal, and biological community and landscape diversity. Restoration Ecology 9(2): 201–208.

Coveney, M. F., D. L. Stites & E. F. Lowe, 2002. Nutrient removal from eutrophic lake water by wetland filtration. Ecological Engineering 19: 141–159.

Cooper, J. R., J. W. Gilliam & R. B. Daniels, 1987. Riparian areas as filters for agricultural sediment. Soil Science Society of American Journal 51(2): 416–420.

Craft, C., 2000. Co-development of wetland soil and benthic invertabrate communities following salt marsh creation. Wetland Ecology and Management 8(2): 197–207.

Grantz, D. A., D. L. Vaughn, R. J. Farber, B. Kim, L. Ashbaugh, T. VanCuren, R. Campbell, D. Bainbridge & T. Zink, 1998. Transplanting native plants to revegetate abandoned farmlands in the western Mojave Desert. Journal of Environmental Quality 27: 960–967.

Hu, Y. & L. G. Jiang, 2004. Returning farmland to lake-restoration of the kidney of Yangtse River. Chinese National Geography 11: 138–146.

Jin, X. S., H. Liu & Q. Y. Tu, 1990. Lake eutrophication in China. Environmental Sciences Press of China (in Chinese).

Kosmas, C., S. T. Gerontidis & M. Marathianou, 2000. The effect of land use change on soil and vegetation over various lithological formations on Lesvos (Greece). Catena 40: 51–68.

Lee, C. S., Y. H. You & G. R. Robinson, 2002. Secondary succession and natural habit restoration in abandoned fields of central Korea. Restoration Ecology 10: 306–314.

Li, Y. Q., 2003. Vegetation dynamics and soil characteristics during grassland succession process in abandoned land. MS degree dissertation thesis, Inner Mongolia Agricultural University, 36 pp.

Naiman, R. J. & H. Décamps, 1997. The ecology of interfaces: riparian zones. Annual Review Ecological Systematics 28: 621–658.

Naiman, R. J., H. Décamps & F. Fournier (eds), 1994. The role of land/inland water ecotones in landscape management and restoration: a proposal for collaborative research. MAB digest 4, UNESCO.

National Bureau of Statistics of China, 2001. Development Report of China in 2001: the ninth five-year plan. Chinese Statistics Press, Beijing, 352 pp (in Chinese).

Peng, S. L., 1995. The restoration and rehabilitation of the degraded ecosystems in the southern part of subtropical China. In Chen, L. Z. & W. L. Chen (eds), Studies on Degraded Ecosystems in China. China's Science and Technology Publishing House, Beijing, 94–113.

Singh, K. P., T. N. Mandal & S. K. Tripathi, 2001. Patterns of restoration of soil physicochemical properties and microbial biomass in different landslide sites in the hilly forest ecosystem of Nepal Himalaya. Ecological Engineering 17: 385–401.

Soil Science Society of America and American Society of Agronomy, 1996. Methods of Soil Analysis: Chemical Methods. SSSA Incorporation and ASA Incorporation, Madison, NJ, USA.

Song, Y. C., 2001. The Species Composition of Vegetation Communities. Vegetation Ecology. East China Normal University Press, 62–89.

Tabacchi, E., D. L. Correll, R. Hauer, G. Pinay, A.-M. Planty-Tabacchi & R. C. Wissmar, 1998.

Development, maintenance and role of riparian vegetation in the river landscape. Freshwater Biology 40: 497–516.

Vollenweider, R. A., 1992. Coastal marine eutrophication: principles and control. In Vollenweider, R. A., R. Marchetti & R. Viviani (eds), Marine Coastal Eutrophication, Elsevier Science Publications, 1–20.

Xiong, Z. P., 2004. The analysis for the causes of floods in China and the counter measures for reducing the damage. China Water Resources, 16 pp.

Yin, C. Q. & Z. W. Lan, 1995. The nutrient retention by ecotone wetlands and their modification for Baiyangdian Lake restoration. Water Science and Technology 32(3): 159–167.

Hydrobiologia (2007) 581:161–165
DOI 10.1007/s10750-006-0518-0

EUTROPHICATION IN LAKES

Quantitative studies on phosphorus transference occuring between *Microcystis aeruginosa* and its attached bacterium (*Pseudomonas* sp.)

Lijuan Jiang · Liuyan Yang · Lin Xiao · Xiaoli Shi · Guang Gao · Boqiang Qin

Abstract Phosphorus release from *Microcystis aeruginosa* and attached bacterium (*Pseudomonas* sp.) isolated from Lake Taihu was examined using a phosphorus isotope tracer in order to investigate the phosphorus transference between the two species. Our results reveal that the amount of phosphorus released form ^{32}P-saturated *M. aeruginosa* is determined by its growth phase and most of phosphorus is assimilated by *Pseudomonas* finally while the amount of phosphorus released from ^{32}P-saturated *Pseudomonas* is also determined by the growth phase of *M. aeruginosa* and most of them are assimilated by *M. aeruginosa*. The results suggest that phosphorus transference occurs between *M. aeruginosa* and its attached *Pseudomonas* . This process makes microenvironment of mucilage of *M. aeruginosa* attached bacteria maintain relative high amounts of phosphorus. Attached bacteria may be a temporary phosphorus bank to the growth of *M. aeruginosa*, and assimilation of phosphorus by *M. aeruginosa* becomes easy when *M. aeruginosa* is in lag growth phase. Thus, the phosphorus exchange between *M. aeruginosa* and attached *Pseudomonas* in microenvironment may be important to microfood web and cyanobacteria bloom.

Keywords *Microcystis aeruginosa* · Attached bacterium · *Pseudomonas* sp. · Phosphorus · Lake Taihu · ^{32}P release

Guest editors: B. Qin, Z. Liu & K. Havens
Eutrophication of shallow lakes with special reference to Lake Taihu, China

L. Jiang · L. Yang (✉) · L. Xiao · X. Shi
State Key Laboratory of Pollution Control and Resource Reuse, School of the Environment, Nanjing University, 22 Hankou Road, Nanjing 210093, P. R. China
e-mail: yangly@nju.edu.cn

L. Yang · X. Shi · G. Gao · B. Qin
Nanjing Institute of Geography and Limnology, CAS, 73 East Beijing Road, Nanjing 210008, P. R. China

Introduction

Microcystis aeruginosa is the dominant cyanobacterium in Lake Taihu, a shallow hypereutrophic lake in eastern China with extensive cyanobacterial algal blooms in summer and autumn (Dokulil et al., 2000). *M. aeruginosa* forms large mucilaginous colonies, usually colonized by the great deal of bacteria. There is a mutualistic relationship between *M. aeruginosa* and attached bacteria (Whitton, 1973). This symbiotic relation may be favorable to *Microcystis* cells (Steppe et al., 1996). *M. aeruginosa* constitute a microhabitat where attached bacteria are sheltered from grazing by embedding in the mucilage. The microenvironment allows attached bacteria to associate with the primary producer of organic carbon for its growth. In this microenvironment, heterotrophic bacteria are important: (1) as contributors of CO_2 and possibly sources of nitrogen, phosphorus and trace elements to *M. aeruginosa*; (2) as primary consumers of algal excretions; and (3) as decomposers. Some bacteria may promote blooms while others have algicidal and lysis effects and are involved in termination and decomposition of blooms (Caiola, 1991; Pellegrini et al., 1997; Lovejoy et al., 1998; Van Hannen et al., 1999; Manage et al., 2000; Manage et al., 2001). Thus, the metabolic relationship of *Microcystis* with attached bacteria is important for understanding bloom dynamics.

Phosphorus is the major limiting nutrient for primary producer, including *M. aeruginosa* in Lake Taihu (Dokulil et al., 2000). Pilot studies were conducted on the phosphorus metabolism of *M. aeruginosa* and its attached bacteria, respectively (Shi et al., 2003; Shi et al., 2004; Zou et al., 2005). The information of interaction of cyanobacterium with its attached bacterium for phosphorus transference was little. In this paper, a phosphorus isotope tracer was used to study phosphorus uptake and release from *M. aeruginosa* and its attached bacterium, *Pseudomonas*, to probe their interactions and phosphorus transformation.

Materials and methods

Organism and cultivation

Microcystis aeruginosa was obtained from Institute of Hydrobiology, Chinese Academy of Sciences. It was grown in the modified MA medium (pH = 8.6) in which inorganic phosphorus was replaced by β-Na_2-glycerophosphate (Oh & Lee, 2000). Cultures were incubated at 25°C under illumination intensity of 2200 lx with a 14:10 h of light:dark cycle.

Pseudomonas sp. was isolated from the mucilage of *Microcystis* in Lake Taihu, China and grown in complex medium: glucose 6.0; NH_4Cl 1.0; NaCl 2.0; $MgSO_4{\cdot}7H_2O$ 0.2; K_2HPO_4 1.0; KH_2PO_4 1.0; yeast extract 0.5 (all in grams per liter of deionized water). pH was adjusted to 7.2–7.5 before the complex medium was autoclaved.

Precultivation

This study was conducted with phosphorus-starved *M. aeruginosa*, ^{32}P-saturated *M. aeruginosa*, phosphorus-starved *Pseudomonas* and ^{32}P-saturated *Pseudomonas*. *M. aeruginosa* and *Pseudomonas* in exponential growth phase were harvested by centrifugation at 10,000*g* for 15 min respectively, washed with sterile deionized water and resuspended in MA medium without phosphorus. Phosphorus-starved cells were directly obtained after incubation for 3 days. Phosphorus-starved cells were collected by centrifugation and transferred to MA medium in which β-Na_2-glycerophosphate was replaced with 12 mg l^{-1} $Na_2H^{32}PO_4$, then were incubated for 3 days to obtain ^{32}P-saturated cells of *M. aeruginosa* and *Pseudomonas* respectively.

Transference of phosphorus from ^{32}P-saturated *M. aeruginosa* to *Pseudomonas*

Attachment of bacteria to *M. aeruginosa* mucilage made it difficult to measure ^{32}P content in bacteria cells or in cyanobacteria cells. Considering this difficulty, a dialysis membrane bag was used to separate *Pseudomonas* from *M. aeruginosa*. However, low molecular weight substances, such as phosphate and small organic compounds, could move across the dialysis membrane.

Phosphorus-starved *Pseudomonas* cells were harvested by centrifugation at 10,000*g* for 15 min, and washed with sterile deionized water and resuspended in a 500 ml beaker containing 200 ml MA medium without phosphorus to 1.0×10^7 cell ml^{-1}. ^{32}P-saturated *Microcystis* cells were harvested by the same way except they were resuspended in the dialysis bag to 5.2×10^6 cell ml^{-1}. This dialysis bag was put in the beaker so then *M. aeruginosa* and *Pseudomonas* cells were separated, while phosphate could migrate across the dialysis membrane. They were incubated at 25°C under a 14:10 h of light:dark cycle.

In order to probe the change of intracellular phosphorus, the cells of *M. aeruginosa* or *Pseudomonas* in certain volume were separated respectively via centrifugation at 10000 *g* for 15 min and rinsed several times with MA medium until rinse solution radioactivity was close to background. These cells were digested with 0.2 ml 60% $HClO_4$ and 0.4 ml 30% H_2O_2 at 80°C for 60 min. Then, the digested liquids and the upper aqueous solution separated by centrifugation were respectively put into scintillation vials containing 5 ml Triton X-100 toluene scintillator to analyze the radioactivity with a Beckman LS9800 liquid scintillation counter. Actual phosphorus concentration was corrected using the radioactive decay rate. All samples had three replicates.

In order to measure the biomass of *M. aeruginosa* and *Pseudomonas*, their optical densities were measured with at 460 nm and 380 nm, respectively. Then, the amounts of ^{32}P in *M. aeruginosa* or *Pseudomonas* were calculated.

Transference of phosphorus from ^{32}P-saturated *Pseudomonas* to *M. aeruginosa*

The experimental procedures were same as described above except that phosphorus-starved *M. aeruginosa* was incubated in a 500 ml beaker, while ^{32}P-saturated *Pseudomonas* was incubated in dialysis bag.

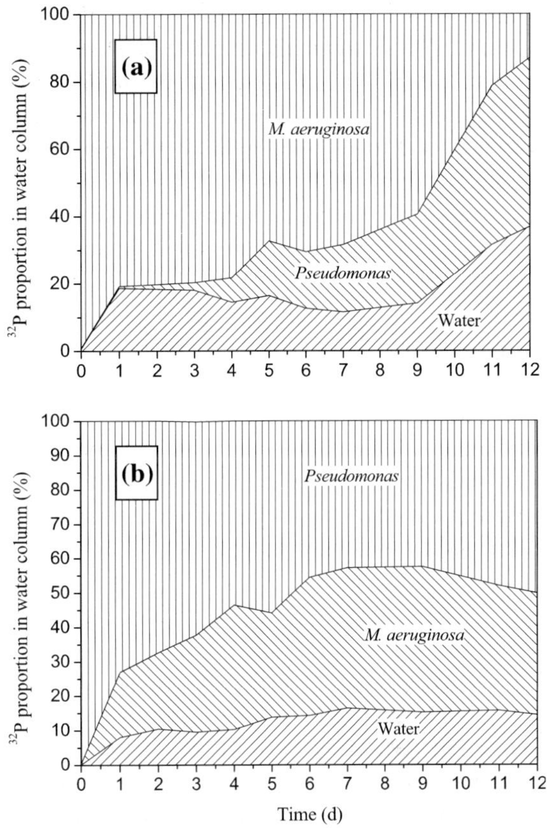

Fig. 1 The transference of ^{32}P released from *M. aeruginosa* (**a**) and from *Pseudomonas* sp. (**b**) among water column.

Results

Transference of phosphorus from ^{32}P-saturated *M. aeruginosa* to *Pseudomonas*

The transference of phosphorus from *M. aeruginosa* to *Pseudomonas* was investigated in a culture system of phosphorus-starved *Pseudomonas* and ^{32}P-saturated *M. aeruginosa* which were separated by a dialysis membrane (Fig. 1a). The growth phases of *M. aeruginosa* comprised four distinct phases of the lag (0–1 day), exponential (1–4 day), stationary (4–6 day) and decline phases (6–12 day) while the growth phases of *Pseudomonas* was in the exponential phase. The total ^{32}P in *Pseudomonas* always increased though the increase was slight for *Pseudomonas* in initial 4-days. Because of the growth of *Pseudomonas* and its incessant assimilation of ^{32}P, the ^{32}P content in *Pseudomonas* increased after *M. aeruginosa* was in the stationary phase. ^{32}P content in *Pseudomonas* increased to 50% of the total ^{32}P on the 12th day when *M. aeruginosa* was in the decline phase.

During the whole experiment period, the intracellular ^{32}P content of *M. aeruginosa* decreased at all times. The ^{32}P concentration in the aqueous solution remained relative steadily from the 1st day to the 9th day, then the sharp increase happened because *M. aeruginosa* in decline growth phase released a large amount of phosphorus to aqueous solution.

Transference of phosphorus from ^{32}P-saturated Pseudomonas to *M. aeruginosa*

Not only phosphorus could move from *M. aeruginosa* to *Pseudomonas*, but also from

Pseudomonas to *M. aeruginosa*. When phosphorus-starved *M. aeruginosa* was present, partitioning of ^{32}P released from ^{32}P-saturated *Pseudomonas* was observed (Fig. 1b). The growth of *M. aeruginosa* exhibited the lag phase (0–2 day), exponential phase (2–9 day) and decline phase (9–12 day) while *Pseudomonas* grew slightly at all times. Most of ^{32}P released from *Pseudomonas* was assimilated by *M. aeruginosa*.

Intracellular ^{32}P content of *M. aeruginosa* increased in the lag and exponential phases, but decreased in decline phase which decrease of ^{32}P content in *M. aeruginosa* was due to the growth of *Pseudomonas* though this decrease was not rapid. ^{32}P concentration in the aqueous solution increased in the first day and then remained relative steadily in the following days.

Discussion

In this experimental system, phosphorus was main factor for the growth of *M. aeruginosa* and *Pseudomonas* and the latter was also controlled by organic carbon. When the intracellular phosphorus content of *M. aeruginosa* was luxury, it grew by utilization of the intracellular phosphorus in its lag and exponential phases (Okada & Sudo, 1982), thus, only 20% of intracellular phosphorus was released form *M. aeruginosa* to aqueous solution and to *Pseudomonas*. However, *M. aeruginosa* released a lot of phosphorus in its stationary and decline phases while it could simultaneously provide *Pseudomonas* with organic carbon and trace elements (Sommaruga & Robarts, 1997; Worm, 1998; Brunberg, 1999). Thus, the growth of *Pseudomonas* was better and the assimilated phosphorus was more when *M. aeruginosa* was in decline phase than that in stationary phase.

After the amount of *M. aeruginosa* was low enough, its second growth became possible. *M. aeruginosa* could use phosphorus in *Pseudomonas* to grow because *Pseudomonas* had a geat deal of phosphorus which was assimilated from overripe *M. aeruginosa* formerly (Fig. 1b). Compared with the phosphorus transference form *M. aeruginosa* to *Pseudomonas*, phosphorus movement form *Pseudomonas* to *M. aeruginosa* was easy and a great deal of ^{32}P was assimilated by *M. aeruginosa* in lag and exponential phases. When *M. aeruginosa* was in the decline phase, its intracellular ^{32}P then turned back to *Pseudomonas* again. This process made soluble phosphorus in aqueous solution stable. Thus, phosphorus released form *M. aeruginosa* could store in attached bacteria and *Pseudomonas* might be a temporary phosphorus bank to *M. aeruginosa* in the microenvironment for its growth.

Conclusion

Phosphorus transference from *Pseudomonas* to *M. aeruginosa* occurs when *M. aeruginosa* is in grow phase while phosphorus transference from *M. aeruginosa* to *Pseudomonas* happens when *M. aeruginosa* is in decline phase. *Pseudomonas* may serve as a temporary phosphorus bank to *M. aeruginosa* in the microenvironment. Thus, phosphorus transference between cyanobacteria and their attached bacteria is decided by their growth phases and phosphorus amount and this exchange could keep an amount of phosphorus in the microenvironment of cyanobacteria and their attached bacteria which is important to microfood web and cyanobacteria bloom.

Acknowledgements We thank F. Wang (Modern Analysis Center, Nanjing University) for their technical assistances. This study was funded by Chinese Academy of Sciences (No. KZCX2-311), "973" Project (No. 2002CB412307) and National Natural Science Foundation of China (No. 40371102, 40501078).

References

Brunberg, A. K., 1999. Contribution of bacteria in the mucilage of *Microcystis* spp. to benthic and pelagic bacterial production in a hypereutrophic lake. FEMS Microbiology Ecology 29: 13–22.

Caiola, M. G., 1991. *Bdellovibrio*-like bacteria in *Microcystis aeruginosa*. Algological Studies 64: 369–376.

Dokulil, M., W. Chen & Q. Cai, 2000. Anthropogenic impacts to large lakes in China: the Tai Hu example. Aquatic Ecosystem Health and Management 3: 81–94.

Lovejoy, C., J. P. Bowman & G. M. Hallegraeff, 1998. Algicidal effects of a novel marine *Pseudoalteromonas* isolate (Class Proteobacteria, Gamma Subdivision) on harmful algal bloom species of the genera

Chattonella, Gymnodinium, and *Heterosigma*. Applied Environmental Microbiology 64: 2806–2813.
Manage, P. A., Z. Kawabata & S. Nakano, 2000. Algicidal effect of the bacterium *Alcaligenes denitrificans* on *Microcystis* spp.. Aquatic Microbial Ecology 22: 111–117.
Manage, P. A., Z. Kawabata & S. Nakano, 2001. Dynamics of cyanophage-like particles and algicidal bacteria causing *Microcystis aeruginosa* mortality. Limnology 2: 73–78.
Oh, H. M. & S. J. Lee, 2000. Microcystin production by *Microcystis aeruginosa* in a phosphorus-limited chemostat. Applied and Environmental Microbiology 66: 176–179.
Okada, M. & R. Sudo, 1982. Phosphorus uptake and growth of bule-green alga, *Microcystis aeruginosa*. Biotechnology and Bioengineering 24: 143–152.
Pellegrini, S., L. Allievi, B. Lolli & M. G. Caiola, 1997. *Bdellovibrio* isolation from the Lake Varese (Italy). Annali Di Microbiologia ED Enzimologa 47: 121–129.
Shi, X., L. Yang, X. Niu & L. Xiao, 2003. Intraintracellular phosphorus metabolism of *Microcystis aeruginosa* under various redox potential in darkness. Micobiological Research 158: 345–352.
Shi, X., L. Yang, F. Wang, L. Xiao, L. Jiang, Z. Kong, G. Gao & B. Qin, 2004. Growth and phosphate uptake kinetics of *Microcystis aeruginosa* under varying environmental conditions. Journal of Environmental Sciences 16: 88–92 (in Chinese).
Sommaruga, R. & R. D. Robarts, 1997. The significance of autotrophic and heterotrophic picoplankton in hypertrophic ecosystems. FEMS Microbiology Ecology 24: 187–200.
Steppe, T. F., J. B. Olson, H. W. Paerl, R. W. Litaker & J. Belnap, 1996. Consortial N2 fixation: a strategy for meeting nitrogen requirements of marine and terrestrial cyanobacterial mats. FEMS Microbiology Ecology 21: 149–156.
Van Hannen, E. J., G. Zwart & H. J. Laanbroek, 1999. Changes in bacterial and eukaryotic community structure after mass lysis of filamentous cyanobacteria associated with viruses. Applied and Environmental Microbiology 65: 795–801.
Whitton, B. A., 1973. Interactions with other organisms. In Carr, N. G. & B. A. Whitton (eds), The Biology of Blue-green Algae. Blackwell, Oxford: 415–433.
Worm, J., 1998. Dynamics of heterotrophic bacteria attached to Microcystis spp. (Cyanobacteria). Aquatic Microbial Ecology 14(1): 19–28.
Zou, D., L. Xiao, L. Yang & Y Wan, 2005. Effects of Phosphorus Sources of different forms on phosphorus metabolism of *Microcystis aeruginosa* and adhesive *Pseudomonas* sp.. Environmental Science 26(3): 118–121 (in Chinese).

Hydrobiologia (2007) 581:167–176
DOI 10.1007/s10750-006-0517-1

EUTROPHICATION IN LAKES

Intracellular phosphorus metabolism and growth of *Microcystis aeruginosa* in dark/light cycles under various redox potential difference conditions

Xiaoli Shi · Liuyan Yang · Lijuan Jiang · Fanxiang Kong · Boqiang Qin · Guang Gao

Abstract Phosphorus metabolism and growth of *M. aeruginosa* were studied under three different conditions of diel fluctuation in redox potential. Redox potential in the culture increased in light and decreased in dark in all treatments except one, when cysteine was added in darkness. Total phosphorus content in *M. aeruginosa* decreased in darkness and increased in light during exponential growth but increased continuously in the stationary phase. Conversely, polyphosphate (PolyP) accumulated in darkness but was lost in the light. Low redox potential in darkness promoted PolyP accumulation. Polyglucose and soluble orthophosphate may provide energy and phosphorus, respectively, for PolyP synthesis. PolyP was important to *M. aeruginosa* survival during poor growth conditions. If the redox potential difference in the dark/light cycle was large, *M. aeruginosa* initially grew faster, but soon lost viability.

Keywords *Microcystis aeruginosa* · Growth · Phosphorus metabolism · Dark/light cycle · Redox potential difference · Lake Taihu

Guest editors: B. Qin, Z. Liu & K. Havens
Eutrophication of shallow lakes with special reference to Lake Taihu, China

X. Shi · L. Yang (✉) · F. Kong · B. Qin · G. Gao
Nanjing Institute of Geography and Limnology, Chinese Academy of Sciences, Nanjing 210008, P.R. China
e-mail: yly@niglas.ac.cn

X. Shi · L. Yang · L. Jiang
State Key Laboratory of Pollution Control and Resource Reuse, School of the Environment, Nanjing University, Nanjing 210093, P.R. China

Introduction

In Lake Taihu, *Microcystis aeruginosa* has formed dense blooms in summer since the 1980s (Chen et al., 2003). Due to oxygen-producing photosynthesis, *Microcystis* usually are associated with aerobic environments, and, consequently, physiological characterization has focused on aerobic metabolism. However, this characterization discounts the possibility that they are found in environments that are permanently anoxic or become anoxic in the dark (Stal & Moezelaar, 1997).

Microcystis is colonial and possesses gas vacuoles, hollow proteinaceous bodies that provide cell buoyancy. In Lake Taihu, when the water column is stable, colonies accumulate at the surface and form surface blooms. Wind may concentrate colonies into dense "scums" on the leeward shore, forming microbial mats, which may experience periodic anoxia. During daytime, oxygenic photosynthesis leads to oxygen supersaturation. In the dark, cyanobacteria continue to respire, and oxygen diffusion into the mat usually is insufficient to prevent anoxia (Revsbech et al., 1983).

Microbial mats exhibit diel oxygen fluctuations due to cyanobacterial metabolism. In light, these organisms grow photoautotrophically, while *Microcystis* switches to chemotrophic metabolism in dark, at the expense of endogenous glycogen accumulated in light (Heyer & Krumbein, 1991). However, a cohesive, overall pattern of dark cyanobacteria metabolism has not emerged (Richardson & Castenholz, 1987).

Phosphorus is a major growth-limiting nutrient in aquatic systems, and, unlike nitrogen, there is no large atmospheric source that is biologically available. All organisms require phosphorus for growth and metabolism (Bostrom et al., 1988; Currie & Kalff, 1984; Howarth, 1988). Phosphorus in microalgae is contained in internal phosphorus pools of polyphosphate (PolyP), soluble orthophosphorus (Pi), structural and other organic phosphorus (John & Flynn, 2000). Cyanobacteria store inorganic phosphate as PolyP granules (Bental et al., 1991). However, little is known about phosphorus regulation in cyanobacteria (Lawrence et al., 1998). Previous studies have found that phosphorus metabolism of *M. aeruginosa* may inhibit continuous anaerobic dark conditions, which are unfavorable for *M. aeruginosa* growth (Shi et al., 2003). *M. aeruginosa* retaining viability under dark anaerobic/light aerobic cycles also may relate to phosphorus metabolism. However, to our knowledge, no studies on phosphorus metabolism of *M. aeruginosa* under dark/light cycle with high redox potential differences have been conducted. Here, we studied phosphorus metabolism and *M. aeruginosa* growth under diel fluctuations in redox potential.

Materials and methods

Organism and cultivation

Axenic *M. aeruginosa* cultures were obtained from the Institute of Hydrobiology of the Chinese Academy of Sciences and grown in batch culture in modified MA medium where inorganic phosphorus was replaced with β-Na_2glycerophosphate (Oh et al., 2000). Media were adjusted to pH 8.6 with 2 mol l^{-1} NaOH and incubated at 25°C under 40 μE m^{-2} s^{-1} light intensity provided by fluorescent lamps with a 14:10 h light/dark cycle. *M. aeruginosa* density was counted on a microscope with a 0.1 mm^3 counter, and about 300 cells were counted. Cell numbers were converted to dry weight using a 1.32×10^{-7} mg dry weight per cell (Johnston et al., 1994).

Experimental methods

Studies were conducted with *M. aeruginosa* cells harvested during exponential growth. Cells were washed twice and resuspended in the medium described above (MA). Aliquots were transferred to serum bottles with screw caps and a butyl rubber inlayer. Bottles were wrapped in black cloth and 0, 0.01, and 0.04% (m/v) cysteine (reducing agent) were introduced through serum caps with a syringe. After one day, all bottles were unwrapped and serum caps were opened. Then, *M. aeruginosa* was incubated under illumination intensity of 40 μE m^{-2} s^{-1} for another day. The cycle then was repeated. Therefore, days 1, 3, 5 and 7 had air-tight dark cultivation periods with different concentrations of cysteine, while days 2, 4, 6 and 8 were open light cultivation. All experiments were carried out with triplicate bottles. Cell density, polyglucose, PolyP and soluble orthophosphate content in *M. aeruginosa*, as well as phosphate concentration in the water, were assayed daily. A microelectrode was used to monitor initial and final redox potential.

Analysis

M. aeruginosa cells were harvested by centrifugation for physiological assay. Samples were digested by autoclaving in 10% persulphate for 30 min at 121°C for total phosphorus measurement (Grillo & Gibson, 1979). Orthophosphate in *M. aeruginosa* was extracted with 10% cold trichloroacetic acid. For PolyP, 40 ml of culture was centrifuged and digested with 1 ml alkaline hypochlorite reagent (5.4%) for 45 min at 25°C. Residue after centrifugation was washed twice, then extracted twice with distilled water. PolyP was precipitated with ethanol and measured after hydrolysis to orthophosphate at

100°C (Harold, 1963; Rao et al., 1985). Later, all phosphorus forms were converted to orthophosphate, and all samples were measured spectrophotometrically with an automated ascorbic acid reduction method (Greenberg, 1985). All phosphorus forms were expressed in milligrams of phosphorus per unit dry weight (DW) for *M. aeruginosa*. Polyglucose was extracted from cells, harvested from 20 ml culture with 1 ml 10% trichloroacetic acid, precipitated by adding three volumes of alcohol, and determined using anthrone reagent (Roe & Dailey, 1966).

Results

Redox potential decreased from 252 to 52 mV with 0.01% cysteine and to –50 mV with 0.04% cysteine at the beginning of the experiment (Fig. 1). Redox potential increased in light and declined in darkness in all treatments except the first dark period. Average redox potential differences were 67 mV, 83.5 mV and 276 mV, respectively, when 0, 0.01 and 0.04% cysteine were added in darkness.

M. aeruginosa cell density increased in all treatments, and high redox potential differences in the diel cycle enhanced *M. aeruginosa* growth, with maximum cell density reaching 7.5×10^6 cell ml^{-1} on the 7th day and 7.8×10^6 cell ml^{-1} on the 5th day when adding 0.01% and 0.04% cysteine in darkness, respectively (Fig. 2). Compared with the control, *M. aeruginosa* had a short growth period and lost viability after reaching maximum cell density.

Phosphate concentration in the aqueous phase increased in darkness and declined in light in all treatments initially, but this pattern ceased on the 5th day, when cysteine concentrations were 0.01% and 0.04% in the dark (Fig. 3). However, total phosphorus in *M. aeruginosa* increased (Fig. 4). There were no significant differences observed among the three treatments ($P > 0.05$), so enlarging the redox difference in the dark/light cycle did not have an obvious effect on phosphorous metabolism.

Soluble orthophosphate in *M. aeruginosa* declined initially, with increasing steepness at high cysteine concentrations, then increased in light and decreased in darkness (Fig. 5). Polyglucose

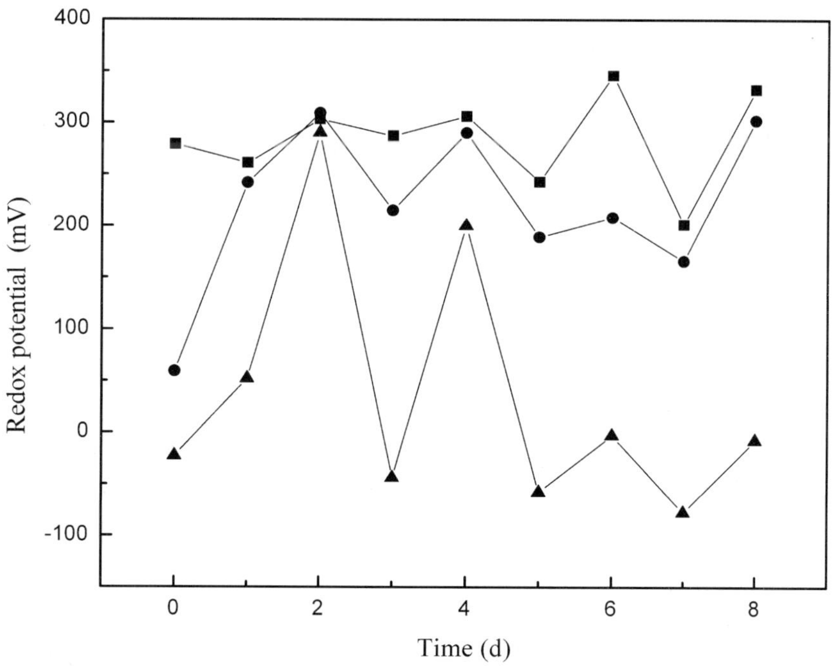

Fig. 1 Redox potential under different treatments. ■, Control; ●, 0.01% of cysteine in darkness; ▲, 0.04% of cysteine in darkness

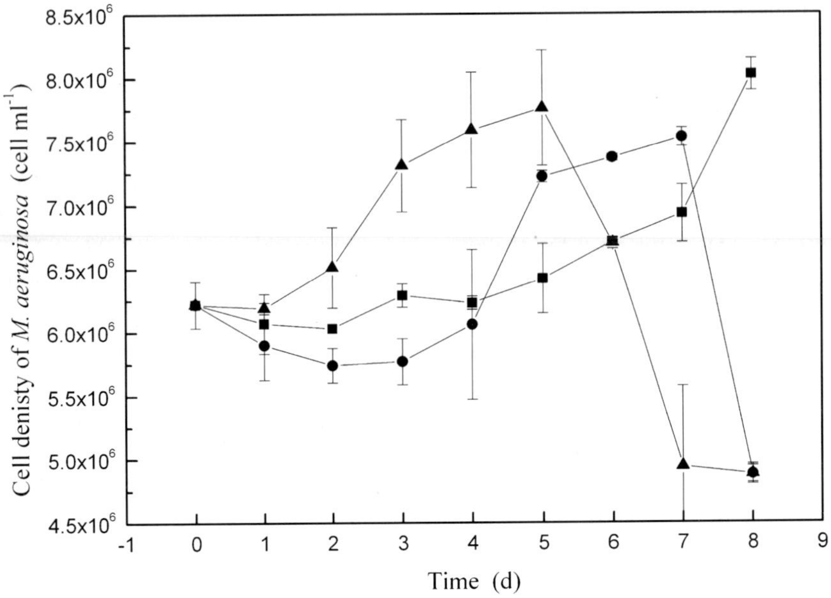

Fig. 2 Growth of *M. aeruginosa* under all the treatments. ■, Control; ●, 0.01% of cysteine in darkness; ▲, 0.04% of cysteine in darkness. Error bars represent standard derivation

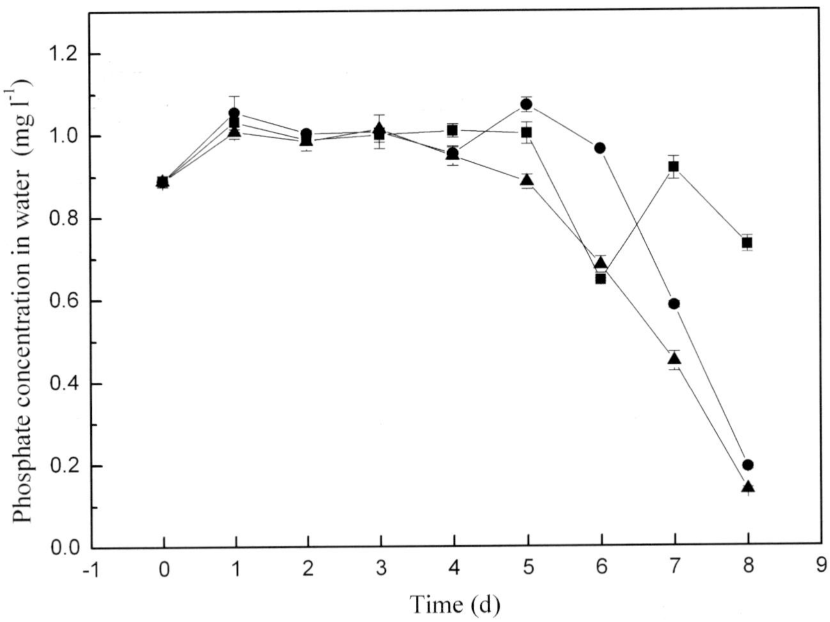

Fig. 3 Variation of phosphate concentration in water. ■, Control; ●, 0.01% of cysteine in darkness; ▲, 0.04% of cysteine in darkness. Error bars represent standard derivation

showed a similar pattern as soluble orthophosphate content (Fig. 6). Adding reducing agent promoted PolyP accumulation in *M. aeruginosa*, with PolyP reaching 0.038 and 0.043 mg g^{-1} DW on the 3rd day when cysteine concentrations were 0.01% and 0.04%, respectively. PolyP decreased in the light and increased in the dark (Fig. 7).

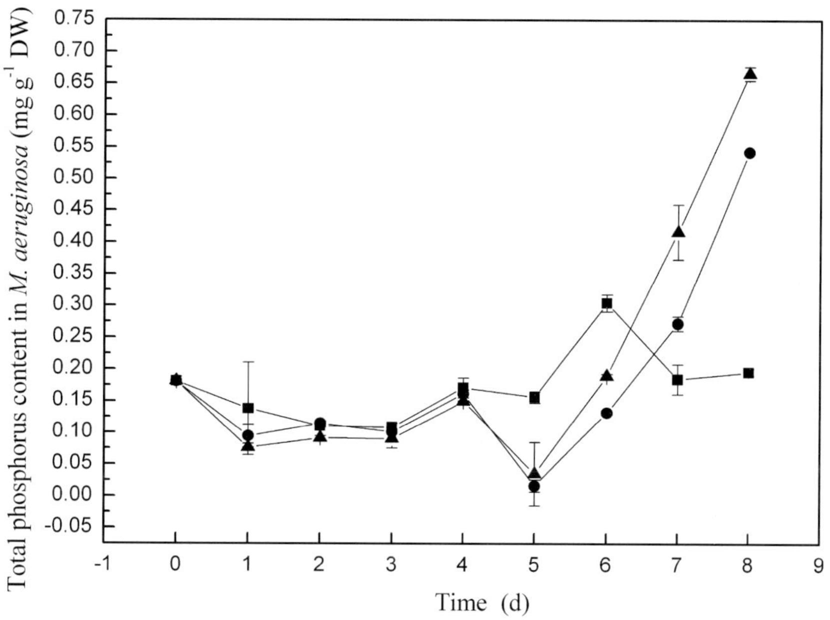

Fig. 4 Total phosphorus content in *M. aeruginosa*. ■, Control; ●, 0.01% of cysteine in darkness; ▲, 0.04% of cysteine in darkness. Error bars represent standard derivation

Discussion

Phosphorus metabolism of *M. aeruginosa* under dark/light cycles with different redox potential difference

M. aeruginosa does not form microbial mats or dense scums, which could cause anoxia, in the laboratory, so phosphorus metabolism of *M. aeruginosa* generally is studied under aerobic cultivation conditions. Phosphate uptake rate is considered a function of ambient phosphate and cell quota, and Droop's equation describes the relationship between cell quota and population growth rate (Fuhs et al., 1972). Phosphorus in *M. aeruginosa* decreased in darkness, increased in light, and increased during the stationary phase when reducing agent was added in darkness (Fig. 4). Thus, *M. aeruginosa* may release phosphorus in darkness and absorb phosphorus in light.

M. aeruginosa may synthesize PolyP when cells are added to fresh medium. PolyP in *M. aeruginosa* increased if redox potential was low. After a short period of growth, PolyP in *M. aeruginosa* increased in darkness and declined in the light (Fig. 7), opposite the trend of TP and polyglucose (Fig. 4). Synthesis of PolyP may help maintain phosphorous balance in the cell. PolyP accumulation coinciding with decreased polyglucose and soluble orthophosphate shows that these compounds may provide energy and phosphorus for PolyP synthesis. *M. aeruginosa* may accumulate polyglucose during daylight and consume it at night, and previous research showed that polyglucose fermentation can provide enough energy for PolyP synthesis (Shi et al., 2003). An experiment with yeast also indicated that endogenous substrate is adequate to supply energy for PolyP synthesis (Schuddemat et al., 1989).

Inorganic polyphosphates are polymers of orthophosphate (Pi) with a phosphoanhydride bond, which has free energy close to that of ATP. The exact physiological function of PolyP remains uncertain, although various biological functions have been demonstrated, including as a reservoir of energy and phosphate, a chelator of divalent cations, a capsule material, and a "channelling" agent in the phenomenon of bacterial transformation (Magrath & Quinn, 2000; Keasling & Hupf, 1996; Kornberg, 1995). In general, PolyP concentration is low during exponential growth

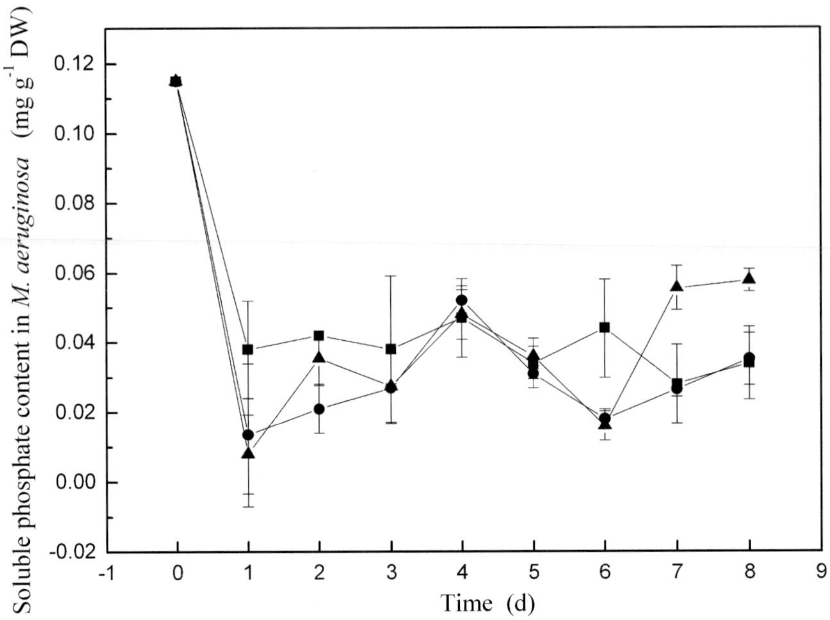

Fig. 5 Soluble phosphate content in *M. aeruginosa* under all the treatments. ■, Control; ●, 0.01% of cysteine in darkness; ▲, 0.04% of cysteine in darkness. Error bars represent standard derivation

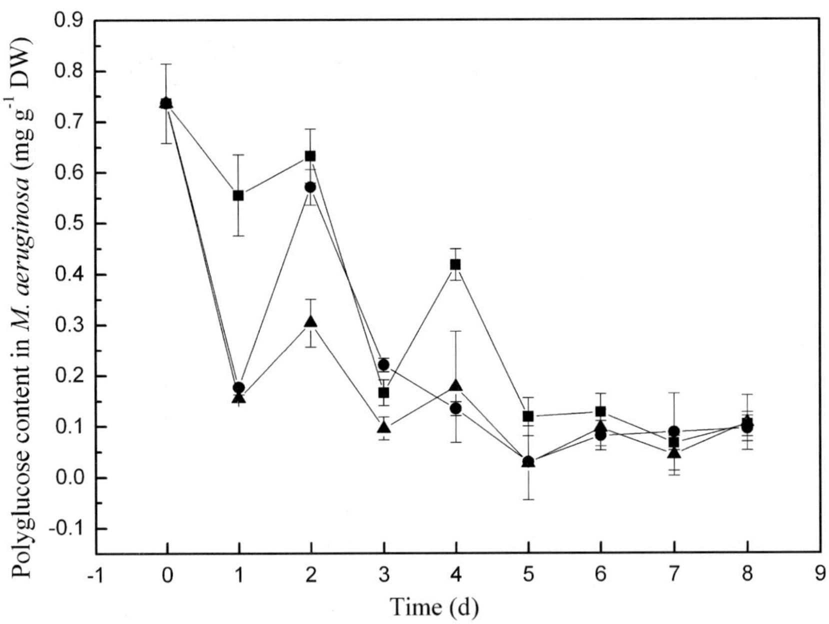

Fig. 6 Polyglucose content in *M. aeruginosa* under all the treatments. ■, Control; ●, 0.01% of cysteine in darkness; ▲, 0.04% of cysteine in darkness. Error bars represent standard derivation

but may increase when the stationary phase begins or when growth is arrested due to nutrient imbalance (Noegel & Gotschlich, 1983). Pi released and accumulated by the cells would be sequestered as PolyP to maintain Pi and H^+ concentration (Harold, 1963). Thus, rapid

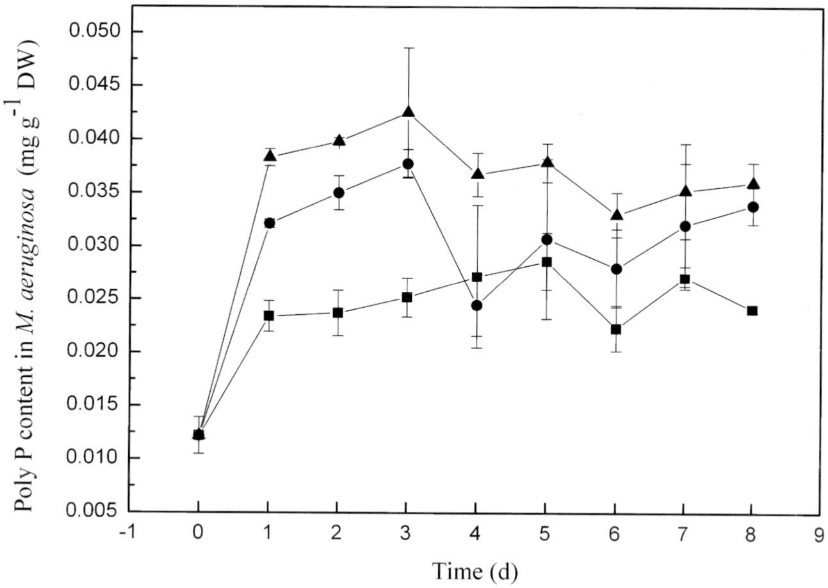

Fig. 7 PolyP content in *M. aeruginosa* under all the treatments. ■,Control; ●, 0.01% of cysteine in darkness; ▲, 0.04% of cysteine in darkness. Error bars represent standard derivation

orthophosphate release may stimulate PolyP synthesis. PolyP is involved in controlling Pi and serves, in effect, as a metabolic buffer.

Prokaryotes are composed of cytoplasm, cell surface, periplasm and plasma membrane. PolyP granules can be found in all of these, and different locations of PolyP may play different roles. Electron microscopy indicated that PolyP granules in *M. aerugnosa* were located in the cytoplasm (Shi et al., 2003). PolyP granules may be used for phosphorus and energy storage in our experiments (Kulaev et al., 1999).

M. aeruginosa growth in the dark/light cycle under different redox potential difference

Some cyanobacterial strains can survive in dark, reducing conditions. *Oscillatoria terebriformis*, a cyanobacterium, moved each night (Richardson & Castenholz, 1987). *O. chalybea* blooms are correlated with lowered partial pressure of O_2 accompanied by increased sulfide at increasing depth in lakes (Leventer & Eren, 1970). The presence of *O. redekei* in the lower strata of some lakes was attributed to reducing conditions of sulfide (Tash, 1967). Thus, *Oscillatoria* can survive in dark, reducing conditions, and some *Oscillatoria* species are dependent on lowered redox potential.

M. aeruginosa can endure low redox potential in darkness, which stimulated slight growth and reduced mortality in darkness (Shi, 2003). Large redox potential difference may promote *M. aeruginosa* growth initially, then cause cell death (Fig. 2). When adding 0.01% and 0.04% cysteine, average redox potential differences between light and dark were 83 mV and 276 mV, respectively. In this case, *M. aeruginosa* lost viability at days 5 and 7, respectively. In previous research, we found that large cell density in culture may result in high diel redox potential difference, and, when initial cell density reached 1.5×10^7, 6×10^7 and 1×10^8 cell ml^{-1}, average diel redox potential differences were 65, 330, and 450 mV, respectively (Fig. 8). During the 18-day experiment, *M. aeruginosa* grew and had no evident mortality (data not shown). Since cysteine had no other chemical effect besides lowing redox potential (Weller et al., 1975), we assumed that, only when *M. aeruginosa* reached a certain cell density, could it survive low redox potential. *M. aeruginsoa* seemed to thrive to

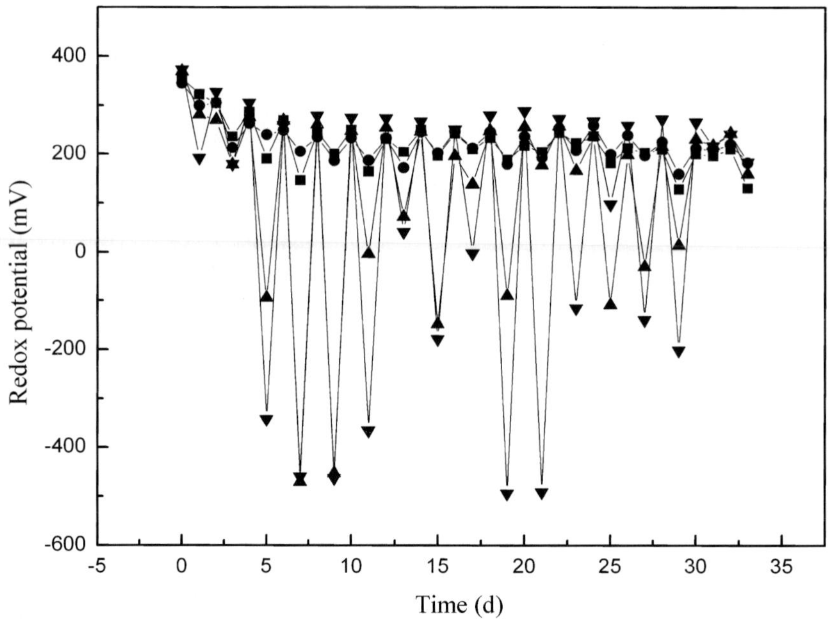

Fig. 8 Redox potential of the medium with different cell density of *M. aeruginosa*. ■, 3×10^6 cell ml^{-1}; ●, 1.5×10^7 cell ml^{-1}; ▲, 6×10^7 cell ml^{-1}; ▼, 1×10^8 cell ml^{-1}

resist the large diel redox potential difference since redox potential increased in the fist darkness period when the reducing agent was added, but could not hold back the designed trend. In our experiment, cell density of 6.2×10^6 cell ml^{-1} was too low to endure 83 and 276 mV diel redox potential difference. In Lake Taihu, large diel redox potential differences are observed in microbial mats or scums, and, in that case, cell density is high enough to adapt to such circumstances. Thus, large diel rodox potential difference may promote algae growth, but shows no negative effect.

In autumn, microbial mats of *Microcystis* sink to the sediment. *M. aeruginosa* colonies remain on the sediment as vegetative cells. These winter colonies are viable and constitute an inoculum for the following year, some reentering the water column in spring (Preston et al., 1980). Redox potential of sediment in Lake Taihu changes with depth from about -100 to + 100 mV (Fig. 9). Sediment suspension often occurs under wind and waves in large shallow lakes. Redox potential at the sediment surface increases at this time and decreases during calm periods. Therefore, *M. aeruginosa* in the sediment regularly experiences variation in redox potential.

Microcystis was a small proportion of biomass in non-stratified enclosures but was dominant in diurnally and continuously stratified enclosures (Nakano et al., 2001). Although thermal stratification was required for *Microcystis* growth, it is unclear why it was important. We hypothesize that stratification may lead to high diel redox potential difference, which is more favorable for *Microcystis* than other species, and, therefore, it is dominant in eutrophic environments.

Anoxia was vital for survival of *O. terebriformi* during dark periods, because fast degradation of glycogen under aerobic conditions would result in exhaustion of energy reserves within a few hours (Moezelaar et al., 1996). However, PolyP accumulation under low redox potential in dark may be important to *M. aeruginosa* for survival in unfavorable conditions. Therefore, the ability to accumulate PolyP in the dark and negative redox potential may be an advantage in low light, organically rich, and low-redox habitats. Finally, the phenomenon reported here may be important in understanding mechanisms of phosphorous

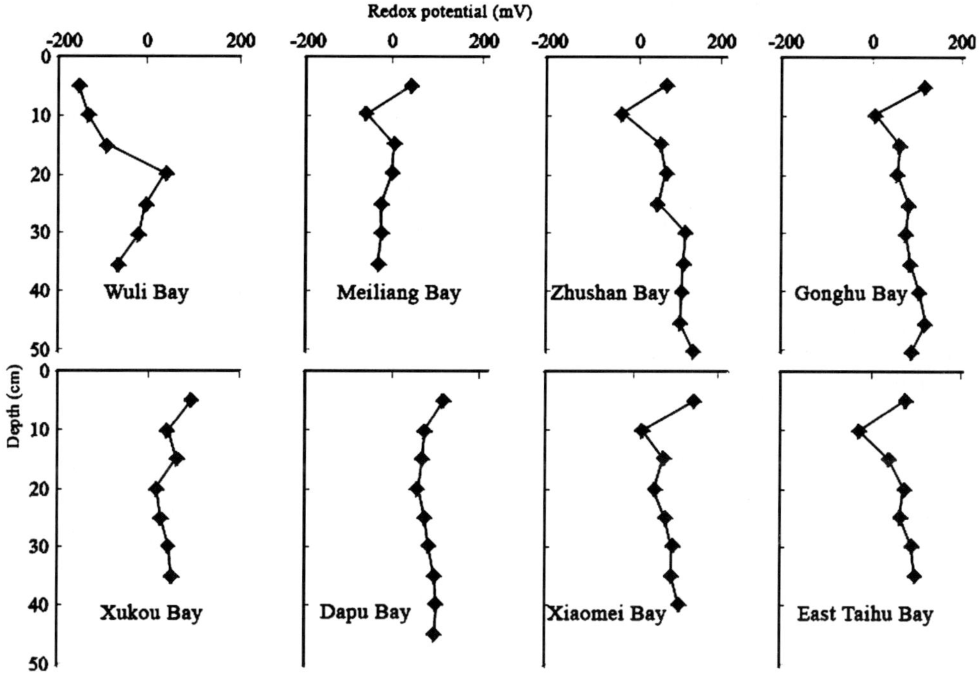

Fig. 9 Redox potential in sediment of main region of Lake Taihu. Data from (Zhu et al., 2004)

metabolism in *Microcystis* during high diel redox potential difference and explain why mats and *Microcystis* in sediment can remain viable for several months in Lake Taihu.

Acknowledgements This research was funded by Chinese Academy of Sciences (KZCX2-403), "973" Project (2002CB412307) and National Natural Science Foundation of China (No. 40371102, 40501078).

References

Bental, M., U. Pick, M. Avron & H. Degani, 1991. Polyphosphate metabolism in the alga *Dunaliella salina* studied by ^{31}P-NMR. Biochimca et Biophysica Acta 1092: 21–28.

Bostrom, B., G. Pertsson & B. Broberg, 1988. Bioavailability of different phosphorus forms in freshwater systems. Hydrobiologia 170: 133–155.

Chen, Y. W., B. Q. Qin, K. Teubner & M. T. Dokulil, 2003. Long-term dynamics of phytoplankton assemblages: Microcystis-domination in Lake Taihu, a large shallow lake in China. Journal of Plankton Research 25: 445–453.

Currie, D. J. & J. Kalff, 1984. The relative importance of bacterioplankton and phytoplankton in phosphorus uptake in freshwater. Limnology and Oceanography 29: 311–321.

Fuhs, G. W., S. D. Demmerle, E. Canelli & M. Chem, 1972. Characterization of phosphorous-limited planktonic algae. In Nutrients and Eutrophication. American Society of Limnology and Oceanography, Special Symposium 1: 113–132.

Greenberg, A. E., 1985. Standard Methods for the Examination of Water and Wastewater. American Public Health Association, Washington DC.

Grillo, J. F. & J. Gibson, 1979. Regulation of phosphate accumulation in the unicellular Cyanobacteria *Synechococcus*. Journal of Bacteriology 140: 508–517.

Harold, F. M., 1963. Inorganic polyphosphate of high molecular weight from aerobacter aerogenes. Journal of Bacteriology 86: 885–887.

Heyer, H. & W. E. Krumbein, 1991. Excretion of fermentation products in dark and anaerobically incubated cyanobacteria. Archives of Microbiology 155: 284–287.

Howarth, R. W., 1988. Nutrient limitation of net primary production in marine ecosystems. Annual Review of Ecology and Systematics 19: 89–110.

John, E. H. & K. J. Flynn, 2000. Modelling phosphate transport and assimilation in microalgae; how much complexity is warranted? Ecological Modelling 125: 145–147.

Johnston, N. A. L., V. S. Campagna, P. R. Hawkins & R. J. Banens, 1994. Response of the eastern rainbowfish (Melanotaenio duboulay) to toxic *Microcystis aeruginosa*. In Jones, G. J. (ed.), Cyanobacterial Research. CSIRO Cataloguing-in-Publication Entry, Australia: 187–193.

Keasling, J. D. & G. A. Hupf, 1996. Genetic manipulation of polyphosphate metabolism affects cadmium tolerance in *Escherichia coli.* Applied and Environmental Microbiology 62: 743–746.

Kornberg, A., 1995. Inorganic polyphosphate: toward making a forgotten polymer unforgettable. Journal of Bacteriology 177: 491–496.

Kulaev, I., V. Vagabov & T. Kulakovskaya, 1999. New aspects of inorganic polyphosphate metabolism and function. Journal of Bioscience and Bioengineering 88: 111–129.

Lawrence, B. A., C. Suarez, A. DePina, E. Click, N. H. Kolodny & M. Allen, 1998. Two internal pools of soluble polyphosphate in the Cyanobacterium Synechocystis sp. Strain PCC 6308: an in vivo ^{31}P NMR spectroscopic study. Archives of Microbiology 169: 195–200.

Leventer, H. & J. Eren, 1970. Taste and odor in the reservoirs of the Israel National Waster System. In Shuval, H. I. (ed.), Developments in Water Quality Research. Ann Arbor Humphrey Science Publishers, Mich: 19–37.

Magrath, J. W. & J. P. Quinn, 2000. Intracellular accumulation of polyphosphate by the yeast *Candida humicola* G-1 in response to acid pH. Applied and Environmental Microbiology 9: 4068–4047.

Mozelaar, R., S. M. Bijvank & L. J. Stal, 1996. Fermentation and sulfur reduction in the mat-building Cyanobacterium *Microcoleus chthonoplastes.* Applied and Environmental Microbiology 62: 1752–1758.

Nakano, S., K. Hayakawa, J. Frenette, T. Nakajima, C. M. Jiao, S. Tsujimura & M. Kumagai, 2001. Cyanobacterial blooms in a shallow lake: a large-scale enclosure assay to test the importance of diurnal stratification. Archiv fur Hydrobiologie 150: 491–509.

Noegel, A. & E. C. Gotschlich, 1983. Isolation of a high molecular weight polyphosphate from *Neisseria gonorrhoeae.* Journal of Experimental Medicine 157: 2049–2060.

Oh, H. M., S. J. Lee, M. H. Jang & B. D. Yoon, 2000. Microcystin production by *Microcystis aeruginosa* in a phosphorus-limited chemostat. Appied and Environmental Microbiology 66: 176–179.

Preston, T., W. D. P. Stewart & C. S. Reynolds, 1980. Bloom-forming cyanobacterium *Microcystis aeruginosa* overwinters on the sediment. Nature 288: 365–367.

Rao, N. N., M. F. Roberts & A. Torriani, 1985. Amount and chain length of polyphosphates in *Escherichia coli* depend on cell growth conditions. Journal of Bacteriology 162: 242–247.

Revsbech, N. P., B. B. Jorgensen, T. H. Blackburn & Y. Cohen, 1983. Microelectrode studies of the photosynthesis and O_2, H_2S and pH profiles of a microbial mat. Limnology and Oceanography 28: 1062–1074.

Richardson, L. L. & R. W. Castenholz, 1987. Enhanced survival of the Cyanobacterium *Oscillatoria terebriformis* in darkness under anaerobic conditions. Applied and Environmental Microbiology 53: 2151–2158.

Roe, J. H. & R. E. Dailey, 1966. Determination of glycogen with the anthrone reagent. Analytical Biochemistry 15: 245–250.

Schuddemat, J., R. De Boo, C. C. M. Van leeuwen, P. J. A. Brock & J. Van Steveninck, 1989. Polyphosphate synthesis in yeast. Biochimica et Biophysica Acta 1010: 191–198.

Shi, X. L., L. Y. Yang, X. J. Niu, L. Xiao, Z. M. Kong, B. Q. Qin & G. Gao, 2003. Intracellular phosphorus metabolism of *Microcystis aeruginosa* under various redox potential in darkness. Microbiological Research 158: 345–352.

Stal, L. J. & R. Moezelaar, 1997. Fermentation in cyanobacteria. FEMS Microbiology Review 21: 179–211.

Tash, C. J., 1967. Summary. In Bartsch, A. F. (ed.), Proceedings of the Symposium on the Environmental Requirements of Blue-green Algae. Federal Water Pollution Control Administration, Oreg: 103–106.

Weller, D., W. Doemel & T. Brock, 1975. Requirement of low oxidation-reduction potential for photosynthesis in a blue-green alga (*Phormidium* sp.). Archives of Microbiology 104: 7–13.

Zhu, G. W. & C. X. Fan, 2004. Physicochemical characters of sediment in Lake Taihu. In Qin B. Q., W. P. Hu & W. M. Chen (eds), Evolution Process and Mechanism of Water Environment of Lake Taihu. Science Publication Company, Beijing, 170–187 (in Chinese).

Hydrobiologia (2007) 581:177–188
DOI 10.1007/s10750-006-0511-7

EUTROPHICATION IN LAKES

The bacterioplankton of Lake Taihu, China: abundance, biomass, and production

Guang Gao · Boqiang Qin · Ruben Sommaruga · Roland Psenner

Abstract Abundance, biomass and production of pelagic bacteria were examined over one year at monthly sampling intervals across a trophic profile in Meiliang Bay, Lake Taihu. With the lowest density in the open lake, the bacterial abundance showed a clear trend in relation to trophic status. The carbon content per cell was higher in autumn and winter, and the opposite was true for bacterial biomass. Bacterial ^{3}[H]-*TdR* and ^{14}[C]-Leu incorporation rates, cell production, turnover times and carbon production varied during the annual cycle at different sites. The ratio of bacterial production to primary production was high, independently of the method used, indicates that the microbial food web in Lake Taihu is an important component of the total food web of the lake and dominated by external inputs.

Keywords Eutrophication · Trophic status · Bacterial biomass · Bacterial production · Lake Taihu

Guest editors: B. Qin, Z. Liu & K. Havens
Eutrophication of shallow lakes with special reference to Lake Taihu, China

Electronic Supplementary Material Supplementary material is available for this article at http://dx.doi.org/10.007/s10750-006-0511-7 and is accessible for authorized users

G. Gao (✉) · B. Qin
Nanjing Institute of Geography and Limnology, Chinese Academy of Sciences, 73 East Beijing Road, Nanjing 210008, P.R. China
e-mail: guanggao@niglas.ac.cn

R. Sommaruga · R. Psenner
Institute of Ecology, University of Innsbruck, Innsbruck, Austria

Introduction

Bacteria play a pivotal role in aquatic ecosystems. They use dissolved organic matter (DOM), which is mineralized and transformed into biomass, forming the base of the microbial food web (Azam et al., 1983). Measurement of bacterial cell numbers and biomass is, thus, a central aspect in the study of plankton.

By using fluorescent dyes and epifluorescence microscopy for direct counting (Hobbie et al., 1977; Porter & Feig, 1980; reviewed by Raymond, 1994), combined with automated image analysis (Psenner, 1991a, b) or electron microscopy (Krambeck et al., 1981) for measuring bacterial size distribution and cell volumes, reliable estimates of bacterial biomass can be made. Using the incorporation of radio-labeled precursors into DNA (Thymidine) and proteins (Leucine) (Fuhrman & Azam, 1980, 1982; Kirchman et al., 1985, 1986), as well as other approaches, e.g., the frequency of dividing cells (Davis & Sieburth,

1984), realistic estimates of in situ bacterioplankton production can be achieved.

Eutrophication and contaminant enrichment can directly and indirectly affect structure and functioning of microbial food webs, because any changes in trophic state of a water body would affect the relative significance of microorganisms and alter the entire pelagic food web (Weisse, 1991; Sommaruga, 1995). Bacterial abundance (Bird & Kalff, 1984) and production (Cole et al., 1988) correlate with phytoplankton biomass and primary production. Therefore, bacterial biomass and production will be enhanced with increased nutrient loading and primary production (Weisse, 1991). Also, the abundance and biomass of other microbial food web components (such as autotrophic picoplankton, small eukaryotic algae, heterotrophic nanoflagellates, ciliates and metazoan microplankton) will increase, which will decrease the relative importance of the microbial food web to carbon flux (Weisse, 1991).

In this paper, we reported a study on selected microbial parameters along a trophic gradient in Lake Taihu, a large shallow lake in eastern China. The major aim of these studies was to evaluate the relationships between the bacterioplankton abundance, cell volume and biomass, production and the trophic state of lake. We tested the hypothesis that bacteria and lake eutrophication are tightly coupled and bacterial growth in lakes largely occurs at the expense of allochthonous carbon sources other than phytoplankton production.

Materials and methods

Sampling sites

Lake Taihu is the third largest fresh water lake in China and is influenced by intensive blooms of blue-green alga during summer. In Meiliang Bay, at the northern end of Lake Taihu at the inflow of the main tributary, a north-south profile across the bay and a transverse profile from west to east were chosen (Fig. 1). These five stations represented different trophic levels, including a highly contaminated area at the river mouth (#0) and several sites in Meiliang Bay with nutrient levels decreasing with distance from the river mouth (3#, #5, and site #2 in front of the Taihu Station). The site with the lowest influence from nutrient inputs by the river is the open area (#8) in the center of the lake.

Sampling

Samples were taken with a 5 l Patalas–Schindler sampler. A 50 ml sub-sample was transferred into a glass bottle (75 ml) containing 2.5 ml of pre-filtered (pore size, 0.2 μm) formaldehyde (final concentration 2%). The glass bottles were washed with acid, rinsed with particle-free distilled water and autoclaved one day before use. After return to the laboratory, samples were stored in a refrigerator at 4°C until slides for enumeration and image analysis were prepared.

Bacterial abundance

For bacteria abundance, a well-mixed sub-sample was diluted with particle-free distilled water (1:40) and stained with 4′, 6′-diamidino-2-phenylindole (DAPI, Sigma) at a final concentration of 2 μg ml^{-1} for 7 min (modified from Porter & Feig, 1980; Raymond et al., 1994; Posch et al., 2001). After staining, samples were filtered onto black polycarbonate filters (0.2 μm pore size, 25 mm diameter, Poretics^{TM}) and embedded in nonfluorescent immersion oil (Cargille type A, Cargille Laboratories, Inc., USA). On each slide, at least 400 bacteria were counted at 1600× with a Zeiss Axiovert 135 M epifluorescence microscope equipped with a HBO 50 W mercury lamp and a filter set for UV excitation (BP 365, FT 395, LP397 nm, Zeiss filter set 01) (Zeiss, Germany).

Bacterial biomass

Bacterial biomass is the product of cell abundance and carbon content per cell. Measurement of cell sizes and carbon contents is described elsewhere. Briefly, cell length and width were measured microscopically using a DAPI concentration of 1 μg ml^{-1} and a sensitive camera coupled to an image analysis system. Carbon

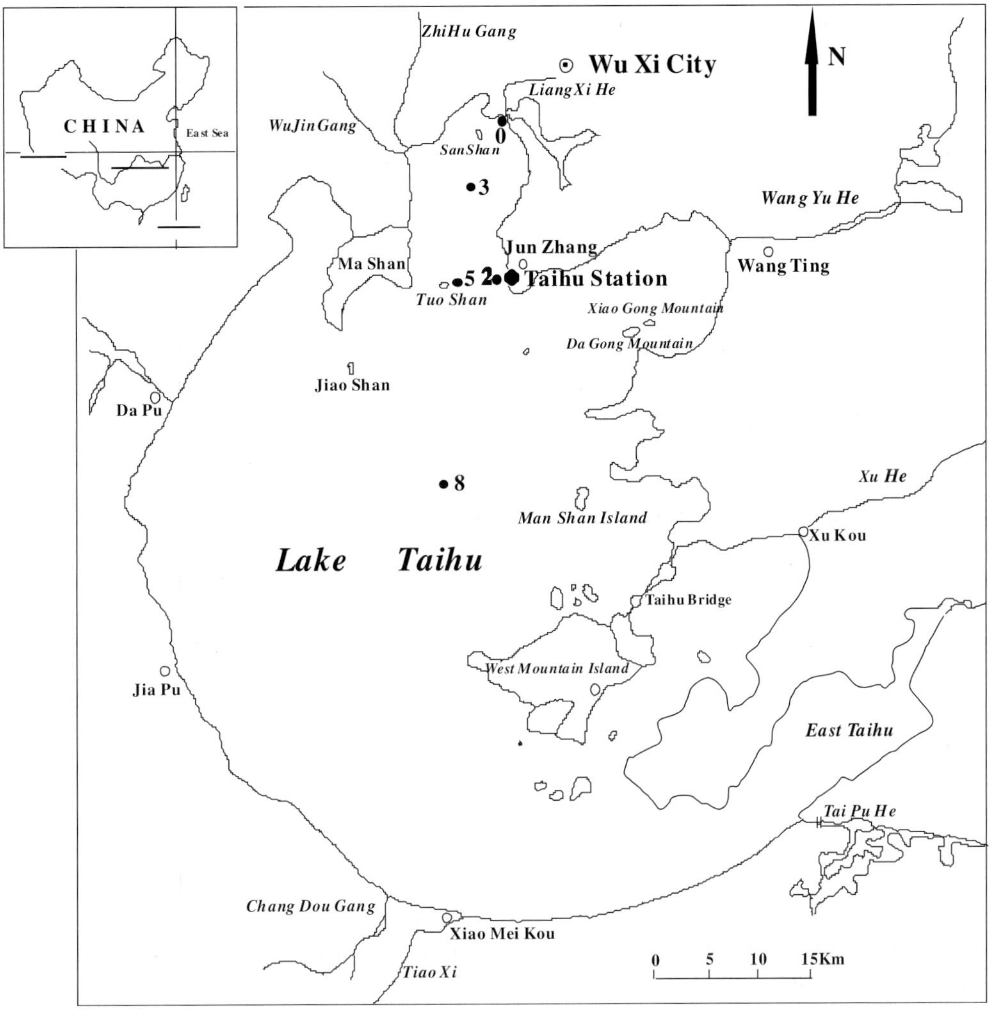

Fig. 1 The location of the sampling sites in Lake Taihu

content was inferred by an empirical formula established under comparable staining procedures (Loferer-Krößbacher et al., 1998; Posch et al., 2001).

Metabolically active bacteria

A 500 ml sub-sample was transferred into a glass bottle prepared as above. After return to the laboratory, 1 ml of 2-(*p*-iodophenyl)-3-(*p*-nitrophenyl)-5-phenyl tetrazolium chloride (INT) stock solution (2 mg ml^{-1}) was added to 10 ml of sample in 20 ml polyethylene vials (INT final conc. ca. 0.02%). All samples were incubated at in situ temperature in a water bath for 1 h without any substrate addition. After incubation, samples were fixed with pre-filtered (pore size, 0.2 μm) formaldehyde (final concentration 2%). Samples were stored in a refrigerator at 4°C at least 1 h and no longer than 24 h before preparation (modified from Zimmermann et al., 1978; Dufour & Colon, 1992; Thom et al., 1993). Other procedures were the same as for bacterial abundance estimations. A blue filter and halogen lamp were used for INT measurements.

Bacterial production (BP)

A dual-label incubation method was used to measure incorporation of thymidine (*TdR*) and leucine (Leu) by adding [^{3}H]-*TdR* and [^{14}C]-Leu to a single sample and collecting cold-TCA-insoluble material (modified from Chin-Leo & Kirchman, 1988; Kirchman & Hoch, 1988). For each sample, triplicates of 10 ml each and one control (already containing 0.5 ml of 0.2 μm pre-filtered formalin at a final concentration of 2%) were used. In each sub-sample, 100 nM (final concentration) of [^{14}C]-Leu (specific activity, 308 mCi mmol^{-1}, Amersham Life Science, England.) was added, and samples were incubated for 40 min in a water bath at in situ temperature. After 40 min, 10 nM (final concentration) of [^{3}H]-*TdR* (specific activity, 81 Ci mmol^{-1}, Amersham Life Science, England.) was added, and samples were incubated for an additional 20 min. The incubation was stopped by adding 0.5 ml formalin at a final concentration of 2%. After incubation, samples were filtered through cellulose nitrate filters (Sartorius^{TM}, 0.2 μm pore size, 25 mm diameter). Then, 5 ml of ice-cold 5% TCA were added for 5 min to extract macromolecular fractions. The filter was rinsed 3 times with 5 ml of ice-cold 5% TCA. Radioactivity incorporated into cellular material was assayed by a Beckman liquid scintillation counter (LS-8000IC).

According to the literature (Simon & Azam 1989, Wetzel & Likens 2000), the following formulas and conversion factors were used to calculate bacterial [^{3}H]-*TdR* and [^{14}C]-Leu incorporation, cell production and carbon production.

$$[^{3}\mathrm{H}] - TdR.\mathrm{incorporation}\ [\mathrm{n\ mol}\ TdR\ \mathrm{l}^{-1}\ \mathrm{h}^{-1}] = \frac{U \times 4.5 \times 10^{-13}}{S \times T \times V}$$

where 4.5×10^{-13} = the number of curies per dpm; U = dpm of the filter; S = Specific activity of the *TdR* [Ci mmol^{-1}]; T = Incubation time [h]; V = Volume of filtered sample [l].

Cell production [cells l^{-1} h^{-1}] = 2×10^{18} cells mol^{-1} *TdR*

Carbon production [μg C l^{-1} h^{-1}] = TI × CP × CC

where TI = *TdR* incorporation [moles*TdR* l^{-1} h^{-1}]; CP = Cell production [cells mol^{-1}]; CC = Cellular carbon content [μg C cell^{-1}].

$$[^{14}\mathrm{C}]\ \text{- Leu. incorporation}\ [\mathrm{n\ mol\ Leu\ l}^{-1}\ \mathrm{h}^{-1}] = \frac{U \times 4.5 \times 10^{-13}}{S \times T \times V}$$

where 4.5×10^{-13} = the number of curies per dpm; U = dpm of the filter; S = Specific activity of the Leu [Ci mmol^{-1}]; T = Incubation time [h]; V = Volume of filtered sample [l].

Cell production [cells l^{-1} h^{-1}] = 0.07×10^{18} cells mol^{-1} Leu.

BCP [μg C l^{-1} h^{-1}] = mol. Leu (inc.) × $(7.3/100)^{-1}$ × 131.2 × ID × (C/protein)

where: 131.2 = molecular weight of leucine

$(7.3/100)^{-1}$ = mole% of leucine in protein, hence 0.073

C/protein = ratio of cellular carbon to protein, best estimate = 0.86

ID = Isotope dilution; here ID = 2

Results

Some physical and chemical parameters in different parts of Lake Taihu during May 1998 to May 1999 are presented in Electronic Supplementary material. Due to the shallowness of the lake and strong winds, the water column remained mixed. However, nutrient concentrations, Chl *a* concentration, and other physical and chemical parameters varied among the sampling sites and showed a clear trophic gradient from the river mouth (hypertrophy), Meliang Bay (eutrophy) to the open lake (meso/eutrophy).

Bacterial abundance

The distribution patterns of bacterial abundance varied significantly along the trophic gradient of this lake (One-way ANOVA, $P = 0.005$). At the hypertrophic river mouth, the highest bacterial abundance was determined. With the mean value of 15.4×10^{9} l^{-1}, the bacterial abundance was ca. 55% higher than the mean in eutrophic Meiliang Bay and almost twice as high as the meso/eutrophic open lake (Fig. 2).

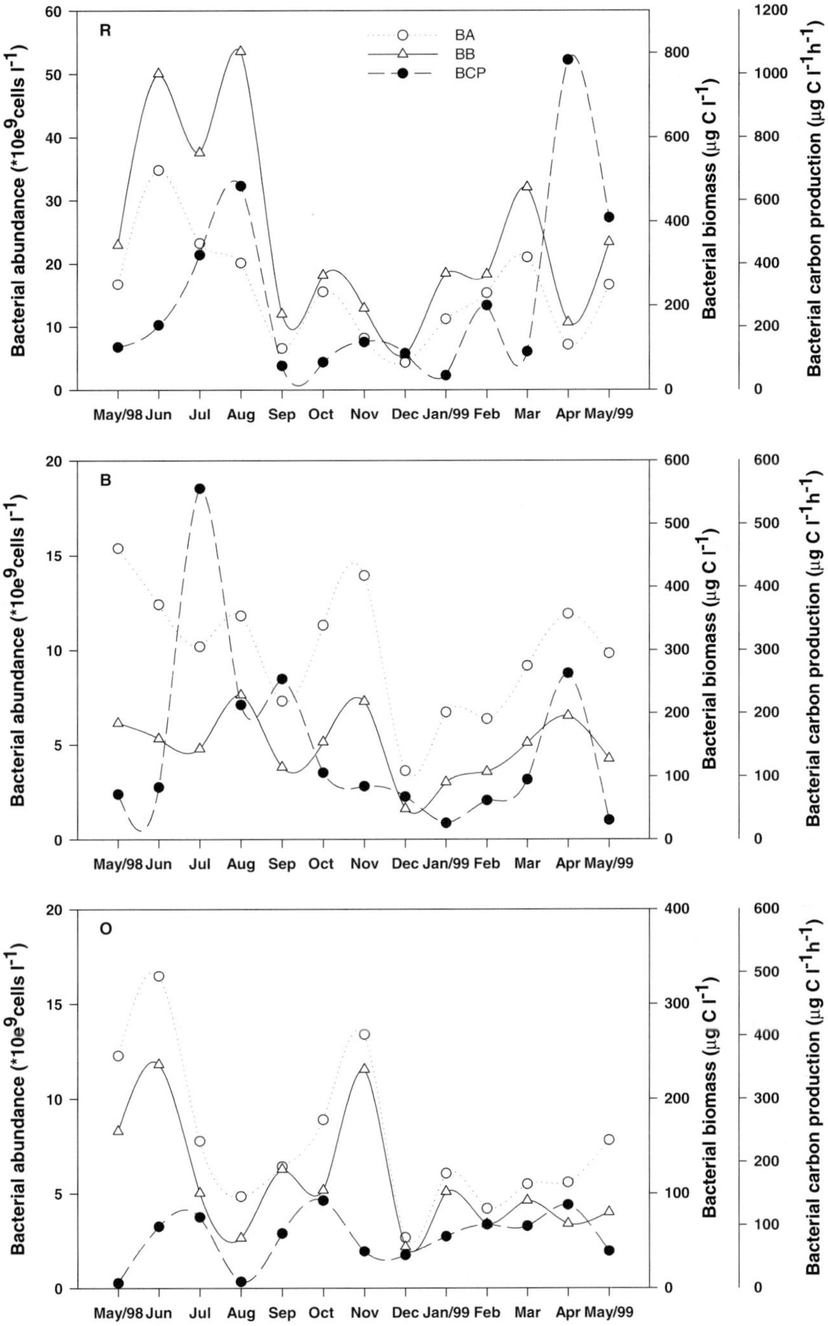

Fig. 2 Seasonal distribution of bacterial abundance, biomass and carbon production at 3 different sites of Lake Taihu from May 1998 to May 1999. R ... river mouth; B ... Meiliang Bay; O ... open lake

Bacterial biomass

The bacterial biomass also markedly increased along the trophic gradient of Lake Taihu (Fig. 2). At the hypertrophic river mouth, the mean bacterial biomass was 364 ± 225 μg C l^{-1}, i.e. ca. 1.5 times and 2.2 times higher than that at eutrophic Meiliang Bay and the open lake, respectively.

The difference between the highest and lowest value was ca. tenfold at the river mouth, while only ca. fivefold were recorded both in the Meiliang Bay and the open lake. No significant differences for seasonality were found in different trophic area of this lake (One-way ANOVA, $P > 0.05$).

Active bacteria

A highly significant positive correlation was found between the number of active bacteria and the total bacterial abundance in all parts of this lake, although this relationship varied among the habitats (Fig. 3.). In the hypertrophic river mouth, the mean ratio of active bacteria versus total bacterial abundance (34%) was slightly higher than that in the eutrophic Meiliang Bay (29%) and the meso/eutrophic open lake (28%). Variation of the percentage of active cells between the different trophic sites was smaller than the variation in abundance and biomass.

Bacterial production

The bacterial [^{3}H]-*TdR* incorporation rate also increased along the trophic gradient, and showed a significantly different over time (One-way ANOVA, $P < 0.0001$, $n = 65$) and space (One-way ANOVA, $P = 0.04$, $n = 65$), although the distribution patterns were similar in all of the sites. At the river mouth, with the maximum value of 14603 pmol l^{-1} h^{-1}, the annual mean value of [^{3}H]-*TdR* incorporation rate was ca. 1.2 times than Meiliang Bay and 2.2 times higher than the open lake, respectively (Table 1).

While the turnover times estimated from cell production were extremely fast and highly variable among the sampling sites (Fig. 3). At the hypertrophic river mouth, the mean turnover times was only ca. 87% less than the open lake, although the cell production was ca. 3 times higher (Fig. 3). Differences of turnover times within the temporal (One-way ANOVA, $P > 0.05$, $n = 65$) and spatial (One-way ANOVA, $P > 0.05$, $n = 65$) distribution were not significant.

Bacterial carbon production at different sites showed a similar distribution pattern as bacterial [^{3}H]-*TdR* incorporation rate, particularly at the river mouth and in Meiliang Bay (Fig. 2). With a mean of 141.8 ± 52.8 μg C l^{-1} h^{-1} at the river mouth, bacterial carbon production was ca. 2.4 and 4.5 times higher than means at Meiliang Bay and the open lake, respectively.

Temporal distribution of bacterial incorporation rates at different sites were similar for both [^{14}C]-Leu and [^{3}H]-*TdR*, even though the value for [^{14}C]-Leu was much higher (Figs. 3 and 4). No significant differences were found between bacterial cell production based on either method (Paired *t*-test, $P > 0.05$). However, turnover times estimated from [^{3}H]-*TdR* and [^{14}C]-Leu at the river mouth (One-way ANOVA, $P = 0.002$, $n = 39$) and Meiliang Bay (One-way ANOVA, $P = 0.001$, $n = 39$) were significantly different, although cell production based on both methods was not significantly different. In the open lake, no significant difference was found (One-way ANOVA, $P > 0.05$, $n = 39$).

While a clear positive correlation between [^{3}H]-*TdR* and [^{14}C]-Leu measurement was found, bacterial carbon production measured by [^{14}C]-Leu was ca. 2 times higher than that of [^{3}H]-*TdR* (Fig. 4). Since both methods are based on measuring some aspect of macromolecular synthesis (DNA for *TdR*, Protein for Leu), it is possible to convert these incorporation rates to rates of macromolecular synthesis and, in turn, to biomass production. But this conversion depends on detailed information about several cellular components, thus bacterial cell size and biomass were measured at the same time (Figs. 3 and 5).

Discussion

Factors affecting bacterial abundance and activity

Bacterioplankton assemblages often are dominated by inactive or dormant cells, and only a small fraction of bacterial cells appear to be metabolically active in aquatic systems (Stevenson, 1978; Mason et al., 1986). However, the fraction of active cells varies in time and space (Zimmerman et al., 1978; Quinn, 1984). Over a broad range of aquatic systems, the total number

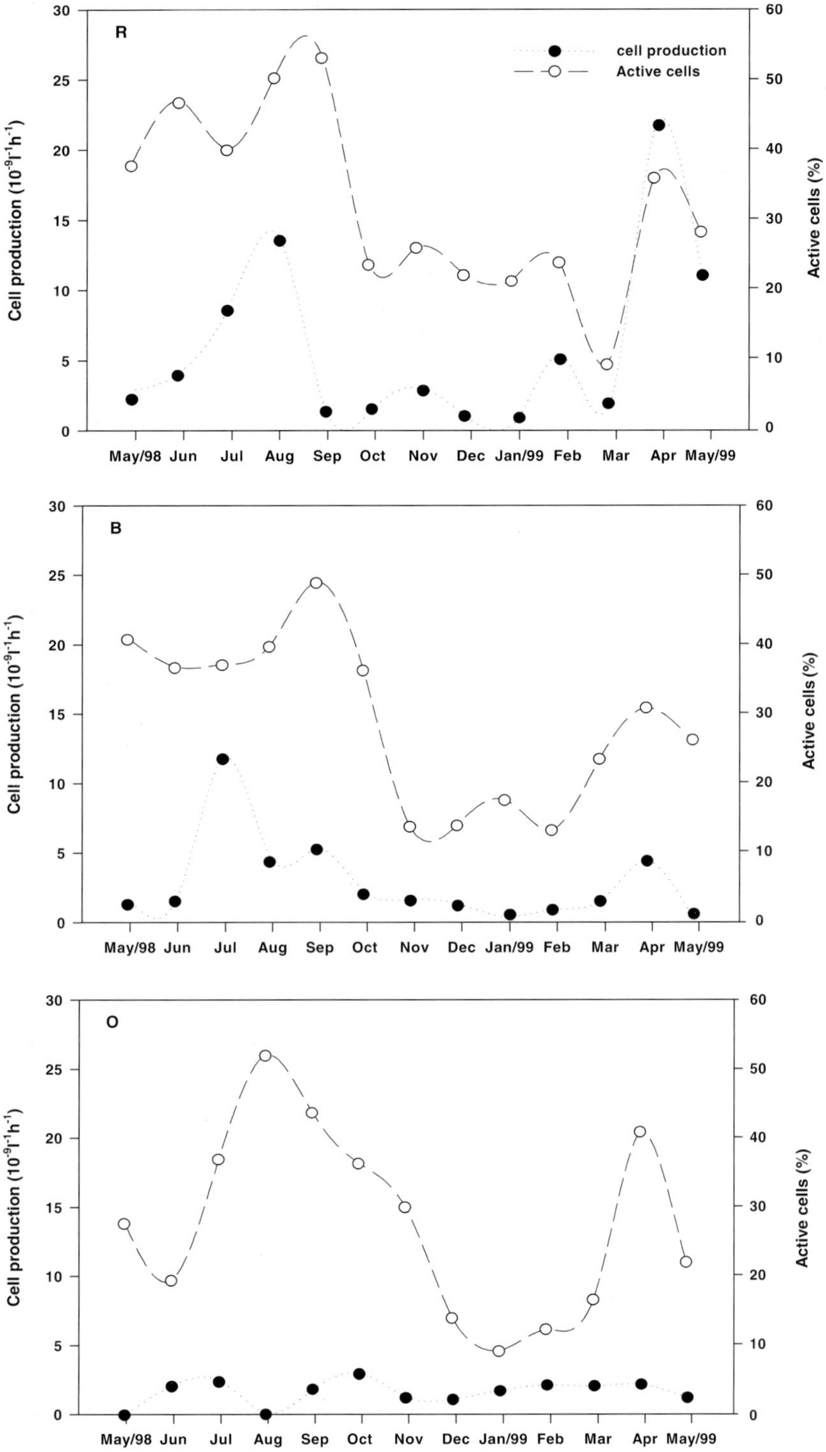

Fig. 3 Seasonal distribution of cell production rates and percentage of active cells at 3 different sites of Lake Taihu from May 1998 to May 1999. R ... river mouth; B ...Meiliang Bay; O ... open lake

Table 1 Leucine and Thymidine incorporation rates (pmol l^{-1} h^{-1}) at different sites of Lake Taihu

		TdR	Leu	*TdR*/Lue
R	Mean ± S.E	2878.8 ± 1069.0	92.2 ± 25.3	31.2
	Range	343.3–14603.2	14.0–323.9	24.5–45.1
B	Mean ± S.E	1338.8 ± 409.3	45.6 ± 12.6	29.4
	Range	195.2–4755.8	8.1–172.4	24.1–27.6
O	Mean ± S.E	886.5 ± 150.0	24.7 ± 3.5	35.9
	Range	68.1–2124.1	2.5–43.0	27.2–49.4

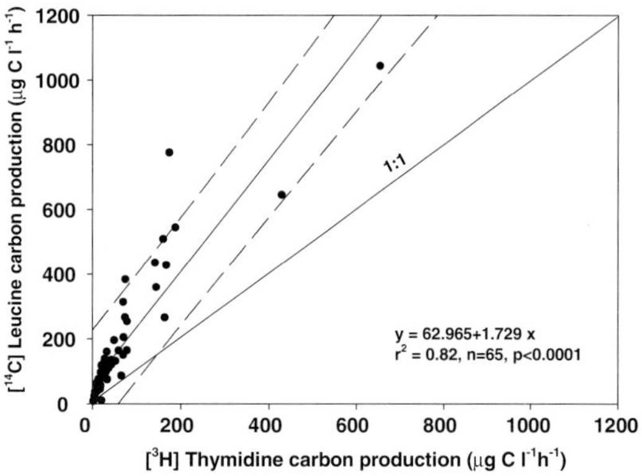

Fig. 4 Correlation of bacterial carbon production measured by [^{3}H] *TdR* and [^{14}C] leu in Lake Taihu from 1998 to 1999. $r^2 = 0.82$, $n = 65$, $p < 0.0001$. The dashed line shows the 95% confidence limits

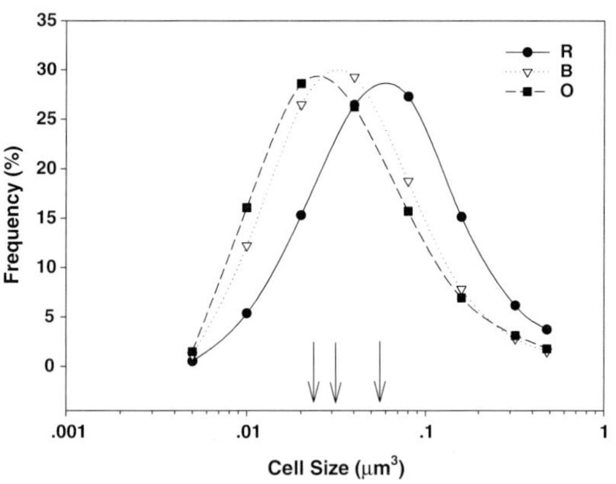

Fig. 5 The distribution of bacterial cell size at 3 different sites of Lake Taihu. R ... river mouth; B ... Meiliang Bay; O ... open lake

of bacteria varied by three orders of magnitude, active bacteria varied by four orders of magnitude, and the proportion of active cells increased systematically along gradients of enrichment from ultra-oligotrophic open ocean areas (<5%) to highly productive estuaries (>50%) (Del Giorgio & Scarborough, 1995).

With lowest bacterial abundance in the open lake, bacterial density showed a clear trend in relation to trophic status in Lake Taihu (Fig. 2), corresponding to the overall trend of abundance increase with system enrichment (Weisse, 1991). Bacterial abundance in the highly contaminated river mouth, however, was only slightly higher and not significantly different from those observed in eutrophic Meiliang Bay. Similar results were obtained by Azam et al. (1983) and Bird & Kalff (1984), who also found that fluctuations of bacterial biomass are small across trophic gradients. The low rate of increase in total bacterial density may imply that total bacterial biomass tends to be more uniform among systems, and their relative contribution to community biomass declines with increasing system productivity (Fuhrman et al., 1989). This indicates that structural and functional characteristics of the 'microbial loop' may be operating differently in stressed versus unstressed ecosystems (Munawar & Weisse, 1989).

With a mean ratio of ca. 34%, 29% and 28% in the river mouth, Meiliang Bay and open lake, respectively, the ratio of metabolically active bacteria to total bacterial abundance does not always correspond to these environmental gradients, although a significant positive correlation between total bacterial abundance and active bacteria existed in all parts of Lake Taihu (Fig. 3). However, data on total and active bacteria from the literature for aquatic systems show that our data are in the range of pelagic systems in general (Del Giorgio & Scarborough, 1995). The estimated average proportion of active cells in this work was similar to the mean for published data on lake bacteria (23%) (Del Giorgio & Scarborough, 1995), although some papers show that this ratio can reach 100% in a hypertrophic lake (Sommaruga, 1995). In most aquatic systems, there is a pool of rapidly growing cells embedded in a larger matrix of inactive bacteria, and the relative size of the active and inactive pools varies systematically along gradients of enrichment (Del Giorgio & Scarborough, 1995). This large variance among systems may also reveal that there are differential control mechanisms for active and inactive bacteria in different ecosystems.

Although Billen (1990) suggested that the lack of simultaneous changes in phytoplankton and bacterioplankton abundance on a seasonal scale is caused, in many cases, by the delayed response of bacteria to phytoplankton development, it is reasonable that the number of active cells is more variable among systems than the total number of bacteria, and that only metabolically active bacterioplankton are responsible for growth, nutrient uptake and transformation of organic carbon substrates (Del Giorgio & Scarborough, 1995). Moreover, this variation may be masked by a rather large and constant pool of inactive bacteria. Thus, the function of bacterioplankton, based on the total number of bacteria, may have to be revised to accommodate large variations in the proportion of metabolically active cells (Del Giorgio & Scarborough, 1995). However, like other planktonic components, bacterioplankton density and production are correlated positively to system productivity, such as nutrient concentration, dissolved organic carbon, chlorophyll *a*, and primary production (Cole et al., 1988).

Leucine and Thymidine incorporation

Although there are still technical problems, incorporation of [^{3}H] thymidine (*TdR*) into DNA and [^{14}C] leucine (Leu) into protein are considered the most reliable methods (Fuhrman & Azam, 1980; Chin-Leo & Kirchman, 1988). However, accurate estimates of bacterial production depend on use of an appropriate conversion factor (Chin-Leo & Kirchman, 1988), which may vary (even by more than a factor of 100) over short time and space scales (Kirchman et al., 1986; Chin-Leo & Kirchman, 1988; Smits & Riemann, 1988). In this case, observed variation in thymidine incorporation could be due to changes in conversion factors rather than bacterial production (Kirchman & Hoch, 1988). So, it is necessary to ensure that observed changes in *TdR* incorporation are due to real changes in bacterial production, not in the conversion factor or some artifact (Chin-Leo & Kirchman, 1988).

When both *TdR* and Leu incorporation are measured in a single incubation by the dual-label approach, agreement between independent measurements made simultaneously should separate real changes from artifacts more effectively than a single-label approach (Fig. 4). Thus, we are confident that observed variations reflect real changes in the metabolism of bacterial assemblages (Kirchman et al., 1986). In addition, incorporation rates determined from dual-label incubations differed only slightly from those obtained from single-label incubations (Chin-Leo & Kirchman, 1988). The dual-label procedure, however, can reduce the time and material needed to perform simultaneous measurements of *TdR* and Leu incorporation, while minimizing errors.

Physico-chemical parameters directly or indirectly affecting bacterial production vary over time and relatively short distances (Kirchman & Hoch, 1988), which may cause methodological problems, especially in attempting to measure bacterial production. For this reason, the dual-label approach was used to measure *TdR* and Leu incorporation. In spite of problems inherent to each method, *TdR* and Leu incorporation rates covaried and gave the same qualitative picture of bacterial production in different parts of Lake Taihu (Table 1, Fig. 4). Although bacterial carbon production data lay within the broad range of the literature, bacterial carbon production measured by Leu was ca. 2 to 3 times higher than by *TdR*. This may indicate that conversion factors in Lake Taihu vary by a factor of two or so and differences may be due to methodology. Servais (1992) suggested that bacterial production measured by *TdR* seems to be systematically underestimated compared with those measured by Leu.

Bacterial production and its coupling with primary production

Bacterial production can be used to estimate average growth rate of bacterial assemblages and as an indicator of bacterial response to fluctuations in environmental conditions (Chin-Leo & Kirchman, 1988). Bacterial abundance and production are correlated positively with phytoplankton biomass and primary production in a broad range of freshwater and marine systems, and bacterial production varied from 2% to >80% of primary production (Bird & Kalff, 1984; Cole et al., 1988). However, bacterial abundance and production do not increase proportionally with chlorophyll concentration as well as primary production (Robarts et al., 1994), although a positive empirical relationship was found between bacterial abundance and chlorophyll concentration in fresh and marine waters (Bird & Kalff, 1984).

Lake Taihu data show that bacterioplankton production was higher at the river mouth than in the open lake (Table 1, Fig. 3). Bacterial production (both measured by *TdR* and Leu) and phytoplankton biomass at the river mouth and in Meiliang Bay were correlated positively, but not in the open lake. The bacterial production to primary production ratio varied among the sampling sites (One-way ANOVA, $P = 0.023$, $n = 65$) and increased along the trophic gradient with a mean value ca. 2 and 4 times higher than in Meiliang Bay and the open lake, respectively. At the same time, no significant correlation between primary production and bacterial secondary production based on [^{3}H]-*TdR* and [^{14}C]-Leu measurement was found in any part of Lake Taihu.

The mean ratio of bacterioplankton production to primary production (Table 2) presented a similar picture to that in humic-rich reservoirs, where bacterioplankton production was usually ca. 2 to 4 times higher than primary production (Balogh & Vörös, 1997). These authors found that lakes receiving large amounts of humic compounds contain larger standing stocks of pelagic bacteria than less humic lakes, suggesting a large non-phytoplankton carbon source. Our results support the hypothesis that bacterial growth in lakes largely occurs at the expense of allochthonous carbon sources rather than phytoplankton production, especially at the river mouth and in Meiliang Bay, where organic matter is entering from the river. Moreover, the ratio of bacterioplankton to primary production decreases with trophic status (Table 2), which is in agreement with other results (Cole et al., 1988), who associated increasing eutrophication and bacterioplankton production, and a decreasing ratio of bacterioplankton biomass and production to Chl *a*. They suggested that the relative importance of bacterioplankton in lake ecosystems seem to decrease markedly with trophic status of the studied ecosystems (Cole et al. 1988).

Conclusions

With the highest value in the hypertrophic river mouth, the abundance, biomass and production of pelagic bacteria markedly increased along the trophic gradient of Lake Taihu (the open lake, Meiliang Bay, and the river mouth). A positive correlation between bacterial production (measured by *TdR* and Leu) and phytoplankton biomass was found in the river mouth and Meiliang Bay, but not in the open lake. The bacterial to primary production ratio increased with trophic status. No significant correlation between primary production and bacterial secondary production was found in any part of Lake Taihu. Our results support the hypothesis that bacteria and lake eutrophication are tightly

Table 2 Comparison of primary production and bacterial secondary production (μg C l^{-1} h^{-1}) at different sites of Lake Taihu from May 1998 to May 1999

		PP	*TdR*	Leu	*TdR*/PP	Leu/PP
R	Mean ± S.E	79.1 ± 8.1	141.8 ± 52.8	297.1 ± 81.5	1.8 ± 0.5	4.0 ± 1.0
	Range	37.0–138.4	17.0–654.5	45.1–1044.1	0.2–5.9	0.5–11.6
B	Mean ± S.E	76.0 ± 9.4	41.2 ± 7.9	147.1 ± 25.6	0.8 ± 0.2	3.1 ± 0.7
	Range	19.8–309.1	4.2–173.8	16.2–775.3	0.1–4.7	0.2–19.7
O	Mean ± S.E	76.1 ± 9.3	25.8 ± 4.7	79.8 ± 11.4	0.5 ± 0.1	1.5 ± 0.4
	Range	18.9–135.6	1.8–65.3	8.2–138.7	0.01–1.6	0.1–5.3

* The primary production (PP) was calculated from Chl *a* concentration by using the empirical formula: PP = –0.63 + 0.197 × Chl *a*, ($r = 0.90$) (Huang et al., 2001), where PP (mg O_2 m^{-2} d^{-1}), Chl *a* (mg m^{-3}), by using the factor of 2.67 to convert the O_2 unit to C unit (Sorokin, 1999)

coupled and bacterial growth in lakes largely occurs at the expense of allochthonous carbon sources other than phytoplankton production (Balogh & Vörös, 1997), especially at the river mouth and in Meiliang Bay. Moreover, the ratio of bacterioplankton to primary production decreases with trophic status (Table 2)

Acknowledgements This study was supported by the key Projects of CAS (No. KZCX1-SW-12, KZCX-SW-01-15). G. Gao was supported by a PhD scholarship of OeAD, the Austrian Exchange Service.

References

Azam, F., T. Fenchel, J. S. Field, J. S. Gray, L. A. Meyer-Reil & F. Thingstad, 1983. The ecological role of water-column microbes in the sea. Marine Ecology Progress Series, 10: 257–263.

Billen, G., 1990. Delayed development of bacterioplankton with respect to phytoplankton: a clue for understanding their trophic relationships. Archiv für Hydrobiologie Beihefte Ergebnisse der Limnologie 34: 191–201.

Bird, D. F. & J. Kalff, 1984. Empirical relationship between bacterial abundance and chlorophyll concentrations in fresh and marine waters. Canadian Journal of Fisheries and Aquatic Sciences 41: 1015–1023.

Chin-Leo, G. & D. L. Kirchman, 1988. Estimating bacterial production in marine water from the simultaneous incorporation of thymidine and leucine. Applied and Environmental Microbiology 54: 1934–1939.

Cole, J. J., S. Findlay & M. L. Pace, 1988. Bacterial production in fresh and saltwater ecosystems: a cross–system overview. Marine Ecology Progress Series 43: 1–10.

Davis, P. G. & J. M. Sieburth, 1984. Estuarine and oceanic microflagellate predation of actively growing bacteria: estimation by frequency of dividing – divided bacteria. Marine Ecology Progress Series 19: 237–246.

Del Giorgio, P. A. & G. Scarborough, 1995. Increase in the proportion of metabolically active bacteria along gradients of enrichment in freshwater and marine plankton: implications for estimates of bacterial growth and production rates. Journal of Plankton Research 17: 1905–1924.

Dufour, P. & M. Colon, 1992. The tetrazolium reduction method for assessing the viability of individual bacterial cells in aquatic environments: improvements, performance and applications. Hydrobiologia 232: 211–218.

Fuhrman, J. A. & F. Azam, 1980. Bacterioplankton secondary production estimates for coastal waters of British Columbia, Antarctica, and California. Applied and Environmental Microbiology 39: 1085–1095.

Fuhrman, J. A. & F. Azam, 1982. Thymidine incorporation as a measure of heterotrophic bacterial production in marine surface waters: evaluation and field results. Marine Biology 66: 109–120.

Fuhrman, J. A., T. D. Sleeter, C. A. Carlson & L. M. Proctor, 1989. Dominance of bacterial biomass in the Sargasso Sea and its ecological implications. Marine Ecology Progress Series 57: 207–217.

Hobbie, J. E., R. J. Daley & S. Jasper, 1977. Use of nucleopore filters for counting bacteria by fluorescence microscopy. Applied and Environmental Microbiology 33: 1225–1228.

Huang, Y. P., C. X. Fan, P. M. Pu, J. F. Jiang & Q. Y. Dai, 2001. The Water Environment and Pollution Control in Lake Taihu. Sciences Press, Beijing (in Chinese).

Kirchman, D. L., E. Knees & R. Hodson, 1985. Leucine incorporation and its potential as a measure of protein synthesis by bacteria in natural aquatic systems. Applied and Environmental Microbiology 49: 599–607.

Kirchman, D. L. & M. P. Hoch, 1988. Bacterial production in the Delaware Bay estuary estimated from thymidine and leucine incorporation rates. Marine Ecology Progress Series 45: 169–178.

Kirchman, D. L., S. Y. Newell & R. E. Hodson, 1986. Incorporation versus biosynthesis of leucine: implications for measuring rates of protein synthesis and biomass production by bacteria in marine systems. Marine Ecology Progress Series 32: 47–59.

Krambeck, C., H. J. Krambeck & J. Overbeck, 1981. Microcomputer assisted biomass determination of plankton bacteria on scanning electron micrographs. Applied and Environmental Microbiology 42: 142–149.

Loferer-Krößbacher, M., K. P. Witzel & R. Psenner, 1998. Determination of bacterial cell dry mass by transmission electron microscopy and densitometric image analysis. Applied and Environmental Microbiology 64: 688–694.

Mason, C. A., G. Hamer & J. D. Bryers, 1986. The death and lysis of microorganisms in environmental processes. FEMS Microbiology Review 39: 373–401.

Munawar, M. & T. Weisse, 1989. Is the 'microbial loop' an early warning indicator of anthropogenic stress? Hydrobiologia 188/189: 163–174.

Porter, K. G. & Y. S. Feig, 1980. The use of DAPI for identifying and counting aquatic microflora. Limnology and Oceanography 25: 943–948.

Posch, T., M. Loferer–Krößbacher, G. Gao, A. Alfreider, J. Pernthaler & R. Psenner, 2001. Precision of bacterioplankton biomass determination: a comparison of two fluorescent dyes, and of allometric and linear volume-to-carbon conversion factors. Aquatic Microbial Ecology 25: 55–63.

Psenner, R., 1991a. Detection and sizing of aquatic bacteria by means of epifluorescence microscopy and image analysis. Microscope Analysis 26: 13–15.

Psenner, R., 1991b. Determination of bacterial cell volumes by image analysis. Verhandlungen Internationale Vereinigung Für Theoretische und Angewandte Limnologie 24: 2605–2608.

Quinn, J. P., 1984. The modification and evaluation of some cytochemical techniques for the enumeration of metabolically active heterotrophic bacteria in the aquatic environment. Journal of Applied Bacteriology 57: 51–57.

Raymond, L., J. R. Kepner & J. R. Pratt, 1994. Use of fluorochromes for direct enumeration of total bacteria in environmental samples: past and present. Microbiology Review 58: 603–615.

Robarts, R. D., M. T. Arts, M. S. Evans & M. J. Waiser, 1994. The coupling of heterotrophic bacterial and phytoplankton production in a hypertrophic, shallow prairie lake. Canadian Journal of Fisheries and Aquatic Sciences 51: 2219–2226.

Servais, P., 1992. Bacterial production measured by ^{3}H-thymidine and ^{3}H-leucine incorporation in various aquatic ecosystems. Archiv für Hydrobiologie Beihefte Ergebnisse der Limnologie 37: 73–81.

Simon, M. & F. Azam, 1989. Protein content and protein synthesis rates of planktonic marine bacteria. Marine Ecology Progress Series 51: 201–213.

Smits, J. D. & B. Riemann, 1988. Calculation of cell production from [^{3}H]thymidine incorporation with freshwater bacteria. Applied and Environmental Microbiology 54: 2213–2219.

Sommaruga, R., 1995. Microbial and classical food web: A visit to a hypertrophic lake. FEMS Microbiology Ecology 17: 257–270.

Sorokin, Y. I., 1999. Aquatic microbial ecology. Backhuys Pbulishers, Leiden, Netherlands.

Stevenson, L. H., 1978. A case for bacterial dormancy in aquatic systems. Microbial Ecology 4: 127–133.

Thom, S. M., R. W. Horobin, E. Seidler & M. R. Barer, 1993. Factors affecting the selection and use of tetrazolium salts as cytochemical indicators of microbial viability and activity. Journal of Applied Bacteriology 74: 433–443.

V.-Balogh, K. & L. Vörös, 1997. High bacterial production in hypertrophic shallow reservois rich in humic substances. Hydrobiogia 342/343: 63–70.

Weisse, T., 1991. The microbial food web and its sensitivity to eutrophication and contaminant enrichment: a cross–system overview. Internationale Revue der gesamten Hydrobiologie 76: 327–337.

Wetzel, R. G. & G. E. Likens, 2000. Limnological Analyses, 3 edn. Springer-Verlag, New York.

Zimmerman, R., R. Iturriaga & J. Becker-Birck, 1978. Simultaneous determination of the total number of bacteria and number thereof involved in respiration. Applied and Environmental Microbiology 36: 926–935.

Hydrobiologia (2007) 581:189–193
DOI 10.1007/s10750-006-0497-1

EUTROPHICATION IN LAKES

Floating-leaved macrophyte (*Trapa quadrispinosa* Roxb) beds have significant effects on sediment resuspension in Lake Taihu, China

Peisheng Huang · Boping Han · Zhengwen Liu

Abstract *Trapa quadrispinosa* Roxb is a common floating-leaved macrophyte in China. In this study, the effects of *Trapa* on sediment resuspension in Lake Taihu, a large, shallow, eutrophic lake in eastern China, were investigated using sediment traps. The study was conducted at stations with and without *Trapa* beds from 7 September to 18 October, 2003. Results showed that sediment resuspension rates at the station without *Trapa* were significantly higher than those at the station with *Trapa*. During the study period (41 days), 10,970 g dw m^{-2} of surface sediment was resuspended within the *Trapa* beds and 29,903 g dw m^{-2} in the pelagic zone. These equate to phosphorus fluxes of 7.4 g m^{-2} to the water column at the station with *Trapa* and 16.1 g m^{-2} at the station without *Trapa*. Floating-leaved macrophytes, such as *Trapa*, may be important in sediment resuspension and internal phosphorus loading in shallow lakes.

Guest editors: B. Qin, Z. Liu & K. Havens
Eutrophication of shallow lakes with special reference to Lake Taihu, China

P. Huang · B. Han · Z. Liu (✉)
Institute of Hydrobiology, Jinan University,
Guangzhou 510632, China
e-mail: zliu@niglas.ac.cn

P. Huang · Z. Liu
Nanjing Institute of Geography and Limnology,
Chinese Academy of Sciences, Nanjing 210008, China

Keywords Lake Taihu · Sediment resuspension · Floating-leaved macrophytes

Introduction

Many shallow lakes have become eutrophic as the result of anthropogenic influences. Eutrophication often results in decreased water transparency and subsequent declines in macrophyte abundance. Macrophytes are important to nutrient cycling in aquatic systems, particularly in shallow lakes, in many ways: nutrient incorporation in plant biomass, inhibiting phosphorus release from sediments, enhancement of sedimentation and reducing sediment resuspension. Sediment resuspension may redistribute settled materials, increase water turbidity and, thus, affect light conditions, nutrient fluxes, photosynthesis, phytoplankton and zooplankton community structure and abundance, fish feeding, and growth of submerged macrophytes (Kristensen et al., 1992; Søndergaard et al., 1992; Carrick et al., 1993; Ogilvie & Mitchell, 1998; Scheffer, 1999; Saulnier & Mucci, 2000; Madsen et al., 2001). Hence, factors controlling sediment resuspension rates have substantial consequences for the lake ecosystem.

Sediment resuspension has direct and indirect effects on aquatic macrophytes. Conversely, aquatic macrophytes can reduce sediment resuspension by attenuating wave and current energy and, thus, increasing net sedimentation rate (Madsen et al., 2001). Effects of submerged macrophytes on sediment resuspension have been the focal point of intensive research (Madsen et al., 2001; Crossley et al., 2002). Effects of emergent and submerged macrophytes on sediment resuspension also have been studied in recent years (Dieter, 1990; Horppila & Nurminen, 2001). However, similar studies have rarely applied to floating-leaved macrophytes, common in shallow lakes. In this study, sediment traps were used to evaluate effects of floating-leaved macrophytes on sediment resuspension in large, shallow, eutrophic Lake Taihu (China).

Methods and materials

The study site was located along the northern coast of Lake Taihu, one of the largest freshwater lakes in China (Qin et al., 2007, Fig. 1). The study was carried out from 7 September to 18 October, 2003, at different stations within a stand of a floating-leaved macrophyte, *Trapa quadrispinosa* Roxb. The station with *Trapa* was located inside the *Trapa* stand ~20 m from the outer edge, and the station without *Trapa* was outside the *Trapa* stand ~30 m from the stand edge. Plant coverage was about 80% of the lake surface within the *Trapa* stand during the study.

Triplicate sediment traps were placed at each station to determine the gross sedimentation rate. The traps were constructed from a cylindrical stainless steel pipe and had an inside diameter of 60 mm and a length: width ratio of 5:1. The tops of the traps were 40 cm above the sediments. Each trap was attached to a fine rope, which was fastened to an anchor and a subsurface buoy.

The traps were recovered after 10–14 days of deployment. In the laboratory, trap contents were removed and dried at 60°C to constant weight. Loss on ignition (LOI) of the entrapped material was determined by ignition at 550°C. On each sampling date, triplicate surface (2 cm) sediment samples were taken from each station with a corer and analyzed for LOI and total phosphorus (TP). For TP analysis air-dry sediment samples were digested with sulphuric acid, then P was determined using the acidic molybdate-ascorbic acid (AMAA) method (Murphy & Riley, 1962). Seston samples were taken with a tube sampler, then filtered onto GF/C filters and analyzed for suspended solids and LOI.

Sediment resuspension rate was calculated using the method of Gasith (1975), which is referred to as the "label approach" by Floderus

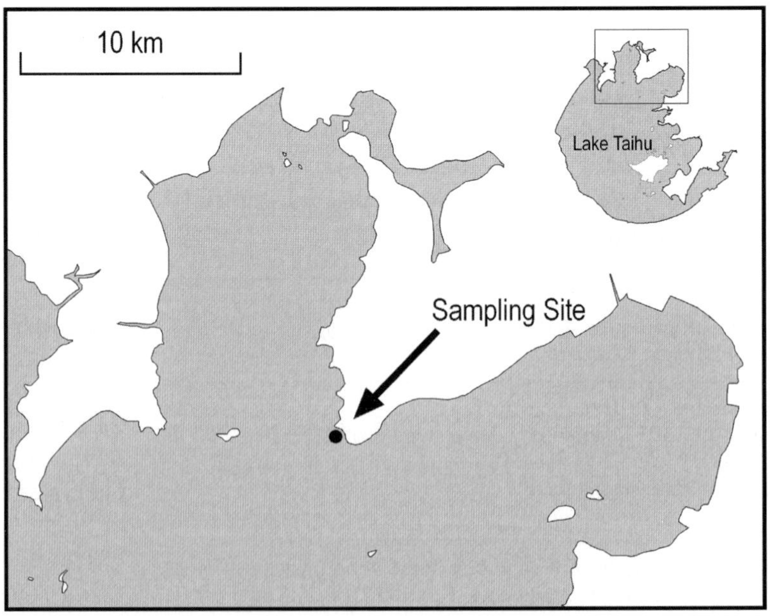

Fig. 1 Lake Taihu and the sampling site

(1989) and has been modified by several authors (Bloesch, 1994). The method uses the equation:

$$R = S\frac{(f_S - f_T)}{(f_R - f_T)}$$

where R is resuspended sediment (g dry weight); S is entrapped settling flux (g dry weight); f_S is the organic fraction (%) of S; f_R is the organic fraction (%) of R and f_T is the organic fraction (%) of seston (T) suspended in the water column. In general, organic matter is lower in resuspended sediments than fresh material (Bloesch, 1994). According to Blomqvist & Hakanson (1981), this method is reliable if the organic content of seston (f_T) is significantly different from surface sediment (f_R). The rate of P resuspension at each station was estimated using the calculated resuspension rate and P concentration in surface sediments.

ANOVA (SPSS 10.0) was used to test for significant differences between the data of the stations with and without *Trapa*.

Results

The organic fraction of suspended seston ranged from 21.3% to 50.3% (Table 1) and was significantly higher than the surface sediments (f_R) ($P < 0.001$), suggesting that the method can be used reliably in this study. The organic fraction of entrapped material was above 12% at the station with *Trapa* and significantly higher than the station without *Trapa* ($P < 0.001$). For surface sediments, the organic fraction at the *Trapa* station was also significantly higher than the no *Trapa* station (Table 1).

Gross sedimentation rates ranged from 128 to 726 and 423–1503 g dw m^{-2} day^{-1} for the station with and without *Trapa*, respectively (Fig. 2). Highest values for both stations were recorded 7–18 October, 2003. The sedimentation rate at the station with *Trapa* was significantly lower than the station without *Trapa* ($P < 0.001$). The resuspension rate at the *Trapa* station ranged from 105 to 566 g dw m^{-2} day^{-1} and was significantly lower than the no *Trapa* station, which ranged from 328 to 1505 g dw m^{-2} day^{-1} (Fig. 2). At both stations the sedimentation rate was related to the sediment resuspension rate ($R^2_{Trapa} = 0.999$, $R^2_{No\ Trapa} = 0.996$).

Total P concentration in surface sediments at the *Trapa* station had a narrow range and was always above 0.64 mg g dw^{-1}, while the station without *Trapa* was always below 0.64 mg g dw^{-1} (Table 2). The resuspension rate for TP ranged from 68 to 362 and 85 to 963 mg day^{-1} m^{-2} at the *Trapa* and no *Trapa* stations, respectively. During the study period (41 days), 10,970 g dw m^{-2}was resuspended within *Trapa* beds and 29,903 g dw m^{-2} in the pelagic zone. With the sediments, 7.4 g P m^{-2} entered the water column at the station with *Trapa* and 16.1 g P m^{-2} for the station without *Trapa*.

Discussion

In shallow waters, the concentration of suspended particles in the water column depends on the continuous processes of sedimentation and

Table 1 The organic fraction (%, ±95% confidence limits) of surface sediment, entrapped material and suspended seston in both sampling station during the study period

	17 September	27 September	7 October	18 October
Sediments				
Station with *Trapa*	8.6 ± 0.2	8.2 ± 1.0	8.0 ± 1.0	7.9 ± 0.0
Station without *Trapa*	6.7 ± 0.5	5.9 ± 0.4	7.0 ± 0.3	7.9 ± 1.0
Entrapped material				
Station with *Trapa*	12.9 ± 0.2	14.4 ± 0.6	12.2 ± 1.0	12.5 ± 2.5
Station without *Trapa*	10.8 ± 0.1	10.3 ± 0.4	8.5 ± 1.5	8.0 ± 0.1
Seston				
Station with *Trapa*	34.6 ± 1.7	32.6 ± 6.4	26.8 ± 1.1	31.3 ± 5.2
Station without *Trapa*	50.3 ± 2.3	22.3 ± 8.9	21.3 ± 3.1	38.1 ± 0.6

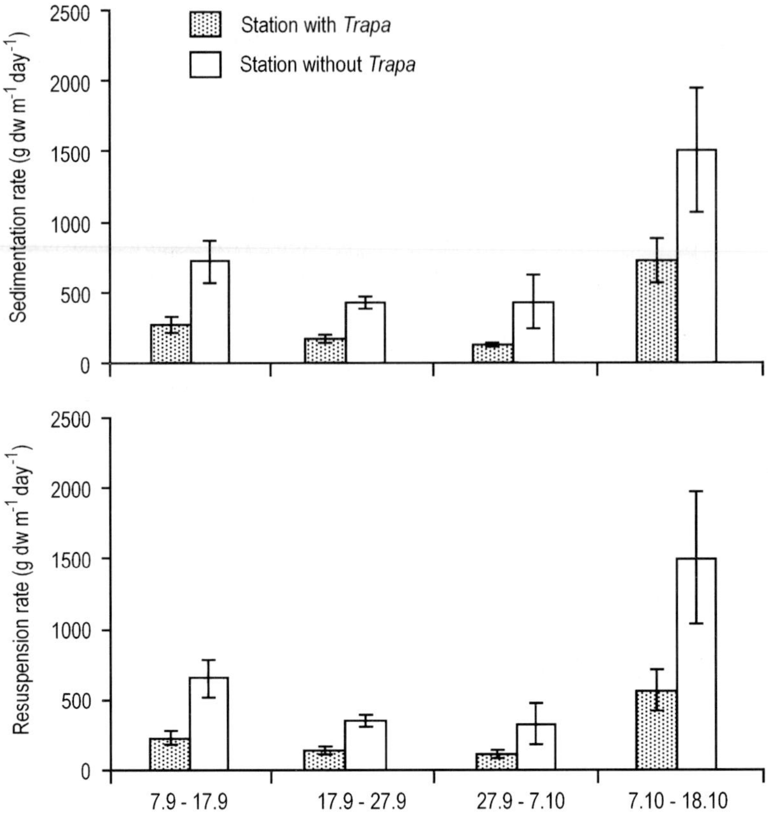

Fig. 2 Sedimentation and resuspension rates at stations with and without a floating-leaved macrophyte *Trapa quadrispinosa* (mean ± SD)

Table 2 Concentration of total phosphorus (mg g dw^{-1}) in the surface sediments (mean ± 95% confidence limits)

	17 September	27 September	7 October	18 October
Station with *Trapa*	0.75 ± 0.07	0.71 ± 0.08	0.65 ± 0.04	0.64 ± 0.01
Station without *Trapa*	0.40 ± 0.18	0.24 ± 0.13	0.63 ± 0.14	0.64 ± 0.09

resuspension. Macrophyte coverage is an important factor influencing sedimentation and resuspension in lakes (Dieter, 1990; Horppila & Nurminen, 2001; Madsen et al., 2001). In Lake Hiidenvesi, sediment resuspension in a stand of emergent *Typha angustifolia* was significantly lower than outside the stand (Horppila & Nurminen, 2001). In the present study, the mean resuspension rate within the *Trapa* stand was 36% of that in the open water, suggesting that this floating-leaved macrophyte attenuates sediment resuspension.

The study area was located along the northern coast of Lake Taihu, which is exposed to frequent southwest wind. *Microcystis* blooms often accumulate in this region and subsequently settle within the *Trapa* beds. This may be responsible in part for the high organic fraction in surface sediments and entrapped material at the station with *Trapa* (Table 1). Rooted macrophytes assimilate P mainly from sediments and, therefore, may reduce sediment P concentration (Chen & Barko, 1988). However, P concentration in sediments at the station with *Trapa* was higher than the station without *Trapa*. Reduced resuspension within the *Trapa* beds and organic matter burial, including settled *Microcystis* cells and macrophyte detritus, may explain this observation.

Large shallow lakes are susceptible to resuspension (Carper & Bachmann, 1984). The high regression coefficient between resuspension and sedimentation rates in Lake Taihu suggests that resuspension supplies particles to the water column. This study demonstrated that *Trapa* reduces resuspension and internal P loading. Re-establishment of macrophytes is a strategy for restoration of eutrophic shallow lakes. However, floating-leaved macrophytes usually are ignored in theoretical study and management. Growth of floating-leaved macrophytes is not dependent on light in the water column and may present an alternative for vegetation restoration in eutrophic, turbid lakes, such as Lake Taihu, at least in some areas.

Acknowledgements The study was supported jointly by the Key Project of CAS (KZCX1-SW-12), "863" Project (2002AA601011-04), a NSFC Project (40371103) and Jinan Science Fundation. The authors are grateful to H. Pan and H. Wang for their assistance in sampling and chemical analyses, to B. Gu and M. McCarthy for his valuable comments and correcting English. The comments of anonymous reviewers were helpful and acknowledged.

References

Bloesch, J., 1994. A review of methods used to measure sediment resuspension. Hydrobiologia 284: 13–18.

Blomqvist, S. & L. Hakanson, 1981. A review on sediment traps in aquatic environments. Archiv für Hydrobiologie 91: 101–132.

Carper, G. L. & R. W. Bachmann, 1984. Wind resuspension of sediments in a prairie lake. Canadian Journal of Fisheries and Aquatic Sciences 41: 763–7.

Carrick, H. J., F. J. Aldridge & C. L. Schelske, 1993. Wind influences phytoplankton biomass and composition in a shallow, productive lake. Limnology and Oceanography 38(6): 1179–1192.

Chen, R. J. & J. W. Barko, 1988. Effects of freshwater macrophytes on sediment chemistry. Journal of Freshwater Ecology 4: 279–289.

Crossley, M. N., W. C. Dennison, R. R. Williams & A. H. Wearing, 2002. The interaction of water flow and nutrients on aquatic plant growth. Hydrobiologia 489: 63–70.

Dieter, C. D., 1990. The importance of emergent vegetation in reducing sediment resuspension in wetlands. Journal of Freshwater Ecology 5: 467–473.

Floderus, S., 1989. The effect of sediment resuspension on nitrogen cycling in the Kattegat–variability in organic matter transport.–PhD Thesis, Dept of Physical Geography, Uppsala University. UNGI Rapport 71. 21 pp.

Gasith, A., 1975. Tripton sedimentation in eutrophic lakes—simple correction for the resuspended matter. Verhandlungen der Internationalen Vereinigung für Limnologie 19: 116–122.

Horppila, J. & L. Nurminen, 2001. The effect of an emergent macrophyte (*Typha angustifolia*) on sediment resuspension in a shallow north temperate lake. Freshwater Biology 46: 1447–1455.

Kristensen, P., M. Søndergaard & E. Jeppesen, 1992. Resuspension in a shallow eutrophic lake. Hydrobiologia 228: 101–109.

Madsen, J. D., P. A. Chambers, W. F. James, E. W. Koch & D. F. Westlake, 2001. The interaction between water movement, sediment dynamics and submersed macrophytes. Hydrobiologia 444: 71–84.

Murphy, J. & J. P. Riley, 1962. A modified single solution method for the determination of phosphate in natural water. Analytica Chimica Acta 27: 31–36.

Ogilvie, B. G. & S. F. Mitchell, 1998. Does sediment resuspension have persistent effects on phytoplankton? Experimental studies in three shallow lakes. Freshwater Biology 40: 51–63.

Qin, B. Q, P. Z. Xu, Q. L. Wu, L. C. Luo & Y. L. Zhang, 2007. Environmental issues of Lake Taihu, China. Hydrobiologia 581: 3–14.

Saulnier, I. & A. Mucci, 2000. Trace metal remobilization following the resuspension of estuarine sediments: Saguenay Fjord, Canada. Applied Geochemistry 15: 203–222.

Scheffer, M., 1999. The effect of aquatic vegetation on turbidity: how important are the filter feeders? Hydrobiologia 408/409: 307–316.

Søndergaard, M., P. Kristensen & E. Jeppesen, 1992. Phosphorus release from resuspended sediment in the shallow and wind-exposed Lake Arresø, Denmark. Hydrobiologia 228: 91–99.

Hydrobiologia (2007) 581:195–207
DOI 10.1007/s10750-006-0496-2

EUTROPHICATION IN LAKES

Nitrogen dynamics and microbial food web structure during a summer cyanobacterial bloom in a subtropical, shallow, well-mixed, eutrophic lake (Lake Taihu, China)

Mark J. McCarthy · Peter J. Lavrentyev · Longyuan Yang · Lu Zhang · Yuwei Chen · Boqiang Qin · Wayne S. Gardner

Abstract Nitrogen dynamics and microbial food web structure were characterized in subtropical, eutrophic, large (2,338 km^2), shallow (1.9 m mean depth), and polymictic Lake Taihu (China) in Sept–Oct 2002 during a cyanobacterial bloom. Population growth and industrialization are factors in trophic status deterioration in Lake Taihu. Sites for investigation were selected along a transect from the Liangxihe River discharge into Meiliang Bay to the main lake. Water column nitrogen and microbial food web measurements were combined with sediment–water interface incubations to characterize and identify important processes related to system nitrogen dynamics. Results indicate a gradient from strong phosphorus limitation at the river discharge to nitrogen limitation or co-limitation in the main lake. Denitrification in Meiliang Bay may drive main lake nitrogen limitation by removing excess nitrogen before physical transport to the main lake. Five times higher nutrient mineralization rates in the water column versus sediments indicate that sediment nutrient transformations were not as important as water column processes for fueling primary production. However, sediments provide a site for denitrification, which, along with nitrogen fixation and other processes, can determine available nutrient ratios. Dissimilatory nitrate reduction to ammonium (DNRA) was important, relative to denitrification, only at the river discharge site, and nitrogen fixation was observed only in the main lake. Reflecting nitrogen cycling patterns, microbial food web structure shifted from autotrophic (phytoplankton dominated) at the river discharge to heterotrophic (bacteria dominated) in and near the main lake.

Keywords Lake Taihu · Eutrophication · Nitrogen · Cyanobacterial blooms · Denitrification

Guest editors: B. Qin, Z. Liu & K. Havens
Eutrophication of shallow lakes with special reference to Lake Taihu, China

M. J. McCarthy (✉) · W. S. Gardner
Marine Science Institute, The University of Texas at Austin, 750 Channel View Drive, Port Aransas, TX 78373, USA
e-mail: markm@utmsi.utexas.edu

P. J. Lavrentyev
Department of Biology, The University of Akron, Akron, OH 44325, USA

L. Yang · L. Zhang · Y. Chen · B. Qin
Nanjing Institute of Geography and Limnology, Chinese Academy of Sciences, 73 East Beijing Road, Nanjing, Jiangsu 210008, P.R. China

Introduction

Nitrogen (N) and phosphorus (P) often limit primary productivity in aquatic systems. Unlike P, dissolved inorganic N (DIN) occurs in different forms and oxidation states. DIN is assimilated and converted to organic N by phytoplankton, other plants, or bacteria. Organic N can be regenerated as ammonium (NH_4^+) or dissolved organic N (DON) compounds, reassimilated by plants or heterotrophs, or oxidized by bacteria.

DIN composition and ratios can affect phytoplankton community structure. For example, NH_4^+ is conducive to cyanobacteria (Blomqvist et al., 1994; Hyenstrand et al., 1998; Dokulil & Teubner, 2000; Jacoby et al., 2000). Unlike diatoms, cyanobacteria are poor competitors for nitrate (NO_3^-; Hyenstrand et al., 1998). Some cyanobacteria can fix atmospheric dinitrogen (N_2), which can be regenerated as bioavailable DIN compounds (Shapiro, 1990; Downing et al., 2001). Cyanobacteria dominance often is regarded as an indicator of eutrophication (Dokulil & Teubner, 2000).

Nitrification coupled with denitrification converts biologically available N forms (NH_4^+ and NO_3^-, respectively) to N_2 gas and may reduce effects of excessive N inputs and eutrophication (Seitzinger, 1988). Nitrate for coupled denitrification derives from organic matter mineralization to NH_4^+ followed by nitrification (Seitzinger, 1988). Dissimilatory NO_3^- reduction to NH_4^+ (DNRA) is an alternative pathway for NO_3^- in sediments (Tobias et al., 2001; An & Gardner, 2002). The relative partitioning between NO_3^- reduction pathways (denitrification versus DNRA) is important since denitrification removes fixed N from the system while DNRA returns it as bioavailable NH_4^+ (Tobias et al., 2001).

The microbial food web (MFW) includes heterotrophic, autotrophic, and mixotrophic prokaryotes and eukaryotes. Multiple cascading trophic interactions within the MFW can affect the biogeochemical N cycle and have feedback effects on trophic conditions in coastal waters (Lavrentyev et al., 1998). MFW structure also is an important factor regulating algal community dynamics in eutrophic, cyanobacteria-dominated systems (Elser, 1999). MFW processes should be included in nutrient-food web models to understand complex processes in aquatic environments (Edwards et al., 2000).

Evaluating internal nutrient cycling and transformations, and conditions enhancing them, in lakes and other aquatic systems help managers address effects of excessive nutrient inputs and eutrophication, such as cyanobacteria dominance. Eutrophication studies in freshwater lakes often focus on P, since it is the most common limiting nutrient. This situation describes the current status of research on Lake Taihu, a large (2,338 km^2), subtropical, shallow (1.9 m mean depth), well-mixed, and eutrophic lake in China (Pu & Yan, 1998). Studies in this lake have been limited to monitoring nutrient sources and dominant compounds (i.e. nutrient concentrations).

Water column NH_4^+ regeneration and uptake, benthic nutrient fluxes and N sinks (i.e. denitrification), water column versus benthic N cycling, and MFW structure are important factors related to trophic status and water quality. These issues have not been addressed in Lake Taihu. The size, depth, trophic status, latitude, and basin land use for Lake Taihu are similar to Lake Okeechobee (Florida, USA; Havens et al., 2001). General information about Lake Taihu geography is presented elsewhere in this issue (Qin et al., 2007).

Lake Taihu was oligotrophic as recently as the 1950s, but increased nutrient inputs related to population and economic growth have led to eutrophication (Cai et al., 1997; Chen et al., 2003). Most pollutants come from rivers discharging into Meiliang Bay and other parts of the lake (Huang, 2000). Meiliang and Wuli Bays (Fig. 1) suffer from severe eutrophication (Huang, 2000). Nutrient concentrations decrease with distance

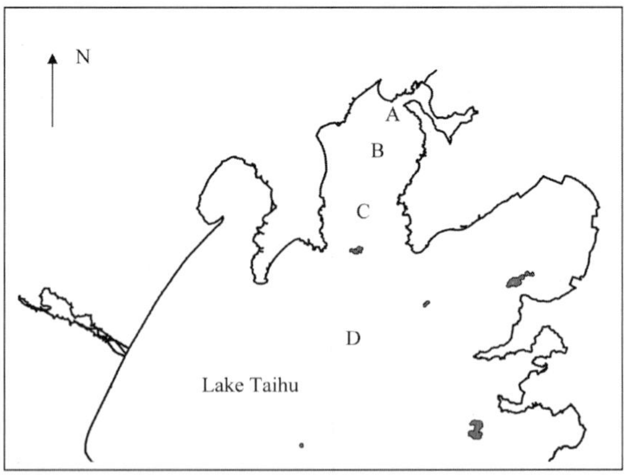

Fig. 1 Map of northern Lake Taihu showing locations of sampling stations in Meiliang Bay. A is the "river" site, B is "inner bay", C is "outer bay", and D is "main lake"

from river outflows (Cai et al., 1997), but whole-lake nutrient concentrations have increased by an order of magnitude (Huang, 2000).

Wind-driven mixing and shallow depth prevent stratification in the lake, and bottom waters remain oxic (Dickman et al., 1998). Phytoplankton diversity has decreased since 1981, but cyanobacteria populations (*Microcystis* and *Anabaena*) have increased (Pu & Yan, 1998) and can comprise 85% of summer phytoplankton biomass (Chen et al., 2003). Annual blooms have clogged intakes at municipal waterworks, interrupted domestic and industrial water supply, disrupted tourism and capture fisheries, and caused cultured fish stock losses (Pu & Yan, 1998). Wind-driven sediment resuspension inhibits photosynthesis below 0.5–1 m depth. However, *Microcystis* use gas vesicles to maintain buoyancy and have a competitive advantage over other phytoplankton by avoiding light limitation (Shapiro, 1990; Chen et al., 2003).

The primary goal of this study was to characterize, for the first time, N dynamics relative to MFW structure in this understudied, eutrophic system. Specific objectives were to: (1) measure water column NH_4^+ regeneration and potential uptake rates; (2) measure sediment N fluxes and transformation rates; (3) compare sediment N transformation rates with water column N cycling rates; and (4) relate microbial plankton composition and distribution to N cycling rates.

Materials and methods

Site description

Four stations were selected along the Liangxihe River discharge gradient from Meiliang Bay to the main lake (Fig. 1). Stations were located at the Liangxihe River discharge into Meiliang Bay (river; "A" on Fig. 1), inner Meiliang Bay (inner bay; "B" on Fig. 1), outer Meiliang Bay (outer bay; "C" on Fig. 1), and the main lake ("D" on Fig. 1). These stations correspond to monitoring stations (0, 1, 3, and 7, respectively) along a larger transect detailed in previous work (Cai et al., 1997; Chen et al., 2003). Sampling occurred on 26 Sept 2002 (outer bay and main lake) and 2 Oct 2002 (river and inner bay). Data from triplicate incubation chambers were averaged for each measured parameter or flux. Standard errors were calculated to determine replicate variation and are reported with means as plus/minus (±) one standard error (SE).

Water column characteristics

Water depth was estimated using a wooden pole and tape measure. Turbidity was estimated using a ~30 cm Secchi disc, and surface water temperature was measured with a mercury thermometer. Dissolved oxygen concentration (DO) was measured using Winkler titration, and chlorophyll *a* concentration (chl) was measured using hot ethanol extraction followed by spectrophotometry (Chen et al., 2003).

Water column nutrient {NO_3^- + nitrite (NO_2^-), NO_2^-, and ortho-phosphate (o-PO_4^{3-})} samples were filtered with 0.2 μm syringe filters (Osmonics) and frozen in 14 ml snap-cap tubes (Falcon). NH_4^+ samples were filtered (0.2 μm syringe filter) and frozen in 8 ml glass vials (Wheaton). Frozen samples were carried to the University of Texas Marine Science Institute (UTMSI) for flow-injection analysis of $NO_3^- + NO_2^-$, NO_2^-, and PO_4^{3-} (Lachat QuikChem 8000) and high performance liquid chromatographic (HPLC) NH_4^+ analysis (Gardner et al., 1995a). Total N (TN) and total P (TP) were determined by persulfate oxidation and spectrophotometry (Jin & Tu, 1990). Total P was oxidized to PO_4^{3-} at 120°C. TN and TP recovery efficiencies were 98.4% and 99.7%, respectively.

Water column NH_4^+ regeneration and uptake

Water from each site was collected in 2 l plastic bottles, returned to the Taihu Laboratory for Lake Ecosystem Research (TLLER), enriched with 99.8% $^{15}NH_4Cl$ (Isotec; 16 μmol l^{-1} final isotope concentration), and partitioned into triplicate light and dark 70 ml tissue culture bottles (Corning). Dark bottles were wrapped with aluminum foil. Initial samples were collected immediately after enrichment and mixing, filtered (0.2 μm syringe filter), and frozen in 8 ml glass vials (Wheaton). Bottles were incubated in a mesh bag suspended from the TLLER pier into surface water. Intermediate samples were collected after 4 (river and inner bay) or 11 h (outer bay and main lake), filtered, and frozen. Intermediate samples were collected to coincide with day/night transition. Final samples were collected after 20 h, filtered, and frozen. All frozen samples were hand-carried to UTMSI for analysis of total NH_4^+ concentration and atom % ^{15}N using HPLC (Gardner et al., 1995a). Ammonium regeneration and potential uptake rates were calculated from these data using the Blackburn/Caperon isotope dilution model (Blackburn, 1979; Caperon et al., 1979). NH_4^+ uptake rates in this study are qualified as "potential", and no instances of complete isotope substrate depletion were observed.

Sediment–water interface incubations

Triplicate sediment cores (depth 15–20 cm) were collected from each site using a coring device allowing acrylic tube (30.5 cm length, 7.6 cm inside diameter, 0.32 cm wall thickness) insertion into the sediment with minimal disturbance and overlying water retention. Sediment cores with overlying water were sealed immediately at both ends with butyl caps and electrical tape. Near-bottom water from each site was collected into two ~20 l carboys using a submersible pump.

Cyanobacteria formed a "scum" covering the water surface at all sampling stations. The water column also was turbid from wind-driven sediment resuspension. Thus, the sediment–water interface was expected to receive little or no light, and intact sediment cores with overlying water were wrapped with aluminum foil to prevent light effects. Wind-driven turbulence and a shallow water column also maintain bottom water normoxia. Therefore, carboys with bottom water were aerated using an aquarium air pump to maintain oxic conditions in overlying water.

Cores were installed into a flow-through incubation system (Lavrentyev et al., 2000; McCarthy & Gardner, 2003) consisting of aerated bottom water, a multi-channel proportioning pump (Technicon), transmission tubing (Teflon), and an acetol plunger with Viton o-ring. The plunger was positioned ~5 cm above the sediment surface to give ~230 ml of overlying water volume (An & Gardner, 2002). This system maintained a gas-tight setting with inflow (flow rate $\cong$ 0.072 l h^{-1}) from the bottom water carboy and outflow from positive displacement of overlying water. The cores were placed in a water bath maintained at in situ temperature and allowed to reestablish steady-state overnight.

After the pre incubation period, discreet inflow (from bottom water carboy) and outflow samples were collected daily for 2 days for nutrient analyses as described above. Nutrient fluxes (in μmol m^{-2} h^{-1}) were calculated by: $(C_o - C_i) \times f/a$, where C_o is the outflow concentration in μmol l^{-1}, C_i is the inflow concentration, f is the flow rate (0.072 l h^{-1}), and a is the sediment surface area (0.0045 m^2).

Dissolved gas samples were collected from core outflow by overflowing 15 ml ground-glass stopper test tubes (Chemglass; 19.8 cm length and 1 cm ID), injecting 200 μl 50% $ZnCl_2$, and inserting the ground-glass stopper quickly while twisting to prevent air bubble entrapment. Parafilm was wrapped around the top of the tubes, which were stored under water in 4 l bottles (Nalgene) to prevent large temperature changes. Samples were hand-carried to UTMSI for dissolved gas (N_2, O_2, and argon) analysis via membrane inlet mass spectrometry (MIMS; Kana et al., 1994; An et al., 2001; An & Gardner, 2002; McCarthy & Gardner, 2003). Net N_2 flux and sediment O_2 demand (SOD; O_2 flux) were calculated from the flux formula above. In the absence of N fixation, net N_2 flux provides a reasonable estimate for "actual" denitrification rates.

Bottom water was enriched with 98% $Na^{15}NO_3^-$ (~100 μmol l^{-1} final isotope concentration) after the second sampling day. Samples were collected for two more days for $^{28}N_2$, $^{29}N_2$, and $^{30}N_2$ analysis via MIMS (An et al., 2001; An & Gardner, 2002). These data allowed isotope pairing (Nielsen, 1992) analysis of potential denitrification (sum of $^{28}N_2$, $^{29}N_2$, and $^{30}N_2$ fluxes and N fixation) driven by ^{14}N and ^{15}N pools and simultaneous evaluation of N fixation. Potential denitrification, ^{14}N and ^{15}N denitrification, and N fixation calculations are detailed in An et al. (2001). Ammonium isotope ratios obtained by HPLC (Gardner et al., 1995a) allowed calculation of $^{15}NH_4^+$ concentrations and, using the above flux formula, $^{15}NH_4^+$ production (potential DNRA; An & Gardner, 2002).

MIMS uses a quadrupole mass spectrometer (QMS) to detect dissolved gases in water. The QMS ion source ionizes gases and produces O^+ ions, which react with N_2 to form nitric oxide (NO; Eyre et al., 2002). This scavenging results in a lower N_2 signal at high O_2 concentrations, and vice versa, and may lead to over- or underestimation of denitrification rates. The error, however, is machine dependant (Eyre et al., 2002), and the extent of this effect was small (0.13%) on the QMS used in this study (McCarthy & Gardner, 2003). Also, sulfide can inhibit the final step in denitrification (nitrous oxide (N_2O) → N_2) and lead to N_2O release (Nedwell & Dong, 2002). Since MIMS uses N_2:Ar to determine N_2 and a liquid N_2 trap to remove interferences (Kana et al., 1994), denitrification leading to N_2O release is not detected and may cause rate underestimation.

Water column MFW structure

Heterotrophic bacteria and phototrophic picoplankton were quantified from 1% formalin-fixed samples using epifluorescence microscopy (EFM; Olympus BX-40). Heterotrophic bacteria were stained with DAPI (Porter & Feig, 1980) and sonicated to count attached bacteria (Velji & Albright, 1993). Attached bacterial abundance was the calculated difference between total cell number (treated samples) and total free-suspended cells (untreated samples). Bacterial cells were measured with a SPOT-2 digital camera and Image Pro 4.5 software. Bacterial and picocyanobacterial biovolumes were converted to carbon following Loferer-Krossbacher et al. (1998) and Menden-Deuer & Lessard (2000), respectively.

Heterotrophic nanoflagellates (HNF) were preserved with 1% formaldehyde, concentrated onto black 0.8 μm polycarbonate membrane filters, and counted using EFM following dual-staining with FITC/DAPI (Sherr et al., 1993). Lugol's iodine preserved microplankton were settled overnight in 20–50 ml chambers and counted with an Olympus IX-70 inverted differential interference contrast (DIC) microscope. Linear dimensions of 30–90 individuals (fewer for less abundant taxa) were measured at 400–600× and converted to volumes using appropriate geometric shapes. Tintinnid volume was determined using the same approach as for aloricate protists, since they were clearly visible under DIC. Phytoplankton and HNF volumes were converted to carbon following Menden-Deuer & Lessard (2000). Ciliate and rotifer volumes were converted following Putt & Stoecker (1989) and Fahnenstiel et al. (1998), respectively.

Results

Water column characteristics

Site data for sampling stations are provided in Table 1. All sites except 'river' were deeper than the mean lake depth. Mean DO (8.4 ± 0.2 mg O_2 l^{-1}) was near saturation (8.6; Colt, 1984) for the mean temperature (22.7 ± 0.4°C). Nutrient concentrations decreased along the transect except PO_4^{3-} concentrations were low at all sites. TN:TP was about eight-fold higher at 'river' versus the other sites.

Water column NH_4^+ regeneration and uptake

Water column NH_4^+ regeneration and potential uptake rates (hereafter "regeneration" and "uptake") were an order of magnitude higher at 'river' than the other sites (Fig. 2). Light/dark uptake differences were observed at all sites. Light/dark differences in regeneration were observed at 'river' and 'outer bay'. Light uptake was higher than light regeneration at all sites.

Sediment–water interface nutrient fluxes

Benthic nutrient fluxes were higher at 'river' than the other sites (Table 2). Mean sediment DIN flux for all sites (129 ± 34 μmol N m^{-2} h^{-1}) was almost two orders of magnitude higher than PO_4^{3-} flux (1.4 ± 0.6 μmol P m^{-2} h^{-1}). Mean NO_2^- (-7.2 ± 3.3 μmol N m^{-2} h^{-1}) and NO_3^- (-13.1 ± 9.6 μmol N m^{-2} h^{-1}) fluxes for all sites were negative due to high fluxes into the sediment at 'river'. Lake Taihu sediments were a source of NH_4^+ (149 ± 47 μmol N m^{-2} h^{-1}).

Table 1 Site characteristics for Lake Taihu sampling stations

Parameter	Units	River	Inner Bay	Outer Bay	Main Lake
Lat (N)	d, m, s	31, 32, 17	31, 29, 37	31, 26, 16	31, 21, 0
Long (E)	d, m, s	120, 13, 17	120, 11, 51	120, 11, 12	120, 10, 58
Depth	m	1.4	2.6	2.9	2.8
Temp	°C	23.8	22.4	22.4	22.0
pH	su	8.65	9.04	9.08	8.55
Secchi	m	0.35	0.35	0.50	0.40
NH_4^+	μmol l^{-1}	185	3.4	2.9	0.07
NO_3^-	μmol l^{-1}	73.3	6.4	0.14	0.35
NO_2^-	μmol l^{-1}	13.8	0.6	0.02	0.06
NH_4^+:NO_x		2.1	0.5	17.9	0.2
PO_4^{3-}	μmol l^{-1}	0.18	0.25	0.22	0.04
TN	μmol l^{-1}	135	20.7	20.6	27.4
TP	μmol l^{-1}	1.5	1.7	2.5	2.0
TN:TP		89	12	8	13
Chl *a*	μg l^{-1}	85.4	4.5	3.9	14.0
DO	Mg l^{-1}	8.3	9.0	8.3	8.0

Lat (N) = north latitude in degrees, minutes, seconds (d, m, s). Long (E) = east longitude. Temp = temperature. NH_4^+ = ammonium. NO_3^- = nitrate. NO_2^- = nitrite. NO_x = NO_3^- + NO_2^-. PO_4^{3-} = ortho-phosphate. TN = total N. TP = total P. Chl *a* = chlorophyll *a*. DO = dissolved oxygen

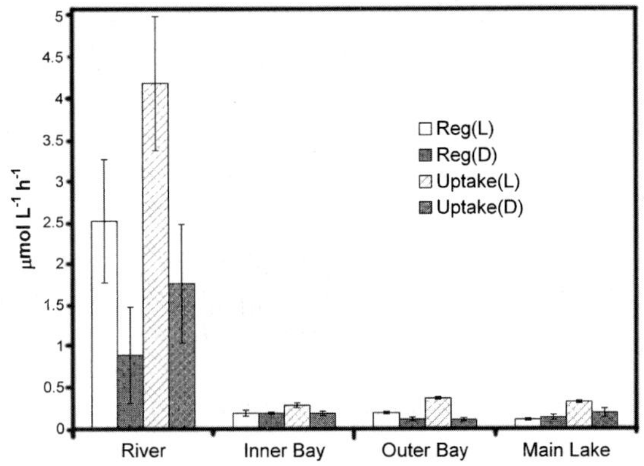

Fig. 2 Light (L) and dark (D) water column ammonium (NH_4^+) regeneration (Reg) and potential uptake rates plus/minus one standard error

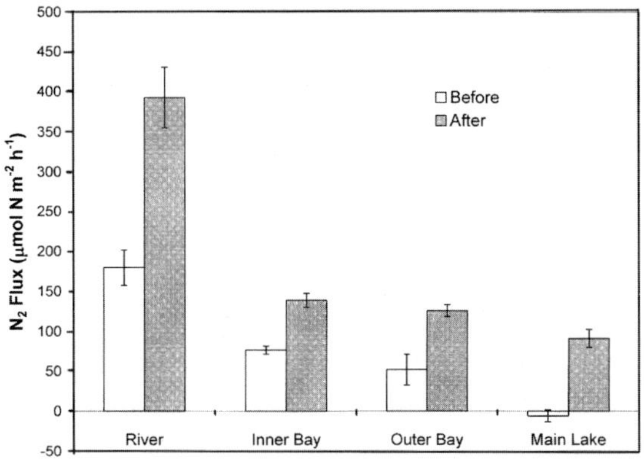

Fig. 3 Sediment–water interface net dinitrogen (N_2) flux before and after isotope ($^{15}NO_3^-$) addition plus/minus one standard error

SOD, denitrification, and N_2 fixation

Before isotope addition, SOD (in μmol O_2 m^{-2} h^{-1}) ranged from 750 ± 0.2 at 'outer bay' to 1350 ± 83 at 'river' and was similar at 'inner bay' (950 ± 32) and 'main lake' (970 ± 23). Net N_2 flux (in μmol N m^{-2} h^{-1}) decreased along the transect before and after $^{15}NO_3^-$ addition (Fig. 3). N fixation (after $^{15}NO_3^-$ addition) was observed only at 'main lake' (14 ± 10 μmol N m^{-2} h^{-1}). Thus, net N_2 fluxes before isotope addition likely reflect "actual" denitrification rates at the other sites.

Potential denitrification was equivalent to net N_2 flux after $^{15}NO_3^-$ addition (Fig. 4) except at 'main lake', where N fixation was observed. Denitrification fueled by $^{15}NO_3^-$ and the ^{14}N pool

Table 2 Lake Taihu sediment–water interface nutrient fluxes (μmol N or P m^{-2} h^{-1}) plus/minus one standard error

Nutrient	River	Inner Bay	Outer Bay	Main Lake
NH_4^+	490 ± 31	40 ± 19	20 ± 0.5	45 ± 1.4
NO_3^-	−83 ± 6.8	5.8 ± 11	12 ± 0.1	13 ± 1.2
NO_2^-	−31 ± 3.8	−2.0 ± 0.4	0.08 ± 0.05	4.2 ± 0.5
DIN	380 ± 27	44 ± 7.2	33 ± 0.6	63 ± 3.1
PO_4^{3-}	5.2 ± 0.2	1.2 ± 0.5	0.8 ± 1.0	−1.5 ± 0.9

See Table 1 legend for nutrient abbreviations

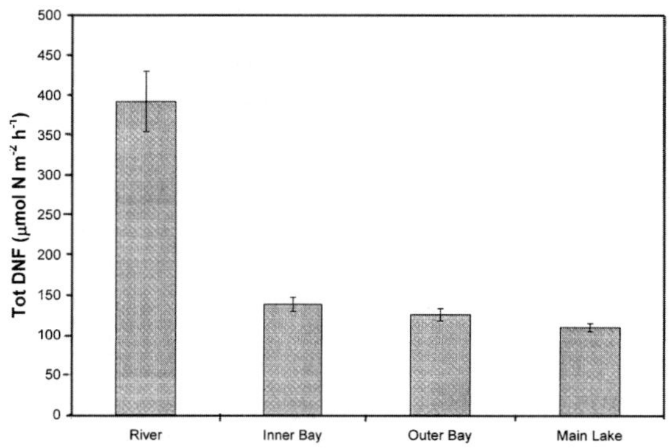

Fig. 4 Sediment–water interface potential (Tot) denitrification (DNF) after isotope ($^{15}NO_3^-$) addition plus/minus one standard error

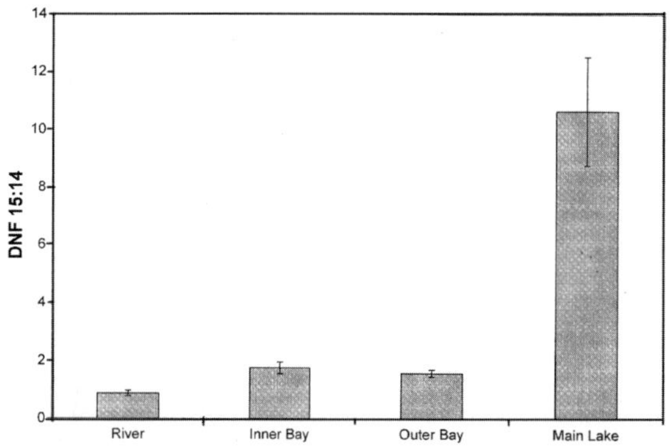

Fig. 5 Ratio of denitrification (DNF) fueled by ^{15}N versus ^{14}N nitrate (15:14) pools plus/minus one standard error

were similar at 'river', but the ^{15}N pool accounted for more potential denitrification than the ^{14}N pool at the other sites (Fig. 5).

Potential DNRA

Potential DNRA (in μmol N m^{-2} h^{-1}) was observed at all sites in Lake Taihu (data not shown). Potential DNRA rates ranged from 0.4 ± 0.1 at 'outer bay' to 250 ± 80 at 'river' and were 9.5 ± 2.4 at 'inner bay' and 3.0 ± 1.9 at 'main lake.

Water column MFW structure

Cyanobacteria were abundant at all sites during sampling (Table 3). Maximum phytoplankton biomass was observed at 'river' and dominated by large (>10,000 μm^3) euglenophytes (~80%; Fig. 6). The "edible" (to microbial grazers) fraction (i.e. < 200 μm^3) was mostly cryptophytes. Bacteria also reached maximum abundance (5×10^7 cells ml^{-1}) at 'river', but >67% were aggregated or attached to particles, and the median cell volume was minimal (0.033 μm^3). Rotifers (mostly *Polyarthra remata*) accounted for >50% of micrograzer biomass (HNF + ciliates + rotifers). The choreotrich ciliate, *Strobilidium* spp., and colorless nano-chysophytes dominated heterotrophic protists.

Meiliang Bay sites (inner bay and outer bay) had lower phytoplankton biomass consisting of green algae, cyanobacteria, and diatoms (Table 3; Fig. 6). Heterotrophic biomass in Meiliang Bay also was lower versus 'river', but to a lesser extent than for bacterioplankton alone, which were represented by large cells (median vol. = 0.085 μm^3). The colonial peritrich ciliate *Epystilis rotans*, chrysophytes, and rotifers *Asplanchna priodonta* and *Brachionus* spp. dominated the microbial grazers.

The main lake site had higher phytoplankton biomass than Meiliang Bay sites and was dominated by *Anabaena flos-aquae* and *Synechococcus*-like picocyanobacteria (>50%; Fig. 6). Bacterial biomass was lowest (cell volume 0.05 μm^3), but micrograzer biomass was high, primarily due to the tintinnid ciliate *Codonella cratera* and a diverse HNF assemblage, including chysophytes, kinetoplastids, and choanoflagellates.

Discussion

Water column nutrients and phytoplankton

Nutrients in Lake Taihu followed the expected pattern of decreasing concentrations with distance from river outflow. Possible explanations include dilution via physical transport (i.e. wind-driven currents), higher nutrient uptake versus regeneration rates along the transect, and nutrient sinks, such as sediment burial and denitrification.

High TN:TP ratios at 'river'(Table 1) support P-limitation observations (Cai et al., 1997; Huang, 2000; Chen et al., 2003). However, TN:TP ratios below Redfield (~16) at the other sites

Table 3 Phytoplankton (Phyto) and microbial food web (MFW) characteristics during Lake Taihu sampling

			River	Inner Bay	Outer Bay	Main Lake
MFW	μg C l^{-1}	Phytoplankton	6930	149	55.4	518
		Bacteria	577	292	263	157
		HNF	20.8	8.94	6.97	15.1
		Ciliates	35.3	13.5	10.3	52.9
		Rotifers	63.8	3.57	5.75	2.62
Phyto.	%phyto	Chlorophytes	4.90	41.8	27.4	33.5
		Cyanobacteria	1.90	28.6	37.2	59.7
		Diatoms	2.00	6.60	16.7	5.87
		Euglenophytes	78.9	0	0	0
		Varia (i.e. cryptophytes)	12.2	23.0	18.7	0.96
Cyanos	%phyto	*Anabaena*	0.14	14.5	6.01	38.9
		Synechococcus	0.2	6.8	29.9	11.7
		Aphanizomenon	0	0	0	3.68
		Total N-fixers	0.33	21.3	35.9	54.3
		Microcystis	0.80	7.33	1.27	0.14
Ratios	1	Bacteria:NH_4^+	0.22	6.13	6.48	160.2
	2	Chl:DIN	0.02	0.03	0.09	2.08
	3	Chl:NH_4^+	0.03	0.09	0.09	14.28
	4	Chl:NO_3^-	0.08	0.05	1.96	2.86
	5	Chl:PO_4^{3-}	15.3	0.58	0.56	11.3
	6	Diatom:NO_3^-	3.18	8.5	216	278
	7	Diatom:PO_4^{3-}	41.8	7.02	4.43	78.5
	8	Heterotrophs:Chl	0.10	2.13	5.16	0.44
	9	MZP:total phyto	0.01	0.11	0.29	0.11
	10	MZP:edible phyto	0.56	0.63	0.58	0.54
	11	Protists:bacteria	0.10	0.08	0.07	0.43
	12	Light uptake:Chl	49.0	63.0	95.3	23.1
	13	Light reg:hetero C	3.61	0.59	0.66	0.48
	14	Dark reg:hetero C	1.28	0.59	0.39	0.60

% phyto = proportion of phytoplankton groups to total phytoplankton; Cyanos = cyanobacteria; C = Carbon; HNF = heterotrophic nanoflagellates; Chl = chlorophyll; MZP = microzooplankton; Reg = regeneration; Hetero = heterotrophic. See Table 1 legend for nutrient abbreviations

indicate surplus P and conditions conducive to cyanobacteria (Jacoby et al., 2000). Nutrient concentrations in Meiliang Bay are lowest in

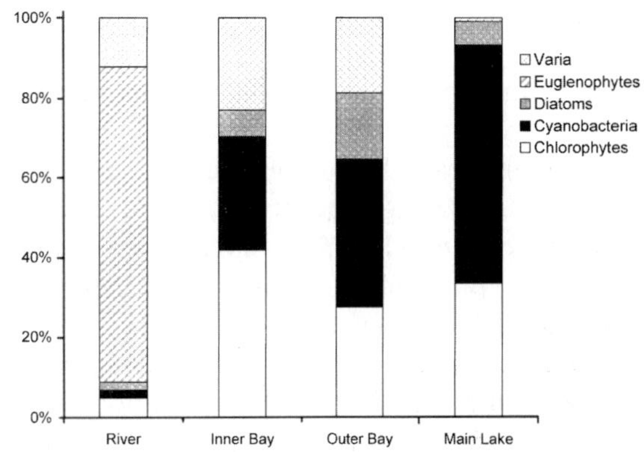

Fig. 6 Proportions of chlorophytes, cyanobacteria, diatoms, euglenophytes, and varia (mostly cryptophytes) in total phytoplankton biomass

summer (wet season) and highest in winter (dry season), when biological activity decreases and inverse water flow occurs (Cai et al., 1997). Thus, nutrient limitation status may change in different seasons.

Nutrient concentrations and algal standing stocks are important factors affecting cyanobacteria dominance (Downing et al., 2001). Transect NH_4^+:NO_x ratios were >1 at 'river' and 'outer bay' and <1 at 'inner bay' and 'main lake' (Table 1). Cyanobacteria percentage biomass, relative to total phytoplankton biomass, was highest at 'main lake' (Table 3), where the NO_3^- proportion peaked and NH_4^+ likely was depleted by cyanobacteria. Diatom percentage biomass peaked at 'outer bay', where highest NH_4^+:NO_x was observed, likely due to NO_3^- uptake by diatoms (Hyenstrand et al., 1998). Percentage cyanobacteria biomass was greater than for diatoms at all sites except 'river'. Phytoplankton biomass at 'river' was dominated (79%) by *Euglena* spp., which are common in hypereutrophic systems (Hewson et al., 2001).

Water column NH_4^+ regeneration and uptake

Water column NH_4^+ cycling rates in Lake Taihu were within ranges observed in other eutrophic systems. High regeneration and uptake (Fig. 2) at 'river' are not surprising given high DIN concentration and phytoplankton biomass. Similar rates were observed in the Mississippi River plume (Bode & Dortch, 1996) and hypereutrophic Lake Maracaibo (Gardner et al., 2000). Uptake in Lake Okeechobee (0.67 μmol l^{-1} h^{-1}; Gu et al., 1997) was lower than the rate at 'river' but in the same range as those from Meiliang Bay and the main lake. Regeneration and uptake at sites other than 'river' were higher than oligotrophic Great Lakes (Gardner et al., 1995b) but lower than south Texas coastal waters (McCarthy et al., unpub. data).

Light/dark uptake differences were observed at all sites (Fig. 2). This result supports the idea that N is not limiting, since phytoplankton uptake depends less on light during N stress (Cochlan et al., 1991). However, light/dark uptake differences are less pronounced in N limited versus N replete systems (Cochlan et al., 1991). Light-to-dark uptake ratio was ~1.7 (less pronounced) at 'inner bay' and 'main lake' and 2.4 and 3.5 (more pronounced) at 'river' and 'outer bay', respectively. This observation suggests possible N limitation at 'inner bay' and 'main lake'. Lack of light/dark regeneration differences at these sites also is consistent with N limitation (Fig. 2; Gardner et al., 2000), since N-starved heterotrophs may retain N for biomass rather than release it. Light/dark uptake differences also may reflect autotrophic versus heterotrophic dominance, especially when those differences are more pronounced ('river' and 'outer bay').

Sediment–water interface nutrient fluxes

Cyanobacteria may acquire sufficient P to sustain all pelagic growth from sediments, thus avoiding competition with other phytoplankton (Blomqvist et al., 1994). Low water column PO_4^{3-} concentrations (Table 1) and PO_4^{3-} flux out of the sediments (Table 2) at 'river' and 'inner bay' indicate rapid P uptake and support P limitation observations at these sites. No sediment P flux at 'outer bay' and negative flux at 'main lake' (Table 2) are consistent with water column TN:TP observations below Redfield and suggest N limitation or co-limitation in outer Meiliang Bay and the main lake.

Positive NH_4^+ flux at all sites (Table 2) indicates high organic matter inputs and rapid remineralization (including DNRA). Cyanobacteria in Lake Taihu may be fueled in part by sediment NH_4^+ flux into the water column, especially at 'main lake' where DIN concentration was below 0.5 μmol l^{-1} (Table 1). Negative NO_3^- flux at 'river' may reflect denitrification and/or DNRA fueled by water column NO_3^-. Positive NO_3^- flux at 'outer bay' and 'main lake' can be explained by sediment nitrification. Small positive NO_2^- fluxes at these sites may result from incomplete nitrification and/or denitrification, since NO_2^- is an intermediate in both processes (Zumft, 1997).

N_2 fixation

Nitrogen fixation was not observed in Meiliang Bay (including 'river'). Before isotope addition, net N_2 flux at 'main lake' was negative, but standard error bars overlapped with zero, indicating that denitrification and N fixation were in balance. This site had the highest water column N-fixer biomass (Table 3). *Anabaena* comprised nearly 40% of total phytoplankton biomass, and 'main lake' is the only site where *Aphanizomenon* was observed. Phytobenthos were not evaluated, so it cannot be assumed that these N-fixers were responsible for observed sediment N fixation. However, high water column N-fixer abundance and sediment N fixation further support possible N limitation at this site (Dokulil & Teubner, 2000).

Denitrification

Sediments provide a site for denitrification, which can drive aquatic systems toward N limitation and regulate nutrient supply ratios (Seitzinger, 1988). Denitrification in Meiliang Bay may be a factor in observed evidence for N limitation in the main lake. If excess N is denitrified before physical transport via wind-driven currents moves it to the main lake, then it follows that organisms in the main lake may have to rely on other N sources, such as fixation. Lack of N fixation at Meiliang Bay sites (including 'river') indicates that net N_2 fluxes before isotope addition (Fig. 3) may reflect "actual" denitrification rates, assuming no N_2O release from incomplete denitrification. Increases in net N_2 flux after $^{15}NO_3^-$ addition (Fig. 3) indicate that potential denitrification may be substrate limited, especially at 'outer bay' and 'main lake', where water column NO_3^- was low (Table 1). Another possible explanation for increased net N_2 flux after isotope addition is that the sediment core incubations eliminated wind-driven turbulence and resuspension, thus allowing settlement of suspended particles and development of a more defined redox gradient. After 2 days of incubation, redox conditions may have become inhibitory for nitrifiers and stimulative for denitrifiers, which could take advantage of increased substrate (as $^{15}NO_3^-$) to replace NO_3^- supply from decreased nitrification.

Decreases in potential denitrification along the transect may reflect diminishing organic matter availability (Fig. 4). Denitrification fueled by $^{15}NO_3^-$ versus the $^{14}NO_3^-$ pool (from water column or coupled to nitrification) shows substrate stimulation (Fig. 5), especially at sites further from river discharge. Decreasing water column NO_3^- concentrations along this gradient may result from lower nitrification and lead to reduced nitrification/denitrification coupling and increased dependence on water column NO_3^-.

DNRA

Since NH_4^+ is a primary factor in cyanobacterial dominance, it is important to characterize N transformations capable of maintaining NH_4^+ in the system. DNRA is associated with sulfidic estuarine or marine sediments (Tiedje, 1988; An & Gardner, 2002) but also can comprise up to 30% of NO_3^- reduction in lake sediments (Tiedje, 1988). Potential denitrification to DNRA partitioning ratios (DNF:DNRA) in Lake Taihu were 1.6 at 'river', 14.6 at 'inner bay', 315 at 'outer bay', and 35 at 'main lake' (data not shown). Since DNRA depends more on organic matter availability than NO_3^- concentration (Tiedje, 1988), higher DNF:DNRA may reflect organic matter limitation of DNRA away from river discharge. While potentially a meaningful process at 'river', DNRA is not as important as fixed N removal via denitrification at the other sites.

Water column versus sediment NH_4^+ regeneration

Depth-averaged water column NH_4^+ regeneration (water column regeneration converted from $\mu mol\ N\ l^{-1}\ h^{-1}$ to $\mu mol\ N\ m^{-2}\ h^{-1}$ given depth) and sediment N recycling rates (sum of denitrification and net DIN flux, which includes DNRA; Table 4) indicates that Lake Taihu nutrient mineralization was four to six times higher in the water column versus sediments. Thus, primary production appears to be driven primarily by

Table 4 Comparison of depth averaged water column (WC) and sediment (Sed) nitrogen regeneration (reg) in Lake Taihu

	River	Inner Bay	Outer Bay	Main Lake
DNF	180	77	52	8
DIN flux	380	44	33	63
Sed reg	560	121	85	71
WC reg	3520	480	560	310
WC:sed	6.3	4.0	6.6	4.4
WC:DIN	9.3	10.9	17.0	4.9

Fluxes in $\mu mol\ N\ m^{-2}\ h^{-1}$. Sediment regeneration determined by adding denitrification (DNF) and net dissolved inorganic N (DIN) flux before $^{15}NO_3^-$ addition. DNF at 'main lake' determined by adding net N_2 flux before $^{15}NO_3^-$ addition to calculated nitrogen fixation rate water column processes. This dependence is more pronounced if denitrification is excluded from the sediment calculation (WC Reg:DIN; Table 4), since denitrification results in N removal from the system.

MFW structure and N cycling

MFW structure shows a general shift from autotrophic to heterotrophic dominance along the river to lake transect. Nutrient concentration and cycling rate comparisons with MFW structure indicate increasing N limitation along the transect (ratio #'s 2, 3, 4, and 6; Table 3). Interestingly, the light uptake-to-chl ratio (#12) was lowest at 'main lake', but N fixers may have accumulated sufficient N and, thus, did not respond to ^{15}N addition. Diatom biomass-to-NO_3^- ratio (#6) indicates severe N limitation at 'outer bay' and 'main lake'. High regeneration-to-heterotrophic biomass ratios (#'s 13 and 14) at 'river' may indicate heterotrophic bacterial NH_4^+ recycling. Recycling at all other sites, particularly 'main lake', likely is driven by grazing, since bacteria may become N limited as indicated by a higher bacteria-to-NH_4^+ ratio (#1).

The microzooplankton (ciliates and rotifers)-to-"edible" phytoplankton biomass ratio (#10) remained remarkably constant along the transect, indicating tight trophic coupling between these groups, despite sharp variation in the microzooplankton-to-total phytoplankton biomass ratio (#9). Strong grazing pressure on nanoplankton may help cyanobacteria by eliminating competitors for limiting nutrients. Grazers often are ineffective at preying on colonial cyanobacteria due to colony size, potential toxicity, and/or gelatinous envelopes (Dokulil & Teubner, 2000). However, grazers prey on competing phytoplankton (i.e. diatoms) and transform organic N to NH_4^+ via excretion (Shapiro, 1990). Although large colonial cyanobacteria and euglenophytes may not be involved in direct herbivorous trophic transfers within the MFW, the latter group are mixotrophs (i.e. can combine autotrophic and heterotrophic modes of nutrition). Therefore, nutrient limitation is less likely to affect these taxa, since they can obtain nutrients by grazing on bacteria and picocyanobacteria.

Low chl:PO_4^{3-} (#5) and diatom:PO_4^{3-} ratios (#7) at 'inner bay' may be due to top-down pressure rather than N limitation. Greater than 1,000 cladocerans l^{-1} were present at this site. Higher heterotrophic C-to-chl ratio (#8), predominance of mesoplankton-sized ciliate colonies (less vulnerable to cladoceran grazing), and lower protist-to-bacteria ratio (#11) also support this suggestion. The latter ratio also indicates strong bacterial grazing pressure at 'river' and 'main lake', particularly if only suspended cells are considered at 'river' (ratio increases to 0.29). Bacteria may dominate plankton biomass in Meiliang Bay due to sediment resuspension introducing benthic bacteria into the water column, indicated by unusually large bacterial cells and balanced dark NH_4^+ regeneration and uptake.

Conclusions

Nitrogen dynamics relative to MFW composition were characterized along the Liangxihe River discharge gradient in and around Meiliang Bay, Lake Taihu, during late summer/early fall cyanobacteria bloom conditions. The observations provide insights into ecological characteristics of a subtropical, shallow, well-mixed, eutrophic lake. The most interesting finding is that N limits or co-limits primary production in and near central Lake Taihu, contrary to the previous paradigm of exclusive P limitation. A gradient was observed from strong P limitation at the river discharge to N limitation or co-limitation in the main lake. Denitrification in Meiliang Bay may drive main lake N limitation by removing excess N before physical forces transport it to the main lake. This result also exemplifies the importance of characterizing N cycling in freshwater systems, where most studies have focused on P dynamics.

Another important finding is the importance of water column N recycling relative to sediment processes. Nutrient flux from Lake Taihu sediments do not appear to drive primary production during late summer/early fall when cyanobacteria bloom conditions are beginning to wane. However, sediments have an important role in nutrient and MFW dynamics by providing a site for denitrification, which, in conjunction with N fixation and other processes, can regulate available nutrient ratios. Reflecting N cycling patterns, MFW structure shifted from autotrophic (phytoplankton dominated) at the river discharge to heterotrophic (bacteria dominated) in and near the main lake. More studies are needed to evaluate seasonal and spatial variability in nutrient limitation status, N transformation rates, and phytoplankton and MFW structure and function in Lake Taihu.

Acknowledgements This project was supported in part by the Chinese Academy of Sciences (contract No. KZCX1-SW-12), the National Science Foundation of China (Grant #40371104), and the Nancy Lee and Perry R. Bass Regents Chair in Marine Science held by Gardner. Several members of the faculty, staff, and graduate students at the Taihu Laboratory for Lake Ecosystem Research contributed field assistance and discussion during this project. Special assistance was provided by Dr. Zhu Guangwei. The authors thank Dr. Dmitri Sobolev for comments on the manuscript and Dr. Hilairy Hartnett for discussion on dissolved gas sample preservation techniques.

References

An, S. & W. S. Gardner, 2002. Dissimilatory nitrate reduction to ammonium (DNRA) as a nitrogen link, versus denitrification as a sink in a shallow estuary (Laguna Madre/Baffin Bay, Texas). Marine Ecology Progress Series 237: 41–50.

An, S., W. S. Gardner & T. M. Kana, 2001. Simultaneous measurement of denitrification and nitrogen fixation using isotope pairing with membrane inlet mass spectrometry analysis. Applied and Environmental Microbiology 67: 1171–1178.

Blackburn, T. H., 1979. Method for measuring rates of NH_4^+ turnover in anoxic marine sediments, using a ^{15}N–NH_4^+ dilution technique. Applied and Environmental Microbiology 37: 760–765.

Blomqvist, P., A. Pettersson & P. Hyenstrand, 1994. Ammonium–nitrogen: a key regulatory factor causing dominance of non-nitrogen-fixing cyanobacteria in aquatic systems. Archiv fur Hydrobiologie 132: 141–164.

Bode, A. & Q. Dortch, 1996. Uptake and regeneration of inorganic nitrogen in coastal waters influenced by the Mississippi River: spatial and seasonal variations. Journal of Plankton Research 18: 2251–2268.

Cai, Q., X. Gao, Y. Chen, S. Ma & M. Dokulil, 1997. Dynamic variations of water quality in Lake Taihu and multivariate analysis of its influential factors. Journal of Chinese Geography 7: 72–82.

Caperon, J., D. Schell, J. Hirota & E. Laws, 1979. Ammonium excretion rates in Kaneohe Bay, Hawaii, measured by a ^{15}N isotope dilution technique. Marine Biology 54: 33–40.

Chen, Y., B. Qin, K. Teubner & M. Dokulil, 2003. Long-term dynamics of phytoplankton assemblages: *Microcystis*-domination in Lake Taihu, a large shallow lake in China. Journal of Plankton Research 25: 445–453.

Cochlan, W. P., N. M. Price & P. J. Harrison, 1991. Effects of irradiance on nitrogen uptake by phytoplankton: comparison of frontal and stratified communities. Marine Ecology Progress Series 69: 103–116.

Colt, J., 1984. Computation of dissolved gas concentrations in water as functions of temperature, salinity, and pressure. American Fisheries Society Special Publication 14.

Dickman, M., P. Pu & C. Zheng, 1998. Taihu Lake: past, present and future. Journal of Lake Sciences (China) 10(suppl.): 75–83.

Dokulil, M. T. & K. Teubner, 2000. Cyanobacterial dominance in lakes. Hydrobiologia 438: 1–12.

Downing, J. A., S. B. Watson & E. McCauley, 2001. Predicting cyanobacteria dominance in lakes. Canadian Journal of Fisheries and Aquatic Sciences 58: 1905–1908.

Edwards, C. A., T. A. Powell & H. P. Batchelder, 2000. The stability of an NZP model subject to realistic levels of vertical mixing. Journal of Marine Research 58: 37–60.

Elser, J. J., 1999. The pathway to noxious cyanobacteria blooms in lakes: the food web as the final turn. Freshwater Biology 42: 537–543.

Eyre, B. D., S. Rysgaard, T. Dalsgaard & P. B. Christensen, 2002. Comparison of isotope pairing and N_2:Ar methods for measuring sediment denitrification—assumptions, modifications, and implications. Estuaries 25: 1077–1087.

Fahnenstiel, G. L., A. E. Krause, M. J. McCormick, H. J. Carrick & C. L. Schelske, 1998. The structure of the planktonic food web in the St. Lawrence Great Lakes. Journal of Great Lakes Research 24: 531–554.

Gardner, W. S., H. A. Bootsma, C. Evans & P. A. St. John, 1995a. Improved chromatographic analysis of ^{15}N:^{14}N ratios in ammonium or nitrate for isotopic addition experiments. Marine Chemistry 48: 271–282.

Gardner, W. S., J. F. Cavaletto, T. H. Johengen, J. R. Johnson, R. T. Heath & J. B. Cotner, 1995b. Effects of the zebra mussel, *Dreissena polymorpha*, on community nitrogen dynamics in Saginaw Bay, Lake Huron. Journal of Great Lakes Research 21: 529–544.

Gardner, W. S., P. J. Lavrentyev, H. A. Bootsma, J. F. Cavaletto, F. Troncone & J. B. Cotner, 2000. Effects of natural light on nitrogen dynamics in diverse aquatic environments. Internationale Vereinigung Fur Theoretische Und Angewandte Limnologie 27: 64–73.

Gu, B., K. E. Havens, C. L. Schelske & B. H. Rosen, 1997. Uptake of dissolved nitrogen by phytoplankton in a eutrophic subtropical lake. Journal of Plankton Research 19: 759–770.

Havens, K. E., T. Fukushima, P. Xie, T. Iwakuma, R. T. James, N. Takamura, T. Hanazato & T. Yamamoto,

2001. Nutrient dynamics and the eutrophication of shallow lakes Kasumigaura (Japan), Donghu (PR China), and Okeechobee (USA). Environmental Pollution 111: 263–272.

Hewson, I., J. M. O'Neil, C. A. Heil, G. Bratbak & W. C. Dennison, 2001. Effects of concentrated viral communities on photosynthesis and community composition of co-occurring benthic microalgae and phytoplankton. Aquatic Microbial Ecology 25: 1–10.

Huang, Y., 2000. The water quality of Lake Taihu and its protection, pp. 239–246. In Nanjing Institute of Geography and Limnology (eds), Research on Lake Sciences. Chinese Academy of Sciences.

Hyenstrand, P., P. Blomqvist & A. Pettersson, 1998. Factors determining cyanobacterial success in aquatic systems: a literature review. Ergebnisse der Limnologie 51: 41–62.

Jacoby, J. M., D. C. Collier, E. B. Welch, F. J. Hardy & M. Crayton, 2000. Environmental factors associated with a toxic bloom of *Microcystis aeruginosa*. Canadian Journal of Fisheries and Aquatic Sciences 57: 231–240.

Jin, X. & Q. Tu (eds), 1990. Standard Methods for Observation and Analysis of Lake Eutrophication (2nd edn.). Chinese Environmental Science Press, Beijing, 240 pp (in Chinese).

Kana, T. M., C. Darkangelo, M. D. Hunt, J. B. Oldham, G. E. Bennett & J. C. Cornwell, 1994. Membrane inlet mass spectrometer for rapid high-precision determination of N_2, O_2, and Ar in environmental water samples. Analytical Chemistry 66: 4166–4170.

Lavrentyev, P. J., H. A. Bootsma, T. H. Johengen, J. F. Cavaletto & W. S. Gardner, 1998. Microbial plankton response to resource limitation: insights from the community structure and seston stoichiometry in Florida Bay, USA. Marine Ecology Progress Series 165: 45–57.

Lavrentyev, P. J., W. S. Gardner & L. Yang, 2000. Effects of the zebra mussel on nitrogen dynamics and the microbial community at the sediment–water interface. Aquatic Microbial Ecology 21: 187–194.

Loferer-Krossbacher, M., J. Klima & R. Psenner, 1998. Determination of bacterial cell dry mass by transmission electron microscopy and densitometric image analysis. Applied and Environmental Microbiology 64: 688–694.

McCarthy, M. J. & W. S. Gardner, 2003. An application of membrane inlet mass spectrometry to measure denitrification in a recirculating mariculture system. Aquaculture 218: 341–355.

Menden-Deuer, S. & E. J. Lessard, 2000. Carbon to volume relationships for dinoflagellates, diatoms, and other protist plankton. Limnology and Oceanography 54: 569–579.

Nedwell, D. B. & L. F. Dong, 2002. Authors' reply to letters to the editor: nitrous oxide formation in the Colne estuary in England: the central role of nitrite. Applied and Environmental Microbiology 68: 5202–5204.

Nielsen, L. P., 1992. Denitrification in sediment determined from nitrogen isotope pairing. FEMS Microbiology Ecology 86: 357–362.

Porter, K. G. & Y. C. Feig, 1980. The use of DAPI for identifying and counting aquatic microflora. Limnology and Oceanography 25: 943–948.

Pu, P. & J. Yan, 1998. Taihu Lake—a large shallow lake in the East China Plain. Journal of Lake Sciences (China) 10(suppl.): 1–12.

Putt, M. & D. K. Stoecker, 1989. An experimentally determined carbon: volume ratio for marine oligotrichous ciliates from estuarine and coastal waters. Limnology and Oceanography 34: 177–183.

Qin, B., P. Xu, Q. Wu, L. Luo & Y. Zhang, 2007. Environmental issues of Lake Taihu, China. Hydrobiologia 581: 3–14.

Seitzinger, S. P., 1988. Denitrification in freshwater and coastal marine ecosystems: ecological and geochemical significance. Limnology and Oceanography 33: 702–724.

Shapiro, J., 1990. Current beliefs regarding dominance by blue-greens: the case for the importance of CO_2 and pH. Internationale Vereinigung Fur Theoretische Und Angewandte Limnologie 24: 38–54.

Sherr, E. B., D. A. Caron & B. F. Sherr, 1993. Staining of heterotrophic protists for visualization via epifluorescence microscopy. In Kemp, P. F., B. F. Sherr, E. B. Sherr & J. J. Cole (eds), Handbook of Methods in Aquatic Microbial Ecology. Lewis, London: 213–228.

Tiedje, J. M., 1988. Ecology of denitrification and dissimilatory nitrate reduction to Ammonium. In Zehnder, A. J. B. (ed.), Biology of Anaerobic Microorganisms. Wiley, New York, 179–244.

Tobias, C. R., I. C. Anderson, E. A. Canuel & S. A. Macko, 2001. Nitrogen cycling through a fringing marsh-aquifer ecotone. Marine Ecology Progress Series 210: 25–39.

Velji, M. I. & L. J. Albright, 1993. Improved sample preparation for enumeration of aggregated aquatic substrate bacteria. In Kemp, P. F., B. F. Sherr, E. B. Sherr & J. J. Cole (eds), Handbook of Methods in Aquatic Microbial Ecology. Lewis, London, 139–142.

Zumft, W. G., 1997. Cell biology and molecular basis of denitrification. Microbiology and Molecular Biology Reviews 61: 533–616.

Hydrobiologia (2007) 581:209–215
DOI 10.1007/s10750-006-0493-5

Different competitive outcomes among four species of cladocerans under different alga combinations of colonial *Microcystis* spp. and green alga *Scenedesmus obliquus*

Feizhou Chen · Ping Xie · Boqiang Qin

Abstract Cyanobacteria blooms (especially *Microcystis* spp.) are thought to alter dominance of large-sized daphnids into small-sized metazoan zooplankton. However, several field investigations show different phenomena. Laboratory experiments were conducted based on the hypothesis that different *Microcystis* spp. concentrations would influence competitive outcomes using two algal combinations of different concentrations and four species of cladocerans. In the algal combination of 50 mg l^{-1} colonial *Microcystis* spp. + 1 mg l^{-1} *Scenedesmus obliquus* (fresh weight), *Daphnia carinata* was absent during the experiment in competition with other cladocerans. Decreasing colonial *Microcystis* spp. concentration (10 mg l^{-1}) resulted in a shift from dominance by small-sized cladocerans to dominance by *D. carinata*. No significant effects of different concentrations of colonial *Microcystis* spp. on competitive outcomes were shown among three small-sized cladocerans. These results support the idea that cyanobacteria concentration affects the dominance status of large-bodied daphnid.

Keywords cladoceran · *Daphnia* · competition · *Microcystis* spp.

Introduction

Cyanobacteria blooms, often occurring in eutrophic systems, can alter zooplankton community structure by disturbing interspecific actions. Generally, large-sized species often are replaced by small-sized species (e.g. Infante & Riehl, 1984). For example, *Ceriodaphnia* and *Bosmina* replaced the large-sized cladoceran *Daphnia ambigua* during a cyanobacteria bloom (Allan, 1977). Other field investigations showed that large-sized cladocerans dominated during blooms (George & Edwards, 1974; Kim et al., 2001). In another case, colonial *Microcystis* did not favor rotifers (*Brachionus calyciflorus*) in competition with *D. ambigua* (Fulton & Paerl, 1988).

Many factors, including concentration, size, *Microcystis* toxin and alternative food abundance

Guest editors: B. Qin, Z. Liu & K. Havens
Eutrophication of shallow lakes with special reference to Lake Taihu, China

F. Chen (✉) · P. Xie
Donghu Experimental Station of Lake Ecosystems, State Key Laboratory for Freshwater Ecology and Biotechnology of China, Institute of Hydrobiology, Chinese Academy of Sciences, Wuhan 430072
P.R. China
e-mail: feizhch@niglas.ac.cn

F. Chen · B. Qin
Nanjing Institute of Geography and Limnology, Chinese Academy of Sciences, Nanjing 210008
P.R. China

during blooms may result in zooplankton shifts (Jarvis et al., 1987; Fulton & Paerl, 1988; Sartonov, 1995; Chen & Xie, 2003, 2004). Different responses of species to these factors result in observed zooplankton community structure.

Since there is little or no evidence for lethal effects of toxic cyanobacteria on crustacean plankton (Haney, 1987) and disregarding effects of planktivorous fish, zooplankton shifts during cyanobacteria blooms may be determined, to a certain extent, by biomass/concentration of cyanobacteria and alternative food. Low concentrations of edible algae favors small-sized cladocerans, while high concentrations favor large-sized cladocerans, such as *Daphnia* (Goulden et al., 1982; Tessier & Goulden, 1987). However, the opposite also has been reported (Nandini et al., 2002). During cyanobacteria blooms, alternative food availability may influence competition outcomes, except feeding inhibition. Although colonial *Microcystis* spp. did not favor *D. ambigua* in competition with *Diaphanosoma brachyurum* and *Bosmina longirostris* in a laboratory experiment (Fulton & Paerl, 1988), field investigation showed daphnid dominance in the cladoceran community. In the present study, we hypothesized that different colonial *Microcystis* spp. concentrations could affect competition outcomes among cladocerans, expecially large-sized daphnids and small-sized cladocerans. Two concentration combinations of colonial *Microcystis* spp. and green alga *S. obliquus* were used to test the hypothesis.

Materials and methods

All stock culture and experimental procedures were conducted at 25°C ± 0.5°C and a 12:12 (hours light:dark) photoperiod. Wild *Daphnia carinata*, *Moina micrura*, *D. brachyurum* and *Ceriodaphnia cornuta* were collected from Lake Donghu and reared through many generations in the laboratory. *S. obliquus*, a green alga, was sourced from cultures maintained by the Institute of Hydrobiology of the Chinese Academy of Sciences. *Sceneesmus obliquus* was cultured in Shuisheng VI medium (Li et al., 1959) with aeration. Colonial *Microcystis* spp. (90–95% *Microcystis aeruginosa*) was collected from a dense bloom in a nearby fishpond, and previous work indicated that this population produced microcystin that was lethal to *Daphnia magna* and *D. carinata* (Carmichael et al., 1988; He et al., 1997; Chen & Xie, 2004). Microcystin concentration was 4.5×10^{-6} μg cell^{-1}. Colonial *Microcystis* between 64–112 μm were used for competitive experiments. A low concentration of this size range of *Microcystis* improves development of *D. carinata* (Chen & Xie, 2003). When colonies were destroyed by ultrasonic waves, toxicity to *D. carinata* increased (Chen & Xie, 2004). Thus, toxicity to cladocerans was less in this study and can be neglected. Colonies were held at 4–6°C in the dark for 24–48 h prior to experiments.

During the cyanobacteria bloom in Lake Donghu, *Microcystis* spp. biomass ranged from 0 to 46 mg l^{-1} (fresh weight) (Tang, 2002). Two algal biomass combinations were used: 10 mg l^{-1} colonial *Microcystis* spp. + 1 mg l^{-1} *S. obliquus* (hereafter 10 M:1 S), and 50 mg l^{-1} colonial *Microcystis* spp. + 1 mg l^{-1} *S. obliquus* (hereafter 50 M:1 S). Algal biomass was fresh weight. Each combination had two replicates, and each replicate included 10 neonates (<12 h old) for *D. carinata*, *M. micrura*, and *D. brachyurum* and 15 neonates for *C. cornuta* considering body size difference. Experiments were conducted in a 250 ml beaker. Experiment length was determined by competitive results, and the maximum was four weeks. The culture medium was changed every two days, and the number of cladocerans was counted.

Maximal population growth rate was calculated as:

$$g = (\ln N_t - \ln N_0)/t$$

where N_0 and N_t are the mean individual number initially and reaching highest density after t days, respectively. For two peaks in certain growth curves, we selected the first peak for calculation of population growth rate. Competition comparison among *M. micrura*, *D. brachyurum* and *C. cornuta* was made using 2-way ANOVA with the index of the maximal population growth rate. All statistical analyses were conducted in Statistica 5.1.

Results

Competitive outcomes were different between the two alga combinations. In the 50 M:1 S combination, *D. carinata* could not compete with *M. micrura, C. cornuta*, and *D. brachyurum*. *D. carinata* was absent within 18 days. Time to extinction of *D. carinata* was shortest (8 days) in the presence of *M. micrura*. However, at 10 M:1 S, an equilibrium was reached in the presence of *M. micrura* and *C. cornuta* without any extinction (Fig. 1). For *D. brachyurum*, extinction occurred within two weeks in competition with *D. carinata* in the 10 M:1 S combination.

Competitions among *M. micrura, C. cornuta* and *D. brachyurum* were less intense than between *D. carinata* and other cladocerans (Fig. 2). No species disappeared within four weeks in the 50 M:1 S combination, while only *C. cornuta* was absent in competition with *D. brachyurum* in the 10 M:1 S concentration. *Moina micrura* and *C. cornuta* had stable density at the end of 50 M:1 S experiments. In 10 M:1 S trials, *M. micrura* dominated over *C. cornuta*. At both *Microcystis* spp. concentrations, *M. micrura* had a higher population growth rate than *D. brachyurum*, but, after 10 days, its density declined gradually along with increasing density of *D. brachyurum*. *C. corunta* dominated in all 50 M:1 S experiments. ANOVA analysis indicted that algal mixtures had no significant effect on the maximal population growth rate of *M. micrura* while species and interaction had. For *D. brachyurum* and *C. cornuta*, algal mixtures, species and interactions had significant

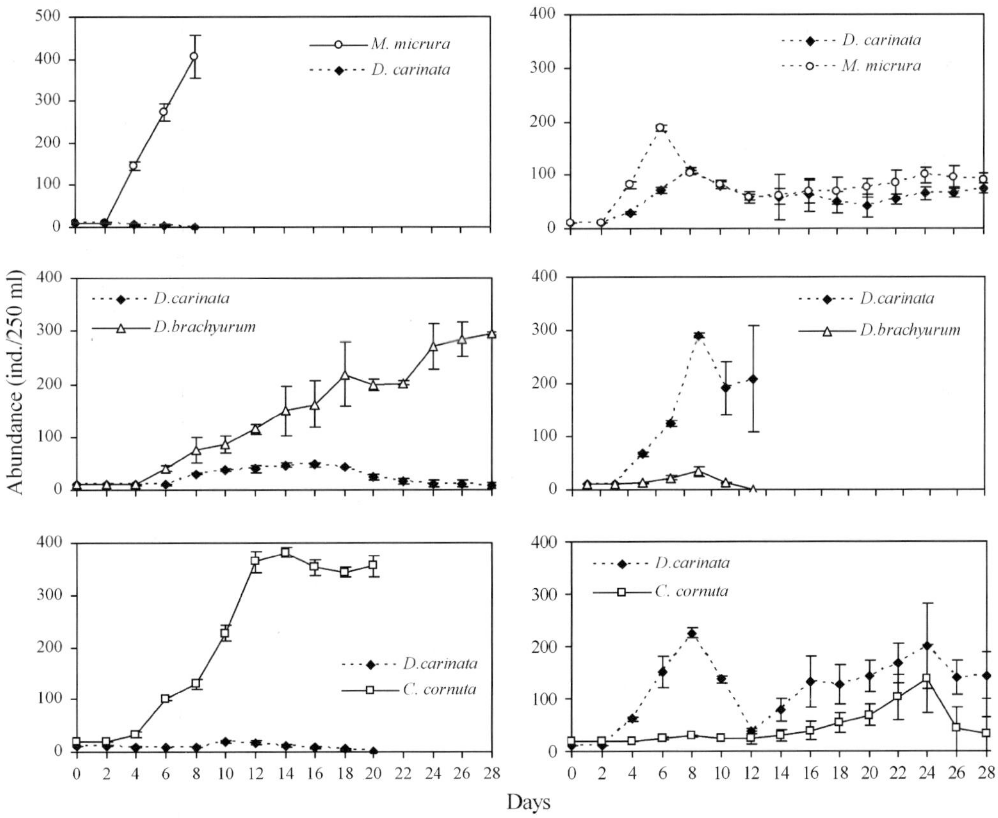

Fig. 1 Cladoceran abundances (±SD) in laboratory competition between *Daphnia carinata* and three small-sized cladoceran at different alga combinations. Ultimate time of experiments is four weeks. Left shows the results at the alga combination of 50 mg l^{-1} colonial *Microcystis* spp. + 1 mg l^{-1} *Scenedesmus obliquus*. Right at the alga combination of 10 mg l^{-1} colonial *Microcystis* spp. + 1 mg l^{-1} *S. obliquus*. Algal biomass is fresh weight. Abundance showed in Y-axis is number in 250 ml

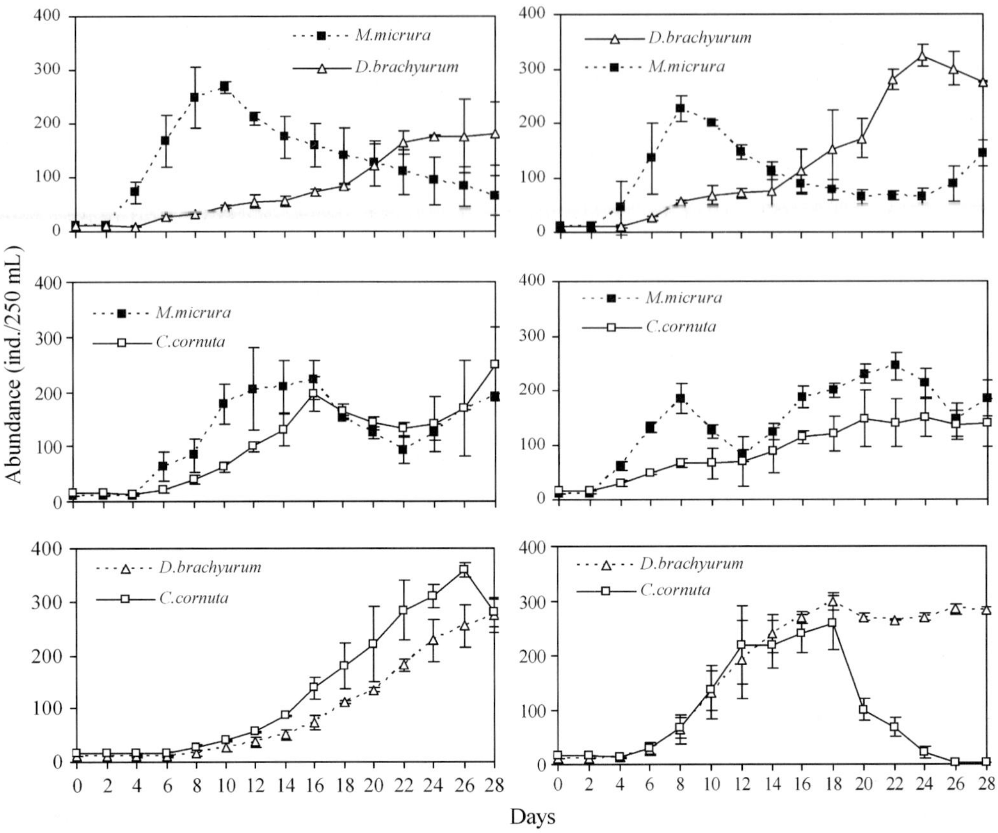

Fig. 2 Cladoceran abundances (±SD) in laboratory competition among *Moina micrura, Diaphanosoma brachyurum* and *Ceriodaphnia cornuta* at different alga combinations. Ultimate time of experiments is four weeks. Left shows results at the alga combination of 50 mg l^{-1} colonial *Microcystis* spp. + 1 mg l^{-1} *Scenedesmus obliquus*. Right at the alga combination of 10 mg l^{-1} colonial *Microcystis* spp. + 1 mg l^{-1} *S. obliquus*. Algal biomass is fresh weight. Abundance in Y-axis is number in 250 ml

effect on their maximal population growth rate (Table 1).

Table 2 shows maximum population growth rate (MGR) and time needed to reach MGR (days) for three cladocerans in competition under two algal combinations. Except for the competition between *M. micrura* and *C. cornuta*, MGR were higher in the 10 M:1 S trials and time reaching MGR shortened correspondingly.

Discussion

Results show that competitive outcomes among cladocerans, especially between large-sized *D. carinata* and small-sized cladocerans, are related to colonial *Microcystis* spp. concentration. Low concentrations of colonial *Microcystis* spp. may not result in competitive extinction of large-sized *D. carinata*, while high *Microcystis* spp. concentration brought on its extinction.

During a cyanobacteria bloom, it is thought that community succession is from larger-sized cladocerans to copepods, rotifers and small-sized cladocerans (e.g. Infante & Riehl, 1984). This has not always been the case in laboratory experiments and field investigations (Mesfin et al., 1988). In fact, concentration, size, *Microcystis* toxin, and alternative food abundance affect zooplankton shifts. Large-sized phytoplankton and high concentrations of colonial algae often inhibit *Daphnia* sp. Feeding and interfere with food availability (Hessen & Donk, 1993). *Daphnia ambigua* dominated without *M. aeruginosa*

Table 1 Results of variance (ANOVA) analysis performed on maximal population growth rate of *M. micrura*, *D. brachyurum* and *C. cornuta* in relation to food mixture and competitive interaction

Source	df	SS	MS	F
M. micrura				
Food combination (A)	1	0.0002	0.0002	7.35^{ns}
With species (B)	1	0.0733	0.0733	2444.68**
Interaction of A·B	1	0.0076	0.0076	252.14**
Error	7	0.0812	0.0116	
D.brachyurum				
Food combination (A)	1	0.0062	0.0062	842.59**
With species (B)	1	0.0018	0.0018	239.94**
Interaction of A·B	1	0.0004	0.0004	58.98**
Error	7	0.0085	0.0012	
C. cornuta				
Food combination (A)	1	0.0008	0.0008	127.99**
With species (B)	1	0.0046	0.0046	737.24**
Interaction of A·B	1	0.0005	0.0005	76.88**
Error	7	0.0059	0.0008	

Ns and **represent significant level of $p > 0.05$ and $p < 0.001$, respectively

and approached extinction with 50% and 90% *M. aeruginosa* (Fulton & Paerl, 1988). Our results show a different competitive shift for *D. carinata* under different concentrations of *Microcystis*. For small-sized cladocerans in the present study, growth rate was related negatively to colonial *M. aeruginosa* concentration with fixed concentration of *S. obliquus*. *D. carinata* obtained the highest growth rate at a biomass level of 10 mg l^{-1} colonial *M. aeruginosa*, indicating its ability to utilize colonial forms (Chen & Xie, 2003). The competitive shift in the present study was from small-sized cladocerans to *D. carinata* when concentration of colonial *Microcystis* spp. was low. It appears that a low concentration of colonial *Microcystis* does not lead to the general successive pattern for zooplankton.

Colony size also affects cladoceran population and their interactions. Only *Daphnia pulex* could feed on 60–100 μm *Microcystis* colonies, while *Ceriodaphnia reticulata*, *M. micrura* and *Diaphanosoma excisum* could not (Jarvis et al., 1987). Previous work showed similar results (Chen & Xie, 2003). The inability of the other three cladocerans to feed on 64–112 μm *Microcystis* colonies determined the status of *D. carinata* in competition. In the present study, larger-sized *M. aeruginosa* (>112 μm) colonies that may have a more negative effect on *D. carinata* were not evaluated. For the selected size of *Microcystis* spp., inhibition of *Daphnia* from low *Microcystis* concentration was weak, and *Daphnia* maintained dominance.

Microcystin and other compounds in *Microcystis* can be lethal to cladocerans (e.g. Jungamnn, 1992; DeMott, 1999). *Keratella cochlearis* was superior in competition with *D. pulex* under toxic *Microcystis* (Sartonov, 1995). However, there is little or no evidence for lethal effects of toxic cyanobacteria on crustacean plankton in the field (Haney, 1987). In fact, many cladocerans can coexist well with *Microcystis* in nature (McNaught et al., 1997; Nandini et al., 2000). Colonial *Microcystis* spp. used in the present study had no obvious toxicity to cladocerans based on the previous study, but it exhibited toxicity when colonies were destroyed (Chen & Xie, 2004). Microcystin released into the water is thought to be inhibited by the mucilaginous structure of the colony (Stangenberg, 1968; Goarant et al., 1994). Once colonial *Microcystis* decomposes, microcystin and other toxic compounds are released (Watanabe et al., 1992). Although toxicity is possible, alternative food availability may weaken toxic effects of microcystin (Ferrão-Filho et al., 2000; Chen & Xie, 2004).

During the bloom, alternative food, including bacteria attached to colonial *Microcystis*, detritus, algae and protozoans also affect population sizes

Table 2 Maximum population growth rate (MGR) and time to reach MGR (days) of three cladocerans among competitions under two types of mixtures of *Microcystis* spp. (M) and *Scenedesmus obliquus* (S)

	50 mg l^{-1} M + 1 mg l^{-1} S			10 mg l^{-1} M + 1 mg l^{-1} S		
	M. micrura	*D. brachyurum*	*C. cornuta*	*M. micrura*	*D. brachyurum*	*C. cornuta*
M. micrura		0.328(10)	0.194(16)		0.390(8)	0.145(22)
D. brachyurum	0.103(28)		0.118(28)	0.145(24)		0.189(18)
C. cornuta	0.089(28)	0.122(26)		0.096(24)	0.158(18)	

of cladocerans. Laboratory experiments showed that growth rates of the four cladocerans used in the present study were correlated positively with *S. obliquus* biomass (Chen & Xie, 2003). Many species, including *Daphnia* sp., can consume colonial *Microcystis* sp. of edible size (Matveev et al., 1994; Nandini et al., 2000). *Daphnia* sp. can consume the mucilaginous part of colonies and associated bacteria and may be more resistant to toxin (Goarant et al., 1994). In the present study, *D. carinata* was not inhibited by low colonial *Microcystis* spp. concentration. *Scenedesmus obliquus*, the mucilaginous part of colonies and associated bacteria may be used by *D. carinata*, which may improve population growth. *Moina micrura* may have a competitive advantage over rotifers in exploiting *Microcystis* as a food source or use other food sources (e.g. bacteria) when *Microcystis* are under stress (Kim et al., 2001). Several genera of *Moina, Diaphanosoma, Bosmina* and *Daphnia* can use cyanobacteria detritus (Hanazato & Yasuno, 1985, 1987; Gulati et al., 2001; Chen & Xie, 2003). This may promote populations of these species in the lake and affect their dominant status in the cladoceran community through interactions.

However, other research shows that low concentrations of edible algae favors small-sized cladocerans, while high concentrations favor large-sized cladocerans, such as *Daphnia* (Goulden et al., 1982; Tessier & Goulden, 1987). During a *Microcystis* bloom, other edible algae may have actual or relative low concentration. Resultant food scarcity may favor small-sized cladocerans. If *Daphnia* are inhibited by colonial *Microcystis* and less alternative food, its population may decline.

No obvious competition outcomes between *M. micrura* and *D. brachyurum* or *C. cornuta* under two alga combinations were observed in the present study. *Diaphanosoma brachyurum* outweighs *M. micrura* later in the experiment. Similar trends appear in the competition between *D. brachyurum* and *C. cornuta*. *D. brachyurum* may utilize bacteria around *Microcystis* spp. and detritus in the beaker more than *M. micrura* and *C. cornuta* (Champ & Pourriot, 1977; Taira, 1989). For these three cladocerans, competition may affect competitive outcomes (Chaneton & Bonsall, 2000; Martínez & Medel, 2002). In nature, this effect should be considered.

In conclusion, our experiments reveal that in 50 mg l^{-1} colonial *Microcystis* spp. + 1 mg l^{-1} *S. obliquus* combinations, *D. carinata* was extinct during the experiment in competition with other cladocerans. Decreasing colonial *Microcystis* spp. concentration (10 mg l^{-1}) resulted in a shift from small-sized cladocerans to *D. carinata*. No obvious effect on competitive outcomes was shown among three small-sized cladocerans. These results suggest that cyanobacteria concentration may influence dominance status of large-bodied daphnids.

Acknowledgements This work was supported by National Natural Science Foundation of China (No. 30600069), a key project of Chinese Academy of Sciences (No. KZCX1-SW-12) and National Natural Science Foundation of China (No. 40501078).

References

Allan, J. D., 1977. An analysis of seasonal dynamics of a mixed population of *Daphnia* and associated cladoceran community. Freshwater Biology 7: 505–512.

Carmichael, W. W., M. J. Yu, Z. R. He, J. W. He & J. L. Yu, 1988. Occurrence of the toxic cyanobacterium (blue-green alga) *Microcystis aeruginosa* in central China. Archiv für Hydrobiologie 114: 21–30.

Champ, P. & R. Pourriot, 1977. The food of freshwater planktonic cladocera (Crustacea). Annee Biology 16: 317–342.

Chaneton, E. J. & M. B. Bonsall, 2000. Enemy-mediated apparent competition: empirical patterns and the evidence. Oikos 88: 380–394.

Chen, F. Z. & P. Xie, 2003. The effects of fresh and decomposed Microcystis aeruginosa on cladocerans from a subtropic Chinese lake. Journal of Freshwater Ecology 18: 97v–104v.

Chen, F. Z. & P. Xie, 2004. The toxicities of single-celled *Microcystis aeruginosa* PCC7820 and liberated *M. aeruginsa* to *Daphnia carinata* in the absence and present of the green alga *Scenedesmus obliquus*. Journal of Freshwater Ecology 19: 539–545.

DeMott, W. R., 1999. Foraging strategies and growth inhibition in five daphnids feeding on mixtures of a toxic cyanobacterium and a green alga. Freshwater Biology 42: 263–274.

Ferrão-Filho, A. S., S. M. F. O. Azevedo & W. R. DeMott, 2000. Effects of toxic and non-toxic cyanbacteria on the life history of tropical and temperature cladocerans. Freshwater Biology 45: 1–19.

Fulton, III R. S. & H. W. Paerl, 1988. Effects of the blue-green alga *Microcystis aeruginosa* on zooplankton competitive relations. Oecologia 76: 383–389.

George, D. G. & R. W. Edwards, 1974. Population dynamics and production of *Daphnia hyalina* in a eutrophic reservoir. Freshwater Biology 4: 445–465.

Goarant, E., G. Prensier & N. Lair, 1994. Specific immunological probes for the identification and tracing of prey in crustacean gut contents. The example of cyanobacteria. Archiv für Hydrobiologie 131: 243–252.

Goulden, C. E., L. L. Henry & A. J. Tessier, 1982. Body size, energy reserves, and competitive ability in three species of cladocera. Ecology 63: 1780–1789.

Gulati, R. D., M. Bronkhorst & E. Van Donk, 2001. Feeding in *Daphnia galeata* on *Oscillatoria limnetica* and on detritus derived from it. Journal of Plankton Research 23: 705–718.

Hanazato, T. & M. Yasuno, 1985. Population dynamics and production of cladoceran zooplankton in the highly eutrophic Lake Kasumigaura. Hydrobiologia 124: 13–22.

Hanazato, T. & M. Yasuno, 1987. Experimental studies on competition between *Bosmina longirotris* and *Bosmina fatalis*. Hydrobiologia 154: 189–199.

Haney, J. F., 1987. Field studies on zooplankton-cyanobacteria interactions. New Zealand Journal of Marine and Freshwater Research 21: 467–475.

He, J. W., Z. R. He & Q. L. Guo, 1997. The toxicity of *Microcystis aeruginosa* to fishes and *Daphnia*. Journal of Lake Sciences 9: 49–56 (in Chinese).

Hessen, D. O. & E. Van Donk, 1993. Morphological changes in *Scenedesmus* induced by substances release from *Daphnia*. Archiv für Hydrobiologie, 127: 129–140.

Infante, A. & W. Riehl, 1984. The effect of cyanophyta upon zooplankton in a eutrophic tropical lake (Lake Valencia, Venezuela). Hydrobiologia 113: 293–298.

Jarvis, A. C., R. C. Hart & S. Combrink, 1987. Zooplankton feeding on size fractionated *Microcystis* colonies and *Chlorella* in a hypertrophic lake (Hartbeespoort Dam, South Africa): Implications to resource utilization and zooplankton succession. Journal of Plankton Research 9: 1231–1249.

Jungmann, D., 1992. Toxic compounds isolated from *Microcystis* PCC7806 that are more active against *Daphnia* than two microcystins. Limnology and Oceanography 37: 1777–1783.

Kim, H. W., G. J. Joo, K. H. Chang & S. J. Hwang, 2001. Zooplankton community dynamics during the summer Mycrocystis bloom in the lower part of the Nakdong River, South Korea. Verhandlungen Internationale Vereinigung für theoretische und angewandte Limnologie 27: 1044–1049.

Li, S. H., H. Zhu, J. Z. Xia, K. E. Lin, K. S. Liu, Z. Y. Le & Y. X. Chen, 1959. Experiment on mass culture of single-celled green alga. Acta Hydrobiological Sinica 4: 462–472 (in Chinese).

Martínez, G. & R. Medel, 2002. Indirect interactions in a microcosm-assembled cladoceran community: implications for apparent competition. Oikos 97: 111–115.

Matveev, V., L. Matveeva & G. Jones, 1994. Study of the ability of *Daphnia carinata* King to control phytoplankton and resist cyanobacterial toxicity: implications of biomanipulation in Australia. Australia Journal of Marine and Freshwater Research 45: 889–904.

McNaught, D. C., W. Chen & H. J. Cai, 1997. Grazing on a *Microcystis* bloom by cladocerans and copepods in Tai Lake. Verhandlungen Internationale Vereinigung für theoretische und angewandte Limnologie 26: 550–552.

Mesfin, M., C. Tudorancea & R. M. Baxter, 1988. Some limnological observations on two Ethiopian hydroelectric reservoirs: Koka (Shewa administrative district) and Finchaa. Hydrobiologia 157: 47–55.

Nandini, S., S. S. S. Sarma & P. Ramirez-Garcia, 2000. Life table demography and population growth of *Daphnia laevisi* (Cladocera, Anomopoda) under different densities of *Chlorella vulgaris* and *Microcystis aeruginosa*. Crustaceana 73: 1273–1286.

Sartonov, A., 1995. Effects of *Microcysits aeruginosa* on interference competition between *Daphnia pulex* and *Keratella cochlearis*. Hydrobiologia 307: 117–126.

Stangenberg, M., 1968. Toxic effects of *Microcystis aeruginosa* Kg. Extracts on *Daphnia longispina* O. F. Müller and *Eucypris virens* Jurine. Hydrobiologia 32: 81–87.

Taira, M., 1989. Species composition and feeding habits of crustacean plankters in the pools of high moor. Japanese Journal of Limnology 50: 333–339.

Tang, H. J., 2002. Ecological studies on phytoplankton of the shallow, eutrophic Lake Donghu. Dotoral dissertation thesis, Institute of Hydrobiology, Chinese Academy of Sciences (in Chinese).

Tessier, A. J. & C. E. Goulden, 1987. Cladoceran juvenile growth. Limnology and Oceanography 32: 680–686.

Watanabe, M. M., K. Kaya & N. Takamura, 1992. Fate of the toxic cyclic heptapeptides, the microcystins, from blooms of *Microcystis* (Cyanobacteria) in a hypertrophic lake. Journal of Phycology 28: 761–767.

Hydrobiologia (2007) 581:217–223
DOI 10.1007/s10750-006-0506-4

EUTROPHICATION IN LAKES

The effect of temperature on growth characteristics and competitions of *Microcystis aeruginosa* and *Oscillatoria mougeotii* in a shallow, eutrophic lake simulator system

Zhaosheng Chu · Xiangcan Jin · Norio Iwami · Yuhei Inamori

Abstract Blue-green algal blooms formed by *Microcystis* and *Oscillatoria* often occur in shallow eutrophic lakes, such as Lake Taihu (China) and Lake Kasumigaura (Japan). Growth characteristics and competitions between *Microcystis aeruginosa* and *Oscillatoria mougeotii* were investigated using lake simulator systems (microcosms) at various temperatures. *Oscillatoria* was the superior competitor, which suppressed *Microcystis*, when temperature was <20°C, whereas the opposite phenomenon occurred at 30°C. *Oscillatoria* had a long exponential phase (20 day) and a low growth rate of 0.22 day^{-1} and 0.20 day^{-1} at 15°C and 20°C, respectively, whereas *Microcystis* had a shorter exponential phase (2–3 days) at 30°C and a higher growth rate (0.86 day^{-1}). Interactions between the algae were stronger and more complex in the lake simulator system than flask systems. Algal growth in the lake simulator system was susceptible to light attenuation and pH change, and algae biomasses were lower than those in flasks. The outcome of competition between *Microcystis* and *Oscillatoria* at different temperatures agrees with field observations of algal communities in Lake Taihu, indicating that temperature is a significant factor affecting competition between *Microcystis* and *Oscillatoria* in shallow, eutrophic lakes.

Keywords *Microcystis* · *Oscillatoria* · Lake simulator · Competition · Lake Taihu

Guest editors: B. Qin, Z. Liu & K. Havens
Eutrophication of shallow lakes with special reference to Lake Taihu, China

Z. Chu · X. Jin (✉) · N. Iwami
Research Center for Lake Ecology & Environments, Chinese Research Academy of Environmental Sciences (CRAES), Beijing 100012, China
e-mail: jinxc@craes.org.cn

Y. Inamori
National Institute for Environmental Studies (NIES), 16-2 Onogawa, Ibaraki 305-0053, Japan

Introduction

Blue-green algal blooms often cause significant problems, such as odors, water deoxygenation and clogging of water supply systems. Furthermore, some blue-green algae produce toxins, which are dangerous to aquatic organisms, domestic animals and human beings. *Microcystis* and *Oscillatoria* are two genera of bloom-forming blue-green algae often present in shallow eutrophic lakes (Reynolds & Walsby, 1975). *Microcystis* usually forms colonies, and *Oscillatoria* grows in filamentous form. In Lake Taihu (China), *Microcystis* often dominates in summer and autumn (Fan, 1996; Dokulil et al., 2000; Yang et al., 2003) whereas *Oscillatoria* often dominates in early spring and later autumn. The climate of Lake Kasumigaura (Japan) is similar to Lake Taihu, and *Microcystis* dominance often occurs in summer (Wei et al., 2001). In recent years, *Microcystis* has been declining and was replaced by filamentous blue-green algae, such as *Oscillatoria* and *Phormidium* (Takamura et al., 1992; Ohkubo et al., 1993; Wei et al., 2001). Despite considerable research on photosynthetic and growth responses of *Microcystis* and *Oscillatoria* in laboratory and field studies (Mur & Schreurs, 1995; Zhang & Prepas, 1996; Fujimoto et al., 1997; Coles and Jones 2000), there is limited information about mechanisms of *Microcystis* and *Oscillatoria* dominance in seasonal dynamics and year-by-year changes in shallow eutrophic lakes.

Generally, there are three categories of studies on competition of bloom-forming algae. The first is large-scale investigations based on field surveys; second, small-scale studies based on laboratory experiments; and, third, intermediate-scale studies applying 'microcosms'. Natural systems are often too complicated to obtain detailed information in a short time or only provide insight into lake-specific circumstances. The second method often oversimplifies natural systems, and information only can be applied in a limited sense. Two kinds of microcosms being used are the 'in-situ microcosm', placed directly into lakes, and the 'laboratory microcosms', constructed in the laboratory. Construction of in situ microcosms depends on weather and space (Xie et al., 2003a, b). Here, a laboratory microcosm, called the lake simulator system, was used. It is a large-scale simulation system where light, temperature, nutrients, water and turbulence can be controlled (Okada & Sudo, 1980). Conveniently, chemical components, such as pH, dissolved oxygen (DO) and oxidation–reduction potential (ORP) can be monitored. Furthermore, the system can be sterilized by steam.

In this study, the lake simulator and a flask system were used to investigate growth and competition of *Microcystis* and *Oscillatoria* at different temperatures, thus increasing understanding of shifts in bloom-forming blue-green algae during seasonal dynamics in shallow eutrophic lakes.

Materials and methods

Algae. Microcystis aeruginosa Kützing and *Oscillatoria mougeotii* Kützing *ex* Lemmermann were isolated from Lake Kasumigaura in 1996 and maintained at 20°C under a 12 h:12 h light:dark cycle in M_{11} medium, which consists of 100 mg $NaNO_3$, 10 mg K_2HPO_4, 75 mg $MgSO_4 \cdot 7H_2O$, 40 mg $CaCl_2 \cdot 2H_2O$, 20 mg Na_2CO_3, 6 mg Fe-citrate and 1 mg $Na_2EDTA \cdot 2H_2O$ in 1:l deionized water. Both algal cultures were axenic.

Lake Simulator System. The lake simulator system consists of two artificial lakes (tanks, 4 m high and 1 m diameter), two artificial suns (Xenon lamp, 5 kW), a sterilization tank, monitoring system, temperature control system, sterilized air supply system and a steam system (an artificial lake and artificial sun is omitted in Fig. 1).

Growth and competition experiments

The lake simulator and small conical flasks (500 ml) were used to investigate growth and competition between *Microcystis* and *Oscillatoria.* M_{11} culture medium was used to cultivate the two algae. Before culture experiments, the simulator was sterilized with steam. The culture medium was sterilized in the sterilization tank and transported to the artificial lake by the pressure of

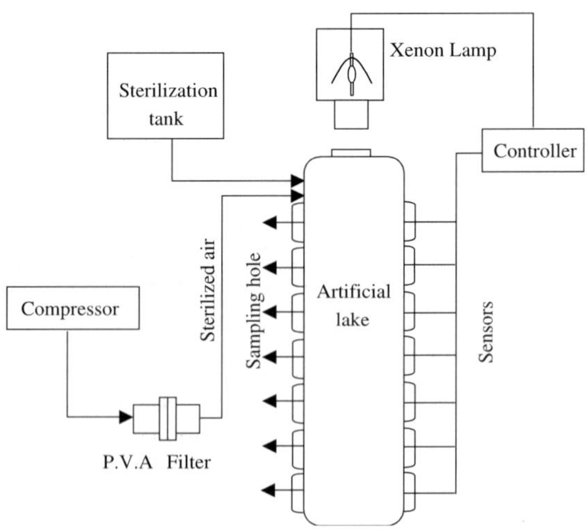

Fig. 1 Schematic diagram of the lake simulator system

sterilized air. Each tank contained 2400 l of culture medium with 3.4 m in depth. To simulate lake turbulence, sterilized air was supplied via the sampling hole, which was 0.7 m above the bottom of the artificial tank, producing 5–10 cm S^{-1} turbulence at the water surface. pH was adjusted to 8.0–8.5 using concentrated HCl. When the medium was saturated with DO, *Microcystis* (in single cell) and *Oscillatoria* were inoculated to the artificial lake resulting in an initial population density of 2.5×10^4 cells ml^{-1} and 5.0 cm ml^{-1} (the length of filaments), respectively. These numbers correspond to 0.40 and 0.20 mg DW (dry weight) l^{-1}. Illumination at the surface water was 100,000 lux (12 h light:12 h dark cycle), similar to natural sunlight at noon at the lake surface (Lake Taihu). 200 ml water was sampled from seven sampling holes and pooled for analysis. Flask cultures (containing 150 ml M_{11} medium) were incubated in light (2,000 lux). The culture medium, the light:dark cycle, temperatures and initial cell densities were the same as those in the lake simulator.

Chlorophyll *a* was extracted with acetone (90%) and determined spectrophotometrically (EPA Method, 1997). pH was determined potentiometrically. Algal density was counted with a haemacytometer (TATAI, Japan) under a microscope. 30–300 cells were counted for *Microcystis*, and 10 μl of algal solution determined for *Oscillatoria*. Counting was carried out three times for each sample. If *Oscillatoria* filaments were longer than 0.25 mm, they were disrupted by ultrasonication before counting. Cell counts were converted to dry weight to determine dominance. The sampling culture medium was filtered through glass fiber filter (Whatman/C) and dried at 90°C for 12 h to determine DW. Conversion factors for DW and cell count were 2.3×10^{-11}, 1.6×10^{-11}, and 1.7×10^{-11} g $cell^{-1}$ for *Microcystis* and 4.1×10^{-8}, 4.0×10^{-8}, and 6.0×10^{-8} g cm^{-1} for *Oscillatoria*, at 15°C, 20°C and 30°C, respectively. Specific growth rates were determined by linear regression of logarithm of biomass and time (Eq. 1) on data from exponential growth.

$$\ln B = \mu^* t + b \qquad (1)$$

where μ is the growth rate, B is dry weight of biomass (mg DW l^{-1}), t is time (day) and b is the ln of inoculation density.

Results

Flask system

Growth curves of uni-algal and mixed cultures of *Microcystis* and *Oscillatoria* in flask experiments consisted of exponential and stationary phases without a lag phase (Figs. 2, 3; since experiments were stopped at the end of the saturation phase, the death phase was not included in the curves). *Oscillatoria* grew very well at 15–30°C (Fig. 3), whereas *Microcystis* did not grow well at 15°C (Fig. 2), whether in uni- or mixed cultures. Growth curves of uni-algal cultures of both algae were similar to those of mixed cultures in flask experiments from 15°C to 30°C (Figs. 2, 3). They had similar growth rates in mixed and uni-algal cultures (Table 1), but maximum biomass in mixed cultures was lower than uni-algal cultures (Table 2).

Growth rates of the two algae increased with temperature (Table 1), and the increase was more pronounced for *Microcystis* than *Oscillatoria*. At 15°C, the growth rate of *Oscillatoria* was higher than *Microcystis*, at 20°C the growth rates were comparable, and at 30°C the growth rate of

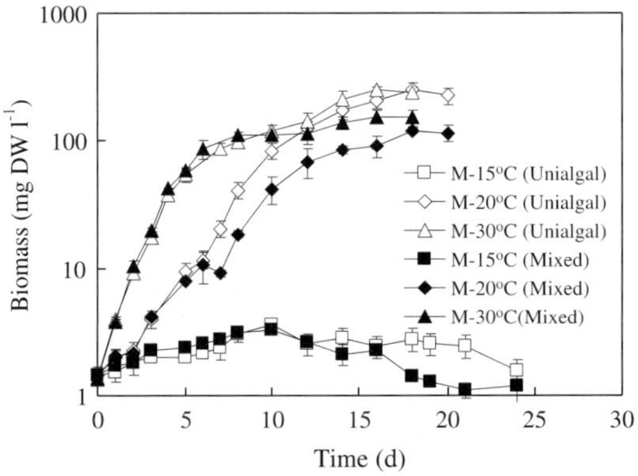

Fig. 2 The growth curves of uni-algal cultures and mixed cultures of *Microcystis* at different temperature in flask experiments

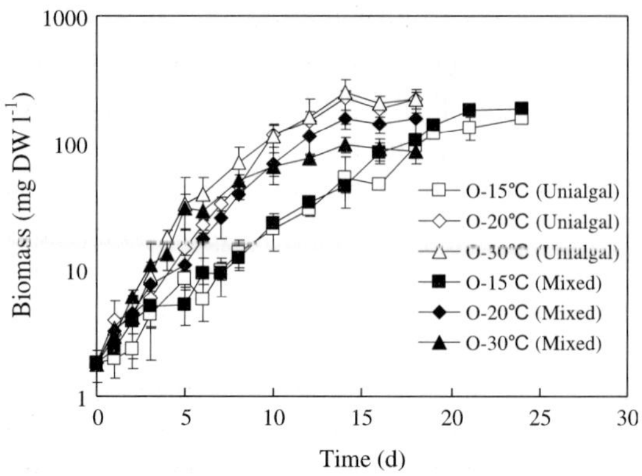

Fig. 3 The growth curves of uni-algal cultures and mixed cultures of *Oscillatoria* at different temperature in flask experiments

Table 2 Maximum biomass (B_{max}) of *Microcystis* and *Oscillatoria* at different temperature in the lake simulator and flasks experiments

		B_{max} (mg DW l^{-1})		
		Flask uni-algal cultures	Flask mixed cultures	Lake simulator cultures
Microcystis	15°C	3.56	3.23	1.14
	20°C	345	155	0.60
	30°C	250	155	28.1
Oscillatoria	15°C	156	188	8.35
	20°C	220	121	19.9
	30°C	254	97.2	0.50

Microcystis was higher. Flask experiments showed that *Oscillatoria* was the superior competitor at 15°C, whereas *Microcystis* was superior at 30°C, and the competition was about equal at 20°C (Tables 1, 2).

Lake simulator system

Growth and competition of *Microcystis and Oscillatoria* in the lake simulator system are shown in Fig. 4, revealing that *Oscillatoria* dominated at 15°C and 20°C, whereas *Microcystis* dominated at 30°C. At 15°C, *Oscillatoria* had a long exponential phase (20 day) and a low growth rate, but *Microcystis* grew little, and, therefore, *Oscillatoria* contributed 91% to the total biomass on day 20. When temperature was increased to 20°C, growth and competition of the two algae changed little, and *Oscillatoria* accounted for 98% of total mass on day 20. When temperature was 30°C, *Microcystis* doubled quickly in the exponential phase, but this phase was very short (about 2–3 days). Afterwards, *Microcystis* biomass increased with time, *Oscillatoria* grew very slowly, and *Microcystis* accounted for 99.7% of total biomass on day 29.

Changes in growth rates of *Microcystis* and *Oscillatoria* in lake simulator experiments did not show the same trends. With increased temperature, *Microcystis* growth rate increased, while that of *Oscillatoria* decreased. Moreover, total maximum biomass increased with temperature: 9.13, 19.9 and 28.1 mg DW l^{-1} at 15°C, 20°C and 30°C, respectively. These biomasses accounted for 4.8, 5.0 and 7.4% of those reached in flasks experiments.

Table 1 Growth rates (μ ± SE) of *Microcystis* and *Oscillatoria* at different temperature in the lake simulator and flasks experiments

		Growth rate μ (day^{-1})		
		Flask uni-algal cultures	Flask mixed cultures	Lake simulator cultures
Microcystis	15°C	0.083 ± 0.009 (0–10 day)	0.079 ± 0.007 (0–10 day)	–
	20°C	0.42 ± 0.02 (0–10 day)	0.33 ± 0.02 (0–10 day)	–
	30°C	0.81 ± 0.04 (0–4 day)	0.85 ± 0.05 (0–4 day)	0.86 (0–2 day)
Oscillatoria	15°C	0.22 ± 0.01 (0–19 day)	0.22 ± 0.01 (0–19 day)	0.22 ± 0.02 (2–20 day)
	20°C	0.40 ± 0.02 (0–10 day)	0.35 ± 0.02 (0–10 day)	0.20 ± 0.01 (0–20 day)
	30°C	0.56 ± 0.05 (0–4 day)	0.55 ± 0.04 (0–5 day)	–

Period of exponential phase is in parentheses

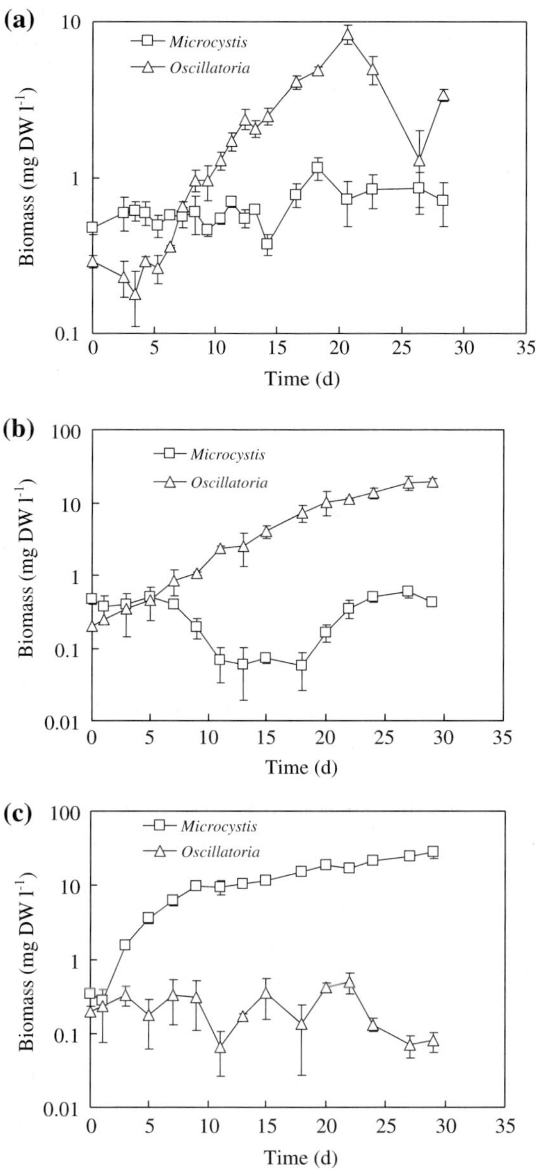

Fig. 4 Competition between *Microcystis* and *Oscillatoria* in lake simulator experiments at 15°C (**a**), 20°C (**b**) and 30°C (**c**)

Discussion

Uni-algal cultures of *Microcystis and Oscillatoria* in flasks showed increased growth rates with increasing temperature. Flasks can be considered a homogeneous system due to small volume, and they are useful for determination of specific growth rates under different conditions. When grown together in flasks, the two algal species grew independently and reached high biomasses. However, the flasks represent a simple system and cannot simulate a lake.

Competition between *Microcystis and Oscillatoria* in the lake simulator experiments showed that *Oscillatoria* dominated when temperature was <20°C, whereas *Microcystis* dominated when temperature was 30°C. Maximum biomasses reached in lake simulator experiments were <10% of those reached in corresponding flasks at all temperatures. Since interactions between the two algae in the lake simulator experiments were stronger than those in flasks, these results show that flask experiments cannot predict competition between algae in a lake system.

Changes in nutrients, light, pH and other factors during algal growth may be important in algal competition. Due to the low biomass in lake simulator experiments, all nutrients except carbon were considered sufficient and could not have limited algal growth rates. Since speciation of total inorganic carbon (TIC) depends on pH, the available inorganic carbon resource is determined by both TIC and pH (Stumm & Morgan, 1996). An increase in pH results in a decrease in the fraction of free CO_2, which is assimilated easily by algae. Low CO_2 conditions decrease growth rate (Thyssen et al., 2001). Indeed, growth rates of the dominant algae decreased when pH increased to 9–9.5 (Fig. 5). This suggests that changes in the growth rate of the dominant alga may be a result of pH changes. Furthermore, the light climate

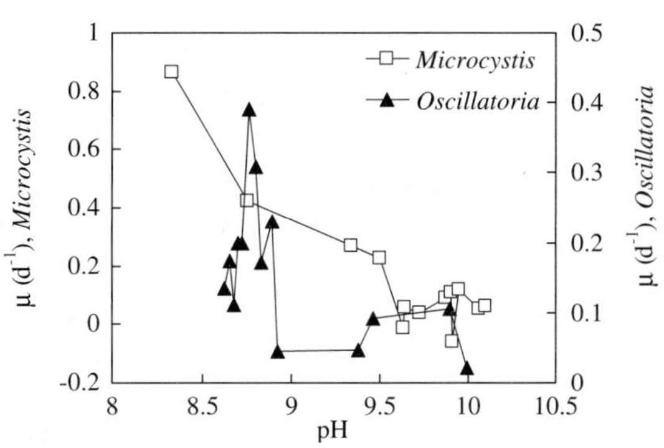

Fig. 5 The effect of pH on the growth rate on the dominant alga (*Microcystis* is the dominant alga at 30°C and *Oscillatoria* is the dominant alga at 20°C)

underwater changed during algal growth. If the wall effect on light is neglected, Eq. 2 (Huisman et al., 2002) can be used to describe changes of light intensity in the lake simulator system.

$$I(z,t) = I_{in} \exp[-\int_0^z k\omega(\sigma,t)d\sigma - K_{bg}z] \quad (2)$$

where I_{in} is incident light intensity, $\varpi(\sigma,t)$ is population density at depth z and time t, σ is an integration variable, k is the specific light attenuation coefficient of algae, and K_{bg} is the background attenuation. As shown by Eq. 2, light intensity attenuates exponentially with biomass and depth. Due to high incident light intensity, increasing algal biomass may alleviate high light stress. However, growth rate did not increase with biomass, suggesting high light stress was not evident in the lake simulator system. Furthermore, increasing algal biomass leads to decreased transparency and development of a steep light gradient in the upper water column. Light limitation caused by self-shading of algae may limit growth when the biomass of dominant alga is significant. At 30°C, a long linear phase following the exponential phase may be the result of light limiting growth of *Microcystis*. At 15°C and 20°C, *Oscillatoria* died quickly after the exponential phase, which may result from increased pH. Due the rapid growth of the superior alga, elevated pH and light shading also may influence growth of inferior algae in the lake simulator system.

As the lake simulator system simulated sunlight, water turbulence, nutrient concentrations and depth in natural, shallow, eutrophic lakes and controlled temperature precisely, results might reflect temperature influences on growth and competition of these two algae in natural lakes. In some areas of Lake Taihu (China), such as Gonghu Bay in the northeast, *Oscillatoria* dominates in spring and late autumn, whereas *Microcystis* dominates in summer and early autumn. Temperatures in natural systems often are between 15 and 25°C, and, therefore, results from the lake simulator experiments can be used to predict cyanobacterial dominance in nature. Although there are many environmental factors affecting the algal community, this study suggests that water temperature is an important factor in the shift of *Microcystis* and *Oscillatoria* dominance in a simulated, shallow, eutrophic lake.

Acknowledgements This work was supported by the Major State Basic Research Development Program of China (No. 2002CB412301), JICA Taihu Water Environment Restoration Model Project and National Natural Science Foundation of China (No. 40501078). We thank Dr. Fengchang Wu, Dr. Shao Yang and three anonymous reviewers for their useful comments and suggestions.

References

Coles, J. F. & R. C. Jones, 2000. Effect of temperature on photosynthesis-light response and growth of four phytoplankton species isolated from a tidal freshwater river. Journal of Phycology 36: 7–16.

Dokulil, M., W. Chen & Q. Cai, 2000. Anthropogenic impacts to large lakes in China: the Tai Hu example. Aquatic Ecosystem Health and Management 3: 81–94.

EPA Method 446.0, 1997. In vitro Determination of Chlorophylls *a*, *b*, $c_1 + c_2$ and Pheopigments in Marine and Freshwater Algae by Visible Spectrophotometry.

Fan, C. X., 1996. Historical evolution of water ecological setting in Taihu Lake. Journal of Lake Sciences 8: 297–304 (in Chinese).

Fujimoto, N., R. Sudo, N. Sugiura & Y. Inamori, 1997. Nutrient-limited growth of *Microcystis aeruginosa* and *Phormidium tenue* and competition under various N:P supply ratios and temperatures. Limnology and Oceanography 42: 250–256.

Huisman, J., M. Arrayas, U. Ebert & B. Sommeijer, 2002. How do sinking phytoplankton species manage to persist? The American Naturalist 159: 245–254.

Mur, L. R. & H. Schreurs, 1995. Light as a selective factor in the distribution of phytoplankton species. Water Science and Technology 32: 25–34.

Ohkubo, N., O. Yagi & M. Okada, 1993. Studies on the succession of blue-green-algae, *Microcystis*, anabaena, *Oscillatoria* and *Phormidium* in lake Kasumigaura. Environmental Technology 14: 433–442.

Okada, M. & R. Sudo, 1980. Microcosm system study of algal bloom formation. Journal WPCF 52: 1029–1033.

Reynolds, C. S & A. E. Walsby, 1975. Water-blooms. Biology Review 50: 437–481.

Stumm, W. & J. J. Morgan, 1996. Aquatic Chemistry: Chemical Equilibria and Rates in Natural Waters, 3rd edn. John Wiley & Sons, New York, 148–205.

Takamura, N., A. Otsuki, M. Aizaki & Y. Nojiri, 1992. Phytoplankton species shift accompanied by transition from nitrogen dependence to phosphorus dependence of primary production in lake Kasumigaura, Japan. Archiv für Hydrobiologie 124: 129–148.

Thyssen, C., R. Schlichting & C. Giersch, 2001. The CO_2-concentrating mechanism in the physiological context: lowering the CO_2 supply diminishes culture growth and economises starch utilisation in Chlamydomonas reinhardtii. Planta 213: 629–639.

Wei, B., N. Sugiura & T. Maekawa, 2001. Use of artificial neural network in the prediction of algal blooms. Water Research 35: 2022–2028.

Xie, L., P. Xie, L. Sixin, H. Tang & H. Liu, 2003a. The low TN:TP ratio, a cause or a result of *Microcystis* blooms? Water Research 37: 2073–2080.

Xie, L., P. Xie & H. J. Tang, 2003b. Enhancement of dissolved phosphorus release from sediment to lake water by *Microcystis* blooms – an enclosure experiment in a hyper-eutrophic, subtropical Chinese lake. Environmental Pollution 122: 391–399.

Yang, Z., W. Shi, L. Chen & Y. Chen, 2003. Ecological environment succession and countermeasure of East Taihu. China Environmental Science 23: 64–68 (in Chinese).

Zhang, Y. & E. E. Prepas, 1996. Regulation of the dominance of planktonic diatoms and cyanobacteria in four eutrophic hardwater lakes by nutrients, water column stability, and temperature. Canadian Journal of Fisheries and Aquatic Sciences 53: 621–633.

Hydrobiologia (2007) 581:225–240
DOI 10.1007/s10750-006-0508-2

EUTROPHICATION IN LAKES

Stability and change of phytoplankton communities in a highly dynamic environment—the case of large, shallow Lake Balaton (Hungary)

Márk Honti · Vera Istvánovics · András Osztoics

Abstract Time series data of key environmental variables (water temperature, global radiation, vertical light attenuation, internal P load) and biomass of four colour classes of photosynthetically active algae were collected during 2003 and 2004 with daily resolution. Using these data, seasonal patterns of phytoplankton were analyzed as a function of the dynamic environment. Abstraction of the environmental state as a point in multi-dimensional space was used to identify habitat templates of bloom-forming groups and derive an indicator of environmental stability/physical disturbance. These templates were synthesized into a simple threshold model that sufficiently simulated development and collapse of various blooms. Blooms were, however, rare events related to specific environments with strong, unidirectional forcing. Tentative quantification of disturbance and compositional stability/community change allowed discriminating disturbance-driven changes and autogenic succession with reasonable success. The two processes were found to be equally important in shaping the composition and biomass of phytoplankton.

Guest editors: B. Qin, Z. Liu & K. Havens
Eutrophication of shallow lakes with special reference to Lake Taihu, China

M. Honti (✉) · V. Istvánovics · A. Osztoics
Department of Civil and Environmental Engineering, Budapest University of Technology and Economics, Műegyetem rkp. 3., Budapest 1111, Hungary
e-mail: mark@vkkt.bme.hu

Keywords Daily time series · Environmental stability · Physical disturbance · Autogenic succession · Predictability · Ground state

Introduction

Phytoplankton communities undergo a seasonal change comparable to the decade-scale succession in terrestrial ecosystems (Reynolds, 1997). Succession is, in a strict sense, an autogenic process during which species selection is driven by environmental changes resulting from metabolism of the community itself. External forcing represents the rest of environmental variability, which phytoplankton cannot influence. Below certain thresholds, external forcing is absorbed efficiently by the community without leaving its recognizable fingerprint on composition or abundance. Beyond these thresholds, it acts as a disturbance upon phytoplankton development. Such events alter community structure, repeatedly distorting the succession pattern and propagating it into altered directions of development. The unpredictable frequency of disturbance sets the time span allowed for autogenic succession and, thus, decreases the chance of attaining stable states.

The intermediate disturbance hypothesis (IDH) of Connell (1978) recognizes that disturbance in a specific range of frequency is a prime factor maintaining species diversity of ecosystems. This range of frequencies must be scaled to the characteristic doubling time of biota making up the examined communities (Reynolds, 1997). Fulfilling this criterion, the IDH recently has been shown to be a successful approach in phytoplankton ecology (Reynolds, 1988; Padisák et al., 1993; Padisák, 1993; Sommer, 1995). However, disturbance is not measured when studying natural phytoplankton but inferred from specific compositional changes (Juhász-Nagy, 1993). Nevertheless, few theoretical studies aimed at resolving the logical problem of this circular argumentation (Reynolds, 1997; Elliott et al., 2001). This is not surprising if one considers the difficulties faced when characterizing environmental stability/disturbance. One point is a priori clear: such a trial will not succeed unless environmental dynamics is captured at a temporal scale matching characteristic period time of phytoplankton. As shown by Fourier spectrum analysis of daily biomass time series from Lake Balaton, the characteristic period time is in the range of a few days (Istvánovics et al., 2005). Consequently, environmental dynamics is to be described at least at a daily scale in this lake.

Large, shallow lakes represent a dynamic environment for phytoplankton, where it is difficult to separate various pathways of community development due to frequent resuspension of sediment. Field work and theoretical analysis revealed the decisive impact of sediment resuspension acting as disturbance on phytoplankton dynamics in Lake Balaton (Padisák et al., 1988, 1990; Padisák, 1993) as well as in a broad assortment of shallow lakes (Padisák et al., 2003).

A high-resolution monitoring program in Lake Balaton yielded time series of key environmental variables and phytoplankton at a daily scale (Istvánovics et al., 2005). The objectives of this work are to identify patterns of community development, which are predictable using available environmental variables, and distinguish compositional changes of phytoplankton driven by disturbance and succession.

Materials and methods

Description of study site

Lake Balaton is a large (593 km^2), shallow (z_{mean} = 3.2 m) lake in Western Hungary (Fig. 1). The long wind fetch from all directions results in almost permanent mixing of the whole water column. Due to frequent resuspension of highly calcareous (50–60% carbonates) sediments and slow sedimentation of precipitating carbonates, the water is always turbid. In the shallowest, westernmost area (z_{mean} = 2.3 m), the coefficient of vertical light attenuation (K_d) averages 4 m^{-1} (Istvánovics et al., 2004). As a consequence, mean euphotic depth is shallower (1.2 m) than mixing depth. Phosphorus desorption from resuspended sediments results in a pulsed internal load (Istvánovics et al., 2004). Mean suspended solids concentration approaches 36 g m^{-3}, and the number of inorganic particles in the ingestible size range exceeds food items of filter-feeding zooplankton by one to three orders of magnitude (G.-Tóth, 1992). Due to strong physical constraints acting upon non-selective cladocerans, the copepod *Eudiaptomus gracilis* (Kiefer, 1932) makes up 80–100% of crustacean plankton. Thus, community grazing rate rarely exceeds 8–10% (G.-Tóth, 2000), and phytoplankton live in a dynamic environment where light, temperature and nutrient supply fluctuate rapidly, but sedimentation and grazing are negligible to moderate.

Data and analysis

An online monitoring station has been maintained since 2001 in the westernmost area of the lake (Fig. 1). The station is equipped with meteorological and turbidity sensors and an online delayed fluorescence (DF) spectrometer that automatically registers composition and biomass of photosynthetically active phytoplankton. DF spectroscopy is a new method (Gerhardt & Bodemer, 2000; Istvánovics et al., 2005) that distinguishes four colour classes of algae: cyanobacteria, chlorophytes (Chlorophyceae, Euglenophyceae, Conjugatophyceae), chromophytes (Bacillariophyceae, Chrysophyceae, Dinophyta,

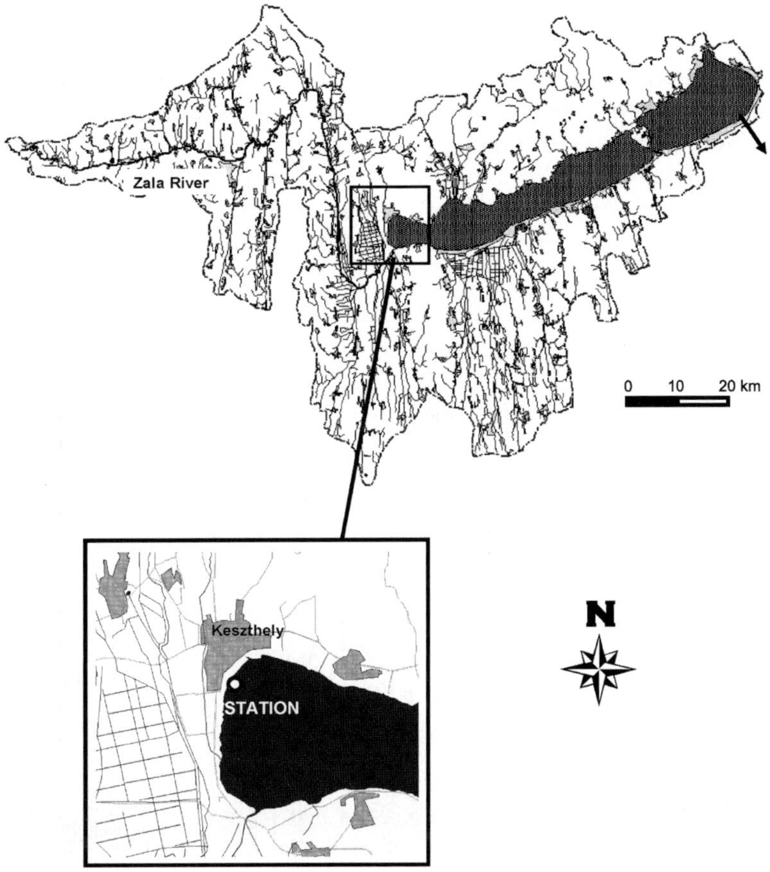

Fig. 1 Lake Balaton and its catchment. (Insertion shows location of the sampling station. Zala River is the largest tributary. Arrow indicates the only outflow)

Xanthophyceae, Haptophyta) and cryptophytes. Phytoplankton data were recorded every 20 min at four depths in a single profile. Meteorological and turbidity data were obtained every 10 and 6 sec, respectively. These data were averaged in 15 min intervals. Turbidity was calibrated against coefficient of vertical light attenuation in the laboratory and using weekly K_d values determined manually. For a more detailed description of the monitoring program and a critical evaluation of the DF method, see Istvánovics et al. (2004, 2005). At the monitoring site, weekly phytoplankton samples were taken and analysed by microscopic counting. (Padisák, pers. comm.)

A database was created consisting of K_d, underwater light, temperature, phytoplankton composition and biomass. Each variable was averaged by day and depth. Time series of dominance was defined with five categories: one for each colour class of algae and an 'uncertain' category. A colour class was considered dominant if its biomass exceeded that of any other class by more than 10%. The uncertain category was reserved for rare cases when this criterion was not satisfied. Since time series collected in 2001 and 2002 were fragmented, we rely on data from 2003 and 2004. Our unique dataset has very high temporal resolution compared to that allowed by traditional methods. Obviously, resolution of phytoplankton along the diversity axis is poor relative to microscopic enumeration.

Daily external loads of various P and N forms are monitored daily at the mouth section of the largest tributary of the lake (database of the West Transdanubian Authority of Water and Environment). This load is estimated to represent 90–95% of total external load of the westernmost area of Lake Balaton (Clement, 2000).

Results and discussion

In 2003, the seasonal pattern of phytoplankton dynamics was typical for shallow, eutrophic lakes (Fig. 2). Three major biomass peaks developed: (i) a spring diatom bloom in late March, (ii) a mixed bloom of *Aphanizomenon flos-aquae* (Lyngbye) Ralfs, *A. klebahnii* Elenkin and *Planktothrix agardhii* Gomont in early July and (iii) a *Cylindrospermopsis raciborskii* (Woloszynska) Seenayya et Subba Raju bloom in mid-August. Development of two successive cyanobacteria blooms is common in Lake Balaton. The high temperature requirement of subtropical *C. raciborskii* is the prime reason to account for the phenomenon (Padisák & Istvánovics, 1987; Padisák & Reynolds, 1998). In addition to the

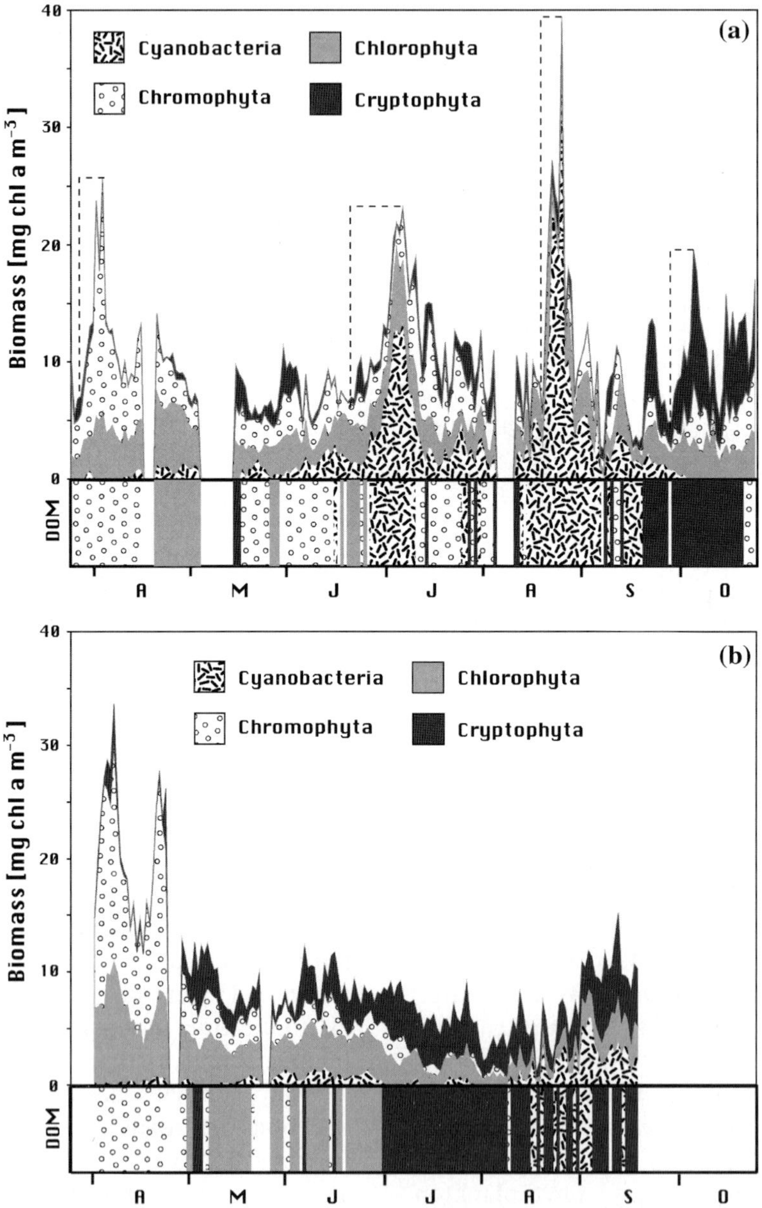

Fig. 2 Biomass and dominance (DOM) of four colour classes of phytoplankton in 2003 (**a**) and 2004 (**b**). Gaps indicate data deficiency due to technical reasons. (Dashed boxes border major blooms used to derive habitat templates for functionally distinct groups; for details see text)

major blooms, smaller peaks of chromo- and cryptophytes occurred during autumn. Since our dataset was recorded at a single point, biomass variations originating from water motion could not be resolved. However, Balaton is well-mixed (Luettich et al., 1990), which allows neglecting spatial heterogeneities.

Environmental conditions were different in 2004 from those in 2003 (Table 1). Higher water discharge resulted in higher external nutrient loads, particularly during spring. Nutrient supply also was more favorable during summer. Water column stability was high throughout the summer and autumn 2004 due to shading and turbulence dampening effects of macrophytes. The latter appeared in moderate densities in 2003, but surface coverage reached about 85% by early summer 2004. Mean water temperature was significantly lower in summer 2004. As a result of the strong influence of macrophytes, phytoplankton dynamics was atypical for Lake Balaton (Fig. 2). A sustained spring bloom of diatoms and the lack of a well-defined summer bloom resembled the seasonal pattern characteristic of mesotrophic areas of the lake. At the same time, there was a long-lasting and virtually unbroken dominance of cryptophytes from early July to the end of the study period (cf. Padisák, 1994).

A few characteristic patterns evolved in dominancy and biomass of phytoplankton: (i) major blooms were separated by periods of low biomass; (ii) each bloom was dominated by species of a single colour class; (iii) between blooms, small peaks of one or the other colour class started but collapsed before attaining significant biomass levels and produced capricious variability in dominance. This pattern suggests that separated blooms may be bound to specific circumstances, while, between blooms, there are chaotic periods of low biomass with various species struggling to outgrow each other.

As a first step to identify environmental conditions selecting for certain species during bloom periods, physical forcing was characterized by two indicators: water temperature and vertical light attenuation. The choice of day- and depth-averaged water temperature is self-explanatory. Daily mean coefficient of vertical light attenuation (K_d) was selected as an indirect indicator of underwater light climate. Although global radiation was measured, shading effect of macrophytes was not quantified. Therefore, time series of incident photon flux density (PFD) was heterogeneous. K_d showed large, irregular fluctuations between about 1 and 15 m^{-1} due to vast sediment resuspension. The corresponding variability in

Table 1 Mean ± standard deviation values of environmental variables (daily data) and nutrient status of phytoplankton (weekly data) in 2003 and 2004

Variable	2003	2004
Spring (March–May)		
Coefficient of vertical light attenuation (m^{-1})	2.9 ± 1.9	2.2 ± 1.8
Water temperature (°C)	14.0 ± 6.3	15.5 ± 2.9
Water discharge (m^3 s^{-1})	4.8 ± 2.4	11.6 ± 4.9
External load of nitrate (mg N m^{-2} day^{-1})	8.3 ± 10.0	10.2 ± 7.4
External load of phosphate (mg P m^{-2} day^{-1})	0.57 ± 0.43	0.70 ± 0.62
Internal load of phosphate (mg P m^{-2} day^{-1})	1.32 ± 0.99	0.78 ± 0.63
Specific surplus P content (μg P [μg chl *a*]$^{-1}$)	0.70 ± 0.23	0.98 ± 0.61
Specific N content (μg N [μg chl *a*]$^{-1}$)	19.1 ± 4.2	22.7 ± 2.9
Late summer (July–August)		
Coefficient of vertical light attenuation (m^{-1})	2.5 ± 1.2	1.1 ± 1.0
Water temperature (°C)	25.9 ± 1.9	23.8 ± 2.3
Water discharge (m^3 s^{-1})	0.4 ± 0.2	2.8 ± 1.5
External load of nitrate (mg N m^{-2} day^{-1})	0.08 ± 0.05	0.80 ± 0.58
External load of phosphate (mg P m^{-2} day^{-1})	0.15 ± 0.10	1.78 ± 0.79
Internal load of phosphate (mg P m^{-2} day^{-1})	1.11 ± 0.50	0.44 ± 0.47
Specific surplus P content (μg P [μg chl *a*]$^{-1}$)	0.53 ± 0.15	1.30 ± 0.36
Specific N content (μg N [μg chl *a*]$^{-1}$)	13.0 ± 4.9	18.6 ± 6.3

underwater light—proportionate to exp(-K_d)—was in the range of 10^6. In comparison, daily variability in incident light (maximum 25 fold) was negligible. Thus, K_d is a prime determinant of underwater PFD and an indicator of sediment resuspension.

Plotting daily values of T and -K_d on a two dimensional graph, the path followed by physical constraints emerged (Fig. 3). Noticeably, seasonality of temperature governed this path into well-expressed directions in early spring and late autumn. In summer, a complex knot pattern was observed with no detectable progress towards any direction.

Dominant colour classes of phytoplankton were inserted into the T versus K_d space (Fig. 3). Their positions agreed with the known physiological requirements of characteristic member species and clearly reflected their different sensitivity to stability of the physical environment (Padisák

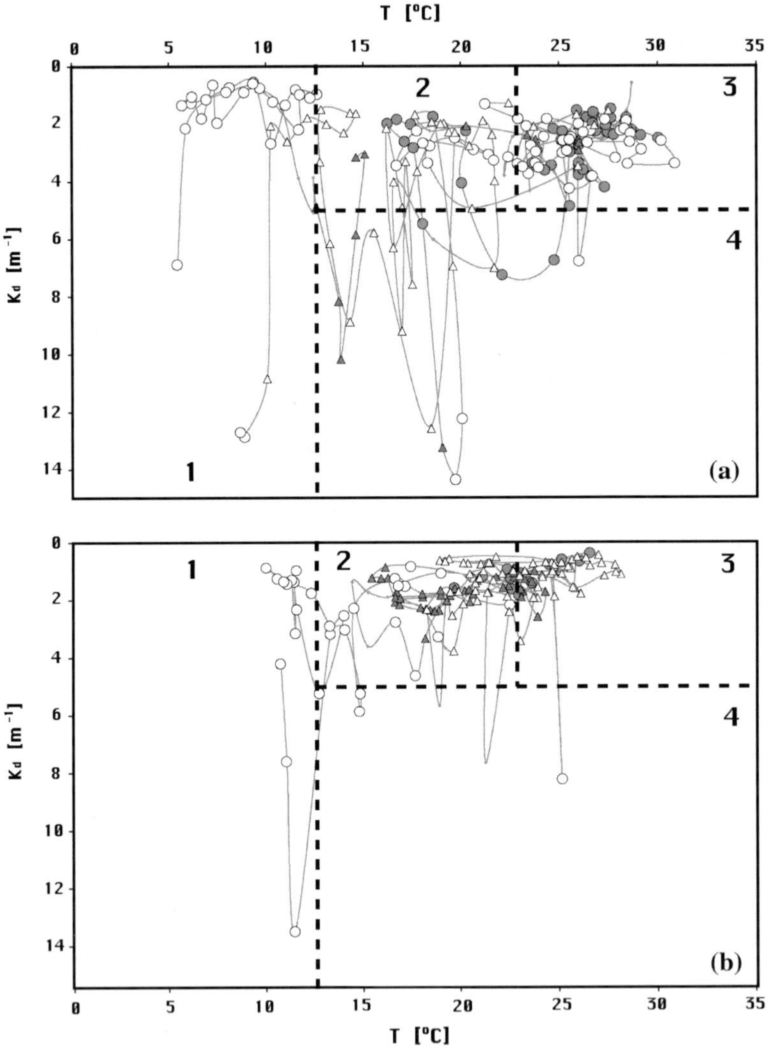

Fig. 3 The path followed by the environment in the space of physical constraints in 2003 (**a**) and 2004 (**b**) and dominant colour classes of phytoplankton placed into this space. (T—vertically averaged daily mean water temperature; K_d—daily mean coefficient of vertical light attenuation. Cyanobacteria—closed circle; chlorophytes—closed triangle; chromophytes—open circle; cryptophytes—open triangle. Numbered areas separated by dashed lines indicate characteristic habitat ranges of various functional groups. 1—acclimating R species, 2—early successional phase, 3—stress tolerant S species, 4—meroplankton). Seasonality means advancing from left to right in spring, staying on the right side in summer and returning to left in autumn

et al., 1988, 1990; Padisák, 1994; Reynolds, 1997). Spring chromophytes (diatoms) are typical acclimating strategists (R species) adapted to low temperature and requiring sufficient turbulence to keep entrained in the water column. Assemblages of early successional stages can be recognized by concentrated occurrence of chlorophytes and chromophytes in the moderate T versus low to moderate K_d sub-space of physical constraints. Summer blooms of cyanobacteria, particularly those of subtropical *C. raciborskii*, are associated with high temperature. Low turbulence allows this group to exploit the advantage of buoyancy regulation. Besides cyanobacteria, stress tolerant chromophytes (*Ceratium hirundinella* (O.F. Müller) Schrank) are typical members of late-summer quasi-equilibria in Lake Balaton. The unusual summer dominance of cryptophytes in 2004 may be explained by metaphytic affiliation of certain species within this group (Borics et al., 2003). Small cryptophytes and several meroplanktonic diatoms benefit from intense wind-induced sediment resuspension, which temporarily extends their habitat into the water column.

Major phytoplankton blooms were identified manually in 2003 (Fig. 2) and placed into the space of physical constraints (Fig. 4). Habitat templates of various colour classes were determined by selecting sequences of days when the bloom-forming class achieved significant and undisturbed growth. The sub-space determined by these days (13 days in 2003) represents physical conditions where the given class has obvious advantage over the others. These regions were separated from each other and occupied margins of the environmental space observed during our measurements (Fig. 4). The marginal position of blooming sub-spaces indicated that strong, unidirectional selective pressure was a prerequisite for bloom formation. Conclusively, blooms are rare events bound to a specific environmental state that provides a few species of a single colour class the opportunity to grow faster than others. Such states are, however, transient and represent the extreme.

A simple threshold model was developed to test the consistency of our observations and hypotheses about seasonal dynamics of phytoplankton. Obviously, the two physical constraints (Fig. 4) did not discriminate habitat templates of various blooms. For example, the path of environmental changes in the T versus K_d space was almost symmetrical during the growth and collapse of the spring diatom bloom in 2003. Thus, additional variables had to be included.

Minimal water temperature in the last two weeks (T_{min}) was chosen because it reflects

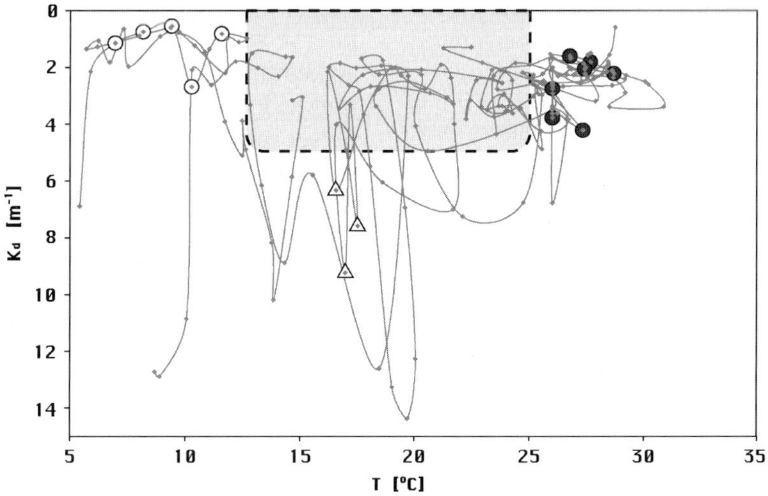

Fig. 4 The path followed by the environment in the space of physical constraints in 2003 and four major biomass peaks placed into this space. (*T*—vertically averaged daily mean water temperature; K_d—daily mean coefficient of vertical light attenuation. Cyanobacteria—closed circle; chromophytes—open circle; cryptophytes—open triangle. The four blooms are those shown in Fig. 2; the two cyanobacterial blooms do not separate in this space. Shaded area indicates the region of "mean" environmental states emphasizing marginal positioning of blooms)

sensitivity of N_2-fixing cyanobacteria to abrupt temperature changes. At the same time, it incorporates the experience that cyanobacteria blooms are preceded regularly by two to three weeks of calm, warm weather (Gorzó, 1987; Padisák, 1994). Compared to *Aphanizomenon* and *Anabaena* species, subtropical *Cylindrospermopsis raciborskii* has an unusually narrow temperature range (22–24°C), where akinetes germinate (Gorzó, 1987), and a high threshold temperature (18°C), where the species starts to grow (Shafik et al., 1997). These differences were expressed in the model by determining different T_{min} values (Table 2).

Nutrient supply of phytoplankton is a discriminative factor during growth periods of various blooms. In Lake Balaton, concentration of silica is always high enough to satisfy requirements of even large diatom blooms (V. Balogh, pers. comm.). Physiological and chemical data (Istvánovics et al., unpublished) showed that carrying capacity might be N determined during spring in dry years without typical spring floods, including 2003. Previous observations also suggest that N deficiency was likely to put an end to spring diatom blooms (Istvánovics & Herodek, 1985, 1995). Thus, area specific external nitrate load was chosen to indicate N supply.

In summer, algal growth can be limited by either N or P, but extreme P deficiency prevails and the carrying capacity usually is P determined (Istvánovics & Herodek, 1985, 1995; Istvánovics & Somlyódy, 2001). This also was the situation in 2003. However, no signs of serious P deficiency were detected in summer 2004, when light availability probably limited algal growth (Table 1). Although daily loads of external and internal P were available (Istvánovics et al., 2004), differential ability of various algal groups to grow on previously acquired and stored P during low external supply prevented incorporation of these variables into the simple threshold model.

Table 2 Discriminative rules for habitat templates and growth rate of the four functionally different blooms, derived from 13 days in 2003

Bloom	Criteria	Growth rate (d^{-1})
Spring chromophytes (diatoms)	$6.5 \le T \le 12.5$, $L_{nitrate-N} \ge 13$	0.74
Early summer cyanobacteria	$T_{min} \ge 23.7$	0.15
Late summer cyanobacteria	$T_{min} \ge 25.5$	0.40
Meroplankton	$T \ge 10$ $K_d \ge 5.6$	–

T, °C is the actual water temperature; T_{min}, °C is the minimum water temperature during the last two weeks; $L_{nitrate}$, mg N m^{-2} d^{-1} is the external load of nitrate; K_d, m^{-1} is the coefficient of vertical light attenuation. Biomass changes of meroplankton were related to frequency of resuspension; see in text

In this way, a four dimensional environmental space was established, including T, K_d, T_{min} and external nitrate load. Actual environmental state is indicated by a point in the space. Considering the seasonal biomass pattern in 2003 (Fig. 2), habitat templates of four functionally distinct blooms could be quantified for modeling purposes: (i) spring diatoms, (ii) early summer cyanobacteria, (iii) late summer cyanobacteria and (iv) meroplankton associated with resuspension. A few simple binary rules were assumed to delimit habitat templates for each functional group. Rules were set up according to ecological knowledge and refined to fit observed data (Table 2). Growth of spring diatoms was dependent on external N supply and water temperature. Early and late summer cyanobacteria grew above a T_{min} of 23.7°C and 25.5°C, respectively. Meroplanktonic biomass peaks were associated with intense sediment resuspension (indicated by high absolute value of K_d) in a certain temperature range. Each resuspension event exceeding the preset criterion was assumed to increase the biomass of this group by 1 mg chl *a* m^{-1}, and the rate of decay was 0.1 d^{-1} in the absence of critical resuspension.

The community most often existed in a "ground state" corresponding to periods between blooms characterized by low biomass and capricious compositional changes (cf. Fig. 2). To ease mathematics, an equal share of biomass was assigned to each group in ground state. As the environment entered a blooming subspace, or fulfilled the preset rules of a habitat template, the appropriate group began to grow at "full speed". Truly planktonic groups (i–iii) were assumed to grow at a fixed exponential rate, while meroplankton (iv) growth was set proportional to frequency of resuspension. As soon as the environment left the template, growth stopped and the bloom started to decay towards ground state. Since growth and collapse of each bloom was roughly symmetrical, loss rates were assumed to equal growth rates, an assumption supported by Sommer (1981).

Biomass simulations fit reasonably to measurements (Fig. 5). Although model parameters were tuned using data from 2003, the model reproduced the lack of cyanobacterial blooms in 2004. This suggested that low mean water temperature during summer 2004 (Table 1) may have been a more influential factor in preventing blooming of cyanobacteria than other differences experienced in this year.

Our simple model reduced the influence of environmental dynamics to a narrow range of conspicuous events: crossing the borders of habitat templates. This oversimplification was useful to demonstrate predictability of bloom formation but could not explain the complexity of phytoplankton dynamics in the ground state, which accounted for about 80% of the study period (Figs. 2, 3, 5). The surprising length of the ground state interestingly demonstrated Reynold's (1997) contention that phytoplankton inhabited an extremely hostile habitat. To get an insight into the effect of environmental dynamics on community composition, the compositional stability was examined as a function of environmental dynamics. For this purpose, appropriate indicators of

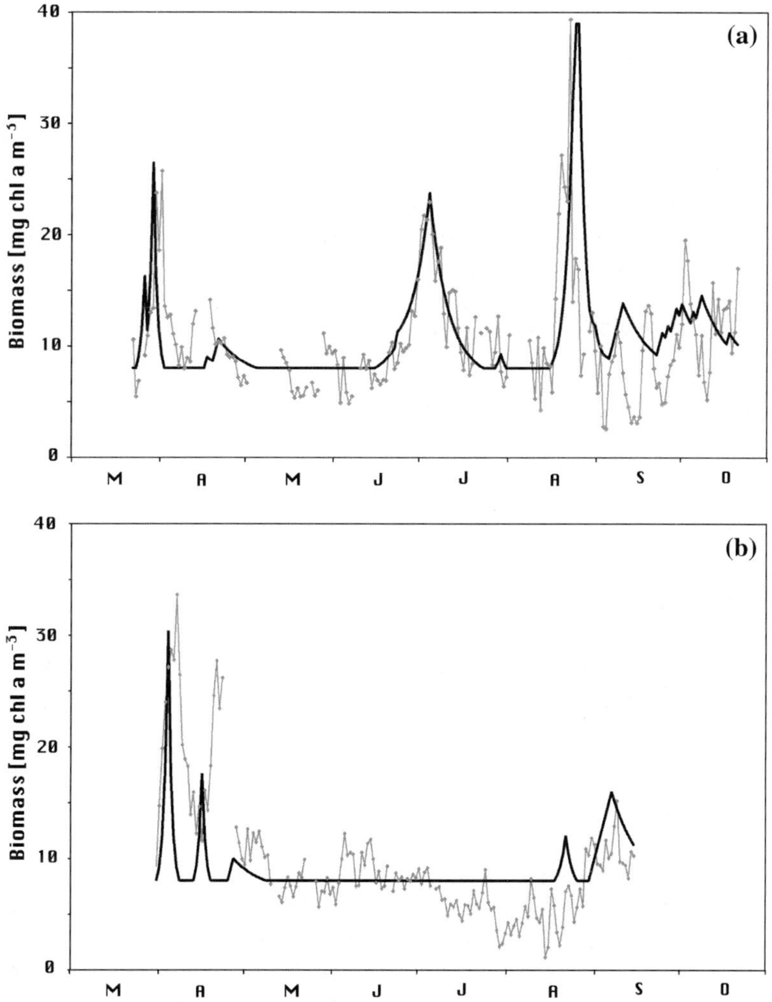

Fig. 5 Observed (thin line) and simulated (thick line) biomass of phytoplankton in 2003 (**a**) and 2004 (**b**)

environmental stability/physical disturbance and community stability/change were identified.

Environmental instability (IS_e) was defined as the step length between the states (e) of the environment in two subsequent days in the T versus K_d space. It can be computed as Euclidean distance:

$$IS_e = \|e_t - e_{t-1}\| = \sqrt{(T_t - T_{t-1})^2 + (Kd_t - Kd_{t-1})^2} \quad (1)$$

where t is the current day. The smaller the step is between two days, the more stable the environment is. Large IS_e is indicative of high instability, which, above some level, can act as physical disturbance. Based on the distribution of step lengths, the upper 15% of events was designated as disturbance (Fig. 6). According to this definition, an environmental change qualifies as a disturbance when $IS_e > 2.5$. As the dimensions of the environmental space have different units

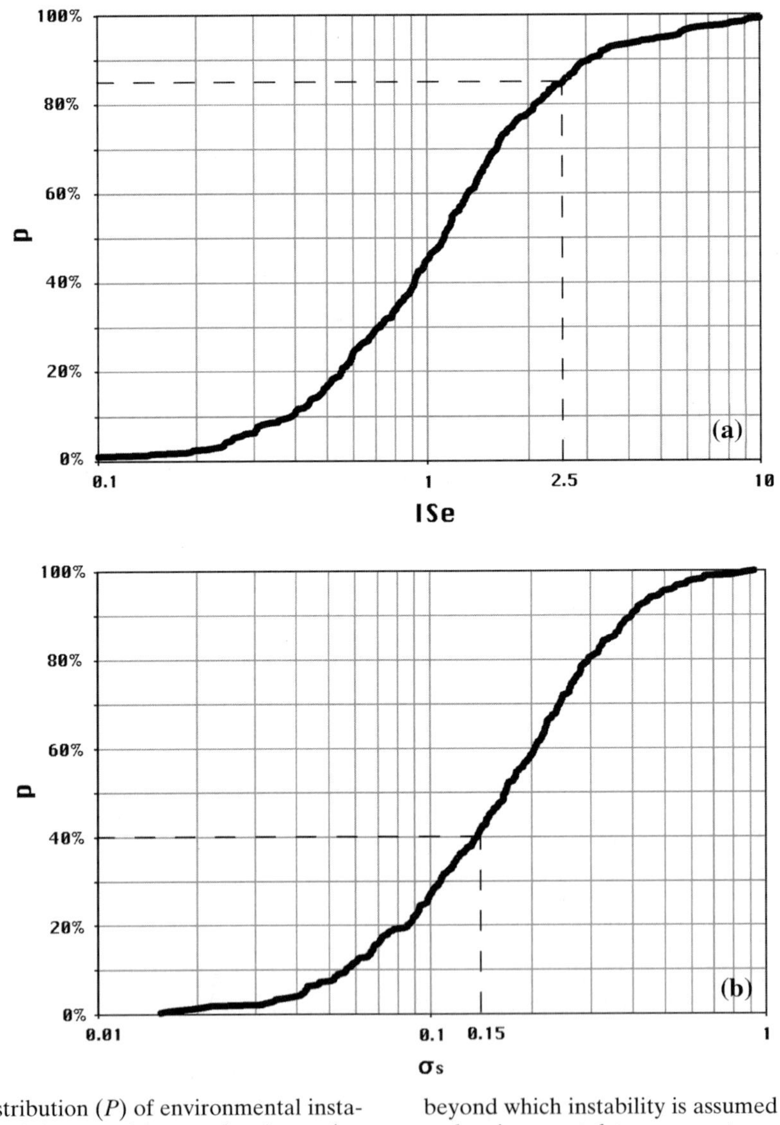

Fig. 6 Probability distribution (P) of environmental instability (IS_e; **a**) and that of the rate of community change (σ_s; **b**). (Dashed lines in **a** and **b** indicate threshold values beyond which instability is assumed to act as a disturbance and σ_s is assumed to represent a real community change, respectively)

(m^{-1} for K_d and °C for T), distances are subject to the weighting used between dimensions. Noticeably, Equation 1 gives equal weights to T and K_d, but the change in the latter variable by 2.5 units is equivalent to a twelve-fold resuspension induced change in underwater PFD. Since, however, a temperature change of 2.5°C represents about 10% of the temperature range covered by the study period, whereas the respective value is 0.001% for the change in underwater PFD, IS_e is far more sensitive to changes in resuspension and PFD than temperature. Disturbances happened throughout 2003 and were less frequent in 2004 (Fig. 7). These events almost each time were coupled with an instantaneous or slightly delayed flip in dominance (Fig. 7).

Compositional changes were characterized using a highly sensitive and more robust indicator

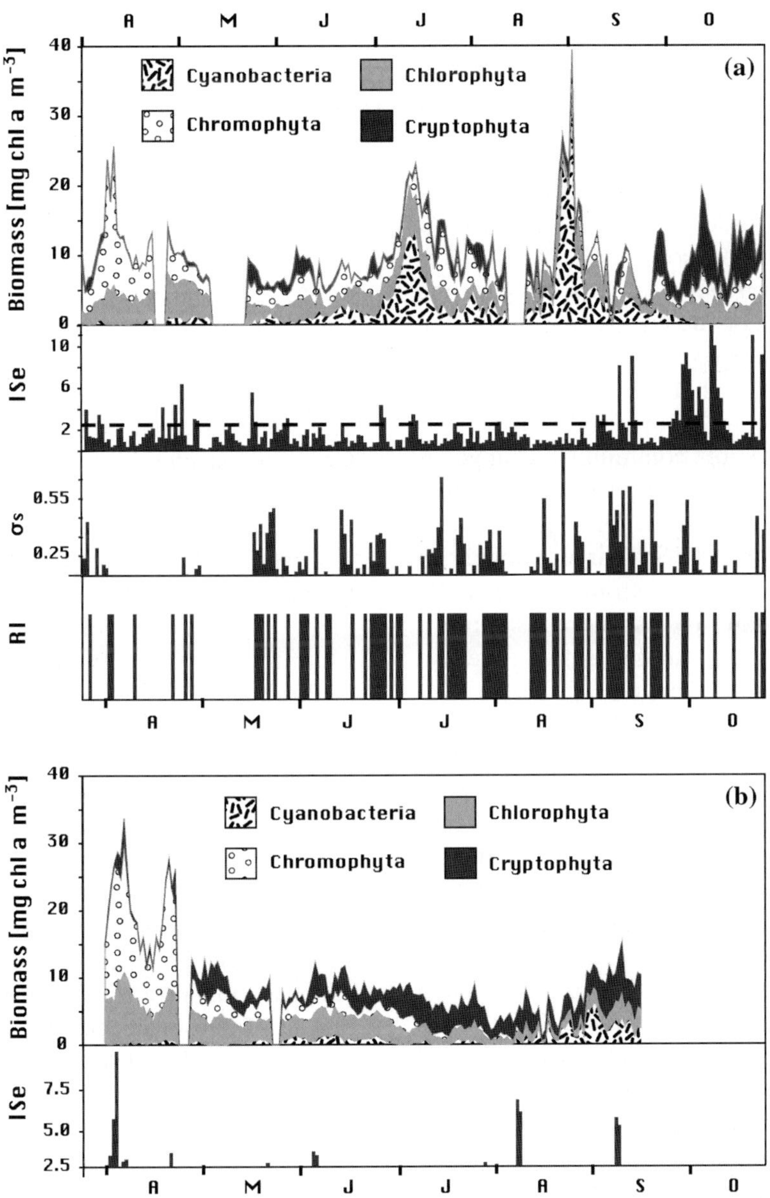

Fig. 7 Seasonal pattern of phytoplankton related to disturbance ($IS_e > 2.5$) in 2003 (**a**) and 2004 (**b**). (In 2003 the rate of community change, σ_s, and community stability, RI, are also indicated)

(Fig. 7). In the former case, the rate of community change (σ_S, d^{-1}) was calculated as the sum of daily change in relative biomass of various colour classes (Reynolds, 1997):

$$\sigma_s = \sum_{i=1}^{4} \left| \frac{b_{t+1}^i}{B_{t+1}} - \frac{b_t^i}{B_t} \right| \quad (2)$$

where B is total biomass (mg chl a m^{-3}), b^i is fractional biomass of the four colour classes (mg chl a m^{-3}), and t is the current day. Taking into account the distribution of σ_s (Fig. 6) and uncertainty of the DF method (Istvánovics et al., 2005), σ_s values below 0.15 d^{-1} were assigned to measurement error.

To derive a less sensitive indicator, the four colour classes were ordered on the basis of their fractional biomass. A positional change was assumed to occur whenever a class jumped one to three ranks up or down compared to the preceding day. The rank indicator (RI) was a simple binary value with RI = 0 when no jump occurred and RI = 1 when any kind of jump was observed. The rate of community change is sensitive to differences in net growth and/or loss rates of various colour classes. These differences, however, are not necessarily large enough to provoke positional changes ensuring lower RI sensitivity. In this way, RI can be considered a measure of community stability rather than an indicator of community change.

To examine the effect of physical disturbance on phytoplankton composition, community change ($\sigma_s > 0.15$ d^{-1}) and stability (RI) were related to IS_e (Figs. 7, 8). We assumed that compositional changes were driven by disturbance when an $IS_e > 2.5$ event was followed by compositional changes within a three day period, including the day of disturbance. Both σ_s and RI showed that disturbance most often induced changes in community composition. In a somewhat larger number of cases, community changes were not related to physical disturbance (Fig. 8). Most of the latter cases occurred during summer (Fig. 7). Although a partial failure of the disturbance indicator could not be excluded, timing of disturbance-independent compositional changes suggested that such changes were driven by autogenic succession of phytoplankton. Certainly, some of the disturbance-independent changes may be associated with selective grazing of zooplankton or other selective loss processes, like viral infection. Although community grazing rate of zooplankton is typically low in Lake Balaton (G.-Tóth, 1992), two pieces of indirect evidence suggested that the situation might be different in 2004. First, dense macrophyte stands provide refuge for zooplankton (Sheffer et al., 1993), where restricted turbulence might favor passive feeders. Second, prolonged dominance of cryptophytes was indicative of sufficient food supply. In spite of these uncertainties, succession and disturbance were equally important in shaping the structure of phytoplankton.

Both σ_s and RI indicated that the most frequent case was communities with stable composition (Fig. 8). Stable composition did not imply that the environment was stable. On the contrary, the environment changed all the time (Figs. 3, 4), but variability usually stayed within the limits where phytoplankton were capable of absorbing external fluctuations. This was indicative of a considerable structural resilience. At the same time, invoked stress usually prevented significant growth.

Cases of stable composition rarely concentrated into long-lasting intervals (Fig. 7). Surprisingly, the longest unbroken period of stable community structure was observed during spring in 2003 and 2004 (Fig. 7). The explanation was twofold. (i) Phytoplankton experienced the most severe and frequent disturbance during autumn, resulting in a number of small biomass peaks with divergent compositions. (ii) The rate of community change increased with water temperature (Fig. 9), indicating that more species were released from stress caused by a sub-optimal level of this variable. Internally driven compositional changes thus were slow in spring compared to the period of autogenic succession.

Fragmented occurrence of stable communities suggested that it was difficult for any colour class to preserve its position for any significant period of time in the dynamic environment of Lake Balaton. This is an important difference compared to deep lakes, where environmental stability associated with thermal stratification supports less disturbed successional sequences, which frequently result in prolonged dominance of one or another K-selected stress tolerant species (Reynolds, 1997). This difference between deep and shallow lakes explain the observation of Padisák et al. (2003) that equilibrium phases sensu Sommer et al. (1993) were relatively infrequent (17 out of 80) in shallow lakes during summer.

Cases when disturbance failed to influence phytoplankton composition (Fig. 8) may reflect measuring errors and/or unsatisfactory definition of disturbance. The number of such cases, however, was low, indicating the merits of the approach. Refinement of this approach deserves attention considering the limited efforts taken to characterize disturbance using independent methods (cf. Juhász-Nagy, 1993; Reynolds, 1997).

Compositional changes were related to fluctuations in the T versus K_d space, considering that stress caused by low nutrient supply was likely a key factor determining phytoplankton dynamics. On one hand, due to lack of top-down control of zooplankton (G.-Tóth, 1992, 2000), autogenic succession is driven by competition for gradually depleting resources, including nutrients. On the other hand, resuspension related disturbance is coupled to increased nutrients, primarily P availability from desorption from sediment particles. The numerous storage and delay factors hidden in the system make it impossible to characterize the

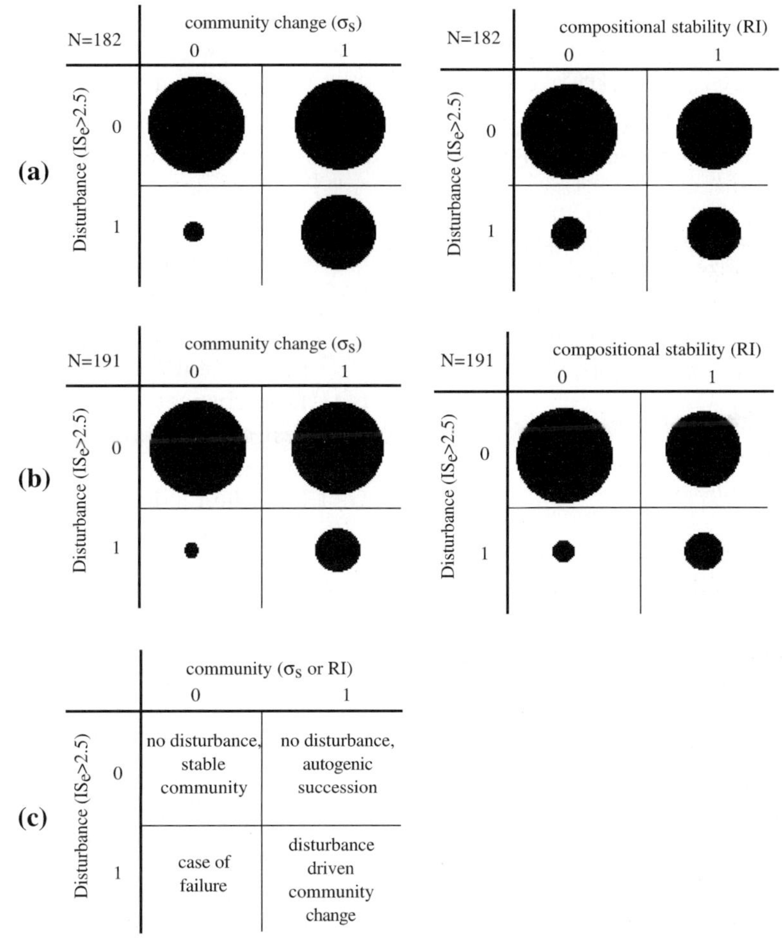

Fig. 8 Compositional stability expressed as either the rate of community change or community stability as a function of disturbance in 2003 (**a**) and in 2004 (**b**). (Disturbance was identified as high environmental instability indicated by $IS_e > 2.5$ values. *N*—total number of cases. Area of symbols is proportionate to number of cases. **c**—interpretation of matrix areas)

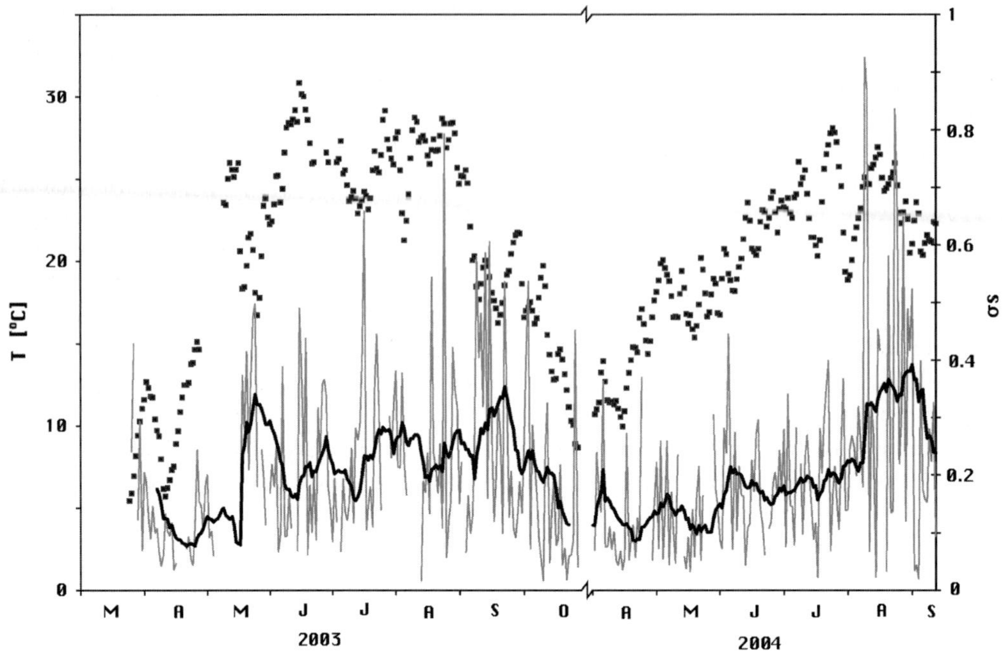

Fig. 9 The rate of community change (σ_s) compared to vertically averaged daily mean water temperature (T). (Temperature—closed symbol; observed σ_s—thin line, moving average on σ_s-thick line)

role of nutrient stress on a daily basis. One should, however, recognize that any change in the phytoplankton community examined in the two dimensional space of physical constraints implicitly reflects variability of nutrient stress.

Conclusions

(1) High-resolution online monitoring systems are required to follow dynamics of the environment and phytoplankton in sufficient detail. Such time series reveal the enormous complexity of ecosystems.

(2) Environmental states providing the opportunity for certain species or functional groups to develop blooms represent extreme margins in the space of environmental constraints with strong unidirectional selective pressure. Correspondingly, blooms are rare, exceptional events and phytoplankton spend nearly 80% of the time in a ground state with low biomass and high diversity.

(3) Growth and collapse of blooms could be described by a simple threshold model based on habitat templates of the bloom-forming functional groups. In contrast to the predictability of major blooms, dominance and biomass was unpredictable during the ground state.

(4) Disturbances could be identified with success using a simple indicator of environmental instability developed here. Most of the identified disturbance events provoked detectable compositional changes in the phytoplankton community, even though time series of phytoplankton consisted of only four colour classes. Autogenic succession and disturbance were found to be equally important in shaping the phytoplankton community.

(5) The highly dynamic environment of shallow lakes is predicted to restrict the occurrence and duration of equilibrium phases.

Acknowledgements This study was supported financially by the "Phytoplankton-on-line" (No. EVK1-CT-1999-00037) project and by the Balaton project of the Office of the Prime Minister (Hungary). We are greatly indebted to Dr. Judit Padisák for microscopic counting results and reading the manuscript as well as to Mr. Norbert Turay, Mr. Gábor Poór, and Mrs. Rozália Kelemen for their invaluable technical assistance with the monitoring program. We thank Dr. János Korponai, Mr. György Süle and Mr. István Nádor from the West Transdanubian Authority of Water and Environment for allowing access to their load data and three anonymous referees for their useful comments.

References

Borics, G., B. Tóthmérész, I. Grigorszky, J. Padisák, G. Várbíró & S. Szabó, 2003. Algal assemblage types of bog-lakes in Hungary and their relation to chemistry, hydrological conditions and habitat diversity. Hydrobiologia 50: 145–155.

Clement, A., 2000. Improving uncertain nutrient load estimates for Lake Balaton. Water Science and Technology 43: 279–286.

Connell, J. H., 1978. Diversity in tropical rain forests and coral reefs. Science 199: 1302–1310.

Elliott, J. A., A. E. Irish & C. S. Reynolds, 2001. The effects of vertical mixing on a phytoplankton community: a modelling approach to the intermediate disturbance hypothesis. Freshwater Biology 46: 1291–1297.

Gerhardt, V. & U. Bodemer, 2000. Delayed fluorescence excitation spectroscopy: a method for determining phytoplankton composition. Archiv für Hydrobiologie Special Issues Advances in Limnology 55: 101–120.

Gorzó, Gy., 1987. The influence of physical and chemical factors on akinete germination of heterocytic cyanobacteria from Lake Balaton. Hidrológiai Közlöny 67: 127–133 (in Hungarian).

G.-Tóth, L., 1992. Limiting effect of abioseston on food ingestion, postembryonic development time and fecundity of daphnids in Lake Balaton (Hungary). Journal of Plankton Research 14: 435–446.

G.-Tóth, L., 2000. Feeding biology of planktonic crustaceans and their role in elimination of phytoplankton in Lake Balaton. D.Sc. Thesis, Balaton Limnological Research Institute, Tihany (in Hungarian).

Istvánovics, V. & S. Herodek, 1985. Orthophosphate uptake of planktonic microorganisms in Lake Balaton. Hydrobiologia 122: 159–166.

Istvánovics, V. & S. Herodek, 1995. Estimation of net uptake and leakage rates of orthophosphate from ^{32}P uptake kinetics by a force-flow model. Limnology and Oceanography 40: 17–32.

Istvánovics, V., M. Honti, A. Osztoics, H. M. Shafik, J. Padisák, Y. Yacobi & W. Eckert, 2005. Continuous monitoring of phytoplankton dynamics in Lake Balaton (Hungary) using on-line delayed fluorescence excitation spectroscopy. Freshwater Biology 50: 1950–1970.

Istvánovics, V., A. Osztoics & M. Honti, 2004. Dynamics and ecological significance of daily internal load of phosphorus in shallow Lake Balaton, Hungary. Freshwater Biology 49: 232–252.

Istvánovics, V. & L. Somlyódy, 2001. Factors influencing lake recovery from eutrophication-the case of Basin 1 of Lake Balaton. Water Research 35: 729–735.

Juhász-Nagy, P., 1993. Notes on compositional diversity. Hydrobiologia 249: 173–182.

Luettich, R. A., D. R. F. Harleman & L. Somlyódy, 1990. Dynamic behavior of suspended sediment concentrations in a shallow lake perturbed by episodic wind events. Limnology and Oceanography 35: 1050–1067.

Padisák, J., 1993. The influence of different disturbance frequencies on the species richness, diversity and equitability of phytoplankton in shallow lakes. Hydrobiologia 249: 135–156.

Padisák, J., 1994. Relationships between short-term and long-term responses of phytoplankton to eutrophication of the largest shallow lake in Central Europe (Balaton, Hungary). In Sund, H., H.-H. Stabel, W. Geller, Y. Xiaogan, Y. Kechang & S. Fengning (eds), Environmental Protection and Lake Ecosystem. Science & Technology Press, Beijing, 419–437.

Padisák, J., G. Borics, G. Fehér, I. Grigorszky, I. Oldal, A. Schmidt & Zs. Zámbóné-Doma, 2003. Dominant species, functional assemblages and frequency of equilibrium phases in late summer phytoplankton assemblages in Hungarian small shallow lakes. Hydrobiologia 502: 157–168.

Padisák, J., L. G.-Tóth & M. Rajczy, 1988. The role of storms in the summer succession of phytoplankton in a shallow lake (Lake Balaton, Hungary). Journal of Plankton Research 10: 249–265.

Padisák, J., L. G.-Tóth & M. Rajczy, 1990. Stir-up effect of wind on a more-or-less stratified shallow lake phytoplankton community, Lake Balaton, Hungary. Hydrobiologia 191: 249–254.

Padisák, J. & V. Istvánovics, 1997. Differential response of blue-green algal groups to phosphorus load reduction in a large shallow lake: Balaton, Hungary. Verhandlungen der Internationale Vereiningung für theoretische und angewandte Limnologie 26: 574–580.

Padisák, J. & C. S. Reynolds, 1998. Selection of phytoplankton associations in Lake Balaton, Hungary, in response to eutrophication and restoration measures, with special reference to the cyanoprokaryotes. Hydrobiologia 384: 41–53.

Padisák, J., C. S. Reynolds & U. Sommer (eds), 1993. The intermediate disturbance hypothesis in phytoplankton ecology. Hydrobiologia 249: 1–199.

Reynolds, C. S., 1988. The concept of ecological succession applied to seasonal periodicity of freshwater phytoplankton. Verhandlungen der Internationale Vereiningung für theoretische und angewandte Limnologie 23: 683–691.

Reynolds, C. S., 1997. Vegetation Processes in the Pelagic: A Model for Ecosystem Theory. Ecology Institute, Oldendorf/Luhe.

Shafik, H. M., L. Vörös, M. Présing, A. Kovács & I. Kóbor, 1997. Growth of Cylindrospermopsis raciborskii in batch and continuous cultures. Hidrológiai Közlöny 77: 17–18 (in Hungarian).

Scheffer, M., S. H. Hosper, M.-L. Meijer, B. Moss & E. Jeppesen, 1993. Alternative equilibria in shallow lakes. Trends in Ecology & Evolution 8: 275–279.

Sommer, U., 1981. The role of r- and K-selection in the succession of phytoplankton in Lake Constance. Acta Oecologica 2: 327–342.

Sommer, U., 1995. An experimental test of the intermediate disturbance hypothesis using cultures of marine phytoplankton. Limnology and Oceanography 40: 1271–1277.

Sommer, U., J. Padisák, C. S. Reynolds & P. Juhász-Nagy, 1993. Hutchinson's heritage: the diversity-disturbance relationship in phytoplankton. Hydrobiologia 249: 1–8.

Hydrobiologia (2007) 581:241–254
DOI 10.1007/s10750-006-0500-x

EUTROPHICATION IN LAKES

Intra-habitat heterogeneity of microbial food web structure under the regime of eutrophication and sediment resuspension in the large subtropical shallow Lake Taihu, China

Qinglong L. Wu · Yuwei Chen · Kuidong Xu · Zhengwen Liu · Martin W. Hahn

Abstract Planktonic microbial community structure and classical food web were investigated in the large shallow eutrophic Lake Taihu ($2338\ km^2$, mean depth 1.9 m) located in subtropical Southeast China. The water column of the lake was sampled biweekly at two sites located 22 km apart over a period of twelve month. Site 1 is under the regime of heavy eutrophication while Site 2 is governed by wind-driven sediment resuspension. Within-lake comparison indicates that phosphorus enrichment resulted in increased abundance of microbial components. However, the coupling between total phosphorus and abundance of microbial components was different between the two sites. Much stronger coupling was observed at Site 1 than at Site 2. The weak coupling at Site 2 was mainly caused by strong sediment resuspension, which limited growth of phytoplankton and, consequently, growth of bacterioplankton and other microbial components. High percentages of attached bacteria, which were strongly correlated with the biomass of phytoplankton, especially *Microcystis* spp., were found at Site 1 during summer and early autumn, but no such correlation was observed at Site 2. This potentially leads to differences in carbon flow through microbial food web at different locations. Overall, significant heterogeneity of microbial food web structure between the two sites was observed. Site-specific differences in nutrient enrichment (i.e. nitrogen and phosphorus) and sediment resuspension were identified as driving forces of the observed intra-habitat differences in food web structure.

Guest editors: B. Qin, Z. Liu & K. Havens
Eutrophication of shallow lakes with special reference to Lake Taihu, China

Q. L. Wu · Y. Chen · Z. Liu
Nanjing Institute of Geography & Limnology,
Chinese Academy of Sciences,
Nanjing 210008, P.R. China

Q. L. Wu · M. W. Hahn
Institute for Limnology,
Austrian Academy of Sciences,
A-5310 Mondsee, Austria

K. Xu
Institute of Oceanography,
Chinese Academy of Sciences, Qingdao
266071, P.R. China

Z. Liu
Institute of Hydrobiology, Jinan University,
Guangzhou 510632, P.R. China

Present Address:
Q. L. Wu (✉)
Division of Environmental Science, Engineering,
Faculty of Engineering, National University
of Singapore, 1 Engineering Drive 2,
Singapore 117576, Singapore
e-mail: wuqinglong@hotmail.com

Keywords Microbial food web · Eutrophication · Resuspension · Shallow lake · Heterogeneity

Introduction

Eutrophication in lakes associated with cyanobacterial blooms, notably the cyanobacteria *Microcystis* spp., is becoming common in many parts of the world as a consequence of excessive input of nutrients and water pollution (Dokulil & Teubner, 2000). Although many studies on pelagic classical food webs have been performed, the microbial food webs of eutrophic lakes in tropical or subtropical zone have received only little attention (Robarts & Wicks, 1990; Sommaruga, 1995). The currently available information on microbial food webs in the pelagic of freshwater systems is mainly about temperate lakes. In these lakes nutrient has significant impacts on microbial food webs (e.g. Weisse, 1991; Berninger et al., 1993). The nutrients enrichment in lakes resulted in increased abundance and biomass of different microbial components (e.g. Beaver & Crisman, 1989; Christoffersen et al., 1990; Berninger et al., 1991). Also under eutrophic and hypertrophic conditions, the microbial food webs may play an important role in the carbon flow and nutrient cycling as they do under oligotrophic conditions (Weisse & Stockner, 1993). On the other hand, the energy transfer efficiency is assumed to decrease with increasing trophic status of lakes (Weisse & Stockner, 1993; Sommaruga, 1995). Along a gradient towards eutrophy, 'bottom-up' control (i.e. supply of nutrients) of the microbial components is increasingly shifted to 'top-down' interactions (Weisse, 1991; Weisse & Stockner, 1993). In eutrophic lakes with high abundance of colonial *Microcystis* spp., the dominance of *Microcystis*-attached or *Microcystis*-associated bacteria may occur (Worm & Søndergaard, 1998). Due to the decreased sensitivity of attached bacteria to grazing by most bacterivorous protists (Jürgens & Güde, 1994), the increased proportion of attached bacteria may have consequences for the structure and energy transfer efficiency of microbial food web in *Microcystis* spp. dominated habitats.

Physical forces may influence the microbial food webs of shallow water habitats. One of the most important processes is wind-driven sediment resuspension and water column mixing. It has been indicated that sediment resuspension may increase the total abundances and may also change the community composition of protists due to passive dispersal of protists from sediment (Rogerson & Laybourn-Parry, 1992; Shimeta & Sisson, 1999; Garstecki et al., 2000) and due to species-specific differences in susceptibility to resuspension (Shimeta & Sisson, 1999; Garstecki et al., 2002). Previous experiments demonstrated that at least some protists are generally well adapted to environmental situations of high sediment load (Boenigk & Novarino, 2004). Field and experimental studies indicate that sediment resuspension may cause release of nutrients and organic matter into the water column (Arfi & Bouvy, 1995). This can result in increased bacterial abundance and biomass (Wainright, 1987, 1990; Ritzrau & Graf, 1992), as well as in increased numbers of bacterivorous protists benefiting from the increased prey availability (Garstecki et al., 2002). The currently available information on the ecological impacts of sediment resuspension was mainly obtained in studies on estuarine and costal systems.

In this study, the structure and dynamics of the planktonic microbial food web in the large subtropical shallow eutrophic Lake Taihu (Qin et al., 2007) were investigated. This lake has been the subject of many limnological studies during the last 50 years (Qin et al., 2004), however, detailed studies on the microbial food web were lacking thus far. For comparison of intra-habitat differences of microbial food web structure two sites located 22 km apart were selected for investigation over a period of twelve month. The two sites differ strongly in their degree of sediment resuspension and eutrophication. In order to reveal possible site-specific differences in the interactions between classical and microbial food web, the dynamics and composition of phytoplankton and zooplankton were also investigated. Attempts to analyse the potential effects of eutrophication and sediment resuspension on microbial food web in this large and shallow lake were made.

Methods

Study sites and sampling scheme

General information of Lake Taihu can be found in the introductory paper (Qin et al., 2007) or elsewhere (Chen et al., 1997; Qin et al., 2004). The lake can be divided into several different regions according to the spatial differences of physical and chemical conditions and phytoplankton community structure (Chen et al., 1997, 2003). One sampling station (Site 1) is located in Meiliang Bay, which was highly enriched with nitrogen and phosphorus (Fig. 1). The bay is located in the northern part of Lake Taihu, has a surface area of ca. 100 km^2, and an average depth of ca. 1.8 m. The eutrophication of the bay was mainly caused by pollution with domestic wastewater discharged from Lujiang and Liangxi River (Fig. 1). The second sampling station (Site 2) is located in the open lake close to its centre (Fig. 1). The open lake is less enriched with nitrogen and phosphorus than most of the bays. The open lake is on average only a few centimetres deeper than Meiliang Bay, but much more exposed to wind.

Twenty-four samplings were performed at both sampling sites from September 30, 2002 to September 30, 2003, at intervals of ca. 15 days. All samples were collected from surface waters (top 50 cm) with a 5-litre Schindler sampler. Sampling was undertaken between 9 and 11 a.m. Fifty ml water samples for determination of numbers of freely suspended bacteria, attached and aggregated bacteria, and heterotrophic nanoflagellates (HNF) were fixed with 2% formalin (final concentration). 250 ml water samples for determination of phytoplankton, ciliates, and nauplii were preserved with 1.5% Lugol's Iodine. Macrozooplankton was collected by filtering 5 litre lake water through a nylon mesh of 30 μm. Samples were fixed with formalin (final concentration 4%). All samples were stored at 4°C in dark until further analysis within 5 months.

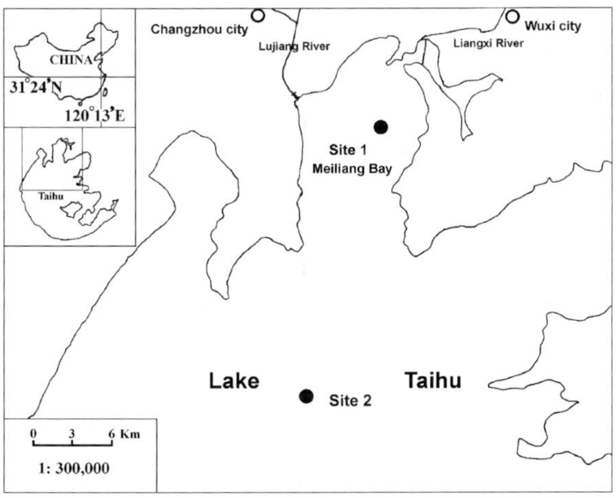

Fig. 1 Map of the northern part of Lake Taihu showing the locations of the sampling sites in Meiliang Bay (Site 1) and the open lake (Site 2)

At each second sampling, secchi-depth (SD), water temperature (WT), pH, conductivity (Cond), dissolved oxygen (DO), total suspended solid (TSS), total nitrogen (TN), total dissolved nitrogen (TDN), ammonium (NH_4^+), nitrite (NO_2^-), nitrate (NO_3^-), total phosphorus (TP), total dissolved phosphorus (TDP), ortho-phosphorus (PO_4^{3-}-P), and chlorophyll *a* (Chl *a*) were measured according to standard methods (Jin & Tu, 1990).

Abundance and biomass determination

Phytoplankton samples were counted using a sedimentation chamber of 1–5 ml at 200× or 400× magnification on an inverted microscope. Phytoplankton species were identified according to Hu et al. (1980). Algal biovolumes were calculated from cell numbers and cell size measurements using appropriate geometric morphs (Hillebrand et al., 1999). Conversion to biomass assumed that 1 mm^3 of biovolume is equivalent to 1 mg of fresh-weight biomass.

Ciliates and nauplii were counted in sedimentation chambers of 5–50 ml at 200× magnification on an inverted microscope. Ciliates were identified mainly according to the keys of Foissner and Berger (1996), and Foissner et al. (1999). Linear dimensions of about 30 individuals of abundant taxa were measured by using the LUCIA G image analysis system (Lucia 4.51, Laboratory Imaging, Prague, Czech Republic), which was connected to the inverted microscope via a video camera. The biovolumes were calculated using appropriate geometric shapes and converted into biomass using a conversion factor of 110 fg C μm^{-3} (Turley et al., 1986). Macrozooplankton were counted in a Bogorov chamber at 45× magnifications, and identified according to Zhang & Huang (1991). For counting of rotifers fixed samples of 0.1 to 0.5 litre were filtered through a 30 μm mesh and poured into a chamber, and counted at 100× magnification following the same approach as for ciliates and nauplii.

For determination of numbers of freely suspended bacteria and numbers of attached bacteria untreated and sonicated subsamples were counted (Velji & Albright, 1993). Subsamples (0.2–1 ml) were diluted to 1 or 2 ml with autoclaved inorganic medium, stained with 4,6-diamidino-2-phenylindole (DAPI) (0.1% wt/vol, final concentration) for 10 min (Porter & Feig, 1980), filtered onto black 0.2 μm pore-size membrane filters (Millipore), and counted at 1250× magnifications by using an epifluorescence microscope (Zeiss, Germany) equipped with a BP 365, FT 395, and LP 397 filter set. Attached bacterial abundances were determined as the differences between total cell numbers (sonicated samples) and total numbers of freely suspended cells (non-sonicated samples). Autotrophic picoplankton cells were identified by checking for the presence of autofluorescence under green excitation and excluded from bacterial counts. Length and width of bacteria were determined with LUCIA G image analysis system as described in Psenner (1993). Cell volumes were calculated with the formula: $V = (w^2 \times \pi/4) \times (l{-}w) + (\pi \times w^3/6)$, where V = Volume (μm^3), w = width (μm) and l = length (μm). The biovolumes were converted into biomass using allometric relationship as suggested by Norland (1993).

Heterotrophic nanoflagellates (HNF) were counted and measured in the same way as for bacteria except sonication of samples. Individual autofluorescence was measured under green excitation using a BP 510–560, FT 580, and LP 590 filter set. Biovolumes were calculated using the formula: $V = (\pi \times w^2/6) \times L$, where V = Volume (μm^3), w = width (μm) and L = length (μm). The biomass of HNF was estimated from mean cell volume using a conversion factor of 220 fg C μm^3 (Borsheim & Bratbak, 1987).

Statistics

Redundancy analysis (RDA) was applied for revealing relationship between the members of planktonic microbial and classical food web and the major environmental variables controlling the dynamics of different members. The biotic variables include the abundances of total bacteria, attached bacteria, HNF, ciliates, rotifers, Cladocerans, *Daphnia* sp., total phytoplankton, and *Microcystis* spp. The tested abiotic environmental variables were water temperature, TP, TN, TSS and COD. All data were log $(x + 1)$ transformed. The RDA was performed with the software CANOCO 4.5 (SCIENTIA Software) by assuming linear species-environment relationships because Detrended Correspondence Analysis (DCA) run on biotic variables indicated that the length of first axis was <2. The significance of first ordination and canonical axes together was assessed in permutation tests with 499 unrestricted Monte Carlo permutations. The statistical analyses were performed separately for data of Site 1 and Site 2. *T*-test was used to test the differences of parameters between the two sites. Log-transformation was used to normalize the data if necessary.

Results

Major ecological differences at the two study sites

Main ecological parameters measured at the two sites of Lake Taihu during the one-year sampling period were summarized in Table 1. Major between-sites differences in chemo-physical parameters were observed for conductivity, secchi depth, concentration of total suspended solid, and concentrations of major algal nutrients (Table 1). No anoxic condition were observed at both sites over the water column and near the bottom, but much lower dissolved oxygen was found at the sediment water interface at Site 1 (unpublished data from Fan et al.). Minimum water temperature (3.8°C) was measured in January and maximum water temperature occurred in July (30°C) (Fig. 2a and b). The concentrations of TSS, TN, NH_4^+, NO_3^-, TP, TDP, and Chl *a* were higher at Site 1 than those at Site 2. Abundance and biomass of components of planktonic community were higher at Site 1 than at Site 2. Little fluctuation of chl *a* value was observed at Site 2

(Fig. 2b). But high seasonal variation of TSS was found at Site 2 relative to Site 1 (Fig. 2e, f). Results of RDA indicated that the most important factors structuring the food web at Site 1 were TP and water temperature (Fig. 3a), while at Site 2 it was only the water temperature (Fig. 3b).

Seasonal changes of microbial components

At Site 1, total bacterial abundance ranged from 2.9×10^6 cells ml^{-1} to 14.5×10^6 cells ml^{-1} with a mean of 7.4×10^6 cells ml^{-1} (Fig. 4a). Bacterial numbers peaked in the middle of summer and showed the lowest values in late winter. The mean of bacterial cell volume ranged from 0.03 to 0.13 μm^3, while the median ranged from 0.02 μm^3 to 0.09 μm^3 (Fig. 4e). Total bacterial biomass ranged from 34.8 μg C l^{-1} to 299.5 μg C l^{-1} with a mean of 98.3 μg C l^{-1} (Fig. 4a). Fluctuation of bacterial biomass was generally in accordance with bacterial abundance. Attached bacterial abundance oscillated between 0 (not measurable) and 6.1×10^6 cells ml^{-1} with a mean of 1.4×10^6 cells ml^{-1} accounting for 0–51.8 % and 14.4% of the total bacterial abundance (Fig. 4c). High numbers of attached bacteria were observed from May to October.

At Site 2, the mean bacterial abundance was 5.7×10^6 cells ml^{-1} with a highest value of 9.2×10^6 cells ml^{-1} in middle of summer and a lowest value of 2.0×10^6 ml^{-1} in early winter (Fig. 4b). The bacterial biomass ranged from 28.9 μg C l^{-1} to 128.4 μg C l^{-1} with an average of 64.9 μg C l^{-1}. Little fluctuation of bacterial abundance and biomass was observed at Site 2. Attached bacteria ranged from 0 (not measurable) to 4.3×10^6 cells ml^{-1} with a mean value of 1.4×10^6 cells ml^{-1}, which account for 0–51.2% and 19.0% of the total bacterial abundance

Table 1 Physical, chemical, and biological parameters determined for the two sampling sites of Lake Taihu in the period from September 2002 to September 2003. The data of the two sampling sites were compared by *t*-test

Parameters	Site 1 Mean (range)	Site 2 Mean (range)	*p* value (*t*-test)
Water depth (m)	2.0–2.7	2.4–3.2	<0.01
Water temperature (°C)	17.6 (4.3–30.0)	17.2 (3.8–30.0)	0.91
Dissolved oxygen (g m^{-3})	9.28 (6.38–12.31)	9.01 (6.82–12.51)	0.91
Conductivity (μs cm^{-1})	457 (380–570)	388 (320–440)	<0.01
PH	8.43 (8.01–8.98)	8.38 (8.19–8.66)	0.59
Secchi-depth (cm)	50 (10–100)	36 (15–50)	0.08
Total suspended substances (g m^{-3})	32.31 (9.56–51.09)	68.18 (20.24–170.12)	0.02
Chemically oxygen demanded (mg l^{-1})	7.6 (4.6–34.2)	5.9 (3.6–5.9)	0.04
Total nitrogen (μmol l^{-1})	282.9 (108.6–780.0)	154.8 (61.4–252.9)	0.03
Total dissolved nitrogen (μmol l^{-1})	182.9 (32.1–311.4)	117.9 (21.4–234.9)	0.09
NH_4^+-N (μmol l^{-1})	40.0 (2.14–110.0)	6.4 (1.4–17.9)	<0.01
NO_2^--N (μmol l^{-1})	5.0 (0.7–19.3)	0.7 (0.1–1.4)	0.03
NO_3^--N (μmol l^{-1})	70 (3.6–132.1)	59.3 (4.3–124.3)	0.58
TP (μmol l^{-1})	3.39 (1.26–6.87)	2.36 (1.45–3.81)	0.06
Total dissolved phosphorus (μmol l^{-1})	1.07 (0.52–2.32)	0.65 (0.32–1.84)	0.04
PO_4^{3-}P (μmol l^{-1})	0.19 (0–1.0)	0.13 (0–0.39)	0.19
Chlorophyll a (μg l^{-1})	20.75 (1.70–52.01)	8.88 (3.83–16.12)	0.04
Total nitrogen: total phosphorus	44.95 (8.82–103.08)	32.35 (12.2–75.8)	0.40
Total bacterial abundance (10^6 cells ml^{-1})	7.4 (2.9–14.5)	5.7 (2.0–9.2)	<0.01
Total bacterial biomass (μg C l^{-1})	98.3 (34.8–299.5)	64.9 (28.9–128.4)	<0.01
HNF abundance (10^7 cells l^{-1})	0.8 (0.1–3.5)	0.5 (0.1–1.6)	0.03
HNF biomass (μg C l^{-1})	127.1 (4.9–699.7)	84.7 (1.6–454.3)	0.02
Ciliate abundance (10^3 cells l^{-1})	9.1 (0.5–29)	8.7 (1.8–62.6)	0.89
Ciliate biomass (μg C l^{-1})	16.2 (0.7–68.6)	10.4 (1.4–37.4)	0.21
Phytoplankton biomass (mg l^{-1})	2.38 (0.06–9.02)	0.28 (0.01–0.86)	0.02
Rotifer abundance (ind. l^{-1})	640 (1–4650)	156 (1–696)	0.02
Cladocerans abundance (ind. l^{-1})	183 (0.2–740)	87.5 (1.4–496)	0.03

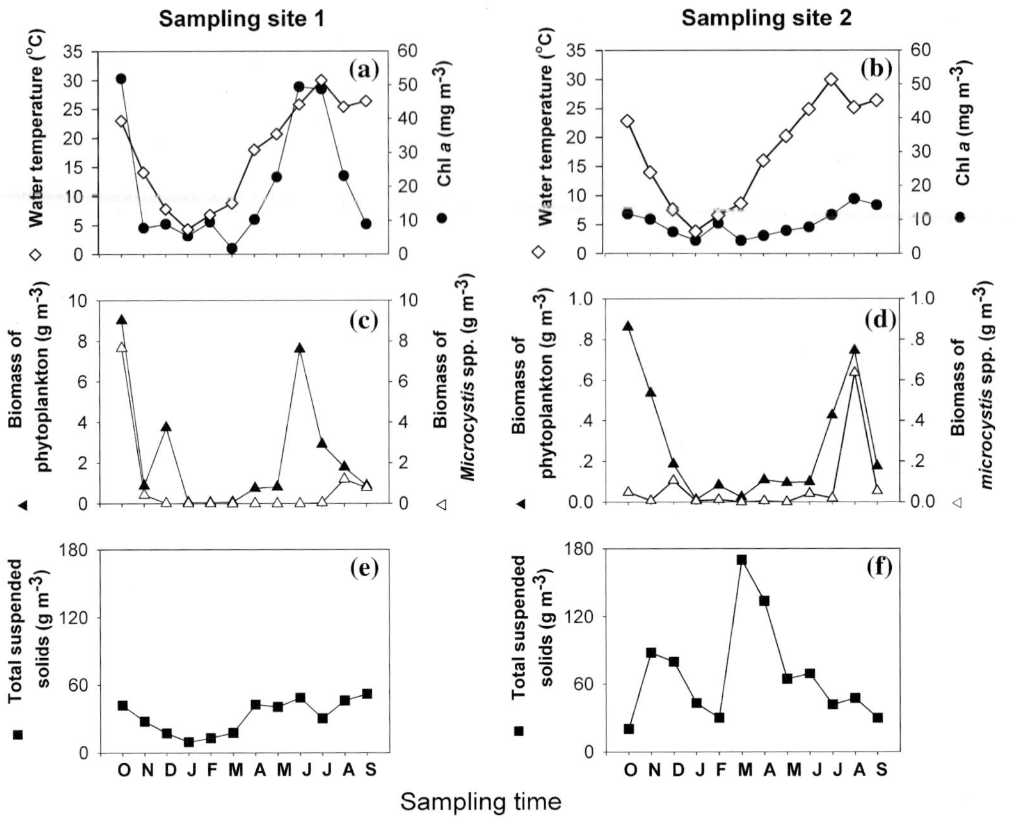

Fig. 2 Seasonal development of Chlorophyll a and water temperature (**a** and **b**), biomass of phytoplankton and biomass of colonial *Microcystis* spp. (**c** and **d**, and total suspended solids (TSS) (**e** and **f**) in the water column at the two sampling stations in Lake Taihu

respectively (Fig. 4d). High abundance and percentage of attached bacteria were found from early spring to late autumn. The mean and median of cell volume fluctuated from 0.03 to 0.07 μm^3 and 0.02 μm^3 to 0.03 μm^3 respectively (Fig. 4f). No correlations between the cell volumes of freely suspended bacteria or attached bacteria, and temperature were found at both sampling stations (linear regression, $P > 0.05$).

The abundance of HNF at Site 1 ranged from 0.1×10^7 cells l^{-1} to 3.5×10^7 cells l^{-1} with a median and mean density of 0.6×10^7 cells l^{-1} and 0.8×10^7 cells l^{-1} respectively (Fig. 5a). HNF in the size class from 5 to 10 μm were the most abundant. HNF biomass ranged from 4.9 μg C l^{-1} to 699.7 μg C l^{-1} with a median and mean biomass of 56.6 μg C l^{-1} and 127.1 μg C l^{-1}, respectively (Fig. 5a). Single peak of HNF abundance and biomass was observed in late autumn, while no strong oscillation was observed during other months (Fig. 5a). At Site 2, the median and average density of HNF was 0.4×10^7 cells l^{-1} and 0.5×10^7 cells l^{-1} with a range from 0.1×10^7 cells l^{-1} to 1.6×10^7 cells l^{-1}, respectively (Fig. 5h). HNF biomass ranged from 1.6 μg C l^{-1} to 454.3 μg C l^{-1} with a median and mean biomass of 57.3 μg C l^{-1} and 84.7 μg C l^{-1} (Fig. 5h). HNF biomass formed one peak in late fall (Fig. 5h). Both HNF abundance and HNF biomass were not correlated with the mean size of total bacteria and freely suspended bacteria at both sites (linear regression, $P > 0.05$).

At Site 1, the abundance of ciliates ranged from 470 cells l^{-1} to 29,000 cells l^{-1} with a mean and median density of 9100 cells l^{-1} and 4000 cells l^{-1} (Fig. 5b). Ciliates in the size class of 20–40 μm were the most abundant in most time of the year (Fig. 5b). Dominant taxa in this size

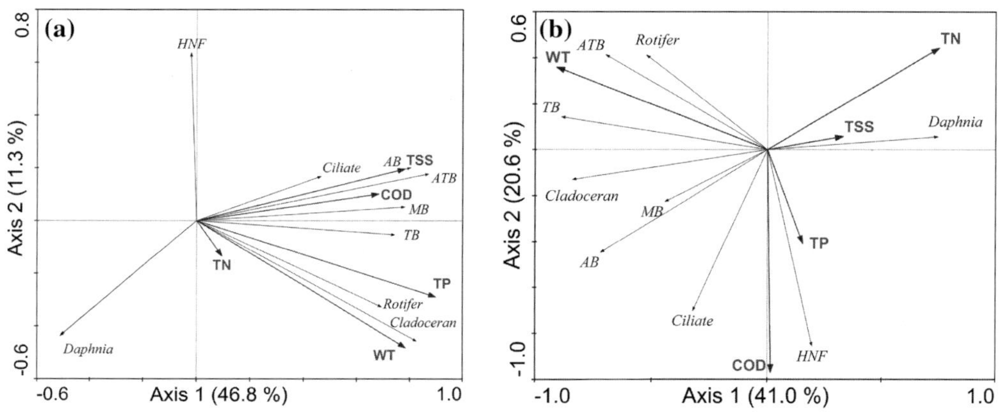

Fig. 3 RDA biplots show different planktonic groups in relation to main environmental factors at Site 1 (**a**) and Site 2 (**b**) in Lake Taihu. TN, TP, TSS, WT and COD refer to total nitrogen, total phosphorus, total suspended substances, water temperature, and chemically oxygen demanded respectively. AB, MB, TB, and ATB refer to abundance of phytoplankton, *Microcystis* spp., total bacteria and attached bacteria, respectively

class were *Strobilidium*-like and unidentified oligotrich small ciliates. The lowest abundance occurred in winter and early spring with exception of 15th of January, while the highest density was observed in summer. Ciliates <20 μm were abundant only from July to September (Fig. 5b)

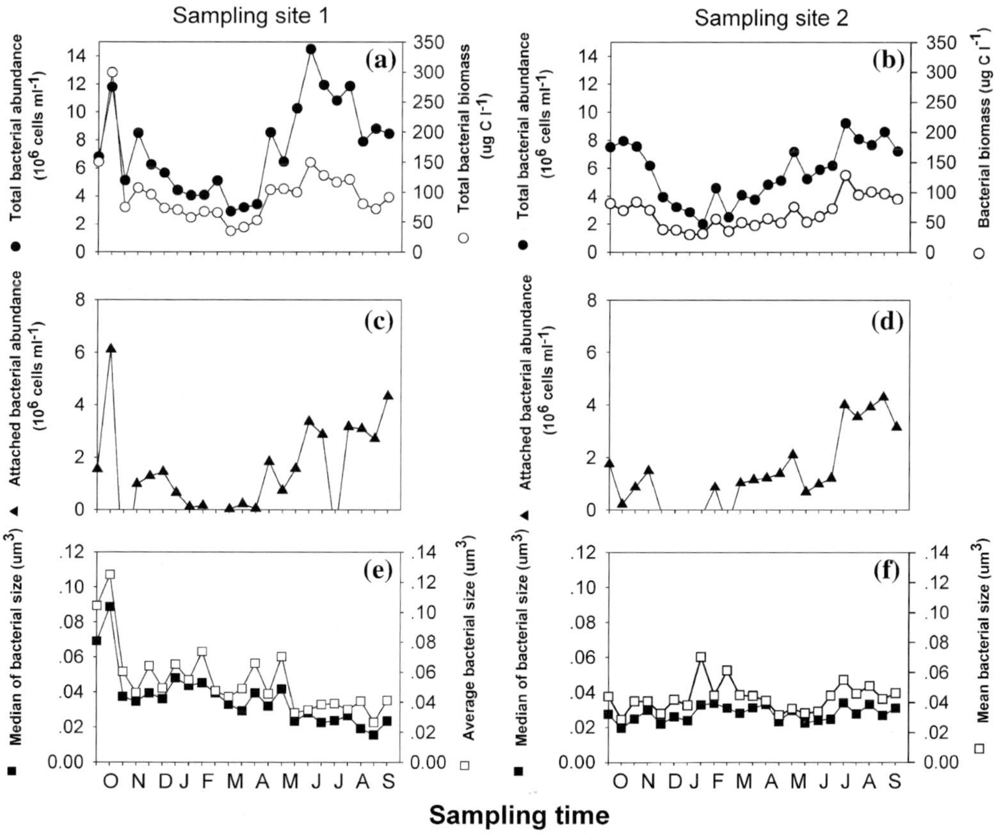

Fig. 4 Seasonal development of total bacterial abundance and total bacterial biomass (**a** and **b**), attached bacterial abundance (**c** and **d**), mean and median of bacterial size (**e** and **f**) at the two sampling sites in Lake Taihu

and dominated by *Urotricha*-like species. In January and June, ciliates >40 μm contributed 50% of total ciliate abundance (Fig. 5b). They were represented by *Vorticella* spp., *Epistylis* spp., *Codenella* sp., and *Tintinnidium* spp. Total ciliate biomass ranged from 0.7 μg C l^{-1} to 68.6 μg C l^{-1} with a mean and median of 16.2 μg C l^{-1} and 10.3 μg C l^{-1} (Fig. 5b). High biomass was observed in January and June, which was mainly contributed by *Vorticella* spp., and *Epistylis* spp. At Site 2, total ciliate abundance averaged 8700 cells l^{-1}, showing two higher values at the end of January with density of 26,100 cells l^{-1} and in midsummer with density of 62,600 cells l^{-1}, while showing lower values during most time of the year (Fig. 5i). During the winter peak, ciliates in the size class of <40 μm were the dominant groups (Fig. 5i). They were represented by an unidentified small ciliate and *Pelagovorticella*-like species, respectively. During the summer peak, *Urotricha*-like ciliates were dominating. During other periods, ciliates were mainly composed of *Codenella* sp., *Tintinnidium* spp., *Stenosemella* sp., and *Vorticella* spp. Total biomass of ciliates ranged from 1.4 μg C l^{-1} to 37.4 μg C l^{-1} with a mean and median of 10.4 μg C l^{-1} and 7.5 μg C l^{-1} (Fig. 5i). Higher biomass was observed from May to August, which was mainly contributed by *Vorticella* spp. and *Codenella* sp.

Community dynamics of phytoplankton and zooplankton

At Site 1, the phytoplankton biomass ranged from 0.06 g m^{-3} in January to 9.02 g m^{-3} in October. The cyanobacteria *Microcystis* spp. dominated in terms of biomass among the cyanobacteria, as well as among the whole phytoplankton assemblages from July to November (Fig. 2a). At Site 2, low phytoplankton biomass was observed with a range from 0.01 g m^{-3} in December to 0.86 g m^{-3} in October. *Microcystis* spp. dominated phytoplankton communities in August and September (Fig. 2b).

At Site 1, rotifer abundance averaged 640 ind. l^{-1} showing the highest value of 4650 ind. l^{-1} in the middle of July and the lowest value of 1 ind. l^{-1} at the end of December (Fig. 5c). The abundant rotifers include *Branchionus* sp., *Keratella valga*, and *Polyarthra* sp., while *Asplanchna priodonta* only contributed to more than 50% of rotifer biomass in middle of July. Nauplii abundance ranged from 6 to 1200 ind. l^{-1} with an average and median of 144 and 27 ind. l^{-1} (Fig. 5d). The highest abundance occurred at the end of August, while much lower abundance was observed during most of other period (Fig. 5d). Abundance of calanoid copepods ranged from 0 to 22 ind. l^{-1} with maximum abundances occurring in middle of fall and spring, and minimum in summer (Fig. 5e). Abundance of cyclopoid copepods averaged 49 ind. l^{-1}, showing highest value of 408 ind. l^{-1} in middle of July and much lower values during most of the other periods (Fig. 5f). The succession of cladocerans can be divided into two main phases (Fig. 5g). In the first phase, *Daphnia* sp. dominated the plankton community from end of February to end of May, with maximum abundance of 96 ind. l^{-1} at the end of March (Fig. 5g). In the second phase, from June to November, *Bosmina* sp. was the most abundant species, showing the highest abundance of 722 ind. l^{-1} at the end of June (Fig. 5g). *Moina* sp. only appeared in considerable numbers in middle of September (Fig. 5g). At Site 2, low rotifer abundances were observed all over the year, ranging from 1 to 696 ind. l^{-1} with an average of 156 ind. l^{-1} (Fig. 5j). Little fluctuation of rotifer abundance was found (Fig. 5j). The abundant rotifers include *Branchionus* sp., *Keratella cochlearis*, and *Polyarthra* sp. Nauplii abundance ranged from 0.6 to 2830 ind. l^{-1} with an average and median of 226 and 17 ind. l^{-1} (Fig. 5k). One nauplii peak appeared at the end of June, while much lower abundance was observed during most of the other periods (Fig. 5k). Higher abundance of calanoid copepods and cyclopoid copepods was observed in later summer while lower abundance appeared during most of other period (Fig. 5l, m). The succession of cladocerans could again be divided into two main phases (Fig. 5n). In the first phase, *Daphnia* sp. also dominated the plankton community at this site from middle of January to

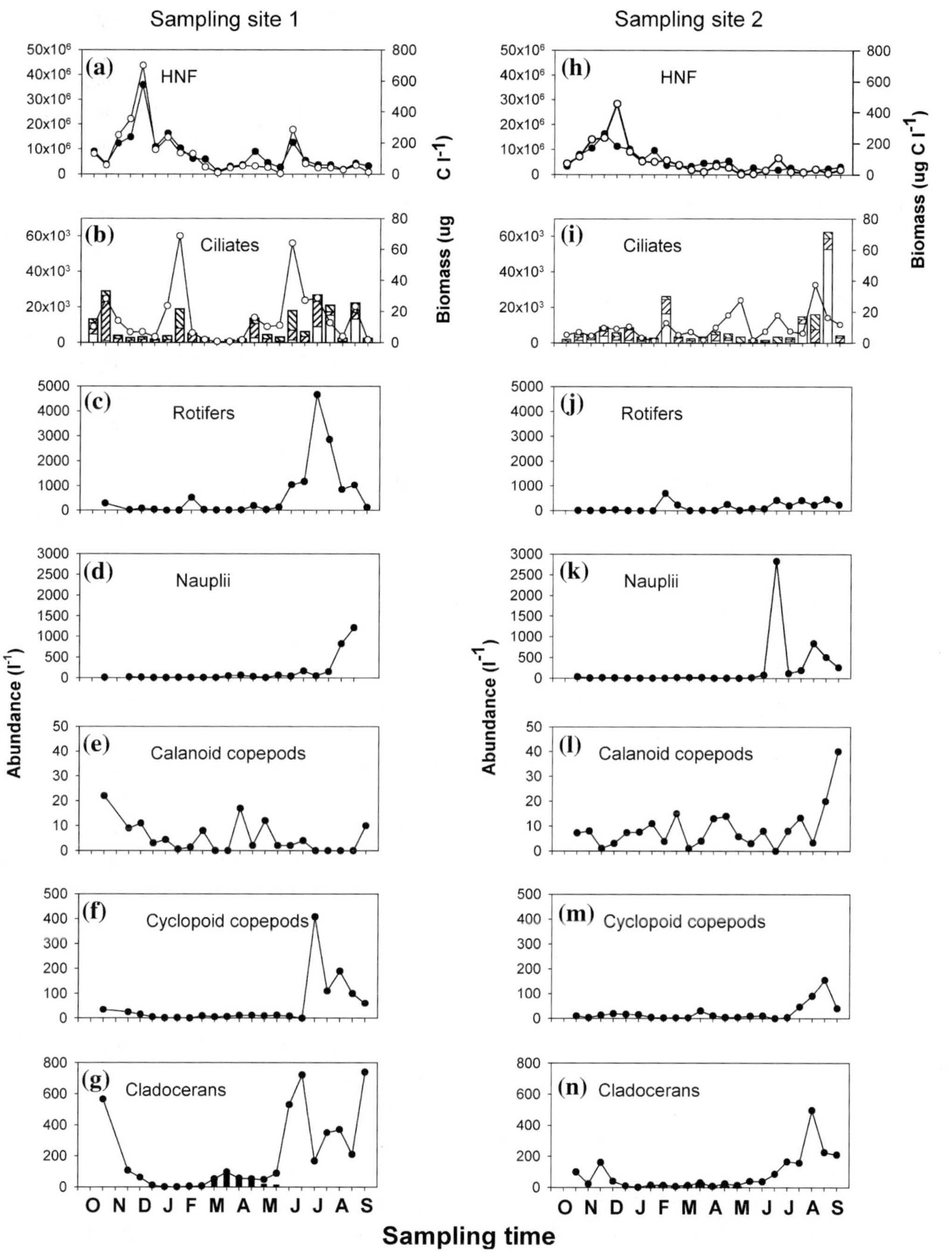

Fig. 5 Seasonal changes in abundance (•) and biomass (○) of HNF (**a** and **h**), ciliates (**b** and **i**), rotifers (**c** and **j**), nauplii (**d** and **k**), calanoid copepods (**e** and **l**), cyclopoid copepods (**f** and **m**), and cladocerans (**g** and **n**). In Fig. **b** and **i**, the vertical bars show abundances of ciliates in three size classes: <20 μm (open part of the bars); 20–40 μm (upward diagonally hatched part of the bars); and > 40 μm (downward diagonally hatched part of the bars). In figure g and n, the vertical black bars show the abundance of *Daphnia*

middle of May, with maximum abundance of 29 ind. l^{-1} at the end of March (Fig. 5n). In the second phase from May to middle of November, *Bosmina* sp. and *Moina* sp. were the most abundant species, showing the highest abundance of 496 ind. l^{-1} in the middle of August (Fig. 5n).

Statistics

RDA model calculated with the data of Site 1 and Site 2 significantly (for first canonical axis, $P < 0.05$) explained the variability of the abundance of members of planktonic food webs at both sites (Figs. 3a and b). At Site 1 (Fig. 3a), the first axis showed high canonical correlation with TP, WT, COD, and TSS, which were positively related to planktonic components except for HNF and *Daphnia*. HNF was negatively related to such micrograzers like rotifer and *Bosmina*-dominated zooplankton. *Daphnia* abundance was negatively related with all microbial components. There was a significant correlation between attached bacteria and TSS and abundance of *Microsystis* spp. At site 2 (Fig. 3b), the first axis showed only high canonical correlation with water temperature. In contrast to Site 1, TN, TSS and TP were even negatively related to abundance of phytoplankton and microbial communities. *Daphnia* abundance was negatively related with all microbial components as the same at Site 1.

Discussion

Influence of sediment resuspension on microbial components

Based on the concentrations of TP and TN (Table 1), Site 1 was more enriched with phosphorus and nitrogen than Site 2. On the other hand, Site 2 showed a higher intensity of sediment resuspension than Site 1, as indicated by higher amount of TSS (Table 1). Furthermore, most planktonic organisms showed on average lower biomass at Site 2.

Sediment resuspension, as a common feature of large shallow lakes, may have positive and negative impacts on microbial food webs. Positive effects on abundance of microbial components could include (i) direct transportation of benthic bacteria and protozoan into the water column (Rogerson & Laybourn-Parry, 1992; Shimeta & Sisson, 1999; Garstecki et al., 2000, 2002); (ii) release of dissolved nutrient and organic matter into water column (Arfi & Bouvy, 1995), and thereby increasing the abundance of bacteria (Wainright, 1987, 1990; Ritzrau & Graf, 1992) and some bacterivorous protists (Garstecki et al., 2002). However, no correlations between the abundance of microbial components (i.e., bacteria, HNF and ciliates) and TSS, the indicator of sediment resuspension, were found at Site 2. The same result was found for Site 1 except the bacterial abundance. Possible explanation for these findings is the different intensity of sediment resuspension, which is dependent on wind speed, water depth, and the size of a lake. Site 1 is located in the Meiliang bay, where only very strong wind can cause a significant resuspension of sediment (Qin et al., 2004). Therefore, concentration of suspended solids in the water column at Site 1 was lower during periods of weak wind intensity or at calm situations. This is confirmed by little variations of TSS and high contribution of phytoplankton biomass to the TSS fraction. This contribution was indicated by the strong positive correlations between phytoplankton biomass and TSS. The microbial organisms in the upper sediment may only contribute to the microbial components in the water column of the bay during periods with high wind intensity. Site 2 is located in the centre of the lake, where moderate wind (3.3–5.0 m s^{-1}) caused strong mixing of the water column and sediment resuspension during almost two thirds of a year (Qin et al., 2004; Luo et al., 2004). Long-lasting sediment accumulation at the lake bottom was not observed at Site 2 (Qin et al., 2004). The bottom (hard clay) water interface at Site 2 can only allow biofilm-forming microorganisms to inhabit this surface, and these biofilm organisms cannot contribute much to the planktonic microbial community in the water column. For the same reason, not much organic carbon can be deposited and later introduced into the water column by resuspension events. Therefore, the bacterial community at Site 2 could, in contrast to Site 1, not benefit from release of organic material from the sediment during resuspension events. This fits to the observation, that higher amount of resuspended sediment did not lead to higher abundances of total bacteria and freely suspended bacteria at Site 2. To reveal the exact contribution of benthic microbial organism to planktonic microbial food web and its controlling factors,

more detailed temporal in situ studies are needed at different locations of the lake.

At Site 2 a negative correlation between TSS and phytoplankton biomass and a relatively low phytoplankton biomass despite relatively high concentration of total phosphorous were observed. Obviously, the phytoplankton was negatively impacted by decrease in light intensities as a result of the frequent resuspension of sediment and the high concentration of TSS (Chen et al., 2003). A negative indirect impact of sediment resuspension on bacterial growth conditions could be caused by the limited growth and production of phytoplankton. Decrease in primary production may lead to a lower supply of dissolved organic carbon for bacterial growth (Chen & Wangersky, 1996). At Site 1, the positive correlations between TSS and the bacterial abundances or phytoplankton biomass suggest no apparent negative impacts.

Impacts of eutrophication on microbial components

Higher abundances of microbial components (i.e. bacteria, HNF) were observed at site 1 than those at Site 2. This within-lake comparison supports the hypothesis that abundances of microbial components are positively related to phosphorus enrichment of a lake (e.g. Weisse, 2003). However, the coupling between TP and abundance of microbial components was much different at the two sites (Fig. 3). Indeed no obvious correlation between TP and abundance of microbial components was observed at Site 2 (Fig. 3b). This weak coupling could be duo to the strong sediment resuspension at Site 2, which limited growth of phytoplankton and, consequently, growth of bacterioplankton and other microbial components (i.e. HNF, ciliate). This was supported by the fact that no correlation between TP and phytoplankton abundance were observed at Site 2 (Fig. 3b), while good correlation between abundance of phytoplankton and bacterioplankton was found at both sites (Figs. 3a and b). On the other hand, on average higher biomass ratios of phytoplankton to bacteria were observed at Site 1 than at Site 2 (Fig. 6). This may indicate different rates of bacterial production at two different locations.

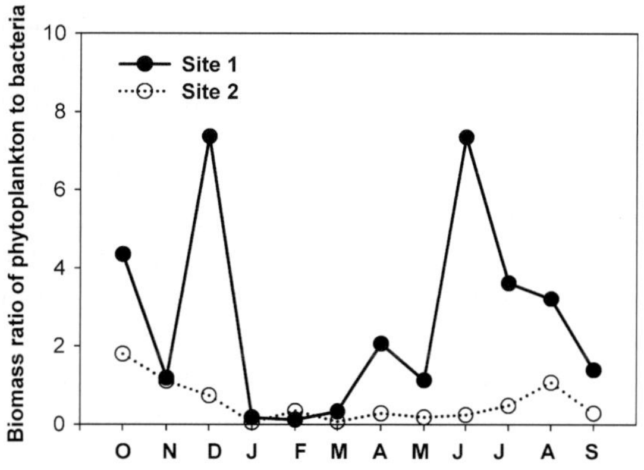

Fig. 6 Biomass ratios of phytoplankton to bacteria at the two sampling sites in Lake Taihu

Along a gradient towards eutrophy, microbial food webs are progressively controlled by 'top-down' forces (Weisse & Stockner, 1993), and zooplankton predation on components of microbial food webs is gaining an increasing importance in linking of planktonic microbial food webs and the classical food chains. Among the zooplankton, *Daphnia*, can strongly suppress the microbial food web by their size-selective feeding abilities (Weisse & Stockner, 1993; Jürgens, 1994, Jürgens & Stolpe, 1995). The data on the seasonal succession of cladocerans, flagellates, bacteria and ciliates at both sites allow insights into the trophic interactions between metazoans and the microbial loop (Figs. 3, 5), and indicate the similar 'top-down' forces of *Daphnia* during spring and early summer in such a subtropical lake. Following the collapse of *Daphnia* population in mid-summer, possibly due to fish predation and blooming of toxic colonial *Microcystis* spp., *Bosmina* sp. and *Moina* sp., dominated the crustacean community.

Significance of attached bacteria

Particles from sediment resuspension, terrestrial input, and large phytoplankton are important for colonization of attached bacteria (Simon et al., 2002). Lake Taihu received a large amount of particles (organic and inorganic) from its watershed due to intensive agriculture and soil erosion (Dokulil et al., 2000). Sediment resuspension especially in the open lake is very significant due to the lake's shallowness (Qin et al., 2004).

Under the process of eutrophication, phytoplankton blooms, notably the colonial *Microcystis* spp. fuel Lake Taihu with high amount of particulate organic matter (Chen et al., 2003). Our investigations indicated that high abundances of attached bacteria occurred from summer to autumn at both sites in this lake. At Site 1, the strong correlations between attached bacterial abundance and biomass of *Microcystis* spp. suggests that much of the attached bacteria were associated with *Microcystis* spp. colonies (Fig. 3a). This fits to the findings in a eutrophic Danish lake (Worm & Søndergaad, 1988). Worm et al. (2001) demonstrated that bacteria attached to phytoplankton blooms hold up to nearly 80% of total bacterial numbers in an enclosure experiment. At Site 2, however, no strong correlations between attached bacterial abundance and phytoplankton biomass, *Microcystis* spp. biomass, and TSS were observed. This might suggest that the communities of attached bacteria differed in their composition at the two investigated sites.

In comparison to freely suspended bacteria, attached bacteria contributed higher proportion to bacterial production (Simon et al., 2002). The dominance of attached bacteria in Lake Taihu from summer to autumn may have consequences for the structure of the microbial food web, as well as for the carbon flow through the entire food web. Since attached bacteria could be less susceptible to predation by protozoan (Jürgens & Güde, 1994), the production by attached bacteria might not be channeled efficiently through the microbial food web to higher trophic levels. On the other hand, planktivorous fish (e.g. *Hypophthalmichthys molitrix*, *Aristichthys nobilis* commonly found in Lake Taihu), which are able to directly graze upon phytoplankton dominated by *Microcystis* spp. (Xie & Liu, 2001), may directly crop a part of the production of attached bacteria. This may lead to a direct channeling of a part of the bacterial production to higher trophic levels.

Conclusion

Significant intra-habitat heterogeneity of microbial food web structure and dynamics was observed in Lake Taihu. Site-specific differences in wind-driven sediment resuspension, as well as differences in intensity of eutrophication were identified as major causes of the observed within habitat heterogeneity. In contrast to previous observations in coastal regions, our investigations indicate that frequent sediment resuspension in large lakes does not contribute to growth of planktonic microbial organisms.

Acknowledgements This work was supported by Chinese Academy of Sciences (No. KZCX1–SW–12), NSFC (No. 30370278, 30200037), and Austrian Academy of Sciences. Jiang Ji and some other staff members at the Taihu Laboratory for Lake Ecosystem Research contributed field assistance. Q. L. Wu was supported by a scholarship from North-South dialogue Program, Ministry of Foreign Affairs, Austria (EZA project 894/01). We thank Thomas Weisse for providing comments on an early version of the manuscript.

References

Alfri, R. & M. Bouvy, 1995. Size, composition and distribution of particles related to wind-induced resuspension in a shallow tropical lagoon. Journal of Plankton Research 17: 557–574.

Beaver, J. R. & T. L. Crisman, 1989. The role of ciliated protozoa in pelagic freshwater ecosystems. Microbial Ecology 17: 111–136.

Berninger, U. G., B. J. Finlay & P. Kuuppo-Leinikki, 1991. Protozoan control of bacterial abundances in freshwater. Limnology and Oceanography 36: 139–146.

Berninger, U. G., S. A. Wickham & B. J. Finlay, 1993. Trophic coupling within the microbial food web: a study with fine temporal resolution in a eutrophic freshwater ecosystem. Freshwater Biology 30: 419–432.

Boenigk, J. & G. Novarino, 2004. Effect of suspended clay on the feeding and growth of bacterivorous flagellates and ciliates. Aquatic Microbial Ecology 34: 181–192.

Borsheim, K. Y. & G. Bratbak, 1987. Cell volume to carbon conversion factors for a bacterivorous Monas sp. enriched from sea water. Marine Ecology Progress Series 36: 171–175.

Chen, W., Y. Chen, X. Gao & I. Yoshida, 1997. Eutrophication of Taihu and its control. Agriculture Engineering Journal 6: 109–120.

Chen, Y., B. Qin, K. Teubner & M. T. Dokulil, 2003. Long-term dynamics of phytoplankton assemblages: *Microcystis*-domination in Lake Taihu, a large shallow lake in China. Journal of Plankton Research 25: 445–453.

Chen, W. & P. Wangersky, 1996. Rates of microbial degradation of dissolved organic carbon from phytoplankton cultures. Journal of Plankton Research 18: 1521–1533.

Christoffersen, K., B. Riemann, L. R. Hansen, A. Klysner & H. B. Sorensen, 1990. Qualitative importance of the microbial food loop and plankton community structure in a eutrophic lake during a bloom of cyanobacteria. Microbial Ecology 20: 253–272.

Dokulil, M. T. & K. Teubner, 2000. Cyanobacterial dominance in lakes. Hydrobiologia 438: 1–12.

Dokulil, M. T., W. Chen & Q. Cai, 2000. Anthropogenic impacts to large lakes in China: the Taihu example. Aquatic Ecosystem Health and Management 3: 81–94.

Foissner, W. & H. Berger, 1996. A user-friendly guide to the ciliates (Protozoa Ciliophora) commonly used by hydrobiologist as bioindicators in rivers, lakes, and waste waters, with notes on their ecology. Freshwater Biology 35: 375–482.

Foissner, W., H. Berger & J. Schaumberg, 1999. Identification and Ecology of Limnetic Plankton Ciliates. Bayer. Landesamt für Wasserwirtschaft, München.

Garstecki, T., R. Verhoeven, S. A. Wickham & H. Arndt, 2000. Benthic-pelagic coupling: a comparison of the community structure of benthic and pelagic heterotrophic protists in shallow inlets of the southern Baltic. Freshwater Biology 45: 147–168.

Garstecki, T., S. A. Wickham & H. Arndt, 2002. Effects of experimental sediment resuspension on a coastal planktonic microbial food web. Estuarine, Coastal and Shelf Science 55: 751–762.

Hillebrand, H., C. D. Dürselen, D. Kirschtel, T. Zohary & U. Pollingher, 1999. Biovolume calculation for pelagic and benthic microalgae. Journal of Phycology 35: 403–424.

Hu, H., Y. Li, Y. Wei, H. Zhu, J. Chen & Z. Shi, 1980. Freshwater Algae in China. Shanghai Science and Technology Press, Shanghai (in Chinese).

Jin, X. & Q. Tu, 1990. The Standard Methods for Observation and Analysis of Lake Eutrophication, 2nd edn. China Environmental Science Press, Beijing (in Chinese).

Jürgens, K. & G. Stolpe, 1995. Seasonal dynamics of crustacean zooplankton, heterotrophic nanoflagellates and bacteria in a shallow eutrophic lake. Freshwater Biology 33: 27–38.

Jürgens, K. & H. Güde, 1994. The potential importance of grazing-resistant bacteria in planktonic systems. Marine Ecology Progress Series 112: 169–188.

Jürgens, K., 1994. The impact of Daphnia on microbial food webs-A review. Marine Microbial Food Webs 8: 295–324.

Luo, L., B. Qin, W. Hu & F. Zhang, 2004. Sediment re-suspension under different hydrodynamic distribution in Lake Taihu. Journal of Lake Science 16: 273–276.

Norland, S., 1993. The relationship between biomass and volume of bacteria. In Kemp, P., B. Sherr, E. Sherr & J. J. Cole (eds), Handbook of Methods in Aquatic Microbial Ecology. Lewis, 303–308.

Porter, K. G. & Y. S. Feig, 1980. The use of DAPI for identifying and counting aquatic microflora. Limnology and Oceanography 25: 943–948.

Psenner, R., 1993. Determination of size and morphology of aquatic bacteria by automated image analysis. In Kemp, P., B. Sherr, E. Sherr & J. J. Cole (eds), Handbook of Methods in Aquatic Microbial Ecology. Lewis, pp. 339–346.

Qin , B., W. Chen & W. Hu, 2004. Succession of Ecological Environment and Its Mechanism in Lake Taihu. China Science Press, Beijing (in Chinese).

Qin, B., P. Z. Xu, Q. L. Wu, L. C. Luo & Y. L. Zhang, 2007. Environmental issues of Lake Taihu, China. Hydrobiologia 581: 3–14.

Ritzrau, W. & G. Graf, 1992. Increase of microbial biomass in the benthic turbidity zone of Kiel Bight after resuspension by a storm event. Limnology and Oceanography 37: 1081–1086.

Robarts, R. D. & R. J. Wickham, 1990. Heterotrophic bacterial production and its dependence on autotrophic production in a hypertrophic African reservoir. Canadian Journal of Fisheries and Aquatic Science 47: 1027–1037.

Rogerson, A. & J. Laybourn-Parry, 1992. The abundance of marine naked amoebae in the water column of the Clyde Estuary. Estuarine, Coastal and Shelf Science 33: 187–196.

Shimeta, J. & J. D. Sisson, 1999. Taxon-specific tidal resuspension of protists into the subtidal benthic boundary layer of a coastal embayment. Marine Ecology Progress Series 177: 51–62.

Simon, M., H. P. Grossart, B. Schweitzer & H. Ploug, 2002. Microbial ecology of organic aggregates in aquatic ecosystems. Aquatic Microbial Ecology 28: 175–211.

Sommaruga, R., 1995. Microbial and classical food webs: A visit to hypertrophic lake. FEMS Microbiology Ecology 17: 257–270.

Turley, C. M., R. C. Newell & D. B. Robins, 1986. Survival strategise of two small marine ciliates and their role on regulating bacterial community structure under experimental conditions. Marine Ecology Progress Series 33: 59-70.

Velji, M. I. & L. J. Albright, 1993. Improved sample preparation for enumeration of aggregated aquatic substrate bacteria. In Kemp, P., B. Sherr, E. Sherr & J. J. Cole (eds), Handbook of Methods in Aquatic Microbial Ecology. Lewis, 139–142.

Wainright, S. C., 1987. Stimulation of heterotrophic microplankton production by resuspended marine sediments. Science 238: 1710–1711.

Wainright, S. C., 1990. Sediment-to-water fluxes of particulate material and microbes by resuspension and their contribution to the planktonic food web. Marine Ecology Progress Series 52: 271–281.

Weisse, T. & J. G. Stockner, 1993. Eutrophication: the role of microbial food webs. Memorie dell'Istituto Italiano di Idrobiologia 52: 133–150.

Weisse, T., 1991. The microbial food web and its sensitivity to eutrophication and contaminant enrichment: a cross-system overview. International Review of Hydrobiology 76: 327–337.

Weisse, T., 2003. Pelagic microbes-Protozoa and the microbial food web. In O'Sulliva, P. E. & C. S. Reynolds (eds), The Lakes Handbook. Blackwell Science Ltd, 417–460.

Worm, J. & M. Søndergaad, 1998. Dynamics of heterotrophic bacteria attached to *Microcystis* spp. (Cyanobacteria). Aquatic Microbial Ecology 14: 19–28.

Worm, J., K. Gustavson, K. Garde, N. H. Borch & M. Søndergaard, 2001. Functional similarity of attached and free-living bacteria during freshwater phytoplankton blooms. Aquatic Microbial Ecology 25: 103–111.

Xie, P. & J. Liu, 2001. Practical success of biomanipulation using filter-feeding fish to control cyanobacteria blooms: a synthesis of decades of research and application in a subtropical hypereutrophic lake. The Scientific World 1: 337–356.

Zhang, Z. S. & Q. F. Huang, 1991. Methods in Freshwater Plankton Study. China Science Press, Beijing (in Chinese).

Hydrobiologia (2007) 581:255–267
DOI 10.1007/s10750-006-0501-9

EUTROPHICATION IN LAKES

Effects of limiting nutrients and N:P ratios on the phytoplankton growth in a shallow hypertrophic reservoir

Ho-Sub Kim · Soon-Jin Hwang · Jae-Ki Shin · Kwang-Guk An · Chun Gyeong Yoon

Abstract The purpose of this study was to evaluate the effects of limiting nutrients and the N:P ratios on the growth of phytoplankton (mainly cyanobacteria) in a shallow hypertrophic reservoir between November 2002 and December 2003. Nutrient enrichment bioassays (NEBs) were conducted, along with analyses of seasonal ambient nutrients and phytoplankton taxa, in the reservoir. The average DIN:TDP and TN:TP mass ratios in the ambient water were 90 (range: 17–187) and 34 (13–60), respectively, during the study period. The dissolved inorganic phosphorus showed seasonal variation, but less than that of inorganic nitrogen. The TN:TP ratios ranged from 13 to 46 (mean: 27 ± 6) during June–December when the cyanobacteria, *Microcystis*, dominated the phytoplankton composition. The NEBs showed that phytoplankton growth was mainly stimulated by the phosphorus (all of total 17 cases), rather than the nitrogen concentration (8 of 17 cases). The rapid growth rate of cyanobacteria was evident with TN:TP ratios less than 30. According to the results of the NEBs with different N concentrations (0.07, 0.7 and 3.5 mg l^{-1}), but the same N:P ratios and when the nitrogen concentration was highest, the cyanobacterial growth reached a maximum at N:P ratios <1. Overall, the response of cyanobacterial growth was a direct function of added phosphorus in the NEBs, and was greater with increased N concentrations. Thus, cyanobacterial blooms favored relatively low N:P ratios in this hypertrophic reservoir system.

Keywords N:P ratio · P-limitation · Phytoplankton growth · Cyanobacteria · Hypertrophic reservoir

Guest editors: B. Qin, Z. Liu & K. Havens
Eutrophication of shallow lakes with special reference to Lake Taihu, China

H.-S. Kim · S.-J. Hwang (✉) · C. G. Yoon
Han River Environmental Research Laboratory, National Institute of Environmental Research, Gyeonggi 476-823, Korea
e-mail: sjhwang@konkuk.ac.kr

J.-K. Shin
Korea Institute of Water and Environment, Korea Water Resources Corporation, Daejon 305-730, Korea

K.-G. An
School of Bioscience and Biotechnology, Chungnam National University, Daejon 305-764, Korea

Present Address:
H.-S. Kim
Watershed Management Research Division, National Institute of Environmental Research, Inchon 404-170, Korea

Introduction

The nitrogen-to-phosphorus ratio (N:P) in lentic systems has been frequently used as a key indicator in predicting algal biomass and compositions, and its seasonal succession (Tilman, 1982; Kilham, 1990). It can also serve as an index that represents the nutrient limitation for the algal growth (Smith, 1983; Fugimoto & Sudo, 1997). The atomic ratio, 16N:1P, which is known as the Redfield ratio, has been used for generally describing the average elemental composition of phytoplankton. Many investigations; however, have shown this ratio to vary considerably, depending on the algal species and surrounding environmental conditions (Rhee, 1978; Forsberg & Ryding, 1980; Geider & La Roche, 2002; Ho et al., 2003), and that optimum nutrient ratios vary interspecifically within the diatoms due to growth competitions (Tilman, 1977, 1978).

A considerable variation in the nutrient stoichiometry of the natural phytoplankton community is a reflection of the type and extent of nutrient limitation and availability. In turn, this is likely related to the supply ratios of the elements loaded to the lake affecting the development of particular phytoplankton communities (Smith, 1983; Smith et al., 1987; Sommer, 1989; Fujimoto & Sudo, 1997; Xie et al., 2003). The hypothesis of "the dominance of cyanobacteria at low N:P ratios" is one of these examples. Smith (1983) pointed out that bloom-forming cyanobacteria had a tendency to dominate the lake when the N:P ratio was less than 29. The significance of the N:P ratio as a critical standard, however, is still controversial, due to variabilities in the other chemical characteristics and phytoplankton compositions within a geographic region.

Unlike Smith's 'N:P ratio rule', which stated that the relative proportion of cyanobacteria to the total algal biomass might decrease when the N:P ratio exceeded 29 (Smith et al., 1987), the bloom of cyanobacteria was affected by increases in the phosphorus concentration rather than simply by a decrease in the N:P ratio (Trimbee & Prepas, 1987; Sheffer et al., 1997; Xie et al., 2003). Moreover, the possibility that the growth response of algae to the limiting nutrient would be inconsistent with the N:P ratio rule in eutrophic systems that had greater amounts of nitrogen and phosphorus than are required for algal growths (Paerl et al., 2001). These results suggest that the nutrient stoichiometry that algae require for growth would vary with the extent of nutrient deficiency and availability.

In this study, we evaluated the response of phytoplankton growth in relation to different ambient concentrations of N and P, and N:P ratios, in a hypertrophic reservoir where the nitrogen background concentrations were high. For these experiments, we conducted nutrient enrichment bioassays (NEBs) and compared the growth rates of phytoplankton among several treatments of the NEBs, along with seasonal analyses of the ambient nutrients and phytoplankton compositions.

Materials and methods

The study system

Singu Reservoir is a small agricultural reservoir, with surface area of 0.1 km^2 and maximum depth of 7.0 m, which has two inflowing streams with channel length of 1.3 and 3.4 km, respectively. The total watershed area is 0.255 km^2, with the water quality mainly influenced by the point source of livestock wastes (KARICO, 2001). The watershed of the reservoir is located in Monsoon climate region, and precipitation is concentrated during June–September. The use for irrigation affects the water level of the reservoir (usually between May and June). Thus, both irrigation and Monsoon rainfall affect both water quality and water level of the reservoir.

Limnological variables

Various limnological variables including temperature, dissolved oxygen (DO) concentration, pH, and electric conductivity (EC) were measured by Hydrolab data logger (RE-232/SDI-12) in situ. Temperature and DO were measured vertically at 20–50 cm intervals. Transparency was measured with a white disk with 20 cm diameter. Water level was measured as the maximum depth at each time. Rainfall data were obtained from the nearest meteorological observation center (Boryung Meteorological Observation Center, Chungnam Province, S. Korea).

Chemical analyses

The surface water was collected biweekly from the reservoir between November 2002 and December 2003. Stream water samples were also collected from an inflowing location prior to entering the reservoir. The discharge volume from the reservoir was estimated through the linear multiplication of the water velocity and the channel area.

Five liter water samples were collected for analyses of nutrients and phytoplankton taxa, using polyethylene bottles, which were kept in a dark ice-box during transportation. On arrival at the laboratory (within 5 h), the water samples were filtered through GF/F filters, and the filtrate used to measure the concentration of chlorophyll *a* (Chl *a*). Phosphorus fractions, such as dissolved inorganic phosphorus (DIP), total dissolved phosphorus (TDP) and total phosphorus (TP), were analyzed in triplicate using the ascorbic acid method from the standard method of the APHA (1995). The NH_4^+–N concentrations were measured using an indolphenol reagent, and those of nitrate–nitrogen (NO_3^-–N) and nitrite–nitrogen (NO_2^-–N) using the cadmium reduction method (APHA, 1995). The total nitrogen (TN) was measured using the cadmium reduction method, following persulfate digestion. Dissolve inorganic nitrogen (DIN) was obtained from the sum of NO_3^-–N, NO_2^-–N and NH_4^+–N. The Chl *a* concentration was analyzed using the methanol extraction method of Maker et al. (1980).

Phytoplankton analysis

Water samples for the identification of phytoplankton species and the measurement of cell densities were collected from a 0.5–1 m depth from the reservoir, stored in Whirl-Pak bags and fixed with Lugol's solution (2%, final conc.). The cell density was measured using the Sedgwick-Rafter counting chamber under microscopic magnification of 200–400×. The phytoplankton composition was identified as containing three major and two minor taxa; Bacillariophyceae, Cyanophyceae and Chlorophyceae, and Dinophyceae and Cryptophyceae, respectively.

Nutrient enrichment bioassays (NEBs)

In addition to the regular sampling, nutrient enrichment bioassays (NEBs) were conducted for the evaluations of the limiting nutrients that regulate phytoplankton growth. The experiments for the evaluation of NEBs were conducted in a shaking incubator at a temperature of 20–25°C and light intensity of 100 ± 5 μE m^{-2} s^{-1}. For the experiments, the water was filtered using 100 μm mesh to minimize the effects associated with zooplankton grazing. Aliquots of filtrate (50 ml) were placed in 250 ml flasks, followed by the addition of KNO_3 and KH_2PO_4 as nutrient sources, to final concentrations of 1 mg l^{-1} for both nutrients. The final volume was made up to 100 ml with the addition of GF/F filtrate. A set of algal bioassay was conducted in triplicate sets for the control (no addition), +N (addition of nitrogen) and N + P (addition of nitrogen and phosphorus), and the responses measured as the difference between the final and initial Chl *a* at 0, 1, 3, 5 and 7 days after incubation. The growth rate (μ: day^{-1}) under each set of treatment conditions was calculated using the equation of the APHA (1995), as follows:

$$\mu\ (\mathrm{day}^{-1}) = \ln\ (X_2/X_1)/(T_2 - T_1)$$

where, X_1 = the concentration of Chl *a* at the initial incubation stage (T_1 h);
X_2 = the concentration T_2 h

Other triplicate experiments to measure the phytoplankton growth rate were conducted with five different P concentrations (+0.05, 0.1, 0.5, 1.0 and 1.5 mg l^{-1}, as KH_2PO_4 spiking), with nitrogen fixed at 1 mg l^{-1}. The medium into which different P concentrations were added was the same GF/F filtered lake water as above. The nitrogen concentrations of the medium used in the NEBs were similar between the treatments categorized with different N:P ratios. The spiking concentration of phosphorus, however, varied within the range 0.05–1.5 mg l^{-1}.

The DIN:TDP mass ratio was calculated using the concentration measurements from the subsamples taken from each treatment prior to the starting of the NEBs. The maximum growth rate (μ_{max}) and half saturation constant (K_u) were calculated as values to the phosphorus concentrations, with the corresponding growth rate (μ) obtained from the model of Monod (1950) employing the Sigma plot program (Version 7.0, SPSS Inc.).

Analysis of cyanobacterial growth under different N:P ratios

Using the same N:P ratio, the cyanobacterial growth rates were determined for different N and P concentrations. The incubation conditions were maintained at 20°C, with a light intensity of 53 ± 2 μE m^{-2} s^{-1} (L:D cycle = 12:12). The light intensities used were values observed during the study period at the water depths from which the samples were taken. The cyanobacteria were collected by vertically pulling up a 30 μm pore size plankton net, with the collected sample then filtered through a 100 μm net to remove the large zooplankton. In order to isolate the cyanobacteria from composite phytoplankton, samples were placed in a large-mouth container for 24 h, and the cyanobacteria that rose to the surface due to their buoyancy were transferred by pipette to a GF/F pre-filtered lake water.

Cyanobacterial cells were starved in MA2 medium with phosphorus and nitrogen excluded. After 2-week incubation, a 2 ml aliquot of the cyanobacteria sample was placed in an acid-washed 250 ml flask, along with MA2 medium that had been modified with different phosphorus and nitrogen concentrations. The N:P mass ratio was controlled by changing the amount of phosphorus with respect to fixed portions of nitrogen (0.07, 0.7 and 3.5 mg l^{-1}). Triplicate treatment sets were incubated for 5 days, and the cell growth determined as the difference between initial Chl *a*, prior to the incubation, and that 5 days later.

Statistical analysis

Statistical analyses were conducted between the variables using Pearson's correlation (SPSS 10.0). The differences in the growth response between the various NEB treatments were analyzed by an ANOVA, and the level of significance identified at $P < 0.05$.

Results

Variation of limnological variables

Rainfall during the study period was concentrated between June and September (41.5 mm in June, 391 mm in July, 79.5 mm in August, and 393 mm in September) (Fig. 1). The use for irrigation between the end of May and mid June decreased the reservoir water level from 7.1 to 5 m. The water level returned to 7 m due to seasonal monsoon rainfall during June–July. Transparency of the reservoir varied seasonally with the peak during May. The time of low transparency was

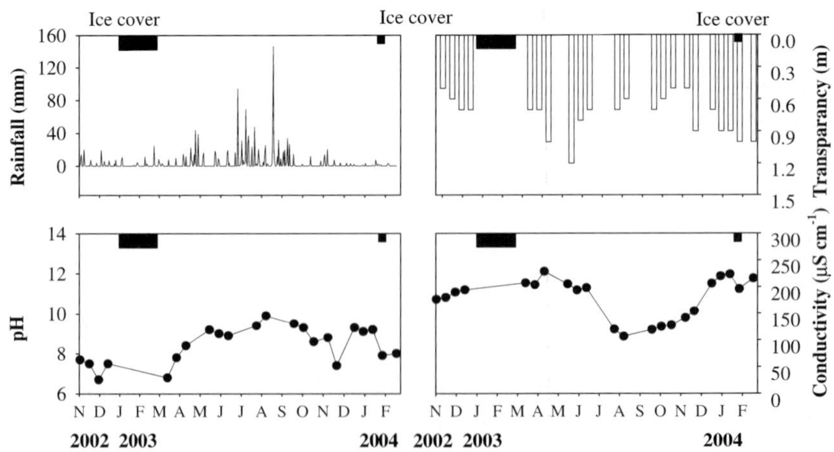

Fig. 1 Seasonal changes of precipitation, Secchi transparency, pH and electric conductivity in Shingu Reservoir during November 2002–February 2004

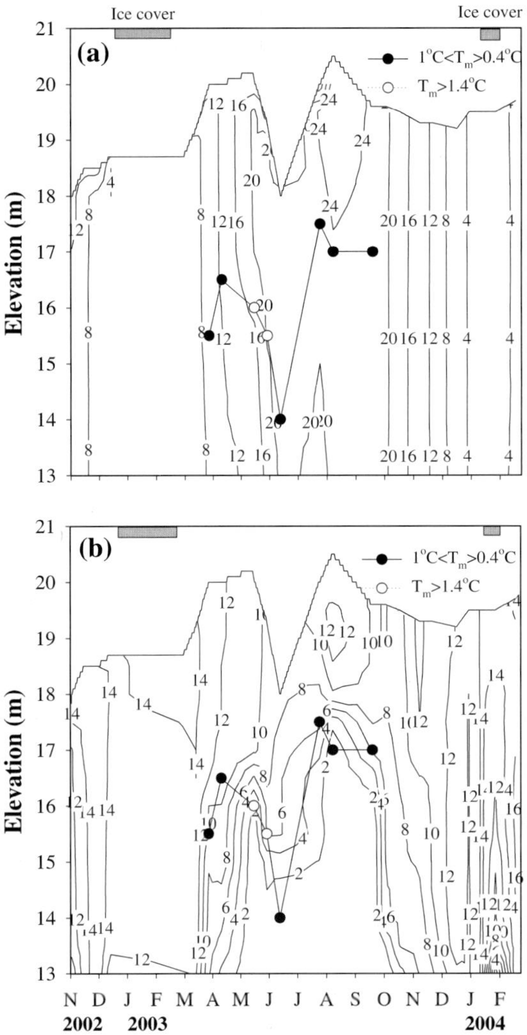

Fig. 2 Temporal and spatial changes of temperature (**a**) and dissolved oxygen concentration (**b**) in Shingu Reservoir during November 2002–February 2004. The line with *black circles* indicates the vertical temperature difference between 0.4 to 1.0°C, and the *open-circled line* indicates the vertical temperature difference greater than 1.0°C

coincided with the period of high pH, but EC variation was almost mirrored with pH. The reservoir had a weak thermal stratification near the hypolimnion, which persisted annually from mid-June to October (Fig. 2). Hypoxia <2 mg O_2 l^{-1} in the hypolimnion was frequently observed from the end of May, and was most intense during mid-August. Hypolimnetic hypoxia was maintained until late October, the beginning of fall.

Variation of nutrients and phytoplankton

During the study period, the TN concentrations in the reservoir were >1.2 mg l^{-1} (1.6–3.2 mg l^{-1}) (Fig. 3). The TN was mainly composed of NO_3^-–N, rather than NH_4-N, with NO_2-N + NO_3-N and NH_4-N ranging from 0.4 to1.9 and 0.01 to 0.92 mg N l^{-1}, respectively (Fig. 3). Dissolved inorganic N peaked in early spring (March–April), then abruptly decreased during May–August. Thus, the variation in the dissolved inorganic N was greater than that of the TN. DIP concentration ranged from 0.6 to 8.8 μg P l^{-1} and varied not much during the study period. DTP ranged from 10.1 to 25.6 μg P l^{-1}, with the peak concentration during August and minimum during December. The average TP was 76.4 μg P l^{-1} and showed large seasonal fluctuations (48.0–126.6 μg P l^{-1}). The TP increased from June, reaching a maximum in October. This reservoir was adjudged to be hypertrophic, based on the conventional criteria of nitrogen and phosphorus (Forsberg & Ryding, 1980). The TN:TP mass ratios in this hypertrophic reservoir were <60, whereas those of DIN:TDP increased to >150 during March–June.

The concentration of Chl *a* varied from 28.8 to 109.7 μg l^{-1} during the study period, and showed a similar seasonal pattern to that of the algal cell density (Fig. 4). The Chl *a* was at a minimum during early spring of 2003, but began to increase from May when cyanobacteria dominated the phytoplankton community. Thus, Chl *a* maxima >90 μg l^{-1} were observed between July and November, which coincided with the peaks in the cell density (Fig. 4). The Chl *a*:TP ratios, the Chl *a* yield for a given phosphorus concentration, were >1.0, indicative of a massive algal bloom. Pearson's correlation analyses showed the Chl *a* concentrations to be positively correlated ($r = 0.66$, $P = 0.002$, $n = 19$) with the TP, but negatively ($r = -0.48$, $P = 0.042$, $n = 18$) with the TN. This result coincided with the inverse relations between Chl *a* and the TN:TP ratio ($r = -0.70$, $P < 0.001$, $n = 18$).

The cell densities (cell number ml^{-1}) in the reservoir showed high seasonal fluctuations. The community structure was characterized by the dominance of cyanobacteria during most of

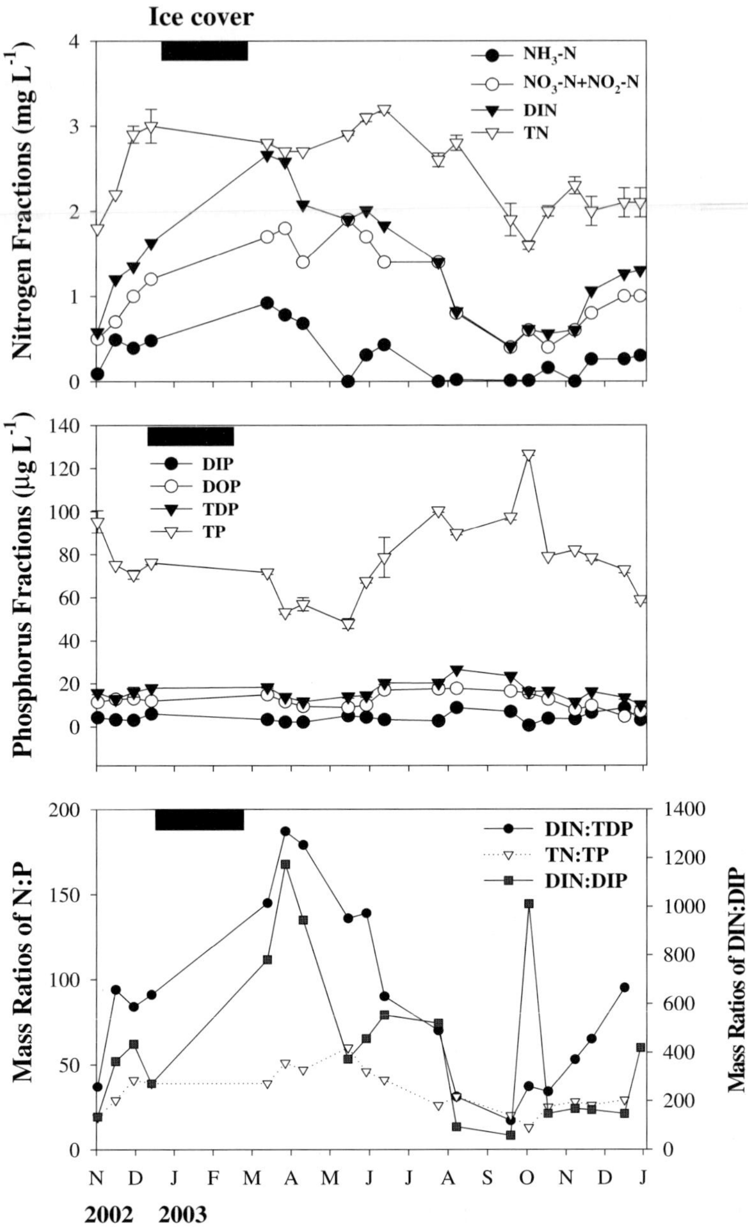

Fig. 3 Seasonal changes in the nitrogen and phosphorus fractions and N:P ratios in Shingu Reservoir during November 2002–December 2003

the year, with the exception of a short period during spring, which was dominated by a few other phytoplankton, such as a diatom (*Melosira varians*) and a green alga (*Dictyosphaerium puchellum*), and cryptomonads (*Rohdomonas* spp.) (Fig. 5). The dominant taxa of cyanobacteria were *Oscillatoria* spp., *Microcystis* spp. and *Aphanizomenon* sp., which dominated from May until the winter ice season. Seasonal succession of the phytoplankton community showed that *Oscillatoria* spp. dominated at the beginning of the study period, which was then followed by *Aphanizomenon* sp. in late spring when the water level of the reservoir abruptly decreased. During the study, the major species causing the algal bloom was identified as *Microcystis* spp. (Fig. 5), and the high Chl *a* >60 $\mu g\ l^{-1}$, continued during July–December 2003 (Fig. 4).

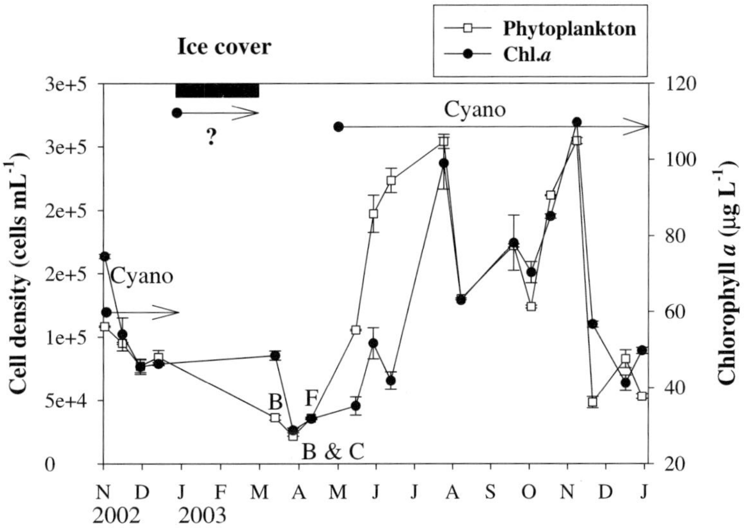

Fig. 4 Seasonal changes in the phytoplankton cell density and chlorophyll-*a* in Shingu Reservoir during November 2002–January 2004. Cyano, B, C and F denote cyanophytes, bacillariophytes, chlorophytes and flagellated algae, respectively

Phytoplankton growth under limiting nutrients and various N:P ratios

Overall, the results of the NEBs experiments showed phosphorus as the key nutrient regulating the phytoplankton growth in the system (Fig. 6, lower panel), but nitrogen was also considered an important nutrient during a short period (Fig. 6, upper panel). Phosphorus enrichment caused a significantly ($P < 0.001$, $n = 17$, ANOVA) greater growth rate than that seen in the control and nitrogen treatments (Fig. 6), indicating P-limitation of the algal growth in the system. No difference was found in the NEBs between the P and N + P additions ($P = 0.229$, $n = 17$, ANOVA). Six of the 17 NEBs experiments (20 July, 8 August, 21 September, 2 October, 16 October, and 1 November) showed significantly ($P < 0.05$, $n = 6$, ANOVA) higher growth rates with the N + P treatments than the P, N treatments and control. This outcome suggests that simultaneous limitation of nitrogen and phosphorus occurred between July and early November 2003. In the mean time, only 3 cases (November 2002, March and April 2003) of the NEBs had greater growth rates in the +P than the N + P treatments (Fig. 6).

The phytoplankton growth was largely determined by the magnitude of P, although the ambient N and P concentrations were different for the same N:P ratio (Fig. 7). The growth rates over a wide range of DIN:TDP ratios decreased with increasing ratio, but the most noticeable decrease was observed at ratios above 30 (Fig. 7). The growth rate consistently increased until the P concentration reached 50 $\mu g\ l^{-1}$ (Fig. 8). The results plotted using the equation of Monod (1950) indicated a maximum growth rate of 0.54 day^{-1} and with a half saturation concentration of 33.1 $\mu g\ l^{-1}$ for phosphorus. The maximum growth rate occurred in those treatments enriched with phosphorus alone or with both phosphorus and nitrogen. The treatments enriched with only nitrogen (+N) had lower growth rate, with the exception of one case during December 2002.

Cyanobacterial growth variation under different N:P ratios

The magnitude of the cyanobacterial growth responses showed differences with respect to the N:P ratios and N concentrations for the same N:P level. The N:P ratios under 20 displayed large differences in the growth rate with different nitrogen concentration (Fig. 9a), with the growth level tending to increase with increasing nitrogen concentration (Fig. 9b). The difference in the growth rates between 0.7 and 3.5 mg N l^{-1} were smaller ($P = 0.86$, ANOVA) than that of the lowest concentration, 0.07 mg N l^{-1} ($P = 0.001$, ANOVA).

At a constant N:P ratio, but with variable concentrations of N, the cyanobacteria showed

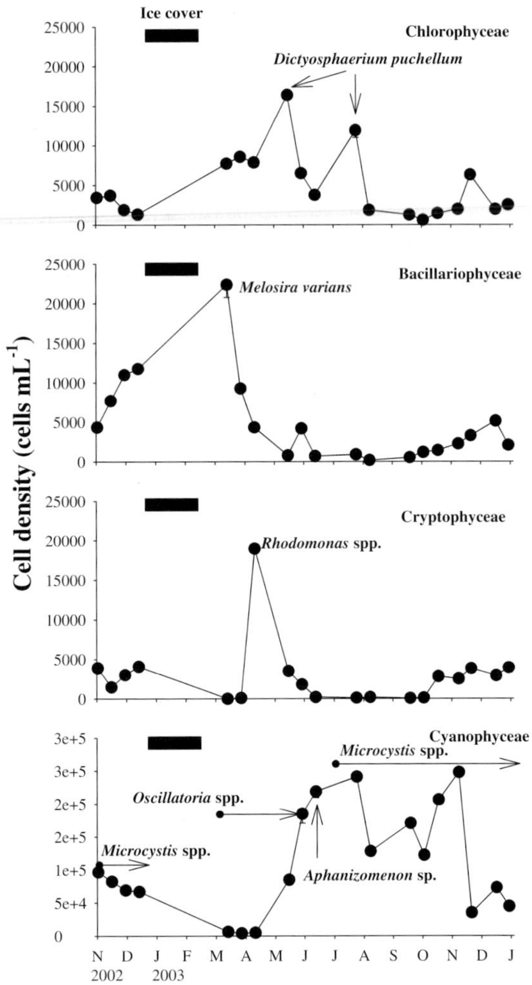

Fig. 5 The seasonal succession of dominant phytoplankton taxa in Shingu Reservoir during November 2002–December 2003

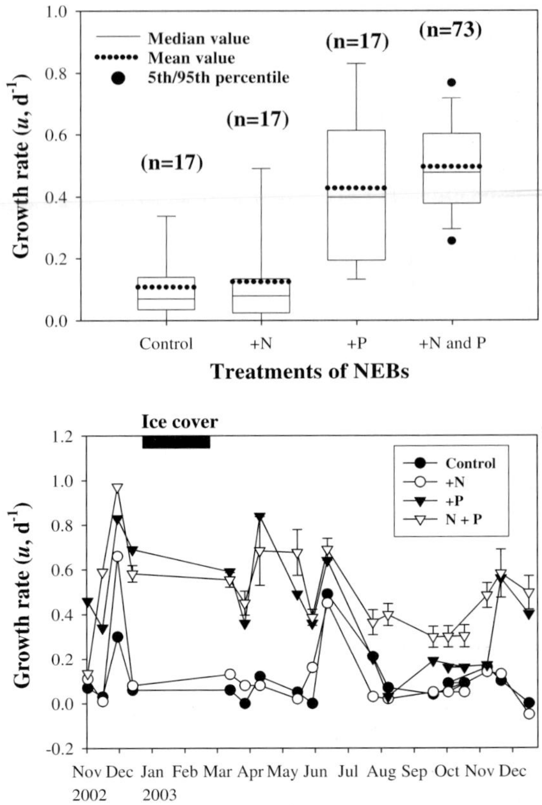

Fig. 6 The growth rates of phytoplankton toward various treatments (control, +N, +P, and N + P) in the NEBs, during November 2002–December 2003 (upper panel), and the seasonal growth response to nutrient additions in the NEBs (lower panel)

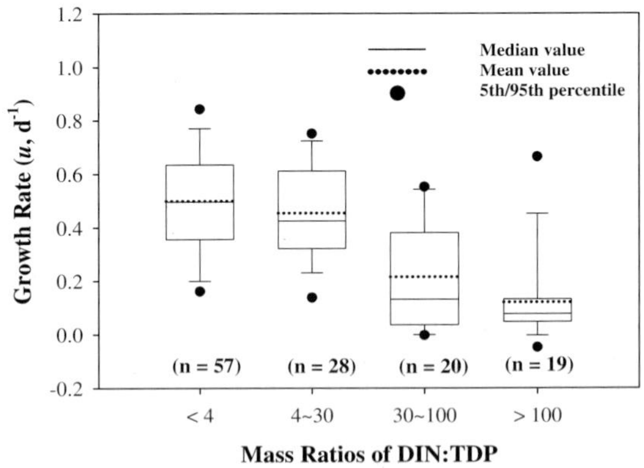

Fig. 7 The growth rates of phytoplankton at various DIN:TDP ratios

different growth patterns in the NEBs (Fig. 9c). The growth response reached a maximum at an N concentration of 3.5 mg l^{-1} when the N:P ratio in the NEBs was 1.0. The growth tended to decrease as the N:P ratio increased ($P < 0.01$, ANOVA) (Fig. 9c). A nitrogen concentration at 0.7 mg N l^{-1} caused the highest algal growth when the N:P ratio was 0.7 ($P < 0.05$, ANOVA). The growth rates were similar within the N:P ratio range 1–10 ($P > 0.5$, ANOVA), while the level of growth tended to decrease for N:P ratios above 10 ($P = 0.009$, ANOVA) (Fig. 9a). Conversely, the lowest nitrogen concentration, 0.07 mg N l^{-1}, showed a much lower level of growth, especially when the N:P ratio was smaller than 1. Ratios >7 tended to give greater phytoplankton growth ($P < 0.004$, ANOVA) (Fig. 9c), compared to ratios <1.

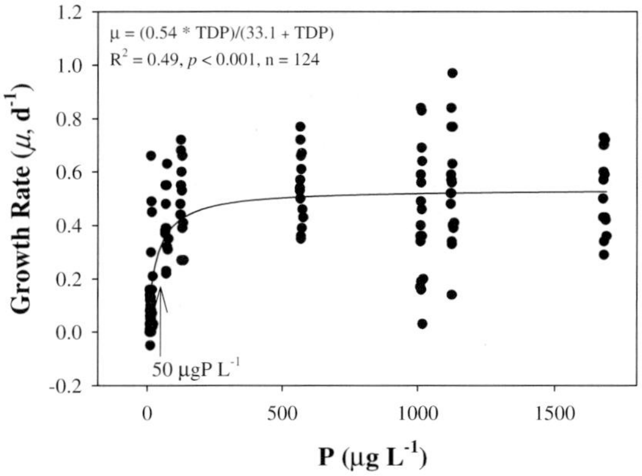

Fig. 8 Phytoplankton growth rate at various P (TDP = ambient + addition) concentrations using natural phytoplankton assemblages during November 2002–December 2003

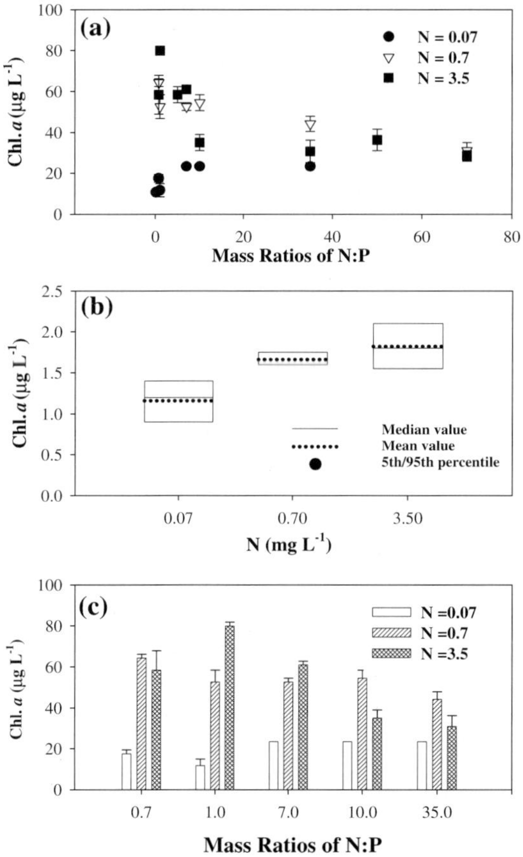

Fig. 9 Response of chlorophyll *a* to the N:P mass ratios (upper panel), and cyanobacterial responses to nitrogen (mid panel) and the N:P ratios (range: 0.7–35.0) at various nitrogen concentrations (lower panel)

Discussion

As important factors for determining the growth of phytoplankton, the ambient nutrient concentrations and nutrient ratios, known as stoichiometry (e.g., Sterner & Elser, 2002), are being widely used to estimate the physico-ecological consequences, such as the algal growth and succession of specific phytoplankton compositions (Goldman et al., 1979; Tilman, 1982; Sommer, 1989). An explanation for the mass conservation in aquatic organisms is that the external nutrient supply determines the cellular element composition of algae, which was supported by the earlier seminal work on stoichiometry (Redfield et al., 1963; Rhee, 1973) in an oligotrophic ocean and culture medium. The rules determining the constraints on stoichiometry; however, have been understood to vary depending on the organism, feeding mode and metabolic process (Geider & La Roche, 2002; Sterner & Elser, 2002; Ho et al., 2003).

A degree of nutrient deficiency and differential nutrients loading to a lake are also constraints on the stoichiometric variation and; thus, cause not only an imbalanced algal growth, but also lead the development of certain algal populations that have adapted to that particular situation (Smith et al., 1987). This implication calls attention to the need for eutrophic systems to be seen as having more variable and complicated limiting nutrients and critical nutrient ratio dynamics (Terry et al., 1985; Leonardos & Geider, 2004). Baring this in mind, our main concern in this study was to elucidate how algal growth (usually cyanobacteria) in a hypertrophic reservoir responded to both variations in ambient nutrient regimes and the N:P ratio in the NEBs in relation to the ambient N:P ratio.

Our major finding was that the phytoplankton (mostly cyanobacteria) in our hypertrophic study reservoir were P-limited and the degree of their growth in P addition increased with higher N concentrations added, indicating that at high nitrogen level, low N:P ratios were favorable for cyanobacterial development. These results suggest that phytoplankton in an N-rich environment appeared to demand a large amount of P to enable the use of a large supply of both N and P simultaneously to meet the great biomass building. These results also suggest that higher nitrogen concentration is likely to induce stronger P-limitation on the phytoplankton growth, and are supported by the inverse relationship between the TN:TP ratio and Chl *a* concentration ($r = -0.7$, $P < 0.001$: Figs. 3, 4). An evident increase in the growth rate was observed with TN:TP ratios <30, which was sustained with increases in TDP up to 50 $\mu g\ l^{-1}$, but reached a plateau thereafter (Figs. 7, 8).

With respect to the seasonal succession of phytoplankton, the variation in the N:P ratio appeared to be related with the cyanobacterial predominance in the reservoir. Due to the characteristic blooms in lentic systems with human-induced eutrophication, there have been numerous studies regarding the development of cyanobacteria and the interspecific changes of phytoplankton within eutrophic systems (Smith, 1983; Smith et al., 1987; Xie et al., 2003). The TN:TP ratios during the cyanobacterial blooms were 27 ± 6, almost coinciding with the criteria of Smith (1983), who suggested a high possibility of cyanobacterial dominance at TN:TP ratios under 29. The phytoplankton growth rate to specific ambient nutrient concentrations and N:P ratios, via the nutrient enrichment experiments, suggests that the level of the nutrient limiting ratios was frequently found near 30 (Fig. 7), even if the criteria are still controversial. P limitation was observed with ratios exceeding this level, but no N-limitation was detected at ratios below 30. The possibility of N-limitation was only detectable under low N concentration conditions, even in the experiment subjected to the cyanobacteria (Fig. 9).

The causes of these results were suspected to be due to the special characteristics of our reservoir, which maintained much higher nitrogen than phosphorus levels, as well as DIN:TN ratios (53%) much higher than those for TDP:TP (23%). The N:P ratio standard for evaluating the possibility of potential P and N limitation would vary in cases where the absolute concentrations were different, even though the N:P ratio would be maintained at the same level. Also, the eutrophic system where the concentration of N is much higher than that of P might create the possibility of N limitation being relatively lower than that of phosphorus, even with a decreasing N:P ratio. This result suggests that any standard for evaluating the limiting nutrient should consider the factors, not only the characteristics of phytoplankton species (Tilman 1977, 1978; Rhee, 1978), but also the absolute concentration of the nutrients existing within the aquatic system. A relative difference in the concentrations of the nutrients is necessary to endorse the growth of phytoplankton, rather than simply by their determination using the relative ratios of the nutrients in various aquatic systems with various nutrient stati (Leonardos & Geider, 2004).

Even though a number of studies have suggested the possibility of nitrogen-fixing cyanobacteria as the dominant species during periods of decreased N:P ratio (Horne, 1979; Tilman, 1982; Howarth et al., 1988; Paerl et al., 2001), our result indicated there was a dominant species shift, from *Aphanozomenon* to *Microsystis*, during periods when the N concentration decreased, and after completing the stratification, it has been suggested that the nitrogen concentration may not be the cause of the cyanobacterial species succession. The level of inorganic N tends to decrease at a higher rate than that of inorganic P during the algal growth period in both oligotrophic and eutrophic lakes and; thus, the growth limiting factor could change from P to N during the stratification period (Hendrey & Welch, 1974). Under this condition, dominant cyanobacteria species could succeed to taxa armed with the ability to fix nitrogen (Horne, 1979; Tilman, 1982; Howarth et al., 1988; Paerl et al., 2001). However, the nitrogen fixing *Apanizomenon* which was observed before establishing the stratification had shifted to non-N fixing *Microcystis* during the period of N:P ratio decrease, which was caused by the drop in the N concentration and the rise in the P concentration after stratification had been established. Therefore, phosphorus seemed to be one of the most substantial factors influencing the dominance of *Microsystis* (Xie et al., 2003). Our result is also supported by Fujimoto & Sudo (1997), who showed that *Microcystis aeruginosa* was a superior competitor under low N:P supply ratio conditions than the filamentous cyanobacterium, *Phormidium tenue*.

A low phytoplankton growth rate was demonstrated, even in the experiments with P or N + P additions between July and October (Fig. 6), which suggests the physiological property of phytoplankton needs to be considered when predicting their growth response, rather than simply relying on the nutrient concentration and N:P ratio (Droop, 1968; Rhee & Gotham, 1981). Due to the luxury of P consumption, cyanobacteria could store enough P in the cell, so their growth would not necessarily rely on the ambient phosphate concentration (Goldman et al., 1987). The DIN:DTP ratio in our reservoir between July and October was in the range 17–37, with the exception of July (DIN:DTP = 70), and this range seemed to be comparable to the optimum DIN:DTP ratio (<30) (Fig. 7) for phytoplankton growth and; thus, both P and N could function as potential limiting nutrients. On the other hand, there could have been an excess amount of external phosphorus supplied to the phytoplankton with the inflowing discharge around this period (Table 1). Moreover, after the end of May, the phosphorus released from the sediment under hypo-oxic conditions (Kim, 2004) could have been supplied to the surface by wind disturbances. Therefore, these phosphorus additions could have resulted in the low growth rate. During August and September especially, when DIP concentration in the water column was at the maximum seasonal concentration, there had been a lot of standard precipitation prior to sampling. The precipitation from June to September accounted for 66% of the total annual precipitation (1,359 mm) and; thus, this concentrated precipitation may have been responsible for increasing the discharge and P loading from the watershed. The increase in the TP concentration with both decreasing N and Chl *a* concentrations during this period may also be reflected by the increase of inorganic particulate matter.

Considering the mass conservation between an algal cell and the external nutrient supply, there should be no problem in using any type of nutrient for evaluating nutrient limitation (e.g., Sterner & Elser, 2002); for example, cellular or external and inorganic or organic forms. However, the factors causing variations in the cellular stoichiometry, such as the physiological status and the property of nutrient storage in phytoplankton species, need to be carefully considered to understand the ecological stoichiometry. The TN:TP ratios have been most frequently used as a measure, or criterion, to evaluate cyanobacterial blooms and limiting nutrients (Forsberg & Ryding, 1980; Smith, 1983; Trimbee & Prepas, 1987; Fujimoto & Sudo, 1997; Sheffer et al., 1997; Xie et al., 2003). Also, the DIN:DIP (Rhee, 1978; Patricia et al., 2000), DIN:OP or DIN:TP (Perkins & Underwood, 2000) ratios have been employed as surrogates of the TN:TP ratio for the assessment of lake conditions. In this study, the TN:TP ratios were closely related to the cyanobacterial density and nutrient ratios, which agreed with the growth response, as based on the soluble fractions of the DIN:TDP ratios. Even though organic N could be used for phytoplankton growth in the long run, the DIN content of our reservoir was 45 times higher than that of the DIP (less than 8.0 $\mu g\ l^{-1}$), and the average DIP

Table 1 Ambient nutrients, chlorophyll-*a* (Chl *a*) and discharge in two inflow streams and the outflow of Shingu Reservoir between May and October, 2003

Date	Inflow 1			Inflow 2			Outflow		
	Discharge $m^3\ day^{-1}$	TN $mg\ l^{-1}$	TP $\mu g\ l^{-1}$	Discharge $m^3\ day^{-1}$	TN $mg\ l^{-1}$	TP $\mu g\ l^{-1}$	TN $mg\ l^{-1}$	TP $\mu g\ l^{-1}$	Chl *a* $\mu g\ l^{-1}$
30 May	–	2.1 ± 0.03	197.7 ± 0.8	–	3.5 ± 0.10	315.8 ± 11.2	–	–	–
13 Jun.	–	6.2 ± 0.17	143.2 ± 2.3	–	7.6 ± 0.08	125.3 ± 9.3	–	–	–
25 Jul.	3,606	5.8 ± 0.04	94.4 ± 0.8	31,548	2.9 ± 0.17	89.8 ± 0.8	3.2 ± 0.10	134.4 ± 3.0	132.0 ± 4.6
8 Aug.	384	1.6 ± 0.04	74.9 ± 0.7	6,150	2.4 ± 0.12	91.2 ± 0.7	2.9 ± 0.13	86.8 ± 0.7	55.6 ± 2.7
19 Sep.	1,051	1.7 ± 0.25	96.7 ± 0.7	7,833	2.5 ± 0.17	105.3 ± 0.7	–	–	–
3 Oct.	303	1.4 ± 0.01	94.6 ± 2.8	2,378	2.5 ± 0.06	68.2 ± 2.1	–	–	–
18 Oct.	370	1.8 ± 0.10	29.1 ± 0.7	2,229	2.5 ± 0.00	84.2 ± 0.7	–	–	–

–: No inflow or outflow water

proportion of the total P was 5.8%. Therefore, we applied the soluble fraction of the TDP, which included the organic form of phosphorus, for phytoplankton to use under the environment with low DIP concentration levels (Islam & Whitton, 1992; Yelloy & Whitton, 1996). The phytoplankton growth in our system had a positive relationship with the dissolved P in the NEBs, which increased as both the TN:TP ($r = 0.596$, $P = 0.012$, $n = 17$) and DIN:DTP ($r = 0.575$, $P = 0.016$, $n = 17$) ratios increased. Thus, our experimental approach of NEBs, based on the algal growth response to the nutrient ratio and ambient nutrients, may be used as a useful tool for the diagnosis of lake eutrophication and efficient managements.

To conclude, cyanobacteria dominated in our hypertrophic study reservoir were P-limited in most times, and thus their growth rate increased with P addition. Also the degree of the cyanobacterial growth in P addition was magnified with more N supplied, reflecting that at high nitrogen level, low N:P ratios were favorable for cyanobacterial dominance. Our results suggest that limiting nutrients and N:P ratio are critical factors to evaluate the cynaobacterial bloom in eutrophic systems. However, cyanobacterial bloom events would be complicated in the ambient condition, so that some other factors may also be important to affect it such as temperature, light, and grazing (Hutchinson, 1957; Reynolds, 1984; Carpenter & Kitchell, 1993). In the companion study (Kim, 2004), we found that temperature had significant effect on the maximum growth rate (U_{max}) of cyanobacteria between 10 and 25°C ($Q_{10} = 2$). Light was also detrimental to their growth rate, but the degree of the growth rate by any single factor of either temperature or light varied with nutrients. As such, we do not deny the potentials of other factors to the cyanobacterial bloom in our reservoir, but, as we found in this study, limiting nutrients and their stochiometry are important factors to understand cyanobacterial dominance in the eutrophic system.

Acknowledgement This paper was supported by Konkuk University in 2000.

References

APHA, 1995. Standard Methods for the Examination of Water and Wastewater, 19th edn. APHA-AWWA-WEF, Washington, DC, USA.

Carpenter, S. R. & J. R. Kitchell, 1993. Cascading trophic interactions and lake productivity. Bioscience 35: 634–639.

Droop, M. R., 1968. Vitamin B_{12} and marine ecology. 4. The kinetics of uptake, growth and inhibition of *Monochrysis lutheri*. Journal of Marine Biology Association, UK 48: 689–733.

Forsberg, C. & S. O. Ryding, 1980. Eutrophication parameters and trophic state indices in 30 Swedish waste-receiving lakes. Archive fur Hydrobiologie 89: 189–207.

Fujimoto, N. & R. Sudo, 1997. Nutrient-limited growth of *Microcystis aeruginosa* and *Phormidium tenue* and competition under various N:P supply ratios and temperatures. Limnology and Oceanography 42: 250–256.

Geider, R. J. & J. La Roche, 2002. Redfield revisited: Variability of C:N:P in marine microalgae and its biochemical basis. European Journal of Phycology 37: 1–17.

Goldman, J. C., J. J. McCarthy & D. G. Peavey, 1979. Growth rate influence on the chemical composition of phytoplankton in oceanic waters. Nature 279: 210–215.

Goldman, J., D. A. Caron & M. R. Dennet, 1987. Nutrient cycling in a microflagellate food chain, IV. Phytoplankton–microflagellate interactions. Marine Ecology Progress Series 38: 75–87.

Hendrey, G. R. & E. B. Welch, 1974. Phytoplankton productivity in Findley Lake. Hydrobiology 45: 45–63.

Ho, T. Y., A. Quigg, Z. V. Finkel, A. J. Milligan, K. Wyman, P. G. Falkowski & F. M. M. Morel, 2003. The elemental composition of some marine phytoplankton. Journal of Phycology 39: 1145–1159.

Horne, A. J., 1979. Management of lakes containing N-fixing blue-green algae. Archive fur Hydrobiologie 13: 133–144.

Howarth, R. W., R. Marino & J. J. Cole, 1988. Nitrogen fixation in freshwater, esturine, and marine ecosystems. 2. Biogeochemical controls. Limnology and Oceanography 33: 669–687.

Hutchinson, G. E., 1957. A Treatise on Limnology. Geography Physics and Chemistry, Vol. 1. Wiley, New York: 1015 pp.

Islam, M. R. & B. A. Whitton, 1992. Retention of P-nitrophenol and 4-methylumbelliferone by marine macroalgae and implications for measurement of alkaline phosphatase activity. Journal of Phycology 32: 819–825.

KARICO, 2001. Report of Water Monitoring in Agricultural Reservoirs, S. Korea. Korea Agriculture and Rural Infrastructure Cooperation (in Korean).

Kilham, P., 1990. Mechanisms controlling the chemical composition of lakes and rivers: Data from Africa. Limnology and Oceanography 35: 80–93.

Kim, H. S., 2004. Study on the growth dynamics and ecotechnological control of algae in reservoirs. Ph. D dissertation thesis. Konkuk University, Seoul, South Korea (in Korean).

Leonardos, N. & R. J. Geider, 2004. Responses of elemental and biochemical composition of *Chaetoceros muelleri* to growth under varying light and nitrate: Phosphate supply ratios and their influence on critical N:P. Limnology and Oceanography 49: 2105–2114.

Marker, A. F. H., E. A. Nusch, I. Rai & B. Riemann, 1980. The measurement of photosynthetic pigments in freshwaters and standardization of methods: Conclusions and recommendations. Archive fur Hydrobiologie 14: 91–106.

Monod, J., 1950. La technique de culture continue: theorie at applications. Annales de l'Institut Pasteur de Lille 79: 390–410.

Paerl, H. W., R. S. Fulton, P. H. Moisander & J. Dyble, 2001. Harmful freshwater algal blooms, with an emphasis on cyanobacteria. The Scientific World Journal 1: 76–113.

Patricia, A. S., W. H. Shaaw & P. A. Bukaveckss, 2000. Differences in nutrient limitation and grazer suppression of phytoplankton in seepage and drainage lakes of the Adirondack region, NY, USA. Freshwater Biology 43: 391–407.

Perkins, R. G. & G. J. C. Underwood, 2000. Gradients of chlorophyll *a* and water chemistry along an eutrophic reservoir with determination of the limiting nutrient by In Situ nutrient addition. Water Research 34: 713–724.

Redfield, A. C., B. H. Ketchum & F. A. Richards, 1963. The influence of organisms on the composition of sea water. In Hill, M. N. (ed.), The Sea. Interscience, New York, 26–27.

Reynolds, C. S., 1984. Phytoplankton periodicity: The interactions of form, function and environmental variability. Freshwater Biology 14: 111–142.

Rhee, G. Y., 1973. A continuous culture study of phosphate uptake, growth rate and polyphosphate in *Scenedesmus* sp. Journal of Phycology 9: 495–506.

Rhee, G. Y., 1978. Effects of N:P atomic ratios and nitrate limitation on algal growth, cell composition, and nitrate uptake. Limnology and Oceanography 23: 10–24.

Rhee, G. Y. & I. J. Gotham, 1981. The effect of environmental factors on phytoplankton growth: Light and the interactions of light with nitrate limitation. Limnology and Oceanography 26: 649–650.

Sheffer, M., S. Rinaldi, A. Grangnani, L. R. Mur & E. H. Nes, 1997. On the dominance of filamentous cyanobacteria in shallow, turbid lakes. Ecology 78: 272–282.

Smith, V. H., 1983. Low nitrogen to phosphorus ratios favors dominance by blue-green algae in lake phytoplankton. Science 221: 669–671.

Smith, V. H., E. Willen & B. Karlsson, 1987. Predicting the summer peak biomass of four species of blue-green algae (cyanphyta/cyanobacteria) in Swedish lakes. Water Research Bulletin 23: 397–402.

Sommer, U., 1989. The role of competition for resources in phytoplankton succession. In Sommer, U. (ed.), Plankton Ecology: Succession in Plankton Communities. Springer-Verlag, Berlin, 57–106.

Sterner, R. W. & J. J. Elser, 2002. Ecological Stoichiometry: The Biology of Elements from Molecules to the Biosphere. Princeton University Press. New Jersey, 439.

Terry, K. L., E. A. Laws & D. J. Burns, 1985. Growth rate variation in the N:P requirement ratio of phytoplankton. Journal of Phycology 21: 323–329.

Tilman, D. 1977. Resource competition between planktonic algae: An experimental and theoretical approach. Ecology 58: 338–348.

Tilman, D. 1978. Ecological competition between algae: Experimental confirmation of resource based competition theory. Science 192: 463–465.

Tilman, D. 1982. Resource competition and community structure. Princeton Monographs in Population Biology 17. Princeton University Press.

Trimbee, A. M. & E. E. Prepas, 1987. Evaluation of total phosphorus as a predictor of the relative biomass of blue-green algae with emphasis on Alberta lakes. Canadian Journal of Fisheries and Aquatic Science 44: 1337–1342.

Xie, L., P. Xie, S. Li, H. Tang & H. Liu, 2003. The low TN:TP ratio, a cause or a result of *Microcystis* blooms. Water Ressearch 37: 2073–2080.

Yelloly, J. M. & B. A. Whitton, 1996. Seasonal changes in ambient phosphate and phosphatase activities of the cyanobacterium *Rivularia atra* in intertidal pools at Tyne Sands, Scotland. Hydrobiologia 325: 201–212.

Hydrobiologia (2007) 581:269–285
DOI 10.1007/s10750-006-0507-3

EUTROPHICATION IN LAKES

Restoration of shallow lakes by nutrient control and biomanipulation—the successful strategy varies with lake size and climate

E. Jeppesen · M. Meerhoff · B. A. Jacobsen · R. S. Hansen · M. Søndergaard · J. P. Jensen · T. L. Lauridsen · N. Mazzeo · C. W. C. Branco

Abstract Major efforts have been made worldwide to improve the ecological quality of shallow lakes by reducing external nutrient loading. These have often resulted in lower in-lake total phosphorus (TP) and decreased chlorophyll *a* levels in surface water, reduced phytoplankton biomass and higher Secchi depth. Internal loading delays recovery, but in north temperate lakes a new equilibrium with respect to TP often is reached after <10–15 years. In comparison, the response time to reduced nitrogen (N) loading is typically <5 years. Also increased top-down control may be important. Fish biomass often declines, and the percentage of piscivores, the zooplankton:phytoplankton biomass ratio, the contribution of *Daphnia* to zooplankton biomass and the cladoceran size all tend to increase. This holds for both small and relatively large lakes, for example, the largest lake in Denmark (40 km^2), shallow Lake Arresø, has responded relatively rapidly to a ca. 76% loading reduction arising from nutrient reduction and top-down control. Some lakes, however, have proven resistant to loading reductions. To accelerate recovery several physico-chemical and biological restoration methods have been developed for north temperate lakes and used with varying degrees of success. Biological measures, such as selective removal of planktivorous fish, stocking of piscivorous fish and implantation or protection of submerged plants, often are cheap versus traditional physico-chemical methods and are therefore attractive. However, their long-term effectiveness is uncertain. It is argued that additional measures beyond loading reduction are less cost-efficient and often not needed in very large lakes.

J. P. Jensen was deceased.

Guest editors: B. Qui, Z. Liu & K. Havens
Eutrophication of shallow lakes with special reference to Lake Taihu, China

E. Jeppesen (✉) · M. Meerhoff · M. Søndergaard · J. P. Jensen · T. L. Lauridsen
Department of Freshwater Ecology, National Environmental Research Institute, Vejlsøvej 25, 8600 Silkeborg, Denmark
e-mail: ej@dmu.dk

E. Jeppesen · M. Meerhoff
Department of Plant Biology, University of Aarhus, Ole Worms Allé 135, 8000 Aarhus, Denmark

M. Meerhoff · N. Mazzeo
Departamento de Ecología, Facultad de Ciencias, Iguá 4225, Montevideo, Uruguay

B. A. Jacobsen · R. S. Hansen
Frederiksborg Amt, Technical Department, Amtsgården, Kongens Vænge 2, 3400 Hillerod, Denmark

C. W. C. Branco
Departamento de Ciências Naturais, Universidade Federal do Estado do Rio de Janeiro, Av. Pasteur 458, Urca, Rio de Janeiro, Brazil 22290-040

Although fewer data are available on tropical lakes these seem to respond to external loading reductions, an example being Lake Paranoá, Brazil (38 km^2). However, differences in biological interactions between cold temperate versus warm temperate-subtropical-tropical lakes make transfer of existing biological restoration methods to warm lakes difficult. Warm lakes often have prolonged growth seasons with a higher risk of long-lasting algal blooms and dense floating plant communities, smaller fish, higher aggregation of fish in vegetation (leading to loss of zooplankton refuge), more annual fish cohorts, more omnivorous feeding by fish and less specialist piscivory. The trophic structures of warm lakes vary markedly, depending on precipitation, continental or coastal regions locations, lake age and temperature. Unfortunately, little is known about trophic dynamics and the role of fish in warm lakes. Since many warm lakes suffer from eutrophication, new insights are needed into trophic interactions and potential lake restoration methods, especially since eutrophication is expected to increase in the future owing to economic development and global warming.

Keywords Lake restoration · Large lakes · Recovery · Nutrient loading reduction · Climate

Introduction

Efforts in the last 20–30 years have attempted to combat eutrophication by reducing external phosphorus (P) loading in Western Europe and North America. Sewage and industrial loading has declined, whereas nutrient input from diffuse sources often remains high, particularly in countries with intensive agriculture (Van der Molen & Portielje, 1999; Kronvang et al., 2005).

Many lakes have responded positively to nutrient loading reductions (Sas, 1989; Marsden, 1989; Jeppesen et al., 2005c). An analysis of European and North American lakes based on 35 case studies, including long time-series data, revealed that, although internal P loading does delay recovery, new equilibria with respect to (total phosphorus) TP were reached in most lakes after <10–15 years, marginally dependent on lake retention time and depth (Jeppesen et al., 2005c). In comparison, response times to reduced N loading are typically <5 years. These findings also apply to shallow Danish lakes after a reduction in mainly external P loading (Søndergaard et al., 2002a, 2005; Jeppesen et al., 2002, 2005a, b). In these lakes, TP has declined during all seasons, particularly in winter, followed by spring and autumn, while weaker effects in late summer can be ascribed to internal P loading. Changes in chlorophyll *a* follow TP, indicating that nutrient constraints are important in determining seasonal behaviour. Diatoms declined precipitously in winter and spring. Non-heterocystous cyanobacteria declined throughout the summer, primarily in early summer, while heterocystous cyanobacteria increased in late summer and autumn. Green algae declined throughout the year but least so in mid-summer. In Barton Broad, a shallow lake in the UK, Phillips et al. (2005) observed a similar pattern with strong effects on diatoms in spring and a decline first in non-heterocystous cyanobacteria and secondly, albeit slower, in heterocystous genera after 15–20 years. Thus, there is clear evidence for enhanced resource control of phytoplankton after an external loading reduction.

However, enhanced top-down control by zooplankton may also play a role. Fish biomass often declined in the Danish lakes and the percentage of piscivores often increased. Similar observations were made in the study of 35 North American and European lakes in recovery (Jeppesen et al., 2005c); in 82% of the lakes for which data were available, fish biomass declined with TP, while the piscivore proportion increased in 80% of the lakes. This is surprising given the high longevity of the dominant species. While there were no changes in zooplankton biomass, the contribution of *Daphnia* to the biomass and mean individual biomass of cladocerans increased most notably during summer. Moreover, the zooplankton:phytoplankton biomass ratio, and probably phytoplankton grazing, rose. This is suggested also by the cladoceran size increase. Higher abundance of large-bodied zooplankton, particularly in summer, may reflect that increased piscivory leads to stronger predation control of young-of-the year (YOY), primarily planktivorous, fish. This reduces predation on zooplankton during summer, when YOY fish are abundant and forage in the pelagic zone. Enhanced abundance of submerged macrophytes or filamentous algae in some lakes may also enhance survival of large-bodied zooplankton as they can serve as a daytime refuge from predation (Burks et al., 2002).

A north temperate example: recovery of Lake Arresø, Denmark

The positive response to external nutrient loading reduction is not restricted to small lakes. Here we elaborate in more detail on the response of Arresø to nutrient loading reduction. Arresø (coordinates) is the largest lake in Denmark (40 km^2). It is shallow (mean depth 3.1 m, max depth 5.9 m) and has a 215 km^2 catchment area. The retention time is long (1.3–10.3 years, mean 1989–2003: 3.5 years) compared with most Danish shallow lakes. The lake is near the coast and exposed to prevailing north-westerly winds and, thus, wave-induced resuspension (Kristensen et al., 1992). Based on intensive sampling (every 2–8 h) during different seasons in 1991, and modelling, resuspension was calculated to occur 50% of the time and to reduce Secchi depth to <0.5 m on average (Kristensen et al., 1992). The lake was loaded heavily with P and N from the catchment, particularly P from the upstream city of Hillerød. TP loading decreased from 0.68 g P m^{-2} $year^{-1}$ in 1989 to 0.16 g P m^{-2} $year^{-1}$ for 1996–2003, and TN decreased from 13 to 11 g N m^{-2} $year^{-1}$ (Jacobsen et al., 2004). The reduction mostly was due to improved sewage treatment, establishment of artificial lakes on the main inlet stream, and possibly to reduced fertilization in the catchment. Nutrient load reduction led to substantial improvements in the lake. TN, TP, chlorophyll *a* and the total biomass (and especially of green algae) of phytoplankton have declined markedly (Fig. 1). Diatom biomass has tended to increase with associated effects on silicate concentration, whereas the trend for cyanobacteria is ambiguous. Algal biomass changes cannot be ascribed entirely to enhanced nutrient control, since there have been marked zooplankton community changes and potential grazing, most likely due to fish community changes. Biomasses of the dominant plankti-benthivorous bream (*Abramis brama*) (Fig. 2) and benthic ruffe (*Gymnocephalus cernua*) are decreasing, whereas zander (*Sander lucioperca*), a potential predator, is increasing, implying enhanced top-down control on prey fish. Accordingly, zooplankton biomass and the share of large-sized cladocerans have increased substantially. This can be explained by reduced predation pressure on zooplankton, since the algal biomass was dominated by edible green algae when biomass was low and the zooplankton were dominated by small forms. With increased zooplankton biomass the grazing pressure on phytoplankton has intensified, and the phytoplankton face simultaneous, enhanced resource control due to lower P and N concentrations. Grazing control is strengthened by the shift to large-sized cladocerans *Daphnia hyalina* and *D. galeata* (Fig. 1), which have broader feeding capabilities than previously dominant *Bosmina* spp. and *Chydorus sphaericus*. Following the increase in zooplankton biomass and cladoceran size, and the zooplankton:phytoplankton ratio, water transparency has improved (Fig. 1), emphasizing the role of enhanced grazer control of phytoplankton.

Seasonal phytoplankton biomass reduction and Secchi depth increases are larger in spring, autumn and winter than in summer (Fig. 3). This also is true for the zooplankton:phytoplankton biomass ratio and cladoceran body mass, suggesting that changes in grazing pressure are important. The minor effect in summer coincides with fish larvae exploiting the pelagic in Danish lakes (Jeppesen et al., 2004). Moreover, during this period, nitrate is low (in all years) (Fig. 3), which likely may have dampened P loading reduction effects.

However, loading still remains too high to promote a clearwater state with macrophytes, which prevailed in the 1900's per paleolimnological analyses (Klein, 1989).

Accelerating the recovery of northern temperate lakes

Positive effects may not occur after nutrient input reduction. Several lakes have been resistant to loading reductions and shown slow recovery (Sas,

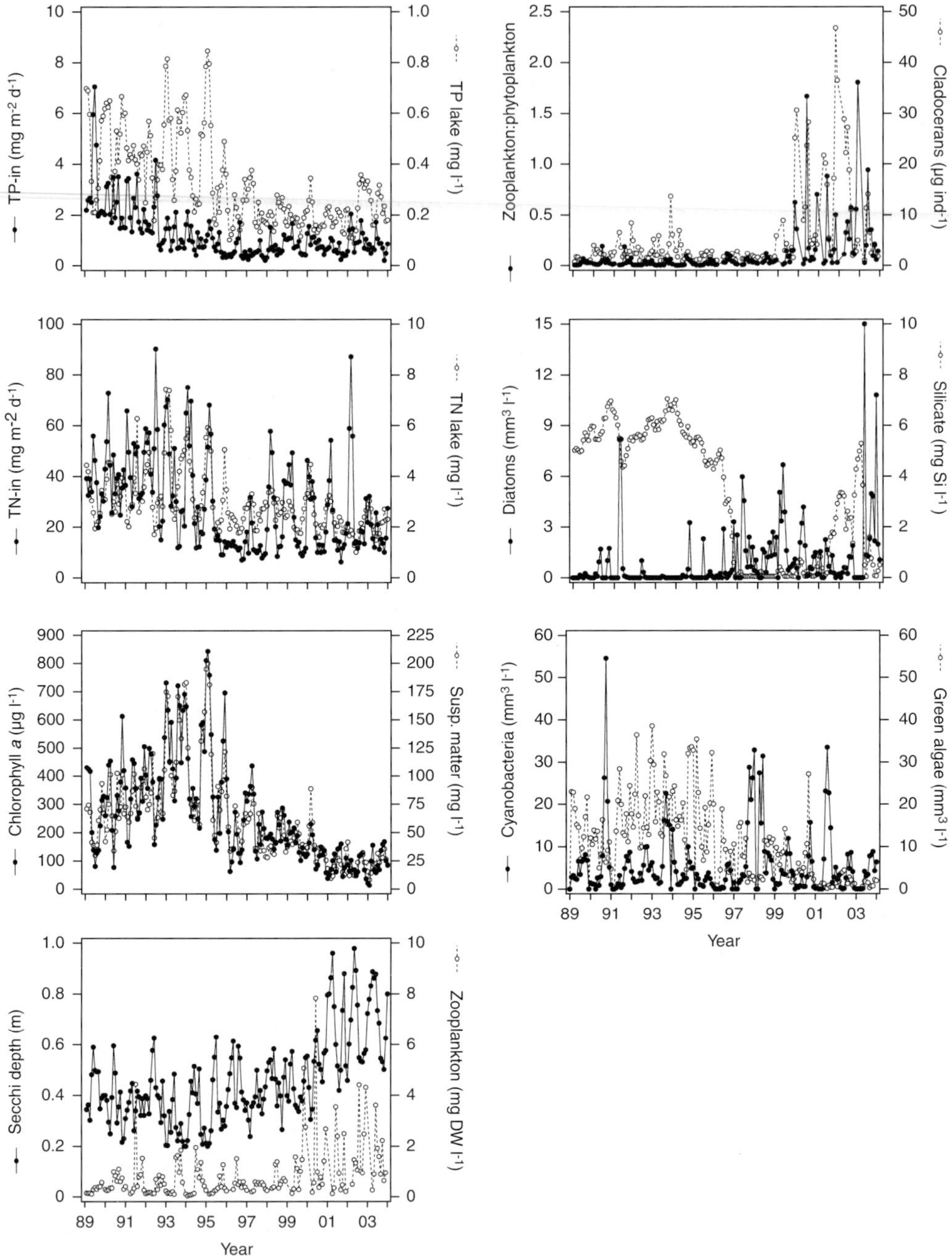

Fig. 1 Monthly mean data on various environmental variables in Arresø. TP_{in} and TN_{in} are the external loading of the lake. Zooplankton:phytoplankton is the biomass (mg DW) ratio

1989; Marsden, 1989). For some lakes, this reflects insufficient nutrient input reduction to promote a clearwater state. For example, significant and sustained changes in the biological community and water transparency of shallow, temperate, freshwater lakes may not appear unless TP is reduced below 0.05–0.1 mg P l^{-1} (Jeppesen et al., 2000) or, for deep lakes, 0.02–0.03 mg P l^{-1} (Sas, 1989). Even when P loading has been reduced sufficiently, resistance to improvement often is observed. This resistance may be "chemical", P concentrations remain high due to release from

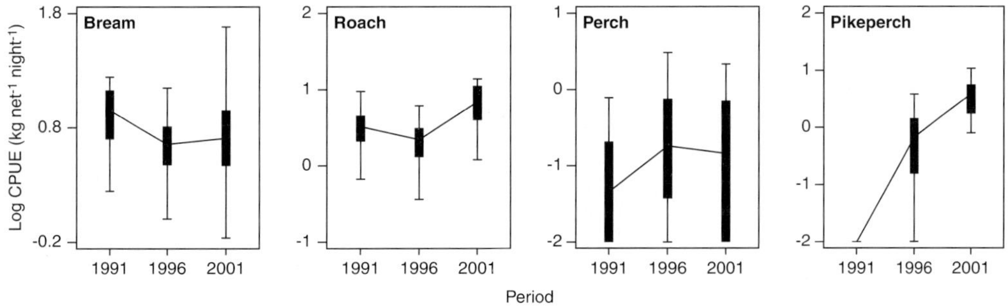

Fig. 2 Box plot (10, 25, 75, 90% percentiles) of log(base e) transformed catch per net per night in Arresø using 52 nets distributed in the lake according to Jeppesen et al. (2004)

the sediment pool accumulated when loading was high (Søndergaard et al., 2002a). Various methods have been used to reduce internal P loading (Cooke et al., 1993; Søndergaard et al., 2002a, b), including sediment removal and chemical treatment with aluminum or iron salts. In stratified lakes, oxygen or nitrate injections to the bottom layer or destabilization of the thermocline have been used (Cooke, 1993). For large, shallow lakes, it is debatable whether sediment removal and chemical treatment are feasible methods. Sediment removal is expensive, particularly for large lakes due to transport and disposal. Moreover, sediment resuspension and redistribution may hamper removal of the upper, nutrient-rich sediment layers. Resuspension also is a problem with chemical methods. P bound to aluminum or iron may return to the water on windy days and loosely bound P made available to phytoplankton. Finally, the sediment P pool often is small in large lakes because resuspension leads to washout of particulate TP (unless hydraulic retention time is long) and organic net sedimentation is low, the latter due to high mineralization. Thus, the usefulness of sediment removal or treatment is more limited in large than small lakes. The fast response of Arresø was attained without sediment removal.

The resistance also may be "biological". Planktivorous and benthivorous fish contribute to biological resistance in shallow eutrophic lakes (Moss, 1990; Scheffer et al., 1993). High fish predation prevents the appearance of large herbivorous zooplankton, which would otherwise clear the water, and diminishes benthic invertebrates, which stabilize and oxidize the sediment. Moreover, nutrient excretion to overlaying waters by benthic-feeding fish or fish bioturbation of surface sediment may play a role (Breukelaar et al., 1994; Persson, 1997). To overcome biological resistance, various fish manipulation methods have been developed (Benndorf, 1995; Drenner & Hambright, 1999; Søndergaard et al., 2001). One method is enhancement of top-down control of phytoplankton by selective removal of planktivorous fish; a method employed extensively in the temperate zone, but only just beginning in the subtropics (Scasso et al., 2001) and other warmer climate zones, such as Mediterranean arid lakes (Beklioglu & Tan, personal communication). Removal of 75–80% of the planktivorous and benthivorous fish stock over 1–2 years is recommended to avoid regrowth and stimulate growth of potentially piscivorous perch (Perrow et al., 1997; Hansson et al., 1998; Meijer et al., 1999). An alternative or supplementary method to fish removal is stocking of 0^+ pike to control newly hatched plankti-benthivorous roach and bream (Prejs et al., 1994; Berg et al., 1997; Skov & Berg, 1999) though the results are ambiguous (C. Skov et al., unpublished data). Others have stocked pikeperch, walleye (*Stizostedion vitreum*) and largemouth bass (*Micropterus salmoides*) (Benndorf, 1995; Lathrop et al., 2002). In contrast to chemical engineering methods, fish manipulation often is cheap (Jeppesen & Sammalkorpi, 2002) and therefore attractive, though its long-term stability is uncertain as a return to a more turbid state has occurred in many of the lakes. The reasons for this remain to be elucidated, but

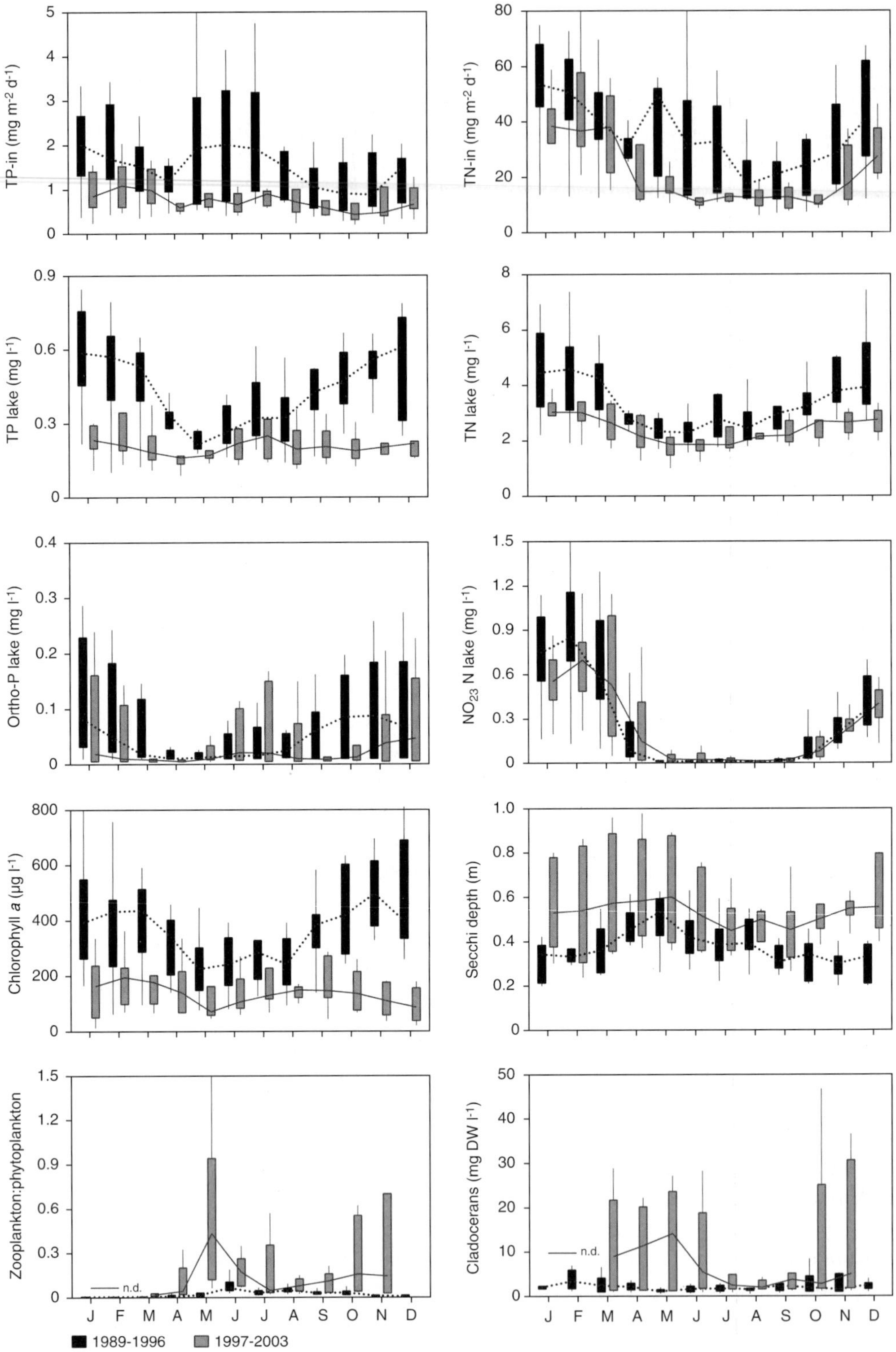

Fig. 3 Box plot (10, 25, 75, 90% percentiles) of selected environmental variables in Arresø during the first (1989–1996) and the last part of the study period (1997–2003). n.d. means no data in Jan–Feb for 1997–2003. See also legend of Fig. 1

factors such as reinforced internal loading of P, return of planktivorous dominance (particularly roach in Danish lakes) and large year-to-year variations in the coverage of the submerged macrophyte community may play a role. However, in Danish lakes suspended matter remains lower and Secchi depth higher likely reflecting an apparent permanent lower bream population (authors unpublished data).

It is unlikely that fish manipulation will be feasible for very large lakes (say >500 km^2). First, it is difficult to obtain a 'sledgehammer' effect on the target fish population in very large lakes. The fish cannot easily be removed fast enough to compensate for their annual reproductive rate. Changes in the fish stock may simply become difficult to discern unless the fishing effort is substantial. Second, recent studies indicate that the fish population responds quickly to a nutrient loading reduction (Jeppesen et al., 2005a, b, c). This may render fish stock intervention less attractive in very large lakes as the natural changes may occur at higher speed than those invoked by biomanipulation in such lakes. Third, fish stocking is expensive for large lakes. However, biomanipulation combined with nutrient loading reduction has yielded a fast response (higher transparency, fewer cyanobacteria and in some cases extensive growth of submerged macrophytes) in medium large lakes, examples being Lake Vesijärvi, Finland (109 km^2), Lake Wolderwijd, Holland (27 km^2), and Lake Christina, USA (16 km^2) (Hanson & Butler, 1990; Peltonen et al., 1999; Meijer et al., 1999).

Delay in recovery and re-establishment of submerged vegetation after nutrient loading reduction have often been noted (Sas, 1989; Cooke et al., 1993; Jeppesen & Sammalkorpi, 2002; Jeppesen et al., 2005c). The delay in plant appearance may be due to lack of seed banks, waterfowl grazing, limited grazing on periphyton or too high turbidity (Søndergaard et al., 1996; Mitchell & Perrow, 1997; Lauridsen et al., 2003). Construction of exclosures to protect macrophytes against waterfowl grazing to improve plant re-establishment has been used as an alternative or supplementary restoration tool to fish manipulation (Cooke, 1993; Jeppesen & Sammalkorpi, 2002). The exclosures enable macrophytes to grow in a grazer-free environment where they can spread seeds, turions or plant fragments and thereby augment colonization. Moreover, they serve as a daytime refuge for zooplankton. Plant refuges as a restoration tool is probably most useful in small lakes and in the littoral zone of other lakes where colonization is restricted to the near shore areas due to light limitation in the deeper main lake. Waterfowl aggregate in littoral emergent zones, which provide shelter and nest sites, and may impart higher grazing pressure on plants. Plants as a refuge have greater effect in nutrient-rich lakes where plant density often is highest and prey fish are confined to the pelagial (Jeppesen et al., 1997; Blindow et al., 2000; Burks et al., 2002). Transplantation or protection of plants (in sheltered beds) and seeds may be useful in large shallow lakes if pre-studies reveal that (re)-colonization is seed-limited (Weisner & Strand, 2002; Ke & Li, 2006). However, it is important to evaluate whether wind (wave) disturbance will prevent colonization. In large lakes effort should be concentrated on sheltered areas where plants can colonize gradually, as seen in large (27 km^2) Lake Veleuwe (Meijer et al., 1999).

Warm lakes

Few long-term lake recovery studies after reduced external loading exist for tropical and subtropical lakes: here we present an example from Brazil. Lake Paranoá (15°48′S, 47°47′W) is a shallow, urban, man-made lake (area 38 km^2, mean depth: 13 m, retention time: 0.9 years). It was created in 1959 for recreational purposes, hydroelectric power production, to increase the humidity of the region, as well as to receive wastewater and run-off from the city of Brasilia. The first plankton study recorded dominance of desmids and large-bodied cladocerans (*Daphnia* and *Ceriodaphnia*) (Oliveira & Krau, 1970). Later, phytoplankton biomass was dominated by *Cylindrospermospsis raciborskii*, which remained dominant in subsequent decades (Cronberg, 1976; Pinto-Coelho & Giani, 1985; Branco & Senna, 1991). *Microcystis aeruginosa* blooms occurred periodically during the end of the dry season

(Branco & Senna, 1994). Besides the permanent bloom of filamentous cyanobacteria (*C. raciborskii*), eutrophic conditions were also indicated by small-bodied zooplankton (mainly rotifers and small cladocerans, such as *Diaphanosoma birgei* and *Bosmina* spp.), low Secchi depth (<50 cm) and high levels of nutrients and chlorophyll *a* (Branco & Senna, 1996; Branco & Cavalcanti, 1999, Fig. 4). Aquatic macrophytes, such as free-floating *Eichhornia crassipes* and *Pistia stratiotes*, also spread in some areas. Eleven fish species colonized the lake from the catchment. In addition, the lake was stocked with exotic fish species, such as bluegill sunfish (*Lepomis macrochirus*), black-bass (*Micropterus salmoides*), carp (*Cyprinus carpio*), Congo (*Tilapia rendalli*) and Nile (*Oreochromis niloticus*) tilapias, and the piscivorous tucunaré (*Cichla ocellaris*) from the Amazon Basin (Ribeiro et al., 2001). In the late 1980s, tilapia dominated and accounted for 50% of total fish biomass. In subsequent years, the lake was stocked with tambaqui (*Colossoma macropomum*) and tamoatá (*Callychthys callychthys*), both from the Amazon Basin and with high commercial value.

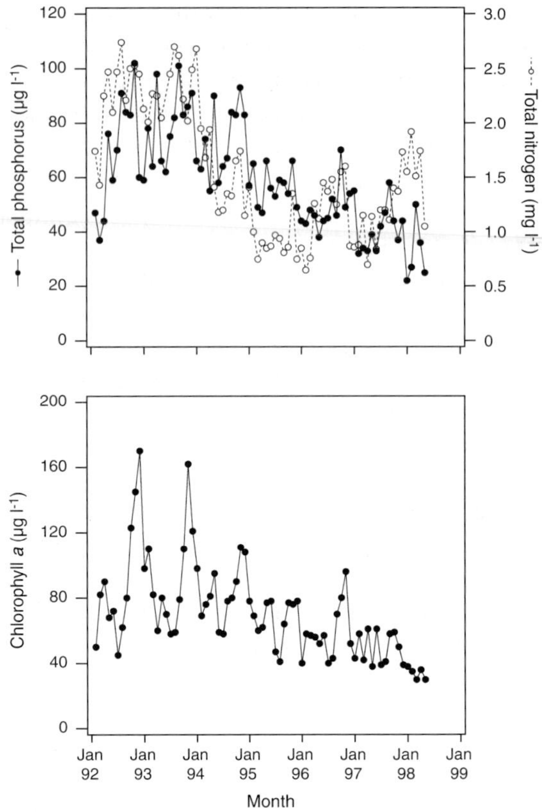

Fig. 4 Total phosphorus, total Kjeldahl nitrogen (TKN) and chlorophyll *a* in Lake Paranoá, Brazil, before and after (1993–1994) a major reduction of nutrient loading

In 1993–1994, two tertiary sewage treatment plants capable of processing 2,400 l s^{-1} were constructed in addition to a complementary sewage collecting system. Resultant decrease in external point-source of TP loading, from 6.3 g m^{-2} $year^{-1}$ in 1989 to 0.48 g m^{-2} $year^{-1}$ in 1998 (Pereira & Cavalcanti, 1997; Burnett et al., 2001), led to decreased mean TP, and also total Kjeldahl nitrogen (TKN) loading and concentration decreased (Fig. 4). Chlorophyll *a* values decreased (Fig. 4), and Secchi depth attained 2.4 m in 1999. Important planktonic community level changes were observed from 1996 to 1999: (1) decreased cyanobacteria dominance and replacement of *C. raciborskii* by green algae and diatoms by late 1998; (2) short-term dominance of colonial floating green algae (*Botryococcus braunii*), replacing *M. aeruginosa* in some hypertrophic areas in 1996; (3) re-appearance of *Ceriodaphnia cornuta*, *Daphnia gessneri*, other small cladocerans, such as *Moina micrura* and *Bosminopsis deitersi* (Elmoor-Loureiro et al., 2004), and *Notodiaptomus cearensis*, an oligotrophic calanoid copepod species (Padovesi-Fonseca et al., 2001). Other strategies, such as reducing the retention time of the lake and fish population management, were used to generate further improvements. An echo-sounding campaign in 1998 revealed a fish stock of 1,500 tons, with tilapia constituting up to 90% of fish biomass. Laboratory studies on fish feeding rates on lake plankton and P excretion rates, plus an assessment of planktivorous fish impacts in enclosures, indicated that tilapia overpopulation control would prevent accumulated P from being available to primary production, especially due to bottom feeding (Starling, 1993a, b; Starling & Rocha, 1990). Tilapia control was initiated in the most eutrophicated region by legalizing commercial cast-net fisheries. Continuous improvement of lake environmental conditions was confirmed in early 2000 by lower chlorophyll *a* values, enhanced water transparency, and decreased fish biomass to 800 tons. Tilapia dominance declined, allowing recovery of a heterogeneous fish population, including native omnivores and piscivores, and some exotic species, all of which contributed to enhancement of large zooplankton populations.

Trophic dynamics and biomanipulation in warm lakes

While results from Lake Paranoá illustrate the importance of reducing external loading to restore subtropical and tropical lakes, it is difficult to evaluate effects of fish manipulations in the lake. It is debatable whether fish manipulation methods used in cold temperate lakes can be employed in warm temperate, subtropical and tropical lakes. Several factors indicate that fish stock manipulation may not have the same positive effects in warm lakes (Moss et al., 2004; Jeppesen et al., 2005d): (a) fish species richness often is higher and many fish show partial niche overlap, which should increase predator control of prey (Lazzaro, 1997; Aguiaro & Caramaschi, 1998); (b) fish stocks often are dominated by omnivores independent of trophic state (Lazzaro, 1997; Branco et al., 1997; Yafe et al., 2002; Blanco et al., 2003; Mazzeo et al., unpublished data) and generally few piscivores are present (Quirós, 1998); (c) fish density, but not necessarily biomass, is higher (1–2 orders of magnitude) (Scasso et al., 2001; Aguiaro et al., 2003; Mazzeo et al., 2003; Meerhoff et al., 2003); (d) fish reproduction occurs throughout the year (Fernando, 1994; Paugy & Lévêque, 1999) and many species are viviparous and/or show parental care (Lorier & Berois, 1995), assuring higher juvenile survival rates. Therefore, top-down control by piscivores probably is weaker in warm lakes. Since small fish are more zooplanktivorous and have higher energy demand per unit biomass (Kalff, 2002), small fish in high abundance leads to higher predation pressure on zooplankton.

Accordingly, zooplankton communities in tropical and subtropical lakes often are dominated by small cladocerans (i.e. *Diaphanosoma*, *Ceriodaphnia* and *Bosmina*), rotifers, and copepod juveniles and nauplii (Dumont, 1994; Lewis, 1996; Branco et al., 2002; Garcia et al., 2002; Lacerot et al., unpublished data). Large *Daphnia* spp. may appear in early spring and autumn in warm-temperate lakes (Romo et al., 2005) or, if fish are absent, for longer periods in subtropical lakes (Mazzeo et al., unpublished data). Omnivorous copepods usually dominate zooplankton biomass in oligo-mesotrophic systems, whereas microzooplankton prevail in eutrophic systems. Therefore, classic control of phytoplankton by large zooplankton usually is not found in warm lakes. However, the numerous microzooplankton in these systems may control nanophytoplankton, thus favouring larger cyanobacteria. Some works demonstrate that cyclopoid copepods can control diatoms, rotifers can graze small green algae, and particulate matter and associated bacteria is an important food resource for zooplankters in tropical lakes (Infante, 1978; González et al., 2005). The zooplanktivorous phantom midge *Chaoborus* also seems to be more abundant in tropical lakes (e.g. Aguilera & Goitia, 1999), perhaps because bottom water anoxia is more common, which may provide *Chaoborus* with a fish predation refuge (Lewis, 1996) and allow strong predation pressure on zooplankton (Arcifa et al., 1992). Finally, shrimps (e.g. *Palaemonetes argentinus* or *Macrobrachium borelli*) also can be important predators on zooplankton considering their high abundance, food preferences and several reproductive events from spring to summer (Boschi, 1981; Collins, 1999; Collins & Paggi, 1998).

Aquatic plants play an important structural role in most freshwater ecosystems (Moss, 1990; Scheffer et al., 1993), also in the tropics (Thomaz & Bini, 2003). In temperate nutrient-rich lakes, submerged plants act as daytime refuges for zooplankton against fish predators (Timms & Moss, 1984; Lauridsen et al., 1996; Burks et al., 2002). At night, large zooplankton move out to the open water to feed, thereby contributing to clearwater conditions in lakes with high macrophyte coverage (Jeppesen et al., 1997). However, in the tropics and subtropics, effects of macrophytes on trophic interactions are more complex, since all forms (i.e. emergent, submerged, floating-leaved and large free-floating species) can be prominent. Macrophytes are extremely important to fish in the tropics and subtropics. Removal of *Chara* spp. beds in an Itaipú reservoir resort

(Santa Helena, Brazil) with the purpose to decrease piranha (*Serrassalmus marginatus*) population led to decreased fish abundance and species richness from 25 to 10 species (Agostinho et al., 2003; Fig. 5). Studies comparing different habitats in tropical and subtropical lakes indicate that the smallest fish species and individuals aggregate in all vegetation forms (Conrow et al., 1990; Meschiatti et al., 2000; Meerhoff et al., 2003; Meerhoff et al. unpublished data). Densities above 140 ind m^{-2} of a single, small-sized species were found in submerged plant beds in a subtropical lake (Meerhoff et al., 2003). Therefore, submerged vegetation in warm lakes may be a poor refuge for large-bodied zooplankton, particularly cladocerans (Meerhoff et al., 2003, 2006). An inverse diel horizontal migration pattern (aggregating in the pelagial during day and among plants at night) was found for *Bosmina longirostris* and *Diaphanosoma birgei* during spring and autumn in a subtropical lake without piscivores (Lake Blanca, Uruguay; Iglesias et al., 2007). The preference of small omnivorous fish for emergent and submerged plant beds during day and the occurrence of *Chaoborus* in the pelagic at night determined the spatial distribution observed. In accordance with the suggestion of a poor refuge effect for zooplankton in warm lakes, no positive effect of plants on water clarity was found in nutrient-rich subtropical lakes in Florida, and nutrient rich lakes with high plant biomass often were turbid (Bachmann et al., 2002).

Considering the higher fish density in tropical systems and lack of refuges for zooplankton among macrophytes, fish biomass thresholds allowing major grazing pressure by zooplankton may be lower and fish reductions therefore have to be very large compared to temperate lakes to allow large-bodied grazers, like *Daphnia* spp, to dominate (Scasso et al., 2001; Mazzeo et al., unpublished data). The efficiency of decreasing planktivorous fish biomass by fish removal is further hampered by continuous and/or viviparous reproduction in many (sub)tropical species (Scasso et al., 2001 and Fig. 6). Therefore, it may be more difficult to provoke and maintain a trophic cascade effect in subtropical and tropical lakes than in temperate lakes. Accordingly, positive, but only short-term, cascading effects of a massive (>80%) fish kill (due to fungal infection) of the dominant planktivorous mosquito fish (*Gambusia affinis*) were found in Lake Naini Tal, India (Nagdali & Gupta, 2002). Zooplankton abundance increased, phytoplankton biomass and productivity declined, as did

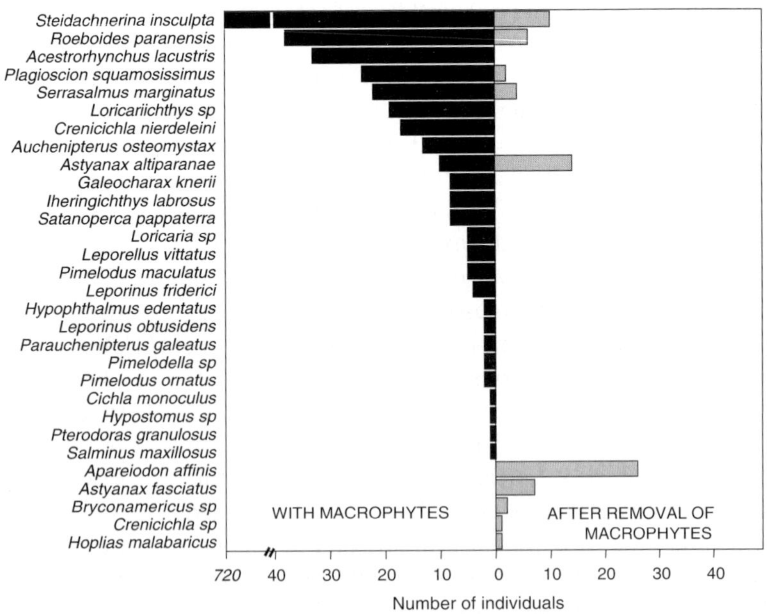

Fig. 5 Specific composition of fish assemblages in Santa Helena, Itaipú Reservoir (Paraná, Brazil) before (1987) and after (1988) a substantial removal of submerged macrophytes. Reproduced with kind permission from Agostinho et al. (2003)

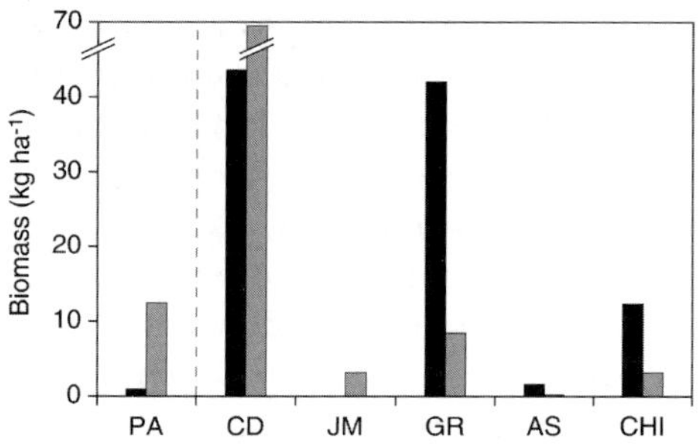

Fig. 6 Changes in the biomass of nekton at the beginning (black) and end (white) of summer 2003 in biomanipulated Lake Rodó (Uruguay) after an intensive programme of planktivorous fish removal. The biomanipulation had no effect on shrimps (*Paleomonetes argentinus*, PA) and viviparous fish, such as *Cnesterodon decemmaculatus* (CD) and *Jenynsia multidentata* (JM), while effective on the short-term for oviparous species: *Gymnogeophagus rhabdotus* (GR), *Astyanas* sp (AS) and *Cheirodon interruptus* (CHI). Moreover, no effect was observed on total fish biomass (from F. Quintans, unpublished data)

nutrient concentrations, resulting in higher water transparency. However, four months later, mosquito fish, plankton and nutrients had returned to previous levels.

In subtropical and tropical lakes, stocking of herbivorous silver carp (*Hypophthalmus molitrix*) has, with some success, been used to combat potentially toxic cyanobacteria (Starling et al., 1998; Datta & Jana, 1998). However, silver carp also consume zooplankton (Wu et al., 1997; Starling et al., 1998; Xu & Zie, 2004), which reduces grazing on small-sized phytoplankton. Therefore, it is unlikely that silver carp stocking, which is a symptom treatment rather than a restoration, will lead to a shift to a clear, macrophyte dominated stage (Matyas et al., 2003). Introduction of strictly piscivorous fish might be the only efficient biomanipulation strategy in (sub)tropical regions to control planktivorous fish. Ongoing mesocosm experiments using the sit-and-wait piscivore tararira (*Hoplias malabaricus*) support this hypothesis (Mazzeo et al., unpublished), but clearly more experiments are needed in this field before any firm conclusions can be drawn.

However, it is important to notice that trophic structure varies among warm lakes, depending on whether they are situated in wet or dry regions and temporal variations in temperature and salinity, and this may have implications on the restoration strategy as well. For example, tropical lakes are particularly sensitive to hydrological changes, which strongly affect their structure and functioning. In Cochabamba, Bolivia (17°S), switches from phytoplankton to macrophyte dominance in Lake Alalay have been associated to an increase in water level (Cadima, 1997). Higher densities and richness of zooplankton and fish were found during the low water season in several shallow lakes in Bolivia (Pouilly et al., 1999; Rejas & Maldonado, 2000). Water level changes also have strong effects on macrophyte abundance and water clarity in Turkish lakes (Beklioglu et al., 2003). These lakes, as well as northern Greek lakes, are, though warm in summer, subjected to cold winters, leading to a mixed fish community, often with low species richness compared to the subtropical and tropical lakes, and dominance of northern species (Beklioglu et al., 2003; Zalidis et al., 2002). These form an intermediate state between the north temperate lakes and winter warm temperate lakes and the subtropical/tropical lakes, which have to be taken into account when planning lake restoration (Beklioglu et al., 2003; Beklioglu & Tan, personal communication). Likewise, saline lakes are typically species poor (Williams, 1998) and, like in the subtropical lakes, submerged macrophytes appear to have little effect on water clarity in saline lakes (Moss, 1994a; Jeppesen et al., 1994) and form special cases.

In many experiments in temperate lakes, improvement in environmental state was observed without a trophic cascade after biomanipulation, i.e. without zooplankton species composition changes towards dominance of large-sized individuals and, consequently, higher phytoplankton grazing (Horppila et al., 1998). However, reductions in cyanobacteria and TP were observed, and water clarity increased. These phenomena may be due to reduced sediment P release and fish foraging following biomanipulation (Horppila et al., 1998). Biomanipulation also may reduce sediment nutrient release in tropical lakes, but the dominance of small fish species and better growth conditions for cyanobacteria suggest that effects may be temporary. Other methods, such as the combined use of aquatic plant harvests (mostly free-floating, if present) and hydraulic management, can be useful, though insufficient, for restoring small, shallow lakes in warm regions by decreasing internal nutrient loads (Rodríguez-Gallego et al., 2004). Sediment removal might be more attractive in such lakes, since P-content per unit volume of sediment exceeds that in the water column and plants (Søndergaard et al., 2003). However, drastic reduction of external nutrient loading seems to be the best method for restoring small or large subtropical and tropical lakes, but the scientific basis (e.g. nutrient threshold levels) to make decisions remains limited. Focus may well have to be on reducing external N loading as nitrogen has often been found to be the limiting nutrient for phytoplankton growth in tropical lakes (Lewis, 1996). Focus on nitrogen, in addition to phosphorus, is also relevant for temperate shallow lakes, as recent studies indicate that submerged macrophyte recovery after loading reduction is more likely to occur at moderately high TP when TN loading is low (Moss et al., 1994b; González Sagrario et al., 2005; James et al., 2005; Jeppesen et al., 2005b).

Conclusions

For small and large shallow lakes, the best strategy for improving the environmental state is external nutrient loading reduction. Like small lakes, large shallow lakes also may respond strongly and quickly to loading reduction. For smaller lakes resistant to nutrient loading reduction, physico-chemical and biological methods may enhance recovery. However, these methods are unlikely to be feasible in large shallow lakes. Sediment removal and chemical treatment, in particular, are less attractive because: (a) sediment transport and storage on land (if removed) will be expensive; (b) there is a large risk of sediment redistribution from resuspension; (c) the sediment P pool often is relatively small in large lakes. Fish manipulation also may be impractical because: (a) the large size makes it difficult to obtain a strong effect on the target fish population; and (b) fish population is among the first to respond to nutrient loading reduction, and natural changes in fish stock become difficult to discern in large lakes. If needed, transplantation/protection of aquatic plants (in sheltered beds) and seeds may be useful in large shallow lakes, the latter if pre-studies reveal that (re)-colonization is seed-limited. However, it is important first to evaluate whether wind (wave) disturbance will prevent colonization where seeds or plants are added.

Differences in biological interactions in cold temperate versus warm temperate-subtropical-tropical lakes make it difficult to apply biological restoration methods to warm lakes. Warm lakes often have prolonged growth seasons with a larger risk of long-lasting algal blooming and dense floating plant communities, higher dominance and abundance of small fish, higher fish aggregation in vegetation (loss of zooplankton refuge), more fish cohorts per year, and more omnivory by fish and less specialist piscivory. However, trophic structure varies among these lakes, depending on whether they are situated in wet or dry regions and temporal variations in temperature and salinity and this have implication for the restoration strategy.

Acknowledgements We thank the Danish counties for access to data from the Survey Programme of Danish Lakes. The work was supported by Eurolimpacs (GOCE-CT-2003-505540) and the Danish Natural Science Research Council funded research project "Consequences of weather and climate changes for marine and freshwater ecosystems. Conceptual and operational forecasting of the aquatic environment – CONWOY" (SWF: 2052-01-0034) and CLEAR (a Villum Kann Rasmussen centre of excellence project). We acknowledge the main Uruguayan Scientific Agencies for funding some of the research projects: CSIC, DINACYT, and PDT. Thanks to Sidinei M. Thomaz and Angelo A. Agostinho (Universidade Estadual de Maringá – Nupélia, Brazil) for their kind permission to use data and figures, and to Federico Quintans for data used in Fig. 6. We also thank Kathe Møgelvang and Anne Mette Poulsen for technical assistance. The two unknown reviewers are thanked for most valuable scientific comments and editorial suggestions and the organisers of the conference about Lake Taihu for inviting this paper.

References

Agostinho, A. A., L. C. Gomes & H. J. Ferreira Jr. 2003. Relações entre macrófitas aquáticas e fauna de peixes. In Thomaz, S. M. & L. M. Bini (eds), Ecologia e manejo de macrófitas aquáticas. EDUEM, Maringá, 261–279.

Aguiaro, T. & E. P. Caramaschi, 1998. Trophic guilds in fish assemblages in three coastal lagoons of Rio de Janeiro State (Brazil). Verhandlungen der Internationale Vereinigung der Limnologie 26: 2166–2169.

Aguiaro, T., C. W. C. Branco, J. R. Verani & E. P. Caramaschi, 2003. Diet of the clupeid fish *Platanichthys platana* (Regan, 1917) in two different Brazilian coastal lagoons. Brazilian Archives of Biology and Technology 46: 215–222.

Aguilera, X. & E. Goitia, 1999. Structure of the zoobenthic community of the Laguna Bufeos (Cochabamba, Bolivia). Revista Boliviana de Ecología 6: 55–64.

Arcifa, M. S., E. A. T. Gomes & A. J. Meschiatti, 1992. Composition and fluctuations of the zooplankton of a tropical Brazilian reservoir. Archiv für Hydrobiologie 123: 479–495.

Bachmann, R. W., C. A. Horsburgh, M. V. Hoyer, L. K. Mataraza & D. E. Canfield Jr. 2002. Relations between trophic state indicators and plant biomass in Florida lakes. Hydrobiologia 470: 219–234.

Beklioglu, M., Ö. Ince & I. Tüzün, 2003. Restoration of eutrophic Lake Eymir, Turkey, by biomanipulation undertaken following a major external nutrient control I. Hydrobiologia 489: 93–105.

Benndorf, J., 1995. Possibilities and limits for controlling eutrophication by biomanipulation. International Revue der Gesamten Hydrobiologie 80: 519–534.

Berg, S., E. Jeppesen, M. Søndergaard, 1997. Pike (*Esox lucius* L.) stocking as a biomanipulation tool. 1. Effects on the fish population in Lake Lyng (Denmark). Hydrobiologia 342/343: 311–318.

Blanco, R., S. Romo, M. J. Villena & S. Martinez, 2003. Fish communities and food web interaction in some shallow Mediterranean lakes. Hydrobiologia 506/509: 473–480.

Blindow, I., A. Hargeby, B. Wagner & G. Andersson, 2000. How important is the crustacean plankton for the maintenance of water clarity in shallow lakes with abundant submerged vegetation? Freshwater Biology 44: 185–197.

Boschi, E., 1981. Fauna de agua dulce de la República Argentina. Vol. XXVI, Decapoda Natantia. FECIC, Buenos Aires, 61 pp.

Branco, C. W. C. & P. A. C. Senna, 1991. The taxonomic elucidation of the Paranoá Lake (Brasília, Brazil) problem: *Cylindrospermopsis raciborskii*. Bulletin du Jardin Botanique National de Belgique 61: 85–91.

Branco, C. W. C. & P. A. C. Senna, 1994. Factors influencing the development of *Cylindrospermopsis raciborskii* and *Microcystis aeruginosa* in Paranoá Reservoir, Brasilia, Brazil. Algological Studies 75: 85–96.

Branco, C. W. C. & P. A. C. Senna, 1996. Plankton studies in Paranoá Reservoir, Brasília, Brazil. I. Relations among heterotrophic bacteria, chlorophyll *a*, total phytoplankton, total zooplankton and physichochemical factors. Hydrobiologia 337: 171–181.

Branco, C. W. C., T. Aguiaro, F. A. Esteves & E. P. Caramaschi, 1997. Food sources of the Teleost *Eucinostomus argenteus* in two coastal lagoons of Brazil. Studies on Neotropical Fauna & Environment 32: 33–40.

Branco, C. W. C. & C. B. Cavalcanti, 1999. A ecologia das Comunidades Planctônicas no Lago Paranoá. Chapter 19. In Henry, R. (ed.), Ecologia de reservatórios: estrutura, função e aspectos sociais. Editora da UNESP, Botucatu, 573–595.

Branco, C. W. C., M.-I. Rocha, G. F. S. Pinto, G. A. Gômara & R. D. Filippo, 2002. Limnological features of Funil Reservoir (Brazil, RJ) and indicator properties of rotifers and cladocerans of the zooplankton community. Lakes and Reservoirs: Research and Management 7: 87–92.

Breukelaar, A. W., E. H. R. R. Lammens, J. P. G. Klein Breteler & I. Tatrai, 1994. Effects of benthivorous bream (*Abramis brama* L.) and carp (*Cyprinus caprio* L.) on sediment resuspension and concentration of nutrients and chlorophyll *a*. Freshwater Biology 32: 113–121.

Burks, R. L., E. Jeppesen, D. M. Lodge & T. Lauridsen, 2002. Diel horizontal migration of zooplankton: costs and benefits of inhabiting littoral zones. Freshwater Biology 47: 343–365.

Burnett, J. A. B., S. P. Mattos & N. M. P. Azzolin, 2001. Intervenções da Companhia de Saneamento. In Olhares sobre o Lago Paranoá. Secretaria de Meio Ambiente e Recursos Hídricos do DF, Brasília, Capítulo IX: 199–211.

Cadima, M. M., 1997. Algae and macrophytes of the lake Alalay (Cochabamba, Bolivia). Revista Boliviana de Ecologia 3: 35–46.

Collins, P, 1999. Feeding of *Palaemonetes argentinus* (Decapoda: Palaemonidae) from an oxbow lake of the Paraná River, Argentina. Journal of Crustacean Biology 19: 485–492.

Collins, P. & J. C. Paggi, 1998. Feeding ecology of *Macrobrachium borelli* (Nobili) (Decapoda: Palaemonidae) in the food valley of the River Paraná, Argentina. Hydrobiologia 363: 21–20.

Conrow, R., A. V. Zale & R. W. Gregory, 1990. Distributions and abundances of early stages of fishes in a Florida lake dominated by aquatic macrophytes. Transactions of the American Fisheries Society 119: 521–528.

Cooke, G. D., 1993. Phosphorus inactivation and sediment oxidation. In Cooke, G. D., E. B. Welch, S. A. Peterson & P. R. Newroth (eds), Restoration and Management of Lakes and Reservoirs, 2nd ed. Lewis Publishers, Boca Raton: 161–209.

Cooke, G. D., E. B. Welch, S. A. Peterson & P. R. Newroth, 1993. Restoration and Management of Lakes and Reservoirs. Boca Raton, Lewis Publishers, Florida.

Cronberg, G., 1976. The Lago Paranoá Restoration Project: Phytoplankton ecology and taxonomy. A preliminary report. Brasília (Project FAHO – World Health Organization, 76/PW/BRA/2000, 24 pp.

Datta, S. & B. B. Jana, 1998. Control of blooom in a tropical lake: grazing efficiency of some herbivorous fishes. Journal of Fish Biology 53: 12–24.

Drenner, R. & D. Hambright, 1999. Review: biomanipulation of fish assemblages as a lake restoration technique. Archiv für Hydrobiologie 146: 129–165.

Dumont, H. J., 1994. On the diversity of the Cladocera in the tropics. Hydrobiologia 272: 27–38.

Elmoor-Loureiro, L. M. A., L. Mendonça-Galvão & C. Padovesi-Fonseca, 2004. New cladocerans from Lake Paranoá, Central Brazil. Brazilian Journal of Biology 64: 415–422.

Fernando, C. H., 1994. Zooplankton, fish and fisheries in tropical freshwaters. Hydrobiologia 272: 105–123.

Garcia, P. R., S. Nandini, S. S. S. Sarma, E. R. Valderrama, I. Cuesta & M. D. Hurtado, 2002. Seasonal variations of zooplankton abundance in the freshwater reservoir Valle de Bravo (Mexico). Hydrobiologia 467: 99–108.

González, Sagrario, M. A., E. Jeppesen, J. Gomà, M. Søndergaard, T. Lauridsen & F. Landkildehus, 2005 Does high nitrogen loading prevent clear-water conditions in shallow lakes at moderately high phosphorus concentrations? Freshwater Biology 50: 27–41.

González, E. J., G. Bernal, H. Hernández, M. L. Matos & C. Penaherrera, 2005. De qué se alimenta el zooplancton de los embalses venezolanos? Memorias del Instituto de Biología Experimental 4: 137–141.

Hanson, M. A. & M. G. Butler, 1990. Early responses of plankton and turbidity to biomanipulation in a shallow prairie lake. Hydrobiologia 200/201: 317–327.

Hansson, L.-A., H. Annadotter, E. Bergman, S. F. Hamrin, E. Jeppesen, T. Kairesalo, E. Luokkanen, P.-Å. Nilsson, M. Søndergaard & J. Strand, 1998. Biomanipulation as an application of food chain theory: constraints, synthesis and recommendations for temperate lakes. Ecosystems 1: 558–574.

Horppila, J., H. Peltonen, T. Malinen, E. Juokkanen & T. Kairesalo, 1998. Top-down or bottom-up effects by fish – issues of concern in biomanipulation of lakes. Restoration Ecology 6: 1–10.

Iglesias, C., G. Goyenola, N. Mazzeo, M. Meerhoff, E. Rodó & E. Jeppesen, 2007. Horizontal dynamics of zooplankton in subtropical Lake Blanca (Uruguay) hosting multiple zooplankton predators and aquatic plant refuges. Hydrobiologia DOI 10.1007/s10750-007-0599-4 (in press).

Infante, A., 1978. Natural food of herbivorous zooplankton of lake Valencia (Venezuela). Archiv Hydrobiologia 82: 347–358.

Jacobsen, B. A., R. S. Hansen, B. Moeslund, E. Jørgensen & G. Jensen, 2004. Arresø-tilstand og udvikling 2003. Report from Frederiksborg Amt, Teknik og Miljø, Denmark, 115 pp.

James, C., J. Fisher, V. Russell, S. Collings & B. Moss, 2005. Nitrate availability and plant species richness: implications for management of freshwater lakes. Freshwater Biology 50: 1049–1063.

Jeppesen, E., M. Søndergaard, E. Kanstrup, B. Petersen, R. B. Henriksen, M. Hammershøj, E. Mortensen, J. P. Jensen & A. Have, 1994. Does the impact of nutrients on the biological structure and function of brackish and freshwater lakes differ? Hydrobiologia 275/276: 15–30.

Jeppesen, E., T. L. Lauridsen, T. Kairesalo & M. Perrow, 1997. Impact of submerged macrophytes on fish-zooplankton relationships in lakes. In Jeppesen, E., Ma. Søndergaard, Mo. Søndergaard & K. Christoffersen (eds), The Structuring Role of Submerged Macrophytes in Lakes. Ecological Studies, Vol. 131, Springer Verlag, 91–115.

Jeppesen, E., J. P. Jensen, M. Søndergaard, T. Lauridsen & F. Landkildehus, 2000. Trophic structure, species richness and biodiversity in Danish lakes: changes along a phosphorus gradient. Freshwater Biology 45: 201–213.

Jeppesen, E. & I. Sammalkorpi, 2002. Lakes. In Perrow, M. & T. Davy (eds), Handbook of Restoration Ecology, Chapter 14, Cambridge University Press, 297–324.

Jeppesen, E., J. P. Jensen & M. Søndergaard, 2002. Response of phytoplankton, zooplankton and fish to re-oligotrophication: an 11-year study of 23 Danish lakes. Aquatic Ecosystem Health & Management 5: 31–43.

Jeppesen, E., J. P. Jensen, M. Søndergaard, M. Fenger-Grøn, K. Sandby, P. Hald, Møller & U. H. Rasmussen, 2004. Does fish predation influence zooplankton community structure and grazing during winter in north-temperate lakes? Freshwater Biology 49: 432–447.

Jeppesen, E., J. P. Jensen, M. Søndergaard & T. Lauridsen, 2005a. Response of fish and plankton to nutrient loading reduction in 8 shallow Danish lakes with special emphasis on seasonal dynamics. Freshwater Biology 50: 1616–1627.

Jeppesen, E., M. Søndergaard, J. P. Jensen, T. Lauridsen, L. Liboriussen, R. Bjerring Hansen, L. S. Johansson & F. Landkildehus, 2005b. The response of north temperate lakes to reduced nutrient loading with special emphasis on shallow Danish lakes. Verhandlungen der Internationale Vereinigung der Limnologie 29: 115–122.

Jeppesen, E., M. Søndergaard, J. P. Jensen, K. Havens, O. Anneville, L. Carvalho, M. F. Coveney, R. Deneke, M. T. Dokulil, B. Foy, D. Gerdeaux, S. E. Hampton, S. Hilt, K. Kangur, J. Köhler, E. Lammens, T. L. Lauridsen, M. Manca, R. Miracle, B. Moss, P. Nõges, G. Persson, G. Phillips, R. Portielje, S. Romo, C. L. Schelske, D. Straile, I. Tatrai, E. Willén & M. Winder, 2005c. Lakes' response to reduced nutrient loading–an analysis of contemporary data from 35 European and North American long term studies. Freshwater Biology 50: 1747–1771.

Jeppesen, E., M. Søndergaard, N. Mazzeo, M. Meerhoff, C. Branco, V. Huszar & F. Scasso, 2005d. Lake restoration and biomanipulation in temperate lakes: relevance for subtropical and tropical lakes. Chapter 11. In Reddy, M. V. (ed.), Tropical Eutrophic Lakes: Their Restoration and Management. Science Publishers, Enfield, 331–359.

Kalff, J., 2002. Limnology – Inland Water Ecosystems. Prentice-Hall Inc.

Ke, X. S. & W. Li, 2006. Germination requirement of *Vallisneria natans* seeds: implications for lake restoration in Chinese Lakes. Hydrobiologia 559: 357–362.

Klein, T., 1989. Søerne i de gode gamle dage-om den "Økologiske baggrundstilstand". Vand & Miljø 5: 211–215.

Kristensen, P., M. Søndergaard & E. Jeppesen, 1992. Resuspension in a shallow hypertrophic lake. Hydrobiologia 228: 101–109.

Kronvang, B., E. Jeppesen, D. J. Conley, M. Søndergaard, S. E. Larsen, N. B. Ovesen & J. Carstensen, 2005. Nutrient pressures and ecological responses to nutrient loading reductions in Danish streams, lakes and coastal waters. Journal of Hydrology 304: 274–288.

Lathrop, R.C., B. M. Johnson, T. B. Johnson, M. T. Vogelsang, S. R. Carpenter, T. R. Hrabik, J. F. Kitchell, J. J. Magnuson, L. G. Rudstam & R. S. Stewart, 2002. Stocking piscivores to improve fishing and water clarity: a synthesis of the Lake Mendota biomanipulation project. Freshwater Biology 47: 2410–2424.

Lauridsen, T. L., L. J. Pedersen, E. Jeppesen & M. Søndergaard, 1996. The importance of macrophyte bed size for cladoceran composition and horizontal migration in a shallow lake. Journal of Plankton Research 18: 2283–2294.

Lauridsen, T., H. Sandsten & P. H. Møller, 2003. The restoration of a shallow lake by introducing *Potamogeton* spp.: the impact of waterfowl grazing. Lakes & Reservoirs: Research & Management 8: 177–187.

Lazzaro, X., 1997. Do the trophic cascade hypothesis and classical biomanipulation approaches apply to tropical lakes and reservoirs? Verhandlungen der Internationale Vereinigung der Limnologie 26: 719–730.

Lewis, W. M. Jr. 1996. Tropical lakes: how latitude makes a different. In Schiemer, F. & K. T. Boland (eds), Perspectives in Tropical Limnology. SPB Academic Publishing B.V., Amsterdam, The Netherlands, 43–64.

Lorier, E. & N. Berois, 1995. Reproducción y nutrición embrionaria en *Cnesterodon decemmaculatus* (Teleoste: Poeciliidae). Revista Brasileira de Biologia 55: 27–44.

Marsden, S., 1989. Lake restoration by reducing external phosphorus loading: the influence of sediment phosphorus release. Freshwater Biology 21: 139–162.

Matyas, K, I. Oldal, J. Korponai, I. Tatrai & G. Paúlovits, 2003. Indirect effect of different fish communities in nutrient chlorophyll relationship in shallow hypertrophic water quality reservoir. Hydrobiologia 504: 231–239.

Mazzeo, N., L. Rodríguez-Gallego, G. Kruk, M. Meerhoff, J. Gorga, G. Lacerot, F. Quintans, M. Loureiro, D. Larrea & F. García-Rodrígues, 2003. Effect of *Egeria densa* Planch. beds on a shallow lake without piscivorous fish. Hydrobiologia 506/509: 591–602.

Meerhoff, M., N. Mazzeo, B. Moss & L. Rodríguez-Gallego, 2003. The structuring role of free-floating versus submerged plants in a subtropical shallow lake. Aquatic Ecology 37: 377–391.

Meerhoff, M., C. Fosalba, C. Bruzzone, N. Mazzeo, W. Noordoven & E. Jeppesen, 2005. An experimental study of habitat choice of Daphnia: plants signal danger more than refuge in subtropical lakes. Freshwater Biology 51: 1320–1330.

Meijer, M.-L., I. de Boois, M. Scheffer, R. Portielje & H. Hosper, 1999. Biomanipulation in the Netherlands: an evaluation of 18 case studies in shallow lakes. Hydrobiologia 408/409: 13–30.

Meschiatti, A. J., M. S. Arcifa & N. Fenerich-Verani, 2000. Fish communities associated with macrophytes in Brazilian floodplain lakes. Environmental Biology of Fishes 58: 133–143.

Mitchell, S. F. & M. Perrow, 1997. Interactons between grazing birds and macrophytes. In Jeppesen, E., Ma. Søndergaard, Mo. Søndergaard, K. Christoffersen (eds), The Structuring Role of Submerged Macrophytes in Lakes. Springer-Verlag, New York, 175–196.

Moss, B., 1990. Engineering and biological approaches to the restoration from eutrophication of shallow lakes in which aquatic plant communities are important components. Hydrobiologia 200/201: 367–377.

Moss, B., 1994a. Brackish and fresh-water shallow lakes – different systems or variations on the same theme? Hydrobiologia 276: 1–14.

Moss, B., S. McGowan & L. Carvalho, 1994b. Determination of phytoplankton crops by top-down and bottom-up mechanisms in a group of English lakes, the West Midland Meres. Limnology and Oceanography 39: 1020–1030.

Moss, B., D. Stephen, D. M. Balayla, E. Bécares, S. E. Collings, C. Fernández-Aláez, M. Fernández-Aláez, C. Ferriol, P. García, J. Gomá, M. Gyllström, L.-A. Hansson, J. Hietala, T. Kairesalo, M. R. Miracle, S. Romo, J. Rueda, V. Russell, A. Ståhl-Delbanco, M. Svensson, K. Vakkilainen, M. Valentín, W. J. Van de Bund, E. Van Donk, E. Vicente & M. J. Villena, 2004. Continental-scale patterns of nutrient and fish effects on shallow lakes: synthesis of a pan-European mesocosm experiment. Freshwater Biology 49: 1633–1649.

Nagdali, S. S. & P. K. Gupta, 2002. Impact of mass mortality of a mosquito fish, *Gambusia affinis* on the ecology of a fresh water eutrophic lake (Lake Naini Tal, India). Hydrobiologia 468: 45–52.

Oliveira, L. P. H. & L. Krau, 1970. Hidrobiologia geral aplicada particularmente a veiculadores de esquistossomos: hipereutrofia, mal moderno das águas. Memórias do Instituto Oswaldo Cruz 68: 89–118.

Padovesi-Fonseca, C., L. Mendonça-Galvão, D. F. Pereira, M. G. Philomeno & D. L. P. Rocha, 2001. O zooplâncton do Lago Paranoá. In Olhares sobre o Lago Paranoá. Secretaria de Meio Ambiente e Recursos Hídricos do DF, Brasília, 115–117.

Paugy, D. & C. Lévêque, 1999. La Reproduction. In Lévêque, C. & D. Paugy (eds), Les poissons des eaux continentales africaines: diversité, écologie, utilisation par ĺhomme. IRD Editions, Paris, 129–152.

Peltonen, H., J. Ruuhijärvi, M. Olin, T. Malinen, J. Horppila & J. Keto, 1999. The effects of food-web management and environmental variations on fish assemblage dynamics in a north-temperate lake. Journal of Fish Biology 55: 54–67.

Pereira, C. E. B. & C. G. P. Cavalcanti, 1997. Lago Paranoá – Rumo à Recuperação. Technical Report, CAESB, 15 pp.

Persson, A., 1997. Phosphorus release by fish in relation to external and internal load in a eutrophic lake. Limnology & Oceanography 43: 577–583.
Perrow, M. R., M.-L. Meijer, P. Dawidowicz & H. Coops, 1997. Biomanipulation in shallow lakes: state of the art. Hydrobiologia 342/343: 355–365.
Phillips, G., A. Kelly, J-A. Pitt, R. Sanderson & E. Taylor, 2005. The recovery of Barton Broad, a very shallow eutrophic lake, 20 years after the control of effluent derived phosphorus. Freshwater Biology 50: 1628–1638.
Pinto-Coelho, R. M. & A. Giani, 1985. Variações sazonais do fitoplâncton e fatores físico-químicos no Reservatório do Paranoá, Brasília. Ciência e Cultura 37: 2000–2006.
Pouilly, M., C. Ibáñez, M. Gutiérrez & T. Yunoki, 1999. Ecological functioning of floodplain lakes of Mamoré river (Beni, Bolivia). Revista Boliviana de Ecologia 6: 41–54.
Prejs, A., A. Martyniak, S. Boroń, P. Hliwa & P. Koperski, 1994. Food web manipulation in a small eutrophic Lake Wirbel, Poland: effect of stocking with juvenile pike on planktivorous fish. Hydrobiologia 275/276: 65–70.
Rejas, D. & M. Maldonado, 2000. Spatial-temporal variations in the composition of the fish community of a shallow lake in the floodplain of the Ichilo river (Cochabamba, Bolivia). Revista Boliviana de Ecologia 7: 37–46.
Quirós, R., 1998. Fish effects on trophic relationships in the pelagic zone of lakes. Hydrobiologia 361: 101–111.
Ribeiro, M. C. L. B., F. L. R. M. Starling, T. Walter & E. M. Farah, 2001. Peixes. In Olhares sobre o Lago Paranoá (ed.), Secretaria de Meio Ambiente e Recursos Hídricos do DF, Brasília, 121–128.
Rodríguez-Gallego, L., N. Mazzeo, M. Meerhoff, J. Clemente, C. Kruk, F. Scasso, G. Lacerot, J. García & F. Quintans, 2004. Effects of a water recirculation system covered by free-floating plants on the restoration of a hypertrophic subtropical lake. Lakes and Reservoirs: Research and Management 9: 205–213.
Romo, S., M.-J. Villena, M. Sahuquillo, J. M. Soria, T. Alfonso, E. Vicente & M. R. Miracle, 2005. Response of a shallow Mediterranean lake to nutrient diversion: does it follow similar patterns as northern shallow lakes? Freshwater Biology 50: 1706–1717.
Sas, H. (ed.), 1989. Lake restoration by reduction of nutrient loading. Expectation, experiences, extrapolation. Acad. Ver. Richardz Gmbh.
Scasso, F., N. Mazzeo, J. Gorga, C. Kruk, G. Lacerot, J. Clemente, D. Fabián & S. Bonilla, 2001. Limnological changes of a subtropical shallow hypertrophic lake during its restoration. Two years of whole-lake experiments. Aquatic Conserversation: Marine Freshwater Ecosystems 11: 31–44.
Scheffer, M., S. H. Hosper, M.-L. Meijer, B. Moss & E. Jeppesen, 1993. Alternative equilibria in shallow lakes. Trends in Ecology and Evolution 8: 275–279.
Skov, C., & S. Berg, 1999. Utilization of natural and artificial habitats by YOY pike in a biomanipulated lake. Hydrobiologia 408/409: 115–122.
Starling, F. L. R. M., 1993a. Control of eutrophication by silver carp (*Hypophthalmus molitrix*) in the tropical Paranoá Reservoir (Brasília, Brazil): a mesocosm experiment. Hydrobiologia 257: 143–152.
Starling, F. L. R. M., 1993b. Análise experimental dos efeitos da tilápia do Congo (*Tilapia rendalli*) e carpa prateada (*Hypophthalmus molitrix*) sobre a estrutura da comunidade planctônica do Lago Paranoá, Brasília (DF). Acta Limnologica Brasileira 4: 144–156.
Starling, F. L. R. M. & A. J. A. Rocha, 1990. Experimental study of the impacts of planktivorous fishes on plankton community and eutrophication of a tropical Brazilian reservoir. In Gulati, R. D., E. H. R. R. Lammens, M. L. Meijer & E. van Donk (eds), Biomanipulation – Tool for Water Management. Developments in Hydrobiology 61. Kluwer Academic Publishers, Dordrecht, 581–591.
Starling, F. L. R. M., M. Beveridge, X. Lazzaro & D. Baird, 1998. Silver carp biomass effectsd on the plankton community in Paranoa Reservoir (Brazil) and an assessment of its potential for improving water quality in lacustrine environments. Internationale Revue der Hydrobiologie 83: 499–507.
Søndergaard, M., L. Olufsen, T. L. Lauridsen, E. Jeppesen, & T. Vindbæk Madsen, 1996. The impact of grazing waterfowl on submerged macrophytes: in situ experiments in a shallow eutrophic lake. Aquatic Botany 53: 73–84.
Søndergaard, M., J. P. Jensen & E. Jeppesen, 2001. Retention and internal loading of phosphorus in shallow, eutrophic lakes. The Scientific World 1: 427–442.
Søndergaard, M., J. P. Jensen, E. Jeppesen & P. H. Møller, 2002a. Seasonal dynamics in the concentrations and retention of phosphorus in shallow Danish lakes after reduced loading. Aquatic Ecosystem Health & Management 5: 19–23.
Søndergaard, M., K.-D. Wolter & W. Ripl, 2002b. Chemical treatment of water and sediments with special reference to lakes. In Perrow, M. & T. Davy (eds), Handbook of Restoration Ecology, Chapter 10. Cambridge University Press, 184–205.
Søndergaard, M., J. P. Jensen & E. Jeppesen, 2005. Seasonal response of nutrients to reduced phosphorus loading in 12 Danish lakes. Freshwater Biology 50: 1605–1615.
Søndergaard, M., J. P. Jensen & E. Jeppesen, 2003. Role of sediment and internal loading of phosphorus in shallow lakes. Hydrobiologia 506/509: 135–145.
Thomaz, S. M. & L. M. Bini, 2003. Ecologia e manejo de macrófitas aquáticas. EDUEM, Maringá, 341 pp.
Timms, R. M. & B. Moss, 1984. Prevention of growth of potentially dense phytoplankton populations by zooplankton grazing in the presence of zooplanktivorous fish, in a shallow wetland ecosystem. Limnology & Oceanography 29: 472–486.
Van der Molen, D.T. R. Portielje, 1999. Multi-lake studies in The Netherlands: trends in eutrophication. Hydrobiologia 409: 359–365.
Weisner, S. E. B. & J. A., Strand, 2002. Ecology and management of plants in aquatic systems. In Perrow,

M. R., & A. J. Davy (eds), Handbook of Ecological Restoration. Vol. 1. Principles of restoration. Cambridge University Press, Cambridge, 242–256.

Williams, W. D., 1998. Salinity as determinant of the structure of biological communities in salt lakes. Hydrobiologia 381: 191–201.

Wu, J., P. Xie, M. Dai & J. Wang, 1997. Effects of silver carp density on zooplankton and water quality: implications for eutrophic lakes in China. Journal of Freshwater Ecology 12: 437–444.

Xu, J. & P. Xie, 2004. Studies on the food web structure of Lake Donghu using stable carbon and nitrogen isotope ratios. Journal of Freshwater Ecology 19: 645–650.

Yafe, A., M. Loureiro, F. Scasso & F. Quintans, 2002. Feeding of two Cichlidae species in a hypertrophic urban lake. Iheringia Série Zoologica, Porto Alegre 92: 73–79.

Zalidis, G. C., T. L. Crisman, & P. A. Gerakis, 2002. Restoration of Mediteranean wetlands. Hellnic Ministry of Environment. Physical Planning and Public Works, Athens and Greek Biotope/Wetland Centre, Thermi, Greece, 237 pp.

Hydrobiologia (2007) 581:287–295
DOI 10.1007/s10750-006-0516-2

Dredging effects on P status and phytoplankton density and composition during winter and spring in Lake Taihu, China

Xiuyun Cao · Chunlei Song · Qingman Li · Yiyong Zhou

Abstract Phytoplankton density and composition, together with phosphorus (P) concentrations and size-fractionated alkaline phosphatase activity (APA), were investigated in dredged and undredged zones in Lake Taihu from January to April 2004. P concentrations were also determined in the corresponding interstitial water. Enzyme Labeled Fluorescence (ELF) was used for localizing extracellular phosphatase on phytoplankton cell membranes in April. The increase in phytoplankton density was paralleled by a significant increase in soluble reactive P (SRP) concentrations in the water column and interstitial water at all sites from January to April, with chlorophyte gradually becoming dominant. In February, at the undredged site, more algae dominated by chlorophyte occurred in overlying water, rather than in the surface, coinciding with higher SRP concentrations in overlying and interstitial water. Therefore, P status in the bottom is important to phytoplankton development in terms of density and composition. Undredged sites had higher SRP concentrations in interstitial water than dredged sites. Furthermore, Higher APA was observed, accompanied by higher dissolved organic P (DOP) and lower total P at the undredged site in February. Enzymatic hydrolysis of DOP may have been an additional source of P for phytoplankton. In April, *Schroederia* sp. was ELF labeled in surface water at the dredged site, which showed markedly lower SRP concentration, but not at the undredged site with higher SRP concentration. Thus, the dredging might regulate algal density and composition in water column by reducing P bioavailability.

Keywords Phytoplankton · Phosphorus · Alkaline phosphatase · ELF · Sediment dredging · Shallow lakes · Eutrophication

Introduction

One symptom of lake eutrophication is an excess phytoplankton growth, which is controlled by various factors. Besides light and temperature, phytoplankton requires a wide range of inorganic nutrients for growth. These include potentially growth-limiting nutrients, such as carbon dioxide, nitrogen, phosphorus (P), and silicon. Phosphorus is the most commonly limiting nutrient in oligotrophic freshwaters (Graham & Wilcox, 2000). However, high phosphorus loading complicates the situation in eutrophicated waters. Lake

Guest editors: B. Qin, Z. Liu & K. Havens
Eutrophication of shallow lakes with special reference to Lake Taihu, China

X. Cao · C. Song · Q. Li · Y. Zhou (✉)
Institute of Hydrobiology, The Chinese Academy of Sciences, Wuhan 430072, China
e-mail: zhouyy@ihb.ac.cn

recovery often is delayed after external loading reductions due to internal P load (Jeppesen et al., 1990). Phosphorus release from the sediment into the lake water may be so intense and persistent that it may substantially increases the bioavailable P pool and consequently also the algal biomass (Knuttila et al., 1994). Dredging to remove surface sediment is widely used to reduce the internal loading, but its effects on phytoplankton are still equivocal. Cell counts and chlorophyll *a* concentrations suggest that after sediment removal, phytoplankton quantity may remain virtually unchanged compared with the period before sediment removal (Poulickova et al., 1998), while the eutrophic Lake Trummen (Sweden) was restored by removal of the upper most 0.5 m nutrient-rich sediment layer by suction dredging, which drastically reduced nutrient concentrations and phytoplankton biomass (Cronberg, 1982).

Phytoplankton can take up only inorganic orthophosphate (Overbeck, 1991). However, an important indirect source is dissolved organic phosphorus (DOP) after hydrolysis with extracellular phosphatase. Phytoplankton may adapt to P limitation by changing species composition and/or increasing alkaline phosphatase activity (APA; Vrba et al., 1993). A method known as ELF (enzyme labeled fluorescence; Gonzáles-Gil et al., 1998), allows for phosphatase labeling and direct microscopic detection. The product of ELF-phosphate hydrolysis can fluorescently tag individual algal cells at or near the fixed enzymatic activity sites (Rengefors et al., 2001).

A comparative study was conducted at dredged and undredged sites in Lake Taihu from January to April 2004. The objective of this study was to investigate the dredging effects on P status and phytoplankton density and composition concerning a large, eutrophic, shallow Chinese lake. We investigated the fluctuations of phytoplankton in terms of density and composition, coupled with phosphorus concentrations and phosphatase activities in the water column. ELF was used to localize extracellular phosphatase on algal cells, and P concentrations were determined in interstitial water.

Materials and methods

Study site and sampling

The common description of Lake Taihu was given elsewhere (Qin et al., 2006). Dredging was carried

Fig. 1 The map of sampling sites

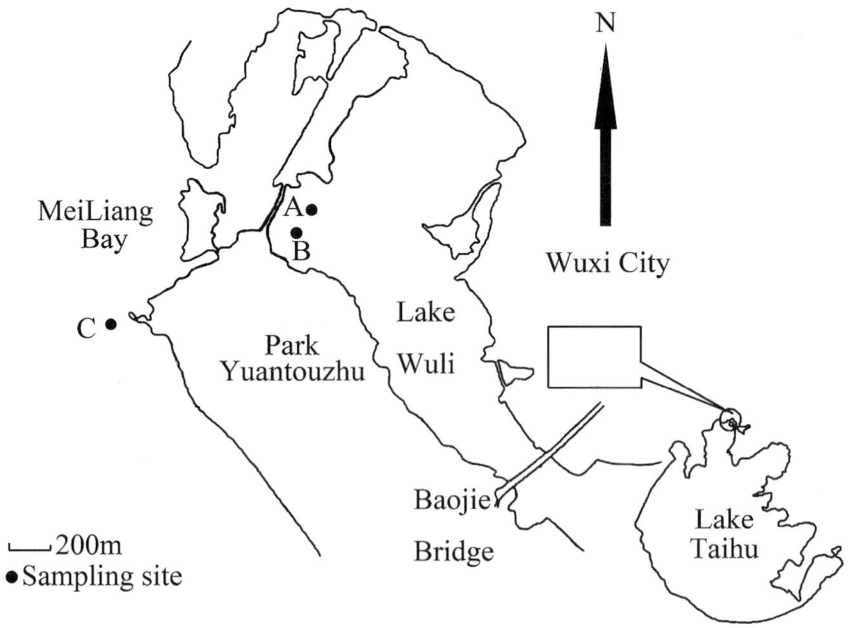

out to remove the surface sediment in order to improve the water quality in 2003–2004. Three sampling sites (Fig. 1), Site A (31°32′4.6″ N, 120°13′22.6″ E) and Site B (31°32′1.7″ N, 120°13′18.7″ E) located in Wuli Bay area and Site C (31°31′32.3″ N, 120°12′37.6″ E) located in the main embayment (Meiliang Bay) were sampled in January, 2004. Site A was dredged before sampling, Site B was sampled before dredging in January and after dredging in February and April, and Site C was undredged. Water samples were collected with a Friedinger sampler at the surface and in the water layer overlying the sediment. Phytoplankton samples (1 l) were preserved immediately with Lugol's iodine solution. Another liter of water was taken for P and phosphatase analysis. Sediment was collected using a Peterson dredge. Interstitial water was extracted from the sediment by centrifugation at 3,000 rpm for 30 min.

Phytoplankton density

Preserved phytoplankton samples in the surface and overlying water were concentrated to 30 ml after sedimentation for one week. After mixing, 0.1 ml concentrated samples were counted directly using a Zeiss microscope at ×640 magnification. Phytoplankton species were identified according to Hu et al. (1979).

P concentration

Total P (TP), dissolved total P (DTP) and soluble reactive phosphorus (SRP) were determined in the surface, overlying and sediment interstitial water. TP was determined following digestion according to Golterman et al. (1978). Phosphorus compounds were hydrolysed by H_2SO_4 and $K_2S_2O_8$ to orthophosphate, which was then determined after Murphy and Riley (1962). 0.5 ml H_2SO_4 and 1 g $K_2S_2O_8$ were added to 50 ml samples. Flasks were covered with an inverted beaker and autoclaved at 120°C for 2 h. DTP was determined using water samples filtered through a 0.45 μm membrane filters (Millipore) using the same procedure. SRP was determined according to Murphy & Riley (1962). DOP was calculated as follows: DOP = DTP-SRP.

Phosphatase activity and microscopic phosphatase detection

Extracellular alkaline phosphatase activity (APA) in the water samples was determined using a procedure modified from Gage and Gorham (1985) and Boon (1989). Triplicate 5 ml water samples were supplied with Tris–HCl buffer (pH = 8.5, final concentration 13 mmol l^{-1}), Na_3N (final concentration 5 mmol l^{-1}), and *p*-nitrophenyl phosphate (*p*NPP, final concentration 0.3 mmol l^{-1}), and incubated at 37°C for 4 h. Absorbency of *p*-nitrophenol was measured spectrophotometrically at 410 nm. APA was determined in unfiltered samples (APA_T) and in samples filtered through 3 μm and 0.45 μm membrane filters ($APA_{<3}$ and $APA_{<0.45}$, respectively). The contribution of APA to the coarser ($APA_{>3}$) and finer ($APA_{0.45-3}$) fractions were calculated as follows: $APA_{>3} = APA_T - APA_{<3}$, $APA_{0.45-3} = APA_{<3} - APA_{<0.45}$ (Chróst et al., 1984).

ELF 97 phosphate (ELFP, Molecular Probes) used for microscopic detection of extracellular PA in the phytoplankton according to the protocol by Štrojsová et al. (2003). Incubations started by adding the ELFP solution (final concentration 27 μmol l^{-1}), and samples were incubated at room temperature for 2–4 h. Each incubation was terminated by transferring the sample to a filter holder (diameter 7 or 10 mm, depending on phytoplankton density) with a membrane filter (Millipore; 0.45 μm pore size) and immediate filtering with a mild vacuum (<20 kPa). The filter with retained plankton was placed on a microscope slide, embedded with anti-fading reagent (Citifluor AF1, London, UK), and covered with a cover slip. To inspect phytoplankton for presence of ELFA precipitates, an epifluorescence microscopy (Carl Zeiss Axioskop 40FL; Olympus AX-70) with UV-excitation was used. For more details, see Štrojsová et al. (2003) and Nedoma et al. (2003).

Results

Variations in P forms

SRP concentration gradually increased both in surface and overlying water from January to April. The same trend was found in interstitial water in February and April (Fig. 2). TP concentration in interstitial water was significantly higher than that in the water column, and, for SRP, interstitial water always showed higher values than overlying water (Figs. 2, 3). In January, Site B (undredged) had significantly ($p < 0.01$) higher SRP concentrations in interstitial water than Site A (dredged). In February and April, the undredged site (Site C) also showed significantly ($p < 0.05$ in February and $p < 0.01$ in April) higher SRP values in both interstitial and overlying waters than dredged sites (Site A and B, which was dredged after January). In February, Site C had the highest TP concentration in interstitial water and DOP concentration in

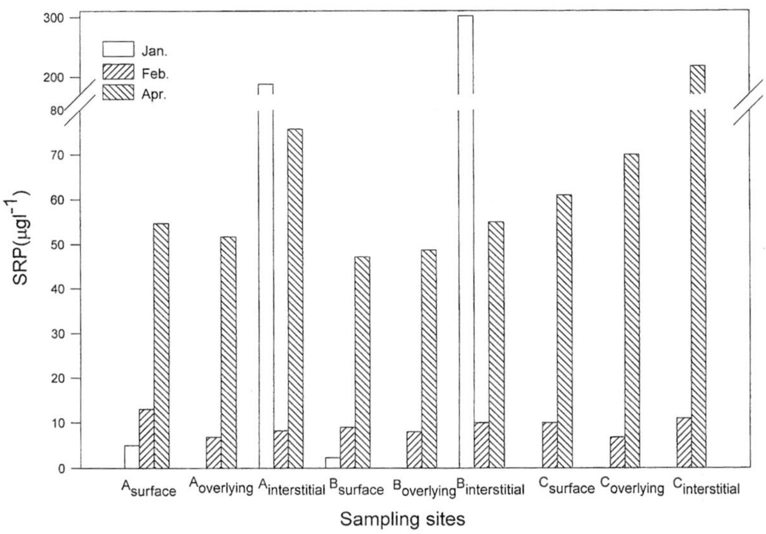

Fig. 2 Seasonal fluctuations of SRP concentrations in surface, overlying and interstitial water of all sampling sites

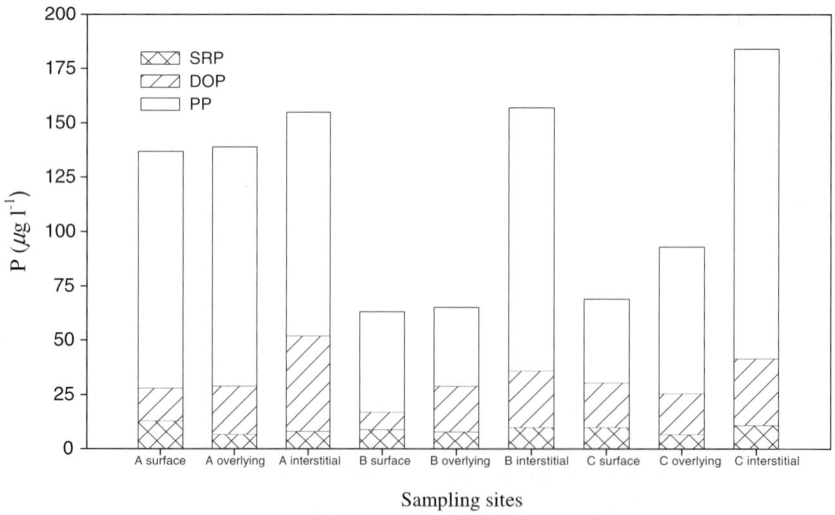

Fig. 3 Concentrations of different forms of P at all sampling sites in February

surface water among all the sites (Fig. 3). Furthermore, Site A showed the highest SRP among the sampling sites in January and February and Site B the highest in April in surface water. For overlying water, Site C had the highest SRP in February and April.

Variations in phytoplankton density and composition

Phytoplankton density gradually increased in surface waters during study period at all sampling sites (Table 1). The same trend was observed in overlying water except at Site C. Algal density was higher in surface water than in overlying water except at Sites B and C in February. Overlying water at Site C had higher phytoplankton density, especially of green algal cells, than surface water in February. Surface water was richer in algal species in April, while it showed similar species numbers to those in overlying water in February. In surface water, Site A showed the highest algal density among the sampling sites in January and February, and Site B had the highest in April. For overlying water, Site C showed the highest value in February and Site B in April, especially in green algae both in density and diversity.

Cryptophytes dominated the phytoplankton assemblage from January to April at Sites A and B, but not at Site C, in surface water. Diatoms were present but were not abundant. Green algae began to appear in February and increased markedly in April at all sites. Few cyanobacteria were found in April at Sites B and C. Similar changes were found in overlying water in February, while more green algae were observed in surface water than in overlying water in April. In overlying water at Site C, green algae (*Coelastrum* sp.) were very abundant.

Extracellular phosphatase activity

The contribution of the coarser ($APA_{>3.0}$) fraction, conventionally taken as "algal" APA to total APA was always over 49% (Fig. 4). No clear trend was found in the relationship between total APA and P concentrations, but an inverse relationship was observed in February. With the highest TP, Site A showed the lowest total APA in surface and overlying water, while the highest total APA was associated with the lowest TP in surface water at Site C. Algae excreted membrane bound phosphatase, as evidenced by ELFA at Sites A and C in surface water in April. A green alga *Schroederia* sp. present, but not dominant at Sites A and C in February and April was ELFA-labeled at Site A. Even so, not all cells were ELF positive. For all the labeled cells, ELFA precipitated on the membrane of the algae in a linear pattern around the whole cell. No labeled cells of this same species were observed at Site C, where SRP concentration in surface water was higher relative to that at Site A.

Discussion

Phytoplankton development is influenced by several factors, e.g., light intensity, water temperature, nutrient concentrations and herbivore grazing (Graham & Wilcox, 2000). Others have reported green algae becoming abundant with increasing water temperature (Robert et al., 2003). In Lake Taihu, algal density increased with temperature, and green algae were abundant in April at all sites. Major shifts in species composition of benthic algae did not result from nutrient additions but seemed to be related to light penetration and *Dreissena* herbivory in Saginaw Bay, Lake Huron (Pillsbury et al., 2002). At all sites in Lake Taihu, algal density

Table 1 Algal density in surface and overlying water during the study period for all sampling sites (10^6 ind l^{-1})

	A surface			A overlying		B surface			B overlying		C surface		C overlying	
	Jan.	Feb.	Apr.	Feb.	Apr.	Jan.	Feb.	Apr.	Feb.	Apr.	Feb.	Apr.	Feb.	Apr.
Density	4.25	5.55	7.3	2.13	4.2	2.65	3.69	15.6	4.09	6.43	4.27	8.06	5.64	3.41

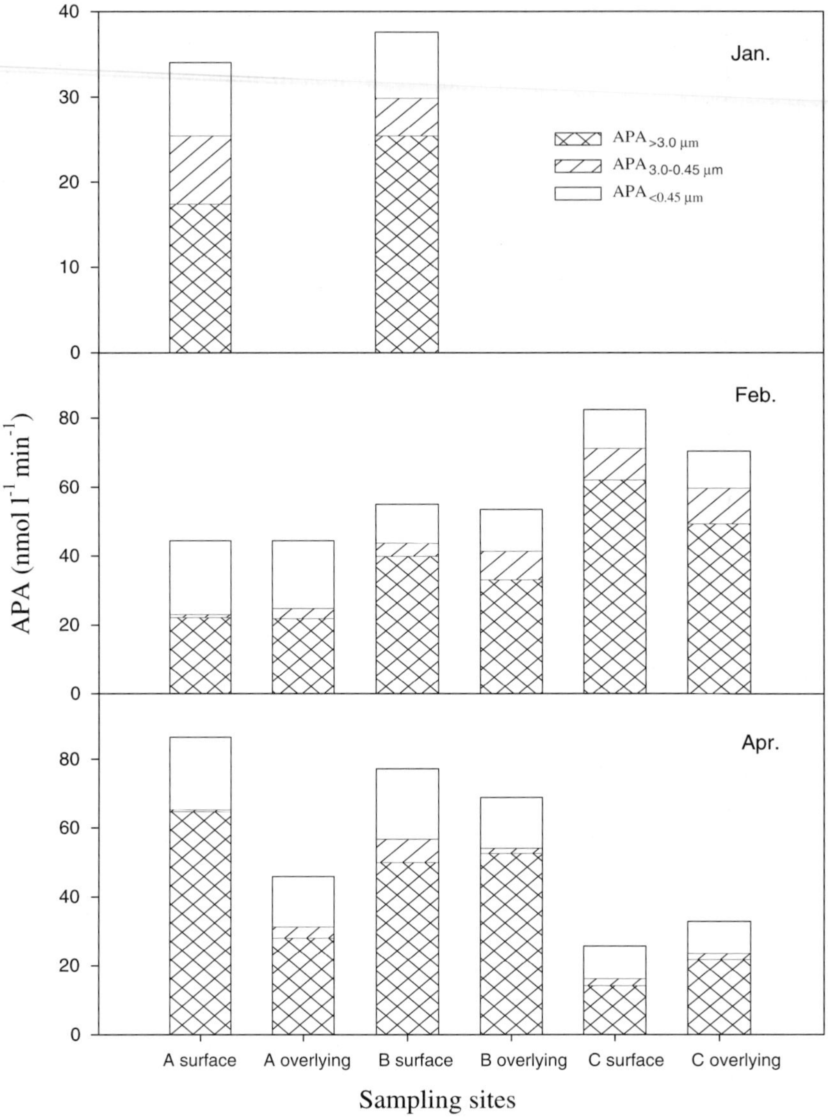

Fig. 4 Size-fractionated APA at all sites studied during experiment period

was higher in surface versus overlying water indicating that light might act as an important factor affecting growth.

SRP appears to have been important in phytoplankton development. From January to April, an increase in bioavailable phosphorus (SRP) was observed at all sites. Following the increase in SRP, water column algae cell counts increased. Higher SRP was accompanied by higher algal density at Site A in January and February and Site B in April. PO_4-P is a major factor in phytoplankton biomass and development elsewhere (Karjalainen et al., 1997–1998; Levine & Whalen, 2001). In a subtropical wetland, algal species composition was more sensitive to P concentration than algal biomass (Pan et al., 2000). In the present study, SRP increased from January to April, and algal composition changed. Green algae occurred in February and became prevalent in April, with cyanobacteria occurring as well. Algal biomass increased after nutrient addition and chlorophytes replaced chrysophytes as the dominant phytoplankton group in another study (Lessmann et al., 2003). In Lake Greifensee

(Switzerland), although total biomass values of phytoplankton did not show any correlation with the long-term decline of phosphorus concentrations, the composition of phytoplankton showed clear trends. During the period with Total P concentrations of >0.25 mg l^{-1}, phytoplankton biomass was dominated by chlorophytes and cryptomonads. With declining P load, chlorophytes and dinoflagellates decreased (Buergi et al., 2003). Overall, in Lake Taihu, higher SRP concentrations in water column was accompanied by higher phytoplankton density, dominated by green algae.

Extracellular phosphatase is important in algal development. For green algae, APA was responsible for making P available, at least partially. For example, *Schroederia* sp. at Site A, responding to low SRP, excreted extracellular phosphatase in April, as illustrated by ELFA-labeling. Total and "algal" APA at Site A was higher than those at Site C (Fig. 4). In a eutrophic Czech reservoir, ELFA-labeling indicated that several green algae produced extracellular phosphatase whenever present (Štrojsová et al., 2003). Green algae can lead to increased phosphatase activity (Sala & Guede, 1995). APA in *Selenastrum capricornutum* was not correlated to molybdate reactive P over a concentration range. The P fraction made available by alkaline phosphatases, enzyme hydrolyzable P, increased the P supply to organisms with high APA (Klotz, 1985). Thus, P fractions mediated by APA, in addition to SRP, are important for P supply to algae. In the present study, higher APA (total or "algal") was accompanied by lower TP and higher DOP in surface water at Site C in February, where green algae were abundant. The alternative method for phytoplankton to obtain P would be use of DOP by APA (Labry et al., 2002). These facts likely reflected the key role of bioavailable phosphorus in the regulation of phytoplankton composition.

From February to April, SRP concentrations in interstitial water paralleled with those in water column. Among all sampling sites, Site C had the highest SRP in interstitial and overlying waters in February. Highest water column phytoplankton density occurred in April, with abundant green algae including *Coelastrum* sp., responding to highest SRP concentrations in interstitial water. In wetland, sediment is of paramount importance, promoting a more diverse phytoplankton assemblage. When sediment or nutrient pulses are absent, phytoplankton is dominated by slow-growing algae (Ortega-Mayagoitial et al., 2003). In Lake Kinneret (Israel), from the bottom to 3 m, green algae (*Pediastrum, Scenedesmus, Coelastrum*, etc.) were observed and changed with increased nutrients from cultural disturbance (Pollingher, 1986). In Canadian lakes, increases at fall turnover consisted mainly of the previous hypolimnetic P and were responsible for fall and winter blooms. These estimates of P were similar in lakes with little internal load, but could differ up to 700% in lakes with high internal load (Nuernberg et al., 1995). In our results, highest algal density was found in overlying water, with green algae (*Coelastrum* sp.) being dominant, which is contrary to other results showing more algae occurring in surface waters. Therefore, the upward flux of phosphorus could support the excess growth of phytoplankton, whereas phytoplankton could adapt for P nutrient by concentrating in bottom as well, which provided further evidence that P in the overlying and interstitial waters is an important factor affecting phytoplankton density and composition.

Compared to the dredged sites, the undredged controls of Lake Taihu showed significantly higher concentration of bioavailable phosphorus (SRP) in interstitial and overlying waters, and considerably larger amounts of algae occurred in bottom water, rather than at the surface, with green algae being dominant. Therefore, in addition to temperature and light, nutrient P may serve as a driving force for development of phytoplankton. A significantly higher density of phytoplankton, dominated by green algae responded to the greatly increased SRP concentrations in interstitial water at all stations in April, strengthening the relation between the density and composition of phytoplankton and P loading from sediments. In general, the undredged sites in Lake Taihu exhibited significantly higher SRP concentrations in interstitial water than the dredged sites. Furthermore, in February, higher total and "algal" APA was accompanied by significantly lower TP and higher DOP concentrations at the undredged sites, where green algae dominated. In

April, *Schroederia* sp. (green algae) excreted extracellular phosphatase as shown by ELF-labeling at a dredged site with relative lower SRP concentration, while, at the dredged site with higher SRP concentration, no such labeling was observed. These facts suggested that organic phosphorus, and its regeneration mediated by phosphatase, constituted an additional way of supplying P for some algae to compensate ambient P deficiency. This was also related to internal loading, and consequently to the sediment dredging.

In summary, the undredged sites in Lake Taihu had significantly higher bioavailable P (SRP) in the interstitial and overlying waters, coinciding with larger amounts of algae in the overlying water, rather than in the surface water, with green algae being dominant. Moreover, higher phytoplankton density, dominated by green algae, responded to higher SRP in the interstitial water at all sites in April. Therefore, in addition to temperature and light, P status in the bottom is important to phytoplankton development in terms of density and composition in Lake Taihu. Additionally, the undredged sites generally had higher SRP in the interstitial water than the dredged sites. Furthermore, in February, higher total and "algal" APA was accompanied by lower TP and higher DOP at the undredged sites, where green algae dominated, suggesting that extracellular phosphatase of chlorophyte mainly mediate P regeneration to improve its bioavailability. In April, *Schroederia* sp. (green algae) excreted extracellular phosphatase as shown by ELF-labeling at a dredged site with lower SRP, while, at the undredged site with higher SRP, no such a labeling was observed. Based on results in an urban lake, Half Moon Lake in Wisconsin, James et al. (2002) found that reduction of internal P sources could substantially reduce by greater than 70% the high concentrations of algae in the lake. Suggested internal P control measures included a sediment chemical treatment to bind P, greater harvesting of *P. crispus* to reduce the macrophyte P pool at the time of senescence, and limiting motor boat activity when the lake is weakly stratified. Our results indicated that the dredging might regulate algal density and composition in water column by reducing P bioavailability in the bottom of eutrophic lakes.

Acknowledgements This work was supported by the National Key Basic Research and Development Program (2002CB412304) and the grant (2002AA601013). We also obliged to the funds from the National Science Foundation of China (20177033). This work was partly supported by Sino-Czech Scientific and Technological Cooperation—project No. 36–20. We would like to thank Jaroslav Vrba, Alena Štrojsová and Feng Weisong for their methodical help and providing facilities to use epifluorescence microscope. Many thanks are given to Li Jianqiu for the sample preparations and to Liu Jingyuan for the review of this manuscript. The thanks also go to Li Lin, Zeng hui for their kind helps in sample collecting.

References

Boon, P. I., 1989. Organic matter degradation and nutrient regeneration in Australian freshwaters: I. Methods for exoenzyme assays in turbid aquatic environments. Archiv Fur Hydrobiologie 115: 339–359.

Buergi, H. R., H. Buehrer & B. Keller, 2003. Long-term changes in functional properties and biodiversity of plankton in Lake Greifensee (Switzerland) in response to phosphorus reduction. Aquatic Ecosystem Health and Management 6: 147–158.

Chróst, R. J., W. Siuda & G. Z. halemejko, 1984. Longterm studies on alkaline phosphatase activity (APA) in a lake with fish-aquaculture in relation to lake eutrophication and phosphorus cycle. Archiv Fur Hydrobiologie 70: 1–32.

Cronberg, G., 1982. Changes in the phytoplankton of lake Trummen induced by restoration. Hydrobiologia 86: 185–193.

Gage, M. A. & E. Gorham, 1985. Alkaline phosphatases activity and cellular phosphorus as an index of the phosphorus status of phytoplankton in Minnesota lakes. Freshwater Biology 15: 227–233.

Golterman, H. L., R. S. Clymo & M. A. M. Ohmstad, 1978. Methods for physical and chemical analysis of fresh waters. Blackwell Scientific Publications, Oxford, Handbook 8, 2nd edn., 111–117.

Gonzáles-Gil, S., B. Keafer, R. V. M. Jovine, A. Aquilera, S. Lu & D. M. Anderson, 1998. Detection and quantification of alkaline phosphatase in single cells of phosphorus-starved marine phytoplankton. Marine Ecology Progress Series 164: 21–35.

Graham, L. E. & L. W. Wilcox, 2000. Algae. Prentice hall, Inc., Upper Saddle River, USA.

Hu, H., R. Li, Y. Wei, C. Zhu, J. Chen & Z. Shi, 1979. Freshwater algae in China. Science Press, Peking (in Chinese).

James, W. F., J. W. Barko, H. L. Eakin & P. W. Sorge, 2002. Phosphorus budget and management strategies for an urban Wisconsin lake. Lake and Reservoir Management 18: 149–163.

Jeppesen, E., P. Kristensen, J. P. Jensen, M. Soendergaard, E. Mortensen & T. Lauridsen, 1990. Recovery resilience following a reduction in external phosphorus loading of shallow, eutrophic Danish lakes: duration, regulating factors and methods for overcoming resilience. 3. Int. Workshop on Ecosystem Research in Freshwater Environment Recovery, 26–29.

Karjalainen, H., S. Seppälä & M. Walls, 1997–1998. Nitrogen, phosphorus and Daphnia grazing in controlling phytoplankton biomass and composition–an experimental study. Hydrobiologia 363: 309–321.

Klotz, R. L., 1985. Influence of light on the alkaline phosphatase activity of Selenastrum capricornutum (Chlorophyceae) in streams. Canadian Journal of Fisheries and Aquatic Sciences 42: 384–388.

Knuttila, S., O.-P. Pietiläinen & L. Kauppi, 1994. Nutrient balances and phytoplankton dynamics in two agriculturally loaded shallow lakes. Hydrobiologia 275/276: 359–369.

Labry, C., A. Herbland & D. Delmas, 2002. The role of phosphorus on planktonic production of the Gironde plume waters in the Bay of Biscay. Journal of Plankton Research 24: 97–117.

Lessmann, D., A. Fyson & B. Nixdorf, 2003. Experimental eutrophication of a shallow acidic mining lake and effects on the phytoplankton. Hydrobiologia 506–509: 753–758.

Levine, M. A. & S. C. Whalen, 2001. Nutrient limitation of phytoplankton production in Alaskan Arctic foothill lakes. Hydrobiologia 455: 189–201.

Murphy, J. & P. Riley, 1962. A modified single solution method of the determination of phosphate in natural waters. Analytica Chimica Acta 27: 1–36.

Nedoma, J., A. Štrojsová, J. Vrba, J. Komárková & K. Šimek, 2003. Extracellular phosphatase activity of natural plankton studied with ELF97 phosphate: fluorescence quantification and labelling kinetics. Environmental Microbiology 5: 462–472.

Nuernberg, G. K, R. W. Bachmann, J. R. Jones, R. H. Peters & D. M. Soballe, 1995. Modeling phosphorus in lakes with possible sediment P release, case studies. Lake and Reservoir Management 11(2):176.

Ortega-Mayagoitial, E., C. Rojo & M. A. Rodrigo, 2003. Controlling factors of phytoplankton assemblages in wetlands: an experimental approach. Hydrobiologia 502: 177–186.

Overbeck, J., 1991. Early studies on ecto- and extracellular enzymes in aquatic environments. In Chróst, R. J. (ed.), Microbial enzymes in aquatic environments. Springer Verlag, New York, 29–59.

Pan, Y. D., R. Stevenson, V. Jan, S. J. Panchabi & C. J. Richardson, 2000. Changes in algal assemblages along observed and experimental phosphorus gradients in a subtropical wetland, USA. Freshwater Biology 44: 339–353.

Pillsbury, R. W., R. L. Lowe, Y. D. Pan & J. L. Greenwood, 2002. Changes in the benthic algal community and nutrient limitation in Saginaw Bay, Lake Huron, during the invasion of the zebra mussel (Dreissena polymorpha). Journal of the North American Benthological Society 21: 238–252.

Pollingher, U., 1986. Non-siliceous algae in a five meter core from Lake Kinneret (Israel). Hydrobiologia 143: 213–216.

Poulickova, A., L. Pechar & M. Kummel, 1998. Influence of sediment removal on fishpond phytoplankton. Archiv Fur Hydrobiologie Supplement 124: 107–120.

Qin, B., P. Xu, Q. Wu, L. Luo & Y. Zhang, 2007. Environmental issues of Lake Taihu, China. Hydrobiologia 581: 3–14.

Rengefors, K., K. Pettersson, T. Blenckner & D. M. Anderson, 2001. Species-specific alkaline phosphatase activity in freshwater spring phytoplankton: application of a novel method. Journal of Plankton Research 23: 435–443.

Roberts, E., J. Kroker, S. Koerner & A. Nicklisch, 2003. The role of periphyton during the re-colonization of a shallow lake with submerged macrophytes. Hydrobiologia 506–509: 525–530.

Sala, M. M. & H. Guede, 1995. Influence of algae and crustacean zooplankton on patterns of microbial hydrolytic enzyme activities—an experimental approach. Aquatic Microbial Ecology 48: 143–154.

Štrojsová, A., J. Vrba, J. Nedoma, J. Komárková & P. Znachor, 2003. Seasonal study on expression of extracellular phosphatases in the phytoplankton of an eutrophic reservoir. European Journal of Phycology 38: 295–306.

Vrba, J., J. Komárková & V. Vyhnálek, 1993. Enhanced activity of alkaline phosphatases—phytoplankton response to epilimnetic phosphorus depletion. Water Science and Technology 28: 15–24.

Hydrobiologia (2007) 581:297–303
DOI 10.1007/s10750-006-0498-0

EUTROPHICATION IN LAKES

Toxicity evaluation of Meiliang Bay, Lake Taihu, China—a drinking water source

Yuangao Chen · Haigang Chen · Yulin Wu · Zhaoli Li · Liwei Sun · Mengmeng Qu · Zhiming Kong

Abstract In this study, potential toxicity in organic pollutants in Meiliang Bay, Lake Taihu, a drinking water source, was investigated using the comet assay and zebrafish embryo test, and two control sites in Lake Changdang and the Changjiang (Yangtze) River were established. For the genotoxicity assay, results showed that organic extracts from water samples induce DNA damage on human lymphocytes and mouse testicular cells. A statistically significant difference ($p < 0.01$) was observed versus the solvent control, as shown by multiple comparisons at a dose of 100 ml tube^{-1}. The degree of DNA damage caused by Meiliang Bay water was most serious (human lymphocytes cells—184 arbitrary units (AU); 234 AU on mouse testicular cells). Organic extracts also affected zebrafish embryo development. Embryo coagulation, axis abnormality, slow absorbability of vitellicle, and multi-edema related to teratogenesis at 96 hpf were observed. In the high dose group, there was obvious edema in the hearts and vitellicles for most dysplastic embryos. Toxic potential in organic pollutants in drinking water sources from the Yangtze River and Lake Changdang were less serious than Meiliang Bay. Therefore, the drinking water source in Meiliang Bay was unsafe compared to the Changjiang River and Lake Changdang, and it is exigent that Meiliang Bay water quality should be ameliorated further.

Keywords Meiliang Bay of Lake Taihu · Drinking water source · Organic pollutant · Toxicity evaluation · Comet assay · Zebrafish embryo test

Guest editors: B. Qin, Z. Liu & K. Havens
Eutrophication of shallow lakes with special reference to Lake Taihu, China

Y. Chen · H. Chen · Y. Wu · Z. Li · L. Sun · M. Qu · Z. Kong (✉)
State Key Laboratory of Pollution Control and Resource Reuse, Environmental School of Nanjing University, Nanjing 210093, China
e-mail: zmkong@nju.edu.cn

Y. Chen
Nanjing Institute of Geography and Limnology, Chinese Academy of Sciences, Nanjing 210008, China

Introduction

There are many methods used to evaluate drinking water toxicity. Application of alkaline single-cell gel electrophoresis (SCGE), or comet assay, in environmental bio-monitoring to clinical applications, DNA repair processes and genetic toxicology, has been explored broadly (Singh et al., 1988; Tice et al., 2000). The comet assay for human lymphocytes has been used to investigate polluted surface water (Zhong et al., 2001; Rajaguru et al., 2002). Results indicate that the comet assay is useful for assessment of aquatic environments. The comet assay is a rapid, simple and sensitive visual technique for detecting potential double-strained DNA breaks in individual cells (Zhu, 1995; Rojas et al., 1999).

Male mice can be used as an animal model for toxicity research in drinking water. Since mice are mammals, results may demonstrate potential adverse effects on humans. Organic extracts may damage male reproductive systems with testis as the target organ (Tian et al., 2002). DNA damage to reproductive cells may induce DNA abnormality and miscarriages as well as developmental abnormalities and cancer susceptibility in offspring (Haines et al., 2001). Since mice spermatid cells are easy to separate and possess particular metabolic activation, they are good for evaluating genotoxicity of chemicals on human reproductive cells (Zhang et al., 2001).

Compared with the conventional acute fish test, the zebrafish embryo test has many advantages, such as low cost, fewer impact factors and high sensitivity. Therefore, it may be used as a standard method. The zebrafish embryo test also may provide information for chemical toxicity studies during different zebrafish embryo growth stages. The zebrafish embryo test has been used to evaluate toxic substance and toxicological mechanisms (Gellert & Heinrichsdorff, 2001; Zhu & Shi, 2002; Strmac et al., 2002).

In this study, toxicity of organic pollutants in Meiliang Bay, Lake Taihu, in Wuxi, Jiangsu province, China, was evaluated using the comet assay, on human peripheral blood lymphocytes and mouse testicular cells, and the zebrafish embryo test. Simultaneously, two control sites in the Yangtze River and Lake Changdang were established. The objective was to develop a simple modeling method for drinking water quality monitoring.

Materials and methods

Materials

Human peripheral blood samples were collected from two volunteers (healthy, non-smoking, adult male) by Nanjing Center Blood Bank (China).

Healthy male mice (Kunming strain) weighing 28–32 g were obtained from Animal Breeding Center of Nanjing University and used after one week of acclimatization. The animals were housed in plastic cages in a temperature-controlled room (25 ± 2°C) with a natural light-dark cycle and were given commercial food and water.

Zebrafish (*Danio rerio*) were obtained from a local pet store. A suitable number of healthy fish (females to males ratio was 1:2) were kept in a 100 l tank filled with water, which was prepared according to ISO (1996) (For per liter water, adding 294 mg $CaCl_2{\cdot}H_2O$, 123 mg $MgSO_4{\cdot}7H_2O$, 65 mg $NaHCO_3$ and 6 mg KCl,). Water was maintained at 26 ± 1°C, DO > 7 mg l^{-1}, with a light/dark cycle of 14/10 h. Fish was fed daily.

Sampling procedures

Three water samples were collected from three sites (Fig. 1), the drinking water source in Meiliang Bay, Lake Taihu, in Wuxi (SL.1), the center of Lake Changdang in Jintan (SL.2), and the Changjiang River in Changzhou (SL.3). Water samples (100 l) were collected from each site at 0.5 m depth in December 2003. After being stored 24 h, each water sample was filtered with gauze and filter paper, to remove suspended material or sediment, and then passed though a column with non-polar neutral resin (XAD-2) to adsorb organic pollutants at a flow rate of 30–40 ml min^{-1}. The column was eluted with carbinol, acetone and dichloromethane. The elution was dried by blowing with nitrogen at 50°C and re-dissolved in 2.0 ml dimethysulphoxide (DMSO).

The comet assay on human peripheral lymphocyte cells

To obtain the peripheral lymphocytes, human blood samples were mixed with 3% glutin solution and subsided at 37°C for 30 min. The layer containing lymphocytes (0.2 ml) were transferred into a test tube with 0.8 ml phosphate-buffered saline (PBS). After centrifuging at 3,000 rpm for 3 min, the supernatant was removed carefully. The pellet containing lymphocytes was re-suspended after being exposed to different solutions containing the organics (i.e., each dosage equals

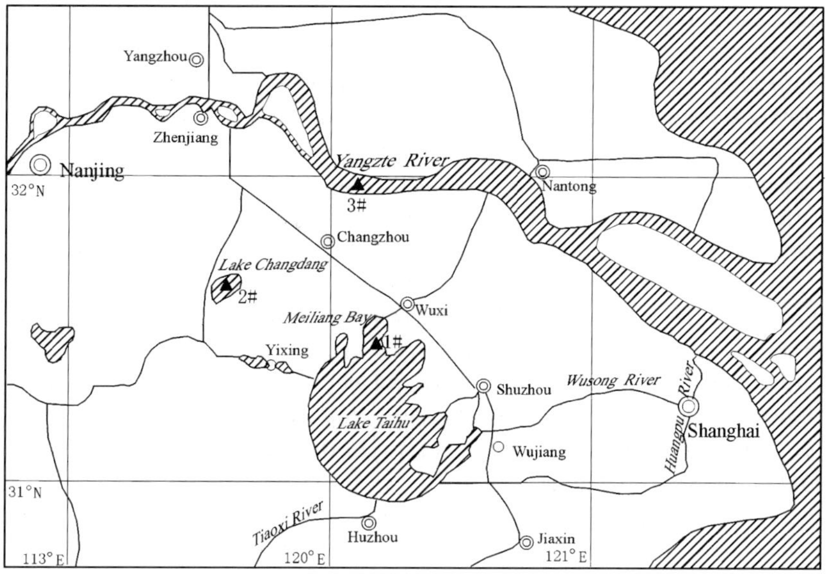

Fig. 1 Water sampling sites

20 ml, 100 ml, and 500 ml source water) in the same buffer including 0.1 ml water sample extract and 0.9 ml PBS incubated for 1 h at 37°C. DMSO was used as a control. Viability of individual cells (>90%) was determined by the Trypan blue exclusion technique.

The comet assay was performed as described by Singh et al. (1988) with some modifications. Briefly, frosted microscope slides, on which cells were embedded in an agarose "sandwich", were submerged in cold lysis buffer (2.5 M NaCl, 100 mM EDTA, 10 mM Tris, adjusted to pH 10.0 with NaOH, 1% Triton X-100 and 10% DMSO both freshly added) at 4°C in the dark for at least 30 min to remove cellular proteins. Electrophoresis was conducted at 18°C with 25 V and 100 mA for 1 h. The experiment was done in red or yellow light to avoid DNA damage. DNA was stained with ethidium bromide, and slides were examined with a fluorescent microscope (BX41, Olympus, Japan). One observer evaluated the slides to minimize variability. Images were analyzed according to Collins et al. (1995). One hundred comets on each slide were scored visually as one of five classes, according to tail intensity, and given a value of 0, 1, 2, 3, or 4 (from undamaged, 0, to maximally damaged, 4). "Arbitrary units" (AU) were used to express DNA damage and was calculated as:

$$\text{Arbitrary units} = \sum_{i=0}^{4} i \times Ni$$

where Ni = frequency in class i; i = DNA damage class (0, 1, 2, 3, 4).

Comparisons of DNA damage induced by control and treated groups were analyzed using a one-way ANOVA using the SPSS software. Since studies have revealed that high dose exposure might damage DNA badly, a very high dose would result in a decline of visual DNA damage for the massive dissociative fraction to be missed in electrophoresis, and a very low dose made differences among groups invisible (Devaux et al., 1997). Therefore, in this study, genotoxicity of water samples was compared at 100 ml tube^{-1}. In addition, differences among groups were analyzed using the Duncan test with a significance level of 0.05.

The comet assay on mouse spermatid cells

Cell suspension from spermaries of adult male mice was performed according to Zhang et al. (2001). Briefly, mouse spermatid was extracted with 90% Dulbecco's Modified Eagle Medium (DMEM) for 1 h at 37°C. The dosage for tested water was the same as the human lymphocytes comet assay. Treated cells were centrifuged, and DMSO was used as a control. Individual cell viability (>90%) was determined by the Try-pan blue exclusion technique.

Zebrafish embryo test

Zebrafish eggs were obtained according to OECD (1998). Eggs were collected in 30 min and washed with tap water. A suitable number of eggs were transferred into prepared chemical concentrations. Eggs were observed and non-fertilized eggs were discarded. The experiment was considered valid when at least 90% of the embryos in the control were viable after 72 h. After hatching, embryos were transferred into wells in a 24-well multi-plate (1 embryo per well). Twenty wells were prepared with 2 ml water sample. Four wells were filled with egg water and were internal controls. DMSO concentration was not more than 0.25%. Tests were replicated at least three times (60 eggs per concentration and 5 concentrations tested for each sample) consecutively. Eggs were incubated at 26 ± 1°C with a light/dark cycle of 12/12 h.

Embryonic development was observed with an inverted microscope (TMD-EF, Nikon, Japan) and recorded daily. The toxicological endpoint was embryo mortality. In addition, differences among treated groups and the blank were analyzed with the χ^2 test.Table 1.

Results

Water samples from SL.1 and SL.2 (500 ml tube^{-1}) damaged DNA so badly that results could not be evaluated. Results from the human peripheral lymphocyte experiment show that all water samples extracts induced DNA damage, and a statistically significant difference ($p < 0.01$) was observed versus the solvent control. Results also showed that the DNA damage score increased with extract concentrations. Comparing genotoxicity of SL.1, 2 and 3 at 100 ml/tube^{-1}, the degree of DNA damage in human peripheral lymphocytes treated with SL.1 water was worst. A typical comet picture is shown in Figs. 2 and 3.

Results from the mouse spermatid experiment are shown in Table 2 and demonstrate that all water sample extracts induced DNA damage. A statistically significant difference ($p < 0.01$) was observed versus the solvent control group. With increasing dose, DNA damage became more extensive. Comparing the genotoxicity among the three water samples at 100 ml tube^{-1}, the order of the degree of DNA damage was SL.1 > SL.2 > SL.3.

Zebrafish embryos were affected by water sample organic extracts (SL.1, SL.2 and SL.3). Among them, SL.1 was the most toxic (Table 3). Compared with the blank control, SL.3 affected the 96 h-hatch-rate dramatically when organics were concentrated by 125. For SL.1 and SL.2, concentrated by 7.8 and 62.5, respectively, toxicity also was significant ($p < 0.01$).

Discussion

Jiangsu province has a developed chemical industry, and the chemical composition of organic

Table 1 DNA damages in human peripheral blood lymphocytes (mean ± SD) exposed to organics from different water samples

		Frequency (%)					
		Class 0	Class 1	Class 2	Class 3	Class 4	
SL.2	20	0.74 ± 1.27	86.61 ± 1.91	11.68 ± 0.17	0.97 ± 0.69	0	112.89 ± 0.66**
	100	0	42.06 ± 6.19	34.93 ± 8.17	20.00 ± 2.36	3.01 ± 1.30	183.95 ± 5.57**
	500	0	1.61 ± 1.21	51.84 ± 3.40	41.35 ± 4.58	5.20 ± 1.81	250.14 ± 1.95**
	20	29.56 ± 4.79	49.18 ± 2.14	21.25 ± 3.73	0	0	91.69 ± 8.32**
	100	0	37.34 ± 12.37	46.46 ± 10.36	16.04 ± 4.04	0.16 ± 0.35	179.01 ± 15.48**
	500	0	14.54 ± 6.86	40.68 ± 7.39	41.32 ± 5.16	3.46 ± 1.50	233.71 ± 9.39**
	20	10.07 ± 10.18	76.74 ± 8.47	13.20 ± 8.63	0	0	103.13 ± 16.87**
	100	0	66.44 ± 13.38	31.43 ± 13.18	2.13 ± 1.07	0	135.69 ± 13.64**
	500	0	33.75 ± 9.84	54.87 ± 12.28	10.21 ± 7.38	1.17 ± 1.43	178.80 ± 12.85**
Solvent control	0	73.02 ± 3.20	25.67 ± 3.16	1.30 ± 0.70	0	0	28.28 ± 3.39

** $p<0.01$

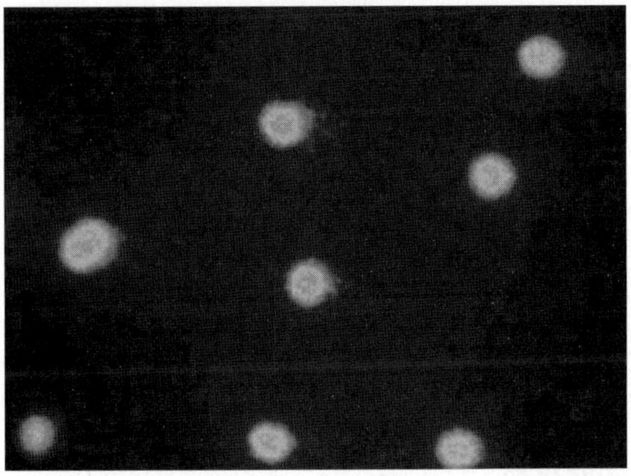

Fig. 2 The comet picture of DMSO on human peripheral blood lymphocytes (×400)

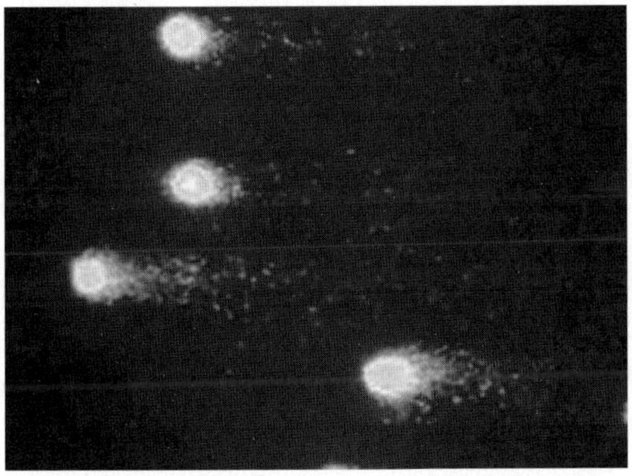

Fig. 3 The comet picture of a sample at the dose of 500 ml/tube^{-1} on human peripheral blood lymphocytes (×400)

extracts from the river and lake was mostly anthropogenic. There were more than 40 priority pollutants (such as pesticides, hydroxybenzenes and heterocyclics). The present study suggests that concentrated organics from water sample extracts were evident in genotoxins and could damage DNA of human peripheral blood lymphocytes and mouse spermatids. Additionally, mouse spermatid DNA damage was so severe in the high dose groups of samples SL.1 and 2 that heads and tails of some comets were separated from each other, which indicates that extracted organic pollutants may have promoted programmed cell death. Compared with human peripheral blood lymphocytes, mouse spermatids may be more sensitive to organic toxins. Mouse germ cells were more sensitive than somatic cells when evaluating potential harmful effects of organic pollutants in tap water (Wu et al., 1999). However, results of the comet assay showed DNA damage in the mouse spermatid control was more serious than human peripheral blood lymphocytes. Therefore, the isolation and culture of mouse spermatids were adjusted.

In the zebrafish embryo test, three aspects were considered when choosing the toxicological endpoints observed, including measurement, obvious toxic reaction and easy operation. The toxicological endpoint was the embryo hatch rate at 96 hpf (hour post fertilization). However, many kinds of dysplasia were observed, such as embryo coagulation and agensis correlated with death, axis abnormality, slow absorption of the vitellicle and

Table 2 DNA damages in mouse spermatid (mean ± SD) exposed to organics from different water samples

		Frequency (%)					
		Class 0	Class 1	Class 2	Class 3	Class 4	
	20	15.34 ± 1.64	48.70 ± 2.52	35.56 ± 3.50	0.40 ± 0.50	0	121.03 ± 4.76**
	100	0	16.93 ± 3.19	39.38 ± 3.55	36.20 ± 4.02	7.49 ± 1.23	234.25 ± 5.09**
	500						
	20	25.93 ± 1.75	45.02 ± 6.02	28.24 ± 6.32	0.80 ± 0.76	0	103.91 ± 8.43**
	100	5.56 ± 1.79	17.24 ± 2.57	37.16 ± 7.27	37.14 ± 3.87	2.91 ± 1.54	214.61 ± 5.89**
	500						
	20	36.44 ± 5.44	51.87 ± 5.81	11.69 ± 1.88	0	0	75.26 ± 5.71**
	100	1.91 ± 1.31	52.94 ± 6.45	41.33 ± 4.03	3.82 ± 2.53	0	147.05 ± 6.26**
	500	0	13.10 ± 3.78	43.39 ± 2.19	38.89 ± 6.50	4.61 ± 2.54	253.00 ± 8.47**
Solvent control	0	51.74 ± 5.25	47.53 ± 5.02	0.74 ± 1.27	0	0	49.00 ± 5.76

** $p<0.01$

Table 3 The 96-h-hatch rate of zebrafish embryo (mean ± SD) affected by different water samples

	Frequency (%)		
	SL.1	SL.2	SL.3
7.8125	66.8 ± 6.2**	93.3 ± 4.7	85.0 ± 14.7
15.625	65.0 ± 7.0**	85.0 ± 7.0	78.3 ± 6.2
31.25	36.7 ± 6.2**	81.7 ± 10.2	78.3 ± 11.8
62.5	0.0**	0.0**	90.0 ± 4.0
125	0.0**	0.0**	38.3 ± 28.9**
Blank control	87.7 ± 9.0		

** $p<0.01$

multi-edema related to teratogenesis at 96 hpf. In the high organics dose from SL.1 and 2, there was obvious edema in the hearts and vitellicles for most dysplastic embryos. The endpoint of teratogenesis was more sensitive than the death endpoint. Some studies showed that minnows had a similar toxicological endpoint for toxins (Strmac et al., 2002). Therefore, the nonspecific reaction to toxin in early life stages, such as acantha distortion, the slow absorption of the vitellicle and formation of edema are worthy of further study.

When exposed to toxin, the zebrafish embryo egg membrane protected the embryo. Selective permeability of the egg membrane changed with growth time and toxin differences. Due to underdevelopment of the embryo metabolizing system, interactions between the toxin and the enzyme system was difficult to observe in the test.

The report of The Environment Status Communique of Jiangsu Province in 2002 indicated that water quality in the Changjiang River was in the secondary category standard for surface water. However, water quality in Lake Taihu was worse than the fifth category standard. Lake Changdang is the third largest lake in Taihu basin, and its water quality approached the third category standard and became worse in recent years. From this study, results of the comet assay (human peripheral blood lymphocytes and mouse spermatid) and the zebrafish embryo test were consistent with the water quality of each water source. The three methods mentioned above could detect biologic toxicity of organic pollutants in water effectively. The research demonstrated that the drinking water source in Meiliang Bay was worst and should be ameliorated further prior to those in Lake Changdang and the Changjiang River.

Acknowledgements We would like to gratefully acknowledge financial support from Ministry of Science and Technology of China (No. 2002AA601013), National Natural Science Foundation of China (No. 40501078) and Chinese Academy of Sciences (No. KZCX1-SW-12, KZCX2-311).

References

Collins, A. R., A. G. Ma & S. J. Duthie, 1995. The kinetics of repair of oxidative DNA damage (strand breaks and oxidised pyrimidines) in human cells. Mutation Research 336: 69–77.

Devaux, A., M. Pesonen & G. Monod, 1997. Alkaline comet assay in rainbow trout hepatocytes. Toxicology in Vitro 11: 71–79.

Gellert, G. & J. Heinrichsdorff, 2001. Effect of age on the susceptibility of zerbrafish eggs to industrial wastewater. Water Research 35: 3754–3757.

Haines, G. A., J. H. Hendry & C. P. Daniel, 2001. Increased levels of comet-detected spermatozoa DNA damage following in vivo isotopic-or X-irradiation of spermatogonia. Mutation Research 495: 21–32.

OECD, 1998. Fish, short-term toxicity test on embryo and sac-fry stages. OECD Guideline for Testing of Chemicals, OECD TG212, Paris.

Rojas, E., M. C. Lopez & M. Valverde, 1999. Single cell gel electrophoresis assay: methodology and applications. Journal of Chromatography B 722: 225–254.

Rajaguru, P., L. Vidya & B. Baskarasethupathi, 2002. Genotoxicity evaluation of polluted ground water in human peripheral blood lymphocytes using the comet assay. Mutation Research 517: 29–37.

Singh, N. P., M. T. McCoy & R. R. Tic, 1988. A simple technique for quantity of low levels of DNA damage in individual cells. Experiment Cell Research 175: 184–191.

Strmac, M., A. Oberemm & T. Braunbeck, 2002. Effects of sediment elutes and extracts from differently polluted small rivers on zebrafish embryos and larvae. Journal of Fish Biology 61: 24–38.

Tian, H. J., D. S. Wu & R. Tao, 2002. Study of male reproductive toxicity induced by organic pollutants in drinking water. Acta Academiae Medicinae Militaris Tertiae 24(9): 1076–1078 (in Chinese).

Tice, R. R., E. Agurell & D. Anderson, 2000. Single cell gel/comet assay: Guidelines for in vitro and in vivo genetic toxicology testing. Environmental and Molecular Mutagenesis 35: 206–221.

Wu, K., C. H. Song & F. Liu, 1999. Evaluation of potential harmful effects of organic pollutants in tap water on mice in vivo. Chinese Journal of Preventive Medicine 33(3): 149–151 (in Chinese).

Zhong, Y., S. L. Feng & Z. M. Kong, 2001. Evaluating the genotoxicity of surface water of Yangzhong city, using the micronucleus test of *Vicia faba* and the comet assay. Bulletin of Environmental Contamination and Toxicology 67(2): 217–224.

Zhu, H. G., 1995. Comprehensive evaluation index of organic mutagens in water. Environment Science of Shanhai 14(10): 44–49 (in Chinese).

Zhang, Z. Z., Z. Z. Heng, Y. Liao & R. Zhao, 2001. Study on DNA damage induced by indirect mutagens using comet assay in testicular cells in vitro. Carcinogenesis, Teratogenesis and Mutagenesis 13(1): 4–7 (in Chinese).

Zhu, L. & S. J. Shi, 2002. Utilization of embryo development technique of zebrafish to evaluating toxicity on various chemicals. Chinese Journal of Application Ecology 13(2): 252–254 (in Chinese).

Hydrobiologia (2007) 581:305–318
DOI 10.1007/s10750-006-0509-1

EUTROPHICATION IN LAKES

Updating water quality targets for shallow Lake Balaton (Hungary), recovering from eutrophication

Vera Istvánovics · Adrienne Clement · László Somlyódy · András Specziár · László G.-Tóth · Judit Padisák

Abstract The paper presents an overview about recovery of shallow Lake Balaton from eutrophication by assessing quantitative and qualitative changes in phytoplankton, zooplankton, and chironomids as a function of load reduction. The aim was to update the present water quality targets. The proposed targeting scheme supplements the existing one with a range of lake-specific ecological criteria. We conclude that simple targets (desired phytoplankton biomass and permissible load) are the best choice during the initial stage of eutrophication management, but more complex schemes including ecological criteria are needed to trace recovery when re-organization of the ecosystem takes place.

Guest editors: B. Qin, Z. Liu & K. Havens
Eutrophication of shallow lakes with special reference to Lake Taihu, China

V. Istvánovics (✉) · A. Clement · L. Somlyódy
Department of Civil and Environmental Engineering, Budapest University of Technology and Economics, Műegyetem rkp. 3., Budapest 1111, Hungary
e-mail: istvera@goliat.eik.bme.hu

A. Specziár · L. G.-Tóth
Balaton Limnological Research Institute, Hungarian Academy of Sciences, Klebelsberg Kuno u. 3., Tihany 8237, Hungary

J. Padisák
Department of Limnology, University of Veszprém, Egyetem u. 10., Veszprem 8200, Hungary

Keywords Phosphorus load · Sediment behavior · Phytoplankton · Cyanobacteria · Zooplankton · Chironomid fauna

Introduction

Water quality targets serve as a measuring scale during eutrophication management that orients managers and the public in questions like "How much can the status of the lake be improved?", "What has been achieved by the implemented measures?", "Which are the main tasks of future management?" Target conditions are usually defined when intolerable economic and ecological consequences of eutrophication call for immediate action that is when the ecosystem had spectacularly transgressed steady state. In the best case, targets are set up using information about past steady state and changes observed during nutrient enrichment. The implicit initial assumption is that apart from some delay, the path of recovery will mirror the trajectory of eutrophication with a resulting steady state identical to the pre-eutrophication status of the lake. Full-scale recovery, however, has not been convincingly confirmed yet, and comprehensive studies of lake recovery suggest hysteretic effects due to both internal P load and biological resilience.

Sas (1989) and coworkers have recognized four stages during lake recovery. No decrease in algal biomass can be expected until external and internal P loads decrease below the level at which P regains its role in determining carrying capacity of the lake (Stage 1). In Stage 2 transparency shows a disproportionately large increase relative to the decrease in area-specific algal biomass due to dispersion of buoyant cyanobacteria and/or motile algae to greater depths with high nutrient concentration. This behavioural response to load reduction can be observed in deep lakes with stable stratification but not in shallow lakes. Once P has become growth limiting, the decrease of phytoplankton biomass is proportionate to the decrease in P supply (Stage 3). During Stage 4, a further reduction in algal biomass is coupled to floristic changes, most conspicuously to the retreat of cyanobacteria. Due to random fluctuations in the biomass and composition of phytoplankton, the various stages have no fixed boundaries. Although not considered in the study of Sas (1989), full recovery of a lake certainly requires one more step (Stage 5): regeneration of former macrovegetation and food web structure (Scheffer et al., 1993). Considering this general pattern of recovery as well as deviations of individual lakes from the "rule", occasional updating of water quality targets is inevitable to maintain the orienting function of these targets over the whole recovery process.

In the present study we overview the recovery of large (596 km^2), shallow (z_{mean} = 3.2 m) Lake Balaton in context of the above general model. Individual processes including the relationship between P load and phytoplankton biomass (Istvánovics & Somlyódy, 2001; Istvánovics et al., 2002), floristic changes of phytoplankton in response to restoration measures (Padisák & Reynolds, 1998), long-term changes in the biomass, composition and production of zooplankton (G.-Tóth, 2000), and the relationship between phytoplankton production and long-term dynamics of the chironomid fauna (Specziár & Vörös, 2001) have been published separately. The present overview was motivated by the practical aim of updating water quality targets considering both the observed response of the ecosystem to eutrophication management and realistic perspectives of further reduction in nutrient loads.

Data and methods

In order to detect trends and clarify patterns, we attempted to use as long data sets as possible. This could be achieved by supplementing the authors own data with critically evaluated historical records from both published sources and various unpublished archives. Whenever quantification was feasible, simple empirical relationships were applied.

Recovery of the lake was traced using one or both of the two independent approaches: (i) analysis of the trajectory of oligotrophication within the same lake area and (ii) comparison of the areas that had previously attained hypertrophy and mesotrophy/slight eutrophy. The advantage of the latter approach is that the influence of random variability caused by physical forcing is partly cancelled (Herodek, 1984). At the same time, poorly explored natural gradients within the lake may impose limitations to the application of this method.

Lake Balaton, its eutrophication and restoration

On the basis of large-scale circulation patterns (Shanahan et al., 1986), Lake Balaton can be divided into four basins (Fig. 1). Mean depth and surface area of the basins increase from west to east with a concomitant decrease in the area of the corresponding sub-watershed (Fig. 1). The largest tributary, the Zala River enters Basin 1, whereas the only outflow is the Sió Canal at the opposite end of the lake. The lake is extremely calcareous with high magnesian calcite making up 50–60% of sediments (Csermák & Máté, 2004). Turbidity is high particularly during summer due to both wind induced sediment resuspension and slow sedimentation of the precipitating carbonates (Istvánovics et al., 2004). Strong wave action restricts macrophyte growth to a narrow littoral zone (during the past 150 years, maximum 10% of lake area; Virág, 1998).

Specific morphometry gives rise to environmental gradients along the longitudinal axis from increasing water retention time to shifts in major ion composition and to decreasing area-specific nutrient loads. One can expect associated

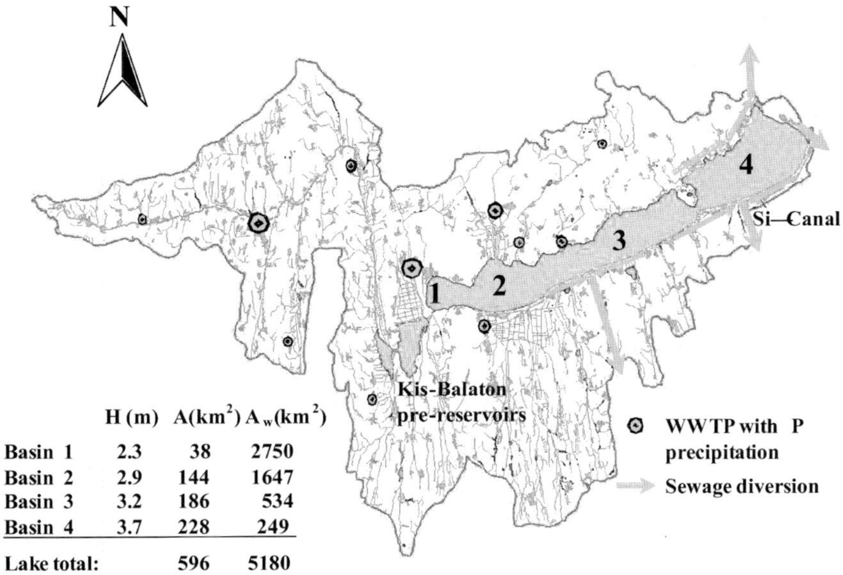

Fig. 1 Lake Balaton and its catchment. (H—average depth, A—surface area, Aw—area of sub-catchment, WWTP—waste water treatment plant)

differentiation in the biotic components under pristine conditions, but the available evidence is faint.

Eutrophication dramatically magnified many of the environmental gradients and particularly the biotic ones. By the end of the 1970s, area-specific P load in Basin 1 exceeded that in Basin 4 by a factor of 8 (Table 1, Somlyódy & Jolánkai, 1986). Primary production has increased 6–8-fold in Basin 1 and doubled in Basin 4 (Herodek, 1986). Blooms of N_2-fixing cyanobacteria developed regularly during summers in the western basins. Four years after its first appearance in 1978 (Padisák, 1997), invasive *Cylindrospermopsis raciborskii* (Woloszynska) Seenayya et Subba Raju became the dominant summer species in each basin (Herodek, 1986; Padisák & Reynolds, 1998; Istvánovics et al., 2002). Significant differences between the western and eastern areas have been reported in composition and production of

Table 1 Multi-annual mean loads (±sd) of total P in tons P year^{-1}

Source	Lake Balaton	Basin 1	Basin 2	Basin 3	Basin 4
Before eutrophication management (1980–1988)					
Geochemical background	18 ± 4	10 ± 2	6 ± 1	2	1
Sewage	61 ± 15	31 ± 7	10 ± 1	5 ± 4	16 ± 10
Urban runoff	56 ± 9	16 ± 4	19 ± 4	10 ± 1	10 ± 2
Agricultural (rural) runoff	137 ± 30	39 ± 17	42 ± 14	30 ± 5	26 ± 4
Atmosphearic	17 ± 2	1	4 ± 1	5 ± 1	7 ± 1
Total	289 ± 42	97 ± 23	80 ± 19	53 ± 4	60 ± 10
Through-flow			20 ± 4	21 ± 7	12 ± 6
After eutrophication management (1989–2002)					
Geochemical background	17 ± 6	9 ± 3	5 ± 2	2 ± 1	1
Sewage[a]	11 ± 3	7 ± 2	3 ± 1	1	0
Urban runoff	34 ± 11	9 ± 3	12 ± 4	6 ± 2	7 ± 3
Agricultural (rural) runoff	74 ± 27	20 ± 11	23 ± 8	14 ± 5	17 ± 6
Atmospheric	10 ± 5	1	2 ± 1	3 ± 2	4 ± 2
Total	145 ± 47	45 ± 19	46 ± 14	26 ± 8	29 ± 9
Through-flow			18 ± 9	21 ± 12	15 ± 11

a—value in 2002

the chironomid fauna (Dévai & Moldován, 1983; Ponyi et al., 1983) as well as in those of fish (Bíró & Vörös, 1990; Bíró, 1997).

In addition to its ecological values, Balaton is an internationally recognized recreational lake. Eutrophication caused significant economic loss in times when two thirds of Hungary's tourist income originated from the Balaton region. An inclusive eutrophication management plan was approved by the government in 1983. The aim was preventing further deterioration in water quality by 1990 and achieving the trophic conditions experienced during the early 1960s in the period 2005–2100 (Láng, 1986). Water quality targets and permissible phosphorus loads were also expressed in quantitative terms.

The most important management measures included (i) sewage diversion from shoreline settlements around Basin 4 and along the southern shore of Basin 3, (ii) introduction of chemical P precipitation at the large waste water treatment plants in the watershed, and (iii) pre-reservoir construction on the Zala River (Fig. 1). In the meantime, Hungarian agriculture has collapsed following the political changes in 1989. From the perspective of eutrophication, the most important consequence was a 90% drop in fertilizer application. Two periods of drought during 1990–1993 and 2000–2003 further decreased diffuse load of nutrients (Fig. 2). Multi-annual mean load of P decreased by about 50% in each basin (Clement et al., in press; Table 1). Considering, however, the additional load originating from through-flow, the reduction was only 30–40% with the exception of Basin 1 (Table 1).

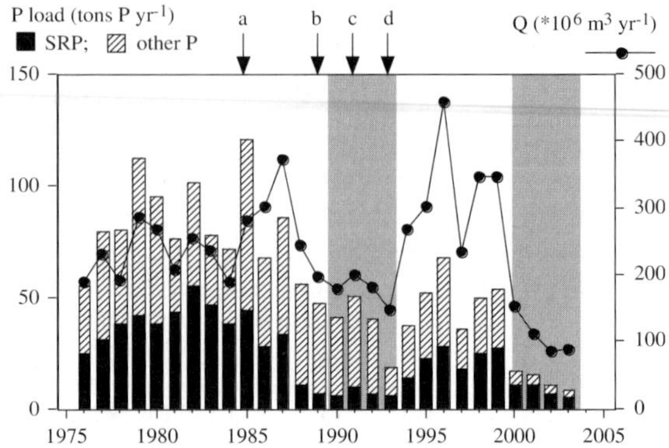

Fig. 2 Loads of soluble reactive P (SRP) and total P (sum of SRP and other P forms), and flow (Q) at the mouth section of the Zala River. (Arrows indicate timing of load reduction (Fig. 1) as follows: a—inundation of the Upper Kis-Balaton reservoir; b—sudden drop in fertilizer application due to the collapse of agriculture; c—introduction of chemical P removal at the sewage treatment plant of Zalaegerszeg, d—inundation of the Lower Kis-Balaton reservoir. Shaded areas indicate two periods of draught)

Results and discussion

The relationship between P load and algal biomass

In the western basins, annual mean biomass of algae decreased significantly after a delay of about one decade (Fig. 3). Although the last few summers were unusually warm and dry, the former adverse conditions have not returned. In the eastern basins, particularly in Basin 3 the improvement was marginal if any. In the period 1995–2003, annual average biomass of phytoplankton exceeded the prescribed range in 5 years (Fig. 3). The mean deviation was the lowest (25%) in Basin 1 and the highest (55%) in

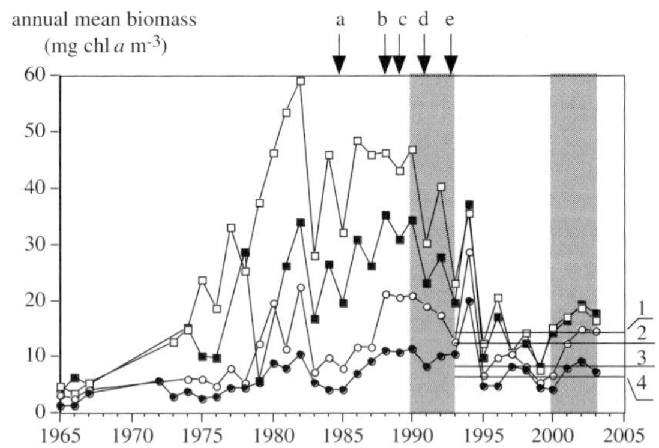

Fig. 3 Annual mean biomass of phytoplankton in the four basins of Lake Balaton. (Basin 1—open square, Basin 2—closed square, Basin 3—open circle, Basin 4—closed circle. Arrows indicate timing of load reduction (Fig. 1) as follows: a—inundation of the Upper Kis-Balaton reservoir; b—sewage diversion from the eastern basins; c—sudden drop in fertilizer application due to the collapse of agriculture; d—introduction of chemical P removal at the sewage treatment plant of Zalaegerszeg, e—inundation of the Lower Kis-Balaton reservoir. Shaded areas indicate two periods of draught. Numbered lines show water quality objectives to be achieved by 2010 in the four basins (cf. Table 2))

Basin 3. The conclusion is that in spite of the conspicuous improvement in the western areas, the lake is still far away from a full recovery.

Due to the large number of small, dynamic inflows (over 50, Fig. 1), annual load estimates are uncertain with the exception of Basin 1 where daily loads through the Zala River have been available from 1976 (Clement, 2000). Consequently, the relationship between algal biomass and phosphorus load can be traced at a higher resolution in this westernmost area than in other basins. The question is then how much the results are transferable to the other basins.

Pre-reservoir construction caused spectacular changes in the loading conditions of Basin 1 in addition to the reduction in absolute terms (Fig. 2). The most important ones were (i) an 80% decrease in the load of suspended solids, (ii) the increase in the calcium load related to total P and suspended solids loads by a factor of 2.5 and 4, respectively, and (iii) the decrease in the mean loading ratio of nitrate-N to phosphate-P below 2 (by weight) during summer. Changes in (i) and (ii) altered the apparent settling velocity of P and improved the composition of the freshly formed sediments (Istvánovics & Somlyódy, 2001), whereas (iii) may favor selection of N_2-fixing cyanobacteria.

Internal load of P is one of the basic processes that determine the path of recovery in shallow lakes (Marsden, 1989; Sas, 1989; Søndergaard et al., 1999). In well-oxygenated Lake Balaton with low concentrations of iron in the sediments (12–20 mg Fe g^{-1} dry mass), resuspension followed by P desoprtion is the predominant mechanism of transport (Boström et al., 1982; Lijklema et al., 1986). At long time scales relevant for recovery, transport can be considered constant provided that no significant changes occur in macrophyte coverage. In such lakes, long-term changes in the internal load depend on the concentration of mobile sediment P. These changes are rarely large enough to detect them directly (Søndergaard et al., 1999). Therefore we estimated long-term trends in mobile sediment P using the mass balance model of Lijklema and coworkers (1986) for the upper sediment layer. Since loads of Ca and suspended solids were available only for the Zala River, additional sources (about 5–10% of the total load) were neglected.

The model (Fig. 4) assumes that the "active" layer is fully mixed to a constant depth of h (m) due primarily to the burrowing activity of

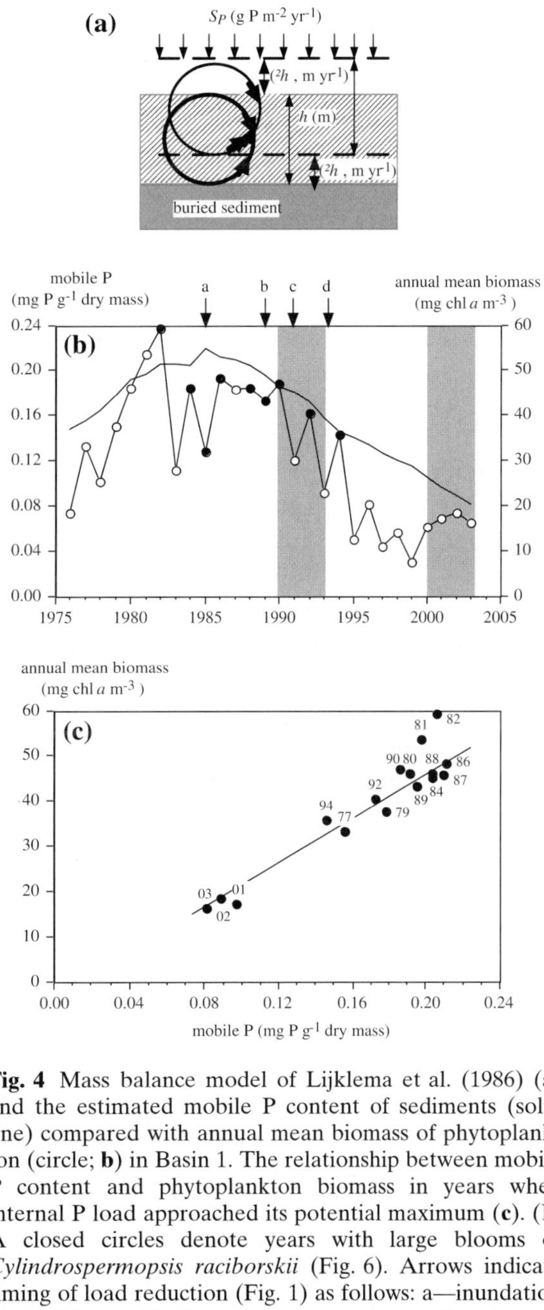

Fig. 4 Mass balance model of Lijklema et al. (1986) (**a**) and the estimated mobile P content of sediments (solid line) compared with annual mean biomass of phytoplankton (circle; **b**) in Basin 1. The relationship between mobile P content and phytoplankton biomass in years when internal P load approached its potential maximum (**c**). (In A closed circles denote years with large blooms of *Cylindrospermopsis raciborskii* (Fig. 6). Arrows indicate timing of load reduction (Fig. 1) as follows: a—inundation of the Upper Kis-Balaton reservoir; b—sudden drop in fertilizer application due to the collapse of agriculture; c—introduction of chemical P removal at the sewage treatment plant of Zalaegerszeg, d—inundation of the Lower Kis-Balaton reservoir. Shaded areas indicate two periods of draught. In B parameters of the regression line are $y = 234.94x$, $r^2 = 0.88$, $n = 16$)

zoobenthos. The rate of sedimentation (Δh, m year^{-1}) can be estimated from the mass balance of suspended solids and calcium. To keep the mixed depth constant, burial must equal the rate of sedimentation. Rate of P sedimentation (S_P, g P m^{-2} year^{-1}) can be derived from the mass balance of total P. The change in the P concentration of the active layer (P, g P m^{-3}) is

$$P = \left(P_0 - \frac{S_P}{\Delta h + kh}\right) \cdot \exp\left(-\frac{\Delta h + kh}{h} t\right) + \frac{S_P}{\Delta h + kh}$$

where P_0 is the initial concentration at $t = 0$ and k (year^{-1}) is the first-order "nutrient decay" rate. For conservative nutrients (like total P) $k = 0$, whereas for non-conservative ones (like total N or mobile P) $k > 0$. Diagenetic processes including re-crystallization of carbonates and exchange of Ca for Mg may result in gradual immobilization of mobile P in the sediments of Lake Balaton. The new equilibrium concentration is $P_{eq} = S_P/(\Delta h + kh)$. The rate of approaching to equilibrium is controlled by the exponent, $\tau = h/(\Delta h + kh)$.

Using the parameters calibrated in 1999 (P_0 = 67 g P m^{-3} = 0.15 mg P g^{-1} dry mass, h = 0.15 m, and k = 0.1 m year^{-1}; Istvánovics & Somlyódy, 2001), the rapid decrease in the mobile P pool is still in progress (Fig. 4). Considering that the concentration of mobile P sets an upper limit to the internal P load, the output of the model can be interpreted as the trend in the potential maximum of internal P load. This maximum has been fully exploited (i) during the initial period after the reduction in the external load, (ii) during the years of large blooms of *C. raciborskii* (there is a partial overlap between (i) and (ii)), and (iii) during the recent dry years when the weir of the Zala River was closed from about June to November. In other years physical constraints, biotic interactions or both might keep the biomass of phytoplankton below the highest achievable level.

In years when internal P load approached to the potential maximum, strong correlation (r^2 = 0.88, n = 16, Fig. 4) was observed between annual mean biomass of phytoplankton and concentration of mobile P. This suggests that the ultimate reason for the fast improvement in the western areas is the favorable long-term behavior of the highly calcareous sediments. The relationship can be used to predict future biomass of phytoplankton in Basin 1 under various loading scenarios.

The OECD approach (Vollenweider & Kerekes, 1982; Fig. 5) was used to explore the response of algal biomass to the reduction in external load of P in the various basins. Mean in-lake concentration of P (P, g P m^{-3}) was calculated from the area-specific P and hydraulic loads (l_{IN}, g P m^{-2} year^{-1} and q, m year^{-1}, respectively) taking into account both inflow and through-flow:

$$P_\lambda = \frac{l_{IN}}{q} \cdot \frac{1}{1 + 2\sqrt{\tau}}$$

where τ (year) is the theoretical water retention time.

Deteriorating water balance of the lake cancelled the effect of reduction in nutrient loads on P in the eastern basins (Fig. 5; Istvánovics et al., 2002). Pre- and post-management status of the various basins seemed to follow a common linear curve suggesting that the behavior of the different areas can be considered fairly similar for managerial purposes.

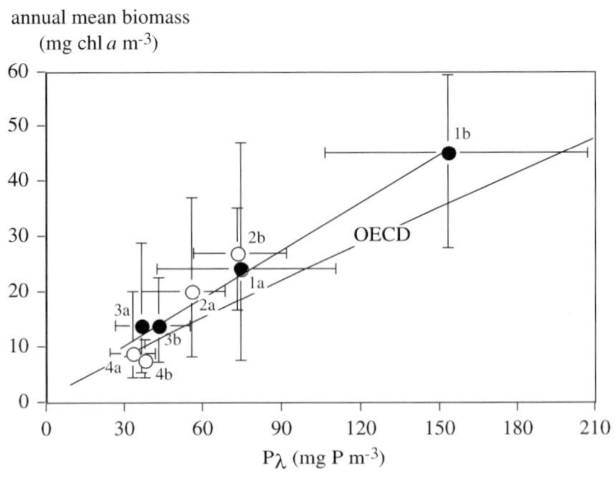

Fig. 5 Multi-annual mean biomass of phytoplankton as a function of the in-lake concentration of P (P_λ). (*X* and *Y* bars denote the minimum and maximum values during the period of averaging. Numbers refer to the four basins, "b" refers to the period before management (1980–1988), "a" refers to the period after management (1989–2003). The OECD model line and the linear trend for the four basins of Lake Balaton are given)

Changes in the composition of phytoplankton

In the eastern basins, phytoplankton composition and dynamics showed little difference during the 1970–1980s from those during the 1950s. *Cyclotella ocellata* Pantocsek and *C. compta* (Ehrenberg) Kützing dominated the spring bloom of diatoms. Main species of the summer maximum were *Ceratium hirundinella* (O.F. Müller) Schrank, meroplanktonic diatoms (*Aulacoseira granulata* (Ehrenberg) Simonsen, *A. granulata* var. *angustissima* (O. Müller) Simonsen and *Snowella lacustris* (Chodat) Komárek et Hindák. *Aphanizomenon klebahnii* Elenkin was also present but never dominant. In contrast to this, monomodal succession with a single large summer bloom characterized the phytoplankton in Basin 1 during the 1970s. Blooms of N_2-fixing cyanobacteria (mainly *Anabaena spiroides* Klebahn and *A. flos-aquae* (Lyngbye) Brébisson) might develop from as early as June and collapse as late as October. New cyanobacterial species were recorded almost each year. Floristic data collected during more than 70 years are clearly indicative of fast structural changes preceding the rapid increase in algal biomass.

Structural resilience could clearly be observed following the reduction in external loads. *Cylindrospermopsis raciborskii* continued to form large blooms in the western areas until 1994 (Fig. 6). A sudden shift occurred in 1995: the share of cyanobacteria has not exceeded 30% from this year and duration of the summer bloom shortened significantly (Fig. 6). Although the seasonal succession was monomodal in many years, the ratio of the summer to spring bloom tended to decrease.

In the eastern areas, the slight increase in annual mean biomass during the early post-management period (Fig. 3) coincided with the increasing share of *C. raciborskii* (Fig. 6). In the period 1988–1996, cyanobacteria contributed to the summer maximum by more than 20% in 7 years, whereas previously this occurred only in 1982 (Padisák, 1994; Padisák & Reynolds, 1998). The possible explanation is that blooms in the

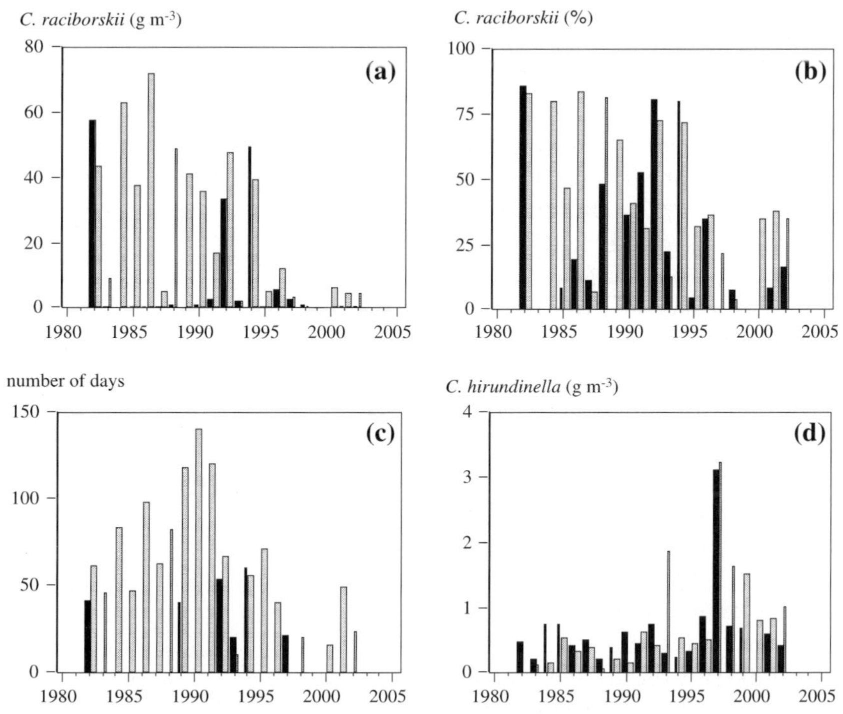

Fig. 6 Maximum biomass (**a**) and contribution to the total biomass during the cyanobacteria bloom (**b**) of *Cylindrospermopsis raciborskii*, number of days when total biomass exceeded 10 g m^{-3} (**c**), and maximum biomass of *Ceratium hirundinella* (**d**) in Basin 1 (dotted bar) and Basin 4 (closed bar). (No data were available in Basin 4 in 2000)

western areas may disperse in 3–4 weeks to inoculate the eastern ones (Istvánovics et al., 2002; Koncsos, 2003).

Apart from *C. raciborskii*, there are two species in Lake Balaton that deserve special attention due to their potential toxicity and/or ability to form excessive blooms. The one is *Aphanizomenon flos-aquae* (Lyngbye) Ralfs, which may form surface scum at the relatively low whole-column biomass of about 4 g m^{-3}. The other one is *Planktothrix agardhii* (Gomont) Anagnostidis et Komárek that bloomed first in the lake in 1989. Its share reached 20% and 50% of the maximum biomass in Basins 1 and 4, respectively. The first bloom might represent one of the unexpected consequences of pre-reservoir construction. By 1988, the upper reservoir (Fig. 1) developed into a hypertrophic pond with yearly mean chlorophyll concentration of 150 mg m^{-3} at the outflow. *P. agardhii* bloomed almost all around the year (Mátyás, 1996), and Basin 1 received a substantial *P. agardhii* "load" between 1988 and 1992. Although the reed-covered lower reservoir has fully eliminated this inoculation from 1993, the cyanobacterium established itself in the western areas of Lake Balaton and made up 30% of the maximum biomass in some years.

Ceratium hirundinella is one of the most successful canopy species of phytoplankton that was historically the prominent summer alga of Lake Balaton (Sebestyén, 1953). Its biomass remained relatively low and stable in Basin 4 (Fig. 6). In contrast to this, biomass of *C. hirundinella* was negatively correlated with that of *Cylindrospermopsis raciborskii* ($r = -0.59$) in Basin 1. The recent increase of *Ceratium hirundinella* (Fig. 6) is indicative of the beginning of structural recovery in the western areas.

In order to exploit the predictive power provided by compositional changes of phytoplankton, Padisák and coworkers (2003) elaborated a novel indicator, the *Q* value. Phytoplankton guilds/associations were identified in accordance with the phyto-sociological classification of Reynolds and coworkers (2002). Considering growth rates, life history, toxicity and distribution of species in 27 guilds, a water quality factor ranging from 2 to 10 was assigned to each functional group. A higher score indicates better quality. Five water quality classes were set up using the biomass-based relative frequency of the guilds weighted by the water quality factor. Presence of potentially toxic or bloom-forming species decreases the value of *Q* even if the total biomass is low. Vice versa, *Q* may attain a high value when the biomass is relatively high but bloom formation is unlikely in the absence of appropriate species.

The *Q* index was tested on phytoplankton time series from Lake Balaton at various temporal scales (Fig. 7). Annual mean biomass of phytoplankton was low in Basin 4 during the period 1983–1988. The *Q* index, however, qualified the area only as moderate or good. Indeed, annual mean biomass increased in the subsequent years

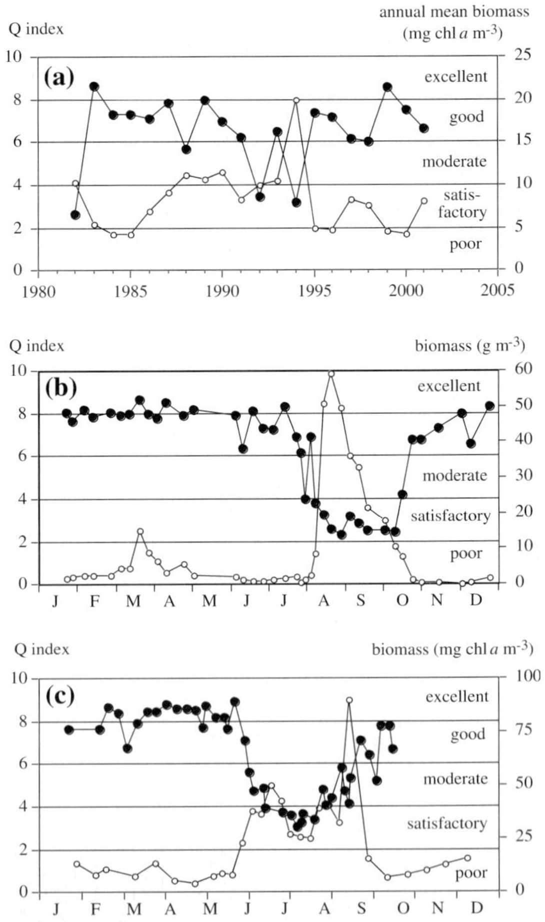

Fig. 7 The value of the *Q* index (closed circle) and the biomass of phytoplankton (open circle) in various years in Basin 4 (**a**), in 1994 in Basin 4 (**b**), and in 2002 in Basin 1 (**c**). (The *Q*-based classification of water quality is shown)

in spite of the reduction in nutrient loads (Fig. 3). At a finer scale, the predictive value of *Q* was even more striking. Low values of *Q* warned before the onset of the summer bloom of cyanobacteria that water quality problems were likely to occur.

Changes in the biomass and composition of crustacean zooplankton

Frequent sediment resuspension and intense carbonate precipitation makes Lake Balaton a highly stressful habitat for filter feeding zooplankton (G.-Tóth, 1992). Before eutrophication, *Eudiaptomus gracilis* (Kiefer, 1932) made up more than 80% of crustacean zooplankton. As an average, 8–10% of primary production was channeled to zooplankton. This is low in comparison to most other temperate lakes (G.-Tóth, 1991). The two important consequences are the prevalence of bottom–up regulation of zooplankton and low efficiency of energy utilization by fish (around 0.01%; Bíró & Vörös, 1982).

The moderate increase in the biomass of crustacean zooplankton by a factor of 1.5–3 was disproportionately low compared to the nearly order of magnitude increase in primary production in the western areas. The reasons of uncoupling might be manifold. (i) The share of inedible algae increased within the phytoplankton. The typically low community clearance rates decreased to negligible during blooms of filamentous cyanobacteria. (ii) Quality of ingestible algae might decrease due to frequent N deficiency. (iii) Elevated primary production further enhanced gross carbonate precipitation and thus counterbalanced the increase in the ratio of food to inorganic particles. In general, the most selective feeders (*Cyclops vicinus* (Uljanin, 1875), *Acanthocyclops vernalis* (Fischer, 1853), *A. vernalis* v. *robustus* (Sars, 1863)) benefited the most from eutrophication. Structural re-organization could be observed after 1996. Number of *E. gracilis* nauplii increased by a factor of 2–3 indicating the improved physiological condition of this species that took back its former leading position in the crustacean zooplankton.

Conclusively, crustacean zooplankton is hardly a suitable group for characterizing trophic status of Lake Balaton due to extreme physical constraints acting upon its composition and biomass.

Changes in the biomass and composition of the chironomid fauna

During the last 150–200 years, *Procladius choreus* (Meigen, 1804), *Tanypus punctipennis* (Meigen, 1818) and *Chironomus balatonicus* (Dévai, Wülker et Scholl, 1983) made up about 90% of the biomass and production of chironomid fauna in the profundal of Lake Balaton (Dévai & Moldován, 1983; Specziár & Vörös, 2001). Of them, carnivorous *P. choreus* was usually present in the highest densities.

Substantial structural and quantitative differences developed along the longitudinal axis of Lake Balaton during the period of eutrophication (Fig. 8; Dévai, 1990; Specziár & Vörös, 2001). The biomass remained low in Basin 4 (0.03–0.6 g dry mass m^{-2} depending on the season and the year), and *P. choreus* preserved its leading position. In the western areas, the biomass increased by a factor of 2–8 (0.2–4.25 g m^{-2})

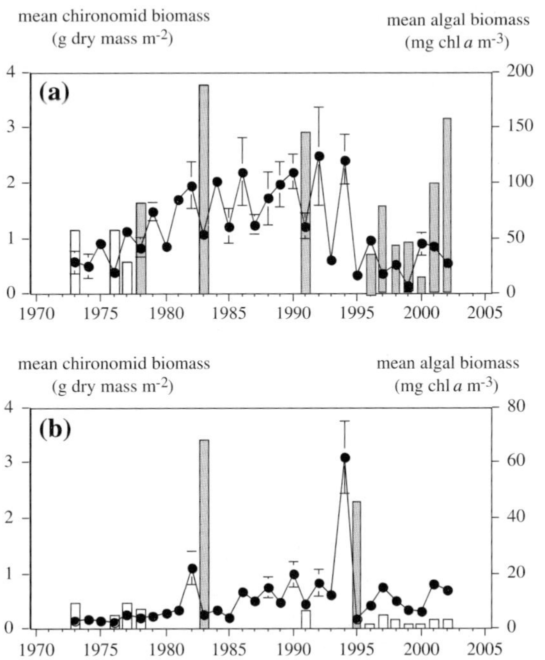

Fig. 8 Annual mean biomass of chironomid larvae (bars) and mean biomass (± sd) of phytoplankton (circle) during summer (July–September) in Basins 1 (**a**) and 4 (**b**)

indicating the relatively strong coupling between algal and chironomid production. The quantitative change was associated with a shift from *P. choreus* dominance to the dominance of phyto- and detritivorous *Ch. balatonicus*.

From the early post-management period, data are available only in 1991. Delayed response of phytoplankton seemed to postpone re-structuring of chironomid assemblages in Basin 1 (Fig. 8). From 1996, mean chironomid biomass tended to decrease but remained higher (0.3–3.1 g m^{-2}) than in the other basins (0.07–0.3 g m^{-2}). Considering the similar biomass of phytoplankton in Basins 1 and 2 (Fig. 3), the difference in the chironomid biomass between the two areas was surprising. The explanation lays in the divergent structure of chironomid assemblages. While *Ch. balatonicus* defended its dominance in Basin 1, *P. choreus* became once again the most abundant species in the other basins. This observation, along with structural changes during eutrophication suggests that high chironomid biomass (>0.5 g m^{-2}) is invariably associated with the dominance of *Ch. balatonicus*. Additional evidence also supports this contention. In Basin 4, two large chironomid peaks were observed following the bloom of *C. raciborskii* in 1982 and 1994. In both cases, *Ch. balatonicus* dominated the chironomid community (*cf.* Figs. 6 and 8).

The above pattern indicates that *Ch. balatonicus* may show a non-linear response to resource availability. To further explore this point, spring biomass of *Ch. balatonicus* was compared to phytoplankton biomass during the previous summer (Fig. 9). Above a threshold chlorophyll concentration of 20–30 mg m^{-3}, abundance of *Ch. balatonicus* exceeded 700–800 ind m^{-2} and the species dominated the chironomid fauna. Below this threshold, its abundance was close to the limit of detectability (≤1 ind m^{-2}). In the latter case, the importance of top–down control by fish increased (Tátrai, 1980; Bíró, 1997).

The shift from *P. choreus* to *Ch. balatonicus* dominance was shown to double (from 0.3–0.6% to 0.7–1.3%) the efficiency of energy transfer from primary producers to chironomids (Specziár & Vörös, 2001; Specziár, 2004). This, in turn, basically influenced population dynamics of several fish species including those that forage primarily for chironomids (*Abramis brama* (Linnaeus, 1758), *Gymnocephalus cernuus* (Linnaeus, 1758)) and those that temporarily supplement their diet with these organisms (*Rutilus rutilus* (Linnaeus, 1758), *Pelecus cultratus*

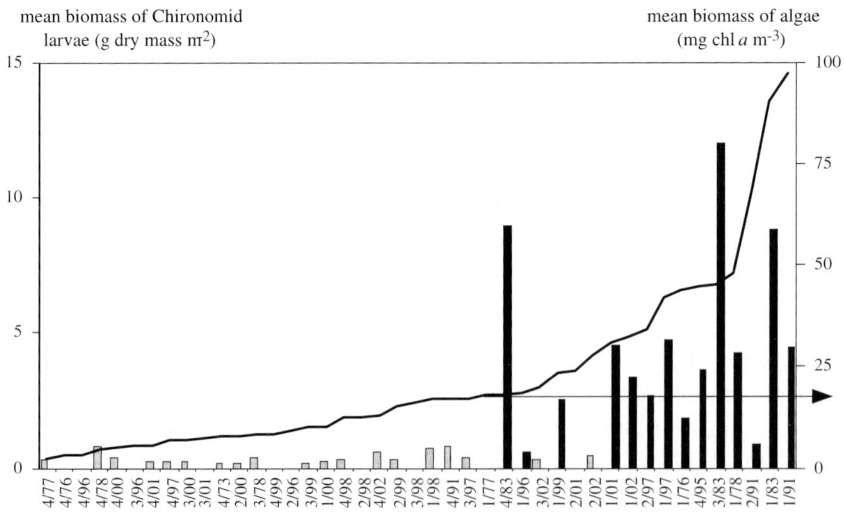

Fig. 9 Mean biomass of chironomid larvae during spring (March–April; bars) and mean biomass of phytoplankton (solid line) during the previous summer (July–September). (In the Z/XY label format of the abscissa Z refers to the basin and XY to the year of the chironomid data. Data were ordered by ascending phytoplankton biomass. Black bars denote years with the dominance of *Chironomus balatonicus*. Arrow indicates the approximate chlorophyll threshold above which *Ch. balatonicus* is likely to be present)

(Linnaeus, 1758), *Alburnus alburnus* (Linnaeus, 1758), *Cyprinus carpio* (Linnaeus, 1758), etc.). For example, growth of *Abramis brama* had been fast during the 1970s with a particularly strong population in Basin 1. In the 1990s, growth slowed down and the bream population decreased dramatically all along the lake (Bíró, 1997).

Updating water quality targets

As a first step, we examine whether the present water quality targets and permissible loads of phosphorus (Table 2) are realistic in the case of Basin 1. Maximum biomass of algae has been predicted for 2010 and 2025 (Table 3) from the relationship between the mobile P pool of the sediments (that is the potential maximum of internal P load) and yearly mean biomass of phytoplankton (Fig. 4). Three distinct load scenarios were defined using the observed loads during the past decade (Table 3). Scenario 1 represented the mean load over the whole period after the latest large management measure taken in 1992. Scenario 2 corresponded to the mean load under average hydro-meteorological conditions. Finally Scenario 3 was the mean load experienced during the draught in 2000–2003.

Table 2 Present water quality targets prescribed for 2010

Basin	Annual mean biomass (mg chl *a* m^{-3})	BAP load (tons P $year^{-1}$)
1	7–14	21
2	6–12	15
3	4–8	8
4	3–6	12

BAP—biologically available P

In order to achieve the present water quality targets until 2010, mean external load of P should be decreased immediately to the level of Scenario 3 (Table 3). We estimate that at the mouth section of the Zala River, 6–8 tons P $year^{-1}$ originate from sewage emission, whereas the diffuse load of biologically available P (BAP) ranges between 10 and 15 tons P $year^{-1}$ in average hydro-meteorological conditions. During the draught, pre-reservoirs released 1–5 tons P $year^{-1}$ of BAP due to increased internal P load. Thus, diffuse load dropped to 0–2 tons P $year^{-1}$, a value below the natural background load in years with average runoff (Table 1). Consequently, no combination of management measures can reduce the external load to the level of Scenario 3. Therefore the present water quality target is unlikely to be attained by 2010.

Taking into consideration planned development in sewerage, introducing an effluent standard of 0.5 g P m^{-3} in the whole watershed will reduce BAP load from point sources by no more than 30–40%. Such reduction would decrease the external load to the level of Scenario 1. The desired water quality could then be achieved by 2025 (Table 3). However, neither Scenario 1 seems to be realistic at the present time because it contains the strong effect of dry years on mean load (Table 3). In this way improved waste water treatment alone does not guarantee the expected improvement in water quality. In order to accomplish our objectives, diffuse load of BAP experienced under mean hydro-meteorological conditions (Scenario 2, Table 3) must be reduced by 3–4 tons P $year^{-1}$ in addition to improved waste water treatment. This reduction is to be

Table 3 Maximal expected value of annual mean algal biomass in Basin 1 assuming three constant load scenarios derived from past observations

Scenario	Period of past observation	TP load (tons P $year^{-1}$)	BAP load (tons P $year^{-1}$)	2010		2025	
				P_{mobil} (mg P g^{-1} dry mass)	Biomass (mg chl *a* m^{-3})	P_{mobil} (Mg P g^{-1} dry mass)	Biomass (mg chl *a* m^{-3})
1	1993–2003	37 ± 22	21 ± 12	0.066	16	0.056	13
2	1993–1999	50 ± 17	27 ± 10	0.077	18	0.074	17
3	2000–2003	15 ± 4	10 ± 2	0.047	11	0.022	5

TP—total phosphorus, BAP—biologically available phosphorus. P_{mobil}—mobil P content of the sediments. Estimates are based on the relationship in Fig. 4b. Loads of Ca and suspended solids were also assumed to be constant realized in a socio-economic environment where the unsustainable low level of fertilizer application is expected to rise, and abandoned agricultural areas are increasingly built in along the shoreline.

Considering the fairly similar behavior of various basins on the OECD plot (Fig. 5), permitted BAP loads can be expected to result in the desired water quality in each area (Table 2) by 2025.

As a second step, present water quality targets were supplemented with ecological criteria (Table 4). Replacing the simple scheme of Table 2 with a more complex classification system (Table 4) demands additional effort from the responsible authorities, which may reluctantly accept the new criteria. Nevertheless, there are at least three good reasons justifying our proposal. (i) The foregoing analysis indicated that recovery of Lake Balaton reached Stages 4 and 5. While mean biomass of phytoplankton may allow properly assessing progress of eutrophication management during Stages 1 and 3, the value of this indicator diminishes when structural changes take place during subsequent stages. (ii) Structural changes precede quantitative changes. Therefore properly elaborated structural indices may have considerable predictive power during Stages 4 and 5. (iii) The European Water Framework Directive urges classification of aquatic ecosystems using ecological indices.

As discussed above, the *Q* index of Padisák and coworkers (2003) is a suitable indicator of structural changes of phytoplankton that has been successfully tested in Lake Balaton at various time scales (Fig. 7). Thus, the index was incorporated as one of the ecological criteria (Table 4).

The long-term relationship between biomass of phytoplankton and that of chironomids (Fig. 8; Specziár & Vörös, 2001) was used to derive appropriate criteria concerning chironomid biomass (Table 4). A structural index, similar to the *Q* index has not been derived for the chironomid fauna yet. At the same time, presence or absence of *Chironomus balatonicus* was shown to basically influence overall efficiency of energy transfer from phytoplankton to fish. In this way abundance of this key species reflects important characteristics of ecosystem functioning. Moreover, among the 14 chironomid species inhabiting Lake Balaton and known to be capable of bloom formation, *Ch. balatonicus* was the single one that produced the actual blooms up to now. Its mass emergence attains an "intolerable" level when the biomass of chironomids exceeds 3–4 g m^{-2}. The threshold behavior of *Ch. balatonicus* (Fig. 9) allowed establishing ecological criteria in relation to the abundance of this species (Table 4).

Although we aimed at setting up the ecological targets in a consistent way, one cannot exclude that various indicators occasionally result in divergent classification. The question arises how an aggregated index of ecological status can be determined. One may accept the worst indication. Alternatively, the aggregated index may be composed from individual indices by appropriate weighting. The final choice must be made after evaluating the

Table 4 The proposed ecological target system to be realized by 2025

Basin	Ecological targets						Permissible TP load[a] (tons P $year^{-1}$)	Load reduction[g] (%)
	Phytoplankton			Chironomidae				
	Annual MB[b]	*Q* index[c]		Maximum MB[e]	*Ch. balatonicus*[f]			
	mg chl *a* m^{-3}	Value	Classification[d]	g dry mass m^{-2}	ind m^{-2}	Classification[d]		
1	14	6–8	Good	2.0	800	Good	37	17
2	12	6–8	Good	1.5	500	Good	21	54
3	8	8–10	Excellent	0.5	<1	Excellent	17	34
4	6	8–10	Excellent	0.5	<1	Excellent	22	24

MB—mean biomass. a—refers to the average of 5 years, yearly variability can reach ±50–60%. b—using the present monitoring frequency. c—in the period July–September. d—scaling: "excellent", "good", "moderate", "satisfactory", "poor". e—in the period March to June. f—maximum density of *Chironomus balatonicus* larvae. g—relative to the present load experienced in years of average runoff experience of experimental introduction of the new targets during the next few years.

Unlike the present scheme that expresses permissible load of P in terms of BAP, we propose to relay on total P loads (Table 4). Several conceptual and methodological problems introduce an additional uncertainty into the estimates of BAP load compared to that of total P load. Undertaking this higher uncertainty has been rational during the initial phase of eutrophication management, when the main effort had to be concentrated on the reduction of loads from point sources (Somlyódy & Jolánkai, 1986). Future management, however, must focus on the reduction of diffuse loads and therefore BAP can be replaced with total P. The proposed total P loads (Table 4) are equivalent to the BAP prescriptions of the present scheme (Table 2). The proposed values refer to loads during the years of average runoff. Depending on the hydrometeorological conditions, the deviation may reach ±50–60%.

Concluding remarks

During the initial stage of eutrophication management, the response of various shallow lakes converges to a single pattern: ones the sum of external and internal P loads reached capacity limiting levels, the phytoplankton biomass decreases in proportion to P supply. Obviously, long-term behavior of sediments determining the magnitude of internal load cause large variability with respect to timing and degree of recovery (Marsden, 1989; Sas, 1989). Nevertheless, a simple targeting scheme that includes the clear objectives (water quality to be restored) and the prerequisites of accomplishing these objectives (permissible loads) is satisfactory to orient both managers and the public. At this early stage, complicated targeting schemes should be avoided since they likely to make confusion.

The situation changes when recovery comes to the stage of re-organization of the aquatic ecosystem. During this stage, the simple targeting system should be replaced with one that integrates specific ecological features of the individual lakes. For example, biomass and composition of zooplankton can be good indicators of recovery in the majority of shallow lakes. In the specific geomorphological environment of Lake Balaton, however, this group fails to properly indicate trophic changes.

As demonstrated in our study, long-term observations must be available about biota in order to choose the most appropriate lake-specific indices and to define a consistent set of ecological criteria. Therefore we recommend including each major group that may a priori be considered to fulfill the role of an indicator in the given lake into the monitoring program from the very beginning of eutrophication management.

Acknowledgements This study was financially supported by the "Phytoplankton-on-line" project (No. EVK1-CT–1999-00037) and by the Ministry for Environment and Water (Hungary). We are greatly indebted to many colleagues who provided their published or unpublished data for this evaluation.

References

Bíró, P., 1997. Temporal variation in Lake Balaton and its fish populations. Ecology of Freshwater Fish 6: 196–216.

Bíró, P. & L. Vörös, 1982. Relationships between phytoplankton and fish yields in Lake Balaton. Hydrobiologia 97: 3–7.

Boström, B., M. Jansson & C. Forsberg, 1982. Phosphorus release from lake sediments. Archiv für Hydrobiologie Beiheft Ergebnisse der Limnologie 18: 5–59.

Clement, A., V. Istvánovics & L. Somlyódy, Evaluation of water quality of Lake Balaton. Vízügyi Közlemények (in Hungarian) (in press) .

Clement, A., 2000. Improving uncertain nutrient load estimates for Lake Balaton. Water Science and Technology 43: 279–286.

Csermák, K. & F. Máté, 2004. Soils of Lake Balaton. V.E. Georgikon Kar, Keszthely, Hungary (in Hungarian).

Dévai, Gy., 1990. Ecological background and importance of the change of chironomid fauna (Diptera: Chironomidae) in shallow Lake Balaton. Hydrobiologia 191: 189–198.

Dévai, Gy. & J. Moldován, 1983. An attempt to trace eutrophication in a shallow lake (Balaton, Hungary) using chironomids. Hydrobiologia 103: 169–175.

G.-Tóth, L., 1991. Respirative Electron Transport System (ETS) Activity and Carbon Metabolism of Net Zooplankton in Lake Balaton. In Bíró, P. (ed.), Hundred Years of Lake Balaton Research. Repro-print, Nemesvámos, 172–180 (in Hungarian).

G.-Tóth, L., 1992. Limiting effect of abioseston on food ingestion, postembryonic development time and

fecundity of daphnids in Lake Balaton (Hungary). Journal of Plankton Research 14: 435–446.
G.-Tóth, L., 2000. Feeding biology of planktonic crustaceans and their role in elimination of phytoplankton in Lake Balaton. DSc Thesis, Balaton Limnological Research Institute, Tihany (in Hungarian).
Herodek, S., 1984. The eutrophication of Lake Balaton: measurements, modeling and management. Internazionale Vereinigung für Theoretische und Angewandte Limnologie 22: 1087–1091.
Herodek, S., 1986. Phytoplankton Changes During Eutrophication and P and N Metabolism. In Somlyódy, L. & G. van Straten (eds), Modeling and Managing Shallow Lake Eutrophication. Springer-Verlag, Berlin, 183–204.
Istvánovics, V., A. Osztoics & M. Honti, 2004. Dynamics and ecological significance of daily internal load of phosphorus in shallow Lake Balaton, Hungary. Freshwater Biology 49: 232–252.
Istvánovics, V. & L. Somlyódy, 2001. Factors influencing lake recovery from eutrophication – the case of Basin 1 of Lake Balaton. Water Research 35: 729–735.
Istvánovics, V., L. Somlyódy & A. Clement, 2002. Cyanobacteria-mediated internal eutrophication in shallow Lake Balaton after load reduction. Water Research 36: 3314–3322.
Koncsos, L., 2003. Evaluation of the Achievement in Water Quality Management. In Mahunka, S. & J. Baczerowski (eds), Results of Lake Balaton Research in 2002. Hungarian Academy of Sciences, Budapest, 79–85 (in Hungarian).
Láng, I., 1986. Impact on Policymaking: Background to a Government Decision. In Somlyódy, L. & G. van Straten (eds), Modeling and Managing Shallow Lake Eutrophication. Springer-Verlag, Berlin, 110–122.
Lijklema, L., P. Gelencsér, F. Szilágyi & L. Somlyódy, 1986. Sediment and its Interaction with Water. In Somlyódy, L. & G. van Straten (eds), Modeling and Managing Shallow Lake Eutrophication. Springer-Verlag, Berlin, 156–183.
Marsden, M. W., 1989. Lake restoration by reducing external phosphorus loading: the influence of sediment phosphorus release. Freshwater Biology 21: 139–162.
Mátyás, K., 1996. Phytoplankton Studies in the Kis-Balaton Water Protection System between 1991 and 1995. In Pomogyi, P. (ed.), 2nd. Kis-Balaton Symposium, ISBN 963 02 9991 7. West-Transdanubian Water Authority, Keszthely, 119–130 (in Hungarian).
Padisák, J., 1994. Relationships between Short-term and Long-term Responses of Phytoplankton to Eutrophication of the Largest Shallow Lake in Central Europe (Balaton, Hungary). In Sund, H., H.-H. Stabel, W. Geller, Y. Xiaogan, Y. Kechang & S. Fengning (eds), Environmental Protection and Lake Ecosystem. Science & Technology Press, Beijing, 419–437.
Padisák, J., 1997. *Cylindrospermopsis raciborskii* (Woloszynska) Seenayya et Subba Raju, an expanding, highly adaptive cyanobacterial species: geographic distribution, population dynamics and autecology. Archive für Hydrobiologie (supplement) 107: 563–593.
Padisák, J. & C. S. Reynolds, 1998. Selection of phytoplankton associations in Lake Balaton, Hungary, in response to eutrophication and restoration measures, with special reference to the cyanoprokaryotes. Hydrobiologia 384: 41–53.
Padisák, J., É. Soróczki Pintér & Zs. Zámbóné Doma, 2003. Diversity and Spatial-temporal Patterns of Phytoplankton in Lake Balaton in 2002. (in Hungarian). In Mahunka, S. & J. Banczerowski (eds), Results of Lake Balaton Research in 2002, ISSN 1419-1075. MTA, Budapest, 35–42.
Ponyi, J., I. Tátrai & A. Frankó, 1983. Quantitative studies on Chironomidae and Oligochaeta in the benthos of Lake Balaton. Archiv für Hidrobiologie 97: 196–207.
Reynolds, C. S., V. Huszar, C. Kurk, L. Naselli-Flores & S. Melo, 2002. Towards a functional classification of the freshwater phytoplankton. Journal of Plankton Research 24: 417–428.
Sas, H., 1989. Lake Restoration and Reduction of Nutrient Loading: Expectations, Experiences, Extrapolations. Academia Verlag Richarz, St Augustin.
Scheffer, M., S. H. Hosper, M.-L. Meijer, B. Moss & E. Jeppesen, 1993. Alternative equilibria in shallow lakes. Trends in Ecology & Evolution 8: 275–279.
Sebestyén, O., 1953. Quantitative plankton studies in Lake Balaton. II. Decadal changes. Annales Instituti Biologici (Tihany) 21: 63–89 (in Hungarian).
Shanahan, P., D. R. F. Harleman & L. Somlyódy, 1986. Wind-induced Water Motion. In Somlyódy, L. & G. van Straten (eds), Modeling and Managing Shallow Lake Eutrophication. Springer-Verlag, Berlin, 204–255.
Somlyódy, L. & G. Jolánkai, 1986. Nutrient Loads. In Somlyódy, L. & G. van Straten (eds), Modeling and Managing Shallow Lake Eutrophication. Springer-Verlag, Berlin, 125–156.
Søndergaard, M., J. P. Jensen & E. Jeppesen, 1999. Internal phosphorus loading in shallow Danish lakes. Hydrobiologia 408/409: 145–152.
Specziár, A., 2004. Long-term Changes in the Chironomid Fauna and Feeding of Fish in Lake Balaton. In Fenyvesi, O. (ed.), Tudományos előadások 2004. MTA Veszprémi Területi Bizottsága, Veszprém, 69–85 (in Hungarian).
Specziár, A. & L. Vörös, 2001. Long-term dynamics of Lake Balaton's chironomid fauna and its dependence on the phytoplankton production. Archiv für Hydrobiologie 152: 119–142.
Tátrai, I., 1980. About feeding conditions of bream, (*Abramis brama* L.) in Lake Balaton. Developments in Hydrobiology 3: 81–86.
Virág, Á., 1998, The Past and the Present of Lake Balaton. (In Hungarian). Egri Nyomda Kft., Eger. ISBN 963 9060 216.
Vollenweider, R. A. & J. J. Kerekes, 1982. Background and Summary Results of the OECD Cooperative Programme on Eutrophication. OECD report, Paris.

Hydrobiologia (2007) 581:319–327
DOI 10.1007/s10750-006-0512-6

EUTROPHICATION IN LAKES

Removal of agricultural non-point source pollutants by ditch wetlands: implications for lake eutrophication control

Cuiling Jiang · Xiaoqiu Fan · Guangbo Cui · Yibin Zhang

Abstract Ditches grown with nature reed (*Phragmites communis* Trin) and wild rice (*Zizania latifolia* Turcz) were selected to study the removal capacity of agricultural non-point source pollutants so as to find a way to alleviate eutrophication in Lake Taihu. Ditches sediment from depths below 40 cm can accumulate organic matter and total nitrogen (TN). TN is correlated positively to organic matter in reed populated sediment and wild rice populated sediment. This suggests that the main composition of TN is organic nitrogen derived from plant decomposition. A significant negative relationship between TN and pH was found in reed and wild rice sediments. Seasonal harvest of helophyte vegetation is an effective method to remove N and P from wetlands. Organic matter and TN concentrations in water and sediments (0–20 cm) in areas where reeds were removed are lower than non-harvested areas (control). Reeds and wild rice have high uptake ability of nitrogen (N) and phosphorus (P). However, the low economic value of these plants will not stimulate voluntary harvest of farmers. *Zizania caduciflora* Turez Hand-mazt is a kind of vegetable widely cultivated in ditches around the lake. It can also absorb N and P effectively. Thus, large scale cultivation of *Z. caduciflora* to replace nature plants may improve water quality.

Keywords Ditch wetland · Helophyte vegetation · Non-point source pollutants

Guest editors: B. Qin, Z. Liu & K. Havens
Eutrophication of shallow lakes with special reference to Lake Taihu, China

C. Jiang (✉) · X. Fan · Y. Zhang
State Key Laboratory of Hydrology-Water Resources and Hydraulic Engineering, Hohai University, 1 Xi Kang Road, Nanjing 210098, China
e-mail: cljianghhu@163.com

C. Jiang · G. Cui
College of Water Resources and Environment, Hohai University, 1 Xi Kang Road, Nanjing 210098, China

Introduction

Lake Taihu is the third largest freshwater lake in China (Qin et al., 2007). Since the 1980s with rapid economic development in the Lake Taihu Basin, increasing nutrient loadings from point and non-point sources have resulted in water quality deterioration. Compared to oligotrophic conditions in the 1950s, Lake Taihu is now eutrophic (Huang et al., 2001). With enhanced control of point-source pollutants, non-point sources are the main influence on water quality of Lake Taihu. Excessive chemical fertilizer application led to large nitrogen (N) and phosphorus (P) increases (Roth & Fox, 1990; Angle et al., 1993; Withers et al., 2001). In Lake Taihu, the amount of fertilizers used (total pure nitrogen, phosphorus and kalium) has doubled from about 400 kg ha^{-1} in the 1980s to 800 kg ha^{-1} at the end of the 1990s (Gao et al., 2002). Uptake efficiency of chemical fertilizers by crops is 25–35% for N (Li & Li, 2001) and 10–20% for P. About 20–25% of N and 20–40% of P applied to crops annually is carried into ditches and channels with runoff and seepage (Xu, 1996; Huang et al., 2001).

Wetlands are effective at trapping and removing N and P by sediment adsorption, decomposition by microorganisms and uptake by helophyte vegetation (Raisin & Mitchell, 1995; Mitsch et al., 2000; Kang et al., 2002; Jordan et al., 2003). In many countries, constructed wetlands are used to treat point or non-point source pollutants (Fleischer et al., 1994; Spieles & Mitsch, 2000; Braskerud, 2002). In general, constructed wetlands are small, making them difficult to develop widely in China. However, there are irrigation and drainage networks that divide croplands in the lake basin. These ditches and channels are the first convergence points for agricultural non-point pollutants and conduits exporting nutrients to rivers and lakes. Wetland plants are flourishing in ditches and channels due to plentiful rainfall and suitable climate. Large emergent plants growing naturally in ditches and channels are reed (*Phragmites communis* Trin) and wild rice (*Zizania latifolia* Turcz), both of which have high capacity for N and P uptake. In the growing season, nutrients are assimilated and accumulated and become plant biomass. Reeds and wild rice have low economic value, and are difficult to harvest. In winter and spring, the plants decompose and release N and P, which may deteriorate water quality of rivers and the lake (Tanner, 2001).

Many studies have focused on the transport and transformation of N and P in constructed wetlands. Much attention was paid to N and P variations in wetland water (including inflow and outflow), but accumulation and transformation processes in sediments were ignored (Arheimer & Wittgren, 1994; Mitsch et al., 2000; Jan, 2002). Removal mechanisms of N and P in ditch wetlands with natural helophyte vegetation are not often discussed, and few, if any, economical methods to prevent secondary pollution have been proposed. The objective of this study is to identify factors affecting N and P transformations in ditches and formulate a plan to remove nutrients to alleviate surface water enrichment.

Materials and methods

Experimental site

The experimental site (32°14′ N, 119°52′ E) was located near Lake Taihu in a ditch wetland 600 m in length and 20–30 m in width (Fig. 1). Reeds and wild rice were growing abundantly in the ditch. The reed area was about 1.4 ha (560 m length × 25 m width) and wild rice was about 0.08 ha (40 m length × 20 m width). There is a concrete weir to keep water in the wetland. Only during flooding does water drain from east to west through the weir. About 1.8 ha of *Zizania caduciflora* Turez Hand-mazt was planted in another nearby ditch. The water table of the wetland in summer was about 40–80 cm deep. Dry and wet conditions alternated with the occurrence of rainfall in winter. During the experiment, only 0.23 ha of cropland was arranged to drain agricultural runoff into the wetland. Wheat and rice were grown in rotation. The wheat growing season was November to May of the next year and June to October for rice. Wheat and rice straws were returned to the soil after harvest. There were 787 kg ha^{-1} urea (46.2% N) and ammonia bicarbonate (17.1% N) and 1275 kg ha^{-1} compound fertilizer (9% N and 7% P) used on the cropland, equaling 607 kg ha^{-1} $year^{-1}$ of N and 90 kg ha^{-1} $year^{-1}$ of P.

Sampling and analyses

Soil samples were taken from cropland, reed and wild rice wetlands in September and December 2002 and March and June 2003 for analysis of organic matter, total nitrogen (TN), NH_4^+-N, NO_3^--N, soil moisture and pH. Two sample stations were selected in cropland and reed wetland, and one in wild rice wetland due to the small area (about 800 m^2). At each station, there were six layers of soil samples (0–10, 10–20,

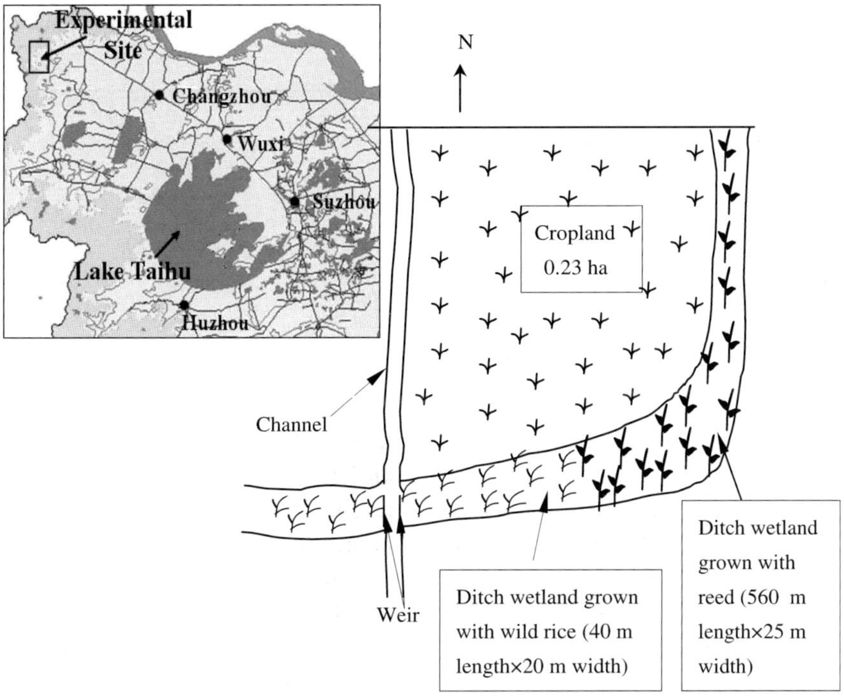

Fig. 1 Schematic illustration of ditch wetland in experimental site

20–40, 40–60, 60–80, 80–100 cm depth) collected using a 5 cm-diameter auger. During replication of sediment sampling in the reed wetland, one station was near the wild rice wetland (in the harvested plot after October) and another was about 150 m from the first (in the non-harvested plot). Two soil sampling stations in cropland were about 8–10 m from the wetland berm. Preliminary investigation showed little variation in nutrient contents between the two stations. Soil nutrient date reported in results is the average of the two stations.

Water samples in wild rice and reed wetlands were collected in September, November and December 2002, and March and June 2003 to analyze organic matter, TN, NH_4^+-N, NO_3^--N, and pH. Two or three water samples in each plot were taken and mixed. In October 2002, about 1/3 of the reed plant area adjacent to wild rice was harvested. Sediment and water samples in the harvested and non-harvested plot (control) were taken. Comparison sampling in the reed wetland was finished in March 2003.

Vegetation sampling was obtained from three randomly selected stations in growing seasons. Plant were sampled and separated into above and below ground fractions after transport to the laboratory. Plant samples were collected in September 2002, and June and September 2003 to analyze N and P in leaf, stem and root fractions. Characteristics of the helophyte plants are shown in Table 1.

Chemical analyses were conducted based on standard methodology (China Standard Press, 1998). Organic matter in water was determined by dichromate oxidation and $FeSO_4$ titration. NH_4^+-N concentration was analyzed using Nessler's reagent colorimetric method, and NO_3^--N concentration was measured using spectrophotometry with phenol disulfonic acid. For TN analyses, sample was acidified with H_2SO_4 to pH < 2 and subjected to Kjeldahl digestion and analysis by colorimetry. pH was measured with a glass electrode. All analyses in water were finished within 48 h.

For analyses of NH_4^+-N and NO_3^--N in soil, fresh samples were collected in pre-weighted aluminum cans and refrigerated immediately to prevent transformation of inorganic nitrogen. NH_4^+-N and NO_3^--N were extracted with 2 M

Table 1 Character of plants (average analyzed result of September 2002, June 2003 and September 2003)

Character	Reed	Wild rice	*Z. caduciflora*	Rice
Height (m)	2.3–3.1	1.47–1.5	2.1–2.2	0.83–0.85
Root length (cm)	27–62	35–47	38–50	10–18
Density (individual m^{-2})	200	184	64	192
Fresh weight of leaf and stem (g $individual^{-1}$)	61.36	21.01	73.34	7.57
Water content of plant (%)	55.56	80.30	76.34	58.12
Fresh weight (kg ha^{-1})	122,720	38,658	46,938	14,534
Dry weight (kg ha^{-1})	54,537	7,615	11,106	6087

KCl by shaking for 1 h with a reciprocal shaker. Determination of NH_4^+-N and NO_3^--N in extracts were the same as above.

Samples for organic matter, TN and pH were collected in plastic bags, air dried for about 1 month in the laboratory, sieved through a 2-mm stainless steel sieve, and kept in brown glass bottles. TN was analyzed using the semi-micro Kjeldahl method, and produced NH_4^+ was collected in a boric acid solution and titrated with H_2SO_4. Samples for pH analysis were extracted with 0.01 M $CaCl_2$ and measured with a glass electrode. Organic matter content was analyzed using potassium dichromate-sulfate oxidation method.

Plant samples, including above and below ground parts, were cleaned and oven-dried at 105°C for 1 h and 80°C for an additional hour. After being cooled, the samples were ground and kept in brown glass bottles. Nutrients in roots, leaves, and stems were analyzed separately. N and P were analyzed via H_2O_2/H_2SO_4 digestion and measured colorimetrically. Biomass was calculated based on fresh and dry weight and surface area of the sampling plot.

Results

Accumulation of organic matter and TN in cropland soil and wetland sediment

Organic matter in cropland soil likely comes from decomposition of crop remains since farmers no longer applied organic fertilizer. When rice and wheat were harvested, only usable parts were removed, and leftover straw was ploughed into soil, which became the organic matter and TN source. Above 60 cm, organic matter content and TN decreased with depth and varied little below 60 cm (Figs. 2, 3). TN was highly correlated with organic matter.

Organic matter and TN also were correlated in wetland sediment (Figs. 4, 5), with $r = 0.97$ ($n = 27$, significant level at $P < 0.001$) in the reed ditch and $r = 0.91$ ($n = 19$, $P < 0.001$) in the wild rice ditch. Therefore, TN in wetlands was composed of organic N from plant decomposition.

Organic matter and TN concentrations from 0 to 100 cm in rice and wheat soil decreased with depth, but, in reed and wild rice sediments, lowest concentrations occurred at 20–40 cm and increased below 40 cm (Figs. 2, 3). Therefore,

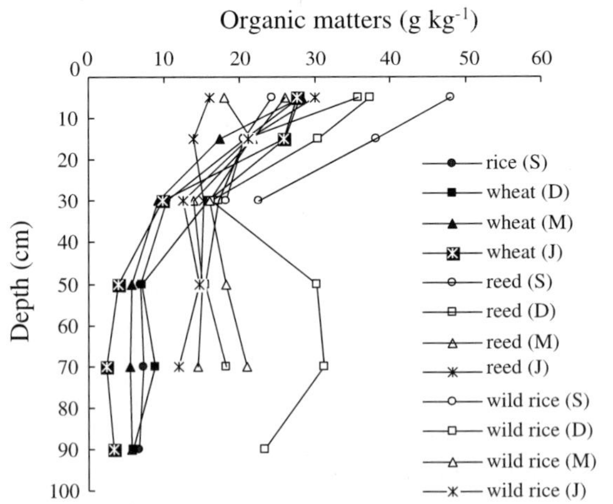

Fig. 2 Contents of organic matter in different layers of field soils (grown with rice and wheat) and wetland sediments (grown with reeds and wild rice) during the period 2002–2003. (S: in September 2002; D: in December 2002; M: in March 2003; J: in June 2003)

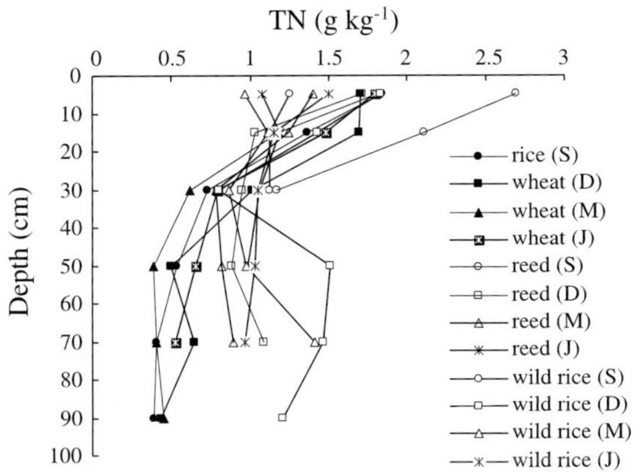

Fig. 3 Contents of TN in different layers of field soils (grown with rice and wheat) and wetland sediments (grown with reeds and wild rice) during the period 2002–2003. (S: in September 2002; D: in December 2002; M: in March 2003; J: in June 2003)

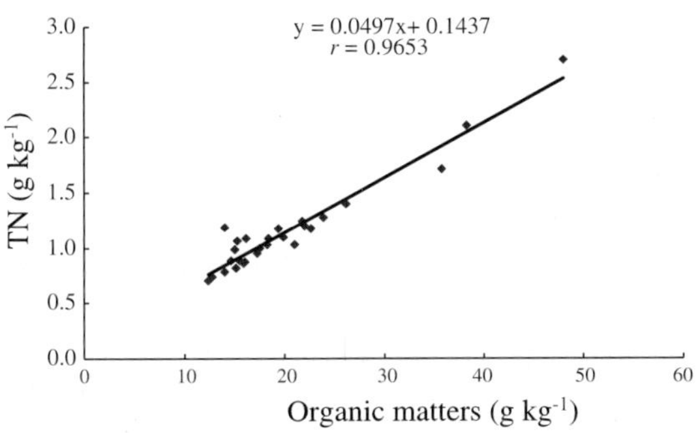

Fig. 4 Correlation between the contents of organic matter and TN in reed wetland sediment

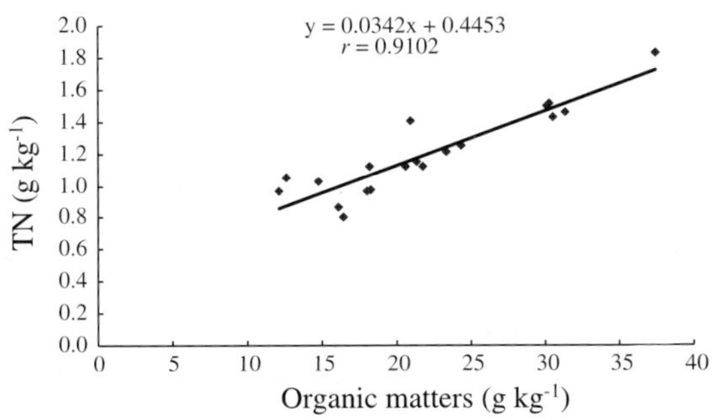

Fig. 5 Correlation between the contents of organic matter and TN in wild rice wetland sediment

organic matter and TN can accumulate in deep layers of wetland sediment.

In sediment above 40 cm, organic matter and TN varied with seasons. Variability in wet sediment was higher than dry soil suggesting that organic matter and TN in wetland surface layers decomposed more quickly due to dry and wet alternation.

Nutrients in harvested plot and in control

The reed wetland adjacent to wild rice was harvested in October 2002. After 20 days, water turbidity in the reaped area was lower than the non-harvested area (control). Organic matter and TN were 21.33 and 1.04 mg l^{-1}, respectively, in water of the reaped area, which were lower than those (28.4 and 1.22 mg l^{-1}, respectively) in the control. In December 2002 and March 2003, sediment organic matter and TN from 0 to 10 and 10 to 20 cm in the harvested area also were lower than the control (Fig. 6). Therefore, harvesting plants in autumn may reduce organic matter and TN accumulation in wetland water and surface sediment. However, NH_4^+-N and NO_3^--N in harvested and non-harvested plots were different from organic matter and TN (Fig. 7). Compared to the non-harvested area, inorganic N from 0 to 10 and 10 to 20 cm in harvested area was high (except low NO_3^--N in 10–20 cm in March 2003). Increased NH_4^+-N and NO_3^--N likely came from organic N decomposition.

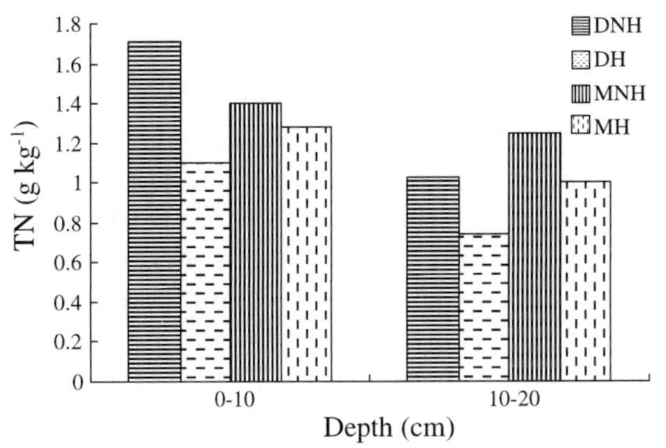

Fig. 6 Contents of TN in the depth of sediment from 0 to 20 cm in harvested and non-harvested plots of reed wetland (DNH: in non-harvested plot in December 2002; MNH: in non-harvested plot in March 2003; DH: in harvested plot in December 2002; MH: in harvested plot in March 2003)

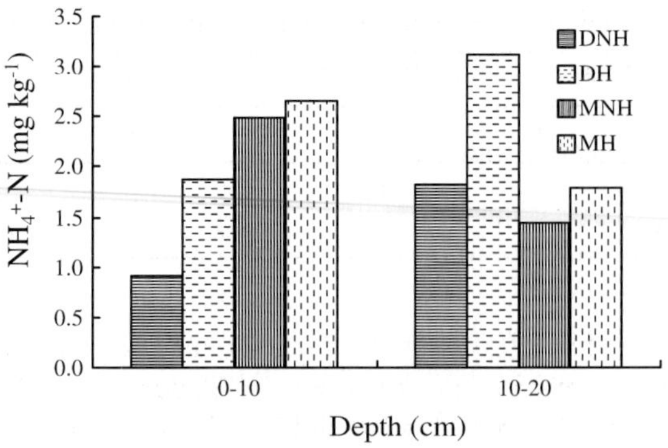

Fig. 7 Contents of NH_4^+-N in the depth of sediment from 0 to 20 cm in harvested and non-harvested plots of reed wetland (DNH: in non-harvested plot in December 2002; MNH: in non-harvested plot in March 2003; DH: in harvested plot in December 2002; MH: in harvested plot in March 2003)

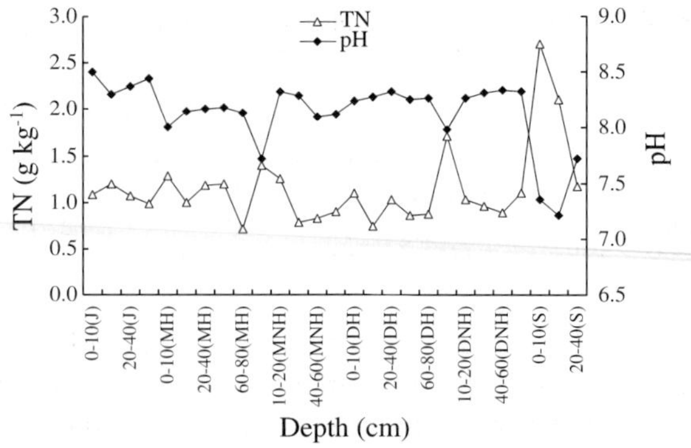

Fig. 8 TN and pH value in different layers of reed wetland sediment during the period 2002–2003. (J: in June 2003; MH: in harvested plot in March 2003; MNH: in non-harvested plot in March 2003; DH: in harvested plot in December 2002; DNH: in non-harvested plot in December 2002; S: in September 2002)

Significant relationships were found among water TN, NH_4^+-N and NO_3^--N in reaped and control plots in the reed wetland. The regression equation for TN and NH_4^+-N is $Y = 0.128X - 0.0338$, $r = 0.90$ ($n = 5$, significant level at $P < 0.05$, n is the number of samples obtained in harvested and non-harvested plots and in different times), and TN and NO_3^--N is $Y = 1.6935X - 0.7413$, $r = 0.96$ ($n = 5$, $P < 0.01$). These relationships suggest that TN mineralization is an important source of NH_4^+-N and NO_3^--N.

Relationship between TN and pH

Nutrient transformation in soil and sediment will cause variation of pH, which may promote chemical reactions. Organic matter and TN in surface soil were high and decreased with depth. In reed and wild rice sediments, organic matter and TN decreased in 0–40 cm, and then increased below 40 cm (Figs. 2, 3). However, pH variation was opposite (Figs. 8, 9). Lower pH coincided with higher organic matter and TN. TN in sediments and soil are negatively correlated with pH. The regression coefficient (r) is –0.78 ($n = 23$, $P < 0.001$) in soil, –0.82 ($n = 27$, $P < 0.001$) in reed sediment, and –0.69 ($n = 19$, $P < 0.01$) in wild rice sediment.

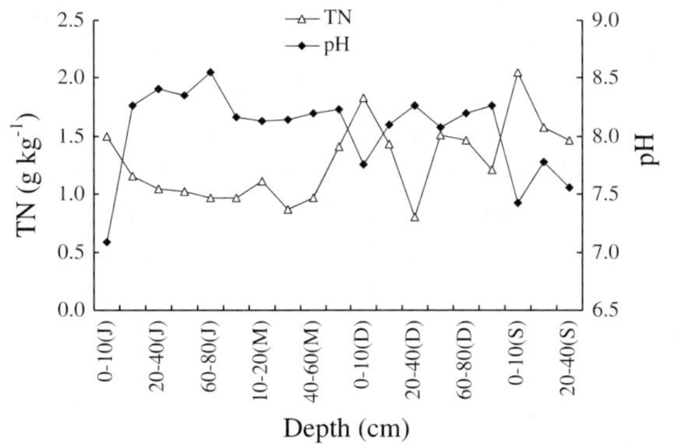

Fig. 9 TN and pH value in different layers of wild rice wetland sediment during the period 2002–2003 (J: in June 2003; M: in March 2003; D: in December 2002; S: in September 2002)

An effective method to remove N and P by plants

Plants are growing naturally in ditch wetlands in the Lake Taihu basin. In late June, reeds and wild rice are fully grown. Plant height, root length, density, fresh weight and water content were measured in September 2002 and June and September 2003 (Table 1). Mean dry reed and wild rice output were 54,537 and 7,615 kg ha^{-1} year^{-1}, respectively. Leaf and stem N and P concentrations are shown in Table 2. There were 818 kg ha^{-1} N and 103.6 kg ha^{-1} P

Table 2 Average contents of N and P in the part of plants above-ground and harvested amount of nutrients in a year

Tissue	*Zizania caduciflora*		Wild rice		Reed		Rice	
	N	P	N	P	N	P	N	P
Leaf and stem (g kg^{-1})	18.0	1.9	17.2	3.8	15.0	1.9	14.0	1.7
Harvested amount (kg ha^{-1})	200	21.1	131	28.9	818	103.6	85.2	10.3

removed after reeds harvest and 131 kg ha^{-1} N and 28.9 kg ha^{-1} P removed by wild rice in autumn. Nutrient uptake by above-ground reed parts is lower than wild rice, but its larger biomass provides high N and P removal capacity.

Zizania caduciflora Turez Hand-mazt is a kind of vegetable and also a perennial aquatic plant similar to wild rice in morphology and ecology. It can assimilate 200 kg ha^{-1} N and 21.1 kg ha^{-1} P in a year (Table 2). Compared with reeds and wild rice, average N content in above-ground parts of *Z. caduciflora* is higher, and P content is similar to reeds but lower than wild rice (Table 2). Therefore, planting *Zizania caduciflora* Turez Hand-mazt to replace natural reeds and wild rice may absorb N accumulated in ditch wetland sediment and alleviate eutrophication in lakes and rivers, but P removal is not as good as natural plants.

Discussion

Inorganic fertilizers used on crops were the primary N and P source in the experimental site. Organic matter and TN in cropland and wetland come from decomposition of plant remains, because farmers no longer use organic fertilizer. In general, inorganic N comprises <5% of TN (Rene et al., 1991; Bai et al., 2002). TN is correlated with organic matter in soil (Yang et al., 2002) and sediments (Figs. 4, 5). However, vertical and spatial distributions of organic matter and TN in cropland soil and wetlands sediment were different.

In rice and wheat soil, organic matter and TN from 0 to 100 cm decreased gradually with depth, but, in reed and wild rice sediments, lowest concentrations occurred in 20–40 cm and increased below 40 cm (Figs. 2, 3). Organic matter and TN decreased from 0 to 40 cm resulted from decomposition by large numbers of microorganisms in the root zone (Peder & Stefan, 1997). Compared with field soil, wetlands sediment contained more organic matter and TN below 40 cm. Therefore, wetlands can accumulate non-point source pollutants.

Organic matter and TN varied with season in surface sediment, more intensively than in soil, because of alternate aerobic and anaerobic conditions (Reddy & Patrick, 1975; Martin et al., 1997), which accelerated decomposition of organic matter and N in wetlands. In addition, cover and removal of plants also affected organic matter and N transformation in surface sediment. When reeds were harvested, improved transparency and aeration accelerated mineralization of organic matter and TN and led to a decrease in sediments from 0 to 20 cm (Fig. 6). Organic N mineralization was correlated with inorganic nitrogen (Tanner et al., 2002). NH_4^+-N and NO_3^--N from 0 to 20 cm in the sediment of harvested area were higher than the control. Significant relationships among TN, NH_4^+-N and NO_3^--N in reed wetland water were found. Therefore, organic nitrogen mineralization in the wetland was an important inorganic N source in addition to cropland runoff (Kovacic et al., 2000; Borin et al., 2001; Berit & Hans, 2002). Mineralization likely increased NH_4^+-N and NO_3^--N concentrations in wetland water temporarily, but TN in sediment decreased permanently. Therefore, harvest of dense plants removed pollutants from wetlands and accelerated organic matter and TN decomposition.

In general, high microorganism biomass occurs with high organic matter content and TN. Organic matter and TN decomposition by microorganisms changes pH. Organic matter produces organic acids during decomposition, and TN release H^+ when oxidized to NO_3^--N by ammonification and nitrification. These reactions result in decreased pH. Nitrification contributes to a pH decrease of two standard units according to DeSimone (1998).

On the basis of 607 kg ha^{-1} N and 90 kg ha^{-1} P fertilizer application in this region, and 20–25% N and 20–40% P loss rates according to Huang et al. (2001), 1 ha of reed wetland will remove N discharged from 5.4 to 8.0 ha of cropland and P discharged from 2.9 to 5.8 ha of field. Wild rice wetland can remove N from 0.9 to 1.3 ha of cropland and P from 0.8 to 1.6 ha of field annually. Therefore, plants are effective at removing agricultural non-point pollutants (Eriksson & Weisner, 1997; Li, 1997; Yang, 1998; Rickerl et al., 2000).

Despite high uptake ability of N and P by reeds and wild rice, wild plants have little economic value and cannot attract harvest of farmers. In winter and spring, the plants die and decompose to release N and P again, which causes secondary pollution. *Z. caduciflora* is a plant with value and is planted in the Lake Taihu basin. It has large biomass and high absorption capacity of N and P (Tables 1, 2). From August to September, *Z. caduciflora* is reproducing and requires high nutrient supply (Qiu et al., 2002). This is the same period for rice harvest, which requires fertilizing and irrigating. High nutrient concentrations in agricultural drainage meet the demand for rapid growth of *Z. caduciflora*. The growth period of *Z. caduciflora* is from March to September, similar to reeds and wild rice, coinciding with the growth period for wheat and rice rotated in crop fields. *Z. caduciflora* can absorb effectively N and P (Liu et al., 2002) discharged from field.

Acknowledgements We wish to express deep appreciation to associate Prof. L. Wang, Prof. A. J. M. Smits and A. Gashaw for valuable comments on the manuscript. We thank Junji Zhang for his practical assistance. The National Natural Science Foundation of China (No.50579018, 50239030) supported this study.

References

Angle, J. S., C. M. Gross, R. L. Hill & M. S. McIntosh, 1993. Soil nitrate concentrations under corn as affected by tillage, manure, and fertilizer application. Journal of Environmental Quality 22: 141–147.

Arheimer, B. & H. B. Wittgren, 1994. Modelling the effects of wetlands on regional nitrogen transport. Ambio 23: 378–386.

Bai, J., W. Deng & Y. Zhang, 2002. Spacial distribution characteristics of soil organic matter and nitrogen in the natural floodplain wetland. Environmental Science 23: 77–81 (in Chinese).

Berit, A. & B. W. Hans, 2002. Modelling nitrogen removal in potential wetlands at the catchment scale. Ecological Engineering 19: 63–80.

Borin, M., G. Bonaiti & L. Giardini, 2001. Controlled drainage and wetlands to reduce agricultural pollution: a lysimetric study. Journal of Environmental Quality 30: 1330–1340.

Braskerud, B. C., 2002. Factors affecting nitrogen retention in small constructed wetlands treating agricultural non-point source pollution. Ecological Engineering 18: 351–370.

China Standard Press, 1998. A Compilation of National Environmental Quality and Pollution Discharge Standards. China Standard Press, Beijing (in Chinese).

DeSimone, L. A., 1998. Nitrogen transport and transformations in a shallow aquifer receiving wastewater discharge: a mass balance approach. Water Resources Research 34: 271–285.

Eriksson, P. G. & S. E. B. Weisner, 1997. Nitrogen removal in a wastewater reservoir: the importance of denitrification by epiphytic biofilms on submersed vegetation. Journal of Environmental Quality 26: 905–910.

Fleischer, S., A. Gustafson, A. Joelsson, J. Pansar & L. Stibe, 1994. Nitrogen removal in created ponds. Ambio 23: 349–357.

Gao, C., J. Zhu & Y. Dou, 2002. Contribution of agricultural non-point source pollution to water quality deterioration in Taihu Lake watershed: recent trends and research priorities. Resources and Environment in the Yangtze Basin 11: 260–263 (in Chinese).

Huang, Y., C. Fang, P. Pu, J. Jiang & Q. Dai, 2001. Water Environment and Pollution Control of Lake Taihu. Science Publishing Company, Beijing, China, 259–264 (in Chinese).

Jan, V., 2002. The use of sub-surface constructed wetlands for wastewater treatment in the Czech Republic: 10 years experience. Ecological Engineering 18: 633–646.

Jordan, T. E., D. F. Whigham, K. H. Hofmockel & M. A. Pittek, 2003. Nutrient and sediment removal by a restored wetland receiving agricultural runoff. Journal of Environmental Quality 32: 1534–1547.

Kang, S., H. Kang, D. Ko & D. Lee, 2002. Nitrogen removal from a riverine wetland: a field survey and simulation study of *Phragmites japonica*. Ecological Engineering 18: 467–475.

Kovacic, D. A., M. B. David, L. E. Gentry, K. M. Starks & R. A. Cooke, 2000. Effectiveness of constructed wetlands in reducing nitrogen and phosphorus export from agricultural tile drainage. Journal of Environmental Quality 29: 1262–1274.

Li, S. & J. Li, 2001. Research progress on losses of fertilizer nitrogen. Agro-Environmental Protection 20: 377–379 (in Chinese).

Li, W., 1997. "Yellow water" in East Taihu Lake caused by *Zizania latifolia* and its prevention. Journal of Lake Sciences 9: 364–368 (in Chinese).

Liu, C., H. Hu & J. Zhang, 2002. Use of new type constructed wetland for treatment of low strength rural sewage. China Water and Wastewater 18: 1–4 (in Chinese).

Martin, H. W., D. B. Ivanoff, D. A. Graetz & K. R. Reddy, 1997. Water table effects on histosol drainage water carbon, nitrogen, and phosphorus. Journal of Environmental Quality 26: 1062–1071.

Mitsch, W. J., A. J. Horne & R. W. Nairn, 2000. Nitrogen and phosphorus retention in wetlands-ecological approaches to solving excess nutrient problems. Ecological Engineering 14: 1–7.

Peder, G. E. & E. B. W. Stefan, 1997. Nitrogen removal in a wastewater reservoir: the importance of denitrification by epiphytic biofilms on submersed vegetation. Journal of Environmental Quality 26: 905–910.

Qin, B. Q, P. Z. Xu, Q. L. Wu, L. C. Luo & Y. L. Zhang, 2007. Environmental issues of Lake Taihu, China. Hydrobiologia 581: 3–14.

Qiu, J., J. Jiang, B. Cao, L. Li, X. Xu & X. Han, 2002. Content and change trend of N, P, K in above ground organs of *Zizania latifolia*. Journal of Hubei Agricultural College 22: 22–125 (in Chinese).

Raisin, G. W. & D. S. Mitchell, 1995. The use of wetlands for the control of non-point source pollution. Water Science and Technology 32: 177–186.

Reddy, K. R. & W. H. Patrick, 1975. Effect of alternate aerobic and anaerobic conditions on redox potentical, organic matter decomposition and nitrogen loss in a flooded soil. Soil Biology & Biochemistry 7: 87–94.

Rene, L., Z. Malgorzata & B. Z. Joy, 1991. Nitrogen assessments in a constructed and a natural salt marsh of San Diego Bay. Ecological Applications 1: 40–51.

Rickerl, D. H., L. L. Janssen, & R. Woodland, 2000. Buffered wetlands in agricultural landscapes in the Prairie Pothole regions: environmental, agronomic, and economic evaluations. Journal of Soil and Water Conservation 55: 220–225.

Roth, G. W. & R. H. Fox, 1990. Soil nitrate accumulations following nitrogen-fertilized corn in Pennsylvania. Journal of Environmental Quality 19: 243–248.

Spieles, D. J. & W. J. Mitsch, 2000. The effects of season and hydrologic and chemical loading on nitrate retention in constructed wetlands: a comparison of low-and high-nutrient riverine systems. Ecological Engineering 14: 77–91.

Tanner, C. C., 2001. Growth and nutrient dynamics of soft-stem bulrush in constructed wetlands treating nutrient-rich wastewaters. Wetlands Ecology and Management 9: 49–73.

Tanner, C. C., R. H. Kadlec, M. M. Gibbs, J. P. S. Sukias & M. L. Nguyen, 2002. Nitrogen processing gradients in subsurface-flow treatment wetlands-influence of wastewater characteristics. Ecological Engineering 18: 499–520.

Withers, P. J. A., S. D. Clay & V. G. Breeze, 2001. Phosphorus transfer in runoff following application of fertilizer, manure, and sewage sludge. Journal of Environmental Quality 30: 180–188.

Xu, Q., 1996. A review on the status of non-point source pollution of chemical fertilizers and pesticides in China. Rural Eco-Environment 12: 39–43 (in Chinese).

Yang, Q., 1998. Ecological functions of aquatic vegetation in East Taihu Lake and its reasonable regulation. Journal of Lake Sciences 10: 76–72 (in Chinese).

Yang, Y., C. Tian, J. Sheng & Q. Wen, 2002. Spatial variability of soil organic matter, total nitrogen, phosphorus and potassium in cotton field. Agricultural Research in the Arid Areas 20: 26–30 (in Chinese).